HAESE MATHEMATICS

Specialists in mathematics education

Mathematics 10

MYP 5 (Standard)

third edition

Michael Haese

Mark Humphries

Ngoc Vo

for use with IB Middle Years Programme

MATHEMATICS 10 MYP 5 (Standard) third edition

Michael Haese	B.Sc.(Hons.), Ph.D.
Mark Humphries	B.Sc.(Hons.)
Ngoc Vo	B.Ma.Sc.

Published by Haese Mathematics
152 Richmond Road, Marleston, SA 5033, AUSTRALIA
Telephone: +61 8 8210 4666
Email: info@haesemathematics.com
Web: www.haesemathematics.com

National Library of Australia Card Number & ISBN 978-1-922416-36-0

First Edition	2014
Reprinted	2015, 2017, 2019
Third Edition	2022
Reprinted	2024

Cartoon artwork by James Hobbs, John Martin, and Ngoc Vo.

Artwork by Brian Houston.

Computer software by Patrick French, Ben Hensley, Huda Kharrufa, Rachel Lee, Linden May, and Han Zong Ng.

Audio recorded by Eloise Quinn-Valentine.

Production work by Hannah Coleman and Michael Mampusti.

Typeset in Australia by Charlotte Frost and Deanne Gallasch. Typeset in Times Roman 10.

Printed in China by Prolong Press Limited.

Acknowledgements: While every attempt has been made to trace and acknowledge copyright, the authors and publishers apologise for any accidental infringement where copyright has proved untraceable. They would be pleased to come to a suitable agreement with the rightful owner.

FOREWORD

Mathematics 10 MYP 5 (Standard) third edition has been designed and written for the International Baccalaureate Middle Years Programme (IB MYP) Mathematics framework, providing complete coverage of the Standard content outlined in the framework.

This book is recommended for students intending to progress to one of the Standard Level courses in the Diploma Programme. Material marked in dark blue is not part of the MYP standard framework, but is useful preparation for the Diploma Programme. Students should be encouraged to complete this material where time permits.

Discussions, Activities, Investigations, and Research exercises are used throughout the chapters to develop conceptual understanding. Material is presented in a clear, easy-to-follow style to aid comprehension and retention, especially for English Language Learners. Each chapter ends with extensive review sets and an online multiple choice quiz.

The associated digital Snowflake subscription supports the textbook content with interactive and engaging resources for students and educators.

The Global Context projects highlight the use of mathematics in understanding history, culture, science, society, and the environment. We have aimed to provide a diverse range of topics and styles to create interest for all students and illustrate the real-world application of mathematics.

We have developed this book in consultation with experienced teachers of IB Mathematics internationally but independent of the International Baccalaureate Organisation (IBO). It is not endorsed by the IBO.

We have endeavoured to publish a stimulating and thorough textbook and digital resource to develop and encourage student understanding and nurture an appreciation of mathematics.

We welcome your feedback.

Email: info@haesemathematics.com

Web: www.haesemathematics.com

ONLINE FEATURES

Each textbook comes with a 12 month subscription to the online edition and its range of interactive features. This can be accessed through the **SNOWFLAKE** online learning platform via a web browser or our offline viewer.

To activate your electronic textbook, please contact Haese Mathematics by emailing info@haesemathematics.com with your proof of purchase such as a copy of your receipt.

For general queries regarding **SNOWFLAKE** and online subscriptions:

- Visit our help page: snowflake.haesemathematics.com.au/help
- Contact Haese Mathematics: info@haesemathematics.com

SELF TUTOR

Self Tutor is an engaging feature that supports students learning independently and in classrooms.

Click on any example box to access a step-by-step animation with teacher voice support providing explanation and understanding.

INTERACTIVE LINKS

The **SNOWFLAKE** icons direct you to interactive tools to enhance learning and teaching.

These features include:

- demonstrations to illustrate and animate concepts
- multiple choice quizzes to test understanding
- games to practise and build skills
- tools for graphing and statistics
- printable pages for use in class.

GRAPHICS CALCULATOR INSTRUCTIONS

Graphics calculator instruction booklets are available for the **Casio fx-CG50**, **TI-84 Plus CE**, **HP Prime,** and the **TI-*n*spire**. Click on the relevant icon below.

CASIO fx-CG50 | TI-84 Plus CE | HP Prime | TI-*n*spire

When additional calculator help may be needed, specific instructions are available from icons within the text.

GRAPHICS CALCULATOR INSTRUCTIONS

GLOBAL CONTEXTS

The International Baccalaureate Middle Years Programme focuses teaching and learning through six Global Contexts:

- Identities and relationships
- Orientation in space and time
- Personal and cultural expression
- Scientific and technical innovation
- Globalisation and sustainability
- Fairness and development

The Global Contexts help students to develop connections between different subject areas in the curriculum.

GLOBAL CONTEXT **THE MINARD MAP**

Global context: Orientation in space and time
Statement of inquiry: Infographics can provide deeper insight into historic events than numbers alone.
Criterion: Communicating

GLOBAL CONTEXT

Each project contains a series of questions, divided into:

- Factual questions (in green)
- Conceptual questions (in blue)
- Debatable questions (in red).

The projects are also accompanied by the general descriptor and a task-specific descriptor for one of the four assessment criteria.

TABLE OF CONTENTS

1

Assumed knowledge

Number

Contents:

CHAPTER

Click on the icon to access this chapter

Assumed knowledge 2

Algebra

Contents:

CHAPTER

Click on the icon to access this chapter

Chapter 1

Exponents

Contents:

OPENING PROBLEM

The **parsec**, the **light-year**, and the **astronomical unit** are all units of length used to measure large distances in space. The length of each unit in metres is shown alongside.

1 parsec $\approx 3.1 \times 10^{16}$ m
1 light-year $\approx 9.5 \times 10^{15}$ m
1 astronomical unit $\approx 1.5 \times 10^{11}$ m

Things to think about:

- **a** Can you write the length of each unit in metres as an ordinary number?
- **b** What advantages are there to writing the lengths in the form given above?
- **c** How much farther is a parsec than a light-year?
- **d** How many astronomical units are there in one parsec?

Exponent notation can be used to represent repeated multiplications of the same number.

For example, we represent $5 \times 5 \times 5$ as 5^3, which reads "5 to the power 3". We say that 5 is the **base**, and 3 is the **exponent**, **power**, or **index**.

If n is a positive integer, then a^n is the product of n factors of a.

$$a^n = \underbrace{a \times a \times a \times a \times \dots \times a}_{n \text{ factors}}$$

A EXPONENT LAWS

In previous years we have seen the following **exponent laws** for **integer exponents**:

If m and n are integers, then:

- $a^m \times a^n = a^{m+n}$ — To **multiply** numbers with the **same base**, keep the base and **add** the exponents.
- $\dfrac{a^m}{a^n} = a^{m-n}$ provided $a \neq 0$ — To **divide** numbers with the **same base**, keep the base and **subtract** the exponents.
- $(a^m)^n = a^{mn}$ — When **raising** a **power** to a **power**, keep the base and **multiply** the exponents.
- $(ab)^n = a^n b^n$ — To **expand** a **product** to a power, raise each factor to the power.
- $\left(\dfrac{a}{b}\right)^n = \dfrac{a^n}{b^n}$ provided $b \neq 0$ — To **expand** a **quotient** to a power, raise both the numerator and denominator to the power.
- $a^0 = 1$ provided $a \neq 0$ — Any non-zero number raised to the power of zero is 1.
- $a^{-n} = \dfrac{1}{a^n}$ — a^n and a^{-n} are **reciprocals** of one another. In particular, $a^{-1} = \dfrac{1}{a}$.

Example 1 — Self Tutor

Simplify using the exponent laws:

a $7^8 \times 7^4$ **b** $\dfrac{x^{11}}{x^k}$ **c** $(5^a)^3$

a $7^8 \times 7^4 = 7^{8+4} = 7^{12}$

b $\dfrac{x^{11}}{x^k} = x^{11-k}$

c $(5^a)^3 = 5^{a\times 3} = 5^{3a}$

EXERCISE 1A

1 Simplify using the exponent laws:

a $3^2 \times 3^5$ **b** $x^6 \times x^3$ **c** $x^5 \times x^n$ **d** $t^3 \times t^4 \times t^5$

e $\dfrac{7^9}{7^5}$ **f** $\dfrac{x^7}{x^3}$ **g** $\dfrac{t^6}{t^x}$ **h** $t^{3m} \div t$

i $(5^3)^2$ **j** $(t^4)^3$ **k** $(y^3)^m$ **l** $(a^{3m})^4$

2 Simplify using the exponent laws:

a $(x^4)^t \times x^5$ **b** $\dfrac{k^2 \times k^m}{k^n}$ **c** $\dfrac{c^5}{(c^x)^4}$ **d** $\dfrac{m^2}{m^3 \times m^t}$

Example 2 — Self Tutor

Write as a power with a prime number base:

a 4×2^p **b** $\dfrac{3^x}{9^y}$ **c** 25^{x-1}

a $4 \times 2^p = 2^2 \times 2^p = 2^{2+p}$

b $\dfrac{3^x}{9^y} = \dfrac{3^x}{(3^2)^y} = \dfrac{3^x}{3^{2y}} = 3^{x-2y}$

c $25^{x-1} = (5^2)^{x-1} = 5^{2(x-1)} = 5^{2x-2}$

3 Write as a power with a prime number base:

a 121 **b** 32 **c** 81 **d** 4^3 **e** 25^2 **f** 9^4

4 Write as a power with a prime number base:

a $2^x \times 2$ **b** $7^t \times 49$ **c** $3^a \div 9$ **d** $8^p \div 4$

e $\dfrac{7^n}{7^{n-2}}$ **f** $\dfrac{9}{3^x}$ **g** $(25^t)^2$ **h** $16^{k-3} \times 2^{-k}$

i $\dfrac{4^a}{2^b}$ **j** $\dfrac{8^x}{16^y}$ **k** $\dfrac{125^x}{5^x}$ **l** $\dfrac{27^{a+2}}{3^a \times 9^a}$

Example 3 — Self Tutor

Expand the brackets then simplify:

a $(2x)^3$ **b** $\left(\dfrac{3c}{b^2}\right)^4$

a $(2x)^3$
$= 2^3 \times x^3$
$= 8x^3$

b $\left(\dfrac{3c}{b^2}\right)^4$
$= \dfrac{3^4 \times c^4}{(b^2)^4}$
$= \dfrac{81c^4}{b^8}$

5 Expand the brackets then simplify:

a $(xy)^2$ **b** $(ab)^3$ **c** $(xyz)^4$ **d** $(3b)^3$
e $(5a)^4$ **f** $(10xy)^5$ **g** $(7p^2q)^2$ **h** $(3a^5b^2)^4$

6 Expand the brackets then simplify:

a $\left(\dfrac{p}{q}\right)^2$ **b** $\left(\dfrac{m}{n}\right)^3$ **c** $\left(\dfrac{x}{3}\right)^4$ **d** $\left(\dfrac{2a}{b}\right)^4$
e $\left(\dfrac{p}{5q}\right)^2$ **f** $\left(\dfrac{3x}{4y}\right)^3$ **g** $\left(\dfrac{m^3}{5n}\right)^2$ **h** $\left(\dfrac{4x}{y^3}\right)^3$

7 Simplify using one or more of the exponent laws:

a $4b^2 \times 2b^3$ **b** $\dfrac{a^6b^3}{a^4b}$ **c** $3ab^2 \times 2a^3$ **d** $(ab)^2 \times 2b^3$
e $(3q)^2 \times (pq)^3$ **f** $\dfrac{5x^3y}{15xy}$ **g** $\left(\dfrac{a^2}{5b}\right)^3$ **h** $\dfrac{24t^2r^4}{15t^2r^2}$
i $\dfrac{(x^2y)^2}{3x^3y}$ **j** $\dfrac{10k^7}{(2k)^5}$ **k** $\dfrac{c^2d}{(2cd^2)^2}$ **l** $\dfrac{(5pq)^3}{(p^2q)^2}$

Example 4 — Self Tutor

Write as a fraction in lowest terms:

a $3^0 - 3^{-1}$ **b** 3^{-2} **c** $\left(\frac{5}{3}\right)^{-2}$

a $3^0 - 3^{-1}$
$= 1 - \frac{1}{3}$
$= \frac{2}{3}$

b 3^{-2}
$= \dfrac{1}{3^2}$
$= \frac{1}{9}$

c $\left(\frac{5}{3}\right)^{-2}$
$= \left(\frac{3}{5}\right)^2$
$= \frac{9}{25}$

8 Evaluate:

a 7^0 **b** $\left(\frac{1}{2}\right)^0$ **c** $\dfrac{5^4}{5^4}$ **d** $2^0 + 2^1$
e $2t^0$ **f** $(2t)^0$ **g** 3×4^0 **h** $5^0 \div 5^3$

9 Write as a fraction in lowest terms:

a 6^{-1} **b** 4^{-1} **c** 13^{-1} **d** 4^{-2}

e 5^{-3} **f** 7^{-2} **g** 10^{-3} **h** 3^{-4}

i $5^0 - 5^{-1}$ **j** $3^0 + 3^1 - 3^{-1}$ **k** $2^{-3} + 2^{-4}$ **l** $12^{-1} - 12^{-2}$

10 Write as an integer or fraction:

a 10^0 **b** 10^{-1} **c** 10^{-2} **d** 10^{-3} **e** 10^{-4} **f** 10^{-5}

11 Write as a power with a negative exponent:

a $\frac{5^3}{5^5}$ **b** $\frac{2^6}{2^{10}}$ **c** $\frac{x^4}{x^9}$ **d** $\frac{y^3}{(y^5)^2}$

12 Write as a fraction in lowest terms:

a $\left(\frac{3}{8}\right)^{-1}$ **b** $\left(\frac{2}{3}\right)^{-1}$ **c** $\left(\frac{1}{5}\right)^{-1}$ **d** $\left(\frac{1}{3}\right)^{-2}$

e $\left(\frac{2}{3}\right)^{-3}$ **f** $\left(\frac{4}{5}\right)^{-2}$ **g** $\left(1\frac{1}{2}\right)^{-3}$ **h** $\left(3\frac{1}{2}\right)^{-2}$

13 Write without brackets or negative exponents:

a $(5x)^{-1}$ **b** $5x^{-1}$ **c** $7a^{-1}$ **d** $(7a)^{-1}$

e $\left(\frac{1}{t}\right)^{-2}$ **f** $\left(\frac{3x}{y}\right)^{-1}$ **g** $(5t)^{-2}$ **h** $(5t^{-2})^{-1}$

i xy^{-1} **j** $(xy)^{-1}$ **k** xy^{-3} **l** $(xy)^{-3}$

m $(3pq)^{-1}$ **n** $3(pq)^{-1}$ **o** $3pq^{-1}$ **p** $\frac{(xy)^3}{y^{-2}}$

q $(5x^{-2}y)^3$ **r** $\left(\frac{c}{2d^3}\right)^{-2}$ **s** $\left(\frac{3r^{-3}}{t}\right)^{-1}$ **t** $\left(\frac{2p}{5q^{-2}}\right)^{-2}$

14 Use the exponent laws to show that, for positive a and b and integer n:

a $\frac{1}{a^{-n}} = a^n$ **b** $\left(\frac{a}{b}\right)^{-n} = \frac{b^n}{a^n}$

15 The units for speed *kilometres per hour* or km/h can be written in exponent form as km h^{-1}. Write these units in exponent form:

a m/s **b** cubic metres/hour

c square centimetres per second **d** cubic centimetres per minute

e grams per second **f** kilogram metres per second

g metres per second per second.

16 Find the smaller of 2^{125} and 3^{75} without using a calculator. **Hint:** $2^{125} = (2^5)^{25}$

17 Order these numbers from smallest to largest: 2^{90}, 3^{60}, 5^{36}, 10^{24}.

GAME

This game can be played by one or two players.

Click on the icon to obtain the printable board and instructions.

GAME

INVESTIGATION RATIONAL EXPONENTS

This Investigation will help you discover the meaning of a number with a rational exponent.

What to do:

1 Notice that $5^{\frac{1}{2}} \times 5^{\frac{1}{2}} = 5^{\frac{1}{2}+\frac{1}{2}} = 5^1 = 5$

and that $\sqrt{5} \times \sqrt{5} = 5$ also.

a Copy and complete:

i $3^{\frac{1}{2}} \times 3^{\frac{1}{2}} = \ldots\ldots = \ldots\ldots = \ldots\ldots$

$\sqrt{3} \times \sqrt{3} = \ldots\ldots$

ii $13^{\frac{1}{2}} \times 13^{\frac{1}{2}} = \ldots\ldots = \ldots\ldots = \ldots\ldots$

$\sqrt{13} \times \sqrt{13} = \ldots\ldots$

iii $a^{\frac{1}{2}} \times a^{\frac{1}{2}} = \ldots\ldots = \ldots\ldots = \ldots\ldots$

$\sqrt{a} \times \sqrt{a} = \ldots\ldots$

b Copy and complete: $a^{\frac{1}{2}} = \ldots\ldots$

2 Notice that $7^{\frac{1}{3}} \times 7^{\frac{1}{3}} \times 7^{\frac{1}{3}} = 7^{\frac{1}{3}+\frac{1}{3}+\frac{1}{3}} = 7^1 = 7$

and that $\sqrt[3]{7} \times \sqrt[3]{7} \times \sqrt[3]{7} = 7$ also.

a Copy and complete:

i $8^{\frac{1}{3}} \times 8^{\frac{1}{3}} \times 8^{\frac{1}{3}} = \ldots\ldots = \ldots\ldots = \ldots\ldots$

$\sqrt[3]{8} \times \sqrt[3]{8} \times \sqrt[3]{8} = \ldots\ldots$

ii $27^{\frac{1}{3}} \times 27^{\frac{1}{3}} \times 27^{\frac{1}{3}} = \ldots\ldots = \ldots\ldots = \ldots\ldots$

$\sqrt[3]{27} \times \sqrt[3]{27} \times \sqrt[3]{27} = \ldots\ldots$

iii $a^{\frac{1}{3}} \times a^{\frac{1}{3}} \times a^{\frac{1}{3}} = \ldots\ldots = \ldots\ldots = \ldots\ldots$

$\sqrt[3]{a} \times \sqrt[3]{a} \times \sqrt[3]{a} = \ldots\ldots$

b Copy and complete: $a^{\frac{1}{3}} = \ldots\ldots$

3 Suggest a rule for the general case: $a^{\frac{1}{n}} = \ldots\ldots$

4 **a** Copy and complete by applying the rule $(a^m)^n = a^{mn}$:

i $(8^2)^{\frac{1}{3}} = 8^{\cdots}$ **ii** $\left(8^{\frac{1}{3}}\right)^2 = 8^{\cdots}$

b Copy and complete by applying the rule you found in **3**.

i $(8^2)^{\frac{1}{3}} = \sqrt[\cdots]{8^2}$ **ii** $\left(8^{\frac{1}{3}}\right)^2 = \left(\sqrt[\cdots]{8}\right)^2$

c Hence copy and complete: $8^{\frac{2}{3}} = \sqrt[\cdots]{8^{\cdots}} = \left(\sqrt[\cdots]{8}\right)^{\cdots}$

5 **a** Use the exponent law $(a^m)^n = a^{mn}$ to simplify:

i $(x^p)^{\frac{1}{q}}$ **ii** $\left(x^{\frac{1}{q}}\right)^p$

b Hence explain why $x^{\frac{p}{q}} = (\sqrt[q]{x})^p = \sqrt[q]{x^p}$.

B STANDARD FORM (SCIENTIFIC NOTATION)

We can use **standard form** or **scientific notation** to write very large and very small numbers in a way that is easier to understand.

Standard form involves writing any positive number as *a number* between 1 inclusive and 10, multiplied by an *integer power of* 10.

The result has the form $a \times 10^k$ where $1 \leqslant a < 10$ and k is an integer.

For example:

- 68 000 000 is written in standard form as 6.8×10^7
- 0.000 471 is written in standard form as 4.71×10^{-4}.

When a number is written in standard form:

- its significant figures are the same and in the same order
- the decimal point is immediately after the first significant figure
- 10^k is the *place value* of the first significant figure in the original number.
  - If the original number is greater than or equal to 10, then k is positive.
  - If the original number is between 1 inclusive and 10, then $k = 0$.
  - If the original number is less than 1, then k is negative.

Example 5 **Self Tutor**

Write in scientific notation:

a 23 600 000 **b** 0.000 023 6

a 23 600 000
$= 2.36 \times 10^7$

b 0.000 023 6
$= 2.36 \div 10^5$
$= 2.36 \times 10^{-5}$

Example 6 **Self Tutor**

Write as an ordinary number:

a 2.57×10^4 **b** 7.853×10^{-3}

a 2.57×10^4
$= 2.5700 \times 10\,000$
$= 25\,700$

b 7.853×10^{-3}
$= 0007.853 \div 10^3$
$= 0.007\,853$

EXERCISE 1B

1 Write in scientific notation:

a 230 **b** 53 900 **c** 0.0361 **d** 0.006 80
e 3.26 **f** 0.5821 **g** 361 000 000 **h** 0.000 001 674

2 Write each quantity in standard form:

a There are approximately 4 million red blood cells in a drop of blood.

b The thickness of a coin is about 0.0008 m.

c A Rubik's Cube has approximately 43 252 000 000 000 000 000 possible arrangements.

d The probability of winning the major prize in a lottery is approximately 0.000 000 003 422.

3 Write as an ordinary number:

a 2.3×10^3 **b** 2.3×10^{-2} **c** 5.64×10^5 **d** 7.931×10^{-4}
e 9.97×10^0 **f** 6.04×10^7 **g** 4.215×10^{-1} **h** 3.621×10^{-8}

4 Write each quantity as an ordinary number:

a The Amazon River is approximately 6.99×10^6 m long.

b A piece of paper is about 1.8×10^{-2} cm thick.

c A test tube holds 3.2×10^7 bacteria.

d A hayfever tablet contains 9.5×10^{-3} g of antihistamine.

Example 7 Self Tutor

Evaluate, giving your answer in scientific notation:

a $(3 \times 10^4) \times (8 \times 10^3)$ **b** $\dfrac{2 \times 10^{-3}}{5 \times 10^{-8}}$ **c** $(8.7 \times 10^8) - (3 \times 10^7)$

a
$$\begin{aligned}&(3 \times 10^4) \times (8 \times 10^3)\\ &= 24 \times 10^{4+3}\\ &= (2.4 \times 10^1) \times 10^7\\ &= 2.4 \times 10^8\end{aligned}$$

b
$$\begin{aligned}&\frac{2 \times 10^{-3}}{5 \times 10^{-8}}\\ &= \tfrac{2}{5} \times 10^{-3-(-8)}\\ &= 0.4 \times 10^5\\ &= (4 \times 10^{-1}) \times 10^5\\ &= 4 \times 10^4\end{aligned}$$

c
$$\begin{aligned}&(8.7 \times 10^8) - (3 \times 10^7)\\ &= 8.7 \times 10^8 - 0.3 \times 10^8\\ &= 8.4 \times 10^8\end{aligned}$$

5 Evaluate, giving your answer in scientific notation:

a $(3 \times 10^3) \times (2 \times 10^7)$ **b** $(4 \times 10^3) \times (7 \times 10^5)$
c $(8 \times 10^{-4}) \times (7 \times 10^{-5})$ **d** $(9 \times 10^{-5}) \times (6 \times 10^{-2})$
e $(3 \times 10^5)^2$ **f** $(4 \times 10^7)^2$
g $(2 \times 10^{-3})^4$ **h** $(5 \times 10^{-3})^3$
i $(6 \times 10^{-1}) \times (4 \times 10^3) \times (5 \times 10^{-4})$ **j** $(6 \times 10^{-3})^2 \times (8 \times 10^{11})$
k $(4 \times 10^3)^{-1}$ **l** $(5 \times 10^{-4})^{-2}$

6 Evaluate, giving your answer in standard form:

a $\dfrac{8\times10^6}{4\times10^3}$ **b** $\dfrac{9\times10^{-3}}{3\times10^{-1}}$ **c** $\dfrac{4\times10^6}{2\times10^{-2}}$

d $\dfrac{3\times10^4}{5\times10^9}$ **e** $\dfrac{2.5\times10^{-4}}{(5\times10^7)^2}$ **f** $\dfrac{(8\times10^{-2})^2}{2\times10^{-6}}$

7 **a** How many times larger is 3×10^{11} than 3×10^5?

b How many times smaller is 5×10^{-21} than 5×10^{-16}?

8 Evaluate, giving your answer in standard form:

To add or subtract numbers in standard form, write the numbers with the same power of 10.

a $(3\times10^8)+(6\times10^8)$

b $(6.4\times10^5)-(5.9\times10^5)$

c $(2\times10^5)+(7\times10^4)$

d $(3.6\times10^{-4})+(1.7\times10^{-5})$

e $(5.81\times10^4)-(1.4\times10^3)$

f $(7\times10^6)+(2.3\times10^8)$

g $(3.54\times10^{-7})+(8.5\times10^{-9})$

h $(1.05\times10^{-5})-(8.2\times10^{-6})$

Example 8 **Self Tutor**

Use your calculator to find:

a $(2.58\times10^7)\times(1.5\times10^6)$ **b** $\dfrac{6.5\times10^{-2}}{1.04\times10^5}$

a $(2.58\times10^7)\times(1.5\times10^6)=3.87\times10^{13}$

b $\dfrac{6.5\times10^{-2}}{1.04\times10^5}=6.25\times10^{-7}$

GRAPHICS CALCULATOR INSTRUCTIONS

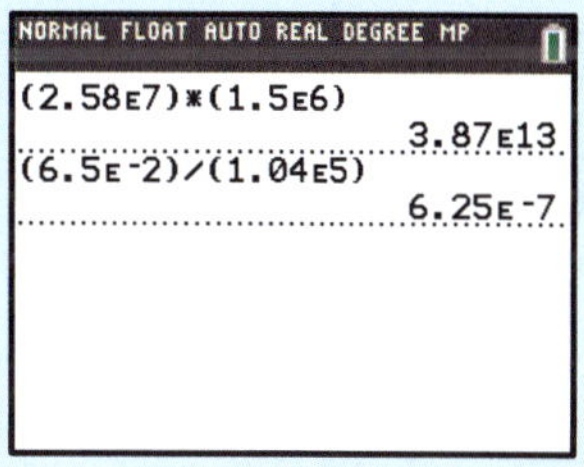

9 Evaluate the following, giving each answer in standard form rounded to 3 significant figures.

a $(4.7\times10^5)\times(8.53\times10^7)$ **b** $(2.7\times10^{-3})\times(9.6\times10^{14})$ **c** $\dfrac{3.4\times10^7}{4.8\times10^{15}}$

d $\dfrac{7.3\times10^{-7}}{1.5\times10^4}$ **e** $(2.83\times10^3)^2$ **f** $(5.96\times10^{-5})^2$

g $\dfrac{(3.56\times10^4)^2}{8.05\times10^{-5}}$ **h** $(1.8\times10^3)^2+(9.85\times10^5)$ **i** $\dfrac{2.9\times10^2}{(7.62\times10^7)^3}$

10 A rocket travels at 4×10^5 km h^{-1}. Assuming 1 year $\approx$ 365.25 days, how far will it travel in:

a 30 days **b** 20 years?

11 A bullet travelling at 2×10^3 km h^{-1} hits a target 500 m away. Find the time of the bullet's flight, in seconds.

12 Mars has volume 1.63×10^{20} m^3. Pluto has volume 7.03×10^{18} m^3. How many times larger is Mars than Pluto?

13 Microbe C has mass 2.63×10^{-5} grams. Microbe D has mass 8×10^{-7} grams.

a Which microbe is heavier?

b How many times heavier is it, than the other microbe?

14 The table shows the land areas of the Canadian provinces (shaded yellow) and territories (shaded green).

a Find the total land area of Canada.

b Place the *provinces* in order from largest to smallest.

c How many times larger is:

i Quebec than Manitoba

ii Nunavut than Prince Edward Island?

d What percentage of the land area of Canada is Nova Scotia?

	Land area (km^2)
Ontario	9.2×10^5
Quebec	1.4×10^6
Nova Scotia	5.3×10^4
New Brunswick	7.1×10^4
Manitoba	5.5×10^5
British Columbia	9.3×10^5
Prince Edward Island	5.7×10^3
Saskatchewan	5.9×10^5
Alberta	6.4×10^5
Newfoundland and Labrador	3.7×10^5
Northwest Territories	1.2×10^6
Yukon	4.7×10^5
Nunavut	1.9×10^6

15 Answer the **Opening Problem** on page **12**.

HISTORICAL NOTE

The ancient Indians explored the concept of expressing very large and very small numbers.

In the *Lalitavistara Sutra*, a Sanskrit text dating from around the 4th century, it is written that the Buddha described the size of an atom in terms of the length of a finger bone as follows:

".... each was the length of seven grains of barley,
each of which was the length of seven mustard seeds,
each of which was the length of seven poppy seeds,
each of which was the length of seven particles of dust stirred up by a cow,
each of which was the length of seven specks of dust disturbed by a ram,
each of which was the length of seven specks of dust stirred up by a hare,
each of which was the length of seven specks of dust carried away by the wind,
each of which was the length of seven tiny specks of dust,
each of which was the length of seven minute specks of dust,
each of which was the length of seven particles of the first atoms."

What to do:

1 Assuming a finger bone is 4 cm long, use the Buddha's description to estimate the length of an atom. Write your answer in metres using standard form.

2 Research the size of a carbon atom. How accurate is the estimate in **1**?

QUICK QUIZ

MULTIPLE CHOICE QUIZ

REVIEW SET 1A

1 Simplify using the exponent laws:

a $k^5 \times k^{3m}$ **b** $\dfrac{p^6}{p}$ **c** $(m^6)^{8t}$

2 Write $\dfrac{49^k}{7^{k-1}}$ as a power with a prime number base.

3 Expand the brackets then simplify:

a $(3w)^2$ **b** $(2x^2y)^3$ **c** $\left(\frac{a}{b}\right)^6$ **d** $\left(\frac{1}{5n}\right)^3$

4 Evaluate:

a 6×6^0 **b** 7^{-1} **c** $2^0 \div 2^4$ **d** $10^0 - 10^{-2}$

5 Write without brackets or negative exponents:

a $(mn)^{-2}$ **b** mn^{-2} **c** $\left(\dfrac{x}{5y^2}\right)^{-3}$

6 Write the unit "square metres per second" in exponent form.

7 Write in standard form:

a 59 000 **b** 0.009 **c** 6 085 000 **d** 0.000 007 71

8 Evaluate, giving your answer in standard form:

a $(6 \times 10^5) \times (3 \times 10^6)$ **b** $\dfrac{9 \times 10^{-5}}{6 \times 10^3}$ **c** $(8 \times 10^9) \times (5 \times 10^{-4})$

d $(7 \times 10^7) + (9 \times 10^7)$ **e** $(8 \times 10^7)^{-1}$ **f** $(2 \times 10^{-4}) - (8 \times 10^{-5})$

9 A probiotic drink contains 6.5×10^9 bacteria, each with mass 4×10^{-13} g. Find the total mass of bacteria in the drink, writing your answer in scientific notation.

10 The speed of light is about 3.00×10^8 m/s. Answer the following in standard form:

a Assuming 1 year $\approx$ 365.25 days, determine how far light travels in:

i 1 hour **ii** 1 day **iii** 1 year.

b How long does it take for light to travel 1 m?

c Sound travels at 343.2 m/s. How many times faster is light than sound?

REVIEW SET 1B

1 Simplify using the exponent laws:

a $\dfrac{x^a}{x^7}$ **b** $(y^{5n})^3$ **c** $\dfrac{m^{15}}{(m^p)^q}$

2 Simplify using one or more of the exponent laws:

a $\dfrac{x^3 \times x^4}{x^n}$ **b** $2t^3 \times 4t^2$ **c** $\dfrac{2a^4b^5}{10a^3b}$

3 Write as a fraction in lowest terms:

a $\left(\frac{4}{3}\right)^{-1}$ **b** $11^0 - 11^{-1}$ **c** $\left(1\frac{3}{4}\right)^{-2}$ **d** $3^{-2} + 3^{-3}$

4 Find the smaller of 2^{60} and 7^{20} without using a calculator.

5 Write $\dfrac{z^2}{z^9}$ as a power with a negative exponent.

6 Write without brackets or negative exponents: **a** $\left(\dfrac{a}{b}\right)^{-2}$ **b** $\left(\dfrac{3a^{-1}}{2b^2}\right)^{-3}$

7 Write as an ordinary number:

a 6.23×10^5 **b** 3.008×10^{-4} **c** 4.597×10^0 **d** 5.06×10^{-5}

8 Evaluate, giving your answer in standard form:

a $(7 \times 10^5) \times (3 \times 10^9)$ **b** $\dfrac{8 \times 10^7}{2 \times 10^{-3}}$ **c** $(4 \times 10^{10}) + (2.5 \times 10^{11})$

9 The human body creates 2.4×10^6 red blood cells per second, and each red blood cell has a life span of approximately 120 days. Estimate, in standard form, the number of red blood cells in an adult at any given time.

10 The table alongside shows the diameters of the planets in the solar system.

a Write, in scientific notation:

i the diameter of Earth in kilometres

ii the radius of Venus.

b Write the planets in order of size, from smallest to largest.

c How much larger is the diameter of Saturn than that of Neptune? Write your answer in scientific notation.

d How many times larger is the diameter of:

i Uranus than Mercury **ii** Jupiter than Mars?

Planet	*Diameter*
Mercury	4.88×10^6 m
Venus	1.21×10^7 m
Earth	1.27×10^7 m
Mars	6.79×10^6 m
Jupiter	1.40×10^8 m
Saturn	1.21×10^8 m
Uranus	5.11×10^7 m
Neptune	4.95×10^7 m

Chapter 2

Algebra: Expansion

Contents:

OPENING PROBLEM

Look at the diagram alongside.

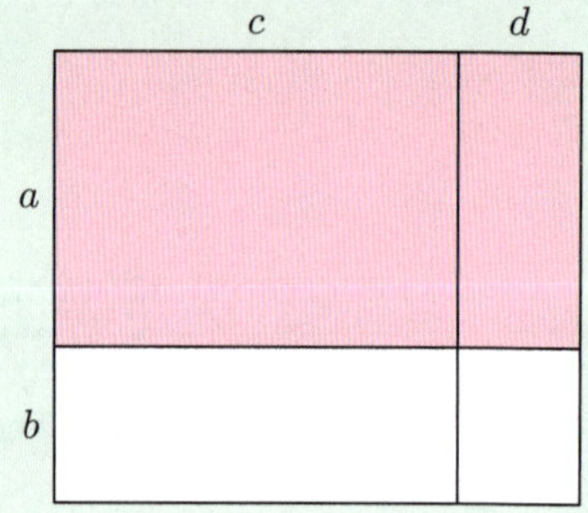

Things to think about:

a Can you write the area of the pink rectangle as:

  i a *sum* of two terms

  ii a *product* of two factors?

b Are the two expressions you wrote in **a** *always* equal?

c Can you use the diagram to explain why $(a+b)(c+d) = ac+ad+bc+bd$?

In mathematics we often want to change the *form* of a mathematical expression.

- If we want to remove brackets so that an expression is written as a sum of terms, we use a process called **expansion**.
- If we want to write an expression as a product of irreducible factors, we use a process called **factorisation**.

Factorisation is the *reverse* of expansion.

A THE DISTRIBUTIVE LAW

In previous years you should have seen the **distributive law** which allows us to multiply a sum of terms by a coefficient.

When a coefficient is multiplied by an expression in brackets, we remove the brackets by multiplying each term inside them by the coefficient, then adding the results.

$$\boldsymbol{a(b+c) = ab+ac}$$

Example 1 **Self Tutor**

Expand:

a $2(3x-1)$ **b** $-3x(x+2)$

a $2(3x-1)$
$= 2\times 3x + 2\times -1$
$= 6x-2$

b $-3x(x+2)$
$= -3x\times x + -3x\times 2$
$= -3x^2-6x$

EXERCISE 2A

1 Expand using the distributive law:

a $2(x+6)$	**b** $7(x-3)$	**c** $4(x-y)$	**d** $3(2a+b)$
e $6(x-3y)$	**f** $x(x+4)$	**g** $3(2x+5)$	**h** $x(3x-2)$
i $4x(x-3)$	**j** $7n(4+2n)$	**k** $2x(x^2-1)$	**l** $x^2(x-3)$

2 Expand using the distributive law:

a $-2(a+4)$ **b** $-5(3-x)$ **c** $-(y+2)$ **d** $-3(x-y)$

e $-2(3+x)$ **f** $-x(x+1)$ **g** $-(4-b^2)$ **h** $-3x(x+y)$

i $-x(1+x)$ **j** $-3x(2x+7)$ **k** $-5n(2-n)$ **l** $-ab(b-a)$

3 Expand using the distributive law:

a $2(3a-5b+1)$ **b** $3x(x-2y+3)$ **c** $-2(a-3b+5c)$

d $6x(x^2+5x-10)$ **e** $-3(2c^2-3c-7)$ **f** $4n(n^2-2n-8)$

Example 2 **Self Tutor**

Expand and simplify: $y(3y-1)-3y(2y-5)$

$$\begin{aligned}&y(3y-1)-3y(2y-5)\\&= y\times 3y + y\times -1 + -3y\times 2y + -3y\times -5\\&= 3y^2 - y - 6y^2 + 15y\\&= 14y - 3y^2\end{aligned}$$

4 Expand then simplify by collecting like terms:

a $2+3(x+1)$ **b** $15-2(x+5)$ **c** $4(x-1)+7$

d $6(x-2)-8x$ **e** $a(a-2)+2a$ **f** $3m(4-m)+m^2$

g $3a(b-a)+4a^2$ **h** $5b-2b(b-2)$ **i** $6a^2-3a(a+4)$

j $4b+(a-b)$ **k** $4b-(a-b)$ **l** $7n-5(3-2n)$

5 Expand then simplify by collecting like terms:

a $2(x-3)+3(x+4)$ **b** $3(x+2)+5(4-x)$

c $6(m-2)-3(2m+1)$ **d** $5(y-x)+6(x-y)$

e $a(a+2)+5(a-3)$ **f** $x(x+5)-3(x-4)$

g $x(x+y)-y(x+y)$ **h** $-3(x-6)-(2-x)$

i $4(3x-2)-(3x+1)$ **j** $2x(x-5)-3x(2-x)$

k $x(x^2+2x)-x^2(2+x)$ **l** $x^2(6-x)+3x(x-4)$

B THE PRODUCT $(a+b)(c+d)$

In the **Opening Problem** we considered the expansion of the product $(a+b)(c+d)$.

We can perform this expansion by applying the distributive law twice:

$$\begin{aligned}&(a+b)(c+d)\\&=(a+b)c+(a+b)d && \{\text{distributive law}\}\\&=c(a+b)+d(a+b)\\&=ac+bc+ad+bd && \{\text{distributive law}\}\\&=ac+ad+bc+bd && \{\text{rearranging terms}\}\end{aligned}$$

The expansion of $(a+b)(c+d)$ is therefore:

$$(a+b)(c+d) = ac + ad + bc + bd$$

The final result contains four terms:

ac is the product of the **F**irst terms of each bracket.
ad is the product of the **O**uter terms of each bracket.
bc is the product of the **I**nner terms of each bracket.
bd is the product of the **L**ast terms of each bracket.

Example 3 — Self Tutor

Expand then simplify by collecting like terms:

a $(x+4)(x-3)$ **b** $(2x-5)(-x+3)$

a $(x+4)(x-3)$
$= x \times x + x \times -3 + 4 \times x + 4 \times -3$
$= x^2 - 3x + 4x - 12$
$= x^2 + x - 12$

b $(2x-5)(-x+3)$
$= 2x \times -x + 2x \times 3 + -5 \times -x + -5 \times 3$
$= -2x^2 + 6x + 5x - 15$
$= -2x^2 + 11x - 15$

EXERCISE 2B

1 Expand then simplify by collecting like terms:

a $(x+2)(x+5)$ **b** $(x+3)(x-4)$ **c** $(x+5)(x-3)$
d $(x-2)(x+4)$ **e** $(x-3)(x+1)$ **f** $(x-2)(x-10)$
g $(1-x)(x-2)$ **h** $(2x+1)(x-3)$ **i** $(3x-4)(2x-5)$

2 Expand then simplify by collecting like terms:

a $(2x+y)(x-y)$ **b** $(x+3)(-2x-1)$ **c** $(x+2y)(-x-1)$
d $(6-x)(2x+3)$ **e** $(4-x)(5-x^2)$ **f** $(1-3x)(2x+1)$
g $(-2-x)(4+5x)$ **h** $(x^2+1)(3-2x)$ **i** $(3+4x)(2-3x)$

3 Expand using $(a+b)(c+d) = ac + ad + bc + bd$, then simplify by collecting like terms:

a $(a+b)(a-b)$ **b** $(a+b)^2$ **c** $(a-b)^2$

4 Expand then simplify:

a $x(x-1)(x+3)$ **b** $-x(x+2)(x-5)$ **c** $3x(2-x)(x+2)$

5 Expand then simplify:

a $(x+3)(x-1)+3(x-5)$

b $(x+7)(x-5)+(x+1)(x+4)$

c $(2x+3)(x-2)-(x+1)(x+6)$

d $(4t-3)(t+1)-(2t-1)(2t+5)$

e $(4x-1)(3-x)+(2x-3)(3x-2)$

f $5(3x-4)(x+2)-(7-x)(8-5x)$

C THE DIFFERENCE BETWEEN TWO SQUARES

In the previous Exercise you should have found the **difference between two squares** expansion:

$$(a+b)(a-b)=a^2-b^2$$

Proof: $(a+b)(a-b)=a^2-\cancel{ab}+\cancel{ab}-b^2$ {FOIL rule}
$=a^2-b^2$

Example 4 Self Tutor

Expand then simplify:

a $(x+4)(x-4)$

b $(3x-2)(3x+2)$

a $(x+4)(x-4)$
$=x^2-4^2$
$=x^2-16$

b $(3x-2)(3x+2)$
$=(3x)^2-2^2$
$=9x^2-4$

EXERCISE 2C

1 Expand then simplify:

a $(x+1)(x-1)$

b $(b+2)(b-2)$

c $(t-6)(t+6)$

d $(x+7)(x\quad 7)$

e $(3+a)(3-a)$

f $(5-x)(5+x)$

g $(x+y)(x-y)$

h $(4+x)(x-4)$

i $(x^2-1)(x^2+1)$

j $(2x+1)(2x-1)$

k $(4-3y)(4+3y)$

l $(3x-4z)(4z+3x)$

2 Expand then simplify:

a $(x+\sqrt{3})(x-\sqrt{3})$

b $(x-\sqrt{2})(x+\sqrt{2})$

c $(\sqrt{7}+x)(\sqrt{7}-x)$

3 By writing two different expressions for the shaded area, show that $(a+b)(a-b)=a^2-b^2$.

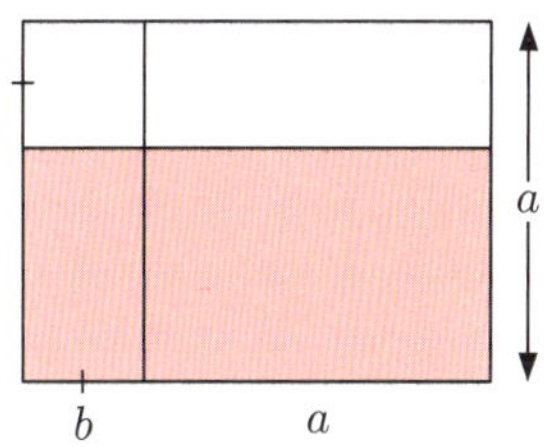

4 Expand then simplify:

a $(x+3)(x-3)-(x+6)(x-6)$

b $(5p-2)(5p+2)-p(3p-1)$

c $(3y-z)(3y+z)-(2y+3z)(2y-3z)$

d $(10-x^2)(10+x^2)-(10-3x^2)(10+3x^2)$

D THE PERFECT SQUARES EXPANSION

In **Exercise 2B** you should have found the **perfect squares expansion**:

$$(a+b)^2 = a^2 + 2ab + b^2$$

Example 5 Self Tutor

Expand then simplify:

a $(2x+1)^2$ **b** $(3-4y)^2$

a $(2x+1)^2$
$= (2x)^2 + 2 \times 2x \times 1 + 1^2$
$= 4x^2 + 4x + 1$

b $(3-4y)^2$
$= 3^2 + 2 \times 3 \times (-4y) + (-4y)^2$
$= 9 - 24y + 16y^2$

EXERCISE 2D

1 Expand then simplify:

a $(x+5)^2$ **b** $(y+6)^2$ **c** $(2+x)^2$
d $(3x+1)^2$ **e** $(4+3x)^2$ **f** $(x+2y)^2$

2 Expand then simplify:

a $(x-4)^2$ **b** $(2-x)^2$ **c** $(a-7)^2$
d $(3x-1)^2$ **e** $(6-5p)^2$ **f** $(2x-3y)^2$

3 Expand then simplify:

a $(2x+3)^2$ **b** $(4a-1)^2$ **c** $(3y+5)^2$
d $(3a-4)^2$ **e** $(2x-7)^2$ **f** $(8+3a)^2$
g $(2+5b)^2$ **h** $(6+5x)^2$ **i** $(4-5y)^2$
j $(5x+3y)^2$ **k** $(2x-9y)^2$ **l** $(1-xy)^2$

4 Use the diagram alongside to show that $(a-b)^2 = a^2 - 2ab + b^2$.

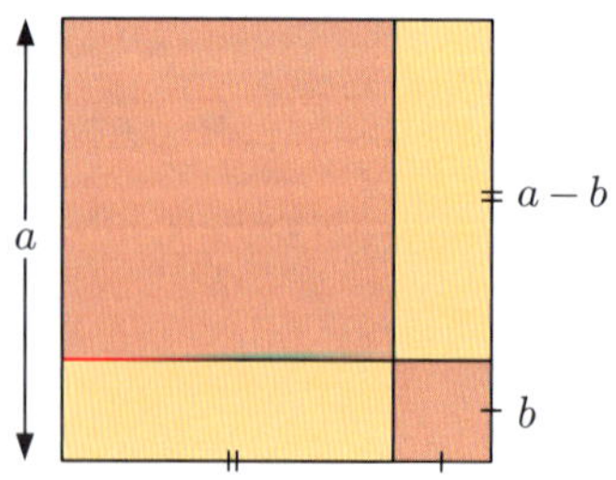

5 Expand then simplify:

a $(x^2+3)^2$ **b** $(y^2-2)^2$ **c** $(2a^2+3)^2$
d $(1-4x^2)^2$ **e** $(a^2+b^2)^2$ **f** $(a^2-b^2)^2$
g $(5x^2+3y)^2$ **h** $(5-x^2)^2$ **i** $(4x-3x^2)^2$
j $(2x^2+4x)^2$ **k** $(7a-3a^2)^2$ **l** $(x^3+8x)^2$

6 Expand then simplify:

a $3x - 2 + (x+5)^2$
b $(x+9)^2 + (x-2)^2$
c $(x+4)(x-4) - (x+1)^2$
d $(1-3x)^2 + (x-2)(x+3)$
e $(3x+1)^2 - (2x-3)^2$
f $(x+8)^2 - (x+2)(x-5)$
g $(5x+2)(5x-2) - (x+2)^2$
h $(3x^2-1)^2 - 4(1-x)^2$

E FURTHER EXPANSION

When expressions containing more than two terms are multiplied together, we can still use the distributive law to expand the brackets. Each term in the first set of brackets is multiplied by each term in the second set of brackets.

So, if there are 2 terms in the first set of brackets and 3 terms in the second set of brackets, there will be $2 \times 3 = 6$ terms in the expansion. However, when we simplify by collecting like terms, the final answer may contain fewer terms.

Example 6 Self Tutor

Expand then simplify: $(2x-1)(x^2-4x+2)$

$(2x-1)(x^2-4x+2)$

$= 2x^3 - 8x^2 + 4x$ {$2x$ × all terms in 2nd bracket}

$\quad - x^2 + 4x - 2$ {-1 × all terms in 2nd bracket}

$= 2x^3 - 9x^2 + 8x - 2$ {collecting like terms}

EXERCISE 2E

1 Expand then simplify:

a $(x+2)(x^2+x+4)$
b $(x-3)(x^2+2x-3)$
c $(x+3)(x^2+2x+1)$
d $(x+1)(2x^2-x-5)$
e $(2x+3)(x^2+2x+1)$
f $(2x-5)(x^2-2x-3)$
g $(x+5)(3x^2-x+4)$
h $(4x-1)(2x^2-3x+1)$
i $(3x^2+4x-1)(3-2x)$
j $(2-5x)(x^2+4x-2)$

Example 7 Self Tutor

Expand then simplify: $(x+1)(x-3)(x+2)$

$(x+1)(x-3)(x+2)$

$= (x^2 - 3x + x - 3)(x+2)$ {expanding first two factors}

$= (x^2 - 2x - 3)(x+2)$ {collecting like terms}

$= x^3 + 2x^2 - 2x^2 - 4x - 3x - 6$ {expanding remaining factors}

$= x^3 - 7x - 6$ {collecting like terms}

2 Expand then simplify:

- **a** $(x+1)(x+2)(x+3)$
- **b** $(x+4)(x+3)(x+2)$
- **c** $(x-3)(x-2)(x+4)$
- **d** $(x-3)(x-2)(x-5)$
- **e** $(2x-3)(x+3)(x-1)$
- **f** $(x-2)(4-x)(3x+2)$
- **g** $(x+3)^2(2x-1)$
- **h** $(3x-2)^2(2-5x)$
- **i** $(x+2)^3$
- **j** $(x-3)^3$

3 State the number of terms you would obtain by expanding:

- **a** $(a+b)(c+d)$
- **b** $(a+b+c)(d+e)$
- **c** $(a+b)(c+d+e)$
- **d** $(a+b+c)(d+e+f)$
- **e** $(a+b)(c+d)(e+f)$
- **f** $(a+b+c)(d+e)(f+g)$

F THE BINOMIAL EXPANSION

The sum $\boldsymbol{a+b}$ is called a **binomial** as it contains two terms.

The **binomial expansion** of $(a+b)^n$ is obtained by writing the expression without brackets.

For example, the binomial expansion of $(a+b)^2$ is the perfect square expansion $(a+b)^2 = a^2+2ab+b^2$.

INVESTIGATION 1 THE BINOMIAL EXPANSION OF $(a+b)^3$

In this Investigation we discover the binomial expansion of $(a+b)^3$.

What to do:

1 Suppose you have a cube 4 cm by 4 cm by 4 cm.
By making 3 cuts parallel to the cube's surfaces, you divide the cube into 8 rectangular prisms, as shown.
How many prisms are:

- **a** 3 cm by 3 cm by 3 cm
- **b** 3 cm by 3 cm by 1 cm
- **c** 3 cm by 1 cm by 1 cm
- **d** 1 cm by 1 cm by 1 cm?

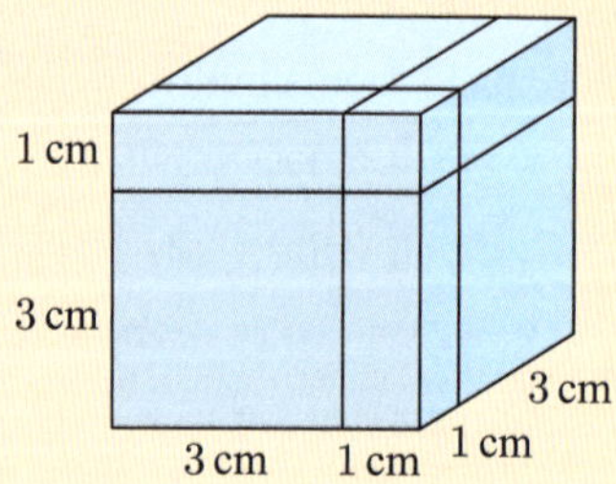

2 Now instead of the $4 \text{ cm} \times 4 \text{ cm} \times 4 \text{ cm}$ cube, suppose you had a cube with edge length $(a+b)$ cm.

- **a** Explain why the volume of the cube is given by $(a+b)^3$ cm^3.
- **b** Suppose you made cuts so each edge was divided into a cm and b cm.
  How many prisms would be:
  - **i** a cm by a cm by a cm
  - **ii** a cm by a cm by b cm
  - **iii** a cm by b cm by b cm
  - **iv** b cm by b cm by b cm?
- **c** By adding the volumes of the 8 rectangular prisms, write an expression for the total volume.
  Hence write down the binomial expansion of $(a+b)^3$.

ANIMATION

Another method of finding the binomial expansion of $(a+b)^3$ is to expand the brackets:

$$\begin{aligned}(a+b)^3 &= (a+b)(a+b)^2 \\ &= (a+b)(a^2+2ab+b^2) \qquad \{\text{perfect square expansion}\} \\ &= a^3+2a^2b+ab^2 \\ &\quad\;\; + a^2b+2ab^2+b^3 \\ &= a^3+3a^2b+3ab^2+b^3 \qquad \{\text{collecting like terms}\}\end{aligned}$$

The **binomial expansion** of $\boldsymbol{(a+b)^3}$ is $\boldsymbol{a^3+3a^2b+3ab^2+b^3}$.

The binomial expansion of $(a+b)^3$ can be used to expand other perfect cubes.

Example 8 **Self Tutor**

Use the binomial expansion $(a+b)^3 = a^3+3a^2b+3ab^2+b^3$ to expand and then simplify:

a $(x+4)^3$ **b** $(3x-2)^3$

a We substitute $a = x$ and $b = 4$.

$$\begin{aligned}\therefore\ (x+4)^3 &= x^3+3\times x^2\times 4+3\times x\times 4^2+4^3 \\ &= x^3+12x^2+48x+64\end{aligned}$$

b We substitute $a=(3x)$ and $b=(-2)$.

$$\begin{aligned}\therefore\ (3x-2)^3 &= (3x)^3+3\times(3x)^2\times(-2)+3\times(3x)\times(-2)^2+(-2)^3 \\ &= 27x^3-54x^2+36x-8\end{aligned}$$

We use brackets around $3x$ and -2 to make sure the whole value is raised to each power.

EXERCISE 2F

1 Use the binomial expansion of $(a+b)^3$ to expand and then simplify:

a $(x+1)^3$ **b** $(x+3)^3$ **c** $(x+5)^3$
d $(x+y)^3$ **e** $(x-1)^3$ **f** $(x-5)^3$
g $(x-4)^3$ **h** $(x-y)^3$ **i** $(2+y)^3$
j $(2x+1)^3$ **k** $(3x+1)^3$ **l** $(2y+3x)^3$
m $(2-y)^3$ **n** $(2x-1)^3$ **o** $(3x-1)^3$

2 By expanding and then simplifying $(a+b)(a+b)^3$, show that $(a+b)^4 = a^4+4a^3b+6a^2b^2+4ab^3+b^4$.

3 Use the binomial expansion $(a+b)^4 = a^4+4a^3b+6a^2b^2+4ab^3+b^4$ to expand and then simplify:

a $(x+y)^4$ **b** $(x+1)^4$ **c** $(x+2)^4$
d $(x+3)^4$ **e** $(x-y)^4$ **f** $(x-1)^4$
g $(x-2)^4$ **h** $(4-x)^4$ **i** $(2x-1)^4$

INVESTIGATION 2 PASCAL'S TRIANGLE

Since $x^0 = 1$ for any $x \neq 0$, we can write the first few binomial expansions as:

$$(a+b)^1 = a+b = 1a^1b^0 + 1a^0b^1$$
$$(a+b)^2 = a^2+2ab+b^2 = 1a^2b^0 + 2a^1b^1 + 1a^0b^2$$
$$(a+b)^3 = a^3+3a^2b+3ab^2+b^3 = 1a^3b^0 + 3a^2b^1 + 3a^1b^2 + 1a^0b^3$$
$$(a+b)^4 = a^4+4a^3b+6a^2b^2+4ab^3+b^4 = 1a^4b^0 + 4a^3b^1 + 6a^2b^2 + 4a^1b^3 + 1a^0b^4$$

Notice how, in each expansion, the terms have been ordered in decreasing powers of a.

What to do:

1 What can you say about the powers of b in the terms of each expansion?

2 When written with the terms in this order, the *coefficients* of the terms form the pattern of numbers called **Pascal's triangle**.

$n = 1$ 1 1

$n = 2$ 1 2 1

$n = 3$ 1 3 3 1

$n = 4$ 1 4 6 4 1

a What can you say about the first and last numbers in each row?

b How can each of the remaining values in a row be determined from the two values diagonally above it?

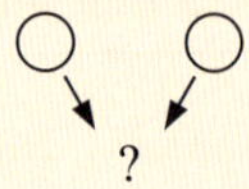

3 **a** Predict the numbers in the 5th and 6th rows of Pascal's triangle.

b Hence predict the binomial expansion of:

i $(a+b)^5$ **ii** $(a+b)^6$

MULTIPLE CHOICE QUIZ

QUICK QUIZ

REVIEW SET 2A

1 Expand then simplify:

a $5(4x-5)$ **b** $-4x(x-3)$ **c** $2(x+6)+x(3x-7)$

2 Expand then simplify:

a $(x+6)(x+1)$ **b** $(x+5)(x-6)$ **c** $(1-4x)(5-3x)$

d $(3x+1)(x-1)-x^2$ **e** $(x+3)(x+2)-(2x-1)(x-6)$

3 Expand then simplify:

a $(t+7)(t-7)$ **b** $(2y+5)(2y-5)$

c $(\sqrt{x}+6)(\sqrt{x}-6)$ **d** $(2y+z)(2y-z)-(y+3z)(y-3z)$

4 Expand then simplify:

a $(k+7)^2$ **b** $(2m-5n)^2$ **c** $(3x-2x^2)^2$

d $(y-4)^2+(4y+1)^2$ **e** $(3x-4)^2+(2x+1)(x-5)$

5 Expand then simplify:

a $x(x-2)-(x+1)(x-3)$ **b** $(k^2-4)^2-(k+2)(k-2)$

c $(x^2+3)(x^2-3)-x(x+5)$

6 Expand then simplify:

a $(x+5)(x-2)(x+1)$ **b** $(2x-3)(x^2+4x+2)$ **c** $(x^2+x-4)(2x^2+5x-2)$

7 **a** Use your calculator to write as a mixed number:

i $\left(1\frac{1}{2}\right)^2$ **ii** $\left(2\frac{1}{2}\right)^2$ **iii** $\left(3\frac{1}{2}\right)^2$ **iv** $\left(4\frac{1}{2}\right)^2$

b What do you suspect is the value of $\left(n+\frac{1}{2}\right)^2$, where n is a positive integer?

c Prove your result by expanding $\left(n+\frac{1}{2}\right)^2$.

d Hence find: **i** $\left(10\frac{1}{2}\right)^2$ **ii** $\left(19\frac{1}{2}\right)^2$

8 Use the binomial expansion of $(a+b)^3$ to expand and then simplify:

a $(6-x)^3$ **b** $(2k+3)^3$ **c** $(r-4t)^3$

REVIEW SET 2B

1 Expand then simplify:

a $6(3-2x)$ **b** $5(3a-2b+7)$ **c** $x(x^2-3)+5(x-4)$

2 Expand then simplify:

a $(2x+5)(3x-1)$ **b** $(x^2+3)(1-4x)$ **c** $(x-4)(x-2)+5(x-1)$

3 Expand then simplify:

a $(2x+3)(2x-3)$ **b** $(x^2-4)(x^2+4)$

c $(\sqrt{10}-\sqrt{m})(\sqrt{10}+\sqrt{m})$ **d** $(a+b)(a-b)-(a+2b)(a-2b)$

4 Expand then simplify:

a $(6-n)^2$ **b** $(5x-3)^2$ **c** $(3x^2-5)^2$

d $3x(x-2)-(x-8)^2$ **e** $2(1-x)^2-(x+1)(2x-5)$

5 Expand then simplify:

a $(x+5)(x-2)-7(x-1)$ **b** $(\sqrt{x}+5)(\sqrt{x}-5)-x(x+2)$

c $(3x-2)^2-(x+2)(2x-1)$ **d** $(4a+b)(4a-b)-(a+2b)^2$

6 Expand then simplify:

a $(x+3)(x^2-2x+2)$ **b** $(2x+1)^2(x-4)$ **c** $(x^2-x+4)(x^2+2x+3)$

7 How many terms would you obtain by expanding $(a+b+c+d)(e+f)(g+h)$?

8 Use the binomial expansion $(a+b)^4 = a^4+4a^3b+6a^2b^2+4ab^3+b^4$ to expand and then simplify:

a $(x-3)^4$ **b** $(2x+1)^4$

9 **a** Write down the binomial expansion of:

i $(a+b)^2$ **ii** $(a+b)^3$ **iii** $(a+b)^4$ **iv** $(a+b)^5$

b In $(a+b)^2 = a^2+2ab+b^2$, the sum of the coefficients of the expansion is $1+2+1=4$. Find the sum of the coefficients in the expansion of:

i $(a+b)^3$ **ii** $(a+b)^4$ **iii** $(a+b)^5$

c What do you suspect is the sum of the coefficients in the expansion of $(a+b)^n$, where n is a positive integer?

d Prove your result by letting $a=b=1$.

Chapter 3

Algebra: Factorisation

Contents:

OPENING PROBLEM

In the diagram alongside, the area of the yellow region is $a^2 - b^2$.

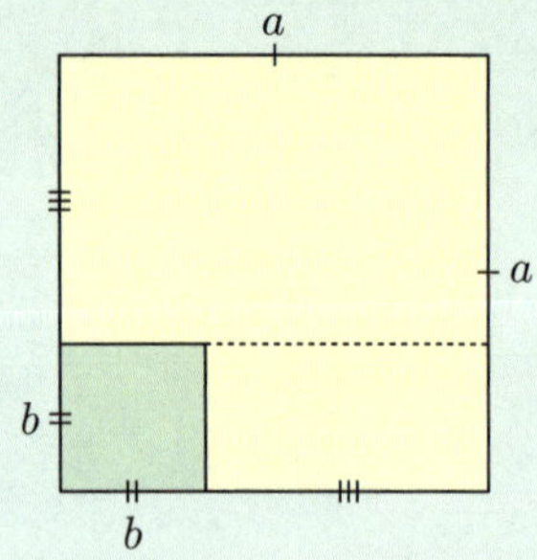

Things to think about:

a If you cut the yellow region along the dashed line, could you rearrange the pieces to show that $a^2 - b^2 = (a+b)(a-b)$?

b Is it possible to write the *sum* of two squares $a^2 + b^2$ as a product of real factors?

Factorisation is the process of writing an expression as a **product** of its **factors**.

Factorisation is the reverse process of expansion. We can therefore perform factorisation by applying the expansion laws in reverse.

An expression is **fully factorised** if none of its algebraic factors can be factorised further.

A COMMON FACTORS

When we apply the distributive law $a(b+c) = ab + ac$, the factor a is a common factor of each term in the expression.

If the HCF of the terms in an expression is not 1, we write it in front of a set of brackets and use the reverse of the distributive law.

$$\boldsymbol{ab + ac = a(b+c)}$$

We use the *highest* common factor in order to *fully* factorise the expression.

Example 1 **Self Tutor**

Fully factorise:

a $2x^2 - 3x$ **b** $6x^2 + 4xy$

a
$$\begin{aligned} &2x^2 - 3x \\ &= 2 \times x \times x - 3 \times x \\ &= x(2x-3) \quad \{\text{HCF} = x\} \end{aligned}$$

b
$$\begin{aligned} &6x^2 + 4xy \\ &= 2 \times 3 \times x \times x + 2 \times 2 \times x \times y \\ &= 2x(3x+2y) \quad \{\text{HCF} = 2x\} \end{aligned}$$

EXERCISE 3A

1 Fully factorise:

a $6x - 2$ **b** $4 + 2x$ **c** $3a + 3b$ **d** $6 - 4x^2$

e $5m + 10n$ **f** $mn - 4n$ **g** $2x - 4y$ **h** $6ab + 3b$

2 Fully factorise:

a $x^2 + 7x$ **b** $3x^2 + 9x$ **c** $4x + x^2$ **d** $8x - 4x^2$

e $2x^2 - 3x$ **f** $5x - 25x^2$ **g** $6x^2 + 15x$ **h** $x^3 + 2x^2$
i $3x^2 - 6x^3$ **j** $8x^3 - x^2$ **k** $x^2y - xy^2$ **l** $2xy^2 + 8x^2y$

3 Fully factorise:

a $x^3 + x^2 + x$ **b** $2x^3 + 11x^2 + 4x$ **c** $4x^2 - 8x + 12$
d $9x^2 - 45x - 27x^3$ **e** $ab + ac + ad$ **f** $xy + x - x^2$

Example 2 **Self Tutor**

Fully factorise: $-3x^2 + 6x$

$-3x^2 + 6x$
$= 6x - 3x^2$ {write with $6x$ first}
$= 2 \times 3 \times x - 3 \times x \times x$
$= 3x(2 - x)$ {HCF $= 3x$}

or

$-3x^2 + 6x$
$= -3 \times x \times x - -3 \times 2 \times x$
$= -3x(x - 2)$ {HCF $= -3x$}

4 Fully factorise:

a $-5a + 10$ **b** $-20 + 12b$ **c** $-8c + 16d$ **d** $-ab + 3a$
e $-p + pq$ **f** $-2x^2 + 9x$ **g** $-4x + 8x^2$ **h** $-4x^2 + 12x$

Example 3 **Self Tutor**

Fully factorise: $-9x^2 - 6x$

$-9x^2 - 6x$
$= -3 \times 3 \times x \times x + -3 \times 2 \times x$
$= -3x(3x + 2)$ {HCF $= -3x$}

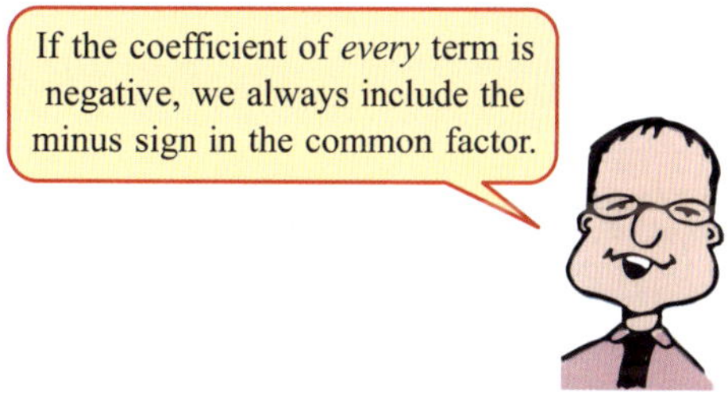

5 Fully factorise:

a $-6x - 24$ **b** $-8 - 8x$ **c** $-10p - 5q$ **d** $-ab - a$
e $-7a - 21a^2$ **f** $-16x^2 - 8x$ **g** $-15x - 10x^2$ **h** $-6x^2 - 16x$

Example 4 **Self Tutor**

Fully factorise:

a $4(a + 1) + (a + 2)(a + 1)$ **b** $-(x + 2)^2 + 2x + 4$

a $4(a + 1) + (a + 2)(a + 1)$
$= (a + 1)[4 + (a + 2)]$ {HCF $= a + 1$}
$= (a + 1)(a + 6)$

b $-(x + 2)^2 + 2x + 4$
$= -(x + 2)(x + 2) + 2(x + 2)$
$= (x + 2)[-(x + 2) + 2]$ {HCF $= x + 2$}
$= (x + 2)(-x - 2 + 2)$
$= -x(x + 2)$

6 Fully factorise:

a $a(a+2)+3(a+2)$ **b** $4(x-1)+x(x-1)$ **c** $5(x+4)-x(x+4)$
d $x(x-9)+(x-9)$ **e** $a(c+d)-b(c+d)$ **f** $x(x-3)-3(x-3)$
g $t(r+s)-(r+s)$ **h** $y(y-6)+y-6$

7 Fully factorise:

a $(x+4)(x+5)+3(x+4)$ **b** $2(x+1)+(x+1)(x-3)$
c $(x-6)(x+2)-2(x+2)$ **d** $(x-y)(y-1)-5(y-1)$
e $3(n+1)(n-3)-n-1$ **f** $n(n-3)-2(n-3)(n+1)$

8 Fully factorise:

a $(x+2)^2-5(x+2)$ **b** $(x-1)^2-3(x-1)$ **c** $-(x+1)^2+2(x+1)$
d $3(x-2)^2-(x-2)$ **e** $2(x+1)^2+x(x+1)$ **f** $(x+1)^3+(x+1)^2$

9 Fully factorise:

a $x+3+(x+3)^2$ **b** $(x-2)^2+3x-6$ **c** $(x+4)^2+8+2x$
d $(x-3)^2-x+3$ **e** $(x+4)^2-2x-8$ **f** $(x-4)^2-5x+20$
g $4(a+b)^2-2a-2b$ **h** $-3(x^2+2)+4+2x^2$

B DIFFERENCE BETWEEN TWO SQUARES FACTORISATION

The **expansion** of $(a+b)(a-b)$ is a^2-b^2, which is the difference between the two squares a^2 and b^2.

$\therefore$ the **factorisation** of a^2-b^2 is: $\boldsymbol{a^2-b^2=(a+b)(a-b)}$

By contrast, the **sum** of two squares does not factorise into real linear factors.

Linear factors have the form $ax+b$.

For example:

- x^2-9 is the difference between two squares, so $x^2-9=(x+3)(x-3)$.
- x^2+9 is the sum of two squares, which does not factorise into real linear factors.

Example 5 Self Tutor

Fully factorise:

a $9-x^2$ **b** $4x^2-25$

a $9-x^2$
$=3^2-x^2$
$=(3+x)(3-x)$

b $4x^2-25$
$=(2x)^2-5^2$
$=(2x+5)(2x-5)$

EXERCISE 3B

1 Factorise, if possible:

a $x^2 - 16$ **b** $4 - x^2$ **c** $x^2 - 81$
d $25 - x^2$ **e** $x^2 + 4$ **f** $64 - a^2$
g $8 + x^2$ **h** $4x^2 - 1$ **i** $9x^2 - 16$
j $4x^2 - 9$ **k** $4x^2 + 1$ **l** $36 - 49x^2$
m $x^2 - 4y^2$ **n** $a^2b^2 - 36$ **o** $16x^2 - 9y^2$

Example 6 Self Tutor

Fully factorise:

a $2x^2 - 8$ **b** $-3x^2 + 48$

a $2x^2 - 8$
$= 2(x^2 - 4)$ {HCF = 2}
$= 2(x^2 - 2^2)$
$= 2(x + 2)(x - 2)$

b $-3x^2 + 48$
$= -3(x^2 - 16)$ {HCF = −3}
$= -3(x^2 - 4^2)$
$= -3(x + 4)(x - 4)$

2 Fully factorise:

a $12 - 3x^2$ **b** $2x^2 - 18$ **c** $3x^2 - 27$
d $-2x^2 + 8$ **e** $3x^2 - 75$ **f** $-5x^2 + 5$
g $8x^2 - 18$ **h** $-27x^2 + 75$ **i** $3x^2 - 12y^2$
j $2 - 18x^2$ **k** $4x - 9x^3$ **l** $x^3y - xy^3$

3 Answer the **Opening Problem** on page **36**.

Example 7 Self Tutor

Factorise using the difference between two squares:

a $(3x + 2)^2 - 9$ **b** $(x + 2)^2 - (x - 1)^2$

a $(3x + 2)^2 - 9$
$= (3x + 2)^2 - 3^2$
$= [(3x + 2) + 3][(3x + 2) - 3]$
$= [3x + 5][3x - 1]$

b $(x + 2)^2 - (x - 1)^2$
$= [(x + 2) + (x - 1)][(x + 2) - (x - 1)]$
$= [x + 2 + x - 1][x + 2 - x + 1]$
$= [2x + 1][3]$
$= 3(2x + 1)$

4 Factorise using the difference between two squares:

a $(x + 1)^2 - 4$ **b** $(2x + 1)^2 - 9$ **c** $(1 - x)^2 - 16$
d $(x + 3)^2 - 4x^2$ **e** $4x^2 - (x + 2)^2$ **f** $9x^2 - (3 - x)^2$
g $(2x + 1)^2 - (x - 2)^2$ **h** $(3x - 1)^2 - (x + 1)^2$ **i** $16x^2 - (2x + 3)^2$

5 Write 391 as the difference between two squares. *Hence* write a factor pair of 391.

C PERFECT SQUARES FACTORISATION

The **expansions** of the perfect squares $(a+b)^2$ and $(a-b)^2$ are

$$(a+b)^2 = a^2 + 2ab + b^2 \quad \text{and} \quad (a-b)^2 = a^2 - 2ab + b^2.$$

$\therefore$ the **factorisations** for these perfect squares are

$$\boldsymbol{a^2 + 2ab + b^2 = (a+b)^2} \quad \text{and} \quad \boldsymbol{a^2 - 2ab + b^2 = (a-b)^2}$$

So, if an expression contains two squares a^2 and b^2, and a middle term $\pm 2ab$, the *sign* of the middle term indicates the *sign* within the perfect square.

Example 8 — Self Tutor

Use the perfect square factorisations to fully factorise:

a $x^2 + 10x + 25$ **b** $x^2 - 14x + 49$

a
$$\begin{aligned} & x^2 + 10x + 25 \\ = \ & x^2 + 2 \times x \times 5 + 5^2 \\ = \ & (x+5)^2 \end{aligned}$$

b
$$\begin{aligned} & x^2 - 14x + 49 \\ = \ & x^2 - 2 \times x \times 7 + 7^2 \\ = \ & (x-7)^2 \end{aligned}$$

EXERCISE 3C

1 Determine whether each expression is a perfect square:

a $x^2 + 4x + 1$ **b** $9 - 6x + x^2$ **c** $x^2 - 2x - 1$

d $x^2 - 4x + 4$ **e** $x^2 + 8x - 16$ **f** $x^2 - 5x + 25$

2 Fully factorise:

a $x^2 + 6x + 9$ **b** $x^2 + 8x + 16$ **c** $x^2 - 6x + 9$

d $x^2 - 8x + 16$ **e** $x^2 + 2x + 1$ **f** $x^2 - 10x + 25$

g $y^2 + 18y + 81$ **h** $m^2 - 20m + 100$ **i** $t^2 + 12t + 36$

Example 9 — Self Tutor

Fully factorise:

a $9x^2 - 6x + 1$ **b** $-8x^2 - 24x - 18$

a
$$\begin{aligned} & 9x^2 - 6x + 1 \\ = \ & (3x)^2 - 2 \times 3x \times 1 + 1^2 \\ = \ & (3x-1)^2 \end{aligned}$$

b
$$\begin{aligned} & -8x^2 - 24x - 18 \\ = \ & -2(4x^2 + 12x + 9) \qquad \{\text{HCF} = -2\} \\ = \ & -2([2x]^2 + 2 \times 2x \times 3 + 3^2) \\ = \ & -2(2x+3)^2 \end{aligned}$$

3 Fully factorise:

a $9x^2 + 6x + 1$ **b** $4x^2 - 4x + 1$ **c** $9x^2 + 12x + 4$

d $25x^2 - 10x + 1$ **e** $16x^2 + 24x + 9$ **f** $25x^2 - 20x + 4$

g $25 - 30x + 9x^2$ **h** $4x^2 + 28x + 49$ **i** $81 - 36x + 4x^2$

4 Fully factorise:

a $x^2 + 4xy + 4y^2$ **b** $9x^2 - 6xy + y^2$ **c** $4x^2 + 12xy + 9y^2$

5 Fully factorise:

a $2x^2 + 4x + 2$ **b** $2x^2 - 12x + 18$ **c** $4x^2 - 32x + 64$

d $-x^2 + 2x - 1$ **e** $-2x^2 - 8x - 8$ **f** $-3x^2 - 30x - 75$

g $3x^2 + 18x + 27$ **h** $-9x^2 + 6x - 1$ **i** $18x^2 + 12x + 2$

6 Explain why:

a $x^2 + 12x + 36$ is never negative **b** $x^2 + 4 \geqslant 4x$ for all real x.

D EXPRESSIONS WITH FOUR TERMS

Some expressions with four terms do not have an overall common factor, but can be factorised by pairing the four terms.

For example, $\underbrace{ab + ac} + \underbrace{bd + cd}$

$= a(b + c) + d(b + c)$ {factorising each pair separately}

$= (b + c)(a + d)$ {removing the common factor $(b + c)$}

Example 10 Self Tutor

Factorise:

a $x^2 + 2x + 5x + 10$ **b** $3x + x^2 - 3 - x$

a $\underbrace{x^2 + 2x} + \underbrace{5x + 10}$

$= x(x + 2) + 5(x + 2)$

$= (x + 2)(x + 5)$

b $3x + x^2 - 3 - x$

$= \underbrace{x^2 + 3x} \underbrace{- x - 3}$

$= x(x + 3) - (x + 3)$

$= x(x + 3) - 1(x + 3)$

$= (x + 3)(x - 1)$

EXERCISE 3D

1 Factorise:

a $2a + 2 + ab + b$ **b** $3x + xy + 3 + y$ **c** $4d + ac + ad + 4c$

d $ab + 6 + 2b + 3a$ **e** $mn + 3p + np + 3m$ **f** $x^2 + 3x + 7x + 21$

g $x^2 + 5x + 4x + 20$ **h** $6x + x + 2x^2 + 3$ **i** $3x^2 + 2x + 12x + 8$

2 Factorise:

a $x^2 - 4x + 5x - 20$ **b** $x^2 - 7x + 2x - 14$ **c** $x^2 - 3x - 2x + 6$

d $x^2 - 5x - 3x + 15$ **e** $x^2 + 7x - 8x - 56$ **f** $x^2 + 2x - 4x - 8$

g $2x^2 + x - 6x - 3$ **h** $3x^2 + 2x - 12x - 8$ **i** $4x^2 - 3x - 8x + 6$

j $2x + 9x^2 - 9x - 2$ **k** $-4x + 3x^2 - 20 + 15x$ **l** $2xy - 5 + 10y - x$

E FACTORISING $x^2 + bx + c$

A **quadratic trinomial** is an algebraic expression of the form $ax^2 + bx + c$ where x is a variable and a, b, c are constants, $a \neq 0$.

In this Section we learn to factorise quadratic trinomials where $a = 1$. They have the form $x^2 + bx + c$.

Consider the expansion

$$\begin{aligned}(x+2)(x+5) &= x^2 + 5x + 2x + 2 \times 5 \quad \{\text{using FOIL}\}\\ &= x^2 + (5+2)x + 2 \times 5\\ &= x^2 + (\textbf{sum} \text{ of 2 and 5})x + (\textbf{product} \text{ of 2 and 5})\\ &= x^2 + 7x + 10\end{aligned}$$

So, to factorise $x^2 + 7x + 10$, we need two numbers whose sum is 7 and whose product is 10. The numbers are 2 and 5, so $x^2 + 7x + 10 = (x+2)(x+5)$.

In general:

$$\boldsymbol{x^2 + bx + c = (x+p)(x+q)} \quad \text{where} \quad p + q = b \quad \text{and} \quad pq = c.$$

We call this the **sum and product method**.

Example 11 — Self Tutor

Fully factorise:

a $x^2 + 5x + 4$ **b** $x^2 - x - 12$

a sum $= 5$ and product $= 4$
$\therefore$ the numbers are 1 and 4.
$\therefore$ $x^2 + 5x + 4 = (x+1)(x+4)$

b sum $= -1$ and product $= -12$
$\therefore$ the numbers are -4 and 3.
$\therefore$ $x^2 - x - 12 = (x-4)(x+3)$

DISCUSSION

What can you say about the *signs* of two numbers if their sum and product are:

- both positive
- both negative
- opposite in sign?

EXERCISE 3E

1 Find two numbers which have:

a sum 8 and product 7
b sum 11 and product 30
c sum 17 and product 60
d sum 2 and product -8
e sum -1 and product -20
f sum -15 and product 56
g sum -22 and product -48
h sum -9 and product -10.

2 Fully factorise:

a $x^2 + 7x + 10$ **b** $x^2 + 14x + 33$ **c** $x^2 + 19x + 34$

3 Fully factorise:

a $x^2 - 9x + 20$ **b** $x^2 - 18x + 56$ **c** $x^2 - 13x + 42$

4 Fully factorise:

a $x^2 - 5x - 6$ **b** $x^2 + 4x - 45$ **c** $x^2 - 7x - 18$
d $x^2 + 11x - 80$ **e** $x^2 - 3x - 10$ **f** $x^2 + 6x - 72$

5 Fully factorise:

a $x^2 + 5x + 6$ **b** $x^2 - x - 6$ **c** $x^2 + 3x - 10$
d $x^2 + 4x - 21$ **e** $x^2 + 3x - 28$ **f** $x^2 - 11x + 24$
g $x^2 + 15x + 44$ **h** $x^2 - x - 56$ **i** $x^2 + 4x - 96$
j $x^2 - 12x + 35$ **k** $x^2 + 17x + 60$ **l** $x^2 - 16x + 39$

Example 12 **Self Tutor**

Fully factorise by first removing a common factor:

a $3x^2 - 9x + 6$ **b** $-2x^2 + 2x + 12$

a $3x^2 - 9x + 6$
$= 3(x^2 - 3x + 2)$ {HCF = 3}
$= 3(x - 2)(x - 1)$ {sum = -3, product = 2
$\therefore$ the numbers are -2 and -1}

b $-2x^2 + 2x + 12$
$= -2(x^2 - x - 6)$ {HCF = -2}
$= -2(x - 3)(x + 2)$ {sum = -1, product = -6
$\therefore$ the numbers are -3 and 2}

6 Fully factorise:

a $2x^2 + 10x + 8$ **b** $3x^2 - 21x + 18$ **c** $2x^2 + 14x + 24$
d $5x^2 - 30x - 80$ **e** $4x^2 - 8x - 12$ **f** $3x^2 - 42x + 99$
g $2x^2 - 2x - 180$ **h** $3x^2 - 6x - 24$ **i** $2x^2 + 18x + 40$
j $4x^2 - 24x + 36$ **k** $7x^2 + 21x - 70$ **l** $x^3 - 7x^2 - 8x$

7 Fully factorise:

a $-x^2 + 4x - 3$ **b** $-x^2 + 4x - 4$ **c** $-x^2 - 2x + 3$
d $-2x^2 + 44x - 240$ **e** $-3x^2 + 30x - 63$ **f** $-5x^2 + 20x + 60$

8 Fully factorise:

a $54 - x^2 - 3x$ **b** $14x + 2x^2 + 20$ **c** $-x^2 - 21 - 10x$
d $4x - 96 + 2x^2$ **e** $18x - 3x^2 - 27$ **f** $-20x + 2x^2 + 42$
g $126 - 2x^2 + 4x$ **h** $20x - 2x^2 - 50$ **i** $x^2 - x^3 + 2x$

9 Given that $x^2 + bx + c = (x + m)(x + n)$, factorise $x^2 - bx + c$.

F MISCELLANEOUS FACTORISATION

Factorisation is an extremely important tool in mathematics. We use it frequently when solving equations, manipulating formulae, and proving results.

By carefully studying the *form* of the expression we are factorising, we can factorise it in the most efficient way.

This summary may help you factorise expressions:

Look for any **common factors** to take out.

If the expression has 2 or 3 terms, look for:

- **difference between two squares**
  $a^2 - b^2 = (a+b)(a-b)$
- **perfect squares**
  $a^2 + 2ab + b^2 = (a+b)^2$ or $a^2 - 2ab + b^2 = (a-b)^2$
- **sum and product type** $\boldsymbol{x^2 + bx + c}$
  $x^2 + bx + c = (x+p)(x+q)$ where $p + q = b$ and $pq = c$

If the expression has 4 terms, look to **group in pairs**.

EXERCISE 3F

1 Fully factorise:

a $7x - 35y$ **b** $m^2 + 3mp$ **c** $-5x^2 - 10x$
d $3x^2 + 2x$ **e** $2p^2 + 8$ **f** $x^2 - 81$
g $x^2 - 8x - 9$ **h** $d^2 + 6d - 7$ **i** $2x^2 - 32$
j $x^2 - 6x + 9$ **k** $x^2 + 22x + 121$ **l** $3x^2 - 108$
m $5a^2 - 5a - 10$ **n** $4t + 8t^2$ **o** $n^4 - 4n^2$
p $2c^2 - 8c + 6$ **q** $x^4 - x^2$ **r** $3b^2 - 75$
s $121 - x^2$ **t** $25y^2 - 16$ **u** $x^3 + 4x^2 + 4x$
v $x^2 - 2xy + y^2$ **w** $49y^2 - 36z^2$ **x** $4d^2 + 28d + 49$
y $4ab^2 - ac^2$ **z** $2\pi R^2 - 2\pi r^2$

2 Fully factorise:

a $ab + ac - 2a$ **b** $a^2b^2 - 2ab$ **c** $18x - 2x^3$
d $x^2 + 14x + 49$ **e** $4a^3 - 4ab^2$ **f** $x^3y - 4xy$
g $4x^4 - 4x^2$ **h** $a^2 + 8a + 15$ **i** $m^2 - 6m + 9$

3 Fully factorise:

a $(x-2)y - (x-2)z$ **b** $(x-y)a + (x-y)$ **c** $x(x+2) + 3(x+2)$
d $(x-1)^2 - 9$ **e** $3x^2 + 3x - 36$ **f** $5x^2 + 5xy - 5x^2y$
g $xy + 2x + 2y + 4$ **h** $4 - (2x+1)^2$ **i** $2x^2 + 10x + x + 5$
j $x^3 + x^2 + x + 1$ **k** $y^2 + 5y - 9y - 45$ **l** $(3x+2)^2 - (x-3)^2$

G FACTORISING $ax^2 + bx + c,\ a \neq 1$

So far we have considered the factorisation of quadratic trinomials of the form $ax^2 + bx + c$ where:

- $a = 1$, for example $x^2 + 5x + 6 = (x+3)(x+2)$
- a is a common factor, for example $2x^2 + 10x + 12 = 2(x^2 + 5x + 6)$
  $= 2(x+3)(x+2)$
- we have a perfect square or difference between two squares,
  for example $4x^2 - 9 = (2x)^2 - 3^2$
  $= (2x+3)(2x-3)$

Quadratic trinomials such as $8x^2 + 22x + 15$ do not fall into any of these categories. We therefore need a method for factorising this type of expression.

FACTORISATION BY "SPLITTING" THE MIDDLE TERM

Using the FOIL rule, we see that

$$\begin{aligned} &(2x+3)(4x+5) \\ &= 8x^2 + 10x + 12x + 15 \\ &= 8x^2 + 22x + 15 \end{aligned}$$

We can *reverse* the process to factorise the quadratic trinomial $8x^2 + 22x + 15$.

$$\begin{aligned} &8x^2 + 22x + 15 \\ &= 8x^2 + 10x + 12x + 15 && \{\text{"splitting" the middle term}\} \\ &= (8x^2 + 10x) + (12x + 15) && \{\text{grouping in pairs}\} \\ &= 2x(4x+5) + 3(4x+5) && \{\text{factorising each pair separately}\} \\ &= (4x+5)(2x+3) && \{\text{completing the factorisation}\} \end{aligned}$$

But how do we know how to correctly "split" the middle term? How do we know that $22x$ should be written as $+\,10x + 12x$ rather than $+\,15x + 7x$ or $+\,20x + 2x$?

INVESTIGATION "SPLITTING" THE MIDDLE TERM

Consider the general quadratic trinomial $ax^2 + bx + c$.

Suppose we "split" the middle term bx into $px + qx$, so $p + q = b$.

What to do:

1. Show that $ax^2 + bx + c = x(ax + p) + (qx + c)$.
2. We can only factorise this expression further if the two terms have a common factor.
   $\therefore\ ax + p = k(qx + c)$ for some k.
   a By equating the coefficients of x, show that $k = \dfrac{a}{q}$.
   b By equating the constant terms, show that $k = \dfrac{p}{c}$.
   c Hence show that $pq = ac$.
3. Summarise the properties of p and q needed so that "splitting" bx into $px + qx$ will enable us to factorise $ax^2 + bx + c$.

From the **Investigation**, you should have concluded that "splitting" the middle term bx into $px + qx$ only works if we can find p and q such that $p + q = b$ and $pq = ac$.

For example, for the quadratic trinomial $8x^2 + 22x + 15$, two numbers with sum $= 22$ and product $= 8 \times 15 = 120$ are 10 and 12. We were therefore successful by "splitting" $22x$ into $+ 10x + 12x$.

To factorise $ax^2 + bx + c$ by "splitting" the middle term:

Step 1: Find two numbers p and q such that $p + q = b$ and $pq = ac$.

Step 2: Replace bx with $px + qx$.

Step 3: Complete the factorisation.

Example 13 — Self Tutor

Factorise by "splitting" the middle term:

a $6x^2 + 19x + 10$ **b** $3x^2 - x - 10$

a In $6x^2 + 19x + 10$, $ac = 60$ and $b = 19$.
We need two numbers with product 60 and sum 19. These are 4 and 15.

$\therefore$ $6x^2 + 19x + 10$
$= 6x^2 + 4x + 15x + 10$ {"splitting" the middle term}
$= 2x(3x + 2) + 5(3x + 2)$ {factorising in pairs}
$= (3x + 2)(2x + 5)$ {taking out the common factor}

b In $3x^2 - x - 10$, $ac = -30$ and $b = -1$.
We need two numbers with product -30 and sum -1. These are 5 and -6.

$\therefore$ $3x^2 - x - 10$
$= 3x^2 + 5x - 6x - 10$ {"splitting" the middle term}
$= x(3x + 5) - 2(3x + 5)$ {factorising in pairs}
$= (3x + 5)(x - 2)$ {taking out the common factor}

EXERCISE 3G

1 Factorise the quadratic trinomial $3x^2 + 7x + 2$ by "splitting" the middle term into:

a $+ 6x + x$ **b** $+ x + 6x$

2 Fully factorise:

a $2x^2 + 5x + 3$ **b** $2x^2 + 13x + 18$ **c** $7x^2 + 9x + 2$
d $3x^2 + 13x + 4$ **e** $3x^2 + 8x + 4$ **f** $3x^2 + 16x + 21$
g $8x^2 + 14x + 3$ **h** $21x^2 + 17x + 2$ **i** $6x^2 + 5x + 1$
j $6x^2 + 19x + 3$ **k** $10x^2 + 17x + 3$ **l** $14x^2 + 37x + 5$

3 Factorise the quadratic trinomial $4x^2 + 4x - 3$ by "splitting" the middle term into:

a $+ 6x - 2x$ **b** $- 2x + 6x$

4 Fully factorise:

a $2x^2 - 9x - 5$ **b** $3x^2 + 5x - 2$ **c** $3x^2 - 5x - 2$
d $2x^2 + 3x - 2$ **e** $2x^2 + 3x - 5$ **f** $5x^2 - 8x + 3$
g $11x^2 - 9x - 2$ **h** $2x^2 - 3x - 9$ **i** $3x^2 - 17x + 10$
j $5x^2 - 13x - 6$ **k** $3x^2 + 10x - 8$ **l** $2x^2 + 17x - 9$
m $2x^2 + 9x - 18$ **n** $15x^2 + x - 2$ **o** $21x^2 - 62x - 3$

Example 14 **Self Tutor**

Fully factorise: $-5x^2 - 7x + 6$

$-5x^2 - 7x + 6$
$= -[5x^2 + 7x - 6]$
$= -[5x^2 + 10x - 3x - 6]$
$= -[5x(x+2) - 3(x+2)]$
$= -[(x+2)(5x-3)]$
$= -(x+2)(5x-3)$

In $5x^2 + 7x - 6$, $ac = -30$ and $b = 7$. The two numbers with product -30 and sum 7 are 10 and -3.

If the coefficient of x^2 is negative, we usually take out -1 as a common factor.

5 Fully factorise by first removing -1 as a common factor:

a $-3x^2 - x + 14$ **b** $-5x^2 + 11x - 2$ **c** $-4x^2 - 9x + 9$
d $-9x^2 + 12x - 4$ **e** $-8x^2 - 14x - 3$ **f** $-12x^2 + 16x + 3$

MULTIPLE CHOICE QUIZ

QUICK QUIZ

REVIEW SET 3A

1 Fully factorise:

a $ab - 5b$ **b** $3x^2 - 12x$ **c** $15x - 6x^2$
d $-4x^2 + 6x$ **e** $12a - 20a^2$ **f** $(x+2)^2 - 3(x+2)$

2 Fully factorise:

a $25 - 9a^2$ **b** $2x^2 - 98$ **c** $(2x+5)^2 - (x-3)^2$

3 If possible, factorise into real linear factors:

a $x^2 - 144$ **b** $x^2 + 16$ **c** $(x-4)^2 - 9$

4 Determine whether each expression is a perfect square:

a $x^2 + 8x + 16$ **b** $x^2 - 2x + 4$

5 Fully factorise:

a $m^2 - 18m + 81$ **b** $16x^2 + 8x + 1$ **c** $3x^2 - 24x + 48$

6 Fully factorise:

a $5x - 5 + xy - y$ **b** $3x + 7 + 6bx + 14b$ **c** $2xy - z - 2xz + y$

7 Fully factorise:

a $x^2 + 14x + 45$ **b** $x^2 - 5x - 36$ **c** $x^2 - 19x + 84$

8 Fully factorise:

a $-x^2 - 7x + 18$ **b** $3x^2 - 9x - 30$ **c** $64 - 2x^2 + 8x$

9 Fully factorise:

a $3t^2 - 27$ **b** $x^2 + 4x - 32$ **c** $4x^2 - 28x + 49$

d $4x^2 + 40x + 100$ **e** $2x^2 + 6x - 20$ **f** $-3x^2 + 24x + 60$

10 Fully factorise:

a $8x^2 + 10x + 3$ **b** $5x^2 - 13x + 6$ **c** $-9x^2 + 3x + 2$

REVIEW SET 3B

1 Fully factorise:

a $6x - xy$ **b** $5ab + 10b^2$ **c** $8m^2 - 10m$

d $-24x - 18x^2$ **e** $(x + 5)^2 - 4(x + 5)$ **f** $(x - 3)^2 - 3x + 9$

2 Fully factorise:

a $x^4 - 9y^2$ **b** $49a^3 - 16a$ **c** $4m^2 - (1 - m)^2$

3 Write 899 as the difference between two squares. *Hence* write a factor pair of 899.

4 Fully factorise:

a $k^2 + 24k + 144$ **b** $9x^2 - 24x + 16$ **c** $a^2 + 6ab + 9b^2$

5 Explain why $x^2 - 6x + 9$ is never negative.

6 Fully factorise:

a $cd + 9 + 3d + 3c$ **b** $2x^2 + 6x - 5x - 15$ **c** $4x - 2x^2 + x^3 - 8$

7 Fully factorise:

a $x^2 - 5x - 66$ **b** $x^2 + 15x + 54$ **c** $2x^2 + 20x - 78$

8 Fully factorise:

a $81x^2 - 4$ **b** $x^2 - 7x - 60$ **c** $4x^2 - 8x - 60$

d $4x^2 + 4x + 1$ **e** $3x^2 - 36x + 108$ **f** $5x^2 - 25x - 70$

9 **a** Evaluate $10^3 + 10^2 + 10 + 1$.

b Fully factorise $x^3 + x^2 + x + 1$.

c Hence write a factor pair of 1111.

10 Fully factorise:

a $5x^2 - 14x - 3$ **b** $4x^2 - 8x - 21$ **c** $-6x^2 - 5x + 50$

Chapter 4

Sets

Contents:

OPENING PROBLEM

A *rational* number is a number which can be written as a *ratio* of integers, or in other words, a fraction with integer numerator and denominator.

Things to think about:

a Is every integer itself a rational number?

b Is every ratio of integers a rational number? If not, can you give an example?

c Is every number on the number line a rational number? If not, can you give an example?

d Are there more integers or rational numbers?

A SETS

A **set** is a collection of numbers or objects.

Each object is called an **element** or **member** of the set.

$\in$ means "is an element of" or "is in".

$\notin$ means "is not an element of" or "is not in".

When we record a set, we write its members within curly brackets, separated by commas.

For example, if P is the set of prime numbers, we write $P = \{2, 3, 5, 7, 11, ...\}$.

We can say $5 \in P$ but $8 \notin P$.

COUNTING ELEMENTS OF SETS

$n(A)$ means "the number of elements in set A".

- A set which contains a finite number of elements is called a **finite set**.
- A set which contains an infinite number of elements is called an **infinite set**.

For example:

- The set of vowels $V = \{\text{a, e, i, o, u}\}$ is a finite set, and $n(V) = 5$.
- The set of even numbers E is an infinite set.

EQUAL SETS

Two sets are **equal** if they contain exactly the same elements.

For example, $\{3, 8, 9, 11, 14\} = \{11, 8, 14, 3, 9\}$.

SUBSETS

Set A is a **subset** of set B if every element of A is also an element of B. We write $A \subseteq B$.

For example, the set $D = \{2, 5, 7\}$ is a subset of the set of prime numbers P. We write $D \subseteq P$.

EMPTY SET

The **empty set** $\varnothing$ or $\{\ \}$ is a set which contains no elements.

The empty set is a subset of all other sets.

EXERCISE 4A

1 **a** List the elements of A, the set of positive two digit whole numbers which contain the digit 4. Hence state $n(A)$.

b List the elements of B, the set of colours in a rainbow. Hence state $n(B)$.

2 Let P be the set of composite factors of 50, and Q be the set of numbers less than 50 which have exactly four factors.

a List the elements of P and Q.

b True or false?

i $10 \in P$ **ii** $18 \in Q$ **iii** $5 \notin P$

c Find:

i $n(P)$ **ii** $n(Q)$

d Is $P \subseteq Q$?

3 For each set:

i List the elements of the set.

ii State whether the set is finite or infinite. If the set is finite, state the number of elements in the set.

a $A = \{\text{factors of } 6\}$ **b** $B = \{\text{multiples of } 6\}$ **c** $C = \{\text{factors of } 17\}$

d $D = \{\text{odd numbers}\}$ **e** $E = \{\text{prime numbers less than } 20\}$

f $F = \{\text{composite numbers between 10 and 30}\}$

4 Decide whether $A \subseteq B$ for:

a $A = \{2, 5, 6\}$ and $B = \{1, 2, 3, 4, 5, 6, 7, 8\}$

b $A = \{4, 8, 11, 12\}$ and $B = \{2, 4, 6, 8, 10, 12, 14, 16\}$

c $A = \varnothing$ and $B = \{1, 4, 7, 10\}$

d $A = \{5, 10, 15, 20, 25, 30\}$ and $B = \{10, 15, 20\}$.

5 Suppose $P = \{\text{prime numbers less than } 10\}$, $Q = \{\text{multiples of 3 less than } 20\}$, $R = \{3, 5, 7\}$, and $S = \{\text{multiples of 6 less than } 20\}$. Decide whether:

a $P \subseteq Q$ **b** $R \subseteq P$ **c** $R \subseteq S$ **d** $S \subseteq Q$

6 Suppose $A \subseteq B$.

a Explain why $n(A) \leqslant n(B)$.

b What can you say if $n(A) = n(B)$?

c What can you conclude about B if A is an infinite set?

7 Let $P = \{\text{multiples of } 10\}$ and $Q = \{\text{multiples of } k\}$, where k is a positive whole number. For what values of k is:

a $P \subseteq Q$ **b** $Q \subseteq P$?

B COMPLEMENT OF A SET

The **universal set** U is the set of all elements we are considering.

For example, if we are only interested in positive whole numbers, then $U = \{1, 2, 3, 4, ...\}$.

The **complement** of a set A is the set of all elements of U that are *not* elements of A.
The complement of A is written A'. We say that A and A' are **complementary**.

For example, if $U = \{1, 2, 3, 4, 5, 6, 7, 8, 9\}$ and $A = \{2, 5, 6, 8\}$, then $A' = \{1, 3, 4, 7, 9\}$.

EXERCISE 4B

1 Let $U = \{1, 2, 3, 4, 5, 6, 7, 8\}$. Find the complement of:

- **a** $A = \{2, 5, 6\}$
- **b** $B = \{\text{odd numbers}\}$
- **c** $C = \{\text{composite numbers}\}$
- **d** $D = \{\text{perfect squares}\}$

2 Suppose $U = \{\text{letters of the English alphabet}\}$. Find the complement of:

- **a** $P = \{\text{C, F, J, M, P, U, Y, Z}\}$
- **b** $Q = \{\text{consonants}\}$
- **c** $R = \{\text{letters in the word HOSPITAL}\}$
- **d** $S = \{\text{letters after J in the alphabet}\}$

3 Let $U = \{1, 2, 3, ..., 30\}$, $A = \{\text{factors of } 30\}$, and $B = \{\text{even numbers}\}$.
List the elements of:

- **a** A
- **b** A'
- **c** B
- **d** B'

4 Suppose U is a finite set. Write a formula for $n(A')$ in terms of $n(A)$ and $n(U)$.

5 Suppose U is a finite set. If $B \subseteq A$, explain why $n(A') \leqslant n(B')$.

6 Suppose U is the set of positive whole numbers.

- **a** Is U finite or infinite?
- **b** Give an example of a set A within U such that:
  - **i** A is finite and A' is infinite
  - **ii** A is infinite and A' is finite
  - **iii** A is infinite and A' is infinite.
- **c** Explain why it is impossible for A and A' to both be finite.

C INTERSECTION AND UNION

INTERSECTION

The **intersection** of two sets A and B is the set of elements that are in **both** set A **and** set B.
The intersection of sets A and B is written $\boldsymbol{A \cap B}$.

For example, if $A = \{1, 3, 4, 6\}$ and $B = \{2, 3, 5, 6, 7\}$, then $A \cap B = \{3, 6\}$.

Two sets A and B are **disjoint** or **mutually exclusive** if they have no elements in common.
In this case $A \cap B = \varnothing$.

UNION

The **union** of two sets A and B is the set of elements that are in **either** set A **or** set B.

The union of sets A and B is written $\boldsymbol{A \cup B}$.

For example, if $A = \{1, 3, 4, 6\}$ and $B = \{2, 3, 5, 6, 7\}$,
then $A \cup B = \{1, 2, 3, 4, 5, 6, 7\}$.

EXERCISE 4C

1 Find $A \cap B$ and $A \cup B$ for:

a $A = \{\text{c, d, j, m, p}\}$ and $B = \{\text{b, d, k, m, t}\}$

b $A = \{2, 5, 8, 11\}$ and $B = \{3, 7, 11, 15, 19\}$

c $A = \{\text{red, pink, green, blue}\}$ and $B = \{\text{purple, blue, green, yellow, red}\}$.

2 Suppose $A = \{\text{factors of } 36\}$ and $B = \{\text{factors of } 63\}$. List the elements of:

a A **b** B **c** $A \cap B$ **d** $A \cup B$

3 Suppose $U = \{1, 2, 3, 4, 5, 6, 7, 8, 9\}$, $A = \{2, 3, 7, 9\}$, and $B = \{\text{multiples of } 3\}$.

a List the elements of:

i A' **ii** B' **iii** $A \cap B$ **iv** $A \cup B$
v $A' \cap B'$ **vi** $A' \cup B'$

b Show that:

i $(A \cup B)' = A' \cap B'$ **ii** $(A \cap B)' = A' \cup B'$

4 Suppose $A = \{\text{factors of } 20\}$ and $B = \{\text{prime numbers less than } 20\}$.

a List the elements of:

i A **ii** B **iii** $A \cap B$ **iv** $A \cup B$

b Show that $n(A \cup B) = n(A) + n(B) - n(A \cap B)$.

5 Suppose $A = \{\text{positive single digit whole numbers}\}$ and $B = \{\text{multiples of 3 less than } 20\}$.

a List the elements of:

i A **ii** B **iii** $A \cap B$ **iv** $A \cup B$

b Show that $n(A \cap B) = n(A) + n(B) - n(A \cup B)$.

6 Let $P = \{\text{factors of } 15\}$, $Q = \{\text{composite numbers less than or equal to } 15\}$, and $R = \{\text{even numbers less than } 15\}$. Which pair of sets is disjoint?

7 Suppose A and B are sets within a universal set U, and $A \subseteq B$. Explain why A and B' are disjoint.

8 Let $U = \{\text{letters of the English alphabet}\}$, $A = \{\text{letters in the word ALLIGATOR}\}$, $B = \{\text{letters in the word PINEAPPLE}\}$, and $C = \{\text{letters in the word MOUNTAIN}\}$.

a List the elements of A, B, and C.

b Find each set and explain what it represents:

i B' **ii** $A \cap C$ **iii** $A \cup B$ **iv** $(A \cup B)'$
v $A \cap B$ **vi** $A \cap C'$ **vii** $A \cap C \cap B'$ **viii** $A \cap B \cap C$

9 Suppose A and B are both infinite sets, and C is finite. If possible, determine whether these sets are finite or infinite:

a $A \cup B$ **b** $A \cap B$ **c** $A \cup C$ **d** $B \cap C$

D SPECIAL NUMBER SETS

The following is a list of some special number sets you should be familiar with. To help refer to them easily, we give them symbols:

- $\mathbb{N} = \{0, 1, 2, 3, 4, 5, 6, 7, ...\}$ is the set of **natural numbers**.

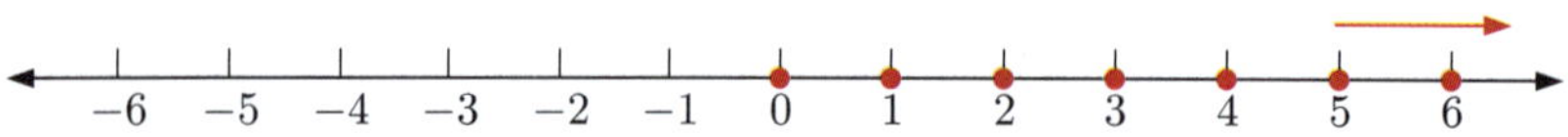

- $\mathbb{Z} = \{0, \pm 1, \pm 2, \pm 3, \pm 4, ...\}$ is the set of **integers**.

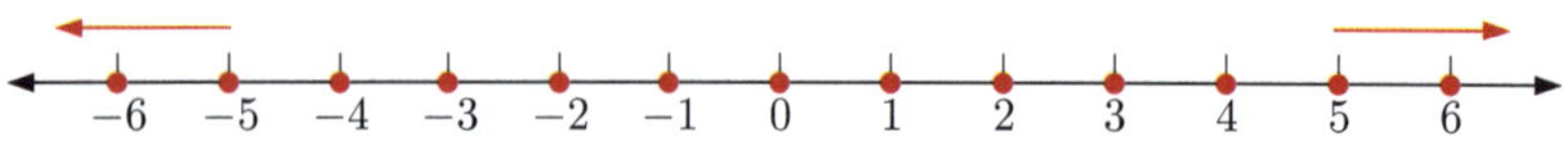

- $\mathbb{Z}^+ = \{1, 2, 3, 4, 5, 6, 7, ...\}$ is the set of **positive integers**.

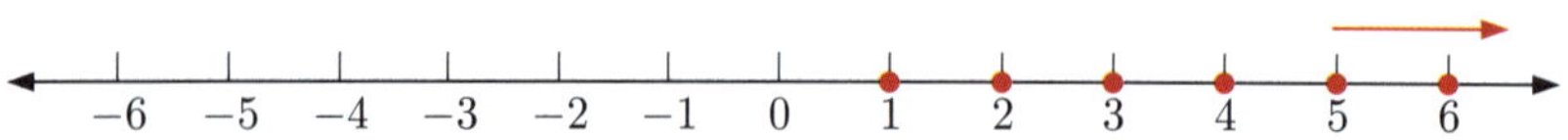

- $\mathbb{Z}^- = \{-1, -2, -3, -4, -5, -6, -7, ...\}$ is the set of **negative integers**.

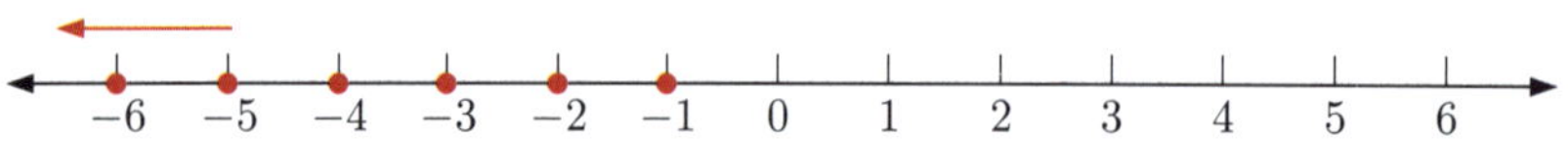

- $\mathbb{Q}$ is the set of **rational numbers**, or numbers which can be written in the form $\frac{p}{q}$ where p and q are integers, $q \neq 0$.

  For example: $\frac{15}{4}$, $10 = \frac{10}{1}$, $0.5 = \frac{1}{2}$, and $-\frac{3}{8}$ are all rational numbers.

- $\mathbb{Q}'$ is the set of **irrational numbers**, or numbers which cannot be written in rational form.

  For example: $\sqrt{3}$, $\sqrt[3]{5}$, and π are all irrational.

- $\mathbb{R}$ is the set of **real numbers**, which are all numbers that can be placed on the number line.

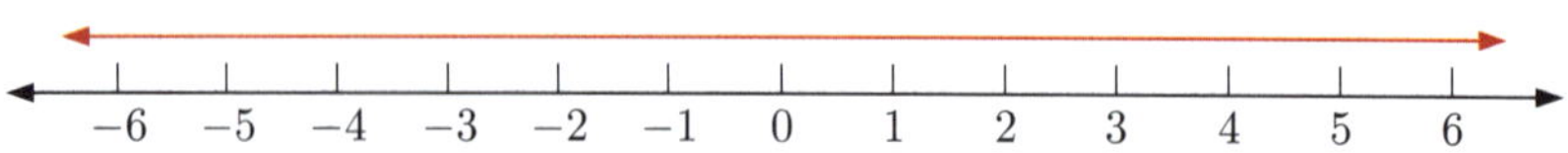

  $\mathbb{R}$ includes all rational and irrational numbers.

  $\frac{3}{0}$ and $\sqrt{-2}$ cannot be placed on a number line, and so are not real numbers.

All of these special number sets are infinite sets.

EXERCISE 4D

1 True or false?

a $3 \in \mathbb{Z}^+$ **b** $6 \in \mathbb{Z}$ **c** $\frac{3}{4} \in \mathbb{Q}$ **d** $\sqrt{2} \notin \mathbb{Q}$

e $-\frac{1}{4} \notin \mathbb{Q}$ **f** $2\frac{1}{3} \in \mathbb{Z}$ **g** $0.3684 \in \mathbb{R}$ **h** $\frac{1}{0.1} \in \mathbb{Z}$

2 True or false?

a $\mathbb{N} \subseteq \mathbb{Z}^-$ **b** $\mathbb{Z}^+ \subseteq \mathbb{Z}$ **c** $\mathbb{Q} \subseteq \mathbb{R}$ **d** $\mathbb{R} \subseteq \mathbb{Q}'$

3 For the universal set $U = \mathbb{N}$, find the complement of $\mathbb{Z}^+$.

4 For the universal set $U = \mathbb{Z}$, find the complement of $\mathbb{Z}^-$.

5 Which of the special number sets is disjoint with $\mathbb{Z}$?

6 Determine whether each of the following numbers is rational, irrational, or neither:

a 8 **b** $-3\frac{1}{4}$ **c** $\sqrt{7}$ **d** $\sqrt{-3}$

e $\sqrt{400}$ **f** 9.176 **g** $\frac{1}{0}$ **h** $\pi - \pi$

7 Explain why 0.527 is a rational number.

8 Give an example to show that each statement is false:

a The sum of two irrational numbers is always irrational.

b The product of two irrational numbers is always irrational.

c The quotient of two irrational numbers is always irrational.

d The product of a rational number and an irrational number is always irrational.

DISCUSSION

The concept of **infinity** or an "infinite number" can be difficult to comprehend.

- Every element in $\mathbb{Z}^+$ and its negative are both in $\mathbb{Z}$. $\mathbb{Z}$ also contains 0. Does it make sense to say that there are about twice as many elements in $\mathbb{Z}$ as in $\mathbb{Z}^+$?
- For any real number greater than 1, its reciprocal is a real number between 0 and 1. Does it make sense to say that there are the same number of real numbers between 0 and 1 as there are greater than 1?

E INTERVAL NOTATION

Consider the set of real numbers between -3 and 2, including 2.

This is an infinite set, so we cannot list all the elements of the set.

Instead we can describe this set using an *interval* on the number line.

Using **interval notation**, we write the set as $\{x \in \mathbb{R} \mid -3 < x \leqslant 2\}$.

We read this as "the set of real numbers x such that x lies between -3 and 2, including 2".

Unless stated otherwise, we *assume* we are dealing with *real* numbers. Thus, the set can also be written as $\{x \mid -3 < x \leqslant 2\}$.

We can also represent the set on a number line.

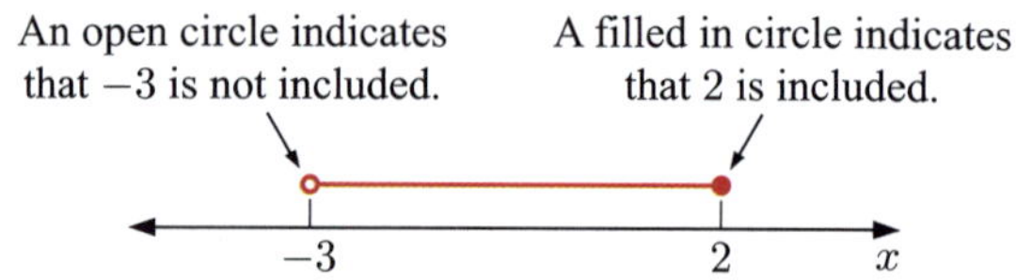

Example 1 **Self Tutor**

Suppose $A = \{x \in \mathbb{Z} \mid 0 < x \leqslant 7\}$.

a Write down the meaning of the interval notation.

b List the elements of A.

c Find $n(A)$.

d Represent A on a number line.

a The set of integers greater than 0 and less than or equal to 7.

b $A = \{1, 2, 3, 4, 5, 6, 7\}$

c $n(A) = 7$

d

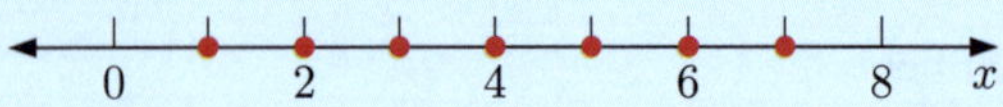

EXERCISE 4E

1 Write down the meaning of the interval notation:

a $\{x \mid x \leqslant 6\}$
b $\{x \in \mathbb{Z} \mid x > -5\}$
c $\{x \in \mathbb{N} \mid x < 10\}$
d $\{x \mid -5 < x < 3\}$
e $\{x \in \mathbb{Z} \mid -6 \leqslant x < 1\}$
f $\{x \mid x < -4 \text{ or } x \geqslant 7\}$

2 List the elements of the set:

a $\{x \in \mathbb{Z} \mid 2 \leqslant x \leqslant 9\}$
b $\{x \in \mathbb{Z}^+ \mid x < 7\}$
c $\{x \in \mathbb{N} \mid -4 < x \leqslant 5\}$
d $\{x \in \mathbb{Z}^- \mid -8 < x \leqslant 2\}$
e $\{x \in \mathbb{Z} \mid x < -3\}$
f $\{x \in \mathbb{N} \mid x > 9\}$

3 Represent on a number line:

a $x < 3$
b $x \geqslant 2$
c $x \leqslant 5$
d $x > -4$
e $-3 < x \leqslant 1$
f $2 \leqslant x < 5$
g $x < 2$ or $x > 4$
h $x \leqslant -1$ or $x > 2$
i $0 < x < 8$
j $3 \leqslant x \leqslant 5$
k $x < 1$ or $x \geqslant 7$
l $x \leqslant 2$ or $x \geqslant 6$

4 Write in interval notation:

a

b

c

d

e

f
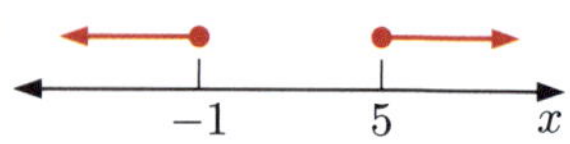

g
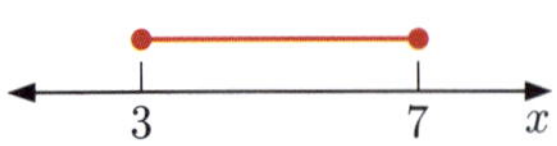

h

i
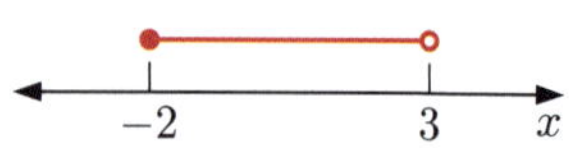

j
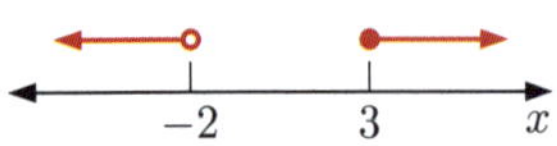

k

l
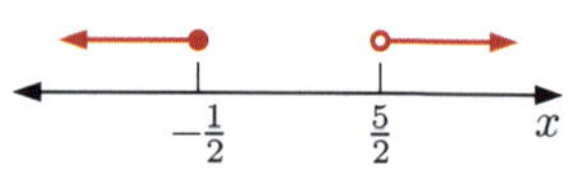

5 Write in interval notation:

a $\{3, 4, 5, 6, 7\}$
b $\{...., -4, -3, -2, -1\}$
c $\{11, 12, 13, 14, ...\}$

d
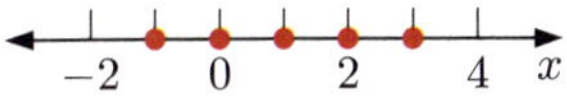

e
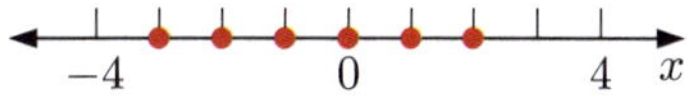

6 Sketch on a number line:

a $\{x \in \mathbb{N} \mid 3 \leqslant x < 10\}$
b $\{x \in \mathbb{Z} \mid -3 \leqslant x < 6\}$
c $\{x \in \mathbb{Z} \mid x \leqslant 2\}$
d $\{x \in \mathbb{N} \mid x < 7\}$
e $\{x \in \mathbb{R} \mid x \geqslant -4\}$
f $\{x \in \mathbb{Z} \mid -1 \leqslant x < 5\}$

7 For each set B:

i Write down the meaning of the interval notation.
ii If possible, list the elements of B.
iii Find $n(B)$.
iv Represent B on a number line.

a $B = \{x \in \mathbb{Z} \mid -2 < x < 5\}$
b $B = \{x \in \mathbb{Z}^+ \mid -6 \leqslant x \leqslant 3\}$
c $B = \{x \mid 1 < x < 2\}$
d $B = \{x \mid 4 \leqslant x < 7\}$

8 Write in interval notation:

a the set of integers between 10 and 100
b the set of real numbers less than 200
c the set of real numbers greater than or equal to -8
d the set of rational numbers between -1 and 1, including 1.

9 Determine whether each set is finite or infinite:

a $\{x \in \mathbb{Z} \mid -10 \leqslant x \leqslant 10\}$
b $\{x \in \mathbb{N} \mid x \geqslant 50\}$
c $\{x \in \mathbb{Q} \mid 4 \leqslant x \leqslant 5\}$
d $\{x \mid x < -1\}$
e $\{x \in \mathbb{Z}^+ \mid x \leqslant 12\}$
f $\{x \mid 0 < x < 1\}$

10 Determine whether $C \subseteq D$ for:

a $C = \{x \mid 0 \leqslant x \leqslant 10\}$ and $D = \{x \mid -10 < x < 20\}$
b $C = \{x \in \mathbb{Z} \mid 2 \leqslant x \leqslant 5\}$ and $D = \{x \mid 2 \leqslant x < 5\}$
c $C = \{x \in \mathbb{Z}^- \mid x > -7\}$ and $D = \{x \in \mathbb{Z} \mid -6 \leqslant x \leqslant 0\}$
d $C = \{x \in \mathbb{Q} \mid 0 < x < 1\}$ and $D = \{x \in \mathbb{Q}' \mid 0 \leqslant x \leqslant 1\}$.

11 Suppose $U = \{x \in \mathbb{Z}^+ \mid x \leqslant 15\}$, $A = \{x \in \mathbb{Z}^+ \mid 5 \leqslant x < 13\}$, and $B = \{x \in \mathbb{Z}^+ \mid 6 < x < 10\}$.

a List the elements of:
i A **ii** A' **iii** B **iv** B'

b True or false?
i $A \subseteq B$ **ii** $B \subseteq A$ **iii** $A' \subseteq B'$ **iv** $B' \subseteq A'$

12 Let $U = \mathbb{R}$, $P = \{x \mid -5 \leqslant x < 10\}$, $Q = \{x \mid 4 < x < 20\}$, and $R = \{x \mid x \leqslant 6\}$.

a Write in interval notation:
i $P \cup Q$ **ii** $Q \cup R$ **iii** $P \cap Q$ **iv** P'

b Represent on a number line:
i $P \cup R$ **ii** Q' **iii** $P \cup Q \cup R$ **iv** $P \cap Q \cap R$

13 **a** Represent on a number line:

i $3 \leqslant x < 9$ **ii** $-1 < x < 4$

b Represent on a number line, and describe using a single interval:

i $\{x \mid 3 \leqslant x < 9 \ \cup \ -1 < x < 4\}$ **ii** $\{x \mid 3 \leqslant x < 9 \ \cap \ -1 < x < 4\}$

QUICK QUIZ

MULTIPLE CHOICE QUIZ

REVIEW SET 4A

1 Let P be the set of all prime numbers between 20 and 40.

a List the elements of P.

b True or false? **i** $37 \in P$ **ii** $27 \in P$

c Find $n(P)$.

2 Decide whether $P \subseteq Q$ for:

a $P = \{5, 6, 7, 8\}$ and $Q = \{1, 2, 3, 4, 5, 6, 7\}$

b $P = \{\text{multiples of 4 between 10 and 30}\}$ and $Q = \{\text{even numbers between 0 and 40}\}$

c $P = \{\text{factors of 100}\}$ and $Q = \{\text{factors of 200}\}$.

3 Suppose $U = \{x \in \mathbb{Z}^+ \mid x \leqslant 10\}$. Find the complement of:

a $A = \{3, 7, 9\}$ **b** $B = \{\text{composite numbers}\}$

4 Suppose P is the set of positive two digit whole numbers which contain the digit 3, and Q is the set of positive two digit whole numbers which contain the digit 8.
List the elements of:

a $P \cup Q$ **b** $P \cap Q$

5 Within an infinite universal set U, A is an infinite set and B is a finite set.
Explain why $A \cap B'$ must be infinite.

6 True or false?

a $-4 \in \mathbb{Z}$ **b** $\mathbb{Q}' \subseteq \mathbb{R}$ **c** $\mathbb{N} \cup \mathbb{Z}^- = \mathbb{Z}$

7 Determine whether each of the following numbers is rational, irrational, or neither:

a $\sqrt{5}$ **b** $\sqrt{4}$ **c** 1.605 **d** $\frac{8}{0}$

8 Write down the meaning of the interval notation:

a $\{x \in \mathbb{Z} \mid -2 \leqslant x \leqslant 7\}$ **b** $\{x \in \mathbb{N} \mid x < 8\}$ **c** $\{x \mid x < -1 \text{ or } x \geqslant 5\}$

9 Write in interval notation:

a
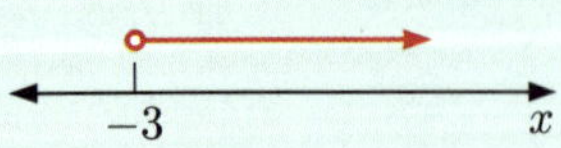

b
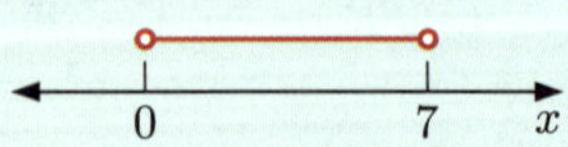

c
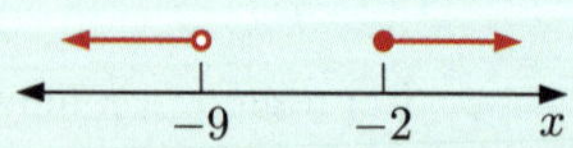

10 Determine whether each set is finite or infinite:

a $\{x \in \mathbb{Z} \mid x \leqslant 7\}$ **b** $\{x \in \mathbb{N} \mid x < 15\}$ **c** $\{x \in \mathbb{Q}' \mid 1 \leqslant x \leqslant 3\}$

11 **a** Represent on a number line:

i $x > -1$ **ii** $-2 < x \leqslant 3$

b Represent on a number line, and describe using a single interval:

i $\{x \mid x > -1 \ \cup\ -2 < x \leqslant 3\}$ **ii** $\{x \mid x > -1 \ \cap\ -2 < x \leqslant 3\}$

12 Suppose $A = \{x \in \mathbb{Z} \mid -3 \leqslant x \leqslant 6\}$.

a Write down the meaning of the interval notation.

b List the elements of A.

c Find $n(A)$.

d Represent A on a number line.

REVIEW SET 4B

1 State whether each set is finite or infinite. If the set is finite, state the number of elements in the set.

a $A = \{\text{factors of } 40\}$ **b** $B = \{\text{multiples of } 10\}$

c $C = \{\text{natural numbers less than } 20\}$

2 Suppose $P = \{\text{factors of } 20\}$ and $Q = \{\text{factors of } k\}$, where k is a positive whole number. For what values of k is $Q \subseteq P$?

3 Let $U = \{\text{positive whole numbers}\}$, $A = \{\text{numbers greater than } 50\}$, and $B = \{\text{numbers between 60 and 100}\}$.

a True or false? **i** $120 \in A'$ **ii** $120 \in B'$

b Is A' finite or infinite?

c Is B' finite or infinite?

d Is $A' \subseteq B'$?

4 Let $A = \{\text{multiples of 3 less than } 40\}$, $B = \{\text{odd numbers less than } 20\}$, and $C = \{\text{square numbers less than } 40\}$. Find:

a $B \cup C$ **b** $A \cap B \cap C$

5 Let $P = \{\text{non-convex polygons}\}$, $Q = \{\text{quadrilaterals}\}$, $R = \{\text{regular polygons}\}$, and $S = \{\text{triangles}\}$.

Which pairs of sets are disjoint?

6 True or false?

a $\frac{1}{\sqrt{2}} \in \mathbb{Q}$ **b** $\mathbb{Z}^- \cup \mathbb{Z}^+ = \mathbb{Z}$ **c** $\mathbb{Q}' \cap \mathbb{Z} = \varnothing$

7 Explain why the following are rational numbers:

a 0.063 **b** $\frac{1}{\sqrt{16}}$

8 Represent on a number line:

a $x \geqslant 5$ **b** $-6 < x \leqslant 1$ **c** $x \leqslant 3$ or $x > 9$

9 Write in interval notation:

a $\{...., -6, -5, -4, -3\}$

b 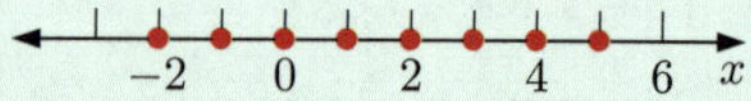

10 Let $A = \{x \mid -4 \leqslant x \leqslant 6\}$, $B = \{x \mid 1 < x < 10\}$, and $C = \{x \mid x \leqslant 3\}$.

a Write in interval notation:

i $A \cup B$ **ii** $A \cap C$

b Represent on a number line:

i $B \cup C$ **ii** $A \cap B \cap C$

11 Suppose $A = \{$even positive numbers less than 15$\}$, $B = \{$factors of 30$\}$, and $C = \{$single digit positive integers$\}$.

a List the elements of:

i A **ii** B **iii** C **iv** $A \cap B$ **v** $B \cap C$

b Show that $A \cap (B \cap C) = (A \cap B) \cap C$.

c Explain why $A \cap (B \cap C) = (A \cap B) \cap C$ for any sets A, B, and C.

12 Suppose $n(A) = 8$, $n(B) = 11$, and $n(U) = 15$.

a Find $n(A')$ and $n(B')$.

b Explain why it is impossible for B to be a subset of A.

c Use interval notation to describe the possible values for $n(A \cup B)$.

d Use a number line to represent the possible values for $n(A \cap B)$.

Chapter 5

Linear equations and inequalities

Contents:

OPENING PROBLEM

Janine has made a 2 kg pot of apricot jam. She plans to keep 200 g of jam for herself, and share the rest equally between her 6 grandchildren.

Things to think about:

a Suppose everything goes to plan, and Janine gives x grams of jam to each grandchild.

- **i** Can you write an *equation* to represent this information?
- **ii** How much jam would each grandchild receive?

b Unfortunately, Janine's cat knocks the pot of jam off the bench. *At least* 300 g of jam is spilt on the floor. Janine decides not to keep any for herself, but to instead share everything she has left between her grandchildren.

Suppose each grandchild receives x grams of jam.

- **i** Can you write an *inequality* to represent this information?
- **ii** How much jam can Janine give to each grandchild?

An **equation** is a mathematical statement which indicates that two expressions have the same value.

The expressions are connected by an **equal sign** $=$.

The **left hand side** (LHS) of an equation is on the left of the $=$ sign.

The **right hand side** (RHS) of an equation is on the right of the $=$ sign.

We can often convert a worded problem into an equation, then follow a formal procedure to **solve** the equation and hence the problem.

In other problems, one quantity we are considering may be greater than or less than another quantity. The problem may then be written as a mathematical **inequality**.

An **inequality** is a mathematical statement which *compares* the values of two expressions. It will contain one of the symbols $>$, $<$, $\geqslant$, or $\leqslant$.

A LINEAR EQUATIONS

Linear equations are equations in which the variable is raised only to the power 1.

All linear equations can be written in the form $ax + b = 0$ where a and b are constants, $a \neq 0$, and x is the variable.

For example, $3x + 4 = 2$, $\frac{2}{3}x + 1 = 6$, and $\frac{x-1}{4} = 8$ are all linear equations

whereas $x^2 + 5x = 7$, $\frac{3}{x} = x^3$, and $\sqrt{x} = 8$ are *not* linear equations.

When we use the $=$ sign between two algebraic expressions, we have an equation which is in balance. Whatever we do to one side of the equation, we must do the same to the other side to **maintain the balance**.

In previous years, we have seen how to solve linear equations algebraically:

Step 1: If necessary, **expand** any **brackets** and **collect like terms**.
Step 2: If necessary, move the unknown to one side of the equation.
Step 3: Use **inverse operations** to **isolate the unknown** and maintain the balance.
Step 4: **Check** your solution by substituting back into the original equation.

Example 1 Self Tutor

Solve for x:

a $3x+7=22$ **b** $\frac{x}{5}+2=-2$

a

$3x+7=22$

$\therefore\ 3x+7-7=22-7$ {$-$ 7 both sides}

$\therefore\ 3x=15$

$\therefore\ \frac{3x}{3}=\frac{15}{3}$ {$\div$ 3 both sides}

$\therefore\ x=5$

Check: $3\times5+7=22$ ✓

b

$\frac{x}{5}+2=-2$

$\therefore\ \frac{x}{5}+2-2=-2-2$ {$-$ 2 both sides}

$\therefore\ \frac{x}{5}=-4$

$\therefore\ \frac{x}{5}\times5=-4\times5$ {$\times$ 5 both sides}

$\therefore\ x=-20$

Check: $\frac{-20}{5}+2=-4+2=-2$ ✓

EXERCISE 5A

1 Solve for x:

a $x+11=0$ **b** $x-4=-10$ **c** $4x=-12$ **d** $\frac{x}{3}=-9$

e $x+7=19$ **f** $x-15=8$ **g** $-2x=16$ **h** $\frac{x}{-12}=-7$

2 Solve for x:

a $5x+35=0$ **b** $4x-5=-17$ **c** $3x-9=18$

d $3x+5=-10$ **e** $13+7x=-1$ **f** $14=3x+5$

g $-3=2x+9$ **h** $8-x=-3$ **i** $3-2x=11$

j $6-4x=-8$ **k** $15=3-2x$ **l** $24-3x=-9$

m $13=-1-7x$ **n** $-21=3-6x$ **o** $23=-4-3x$

3 Solve for x:

a $\frac{2x}{5}=-6$ **b** $\frac{x}{2}+3=-5$ **c** $\frac{x}{4}-2=-5$

d $\frac{x-1}{3}=6$ **e** $\frac{x+5}{6}=-1$ **f** $4=\frac{2+x}{3}$

g $-1+\frac{x}{3}=7$ **h** $6-\frac{x}{8}=10$ **i** $\frac{3}{5}=2-\frac{x}{5}$

Example 2 Self Tutor

Solve for x: $\dfrac{4x+3}{5} = -2$

$$\frac{4x+3}{5} = -2$$

$$\therefore\ 5 \times \frac{(4x+3)}{5} = -2 \times 5 \qquad \{\times 5 \text{ both sides}\}$$

$$\therefore\ 4x+3 = -10$$

$$\therefore\ 4x+3-3 = -10-3 \qquad \{-3 \text{ both sides}\}$$

$$\therefore\ 4x = -13$$

$$\therefore\ \frac{4x}{4} = \frac{-13}{4} \qquad \{\div 4 \text{ both sides}\}$$

$$\therefore\ x = -\tfrac{13}{4}$$

Check: $\dfrac{4 \times -\frac{13}{4} + 3}{5} = \dfrac{-13+3}{5} = \dfrac{-10}{5} = -2$ ✓

4 Solve for x:

a $\dfrac{2x+11}{3} = 0$ **b** $\frac{1}{2}(3x+1) = -4$ **c** $\dfrac{1+2x}{5} = 7$

d $\dfrac{1-2x}{5} = 3$ **e** $\frac{1}{4}(1-3x) = -2$ **f** $\frac{1}{4}(5-2x) = -3$

5 Solve for x:

a $-6(4-5x)+2 = -40$ **b** $\frac{1}{4}(9x-7)-11 = 11$ **c** $10 - \dfrac{2-3x}{6} = -2$

Example 3 Self Tutor

Solve for x: $5(2x+3) - 4(x-2) = -1$

$$5(2x+3) - 4(x-2) = -1$$

$$\therefore\ 10x + 15 - 4x + 8 = -1 \qquad \{\text{distributive law}\}$$

$$\therefore\ 6x + 23 = -1 \qquad \{\text{collecting like terms}\}$$

$$\therefore\ 6x + 23 - 23 = -1 - 23 \qquad \{-23 \text{ both sides}\}$$

$$\therefore\ 6x = -24$$

$$\therefore\ \frac{6x}{6} = \frac{-24}{6} \qquad \{\div 6 \text{ both sides}\}$$

$$\therefore\ x = -4$$

Check: $5(2 \times -4 + 3) - 4(-4-2) = 5 \times (-5) - 4 \times (-6) = -25 + 24 = -1$ ✓

6 Solve for x:

a $4x + 10 + 3x = 24$ **b** $8x - 5 - 3x = -40$

c $3(x-2) - x = 12$ **d** $2(3x+2) - x = -6$

e $-2(4x+3) + 2x = 12$ **f** $3(x+2) + 2(x+4) = -1$

g $4(x-3) - 2(x-1) = -6$ **h** $4(5x-3) - 3(2x-5) = 17$

Example 4 Self Tutor

Solve for x: $5x + 2 = 3x - 5$

$$
\begin{aligned}
5x + 2 &= 3x - 5 \\
\therefore\ 5x + 2 - 3x &= 3x - 5 - 3x \quad \{-\,3x \text{ both sides}\} \\
\therefore\ 2x + 2 &= -5 \\
\therefore\ 2x + 2 - 2 &= -5 - 2 \quad \{-\,2 \text{ both sides}\} \\
\therefore\ 2x &= -7 \\
\therefore\ \frac{2x}{2} &= \frac{-7}{2} \quad \{\div\, 2 \text{ both sides}\} \\
\therefore\ x &= -\tfrac{7}{2}
\end{aligned}
$$

Check: LHS $= 5 \times -\frac{7}{2} + 2 = -\frac{35}{2} + \frac{4}{2} = -\frac{31}{2}$,
RHS $= 3 \times -\frac{7}{2} - 5 = -\frac{21}{2} - \frac{10}{2} = -\frac{31}{2}$ ✓

7 Solve for x:

- **a** $8x + 7 = 4x - 5$
- **b** $7x + 3 = 2x + 9$
- **c** $3x - 8 = 5x - 2$
- **d** $x - 3 = 5x + 11$
- **e** $6 + 2x = 15 - x$
- **f** $3x + 7 = 15 - x$
- **g** $17 - 3x = 4 - x$
- **h** $8 - x = x + 6$
- **i** $9x - 4 = x + 13 - 7x$

8 Solve for x:

- **a** $5(2 - 3x) = -8 - 6x$
- **b** $4(2x - 1) + 9 = 3x$
- **c** $8 - (2 - x) = 2x$
- **d** $5x - 4(4 - x) = x + 12$
- **e** $4(x - 1) = 1 - (3 - x)$
- **f** $5(3 - 2x) + x = 3(2x + 1) - 8$
- **g** $9 - 7(3x + 5) = x - 2(x + 4)$
- **h** $2(4x - 1) - 5(x - 3) = 1 - 4x$
- **i** $5(3x + 2) - 7(1 - x) = 10(x - 3)$
- **j** $6(x - 5) - 5(3 - 2x) = 3(9 - 11x)$

LEARNING ALGEBRA

B PROBLEM SOLVING WITH EQUATIONS

Many problems can be translated into **algebraic equations**.

To solve problems using algebra, we follow these steps:

Step 1: Decide on the unknown quantity and allocate it a variable such as x.

Step 2: Translate the problem into an equation.

Step 3: Solve the equation by isolating the variable.

Step 4: Check that the solution satisfies the original problem.

Step 5: Write the answer in sentence form, describing how the solution relates to the original problem.

Example 5 Self Tutor

The sum of three consecutive even integers is 132. Find the smallest integer.

Let x be the smallest even integer.

$\therefore$ the next is $x+2$, and the largest is $x+4$.

The sum of the numbers is 132.

$\therefore\ x+(x+2)+(x+4)=132$

$\therefore\ 3x+6=132$ {collecting like terms}

$\therefore\ 3x=126$ {$-$ 6 both sides}

$\therefore\ x=42$ {$\div$ 3 both sides}

So, the smallest integer is 42.

EXERCISE 5B

1. When a number is trebled and then decreased by 5, the result is 19. Find the number.
2. Two consecutive integers have a sum of 173. Find the integers.
3. Three consecutive integers add to 108. Find the smallest integer.
4. A number is decreased by 1 and the result is doubled. The final result is 34. Find the number.
5. A number is increased by 4 and the result is halved. The final result is equal to the original number. Find the number.
6. Three times a number is equal to 17 minus the number. Find the number.
7. Five times a number is 2 more than the number. Find the number.
8. Tony made some small pizzas for his birthday party. He cut 4 of them in halves, and the rest in thirds. There were 29 pieces of pizza in total. How many pizzas did Tony make?
9. An art exhibition was held from Friday to Sunday. 50 more people attended on Saturday than on Friday, and 80 more people attended on Sunday than on Saturday. A total of 540 people attended the exhibition. How many people attended on Friday?
10. Dianne used a \$50 note to pay for a tour of the Royal Canadian Mint. The change she received was \$30 more than the cost of the tour. Find the cost of the tour.
11. Marcia bought a 500 g packet of flour. She used some flour to bake a cake, and she used two thirds of the remaining flour to make some scones. She used 60 g more flour for the cake than for the scones. How much flour did Marcia use for the cake?
12. Of the lambs born on a farm one springtime, there were 50 fewer males than females. If $\frac{9}{20}$ of the lambs were male, find the number of lambs born that springtime.

13. In a game of 10-pin bowling, Emma scored 15 points less than Alex, and Alex scored 30% more points than Toni. Their average score was 127. How many points did each person score?

Example 6 **Self Tutor**

Britney has only 10 cent and 50 cent stamps. There are two more 50 cent stamps than there are 10 cent stamps, and their total value is \$6.40. How many 10 cent stamps are there?

Let x be the number of 10 cent stamps.

$\therefore$ there are $(x+2)$ 50 cent stamps.

The total value of the stamps is \$6.40 or 640 cents.

Type	*Number*	*Value*
10 cent	x	$10x$ cents
50 cent	$x+2$	$50(x+2)$ cents

$\therefore\ 10x + 50(x+2) = 640$

$\therefore\ 10x + 50x + 100 = 640$ {distributive law}

$\therefore\ 60x + 100 = 640$ {collecting like terms}

$\therefore\ 60x = 540$ $\{-100$ both sides$\}$

$\therefore\ x = 9$ $\{\div 60$ both sides$\}$

So, there are nine 10 cent stamps.

14 Lynn has some 20 cent coins and 50 cent coins in her wallet. There are five more 20 cent coins than 50 cent coins, and their total value is \$5.90. How many 50 cent coins are there?

15 250 g cans of soup cost \$3 each, and 500 g cans cost \$5.50 each. I bought a total of 12 cans, and they cost me \$48.50. How many 500 g cans did I buy?

16 A corner shop has an equal number of 600 mL and 1 L cartons of milk, and double that number of 2 L bottles of milk. If there are 84 L of milk in total, how many 600 mL cartons does the shop have?

17 Inka invested three times as much money in mining shares as she invested in telecommunications shares. Mining shares earn 9% yearly, and telecommunications shares earn 7% yearly. Inka's yearly income from the shares is \$816. How much money did Inka invest in mining shares?

18 Bruce collects stamps. He has six times as many 10 cent stamps as 5 cent stamps, and he has some 20 cent stamps as well. Bruce has 72 stamps with total value \$8.40. How many of each stamp does he have?

Example 7 **Self Tutor**

Show that $0.\overline{36}$ is a rational number.

Let $x = 0.\overline{36} = 0.363\,636\,36....$

$\therefore\ 100x = 36.363\,636.... = 36 + x$

$\therefore\ 99x = 36$

$\therefore\ x = \frac{36}{99} = \frac{4}{11}$

So, $0.\overline{36}$ is actually the rational number $\frac{4}{11}$.

19 Show that each number is rational:

a $0.\overline{7}$ **b** $0.\overline{41}$ **c** $0.\overline{324}$ **d** $0.0\overline{69}$

C LINEAR INEQUALITIES

An **inequality** is a mathematical statement which compares the values of the two expressions. It will contain one of the symbols $>$, $<$, $\geqslant$, or $\leqslant$.

Linear inequalities take the same form as linear equations, except they contain an inequality sign instead of an "equals" sign. For example, $4 - 3x < 0$ and $5x + 2 \geqslant 7$ are linear inequalities.

RULES FOR SOLVING LINEAR INEQUALITIES

- If we **swap** the LHS and RHS, we **reverse** the inequality sign.
- If we **add** or **subtract** both sides by the same number, we **keep** the inequality sign.
- If we **multiply** or **divide** both sides by:
  - a **positive** number we **keep** the inequality sign
  - a **negative** number we **reverse** the inequality sign.

For example, starting with the inequality $5 \geqslant 3$, if we:

- swap the LHS and RHS, we have $3 \leqslant 5$
- add 2 to both sides, we have $7 \geqslant 5$
- multiply both sides by 3, we have $15 \geqslant 9$
- divide both sides by -1, we have $-5 \leqslant -3$.

We should not multiply or divide both sides of an inequality by an expression involving the variable, unless we are certain that the expression is always positive or always negative.

Example 8 Self Tutor

Solve for x and graph the solution:

a $3x - 4 \leqslant 2$ **b** $3 - \frac{x}{2} < 7$

a $3x - 4 \leqslant 2$

$\therefore\ 3x \leqslant 6$ {+ 4 both sides}

$\therefore\ x \leqslant 2$ {÷ 3 both sides}

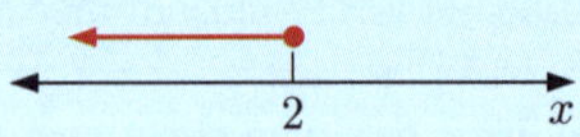

Check: If $x = 1$ then $3x - 4 = 3 \times 1 - 4 = -1$ and $-1 \leqslant 2$ ✓

b $3 - \frac{x}{2} < 7$

$\therefore\ -\frac{x}{2} < 4$ {− 3 both sides}

$\therefore\ x > -8$ {× −2 both sides}

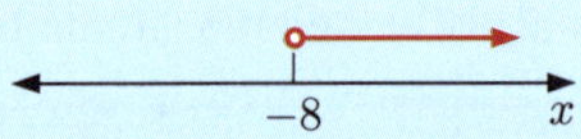

Check: If $x = 0$ then $3 - \frac{0}{2} = 3$ and $3 < 7$ ✓

EXERCISE 5C.1

1 Solve the inequality and graph its solution:

a $x+5<9$ **b** $6a \geqslant 12$ **c** $k-3 \leqslant -7$ **d** $\frac{x}{4} \geqslant 8$

e $-3x<15$ **f** $7+y>0$ **g** $10a \leqslant -80$ **h** $-\frac{c}{7} \geqslant 6$

2 Solve the inequality and graph its solution:

a $3x+2<0$ **b** $5a-7>2$ **c** $2-3x \geqslant 1$

d $5-2x \leqslant 11$ **e** $6x-2<4$ **f** $-x+8 \leqslant 5$

g $5x-3<-8$ **h** $2(3x-1)<4$ **i** $5(1-3x) \geqslant 8$

3 Solve for x and graph the solution:

a $\frac{x}{2}-3>5$ **b** $2-\frac{x}{3} \leqslant 9$ **c** $\frac{x}{4}+4 \geqslant -5$

d $\frac{2x}{3}-2 \leqslant 4$ **e** $8-\frac{3x}{2}>20$ **f** $1-\frac{2x}{5} \geqslant -6$

g $\frac{x+5}{4} \geqslant 1$ **h** $\frac{4-x}{2}<5$ **i** $\frac{3-2x}{-3} \geqslant -1$

4 Solve by first swapping the LHS and RHS:

a $7 \geqslant 2x-1$ **b** $-13<3x+2$ **c** $20>-5x$

d $-3 \geqslant 4-3x$ **e** $1<\frac{5-2x}{3}$ **f** $2 \leqslant 5(1-x)$

Example 9 — Self Tutor

Solve for x and graph the solution: $3-5x \geqslant 2x+7$

$3-5x \geqslant 2x+7$

$\therefore\ 3-7x \geqslant 7$ {$-\ 2x$ both sides}

$\therefore\ -7x \geqslant 4$ {$-\ 3$ both sides}

$\therefore\ x \leqslant -\frac{4}{7}$ {$\div -7$ both sides}

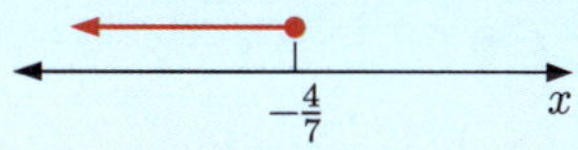

5 Solve for x and graph the solution:

a $3x+2>x-5$ **b** $2x-3<5x-7$ **c** $5-2x \geqslant x+4$

d $7-3x \leqslant 5-x$ **e** $6-4x>x-5$ **f** $1-7x<\frac{5-2x}{3}$

6 Solve for x:

a $4(x-1) \leqslant x+14$ **b** $3x-2>2(x-1)+5x$ **c** $1-(x-3) \geqslant 2(x+5)-1$

d $2(x-5)>\frac{4x-3}{3}$ **e** $6(1-x) \leqslant \frac{3x-1}{2}$ **f** $3(2-3x) \geqslant \frac{2x}{5}-2$

DOUBLE LINEAR INEQUALITIES

A **double inequality** is a statement which contains *two* inequality signs.

You have already seen double inequalities when using interval notation.

For example, the double inequality $5<x<8$ means that $x>5$ *and* $x<8$. So, x can take any value between 5 and 8.

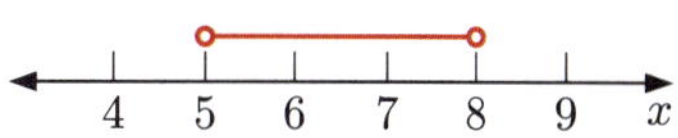

We solve double linear inequalities in the same way we solve linear inequalities. However, we must be careful to apply the same inverse operations to *every* expression in the inequality.

In our final answer, we write the expressions from smallest to largest. For example, we write $5 < x < 8$ rather than $8 > x > 5$.

Example 10 — Self Tutor

Solve for x and graph the solution:

a $5 < 3x - 4 < 17$ **b** $1 \leqslant \frac{1-2x}{3} < 4$

a $5 < 3x - 4 < 17$

$\therefore\ 9 < 3x < 21$ {+ 4 each expression}

$\therefore\ 3 < x < 7$ {÷ 3 each expression}

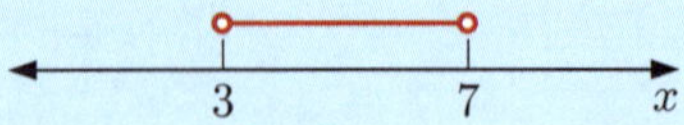

b $1 \leqslant \frac{1-2x}{3} < 4$

$\therefore\ 3 \leqslant 1 - 2x < 12$ {× 3 each expression}

$\therefore\ 2 \leqslant -2x < 11$ {− 1 each expression}

$\therefore\ -1 \geqslant x > -\frac{11}{2}$ {÷ −2 each expression}

$\therefore\ -\frac{11}{2} < x \leqslant -1$ {reversing the inequality}

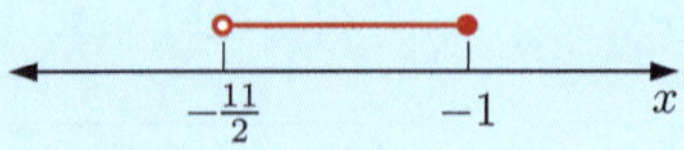

EXERCISE 5C.2

1 Solve the double inequality and graph its solution:

a $4 < x + 2 < 7$ **b** $-10 \leqslant 5a \leqslant 30$ **c** $0 \leqslant x - 4 < 9$

d $6 \leqslant -2x \leqslant 10$ **e** $-8 < 4t \leqslant -1$ **f** $-1 < \frac{m}{-3} \leqslant 2$

2 Solve for x and graph the solution:

a $1 \leqslant 2x - 3 \leqslant 9$ **b** $-3 < 5x + 1 \leqslant 21$ **c** $6 < 2 - x < 11$

d $0 \leqslant 7 - 3x < 25$ **e** $3 < \frac{x}{4} - 1 < 5$ **f** $-10 < \frac{x}{2} + 2 \leqslant -2$

g $-4 \leqslant 1 - \frac{x}{3} \leqslant 1$ **h** $6 < 5 - \frac{x}{5} \leqslant 11$ **i** $2 < \frac{x-1}{6} \leqslant 3$

j $-5 \leqslant \frac{x+3}{8} \leqslant -1$ **k** $0 < \frac{1-x}{4} < 6$ **l** $-2 \leqslant \frac{4-x}{3} < 5$

3 Solve for x:

a $-3 \leqslant \frac{2x}{3} - 5 < 1$ **b** $2 < 4 - \frac{5x}{2} < 3$ **c** $-5 \leqslant 1 - \frac{3x}{4} < 8$

d $4 < \frac{3x-5}{2} < 11$ **e** $-8 \leqslant \frac{1-2x}{3} \leqslant -3$ **f** $-9 < \frac{6-5x}{4} < 2$

DISCUSSION

How could you solve a double inequality where the unknown appears in more than one expression?
For example, how would you solve $x + 1 \leqslant 3x - 5 < 10$?

D PROBLEM SOLVING WITH INEQUALITIES

When solving the following worded problems, you must first convert the problem into an inequality. It is important to read the question carefully to determine what type of inequality is required.

As with solving worded problems with equations, you should always answer the question in sentence form.

Example 11 Self Tutor

A 50 hectare property is divided between Fiona, Greg, and Holly.
Greg's share is twice the size of Fiona's share, but it is smaller than Holly's share.
Discuss the possible size of Fiona's share.

Let Fiona's share be x hectares.

$\therefore$ Greg's share is $2x$ hectares, and Holly's share is $50 - x - 2x = 50 - 3x$ hectares.

Now $2x < 50 - 3x$ {Greg's share is smaller than Holly's share}

$\therefore$ $5x < 50$ {$+ 3x$ both sides}

$\therefore$ $x < 10$ {$\div 5$ both sides}

So, Fiona's share must be less than 10 hectares in size.

EXERCISE 5D

1. Jenny was told to think of a number, treble it, and then add 15. The result was greater than 60. Find the possible values of Jenny's original number.

2. As part of a special promotion, a furniture store is offering customers the choice of either \$60 off or else 15% off the purchase price.
Determine the purchase prices for which customers would be better off choosing the 15% discount.

3. The cost of sending a parcel depends on its weight. The amounts charged by two delivery services are:
Swift Delivery: \$8 flat fee plus \$1.30 per kg
Lightning Pace: \$6 flat fee plus \$1.50 per kg
Determine the weights for which it is cheaper to use *Lightning Pace*.

4. To convert a temperature from degrees Celsius (°C) to degrees Fahrenheit (°F), we multiply by $\frac{9}{5}$, then add 32.
   a. Find the range of temperatures in °F for which water is a liquid at sea level atmospheric pressure.
   b. Find the range of temperatures in °C, such that converting from °C to °F results in a lower number.

Example 12 — Self Tutor

Marty took two \$50 notes to go clothes shopping. He bought a \$20 pair of jeans and 4 identical shirts. He used both \$50 notes to pay for the clothes.

Find the range of possible prices for each shirt.

Let the price of each shirt be x dollars.

$\therefore$ the total cost of the clothes is $4x + 20$ dollars.

Now $50 < 4x + 20 \leqslant 100$ {Marty used both \$50 notes}

$\therefore 30 < 4x \leqslant 80$ {$-$ 20 each expression}

$\therefore \frac{30}{4} < x \leqslant 20$ {$\div$ 4 each expression}

$\therefore \frac{15}{2} < x \leqslant 20$

So, the price of each shirt must be more than \$7.50, but less than or equal to \$20.

5 The cost of parking in a car park is shown in the table.
Amanda parked in the car park, walked to a restaurant, spent 70 minutes at the restaurant, then walked back to her car. Her parking fee was \$16.
Find the range of times it could have taken Amanda to walk between the car park and the restaurant.

Time (t minutes)	*Cost*
$0 < t \leqslant 30$	\$5
$30 < t \leqslant 60$	\$9
$60 < t \leqslant 90$	\$13
$90 < t \leqslant 120$	\$16
$120 < t \leqslant 150$	\$19

6 The largest angle of a triangle is twice the size of the smallest angle. Find the possible angle sizes of the third angle.

7 Two angles of a quadrilateral are equal in size. The third angle is 30° larger than the equal angles, and the remaining angle is obtuse. Find the possible angle sizes of the equal angles.

8 Pierre took three days to cycle 192 km from Toulouse to Andorra. He travelled 24 km farther on the second day than on the first day.
Given that Pierre travelled between 45 km and 65 km on the third day, find the range of distances he could have travelled on the first day.

PUZZLE

In an experiment, 100 people were located on each of the islands A, B, and C. The people were allowed to move between the islands, but no one could enter or leave the set of islands.

After one month, the people on each island were asked if they were living on the same island where they were initially.

On Island A there were 80 "Yes" and 30 "No" responses.

On Island B there were 85 "Yes" and 5 "No" responses.

What can you say about the number of "Yes" responses on Island C?

QUICK QUIZ

MULTIPLE CHOICE QUIZ

REVIEW SET 5A

1 Solve for x:

a $9+2x=-11$ **b** $\frac{x}{3}-4=5$ **c** $\frac{3-2x}{7}=-5$

2 Solve for x:

a $7x+4-3x=-9$ **b** $2(x+5)-4x=9$

c $3(5-x)-2(x-1)=-7(x+3)$

3 The sum of three consecutive integers is 63. Find the smallest integer.

4 A school has 200 students. One third of the left-handed students have blue eyes, and one quarter of the right-handed students have blue eyes. A total of 52 students have blue eyes. How many left-handed students attend the school?

5 I have 25 coins consisting of 5 cent and 50 cent pieces. The total value of the coins is \$7.10. How many 5 cent coins do I have?

6 Solve for x:

a $3x+1\leqslant 16$ **b** $-\frac{x}{3}>7$ **c** $\frac{x}{4}-7\geqslant -2$

7 Solve for x and graph the solution:

a $4x-5<x-11$ **b** $7-5(x-2)\geqslant 2(x-4)$

8 Solve the double inequality and graph its solution:

a $-4\leqslant m-3<9$ **b** $1<5-\frac{x}{2}\leqslant 4$ **c** $-2\leqslant \frac{3k+1}{4}\leqslant 3$

9 Alona and Bethany have different phone plans for international calls:

Alona: \$0.20 flat fee plus \$1.10 per minute

Bethany: \$0.60 flat fee plus \$1.05 per minute

For what call times is it cheaper to use Bethany's phone?

10 The cost of sending a parcel overseas is shown in the table. Harsha sent 5 identical books packed in a 0.6 kg box. The shipping cost was \$50.
Find the possible values for the weight of each book.

Weight (w kg)	*Cost*
$0<w\leqslant 1$	\$20
$1<w\leqslant 2$	\$35
$2<w\leqslant 3$	\$50
$3<w\leqslant 4$	\$60

11 Answer the **Opening Problem** on page **62**.

REVIEW SET 5B

1 Solve for x:

a $15 - 4x = -21$ **b** $3(2x - 5) = 21$ **c** $\frac{2x}{3} + 11 = -2$

2 Solve for x:

a $10 - 7x = 8 + 5x$ **b** $3(2x - 1) + 9 = 2(x + 7)$

c $2(x + 3) - (3x - 1) = 4(4 - x)$

3 Glasses A, B, and C contain a total of 1 litre of water. Glass B contains three times as much water as glass A, and half as much water as glass C. How much water is in glass B?

4 Laura won 12 more tennis matches than she lost last year. If Laura won $\frac{7}{12}$ of her matches, how many matches did she play last year?

5 Show that $0.\overline{18}$ is a rational number.

6 Writing pads cost \$1.35 each and pens cost \$0.85 each. I bought twice as many pens as pads, and the total cost was \$18.30. How many pads did I buy?

7 Solve for x:

a $8 - x < -1$ **b** $\frac{5x}{4} + 3 \geqslant -8$ **c** $\frac{1 - x}{4} < -6$

8 Solve for x and graph the solution:

a $4x - 3 \geqslant x + 1$ **b** $\frac{x - 5}{3} \leqslant 1 - 2x$

9 Solve for x and graph the solution:

a $3 \leqslant \frac{x + 5}{4} < 7$ **b** $-2 < \frac{6 - x}{3} < 4$

10 Roger was told to think of a number, add 20, then divide the result by 3. The final result was less than Roger's original number. Find the possible values of Roger's original number.

11 A pet shelter receives \$150 for each dog, \$100 for each cat, and \$60 for each rabbit they sell.

a In one week, the shelter sells the same number of cats and dogs, and 5 fewer rabbits. The shelter sold 28 animals in total.

i How many of each animal did the shelter sell?

ii How much money did the shelter receive?

b In the next week, the shelter sold twice as many cats as dogs, and 3 fewer rabbits than dogs. If the shelter received \$3100, how many dogs did they sell?

12 Sue is told to think of a number, subtract it from 20, and then divide the result by 3.

a Find the possible values of Sue's original number if her final result is:

i 4 **ii** less than or equal to 2 **iii** between -3 and 3.

b Graph each solution set in **a** on the same number line.

Chapter 6

Venn diagrams

Contents:

OPENING PROBLEM

A city has two newspapers, The Herald and The Prophet. 56% of the people read The Herald and 71% of the people read The Prophet. 18% read neither newspaper.

Things to think about:

a How can we represent this information on a diagram?

b What percentage of the people read:

- **i** both newspapers
- **ii** at least one of the newspapers
- **iii** The Herald but not The Prophet
- **iv** exactly one of the newspapers?

In **Chapter 4** we used set notation and interval notation to *describe* the elements of a set. In this Chapter we will use **Venn diagrams** to *illustrate* sets. Venn diagrams allow us to quickly observe the relationship between different sets.

A VENN DIAGRAMS

A **Venn diagram** consists of a universal set U represented by a rectangle, and sets within it that are usually represented by circles.

The Venn diagram alongside shows the set A within the universal set U.

The **complement** of A is the region outside the circle.

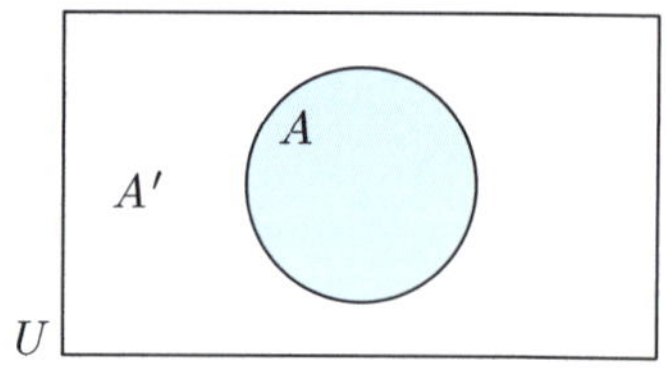

VENN DIAGRAMS FOR TWO SETS

If we are representing two sets A and B on the same Venn diagram, there are several layouts we can choose from.

- If A and B have elements in common, but neither is a subset of the other, the circles for the sets overlap.

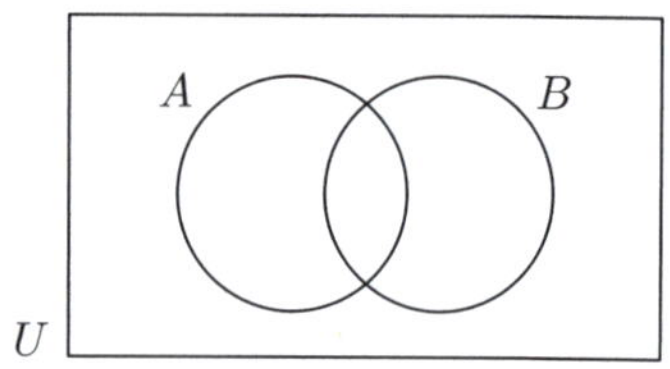

Notice that:

- The **intersection** of A and B is the region which is in *both* circles.
  $A \cap B = \{x \mid x \in A \textbf{ and } x \in B\}$

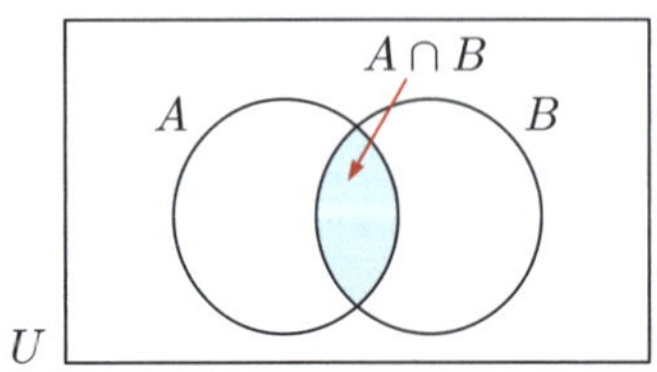

- The **union** of A and B is the region which is in *either* circle.
  $A \cup B = \{x \mid x \in A \textbf{ or } x \in B\}$

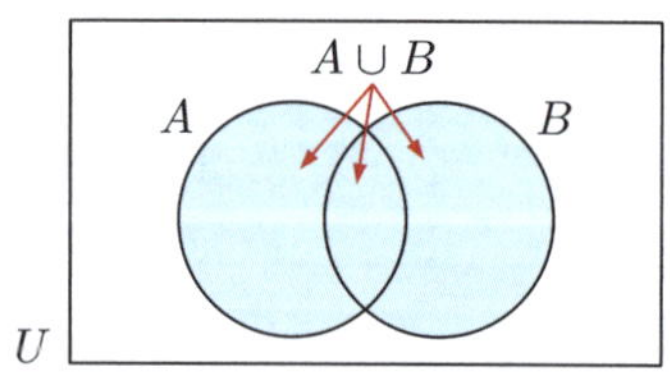

- If A is a **subset** of B, then every element of A is also an element of B.
  We place a circle representing A completely within the circle representing B.

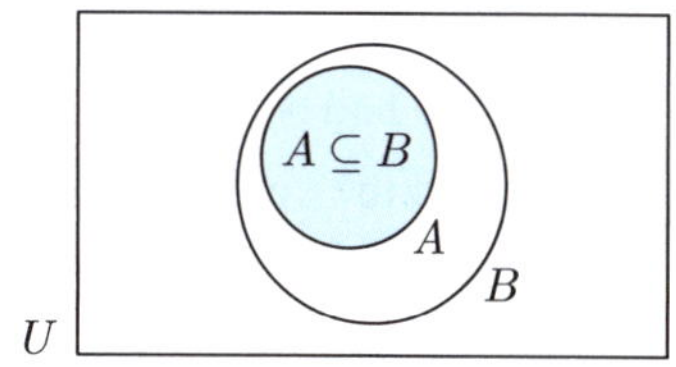

- If A and B are **disjoint** or **mutually exclusive** then there are no elements in both A and B.
  We represent A and B using circles which do not overlap.

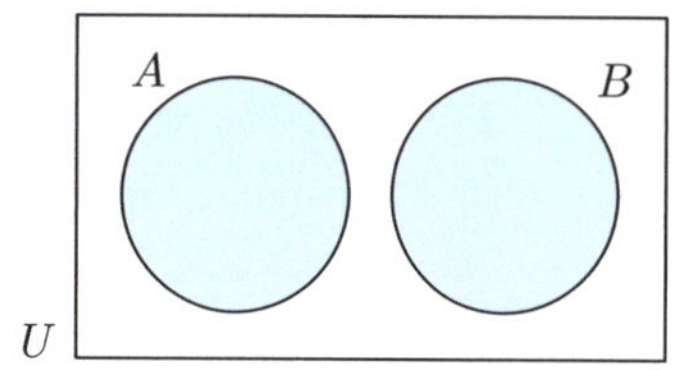

Example 1 — Self Tutor

Let $U = \{x \in \mathbb{Z} \mid 0 \leqslant x \leqslant 10\}$, $A = \{2, 3, 5, 7\}$, and $B = \{1, 3, 6, 7, 8\}$.

a Find $A \cap B$ and $A \cup B$. **b** Illustrate A and B on a Venn diagram.

a $A \cap B = \{3, 7\}$
$A \cup B = \{1, 2, 3, 5, 6, 7, 8\}$

b

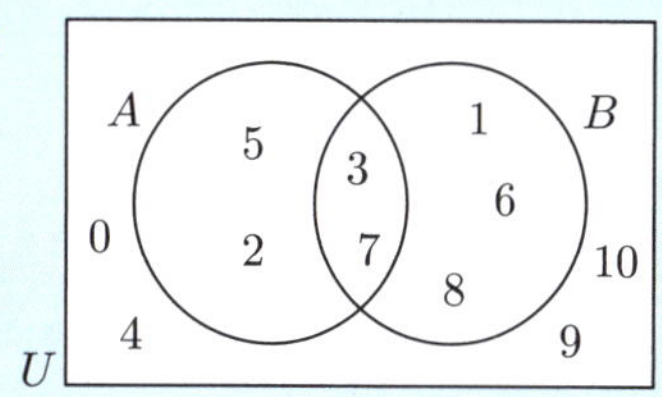

EXERCISE 6A

1 Suppose $U = \{x \in \mathbb{Z}^+ \mid x \leqslant 8\}$ and $A = \{\text{prime numbers}\}$.

a List the elements of A and A'. **b** Illustrate A on a Venn diagram.

c Find $n(A)$ and $n(A')$.

2

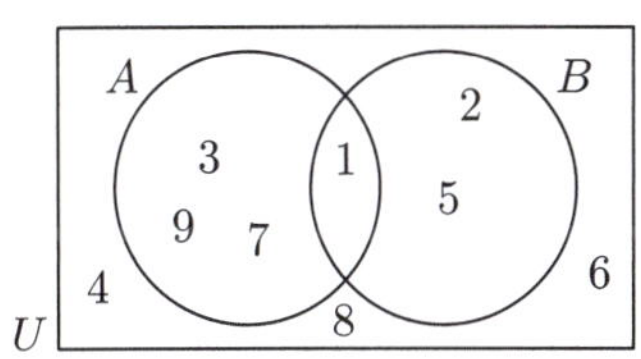

List the elements of:

a A **b** B **c** A' **d** B'

e $A \cap B$ **f** $A \cup B$ **g** U

3 Let $U = \{1, 2, 3, 4, 5, 6, 7, 8, 9, 10\}$. Draw a Venn diagram to represent:

a $A = \{1, 2, 3, 4\}$, $B = \{3, 4, 5, 6\}$ **b** $A = \{6, 7, 9, 10\}$, $B = \{2, 5, 6, 8, 9\}$

c $A = \{3, 5, 7, 9\}$, $B = \{4, 6, 8\}$ **d** $A = \{1, 4, 7, 8, 10\}$, $B = \{1, 7, 10\}$

4 Suppose $U = \{x \in \mathbb{Z}^+ \mid x \leqslant 30\}$, $A = \{\text{factors of } 30\}$, and $B = \{\text{prime numbers}\}$.

a List the elements of: **i** A **ii** B **iii** $A \cap B$ **iv** $A \cup B$

b Find: **i** $n(A)$ **ii** $n(B)$ **iii** $n(A \cap B)$ **iv** $n(A \cup B)$

c Show that $n(A \cup B) = n(A) + n(B) - n(A \cap B)$.

d Illustrate A and B on a Venn diagram.

5 Suppose $U = \{x \in \mathbb{Z}^+ \mid x \leqslant 25\}$, $P = \{\text{multiples of } 3\}$, and $Q = \{\text{multiples of } k\}$, where k is the smallest positive integer such that P and Q are disjoint.

- **a** Find the value of k.
- **b** List the elements of P and Q.
- **c** Illustrate P and Q on a Venn diagram.

Example 2 — **Self Tutor**

Illustrate the sets $\mathbb{N}$, $\mathbb{Z}$, and $\mathbb{Q}$ on a Venn diagram with $U = \mathbb{R}$.

Place these numbers on the Venn diagram: $\sqrt{3}$, $8\frac{1}{2}$, -2, 7.1, 16, and 0.115.

$\mathbb{N} \subseteq \mathbb{Z} \subseteq \mathbb{Q} \subseteq \mathbb{R}$

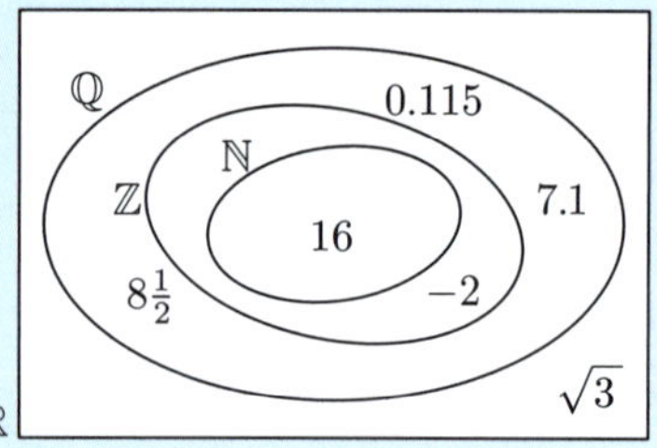

6 Consider the universal set $U = \mathbb{R}$.

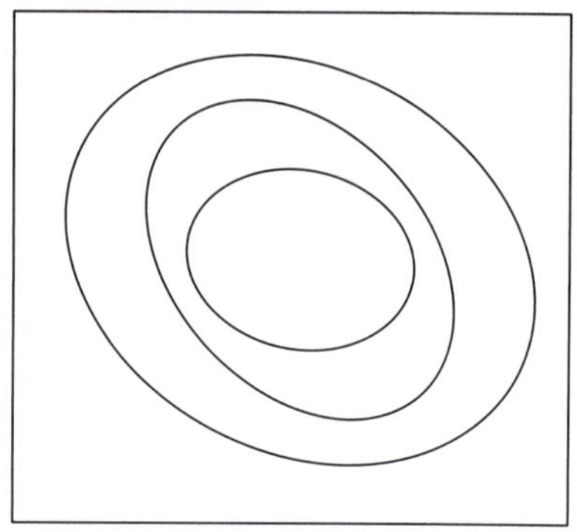

- **a** Copy the Venn diagram and label the sets $\mathbb{R}$, $\mathbb{Q}$, $\mathbb{Z}$, and $\mathbb{N}$. Place these numbers on the Venn diagram: $\frac{1}{2}$, $\sqrt{2}$, $0.\overline{3}$, -5, $5\frac{1}{4}$, 0, 10, and $0.213\,700\,561\,8....$ which does not terminate nor recur.
- **b** True or false?
  - **i** $\mathbb{N} \subseteq \mathbb{Z}$
  - **ii** $\mathbb{Z} \subseteq \mathbb{Q}$
  - **iii** $\mathbb{N} \subseteq \mathbb{Q}$
- **c** Shade the region representing the set of irrationals $\mathbb{Q}'$.

7 Illustrate on a Venn diagram:

- **a** $U = \{\text{triangles}\}$, $E = \{\text{equilateral triangles}\}$, $I = \{\text{isosceles triangles}\}$
- **b** $U = \{\text{quadrilaterals}\}$, $P = \{\text{parallelograms}\}$, $R = \{\text{rectangles}\}$
- **c** $U = \{\text{quadrilaterals}\}$, $P = \{\text{parallelograms}\}$, $R = \{\text{rectangles}\}$, $H = \{\text{rhombuses}\}$
- **d** $U = \{\text{quadrilaterals}\}$, $P = \{\text{parallelograms}\}$, $N = \{\text{non-convex quadrilaterals}\}$
- **e** $U = \{\text{triangles}\}$, $I = \{\text{isosceles triangles}\}$, $R = \{\text{right angled triangles}\}$, $E = \{\text{equilateral triangles}\}$

8 **a** List the elements of:

- **i** A
- **ii** B
- **iii** C
- **iv** $A \cap B$
- **v** $B \cup C$
- **vi** $A \cap C'$

b Find:

- **i** $n(A \cup C)$
- **ii** $n(A' \cap B')$
- **iii** $n(A \cap B \cap C)$

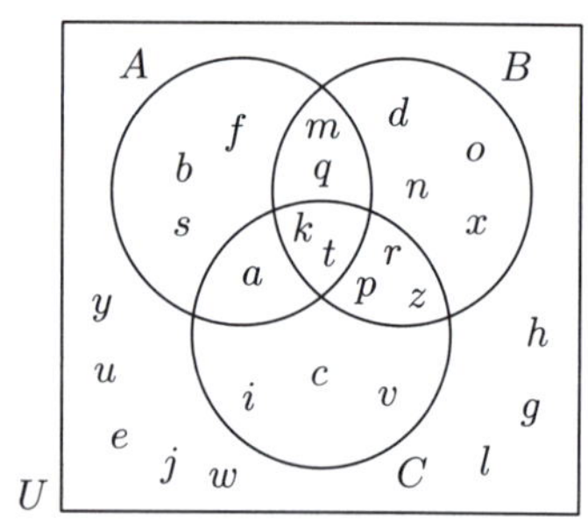

9 Suppose $U = \{x \in \mathbb{Z}^+ \mid x \leqslant 30\}$, $A = \{\text{prime numbers}\}$, $B = \{\text{multiples of } 5\}$, and $C = \{\text{odd numbers}\}$.

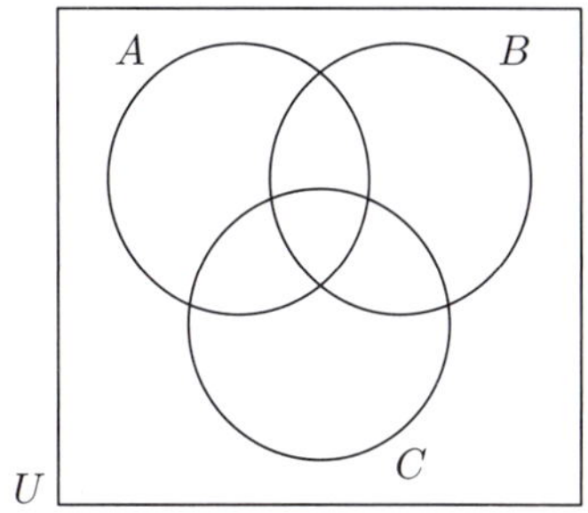

a Represent A, B, and C on a Venn diagram.

b List the elements which are in:

i both B and C **ii** C but not A
iii A and C but not B **iv** A, B, and C
v none of the sets.

B VENN DIAGRAM REGIONS

We can use Venn diagrams to help illustrate regions such as the union or intersection of sets.

The regions of a Venn diagram can be used to verify **set identities**. These are equations involving sets which are true for *all* sets.

In previous years we have seen set identities including:

- $A \cup A' = U$
- $A \cap A' = \varnothing$
- $(A \cup B)' = A' \cap B'$
- $(A \cap B)' = A' \cup B'$

Example 3 Self Tutor

On separate Venn diagrams, shade these regions for two intersecting sets A and B:

a in either A or B, but not both **b** $A \cap B'$

a

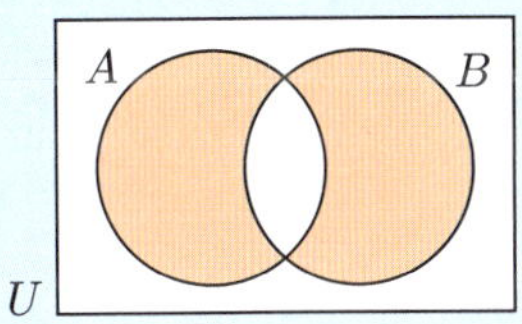

b

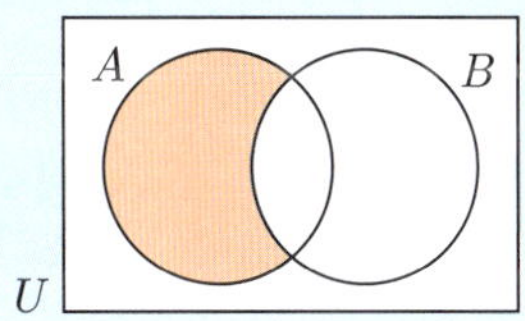

Set identities can be verified by checking that each side of the equation occupies the same region on a Venn diagram.

EXERCISE 6B

1

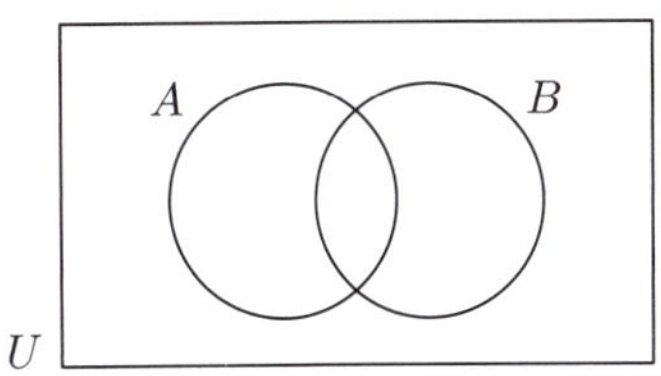

On separate Venn diagrams, shade these regions:

a in B but not A **b** not in B
c $A' \cup B$

PRINTABLE VENN DIAGRAMS

2 Describe in words, the shaded region:

a

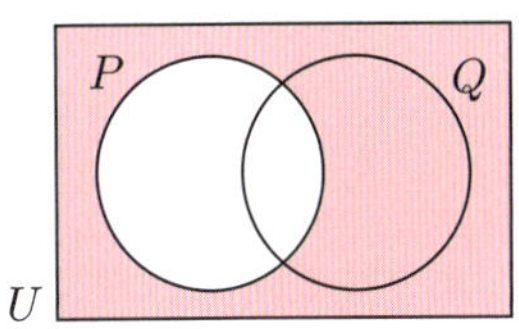

b

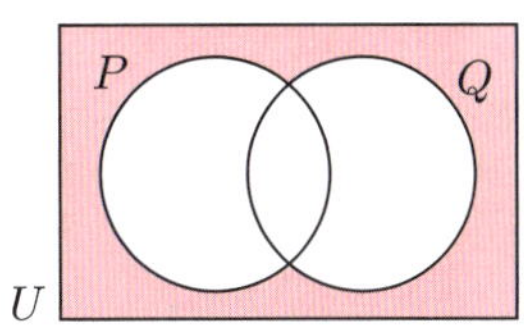

c

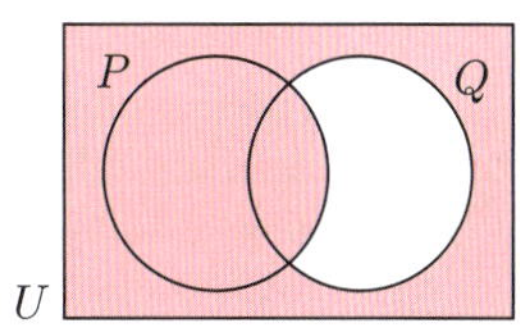

3 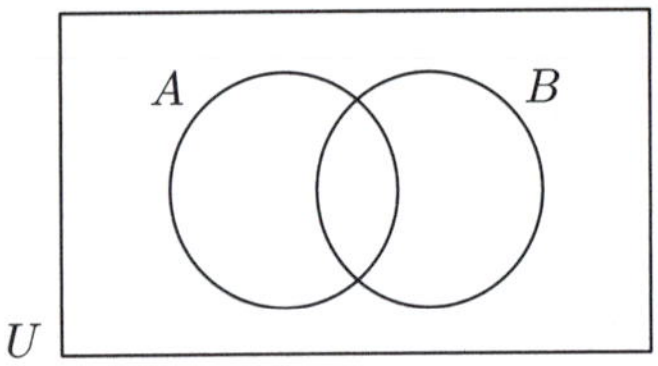

a On separate Venn diagrams, shade:

i A' **ii** B' **iii** $A \cup B$ **iv** $A \cap B$

b Verify that:

i $(A \cup B)' = A' \cap B'$ **ii** $(A \cap B)' = A' \cup B'$

4 Verify that, for $B \subseteq A$:

a $A \cap B = B$ **b** $A \cup B = A$

c $A' \cap B' = A'$ **d** $A' \cup B' = B'$

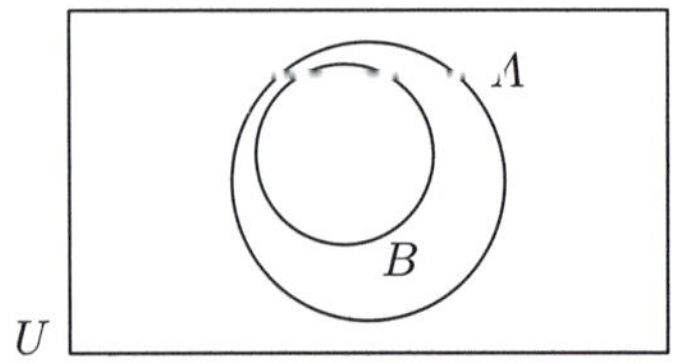

5 Verify that, for disjoint sets A and B:

a $(A \cap B)' = U$ **b** $A' \cup B' = U$

c $A' \cap B = B$

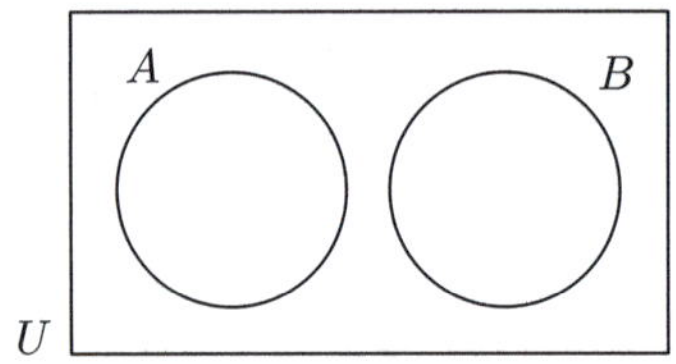

6 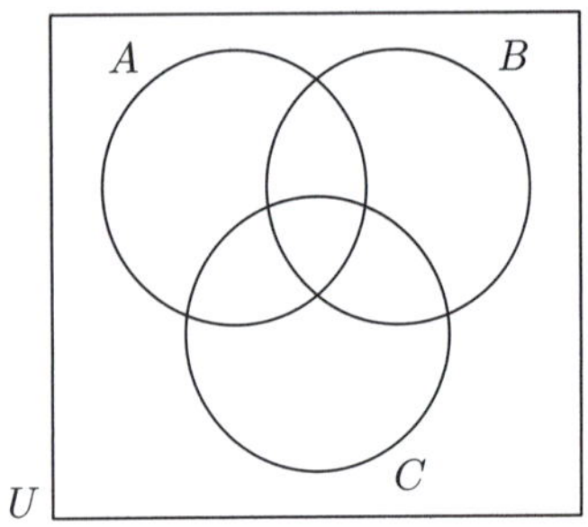

a On separate Venn diagrams, shade:

i A **ii** B'

iii $A \cap B$ **iv** $B \cap C$

v $A \cap C$ **vi** $A \cup B$

vii $B \cup C$ **viii** $A \cup C$

ix $A \cap B \cap C$ **x** $A \cup B \cup C$

xi $(A \cap B \cap C)'$ **xii** $(A \cup B) \cap C$

b By shading the relevant regions, verify that:

i $A \cup (B \cap C) = (A \cup B) \cap (A \cup C)$

ii $A \cap (B \cup C) = (A \cap B) \cup (A \cap C)$

ACTIVITY

Click on the icon to practise shading regions representing various subsets. The software includes two and three intersecting sets.

VENN DIAGRAM REGIONS

C NUMBERS IN REGIONS

To indicate the **number of elements** in a region of a Venn diagram, we write the number in brackets.

For example, the Venn diagram alongside indicates that $A \cap B$ contains 6 elements, A contains $4 + 6 = 10$ elements, and $A \cup B$ contains $4 + 6 + 3 = 13$ elements.

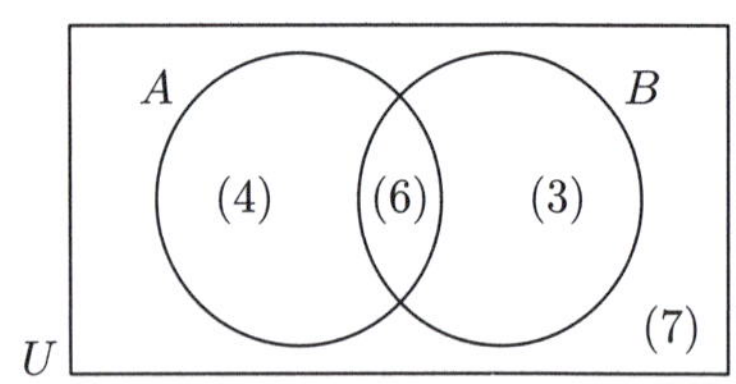

Example 4 **Self Tutor**

Suppose $n(U) = 25$, $n(P) = 10$, $n(Q) = 12$, and $n(P \cap Q) = 3$.
Find the number of elements in:

a $P \cup Q$ **b** P but not Q.

We are given $n(P \cap Q) = 3$

$\therefore$ $n(P \cap Q') = 10 - 3 = 7$

and $n(Q \cap P') = 12 - 3 = 9$

$\therefore$ $n(P' \cap Q') = 25 - 3 - 7 - 9 = 6$

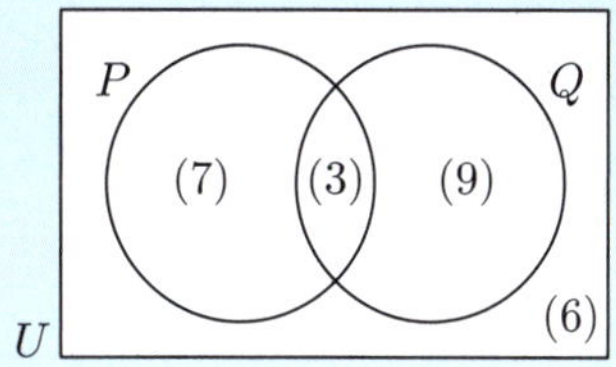

a $n(P \cup Q) = 7 + 3 + 9 = 19$,
so there are 19 elements in $P \cup Q$.

b $n(P \cap Q') = 7$, so there are 7 elements in P but not Q.

EXERCISE 6C

1

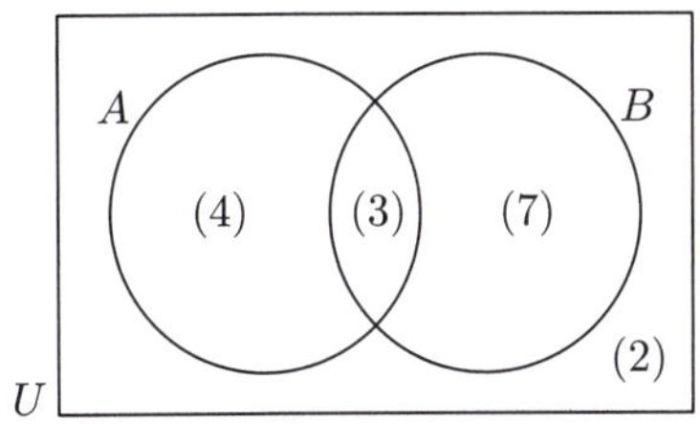

State the number of elements in:

a B **b** A'
c $A \cup B$ **d** A but not B
e B but not A **f** neither A nor B.

2 **a** State the number of elements in:

i $P \cap Q$ **ii** Q
iii $P' \cap Q$ **iv** $P \cup Q$

b Given that $n(U) = 54$, how many elements are in neither P nor Q?

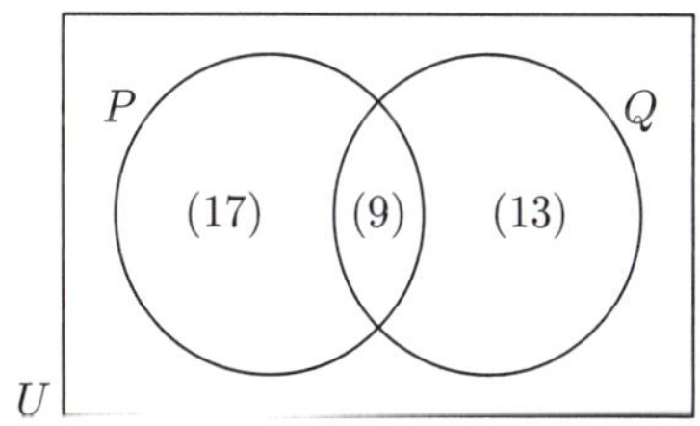

3

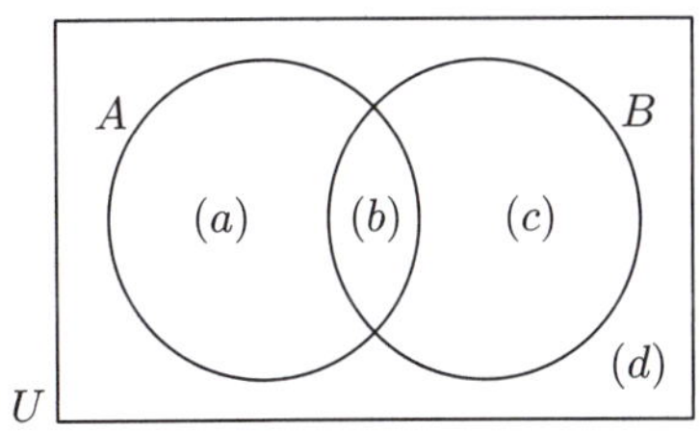

a Write an expression for:

i $n(A)$ **ii** $n(B)$
iii $n(A \cap B')$ **iv** $n(A \cap B)$
v $n(A \cup B)$ **vi** $n(U)$

b Show that:

i $n(A \cap B) = n(A) + n(B) - n(A \cup B)$
ii $n(A \cap B') = n(A) - n(A \cap B)$
iii $n(A \cup B') = n(U) - n(A' \cap B)$

4 Suppose $n(U) = 20$, $n(A) = 12$, $n(B) = 13$, and $n(A \cap B) = 8$.

a Use this information to construct a Venn diagram for sets A and B.

b Hence find the number of elements in:

i $A \cup B$ **ii** B but not A.

5 Suppose $n(U) = 28$, $n(M) = 14$, $n(M \cap N) = 3$, and $n(M \cup N) = 18$. Find:

a $n(N)$ **b** $n((M \cup N)')$ **c** $n(M' \cap N)$

Example 5 — Self Tutor

Suppose $n(U) = 30$, $n(A) = 13$, $n(B) = 15$, and $n(A' \cap B') = 7$.

Find the number of elements in: **a** $A \cap B$ **b** B but not A.

Let $n(A \cap B) = x$

$\therefore\ n(A \cap B') = 13 - x$ and $n(A' \cap B) = 15 - x$

Since $n(U) = 30$, $(13 - x) + x + (15 - x) + 7 = 30$

$\therefore\ 35 - x = 30$

$\therefore\ x = 5$

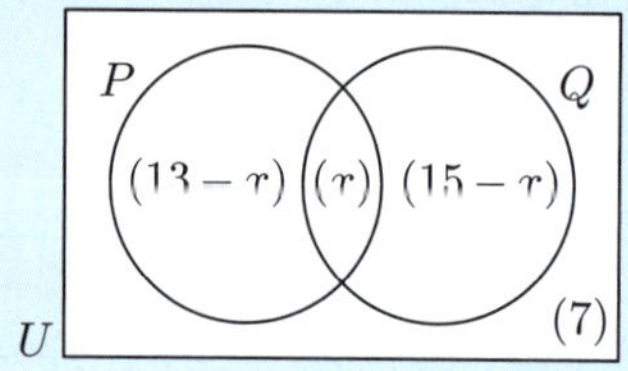

a $n(A \cap B) = 5$, so there are 5 elements in $A \cap B$.

b $n(A' \cap B) = 15 - 5 = 10$, so there are 10 elements in B but not A.

6 Suppose $n(U) = 40$, $n(A) = 17$, $n(B) = 25$, and $n(A' \cap B') = 9$.

a Represent this information on a Venn diagram.

b Find the number of elements in: **i** $A \cap B$ **ii** A or B but not both.

7 Suppose $n(P) = 28$, $n(Q) = 32$, $n(P' \cap Q') = 12$, and $n(U) = 59$.

Find the number of elements in: **a** $P \cap Q$ **b** Q but not P.

8 Suppose $n(A) = 22$, $n(A') = 11$, $n(B') = 9$, and $n(A' \cup B') = 18$.

a Represent this information on a Venn diagram.

b Find the number of elements in: **i** $A \cup B$ **ii** B but not A.

9

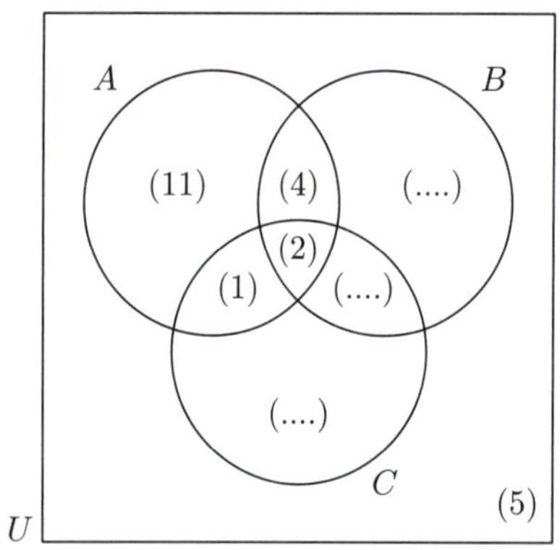

In the given Venn diagram, $n(B \cap C) = 5$, $n(B) = 25$, and $n(U) = 60$.

a Find:

i $n(A \cap B \cap C)$ **ii** $n(A)$

b Copy and complete the Venn diagram.

c Find the number of elements in:

i C **ii** A and C but not B

iii neither A nor B

iv $(A \cup B) \cap C$

v exactly two of the sets.

D PROBLEM SOLVING WITH VENN DIAGRAMS

By considering the number of elements in different regions, we can use Venn diagrams to solve problems.

Example 6 — Self Tutor

There are 54 countries in Africa. English is spoken in 24 of them, and French is spoken in 21 of them. Both languages are spoken in 4 countries.

a Place this information on a Venn diagram.

b In how many countries is:

i English but not French spoken **ii** neither language spoken?

a Let E be the set of countries where English is spoken, and F be the set of countries where French is spoken.

$n(E \cap F) = 4$

$\therefore\ n(E \cap F') = 24 - 4 = 20$

and $n(E' \cap F) = 21 - 4 = 17$

$\therefore\ n(E' \cap F') = 54 - 4 - 20 - 17 = 13$

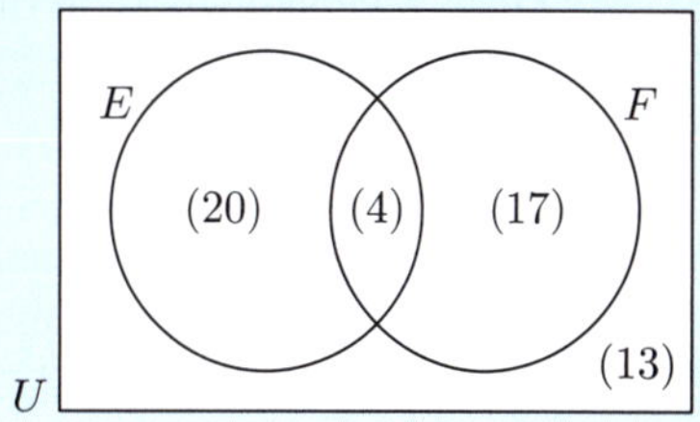

b **i** English but not French is spoken in 20 countries.

ii Neither language is spoken in 13 countries.

EXERCISE 6D

1 In a survey at a resort, guests were asked whether they went sailing (S) or fishing (F) during their stay.
Determine the number of guests who:

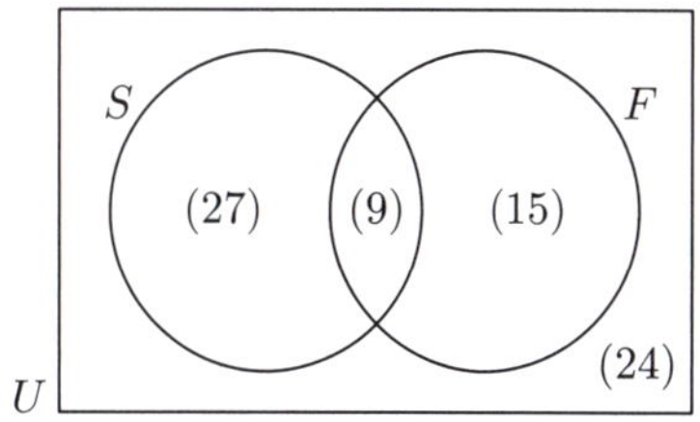

- **a** were surveyed
- **b** did both activities
- **c** did neither activity
- **d** did exactly one of the activities.

2 Tess played 22 cricket games last season. She batted in 17 of the games, and bowled in 11 of the games. In 8 games she both batted and bowled.

- **a** Place this information on a Venn diagram.
- **b** In how many games did Tess:
  - **i** bowl but not bat
  - **ii** neither bat nor bowl?

3 In a class of 40 students, 27 like bananas, 22 like pineapples, and 17 like both fruits. Find the number of students who:

- **a** dislike both fruits
- **b** like exactly one fruit.

Example 7 **Self Tutor**

There are 42 children at a childcare centre. 15 have fair hair (F), 12 have blue eyes (B), and 22 have neither fair hair nor blue eyes.

Find the number of children with fair hair but not blue eyes.

Let $n(F \cap B) = x$

$\therefore\ n(F \cap B') = 15 - x$ and $n(F' \cap B) = 12 - x$

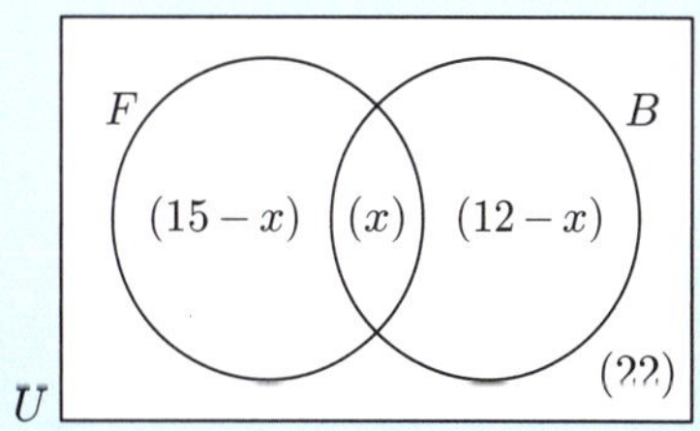

We know that 22 have neither fair hair nor blue eyes.

Since there are 42 children in total,

$(15 - x) + x + (12 - x) + 22 = 42$

$\therefore\ 49 - x = 42$

$\therefore\ x = 7$

$n(F \cap B') = 15 - 7 = 8$, so 8 children have fair hair but not blue eyes.

4 26 teenagers were surveyed to see how many like going to the cinema and ice-skating. 20 like going to the cinema, 19 like ice-skating, and 1 likes neither activity.

- **a** Draw a Venn diagram to display this information.
- **b** How many teenagers:
  - **i** like both activities
  - **ii** only like ice-skating?

5 A service club has 65 members. 38 members attend the Monday meetings, 35 attend the Thursday meetings, and 19 attend neither meeting.
How many members attend:

- **a** both the Monday and Thursday meetings
- **b** only the Monday meetings?

6 At a dance school, 95% of the members study at least one of classical ballet or modern dance. 72% study classical ballet. 35% study modern dance. What percentage of the members study both classical ballet and modern dance?

7 Joe owns an automotive garage which does car services and mechanical repairs. In one week 9 cars had services, 5 had both services and repairs, and 13 had either a service or a repair but not both. How many cars did the garage attend to in total?

8 Of the first 1000 positive integers, 272 contain the digit 1, 819 do *not* contain the digit 0, and 584 contain neither a 0 nor a 1.
How many of the first 1000 positive integers contain:

- **a** both a 1 and a 0
- **b** either a 1 or a 0, but not both?

Example 8 **Self Tutor**

In a senior class of 35 students, the school records show that 15 take biology, 15 take chemistry, and 18 take physics. 10 students take biology and chemistry, 7 take biology and physics, and 5 take chemistry and physics. 3 students take all three subjects.

How many students take none of these three subjects?

Let B, C, and P represent the students taking biology, chemistry, and physics respectively.

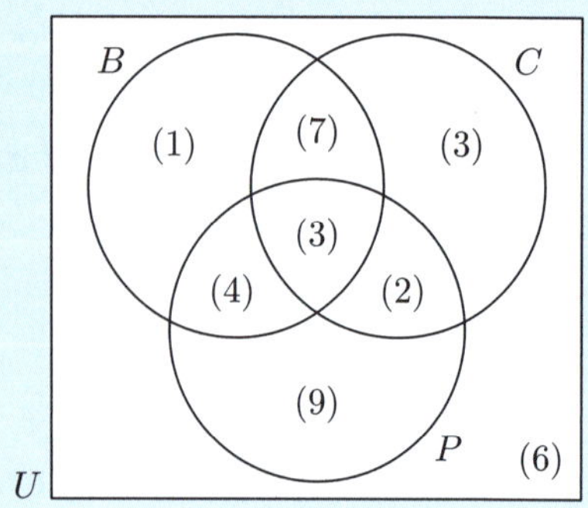

$$n(B \cap C \cap P) = 3$$
$$\therefore\ n(B \cap C \cap P') = 10 - 3 = 7$$
$$n(B \cap C' \cap P) = 7 - 3 = 4$$
$$\text{and}\ \ n(B' \cap C \cap P) = 5 - 3 = 2$$
$$\therefore\ n(B \cap C' \cap P') = 15 - 7 - 3 - 4 = 1$$
$$n(B' \cap C \cap P') = 15 - 7 - 3 - 2 = 3$$
$$\text{and}\ \ n(B' \cap C' \cap P) = 18 - 4 - 3 - 2 = 9$$

We know $n(U) = 35$

$$\therefore\ n(B' \cap C' \cap P') = 35 - 1 - 7 - 3 - 4 - 3 - 2 - 9 = 6$$

So, 6 students take none of these three subjects.

9 92 countries competed in the 2018 Winter Olympics. Of these countries, 22 won gold medals, 21 won silver medals, and 25 won bronze medals.

18 countries won gold and silver medals, 18 won silver and bronze medals, and 18 won gold and bronze medals.

16 countries won all three types of medal.

a Display this information on a Venn diagram.

b How many of the countries won:

i no medals

ii silver and bronze, but not gold medals

iii exactly one type of medal?

10 In a small country town there are three restaurants. 42% of the residents eat at A, 45% at B, and 41% at C. 15% eat at both A and B, 9% at A and C, and 17% at B and C. 4% eat at all three restaurants.

a Display this information on a Venn diagram.

b Find the percentage of the residents who eat at:

i none of the restaurants

ii at least one of the restaurants

iii exactly one of the restaurants

iv either A or B

v C only.

11 In a school of 163 students, each was invited to choose activities from sport, music, and drama. 76 students chose sport, 64 students chose music, and 40 students chose drama. 12 students chose both sport and music, 7 chose both sport and drama, and 21 chose both music and drama. 5 students chose all three activities. How many students chose no activities?

PUZZLE

In the Venn diagram alongside, the universal set contains the smallest positive integers such that each region of the diagram contains exactly one element.

Find the sum of all numbers in the universal set.

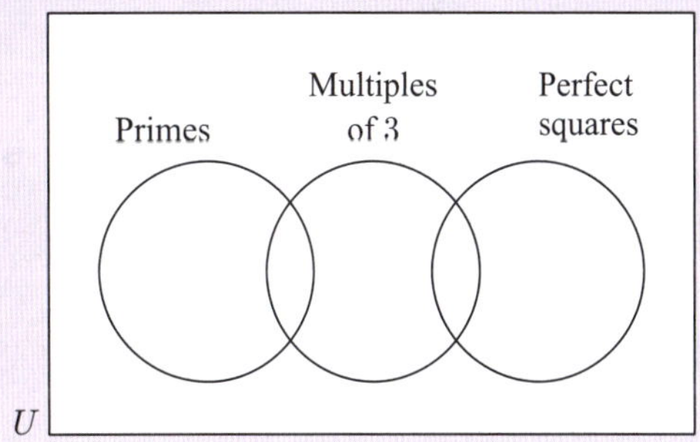

MULTIPLE CHOICE QUIZ

QUICK QUIZ

REVIEW SET 6A

1 Let $U = \{1, 2, 3, 4, 5, 6, 7, 8, 9, 10, 11, 12\}$. Draw a Venn diagram to represent:

a $P = \{2, 5, 6, 9\}$ and $Q = \{1, 5, 8, 9, 11\}$

b $P = \{1, 3, 6, 10, 11\}$ and $Q = \{2, 7, 12\}$

c $P = \{\text{factors of } 6\}$ and $Q = \{\text{factors of } 60\}$.

2 Consider the universal set $U = \mathbb{R}$.
Copy the Venn diagram and label the sets $\mathbb{R}$, $\mathbb{Q}$, $\mathbb{Z}$, and $\mathbb{N}$. Place these numbers on the Venn diagram:
-1, $\sqrt{2}$, 2, 3.1, $\frac{\pi}{2}$, and $4.\overline{2}$.

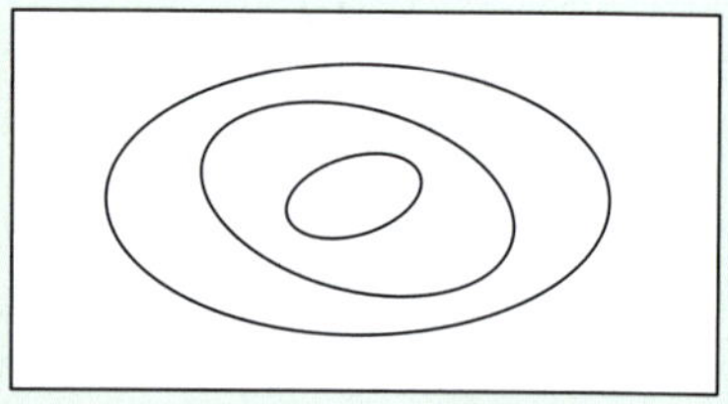

3 Let $U = \{x \in \mathbb{Z}^+ \mid x \leqslant 20\}$, $A = \{\text{factors of } 18\}$, and $B = \{\text{prime numbers}\}$.

a Illustrate A and B on a Venn diagram.

b Find $n(A \cup B)$.

4 Describe in words, the shaded region:

a

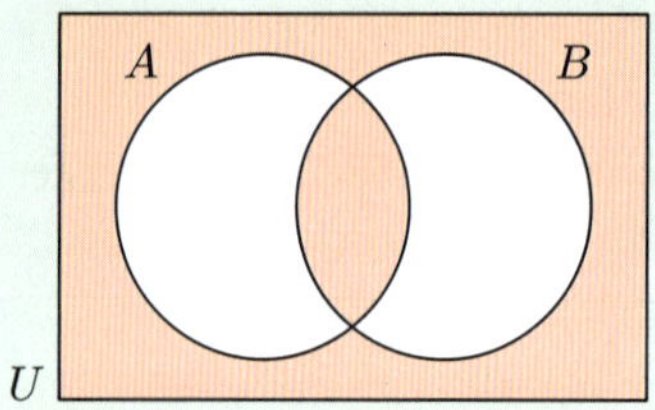

b

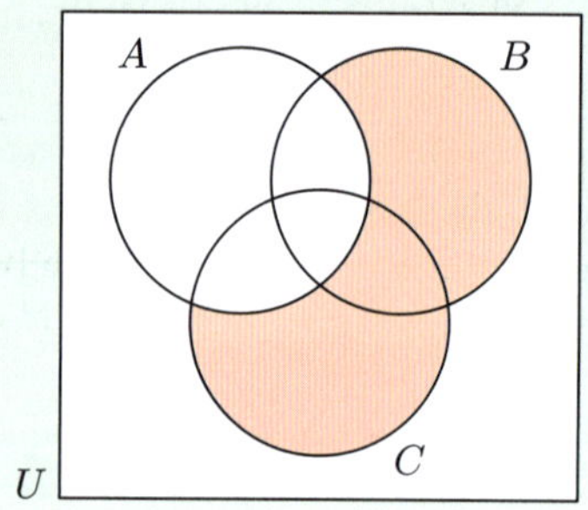

5

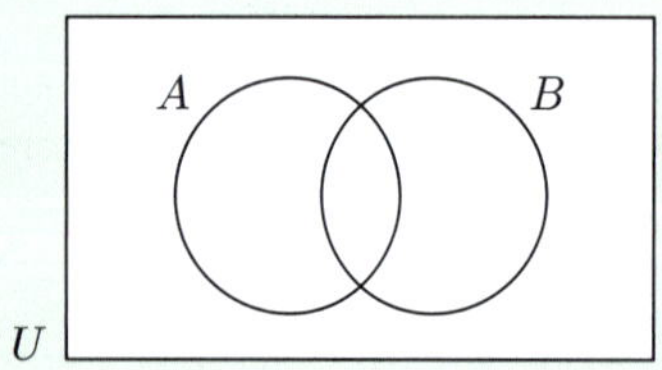

a On separate Venn diagrams, shade:

i $A \cap B'$ **ii** $A' \cup B$

b Verify that $(A' \cup B)' = A \cap B'$.

6 Suppose $n(A) = 21$, $n(B) = 25$, $n(A' \cap B') = 5$, and $n(U) = 44$.
Find the number of elements in:

a $A \cap B$

b A but not B.

7 There are 50 stalls at a market. 28 stalls sell food, 23 stalls sell craft, and 7 stalls sell both food and craft. How many stalls sell:

a neither food nor craft

b food or craft, but not both?

8 A tennis team played 35 matches in a season. 16 matches were played on their home court. Of the 15 matches the team lost, 11 were played away from their home court.
How many matches did the team win on their home court?

9 In a survey at an English airport, 55 travellers said that they had been to Spain, 53 had been to France, and 49 had been to Germany. 18 had been to Spain and France, 15 had been to Spain and Germany, and 25 had been to France and Germany. 10 had been to all three countries, and 9 had been to none of the countries.

a Display this information on a Venn diagram.

b Find how many travellers:

i took part in the survey **ii** had been to exactly two of the countries.

REVIEW SET 6B

1 Suppose $U = \{x \in \mathbb{Z}^+ \mid x \leqslant 10\}$, $P = \{\text{prime numbers}\}$, and $Q = \{\text{even numbers}\}$.

a Illustrate P and Q on a Venn diagram.

b Find: **i** $n(P')$ **ii** $n(P \cap Q)$

c Is $P \cap Q \subseteq P$?

2 Illustrate $U = \{\text{positive integers}\}$, $A = \{\text{multiples of 2}\}$, $B = \{\text{multiples of 3}\}$, and $C = \{\text{multiples of 4}\}$ on a Venn diagram.

3 State the number of elements in:

a A **b** B

c $A \cup B$ **d** neither A nor B.

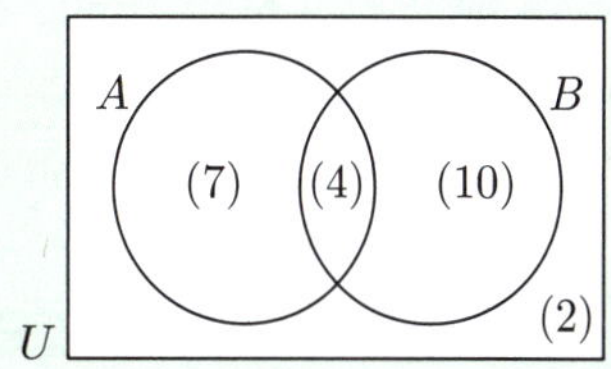

4 **a** On separate Venn diagrams, shade:

i $P \cup Q$ **ii** $Q \cap R$

iii $Q \cup R'$ **iv** $P' \cap Q$

b By shading the relevant regions, verify that $(P \cup Q)' \cap R = P' \cap (Q' \cap R)$.

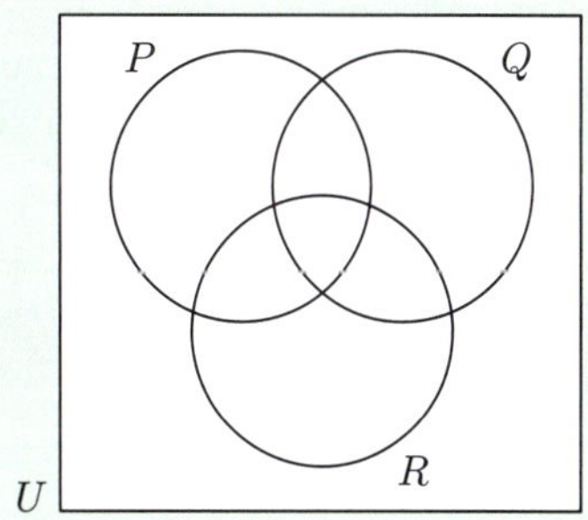

5 Suppose $n(A) = 16$, $n(B) = 19$, $n(A \cap B) = 11$, and $n(U) = 28$.

a Use this information to construct a Venn diagram for sets A and B.

b Hence find the number of elements in:

i $A \cup B$ **ii** A but not B.

6 100 families were surveyed. It was found that 60 had a TV set, and 72 had a computer. 38 families had both items. How many families had:

a a computer but not a TV set **b** neither item?

7 There are 60 houses on a street. 34 houses put their rubbish bin out for collection this morning, and 42 houses put their recycling bins out. 3 houses put neither bin out.
How many houses put both bins out?

8 Answer the **Opening Problem** on page **76**.

9 Suppose $U = \{x \in \mathbb{Z}^+ \mid x \leqslant 20\}$, $A = \{\text{factors of } 20\}$, $B = \{\text{even numbers}\}$, and $C = \{\text{composite numbers}\}$.

a List the elements of A, B, and C.

b Represent A, B, and C on a Venn diagram.

c Find:

i $n(A \cup B)$ **ii** $n(A \cap B \cap C)$

d Show that $(B \cup C) \cap A' = A' \cap C$.

e Is $(B \cup C) \cap A' = A' \cap C$ a set identity? Explain your answer.

10 In a game of Bingo, numbers from 1 to 75 are selected at random. So far, 30 different numbers have been selected.

16 of these numbers are multiples of 2, 11 are multiples of 3, and 6 are multiples of 5. 6 numbers are multiples of 2 and 3, 3 numbers are multiples of 3 and 5, and 5 numbers are multiples of 2 and 5. Two of the numbers are multiples of 2, 3, and 5.

Let U be the set of numbers which have been selected,

A be the set of multiples of 2,

B be the set of multiples of 3,

and C be the set of multiples of 5.

a Represent this information on a Venn diagram like the one alongside.

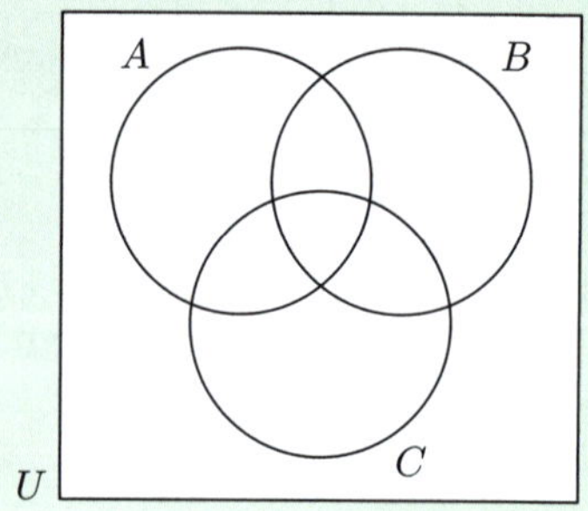

b How many selected numbers are:

i a multiple of 2 or 5

ii a multiple of 3, but not 5?

c Determine whether the numbers 35 and 60 have been selected. Explain your reasoning.

d The next 5 numbers selected are 10, 51, 25, 11, and 48. Update your Venn diagram to include this information.

Chapter 7

Surds and other radicals

Contents:

OPENING PROBLEM

Look at the figure alongside.

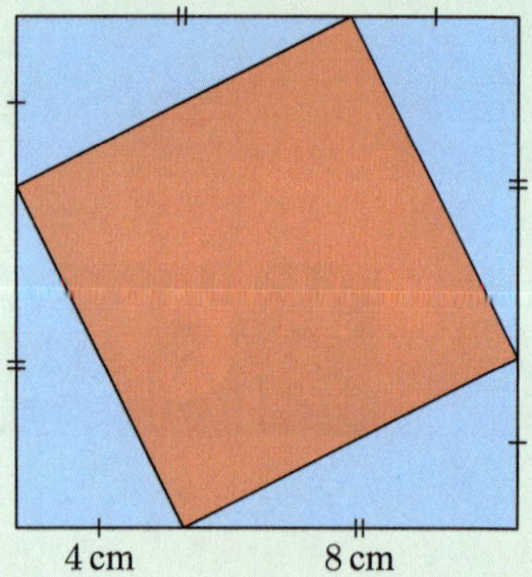

Things to think about:

- **a** Can you use the area of the blue square and the areas of the four triangles to find the area of the red square?
- **b** How can we write *exactly*:
  - **i** the side length of the red square
  - **ii** the perimeter of each triangle?
- **c**
  - **i** What *fraction* of the side length of the blue square is the side length of the red square?
  - **ii** How can this fraction be written in the most simple form?

A RADICALS

A **radical** is a number that is written using the radical sign $\sqrt{}$.

$\sqrt{3}$, $\sqrt{\frac{1}{5}}$, and $\sqrt[3]{8}$ are examples of radicals.

In previous years, you should have seen that:

- The **square root of a**, written $\sqrt{a}$, is the non-negative number which, when squared, gives a.
$$\sqrt{a} \times \sqrt{a} = a$$
$\sqrt{a}$ is a real number only if $a \geqslant 0$.
- The **cube root of a**, written $\sqrt[3]{a}$, is the number which, when cubed, gives a.
$$\sqrt[3]{a} \times \sqrt[3]{a} \times \sqrt[3]{a} = a$$
  - If $a > 0$ then $\sqrt[3]{a} > 0$.
  - If $a < 0$ then $\sqrt[3]{a} < 0$.
- The **nth root of a**, written $\sqrt[n]{a}$, is the number such that $\left(\sqrt[n]{a}\right)^n = a$.
If n is even, then we define $\sqrt[n]{a}$ as the *non-negative* number with this property, and $\sqrt[n]{a}$ is only real for $a \geqslant 0$.

Alongside the exponent laws, these definitions enable us to evaluate or simplify many expressions.

Example 1 — Self Tutor

Simplify: **a** $\left(\sqrt{2}\right)^3$ **b** $\left(\frac{4}{\sqrt{2}}\right)^2$

a
$$\begin{aligned}&\left(\sqrt{2}\right)^3\\ &= \sqrt{2} \times \sqrt{2} \times \sqrt{2}\\ &= 2\sqrt{2}\end{aligned}$$

b
$$\begin{aligned}\left(\frac{4}{\sqrt{2}}\right)^2 &= \frac{4^2}{\left(\sqrt{2}\right)^2}\\ &= \frac{16}{2}\\ &= 8\end{aligned}$$

SURDS

A **surd** is an irrational radical.

For example:
- $\sqrt{2}$, $\sqrt{3}$, and $\sqrt{5}$ are surds.
- $\sqrt{4}$ is not a surd, since $\sqrt{4} = \sqrt{2^2} = 2$.
- $\sqrt{\frac{1}{9}}$ is not a surd, since $\sqrt{\frac{1}{9}} = \sqrt{\left(\frac{1}{3}\right)^2} = \frac{1}{3}$.

HISTORICAL NOTE

Many centuries after Pythagoras, when the Golden Age of the Greeks was past, the writings of the Greeks were preserved, translated, and extended by Arab mathematicians.

The Arabs thought of a square number as growing out of its roots. The roots had to be extracted. The Latin word for "root" is *radix*, from which we get the words "radical" and "radish"! The printed symbol for radix was first R, then r, which was copied by hand as $\sqrt{\ }$.

In the 9th century AD, the Arab mathematician **Al-Khwarizmi** made a translational error. The Greek word *a-logos* means "irrational" but also means "deaf". So, the Greek *a-logos* was interpreted as "deaf" which in Latin is *surdus*.

Hence to this day, irrational radicals like $\sqrt{2}$ are called *surds*.

EXERCISE 7A

1 Determine whether each number is a surd. If it is a surd, find its value correct to 4 decimal places.

a $\sqrt{5}$ **b** $\sqrt{36}$ **c** $\sqrt{72}$ **d** $\sqrt{81}$

e $\sqrt{200}$ **f** $\sqrt{\frac{1}{10}}$ **g** $\sqrt{\frac{1}{64}}$ **h** $\sqrt{\frac{3}{14}}$

2 Match each surd to its corresponding point on the number line:

a $\sqrt{3}$ **b** $-\sqrt{5}$ **c** $\sqrt{\frac{1}{2}}$ **d** $\sqrt{11}$ **e** $-\sqrt{\frac{1}{3}}$

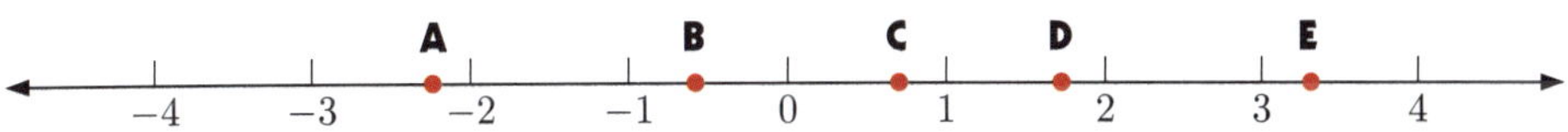

3 Consider the 1000 numbers $\sqrt{1}, \sqrt{2}, \sqrt{3}, ..., \sqrt{1000}$. How many of these numbers are surds?

4 Simplify:

a $(\sqrt{3})^2$ **b** $(\sqrt{3})^3$ **c** $(\sqrt{3})^5$ **d** $\left(\frac{1}{\sqrt{3}}\right)^2$

e $(\sqrt{5})^2$ **f** $(\sqrt{5})^4$ **g** $\left(\frac{5}{\sqrt{5}}\right)^2$ **h** $\left(\frac{10}{\sqrt{5}}\right)^2$

i $\left(\frac{1}{\sqrt{2}}\right)^2$ **j** $\left(\frac{2}{\sqrt{11}}\right)^2$ **k** $\left(\frac{\sqrt{3}}{\sqrt{17}}\right)^2$ **l** $\left(\sqrt{\frac{2}{23}}\right)^2$

Example 2 — Self Tutor

Simplify: **a** $(3\sqrt{2})^2$ **b** $3\sqrt{3}\times(-2\sqrt{3})$

a $(3\sqrt{2})^2$
$= 3\sqrt{2}\times 3\sqrt{2}$
$= 9\times 2$
$= 18$

b $3\sqrt{3}\times(-2\sqrt{3})$
$= 3\times -2\times\sqrt{3}\times\sqrt{3}$
$= -6\times 3$
$= -18$

5 Simplify:

a $(2\sqrt{2})^2$ **b** $(4\sqrt{2})^2$ **c** $(2\sqrt{3})^2$ **d** $(2\sqrt{5})^2$
e $(3\sqrt{5})^2$ **f** $(2\sqrt{7})^2$ **g** $(2\sqrt{10})^2$ **h** $(7\sqrt{10})^2$

6 Simplify:

a $3\sqrt{2}\times 4\sqrt{2}$ **b** $5\sqrt{3}\times 2\sqrt{3}$ **c** $7\sqrt{2}\times 5\sqrt{2}$
d $(-4\sqrt{2})^2$ **e** $(-7\sqrt{3})^2$ **f** $\sqrt{2}\times(-3\sqrt{2})$
g $(-2\sqrt{3})\times(-5\sqrt{3})$ **h** $(-2\sqrt{7})\times 3\sqrt{7}$ **i** $\sqrt{11}\times(-2\sqrt{11})$
j $(4\sqrt{3})^3$ **k** $\sqrt{5}\times(3\sqrt{5})^2$ **l** $(2\sqrt{2})^3\times(-7\sqrt{2})$

7 Simplify:

a $(\sqrt[3]{2})^3$ **b** $(\sqrt[3]{-5})^3$ **c** $\left(\frac{1}{\sqrt[3]{5}}\right)^3$ **d** $(\sqrt[4]{7})^4$

B PROPERTIES OF RADICALS

In previous years you should have discovered that:

For positive numbers a and b, $\sqrt{a}\times\sqrt{b}=\sqrt{ab}$ and $\dfrac{\sqrt{a}}{\sqrt{b}}=\sqrt{\dfrac{a}{b}}$.

Example 3 — Self Tutor

Simplify using the properties of radicals:

a $3\sqrt{2}\times 4\sqrt{11}$ **b** $\dfrac{\sqrt{75}}{\sqrt{3}}$ **c** $\sqrt{\frac{9}{49}}$

a $3\sqrt{2}\times 4\sqrt{11}$
$= 3\times 4\times\sqrt{2}\times\sqrt{11}$
$= 12\times\sqrt{2\times 11}$
$= 12\sqrt{22}$

b $\dfrac{\sqrt{75}}{\sqrt{3}}$
$= \sqrt{\frac{75}{3}}$
$= \sqrt{25}$
$= 5$

c $\sqrt{\frac{9}{49}}$
$= \dfrac{\sqrt{9}}{\sqrt{49}}$
$= \frac{3}{7}$

EXERCISE 7B

1 Simplify:

a $\sqrt{2}\times\sqrt{3}$ **b** $\sqrt{2}\times\sqrt{17}$ **c** $\sqrt{7}\times\sqrt{3}$
d $\sqrt{3}\times\sqrt{2}\times\sqrt{11}$ **e** $2\sqrt{2}\times 5\sqrt{3}$ **f** $5\sqrt{2}\times\sqrt{7}$
g $-5\sqrt{2}\times 2\sqrt{7}$ **h** $\sqrt{5}\times 2\sqrt{2}\times\sqrt{3}$ **i** $(-\sqrt{7})\times(-2\sqrt{3})$
j $(\sqrt{3})^3\times\sqrt{5}$ **k** $(2\sqrt{2})^3\times 5\sqrt{3}$ **l** $(3\sqrt{2})^3\times(\sqrt{3})^3$

2 Simplify:

a $\frac{\sqrt{8}}{\sqrt{2}}$ **b** $\frac{\sqrt{3}}{\sqrt{27}}$ **c** $\frac{\sqrt{18}}{\sqrt{3}}$ **d** $\frac{\sqrt{2}}{\sqrt{50}}$
e $\frac{\sqrt{75}}{\sqrt{5}}$ **f** $\frac{\sqrt{5}}{\sqrt{75}}$ **g** $\frac{\sqrt{18}}{\sqrt{2}}$ **h** $\frac{\sqrt{3}}{\sqrt{48}}$
i $\frac{\sqrt{6}}{\sqrt{2}}$ **j** $\frac{3\sqrt{2}}{\sqrt{18}}$ **k** $\frac{2\sqrt{6}}{\sqrt{24}}$ **l** $\frac{5\sqrt{75}}{2\sqrt{3}}$

3 Simplify:

a $\sqrt{\frac{1}{16}}$ **b** $\sqrt{\frac{81}{144}}$ **c** $\sqrt{\frac{49}{64}}$ **d** $\sqrt{\frac{121}{9}}$
e $\sqrt{6\frac{1}{4}}$ **f** $\sqrt{1\frac{7}{9}}$ **g** $\sqrt{2\frac{14}{25}}$ **h** $\sqrt{7\frac{1}{9}}$

Example 4 — Self Tutor

Write $\sqrt{32}$ in the form $k\sqrt{2}$.

$\sqrt{32} = \sqrt{16\times 2}$
$= \sqrt{16}\times\sqrt{2}$ {using $\sqrt{ab}=\sqrt{a}\times\sqrt{b}$}
$= 4\sqrt{2}$

4 Write in the form $k\sqrt{2}$:

a $\sqrt{8}$ **b** $\sqrt{18}$ **c** $\sqrt{50}$ **d** $\sqrt{98}$
e $\sqrt{200}$ **f** $\sqrt{288}$ **g** $\sqrt{20\,000}$

5 Write in the form $k\sqrt{3}$:

a $\sqrt{12}$ **b** $\sqrt{27}$ **c** $\sqrt{75}$

6 Write in the form $k\sqrt{5}$:

a $\sqrt{20}$ **b** $\sqrt{45}$ **c** $\sqrt{125}$

7 Use the expansions of $(\sqrt{a}\sqrt{b})^2$ and $(\sqrt{ab})^2$ to prove that $\sqrt{a}\sqrt{b}=\sqrt{ab}$ for all positive a and b.

8 Use the expansions of $\left(\frac{\sqrt{a}}{\sqrt{b}}\right)^2$ and $\left(\sqrt{\frac{a}{b}}\right)^2$ to prove that $\frac{\sqrt{a}}{\sqrt{b}}=\sqrt{\frac{a}{b}}$ for all positive a and b.

9 By considering the expansions of $(\sqrt{a}+\sqrt{b})^2$ and $(\sqrt{a+b})^2$, explain why in general, $\sqrt{a}+\sqrt{b}\neq\sqrt{a+b}$.

C SIMPLEST SURD FORM

A radical is in **simplest surd form** when the number under the radical sign is the smallest possible integer.

Example 5 Self Tutor

Write $\sqrt{432}$ in simplest surd form.

$$\begin{aligned}\sqrt{432} &= \sqrt{2^4 \times 3^3}\\ &= \sqrt{2^4} \times \sqrt{3^3}\\ &= 4 \times 3\sqrt{3}\\ &= 12\sqrt{3}\end{aligned}$$

EXERCISE 7C

1 Write in simplest surd form:

a $\sqrt{12}$ **b** $\sqrt{28}$ **c** $\sqrt{54}$ **d** $\sqrt{60}$
e $\sqrt{99}$ **f** $\sqrt{52}$ **g** $\sqrt{40}$ **h** $\sqrt{63}$
i $\sqrt{48}$ **j** $\sqrt{125}$ **k** $\sqrt{147}$ **l** $\sqrt{175}$
m $\sqrt{176}$ **n** $\sqrt{150}$ **o** $\sqrt{2000}$

2 Write in simplest surd form:

a $\sqrt{\frac{5}{9}}$ **b** $\sqrt{\frac{18}{4}}$ **c** $\sqrt{\frac{12}{16}}$ **d** $\sqrt{\frac{75}{36}}$

3 Write in simplest surd form $a + b\sqrt{n}$ where $a, b \in \mathbb{Q}$, $n \in \mathbb{Z}$:

a $\frac{4+\sqrt{8}}{2}$ **b** $\frac{6-\sqrt{12}}{2}$ **c** $\frac{4+\sqrt{20}}{4}$
d $\frac{8-\sqrt{32}}{4}$ **e** $\frac{12+\sqrt{72}}{6}$ **f** $\frac{18+\sqrt{27}}{6}$
g $\frac{14-\sqrt{60}}{8}$ **h** $\frac{5-\sqrt{200}}{10}$

D POWER EQUATIONS

A **power equation** is an equation of the form $x^n = k$ where $n \neq 0$.

You should have seen previously that:

If $x^2 = k$ then $\begin{cases} x = \pm\sqrt{k} & \text{if } k > 0 \\ x = 0 & \text{if } k = 0 \\ x \text{ has no real solutions if } k < 0. \end{cases}$

More generally, for any $n \in \mathbb{Z}^+$, $n > 2$:

- if $x^n = k$ and n is odd, then $x = \sqrt[n]{k}$
- if $x^n = k$ and n is even, then $\begin{cases} x = \pm\sqrt[n]{k} & \text{if } k > 0 \\ x = 0 & \text{if } k = 0 \\ x \text{ has no real solutions if } k < 0. \end{cases}$

Example 6 — Self Tutor

Solve for x, giving your answer correct to 3 significant figures:

a $x^2 = 6$ **b** $x^2 = 13$, $x > 0$ **c** $x^3 = 31$

a $x^2 = 6$
$\therefore\ x = \pm\sqrt{6}$
$\therefore\ x \approx \pm 2.45$

b $x^2 = 13$, $x > 0$
$\therefore\ x = \sqrt{13}$
$\therefore\ x \approx 3.61$

c $x^3 = 31$
$\therefore\ x = \sqrt[3]{31}$
$\therefore\ x \approx 3.14$

EXERCISE 7D

1 Solve for x, giving your answer correct to 3 significant figures where appropriate:

a $x^2 = 11$ **b** $x^2 = -5$ **c** $x^2 = 71$ **d** $x^2 = 89$, $x > 0$
e $x^3 = 8$ **f** $x^3 = 11$ **g** $x^3 = -11$ **h** $x^4 = 81$
i $x^4 = -1$ **j** $x^5 = 23$ **k** $x^5 = -113$ **l** $x^6 = 39.2$, $x > 0$

2 Solve for x, giving your answer in simplest surd form:

a $x^2 = 8$ **b** $x^2 = 72$ **c** $x^2 = 48$ **d** $x^2 = 162$

3 Solve for x:

a $x^2 - 5 = 11$ **b** $13 + x^2 = 49$ **c** $x^2 + 19 = 79$
d $2x^2 = 18$ **e** $3x^2 = -72$ **f** $29 - 2x^2 = 21$

4 Solve for x:

a $64 - x^3 = 0$ **b** $7 + 2x^3 = -9$ **c** $39 + x^4 = 120$

Example 7 — Self Tutor

Solve for x: $\dfrac{x}{2} = \dfrac{5}{x}$

$\dfrac{x}{2} = \dfrac{5}{x}$

$\therefore\ \dfrac{x \times x}{2 \times x} = \dfrac{5 \times 2}{x \times 2}$ {LCD $= 2x$}

$\therefore\ x^2 = 10$ {equating numerators}

$\therefore\ x = \pm\sqrt{10}$

5 Solve for x:

a $\dfrac{x}{3} = \dfrac{4}{x}$ **b** $\dfrac{x}{6} = \dfrac{6}{x}$ **c** $\dfrac{1}{x} = \dfrac{x}{3}$ **d** $\dfrac{x}{7} = \dfrac{7}{x}$
e $\dfrac{2}{x} = \dfrac{x}{5}$ **f** $\dfrac{7}{x} = \dfrac{x}{5}$ **g** $\dfrac{x}{2} = \dfrac{8}{x}$ **h** $\dfrac{x}{5} = \dfrac{-2}{x}$

Example 8 **Self Tutor**

Solve for x: $\sqrt{x} = 6$

$$\sqrt{x} = 6$$
$$\therefore \left(\sqrt{x}\right)^2 = 6^2 \quad \{\text{squaring both sides}\}$$
$$\therefore x = 36$$

6 Solve for x:

a $\sqrt{x} = 2$ **b** $\sqrt{x} = 9$ **c** $\sqrt{x} = 0$ **d** $\sqrt{x} = \sqrt{7}$

e $\sqrt{x} = 2\sqrt{3}$ **f** $\sqrt{x} - 3 = 5$ **g** $7 + \sqrt{x} = 18$ **h** $3\sqrt{x} = 30$

7 Explain why $\sqrt{x} = -5$ has no real solutions.

DISCUSSION

Consider what happens to each of these equations when both sides are squared:

- $\sqrt{x} = 3$
- $\sqrt{x} = -3$
- $x = 3$
- $x = -3$

1 Is the resulting equation always consistent with the original equation?

2 Is it possible to introduce extra solutions to an equation by squaring both sides?

3 If we need to square both sides of an equation, how do we avoid errors?

E OPERATIONS WITH RADICALS

We can perform operations with radicals just as we do with ordinary numbers. In particular:

- We add and subtract "like" radicals in the same way that we add and subtract "like" algebraic terms.
- We can use the usual rules for expanding brackets.

Example 9 **Self Tutor**

Simplify:

a $3\sqrt{7} - 4\sqrt{7}$ **b** $5\sqrt{3} + 2\sqrt{5} - \sqrt{3} + 7\sqrt{5}$ **c** $5\sqrt{3} - 2\sqrt{12}$

a
$$3\sqrt{7} - 4\sqrt{7}$$
$$= -1\sqrt{7}$$
$$= -\sqrt{7}$$

b
"like" radicals
$$5\sqrt{3} + 2\sqrt{5} - \sqrt{3} + 7\sqrt{5}$$
"like" radicals
$$= 4\sqrt{3} + 9\sqrt{5}$$

c
$$5\sqrt{3} - 2\sqrt{12}$$
$$= 5\sqrt{3} - 2\sqrt{4 \times 3}$$
$$= 5\sqrt{3} - 4\sqrt{3}$$
$$= \sqrt{3}$$

EXERCISE 7E

1 Simplify:

a $\sqrt{5} + \sqrt{5}$ **b** $\sqrt{7} - \sqrt{7}$ **c** $4\sqrt{3} + 2\sqrt{3}$

d $3\sqrt{2} + 7\sqrt{2}$ **e** $11\sqrt{3} - 8\sqrt{3}$ **f** $6\sqrt{5} - 7\sqrt{5}$

g $-\sqrt{10}+2\sqrt{10}$ **h** $\sqrt{6}+7\sqrt{6}-3\sqrt{6}$ **i** $9\sqrt{15}-4\sqrt{15}-11\sqrt{15}$

2 Simplify:

a $5\sqrt{2}-\sqrt{3}+\sqrt{2}-\sqrt{3}$ **b** $4\sqrt{7}+5\sqrt{6}+3\sqrt{7}-2\sqrt{6}$
c $9\sqrt{10}-5\sqrt{5}+8\sqrt{5}-\sqrt{10}$ **d** $11\sqrt{3}-8\sqrt{13}+\sqrt{13}-13\sqrt{3}$
e $\sqrt{7}+5\sqrt{11}+9\sqrt{7}-4\sqrt{7}$ **f** $-8-\sqrt{15}+7\sqrt{15}-13$

3 Simplify:

a $3\sqrt{2}+2\sqrt{18}$ **b** $2\sqrt{7}-5\sqrt{28}$ **c** $13\sqrt{6}+3\sqrt{150}$

4 Show that:

a $\sqrt{20}+\sqrt{5}=\sqrt{45}$ **b** $\sqrt{147}-\sqrt{75}=\sqrt{12}$ **c** $\sqrt{175}-\sqrt{63}=\sqrt{28}$

Example 10 **Self Tutor**

Simplify: $\frac{\sqrt{7}}{2}+\frac{\sqrt{7}}{3}$

$$\frac{\sqrt{7}}{2}+\frac{\sqrt{7}}{3}$$
$$=\frac{\sqrt{7}\times 3}{2\times 3}+\frac{\sqrt{7}\times 2}{3\times 2} \quad \{\text{LCD}=6\}$$
$$=\frac{3\sqrt{7}}{6}+\frac{2\sqrt{7}}{6}$$
$$=\frac{3\sqrt{7}+2\sqrt{7}}{6}$$
$$=\frac{5\sqrt{7}}{6}$$

5 Simplify:

a $\frac{\sqrt{5}}{3}+\frac{\sqrt{5}}{4}$ **b** $\frac{\sqrt{6}}{2}-\frac{\sqrt{6}}{7}$ **c** $\frac{5\sqrt{3}}{6}+\frac{\sqrt{3}}{8}$
d $\frac{2\sqrt{11}}{9}-\frac{8\sqrt{11}}{15}$ **e** $\frac{\sqrt{10}}{2}+\frac{\sqrt{10}}{3}+\frac{\sqrt{10}}{4}$ **f** $\sqrt{2}+\frac{5\sqrt{2}}{14}-\frac{7\sqrt{2}}{4}$

Example 11 **Self Tutor**

Expand and simplify:

a $\sqrt{2}\left(\sqrt{2}+\sqrt{3}\right)$ **b** $\sqrt{3}\left(6-2\sqrt{3}\right)$

a $\sqrt{2}\left(\sqrt{2}+\sqrt{3}\right)$
$=\sqrt{2}\times\sqrt{2}+\sqrt{2}\times\sqrt{3}$
$=2+\sqrt{6}$

b $\sqrt{3}\left(6-2\sqrt{3}\right)$
$=\sqrt{3}\times 6+\sqrt{3}\times(-2\sqrt{3})$
$=6\sqrt{3}-6$

6 Expand and simplify:

a $\sqrt{2}\left(\sqrt{5}+\sqrt{2}\right)$ **b** $\sqrt{2}\left(3-\sqrt{2}\right)$ **c** $\sqrt{3}\left(\sqrt{3}+1\right)$
d $\sqrt{3}\left(1-\sqrt{3}\right)$ **e** $\sqrt{7}\left(7-\sqrt{7}\right)$ **f** $\sqrt{5}\left(2-\sqrt{5}\right)$
g $\sqrt{11}\left(2\sqrt{11}-1\right)$ **h** $\sqrt{6}\left(1-2\sqrt{6}\right)$ **i** $\sqrt{3}\left(\sqrt{3}+\sqrt{2}-1\right)$
j $2\sqrt{3}\left(\sqrt{3}-\sqrt{5}\right)$ **k** $2\sqrt{5}\left(3-\sqrt{5}\right)$ **l** $3\sqrt{5}\left(2\sqrt{5}+\sqrt{2}\right)$

Example 12 Self Tutor

Expand and simplify:

a $-\sqrt{2}\,(\sqrt{2}+3)$

b $-\sqrt{3}\,(7-2\sqrt{3})$

a
$$\begin{aligned} &-\sqrt{2}\,(\sqrt{2}+3)\\ &=(-\sqrt{2})\times\sqrt{2}+(-\sqrt{2})\times 3\\ &=-2-3\sqrt{2} \end{aligned}$$

b
$$\begin{aligned} &-\sqrt{3}\,(7-2\sqrt{3})\\ &=(-\sqrt{3})\times 7+(-\sqrt{3})\times(-2\sqrt{3})\\ &=-7\sqrt{3}+6 \end{aligned}$$

7 Expand and simplify:

a $-\sqrt{2}\,(3-\sqrt{2})$
b $-\sqrt{2}\,(\sqrt{2}+\sqrt{3})$
c $-\sqrt{2}\,(4-\sqrt{2})$
d $-\sqrt{3}\,(1+\sqrt{3})$
e $-\sqrt{3}\,(\sqrt{3}+2)$
f $-\sqrt{5}\,(2+\sqrt{5})$
g $-\sqrt{5}\,(\sqrt{5}-4)$
h $-(\sqrt{3}-\sqrt{7})$
i $-2\sqrt{2}\,(1-\sqrt{2})$
j $-3\sqrt{3}\,(5-\sqrt{3})$
k $-7\sqrt{2}\,(\sqrt{2}+\sqrt{6})$
l $(-\sqrt{2})^3\,(3-\sqrt{2})$

Example 13 Self Tutor

Expand and simplify: $(3-\sqrt{2})\,(4+2\sqrt{2})$

$$\begin{aligned} &(3-\sqrt{2})\,(4+2\sqrt{2})\\ &=12+3\times 2\sqrt{2}+(-\sqrt{2})\times 4+(-\sqrt{2})\times 2\sqrt{2}\\ &=12+6\sqrt{2}-4\sqrt{2}-4\\ &=8+2\sqrt{2} \end{aligned}$$

8 Expand and simplify:

a $(1+\sqrt{2})\,(2+\sqrt{2})$
b $(2+\sqrt{3})\,(2+\sqrt{3})$
c $(\sqrt{3}+2)\,(\sqrt{3}-1)$
d $(4-\sqrt{2})\,(3+\sqrt{2})$
e $(1+\sqrt{3})\,(1-\sqrt{3})$
f $(5+\sqrt{7})\,(2-\sqrt{7})$
g $(\sqrt{5}+2)\,(\sqrt{5}-3)$
h $(2\sqrt{2}+\sqrt{3})\,(\sqrt{2}-\sqrt{3})$
i $(4-\sqrt{2})\,(3-\sqrt{2})$

Example 14 Self Tutor

Expand and simplify:

a $(\sqrt{3}+2)^2$

b $(\sqrt{3}-\sqrt{7})^2$

a
$$\begin{aligned} &(\sqrt{3}+2)^2\\ &=(\sqrt{3})^2\ +\ 2\times\sqrt{3}\times 2\ +\ 2^2\\ &=3+4\sqrt{3}+4\\ &=7+4\sqrt{3} \end{aligned}$$

b
$$\begin{aligned} &(\sqrt{3}-\sqrt{7})^2\\ &=(\sqrt{3})^2\ +\ 2\times\sqrt{3}\times(-\sqrt{7})\ +\ (-\sqrt{7})^2\\ &=3-2\sqrt{21}+7\\ &=10-2\sqrt{21} \end{aligned}$$

9 Expand and simplify:

a $(1+\sqrt{2})^2$	**b** $(2-\sqrt{3})^2$	**c** $(\sqrt{3}+2)^2$
d $(\sqrt{2}-\sqrt{3})^2$	**e** $(\sqrt{2}+\sqrt{7})^2$	**f** $(4-\sqrt{6})^2$
g $(\sqrt{6}-\sqrt{2})^2$	**h** $(\sqrt{5}+2\sqrt{2})^2$	**i** $(\sqrt{5}-2\sqrt{2})^2$
j $(6+\sqrt{8})^2$	**k** $(5\sqrt{2}-1)^2$	**l** $(1+3\sqrt{2})^2$

Example 15 **Self Tutor**

Expand and simplify:

a $(3+\sqrt{2})(3-\sqrt{2})$ **b** $(2\sqrt{3}-5)(2\sqrt{3}+5)$

a $(3+\sqrt{2})(3-\sqrt{2})$
$= 3^2-(\sqrt{2})^2$
$= 9-2$
$= 7$

b $(2\sqrt{3}-5)(2\sqrt{3}+5)$
$= (2\sqrt{3})^2-5^2$
$= 4\times 3-25$
$= 12-25$
$= -13$

10 Expand and simplify:

a $(4+\sqrt{3})(4-\sqrt{3})$	**b** $(5-\sqrt{2})(5+\sqrt{2})$
c $(\sqrt{5}-2)(\sqrt{5}+2)$	**d** $(\sqrt{7}+4)(\sqrt{7}-4)$
e $(3\sqrt{2}+2)(3\sqrt{2}-2)$	**f** $(2\sqrt{5}-1)(2\sqrt{5}+1)$
g $(2-4\sqrt{2})(2+4\sqrt{2})$	**h** $(1+5\sqrt{7})(1-5\sqrt{7})$
i $(\sqrt{3}+\sqrt{2})(\sqrt{3}-\sqrt{2})$	**j** $(\sqrt{7}-\sqrt{11})(\sqrt{7}+\sqrt{11})$
k $(3\sqrt{2}+\sqrt{3})(3\sqrt{2}-\sqrt{3})$	**l** $(2\sqrt{5}-3\sqrt{2})(3\sqrt{2}+2\sqrt{5})$

11 Use the binomial expansion for $(a+b)^3$ to write the following in simplest surd form:

a $(3+\sqrt{7})^3$ **b** $(\sqrt{3}-\sqrt{2})^3$ **c** $(2\sqrt{3}-\sqrt{2})^3$

F DIVISION WITH SURDS

When an expression involves division by a surd, we can write the expression with an **integer denominator** which does *not* contain surds. This process is called **rationalising the denominator**.

If the denominator has only one term, we can remove the surd by using the definition $\sqrt{a}\times\sqrt{a}=a$.

For any fraction of the form $\dfrac{b}{c\sqrt{a}}$, $a>0$, we can remove the radical from the denominator by multiplying by $\dfrac{\sqrt{a}}{\sqrt{a}}$.

Example 16 **Self Tutor**

Write with an integer denominator:

a $\dfrac{7}{\sqrt{3}}$ **b** $\dfrac{10}{\sqrt{5}}$ **c** $\dfrac{10}{2\sqrt{2}}$

a $\dfrac{7}{\sqrt{3}} = \dfrac{7}{\sqrt{3}} \times \dfrac{\sqrt{3}}{\sqrt{3}}$
$= \dfrac{7\sqrt{3}}{3}$

b $\dfrac{10}{\sqrt{5}} = \dfrac{10}{\sqrt{5}} \times \dfrac{\sqrt{5}}{\sqrt{5}}$
$= \dfrac{10\sqrt{5}}{5}$
$= 2\sqrt{5}$

c $\dfrac{10}{2\sqrt{2}} = \dfrac{10}{2\sqrt{2}} \times \dfrac{\sqrt{2}}{\sqrt{2}}$
$= \dfrac{10\sqrt{2}}{4}$
$= \dfrac{5\sqrt{2}}{2}$

EXERCISE 7F

1 Write with an integer denominator:

a $\dfrac{1}{\sqrt{5}}$ **b** $\dfrac{2}{\sqrt{2}}$ **c** $\dfrac{5}{\sqrt{3}}$ **d** $\dfrac{1}{\sqrt{8}}$

e $\dfrac{4}{\sqrt{7}}$ **f** $\dfrac{3}{\sqrt{6}}$ **g** $\dfrac{\sqrt{2}}{\sqrt{3}}$ **h** $\dfrac{\sqrt{3}}{\sqrt{7}}$

2 Rationalise the denominator:

a $\dfrac{1}{2\sqrt{3}}$ **b** $\dfrac{4}{3\sqrt{2}}$ **c** $\dfrac{\sqrt{3}}{3\sqrt{5}}$ **d** $\dfrac{2\sqrt{5}}{5\sqrt{3}}$

e $\dfrac{\sqrt{7}}{3\sqrt{2}}$ **f** $\dfrac{22}{4\sqrt{11}}$ **g** $\dfrac{9\sqrt{2}}{3\sqrt{5}}$ **h** $\dfrac{3\sqrt{2}}{5\sqrt{6}}$

3 Evaluate, writing your answer with an integer denominator:

a $2\sqrt{3} - \dfrac{1}{\sqrt{3}}$ **b** $\dfrac{3}{\sqrt{2}} + 2\sqrt{2}$ **c** $\left(\dfrac{1}{\sqrt{3}}\right)^3$ **d** $\dfrac{5}{\sqrt{2}} - \sqrt{8}$

INVESTIGATION RADICAL CONJUGATES

The radical expressions $3 + \sqrt{2}$ and $3 - \sqrt{2}$ are identical except for opposite signs in the middle. We say that these expressions are *radical conjugates*.

If an expression has two terms and includes a radical, its **radical conjugate** is identical except the sign in the middle is opposite.

What to do:

1 Write down the radical conjugate of:

a $1 + \sqrt{3}$ **b** $\sqrt{2} - 1$ **c** $\sqrt{3} - \sqrt{2}$ **d** $3 + 2\sqrt{2}$

2 Expand and simplify:

a $(a + \sqrt{b})(a - \sqrt{b})$ **b** $(\sqrt{a} + b)(\sqrt{a} - b)$

c $(\sqrt{a} + \sqrt{b})(\sqrt{a} - \sqrt{b})$ **d** $(a + c\sqrt{b})(a - c\sqrt{b})$

3 Describe what you should multiply a fraction by to rationalise the denominator, if the denominator has the form:

a $a+\sqrt{b}$ **b** $\sqrt{a}-b$ **c** $\sqrt{a}+\sqrt{b}$ **d** $a-c\sqrt{b}$

4 Copy and complete:
To remove the radicals from the denominator of a fraction, we multiply both the numerator *and* the denominator by the .... of the denominator.

QUICK QUIZ

MULTIPLE CHOICE QUIZ

REVIEW SET 7A

1 How many of the numbers $\sqrt{1}, \sqrt{2}, \sqrt{3}, ..., \sqrt{100}$ are surds?

2 Simplify:

a $(2\sqrt{3})^2$ **b** $\left(\frac{8}{\sqrt{2}}\right)^2$ **c** $3\sqrt{2}\times 6\sqrt{2}$ **d** $(\sqrt[3]{-10})^3$

3 Simplify:

a $\sqrt{7}\times\sqrt{11}$ **b** $5\sqrt{3}\times 6\sqrt{5}$ **c** $\frac{\sqrt{35}}{\sqrt{5}}$ **d** $\frac{\sqrt{2}}{\sqrt{50}}$

4 Write in the form $k\sqrt{7}$:

a $\sqrt{63}$ **b** $\sqrt{252}$

5 Write in simplest surd form:

a $\sqrt{44}$ **b** $\sqrt{90}$ **c** $\sqrt{\frac{18}{25}}$

6 Solve for x, giving your answer in simplest surd form:

a $x^2=24$ **b** $x^2=150$ **c** $x^2+4=100$

7 Solve for x:

a $3x^2=147$ **b** $3-x^3=11$ **c** $\frac{x}{7}=\frac{3}{x}$ **d** $\sqrt{x}=8$

8 Simplify:

a $2\sqrt{7}-9\sqrt{7}+3\sqrt{7}$ **b** $\sqrt{6}-4\sqrt{3}+8\sqrt{6}+\sqrt{3}$ **c** $\frac{2\sqrt{7}}{3}-\frac{3\sqrt{7}}{10}$

9 Expand and simplify:

a $2\sqrt{3}(4-\sqrt{3})$ **b** $(3-\sqrt{7})^2$ **c** $(2-\sqrt{3})(2+\sqrt{3})$

d $(3+2\sqrt{5})(2-\sqrt{5})$ **e** $(4-\sqrt{2})(3+2\sqrt{2})$ **f** $(\sqrt{7}+3\sqrt{8})^2$

10 Write with an integer denominator:

a $\frac{15}{\sqrt{3}}$ **b** $\frac{\sqrt{2}}{2\sqrt{5}}$

11 Evaluate, writing your answer with an integer denominator:

a $\left(\frac{3}{\sqrt{2}}\right)^3$ **b** $\frac{9}{\sqrt{3}} + \sqrt{27}$

REVIEW SET 7B

1 Simplify:

a $\left(\frac{3}{\sqrt{10}}\right)^2$ **b** $(3\sqrt{6})^2$ **c** $\sqrt{2} \times (5\sqrt{2})^2$

2 Simplify:

a $\sqrt{5} \times \sqrt{13}$ **b** $(2\sqrt{5})^2 \times \sqrt{3}$ **c** $\frac{\sqrt{56}}{\sqrt{14}}$ **d** $\sqrt{5\frac{4}{9}}$

3 Write $\sqrt{80}$ in simplest surd form.

4 Write in simplest surd form $a + b\sqrt{n}$ where $a, b \in \mathbb{Q}$, $n \in \mathbb{Z}$:

a $\frac{8+\sqrt{12}}{2}$ **b** $\frac{12-\sqrt{45}}{4}$ **c** $\frac{3+\sqrt{72}}{15}$

5 Solve for x, giving your answer correct to 3 significant figures:

a $x^2 = 29$ **b** $x^4 = 90, \quad x > 0$ **c** $x^3 = -40$

6 Solve for x:

a $3 - 2x^2 = 1$ **b** $11 + 2x^3 = 65$ **c** $\sqrt{x} = 2\sqrt{5}$

7 Simplify:

a $5\sqrt{3} + 2\sqrt{48}$ **b** $\frac{\sqrt{13}}{12} - \frac{3\sqrt{13}}{10}$ **c** $\frac{\sqrt{6}}{3} - \frac{\sqrt{6}}{4} + \frac{2\sqrt{6}}{5}$

8 Show that:

a $\sqrt{54} - \sqrt{24} = \sqrt{6}$ **b** $\sqrt{18} + \sqrt{32} = \sqrt{98}$

9 Expand and simplify:

a $2\sqrt{3}\,(\sqrt{3} - 1)$ **b** $-\sqrt{2}\,(2 - \sqrt{2})$ **c** $(5 - \sqrt{3})\,(5 + \sqrt{3})$

d $(2 - \sqrt{5})^2$ **e** $(2\sqrt{2} - 5)\,(1 - \sqrt{2})$ **f** $(\sqrt{5} + \sqrt{7})\,(\sqrt{7} - \sqrt{5})$

10 Write with an integer denominator:

a $\frac{14}{\sqrt{2}}$ **b** $\frac{5}{2\sqrt{3}}$

11 Answer the **Opening Problem** on page **90**.

Chapter 8

Pythagoras' theorem

Contents:

OPENING PROBLEM

Sven has challenged Jill to a swimming race in a lake. Both swimmers will swim from A to B, but each needs to touch a buoy along the way. Sven thinks he is a very good swimmer, so he will swim to buoy C on his way to B. Jill will swim to buoy D, as shown.

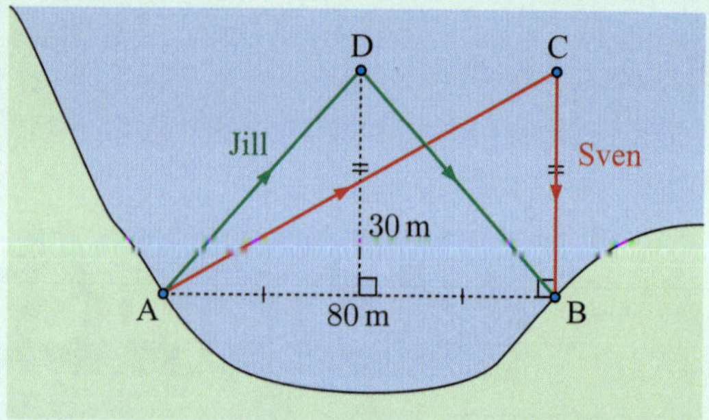

Things to think about:

a How can you calculate the distance each swimmer will swim?

b How far will Jill swim?

c How much farther will Sven swim than Jill?

A PYTHAGORAS' THEOREM

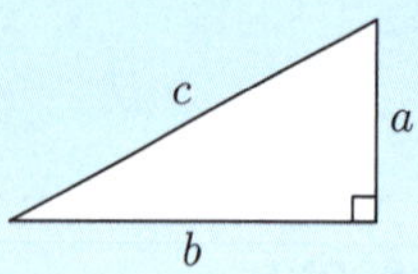

In a right angled triangle with hypotenuse of length c, and other sides of length a and b,

$$c^2 = a^2 + b^2.$$

In geometric form, Pythagoras' theorem states:

In any right angled triangle, the area of the square on the hypotenuse is equal to the sum of the areas of the squares on the other two sides.

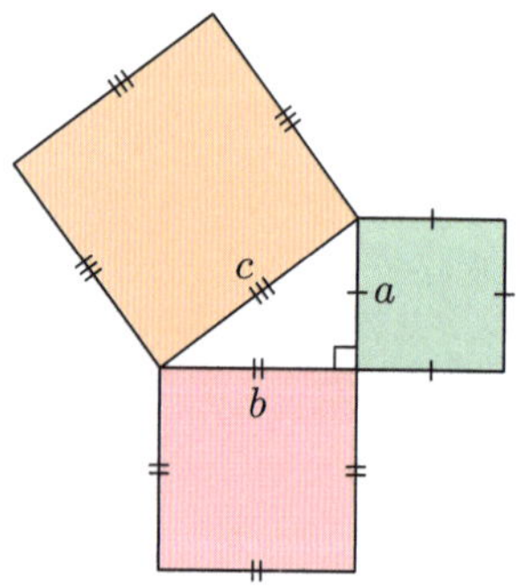

PUZZLE

This tiling pattern is called the Pythagorean tiling because it has been used by several mathematicians to prove Pythagoras' theorem.

In the 9th century, the Persian mathematicians **Al-Nayrīzī** and **Thābit ibn Qurra** used the marked square divided into five pieces as the basis for their proof.

Use the same "five-piece dissection" to prove Pythagoras' theorem for yourself.

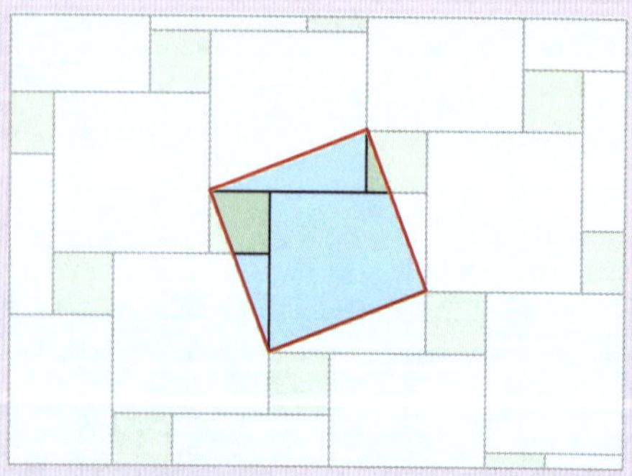

To find side lengths in triangles using Pythagoras' theorem, we need to solve a power equation. Since side lengths must be positive, we can always reject the negative solution.

Example 1 **Self Tutor**

Find the length of the unknown side, giving your answer in simplest surd form if necessary.

a

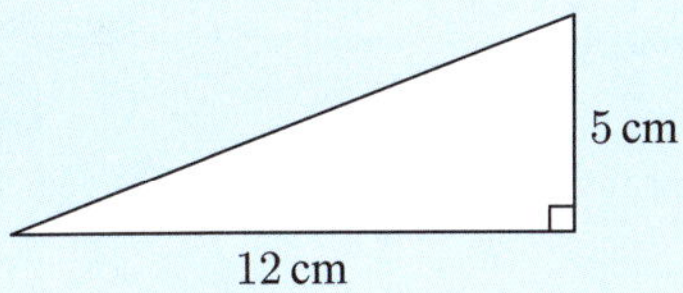

b

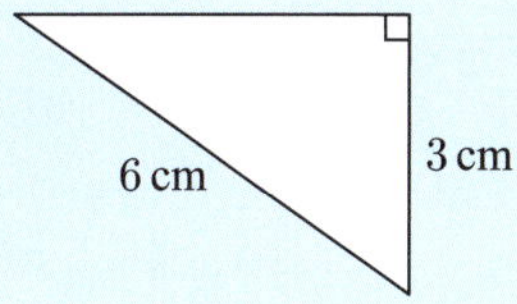

a Let the hypotenuse have length x cm

$\therefore\ x^2 = 12^2 + 5^2$ {Pythagoras}

$\therefore\ x^2 = 144 + 25$

$\therefore\ x^2 = 169$

$\therefore\ x = \sqrt{169}$ {as $x > 0$}

$\therefore\ x = 13$

The hypotenuse is 13 cm long.

b Let the unknown side have length x cm

$\therefore\ x^2 + 3^2 = 6^2$ {Pythagoras}

$\therefore\ x^2 + 9 = 36$

$\therefore\ x^2 = 27$

$\therefore\ x = \sqrt{27}$ {as $x > 0$}

$\therefore\ x = 3\sqrt{3}$

The third side is $3\sqrt{3}$ cm long.

EXERCISE 8A

1 Find the length of the hypotenuse in each triangle. Give your answer in simplest surd form where appropriate.

a

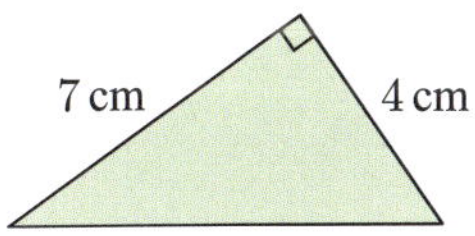

b

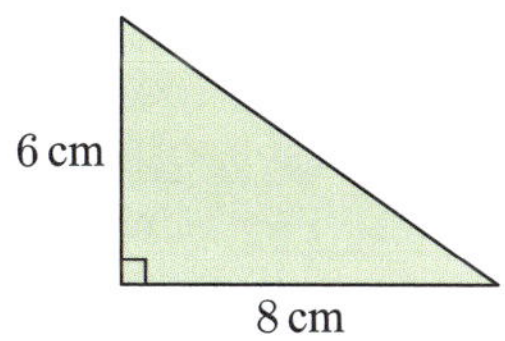

c

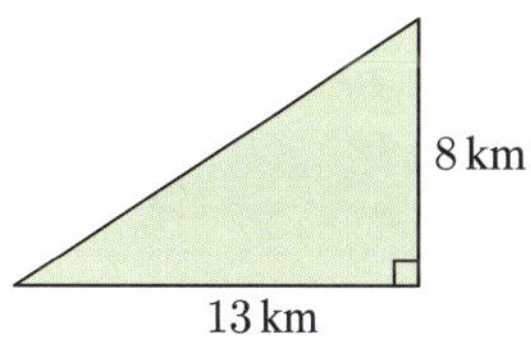

d

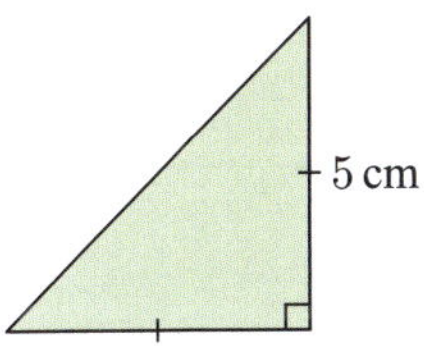

e

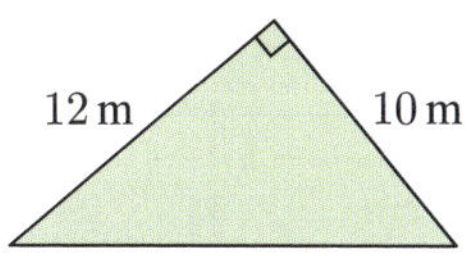

f

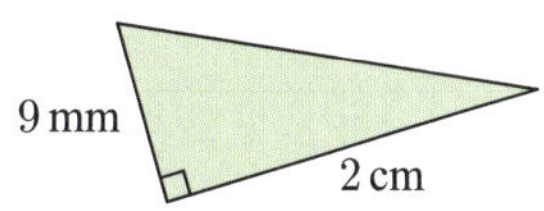

2 Find the length of the unknown side in each triangle. Give your answer in simplest surd form where appropriate.

a

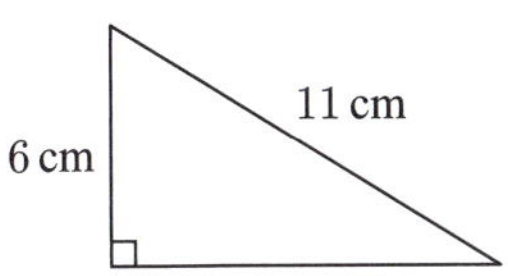

b

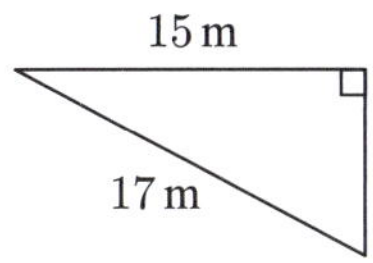

c

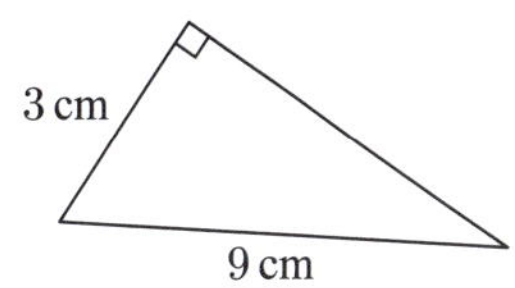

d

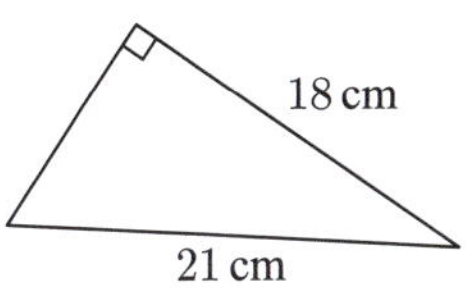

e

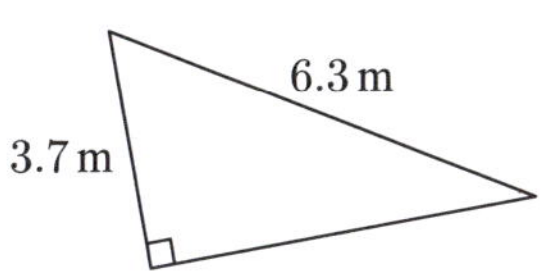

f

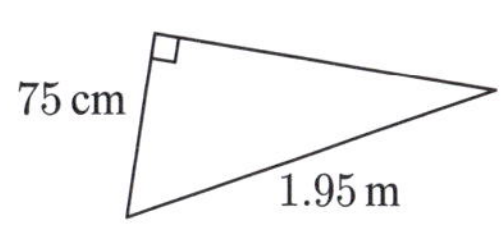

3 Find the length of the unknown side in each triangle. Give your answer to 1 decimal place where appropriate.

a

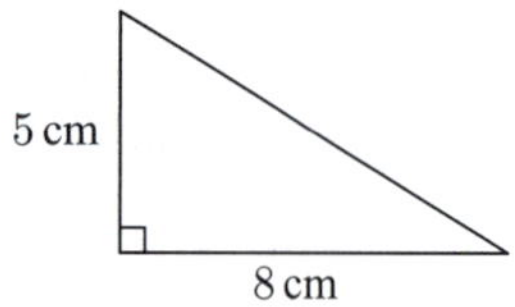

b

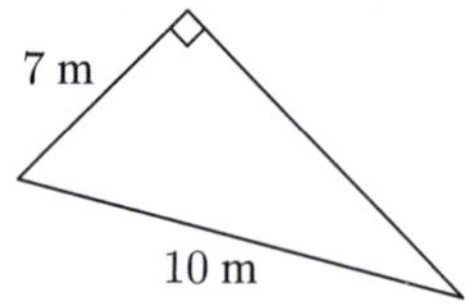

c

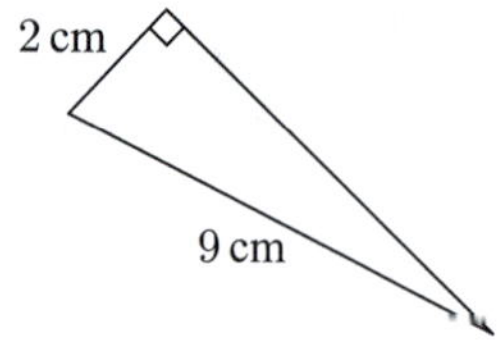

d

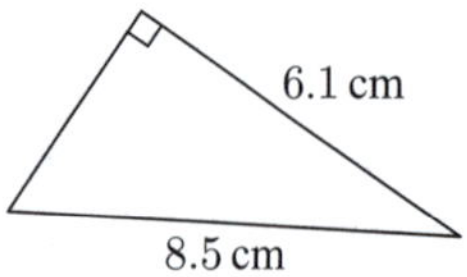

e

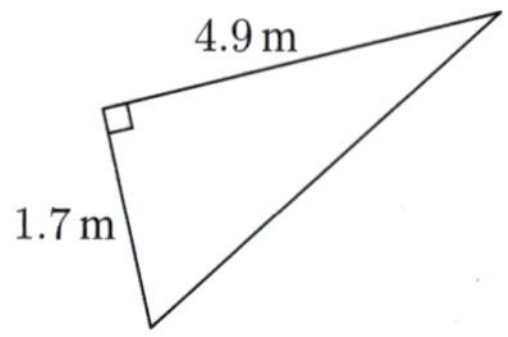

f

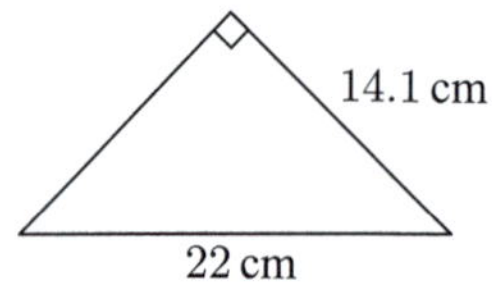

4 Find the value of x:

a

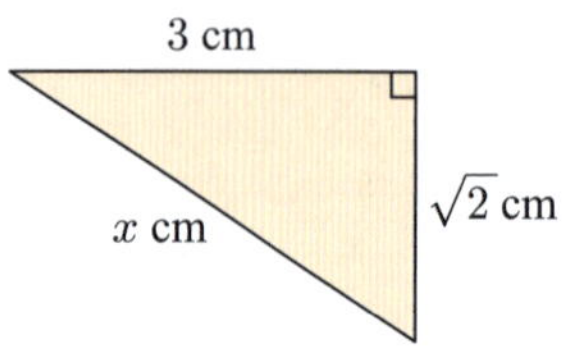

b

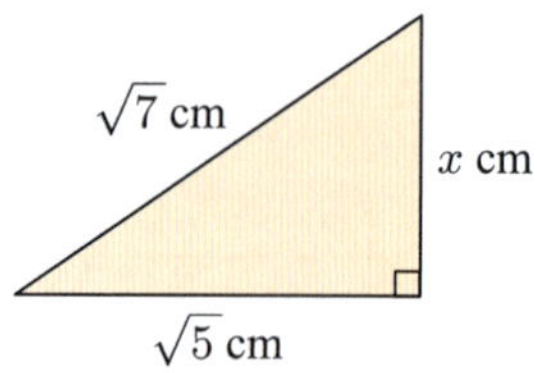

c

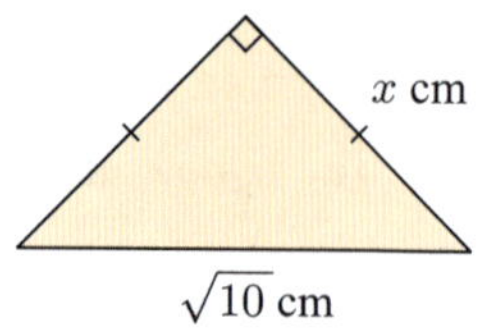

d

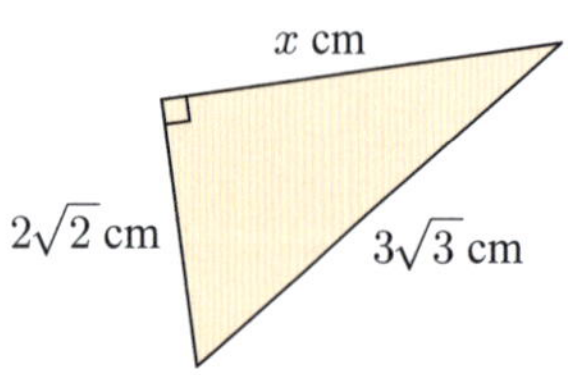

e

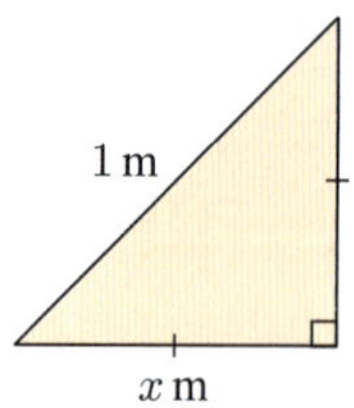

f

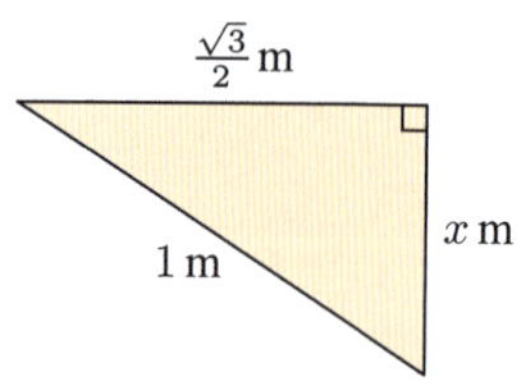

Example 2 — Self Tutor

Find the value of x:

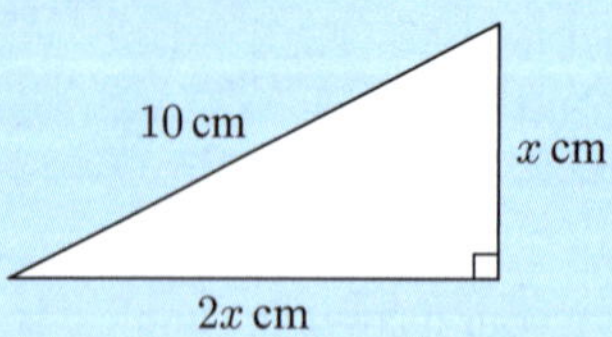

$$x^2 + (2x)^2 = 10^2 \quad \{\text{Pythagoras}\}$$
$$\therefore \; x^2 + 4x^2 = 100$$
$$\therefore \; 5x^2 = 100$$
$$\therefore \; x^2 = 20$$
$$\therefore \; x = \sqrt{20} \quad \{\text{as } x > 0\}$$
$$\therefore \; x = 2\sqrt{5}$$

5 Find the value of x, rounding your answer to 3 significant figures where appropriate:

a

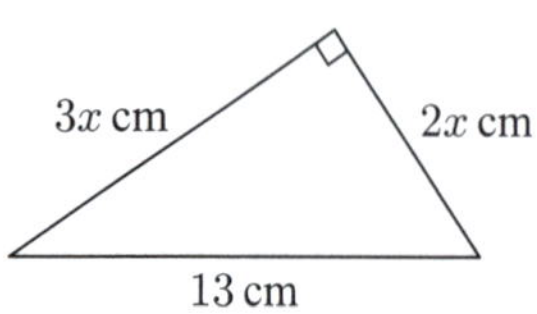

b

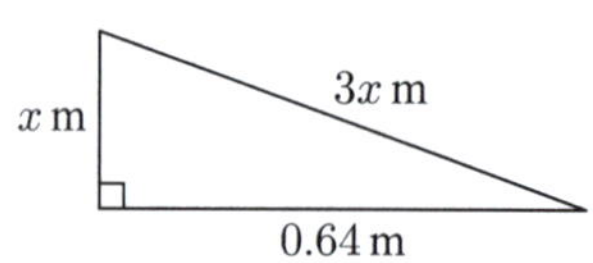

c

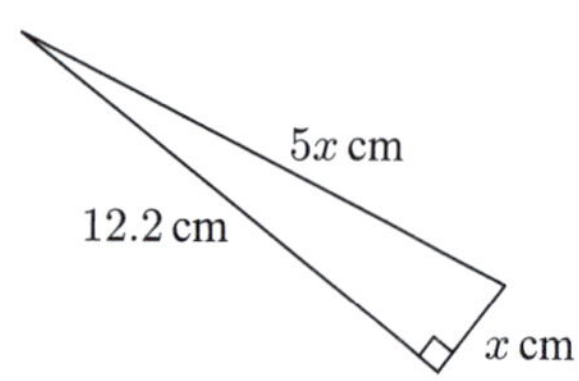

Example 3 — Self Tutor

Find the values of the unknowns:

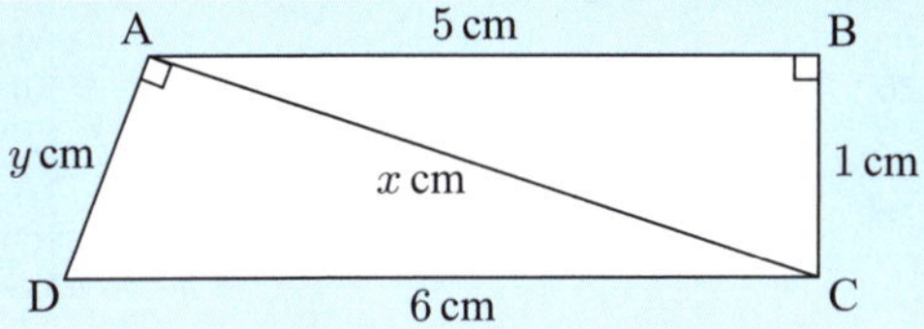

In triangle ABC, the hypotenuse is x cm.

$\therefore\ x^2 = 5^2 + 1^2$ {Pythagoras}

$\therefore\ x^2 = 26$

$\therefore\ x = \sqrt{26}$ {as $x > 0$}

In triangle ACD, the hypotenuse is 6 cm.

$\therefore\ y^2 + (\sqrt{26})^2 = 6^2$ {Pythagoras}

$\therefore\ y^2 + 26 = 36$

$\therefore\ y^2 = 10$

$\therefore\ y = \sqrt{10}$ {as $y > 0$}

6 Find the values of the unknowns:

a

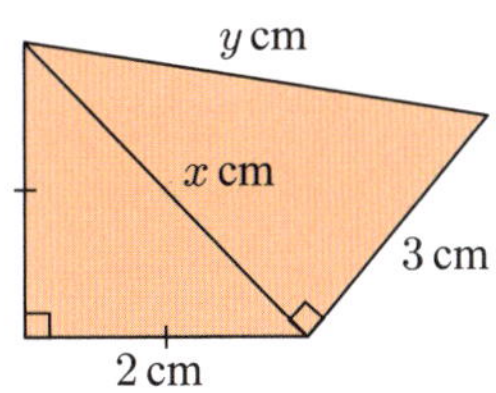

b

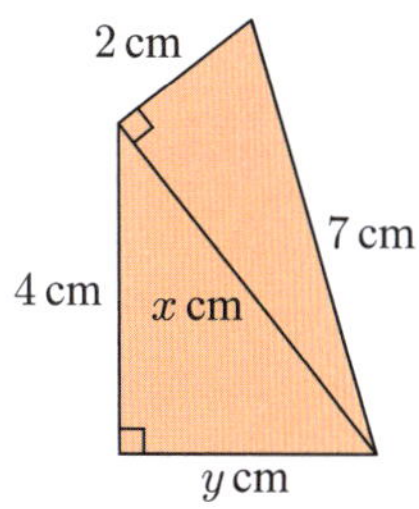

c

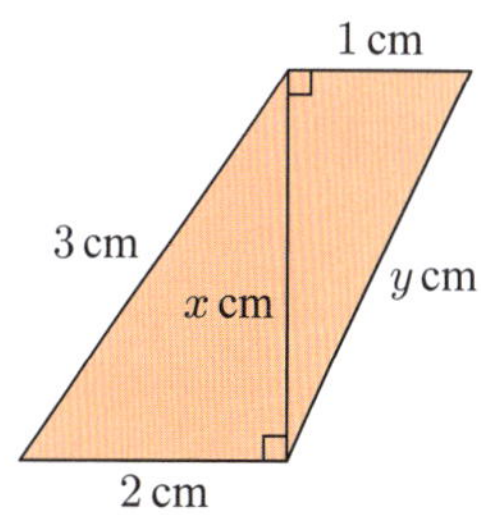

7 Find the value of x, giving your answer in simplest surd form where appropriate:

a

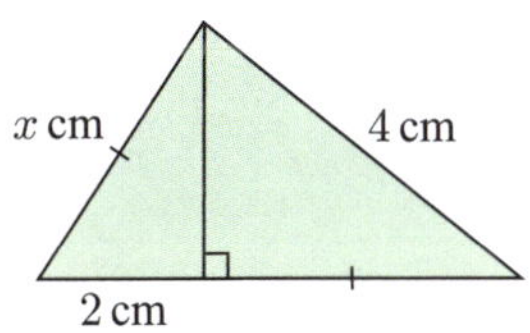

b

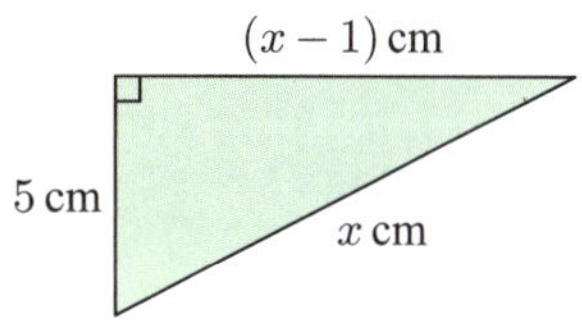

c

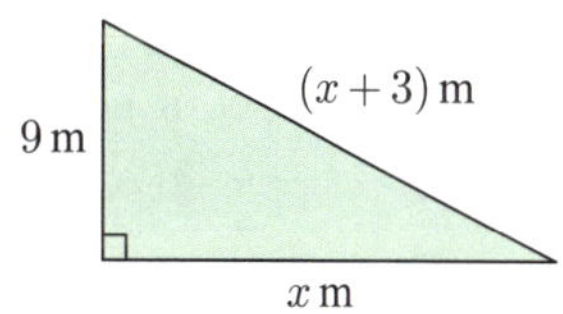

8 Find the length AC:

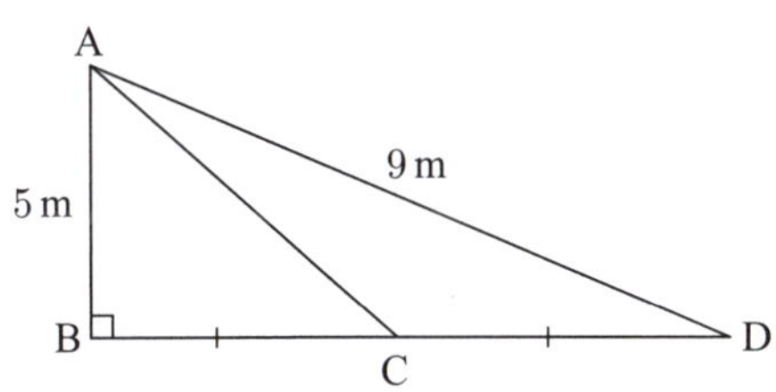

9 Use the figure below to show that $\sqrt{2} + \sqrt{8} = \sqrt{18}$.

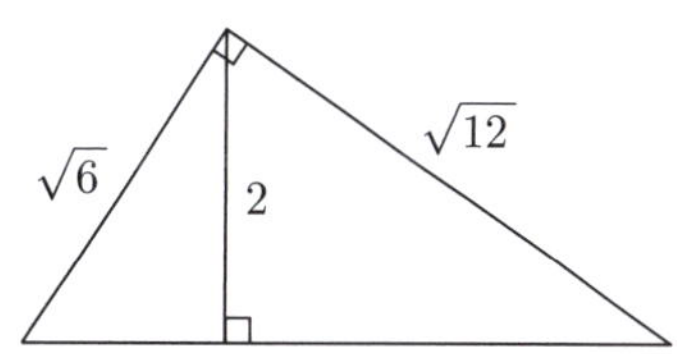

10 Find the length AB:

a

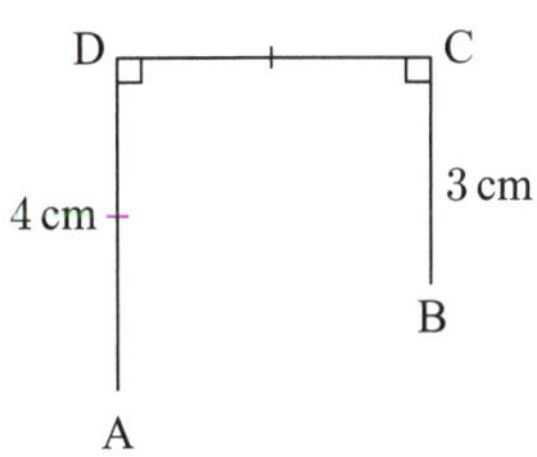

b

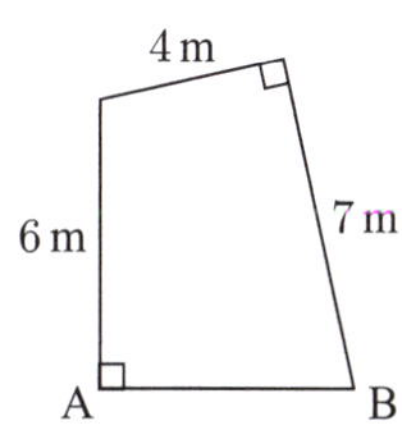

c

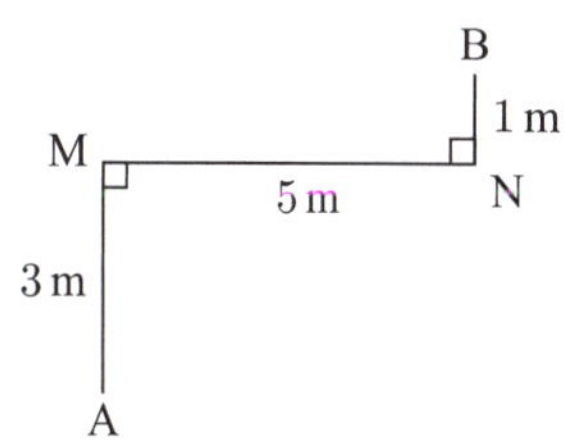

ACTIVITY — PRESIDENT GARFIELD'S PROOF

Prior to being President of the United States, **James Garfield** used the diagram alongside to prove Pythagoras' theorem.

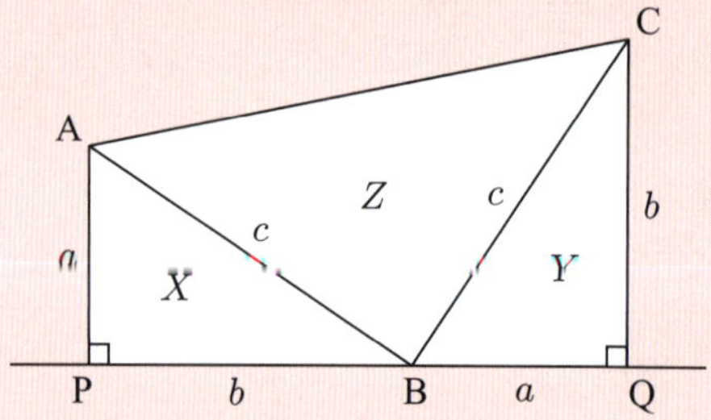

Two identical right angled triangles, ABP and BCQ, are placed on (PQ).

What to do:

1 Find the size of $\widehat{ABC}$. Explain your answer.

2 Find the areas of triangles X, Y, and Z.
Hence write an expression for area X + area Y + area Z in simplest form.

3 Find the area of trapezium ACQP.

4 Use your results from **2** and **3** to find a relationship between a, b, and c.

B PYTHAGOREAN TRIPLES

The simplest right angled triangle with sides of integer length is the **3-4-5 triangle**.

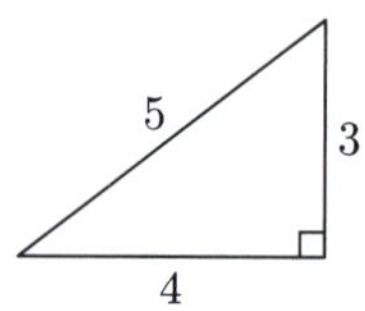

The numbers 3, 4, and 5 satisfy the rule $3^2 + 4^2 = 5^2$.

The set of positive integers $\{a, b, c\}$ is a **Pythagorean triple** if it obeys the rule $a^2 + b^2 = c^2$.

When we write down a Pythagorean triple, we list the numbers in ascending order.

Example 4 — Self Tutor

Find k given that $\{9, k, 15\}$ is a Pythagorean triple.

$$9^2 + k^2 = 15^2$$
$$\therefore\ 81 + k^2 = 225$$
$$\therefore\ k^2 = 144$$
$$\therefore\ k = 12 \quad \{\text{as } k > 0\}$$

EXERCISE 8B

1 Determine whether the following are Pythagorean triples:

a $\{15, 20, 25\}$ **b** $\{8, 17, 19\}$ **c** $\{7, 24, 25\}$

2 Explain why the following are *not* Pythagorean triples:

a $\{5, 7, 9\}$ **b** $\{-5, 12, 13\}$ **c** $\{1, 2, \sqrt{5}\}$

3 Find k given that the following are Pythagorean triples:

a $\{12, 16, k\}$ **b** $\{11, 60, k\}$ **c** $\{24, k, 51\}$
d $\{8, k, k+2\}$ **e** $\{20, k, k+8\}$ **f** $\{k, 60, k+50\}$

4 **a** Given that $\{a, b, c\}$ and $\{d, e, f\}$ are Pythagorean triples, show that $\{be-ad, bd+ae, cf\}$ is also a Pythagorean triple.

b Given that $\{3, 4, 5\}$ and $\{8, 15, 17\}$ are Pythagorean triples, use **a** to construct a new Pythagorean triple.

5 Show that $\{2n+1, 2n^2+2n, 2n^2+2n+1\}$ is a Pythagorean triple for:

a $n=1$ **b** $n=2$ **c** $n=3$ **d** $n=4$

6 **a** Prove that if $m, n \in \mathbb{Z}^+$, $m>n$, then $\{m^2-n^2, 2mn, m^2+n^2\}$ is a Pythagorean triple.

b Generate a Pythagorean triple using $m=5$, $n=2$.

PUZZLE — PYTHAGOREAN TRIPLE SEQUENCES

Consider a sequence of numbers such that any two consecutive numbers are members of the same Pythagorean triple.

For example, one such sequence is 6, 10, 24, 18.

(6, 10 from $\{6, 8, 10\}$; 10, 24 from $\{10, 24, 26\}$; 24, 18 from $\{18, 24, 30\}$)

What to do:

1 Find the missing number in the sequence:

a 3, 5, □, 9, 40 **b** 8, 10, □, 168, 126

2 Create a sequence of 6 numbers with this property, which starts with 3 and ends with:

a 6 **b** 50 **c** 75

3 Find the shortest possible sequence with this property which starts with 3 and ends with 1000.

C PROBLEM SOLVING

Right angled triangles occur in many practical problems. In these situations we can apply Pythagoras' theorem to help find unknown side lengths.

The problem solving approach involves the following steps:

Step 1:	Draw a neat, clear diagram of the situation.
Step 2:	Mark known lengths and right angles on the diagram.
Step 3:	Use a variable such as x to represent the unknown length.
Step 4:	Write down Pythagoras' theorem for the given information.
Step 5:	Solve the equation.
Step 6:	Where necessary, write your answer in sentence form.

Example 5 — Self Tutor

A rhombus has diagonals of length 6 cm and 8 cm. Find the length of its sides.

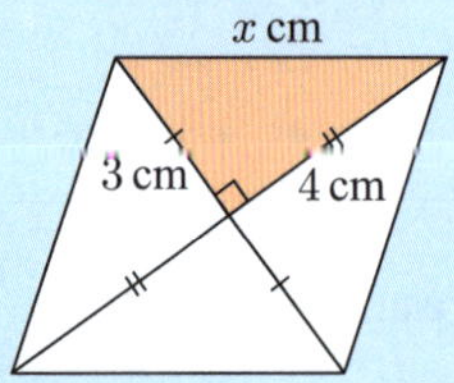

The diagonals of a rhombus bisect at right angles.

Let each side of the rhombus have length x cm.

$\therefore\ x^2 = 3^2 + 4^2$ {Pythagoras}

$\therefore\ x^2 = 25$

$\therefore\ x = \sqrt{25}$ {as $x > 0$}

$\therefore\ x = 5$

The sides are 5 cm in length.

Example 6 — Self Tutor

A 50 m rope is attached inside an empty cylindrical wheat silo of diameter 12 m as shown. How high is the silo?

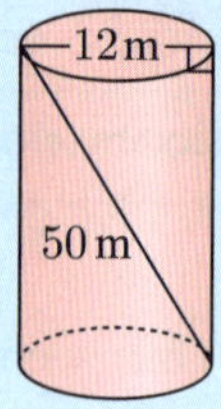

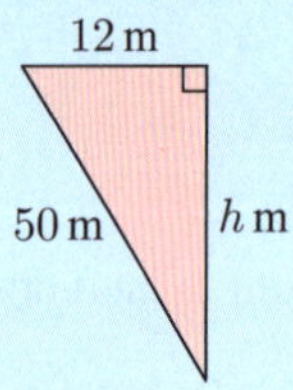

Let the height be h m.

$\therefore\ h^2 + 12^2 = 50^2$ {Pythagoras}

$\therefore\ h^2 + 144 = 2500$

$\therefore\ h^2 = 2356$

$\therefore\ h \approx 48.5$ {as $h > 0$}

The wheat silo is about 48.5 m high.

EXERCISE 8C

1 A rectangle has sides of length 8 cm and 3 cm. Find the length of its diagonals.

2 When the feet of a fireman's ladder are placed 1.6 m from a wall, the ladder just reaches a window which is 4.3 m from the ground. How long is the ladder?

3 A rhombus has sides of length 6 cm. One of its diagonals is 10 cm long. Find the length of the other diagonal.

4 A square has diagonals of length 10 cm. Find the length of its sides.

5 Pirate Captain William Hawk left his hat on Treasure Island. He sailed 18 km northeast through the Forbidden Strait, then 11 km southeast to his home before realising it was missing. He sent his parrot to fetch the hat and return it to the boat. How far did the parrot need to fly?

6 The longer side of a rectangle is three times the length of the shorter side. Each diagonal of the rectangle is 10 cm long. Find the dimensions of the rectangle.

7 An isosceles triangle has equal sides measuring 10 cm, and a base which is 12 cm long. Find the length of the altitude of the triangle from the apex to the base.

8 Find the length of the truss [AB] in the roof structure shown:

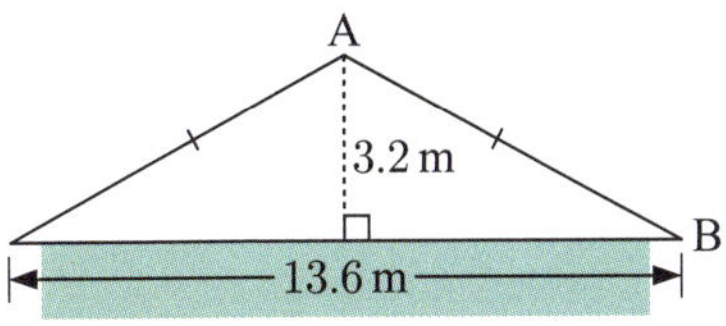

9 How high is the roof above the walls in the roof structure shown?

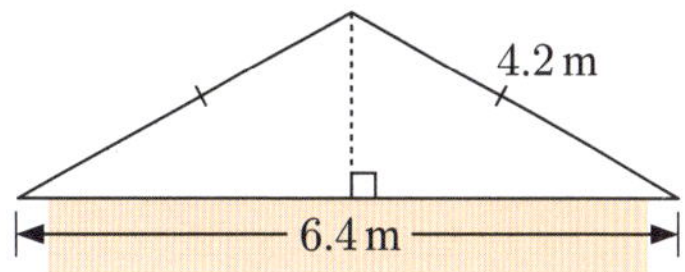

10 An equilateral triangle has an altitude of length 16 cm. Find the length of each side in simplest surd form.

11 Consider the **Opening Problem** on page **104**.

a Find the distance each swimmer will swim.

b How much farther will Sven swim than Jill?

12 A yacht sails 5 km due west at 15 km/h, due south for a while at 12 km/h, and then 10 km directly back to its starting point at 8 km/h.

a Find the total distance sailed.

b Find the total time for the voyage.

13 Ella's house is 2.5 km in a straight line from her school gate. To get to school, Ella walks down Bernard Street, then turns 90° and walks down Thompson Road until she reaches the school gate. She walks twice as far along Bernard Street as she does along Thompson Road. How far does Ella walk along Bernard Street?

14 An extension ladder rests 4 m up a wall. If the ladder is extended a further 0.8 m without moving its feet, then it will now rest 1 m further up the wall. How long is the extended ladder?

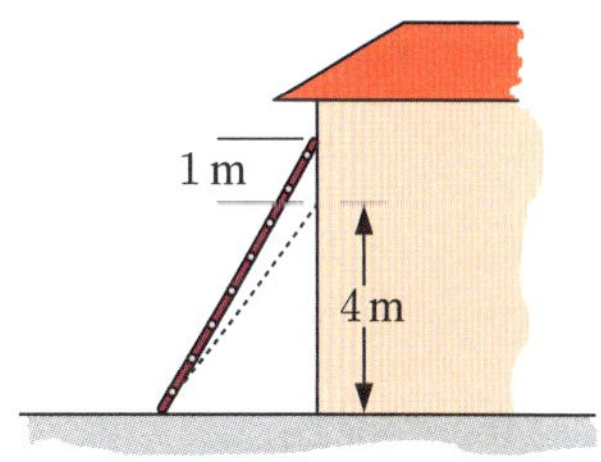

15 A highway runs east-west between two towns B and C that are 25 km apart. Town A lies 15 km directly north of B. A straight road is built from A to meet the highway at D. Given that D is equidistant from A and C, find the position of D on the highway.

16 A cone has a slant height of 17 cm, and a base radius of 8 cm. How high is the cone?

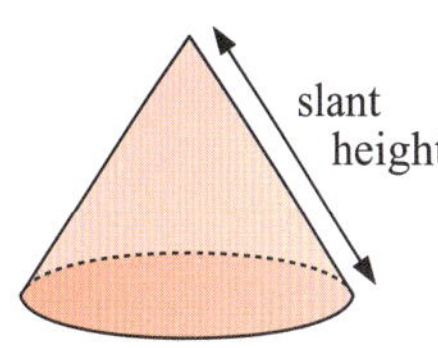

17 A cylindrical drinking glass has radius 3 cm and height 10 cm. Can a 12 cm long thin stirrer be placed in the glass so that it will stay entirely within the glass?

18 A 20 cm nail just fits inside a cylindrical can with base diameter to height ratio $1 : 3$. Find the radius of the can.

Example 7 **Self Tutor**

A pyramid of height 40 m has a square base with edges of length 50 m.
Determine the length of the slant edges.

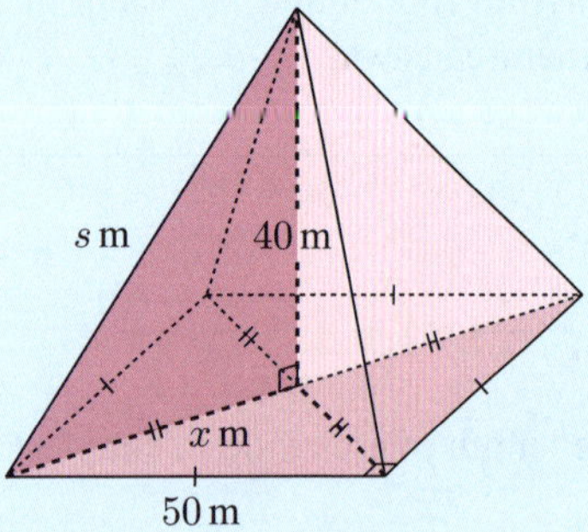

Let a slant edge have length s m.
Let half of a diagonal have length x m.

Using

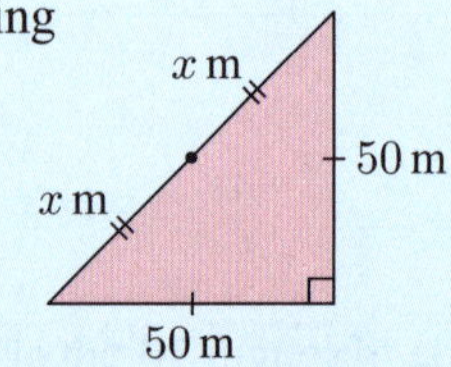

$$(2x)^2 = 50^2 + 50^2 \quad \{\text{Pythagoras}\}$$
$$\therefore \; 4x^2 = 5000$$
$$\therefore \; x^2 = 1250$$

Using

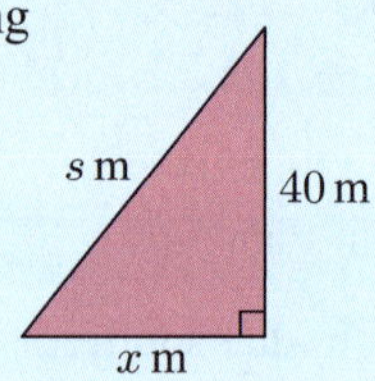

$$s^2 = x^2 + 40^2 \quad \{\text{Pythagoras}\}$$
$$\therefore \; s^2 = 1250 + 1600$$
$$\therefore \; s^2 = 2850$$
$$\therefore \; s \approx 53.4 \quad \{\text{as } s > 0\}$$

Each slant edge is about 53.4 m long.

19 A cube has sides of length 3 cm. Find the length of a diagonal of the cube, giving your answer in simplest surd form.

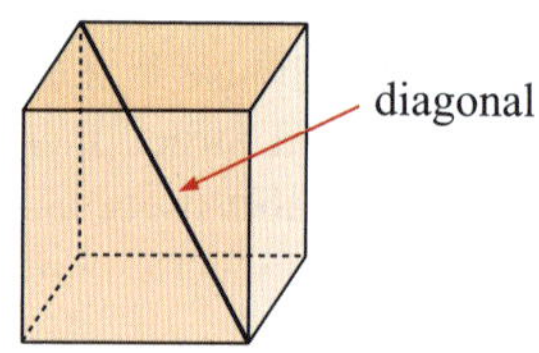

20 A room has floor dimensions 5 m by 3 m, and is 3.5 m high. Find the distance from a corner point on the floor to the opposite corner of the ceiling.

21 In the office building alongside, the entrance is at E. A radio antenna at A, the centre of the roof, receives security signals from the door E.
Find the direct distance between the entrance and the antenna.

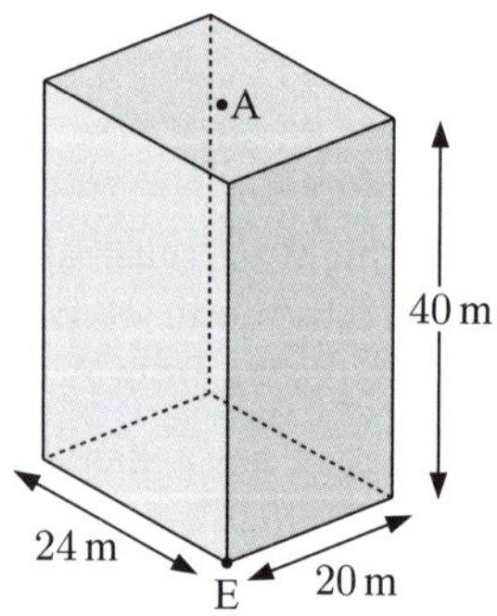

22 An Egyptian Pharaoh wishes to build a square-based pyramid with all edges of length 100 m. Its apex will be directly above the centre of its base. How high, to the nearest metre, will the pyramid reach above the desert sands?

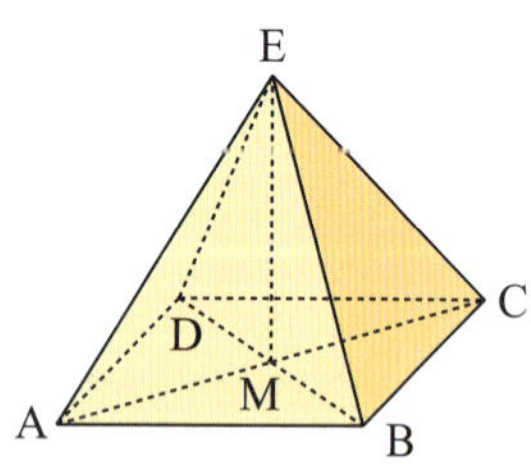

D THE CONVERSE OF PYTHAGORAS' THEOREM

If we are given the three side lengths of a triangle, the **converse of Pythagoras' theorem** gives us a simple **test** to determine whether the triangle is right angled.

THE CONVERSE OF PYTHAGORAS' THEOREM

If a triangle has sides of length a, b, and c units where $a^2 + b^2 = c^2$, then the triangle is right angled.

PYTHAGORAS SIMULATION

Example 8 Self Tutor

Is a triangle with sides 8 cm, 9 cm, and 12 cm right angled?

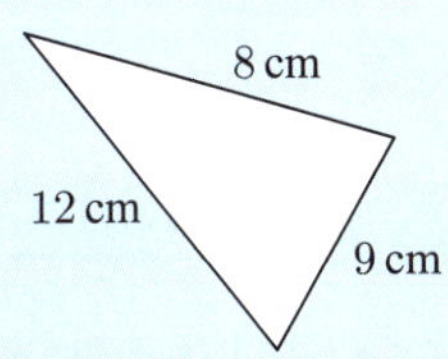

The two shorter sides have lengths 8 cm and 9 cm.

Now $8^2 + 9^2 = 64 + 81 = 145$

and $12^2 = 144$

$\therefore\ 8^2 + 9^2 \neq 12^2$

$\therefore$ the triangle is not right angled.

EXERCISE 8D

1 The following figures are not drawn to scale. Which of the triangles are right angled?

a

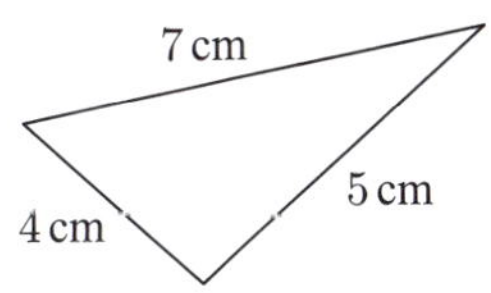

b

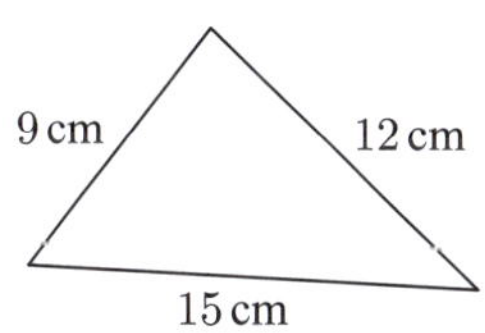

c

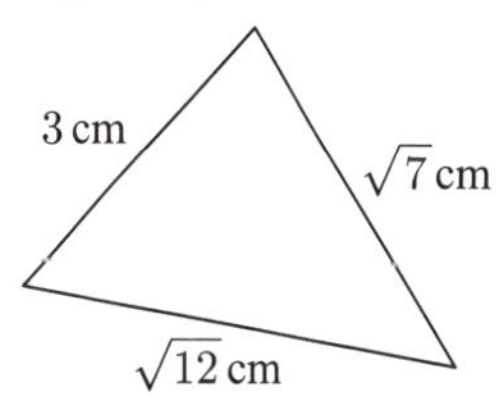

2 The following triangles are not drawn to scale. Determine whether each triangle is right angled. If so, identify the right angle.

a

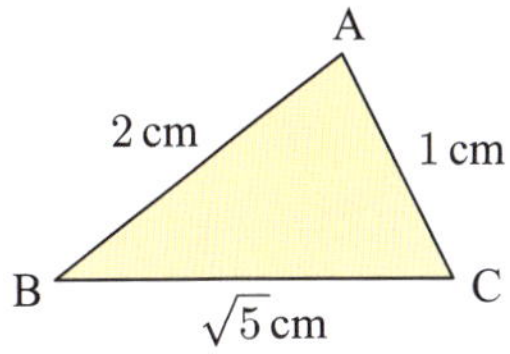

b

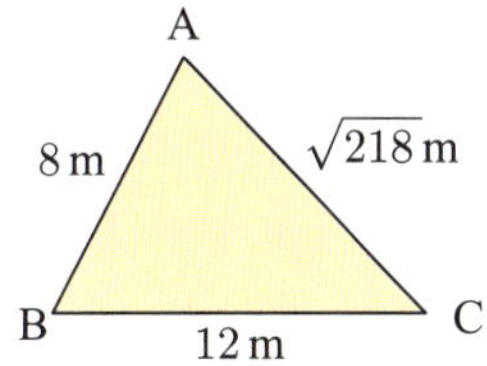

c

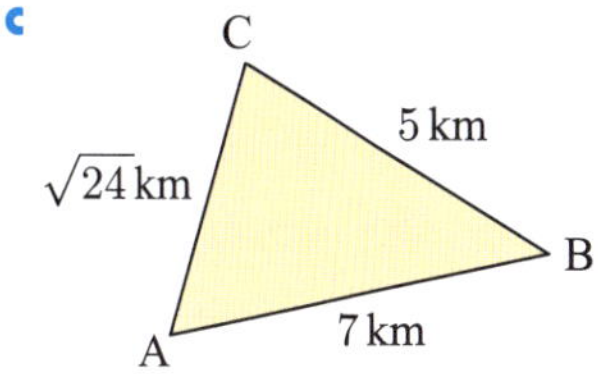

d

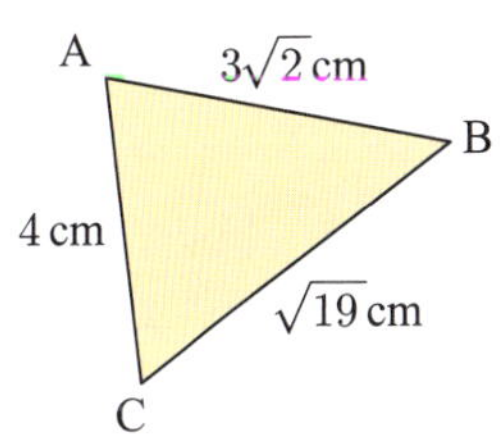

e

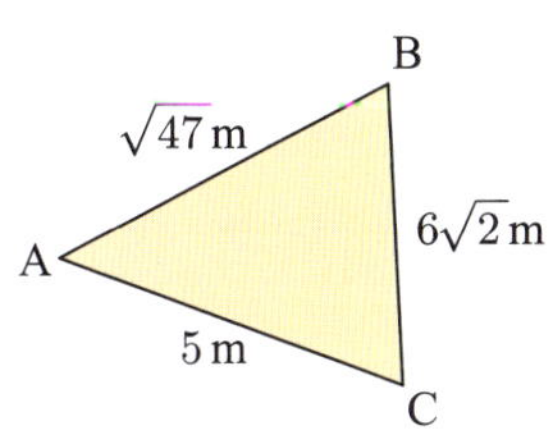

f

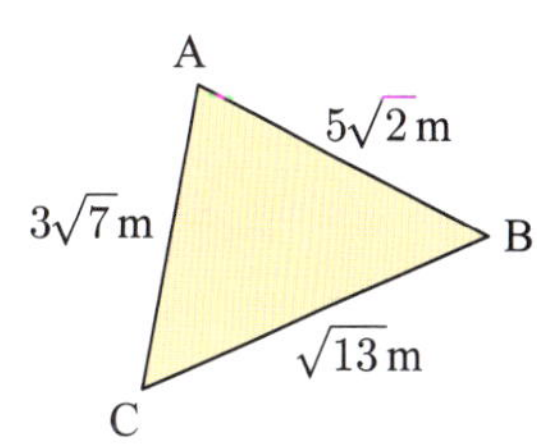

3 To check that his triangular set square is right angled, Roger measures its sides. The two shorter sides are 8 cm and 11.55 cm long, and the longest side is 14.05 cm long.
Is the set square right angled?

4 Meredith is trying to make a scarf by cutting some cloth into a right angled triangle. The triangle she makes has side lengths 40 cm, 42 cm, and 58 cm. Is Meredith's triangle right angled?

5 Is triangle ABC right angled at B?

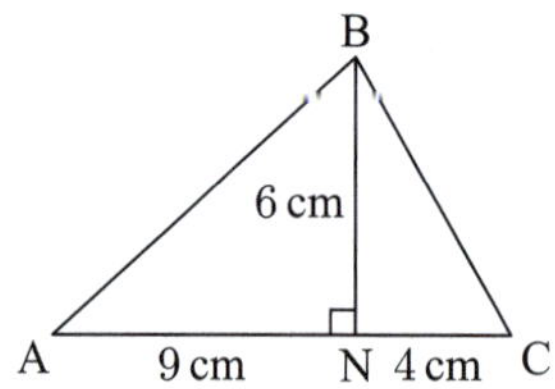

MULTIPLE CHOICE QUIZ

QUICK QUIZ

REVIEW SET 8A

1 Find the value of x, writing your answer in simplest surd form where necessary:

a

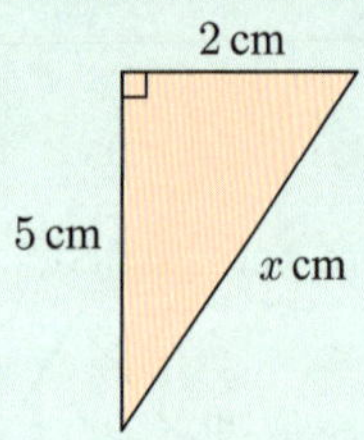

b

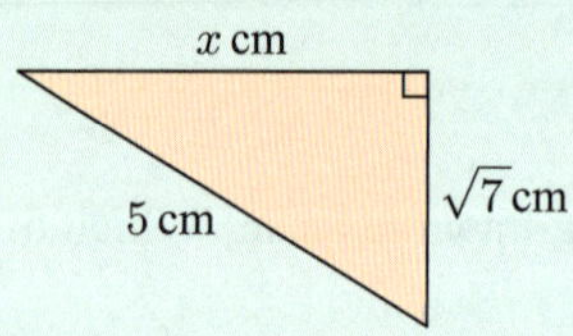

c

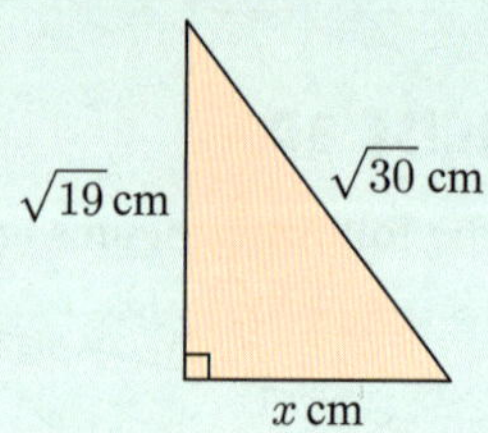

d

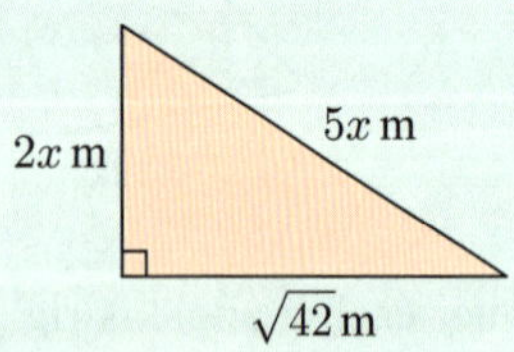

e

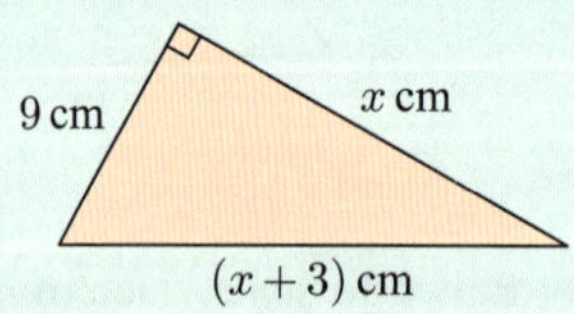

2 Use the figure alongside to show that $\sqrt{2}+\sqrt{18}=\sqrt{32}$.

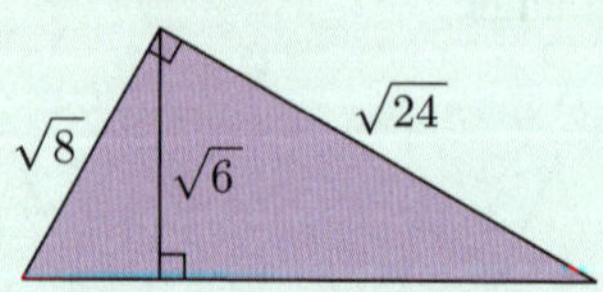

3 Find the values of the unknowns:

a

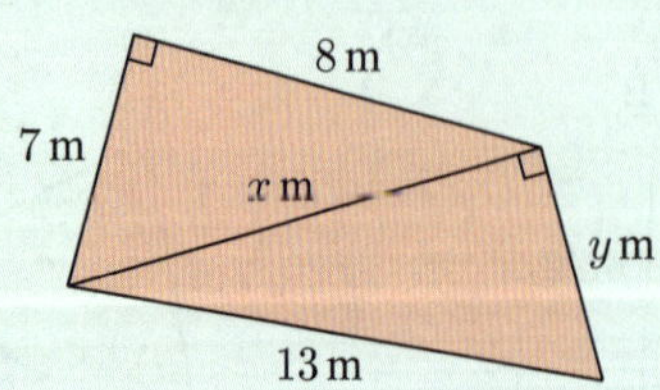

b

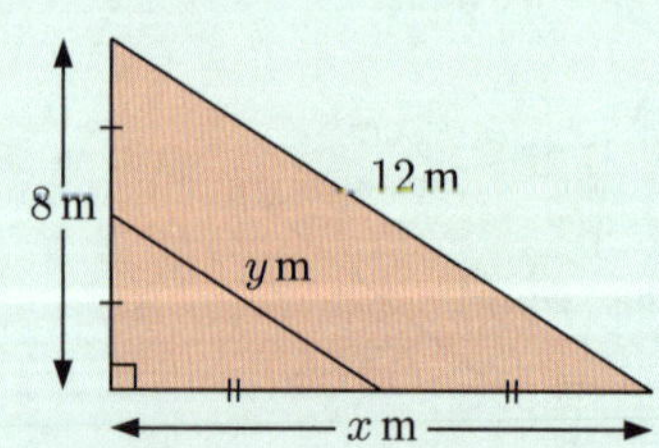

4 Determine whether $\{5, 11, 13\}$ is a Pythagorean triple.

5 Find k given that the following are Pythagorean triples:

a $\{16, k, 34\}$ **b** $\{9, 40, k\}$ **c** $\{25, k, k+5\}$

6 A rectangle is 11 cm long, and its diagonals have length 15 cm. Find the width of the rectangle.

7 A helicopter flies 30 km due north at 40 km/h, and then 25 km due west at 30 km/h. It then flies directly back to its starting point at 35 km/h. Find:

a the total distance travelled **b** the total time for the flight.

8

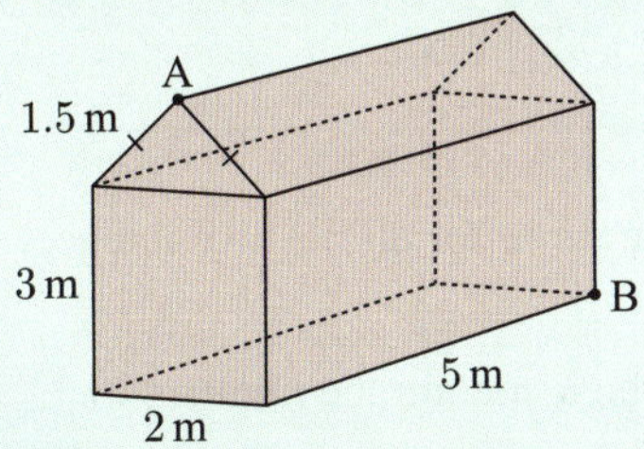

A shed has the dimensions given. Find the shortest distance from A to B.

9 Is this triangle right angled? Explain your answer.

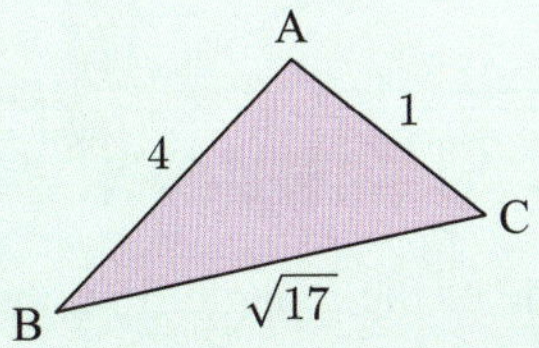

10 Is triangle ABC right angled at B?

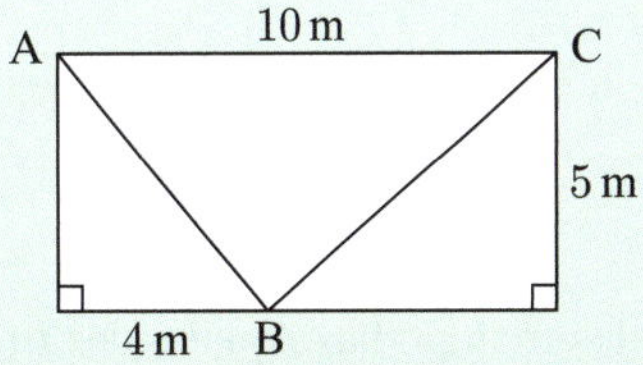

11 In the given figure, AB = 1 cm and AC = 3 cm. Find the radius of the circle.

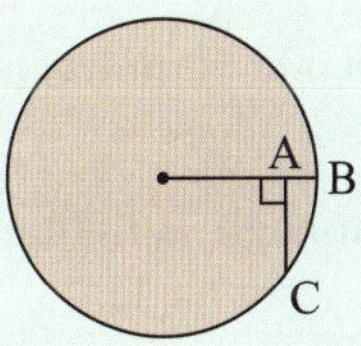

REVIEW SET 8B

1 Find the value of x, rounding your answer to 3 significant figures where appropriate:

a

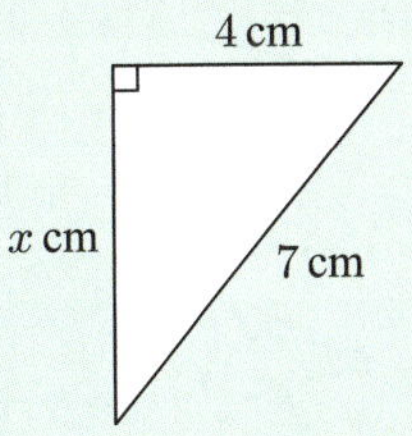

b

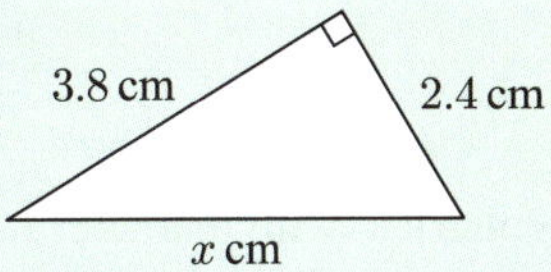

c

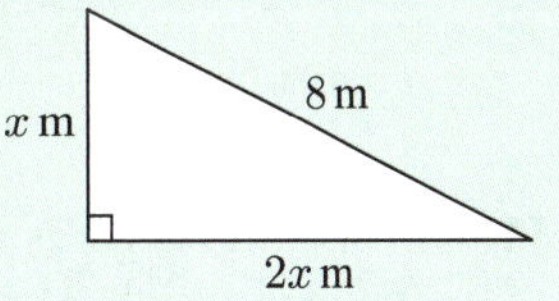

d

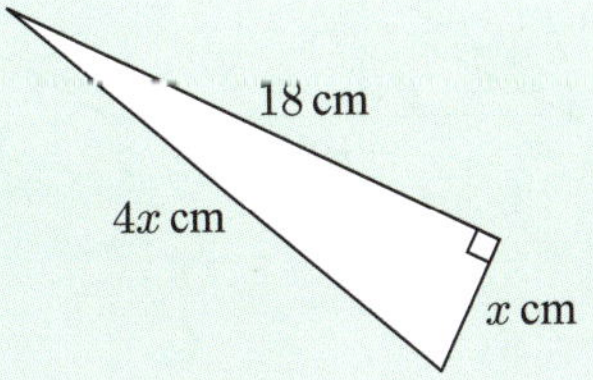

e

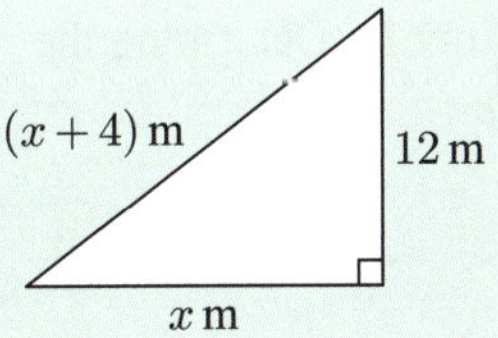

2 Find the distance AB in each figure:

a

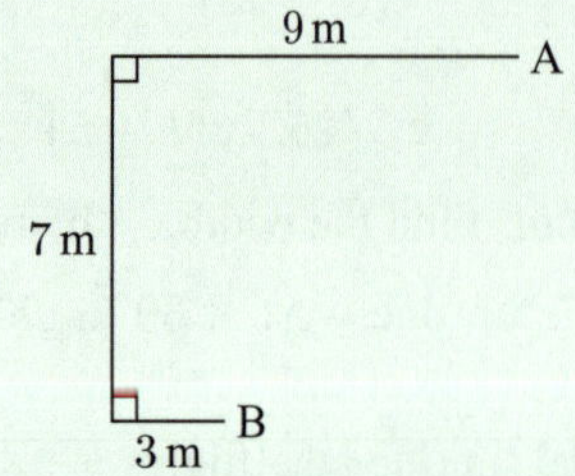

b

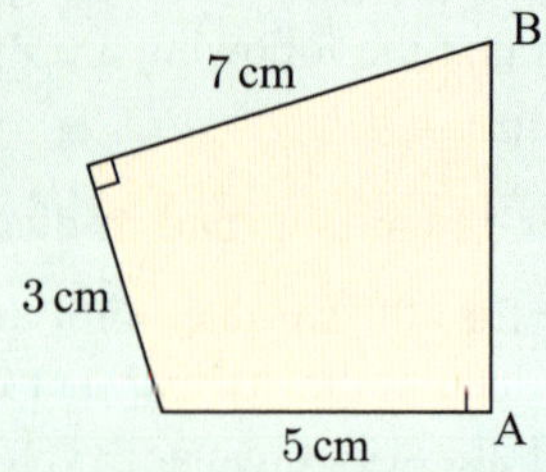

3 Explain why $\{-3, 4, 5\}$ is *not* a Pythagorean triple.

4 Find k given that $\{12, k, 37\}$ is a Pythagorean triple.

5 **a** Prove that $\{4k,\ k^2-4,\ k^2+4\}$ is a Pythagorean triple for all positive integers $k > 2$.

b Find the Pythagorean triple which results when $k = 5$.

c Write down two Pythagorean triples of this form which contain the number 40.

6 A rectangular gate is twice as wide as it is high. It is held in shape by a diagonal strut 3.2 m long.
Find the height of the gate to the nearest millimetre.

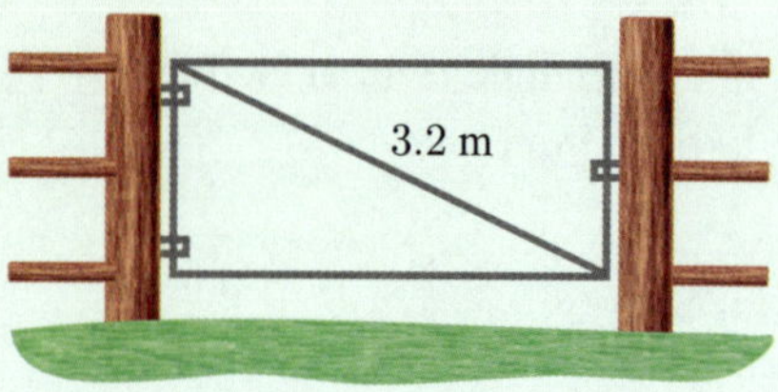

7 An isosceles triangle has equal sides of length 7 cm. The altitude of the triangle from the apex to the base is 3 cm. Find the length of the base of the triangle in simplest surd form.

8 A pyramid of height 50 m has a rectangular base which is 40 m long and 30 m wide. Find the length of the slant edges.

9 Show that this triangle is right angled, and identify the right angle.

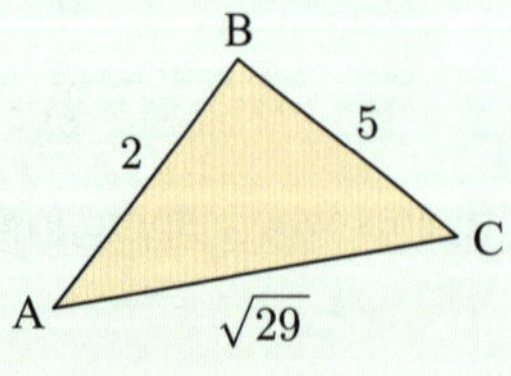

10 In the quadrilateral ABCD below, $AC = 12$ cm and $BD = 13$ cm. Identify the right angle.

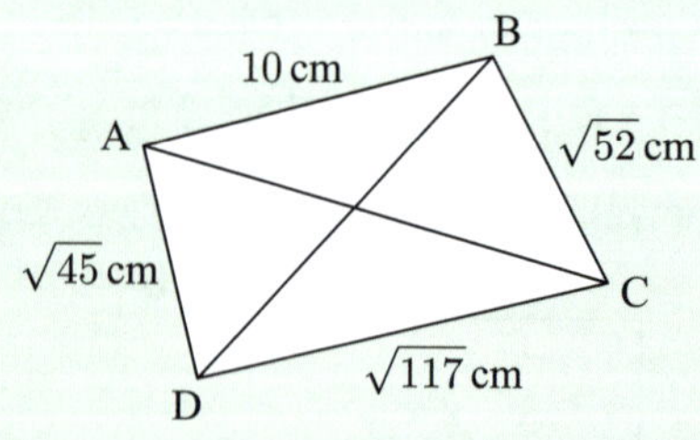

11 An aircraft hanger is semi-cylindrical with diameter 40 m and length 50 m. A helicopter places a cable across the top of the hangar, and one end is pinned to the corner at A. The cable is then pulled tight and pinned at the opposite corner B. Determine the length of the cable, to the nearest cm.

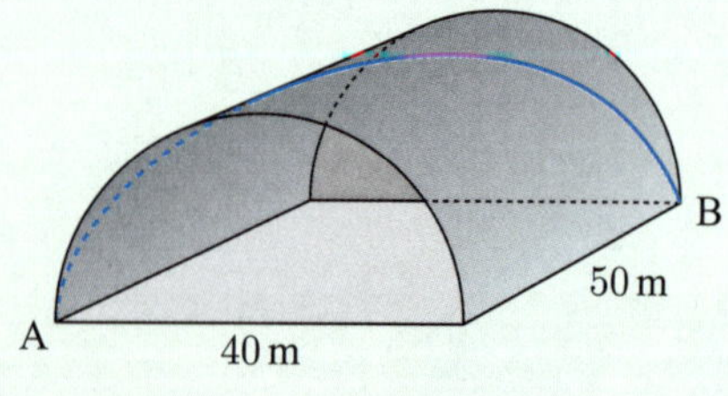

Chapter 9

Algebraic fractions

Contents:

OPENING PROBLEM

To participate in a tennis league, each team must pay \$150 in court fees plus a \$60 team registration fee.

The court fees are shared equally between the players in the team.

The registration fee is shared equally between the players and the team coach.

Suppose a team has x players.

Things to think about:

- **a** Can you write the amount that each player must pay as a single fraction involving x?
- **b** How much must each player pay if the team has:
  - **i** 4 players
  - **ii** 5 players?

Algebraic fractions are fractions which contain at least one variable or unknown.

The variable may be in the numerator, the denominator, or both the numerator and denominator.

For example, $\dfrac{x}{7}$, $\dfrac{-2}{5-y}$, and $\dfrac{x+2y}{1-y}$ are all algebraic fractions.

Algebraic fractions are sometimes called **rational expressions**.

In this Chapter we will see how we can simplify, add, subtract, multiply, and divide algebraic fractions in the same way as we do ordinary fractions. We will then apply our knowledge to solve equations involving algebraic fractions.

A EVALUATING ALGEBRAIC FRACTIONS

To **evaluate** an algebraic expression, we replace the variables with their known values. We then give our answer in lowest terms.

Example 1 Self Tutor

If $x = 3$, $y = -2$, and $z = 4$, evaluate:

a $\dfrac{y}{z^2}$ **b** $\dfrac{2x - z}{y}$

a
$$\frac{y}{z^2} = \frac{(-2)}{4^2} = -\tfrac{2}{16} = -\tfrac{1}{8}$$

b
$$\frac{2x-z}{y} = \frac{2(3)-4}{(-2)} = \frac{6-4}{-2} = \frac{2}{-2} = -1$$

EXERCISE 9A

1 If $a = 2,\ b = 3,\ c = 5,$ evaluate:

a $\dfrac{a}{2}$ **b** $\dfrac{b}{6}$ **c** $\dfrac{b+1}{a}$ **d** $\dfrac{a+2b}{c}$

e $\dfrac{c-b}{a}$ **f** $\dfrac{3c-a}{b}$ **g** $\dfrac{a^2}{b-1}$ **h** $\dfrac{ab}{-1-c}$

i $\dfrac{2a^2}{b+1}$ **j** $\dfrac{ac}{b^2+1}$ **k** $\dfrac{b}{\sqrt{2a+c}}$ **l** $\dfrac{\sqrt{a^2-b}}{c-b}$

2 If $x = -1,\ y = 2,\ z = -3,$ evaluate:

a $\dfrac{y}{x}$ **b** $\dfrac{z}{x}$ **c** $-\dfrac{z}{y+1}$ **d** $\dfrac{y^2}{-4}$

e $\dfrac{z-y}{x}$ **f** $\dfrac{x-y}{z}$ **g** $\dfrac{y}{x+z}$ **h** $\dfrac{x+y}{x-y}$

i $\dfrac{x+z}{y^2}$ **j** $\dfrac{x^2y}{z-1}$ **k** $\dfrac{x^2+y^2}{10}$ **l** $\dfrac{x^2+z^2}{y}$

m $\dfrac{x-z^2}{y}$ **n** $\dfrac{y^2-z^2}{-2y}$ **o** $\dfrac{3x-2y}{z^2}$ **p** $\dfrac{z-xy}{\sqrt{4y+x^2}}$

B SIMPLIFYING ALGEBRAIC FRACTIONS

To write a fraction with integer numerator and denominator in **lowest terms**, we divide the numerator and denominator by their **highest common factor**.

In practice, we often achieve this by **cancelling common factors**.

For example, $\dfrac{15}{35} = \dfrac{3 \times \cancel{5}^1}{7 \times \cancel{5}_1}$ {cancelling the HCF 5}

$= \dfrac{3}{7}$

The same principle can be applied to simplify algebraic fractions.

For example, $\dfrac{6bc}{9c} = \dfrac{2 \times {}^1\cancel{3} \times b \times \cancel{c}^1}{3 \times \cancel{3}_1 \times \cancel{c}_1}$ {cancelling the HCF $3c$}

$= \dfrac{2b}{3}$

Fractions such as $\dfrac{3xy}{7z}$ cannot be simplified since the numerator and denominator do not have any common factors.

ERRORS IN CANCELLING

A common **error** in cancelling is to cancel terms, rather than common factors.

For example, this cancelling is **incorrect**: $\dfrac{x+\cancel{2}^1}{\cancel{6}_3} = \dfrac{x+1}{3}$

The numerator $x+2$ cannot be written as a product of factors other than $1 \times (x+2)$. x and 2 are *terms* of the expression, not factors.

Example 2 Self Tutor

Simplify, if possible:

a $\frac{a^2}{2a}$ b $\frac{6a^2b}{3b}$ c $\frac{a+b}{a}$

a $\frac{a^2}{2a} = \frac{a \times \cancel{a}^1}{2 \times \cancel{a}_1}$
$= \frac{a}{2}$

b $\frac{6a^2b}{3b} = \frac{^2\cancel{6} \times a \times a \times \cancel{b}^1}{_1\cancel{3} \times \cancel{b}_1}$
$= \frac{2 \times a \times a}{1}$
$= 2a^2$

c $\frac{a+b}{a}$ cannot be simplified as there are no common factors.

EXERCISE 9B

1 Simplify:

a $\frac{2a}{4}$ b $\frac{4m}{2}$ c $\frac{6a}{a}$ d $\frac{6a}{2a}$ e $\frac{2a^2}{a}$

f $\frac{2x^3}{2x}$ g $\frac{2x^3}{x^2}$ h $\frac{2x^3}{x^3}$ i $\frac{2a^2}{4a^3}$ j $\frac{8m^2}{4m}$

k $\frac{4a^2}{a^2}$ l $\frac{6t}{3t^2}$ m $\frac{4d^2}{2d}$ n $\frac{ab^2}{2ab}$ o $\frac{4ab^2}{6a^2b}$

2 Simplify:

a $\frac{-m}{3m^2}$ b $\frac{2m^2}{-4m}$ c $\frac{-m}{-3m^3}$ d $\frac{-5m^2}{m^2}$

e $\frac{-9n^2}{-3mn}$ f $\frac{-8n}{4m^2n}$ g $\frac{-mn^2}{2mn}$ h $\frac{12m^2n}{-4mn^2}$

3 If possible, simplify by cancelling common factors:

a $\frac{m+n}{m}$ b $\frac{mn}{m}$ c $\frac{m-n}{n}$ d $\frac{m+n}{m+n}$ e $\frac{n}{m+n}$

Example 3 Self Tutor

Simplify: a $\frac{(-4b)^2}{2b}$ b $\frac{18}{3(c-1)}$

a $\frac{(-4b)^2}{2b} = \frac{(-4b) \times (-4b)}{2 \times b}$
$= \frac{^8\cancel{16} \times b \times \cancel{b}^1}{_1\cancel{2} \times \cancel{b}_1}$
$= 8b$

b $\frac{18}{3(c-1)} = \frac{\cancel{18}^6}{_1\cancel{3}(c-1)}$
$= \frac{6}{c-1}$

4 Simplify:

a $\frac{(2a)^2}{a^2}$ b $\frac{(4n)^2}{8n}$ c $\frac{(-a)^2}{a}$ d $\frac{a^2}{(-a)^2}$

e $\frac{(-2a)^2}{4}$ f $\frac{(-3n)^2}{6n}$ g $\frac{2b}{(2b^2)^2}$ h $\frac{(3k^2)^2}{18a^3}$

5 Simplify:

a $\dfrac{4(x+5)}{2}$ **b** $\dfrac{2(n+5)}{12}$ **c** $\dfrac{7(b+2)}{14}$ **d** $\dfrac{6(k-2)}{8}$

e $\dfrac{15}{3(t-1)}$ **f** $\dfrac{10}{25(k+4)}$ **g** $\dfrac{4}{12(x-3)}$ **h** $\dfrac{20(p+4)}{12}$

Example 4 **Self Tutor**

Simplify:

a $\dfrac{(2x+3)(x+4)}{5(2x+3)}$ **b** $\dfrac{12(x+4)^2}{3(x+4)}$

a $\dfrac{{}^{1}\cancel{(2x+3)}(x+4)}{5\cancel{(2x+3)}_{1}} = \dfrac{(x+4)}{5}$

$= \dfrac{x+4}{5}$

b $\dfrac{12(x+4)^2}{3(x+4)} = \dfrac{{}^{4}\cancel{12}(x+4)\cancel{(x+4)}^{1}}{{}_{1}\cancel{3}\cancel{(x+4)}_{1}}$

$= 4(x+4)$

6 Simplify:

a $\dfrac{(x+4)(x+2)}{9(x+4)}$ **b** $\dfrac{12(a-3)}{(a-3)(a+1)}$ **c** $\dfrac{(x+y)(x-y)}{3(x-y)}$

d $\dfrac{(x+y)^2}{x+y}$ **e** $\dfrac{2(x+2)}{(x+2)^2}$ **f** $\dfrac{(a+5)^2}{3(a+5)}$

g $\dfrac{2xy(x-y)}{6x(x-y)}$ **h** $\dfrac{5(y+2)(y-3)}{15(y+2)}$ **i** $\dfrac{x(x+1)(x+2)}{3x(x+2)}$

j $\dfrac{3(b-4)}{6(b-4)^2}$ **k** $\dfrac{8(p+q)^2}{12(p+q)}$ **l** $\dfrac{24(r-2)}{15(r-2)^2}$

m $\dfrac{x^2(x-2)}{x(x-2)(x-1)}$ **n** $\dfrac{(x+2)^2(x+1)}{4(x+2)}$ **o** $\dfrac{2(x+3)(x-1)^2}{8x(x+3)^2}$

Example 5 **Self Tutor**

Factorise, then simplify:

a $\dfrac{3a+9}{3}$ **b** $\dfrac{4a+12}{8}$

a $\dfrac{3a+9}{3} = \dfrac{{}^{1}\cancel{3}(a+3)}{\cancel{3}_{1}}$

$= a+3$

b $\dfrac{4a+12}{8} = \dfrac{{}^{1}\cancel{4}(a+3)}{\cancel{8}_{2}}$

$= \dfrac{a+3}{2}$

7 Factorise, then simplify if possible:

a $\dfrac{2x+4}{2}$ **b** $\dfrac{3x-6}{3}$ **c** $\dfrac{3x+6}{6}$ **d** $\dfrac{4x-20}{8}$

e $\dfrac{12}{4y+12}$ **f** $\dfrac{4}{6x-30}$ **g** $\dfrac{4x+6}{6}$ **h** $\dfrac{4x+6}{5}$

i $\dfrac{6a-3}{2}$ **j** $\dfrac{6a-3}{3}$ **k** $\dfrac{4}{6a+2}$ **l** $\dfrac{3b+9}{2}$

m $\dfrac{3b+9}{6}$ **n** $\dfrac{6}{8b-12}$ **o** $\dfrac{ax+bx}{x}$ **p** $\dfrac{ax+a}{x}$

Example 6 — Self Tutor

Factorise, then simplify:

a $\dfrac{ab+ac}{b+c}$ **b** $\dfrac{6x^2-6xy}{3x-3y}$

a $\dfrac{ab+ac}{b+c}$ ← HCF is a

$= \dfrac{a(b+c)}{b+c}$

$= \dfrac{a\cancel{(b+c)}^{1}}{\cancel{(b+c)}_{1}}$

$= a$

b $\dfrac{6x^2-6xy}{3x-3y}$ ← HCF is $6x$ / ← HCF is 3

$= \dfrac{6x(x-y)}{3(x-y)}$

$= \dfrac{{}^{2}\cancel{6} \times x \times \cancel{(x-y)}^{1}}{{}_{1}\cancel{3} \times \cancel{(x-y)}_{1}}$

$= 2x$

8 Factorise, then simplify:

a $\dfrac{3x+6}{4x+8}$ **b** $\dfrac{5x-15}{3x-9}$ **c** $\dfrac{2x+12}{5x+30}$ **d** $\dfrac{16x-8}{20x-10}$

e $\dfrac{6x+9}{2x+3}$ **f** $\dfrac{3x-4}{6x-8}$ **g** $\dfrac{4x^2+8x}{x+2}$ **h** $\dfrac{3x^2+9x}{x+3}$

i $\dfrac{5x^2-5xy}{7x-7y}$ **j** $\dfrac{9b^2-9ab}{12b-12a}$ **k** $\dfrac{6x^2-18x}{9x-27}$ **l** $\dfrac{6a+6b}{8a^3+4a^2b}$

9 Factorise, then simplify:

a $\dfrac{ax+bx}{a+b}$ **b** $\dfrac{a+b}{ay+by}$ **c** $\dfrac{ax+bx}{ay+by}$ **d** $\dfrac{ax+bx}{cx+dx}$

Example 7 — Self Tutor

Factorise, then simplify:

a $\dfrac{6a-6b}{b-a}$ **b** $\dfrac{xy^2-xy}{1-y}$

a $\dfrac{6a-6b}{b-a}$

$= \dfrac{6\cancel{(a-b)}^{1}}{-\cancel{(a-b)}_{1}}$

$= -6$

b $\dfrac{xy^2-xy}{1-y}$

$= \dfrac{xy\cancel{(y-1)}^{1}}{-\cancel{(y-1)}_{1}}$

$= -xy$

10 Factorise, then simplify:

a $\dfrac{2x-2y}{y-x}$ **b** $\dfrac{3x-3y}{2y-2x}$ **c** $\dfrac{m-n}{n-m}$ **d** $\dfrac{r-2s}{4s-2r}$

e $\dfrac{3r-6s}{2s-r}$ **f** $\dfrac{2x-2}{x-x^2}$ **g** $\dfrac{ab^2-ab}{2-2b}$ **h** $\dfrac{4x^2-4x}{2-2x}$

Example 8 **Self Tutor**

Factorise, then simplify: $\dfrac{x^2-1}{x^2+3x+2}$

$$\frac{x^2-1}{x^2+3x+2} = \frac{(x-1)(x+1)^1}{(x+2)(x+1)_1}$$
$$= \frac{x-1}{x+2}$$

11 Factorise, then simplify:

a $\dfrac{x^2-1}{x-1}$ b $\dfrac{x^2-1}{x+1}$ c $\dfrac{x^2-1}{1-x}$ d $\dfrac{x+2}{x^2-4}$

e $\dfrac{a^2-b^2}{a+b}$ f $\dfrac{a^2-b^2}{b-a}$ g $\dfrac{2x+2}{x^2-1}$ h $\dfrac{9-x^2}{3x-x^2}$

i $\dfrac{3x^2-3y^2}{2xy-2y^2}$ j $\dfrac{2b^2-2a^2}{a^2-ab}$ k $\dfrac{4xy-y^2}{16x^2-y^2}$ l $\dfrac{4x(x-4)}{16-x^2}$

12 Factorise, then simplify:

a $\dfrac{x^2-x-2}{x-2}$ b $\dfrac{x+4}{x^2+x-12}$ c $\dfrac{a+3}{a^2-2a-15}$ d $\dfrac{2x^2+2x}{x^2-4x-5}$

e $\dfrac{x^2-4}{x^2+4x+4}$ f $\dfrac{x^2-x-12}{x^2-5x+4}$ g $\dfrac{x^2+2x+1}{1-x^2}$ h $\dfrac{n^2-n-20}{n^2+7n+12}$

Example 9 **Self Tutor**

Factorise, then simplify: $\dfrac{6x^2+10x-4}{18x^2+3x-3}$

$$\frac{6x^2+10x-4}{18x^2+3x-3} = \frac{2(3x^2+5x-2)}{3(6x^2+x-1)}$$
$$= \frac{2(3x-1)^1(x+2)}{3(3x-1)_1(2x+1)}$$
$$= \frac{2(x+2)}{3(2x+1)}$$

13 Factorise, then simplify:

a $\dfrac{2x^2+3x-2}{x^2+2x}$ b $\dfrac{2x^2+5x+2}{2x^2+7x+3}$ c $\dfrac{3x^2+7x+2}{6x^2-x-1}$

d $\dfrac{8x^2+2x-1}{4x^2-5x+1}$ e $\dfrac{12x^2-5x-3}{6x^2+5x+1}$ f $\dfrac{15x^2+17x-4}{5x^2+9x-2}$

14 Given that $k \in \mathbb{Z}$ and $k \neq 5$, show that $\dfrac{2k^2-7k-15}{k-5}$ is an odd integer.

LEARNING ALGEBRA

15 Find the value(s) of n such that $\dfrac{6x^2+11x+3}{nx+1}$ is an integer for all $x \in \mathbb{Z}$.

PUZZLE

Click on the icon to load a dynamic puzzle for algebraic fractions.

RATIONAL MAZE

C MULTIPLYING ALGEBRAIC FRACTIONS

Variables are used in algebraic fractions to represent unknown numbers. We can perform operations with algebraic fractions in the same way as numerical fractions, since they are in fact *representing* numerical fractions.

To **multiply** two or more fractions, we multiply the numerators together and multiply the denominators together.

$$\frac{a}{b} \times \frac{c}{d} = \frac{a \times c}{b \times d}$$

To write the result in lowest terms, we cancel any common factors in the numerator and denominator.

Example 10 Self Tutor

Simplify: **a** $\frac{3}{m} \times \frac{m}{6}$ **b** $\frac{3}{m} \times m^2$

a $\frac{{}^1\cancel{3}}{{}_1\cancel{m}} \times \frac{\cancel{m}^1}{\cancel{6}_2} = \frac{1}{2}$

b $\frac{3}{m} \times m^2 = \frac{3}{{}_1\cancel{m}} \times \frac{m^{\cancel{2}}}{1}$
$= 3m$

EXERCISE 9C

1 Simplify:

a $\frac{x}{2} \times \frac{y}{5}$ **b** $\frac{a}{2} \times \frac{3}{a}$ **c** $\frac{a}{2} \times a$ **d** $\frac{a}{4} \times \frac{2}{3a}$

e $\frac{c}{5} \times \frac{1}{c}$ **f** $\frac{c}{5} \times \frac{c}{2}$ **g** $\frac{a}{b} \times \frac{c}{d}$ **h** $\frac{a}{b} \times \frac{b}{a}$

i $\frac{1}{m^2} \times \frac{m}{2}$ **j** $\frac{m}{2} \times \frac{4}{m}$ **k** $\frac{a}{x} \times \frac{x}{b}$ **l** $m \times \frac{4}{m}$

m $\frac{3}{m^2} \times m$ **n** $\left(\frac{a}{b}\right)^2$ **o** $\left(\frac{2}{x}\right)^2$ **p** $\frac{1}{a} \times \frac{a}{b} \times \frac{b}{c}$

2 Simplify:

a $\frac{4x^2}{y} \times \frac{y}{2}$ **b** $\frac{2}{xy} \times \frac{3x}{y}$ **c** $\frac{4y}{x^2} \times \frac{xy}{2}$ **d** $\frac{x}{2y^2} \times \frac{6y}{x^2}$

Example 11 Self Tutor

Factorise, then simplify: $\frac{y^2 - y}{y - 2} \times \frac{3y - 6}{4y - 4}$

$\frac{y^2 - y}{y - 2} \times \frac{3y - 6}{4y - 4} = \frac{y\cancel{(y-1)}^1}{{}_1\cancel{y-2}} \times \frac{3\cancel{(y-2)}^1}{4\cancel{(y-1)}_1}$
$= \frac{3y}{4}$

3 Factorise, then simplify:

a $\frac{x^2 + 3x}{x - 2} \times \frac{5}{2x + 6}$ **b** $\frac{t - 5}{t^2 + t} \times \frac{4t + 4}{3t - 15}$ **c** $\frac{6k - 2}{k + 2} \times \frac{2k^2 + 4k}{9k - 3}$

D DIVIDING ALGEBRAIC FRACTIONS

To **divide** by a fraction, we multiply by the **reciprocal** of that fraction.

$$\frac{a}{b} \div \frac{c}{d} = \frac{a}{b} \times \frac{d}{c}$$

Example 12 — Self Tutor

Simplify: **a** $\frac{4}{n} \div \frac{2}{n^2}$ **b** $\frac{3}{a} \div 2$

a $\frac{4}{n} \div \frac{2}{n^2} = \frac{4}{n} \times \frac{n^2}{2}$

$= \frac{{}^{2}\cancel{4} \times n \times \cancel{n}^{1}}{{}_{1}\cancel{n} \times \cancel{2}_{1}}$

$= 2n$

b $\frac{3}{a} \div 2 = \frac{3}{a} \times \frac{1}{2}$

$= \frac{3 \times 1}{a \times 2}$

$= \frac{3}{2a}$

EXERCISE 9D

1 Simplify:

a $\frac{a}{2} \div \frac{a}{3}$ **b** $\frac{2}{a} \div \frac{2}{3}$ **c** $\frac{3}{4} \div \frac{4}{x}$ **d** $\frac{3}{x} \div \frac{4}{x}$

e $\frac{2}{n} \div \frac{1}{n}$ **f** $\frac{c}{5} \div 5$ **g** $\frac{c}{5} \div c$ **h** $m \div \frac{2}{m}$

i $m \div \frac{m}{2}$ **j** $1 \div \frac{m}{n}$ **k** $\frac{3}{g} \div 4$ **l** $\frac{3}{g} \div \frac{9}{g^2}$

m $\frac{4}{x} \div \frac{x^2}{2}$ **n** $\frac{2}{x} \div \frac{6}{x^3}$ **o** $\frac{a}{b} \div \frac{a^2}{b}$ **p** $\frac{a^2}{5} \div \frac{a}{3}$

2 Simplify:

a $\frac{2x^2}{3y} \div \frac{2}{x}$ **b** $\frac{5}{y} \div \frac{y}{2z^2}$ **c** $\frac{7a^2}{b} \div \frac{3b^2}{a}$ **d** $\frac{12x}{y^2} \div \frac{8xy}{3}$

Example 13 — Self Tutor

Simplify by factorising: $\frac{5m-20}{4} \div \frac{2m-8}{3}$

$\frac{5m-20}{4} \div \frac{2m-8}{3} = \frac{5m-20}{4} \times \frac{3}{2m-8}$

$= \frac{5\cancel{(m-4)}^{1}}{4} \times \frac{3}{2\cancel{(m-4)}_{1}}$

$= \frac{15}{8}$

3 Simplify by factorising:

a $\frac{4a-28}{a} \div \frac{a-7}{5}$ **b** $\frac{x}{3-9x} \div \frac{x^2}{3x-1}$ **c** $\frac{m^2-4m}{10m-2} \div \frac{m^2-16}{3m+12}$

E ADDING AND SUBTRACTING ALGEBRAIC FRACTIONS

The rules for addition and subtraction of algebraic fractions are identical to those for numerical fractions.

To **add** two or more fractions, we obtain the *lowest common denominator* and then add the resulting numerators. $\dfrac{a}{c}+\dfrac{b}{c}=\dfrac{a+b}{c}$

To **subtract** two or more fractions, we obtain the *lowest common denominator* and then subtract the resulting numerators. $\dfrac{a}{c}-\dfrac{d}{c}=\dfrac{a-d}{c}$

To find the lowest common denominator of numerical fractions, we look for the **lowest common multiple of the denominators**.

For example, when finding $\frac{5}{6}+\frac{4}{9}$, the lowest common denominator is 18.

The same method is used when there are variables in the denominator.

For example:

- when finding $\frac{2}{a}+\frac{3}{b}$, the lowest common denominator is ab
- when finding $\frac{5}{6x}+\frac{4}{9y}$, the lowest common denominator is $18xy$.

Example 14 Self Tutor

Simplify by writing as a single fraction:

a $\dfrac{x}{2}+\dfrac{3x}{4}$ **b** $\dfrac{a}{3}-\dfrac{2a}{5}$

a $\dfrac{x}{2}+\dfrac{3x}{4}$ {LCD = 4}

$=\dfrac{x\times 2}{2\times 2}+\dfrac{3x}{4}$

$=\dfrac{2x}{4}+\dfrac{3x}{4}$

$=\dfrac{2x+3x}{4}$

$=\dfrac{5x}{4}$

b $\dfrac{a}{3}-\dfrac{2a}{5}$ {LCD = 15}

$=\dfrac{a\times 5}{3\times 5}-\dfrac{2a\times 3}{5\times 3}$

$=\dfrac{5a}{15}-\dfrac{6a}{15}$

$=\dfrac{5a-6a}{15}$

$=-\dfrac{a}{15}$

EXERCISE 9E

1 Simplify by writing as a single fraction:

a $\dfrac{a}{2}+\dfrac{a}{3}$ **b** $\dfrac{b}{5}-\dfrac{b}{10}$ **c** $\dfrac{c}{4}+\dfrac{3c}{2}$ **d** $\dfrac{d}{2}-\dfrac{3}{5}$

e $\dfrac{x}{7}-\dfrac{x}{2}$ **f** $\dfrac{a}{3}+\dfrac{b}{4}$ **g** $\dfrac{t}{3}-\dfrac{5t}{9}$ **h** $\dfrac{m}{7}+\dfrac{2m}{21}$

i $\dfrac{5d}{6}-\dfrac{d}{3}$ **j** $\dfrac{3p}{5}-\dfrac{2p}{7}$ **k** $\dfrac{2t}{9}+\dfrac{4t}{15}$ **l** $\dfrac{7k}{8}-\dfrac{11k}{18}$

m $\frac{m}{2} + \frac{m}{3} + \frac{m}{6}$ **n** $\frac{a}{2} - \frac{a}{3} + \frac{a}{4}$ **o** $\frac{x}{4} - \frac{x}{3} + \frac{x}{6}$ **p** $2q - \frac{q}{3} + \frac{2q}{7}$

Example 15 Self Tutor

Simplify by writing as a single fraction:

a $\frac{b}{3} + 1$ **b** $\frac{a}{4} - a$

a $\frac{b}{3} + 1 = \frac{b}{3} + \frac{3}{3}$
$= \frac{b+3}{3}$

b $\frac{a}{4} - a = \frac{a}{4} - \frac{a \times 4}{1 \times 4}$
$= \frac{a}{4} - \frac{4a}{4}$
$= \frac{-3a}{4}$
$= -\frac{3a}{4}$

2 Simplify by writing as a single fraction:

a $\frac{x}{2} + 1$ **b** $\frac{b}{4} - 3$ **c** $2 + \frac{a}{3}$ **d** $\frac{y}{3} - 1$

e $\frac{a}{2} + a$ **f** $4 - \frac{x}{2}$ **g** $x - \frac{x}{5}$ **h** $2 + \frac{1}{x}$

i $5 - \frac{2}{x}$ **j** $a + \frac{2}{a}$ **k** $\frac{3}{b} + b$ **l** $\frac{1}{x^2} - 2x$

Example 16 Self Tutor

Simplify by writing as a single fraction:

a $\frac{4}{a} + \frac{3}{b}$ **b** $\frac{5}{x} - \frac{4}{3x}$

a $\frac{4}{a} + \frac{3}{b}$ {LCD = ab}
$= \frac{4 \times b}{a \times b} + \frac{3 \times a}{b \times a}$
$= \frac{4b}{ab} + \frac{3a}{ab}$
$= \frac{4b + 3a}{ab}$

b $\frac{5}{x} - \frac{4}{3x}$ {LCD = $3x$}
$= \frac{5 \times 3}{x \times 3} - \frac{4}{3x}$
$= \frac{15}{3x} - \frac{4}{3x}$
$= \frac{15 - 4}{3x}$
$= \frac{11}{3x}$

3 Simplify by writing as a single fraction:

a $\frac{7}{a} + \frac{3}{b}$ **b** $\frac{3}{a} + \frac{2}{c}$ **c** $\frac{4}{3} + \frac{5}{a}$ **d** $\frac{2a}{m} - \frac{a}{n}$

e $\frac{3}{a} - \frac{1}{2a}$ **f** $\frac{a}{x} + \frac{3}{2x}$ **g** $\frac{4}{x} - \frac{1}{xy}$ **h** $\frac{5}{x} + \frac{6}{5x}$

i $\frac{11}{3z} - \frac{3}{4z}$ **j** $\frac{3}{a} + \frac{a}{2}$ **k** $\frac{x}{y} - \frac{2}{3}$ **l** $\frac{x}{6y} + \frac{2x}{9y}$

m $\frac{1}{8t} - \frac{3}{5t}$ **n** $\frac{1}{x} + \frac{1}{x^2}$ **o** $\frac{1}{x} - \frac{2}{x^2}$ **p** $\frac{5}{2x} + \frac{3}{x^2}$

4 Simplify by writing as a single fraction:

a $\dfrac{1}{x^2y}+\dfrac{2}{xy}$ **b** $\dfrac{2}{x}-\dfrac{1}{xy^2}$ **c** $\dfrac{3}{2x^2}-\dfrac{4}{x}$

d $\dfrac{q}{p^2}-\dfrac{1}{pq}$ **e** $\dfrac{3}{rs}+\dfrac{2r}{s^2}$ **f** $\dfrac{3}{x^2y}-\dfrac{1}{2y^2}$

Example 17 Self Tutor

Write as a single fraction, and hence simplify:

a $\dfrac{x}{6}+\dfrac{x-2}{3}$ **b** $\dfrac{x+1}{2}-\dfrac{x-2}{3}$

a $\dfrac{x}{6}+\dfrac{x-2}{3}$ {LCD = 6}

$=\dfrac{x}{6}+\dfrac{2}{2}\times\left(\dfrac{x-2}{3}\right)$

$=\dfrac{x}{6}+\dfrac{2(x-2)}{6}$

$=\dfrac{x+2(x-2)}{6}$

$=\dfrac{x+2x-4}{6}$

$=\dfrac{3x-4}{6}$

b $\dfrac{x+1}{2}-\dfrac{x-2}{3}$ {LCD = 6}

$=\dfrac{3}{3}\times\left(\dfrac{x+1}{2}\right)-\dfrac{2}{2}\times\left(\dfrac{x-2}{3}\right)$

$=\dfrac{3(x+1)}{6}-\dfrac{2(x-2)}{6}$

$=\dfrac{3(x+1)-2(x-2)}{6}$

$=\dfrac{3x+3-2x+4}{6}$

$=\dfrac{x+7}{6}$

5 Write as a single fraction, and hence simplify:

a $\dfrac{x}{2}+\dfrac{x+1}{3}$ **b** $\dfrac{x-1}{4}-\dfrac{x}{2}$ **c** $\dfrac{2x}{3}+\dfrac{x+3}{4}$ **d** $\dfrac{x+1}{2}-\dfrac{x-1}{3}$

e $\dfrac{x-1}{3}+\dfrac{1-2x}{4}$ **f** $\dfrac{2x+3}{2}+\dfrac{2x-3}{3}$ **g** $\dfrac{x}{3}-\dfrac{x+1}{4}$ **h** $\dfrac{x}{6}+\dfrac{3x-1}{5}$

i $\dfrac{a+b}{3}-\dfrac{b-a}{2}$ **j** $\dfrac{x+1}{5}+\dfrac{2x-1}{4}$ **k** $\dfrac{x}{6}-\dfrac{2-x}{5}$ **l** $\dfrac{2x-1}{5}-\dfrac{x}{4}$

Example 18 Self Tutor

Write as a single fraction, and hence simplify:

a $\dfrac{1}{x}+\dfrac{2}{x-1}$ **b** $\dfrac{2}{x-1}-\dfrac{3}{x+1}$

a $\dfrac{1}{x}+\dfrac{2}{x-1}$ {LCD = $x(x-1)$}

$=\dfrac{1}{x}\times\left(\dfrac{x-1}{x-1}\right)+\left(\dfrac{2}{x-1}\right)\times\dfrac{x}{x}$

$=\dfrac{1(x-1)+2x}{x(x-1)}$

$=\dfrac{x-1+2x}{x(x-1)}$

$=\dfrac{3x-1}{x(x-1)}$

b $\dfrac{2}{x-1}-\dfrac{3}{x+1}$ {LCD = $(x-1)(x+1)$}

$=\left(\dfrac{2}{x-1}\right)\times\left(\dfrac{x+1}{x+1}\right)-\left(\dfrac{3}{x+1}\right)\times\left(\dfrac{x-1}{x-1}\right)$

$=\dfrac{2(x+1)-3(x-1)}{(x-1)(x+1)}$

$=\dfrac{2x+2-3x+3}{(x-1)(x+1)}$

$=\dfrac{-x+5}{(x-1)(x+1)}$ or $\dfrac{5-x}{(x-1)(x+1)}$

6 Write as a single fraction, and hence simplify:

a $\dfrac{3}{x}+\dfrac{4}{x+1}$ **b** $\dfrac{5}{x+2}-\dfrac{3}{x}$ **c** $\dfrac{4}{x+1}-\dfrac{3}{x-1}$

d $3+\dfrac{1}{x+2}$ **e** $\dfrac{1}{x}+\dfrac{4}{x-4}$ **f** $\dfrac{2}{x+3}-4$

g $\dfrac{x+1}{x-1}+\dfrac{x}{x+1}$ **h** $\dfrac{5}{x}+\dfrac{6}{x-2}$ **i** $\dfrac{2}{x+2}-\dfrac{4}{x+1}$

j $\dfrac{x}{x-1}+\dfrac{4}{2x+1}$ **k** $\dfrac{5}{x-1}+\dfrac{x-2}{x+3}$ **l** $\dfrac{x}{x+5}-\dfrac{x}{x-3}$

7 Answer the **Opening Problem** on page **118**.

8 **a** Write $\dfrac{a+2}{a+1}-\dfrac{a+1}{a}$ as a single fraction.

b Hence show that if a is positive, then $\dfrac{a+2}{a+1}-\dfrac{a+1}{a}$ is negative.

9 Write as a single fraction, and hence simplify:

a $\dfrac{2}{x(x+1)}+\dfrac{1}{x+1}$ **b** $\dfrac{2x}{x-3}+\dfrac{4}{(x+2)(x-3)}$ **c** $\dfrac{3}{(x-2)(x+3)}+\dfrac{x}{x+3}$

10 A *unit fraction* is a fraction with numerator 1, for example $\frac{1}{4}$ or $\frac{1}{7}$.

a Show that $\dfrac{1}{n}=\dfrac{1}{n+1}+\dfrac{1}{n(n+1)}$ for $n\neq-1, 0$.

b Hence write the following unit fractions as the sum of two other unit fractions:

i $\frac{1}{2}$ **ii** $\frac{1}{5}$ **iii** $\frac{1}{9}$

F EQUATIONS WITH ALGEBRAIC FRACTIONS

Some equations involving algebraic fractions can be rewritten in linear form, even if the variable started in the denominator. In these cases, we:

- write all fractions with the same **lowest common denominator (LCD)**, and then
- **equate numerators**.

Example 19 **Self Tutor**

Solve for x: $\dfrac{2x+3}{4}=\dfrac{x-2}{3}$

$$\dfrac{2x+3}{4}=\dfrac{x-2}{3}$$

$\therefore\ \dfrac{3}{3}\times\left(\dfrac{2x+3}{4}\right)=\dfrac{4}{4}\times\left(\dfrac{x-2}{3}\right)$ {LCD = 12}

$\therefore\ 3(2x+3)=4(x-2)$ {equating numerators}

$\therefore\ 6x+9=4x-8$ {distributive law}

$\therefore\ 2x+9=-8$ {$-\,4x$ both sides}

$\therefore\ 2x=-17$ {$-\,9$ both sides}

$\therefore\ x=-\frac{17}{2}$ {$\div\,2$ both sides}

EXERCISE 9F

1 Solve for x:

a $\frac{x-2}{4} = \frac{3}{2}$ **b** $\frac{3+2x}{5} = \frac{1}{2}$ **c** $\frac{x}{5} = \frac{x+3}{10}$

d $\frac{x+6}{2} = \frac{x}{3}$ **e** $\frac{2x-11}{7} = \frac{3x}{5}$ **f** $\frac{3x+2}{2} = \frac{x-1}{4}$

g $\frac{1-x}{2} = \frac{x+2}{3}$ **h** $\frac{x+5}{2} = 1 - x$ **i** $\frac{2x+7}{3} = x + 4$

2 Solve for x:

a $\frac{x}{3} + \frac{x+1}{6} = 4$ **b** $\frac{x-4}{2} - \frac{3x}{4} = -5$ **c** $\frac{2x}{3} + \frac{x+1}{6} = \frac{7}{2}$

d $\frac{x+1}{12} - \frac{x}{3} = \frac{x-3}{4}$ **e** $6 - \frac{x+7}{2} = \frac{2x}{5}$ **f** $\frac{2x-1}{3} + x = \frac{8}{9}$

3 Solve for x:

a $\frac{8x+5}{3} - \frac{12x}{5} = -5$ **b** $\frac{x}{2} + \frac{2-x}{3} = \frac{x+3}{5}$

c $\frac{2x-1}{3} - \frac{1}{4} = \frac{3x-1}{5}$ **d** $\frac{x-1}{8} + \frac{2x}{3} = \frac{3x+2}{4}$

e $\frac{x+5}{3} - \frac{4-x}{2} = \frac{4x-1}{9}$ **f** $\frac{3x+20}{10} - \frac{7x}{15} = \frac{8-3x}{12}$

Example 20 Self Tutor

Solve for x: $\frac{4}{3x} = -\frac{5}{6}$

$$\frac{4}{3x} = -\frac{5}{6}$$

$\therefore \frac{4}{3x} \times \frac{2}{2} = -\frac{5}{6} \times \frac{x}{x}$ {LCD = $6x$}

$\therefore 8 = -5x$ {equating numerators}

$\therefore x = -\frac{8}{5}$ {$\div -5$ both sides}

4 Solve for x:

a $\frac{3}{x} = \frac{1}{5}$ **b** $\frac{3}{x} = \frac{2}{3}$ **c** $\frac{2}{7} = \frac{5}{2x}$

d $\frac{4}{9} = -\frac{1}{6x}$ **e** $\frac{1}{2x} = \frac{4}{3}$ **f** $\frac{7}{3x} = -\frac{4}{5}$

g $\frac{4}{5x} = 3$ **h** $-5 = \frac{2}{3x}$ **i** $\frac{x+2}{x} = \frac{5}{4}$

j $\frac{x-3}{x} = \frac{3}{5}$ **k** $\frac{x+5}{2x} = -\frac{8}{5}$ **l** $\frac{x-1}{4x} = \frac{5}{6}$

Example 21 Self Tutor

Solve for x: $\frac{5}{x+2} = \frac{2}{x-1}$

$$\frac{5}{x+2} = \frac{2}{x-1}$$

$$\therefore \left(\frac{5}{x+2}\right) \times \left(\frac{x-1}{x-1}\right) = \left(\frac{2}{x-1}\right) \times \left(\frac{x+2}{x+2}\right) \qquad \{\text{LCD} = (x+2)(x-1)\}$$

$$\therefore 5(x-1) = 2(x+2) \qquad \{\text{equating numerators}\}$$

$$\therefore 5x - 5 = 2x + 4 \qquad \{\text{distributive law}\}$$

$$\therefore 3x - 5 = 4 \qquad \{-\,2x \text{ both sides}\}$$

$$\therefore 3x = 9 \qquad \{+\,5 \text{ both sides}\}$$

$$\therefore x = 3 \qquad \{\div\,3 \text{ both sides}\}$$

5 Solve for x:

a $\dfrac{4}{x+1} = \dfrac{1}{2}$ **b** $\dfrac{5}{2-x} = -2$ **c** $\dfrac{x}{x+3} = 4$

d $\dfrac{2x+1}{x+1} = \dfrac{2}{5}$ **e** $\dfrac{x+4}{3x-1} = \dfrac{1}{2}$ **f** $\dfrac{5x-1}{1-2x} = -6$

6 Solve for x:

a $\dfrac{2}{x} = \dfrac{8}{x+6}$ **b** $\dfrac{3}{x} = \dfrac{5}{x-6}$ **c** $\dfrac{1}{x+1} = \dfrac{7}{x-1}$

d $\dfrac{10}{x-3} = \dfrac{7}{x+6}$ **e** $\dfrac{8}{2x+3} = \dfrac{9}{x-1}$ **f** $\dfrac{5}{3x-2} = \dfrac{4}{x-3}$

Example 22 **Self Tutor**

Solve for x: $\dfrac{1}{x} - \dfrac{2}{x+1} = \dfrac{x+4}{x(x+1)}$

$$\frac{1}{x} - \frac{2}{x+1} = \frac{x+4}{x(x+1)}$$

$$\therefore \frac{1}{x} \times \left(\frac{x+1}{x+1}\right) - \frac{x}{x} \times \left(\frac{2}{x+1}\right) = \frac{x+4}{x(x+1)} \qquad \{\text{LCD} = x(x+1)\}$$

$$\therefore (x+1) - 2x = x + 4 \qquad \{\text{equating numerators}\}$$

$$\therefore 1 - x = x + 4 \qquad \{\text{collecting like terms}\}$$

$$\therefore 1 = 2x + 4 \qquad \{+\,x \text{ both sides}\}$$

$$\therefore -3 = 2x \qquad \{-\,4 \text{ both sides}\}$$

$$\therefore x = -\tfrac{3}{2} \qquad \{\div\,2 \text{ both sides}\}$$

7 Solve for x:

a $\dfrac{6}{x} - \dfrac{4}{x-2} = \dfrac{x+1}{x(x-2)}$ **b** $\dfrac{3}{x} + \dfrac{2}{x+4} = \dfrac{x-8}{x(x+4)}$

c $\dfrac{1}{x+3} - \dfrac{4}{x-2} = \dfrac{x}{(x+3)(x-2)}$ **d** $\dfrac{9}{x-3} + \dfrac{2}{x} = \dfrac{11}{x(x-3)}$

e $\dfrac{2}{x-1} + \dfrac{5}{x+3} = \dfrac{2-x}{(x-1)(x+3)}$ **f** $\dfrac{x}{x+1} + \dfrac{6}{x+2} = \dfrac{x^2}{(x+1)(x+2)}$

g $\dfrac{x}{x+2} + \dfrac{x}{x-3} = 2$ **h** $\dfrac{x}{x-5} + \dfrac{2x}{x+1} = 3$

Example 23 **Self Tutor**

The denominator of a fraction is 20 more than the numerator, and the value of the fraction is $\frac{3}{8}$. Find the numerator of the fraction.

Let x be the numerator of the fraction, so the denominator is $x+20$.

$$\therefore \quad \frac{x}{x+20} = \frac{3}{8}$$

$$\therefore \quad \frac{8}{8} \times \left(\frac{x}{x+20}\right) = \frac{3}{8} \times \left(\frac{x+20}{x+20}\right) \qquad \{\text{LCD} = 8(x+20)\}$$

$$\therefore \quad 8x = 3(x+20) \qquad \{\text{equating numerators}\}$$

$$\therefore \quad 8x = 3x + 60 \qquad \{\text{distributive law}\}$$

$$\therefore \quad 5x = 60 \qquad \{-\ 3x \text{ both sides}\}$$

$$\therefore \quad x = 12 \qquad \{\div\ 5 \text{ both sides}\}$$

So, the numerator of the fraction is 12.

8 The denominator of a fraction is 8 less than the numerator, and the value of the fraction is $\frac{5}{3}$. Find the numerator of the fraction.

9 What number must be subtracted from both the numerator and the denominator of $\frac{3}{5}$ to obtain a fraction equal to $\frac{1}{4}$?

10 Write a fraction for which the sum of the numerator and denominator is 35, and the value of the fraction is $\frac{3}{4}$.

11 How much water must be added to 15 mL of 30% vinegar mixture so that the resulting mixture is 10% vinegar?

12 Steve mixed 100 mL of oil with 2 L of petrol for use in his chainsaw. He realised he had made a mistake, and that actually he needed to mix oil with petrol in the ratio $2 : 75$. How much petrol does Steve need to add to his current mixture?

MULTIPLE CHOICE QUIZ

QUICK QUIZ

REVIEW SET 9A

1 If $p = 5$, $q = -3$, and $r = 6$, evaluate:

a $\dfrac{p-q}{p+q}$ **b** $\dfrac{\sqrt{p^2-16}}{r-q}$ **c** $\dfrac{3q-2}{r^2-p^2}$

2 Simplify:

a $\dfrac{a^2b}{3a}$ **b** $\dfrac{(2t)^2}{6t}$ **c** $\dfrac{x(x-4)}{3(x-4)}$

3 Factorise, then simplify:

a $\dfrac{8}{4x+8}$ **b** $\dfrac{2x+6}{x^2-9}$ **c** $\dfrac{16a+8b}{6a+3b}$

d $\dfrac{16-x^2}{2x-8}$ **e** $\dfrac{a^2+a-12}{a^2+10a+24}$ **f** $\dfrac{3x^2-6x}{3x^2-5x-2}$

4 Simplify:

a $\dfrac{m}{n}\times\dfrac{2}{n}$ **b** $\dfrac{m}{n}\div\dfrac{2}{n}$ **c** $\dfrac{3x}{2y^2}\times\dfrac{4y}{9x^2}$

d $\dfrac{y^2-5y}{y+2}\times\dfrac{3}{2y-10}$ **e** $\dfrac{9-3x}{4x+16}\div\dfrac{x^2-3x}{x+4}$

5 Simplify by writing as a single fraction:

a $\dfrac{2x}{3}+\dfrac{x}{4}$ **b** $2+\dfrac{x}{7}$ **c** $\dfrac{x}{4}-1$

d $\dfrac{x}{2}+\dfrac{x}{4}-\dfrac{x}{3}$ **e** $\dfrac{5}{a}+\dfrac{9}{b}$ **f** $\dfrac{6}{x}-\dfrac{7}{4x}$

6 Write as a single fraction, and hence simplify:

a $\dfrac{2x+1}{5}-\dfrac{x-1}{10}$ **b** $\dfrac{1}{x+1}+\dfrac{2}{x-2}$ **c** $\dfrac{3}{(x+1)(x+2)}+\dfrac{4x}{x+2}$

7 Solve for x:

a $\dfrac{4x+5}{3}=\dfrac{x}{2}$ **b** $\dfrac{4-3x}{5}=x-6$ **c** $\dfrac{1}{3x}=5$

d $\dfrac{x+6}{3-2x}=-1$ **e** $\dfrac{5}{x}=\dfrac{11}{x+1}$ **f** $\dfrac{2}{x}-\dfrac{3}{x+1}=\dfrac{7}{x(x+1)}$

8 Write a fraction for which the difference between the numerator and denominator is 24, and the fraction has value $\frac{3}{2}$.

9 Ivan and Rico play in the same soccer team. Last season, Ivan scored half of his goals from penalties. Rico scored 8 more goals than Ivan, and scored one fifth of his goals from penalties. Suppose Ivan scored x goals for the season.

a Write the total number of penalty goals scored by Ivan and Rico as a single fraction in terms of x.

b How many penalty goals did they score between them, if Ivan scored:

i 12 goals for the season **ii** 22 goals for the season?

REVIEW SET 9B

1 If $m=-4$, $n=3$, and $p=6$, evaluate:

a $\dfrac{p}{m+n}$ **b** $\dfrac{p-2n}{m+n}$ **c** $\dfrac{p-m}{\sqrt{m^2+n^2}}$

2 Factorise, then simplify:

a $\dfrac{a+b}{3b+3a}$ **b** $\dfrac{2a-2b}{b-a}$ **c** $\dfrac{2x^2-8}{x+2}$

d $\dfrac{5x-15}{3x-x^2}$ **e** $\dfrac{x^2+4x+4}{x^2+2x}$ **f** $\dfrac{x^2-6x+9}{2x^2-5x-3}$

3 Simplify:

a $\frac{4}{k} \times \frac{k^2}{12}$ **b** $\frac{9a}{b^2} \times \frac{ab}{3}$ **c** $\frac{t^2-3t}{6t+6} \times \frac{t+1}{4t-12}$

4 Simplify:

a $\frac{9}{n} \div 6$ **b** $\frac{x^2}{4y} \div \frac{3}{x}$ **c** $\frac{7}{3x-6} \div \frac{x+5}{x^2-2x}$

5 Simplify by writing as a single fraction:

a $\frac{3}{x} + \frac{5}{2x}$ **b** $\frac{6}{y} - \frac{a}{b}$ **c** $\frac{3x}{7} - \frac{x}{14}$

d $\frac{4}{3x} + \frac{3}{x^2}$ **e** $3 - \frac{y}{x}$ **f** $1 + \frac{x}{2} + \frac{y}{3}$

6 Write as a single fraction, and hence simplify:

a $\frac{x}{4} - \frac{2-x}{8}$ **b** $\frac{2}{x-1} - \frac{3}{x+2}$ **c** $\frac{x}{x+2} - \frac{2}{x} + \frac{x}{x+3}$

7 If m is an even integer and $m \neq 10$, show that $\frac{m^2-3m-70}{m-10}$ is an odd integer.

8 Solve for x:

a $\frac{1-3x}{4} = \frac{x-2}{2}$ **b** $\frac{2x+1}{3} - \frac{4-x}{6} = -2$ **c** $\frac{18}{x+2} = 3$

d $\frac{3}{x+4} = \frac{8}{x-1}$ **e** $\frac{x+2}{4x} = \frac{2}{3}$ **f** $\frac{x}{x+3} + \frac{x}{x+1} = 2$

9 How much water must be added to 250 mL of 20% cordial mixture so that the resulting mixture is 16% cordial?

10 **a** Determine which sum is larger:

i $\frac{1}{3} + \frac{1}{3}$ or $\frac{1}{2} + \frac{1}{4}$ **ii** $\frac{1}{5} + \frac{1}{5}$ or $\frac{1}{4} + \frac{1}{6}$ **iii** $\frac{1}{9} + \frac{1}{9}$ or $\frac{1}{8} + \frac{1}{10}$

b For any positive integer $x > 1$, which do you think is larger:

$\frac{1}{x} + \frac{1}{x}$ or $\frac{1}{x-1} + \frac{1}{x+1}$?

c Prove your answer to **b** by simplifying $\frac{1}{x-1} + \frac{1}{x+1} - \frac{2}{x}$.

Chapter 10

Formulae

Contents:

OPENING PROBLEM

An architect designing a house wants to equally space n windows along a wall.

Each window has width w, and the wall has length l. The distance between windows, and the distance between the end windows and the ends of the wall, is d.

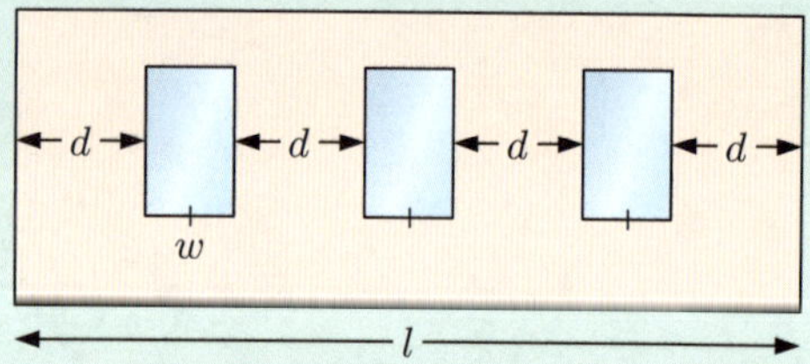

Things to think about:

- **a** Can you explain why $nw + (n+1)d = l$?
- **b** Can you *rearrange* this formula to make d the subject?
- **c** Suppose a wall has length 8 m, and the architect places 4 windows of width 1.2 m in the wall. What will be the distance between the windows?

A **formula** is an equation which connects two or more variables.

For example, the formula $s = ut + \frac{1}{2}at^2$ relates the variables *distance travelled* (s), *initial speed* (u), *acceleration* (a), and *time* (t).

We usually write a formula with one variable on its own on the left hand side. The other variable(s) and constants are written on the right hand side.

The variable on its own is called the **subject** of the formula. We say this variable is written *in terms of* the other variables.

A FORMULA CONSTRUCTION

To **construct** or **derive** a formula to connect related variables, we often start with numerical examples. Once we understand the specific situation, we can **generalise** the result.

Example 1 — Self Tutor

Write a formula for the amount $\$A$ in a person's bank account if initially the balance was:

- **a** \$5000, and \$200 was withdrawn each week for 10 weeks
- **b** \$5000, and \$200 was withdrawn each week for w weeks
- **c** \$5000, and $\$x$ was withdrawn each week for w weeks
- **d** $\$B$, and $\$x$ was withdrawn each week for w weeks.

a $A = 5000 - 200 \times 10$

b $A = 5000 - 200 \times w$
$\therefore\ A = 5000 - 200w$

c $A = 5000 - x \times w$
$\therefore\ A = 5000 - xw$

d $A = B - x \times w$
$\therefore\ A = B - xw$

EXERCISE 10A

1 Write a formula for the total cost $\$C$ of:

a 4 mugs costing $5 each **b** 4 mugs costing $\$x$ each **c** m mugs costing $\$x$ each.

2 Write a formula for the amount $\$A$ earnt for working:

a 5 hours at $15 per hour **b** 5 hours at $\$p$ per hour **c** t hours at $\$p$ per hour.

3 Write a formula for the amount $\$A$ in a bank account if the initial balance was:

a $2000, and then $150 was deposited each week for 8 weeks

b $2000, and then $150 was deposited each week for w weeks

c $2000, and then $\$d$ was deposited each week for w weeks

d $\$P$, and then $\$d$ was deposited each week for w weeks.

4 Write a formula for the total cost €C of hiring a plumber given a fixed call-out fee of:

a €40, plus €80 per hour for 5 hours of work

b €40, plus €80 per hour for t hours of work

c €40, plus €x per hour for t hours of work

d €F, plus €x per hour for t hours of work.

5 In a multiple choice mathematics competition, students are awarded 3 points for each question answered correctly, and penalised 1 point for each question answered incorrectly. Write a formula for the number of points P scored by a student who:

a answers 15 questions and gets 10 of them correct

b answers 20 questions and gets c of them correct

c answers a questions and gets c of them correct.

6 A music recital consists of performances by a number of musicians, with a short break between each performance. Write a formula for the duration D minutes of a recital consisting of:

a 4 performances of 6 minutes each, with a 2 minute break between performances

b p performances of m minutes each, with a b minute break between performances.

7 A rectangular paddock is fenced into a rectangular array of yards so that each yard is connected by a gate to each adjacent yard. A 3×2 arrangement of yards is shown alongside.
Write a formula for the number of gates G for:

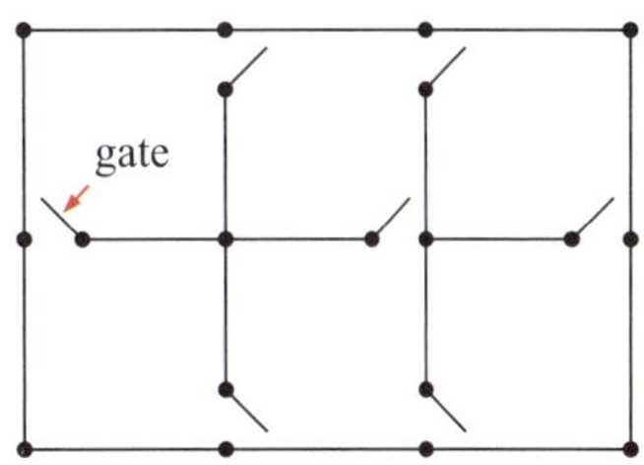

a a 3×2 arrangement **b** a 5×3 arrangement

c a 4×4 arrangement **d** an $m \times n$ arrangement.

ACTIVITY DIMENSIONAL ANALYSIS

When we construct a formula, the units of each term can be used to check whether the formula is reasonable.

Click on the icon to obtain this Activity.

ACTIVITY

B SUBSTITUTING INTO FORMULAE

Suppose a formula contains two or more variables, and we know the value of all but one of them. We can **substitute** the known values into the formula, leaving an equation which can be solved to find the value of the remaining variable.

Example 2 Self Tutor

When a stone is dropped from a cliff, the total distance fallen after t seconds is given by the formula $D = \frac{1}{2}gt^2$ metres, where $g = 9.8\ \text{m/s}^2$. Find:

a the distance fallen after 4 seconds

b the time, to the nearest $\frac{1}{100}$th second, taken for the stone to fall 200 metres.

a $D = \frac{1}{2}gt^2$ where $g = 9.8$ and $t = 4$

$\therefore\ D = \frac{1}{2} \times 9.8 \times 4^2 = 78.4$

$\therefore$ the stone has fallen 78.4 metres.

b $D = \frac{1}{2}gt^2$ where $D = 200$ and $g = 9.8$

$\therefore\ \frac{1}{2} \times 9.8 \times t^2 = 200$

$\therefore\ 4.9t^2 = 200$

$\therefore\ t^2 = \frac{200}{4.9}$ {÷ 4.9 both sides}

$\therefore\ t = \sqrt{\frac{200}{4.9}}$ {t must be positive}

$\therefore\ t \approx 6.39$

$\therefore$ the time taken is about 6.39 seconds.

EXERCISE 10B

1 When a car travels a distance d kilometres in time t hours, the average speed for the journey is given by the formula $s = \dfrac{d}{t}$ km/h. Find:

   a the average speed of a car which travels 250 km in $3\frac{1}{2}$ hours

   b the distance travelled by a car in $2\frac{3}{4}$ hours if its average speed is 80 km/h

   c the time taken, to the nearest minute, for a car to travel 790 km at an average speed of 95 km/h.

2 When a stone is dropped from the top of a cliff, the distance fallen after t seconds is given by the formula $D = \frac{1}{2}gt^2$ metres, where $g = 9.8\ \text{m/s}^2$. Find:

   a the distance fallen in the first 2 seconds

   b the time taken for the stone to fall 100 metres.

3 The *period* or time taken for a simple pendulum to complete one swing is given approximately by $T = \frac{1}{5}\sqrt{l}$ seconds, where l is the length of the pendulum in centimetres. Find:

   a the time for one complete swing of a pendulum with length 45 cm

   b the length of a pendulum which has period 1.8 seconds.

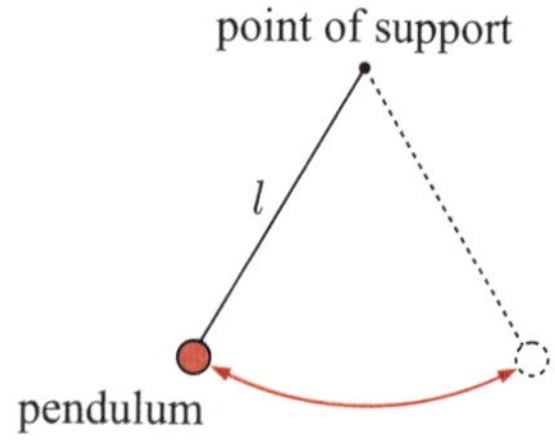

4 n points are equally spaced around a circle.
When every point is connected to every other point with a straight line segment, there are $L = \frac{1}{2}n(n-1)$ lines in total. Find the number of line segments if there are:

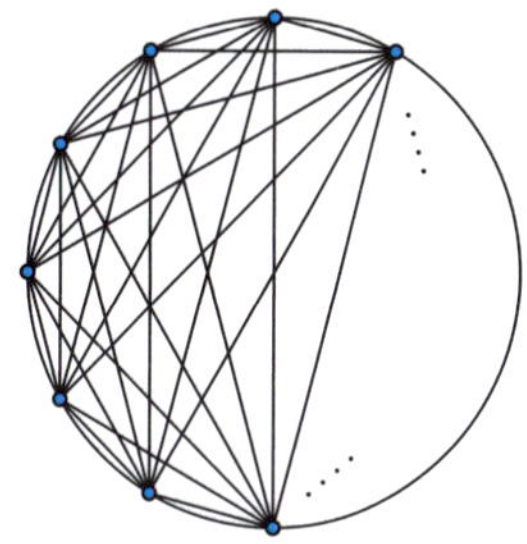

- **a** 12 points around the circle
- **b** 25 points around the circle.

5 The formula $D = 3.56\sqrt{h}$ km gives the approximate distance to the horizon which can be seen by a person with eye level h metres above sea level. Find:

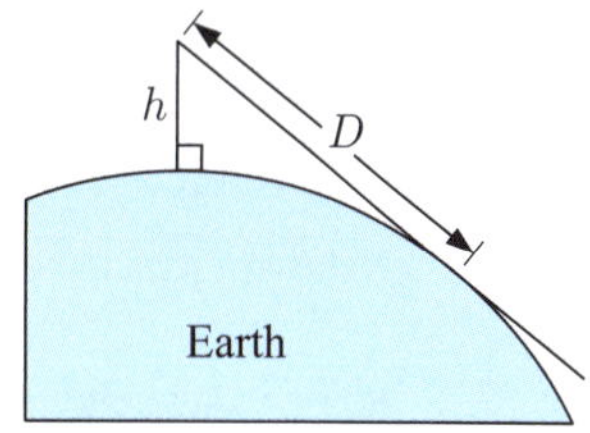

- **a** the distance to the horizon for a person whose eye level is 20 m above sea level
- **b** how far above sea level a person's eye must be, for the person to see for 25 km.

6 The kinetic energy of a non-rotating object with mass m kg and speed v m/s is given by $E = \frac{1}{2}mv^2$ J. Find:

- **a** the kinetic energy of a 0.8 kg ball thrown at 20 m/s
- **b** the mass of a marble with kinetic energy 0.156 J and speed 1.4 m/s
- **c** the speed of an aircraft with kinetic energy 9.26×10^7 J and mass 5360 kg.

C REARRANGING FORMULAE

We can **rearrange** a formula to make one of the other variables the subject using **inverse operations** on both sides. This ensures the formula is still true.

Example 3 Self Tutor

Make y the subject of:

a $2x + 3y = 12$ **b** $x = 5 - cy$

a $2x + 3y = 12$

$\therefore\ 3y = 12 - 2x$ $\{-2x$ both sides$\}$

$\therefore\ y = \dfrac{12 - 2x}{3}$ $\{\div 3$ both sides$\}$

b $x = 5 - cy$

$\therefore\ x + cy = 5$ $\{+cy$ both sides$\}$

$\therefore\ cy = 5 - x$ $\{-x$ both sides$\}$

$\therefore\ y = \dfrac{5 - x}{c}$ $\{\div c$ both sides$\}$

EXERCISE 10C

1 Make y the subject of:

a $x + y = 7$	**b** $x - y = 3$	**c** $x + 2y = 1$
d $x - 2y = 7$	**e** $2x + 5y = 10$	**f** $3x - 5y = 18$
g $3x + 4y = 20$	**h** $2x - 3y = -12$	**i** $5x + 2y = 20$

2 Make x the subject of:

a $p+x=r$ **b** $xy=z$ **c** $-3x+a=d$

d $\dfrac{5-x}{2}=y$ **e** $ax+by=p$ **f** $y=mx+c$

3 Make x the subject of each formula, given $x>0$:

a $x^2=k$ **b** $M=4x^2$ **c** $N=\dfrac{x^2}{a}$

d $m-x^2=n$ **e** $b=\dfrac{a+x^2}{3}$ **f** $D=3x^3-y$

4 Make:

a r the subject of $A=\pi r^2$, $r>0$ **b** x the subject of $y=4x^2-7$, $x>0$

c b the subject of $a^2+b^2=c^2$, $b>0$ **d** r the subject of $V=\frac{4}{3}\pi r^3$.

Example 4 Self Tutor

Make y the subject of $c=\dfrac{m}{y}$.

$$c=\frac{m}{y}$$

$\therefore\ cy=m$ {$\times\, y$ both sides}

$\therefore\ y=\dfrac{m}{c}$ {$\div\, c$ both sides}

5 Make z the subject of:

a $az=\dfrac{b}{c}$ **b** $\dfrac{p}{z}=-\dfrac{q}{t}$ **c** $\dfrac{3}{d}=\dfrac{2}{z}$

d $\dfrac{z}{2}=\dfrac{a}{z}$, $z>0$ **e** $\dfrac{b}{z^2}=\dfrac{z}{n}$ **f** $\dfrac{m}{z}=\dfrac{z}{a-b}$, $z<0$

6 Henry has mass h kg. His twin sisters Jessica and Jacinta both have mass j kg. The average mass of the three siblings is given by $M=\dfrac{h+2j}{3}$ kg. Rearrange this formula to make:

a h the subject **b** j the subject.

Example 5 Self Tutor

Make x the subject of $T=\dfrac{a}{\sqrt{x}}$.

$$T=\frac{a}{\sqrt{x}}$$

$\therefore\ T^2=\left(\dfrac{a}{\sqrt{x}}\right)^2$ {squaring both sides}

$\therefore\ T^2=\dfrac{a^2}{x}$

$\therefore\ T^2x=a^2$ {$\times\, x$ both sides}

$\therefore\ x=\dfrac{a^2}{T^2}$ {$\div\, T^2$ both sides}

7 Make:

a a the subject of $d=\dfrac{\sqrt{a}}{n}$ **b** a the subject of $c=\sqrt{a^2-b^2}$, $a>0$

c l the subject of $T = 2\pi\sqrt{\frac{l}{g}}$

d b the subject of $A = 4\sqrt{\frac{a}{b}}$.

Example 6 — Self Tutor

Make x the subject of $T = \frac{a}{x-b}$.

$$T = \frac{a}{x-b}$$

$\therefore\ T(x-b) = a$ $\quad\{\times\,(x-b)$ both sides$\}$

$\therefore\ Tx - Tb = a$

$\therefore\ Tx = a + Tb$ $\quad\{+Tb$ both sides$\}$

$\therefore\ x = \frac{a+Tb}{T}$ $\quad\{\div\,T$ both sides$\}$

8 Make:

a a the subject of $P = \frac{2}{a+b}$

b q the subject of $A = \frac{B}{p-q}$

c x the subject of $A = \frac{3}{2x+y}$

d y the subject of $M = \frac{4}{x^2+y^2}$, $y > 0$.

9 Make x the subject of:

a $y = 1 + \frac{2}{x-3}$

b $y = -2 + \frac{5}{x+4}$

c $y = -3 - \frac{6}{x-2}$

Example 7 — Self Tutor

Make x the subject of $ax + 3 = bx + d$.

$ax + 3 = bx + d$

$\therefore\ ax - bx = d - 3$ $\quad\{$writing terms containing x on the LHS$\}$

$\therefore\ x(a-b) = d - 3$ $\quad\{$factorising LHS$\}$

$\therefore\ x = \frac{d-3}{a-b}$ $\quad\{\div\,(a-b)$ both sides$\}$

10 Make x the subject of:

a $3x + a = bx + c$

b $ax = c - bx$

c $mx + a = nx - 2$

d $8x + a = -bx$

e $a - x = b - cx$

f $rx + d = e - sx$

g $4(x+y) = x + 1$

h $a(x+5) = b(x-1)$

i $3(xy+2) = 2x$

Example 8 — Self Tutor

Make x the subject of $y = \frac{3x+2}{x-1}$.

$$y = \frac{3x+2}{x-1}$$

$\therefore\ y(x-1) = 3x + 2$ $\quad\{\times\,(x-1)$ both sides$\}$

$\therefore\ xy - y = 3x + 2$

$\therefore\ xy - 3x = y + 2$ $\quad\{$writing terms containing x on LHS$\}$

$\therefore\ x(y-3) = y + 2$ $\quad\{$factorising LHS$\}$

$\therefore\ x = \frac{y+2}{y-3}$ $\quad\{\div\,(y-3)$ both sides$\}$

11 Make x the subject of:

a $y = \dfrac{x}{x+1}$ **b** $y = \dfrac{x-3}{x+2}$ **c** $y = \dfrac{3x-1}{x+3}$

d $y = \dfrac{5x-2}{x-1}$ **e** $y = \dfrac{4x-1}{2-x}$ **f** $y = \dfrac{3x+7}{3-2x}$

D REARRANGEMENT AND SUBSTITUTION

In the Section on formula substitution, the known variables were replaced by numbers, and we then solved an equation to find the unknown.

In situations when we need to perform this process several times, it is quicker to **rearrange** the formula first, and then **substitute**.

Example 9 **Self Tutor**

The **power** of a wind turbine is a measure of the rate at which it produces energy. It can be calculated using the formula $P = \frac{1}{2}\pi r^2 \rho v^3 c_p$ where:

P is the power in watts (W)
r is the radius of the rotors in m
ρ is the density of air ≈ 1.23 kg/m^3 at sea level
v is the wind speed in m/s
c_p is the power coefficient of the particular wind turbine.

a Rearrange the formula to make v the subject.

b Hence find the wind speed required to produce 32 000 W of power using a wind turbine with rotor radius 5.2 m and power coefficient 0.326 .

a $P = \frac{1}{2}\pi r^2 \rho v^3 c_p$

$\therefore \pi r^2 \rho v^3 c_p = 2P$ $\{\times 2 \text{ both sides}\}$

$\therefore v^3 = \dfrac{2P}{\pi r^2 \rho c_p}$ $\{\div \pi r^2 \rho c_p \text{ both sides}\}$

$\therefore v = \sqrt[3]{\dfrac{2P}{\pi r^2 \rho c_p}}$ {cube root of both sides}

b When $P = 32\,000$, $r = 5.2$, $\rho \approx 1.23$, and $c_p = 0.326$,

$v \approx \sqrt[3]{\dfrac{2 \times 32\,000}{\pi \times 5.2^2 \times 1.23 \times 0.326}} \approx 12.3$

$\therefore$ a wind speed of about 12.3 m/s is required.

EXERCISE 10D

1 The *simple interest formula* is $I = Pin$ where I is the interest earned over n years by an investment P at an interest rate i expressed as a decimal.

a Make P the subject of the formula.

b Hence find the amount that must be invested at 3.6% p.a. to earn \$1800 interest over 5 years.

2 **a** Make a the subject of the formula $K = \dfrac{d^2}{2ab}$.

b Find the value of a when:

i $K = 112$, $d = 24$, $b = 2$ **ii** $K = 400$, $d = 72$, $b = 0.4$

3 The height of a bush after t years is given by the formula $H = 1 + \sqrt{t}$ metres.

a Rearrange this formula to make t the subject.

b How long will it take for the bush to reach a height of:

i 2 m **ii** 3 m **iii** 3.5 m?

4 An object with constant acceleration a m/s^2 travels s m. Its initial speed is u m/s and its final speed is v m/s. The variables are connected by the formula $v^2 - u^2 = 2as$.

a Given that $v \geqslant 0$, rearrange the formula to make v the subject.

b Find the final speed of an object which travels:

i 100 m with initial speed 5 m/s and constant acceleration 2 m/s^2

ii 1.5 km with initial speed 10 m/s and constant acceleration 0.9 m/s^2.

5 Consider the wind turbine power formula in **Example 9**.

a Rearrange the formula to make r the subject.

b Find the rotor radius required to produce 46 000 W of power if:

i $c_p = 0.41$ and $v = 9.2$ m/s **ii** $c_p = 0.369$ and $v = 16.4$ m/s.

6 The *winning percentage* of a tennis player who has won w matches and lost l matches is given by the formula $P = \dfrac{w}{w+l} \times 100\%$.

a Find the winning percentage of a player who has won 10 matches and lost 7 matches.

b Rearrange the formula to make w the subject.

c This year Mary has lost 15 matches, and her winning percentage is 37.5%. How many matches has she won?

d Over his career, Claude has won 84 matches and lost 49 matches. His aim is to increase his winning percentage to 65%. How many consecutive matches must he win to reach his target?

7 Consider two objects with masses m_1 kg and m_2 kg, which are d m apart.
The gravitational force between the objects is given by $F = G\dfrac{m_1 m_2}{d^2}$ newtons where $G \approx 6.67 \times 10^{-11}$ is the universal gravitational constant.

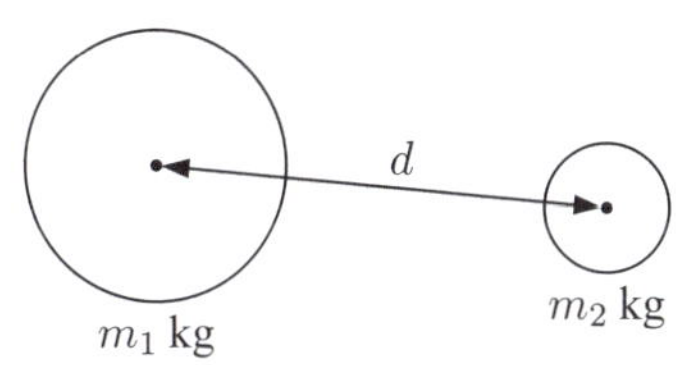

a The Earth has mass 5.97×10^{24} kg, the Moon has mass 7.35×10^{22} kg, and the Sun has mass 1.99×10^{30} kg.
Find the gravitational force between the Earth and the Moon at a time when they are 3.82×10^8 m apart. Give your answer in standard form.

b Rearrange the formula to make d the subject.

c **i** If the gravitational force between the Sun and the Earth is 3.54×10^{22} newtons, find the distance between them.

ii Two planets each have mass 2.32×10^{26} kg and the gravitational force between them is 1.76×10^{14} newtons. Find the distance between the planets.

8 The *thin lens equation* $\frac{1}{o}+\frac{1}{i}=\frac{1}{f}$ describes the relationship between the distance o of an object from the lens, the distance i of its image from the lens, and the focal length f of the lens.

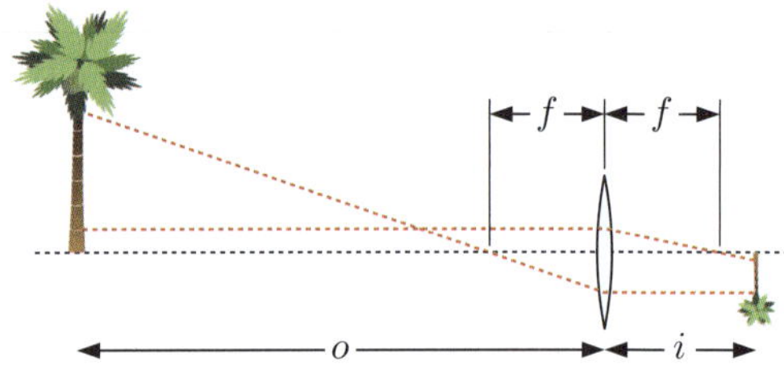

a Rearrange the equation to make o the subject.

b Hence find the distance of an object from the lens if:

i the focal length $f=25$ cm and the image distance $i=36$ cm

ii the focal length $f=2.8$ cm and the image distance $i=14.3$ cm.

GLOBAL CONTEXT **THE US HOUSE OF REPRESENTATIVES**

Global context: Fairness and development

Statement of inquiry: Mathematics can be used to define and identify fair methods of allocating resources.

Criterion: Communicating

GLOBAL CONTEXT

E PREDICTING FORMULAE

We can often predict formulae by examining simple cases and looking for a pattern.

For example, the sequence of even numbers is 2, 4, 6, 8, 10, 12, ...

We observe that: the 1st term is 2×1
the 2nd term is 2×2
the 3rd term is 2×3, and so on.

Assuming the pattern continues, the 13th term will be 2×13.

So, we generalise by saying that "the nth even number is $2n$".

The coefficient 2 in $2n$ causes the terms to increase by 2 each time n is increased by 1.

Example 10 **Self Tutor**

Consider the pattern:

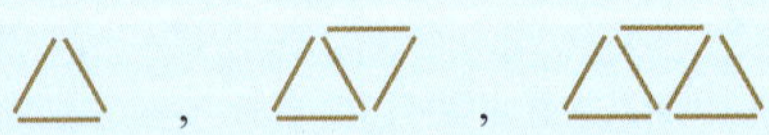

, , ...

a How many matchsticks are needed to make the:

i 1st diagram **ii** 2nd diagram **iii** 3rd diagram **iv** 4th diagram?

b Assuming the pattern continues, write a formula for the number of matchsticks M in the nth diagram.

---

a **i** 3 matches **ii** 5 matches **iii** 7 matches

iv The 4th diagram is

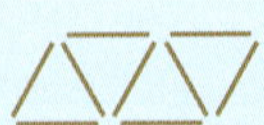

, which contains 9 matches.

b So far, the sequence is 3, 5, 7, 9, ...
We are adding 2 matches each time, so the formula must involve $2n$.

The expression $2n$ generates the sequence 2, 4, 6, 8, ..., whereas the sequence we need is always 1 more than these values.

$\therefore$ there are $M = 2n + 1$ matches in the nth diagram.

Check: If $n = 1$, $M = 2(1) + 1 = 3$ ✓
If $n = 2$, $M = 2(2) + 1 = 5$ ✓
If $n = 3$, $M = 2(3) + 1 = 7$ ✓

EXERCISE 10E

1 Consider the pattern:

a Copy and complete:

Diagram	1	2	3	4
Number of matchsticks				

b Assuming the pattern continues, write a formula for the number of matchsticks M in the nth diagram.

c Hence predict the number of matchsticks in the 20th diagram.

2 Consider the pattern:

a Copy and complete:

Diagram	1	2	3	4
Number of matchsticks				

b Assuming the pattern continues, write a formula for the number of matchsticks M in the nth diagram.

c Hence predict the number of matchsticks in the 20th diagram.

3 Let S_n be the sum of the first n positive odd numbers, so $S_2 = 1 + 3$, $S_3 = 1 + 3 + 5$, and so on.

a Find the value of:

i S_2 **ii** S_3 **iii** S_4 **iv** S_5

b Assuming the pattern continues, write down a formula for S_n.

4 **a** Write down the value of:

i 2^0 **ii** 2^1 **iii** 2^2 **iv** 2^3 **v** 2^4 **vi** 2^5

b Find:

i $1 + 2$ **ii** $1 + 2 + 4$ **iii** $1 + 2 + 4 + 8$ **iv** $1 + 2 + 4 + 8 + 16$

c Write down a formula for the sum S_n of the first n terms in the sequence 1, 2, 4, 8, 16, ...

5 Consider the pattern: $S_1 = \frac{1}{1 \times 2}$, $S_2 = \frac{1}{1 \times 2} + \frac{1}{2 \times 3}$, $S_3 = \frac{1}{1 \times 2} + \frac{1}{2 \times 3} + \frac{1}{3 \times 4}$, ...

a Find the value of:

i S_1 **ii** S_2 **iii** S_3 **iv** S_4

b Assuming the pattern continues, write down a formula for S_n.

c Hence predict the value of S_{99}.

6 Consider the pattern: $S_1 = 1^2$, $S_2 = 1^2 + 2^2$, $S_3 = 1^2 + 2^2 + 3^2$, ...

a Verify that the formula $S_n = \frac{n(n+1)(2n+1)}{6}$ is true for $n = 1$, 2, 3, and 4.

b Assuming the formula in **a** is always true, evaluate $1^2 + 2^2 + 3^2 + 4^2 + 5^2 + ... + 100^2$.

QUICK QUIZ

MULTIPLE CHOICE QUIZ

REVIEW SET 10A

1 Write a formula for the bill $\$B$ at a restaurant if there is a charge of:

a \$15 for service, plus \$25 per person for 5 people

b $\$s$ for service, plus \$25 per person for p people

c $\$s$ for service, plus $\$m$ per person for p people.

2 The formula for the density D of a substance with mass M and volume V is $D = \frac{M}{V}$. Find:

a the density of lead given that 30 cm^3 of lead has mass 340 g

b the mass of a lump of uranium with density 18.97 g/cm^3 and volume 2 cm^3

c the volume of a length of timber with mass 6 kg and density 0.65 g/cm^3.

3 Make x the subject of:

a $mx + n = 3p$

b $\frac{x-4}{3} = y$

c $\frac{7}{y} = \frac{5}{x}$

d $F = \frac{c}{x-2}$

e $4x + a = bx + 5$

f $y = \frac{2x-3}{x-2}$

4 Make h the subject of:

a $T = \sqrt{h - l^2}$

b $G = \sqrt{\frac{5}{h+1}}$

5 Amy is trying to find pairs of numbers which have the same sum and product. In other words, she is looking for number pairs a and b such that $ab = a + b$.

a Rearrange this formula to make b the subject.

b Find b given $a = 3$. Check your answer by finding the sum and product of the numbers.

6 Consider the pattern:

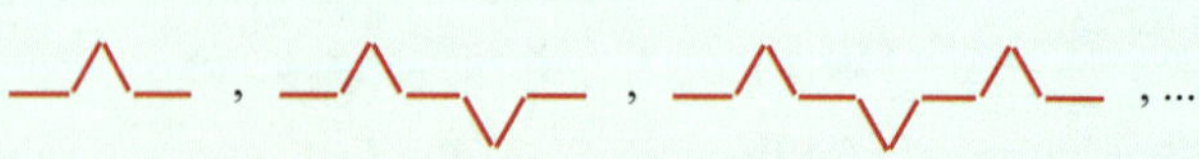

a Copy and complete:

Diagram	1	2	3	4
Number of matchsticks				

b Assuming the pattern continues, write a formula for the number of matchsticks M in the nth diagram.

c Hence predict the number of matchsticks in the 50th diagram.

7 The *coefficient of restitution* of a ball gives a measure of how "bouncy" the ball is.

It is given by the formula $C = \sqrt{\frac{h}{H}}$ where H is the drop height and h is the bounce height.

a Find the coefficient of restitution of each ball:

i

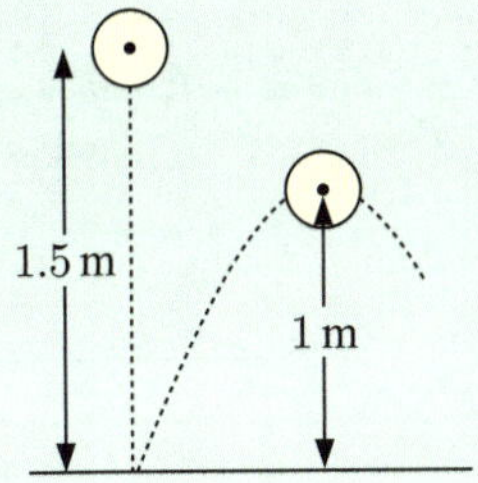

ii

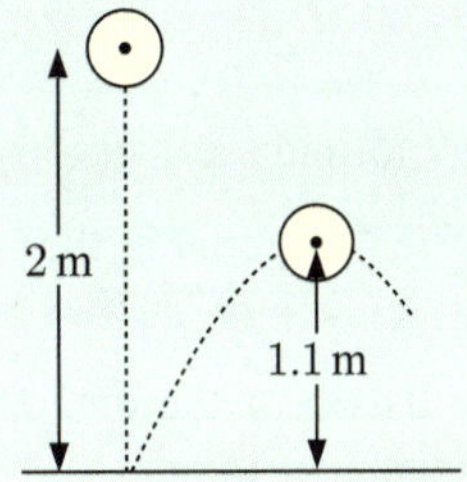

b Rearrange the formula to make h the subject.

c The coefficients of restitution of several balls dropped onto pavement are shown alongside.
Find how high each of these balls would bounce when dropped onto pavement from a height of 5 metres.

Tennis ball	0.728
Golf ball	0.858
Marble	0.657
Superball	0.893

8 Consider the pattern: $S_1 = \frac{1}{2^1}$, $S_2 = \frac{1}{2^1} + \frac{1}{2^2}$, $S_3 = \frac{1}{2^1} + \frac{1}{2^2} + \frac{1}{2^3}$, ...

a Find the value of:

i S_1 **ii** S_2 **iii** S_3 **iv** S_4

b Assuming the pattern continues, write down a formula for S_n.

c Hence predict the value of S_{10}.

REVIEW SET 10B

1 Write a formula for the number of edge pieces E (excluding corner pieces) in a:

a 3×5 jigsaw puzzle

b 4×8 jigsaw puzzle

c $m \times n$ jigsaw puzzle.

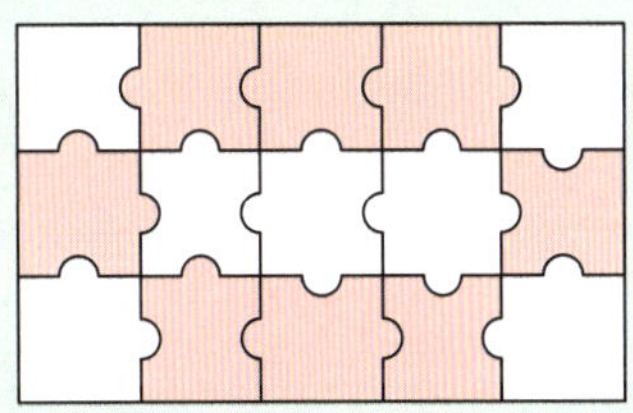

2

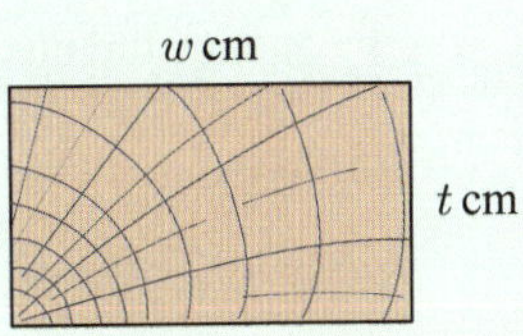

The strength S of a wooden beam is given by $S = 200w^2t$ units, where w cm is its width and t cm is its thickness. Find:

a the strength of a beam of width 16 cm and thickness 4 cm

b the width of a 5 cm thick beam of strength 60 000 units.

3 Make x the subject of:

a $B = dx - f$

b $\frac{Q}{\sqrt{x}} = \frac{t}{3}$

c $2(xy - 1) = 5x$

d $y = \frac{4x - 3}{3x + 2}$

e $P = 2x^2 - r, \; x < 0$

4 In a sideshow game, William is given two attempts to hole a golf putt. If he succeeds with either attempt then he wins a prize. Given that William has probability p of missing each putt, the expected number of games he needs to play in order to win a prize is $N = \dfrac{1}{1-p^2}$.

Rearrange this formula to make p the subject.

5 The length of the diagonal in the rectangular prism shown is given by $L = \sqrt{a^2+b^2+c^2}$.

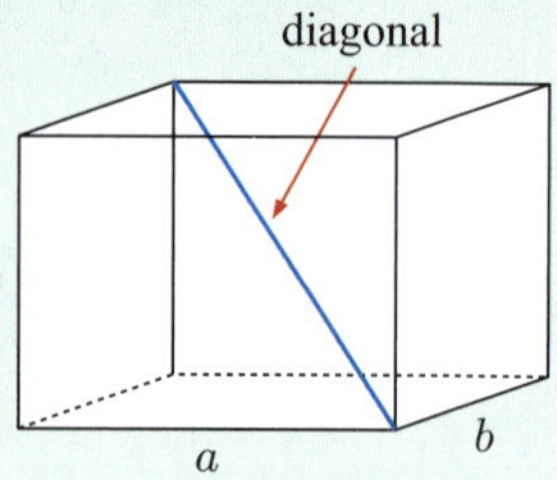

a Find the length of the diagonal in a 2 cm by 6 cm by 9 cm rectangular prism.

b Rearrange the formula to make c the subject.

c A rectangular prism has an 8 cm by 12 cm base, and a diagonal of length 17 cm. Find the height of the rectangular prism.

6 The electric current in a circuit with voltage E volts, resistance r ohms, and load resistance R ohms, is given by the formula $I = \dfrac{E}{r+R}$ amperes.

a Find the current in a circuit with voltage 24 volts, resistance 0.5 ohms, and load resistance 2.5 ohms.

b Rearrange the formula to make r the subject.

c Find the resistance of a circuit with current 1.5 amperes, voltage 7.725 volts, and load resistance 5 ohms.

7 Consider an object orbiting in a circle with radius r m about a much larger object with mass M kg.
Kepler's third law of planetary motion says that the *period* of this orbit is given by

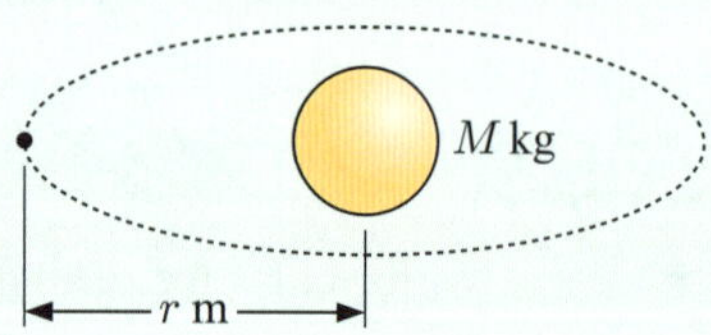

$$T = 2\pi\sqrt{\frac{r^3}{GM}} \text{ seconds}$$

where $G \approx 6.67 \times 10^{-11}$ is the universal gravitational constant.

a Given the average orbit radius of Mars is 2.28×10^{11} m and the mass of the Sun is about 1.99×10^{30} kg, find the period of Mars' orbit.

b Rearrange the formula to make r the subject.

c Write 1 year in seconds using scientific notation. Assume that 1 year $= 365.25$ days.

d 1 astronomical unit (AU) is defined as the average radius of Earth's orbit.
Show that 1 AU $\approx 1.50 \times 10^{11}$ m.

8 **a** Find:

i $1+2$ **ii** $1+2+3$ **iii** $1+2+3+4$

b Find:

i 1^3+2^3 **ii** $1^3+2^3+3^3$ **iii** $1^3+2^3+3^3+4^3$

c Suppose S_n is the sum of the first n positive integers, and C_n is the sum of the first n positive perfect cubes. Write a formula for C_n in terms of S_n.

Chapter 11

Measurement

Contents:

OPENING PROBLEM

Jacob wants to throw a water balloon at his brother. It is a sphere containing 1 litre of water.

Things to think about:

a What is the *volume* of the filled balloon?

b What is the *radius* of the water balloon?

c What is the *circumference* of the water balloon?

d What is the outer *surface area* of the water balloon?

A LENGTH AND PERIMETER

The base unit of **length** in the metric system is the **metre** (m). Other units of length based on the metre are connected as follows:

1 kilometre (km) = 1000 metres (m)
1 metre (m) = 100 centimetres (cm)
1 centimetre (cm) = 10 millimetres (mm)

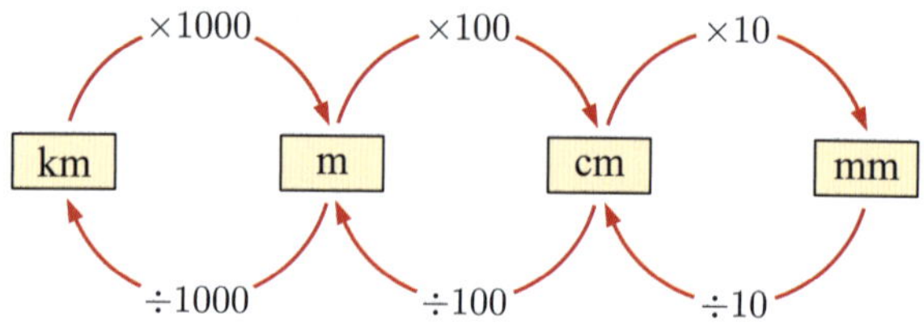

Example 1 Self Tutor

Convert:

a 6.32 km into m **b** 2350 cm into m **c** 32.8 m into mm.

a 6.32 km
$= (6.32 \times 1000)$ m
$= 6320$ m

b 2350 cm
$= (2350 \div 100)$ m
$= 23.5$ m

c 32.8 m
$= (32.8 \times 100 \times 10)$ mm
$= 32\,800$ mm

PERIMETER

The **perimeter** of a closed figure is the total distance around its **boundary**.

POLYGONS

The perimeter of a polygon is the sum of the lengths of its sides.

CIRCLES

In the case of the circle, the perimeter is given the special name **circumference**.

Circumference $C = 2\pi r$
or $C = \pi d$

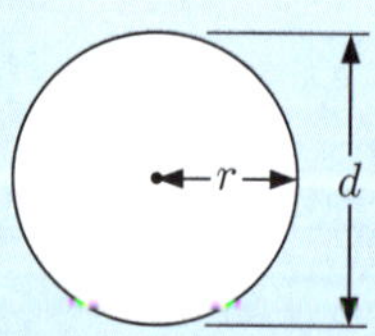

ARC LENGTH

The length of an arc is a fraction of the circumference of the circle.

Arc length $l = \frac{\theta}{360} \times 2\pi r$

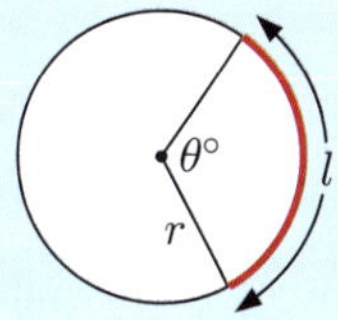

Example 2 | Self Tutor

Find the perimeter of each figure:

a

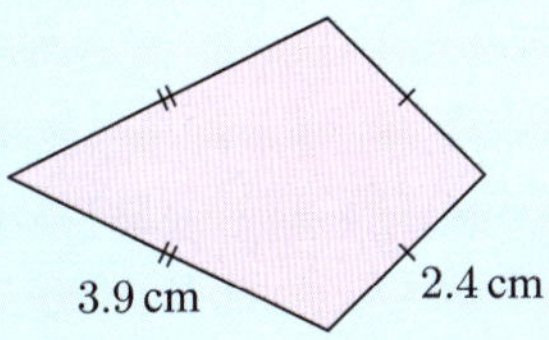

b

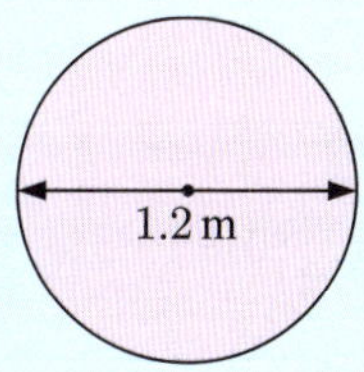

c

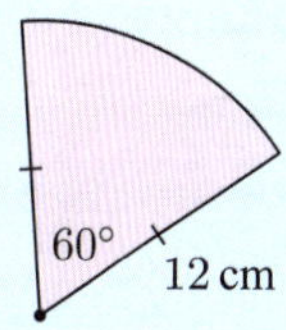

a $\text{Perimeter} = (2 \times 3.9 + 2 \times 2.4) \text{ cm}$
$= (7.8 + 4.8) \text{ cm}$
$= 12.6 \text{ cm}$

b $\text{Circumference} = \pi \times 1.2 \text{ m}$
$\approx 3.77 \text{ m}$

c $\text{Perimeter} = (12 + 12) \text{ cm} + \text{length of arc}$
$= 24 \text{ cm} + \left(\frac{60}{360}\right) \times 2 \times \pi \times 12 \text{ cm}$
$\approx 36.6 \text{ cm}$

EXERCISE 11A

1 Convert:

a 2.61 km into m
b 4300 mm into m
c 865 cm into m
d 700 mm into cm
e 11 500 m into km
f 3.675 m into cm
g 381 mm into m
h 68.2 cm into mm
i 5.67 km into cm
j 2860 cm into m
k 24 300 mm into m
l 0.328 km into cm.

2 Find the perimeter of each figure:

a

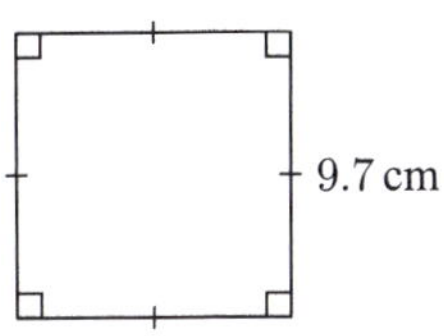

b

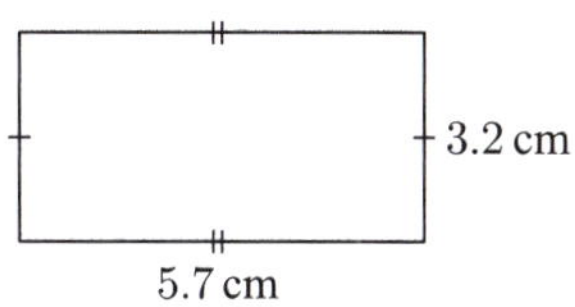

c

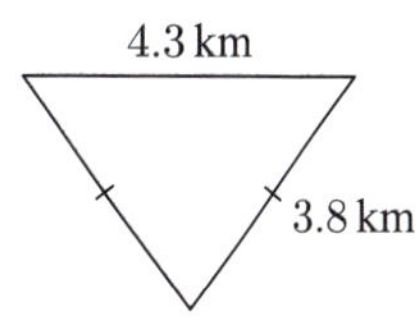

d

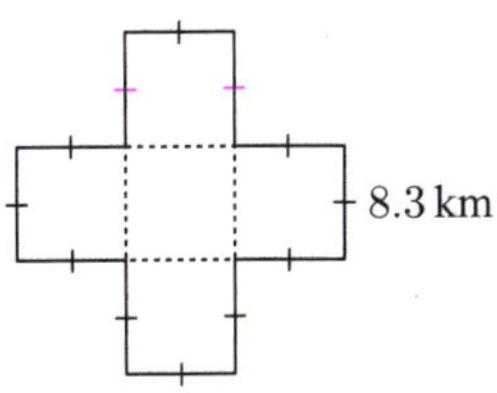

e

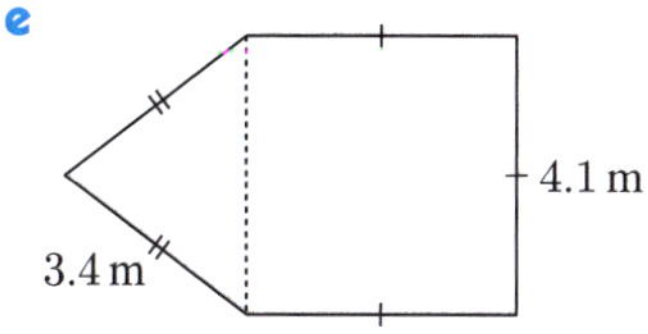

f

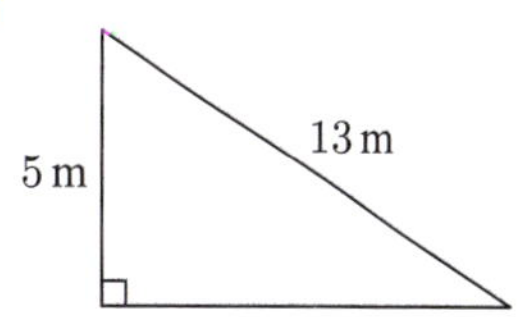

3 An athlete runs around a rectangular housing estate 10 times. The estate is 1.08 km by 420 m. How far has the athlete run?

4 An isosceles triangle has perimeter 30 cm. If the base is 7.8 cm long, find the length of the equal sides.

5 A rectangular swimming pool is 50 m long and 20 m wide. It has brick paving 3 m wide around it. Find the outer perimeter of the brick paving.

6 A rectangle is 16.4 cm by 11.8 cm and has the same perimeter as an equilateral triangle. Find the length of the sides of the triangle.

7 Calculate, correct to 3 significant figures, the circumference of a circle with:

a diameter 11 cm **b** radius 0.54 m.

8 Find the perimeter of:

a

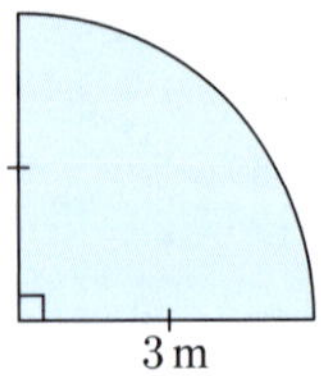

b

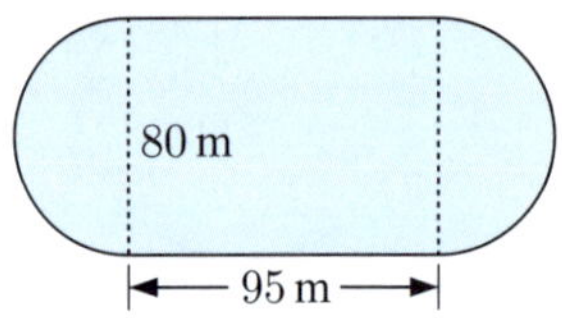

c

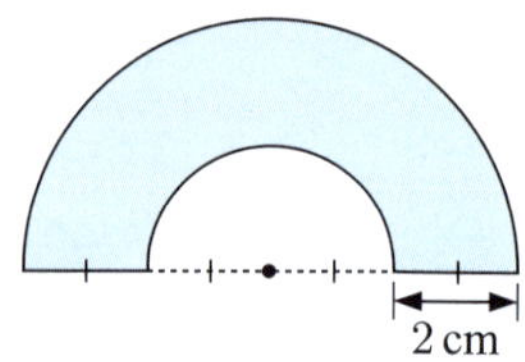

d

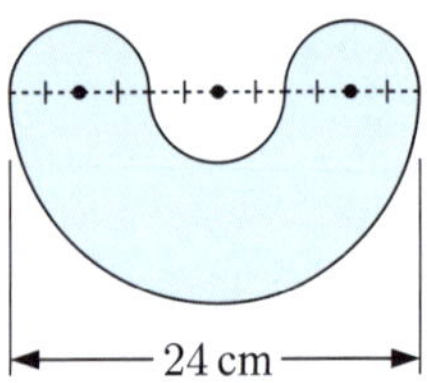

e

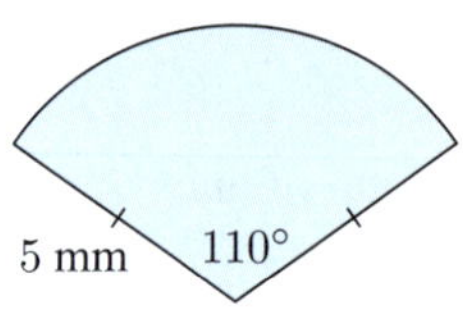

f

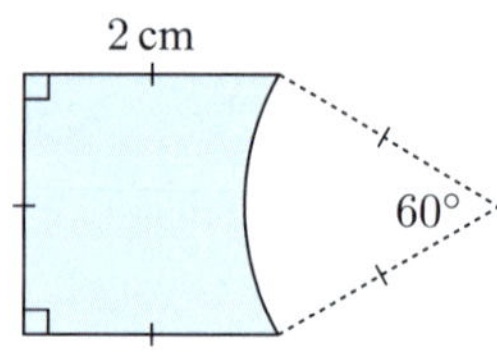

9 Find the diameter of a circle with circumference 400 metres.

10 Calculate the length of the arc of a circle if the radius is 12.5 cm and the angle at the centre is 60°.

11 Write a formula for the perimeter P of:

a

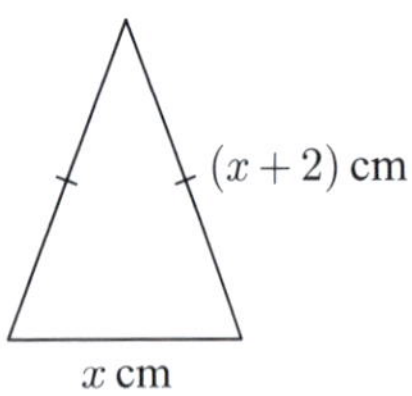

b

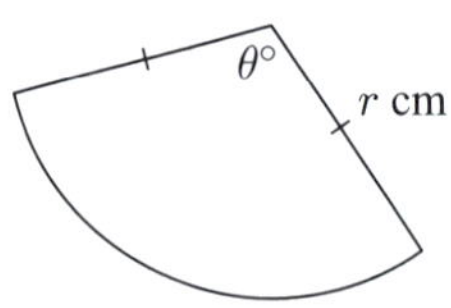

c

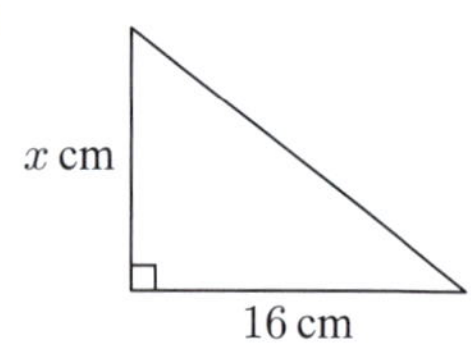

12 The second hand of a clock is 10 cm long. How far does the tip of the second hand travel in 20 seconds?

13 A rhombus has diagonals of length 8 cm and 10 cm. Find its perimeter.

14

Find the total length of ribbon used to tie the box given 25 cm of ribbon is required for the bow.

15 A tennis court has the dimensions shown.

a Find the perimeter of the court.

b Find the total length of all the marked lines, not including the net.

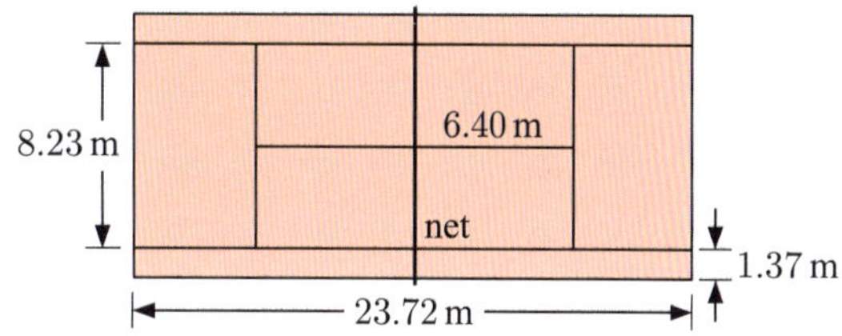

16

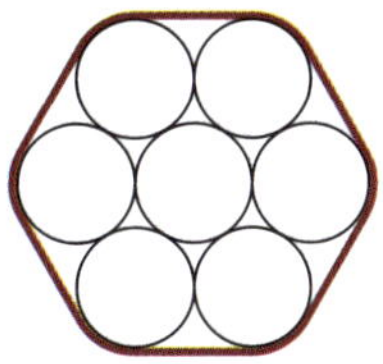

A thin plastic band is used to tie the seven plastic pipes together. If the band is 50 cm long, what is the radius of each pipe?

17 At Bushby Park there is a 5 m diameter circular pond which is surrounded by a 1 m wide garden bed, and then a 3 m wide lawn. A safety fence is placed around the lawn with posts every 3 m and a gateway 1.84 m wide.

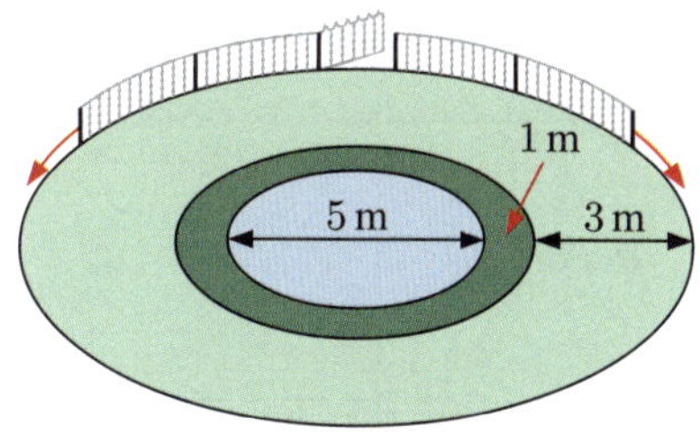

a Find the length of safety fence required. Do not include the gateway.

b How many posts are needed?

c If the posts cost \$15.75 and the safety fence costs \$18.35 per metre, calculate the total cost of the fence, excluding the gate.

GLOBAL CONTEXT — THE MINARD MAP

Global context: Orientation in space and time

Statement of inquiry: Infographics can provide deeper insight into historic events than numbers alone.

Criterion: Communicating

GLOBAL CONTEXT

B AREA

Area is the measurement of the size of a surface.

In the metric system, area is measured using the *squares* of the length units. In addition, we use another unit called a **hectare** (ha).

$$1 \text{ mm}^2 = 1 \text{ mm} \times 1 \text{ mm}$$
$$1 \text{ cm}^2 = 10 \text{ mm} \times 10 \text{ mm} = 100 \text{ mm}^2$$
$$1 \text{ m}^2 = 100 \text{ cm} \times 100 \text{ cm} = 10\,000 \text{ cm}^2$$
$$1 \text{ ha} = 100 \text{ m} \times 100 \text{ m} = 10\,000 \text{ m}^2$$
$$1 \text{ km}^2 = 1000 \text{ m} \times 1000 \text{ m} = 1\,000\,000 \text{ m}^2 \text{ or } 100 \text{ ha}$$

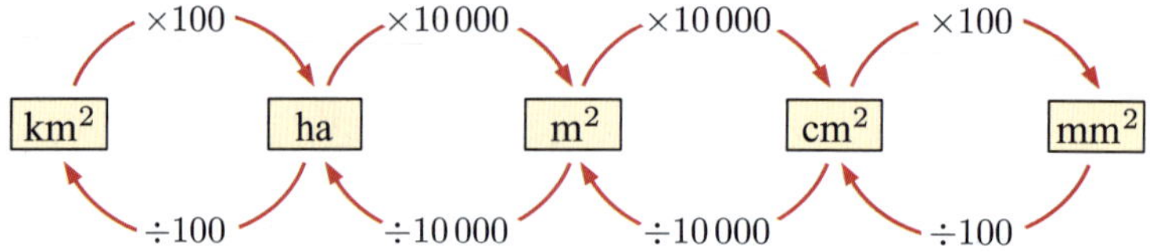

Example 3 — Self Tutor

Convert: **a** 4.8 m^2 into cm^2 **b** 4500 m^2 into ha.

a $4.8\text{ m}^2 = (4.8 \times 10\,000)\text{ cm}^2$
$= 48\,000\text{ cm}^2$

b $4500\text{ m}^2 = (4500 \div 10\,000)\text{ ha}$
$= 0.45\text{ ha}$

In previous years you should have seen the following area formulae:

Shape	*Figure*	*Formula*
Rectangle	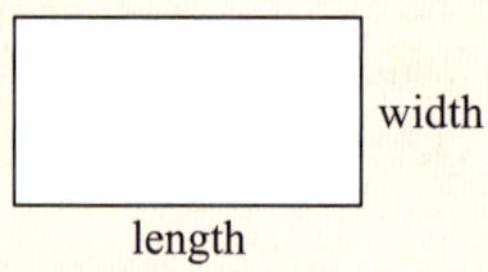	**Area = length × width**
Triangle	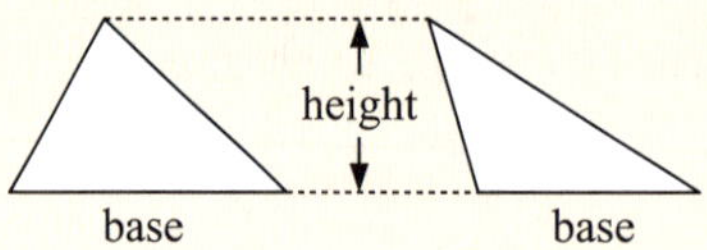	**Area =** $\frac{1}{2}$ **× base × height**
Parallelogram	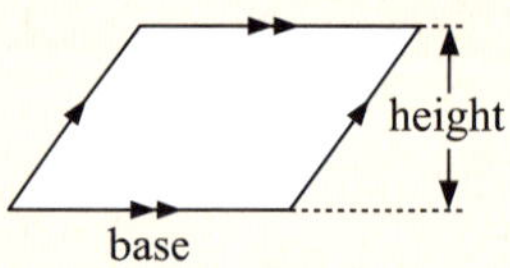	**Area = base × height**
Trapezium	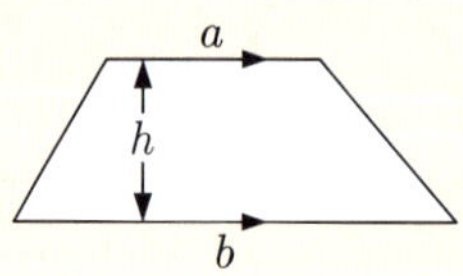	**Area =** $\left(\frac{a+b}{2}\right) \times h$
Kite	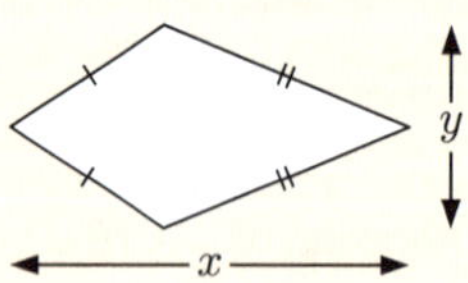	**Area =** $\frac{1}{2} \times x \times y$
Circle	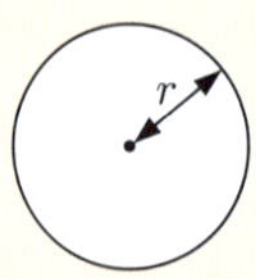	**Area =** πr^2
Sector	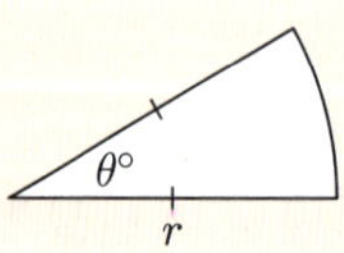	**Area =** $\frac{\theta}{360} \times \pi r^2$

DEMO

DEMO

DEMO

DEMO

Example 4 — Self Tutor

Find the area of each figure:

a
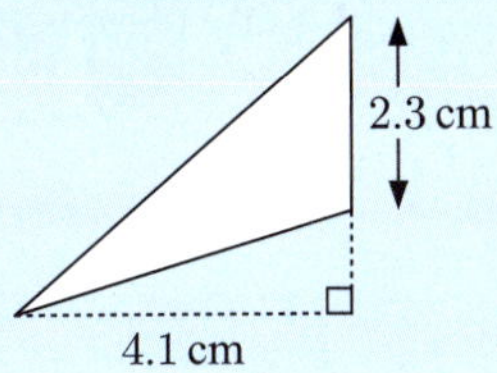

b
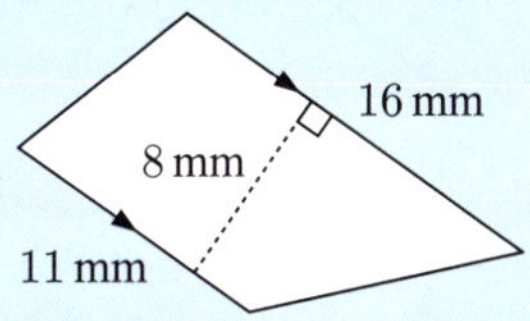

c
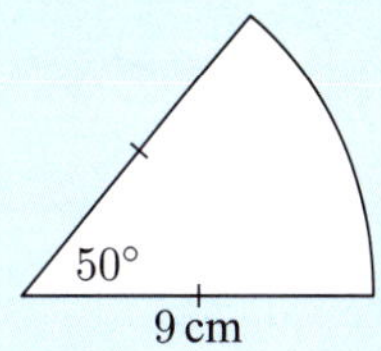

a Area
$= \frac{1}{2} \times \text{base} \times \text{height}$
$= \frac{1}{2} \times 2.3 \times 4.1 \text{ cm}^2$
$= 4.715 \text{ cm}^2$

b Area
$= \left(\frac{a+b}{2}\right) \times h$
$= \left(\frac{11+16}{2}\right) \times 8 \text{ mm}^2$
$= 108 \text{ mm}^2$

c Area
$= \frac{\theta}{360} \times \pi r^2$
$= \frac{50}{360} \times \pi \times 9^2 \text{ cm}^2$
$\approx 35.3 \text{ cm}^2$

EXERCISE 11B

1 Convert:

a 38 400 m^2 into ha
b 25.7 ha into m^2
c 3.5 m^2 into cm^2
d 500 000 cm^2 into m^2
e 18 cm^2 into mm^2
f 460 mm^2 into cm^2
g 9 km^2 into m^2
h 35 km^2 into ha
i 500 ha into km^2.

2 Find the area of each figure:

a
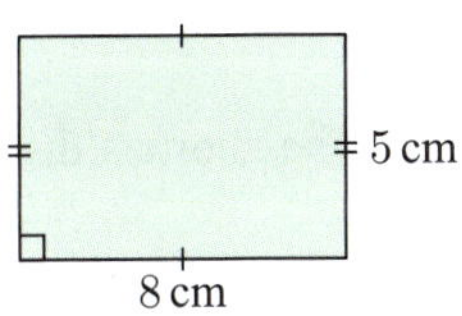

b
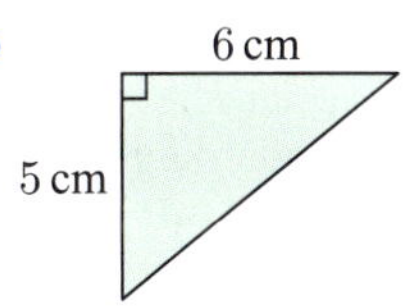

c
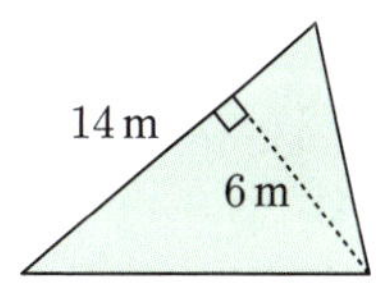

d
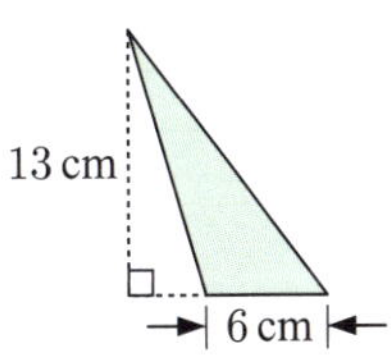

e
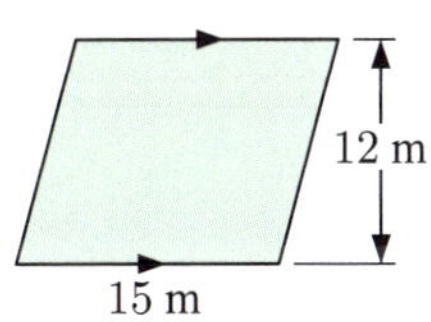

f
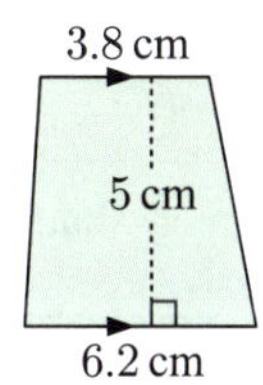

g
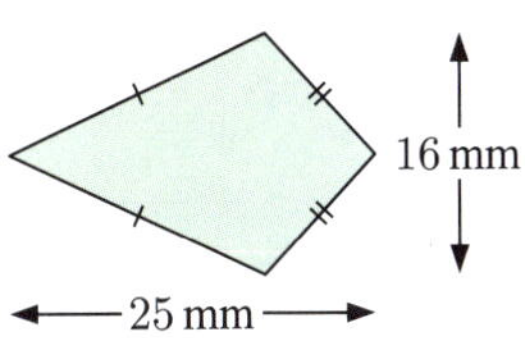

h
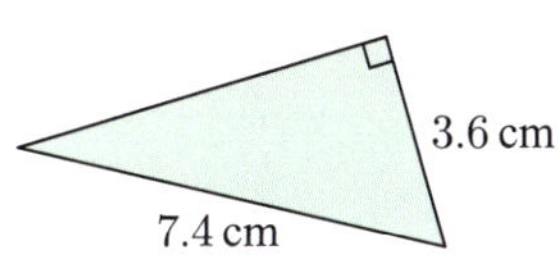

i
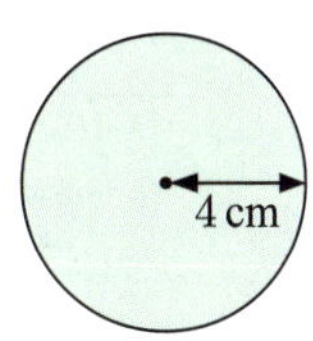

j
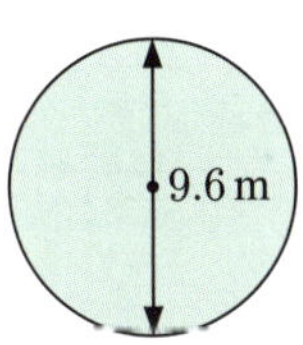

k
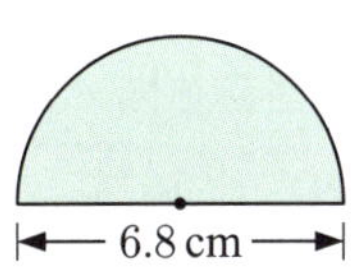

l
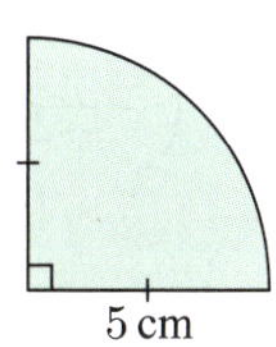

3 Find the area of each sector:

a
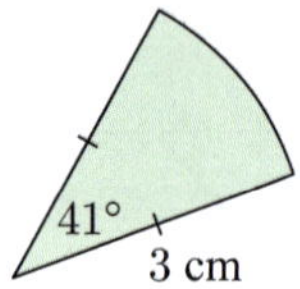

b
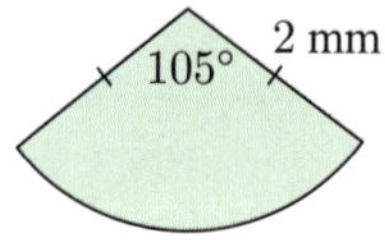

c
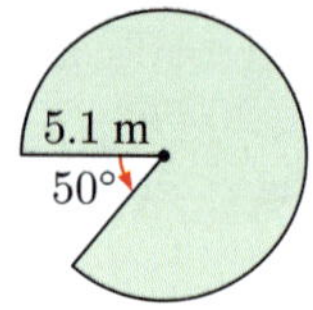

4 Find the value of x:

a
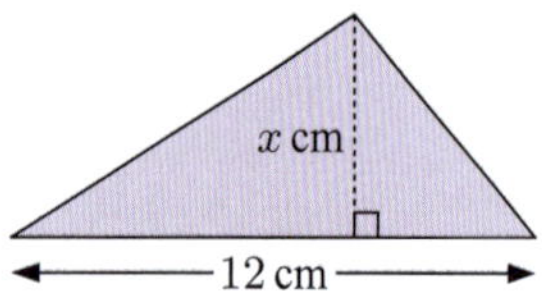

Area $= 102$ cm^2

b
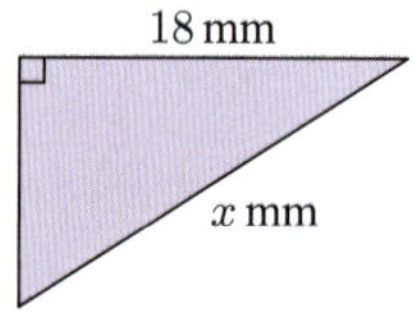

Area $= 99$ mm^2

c
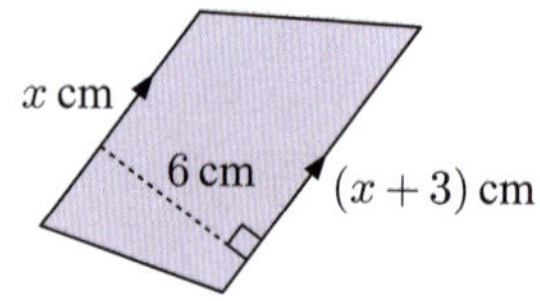

Area $= 57$ cm^2

5 An isosceles triangle has equal sides of length 8 cm and a base of length 6 cm. Find the area of the triangle.

6 Find the area of glass in a circular window with diameter 20 cm.

7 A rectangle with diagonals of length 20 cm has sides in the ratio $2:1$. Find the rectangle's:

a perimeter **b** area.

8 A cropduster can carry 1.2 tonnes of fertiliser at a time. A farmer needs to spread 28 kg of fertiliser per hectare over his 3.4 km by 2.1 km rectangular property. How many flights are needed?

9 Kite ABCD has $\text{AB} = \text{AD} = 6$ cm and $\text{BC} = \text{DC} = 9$ cm. The diagonal [BD] is 8 cm long. Find the area of the kite.

10 Parallelogram ABCD has $\text{AB} = 10$ cm and diagonal $\text{DB} = 15$ cm. The shortest distance from C to [AB] is 8 cm. Find the shortest distance from A to [DB].

11 An equilateral triangle has area $16\sqrt{3}$ cm^2. Find the length of its sides.

Example 5 **Self Tutor**

Find the green shaded area in each figure:

a
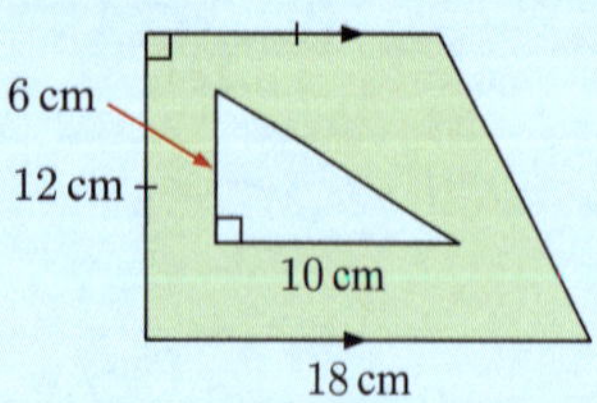

b
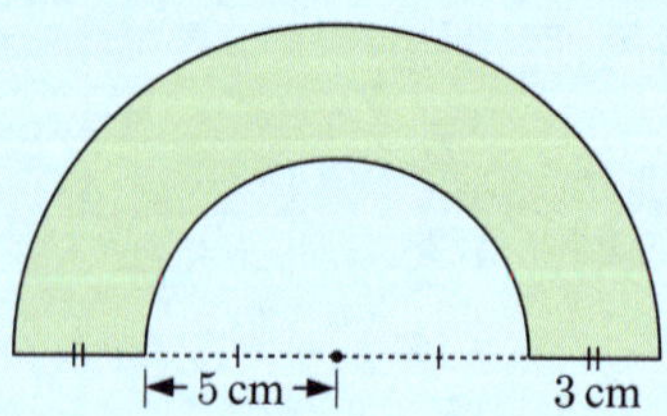

a Area = area of trapezium − area of triangle

$$= \left(\frac{12+18}{2}\right) \times 12 \text{ cm}^2 - \tfrac{1}{2} \times 10 \times 6 \text{ cm}^2$$

$$= (15 \times 12 - 5 \times 6) \text{ cm}^2$$

$$= 150 \text{ cm}^2$$

b Area = area of large semi-circle − area of small semi-circle

$= \frac{1}{2} \times (\pi \times 8^2)\ \text{cm}^2 - \frac{1}{2} \times (\pi \times 5^2)\ \text{cm}^2$

$\approx 61.3\ \text{cm}^2$

12 Find the shaded area:

a

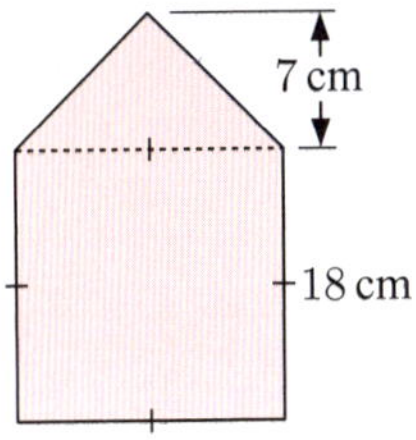

b

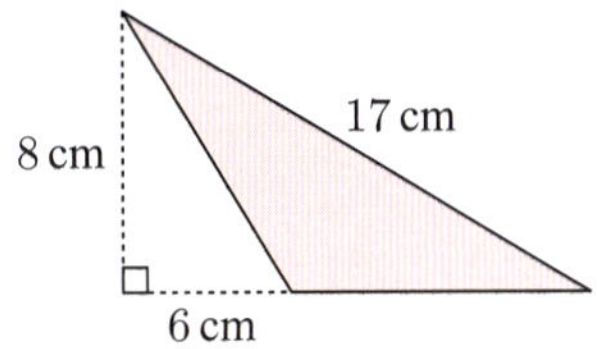

c

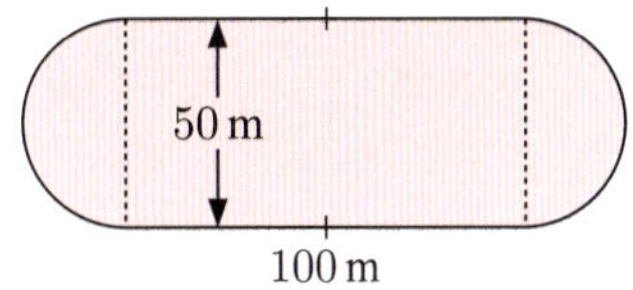

13 A manufacturer produces metal washers with external diameter 2 cm and internal diameter 1 cm.

a How many washers can be cut from a sheet of steel 3 m by 1 m?

b What percentage of the metal is left over?

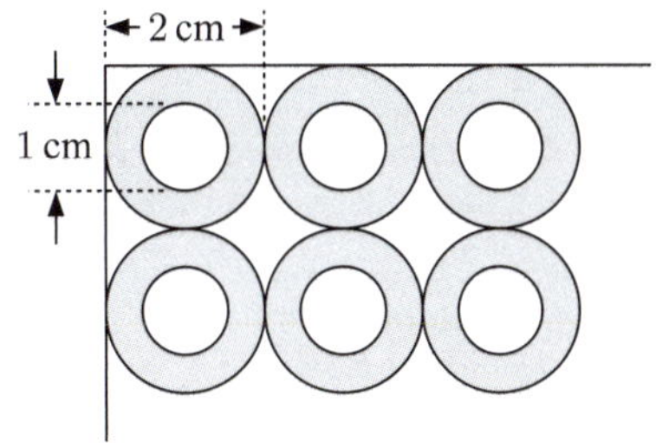

14 Find the perimeter and area of each figure:

a

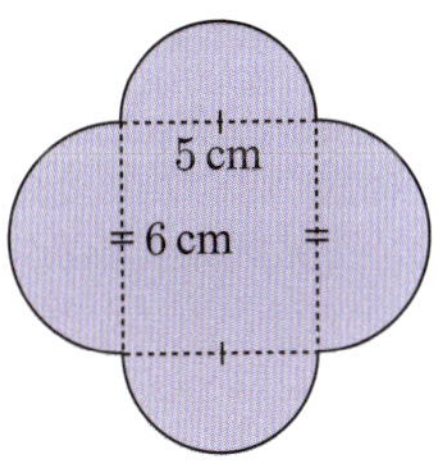

b

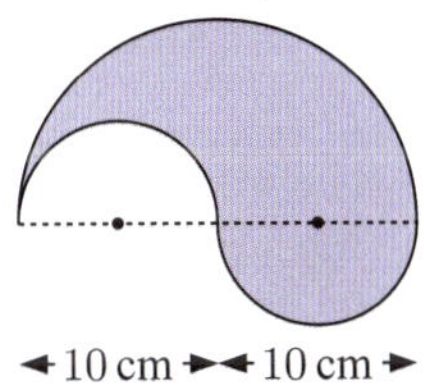

c

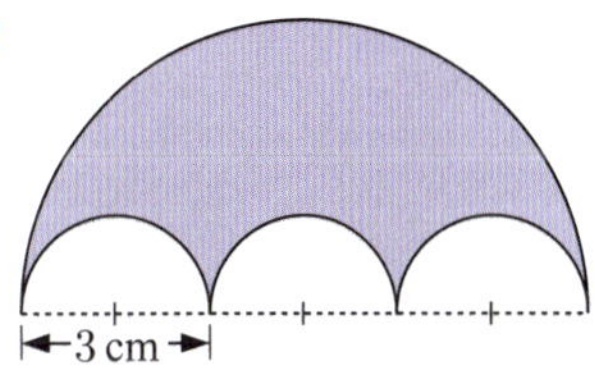

15 Write a formula for the area A of each figure:

a

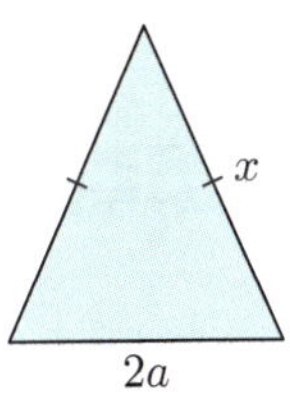

b

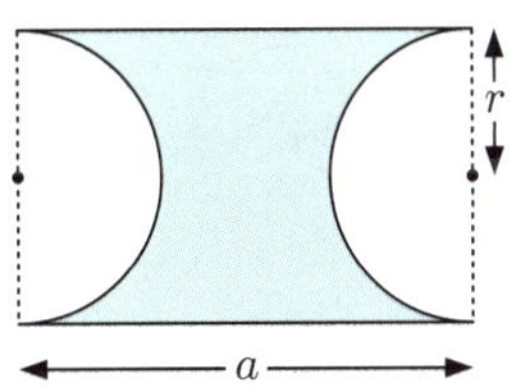

c

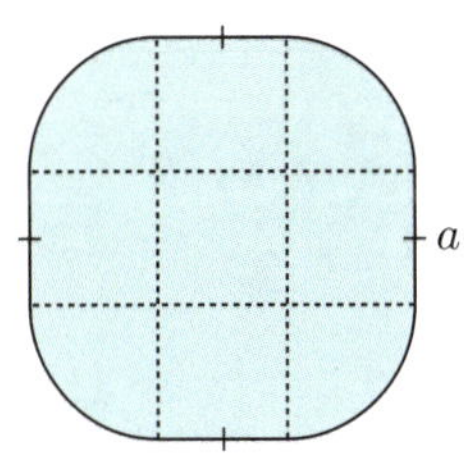

d

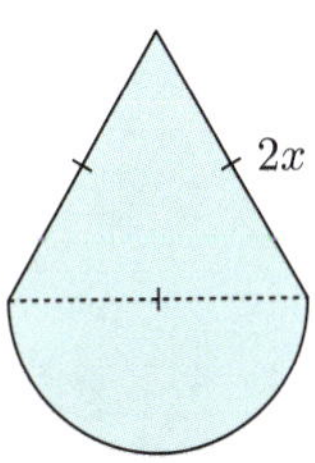

e

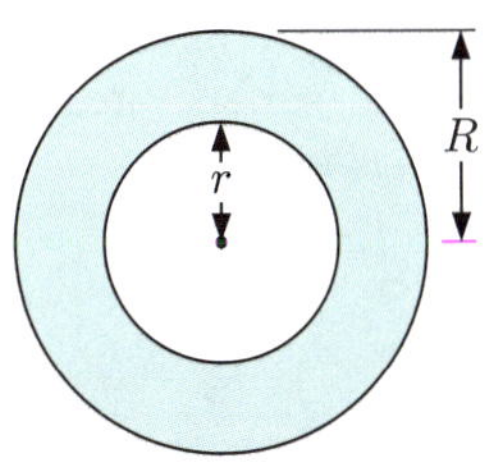

f

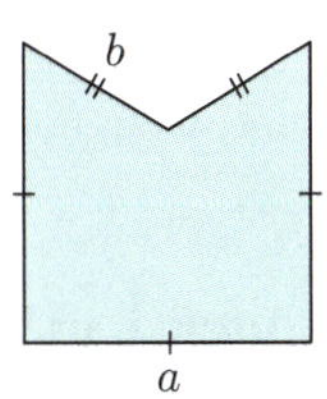

INVESTIGATION — HERON'S FORMULA

Heron or **Hero** of Alexandria was an important geometer and engineer of the first century AD. He invented machines such as the steam turbine, but is most famous for devising a formula for calculating the area of a triangle given the lengths of its three sides:

Area $A = \sqrt{s(s-a)(s-b)(s-c)}$

where $s = \dfrac{a+b+c}{2}$ is the **semi-perimeter** of the triangle.

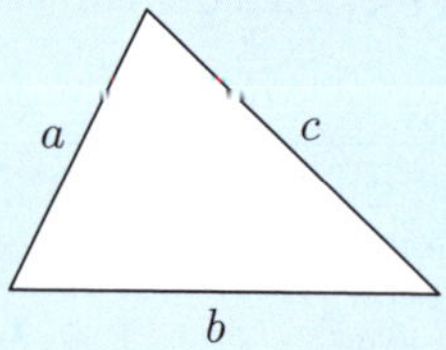

What to do:

1. Find the area of the triangle using:
   a. the formula $A = \frac{1}{2} \times$ base $\times$ height
   b. Heron's formula.

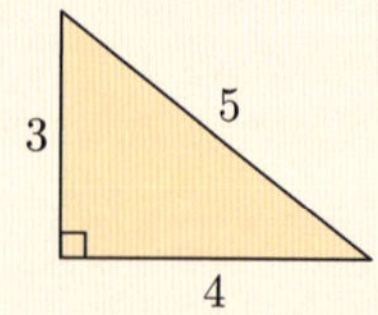

2. Use Heron's formula to find, correct to 1 decimal place, the area of:

a

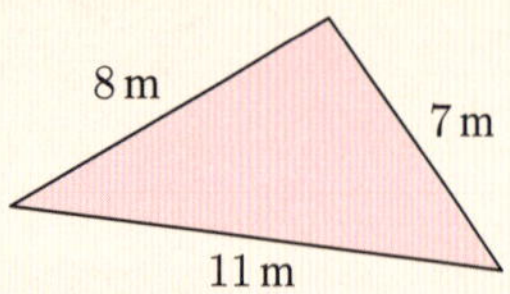

b

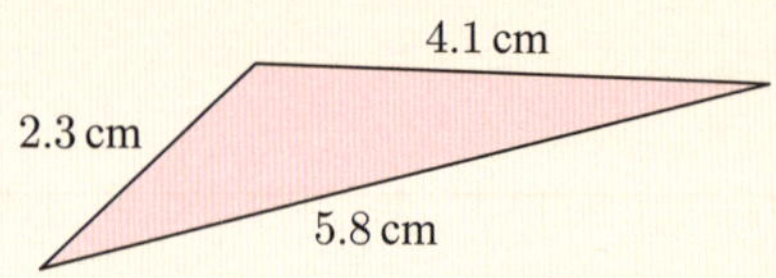

3. Use Heron's formula to find the area of a "triangle" with side lengths 5 m, 3 m, and 8 m. Comment on your result.

4. Find the area of an equilateral triangle with sides of length a cm using:
   a. the perpendicular height of the triangle
   b. Heron's formula.

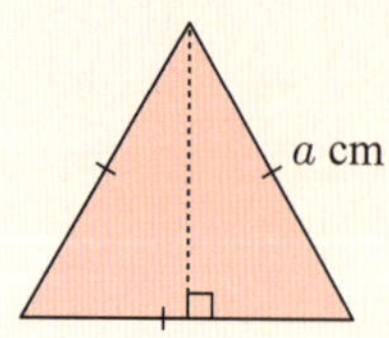

C SURFACE AREA

The **surface area** of a solid is the total area of its surfaces.

In general, we *assume* this definition refers to *exterior* surfaces. So, if we are talking about the surface area of a hollow pipe, we ignore the exposed *interior* surface.

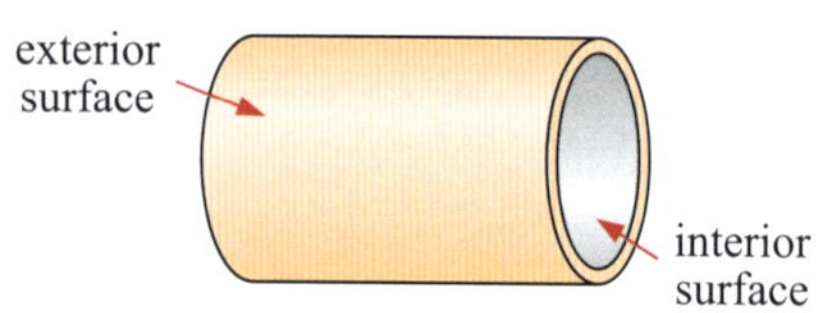

SOLIDS WITH PLANAR FACES

To find the surface area of a solid with flat or **planar** faces, we add the areas of the faces.

Example 6 — Self Tutor

Find the surface area of:

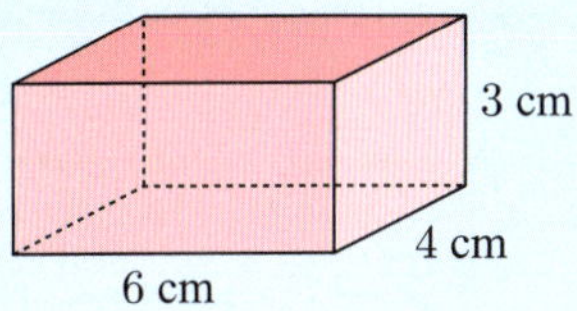

The figure has 2 faces which are $6 \text{ cm} \times 4 \text{ cm}$,
2 faces which are $6 \text{ cm} \times 3 \text{ cm}$,
and 2 faces which are $3 \text{ cm} \times 4 \text{ cm}$.

Surface area
$= (2 \times 6 \times 4 + 2 \times 6 \times 3 + 2 \times 3 \times 4) \text{ cm}^2$
$= (48 + 36 + 24) \text{ cm}^2$
$= 108 \text{ cm}^2$

EXERCISE 11C.1

1 Find the surface area of:

a

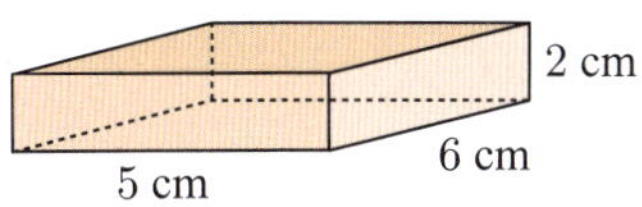

b

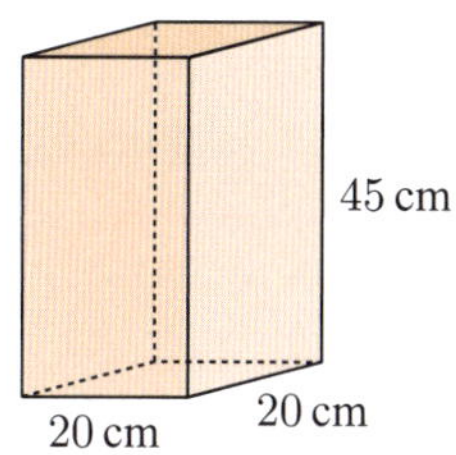

c

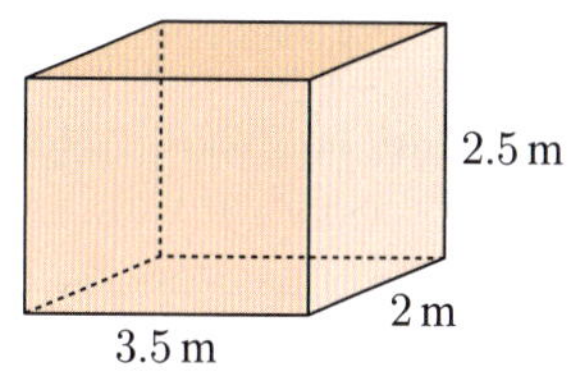

Example 7 — Self Tutor

Find the surface area of:

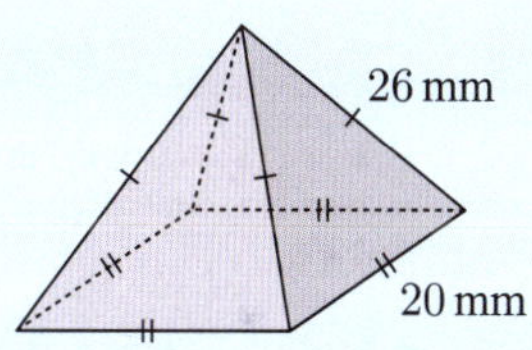

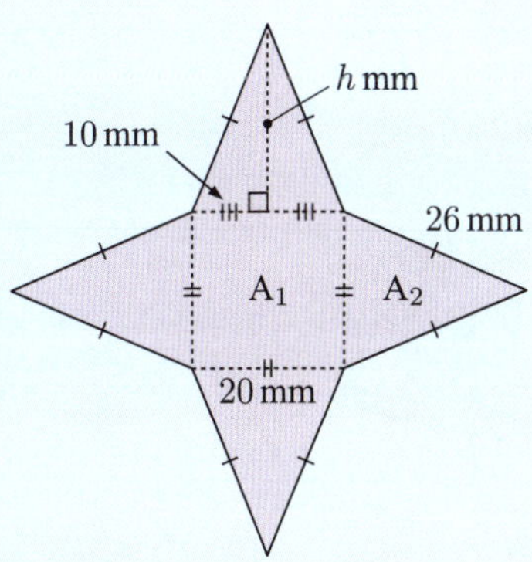

Let the height of the triangular faces be h mm.

$h^2 + 10^2 = 26^2$ {Pythagoras}

$\therefore h^2 + 100 = 676$

$\therefore h^2 = 576$

$\therefore h = 24$ {since $h > 0$}

Surface area $= A_1 + 4 \times A_2$
$= \left(20 \times 20 + 4 \times \frac{1}{2} \times 20 \times 24\right) \text{ mm}^2$
$= (400 + 960) \text{ mm}^2$
$= 1360 \text{ mm}^2$

2 Find the surface area of:

a

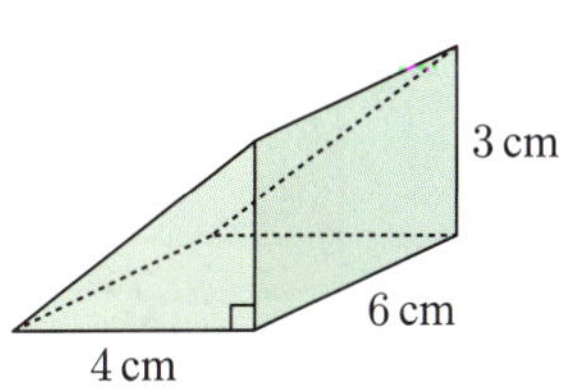

b

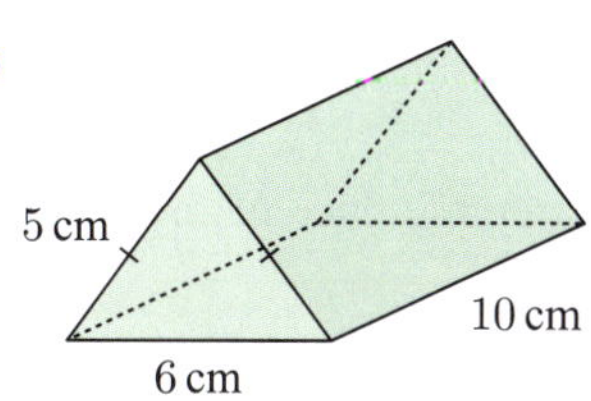

c

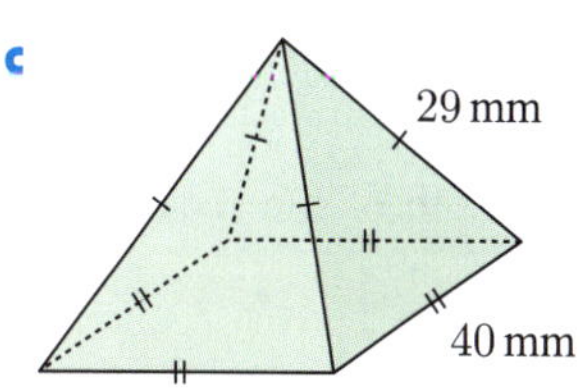

3 Find the surface area of:

a

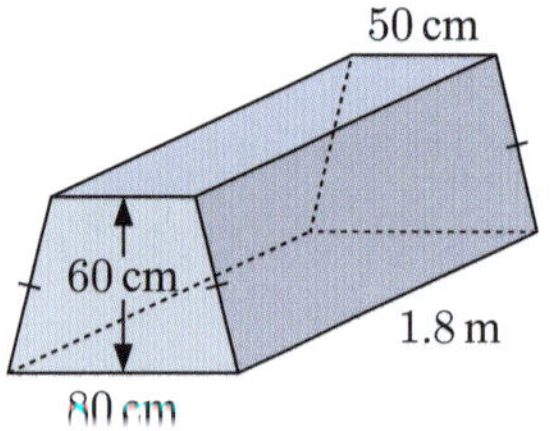

b

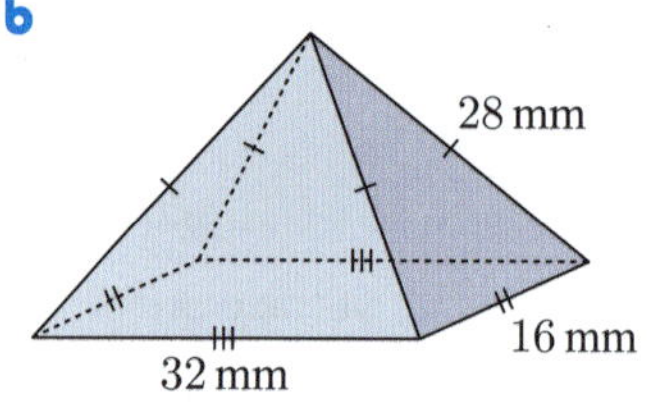

c

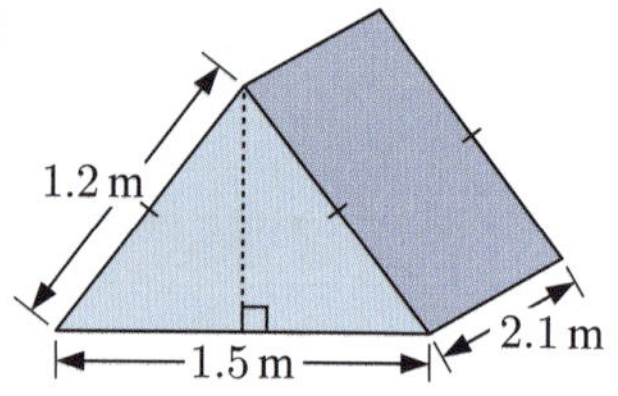

4 Write a formula for the surface area A of:

a

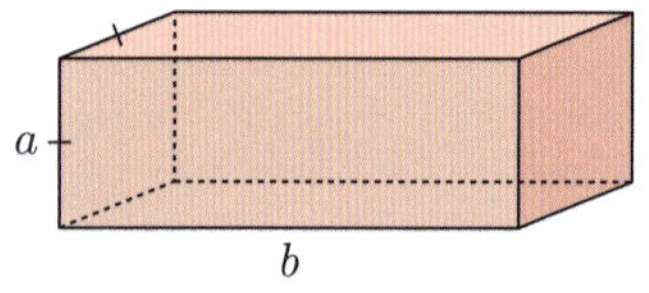

b

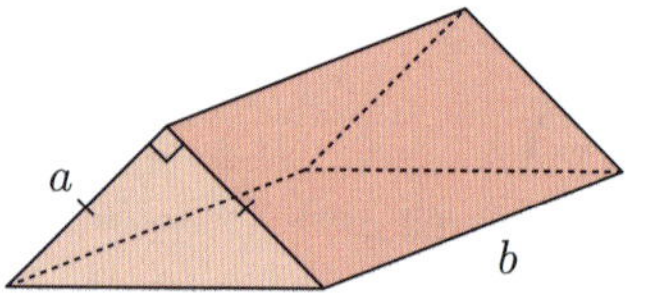

5 Mark has made a cardboard box to post a painting overseas. The box is 60 cm wide and 10 cm deep.

Given that Mark used 1.24 m^2 of cardboard, find the height of the box.

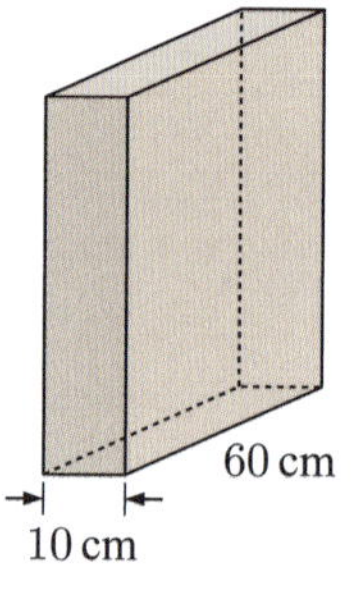

6

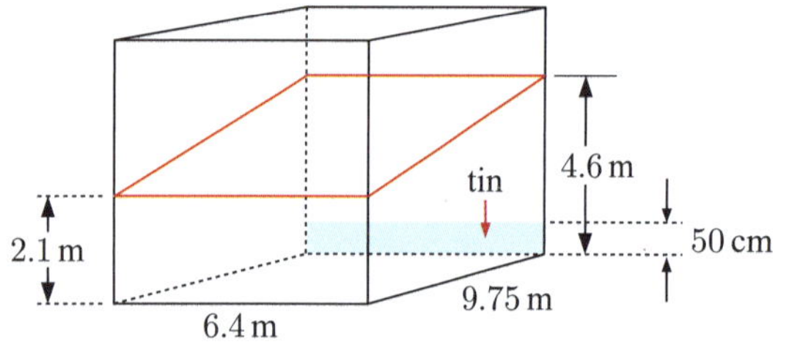

A squash court has the dimensions given. Each shot may strike the floor, or the walls below the red line, except for the tin on the front wall. Calculate the total surface playing area of the walls and the floor.

7 All of the edges of an eight-sided die are 2 cm long. Find the surface area of the die.

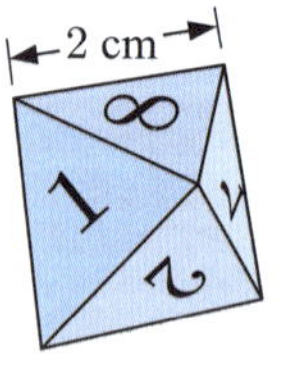

CYLINDERS

This cylinder with radius r and height h has no top or bottom.

If the cylinder is cut, opened out, and flattened, it takes the shape of a rectangle.

The length of the rectangle is the circumference of the cylinder, which is $2\pi r$.

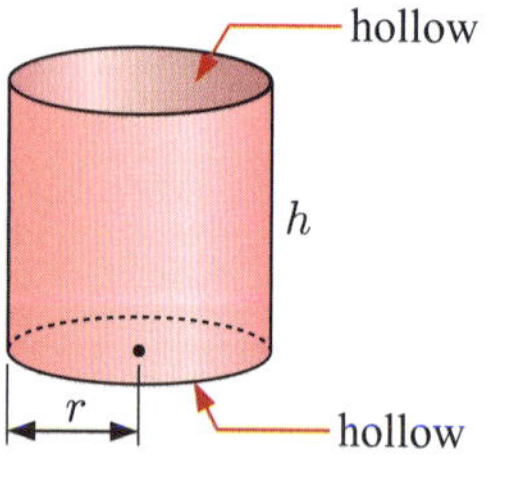

So, the area of the *curved surface* of a cylinder is

$$\begin{aligned} A &= \text{area of rectangle} \\ &= \text{length} \times \text{width} \\ &= 2\pi r h \end{aligned}$$

DEMO

Hollow cylinder (no ends)	**Open cylinder** (one end)	**Solid cylinder** (two ends)
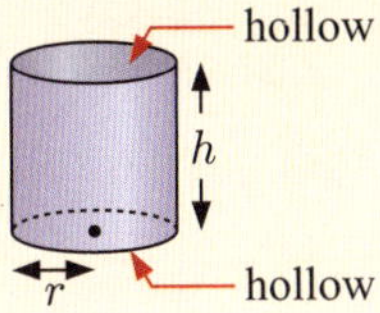	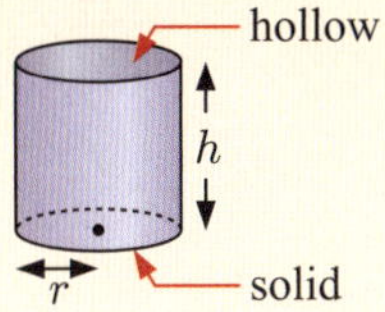	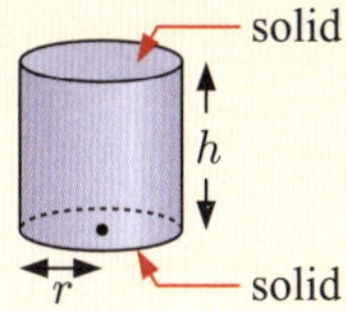
$A = 2\pi r h$	$A = 2\pi r h + \pi r^2$	$A = 2\pi r h + 2\pi r^2$

SPHERES

Archimedes of Syracuse (287 BC - 212 BC) was first to prove a formula for the surface area of a sphere. He proved that the surface area of a sphere is equal to the area of the curved surface of the cylinder that exactly encloses it.

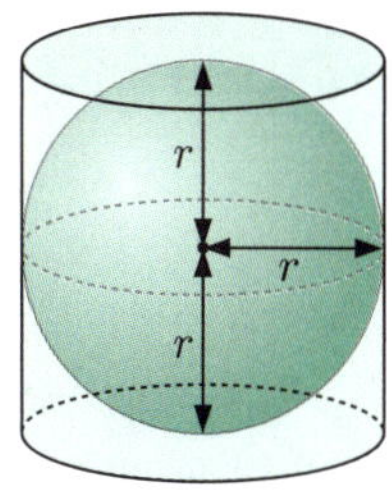

The cylinder exactly enclosing this sphere has radius r and height $2r$, so its surface area $= 2\pi r \times 2r$
$= 4\pi r^2$

$$A = 4\pi r^2$$

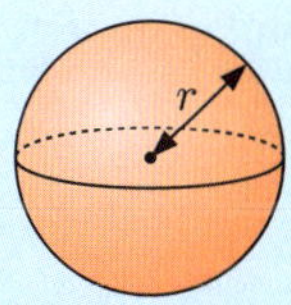

Example 8 **Self Tutor**

Find the surface area of each solid:

a

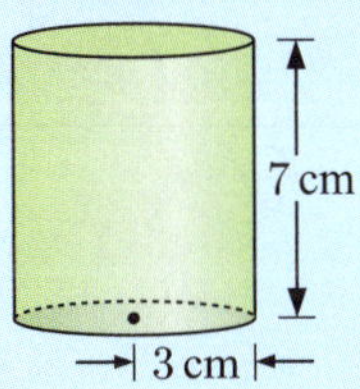

b

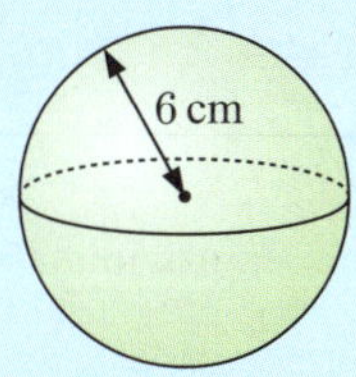

a Surface area
$= 2\pi r h + 2\pi r^2$
$= (2 \times \pi \times 3 \times 7 + 2 \times \pi \times 3^2)\ \text{cm}^2$
$\approx 188\ \text{cm}^2$

b Surface area
$= 4\pi r^2$
$= 4 \times \pi \times 6^2\ \text{cm}^2$
$\approx 452\ \text{cm}^2$

CONES

The curved surface of a cone is made from a sector of a circle. The radius of the sector is equal to the slant height s of the cone. The arc length AB of the sector is equal to the circumference of the base.

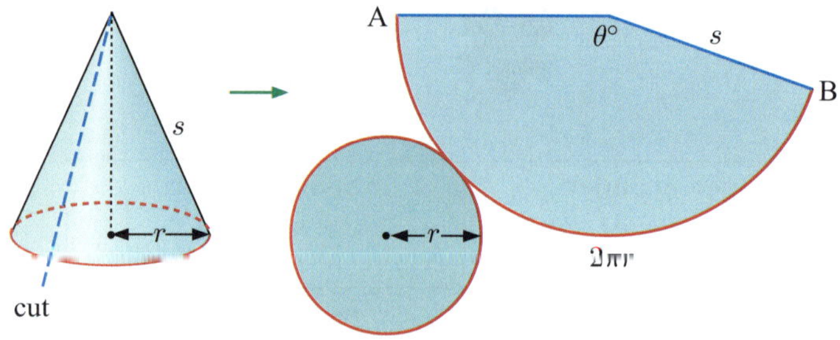

$$\text{arc AB} = 2\pi r$$

$$\therefore \left(\frac{\theta}{360}\right) 2\pi s = 2\pi r$$

$$\therefore \frac{\theta}{360} = \frac{r}{s}$$

$\therefore$ the area of the curved surface

$= \text{the area of the sector}$

$= \left(\frac{\theta}{360}\right) \pi s^2$

$= \frac{r}{s} \times \pi s^2$

$= \pi r s$

Hollow cone
(no end)

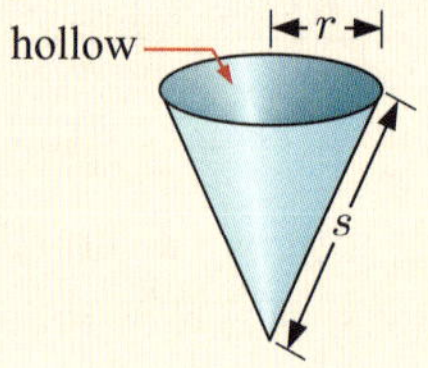

$\boldsymbol{A = \pi r s}$

Solid cone
(closed end)

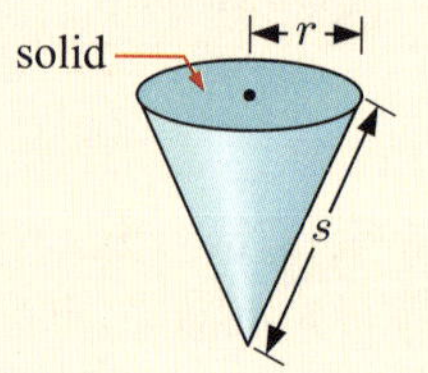

$\boldsymbol{A = \pi r s + \pi r^2}$

Example 9 — Self Tutor

Find the surface area of a solid cone with base radius 5 cm and height 12 cm.

Let the slant height be s cm.

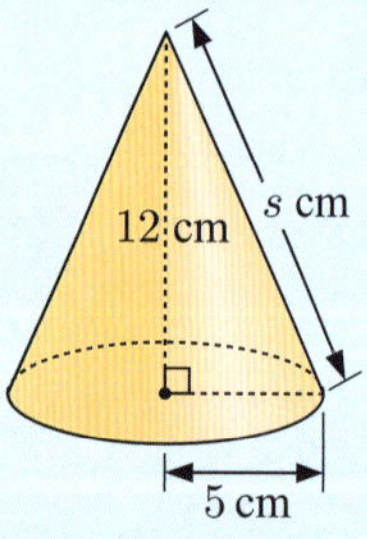

$s^2 = 5^2 + 12^2$ {Pythagoras}

$\therefore s^2 = 169$

$\therefore s = \sqrt{169} = 13$ {since $s > 0$}

Now $A = \pi r^2 + \pi r s$

$\therefore A = \pi \times 5^2 + \pi \times 5 \times 13$

$\therefore A = 90\pi$

$\therefore A \approx 283$

The surface area is approximately 283 cm^2.

EXERCISE 11C.2

1 Find the surface area of:

a hollow pipe

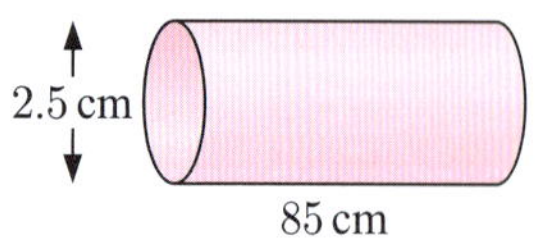

b open top

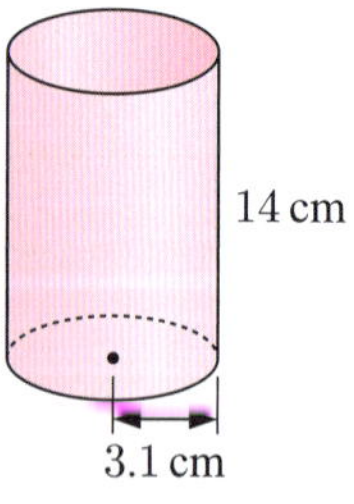

c solid

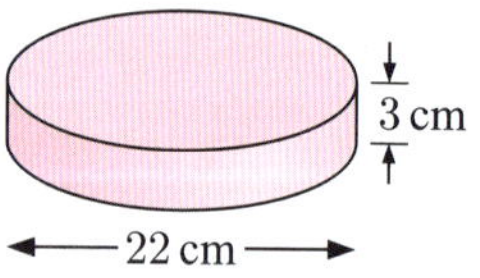

2 Find the surface area of:

a

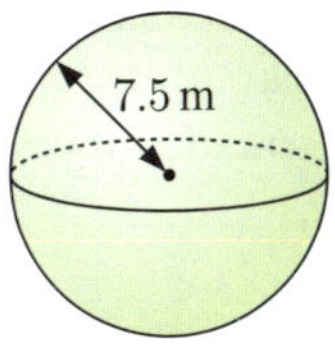

b

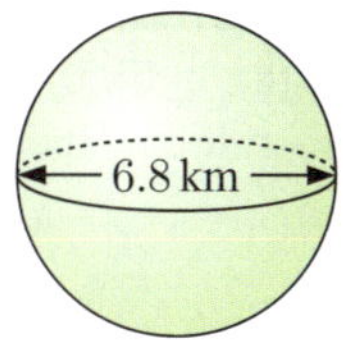

c

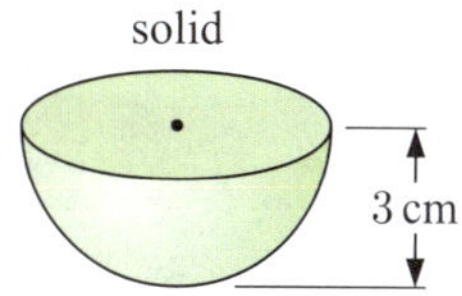

3 Find the surface area of:

a solid

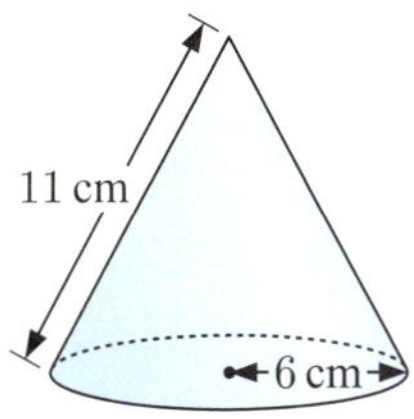

b hollow

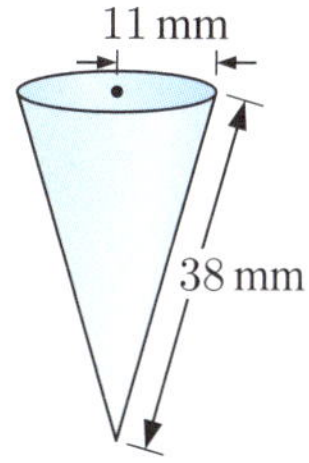

c solid

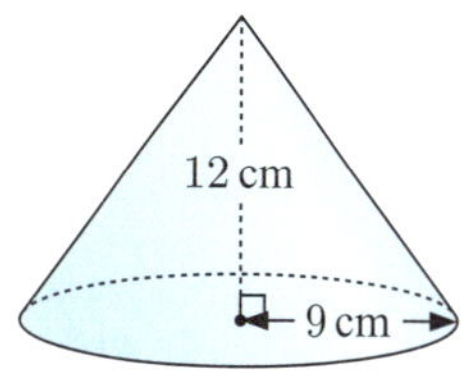

4 Find the total surface area of:

a a sphere with radius 6 cm

b a hemisphere with radius 10 m

c a solid cylinder with base radius 9 cm and height 20 cm

d a solid cone with base radius and perpendicular height both 10 cm.

5 A conical piece of filter paper has base radius 2 cm and height 5 cm.

a Find the slant height s.

b Hence find the outer surface area of the filter paper.

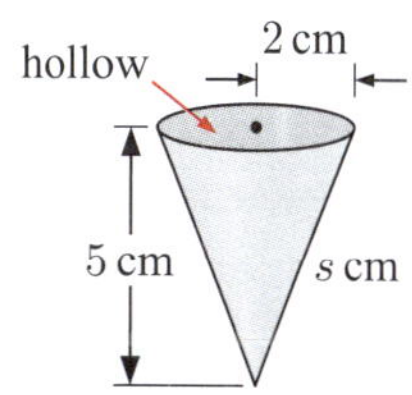

6 A cylindrical tank is 6 m high and has diameter 8 m. Its outer surface (including both ends) is to be painted bright red. Each litre of paint covers 5 m^2. It is purchased in 5 litre cans costing $52.50 each.

a Find the surface area to be painted.

b Find the number of cans of paint which must be purchased.

c Find the cost of the paint.

7 Find:

a the radius of a sphere with surface area 400 m^2

b the height of a solid cylinder with radius 10 cm and surface area 2000 cm^2

c the slant height of a solid cone with base radius 8 m and surface area 850 m^2.

8 We commonly use a sphere to model the Earth, even though it is not a *perfect* sphere. The Earth has a radius of approximately 6400 km.

a Estimate the surface area of the Earth.

b 71% of the Earth's surface is covered by water. Estimate this area.

c China has a land area of 9 706 961 km^2.

i What percentage of the surface area of the Earth is China?

ii What percentage of the *land* area of the Earth is China?

9 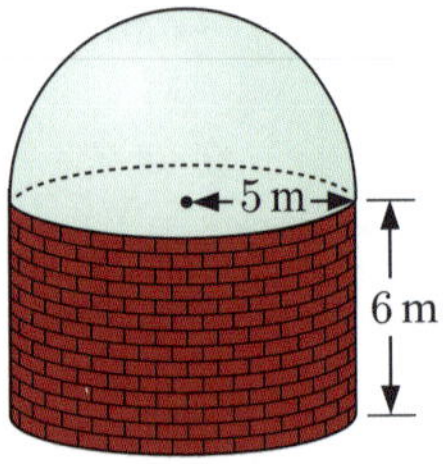

The outside of this observatory needs cleaning. The cost of cleaning the hemispherical glass roof is \$4 per square metre, and the cost of cleaning the cylindrical walls is \$2.50 per square metre. Find, to the nearest dollar, the total cost of cleaning the observatory.

10 Write a formula for the surface area A of each solid:

a 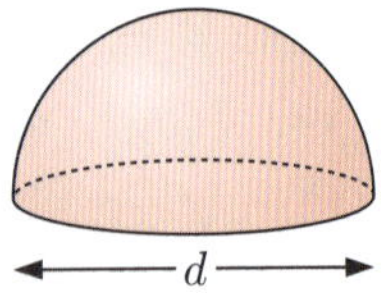

b 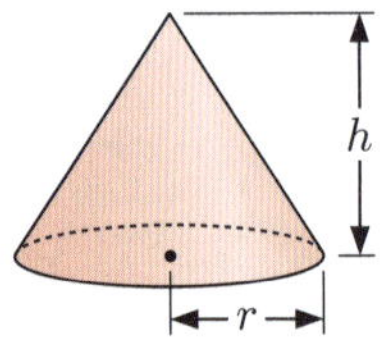

c 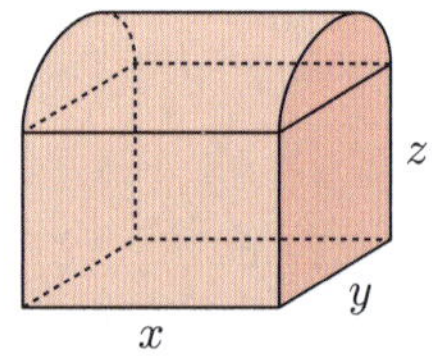

11 A wedge with angle 20° is cut from the centre of a cylindrical cake with radius 13 cm and height 6 cm. Each side of the wedge, excluding the bottom, is to be covered with icing.
Find the total surface area of cake to be iced.

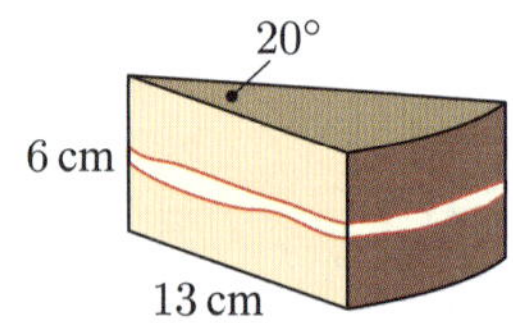

D VOLUME

The **volume** of a solid is the amount of space it occupies.

In the metric system, volume is measured using the *cubes* of the length units.

- 1 cm^3
  $= 10\text{ mm} \times 10\text{ mm} \times 10\text{ mm}$
  $= 1000\text{ mm}^3$
- 1 m^3
  $= 100\text{ cm} \times 100\text{ cm} \times 100\text{ cm}$
  $= 1\,000\,000\text{ cm}^3$

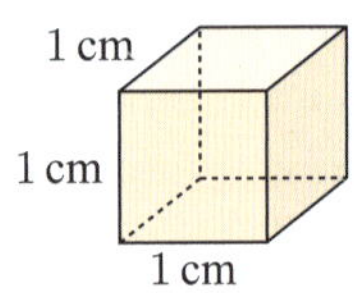

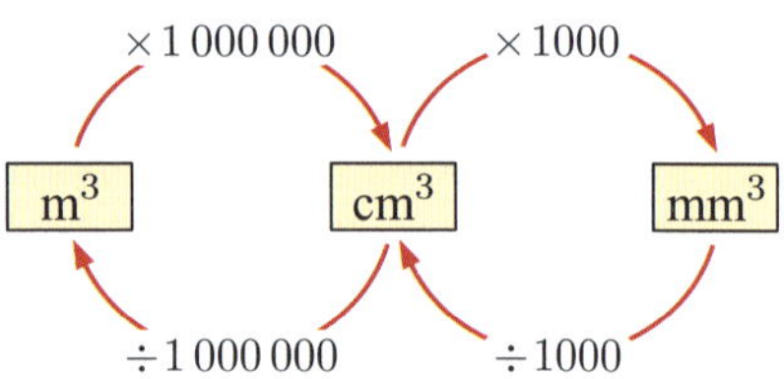

Example 10 **Self Tutor**

Convert: **a** 5 m^3 into cm^3 **b** $25\,000\text{ mm}^3$ into cm^3.

a 5 m^3
$= (5 \times 1\,000\,000)\text{ cm}^3$
$= 5\,000\,000\text{ cm}^3$

b $25\,000\text{ mm}^3$
$= (25\,000 \div 1000)\text{ cm}^3$
$= 25\text{ cm}^3$

EXERCISE 11D.1

1 Convert:

a 8.65 cm^3 into mm^3 **b** $86\,000\text{ mm}^3$ into cm^3 **c** $300\,000\text{ cm}^3$ into m^3

d 124 cm^3 into mm^3 **e** 300 mm^3 into cm^3 **f** 3.7 m^3 into cm^3.

2 A manufacturer of erasers has 2.4 m^3 of rubber. Each eraser has volume 8 cm^3. How many erasers can be made?

3 A manufacturer of lead fishing sinkers has 0.237 m^3 of lead. Each sinker has a volume of 5 cm^3. How many sinkers can be made?

SOLIDS OF UNIFORM CROSS-SECTION

All solids of uniform cross-section contain two identical end faces.

Volume of solid of uniform cross-section
= area of end × length

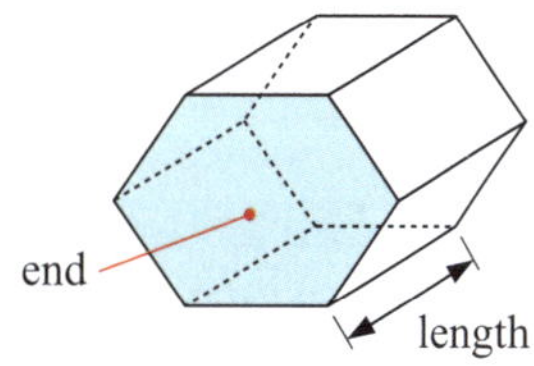

We can use more specific formulae for some special solids of uniform cross-section.

- **Volume of rectangular prism = length × width × height**

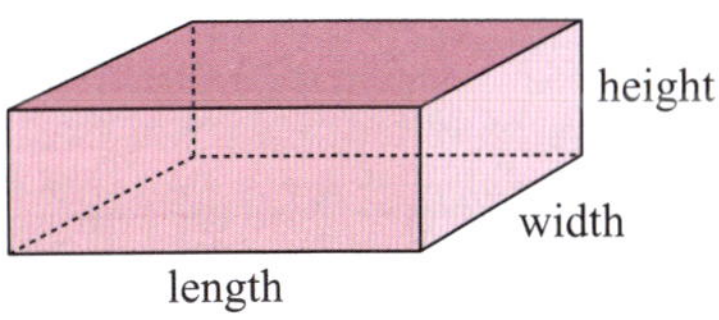

- **Volume of cylinder** $V = \pi r^2 l$ or $V = \pi r^2 h$

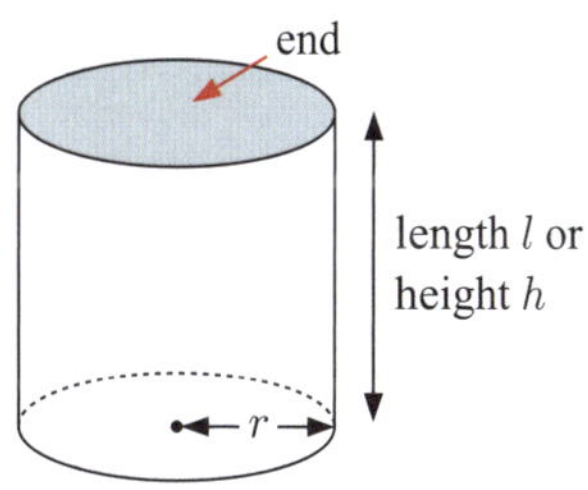

Example 11 **Self Tutor**

Find the volume of each solid:

a

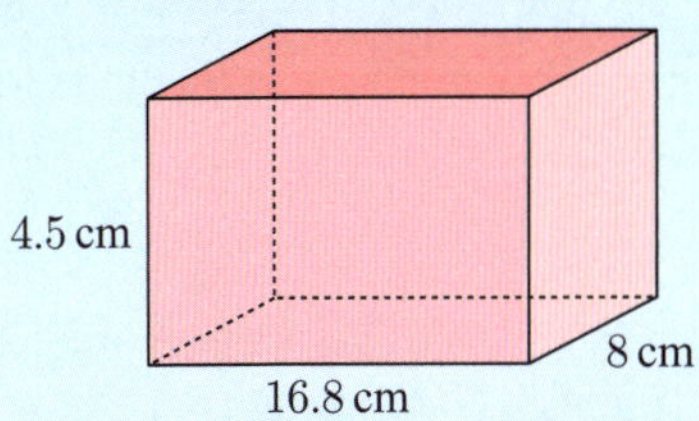

b

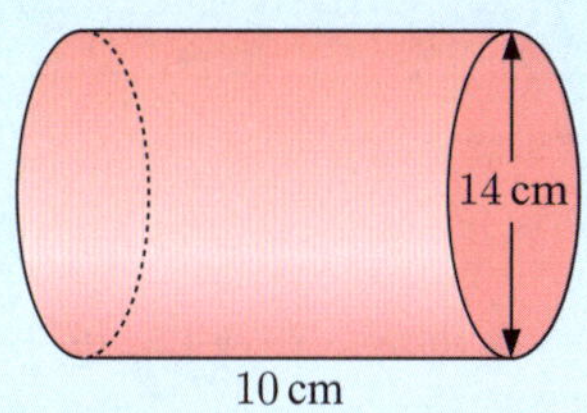

a Volume = length × width × height
$= 16.8 \times 8 \times 4.5\ cm^3$
$= 604.8\ cm^3$

b Volume $= \pi r^2 h$
$= \pi \times 7^2 \times 10\ cm^3$
$\approx 1540\ cm^3$

EXERCISE 11D.2

1 Find the volume of each solid:

a
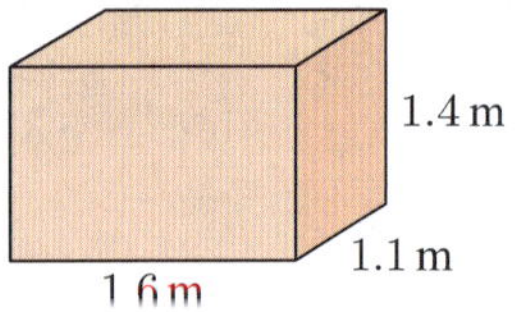

b
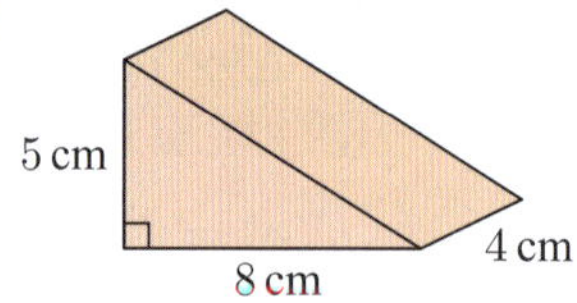

c
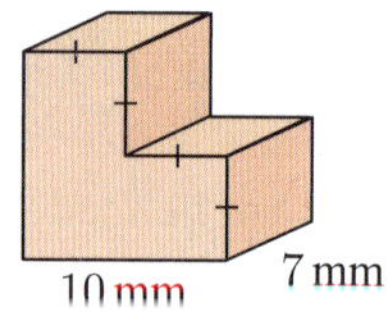

d
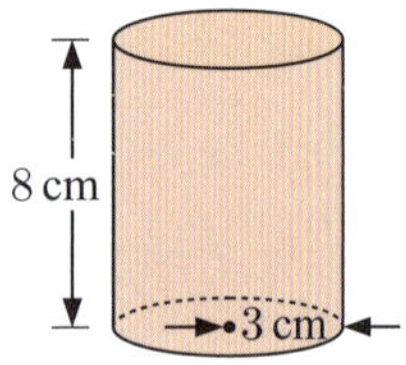

e
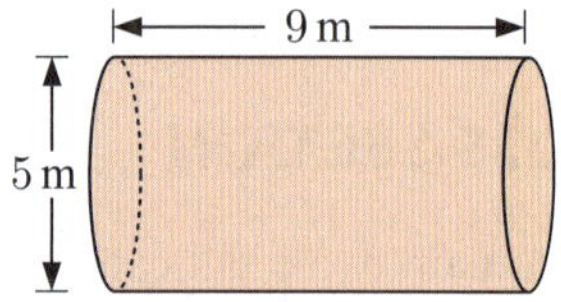

f
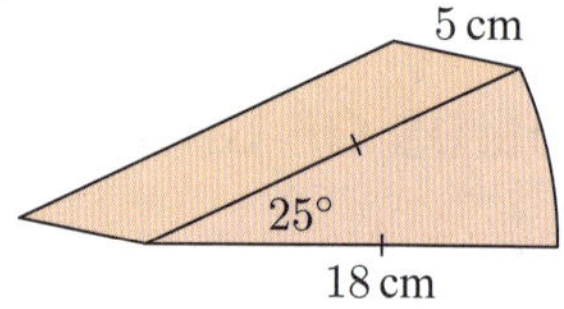

2 **a** Make r the subject of the formula $V = \pi r^2 h$.

b Hence find the radius of a steel rod with volume 470 cm^3 and length 2 m.

3
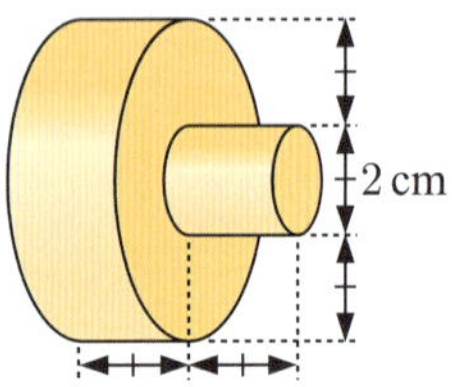

1 m^3 of brass is melted down and cast into solid door handles with the dimensions shown. How many handles can be cast?

4 A cedar table is made from 9 pieces of timber as shown. Its density is 510 kg/m^3.

a Find the volume of wood used to make the table.

b Find the mass of the table.

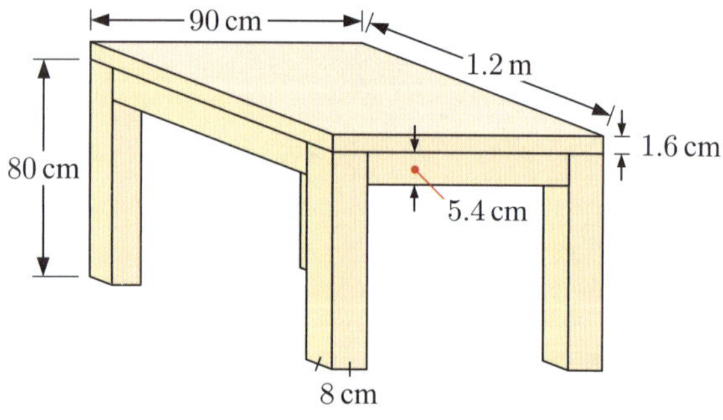

5 In the town square, there is a fountain in the middle of a circular pond. The pond is 6 metres in diameter. A concrete wall 30 cm wide and 60 cm high is built around the edge of the pond.

a Find the area of the top of the wall.

b Find the volume of concrete required for the wall.

6 Calculate the volume of wood in this model train tunnel.

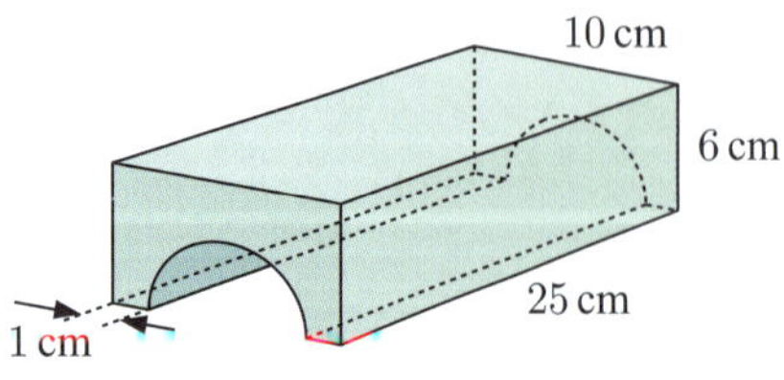

7 A cylindrical pipe has outside radius R, inside radius r, and length l. Show that the volume of concrete used to make the pipe is given by $V = \pi l(R^2 - r^2)$.

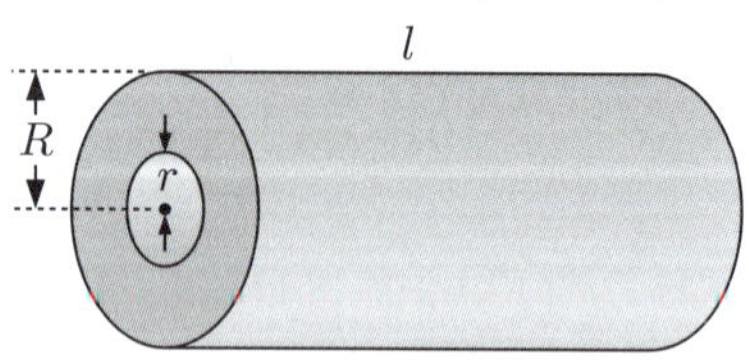

TAPERED SOLIDS

Pyramids and cones are known as **tapered solids**.

The volume of a tapered solid is a third of the volume of the solid of uniform cross-section with the same base and height.

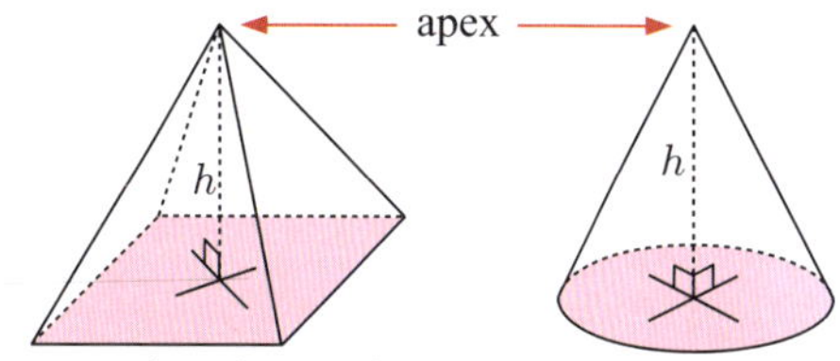

Volume of a tapered solid $= \frac{1}{3} \times$ area of base $\times$ height

Example 12 — Self Tutor

Find the volume of each solid:

a

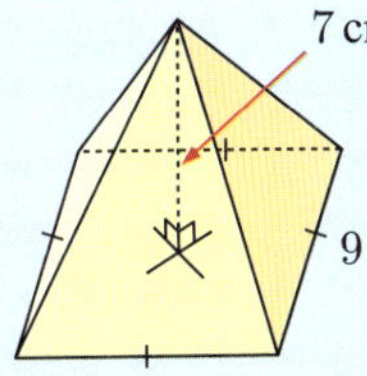

b

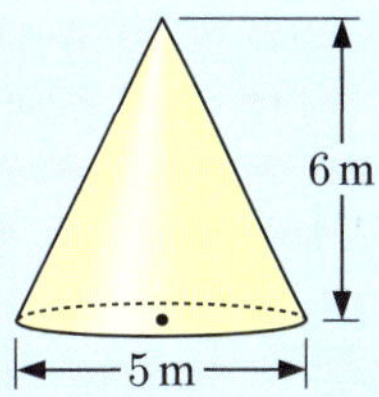

a Volume
$= \frac{1}{3} \times \text{area of base} \times \text{height}$
$= \frac{1}{3} \times 9 \times 9 \times 7$
$= 189 \text{ cm}^3$

b Volume
$= \frac{1}{3} \times \text{area of base} \times \text{height}$
$= \frac{1}{3} \times \pi \times 2.5^2 \times 6$
$\approx 39.3 \text{ m}^3$

SPHERES

Another discovery of **Archimedes of Syracuse** is that the volume of a sphere is two thirds of the volume of the smallest cylinder which encloses it.

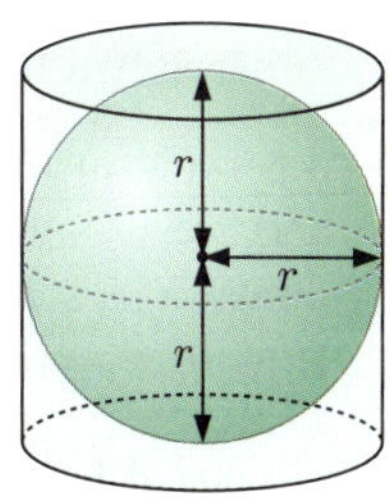

Volume of a sphere $= \frac{4}{3}\pi r^3$

Example 13 — Self Tutor

Find the volume of this sphere:

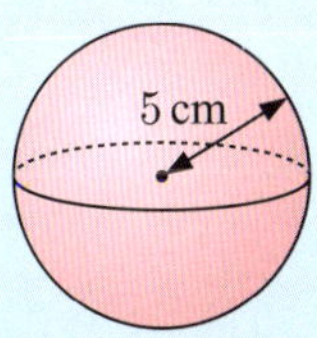

Volume $= \frac{4}{3}\pi r^3$
$= \frac{4}{3} \times \pi \times 5^3$
$\approx 524 \text{ cm}^3$

EXERCISE 11D.3

1 Find the volume of each solid:

a
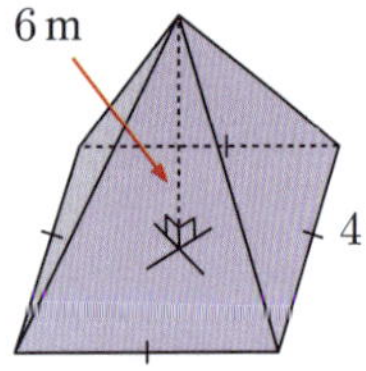

b
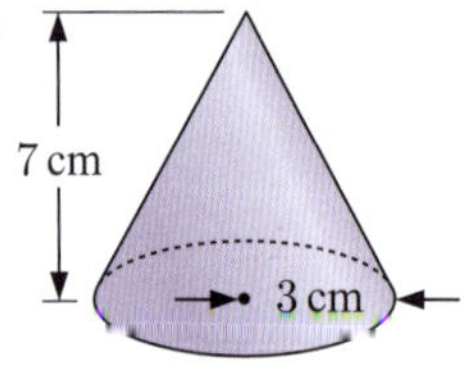

c
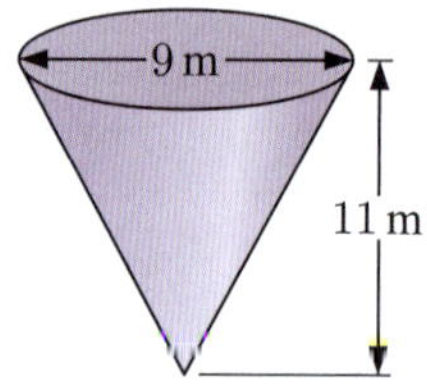

d
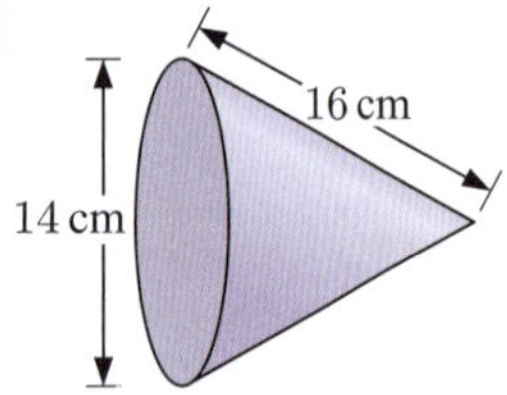

e
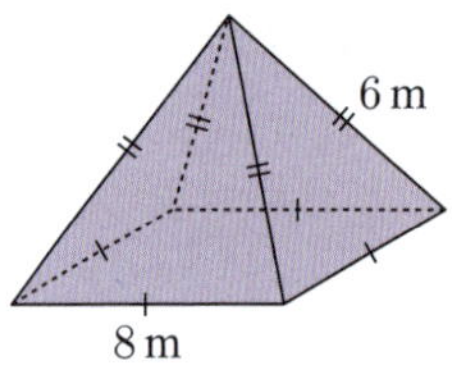

f
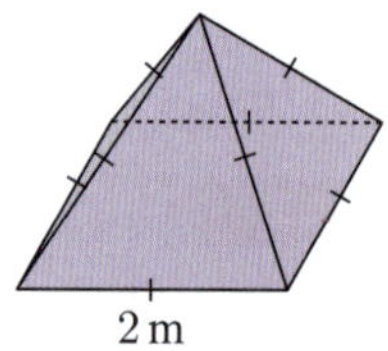

2 Find the volume of each solid:

a
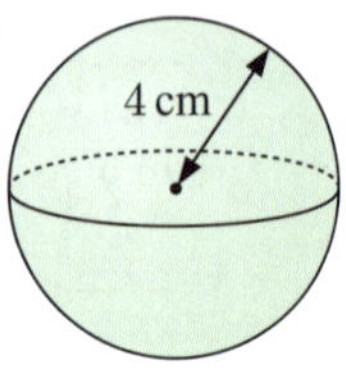

b
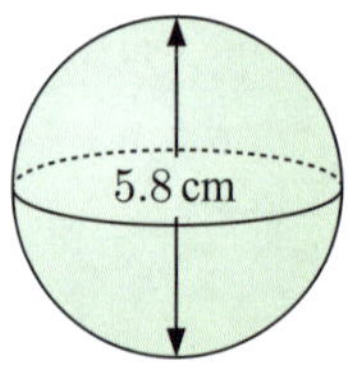

c
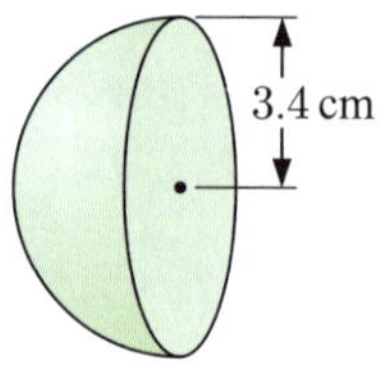

d
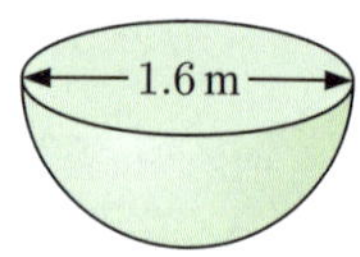

3
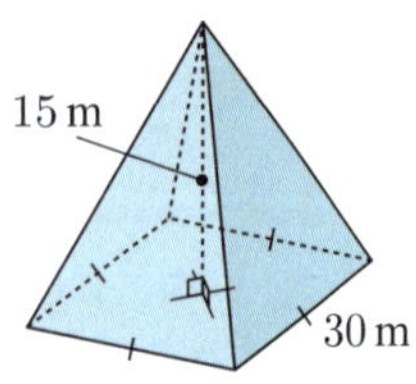

The conservatory for tropical plants at the Botanic gardens is a square-based pyramid with sides 30 metres long and height 15 metres. Calculate the volume of air in this building.

4 A conical heap of gravel is 1.4 m high and has diameter 2.6 m. Find the volume of the heap.

5 A beach ball has diameter 1.2 m. Find the volume of air inside the ball.

6 0.6 cubic metres of metal is melted down and cast into garden stakes with the dimensions shown.
How many garden stakes can be made?

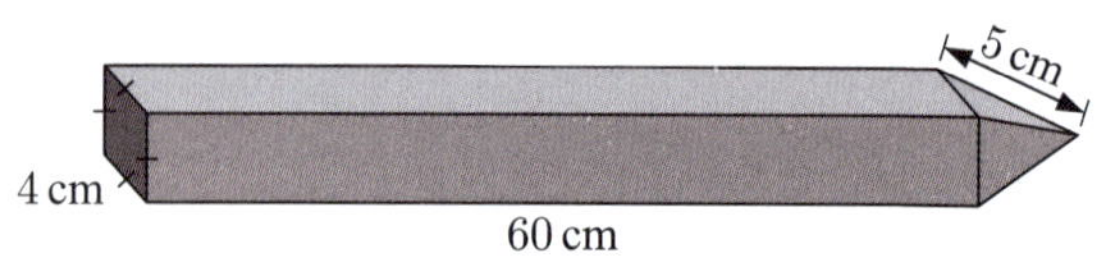

7 A conical heap of garden soil is dumped on a flat surface. The diameter of the heap equals its height, and its volume is 1.5 m^3. How high is the heap?

8 A cone has height equal to its base diameter d. Write a formula for its volume V.

9 Find the volume of this capsule:

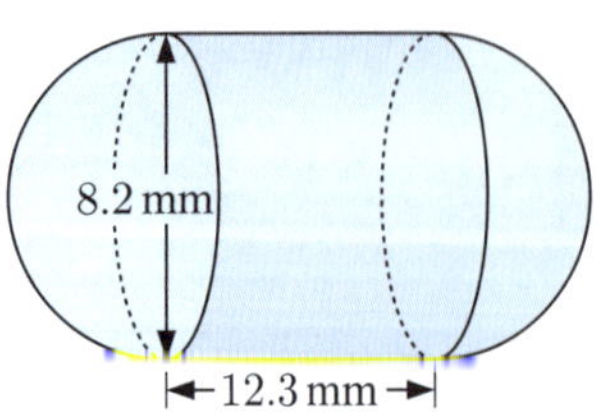

10 Write a formula for the volume V of:

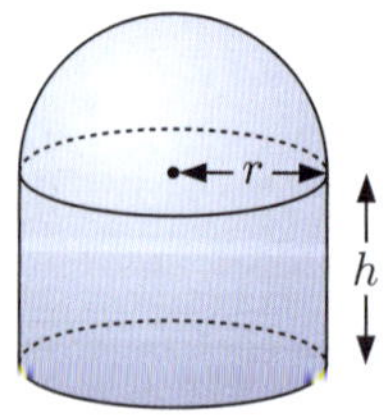

11 A garden ornament is shaped like a mushroom. The base is a cylinder 8 cm high and 6 cm in diameter. The top of the mushroom is 20 cm in diameter.
What volume of concrete, in m^3, is needed to make 50 mushrooms?

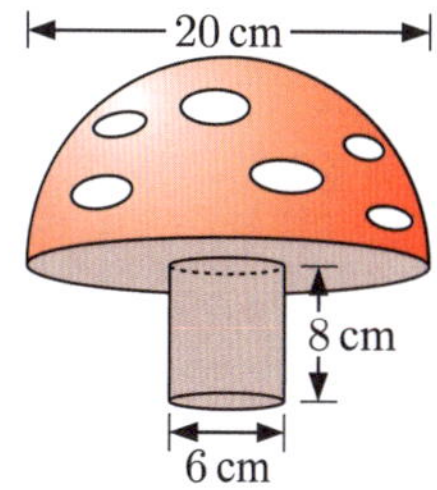

12 In the diagram alongside, the cylinder exactly encloses both the sphere and the cone.
Show that the volume of the cylinder is equal to the sum of the volumes of the other two solids.

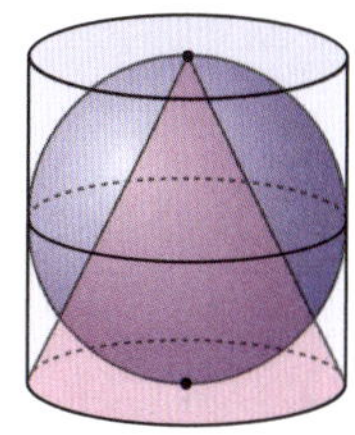

E CAPACITY

The **capacity** of a container is the amount of fluid it can contain.

CAPACITY CONVERSIONS

1 ML = 1000 kL
1 kL = 1000 L
1 L = 1000 mL

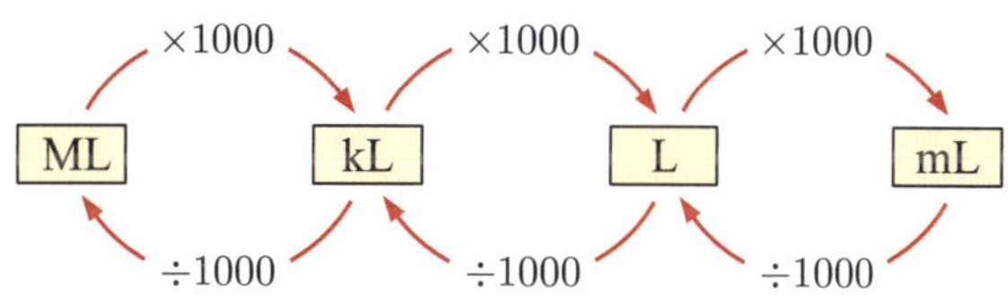

Example 14 Self Tutor

Convert:
a 4.2 L into mL **b** 36 800 L into kL.

a 4.2 L
= (4.2 × 1000) mL
= 4200 mL

b 36 800 L
= (36 800 ÷ 1000) kL
= 36.8 kL

CONNECTING VOLUME AND CAPACITY

You should remember that the units of volume and capacity are connected:

$1 \text{ mL} \equiv 1 \text{ cm}^3$
$1 \text{ L} \equiv 1000 \text{ cm}^3$
$1 \text{ kL} \equiv 1 \text{ m}^3$

Example 15 **Self Tutor**

Find the capacity of a 3 m by 2.4 m by 1.8 m tank.

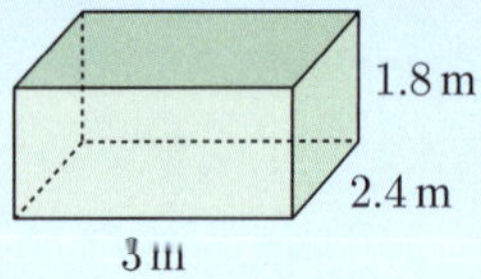

Volume $V = 3 \times 2.4 \times 1.8 \text{ m}^3$
$= 12.96 \text{ m}^3$

$\therefore$ the capacity of the tank = 12.96 kL
= 12 960 L

EXERCISE 11E

1 Convert:

a 5.6 L into mL
b 524 kL into L
c 280 mL into L
d 3.5 ML into kL
e 743 000 L into ML
f 0.84 kL into mL.

2 Find the capacity of each container:

a

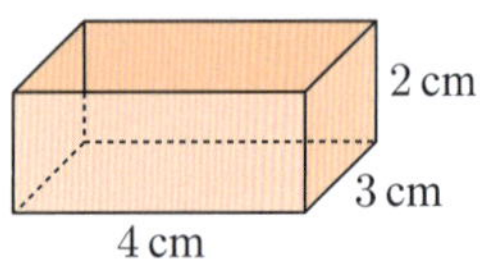

b

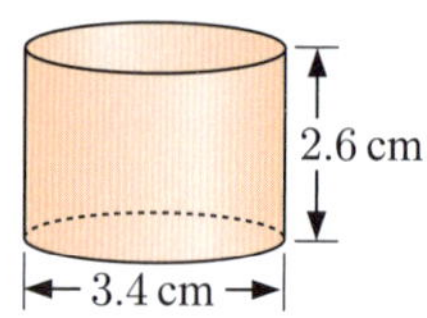

c

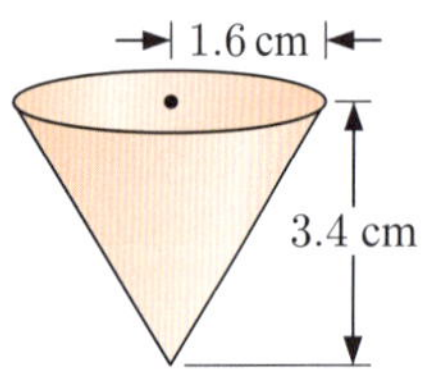

d

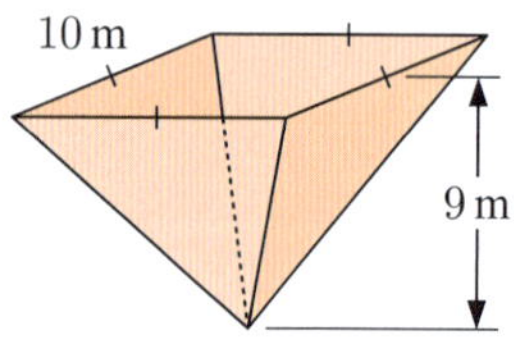

3 How many cylindrical bottles 12 cm high and with 6 cm diameter could be filled from a tank containing 125 litres of detergent?

4 A bottle of perfume is a sphere with diameter 36 mm. Calculate the capacity of the bottle in mL.

5 A car has a rectangular prism petrol tank with dimensions 48 cm by 56 cm by 20 cm.

a Find the capacity of the petrol tank in litres.

b The car consumes petrol at an average rate of 8.7 litres per 100 km. How far could it travel on a full tank of petrol?

6 A spherical snow globe is made of glass 2 mm thick. If the outer diameter of the globe is 12 cm, find the capacity of its interior.

7 A 1 litre cylindrical can of paint has base radius 5 cm. Find the height of the can.

8 The average depth of water in a lake is 4.7 m. It is estimated that the total surface area of the lake is 1.35 ha. How many litres of water does the lake contain? Write your answer using standard form.

9 A castle is surrounded by a circular moat 5 m wide and 2 m deep. The diameter of the outer edge of the moat is 50 m. Find the quantity of water in the moat in kL.

10 The inside of a glass vase consists of a cylinder with diameter 8 cm and depth 15 cm, with a hemisphere below it. Find the capacity of the vase in litres.

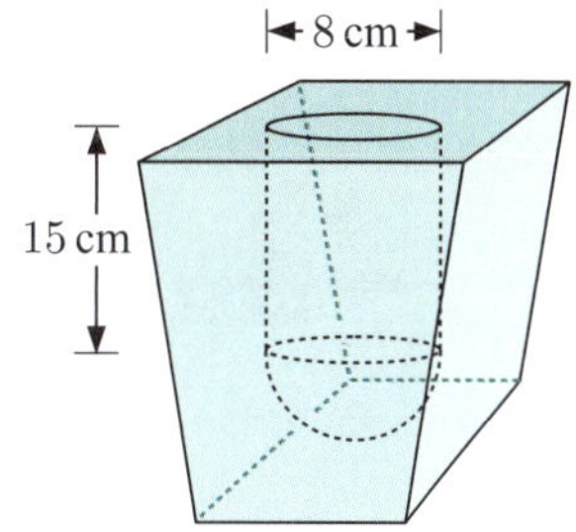

11 Answer the **Opening Problem** on page **150**.

Example 16 — Self Tutor

14.4 kL of water is pumped into an empty cylindrical tank with base radius 2 m. How high up the tank will the water level rise?

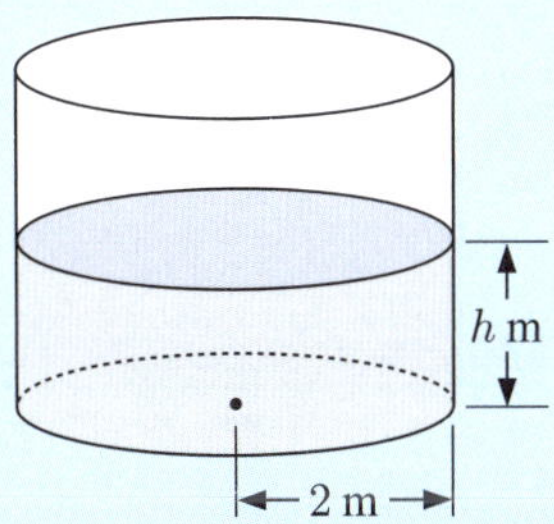

Since $1 \text{ kL} \equiv 1 \text{ m}^3$, the volume of water pumped into the tank is 14.4 m^3.

If the water in the cylindrical tank is h m deep,

its volume $= \pi \times 2^2 \times h$ $\quad \{V = \pi r^2 h\}$

$= 4\pi h \text{ m}^3$

$\therefore \ 4\pi h = 14.4$ $\quad$ {equating volumes}

$\therefore \ h = \dfrac{14.4}{4\pi} \approx 1.15$

$\therefore$ the water level will rise to 1.15 m.

12 20 kL of water is pumped into an empty cylindrical tank with base radius 3 m. How high up the tank will the water level rise?

13 Water enters a cylindrical tank at 80 litres per minute. The base diameter of the tank is 2.4 m and the height is 4 m.

- **a** Find the capacity of the full tank.
- **b** How high up the tank will the water level be after 10 minutes?

14 Consider these containers:

A

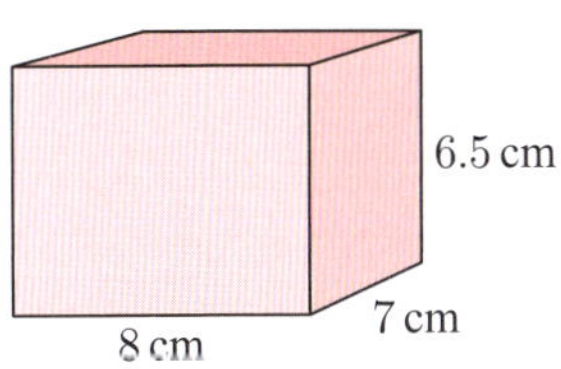

B

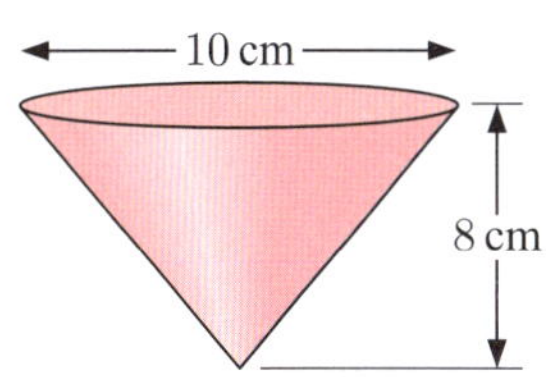

C

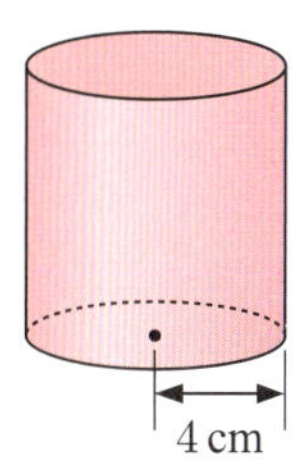

Container A is filled with water. The water from container A is poured into container B until it is full, and the remainder is poured into container C. Find the height of the water in container C.

15 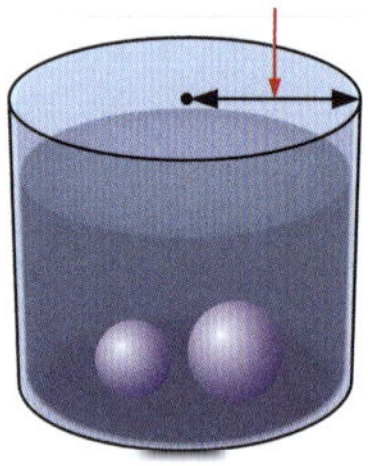

A cylindrical container with radius 12 cm is partly filled with water. Two spheres with radii 3 cm and 4 cm are dropped into the water and are fully submerged. Find the *exact* increase in height of water in the cylinder.

MULTIPLE CHOICE QUIZ

QUICK QUIZ

REVIEW SET 11A

1 Find the perimeter and area of each figure:

a 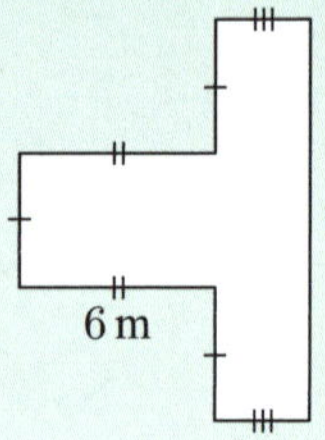

b 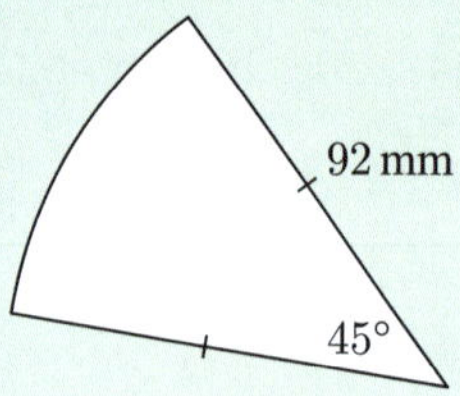

c 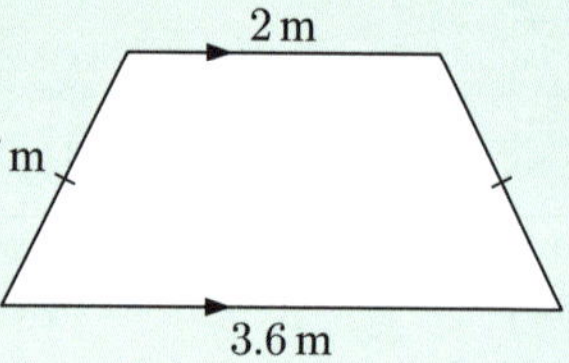

2 The diameter of a car tyre is 50 cm. How many revolutions does the tyre complete if the car travels 2 km?

3 Write a formula for the perimeter P of each figure:

a 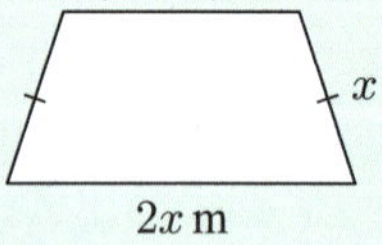

b 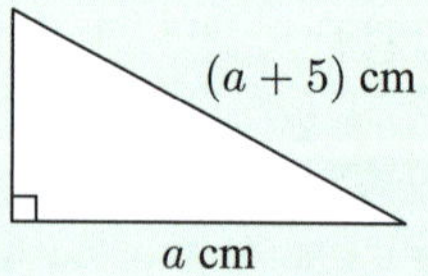

4 A circle has circumference 40.8 m. Find its area.

5 Find the surface area of each solid:

a 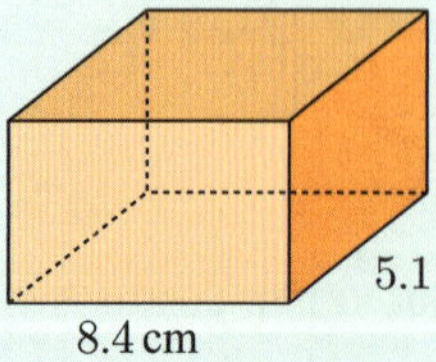

b open top 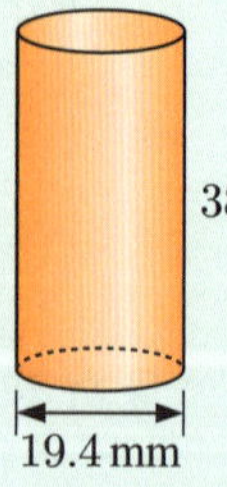

c 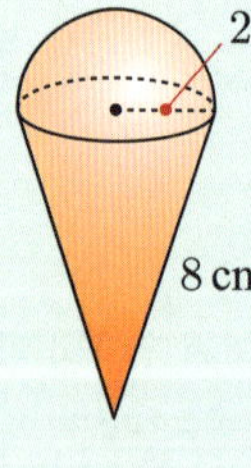

6 What area of leather is required to manufacture 10 dozen footballs, each with diameter 24 cm?

7 Find the volume of each solid:

a

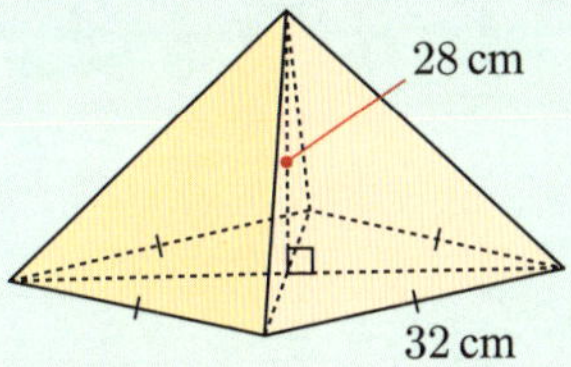

b

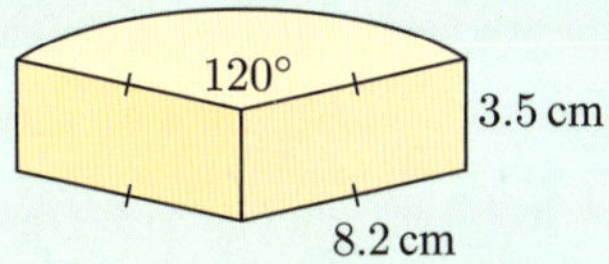

c

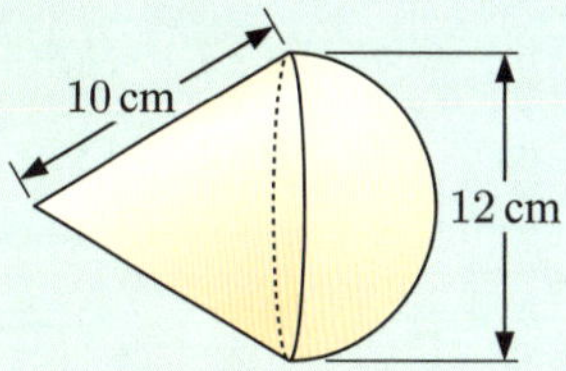

8 A solid cone has radius 5 cm and perpendicular height 8 cm. Find its:

a volume **b** surface area.

9 Determine the total volume of steel required to construct 450 m of steel piping with internal diameter 1.46 m and external diameter 1.50 m.

10 Find the capacity of each container:

a

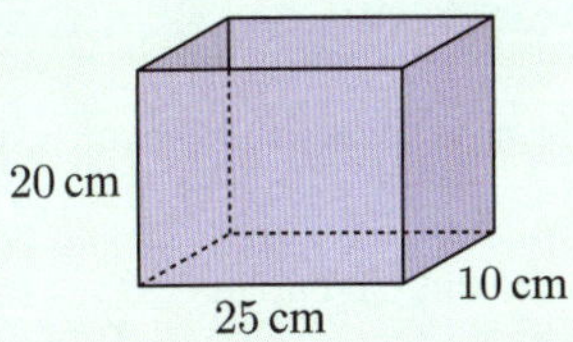

b

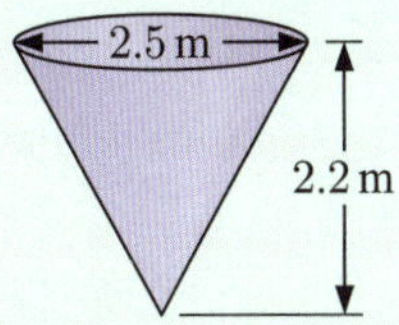

c

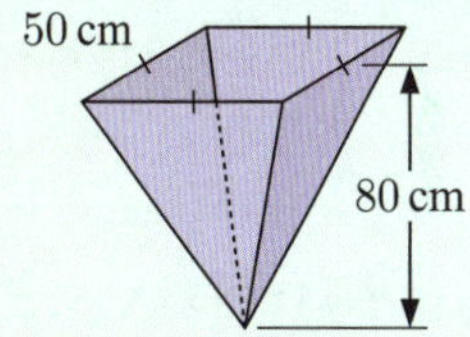

11

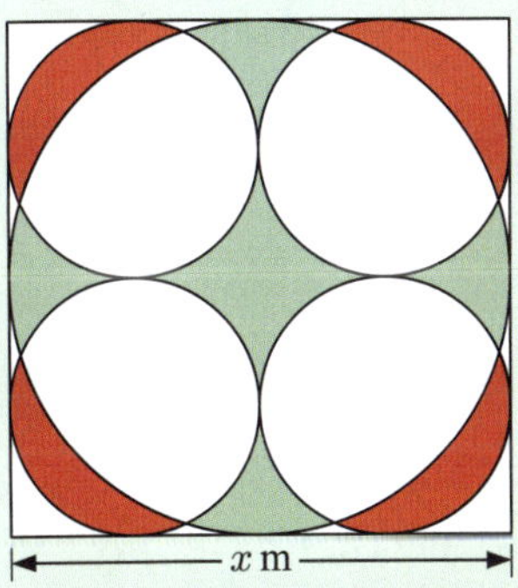

One large circle and four smaller circles are inscribed as shown inside a square with side length x m.

a Find, in terms of x, the radius of:

i the large circle

ii the smaller circles.

b Show that the area of the large circle equals the area of the four smaller circles.

c Hence show that the green shaded area is equal to the red shaded area.

REVIEW SET 11B

1 Find the perimeter and area of each figure:

a

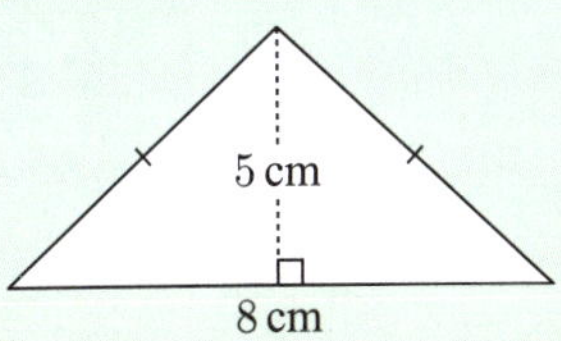

b

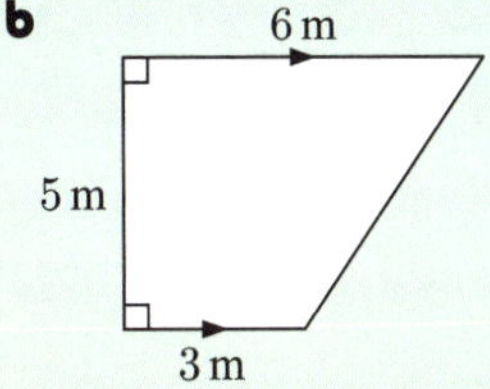

c

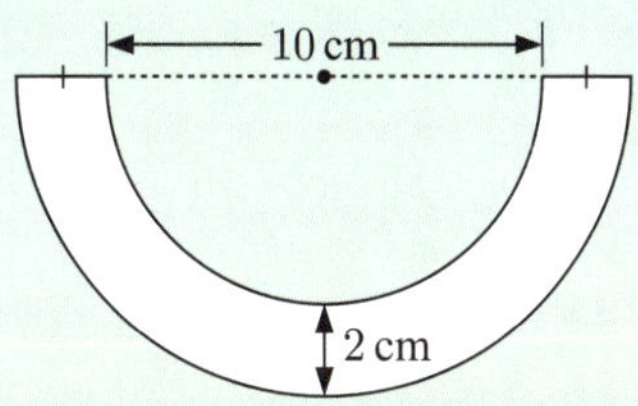

2 A rhombus has diagonals of length 12 cm and 20 cm. Find its perimeter.

3 An isosceles triangle has area 30 cm^2 and base of length 12 cm. Find the length of the equal sides.

4 Write a formula for the area A of each figure:

a

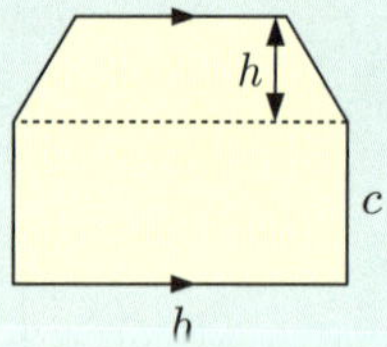

b

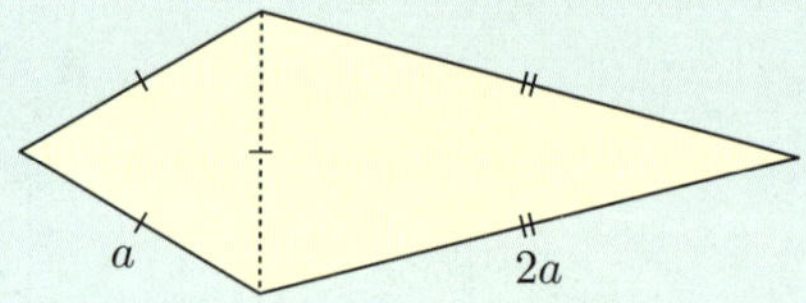

5 Find the surface area and volume of each solid:

a

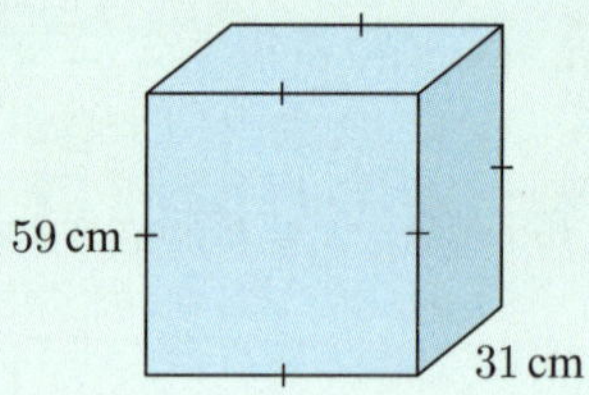

b

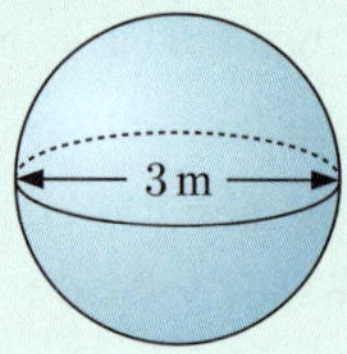

c

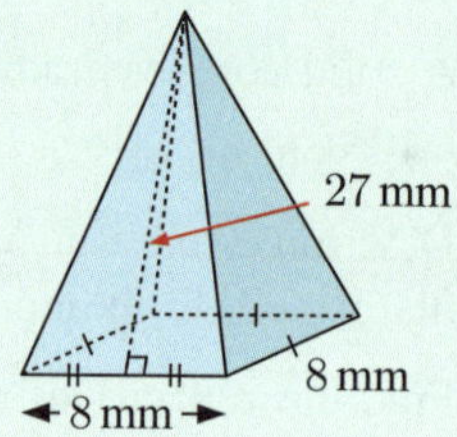

6 A solid cone has base radius 2 m and surface area 10π m^2. Find the slant height of the cone.

7 25 m^3 of concrete is used to make cylindrical bollards which are 90 cm high and have radius 12 cm. How many bollards are made?

8 A sphere fits exactly inside a cubic box. What percentage of the cube's capacity does the sphere occupy?

9 A cylindrical bucket is 20 cm wide and 30 cm high. The bucket is filled with 6 litres of water. How far from the top of the bucket does the water reach?

10

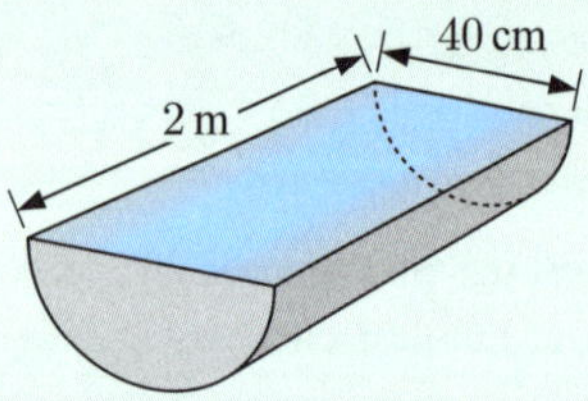

A horse's drinking trough has the dimensions shown. Water flows into the empty trough at 12 litres per minute for 10 minutes. Will the trough overflow?

11 The Pantheon in Rome was built during the reign of emperor Augustus around 27 BC, and rebuilt by the emperor Hadrian around 126 AD. Its centre is designed as a hemisphere on a cylinder so that a sphere with diameter 43.3 m will fit exactly inside it.
Find the capacity of this chamber in kL.

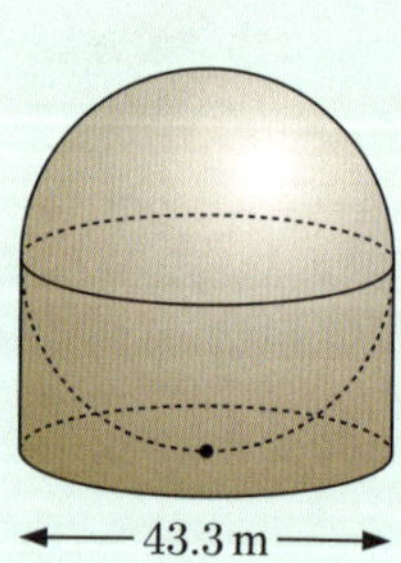

12 An ice block with uniform cross-section has the dimensions shown.

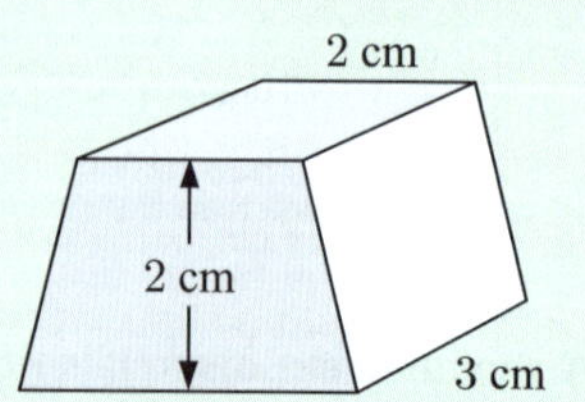

- **a** Find the surface area of the ice block.
- **b** Find the volume of the ice block.
- **c** The density of ice is 91.7% of the density of liquid water.
  - **i** Find the amount of liquid water, in mL, which must be frozen to form the ice block.
  - **ii** How many complete ice blocks can be made from a 2 L jug of liquid water?

Chapter 12

Quadratic equations

Contents:

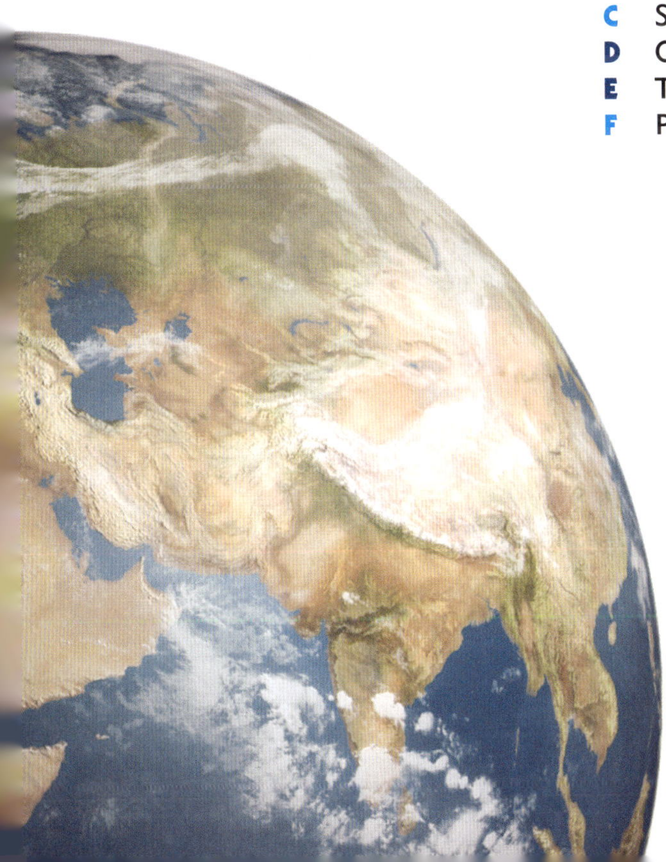

OPENING PROBLEM

In a formal garden there is a $16 \text{ m} \times 16 \text{ m}$ square. The circle of red flowers in its centre has radius x m.

The total area of the quarter circles of yellow flowers is nine times the area of red flowers.

The remainder of the square is a gravel path.

The width of the path at each entrance is $2x$ m.

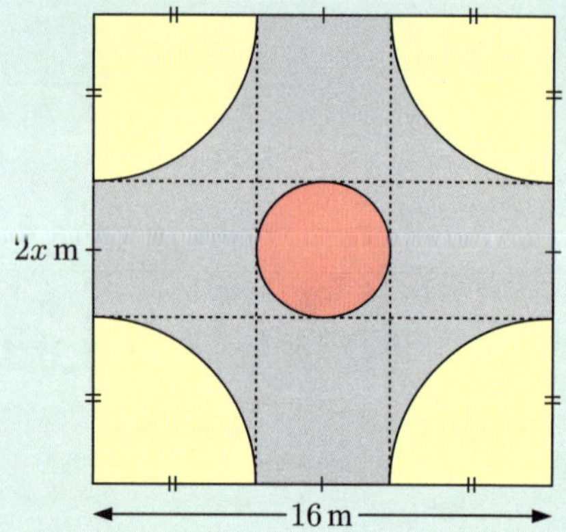

Things to think about:

- **a** Can you explain why the radius of each quarter circle is $(8 - x)$ m?
- **b** Can you explain why $(8 - x)^2 = 9x^2$?
- **c** Can you write this equation in the form $ax^2 + bx + c = 0$?
- **d** Can you *solve* this equation? In which form is it easier to solve?
- **e** Are both of the solutions to the equation possible in the context of the problem?
- **f** What is the radius of the circle of red flowers?

We have previously solved many *linear* equations which can be written in the form $ax + b = 0$, $a \neq 0$. Linear equations have only *one* solution.

The equation in the **Opening Problem** is *not* a linear equation. However, it can be written in the form $ax^2 + bx + c = 0$, $a \neq 0$.

> A **quadratic equation** is an equation which can be written in the form $ax^2 + bx + c = 0$, where a, b, and c are constants, $a \neq 0$.
>
> A quadratic equation may have *two*, *one*, or *zero* real solutions.

For example:

- $x^2 - 2x - 8 = 0$ has two solutions $x = 4$ and $x = -2$
- $x^2 + 2x + 1 = 0$ has only one solution $x = -1$
- $x^2 + 1 = 0$ has no real solutions since $x^2 \geqslant 0$ for all $x \in \mathbb{R}$.

We can verify that the solutions to a quadratic equation are correct by substituting them into the equation.

For example, for the equation $x^2 - 2x - 8 = 0$:

If $x = 4$,
$$\begin{aligned} & x^2 - 2x - 8 \\ &= 4^2 - 2 \times 4 - 8 \\ &= 16 - 8 - 8 \\ &= 0 \quad \checkmark \end{aligned}$$

If $x = -2$,
$$\begin{aligned} & x^2 - 2x - 8 \\ &= (-2)^2 - 2 \times (-2) - 8 \\ &= 4 + 4 - 8 \\ &= 0 \quad \checkmark \end{aligned}$$

A EQUATIONS OF THE FORM $x^2 = k$

In **Chapter 7** we studied **power equations** which can be written in the form $x^n = k$ where $n \neq 0$.

You should recognise that when $n = 2$, we also have a *quadratic* equation.

$$\text{If } x^2 = k \text{ then } \begin{cases} x = \pm\sqrt{k} & \text{if } k > 0 \\ x = 0 & \text{if } k = 0 \\ x \text{ has no real solutions if } k < 0. \end{cases}$$

This principle can be extended to other perfect squares, allowing us to solve equations of the form $(ax+b)^2 = k$ without having to expand the LHS.

Example 1 — Self Tutor

Solve for x:

a $2x^2 + 1 = 15$ **b** $(x-3)^2 = 16$ **c** $(x+2)^2 = 11$

a $2x^2 + 1 = 15$
$\therefore\ 2x^2 = 14$
$\therefore\ x^2 = 7$
$\therefore\ x = \pm\sqrt{7}$

b $(x-3)^2 = 16$
$\therefore\ x - 3 = \pm\sqrt{16}$
$\therefore\ x - 3 = \pm 4$
$\therefore\ x = 3 \pm 4$
$\therefore\ x = 7$ or -1

c $(x+2)^2 = 11$
$\therefore\ x + 2 = \pm\sqrt{11}$
$\therefore\ x = -2 \pm \sqrt{11}$

In **c**, the two solutions are $x = -2 + \sqrt{11}$ and $x = -2 - \sqrt{11}$.

EXERCISE 12A

1 Solve for x:

a $x^2 - 10 = 90$ **b** $2x^2 = 50$ **c** $5x^2 = 20$
d $6x^2 = 54$ **e** $5x^2 = -45$ **f** $7x^2 = 0$
g $3x^2 - 2 = 25$ **h** $4 - 2x^2 = 12$ **i** $4x^2 + 2 = 10$

2 Solve for x:

a $(x-1)^2 = 9$ **b** $(x+4)^2 = 16$ **c** $(x+2)^2 = -1$
d $(x-4)^2 = 5$ **e** $(x-6)^2 = -4$ **f** $(x+2)^2 = 0$
g $(2x-5)^2 = 0$ **h** $(3x+2)^2 = 4$ **i** $(3x+1)^2 = 81$
j $(2x+1)^2 = 48$ **k** $(3-2x)^2 = 7$ **l** $\frac{1}{3}(2x+3)^2 = 2$

3 Solve for x by first eliminating the algebraic fractions:

a $\dfrac{x}{4} = \dfrac{1}{x}$ **b** $\dfrac{5}{x} = \dfrac{x}{2}$ **c** $\dfrac{x}{8} = \dfrac{2}{x}$

B THE NULL FACTOR LAW

The **null factor law** states that:

> If the product of two or more numbers is zero, then at least one of them must be zero.
>
> So, if $ab = 0$ then $a = 0$ or $b = 0$.

If we are given an equation where the LHS is fully factorised and the RHS is zero, we can use the null factor law to find its solutions.

Example 2 **Self Tutor**

Solve for x:

a $2x(x-4)=0$ **b** $(x+3)(2x-5)=0$

a $2x(x-4)=0$

$\therefore\ 2x=0$ or $x-4=0$

$\therefore\ x=0$ or 4

b $(x+3)(2x-5)=0$

$\therefore\ x+3=0$ or $2x-5=0$

$\therefore\ x=-3$ or $2x=5$

$\therefore\ x=-3$ or $\frac{5}{2}$

EXERCISE 12B

1 Explain what can be deduced from:

a $pq=0$ **b** $2a=0$ **c** $2xy=0$ **d** $wx^2=0$

2 Solve for x:

a $x(x+3)=0$ **b** $2x(x-5)=0$ **c** $4x(2-x)=0$

d $-3x(2x+1)=0$ **e** $2x(3x-5)=0$ **f** $-3x(6-5x)=0$

3 Solve for x:

a $(x-1)(x-3)=0$ **b** $(x+5)(x-2)=0$ **c** $5(x+1)^2=0$

d $-2(x-5)(x+4)=0$ **e** $11(x+2)(x-7)=0$ **f** $2(x-3)^2=0$

g $5(x+2)(2x-1)=0$ **h** $8(3x+1)^2=0$ **i** $-6(x-5)(3x+2)=0$

j $-4(2x+1)^2=0$ **k** $(2x+3)(2x+1)=0$ **l** $(3x+1)(2x-5)=0$

4 Solve for x:

a $ax(x-b)=0$ **b** $a(x-b)(x-c)=0$ **c** $a(x-b)^2=0$

d $ax(bx-c)=0$ **e** $a(x-b)(cx-d)=0$

C SOLVING BY FACTORISATION

We have seen that if we can write an equation with its LHS fully factorised and RHS zero, then we can solve it using the null factor law.

This gives us a method for solving quadratic equations:

Step 1: If necessary, rearrange the equation so one side is **zero**.

Step 2: **Fully factorise** the other side (usually the LHS).

Step 3: Apply the **null factor law**.

Step 4: **Solve** the resulting linear equations.

COMMON FACTORS

When we factorise an expression, the first thing we look for are common factors.

Example 3 — Self Tutor

Solve for x: $x^2 = 3x$

$$x^2 = 3x$$
$$\therefore\ x^2 - 3x = 0 \quad \{\text{making the RHS} = 0\}$$
$$\therefore\ x(x-3) = 0 \quad \{\text{factorising the LHS}\}$$
$$\therefore\ x = 0 \text{ or } x - 3 = 0 \quad \{\text{null factor law}\}$$
$$\therefore\ x = 0 \text{ or } 3$$

WARNING ON INCORRECT CANCELLING

In the equation $x^2 = 3x$ from **Example 3**, we may be tempted to divide both sides by the common factor x, giving $x = 3$.

While $x = 3$ is a solution to the equation, we have lost the solution $x = 0$.

From this example we conclude that:

> We must never cancel a common factor involving a variable from both sides of an equation unless we are *sure* that this factor cannot be zero.

EXERCISE 12C.1

1 Solve for x:

- **a** $x^2 - 7x = 0$
- **b** $x^2 + 11x = 0$
- **c** $x^2 - 5x = 0$
- **d** $3x^2 + 6x = 0$
- **e** $2x^2 + 5x = 0$
- **f** $4x^2 - 3x = 0$

2 Solve for x:

- **a** $x^2 = 8x$
- **b** $x^2 = -4x$
- **c** $4x^2 = 5x$
- **d** $9x = -3x^2$
- **e** $x - x^2 = 2x$
- **f** $2x^2 + 3x = 4x - x^2$

3 Solve for x:

- **a** $x(x+5) = 4(x+5)$
- **b** $3(x \quad 2) = x(2-x)$
- **c** $3 + x = x(x+3)$

THE DIFFERENCE BETWEEN TWO SQUARES

Equations such as $9x^2 - 49 = 0$ can be rearranged into the form $x^2 = k$ and then solved.

However, we can also solve them by factorising the LHS as the difference between two squares.

Example 4 — Self Tutor

Solve for x: $9x^2 - 49 = 0$

$$9x^2 - 49 = 0$$
$$\therefore\ (3x)^2 - 7^2 = 0$$
$$\therefore\ (3x+7)(3x-7) = 0$$
$$\therefore\ 3x + 7 = 0 \text{ or } 3x - 7 = 0$$
$$\therefore\ 3x = -7 \text{ or } 3x = 7$$
$$\therefore\ x = \pm\tfrac{7}{3}$$

EXERCISE 12C.2

1 Solve for x:

a $x^2 - 25 = 0$ **b** $64 - x^2 = 0$ **c** $x^2 + 1 = 0$

d $2x^2 - 18 = 0$ **e** $27 - 3x^2 = 0$ **f** $-2x^2 + 72 = 0$

2 Solve for x:

a $4x^2 - 25 = 0$ **b** $49 - 4x^2 = 0$ **c** $9x^2 - 16 = 0$

d $9 - 16x^2 = 0$ **e** $100x^2 - 25 = 0$ **f** $8 - 18x^2 = 0$

3 To solve the equation $4x^2 = (x+1)^2$, Jack notices the LHS can be written as $(2x)^2$.

He argues:
$$4x^2 = (x+1)^2$$
$$\therefore (2x)^2 = (x+1)^2$$
$$\therefore 2x = x + 1$$
$$\therefore x = 1$$

a Verify that $x = 1$ is indeed a solution to the equation.

b Verify that $x = -\frac{1}{3}$ is also a solution to the equation.

c Explain what Jack did wrong which caused him to lose a solution.

d Use the difference between two squares factorisation to correctly solve the equation.

4 Solve for x:

a $4x^2 - (1-x)^2 = 0$ **b** $(x+2)^2 - x^2 = 0$

c $(x+1)^2 = (1-3x)^2$ **d** $(2x-1)^2 = (2-3x)^2$

OTHER FACTORISATIONS

In addition to common factors and the difference between two squares, we also look to factorise using **perfect squares**, **sum and product** factorisation, and **splitting the middle term**.

Example 5 Self Tutor

Solve for x: $x^2 + 3x = 28$

$$x^2 + 3x = 28$$
$$\therefore x^2 + 3x - 28 = 0 \quad \{\text{making the RHS} = 0\}$$
$$\therefore (x+7)(x-4) = 0 \quad \{\text{the numbers } +7 \text{ and } -4 \text{ have sum 3 and product } -28\}$$
$$\therefore x + 7 = 0 \text{ or } x - 4 = 0$$
$$\therefore x = -7 \text{ or } 4$$

EXERCISE 12C.3

1 Solve for x:

a $x^2 + 3x + 2 = 0$ **b** $x^2 - 3x + 2 = 0$ **c** $x^2 - 10x + 25 = 0$

d $x^2 + 5x + 6 = 0$ **e** $x^2 + 6 = 5x$ **f** $x^2 + 7x = -6$

g $x^2 + 14 = -9x$ **h** $x^2 + 11x = -30$ **i** $x^2 = 15 - 2x$

j $x^2 + 4x = 12$ **k** $x^2 = 11x - 24$ **l** $x^2 = 14x - 49$

2 Solve for x:

a $2x^2 + 18x + 36 = 0$ **b** $-x^2 - 11x - 28 = 0$ **c** $3x^2 + 6x = 24$

d $2x^2 + 2x = 24$ **e** $4x^2 + 24 = 20x$ **f** $5x^2 + 20 = 20x$

g $3x^2 = 3x + 18$ **h** $-x^2 = 7x - 60$ **i** $140 + 6x = 2x^2$

3 Solve for x:

a $x(x+5) + 2(x+6) = 0$ **b** $x(1+x) + x = 3$ **c** $(x-1)(x+9) = 8x$

d $x(8-x) + 20 = -13$ **e** $4x(x+1) = -1$ **f** $(x+15)(x-2) = x(9-x)$

Example 6 Self Tutor

Solve for x: $\dfrac{x-2}{x} = \dfrac{6+x}{2}$

$$\frac{x-2}{x} = \frac{6+x}{2}$$

$$\therefore \left(\frac{x-2}{x}\right) \times \frac{2}{2} = \left(\frac{6+x}{2}\right) \times \frac{x}{x} \qquad \{\text{LCD} = 2x\}$$

$$\therefore 2(x-2) = x(6+x) \qquad \{\text{equating numerators}\}$$

$$\therefore 2x - 4 = 6x + x^2$$

$$\therefore x^2 + 4x + 4 = 0$$

$$\therefore (x+2)^2 = 0$$

$$\therefore x + 2 = 0$$

$$\therefore x = -2$$

4 Solve for x:

a $x + 4 = \dfrac{21}{x}$ **b** $\dfrac{x+1}{4} = \dfrac{1}{2x}$ **c** $\dfrac{x+4}{2} = \dfrac{6}{x}$

d $\dfrac{x+2}{x} = x$ **e** $\dfrac{x-1}{x+2} = \dfrac{2}{x}$ **f** $\dfrac{3x}{6x-1} = \dfrac{1}{3x}$

Example 7 Self Tutor

Solve for x: $\dfrac{1}{x} + \dfrac{4}{x+6} = 1$

$$\frac{1}{x} + \frac{4}{x+6} = 1$$

$$\therefore \frac{1}{x} \times \frac{(x+6)}{(x+6)} + \left(\frac{4}{x+6}\right) \times \frac{x}{x} = 1 \times \frac{x(x+6)}{x(x+6)} \qquad \{\text{LCD} = x(x+6)\}$$

$$\therefore x + 6 + 4x = x(x+6) \qquad \{\text{equating numerators}\}$$

$$\therefore 5x + 6 = x^2 + 6x$$

$$\therefore 0 = x^2 + x - 6$$

$$\therefore (x+3)(x-2) = 0$$

$$\therefore x = -3 \text{ or } 2$$

5 Solve for x:

a $\dfrac{2}{x} + \dfrac{3}{x+2} = 1$ **b** $\dfrac{5}{x} + \dfrac{x}{x-6} = 2$ **c** $\dfrac{4}{x+1} + \dfrac{3}{x+3} = -1$

d $\dfrac{1}{x+3} + \dfrac{1}{x-3} = x$ **e** $\dfrac{4}{x+2} - \dfrac{2}{x+1} = x$

Example 8 Self Tutor

Solve for x: $5x^2 = 3x + 2$

$5x^2 = 3x + 2$

$\therefore\ 5x^2 - 3x - 2 = 0$ {making the RHS = 0}

$\therefore\ 5x^2 - 5x + 2x - 2 = 0$ {the numbers -5 and 2 have sum -3 and product -10}

$\therefore\ 5x(x-1) + 2(x-1) = 0$

$\therefore\ (x-1)(5x+2) = 0$

$\therefore\ x - 1 = 0$ or $5x + 2 = 0$

$\therefore\ x = 1$ or $-\frac{2}{5}$

6 Solve for x:

a $2x^2 - 5x + 2 = 0$ **b** $3x^2 + 8x - 3 = 0$ **c** $3x^2 + 17x + 20 = 0$

d $2x^2 + 5x = 3$ **e** $2x^2 + 5 = 11x$ **f** $2x^2 + 7x + 5 = 0$

g $3x^2 + 13x + 4 = 0$ **h** $5x^2 - 6 = 13x$ **i** $2x^2 + 17x = 9$

j $2x^2 = 3x + 5$ **k** $3x^2 = 8 - 2x$ **l** $2x^2 = 18 - 9x$

m $-6x^2 + 17x + 3 = 0$ **n** $-2x^2 - 5x + 12 = 0$ **o** $12x^2 - 22x + 6 = 0$

p $9x^2 + 6x - 48 = 0$ **q** $28x^2 - 8 = 2x$ **r** $36x^2 + 39x = 12$

7 Solve for x:

a $3x(x+2) - 5(x-3) = 17$ **b** $2x(x-6) = x - 20$ **c** $(8x+1)(x-2) = 13 - x$

d $\dfrac{x}{1+2x} = \dfrac{1}{3x}$ **e** $\dfrac{3x+1}{2x} = x + 2$ **f** $\dfrac{5}{x+2} - \dfrac{6}{x-1} = -2$

g $\dfrac{4}{x-2} + \dfrac{3}{x-1} = 3$ **h** $\dfrac{x}{x-2} - \dfrac{x+3}{x} = -5$

8 Solve for x: $\sqrt{x+2} = x$

9 Solve for x: $x^4 - 5x^2 + 4 = 0$
Hint: Let $z = x^2$ and solve for z first.

D COMPLETING THE SQUARE

Many quadratics cannot be easily factorised using the methods we have seen previously.

For example, we cannot easily factorise the LHS of $x^2 + 4x - 7 = 0$ since we cannot find two integers with sum 4 and product -7.

An alternative method is to rewrite the equation in the form $(x-h)^2 = k$. We call this **completed square form** and we call the process which generates it **completing the square**.

For the equation $x^2 + 4x - 7 = 0$ we use this procedure:

$$\begin{aligned} x^2 + 4x - 7 &= 0 \\ \therefore\ x^2 + 4x \quad &= 7 && \{\text{moving constant to RHS}\} \\ \therefore\ x^2 + 4x + 4 &= 7 + 4 && \{\text{completing a perfect square on LHS}\} \\ \therefore\ (x+2)^2 &= 11 && \{\text{factorising LHS}\} \\ \therefore\ x + 2 &= \pm\sqrt{11} && \{\text{If } a^2 = k,\ k > 0 \text{ then } a = \pm\sqrt{k}\} \\ \therefore\ x &= -2 \pm \sqrt{11} \end{aligned}$$

In order to complete the square, we need to know what number to add to both sides.

The perfect square expansion law tells us that $(x+a)^2 = x^2 + 2ax + a^2$.

> To complete a perfect square, the number we must add to both sides is found by **halving the coefficient of x, then squaring this value.**

Example 9 — Self Tutor

For each equation:

- **i** What number must be added to both sides to complete a perfect square on the LHS?
- **ii** Hence write the equation in completed square form.

a $x^2 + 8x = -5$ **b** $x^2 - 6x = 13$

a **i** In $x^2 + 8x = -5$, half the coefficient of x is $\frac{8}{2} = 4$.
$\therefore$ we add 4^2 to both sides.

ii The equation becomes $x^2 + 8x + 4^2 = -5 + 4^2$
$\therefore\ (x+4)^2 = -5 + 16$
$\therefore\ (x+4)^2 = 11$

b **i** In $x^2 - 6x = 13$, half the coefficient of x is $\frac{-6}{2} = -3$.
$\therefore$ we add $(-3)^2 = 3^2$ to both sides.

ii The equation becomes $x^2 - 6x + 3^2 = 13 + 3^2$
$\therefore\ (x-3)^2 = 13 + 9$
$\therefore\ (x-3)^2 = 22$

EXERCISE 12D

1 For each equation:

- **i** What number must be added to both sides to complete a perfect square on the LHS?
- **ii** Hence write the equation in completed square form.

a $x^2 + 2x = 5$	**b** $x^2 - 2x = -7$	**c** $x^2 + 6x = 2$
d $x^2 - 6x = -3$	**e** $x^2 + 10x = 1$	**f** $x^2 - 8x = 5$
g $x^2 + 12x = 13$	**h** $x^2 + 5x = -2$	**i** $x^2 - 7x = 4$

Example 10 Self Tutor

Solve for x by completing the square, giving your answer in simplest surd form:

a $x^2 + 2x - 2 = 0$ **b** $x^2 - 5x + 3 = 0$

a
$$x^2 + 2x - 2 = 0$$
$$\therefore\ x^2 + 2x = 2 \quad \{\text{moving constant to RHS}\}$$
$$\therefore\ x^2 + 2x + 1^2 = 2 + 1^2 \quad \{\text{adding } \left(\tfrac{2}{2}\right)^2 = 1^2 \text{ to both sides}\}$$
$$\therefore\ (x+1)^2 = 3$$
$$\therefore\ x + 1 = \pm\sqrt{3}$$
$$\therefore\ x = -1 \pm \sqrt{3}$$

b
$$x^2 - 5x + 3 = 0$$
$$\therefore\ x^2 - 5x = -3 \quad \{\text{moving constant to RHS}\}$$
$$\therefore\ x^2 - 5x + \left(\tfrac{5}{2}\right)^2 = -3 + \left(\tfrac{5}{2}\right)^2 \quad \{\text{adding } \left(\tfrac{-5}{2}\right)^2 = \left(\tfrac{5}{2}\right)^2 \text{ to both sides}\}$$
$$\therefore\ \left(x - \tfrac{5}{2}\right)^2 = -3 + \tfrac{25}{4}$$
$$\therefore\ \left(x - \tfrac{5}{2}\right)^2 = \tfrac{13}{4}$$
$$\therefore\ x - \tfrac{5}{2} = \pm\sqrt{\tfrac{13}{4}}$$
$$\therefore\ x = \tfrac{5}{2} \pm \tfrac{\sqrt{13}}{2}$$
$$\therefore\ x = \frac{5 \pm \sqrt{13}}{2}$$

2 Solve for x by completing the square, giving your answer in simplest surd form:

a $x^2 - 4x + 1 = 0$ **b** $x^2 - 2x - 2 = 0$ **c** $x^2 - 4x - 3 = 0$
d $x^2 + 2x - 1 = 0$ **e** $x^2 + 4x + 1 = 0$ **f** $x^2 + 6x + 3 = 0$
g $x^2 + 3x + 2 = 0$ **h** $x^2 + 8x + 14 = 0$ **i** $x^2 - 3x - 1 = 0$

Example 11 Self Tutor

Solve for x by completing the square: $x^2 - 4x + 6 = 0$

$$x^2 - 4x + 6 = 0$$
$$\therefore\ x^2 - 4x = -6 \quad \{\text{moving constant to RHS}\}$$
$$\therefore\ x^2 - 4x + 2^2 = -6 + 2^2 \quad \{\text{adding } \left(\tfrac{-4}{2}\right)^2 = 2^2 \text{ to both sides}\}$$
$$\therefore\ (x-2)^2 = -2$$

$\therefore$ no real solutions exist.

If $(x - a)^2 < 0$, then there are no real solutions.

3 If possible, solve for x by completing the square:

a $x^2 + 2x + 4 = 0$ **b** $x^2 - 5x + 6 = 0$ **c** $x^2 - 6x + 11 = 0$
d $x^2 + x - 1 = 0$ **e** $x^2 + 5x - 2 = 0$ **f** $x^2 - 7x + 13 = 0$

4 **a** Solve by completing the square: $x^2 + bx + c = 0$

b Under what conditions does the equation $x^2 + bx + c = 0$ have:

i two real solutions **ii** one real solution **iii** no real solutions?

Example 12 Self Tutor

Solve by completing the square: $3x^2 + 6x - 2 = 0$

$$\begin{aligned} 3x^2 + 6x - 2 &= 0 \\ \therefore\ x^2 + 2x - \tfrac{2}{3} &= 0 \\ \therefore\ x^2 + 2x \quad &= \tfrac{2}{3} \\ \therefore\ x^2 + 2x + 1^2 &= \tfrac{2}{3} + 1^2 \\ \therefore\ (x+1)^2 &= \tfrac{5}{3} \\ \therefore\ x + 1 &= \pm\sqrt{\tfrac{5}{3}} \\ \therefore\ x &= -1 \pm \sqrt{\tfrac{5}{3}} \end{aligned}$$

5 Solve by completing the square:

a $2x^2 + 4x - 1 = 0$ **b** $3x^2 - 12x + 7 = 0$ **c** $5x^2 - 10x + 3 = 0$

d $2x^2 + 12x - 5 = 0$ **e** $2x^2 - 2x + 3 = 0$ **f** $3x^2 + 3x - 1 = 0$

E THE QUADRATIC FORMULA

Since completing the square is tedious, we can perform it on the general quadratic equation to generate a formula.

Suppose $ax^2 + bx + c = 0$

$$\therefore\ x^2 + \frac{b}{a}x + \frac{c}{a} = 0 \qquad \{\div a \text{ since } a \neq 0\}$$

$$\therefore\ x^2 + \frac{b}{a}x \quad = -\frac{c}{a}$$

$$\therefore\ x^2 + \frac{b}{a}x + \left(\frac{b}{2a}\right)^2 = -\frac{c}{a} + \left(\frac{b}{2a}\right)^2 \qquad \{\text{completing the square}\}$$

$$\therefore\ \left(x + \frac{b}{2a}\right)^2 = -\frac{c}{a}\left(\frac{4a}{4a}\right) + \frac{b^2}{4a^2} \qquad \{\text{factorising}\}$$

$$\therefore\ \left(x + \frac{b}{2a}\right)^2 = \frac{b^2 - 4ac}{4a^2}$$

$$\therefore\ x + \frac{b}{2a} = \pm\sqrt{\frac{b^2 - 4ac}{4a^2}}$$

$$\therefore\ x = -\frac{b}{2a} \pm \frac{\sqrt{b^2 - 4ac}}{2a}$$

$$\therefore\ x = \frac{-b \pm \sqrt{b^2 - 4ac}}{2a}$$

The result is called the **quadratic formula**:

$$\text{If } ax^2 + bx + c = 0, \quad a \neq 0, \quad \text{then } x = \frac{-b \pm \sqrt{b^2 - 4ac}}{2a}.$$

Example 13 **Self Tutor**

Solve for x: **a** $x^2 - 2x - 2 = 0$ **b** $2x^2 + 3x - 4 = 0$

a $x^2 - 2x - 2 = 0$ has $a = 1$, $b = -2$, $c = -2$

$$\therefore\ x = \frac{-(-2) \pm \sqrt{(-2)^2 - 4(1)(-2)}}{2(1)}$$

$$\therefore\ x = \frac{2 \pm \sqrt{4+8}}{2}$$

$$\therefore\ x = \frac{2 \pm \sqrt{12}}{2}$$

$$\therefore\ x = \frac{2 \pm 2\sqrt{3}}{2}$$

$$\therefore\ x = 1 \pm \sqrt{3}$$

b $2x^2 + 3x - 4 = 0$ has $a = 2$, $b = 3$, $c = -4$

$$\therefore\ x = \frac{-3 \pm \sqrt{3^2 - 4(2)(-4)}}{2(2)}$$

$$\therefore\ x = \frac{-3 \pm \sqrt{9+32}}{4}$$

$$\therefore\ x = \frac{-3 \pm \sqrt{41}}{4}$$

THE DISCRIMINANT Δ

The quantity $b^2 - 4ac$ under the square root sign is called the **discriminant**.

The symbol **delta** Δ is used to represent the discriminant, so $\Delta = b^2 - 4ac$.

- If $\Delta > 0$, the quadratic equation has two distinct real solutions.
- If $\Delta = 0$, the quadratic equation has one real solution which is $x = -\frac{b}{2a}$.
- If $\Delta < 0$, there are no real solutions.

EXERCISE 12E

1 Solve the following equations using: **i** factorisation **ii** the quadratic formula.

a $x^2 + 6x + 8 = 0$ **b** $x^2 - 10x + 25 = 0$ **c** $3x^2 - 7x - 6 = 0$

2 Use the quadratic formula to solve for x, giving exact answers:

a $x^2 + x - 5 = 0$ **b** $x^2 - 5x + 5 = 0$ **c** $x^2 - 4x - 1 = 0$
d $x^2 + 1 = 3x$ **e** $2x^2 = 2x + 3$ **f** $9x^2 = 6x + 1$

3 Use the quadratic formula to solve for x, rounding your answers to 2 decimal places:

a $3x^2 + 5x - 1 = 0$ **b** $-2x^2 + x + 7 = 0$ **c** $5x^2 - 8x + 1 = 0$
d $7x^2 = 5x + 1$ **e** $3x^2 + 2x = 2$ **f** $25x^2 + 1 = 20x$

4 Use the discriminant Δ to show that the following quadratic equations have no real solutions:

a $x^2 - 3x + 12 = 0$ **b** $x^2 + 2x + 4 = 0$ **c** $-2x^2 + x - 1 = 0$

5 Use the discriminant Δ to determine the values of k for which $3x^2 - 5x + k = 0$ has:

a two distinct real solutions **b** one real solution **c** no real solutions.

6 Solve for x, where possible:

a $x^2 - 4x + 5 = 0$ **b** $x^2 - 4x - 5 = 0$ **c** $x^2 - 10x + 29 = 0$

d $x^2 + 6x + 25 = 0$ **e** $2x^2 - 6x - 5 = 0$ **f** $2x^2 + x - 2 = 0$

7 Use the quadratic formula to solve for x:

a $(x+2)(x-1) = 5$ **b** $(x+1)^2 = 3 - x^2$ **c** $\dfrac{x-1}{x} = \dfrac{x}{3}$

HISTORICAL NOTE — BABYLONIAN ALGEBRA

The mathematics used by the **Babylonians** was recorded on clay tablets in cuneiform. One tablet that has been preserved is called *Plimpton 322*, written around 1600 BC.

Plimpton 322

The Ancient Babylonians were able to solve difficult equations using the rules we use today, such as transposing terms, multiplying both sides by like quantities to remove fractions, and factorisation.

This was all achieved without the use of letters for unknown quantities. Instead, they often used words for the unknown.

Consider the following example from about 4000 years ago:

Problem: *"I have subtracted the side of my square from the area and the result is 870. What is the side of the square?"*

Solution: Take half of 1, which is $\frac{1}{2}$, and multiply $\frac{1}{2}$ by $\frac{1}{2}$ which is $\frac{1}{4}$.
Add this to 870 to get $870\frac{1}{4}$. This is the square of $29\frac{1}{2}$.
Now add $\frac{1}{2}$ to $29\frac{1}{2}$ and the result is 30, the side of the square.

Using our modern symbols, the equation is $x^2 - x = 870$, and one of the solutions is $x = \sqrt{(\frac{1}{2})^2 + 870} + \frac{1}{2} = 30$.

Can you see how the Babylonian solution is related in form to the quadratic formula?

F PROBLEM SOLVING

To solve practical problems involving a quadratic equation, follow these steps:

Step 1: Carefully **read the question** until you understand the problem. A **sketch** may be useful.

Step 2: Decide on the **unknown** quantity and label it with a variable such as x.

Step 3: Use the information given to write an **equation**.

Step 4: **Solve** the equation.

Step 5: **Check** that any solutions satisfy the equation and are reasonable.

Step 6: Write your answer to the question in **sentence form**.

Example 14 — Self Tutor

The sum of a number and its square is 42. Find the number.

Let the number be x, so its square is x^2.

$$\therefore\ x + x^2 = 42$$
$$\therefore\ x^2 + x - 42 = 0 \qquad \{\text{making RHS} = 0\}$$
$$\therefore\ (x+7)(x-6) = 0 \qquad \{\text{factorising}\}$$
$$\therefore\ x = -7 \text{ or } x = 6$$

So, the number is -7 or 6.

Check: The sum of -7 and its square is $-7 + (-7)^2 = -7 + 49 = 42$ ✓
The sum of 6 and its square is $6 + 6^2 = 6 + 36 = 42$ ✓

EXERCISE 12F

1. The sum of a number and its square is 110. Find the number.
2. When 24 is subtracted from the square of a number, the result is five times the original number. Find the number.
3. The sum of two numbers is 9, and the sum of their squares is 45. Find the numbers.
4. Two numbers differ by 7, and the sum of their squares is 29. Find the numbers.

Example 15 — Self Tutor

A rectangle has length 5 cm greater than its width, and its area is 84 cm^2. Find the dimensions of the rectangle.

If the rectangle has width x cm, then its length is $(x+5)$ cm.

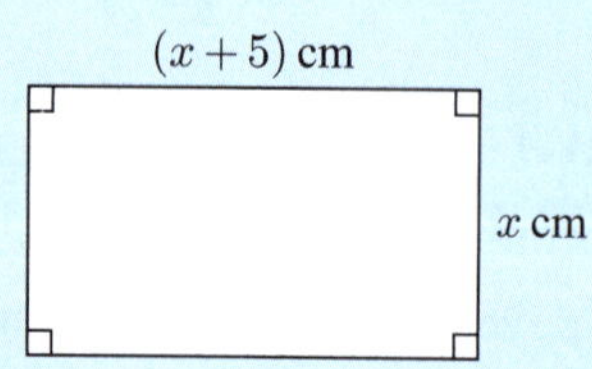

$$\therefore\ x(x+5) = 84$$
$$\therefore\ x^2 + 5x = 84$$
$$\therefore\ x^2 + 5x - 84 = 0$$
$$\therefore\ (x+12)(x-7) = 0$$
$$\therefore\ x = 7 \qquad \{\text{lengths must be positive}\}$$

$\therefore$ the rectangle is 7 cm $\times$ 12 cm.

5. A rectangle has length 4 cm greater than its width, and its area is 96 cm^2. Find the width of the rectangle.
6. The base of a triangle is 4 m longer than its altitude. The area of the triangle is 70 m^2. Find the triangle's altitude.
7. Use Pythagoras' theorem to find the value of x:

a
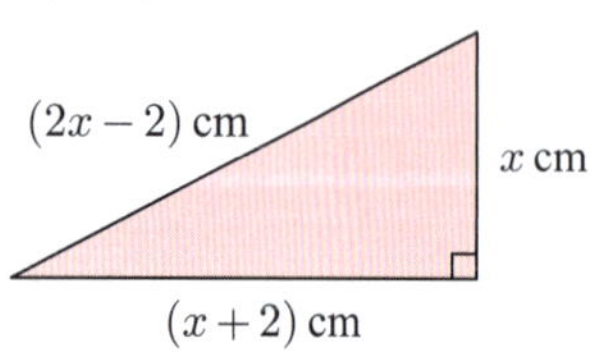

b
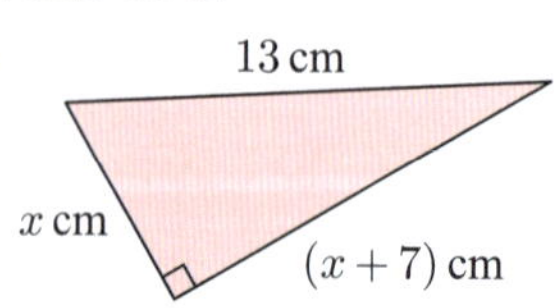

c
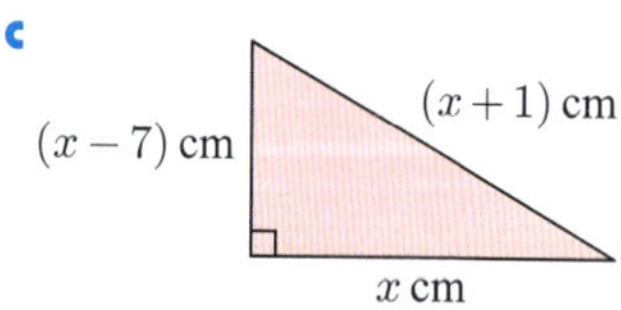

8 A rectangular enclosure is made from 60 m of fencing. The area enclosed is 216 m^2.

a Let the width of the enclosure be x m. Write the length of the enclosure in terms of x.

b Hence find the dimensions of the enclosure.

9 A right angled triangle has sides 1 cm and 2 cm respectively less than its hypotenuse. Find the length of the hypotenuse.

10 A rectangular garden bed was built against an existing brick wall. 22 m of edging was used to enclose 60 m^2. Find the dimensions of the garden bed.

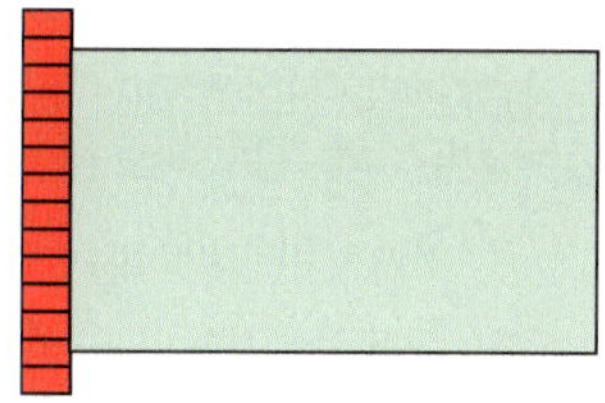

11 ABCD is a rectangle in which AB = 21 cm.
The square AXYD is removed and the remaining rectangle has area 80 cm^2.
Find the length BC.

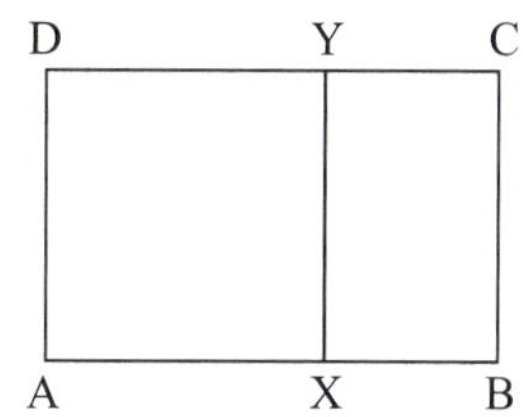

12 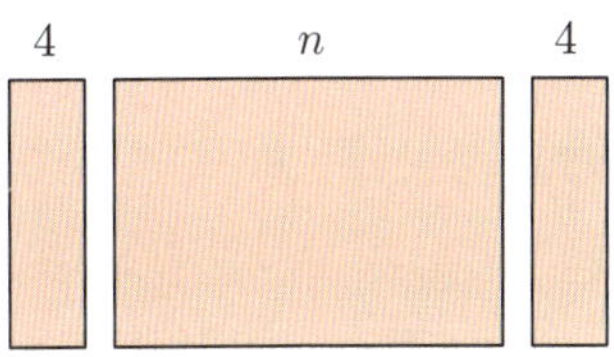

A theatre contains a central block of seats with n seats per row. Blocks on either side contain 4 seats per row. The number of rows is 5 less than the total number of seats per row. In total there are 126 seats in the theatre. Find the value of n.

13 The numerator of a fraction is 3 less than the denominator. If the numerator is increased by 6 and the denominator is increased by 5, the fraction is doubled in value. Find the original fraction.

14 At a fruit market, John bought some oranges for a total of \$20. When Jenny visited a different stall, she bought 10 more oranges than John for the same total amount. Given that the difference in price per orange was 10 cents, how many oranges did John buy?

15 Answer the **Opening Problem** on page **176**.

16 A circular magnet has inner radius x cm, outer radius 2 cm larger, and its depth is the same as the inner radius. The total volume of the magnet is 120π cm^3.
Find the value of x.

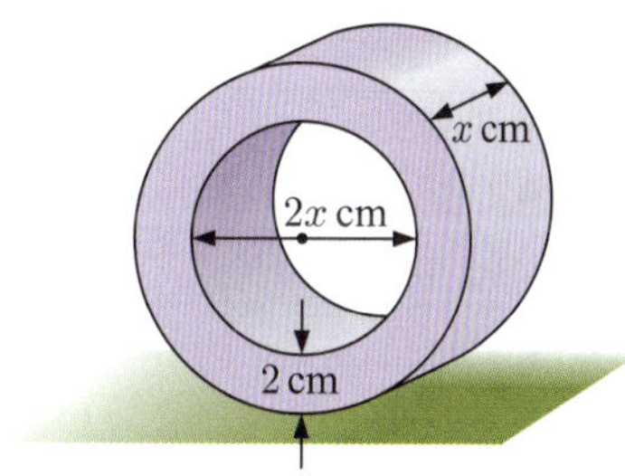

17 The sum of a number and its reciprocal is $2\frac{1}{12}$. Find the number.

18 Two numbers have a sum of 4, and the sum of their reciprocals is 8. Find the numbers.

19 A sheet of cardboard is 15 cm long and 10 cm wide. It is to be made into an open box with base area 66 cm^2 by cutting out equal squares from the four corners and then bending the edges upwards.
Find the size of the squares to be cut out.

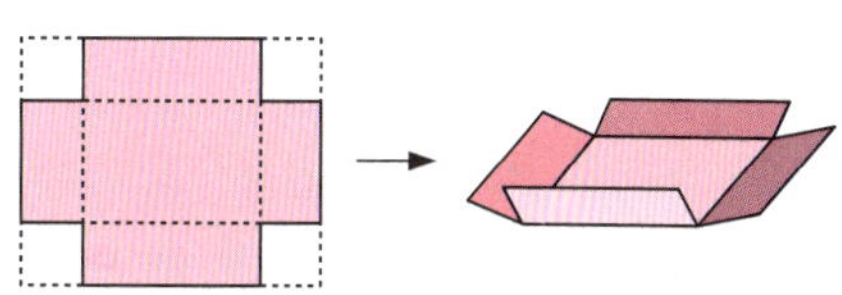

20 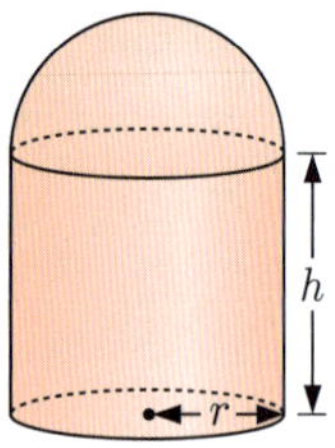

A takeaway milkshake container is cylindrical, with a hemispherical lid on top.
The height h of the container is 7 cm greater than its base radius r.
The surface area of the container and lid is 96π cm^2.
Find the base radius of the container.

21 A rectangular swimming pool is 12 m long by 6 m wide. It is surrounded by a pavement of uniform width x m. The area of the pavement is $\frac{7}{8}$ of the area of the pool.

- **a** Show that the area of the pavement is $(4x^2 + 36x)$ m^2.
- **b** Hence show that $4x^2 + 36x - 63 = 0$.
- **c** How wide is the pavement?

QUICK QUIZ

MULTIPLE CHOICE QUIZ

REVIEW SET 12A

1 Solve for x:

a $-x^2 + 11 = 0$ **b** $7 - 2x^2 = -25$ **c** $3(x-6)(3x+2) = 0$

2 Solve for x:

a $(x-4)^2 = 25$ **b** $(2x+1)^2 - 1 = 0$ **c** $\frac{x}{9} = \frac{4}{x}$

3 Solve for x:

a $2x^2 - 10x = 0$ **b** $7x - x^2 = x + 4x^2$ **c** $5(x-1) = x(1-x)$

4 Solve for x:

a $x^2 - 4x - 21 = 0$ **b** $x^2 + 13x = 30$ **c** $(x+6)(x-3) = 10x$

5 Solve for x:

a $81 - 16x^2 = 0$ **b** $5x^2 + 30x + 45 = 0$ **c** $4x^2 + 1 = 4x$

6 Solve for x:

a $\frac{x-2}{3} = \frac{5}{x}$ **b** $\frac{x}{x+3} = \frac{18}{x+15}$

7 Solve for x:

a $2x^2 - x - 6 = 0$ **b** $10x^2 - x - 2 = 0$ **c** $\frac{x+3}{2} = \frac{x+5}{3x}$

8 Solve for x by completing the square:

a $x^2 - 8x = 2$ **b** $x^2 + 6x + 4 = 0$ **c** $2x^2 + 2x = 48$

9 Use the quadratic formula to solve for x:

a $2x^2 - 3x - 2 = 0$ **b** $3x^2 + 4x - 5 = 0$ **c** $\frac{x+4}{x-2} = \frac{5x}{x-1}$

10 The sum of two numbers is 13, and the sum of their squares is 89. Find the two numbers.

11 Find the value of x:

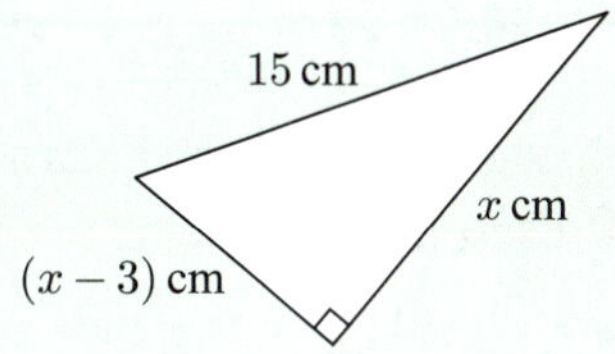

12 A group of friends hires a bus for a day for \$480, agreeing to share the cost equally. At the last minute, two more people decide to go on the trip. As a result, each person pays \$8 less. How many people go on the trip, and how much does each person pay?

13 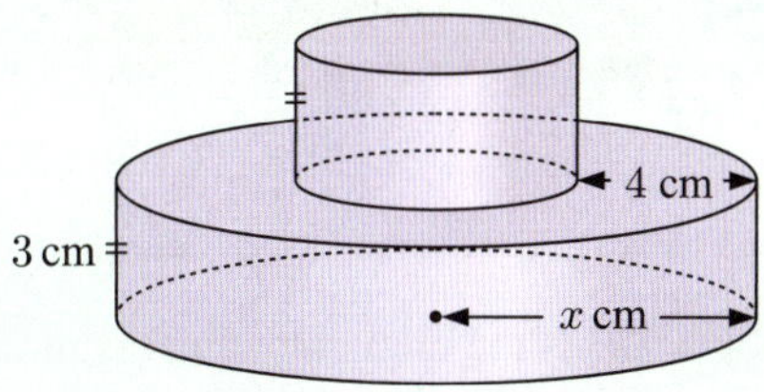

The volume of this solid is 174π cm^3. Find the value of x.

14 The sum of a number and its reciprocal is $2\frac{1}{6}$. Find the number.

15 In a right angled triangle, the second longest side is 5 cm longer than the shortest side, and the hypotenuse is three times the length of the shortest side. Find the exact length of the hypotenuse.

REVIEW SET 12B

1 Solve for x:

a $8-3x^2=-10$ **b** $2x^2-5=-3$ **c** $(4-x)(3x+7)=0$

2 Solve for x:

a $(x+3)^2-19=0$ **b** $5(2x-3)^2=20$ **c** $(3x-1)^2=17$

3 Solve using the difference between two squares factorisation:

a $144-x^2=0$ **b** $18x^2-50=0$ **c** $(3x+1)^2=(2x-5)^2$

4 Solve for x:

a $x^2-8x-33=0$ **b** $x^2-45=4x$ **c** $x(2x-9)=(x-5)(x+12)$

5 Solve for x:

a $3x^2-3x-6=0$ **b** $8x+10=2x^2$ **c** $9x^2+6x=-1$

6 Solve for x:

a $\dfrac{2x+3}{x+1}=\dfrac{3}{x}$ **b** $\dfrac{x}{3-2x}=\dfrac{x+1}{3-x}$

7 Solve for x:

a $6x^2-x-2=0$ **b** $2(5x^2-3)=11x$ **c** $\dfrac{7}{x-2}-\dfrac{2}{x+1}=2$

8 Solve for x by completing the square:

a $x^2-2x=100$ **b** $x^2+7x=\frac{3}{4}$ **c** $-2x^2+12x=-16$

9 Use the quadratic formula to solve for x:

a $2x^2 + 2x - 1 = 0$ **b** $6x^2 + 26x = 20$ **c** $\dfrac{1}{x} - \dfrac{1}{1-x} = 2$

10 Use the discriminant Δ to determine the values of k for which $2x^2 + 3x + k = 0$ has:

a two distinct real solutions **b** one real solution **c** no real solutions.

11 A rectangle has length 5 mm greater than its width, and its area is 84 mm^2. Find the length of the rectangle.

12 In the flag alongside, the area of the red stripes is 700 cm^2. Find the width of the red stripes.

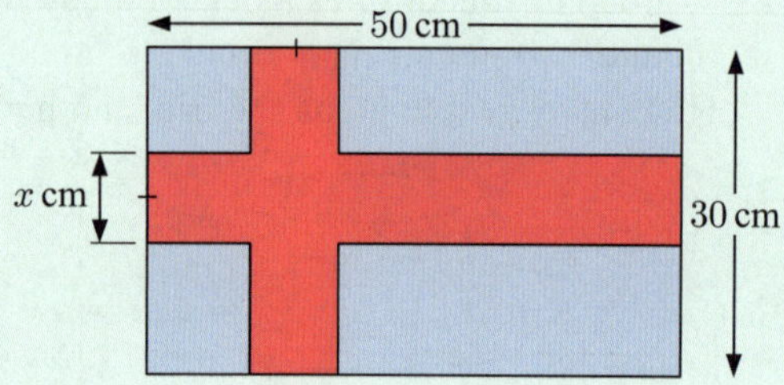

13 The sum of the numerator and denominator of a fraction is 10. If the numerator and denominator are each increased by 4, the fraction is doubled in value. Find the original fraction.

14 The sum of a number and five times its square is equal to four. Find the number.

15 A 20 cm length of wire is bent at right angles to form the two shorter sides of a right angled triangle.

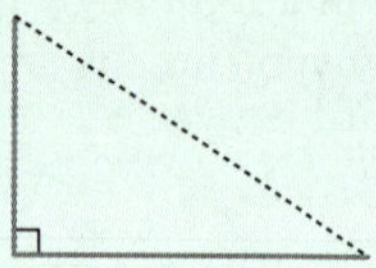

a If the triangle's area is 30 cm^2, find:

i the length of each side

ii the perimeter of the triangle.

b Is it possible for a right angled triangle whose shorter sides are made from a 20 cm length of wire to have area 51 cm^2? Explain your answer.

Chapter 13

Coordinate geometry

Contents:

OPENING PROBLEM

A city has three ambulance stations, located at A(1, 6), B(3, −4), and C(−5, −2). The grid lines are 1 km apart.

This diagram, called a **Voronoi diagram**, shows the regions of the city that are closest to each ambulance station.

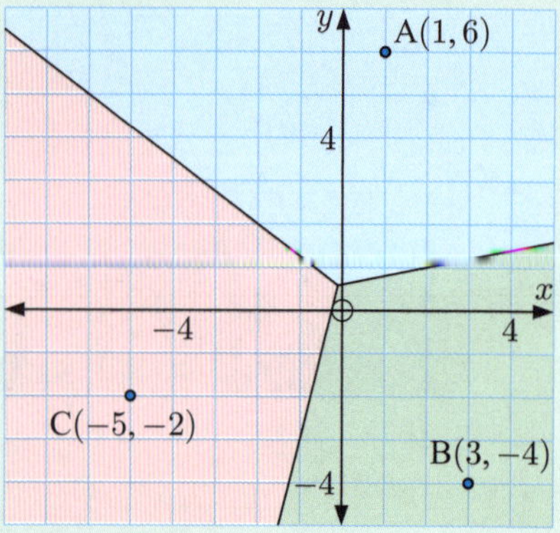

When an accident occurs, an ambulance from the closest station will attend.

Things to think about:

- **a** How far is it between stations A and B?
- **b** Can you describe how the boundaries of the regions are constructed?
- **c** What is the equation of each boundary between the regions?
- **d** Suppose an accident occurs at (1, 0). Which ambulance station should respond?
- **e** Can you show that (−6, 5) lies on one of the edges of the Voronoi diagram? What does this tell us about the closest ambulance station to this location?

HISTORICAL NOTE

The two Frenchmen **René Descartes** and **Pierre de Fermat** arrived at the idea of **analytical geometry** at about the same time. Descartes' work *La Géométrie* was published first, in 1637, while Fermat's *Introduction to Loci* was not published until after his death.

Today, they are considered the co-founders of this important branch of mathematics which links algebra and geometry.

The initial approaches used by these mathematicians were quite opposite. Descartes began with a line or curve and then found the equation which described it. Fermat started with an equation and investigated the shape of the curve it described.

René Descartes

Analytical geometry and its use of coordinates enabled **Isaac Newton** to develop another important branch of mathematics called **calculus**. Newton humbly stated: *"If I have seen further than Descartes, it is because I have stood on the shoulders of giants."*

The **number plane** or **Cartesian plane** consists of two perpendicular axes which intersect at the **origin**, O.

The **x-axis** is horizontal and the **y-axis** is vertical.

The axes divide the number plane into four **quadrants**.

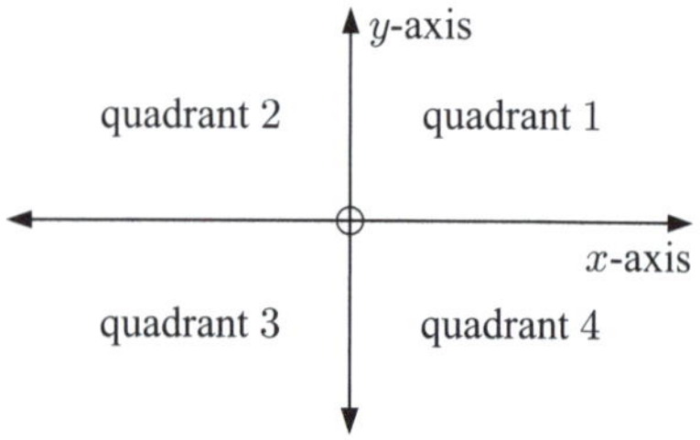

We can describe any point on the Cartesian plane using an **ordered pair** of numbers called **coordinates**.

DEMO

- The coordinates of a point are written (x-coordinate, y-coordinate).
- The x-coordinate gives the horizontal position along the x-axis.
- The y-coordinate gives the vertical position along the y-axis.

A THE DISTANCE BETWEEN TWO POINTS

To find the distance d between the points A$(-1, 5)$ and B$(4, 1)$, we use the horizontal and vertical grid lines to help construct a right angled triangle with hypotenuse [AB].

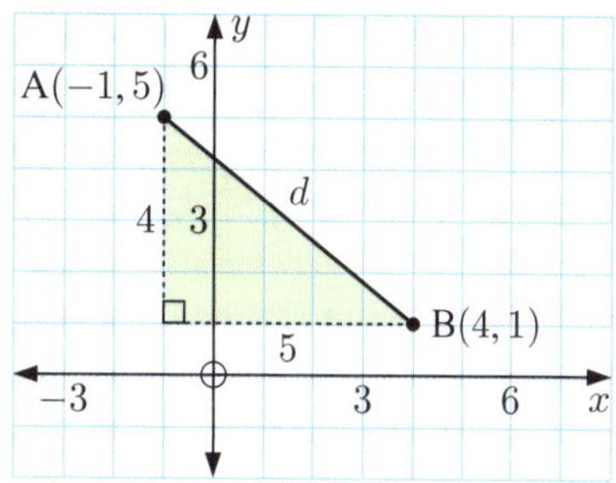

The horizontal distance or ***x*-step** between the points is 5 units. The vertical distance or ***y*-step** between the points is 4 units.

$\therefore\ d^2 = 5^2 + 4^2$ {Pythagoras}

$\therefore\ d^2 = 41$

$\therefore\ d = \sqrt{41}$ {since $d > 0$}

So, the distance between A and B is $\sqrt{41}$ units.

THE DISTANCE FORMULA

To make the process quicker, we can develop a **distance formula** for the distance between two general points (x_1, y_1) and (x_2, y_2).

To go from (x_1, y_1) to (x_2, y_2), we find the

x-step $= x_2 - x_1$

and y-step $= y_2 - y_1$.

$\therefore\ d^2 = (x\text{-step})^2 + (y\text{-step})^2$ {Pythagoras}

$\therefore\ d = \sqrt{(x\text{-step})^2 + (y\text{-step})^2}$

$\therefore\ d = \sqrt{(x_2 - x_1)^2 + (y_2 - y_1)^2}$

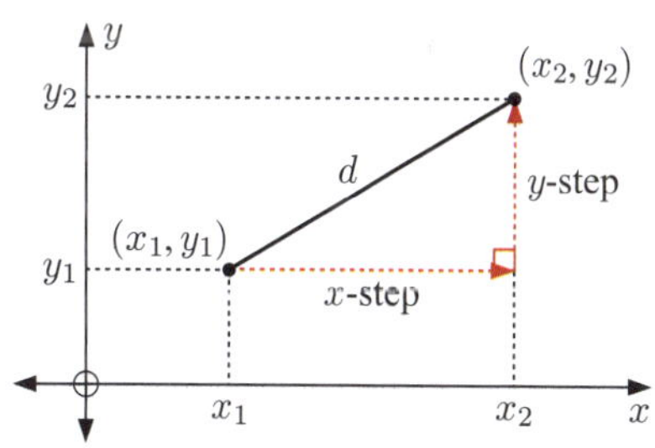

The distance d between two points (x_1, y_1) and (x_2, y_2) is given by

$$\boldsymbol{d = \sqrt{(x_2 - x_1)^2 + (y_2 - y_1)^2}}.$$

Example 1 **Self Tutor**

Find the distance between A$(-2, 1)$ and B$(3, 4)$.

A$(-2, 1)$ with $x_1 = -2$, $y_1 = 1$; B$(3, 4)$ with $x_2 = 3$, $y_2 = 4$

$$\begin{aligned} AB &= \sqrt{(3 - -2)^2 + (4 - 1)^2} \\ &= \sqrt{5^2 + 3^2} \\ &= \sqrt{25 + 9} \\ &= \sqrt{34} \text{ units} \end{aligned}$$

EXERCISE 13A

1 Find the distance between:

a A(3, 1) and B(5, 3) **b** C(−1, 2) and D(6, 2) **c** O(0, 0) and P(−2, 4)

d E(8, 0) and F(2, −3) **e** G(0, −2) and H(0, 5) **f** I(2, 0) and J(0, −1)

g P(1, 2) and Q(−2, 3) **h** R(1, −1) and S($\frac{1}{2}$, −2) **i** W(1, $\sqrt{3}$) and Z(2, $-\sqrt{3}$).

2 The grid lines on this map are 10 km apart.

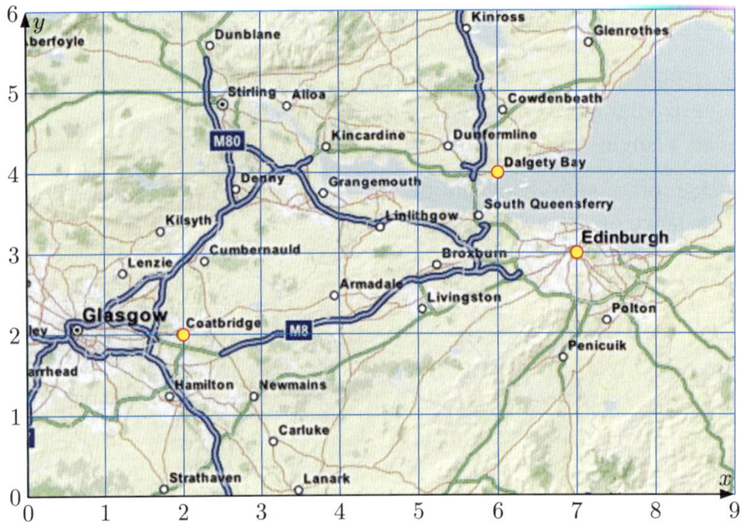

Find the direct distance between:

a Dalgety Bay and Edinburgh **b** Coatbridge and Dalgety Bay

c Coatbridge and Edinburgh.

Example 2 **Self Tutor**

Consider the triangle formed by the points A(1, 2), B(2, 5), and C(4, 1).

a Classify the triangle as equilateral, isosceles, or scalene.

b Determine whether the triangle is right angled.

a $AB = \sqrt{(2-1)^2 + (5-2)^2}$
$= \sqrt{1^2 + 3^2}$
$= \sqrt{10}$ units

$AC = \sqrt{(4-1)^2 + (1-2)^2}$
$= \sqrt{3^2 + (-1)^2}$
$= \sqrt{10}$ units

$BC = \sqrt{(4-2)^2 + (1-5)^2}$
$= \sqrt{2^2 + (-4)^2}$
$= \sqrt{20}$ units

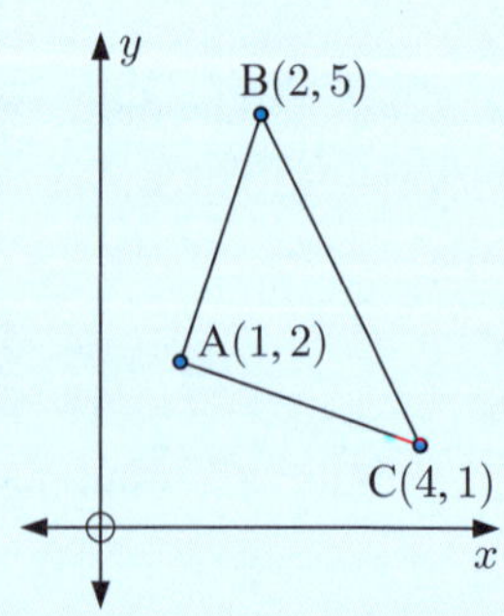

Since AB = AC, the triangle is isosceles.

b The shortest sides are [AB] and [AC].

Now $AB^2 + AC^2 = 10 + 10 = 20 = BC^2$

Using the converse of Pythagoras' theorem, triangle ABC is right angled. The right angle is at A, opposite the longest side.

3 Classify triangle ABC as equilateral, isosceles, or scalene:

a A(3, −1), B(1, 8), C(−6, 1)
b A(1, 0), B(3, 1), C(4, 5)
c A(−1, 0), B(2, −2), C(4, 1)
d A($\sqrt{2}$, 0), B($-\sqrt{2}$, 0), C(0, $-\sqrt{5}$)
e A($\sqrt{3}$, 1), B($-\sqrt{3}$, 1), C(0, −2)
f A(a, b), B($-a$, b), C(0, 2)

4 Determine whether triangle ABC is right angled. If there is a right angle, state the vertex where it occurs.

a A(−2, −1), B(3, −1), C(3, 3)
b A(−1, 2), B(4, 1), C(4, −5)
c A(1, −2), B(3, 0), C(−3, 2)
d A(3, −4), B(−2, −5), C(−1, 1)

5 Find, to 1 decimal place, the perimeter of the triangle with vertices P(−1, 4), Q(2, 1), and R(−2, −1).

6 Classify the quadrilateral with vertices A(1, 0), B(4, −2), C(2, 1), and D(−1, 3).

Example 3 — Self Tutor

Find b given that A(3, −2) and B(b, 1) are $\sqrt{13}$ units apart. Explain your result using a diagram.

From A to B, the x-step $= b - 3$
and y-step $= 1 - -2 = 3$.

$\therefore \sqrt{(b-3)^2 + 3^2} = \sqrt{13}$
$\therefore (b-3)^2 + 9 = 13$ {squaring both sides}
$\therefore (b-3)^2 = 4$
$\therefore b - 3 = \pm 2$
$\therefore b = 3 \pm 2$
$\therefore b = 5$ or 1

Point B could be at two possible locations: (5, 1) or (1, 1).

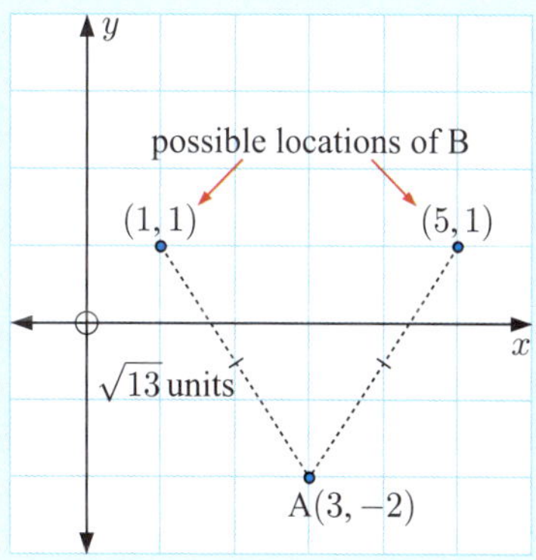

7 Find a and explain the result using a diagram, given:

a P(2, 3) and Q(a, −1) are 4 units apart
b P(−1, 1) and Q(a, −2) are 5 units apart
c P(3, a) and Q(a, 5) are $\sqrt{10}$ units apart
d X(a, a) is $\sqrt{8}$ units from the origin
e A(0, a) is equidistant from P(3, −3) and Q(−2, 2).

8 **a** Find the relationship between x and y if the point P(x, y) is always:

i 3 units from O(0, 0)
ii 2 units from A(1, 3).

b Illustrate and describe the set of points (x, y) such that $x^2 + y^2 = 1$.

c Find the equation of a circle with centre (a, b) and radius r.

B MIDPOINTS

The **midpoint** of line segment [AB] is the point on [AB] which is equidistant from A and B.

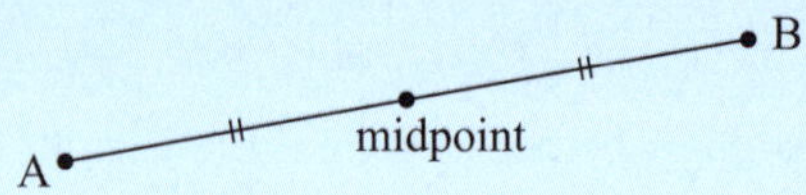

Consider the points A(−1, −2) and B(6, 4).

From the diagram we see that the midpoint of [AB] is M($2\frac{1}{2}$, 1).

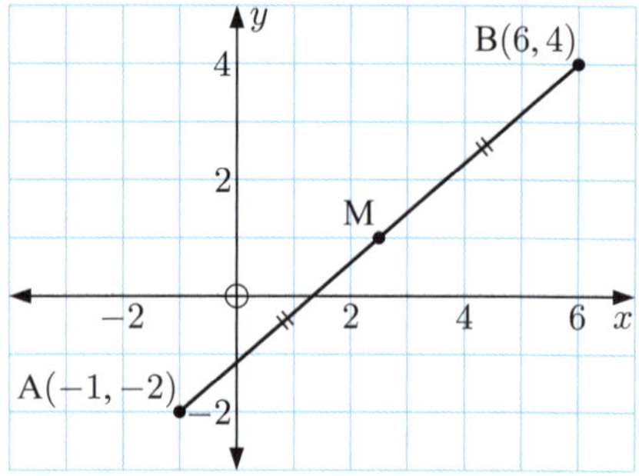

The x-coordinate of M is the *average* of the x-coordinates of A and B.

$\therefore$ the x-coordinate of M $= \dfrac{-1+6}{2} = \frac{5}{2} = 2\frac{1}{2}$

The y-coordinate of M is the *average* of the y-coordinates of A and B.

$\therefore$ the y-coordinate of M $= \dfrac{-2+4}{2} = 1$

If A(x_1, y_1) and B(x_2, y_2) are two points, then the **midpoint** of [AB] has coordinates $\left(\dfrac{x_1+x_2}{2}, \dfrac{y_1+y_2}{2}\right)$.

DEMO

Example 4 Self Tutor

Find the midpoint M of [AB] given A(−1, 3) and B(4, 7).

A(−1, 3): x_1 y_1 B(4, 7): x_2 y_2

The x-coordinate of M

$= \dfrac{x_1+x_2}{2}$

$= \dfrac{-1+4}{2}$

$= \frac{3}{2} = 1\frac{1}{2}$

The y-coordinate of M

$= \dfrac{y_1+y_2}{2}$

$= \dfrac{3+7}{2}$

$= 5$

So, the midpoint is M($1\frac{1}{2}$, 5).

EXERCISE 13B

1 Find the midpoint M of [AB] given:

a A(8, 1) and B(2, 5)
b A(2, −3) and B(0, 1)
c A(3, 0) and B(0, 6)
d A(−1, 4) and B(1, 4)
e A(5, −3) and B(−1, 0)
f A(5, 9) and B(−3, −4)
g A(3, a) and B(−1, $a-2$)
h A(−2, a) and B(4, $-a$).

2 **a** Find the midpoint of:

i [AC] **ii** [BD]

b Hence classify quadrilateral ABCD. Explain your answer.

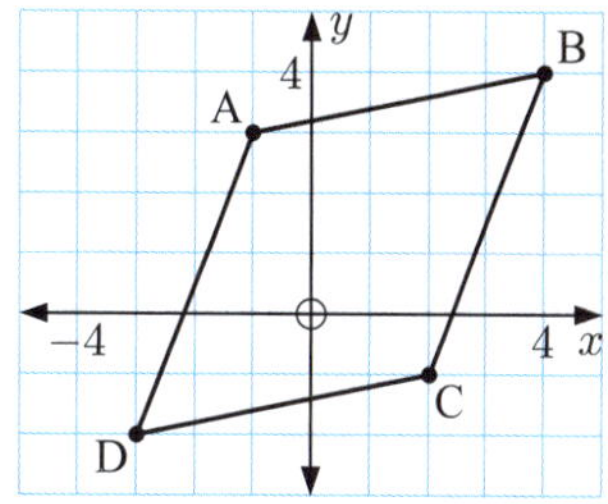

Example 5 **Self Tutor**

Suppose M is the midpoint of [AB] where A is $(1, 3)$ and M is $(4, -2)$.
Find the coordinates of B using:

a the midpoint formula **b** equal steps.

a Suppose B has coordinates (a, b).

$\therefore \frac{a+1}{2} = 4$ and $\frac{b+3}{2} = -2$

$\therefore a + 1 = 8$ and $b + 3 = -4$

$\therefore a = 7$ and $b = -7$

So, B is $(7, -7)$.

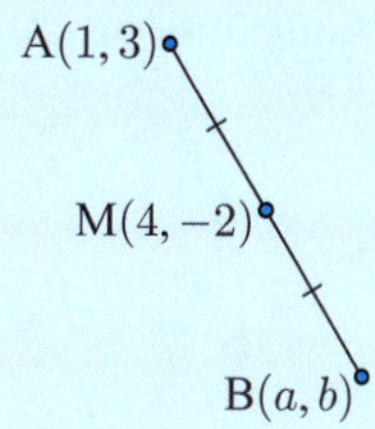

b x-step: $1 \xrightarrow{+3} 4 \xrightarrow{+3} 7$

y-step: $3 \xrightarrow{-5} -2 \xrightarrow{-5} -7$

So, B is $(7, -7)$.

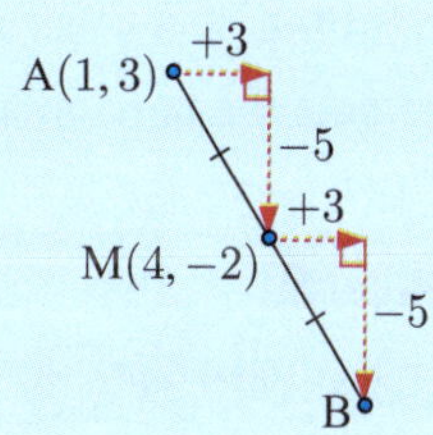

3 Suppose M is the midpoint of [AB]. Find the coordinates of B given:

a A(6, 4) and M(3, −1) **b** A(−5, 0) and M(0, −1)

c A(3, −2) and M($1\frac{1}{2}$, 2) **d** A(−1, −2) and M($-\frac{1}{2}$, $2\frac{1}{2}$)

e A(7, −3) and M(0, 0) **f** A(3, −1) and M(0, $-\frac{1}{2}$).

4 [AB] is a diameter of a circle with centre C. Given A is $(3, -2)$ and B is $(-1, -4)$, find the coordinates of C.

5 [PQ] is a diameter of a circle with centre $(3, -\frac{1}{2})$. Given Q is $(-1, 2)$, find the coordinates of P.

6 Triangle ABC has vertices $A(-1, 3)$, $B(1, -1)$, and $C(5, 2)$. Find the length of the line segment from A to the midpoint of [BC].

7 ABCD is a parallelogram. A is $(-1, -1)$ and B is $(-4, 3)$. The diagonals of the parallelogram intersect at $X(0, 1)$.

a Find the coordinates of: **i** C **ii** D

b Find the length of: **i** [AB] **ii** [BC]

c Hence classify ABCD more completely.

C GRADIENT

The **gradient** of a line is a measure of its steepness.

We can calculate the rate at which a line rises or falls by choosing two points on the line, then dividing the vertical step between them by the horizontal step between them.

gradient $= \dfrac{\textbf{vertical step}}{\textbf{horizontal step}}$ or $\dfrac{y\textbf{-step}}{x\textbf{-step}}$

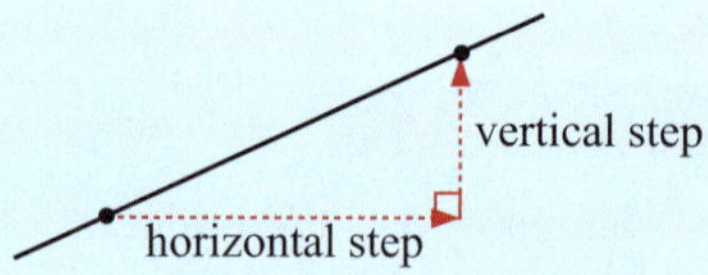

In general:

- An **upward sloping line** has a **positive gradient**.

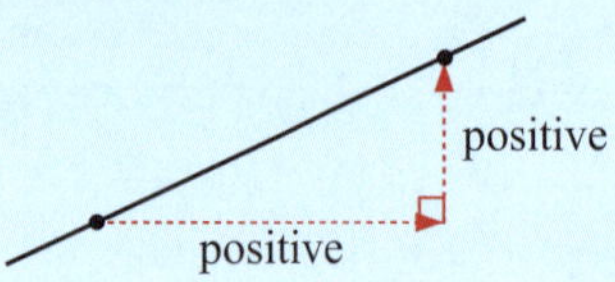

- A **downward sloping line** has a **negative gradient**.

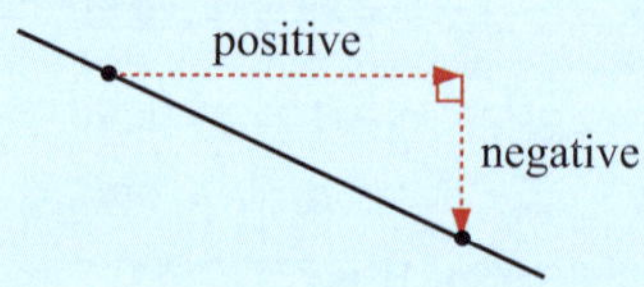

- For a **horizontal line**, the vertical step is 0, so the **gradient of a horizontal line is 0**.
- For a **vertical line**, the horizontal step is 0, so the **gradient of a vertical line is undefined**.

Example 6 Self Tutor

Draw lines with gradients $\frac{3}{4}$ and -2 through the point $(2, 1)$.

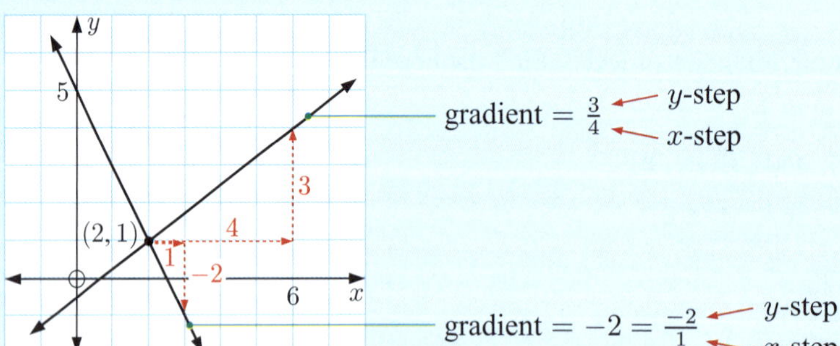

EXERCISE 13C.1

1 Find the gradient of each line segment:

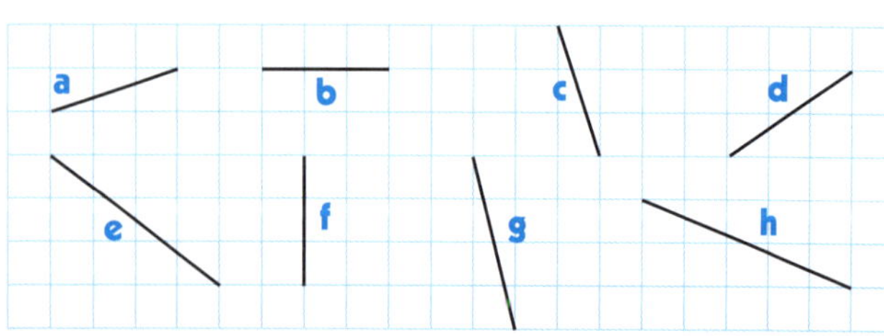

2 On grid paper, draw a line segment with gradient:

a $\frac{3}{4}$ **b** $-\frac{1}{2}$ **c** -4 **d** $\frac{5}{2}$ **e** 0 **f** $-\frac{2}{5}$

3 Draw a line through the point:

a $(3, -1)$ with gradient $\frac{1}{2}$ **b** $(-1, 3)$ with gradient $-\frac{3}{4}$.

4 On the same set of axes, draw lines through $(2, 3)$ with gradients $\frac{1}{3}$, $\frac{3}{4}$, 2, and 4.

5 On the same set of axes, draw lines through $(-1, 2)$ with gradients 0, $-\frac{2}{5}$, -2, and -5.

THE GRADIENT FORMULA

Although we can find gradients using steps on a diagram, it is often quicker to use a formula.

For points $A(x_1, y_1)$ and $B(x_2, y_2)$, the vertical step is $y_2 - y_1$, and the horizontal step is $x_2 - x_1$.

$\therefore$ the gradient is $\dfrac{y_2 - y_1}{x_2 - x_1}$.

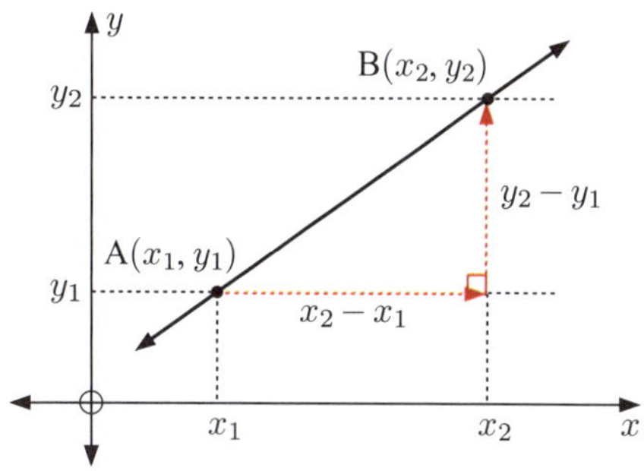

The **gradient** of the line through (x_1, y_1) and (x_2, y_2) is $\dfrac{y_2 - y_1}{x_2 - x_1}$.

Example 7 **Self Tutor**

Find the gradient of the line through $(3, -2)$ and $(6, 4)$.

$(3, -2)$: x_1, y_1 $(6, 4)$: x_2, y_2

$$\text{gradient} = \frac{y_2 - y_1}{x_2 - x_1} = \frac{4 - -2}{6 - 3} = \frac{6}{3} = 2$$

EXERCISE 13C.2

1 Find the gradient of the line segment joining:

a $O(0, 0)$ and $P(6, -2)$ **b** $O(0, 0)$ and $P(-3, -4)$ **c** $O(0, 0)$ and $P(a, b)$.

2 Use the gradient formula to find the gradient of the line through $A(-2, -3)$ and $B(5, 1)$. Plot [AB] on a set of axes to illustrate your answer.

3 Find the gradient of the line through:

a $(2, 3)$ and $(7, 4)$ **b** $(5, 7)$ and $(1, 6)$ **c** $(1, -2)$ and $(3, 6)$

d $(5, 5)$ and $(-1, 5)$ **e** $(3, -1)$ and $(3, -4)$ **f** $(5, -1)$ and $(-2, -3)$

g $(-5, 2)$ and $(2, 0)$ **h** $(0, -1)$ and $(-2, -3)$ **i** $(-1, 7)$ and $(11, -9)$.

Example 8 **Self Tutor**

Find t given that the line segment joining $(5,\ -2)$ and $(9,\ t)$ has gradient $\frac{2}{3}$.

The line segment joining $(5,\ -2)$ and $(9,\ t)$ has gradient $= \dfrac{t - -2}{9-5} = \dfrac{t+2}{4}$.

$$\therefore\ \frac{t+2}{4} = \frac{2}{3}$$

$$\therefore\ t+2 = \frac{8}{3}$$

$$\therefore\ t = \tfrac{2}{3}$$

4 Find t given that the line segment joining:

a $(-3,\ 5)$ and $(4,\ t)$ has gradient 2

b $(5,\ t)$ and $(10,\ 12)$ has gradient $-\frac{1}{2}$

c $(3,\ -6)$ and $(t,\ -2)$ has gradient 3

d $(t,\ 9)$ and $(4,\ 7)$ has gradient $-\frac{3}{5}$

e $(2,\ 5)$ and $(t,\ t)$ has gradient $\frac{4}{7}$

f $(t,\ 2t)$ and $(-3,\ 12)$ has gradient $-\frac{1}{4}$.

5

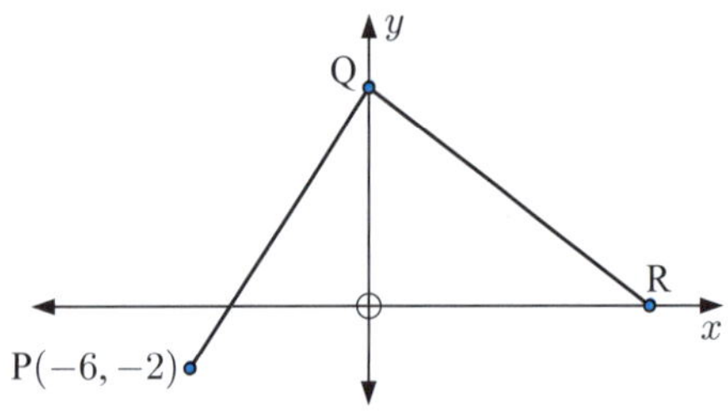

The gradient of [PQ] is $\frac{3}{2}$, and the gradient of [QR] is $-\frac{3}{4}$.

Find the coordinates of R.

D PARALLEL AND PERPENDICULAR LINES

Two lines are **parallel** if there is always the same distance between them.

Two lines are **perpendicular** if they are at right angles to one another.

PARALLEL LINES

The given lines are parallel, and both of them have a gradient of 3.

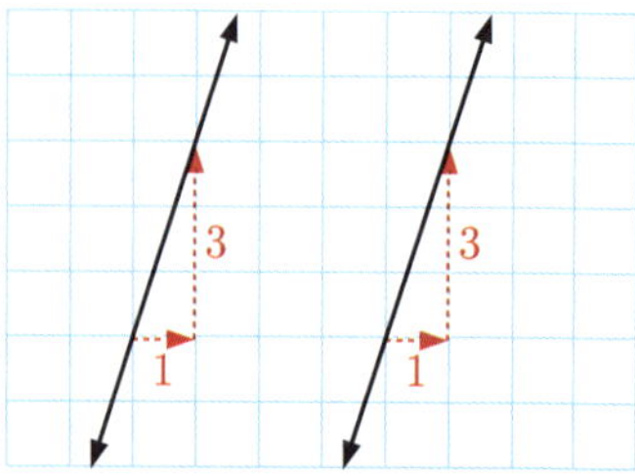

For lines which are neither horizontal nor vertical, the lines are **parallel** if and only if their gradients are **equal**.

PERPENDICULAR LINES

INVESTIGATION 1 PERPENDICULAR LINES

Consider two lines l_1 and l_2 which are neither horizontal nor vertical.

Suppose l_1 has positive gradient $\frac{b}{a}$ and l_2 has negative gradient $-\frac{a}{b}$ where $a, b > 0$.

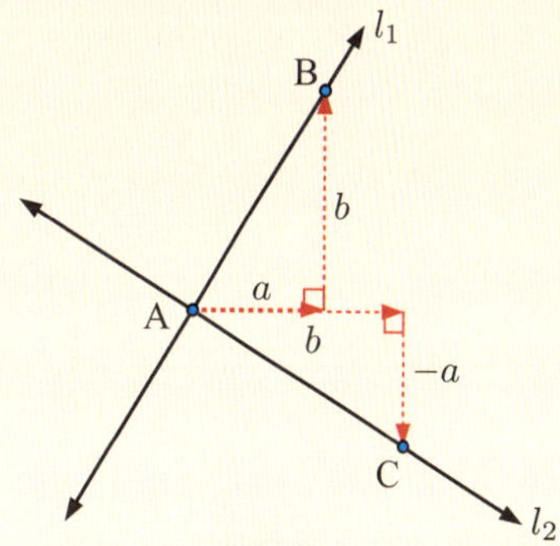

What to do:

1. Consider the points A, B, and C marked. Write, in terms of a and b:
   a AB **b** AC **c** BC
2. Hence show that $AB^2 + AC^2 = BC^2$.
3. Hence find $B\widehat{A}C$.
4. Copy and complete:
   For lines which are neither horizontal nor vertical, if their gradients are negative reciprocals, then the lines are ......

For lines which are neither horizontal nor vertical, the lines are **perpendicular** if and only if their gradients are **negative reciprocals**.

If two perpendicular lines have gradients m_1 and m_2, then $m_1 \times m_2 = -1$.

DISCUSSION

Discuss the cases of parallel and perpendicular lines when they are horizontal or vertical. What can we say about the gradients of the lines in these cases?

Example 9 **Self Tutor**

Find the gradient of all lines perpendicular to a line with a gradient of:

a $\frac{2}{7}$ **b** -5

a The negative reciprocal of $\frac{2}{7}$ is $-\frac{7}{2}$.

$\therefore$ the gradient of any perpendicular line is $-\frac{7}{2}$.

b The negative reciprocal of $-5 = \frac{-5}{1}$ is $\frac{1}{5}$.

$\therefore$ the gradient of any perpendicular line is $\frac{1}{5}$.

EXERCISE 13D

1 Find the gradient of all lines parallel to a line with gradient:

a 2 **b** -4 **c** $\frac{1}{3}$ **d** $-\frac{3}{4}$

2 Find the gradient of all lines perpendicular to a line with gradient:

a $\frac{1}{2}$ **b** $\frac{2}{5}$ **c** 3 **d** 7

e $-\frac{2}{5}$ **f** $-\frac{7}{2}$ **g** $-1\frac{1}{3}$ **h** -1

3 If two lines have these gradients, are they perpendicular?

a $\frac{1}{3}, 3$ **b** $5, -5$ **c** $\frac{3}{7}, -2\frac{1}{3}$ **d** $4, -\frac{1}{4}$

e $6, -\frac{5}{6}$ **f** $\frac{2}{3}, -\frac{3}{2}$ **g** $\frac{p}{q}, \frac{q}{p}$ **h** $\frac{a}{b}, -\frac{b}{a}$

4 Find the gradient of all lines perpendicular to a line whose gradient is undefined.

5 **a** Calculate the gradient of each side of the hexagon.

b Which sides are:

i parallel **ii** perpendicular?

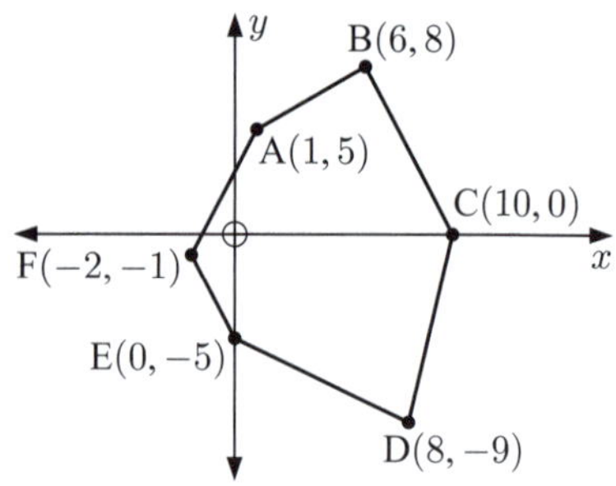

Example 10 — Self Tutor

Find t given that the line joining D$(-1, -3)$ to C$(1, t)$ is perpendicular to a line with gradient 2.

Gradient of [DC] $= -\frac{1}{2}$ {perpendicular to line with gradient 2}

$\therefore \frac{t - -3}{1 - -1} = -\frac{1}{2}$

$\therefore \frac{t+3}{2} = \frac{-1}{2}$

$\therefore t + 3 = -1$ {equating numerators}

$\therefore t = -4$

6 Find a given that the line joining:

a A$(1, 3)$ to B$(3, a)$ is parallel to a line with gradient 3

b P$(a, -3)$ to Q$(4, -2)$ is parallel to a line with gradient $\frac{1}{3}$.

7 Find t given that the line joining:

a A$(2, -3)$ to B$(-2, t)$ is perpendicular to a line with gradient $1\frac{1}{4}$

b C$(t, -2)$ to D$(1, 4)$ is perpendicular to a line with gradient $\frac{2}{3}$.

8 Consider the points A(1, 4), B(−1, 0), C(6, 3), and D(t, −1). Find t such that:

- **a** [AB] is parallel to [CD]
- **b** [AC] is parallel to [DB]
- **c** [AB] is perpendicular to [CD]
- **d** [AD] is perpendicular to [BC].

9 Find the value of k:

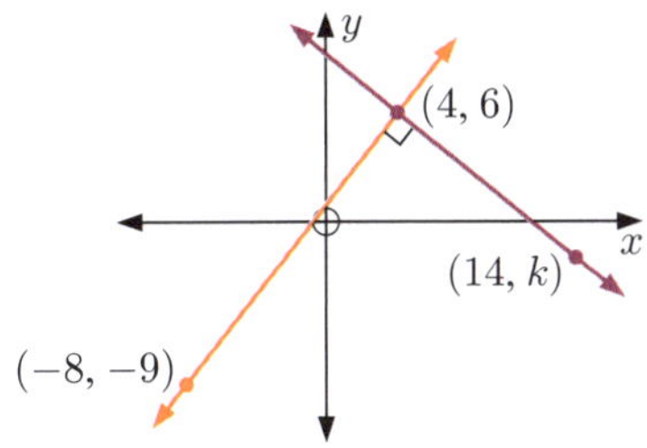

Example 11 — Self Tutor

Show that the points A(1, −1), B(6, 9), and C(3, 3) are collinear.

Gradient of [AB] $= \frac{9 - -1}{6 - 1} = \frac{10}{5} = 2.$ Gradient of [BC] $= \frac{3 - 9}{3 - 6} = \frac{-6}{-3} = 2.$

$\therefore$ [AB] is parallel to [BC], and point B is common to both line segments.

$\therefore$ A, B, and C are collinear.

10 Determine whether each set of points are collinear:

- **a** A(1, 2), B(4, 6), and C(−4, −4)
- **b** P(−6, −6), Q(−1, 0), and R(4, 6)
- **c** R(5, 2), S(−6, 5), and T(0, −4)
- **d** A(0, −2), B(−1, −5), and C(3, 7).

11 Find c given that each set of points are collinear:

- **a** A(−4, −2), B(0, 2), and C(c, 5)
- **b** P(3, −2), Q(4, c), and R(−1, 10).

PUZZLE — THE MISSING SQUARE

Stephanie presents the following puzzle to her friend Courtney:

"I can arrange these four shapes to form a right angled triangle 13 units long and 5 units high."

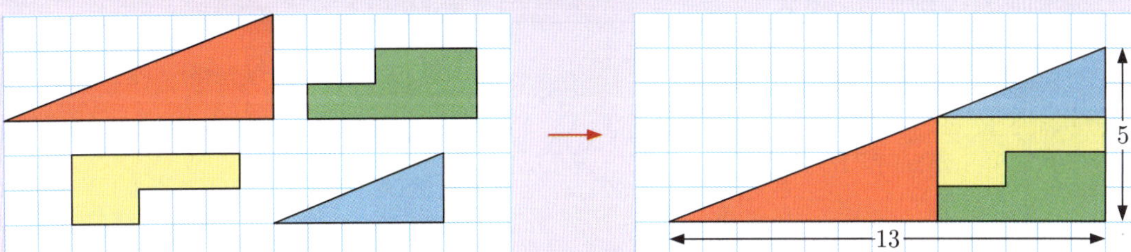

"I can then rearrange the shapes to form a triangle of the same size. However, there is now a missing square."

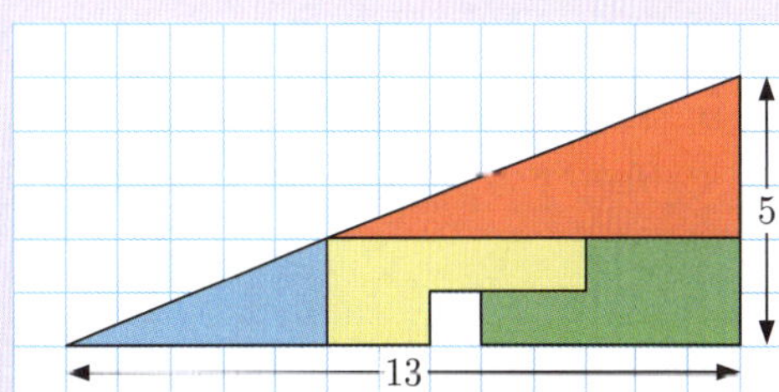

Can you explain why there is a missing square?

E USING COORDINATE GEOMETRY

Coordinate geometry is a powerful tool which can be used to:

- **check** the truth of a geometrical fact
- **prove** geometrical facts using general cases.

Example 12 — Self Tutor

P(3, −1), Q(1, 7), and R(−1, 5) are the vertices of triangle PQR.
M is the midpoint of [PQ]. N is the midpoint of [PR].

a Find the coordinates of M and N.
b Find the gradients of [MN] and [QR].
c Find distances MN and QR exactly.
d Explain the significance of your results.

a M is $\left(\frac{3+1}{2}, \frac{-1+7}{2}\right)$ which is (2, 3). N is $\left(\frac{3+-1}{2}, \frac{-1+5}{2}\right)$ which is (1, 2).

b gradient of [MN] $= \frac{2-3}{1-2}$
$= 1$

gradient of [QR] $= \frac{5-7}{-1-1}$
$= 1$

c MN $= \sqrt{(1-2)^2+(2-3)^2}$
$= \sqrt{1+1}$
$= \sqrt{2}$

QR $= \sqrt{(-1-1)^2+(5-7)^2}$
$= \sqrt{4+4}$
$= \sqrt{8}$
$= 2\sqrt{2}$

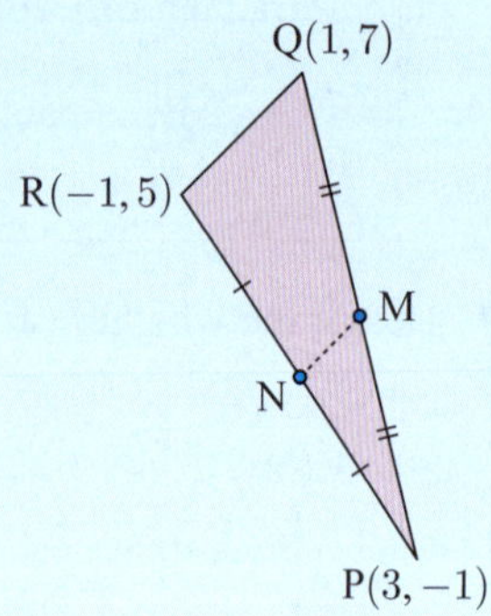

d The line segment between the midpoints of two sides of this triangle is parallel to the third side and half its length.

EXERCISE 13E

1 Consider the points P(1, 5), Q(5, 7), and R(3, 1).
a Show that triangle PQR is isosceles.
b Find the midpoint M of [QR].
c Show that [PM] is perpendicular to [QR].
d Illustrate what you have found.

2 Consider the points A(−1, 1), B(1, 5), and C(5, 1). M is the midpoint of [AB]. N is the midpoint of [BC]. Show that [MN] is parallel to [AC] and half its length.

3 The points A(−2, −7), B(0, −3), C(6, 1), and D(2, −5) form a kite.
a Find the midpoint M of [BD].
b Show that A, M, and C are collinear.
c Show that [AC] is perpendicular to [BD].

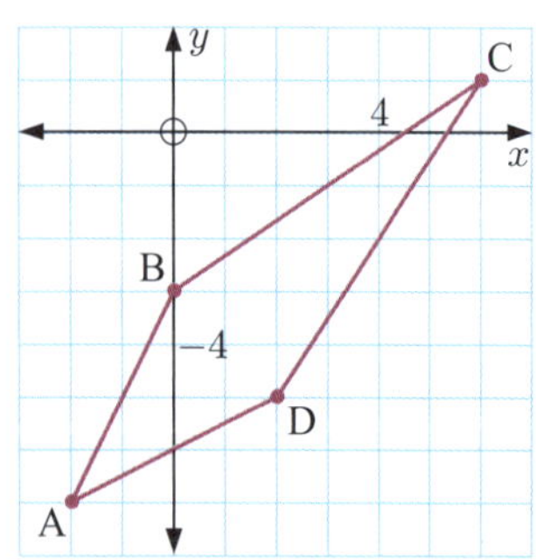

4 Consider the points A(1, 3), B(6, 3), C(3, −1), and D(−2, −1).

- **a** Use the distance formula to show that ABCD is a rhombus.
- **b** Find the midpoints of [AC] and [BD].
- **c** Show that [AC] and [BD] are perpendicular.
- **d** Illustrate what you have found.

5

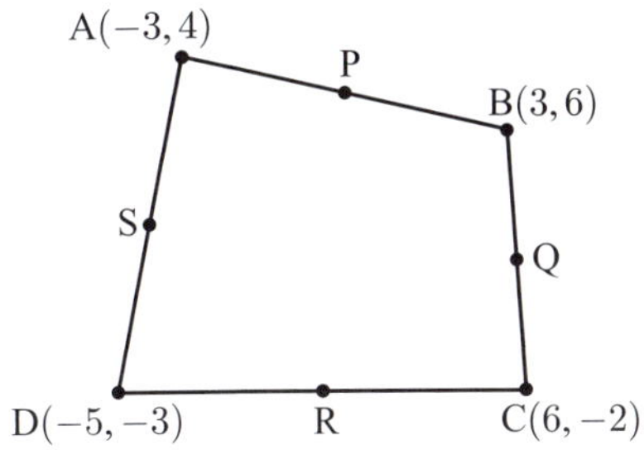

The sketch of quadrilateral ABCD is not drawn to scale. P, Q, R, and S are the midpoints of [AB], [BC], [CD], and [DA] respectively.

- **a** Find the coordinates of:
  - **i** P **ii** Q **iii** R **iv** S
- **b** Find the gradient of:
  - **i** [PQ] **ii** [QR] **iii** [RS] **iv** [SP]
- **c** What can be deduced about quadrilateral PQRS?

6 S(s, 8) lies on the semi-circle shown.

- **a** Find s.
- **b** Find the gradient of:
  - **i** [PS] **ii** [SQ]
- **c** Hence show that $\widehat{PSQ}$ is a right angle.

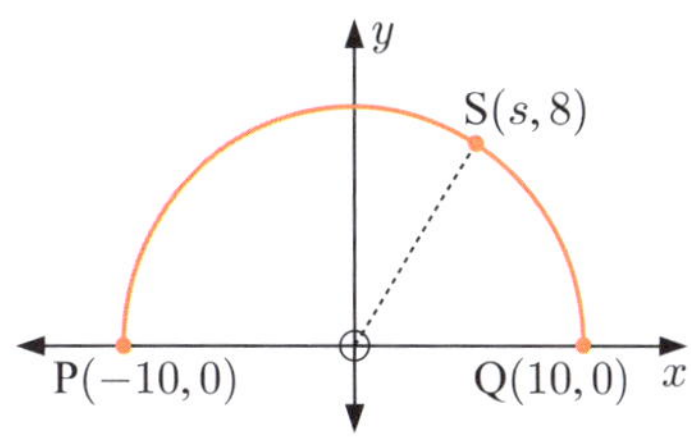

7

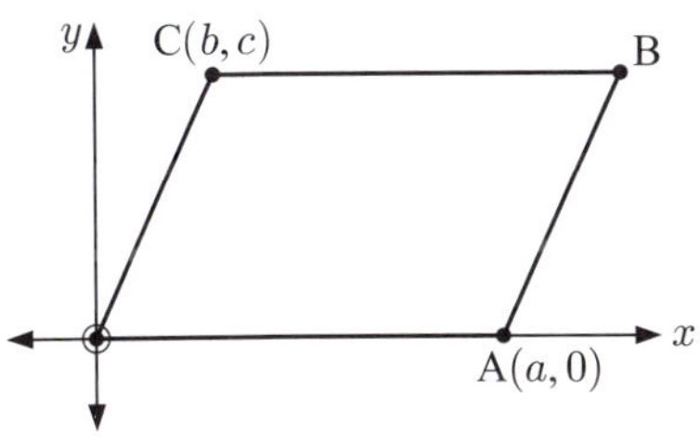

OABC is a parallelogram. You may assume that the opposite sides of the parallelogram are equal in length.

- **a** Find the coordinates of B.
- **b** Find the midpoints of [AC] and [OB].
- **c** What property of parallelograms has been deduced in **b**?

INVESTIGATION 2 THE ANGLE IN A SEMI-CIRCLE

Given any semi-circle, we can *choose* to draw a set of axes with the horizontal axis along the diameter, the semi-circle being above the x-axis, and with scale so the radius is 1 unit.

Let P(X, Y) be a point on the semi-circle.

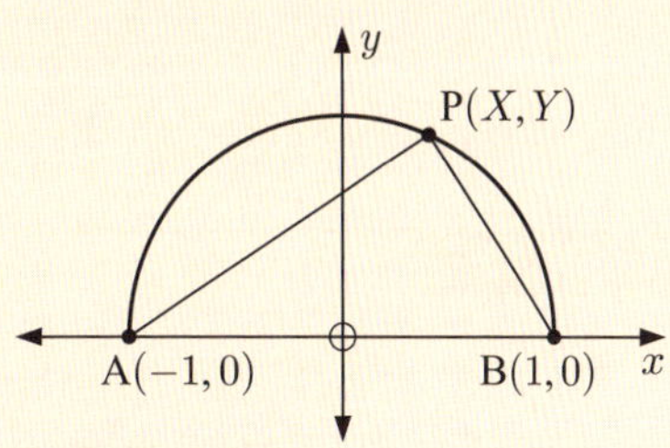

What to do:

1. Use Pythagoras' theorem to show that $Y = \sqrt{1 - X^2}$.
2. Write an expression involving X only, for the gradient of:
   - **a** [AP] **b** [BP]
3. Show that the product of the gradients of [AP] and [BP] is always −1.
4. Copy and complete:
   The angle in a semi-circle is always ......

F THE EQUATION OF A LINE

In previous years you should have seen that:

- The **x-intercept** of a line is the value of x where the line cuts the x-axis.
- The **y-intercept** of a line is the value of y where the line cuts the y-axis.
- These values are together called the **axes intercepts** of a line.
- The **equation of a line** is a rule which connects the x and y coordinates of *all* points on the line.

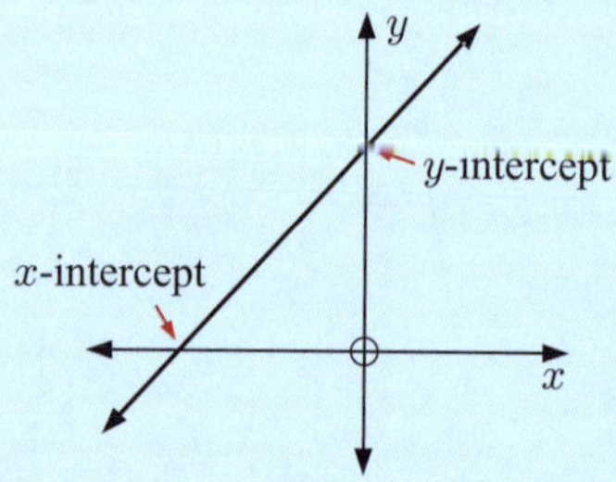

Consider a line with gradient m which passes through the point (x_1, y_1).

Suppose (x, y) is a general point on the line.

Applying the gradient formula, $\dfrac{y-y_1}{x-x_1} = m$

$\therefore \ y - y_1 = m(x - x_1)$

We call this the **point-gradient** form of the equation of a line.

We can then rearrange the equation into other forms:

- The **gradient-intercept form** of the equation of the line with gradient m and y-intercept c is $\boldsymbol{y = mx + c}$.
- The **general form** of the equation of a line is $\boldsymbol{ax + by = d}$ where a, b, and d are constants.

The general form allows us to write the equations of vertical lines, for which the gradient is undefined.

For example, this vertical line is $x = 2$. This is written in general form with $a = 1$, $b = 0$, and $d = 2$.

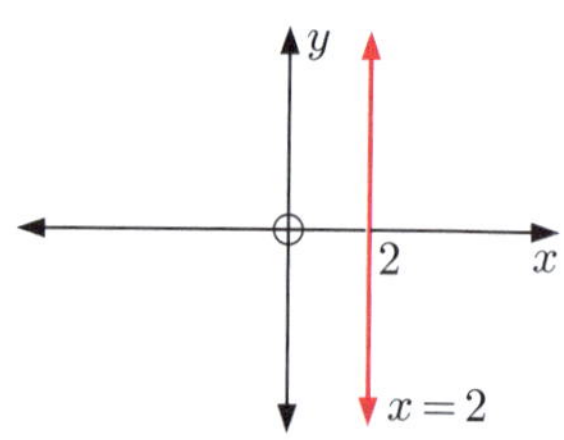

Equations in general form are usually written with a positive coefficient of x.

Example 13 — Self Tutor

Write the equation:

a $y = \frac{1}{2}x - 3$ in general form

b $3x - y = 11$ in gradient-intercept form.

a
$y = \frac{1}{2}x - 3$
$\therefore \ -\frac{1}{2}x + y = -3$ $\{-\frac{1}{2}x$ both sides$\}$
$\therefore \ x - 2y = 6$ $\{\times -2$ both sides$\}$

b
$3x - y = 11$
$\therefore \ 3x = y + 11$ $\{+y$ both sides$\}$
$\therefore \ y = 3x - 11$ $\{-11$ both sides$\}$

EXERCISE 13F

1 State the gradient and y-intercept of the line with equation:

a $y = 2x + 5$ **b** $y = -6x - 1$ **c** $y = \frac{3}{5}x - \frac{2}{5}$

d $y = 10 - x$ **e** $y = 1 - 4x$ **f** $y = \frac{5}{4} - \frac{7}{4}x$

g $y = \dfrac{3x+1}{7}$ **h** $y = \dfrac{4x-5}{8}$ **i** $y = \dfrac{6-7x}{9}$

2 Write in general form:

a $y = x - 7$ **b** $y = -x + 4$ **c** $y = \frac{1}{2}x - 5$

d $y = -3x + 2$ **e** $y = \frac{3}{4}x + \frac{1}{2}$ **f** $y = -\frac{5}{2}x - \frac{3}{2}$

3 Write in gradient-intercept form:

a $2x - y = -1$ **b** $3x + y = 3$ **c** $x - 4y = -8$

d $5x + 2y = 20$ **e** $2x - 6y = 7$ **f** $7x + 2y = -28$

4 $2x - 3y = 6$ and $6x + ky = 4$ are two straight lines.

a Find the gradient of each line.

b Find the value of k such that the lines are: **i** parallel **ii** perpendicular.

Example 14 **Self Tutor**

Determine whether:

a $(1, -4)$ lies on the line with equation $y = 3x - 8$

b $(2, 5)$ lies on the line with equation $3x + 4y = 26$.

a When $x = 1$, we have
$y = 3(1) - 8$
$= 3 - 8$
$= -5$ ✗

So, $(1, -4)$ does not lie on the line.

b Substituting $x = 2$ and $y = 5$ into the LHS gives
$3(2) + 4(5)$
$= 6 + 20$
$= 26$ ✓

So, $(2, 5)$ lies on the line.

A point lies on a line if it satisfies the line's equation.

5 Determine whether:

a $(5, 13)$ lies on the line $y = 3x - 2$ **b** $(-2, -9)$ lies on the line $y = \frac{1}{2}x - 10$

c $(\frac{1}{2}, \frac{3}{2})$ lies on the line $y = 5x - 1$ **d** $(-\frac{1}{3}, -1)$ lies on the line $y = -\frac{1}{2}x - \frac{7}{6}$.

6 Determine whether:

a $(6, 5)$ lies on the line $3x + y = 23$ **b** $(-2, 3)$ lies on the line $4x - 3y = 1$

c $(4, \frac{1}{2})$ lies on the line $3x + 8y = 14$ **d** $(-\frac{2}{3}, \frac{1}{3})$ lies on the line $6x + 9y = -1$.

7 **a** Find c given that $(3, 10)$ lies on the line $y = 2x + c$.

b Find m given that $(-5, \frac{1}{2})$ lies on the line $y = mx + \frac{3}{2}$.

c Find a given that $(a, -\frac{3}{4})$ lies on the line $y = \frac{1}{2}x + \frac{1}{4}$.

d Find k given that (k, k) lies on the line $y = 3x - \frac{15}{2}$.

8 Find k given that:

a $(5, -1)$ lies on the line $3x + 4y = k$
b $(-2, -7)$ lies on the line $x - 3y = k$
c $(3, k)$ lies on the line $5x + 6y = -9$
d $(k, 0)$ lies on the line $2x + 5y = 20$.

9 Find the x-intercept of:

a $y = 2x + 4$
b $y = -x + 5$
c $y = -3x + 15$
d $y = 5x - 10$
e $y = \frac{1}{2}x - 3$
f $y = -\frac{2}{3}x + 6$
g $3x - y = -9$
h $2x + 5y = 7$
i $4x - 9y = -9$

10 Consider the general equation $ax + by = d$.

a Find the y-intercept of the line, stating the condition necessary for the y-intercept to exist. Explain the condition geometrically.

b Find the x-intercept of the line, stating the condition necessary for the x-intercept to exist. Explain the condition geometrically.

c By rearranging the equation into gradient-intercept form, find the gradient of the line. State the condition necessary for the gradient to be defined.

d *Hence* find the gradient of the line with equation:

i $x + 3y = 2$
ii $2x - y = -1$
iii $4x + y = 7$
iv $3x - 2y = -3$
v $x + 5y = 4$
vi $4x - 3y = 12$

e Explain why:

i any line parallel to $3x + 5y = 2$ can be written in the form $3x + 5y = d$
ii any line perpendicular to $3x + 5y = 2$ can be written in the form $5x - 3y = d$.

G GRAPHING STRAIGHT LINES

GRAPHING LINES IN GRADIENT-INTERCEPT FORM

To draw the graph of $y = mx + c$:

- Use the y-intercept c to plot the point $(0, c)$.
- Starting from $(0, c)$, use horizontal and vertical steps from the gradient m to locate another point on the line.
- Join the two points and extend the line in both directions.

Example 15 **Self Tutor**

Draw the graph of $y = \frac{4}{3}x - 5$.

For $y = \frac{4}{3}x - 5$:

- the y-intercept is $c = -5$
- the gradient is $m = \frac{4}{3}$ ← y-step, ← x-step

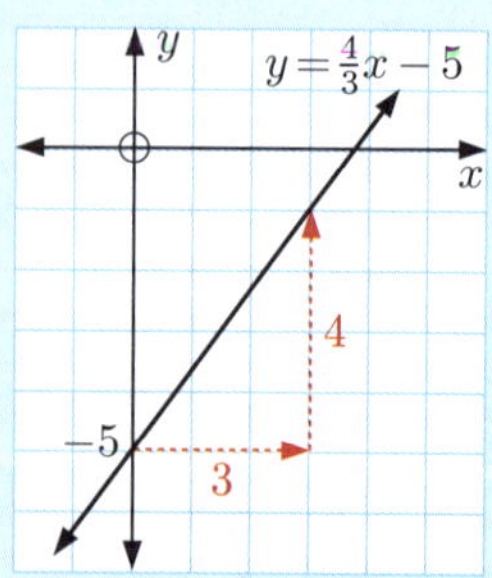

GRAPHING LINES IN GENERAL FORM

To draw the graph of $ax + by = d$:

- Find the y-intercept by letting $x = 0$.
- Find the x-intercept by letting $y = 0$.
- Join the points where the line cuts the axes and extend the line in both directions.

Example 16 — Self Tutor

Draw the graph of $2x + 3y = 12$.

When $x = 0$, $3y = 12$

$\therefore$ $y = 4$

So, the y-intercept is 4.

When $y = 0$, $2x = 12$

$\therefore$ $x = 6$

So, the x-intercept is 6.

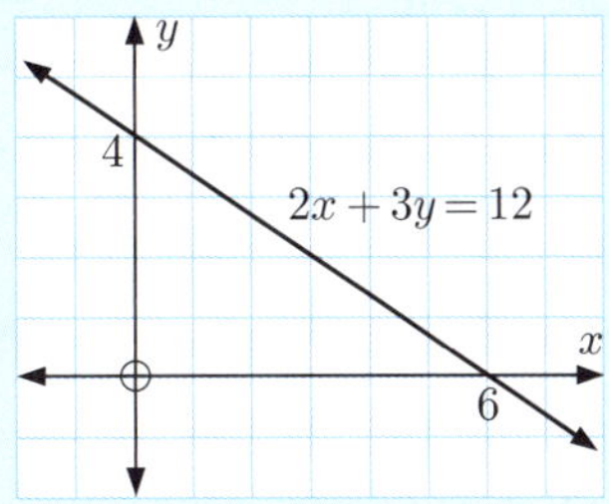

In previous years you should have seen that:

All **vertical** lines have equations of the form $x = a$.
All **horizontal** lines have equations of the form $y = c$.

EXERCISE 13G

1 Draw the graph of:

- **a** $y = 2x - 1$
- **b** $y = \frac{1}{3}x$
- **c** $y = -x + 3$
- **d** $y = \frac{1}{2}x - 3$
- **e** $y = -2x + 4$
- **f** $y = 3x + 1$
- **g** $y = -\frac{2}{5}x$
- **h** $y = \frac{5}{3}x - 5$
- **i** $y = 2 - \frac{1}{3}x$

2 Draw the graph of:

- **a** $x + y = 7$
- **b** $5x + 2y = 10$
- **c** $3x - 4y = 12$
- **d** $2x + 6y = -6$
- **e** $4x + y = -4$
- **f** $3x - 2y = 18$
- **g** $x - 4y = -8$
- **h** $4x + 3y = 18$
- **i** $2x - 5y = 15$

3 Draw the graph of:

- **a** $x = 4$
- **b** $y = -2$
- **c** $x = -1$
- **d** $y = 3$

4 On the same set of axes, draw the graphs of:

- **a** $y = 3x - 1$ and $y = -2x + 6$
- **b** $y = -\frac{1}{2}x + 6$ and $3x - y = 3$
- **c** $4x + 5y = 20$ and $y = 3x$
- **d** $2x - 3y = -18$ and $3x + y = 3$
- **e** $x = 0$, $y = 0$, $x = -3$, and $y = 4$.

5 A landscape company sells sand at \$39 per tonne plus a fixed delivery fee of \$95.

- **a** Write a formula for the total cost C dollars if x tonnes of sand is purchased.
- **b** Draw the graph of C against x for $x \geqslant 0$.
- **c** State and interpret the gradient and C-intercept of the graph.

6 Belinda buys x skipping ropes and y hoops for her PE class. The skipping ropes cost \$12 each and the hoops cost \$16 each. In total she spends \$528.

a Explain why $12x + 16y = 528$.

b Draw the graph of $12x + 16y = 528$.

c Explain the significance of:

i the x-intercept **ii** the y-intercept **iii** points on the line for which $x, y \in \mathbb{N}$.

H FINDING THE EQUATION OF A LINE

To determine the equation of a line, we need to know either:
- its gradient and at least one point which lies on the line, *or*
- two points which lie on the line.

If we are given two points but not the gradient, our first step is always to find the gradient of the line.

What we do next depends on the information we have and our end goal:

- If a line has gradient m and y-intercept c, its equation is $y = mx + c$.
- If a line has gradient m and passes through (x_1, y_1), its equation is $y - y_1 = m(x - x_1)$. We can quickly rearrange the equation into either gradient-intercept or general form.
- We can use the property that a line with gradient $-\frac{a}{b}$ has an equation in the general form $ax + by = d$. The constant d is found by substituting the coordinates of a known point.

Example 17 **Self Tutor**

Find, in gradient-intercept form, the equation of the line which passes through A(1, 3) and B(−2, 5).

The line has gradient $= \dfrac{5-3}{-2-1} = \dfrac{2}{-3} = -\dfrac{2}{3}$.

Since it passes through the point A(1, 3), the equation of the line is

$$y - 3 = -\tfrac{2}{3}(x - 1)$$
$$\therefore\ y - 3 = -\tfrac{2}{3}x + \tfrac{2}{3}$$
$$\therefore\ y = -\tfrac{2}{3}x + \tfrac{11}{3}$$

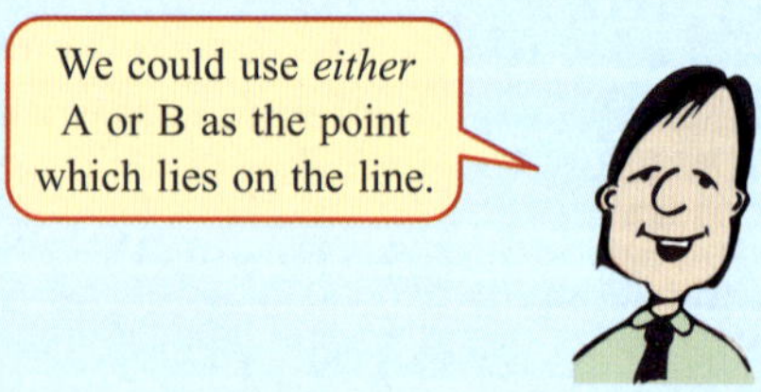

EXERCISE 13H

1 Find, in gradient-intercept form, the equation of the line with:

a gradient 2 and y-intercept 4

b gradient −3 and y-intercept −2

c gradient $\frac{1}{2}$ and y-intercept $\frac{1}{4}$

d gradient $-\frac{1}{3}$ and y-intercept 0.

2 Find, in general form, the equation of the line with:

a gradient $\frac{3}{4}$ and y-intercept 2

b gradient $-\frac{2}{3}$ and y-intercept −3.

3 A straight line has gradient $-\frac{2}{3}$ and y-intercept 5.

a Find the equation of the line. **b** Find the x-intercept of the line.

4 Find the equation of each line in:

i gradient-intercept form **ii** general form.

a

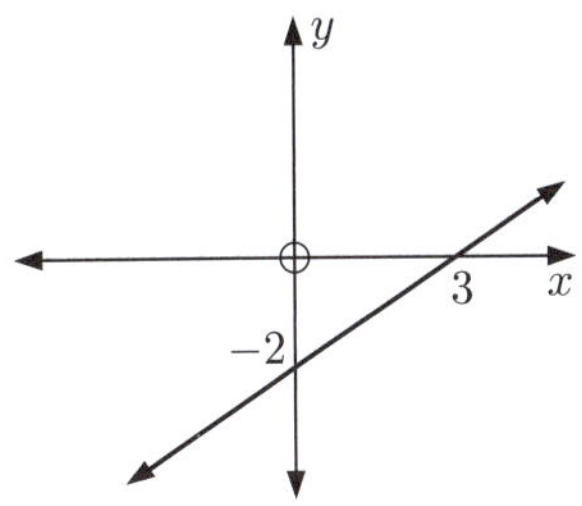

b

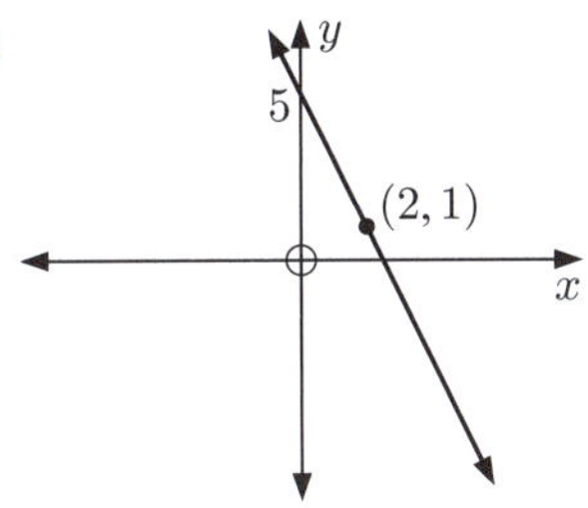

c

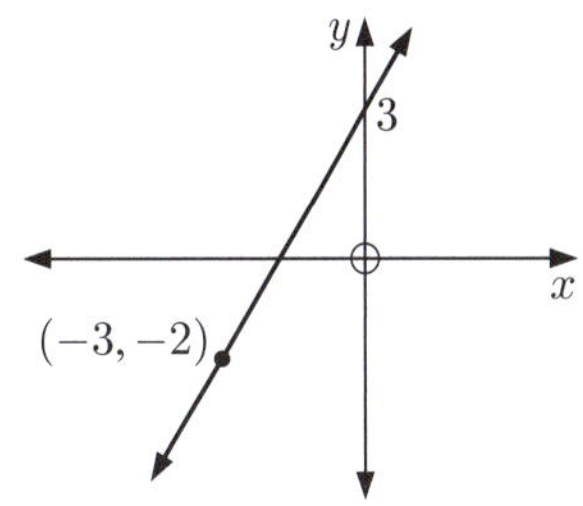

5 Find, in gradient-intercept form, the equation of the line with gradient:

a 2 which passes through $(1, 3)$ **b** -1 which passes through $(-1, 2)$

c $\frac{2}{3}$ which passes through $(-3, 1)$ **d** $-\frac{4}{5}$ which passes through $(4, -2)$.

6 Find, in gradient-intercept form, the equation of the line which passes through:

a $(8, 4)$ and $(5, 1)$ **b** $(5, -1)$ and $(4, 0)$ **c** $(-2, 4)$ and $(-3, -2)$

d $(-4, 6)$ and $(2, 9)$ **e** $(-1, -2)$ and $(5, -4)$ **f** $(2, -4)$ and $(7, -7)$.

Example 18 — Self Tutor

Find, in general form, the equation of the line:

a with gradient $\frac{3}{4}$ which passes through $(5, -2)$

b which passes through $(1, 7)$ and $(-3, 10)$.

a The equation is $3x - 4y = 3(5) - 4(-2)$

$\therefore\ 3x - 4y = 23$

b The line has gradient $= \dfrac{10-7}{-3-1} = \dfrac{3}{-4} = -\dfrac{3}{4}$

$\therefore$ the equation is $3x + 4y = 3(1) + 4(7)$

$\therefore\ 3x + 4y = 31$

7 Find, in general form, the equation of the line with gradient:

a 4 which passes through $(3, 5)$ **b** $-\frac{3}{5}$ which passes through $(-2, 1)$

c $\frac{1}{3}$ which passes through $(1, 4)$ **d** $\frac{2}{7}$ which passes through $(-5, -5)$.

8 Find, in general form, the equation of the line which passes through:

a $(-2, -5)$ and $(1, 4)$ **b** $(-1, 5)$ and $(4, -5)$ **c** $(-4, -3)$ and $(5, -3)$.

9 **a** Find, in general form, the equation of the line through $A(-3, 5)$ and $B(2, 1)$.

b Show that the point $C(12, -7)$ also lies on this line.

10 Find, in general form, the equation of each line:

a

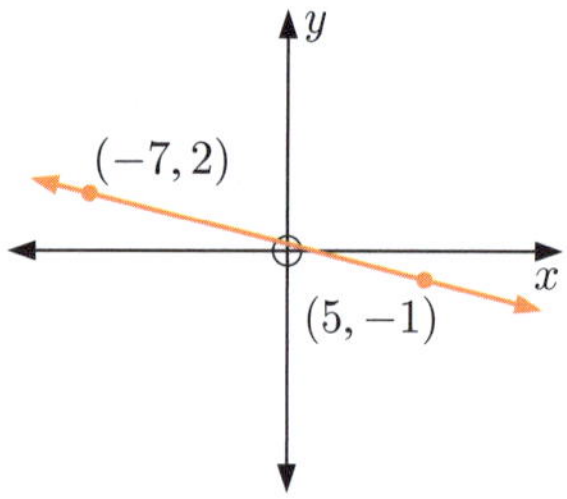

b

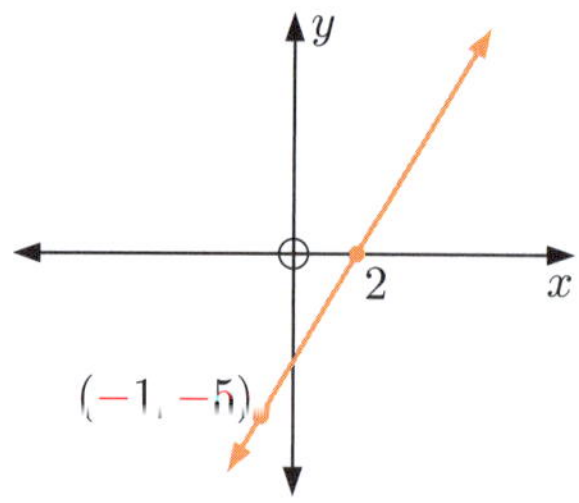

c

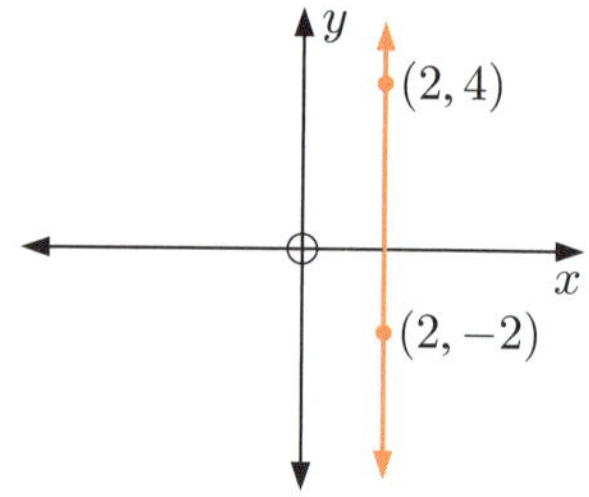

11 Find the equation of the line which:

a is parallel to the line $y = 2x - 2$ and passes through $(-1, 4)$

b is perpendicular to a line with gradient $\frac{3}{4}$, and cuts the x-axis at 5.

12 Line 1 has equation $y = \frac{2}{3}x + 2$.
Find the y-intercept of line 2.

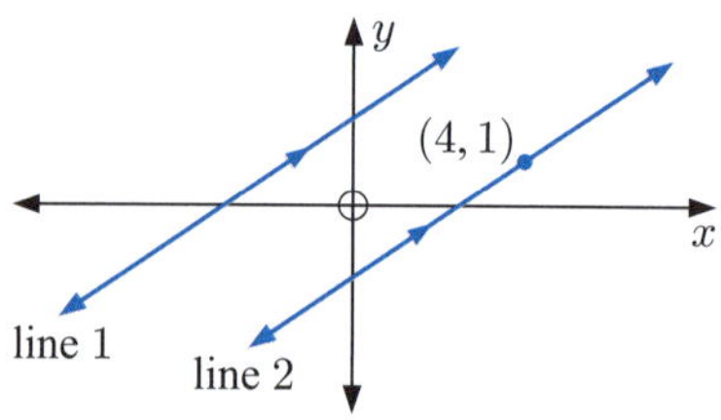

13 Lines l_1 and l_2 intersect at right angles at $(-2, 5)$. The equation of l_1 is $y = 3x + 11$. Find, in general form, the equation of l_2.

14

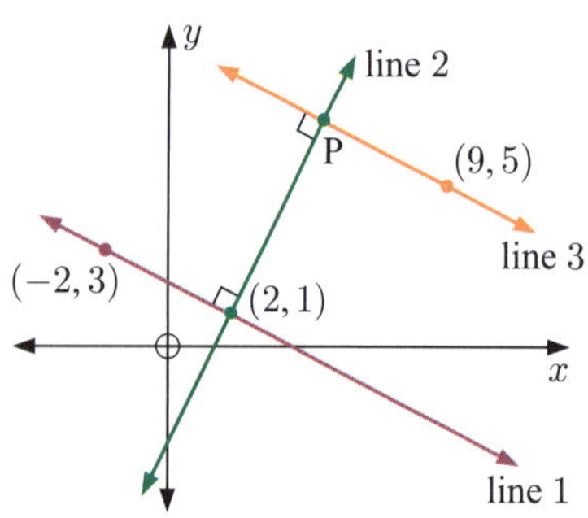

a Find, in gradient-intercept form, the equation of:
i line 1 **ii** line 2 **iii** line 3.

b Show that the coordinates of P are $(5, 7)$.

I PERPENDICULAR BISECTORS

The **perpendicular bisector** of a line segment [AB] is the line perpendicular to [AB] which passes through its midpoint.

Notice that:

- Points on the perpendicular bisector are equidistant from A and B.
- The perpendicular bisector divides the number plane into two regions. On one side of the line are points that are closer to A than to B, and on the other side are points that are closer to B than to A.

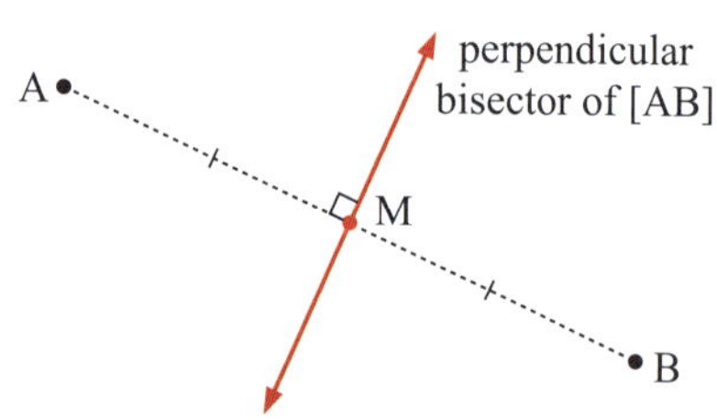

Example 19 **Self Tutor**

Given A$(-1, 2)$ and B$(3, 4)$, find the equation of the perpendicular bisector of [AB].

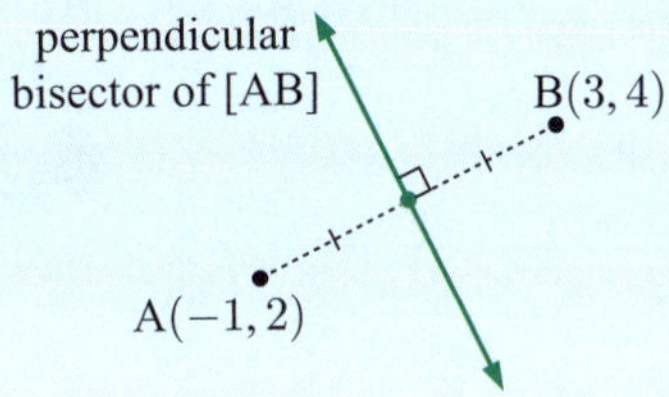

The midpoint M of [AB] is $\left(\frac{-1+3}{2}, \frac{2+4}{2}\right)$ or $(1, 3)$.

The gradient of [AB] is $\frac{4-2}{3--1} = \frac{2}{4} = \frac{1}{2}$

$\therefore$ the gradient of the perpendicular bisector is $-\frac{2}{1}$.

$\therefore$ the equation of the perpendicular bisector is $2x + y = 2(1) + (3)$
which is $2x + y = 5$.

EXERCISE 13I

1. Consider the points A$(7, 2)$ and B$(1, 6)$.
   a. Find the midpoint of [AB].
   b. Find the gradient of [AB].
   c. Hence state the gradient of the perpendicular bisector.
   d. Find the equation of the perpendicular bisector.

2. Find the equation of the perpendicular bisector of [AB] given:
   a. A$(3, -3)$, B$(1, -1)$
   b. A$(1, 3)$, B$(-3, 5)$
   c. A$(3, 1)$, B$(-3, 6)$

3. Suppose A is $(-1, -4)$ and B is $(3, 2)$.
   a. Find the equation of the perpendicular bisector of [AB].
   b. Show that C$(-5, 3)$ lies on the perpendicular bisector.
   c. Show that C is equidistant from A and B.

4. Two post offices are located at P$(3, 8)$ and Q$(7, 2)$ on a council map. Find the equation of the line which should form the boundary between the two regions serviced by the post offices.

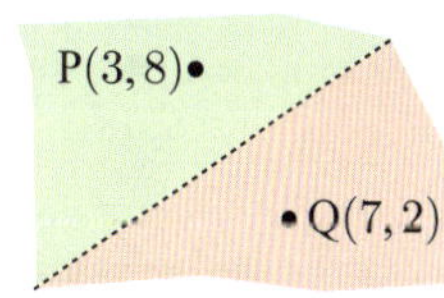

5. Two hospitals are located at A$(-3, 11)$ and B$(1, 1)$.
   a. Find the equation of the line which forms the boundary between the two regions serviced by the hospitals.
   b. Draw a graph on grid paper showing A, B, and the boundary between the regions.
   c. Clara lives at $(-7, 3)$. Which hospital is she closest to?

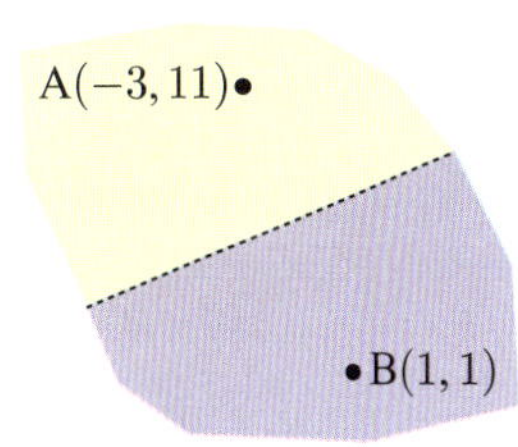

6. Consider the quadrilateral ABCD.
   a. Use side lengths to show that ABCD is a kite.
   b. Find the equation of the perpendicular bisector of [BD].
   c. Show that A and C both lie on this perpendicular bisector.

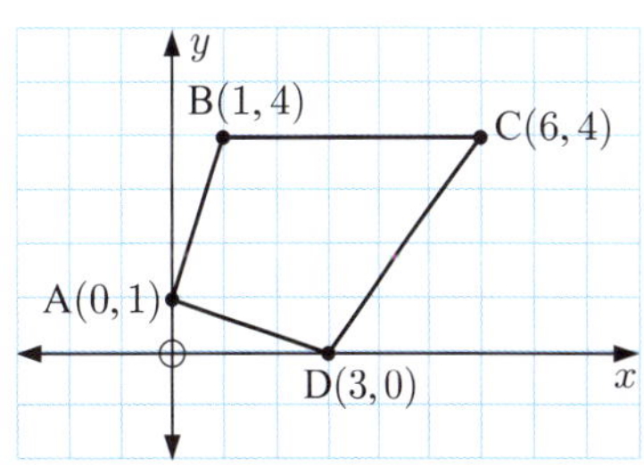

7. Answer the **Opening Problem** on page **194**.

J 3-DIMENSIONAL COORDINATE GEOMETRY

In 3-dimensional coordinate geometry, we specify an origin O and three mutually perpendicular axes called the X-axis, Y-axis, and Z-axis.

3D-POINT PLOTTER

Any point in space can then be specified using an ordered triple in the form (x, y, z).

We generally suppose that the Y and Z-axes are in the plane of the page, and the X-axis is coming out of the page as shown.

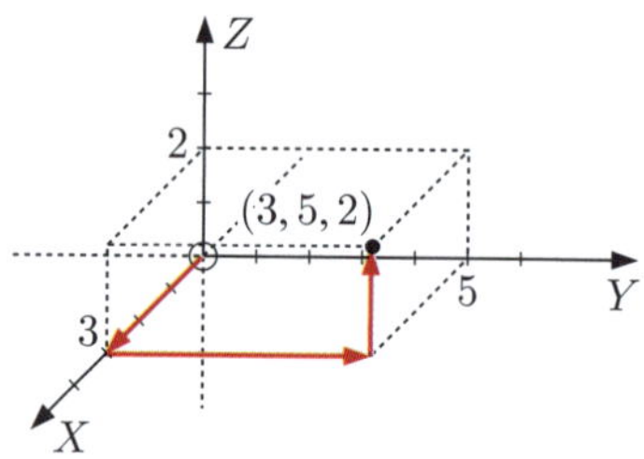

The point $(3, 5, 2)$ has X-coordinate 3, Y-coordinate 5, and Z-coordinate 2.

We see that $(3, 5, 2)$ is located on the corner of a rectangular prism opposite O.

Consider the rectangular prism illustrated, in which A is opposite B.

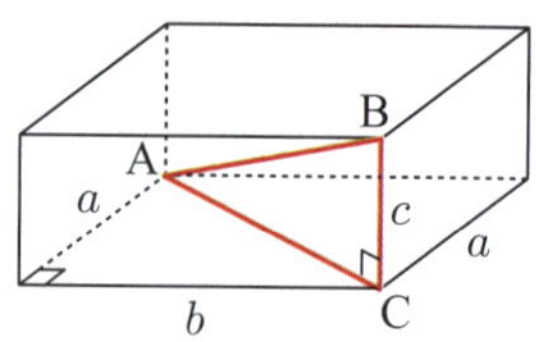

$$\begin{aligned} \text{AC}^2 &= a^2 + b^2 && \{\text{Pythagoras}\} \\ \text{and} \quad \text{AB}^2 &= \text{AC}^2 + c^2 && \{\text{Pythagoras}\} \\ \therefore \quad \text{AB}^2 &= a^2 + b^2 + c^2 \\ \therefore \quad \text{AB} &= \sqrt{a^2 + b^2 + c^2} && \{\text{AB} > 0\} \end{aligned}$$

If A is (x_1, y_1, z_1) and B is (x_2, y_2, z_2), then:

- the **distance** $\text{AB} = \sqrt{(x_2 - x_1)^2 + (y_2 - y_1)^2 + (z_2 - z_1)^2}$
- the **midpoint** of [AB] is $\left(\dfrac{x_1 + x_2}{2}, \dfrac{y_1 + y_2}{2}, \dfrac{z_1 + z_2}{2}\right)$.

Example 20 **Self Tutor**

Consider A$(1, -2, 4)$ and B$(3, -1, 2)$. Find:

a the distance AB **b** the midpoint of [AB].

a $\text{AB} = \sqrt{(3-1)^2 + (-1 - -2)^2 + (2-4)^2}$
$= \sqrt{2^2 + 1^2 + (-2)^2}$
$= \sqrt{4 + 1 + 4}$
$= \sqrt{9}$
$= 3$ units

b The midpoint is $\left(\dfrac{1+3}{2}, \dfrac{-2 + -1}{2}, \dfrac{4+2}{2}\right)$,
which is $(2, -\frac{3}{2}, 3)$.

EXERCISE 13J

PRINTABLE 3D PLOTTER PAPER

1. On separate axes, plot the points:
   - **a** (2, 0, 0) **b** (0, 4, 0) **c** (0, 0, −1)
   - **d** (2, 0, 3) **e** (1, 3, 0) **f** (0, −1, 2)
   - **g** (1, 4, 3) **h** (−2, 3, −1) **i** (2, −3, 1)
   - **j** (−1, 2, −2) **k** (1, 3, −3) **l** (3, −2, 4)
2. Write a formula for the distance d between the origin O(0, 0, 0) and P(a, b, c).
3. For each pair of points, find:
   i the distance AB **ii** the midpoint of [AB].
   - **a** A(3, −1, 1) and B(2, 0, −2) **b** A(2, −1, 4) and B(0, 0, 2)
   - **c** A(3, −2, 0) and B(5, 0, −3) **d** A(−2, 1, 5) and B(1, −3, 2)
4. Classify triangle ABC given points:
   - **a** A(3, −3, 6), B(6, 2, 4), C(4, −1, 3) **b** A(−1, 2, −3), B(3, 1, 1), C(2, 5, 3)
5. The distance from P(1, 2, 3) to Q(k, 1, −1) is 6 units. Find the possible values of k.
6. Find the relationship between x, y, and z if the point P(x, y, z) is always 2 units from O(0, 0, 0). Describe the set of points geometrically.

MULTIPLE CHOICE QUIZ

QUICK QUIZ

REVIEW SET 13A

1. Find the distance between:
   - **a** P(4, 0) and Q(0, −3) **b** R(2, −5) and S(−1, −3).
2. K(−3, 2) and L(3, m) are $\sqrt{52}$ units apart. Find the possible values of m.
3. Find the midpoint of the line segment joining:
   - **a** A(8, −3) and B(2, 1) **b** A(−2, 3) and B(−9, 10).
4. [AB] and [CD] are both diameters of the circle. Find:
   - **a** the coordinates of D
   - **b** the radius of the circle.

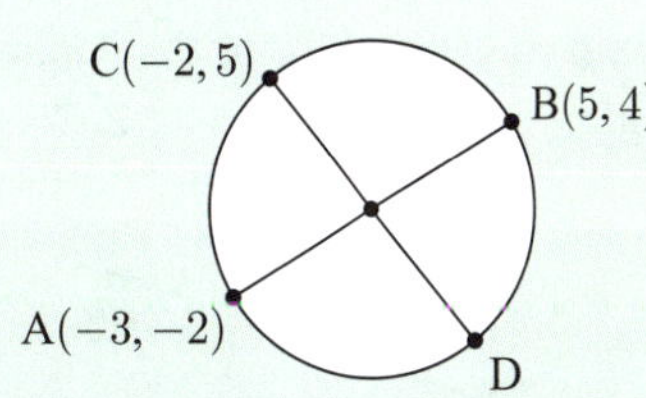

5. On the same set of axes, draw lines through (2, −1) with gradients 0, $\frac{1}{2}$, $\frac{3}{2}$, and 3.

6 Find the gradient of the line through:

a $(-1, 3)$ and $(3, 11)$ **b** $(-2, -3)$ and $(5, -5)$.

7 Find the gradient of all lines perpendicular to a line with gradient $3\frac{1}{3}$.

8 $P(-1, 5)$, $Q(3, 3)$, $R(1, -2)$, and $S(-5, -5)$ are the vertices of trapezium PQRS. Which pair of sides are parallel?

9 Consider the points $A(-3, 1)$, $B(1, 4)$, and $C(4, 0)$.

a Show that triangle ABC is right angled and isosceles.

b Find the midpoint X of [AC].

c Show that [BX] is perpendicular to [AC].

10 Show that $A(1, -2)$, $B(4, 4)$, and $C(5, 6)$ are collinear.

11 $A(-3, 2)$, $B(2, 3)$, $C(4, -1)$, and $D(-1, -2)$ are the vertices of quadrilateral ABCD.

a Find the gradient of each side of the quadrilateral. What can you deduce from your answers?

b Find the midpoint of the diagonals [AC] and [BD]. What can you deduce from your answers?

c **i** Find the gradients of the diagonals.

ii What does the product of these gradients tell us about the quadrilateral?

12 State the gradient and y-intercept of the line with equation:

a $y = 6x - 4$ **b** $y = 3 - \frac{2}{5}x$ **c** $y = \frac{4x - 7}{6}$

13 $3x + 4y = 24$ and $6x + ky = 12$ are two straight lines.

a Write down the gradient of each line.

b Find the value of k such that the lines are:

i parallel **ii** perpendicular.

14 Determine whether:

a $(\frac{1}{2}, 3)$ lies on the line $y = 4x - 1$ **b** $(\frac{1}{3}, -1)$ lies on the line $9x - 2y = 5$.

15 Draw the graph of:

a $y = 4 - 3x$ **b** $y = \frac{2}{3}x - 4$ **c** $5x - 3y = -15$

16 Find, in gradient-intercept form, the equation of the line:

a with gradient -2 and y-intercept 7

b which passes through $(-1, 3)$ and $(2, 1)$.

17 Find, in general form, the equation of each line:

a

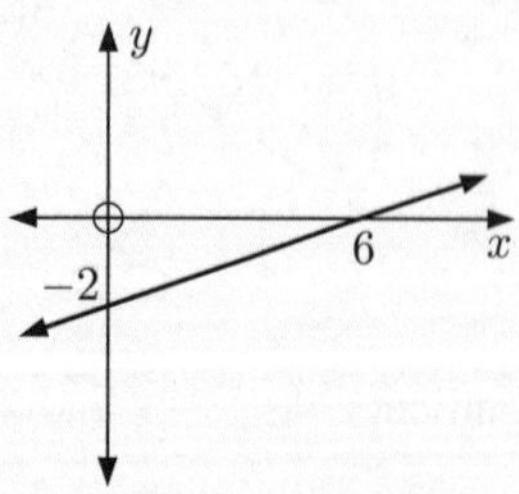

b

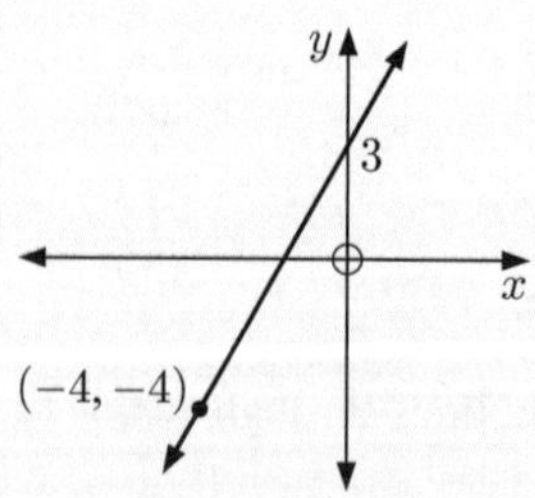

c

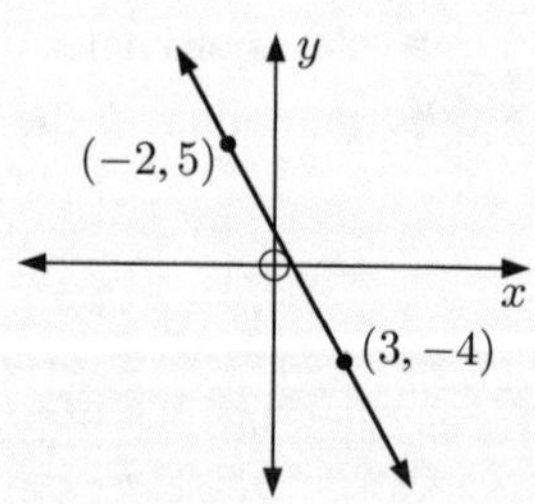

18 Find the equation of the perpendicular bisector of [PQ] given P$(-6, 3)$ and Q$(2, -7)$.

19 Suppose A is $(-5, 1)$ and B is $(3, 0)$.

a Find the equation of the perpendicular bisector of [AB].

b Find c such that $(-2, c)$ lies on the perpendicular bisector.

20 Consider the triangle PQR.

a Use side lengths to show that triangle PQR is isosceles.

b Find the equation of the perpendicular bisector of [PR].

c Show that Q lies on this perpendicular bisector.

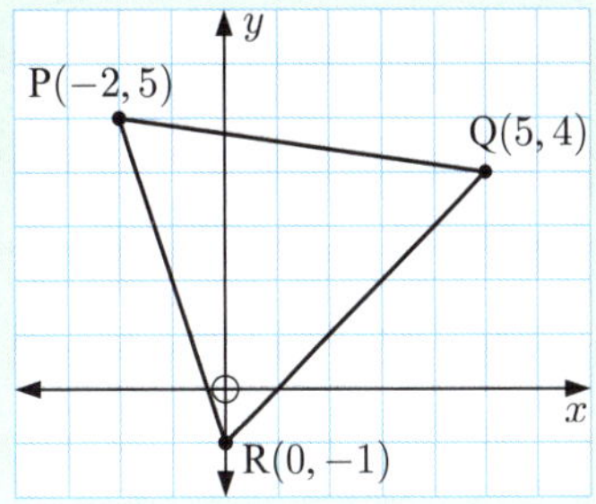

21 For P$(-1, 2, 3)$ and Q$(1, -2, -3)$, find:

a the distance PQ

b the midpoint of [PQ].

REVIEW SET 13B

1 Consider the points S$(7, -2)$ and T$(-1, 1)$. Find:

a the distance ST

b the midpoint of [ST].

2 X$(-2, 3)$ and Y$(a, -1)$ are $\sqrt{17}$ units apart. Find the possible values of a.

3 M$(1, -1)$ is the midpoint of [AB], and A is $(-3, 2)$. Find the coordinates of B.

4 A$(2, -1)$, B$(-5, 3)$, and C$(3, 4)$ are the vertices of triangle ABC.

a Classify triangle ABC as equilateral, isosceles, or scalene.

b Determine whether triangle ABC is right angled.

5 On the same set of axes, draw lines through $(-3, 1)$ with gradients $-\frac{1}{3}$, $-\frac{3}{4}$, -2, and -5.

6 Find the gradient of the line segment joining:

a $(5, -1)$ and $(-2, 6)$

b $(5, 0)$ and $(5, -2)$

c $(4, 7)$ and $(-2, -2)$.

7 Find k given that the line joining X$(2, -3)$ and Y$(-1, k)$ is:

a parallel to a line with gradient $\frac{1}{2}$

b perpendicular to a line with gradient $-\frac{1}{4}$.

8 On the same set of axes, draw the graphs of $y = 2x - 4$ and $2x + 5y = -5$.

9 Find the value of a:

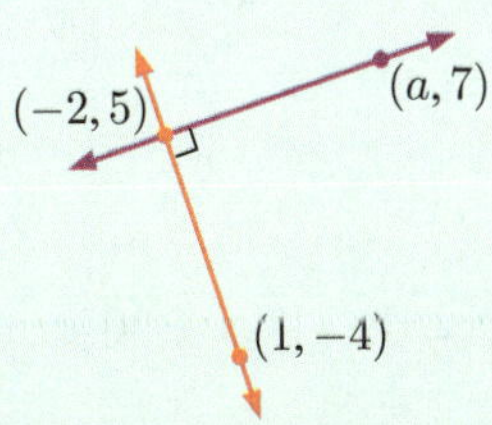

10 Find b given that A$(-6, 2)$, B$(b, 0)$, and C$(3, -4)$ are collinear.

11 Consider the points $P(2, 7)$, $Q(7, 4)$, and $R(-4, -3)$.

- **a** Find the gradient of:
  - **i** [PQ] **ii** [QR] **iii** [PR]
- **b** Hence show that triangle PQR is right angled.
- **c** Find the area of triangle PQR.

12 **a** Write the equation $4x - 3y = 2$ in gradient-intercept form.

- **b** State the gradient of the line.

13 Write $y = -3x + 5$ in general form.

14 Find k given that:

- **a** $(k, -2)$ lies on the line $y = 7 - 4x$
- **b** $(3, -8)$ lies on the line $5x - 3y = k$.

15 Draw the graph of:

- **a** $y = \frac{1}{3}x + 2$
- **b** $x + 4y = 8$
- **c** $4x - 3y = -24$

16 Find, in general form, the equation of the line:

- **a** with gradient 3 which passes through $(-4, -5)$
- **b** which passes through $(-3, 2)$ and $(3, -1)$.

17 The cost of hiring a storage locker is \$15 per week, plus a fixed reservation fee of \$50.

- **a** Write a formula for the total cost C dollars of hiring a storage locker for t weeks.
- **b** Draw the graph of C against t for $t \geqslant 0$.
- **c** State and interpret the gradient and C-intercept of the graph.

18 Find the equation of the line through $(-3, 1)$, which is perpendicular to the line through $(-1, 4)$ and $(2, -2)$.

19 This Voronoi diagram shows the location of Somchai's Thai restaurants.
Home delivery orders are directed to the nearest restaurant.

- **a** Find the equation of each boundary between the regions.
- **b** Susan lives at $(1, 3)$. From which restaurant will her order be delivered?
- **c** Verify that the point $(-1, -2)$ satisfies *all* of the boundary equations. Explain the significance of this location.

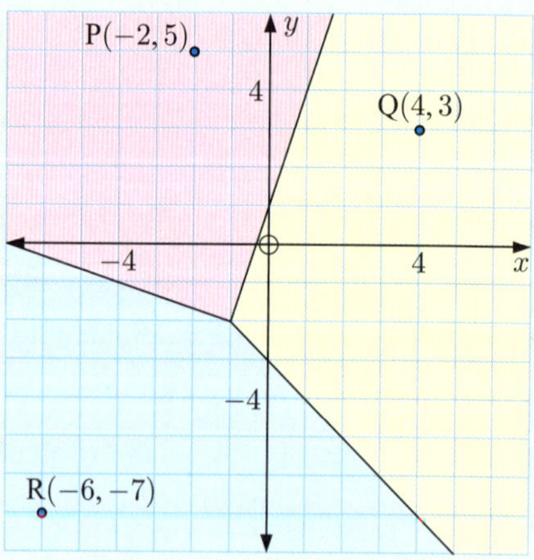

20 How far is $A(-1, -2, 5)$ from the origin O?

21 The distance between $P(1, 3, -1)$ and $Q(2, 1, k)$ is $\sqrt{30}$ units. Find the possible values of k.

Chapter 14

Simultaneous equations

Contents:

OPENING PROBLEM

Amir wants to buy a Ramly burger, but only has RM 4. He sees his friend Hafiz leaving the canteen with a Ramly burger, and asks him how much it cost.

"I'm not sure," said Hafiz, "I bought 3 Ramly burgers and 2 satay sticks. They cost me RM 12.50 altogether."

Amir sees another friend, Ismail, with a Ramly burger, and asks him how much it cost.

"Well, I bought a Ramly burger and 3 satay sticks, and they cost me RM 6.50 altogether."

Things to think about:

a Can Amir determine the cost of a Ramly burger using the information from only one of his friends?

b Suppose a Ramly burger costs RM x and a satay stick costs RM y.
Can you explain why:
i $3x + 2y = 12.5$ **ii** $x + 3y = 6.5$?

c Can you find values for x and y which make *both* equations true?

d Does Amir have enough money to buy a Ramly burger?

To answer the **Opening Problem**, we need values for x and y which satisfy both $3x + 2y = 12.5$ and $x + 3y = 6.5$ *at the same time*. We need to solve the equations *simultaneously*.

We say that $\begin{cases} 3x + 2y = 12.5 \\ x + 3y = 6.5 \end{cases}$ is a set of **simultaneous equations**.

Notice that if $x = 3.5$ and $y = 1$, then $3x + 2y = 3(3.5) + 2(1) = 12.5$,
and $x + 3y = 3.5 + 3(1) = 6.5$.

So, $x = 3.5$, $y = 1$ is a **solution** to the set of simultaneous equations $\begin{cases} 3x + 2y = 12.5 \\ x + 3y = 6.5. \end{cases}$

A GRAPHICAL SOLUTION

Suppose we are given two linear equations involving x and y. The solutions to each equation will form a straight line. If we graph the two equations on the same set of axes, any **point of intersection** corresponds to a **simultaneous solution** of the equations.

- If the lines are not parallel, the lines will meet at exactly one point. The simultaneous equations have *exactly one solution.*

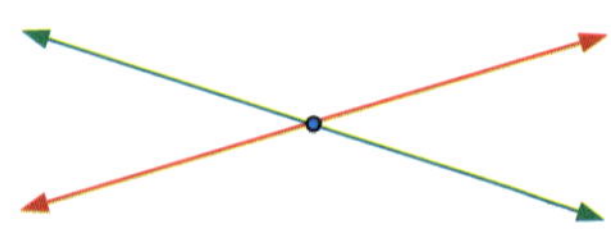

- If the lines are parallel but not identical, the lines will never meet. The simultaneous equations have *no solutions.*

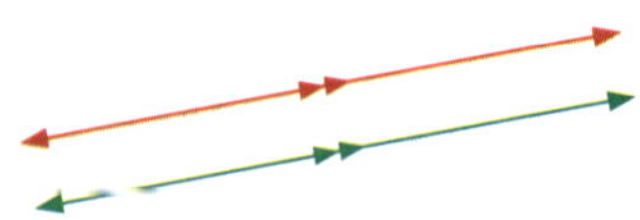

- If the lines are identical, there will be infinitely many points of intersection. The simultaneous equations have *infinitely many solutions.*

Example 1 — **Self Tutor**

Solve graphically: $\begin{cases} y = 2x - 4 \\ 2x + 3y = 12 \end{cases}$

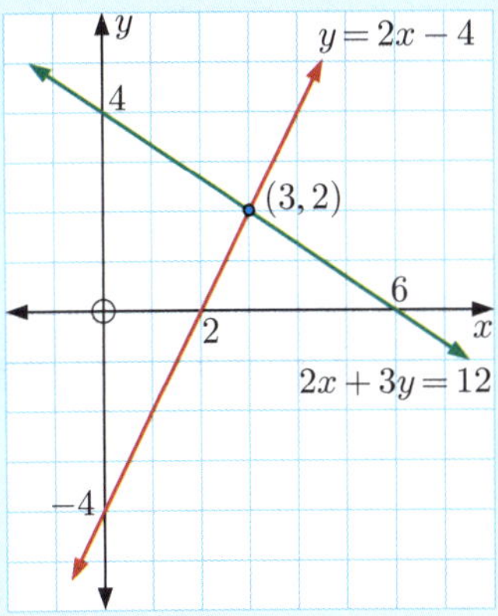

We draw the graphs of $y = 2x - 4$ and $2x + 3y = 12$ on the same set of axes.

The graphs meet at the point $(3, 2)$.

$\therefore$ the solution is $x = 3$, $y = 2$.

Check:

Substituting these values into:

- $y = 2x - 4$ gives $2 = 2(3) - 4$ ✓
- $2x + 3y = 12$ gives $2(3) + 3(2) = 12$ ✓

EXERCISE 14A.1

1 Solve graphically:

a $\begin{cases} y = 4x - 1 \\ y = 2x - 3 \end{cases}$ **b** $\begin{cases} y = 3x \\ y = -2x + 5 \end{cases}$ **c** $\begin{cases} y = x + 2 \\ y = -3x - 6 \end{cases}$

2 Solve graphically:

a $\begin{cases} y = 2x - 8 \\ 2x + 5y = 20 \end{cases}$ **b** $\begin{cases} 4x + y = 8 \\ 2x - 3y = 18 \end{cases}$ **c** $\begin{cases} 3x - y = -6 \\ 3x + 4y = -36 \end{cases}$

3 Try to solve the following simultaneous equations graphically. State the number of solutions in each case.

a $\begin{cases} y = 4x + 1 \\ y = 4x - 3 \end{cases}$ **b** $\begin{cases} 4x - 2y = -8 \\ y = 2x + 4 \end{cases}$

USING TECHNOLOGY

If the solutions to a set of simultaneous equations are not integers, it may be difficult to find them on a graph drawn by hand. In these situations, we can use technology to graph the equations and find the solution.

When using your **graphics calculator**, any equations in the general form $ax + by = d$ will need to be rearranged to make y the subject.

GRAPHING PACKAGE

GRAPHICS CALCULATOR INSTRUCTIONS

Example 2 | **Self Tutor**

Use technology to solve: $\begin{cases} y = 4x + 13 \\ x + y = 7 \end{cases}$

$x + y = 7$

$\therefore \quad y = 7 - x$

We graph $y = 4x + 13$ and $y = 7 - x$ on the same set of axes.

Casio fx-CG50 | **TI-84 Plus CE** | **HP Prime**

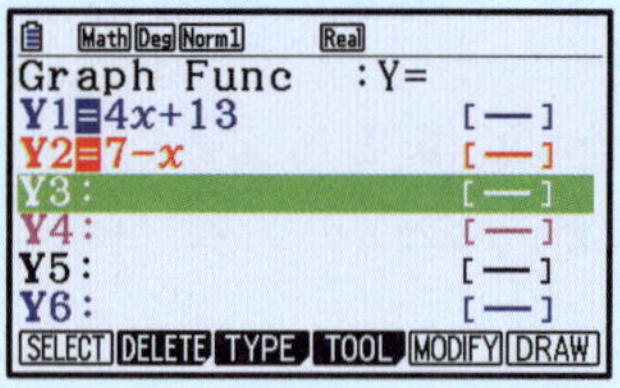

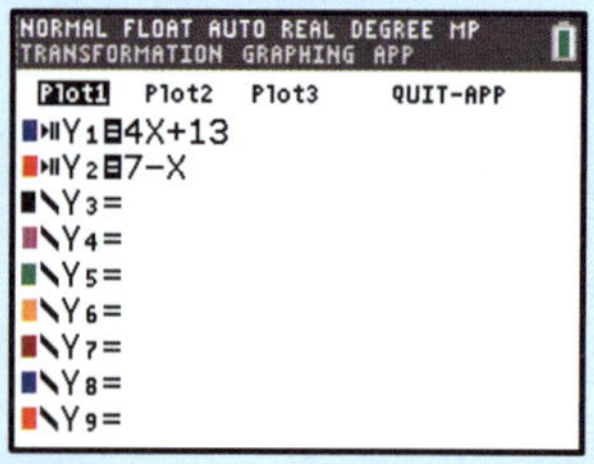

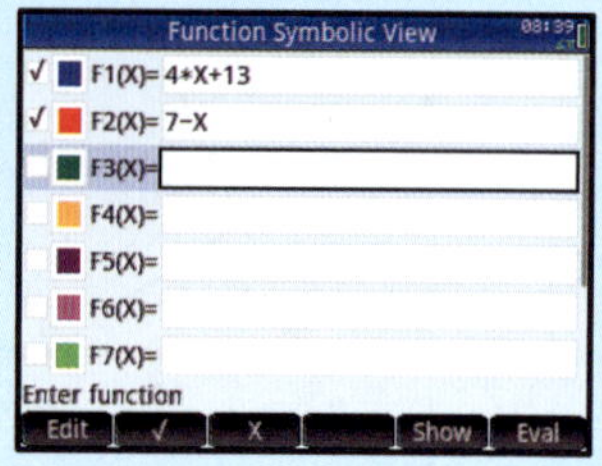

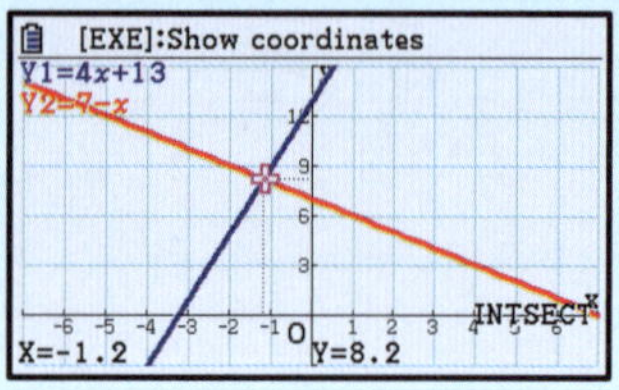

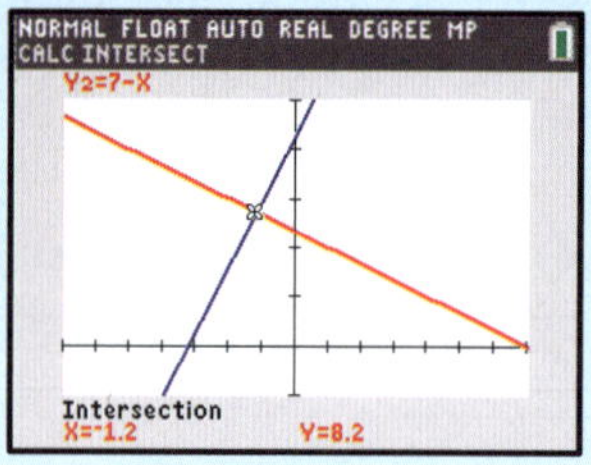

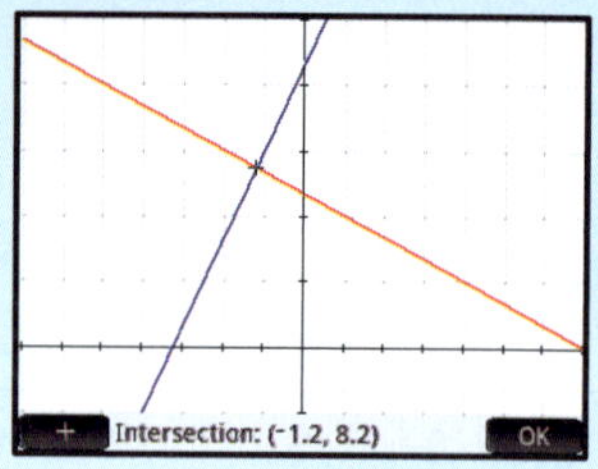

The solution is $x = -1.2$, $y = 8.2$.

EXERCISE 14A.2

1 Use technology to solve:

a $\begin{cases} y = 5x - 2 \\ y = x + 1 \end{cases}$ **b** $\begin{cases} y = 3x - 6 \\ y = 1 - 2x \end{cases}$ **c** $\begin{cases} y = 3x - 4 \\ x + y = -5 \end{cases}$

d $\begin{cases} x + y = 6 \\ 6x + y = 15 \end{cases}$ **e** $\begin{cases} x + 2y = -8 \\ 7x + 3y = -5 \end{cases}$ **f** $\begin{cases} 3x - 4y = -2 \\ 2x + 3y = -10 \end{cases}$

2 Try to solve the following simultaneous equations using technology. State the number of solutions in each case.

a $\begin{cases} y = 1.5x + 2 \\ 3x - 2y = -4 \end{cases}$ **b** $\begin{cases} 4x + 5y = -3 \\ y = -0.8x + 1 \end{cases}$ **c** $\begin{cases} x - 4y = 7 \\ 2x - 8y = -1 \end{cases}$

B SOLUTION BY SUBSTITUTION

We now consider some **algebraic** methods for solving simultaneous equations.

The method of **substitution** is used when one variable is given as the *subject* of one of the equations. We substitute the expression for this variable into the other equation.

Example 3 **Self Tutor**

Solve simultaneously by substitution: $\begin{cases} y = x + 5 \\ 3x - y = 1 \end{cases}$

$y = x + 5$ .... (1)
$3x - y = 1$ .... (2)

Substituting (1) into (2) gives $3x - (x + 5) = 1$
$\therefore\ 3x - x - 5 = 1$
$\therefore\ 2x = 6$
$\therefore\ x = 3$

Substituting $x = 3$ into (1) gives $y = 3 + 5$
$\therefore\ y = 8$

The solution is $x = 3$, $y = 8$.

Check: In (2), $3x - y = 3(3) - 8 = 9 - 8 = 1$ ✓

In equation (2) we replace y with $(x + 5)$. The brackets ensure that *all of* the expression is multiplied by -1.

EXERCISE 14B

1 Solve simultaneously by substitution:

a $\begin{cases} y = x - 3 \\ 2x + 3y = 26 \end{cases}$ **b** $\begin{cases} y = 3x - 2 \\ 5x - 2y = 5 \end{cases}$ **c** $\begin{cases} 7x + 4y = -7 \\ y = 8 - 5x \end{cases}$

d $\begin{cases} y = 2x - 12 \\ y = 13 - 3x \end{cases}$ **e** $\begin{cases} y = 3x + 4 \\ 5x + 3y = 5 \end{cases}$ **f** $\begin{cases} 5x - 6y = 2 \\ y = 3 - 7x \end{cases}$

2 Solve simultaneously by substitution:

a $\begin{cases} x = y + 9 \\ 5x + 2y = 10 \end{cases}$ **b** $\begin{cases} 7x - 5y = 1 \\ x = 2y + 4 \end{cases}$ **c** $\begin{cases} x = 2y - 7 \\ x = 13 - 6y \end{cases}$

d $\begin{cases} y = 2x - 2 \\ x = 5y + 7 \end{cases}$ **e** $\begin{cases} 2x + 4y = 1 \\ x = 10y - 7 \end{cases}$ **f** $\begin{cases} x = -4y - 3 \\ 9x + 5y = 2 \end{cases}$

3 Solve simultaneously by substitution:

a $\begin{cases} y = \frac{1}{2}x - 2 \\ 3x - 8y = 11 \end{cases}$ **b** $\begin{cases} x = -\frac{2}{3}y \\ 6x + 7y = 6 \end{cases}$ **c** $\begin{cases} 5x - 12y = 18 \\ y = \frac{1}{4}x - 1 \end{cases}$

d $\begin{cases} y = -\frac{1}{2}x - 1 \\ 4x + 3y = -2 \end{cases}$ **e** $\begin{cases} 2x - 3y = 3 \\ x = \frac{11}{5}y + 2 \end{cases}$ **f** $\begin{cases} 14x + 15y = 25 \\ y = \frac{2}{5}x - 3 \end{cases}$

4 Try to solve by substitution:

a $\begin{cases} y = 3x + 1 \\ 9x - 3y = 5 \end{cases}$ **b** $\begin{cases} y = 3x + 1 \\ 6x - 2y = -2 \end{cases}$

Discuss your results.

5 Consider the simultaneous equations: $\begin{cases} 4x + 5y = -4 & .... (1) \\ 3x + 2y = -5 & .... (2) \end{cases}$

a Rearrange equation (1) so that y is the subject.

b *Hence* solve the simultaneous equations by substitution.

C SOLUTION BY ELIMINATION

If neither variable is given as the subject of an equation, then solution by substitution is tedious. We instead use the method of **elimination**.

In this method, we make the coefficients of x (or y) the **same size** but **opposite in sign**. We then **add** the equations to **eliminate** one of the variables.

Example 4 Self Tutor

Solve simultaneously by elimination: $\begin{cases} 4x + 3y = 2 \\ x - 3y = 8 \end{cases}$

The coefficients of y are the same size but opposite in sign.

$\therefore$ we **add** the LHSs and the RHSs to eliminate y.

$$\begin{aligned} 4x + 3y &= 2 \quad .... (1) \\ x - 3y &= 8 \quad .... (2) \end{aligned}$$

Adding, $5x = 10$

$\therefore x = 2$

Substituting $x = 2$ into (1) gives $4(2) + 3y = 2$

$\therefore 8 + 3y = 2$

$\therefore 3y = -6$

$\therefore y = -2$

The solution is $x = 2$, $y = -2$.

Check: In (2), $x - 3y = 2 - 3(-2) = 2 + 6 = 8$ ✓

If the coefficients of x (or y) are *not* the same size and opposite in sign, we first need to **multiply** one or both equations by a constant.

Example 5 Self Tutor

Solve simultaneously by elimination: $\begin{cases} 3x + 2y = 7 \\ 2x - 5y = 11 \end{cases}$

$3x + 2y = 7 \quad .... (1)$
$2x - 5y = 11 \quad .... (2)$

Since the LCM of 3 and 2 is 6, we eliminate x by multiplying (1) by 2 and (2) by -3.

$$\begin{aligned} 6x + 4y &= 14 \quad \{(1) \times 2\} \\ -6x + 15y &= -33 \quad \{(2) \times -3\} \end{aligned}$$

Adding, $19y = -19$

$\therefore y = -1$

Substituting $y = -1$ into (1) gives $3x + 2(-1) = 7$

$\therefore\ 3x - 2 = 7$

$\therefore\ 3x = 9$

$\therefore\ x = 3$

The solution is $x = 3,\ y = -1$.

Check: In (2), $2x - 5y = 2(3) - 5(-1) = 6 + 5 = 11$ ✓

We can choose to eliminate either x or y. For example, in **Example 5** we could have eliminated y by multiplying (1) by 5 and (2) by 2.

In general, we choose the variable which is easiest to eliminate.

EXERCISE 14C

1 What equation results when the following are added vertically?

a $\begin{cases} 3x + 4y = 6 \\ 8x - 4y = 5 \end{cases}$ **b** $\begin{cases} 2x - y = 7 \\ -2x + 5y = 5 \end{cases}$ **c** $\begin{cases} 7x - 3y = 2 \\ 2x + 3y = 7 \end{cases}$

d $\begin{cases} 6x - 11y = 12 \\ 3x + 11y = -6 \end{cases}$ **e** $\begin{cases} -7x + 2y = 5 \\ 7x - 3y = 6 \end{cases}$ **f** $\begin{cases} 2x - 3y = -7 \\ -2x - 8y = -4 \end{cases}$

2 Solve simultaneously by elimination:

a $\begin{cases} 5x - y = 4 \\ 2x + y = 10 \end{cases}$ **b** $\begin{cases} 3x - 2y = 7 \\ 3x + 2y = -1 \end{cases}$ **c** $\begin{cases} -5x - 3y = 14 \\ 5x + 8y = -29 \end{cases}$

d $\begin{cases} 4x + 3y = -11 \\ -4x - 2y = 6 \end{cases}$ **e** $\begin{cases} 2x - 5y = 14 \\ 4x + 5y = -2 \end{cases}$ **f** $\begin{cases} -6x - y = 23 \\ 6x + 5y = -13 \end{cases}$

3 Give the equation that results when both sides of the equation:

a $2x + 5y = 1$ are multiplied by 5

b $3x - y = 4$ are multiplied by -1

c $x - 7y = 8$ are multiplied by 3

d $5x + 4y = 9$ are multiplied by -2

e $-3x - 2y = 2$ are multiplied by 6

f $4x - 2y = 3$ are multiplied by -4.

4 Solve simultaneously by elimination:

a $\begin{cases} 2x + y = 8 \\ x - 3y = 11 \end{cases}$ **b** $\begin{cases} 3x + 2y = 7 \\ x + 3y = 7 \end{cases}$ **c** $\begin{cases} 5x - 2y = 17 \\ 3x - y = 9 \end{cases}$

d $\begin{cases} 2x + 5y = -14 \\ -6x + 2y = -9 \end{cases}$ **e** $\begin{cases} 7x + 3y = 5 \\ 5x - 6y = 9 \end{cases}$ **f** $\begin{cases} 4x + 9y = 24 \\ 12x - 7y = -30 \end{cases}$

5 Solve simultaneously by elimination:

a $\begin{cases} 2x + 3y = 13 \\ 3x + 2y = 17 \end{cases}$ **b** $\begin{cases} 4x - 3y = 1 \\ 2x + 5y = 7 \end{cases}$ **c** $\begin{cases} 2x + 5y = 14 \\ 5x - 3y + 27 = 0 \end{cases}$

d $\begin{cases} 5x + 3y = 1 \\ 3x - 2y = 12 \end{cases}$ **e** $\begin{cases} 4x - 7y = -14 \\ 3x + 2y = -25 \end{cases}$ **f** $\begin{cases} 12x + 9y = 7 \\ 7x - 2y = -8 \end{cases}$

g $\begin{cases} 7x + 2y = 20 \\ 13x + 3y = 34 \end{cases}$ **h** $\begin{cases} 3x + 7y = 5 \\ 5x + 11y = 10 \end{cases}$ **i** $\begin{cases} 5x - 7y - 9 = 0 \\ 4x - 5y - 5 = 0 \end{cases}$

6 Try to solve by elimination:

a $\begin{cases} 2x - y = 3 \\ 4x - 2y = 6 \end{cases}$ **b** $\begin{cases} 3x + 4y = 6 \\ 6x + 8y = 7 \end{cases}$

Discuss your results.

D PROBLEM SOLVING

When given a problem in words, we need to interpret the information and use it to write equations involving the unknowns. We can then choose the most appropriate technique to solve the equations simultaneously, and hence answer the original problem.

Example 6 **Self Tutor**

Two numbers have a difference of 7 and an average of 4. Find the numbers.

Let x be the larger number and y be the smaller number.

The difference between x and y is $x - y = 7$ .... (1)

The average of x and y is $\dfrac{x+y}{2} = 4$ .... (2)

$$\begin{aligned} x - y &= 7 \quad \{(1)\} \\ x + y &= 8 \quad \{(2) \times 2\} \\ \hline \text{Adding,} \quad 2x &= 15 \\ \therefore \ x &= \tfrac{15}{2} \end{aligned}$$

Substituting $x = \frac{15}{2}$ into (1) gives $\frac{15}{2} - y = 7$

$\therefore \ y = \frac{15}{2} - 7$

$\therefore \ y = \frac{1}{2}$

The numbers are $\frac{1}{2}$ and $\frac{15}{2}$.

Check: (1) $\frac{15}{2} - \frac{1}{2} = 7$ ✓ (2) $\dfrac{\frac{15}{2} + \frac{1}{2}}{2} = 4$ ✓

EXERCISE 14D

1 The sum of two numbers is 72 and their difference is 40. Find the numbers.

2 Two numbers have a sum of 30. Half of their difference is 7. Find the numbers.

3 The larger of two numbers is 5 more than twice the smaller number, and their difference is 22. Find the numbers.

4 Answer the **Opening Problem** on page **222**.

5 At a garden nursery, 2 geraniums and 3 marigolds cost \$25, whereas 5 geraniums and 6 marigolds cost \$58. Find the cost of each type of plant.

6 At a food van, 4 bottles of water and 2 bottles of soft drink cost \$26, whereas 3 bottles of water and 4 bottles of soft drink cost \$32. Find the total cost of 2 bottles of water and 3 bottles of soft drink.

Example 7 **Self Tutor**

When shopping in Jamaica, 5 coconuts and 14 bananas cost \$8.70, and 8 coconuts and 9 bananas cost \$9.90.

Find the cost of each coconut and each banana.

Let each coconut cost x cents, and each banana cost y cents.

5 coconuts and 14 bananas cost \$8.70, so $5x + 14y = 870$ .... (1)
8 coconuts and 9 bananas cost \$9.90, so $8x + 9y = 990$ .... (2)

Since the LCM of 5 and 8 is 40, we eliminate x by multiplying (1) by 8 and (2) by -5.

$$40x + 112y = 6960 \quad \{(1) \times 8\}$$
$$-40x - 45y = -4950 \quad \{(2) \times -5\}$$

Adding, $$67y = 2010$$
$$\therefore \ y = 30$$

Substituting $y = 30$ in (2) gives $8x + 9(30) = 990$
$$\therefore \ 8x + 270 = 990$$
$$\therefore \ 8x = 720$$
$$\therefore \ x = 90$$

The coconuts cost 90 cents each, and the bananas cost 30 cents each.

7 Three pieces of fish and two serves of chips cost \$8.10. Five pieces of fish and three serves of chips cost \$13.25. Find the cost of each piece of fish and each serve of chips.

8 Seven cups of coffee and four muffins cost \$25.30. Two cups of coffee and three muffins cost \$9.55. Find the cost of each item.

9 Juice is sold in either 600 mL or 1 L bottles. A supermarket manager ordered 79.8 litres of juice and received 93 bottles. How many 600 mL bottles did the manager receive?

10 The triangle alongside is equilateral.
Find the side length of the triangle.

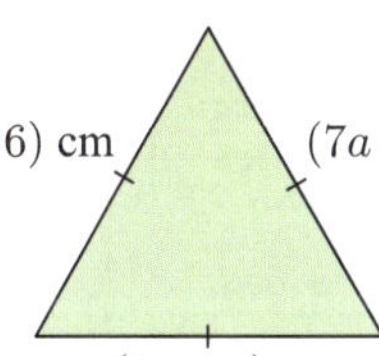

11 A car park charges a certain rate per hour before 3 pm, and a higher rate per hour after 3 pm.
Parking from 11 am to 5 pm costs \$18, and parking from 9 am to 7 pm costs \$31.
Find the rates charged before and after 3 pm.

12 An art store sells two different painting sets. The small sets contain 2 brushes and 3 tubes of paint, and the large sets contain 5 brushes and 8 tubes of paint. Leanne bought some of each set to use as gifts. In total she received 37 brushes and 58 tubes of paint. How many of each set did she buy?

13 Let P be $(1, 13)$ and Q be $(-2, -11)$. The gradient of [PR] is -2 and the gradient of [QR] is 4.
Find the coordinates of R.

14 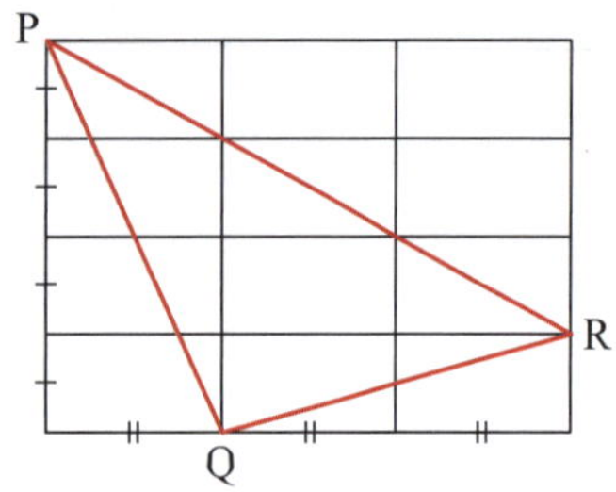

[PQ] has length $\sqrt{175}$ units and [QR] has length $\sqrt{70}$ units. Find the exact length of [PR].

ACTIVITY — LARGER SYSTEMS OF EQUATIONS

Simultaneous equations like $\begin{cases} 2x+3y=11 \\ 3x-2y=10 \end{cases}$ are called "2 by 2" systems because there are 2 equations involving 2 unknowns.

LARGER SYSTEMS OF EQUATIONS

In this online Activity we consider larger systems of equations.

INVESTIGATION — DISTANCE FROM A POINT TO A LINE

When we talk about the distance from a point to a line, we actually mean the *shortest* distance from the point to the line.

Suppose N is the foot of the perpendicular from P to the line l.

If M is any point on the line other than at N, then triangle MNP is right angled with hypotenuse [MP], and so $\text{MP} \geqslant \text{NP}$.

Hence NP is the shortest distance from P to line l.

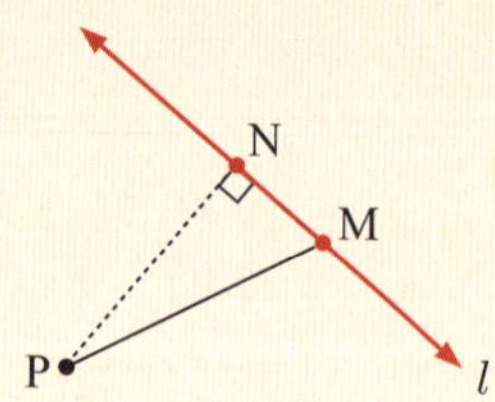

What to do:

1 Consider the point P(7, -4) and the line l with equation $2x + y = 5$.
Let N be the foot of the perpendicular from P to l.

a Find the gradient of: **i** l **ii** [NP]

b Find the equation of (NP).

c Explain why the coordinates of N are found by solving the system of simultaneous equations: $\begin{cases} 2x+y=5 \\ x-2y=15\,. \end{cases}$

d Find the coordinates of N.

e Hence find the shortest distance from P to l.

2 Find the shortest distance from:

a $(8, -5)$ to $y = -2x - 4$ **b** $(1, 7)$ to $4x - 3y = 8$.

3 Find the distance between the parallel lines $y = 3x + 2$ and $y = 3x - 8$ by choosing any point on one line, then finding the shortest distance to the other line.

4 A straight water pipeline passes through two points with coordinates $(3, 2)$ and $(7, -1)$. The shortest spur pipe from the pipeline to the farm at P(9, 7) is [NP]. 1 grid unit = 100 m.

a Find the coordinates of N. **b** Hence find the length of the pipeline [NP].

E NON-LINEAR SIMULTANEOUS EQUATIONS

The technique of substitution can be used to solve non-linear simultaneous equations, especially if the result of the substitution is a quadratic equation. In such cases, the simultaneous equations may have zero, one, or two solutions.

Example 8 Self Tutor

Solve simultaneously by substitution: $\begin{cases} y = x^2 + 4x + 2 \\ 5x + y = 12 \end{cases}$

$y = x^2 + 4x + 2$ .... (1)
$5x + y = 12$ .... (2)

Substituting (1) into (2) gives $5x + (x^2 + 4x + 2) = 12$
$\therefore\ x^2 + 9x - 10 = 0$
$\therefore\ (x - 1)(x + 10) = 0$
$\therefore\ x = 1$ or -10

Substituting $x = 1$ into (1) gives $y = (1)^2 + 4(1) + 2 = 7$
Substituting $x = -10$ into (1) gives $y = (-10)^2 + 4(-10) + 2 = 62$
The solutions are $x = 1,\ y = 7$ and $x = -10,\ y = 62$.

EXERCISE 14E

1 Solve simultaneously by substitution:

a $\begin{cases} y = x^2 - 2x + 5 \\ 3x + y = 11 \end{cases}$
b $\begin{cases} 2x + y = 17 \\ y = x^2 + 5x - 1 \end{cases}$
c $\begin{cases} y = 2x^2 - x + 3 \\ 5x + y = 9 \end{cases}$
d $\begin{cases} y = 3x^2 - 5x - 4 \\ 4x - y = -26 \end{cases}$
e $\begin{cases} y = 2x^2 + x - 5 \\ y = x^2 + 5x + 7 \end{cases}$
f $\begin{cases} y = x^2 - 7x - 11 \\ y = 3x^2 + 5x + 7 \end{cases}$

Example 9 Self Tutor

Solve simultaneously by substitution: $\begin{cases} xy = 12 \\ y = 2x + 2 \end{cases}$

$xy = 12$ .... (1)
$y = 2x + 2$ .... (2)

Substituting (2) into (1) gives $x(2x + 2) = 12$
$\therefore\ 2x^2 + 2x = 12$
$\therefore\ 2(x^2 + x - 6) = 0$
$\therefore\ 2(x + 3)(x - 2) = 0$
$\therefore\ x = -3$ or 2

Substituting $x = -3$ into (2) gives $y = 2(-3) + 2 = -4$
Substituting $x = 2$ into (2) gives $y = 2(2) + 2 = 6$
The solutions are $x = -3,\ y = -4$ and $x = 2,\ y = 6$.

2 Solve simultaneously by substitution:

a $\begin{cases} xy = 20 \\ y = x + 1 \end{cases}$ **b** $\begin{cases} y = 16 - 2x \\ xy = 32 \end{cases}$ **c** $\begin{cases} xy = 5 \\ y = 6x + 7 \end{cases}$

3 Use substitution and the quadratic formula to try to solve:

a $\begin{cases} y = x^2 - 4x + 10 \\ 3x + 2y = 7 \end{cases}$ **b** $\begin{cases} xy = 6 \\ y = x + 4 \end{cases}$ **c** $\begin{cases} y = 5x - 7 \\ xy = -4 \end{cases}$

4 Show algebraically that it is impossible for two real numbers x and y to have a sum of 10 and a product of 30.

5 A and B are two points such that $\text{AB} = \sqrt{60}$ units and the gradient of [AB] is 3. Given that A is (1, 2), find the possible coordinates of B.

MULTIPLE CHOICE QUIZ

QUICK QUIZ

REVIEW SET 14A

1 Solve graphically without using technology:

a $\begin{cases} y = 5x + 2 \\ y = 2x - 7 \end{cases}$ **b** $\begin{cases} 2x - 3y = 18 \\ x + y = 4 \end{cases}$

2 Use technology to solve simultaneously:

a $\begin{cases} y = 3x - 4 \\ y = 9 - x \end{cases}$ **b** $\begin{cases} 3x - 4y = 3 \\ 7x - 11y = 8 \end{cases}$

3 Solve simultaneously by substitution:

a $\begin{cases} x = 8 - 3y \\ 2x - 7y = -10 \end{cases}$ **b** $\begin{cases} y = 7 - 2x \\ 4x + 5y = 5 \end{cases}$

4 **a** Try to solve $\begin{cases} y = 4x - 2 \\ 8x - 2y = 4 \end{cases}$ by substitution.

b How many solutions do these simultaneous equations have?

5 What equation results when the following are added vertically?

a $\begin{cases} 4x - 3y = 8 \\ 2x + 3y = -14 \end{cases}$ **b** $\begin{cases} 5x - y = 7 \\ -5x - 3y = 33 \end{cases}$

6 Solve simultaneously by elimination:

a $\begin{cases} -2x + 7y = 13 \\ 2x - 3y = -9 \end{cases}$ **b** $\begin{cases} 2x - 5y = -2 \\ 8x - 9y = 30 \end{cases}$ **c** $\begin{cases} 4x - 6y = 21 \\ 3x + 5y = -8 \end{cases}$

7 6 adult tickets and 5 student tickets for the cinema cost \$168. 2 adult tickets and 3 student tickets cost \$72. Find the cost of each type of ticket.

8 Carmel has some fifty cent coins and some twenty cent coins in her purse. She has 21 coins in total, and their total value is \$8.40. How many of each type of coin does Carmel have?

9 Find the perimeter of this rectangle:

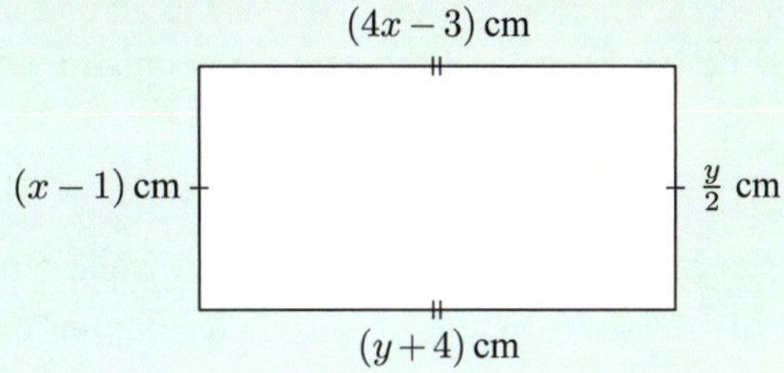

10 Solve simultaneously by substitution:

a $\begin{cases} y = x^2 - 4x - 5 \\ 2x + y = 10 \end{cases}$ **b** $\begin{cases} xy = -8 \\ y = 2x - 17 \end{cases}$

REVIEW SET 14B

1 Solve graphically without using technology:

a $\begin{cases} y = 3x - 8 \\ y = -2x + 7 \end{cases}$ **b** $\begin{cases} 3x - 2y = 18 \\ y = x - 7 \end{cases}$

2 Try to solve $\begin{cases} y = \frac{3}{5}x - 1 \\ 3x - 5y = 5 \end{cases}$ using technology.

State how many solutions the simultaneous equations have.

3 Solve by substitution:

a $\begin{cases} y = 4x - 7 \\ x = 9 - 2y \end{cases}$ **b** $\begin{cases} y = 5x + 3 \\ 4x + y = 1 \end{cases}$

4 **a** Try to solve $\begin{cases} 4x - 2y = 3 \\ y = 2x - 5 \end{cases}$ by substitution.

b How many solutions do these simultaneous equations have?

5 Give the equation that results when both sides of $2x - 3y = -4$ are multiplied by -3.

6 Solve simultaneously by elimination:

a $\begin{cases} 3x - 2y = 17 \\ -3x + y = -13 \end{cases}$ **b** $\begin{cases} 7x + 6y = -16 \\ 3x - 2y = 16 \end{cases}$ **c** $\begin{cases} 5x - 7y = 2 \\ -3x + 4y = 5 \end{cases}$

7 In an egg and spoon race, children must transport eggs from one end of a course to the other. Points are awarded for each egg successfully transported, and points are deducted for each egg which drops and breaks.

Jenny transported 5 eggs and dropped 2 eggs, for a score of 11 points. Heath transported 8 eggs and dropped 5 eggs, for a score of 5 points.

How many points are deducted for dropping an egg?

8 A sports store sells packs of 3 tennis balls for \$5, and packs of 8 tennis balls for \$12. In one day the store sold 97 tennis balls for \$151 in total. How many packs of 3 balls did they sell?

9 Let A be the fraction of the square occupied by the green region, and B be the fraction of the square occupied by the red region.

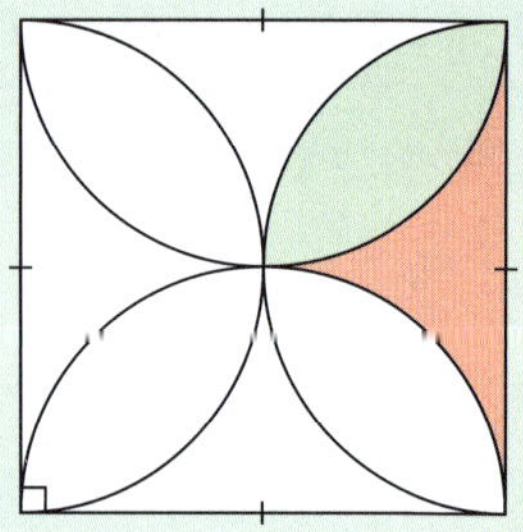

a Explain why:

i $A + B = \frac{1}{4}$ **ii** $2A + B = \frac{\pi}{8}$

b Hence find the exact fraction of the square occupied by:

i the green region **ii** the red region.

c Use the fact that $\pi > 3$ to determine which region is larger.

10 Solve by substitution:

a $\begin{cases} y = 3x^2 - 4x + 20 \\ y = 2x^2 + 6x - 5 \end{cases}$ **b** $\begin{cases} xy = 36 \\ y = 2x + 6 \end{cases}$

Chapter 15

Congruence and similarity

Contents:

OPENING PROBLEM

Nanna's Treat is a start-up company manufacturing homestyle jams and marmalade. They will sell their products in cylindrical glass jars in three sizes: 300 g, 500 g, and 750 g.

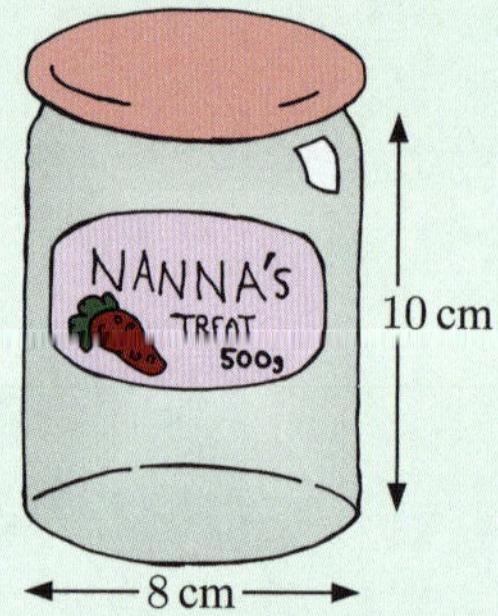

The dimensions of the jars will be *in proportion*.

Things to think about:

- **a** What does it mean that the dimensions are *in proportion*?
- **b** Can you write down the ratio of the jars':
  - **i** capacities **ii** heights **iii** surface areas?
- **c** Can you find:
  - **i** the height of the 300 g jar
  - **ii** the base radius of the 750 g jar
  - **iii** the surface area of the 300 g jar, excluding the lid?

In previous years you should have seen that:

- **Congruent figures** are identical in size and shape. They do not need to have the same position or orientation.
  When we perform a translation, rotation, or reflection, the object and image are congruent.
- **Similar figures** have the same shape but different size. One figure is an enlargement of the other.

A CONGRUENT TRIANGLES

These triangles are congruent because they have the same size and shape.

Their corresponding angles and side lengths are identical.

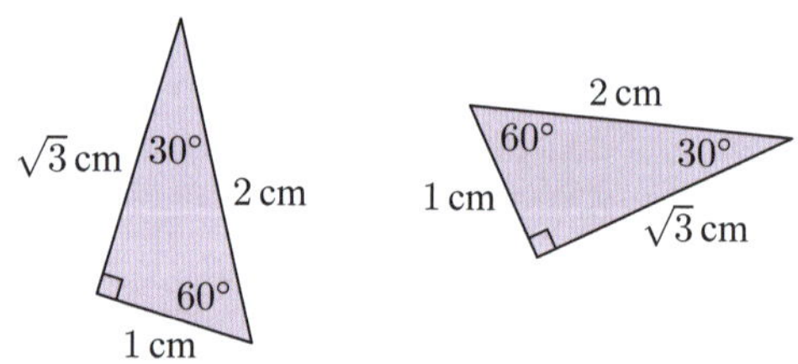

If we are given sufficient information about a triangle, there will be only one way in which it can be drawn. Any two triangles which have this information in common must be congruent.

Two triangles are **congruent** if any one of the following is true:

- All corresponding sides are equal in length. **(SSS)**

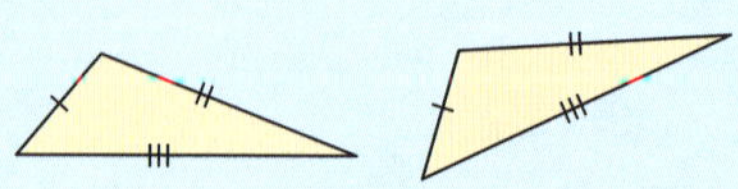

- Two sides and the **included angle** are equal. **(SAS)**

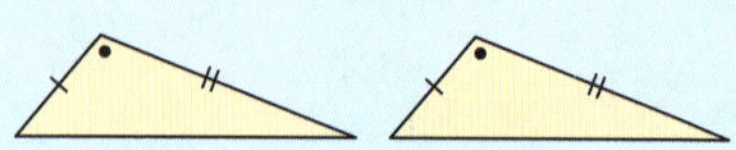

- Two angles and a pair of **corresponding sides** are equal. (**AAcorS**)

- For right angled triangles, the hypotenuses and one pair of sides are equal. **(RHS)**

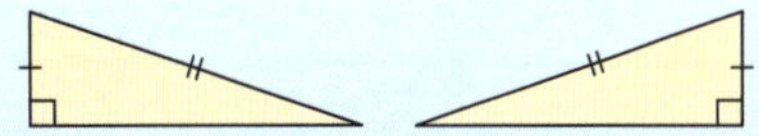

When we consider whether two figures are congruent, we assume the diagrams are *not* drawn to scale, but that information marked on the diagrams is correct.

If we conclude that two triangles are congruent, we:

- use the symbol $\cong$ to mean "is congruent to"
- record the vertices that are in corresponding positions in the same order
- use the appropriate abbreviation to record our reason.

Example 1

Self Tutor

Determine whether each pair of triangles is congruent. Give reasons for your answers.

a

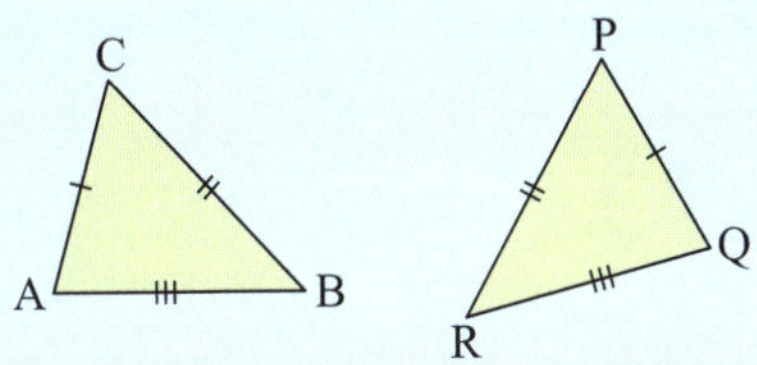

b

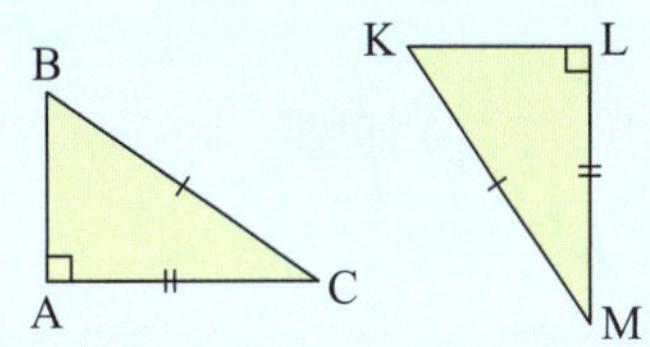

c

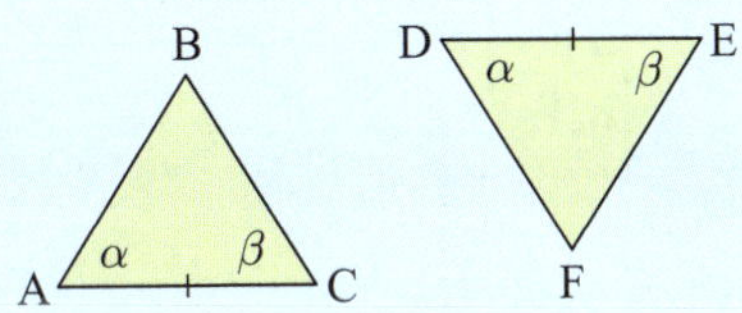

d

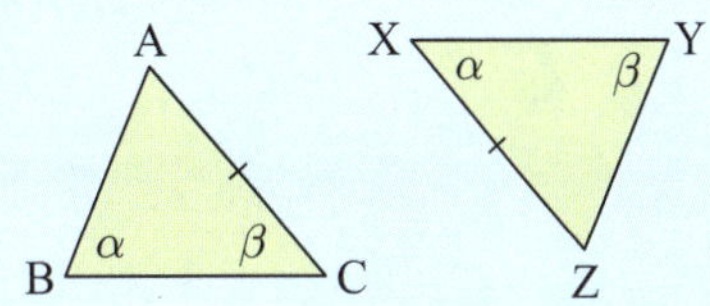

a $\triangle ABC \cong \triangle QRP$ {SSS}

b $\triangle ABC \cong \triangle LKM$ {RHS}

c $\triangle ABC \cong \triangle DFE$ {AAcorS}

d This is *not* AAcorS since the equal sides are not in corresponding positions.
[AC] is opposite α whereas [XZ] is opposite β.
$\therefore$ we cannot conclude that the triangles are congruent.

EXERCISE 15A

1 Determine whether each pair of triangles is congruent, giving reasons for your answers.

a

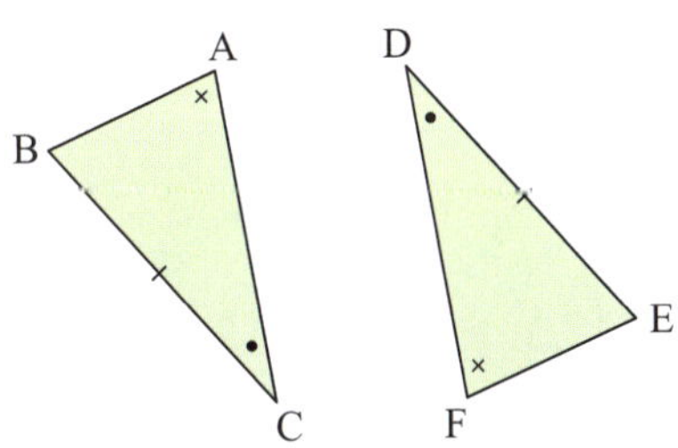

b

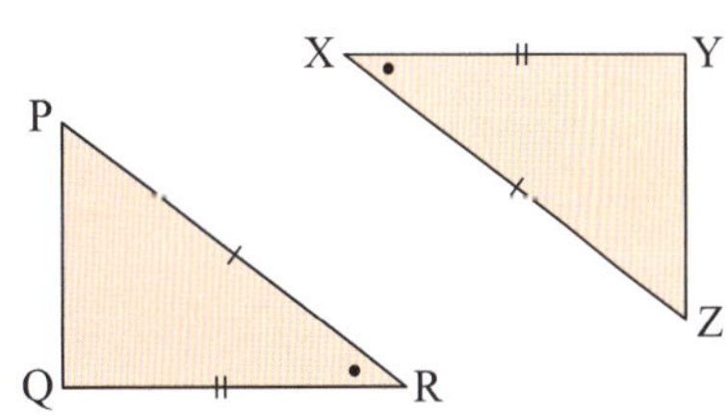

c

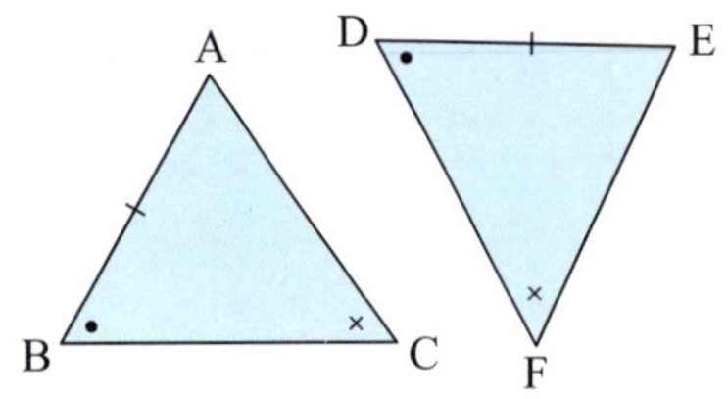

d

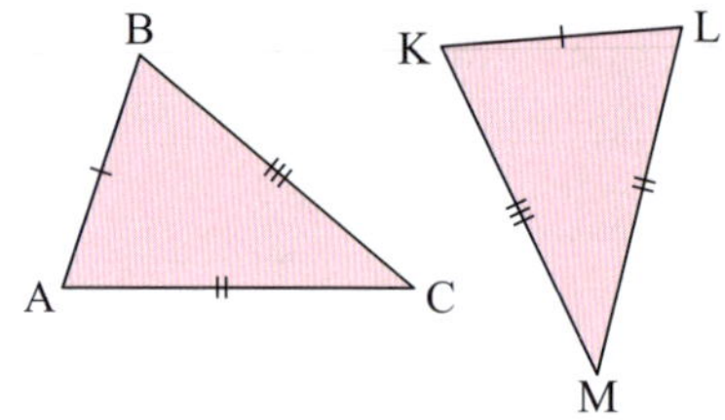

e

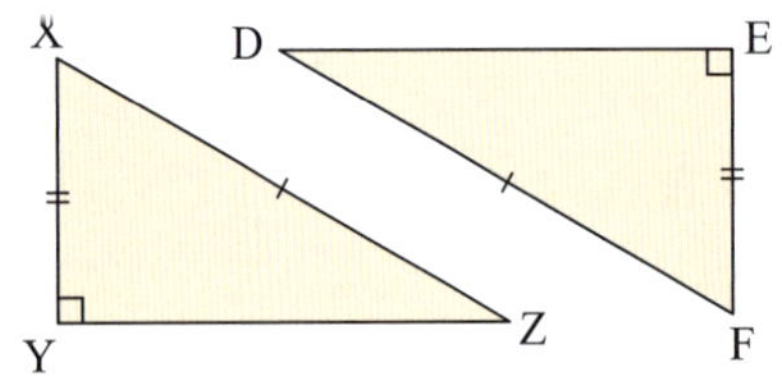

f

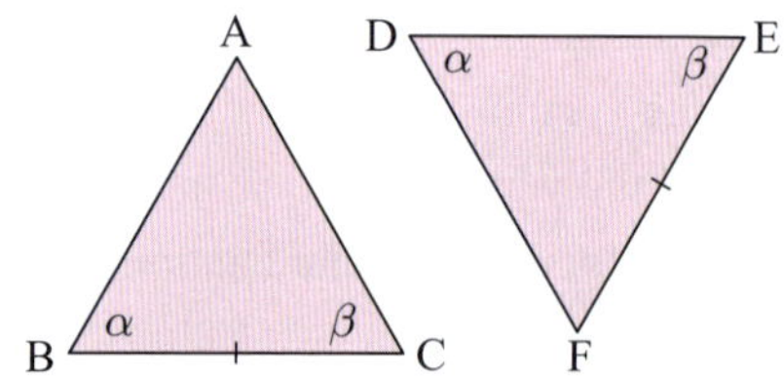

g

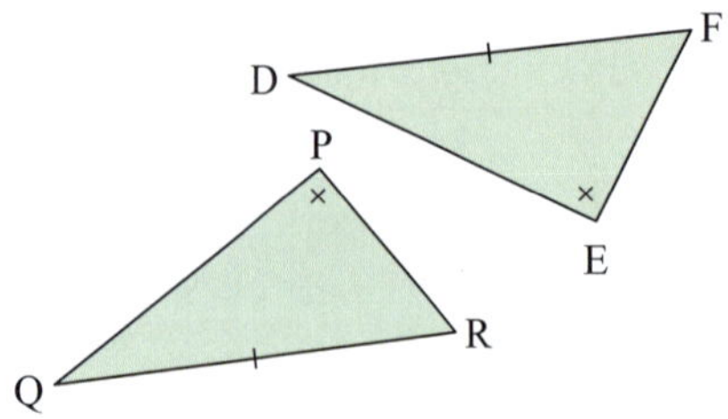

h

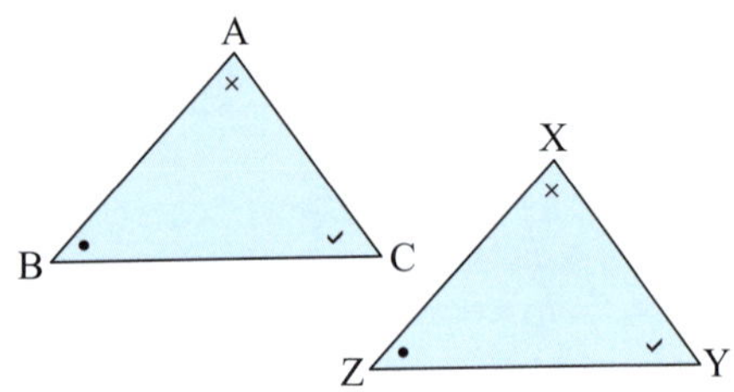

i

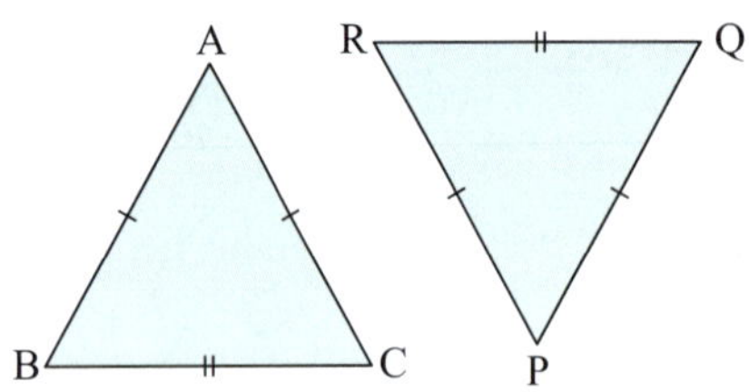

j

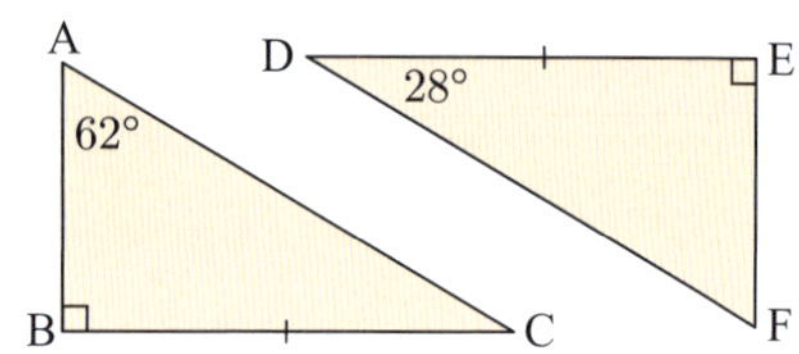

2 In each set of triangles, two are congruent. The triangles are *not* drawn to scale, but contain correct information. State which pair of triangles is congruent, giving a reason for your answer.

a

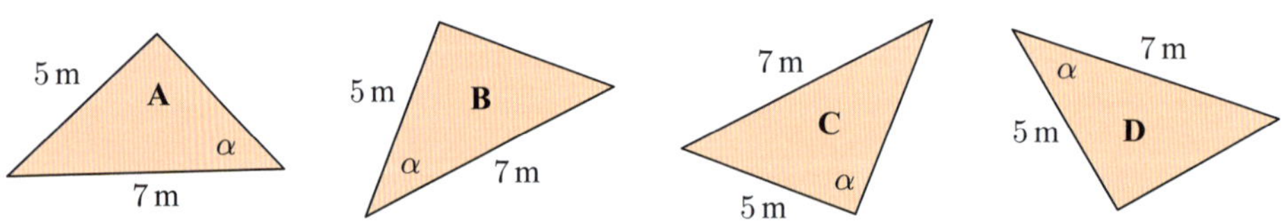

b

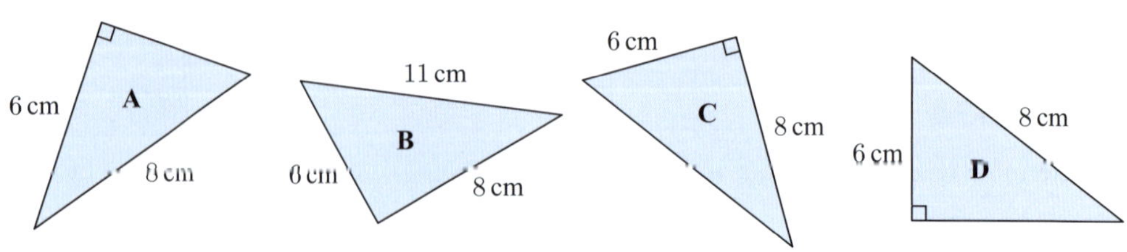

c

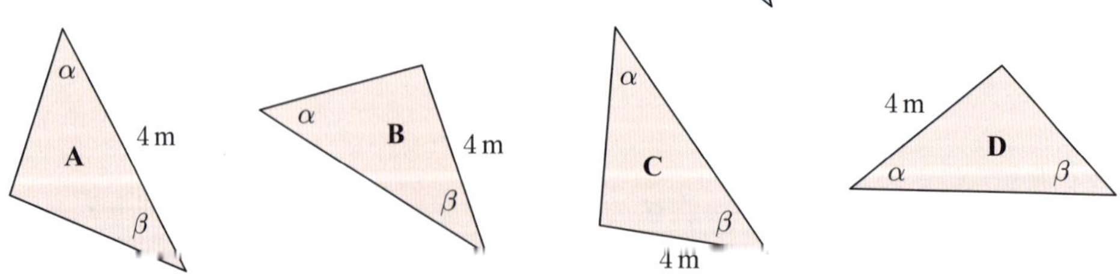

d

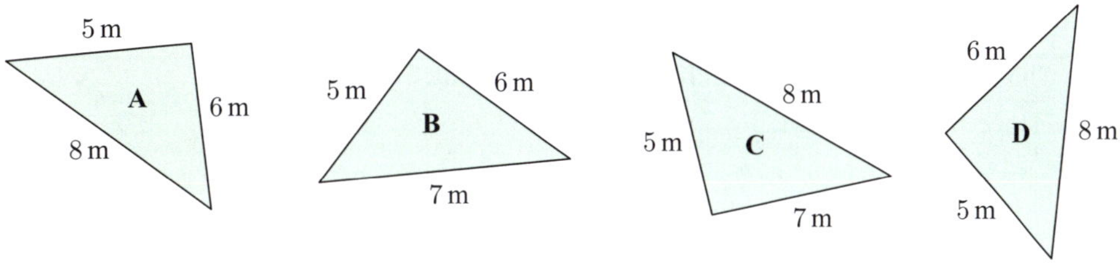

3

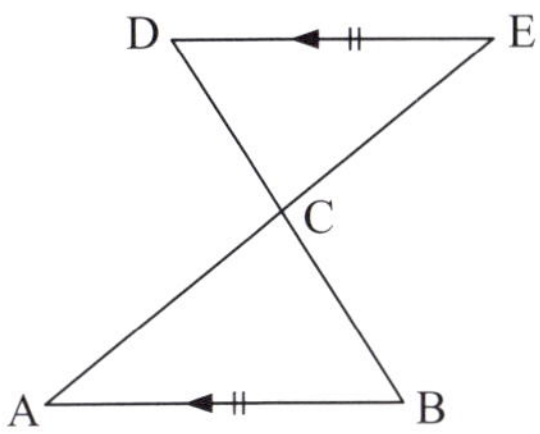

In the given figure, [DE] is parallel to [AB] and DE = AB. Prove that the triangles are congruent.

4

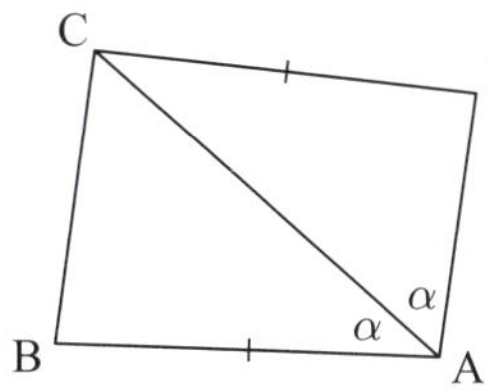

Is there sufficient evidence to conclude that triangles ABC and ADC are congruent? Explain your answer.

B PROOF USING CONGRUENCE

Once we have established that two triangles are congruent, we can deduce that the remaining corresponding sides and angles of the triangles are equal. We can therefore use congruence to prove facts about geometric figures.

Example 2 Self Tutor

Show that [AB] is parallel to [DE].

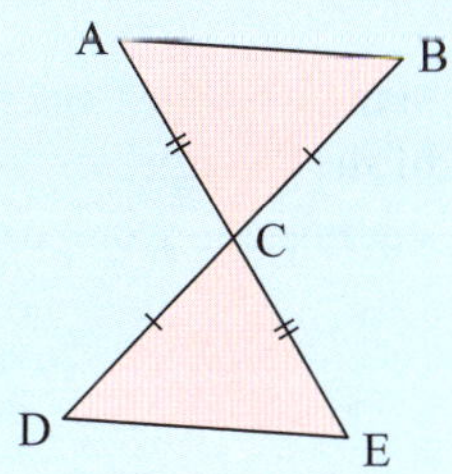

In $\triangle$s ABC and EDC:

- BC = DC {given}
- AC = EC {given}
- $A\widehat{C}B = E\widehat{C}D$ {vertically opposite angles}

$\therefore$ $\triangle ABC \cong \triangle EDC$ {SAS}

$\therefore$ $A\widehat{B}C = E\widehat{D}C$ {equating angles}

$\therefore$ [AB] $\parallel$ [DE] {equal alternate angles}

EXERCISE 15B

1 **a** Show that triangles ABD and CBD are congruent.

b Hence find the size of:

i $C\widehat{B}D$ **ii** $C\widehat{D}B$

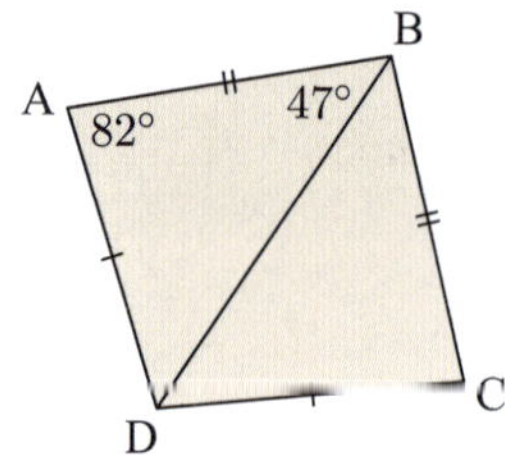

2 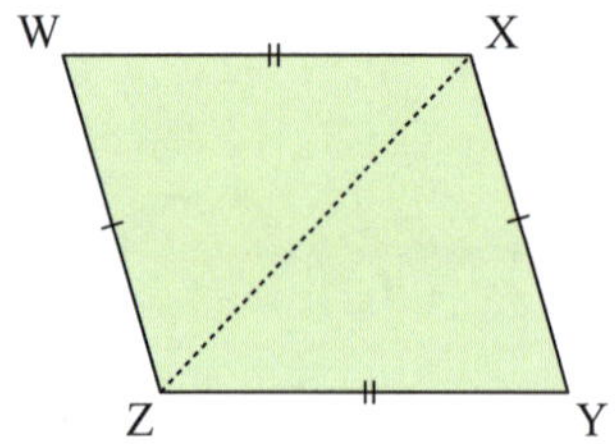

WXYZ is a quadrilateral with opposite sides equal.

a If [XZ] is added to the figure, show that the two triangles created are congruent.

b Hence deduce that WXYZ is a parallelogram.

3 Consider the quadrilateral ABCD alongside.

a Use congruence to show that the opposite sides are equal in length.

b Hence show that the diagonals of a parallelogram bisect each other.

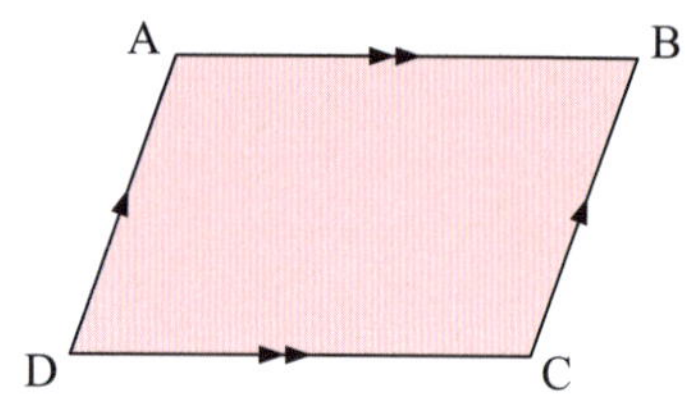

4 Triangle ABC is isosceles with AB = AC.
The angle bisectors of B and C are drawn, meeting the triangle at P and Q respectively.
Prove that AP = AQ.

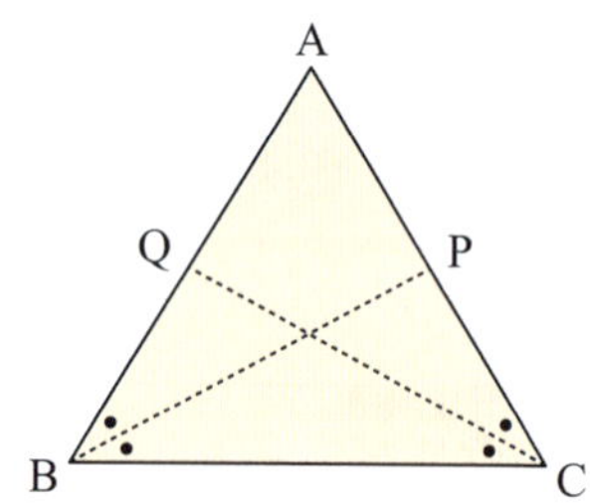

5 The perpendicular bisectors of a triangle's edges meet at a point called the **circumcentre** of the triangle.
Prove that the circumcentre is equidistant from each vertex of the triangle.

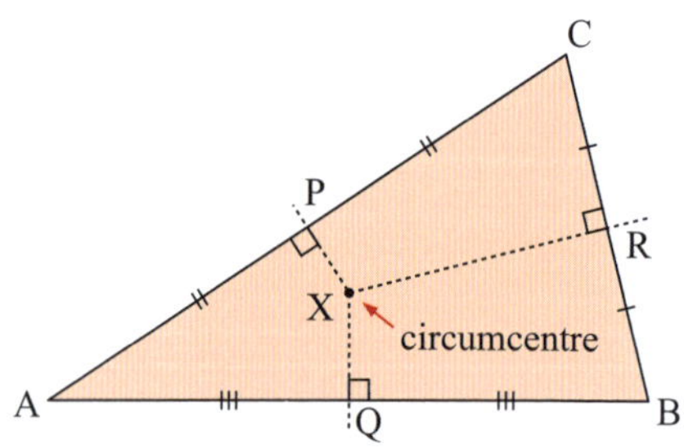

6 ABCD is a rectangle. Equilateral triangles are drawn from each side of the rectangle, with apexes W, X, Y, and Z.
Show that WXYZ is a rhombus.

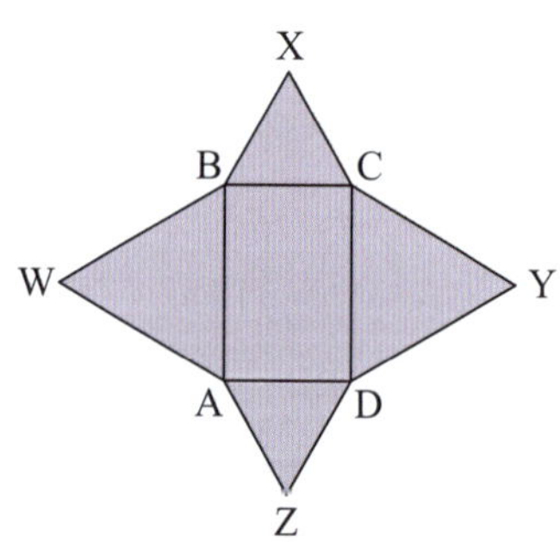

INVESTIGATION 1 THE ISOSCELES TRIANGLE THEOREM

An **isosceles triangle** is a triangle in which two side lengths are equal.

The **isosceles triangle theorem** states that:

In any isosceles triangle:
- the base angles are equal
- the line joining the apex to the midpoint of the base bisects the vertical angle and meets the base at right angles.

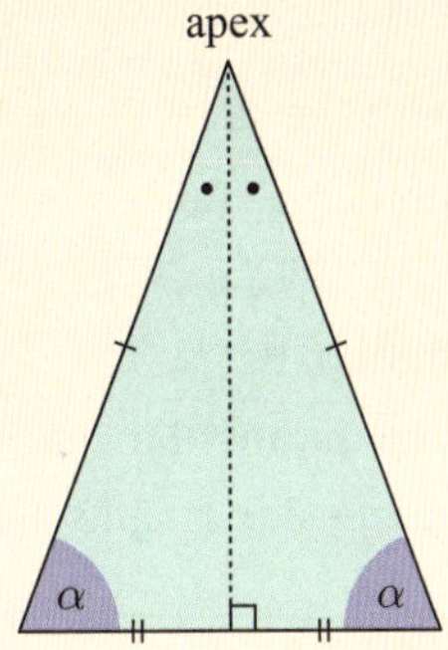

GEOMETRY PACKAGE

The isosceles triangle theorem has these **converses**:

Converse 1: If a triangle has two equal angles then it is isosceles.

Converse 2: The angle bisector of the apex of an isosceles triangle bisects the base at right angles.

Converse 3: The perpendicular bisector of the base of an isosceles triangle passes through its apex.

In this Investigation you will *prove* the three converses using congruence.

What to do:

1 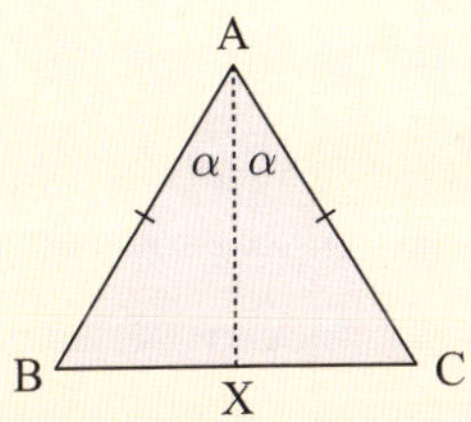

Triangle ABC is isosceles. The angle bisector at A meets [BC] at X.
Prove *Converse 2* by using congruence to show that [AX] is perpendicular to [BC].

2 Sam wants to prove *Converse 1*.

a Suppose Sam draws a line from the apex to the midpoint of the base. Explain why we cannot conclude that the triangles formed are congruent.

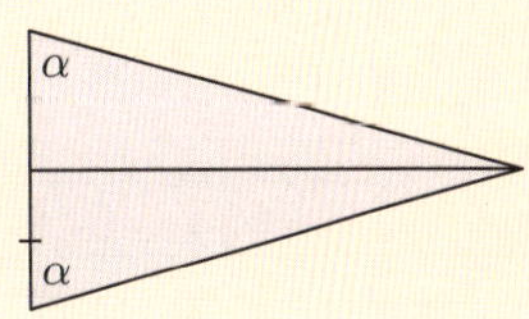

b Suppose that Sam instead draws the perpendicular from the apex to the base. Can he now use congruence to prove *Converse 1*?

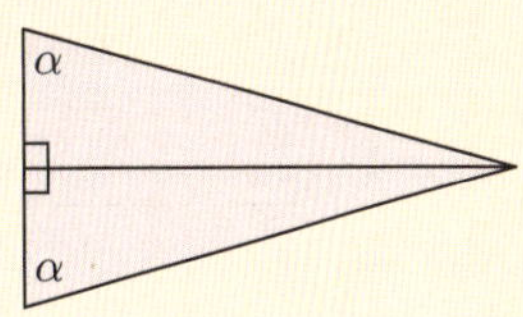

3 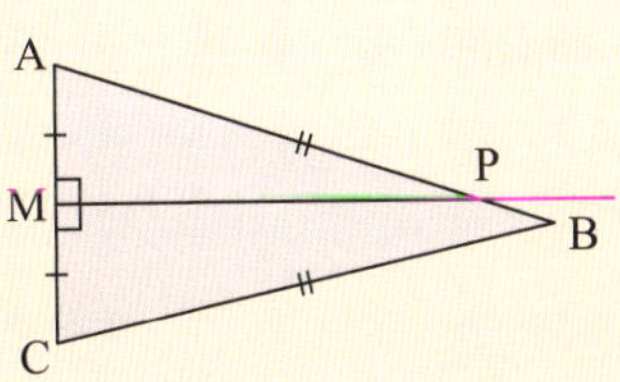

In the diagram alongside, the perpendicular bisector of the base is drawn so that it does not pass through vertex B, but instead meets [AB] at some other point P.
By joining [CP], complete the proof of *Converse 3*.

C SIMILAR TRIANGLES

Similar figures have the same *shape* but different *size*.

Two figures are similar if one can be placed exactly on top of the other using an enlargement or reduction, and if necessary a combination of a translation, a rotation, and a reflection.

Two figures are **similar** if:

- the figures are **equiangular** *and*
- the corresponding side lengths are in the **same ratio**.

SIMILAR TRIANGLES

If two *triangles* are equiangular then their corresponding sides *must* be in the same ratio, and vice versa. So, to show that two triangles are similar, we only need to show that *one* of these properties is true.

Two *triangles* are **similar** if ***either***:

- they are **equiangular** ***or***
- their side lengths are in the **same ratio**.

Since the angles of any triangle sum to 180°, if two angles of one triangle are equal to two angles of another triangle, then the remaining angles of the triangles must also be equal.

Example 3 — Self Tutor

Show that the following figures possess similar triangles:

a

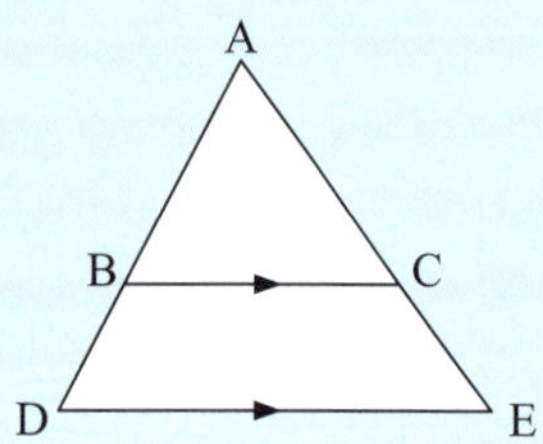

b

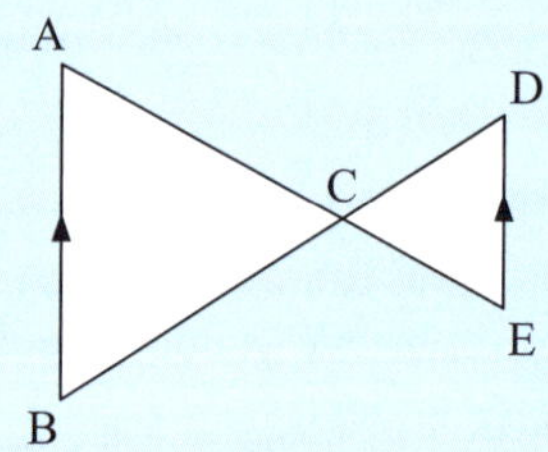

a

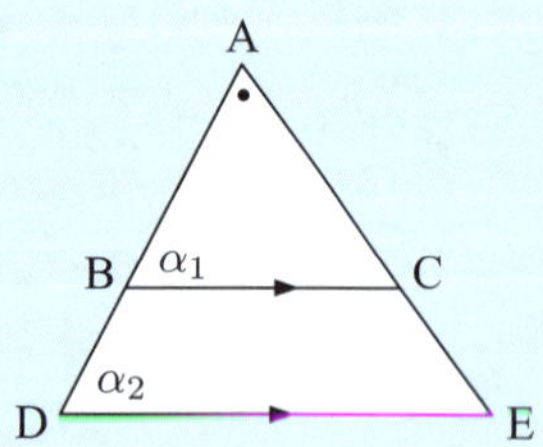

$\triangle$s ABC and ADE are equiangular as:

- $\alpha_1 = \alpha_2$ {equal corresponding angles}
- the angle at A is common to both triangles

$\therefore$ the triangles are similar.

b

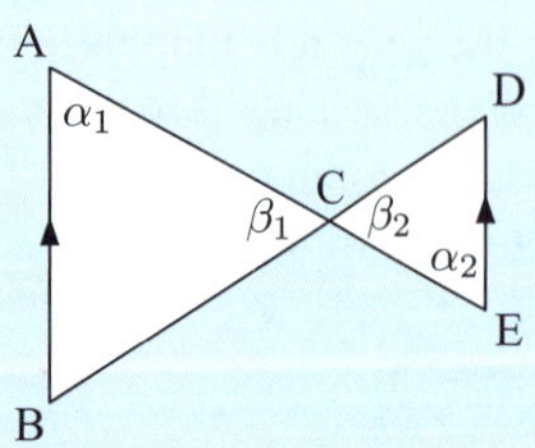

$\triangle$s ABC and EDC are equiangular as:

- $\alpha_1 = \alpha_2$ {equal alternate angles}
- $\beta_1 = \beta_2$ {vertically opposite angles}

$\therefore$ the triangles are similar.

EXERCISE 15C

1 Show that these figures possess similar triangles:

a
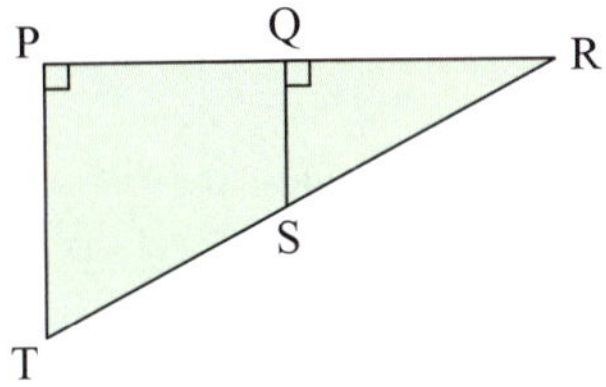

b
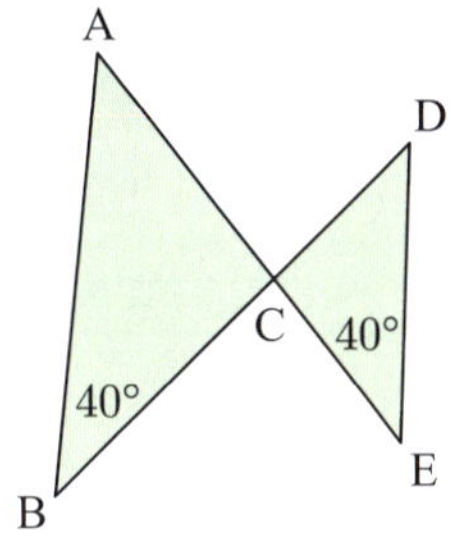

c
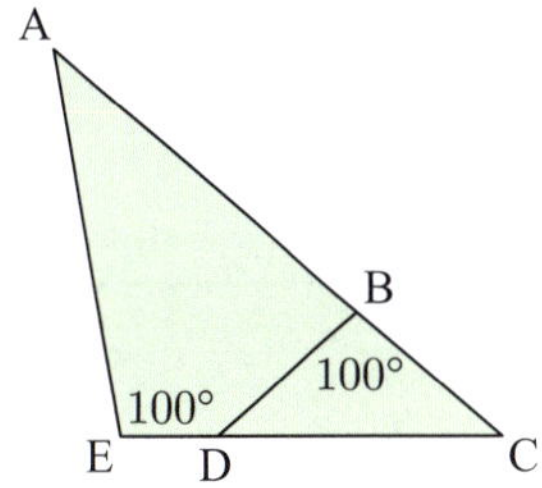

d
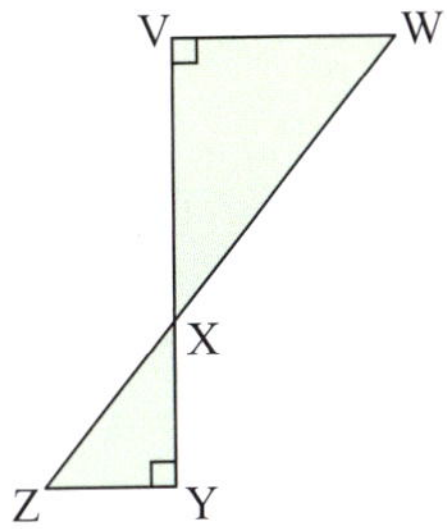

e
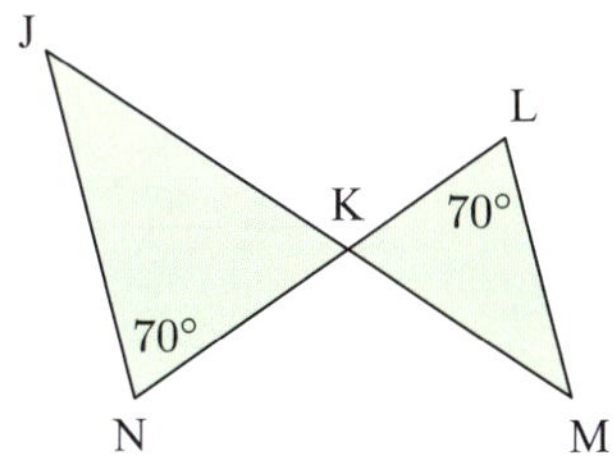

f
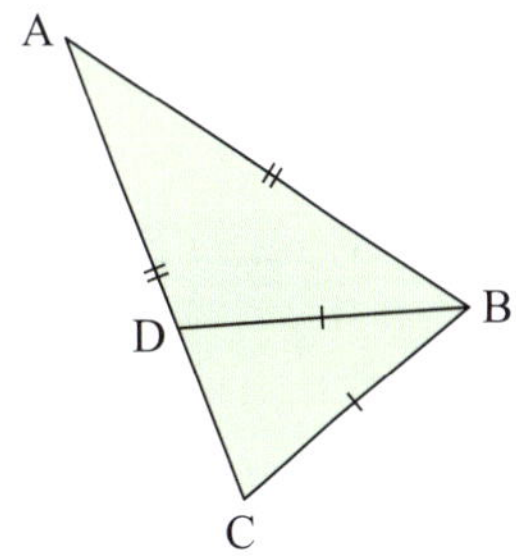

Example 4 — Self Tutor

Show that a pair of triangles is similar, and hence find the value of x:

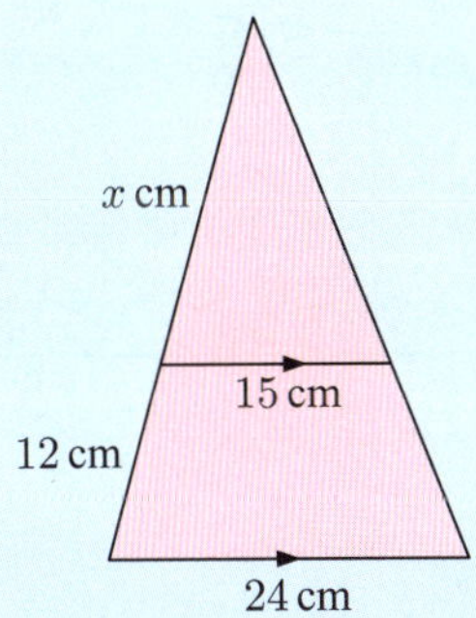

We label the vertices and angles of the figure so that we can easily refer to them.

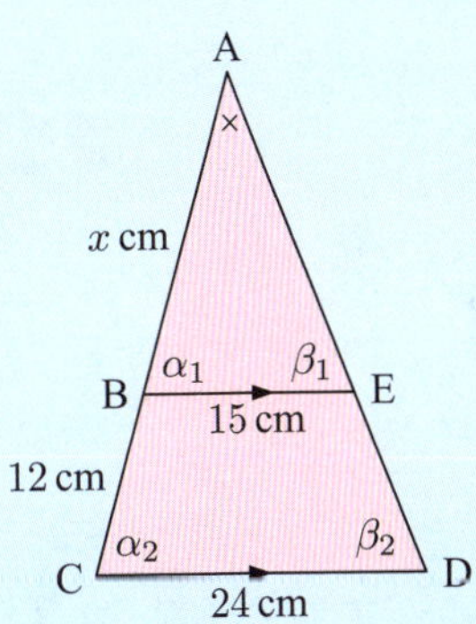

$\triangle$s ABE and ACD are equiangular as:

- $\alpha_1 = \alpha_2$ {equal corresponding angles}
- $\beta_1 = \beta_2$ {equal corresponding angles}

$\therefore$ $\triangle$s ABE and ACD are similar.

$\therefore \dfrac{AC}{AB} = \dfrac{CD}{BE}$ {same ratio}

$\therefore \dfrac{x+12}{x} = \dfrac{24}{15}$

$\therefore 15(x+12) = 24x$

$\therefore 15x + 180 = 24x$

$\therefore 180 = 9x$

$\therefore x = 20$

2 In each figure, show that a pair of triangles is similar. Hence find the value of x:

a

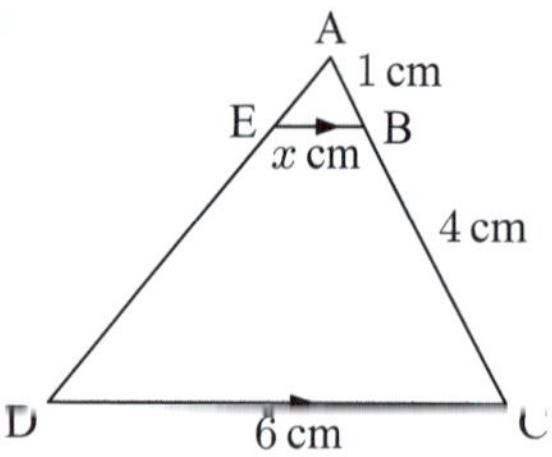

b

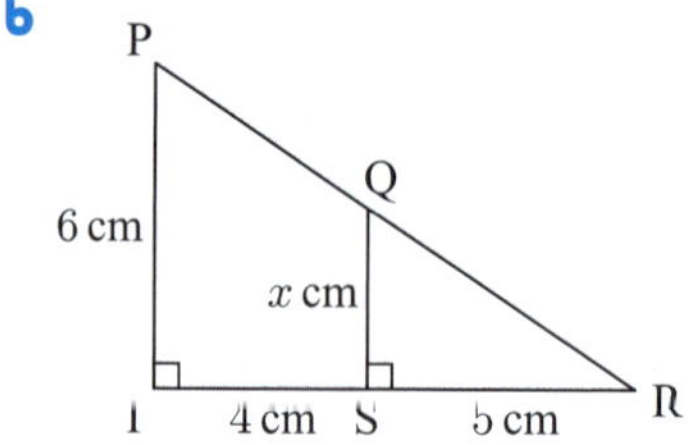

c

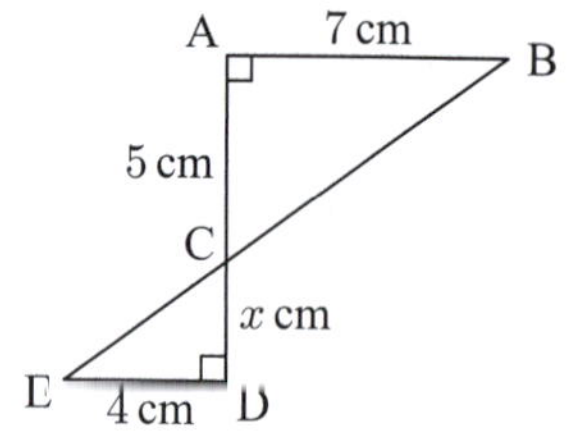

d

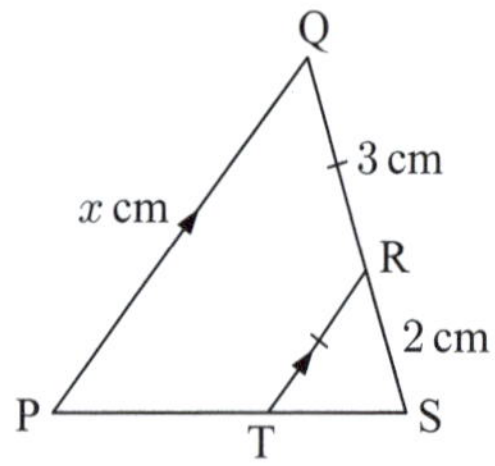

e

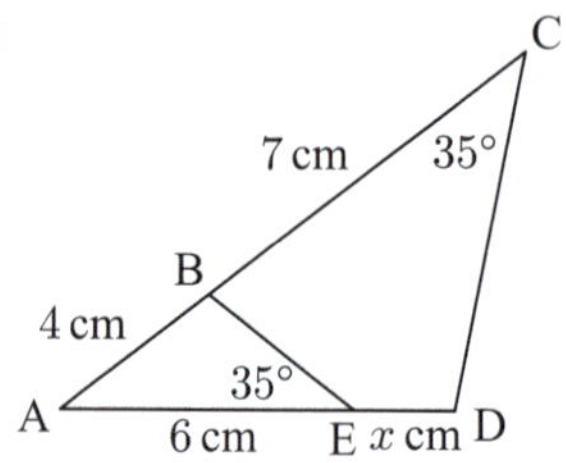

f

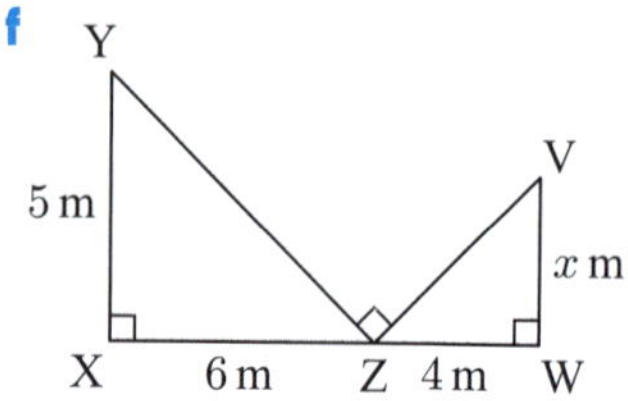

g

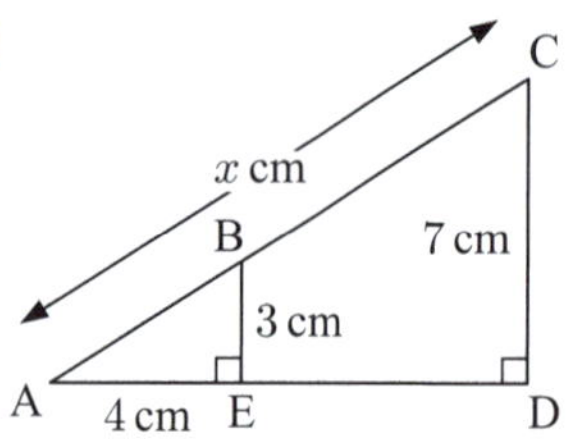

h

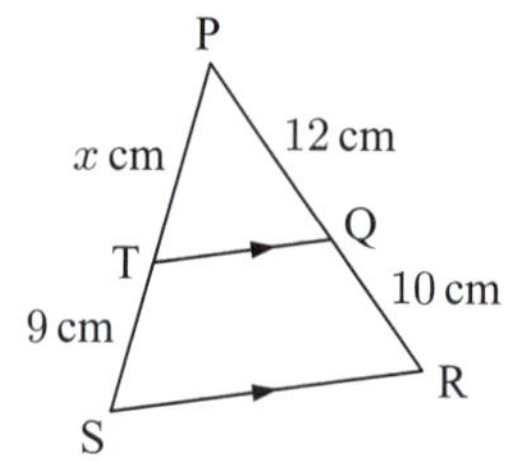

i

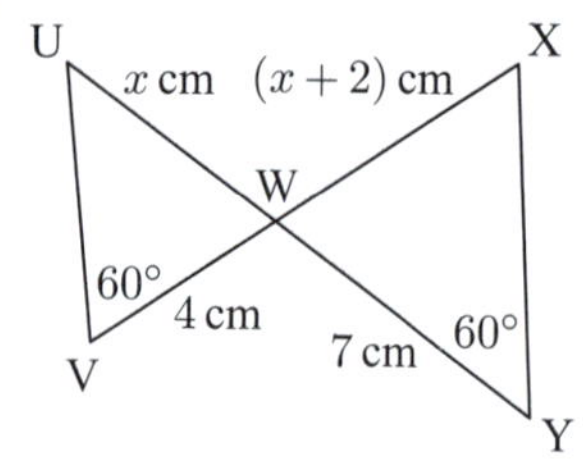

j

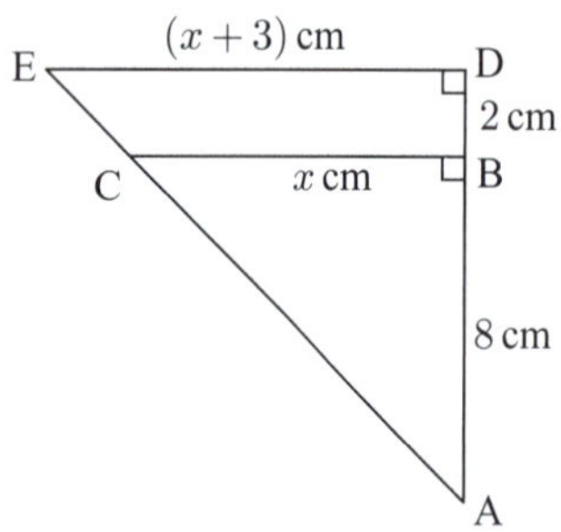

k

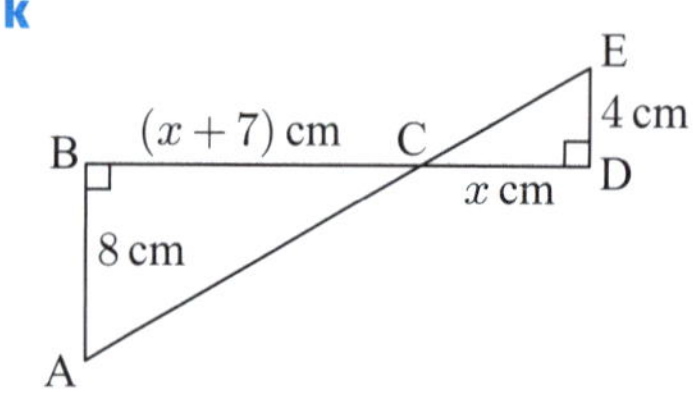

l

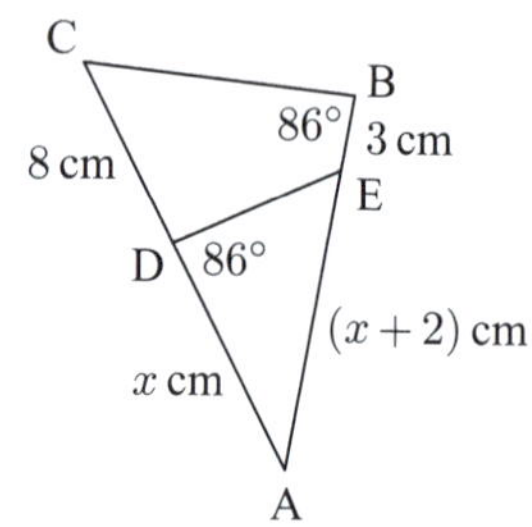

Example 5 **Self Tutor**

An electric light post E is directly opposite a mail box M on the other side of a straight road. Taj walks 30 metres along the road away from E to point T.

Kanvar is 4 metres away from M at point S, so that E, M, and S are collinear.

Kanvar walks 6 metres parallel to the road in the opposite direction to Taj, to point K. Now T, M, and K are collinear.

Find the width of the road.

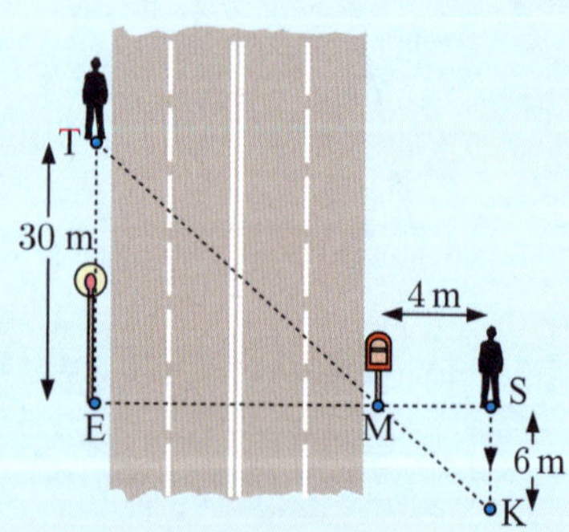

Let the width of the road be x m.

$\triangle$s TEM and KSM are equiangular as:

- $T\widehat{E}M = K\widehat{S}M = 90°$
- $E\widehat{M}T = S\widehat{M}K$ {vertically opposite angles}

$\therefore$ the triangles are similar.

$\therefore \dfrac{EM}{SM} = \dfrac{TE}{KS}$ {same ratio}

$\therefore \dfrac{x}{4} = \dfrac{30}{6}$

$\therefore x = 5 \times 4 = 20$

So, the road is 20 metres wide.

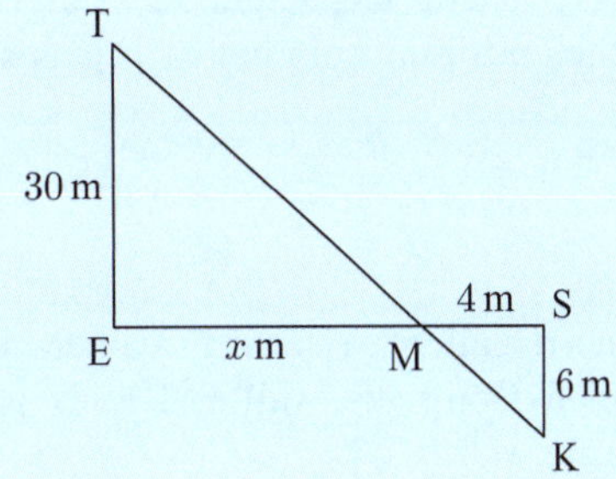

3 A boy 1.2 m tall stands 8.1 m from the base of an electric light pole. He casts a shadow 2.4 m long. How high above the ground is the light globe?

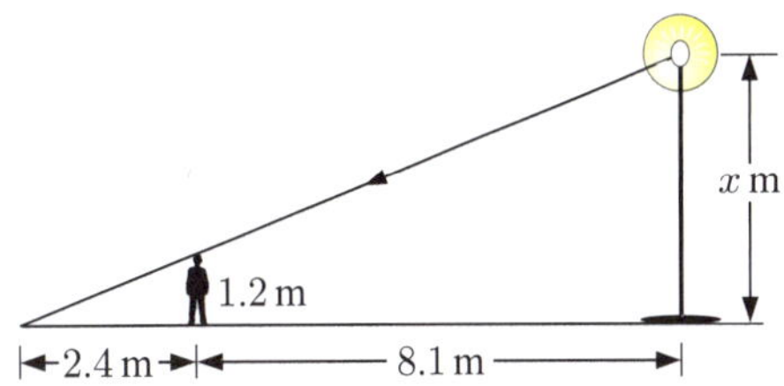

4 A 3.5 m ladder rests on both the top of a fence and the wall behind it as shown. The fence is 1.8 m tall, and the ladder reaches 2.9 m up the wall.
How far is the bottom of the ladder from the fence?

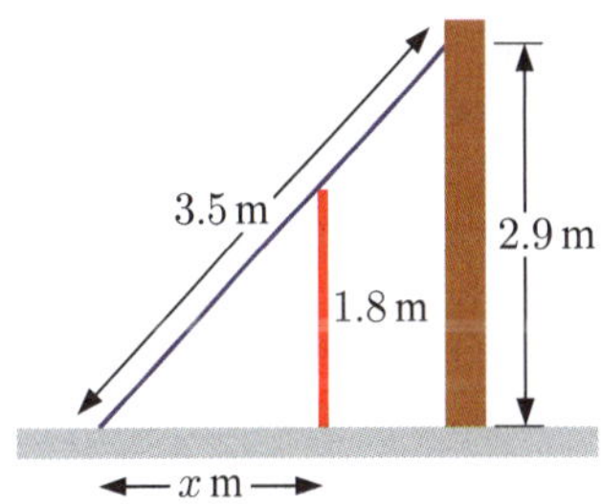

5 A hospital H receives a report about a serious road accident at C. An ambulance reaches the scene by travelling 5 km north, 15 km east, then 3 km north.
A helicopter travels directly from H to C. Their paths intersect at X.
When the helicopter reaches X, how far does it have left to fly?

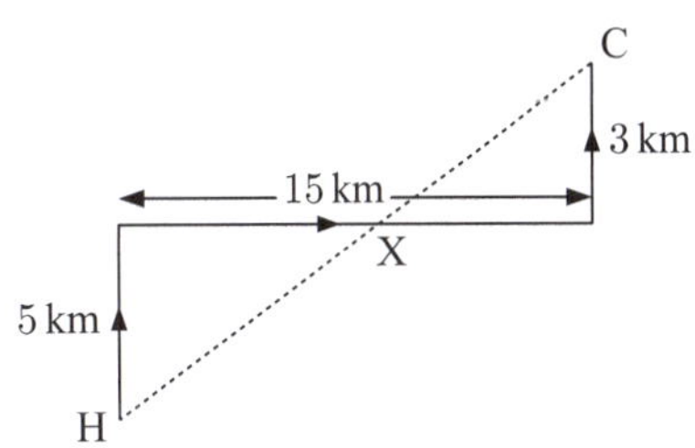

6 Two surveyors estimate the height of a hill. One stands 5 m away from the other on horizontal ground holding a measuring stick vertically. The "line of sight" to the top of the hill passes the vertical stick at a height of 2.4 m. Using laser equipment, they measure the distance from the stick to the top of the hill to be 1500 m.
Estimate the height of the hill, to the nearest metre.

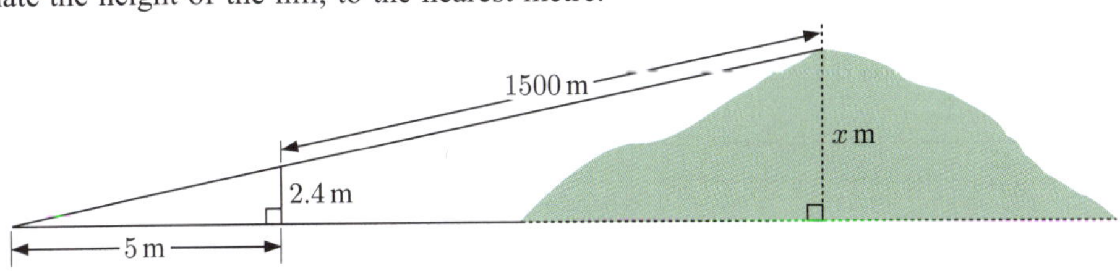

7 Mitchell pushes a coin with diameter 3 cm into a cone with diameter 9 cm and height 12 cm. How far into the cone can Mitchell push the coin before it gets stuck?

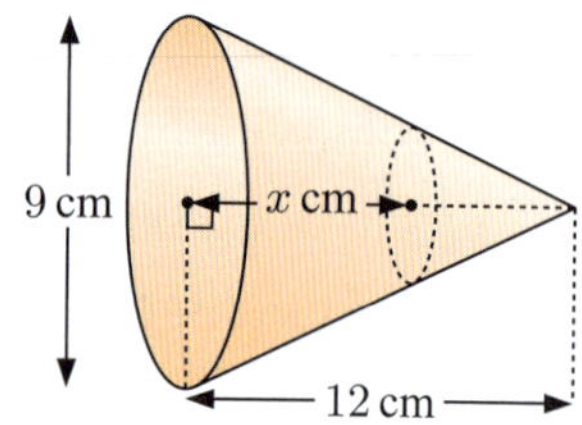

8 In an art gallery, a security camera is being installed along the wall [AB], to view the opposite wall [CD].

Peter is wondering which location for the camera along [AB] maximises the viewing region on the opposite wall.

"It doesn't matter," says Linda. "No matter where the camera is placed, the size of the viewing region will be the same."

Determine whether Linda is correct. Explain your answer.

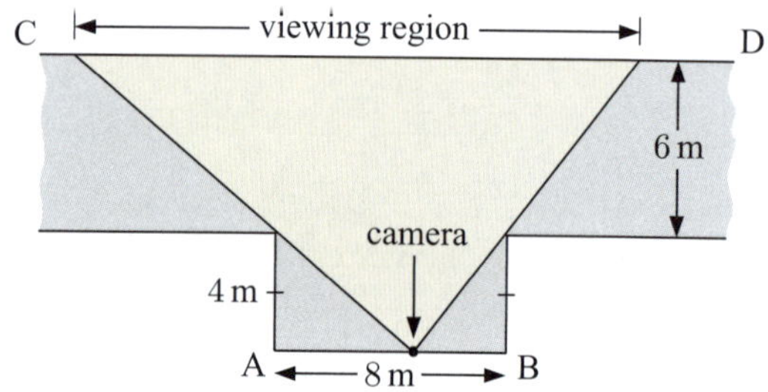

9

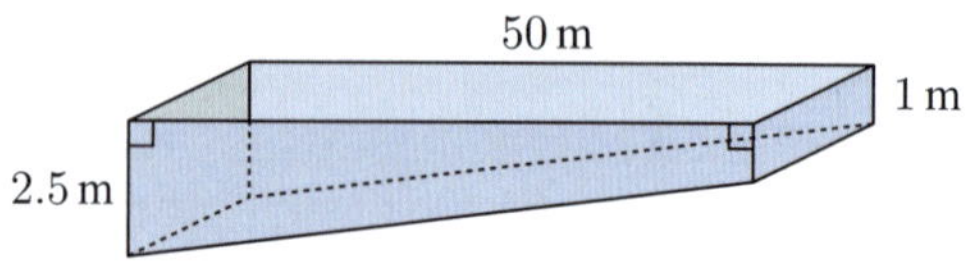

A swimming pool is 1 m deep at one end and 2.5 m deep at the other end. The pool is 50 m long. Bert jumps into the pool 20 m from the shallow end. How deep is the pool at this point?

10 It is safe to let go of the zip-line shown when you are 3 m above the ground.

How far must you travel along the zip-line before it is safe to let go?

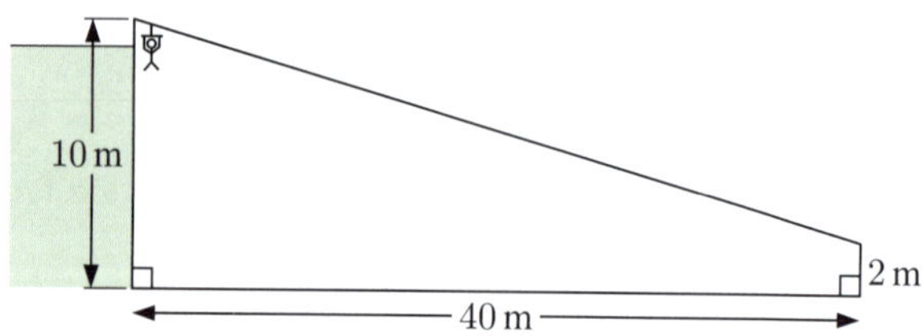

11

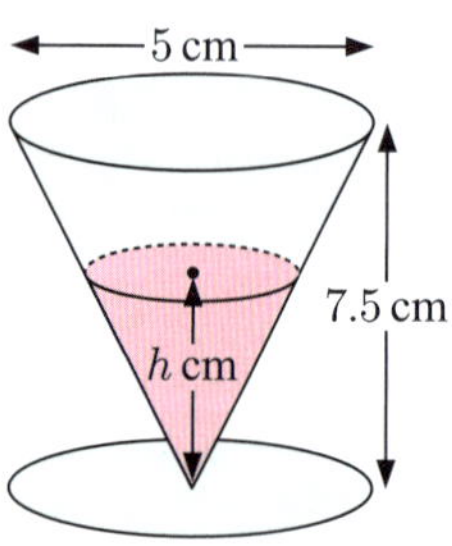

This medicine glass is filled with 20 mL of medicine.

To what height does the medicine level reach?

Example 6 — Self Tutor

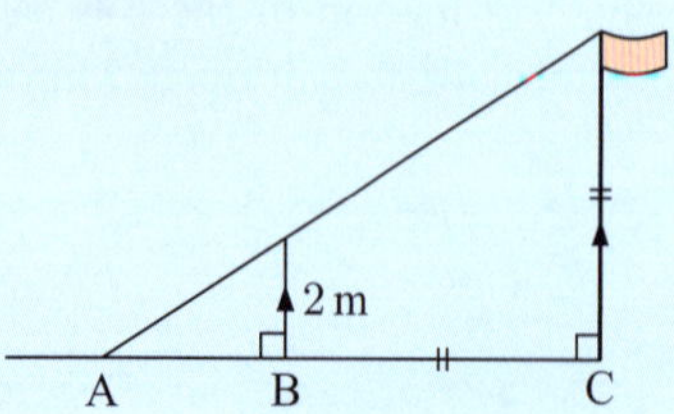

Given that [AB] is 3 m shorter than [BC], find the height of the flagpole.

Let the height of the flagpole be x m.

$\therefore$ BC $= x$ m and AB $= (x - 3)$ m

$\triangle$s ABE and ACD are equiangular as:

- $\widehat{ABE} = \widehat{ACD}$
- the angle at A is common

$\therefore$ the triangles are similar.

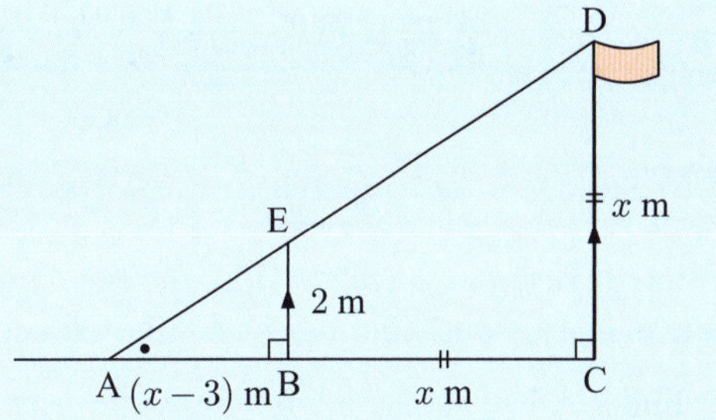

$$\therefore \frac{x}{2} = \frac{(x-3)+x}{x-3} \quad \{\text{same ratio}\}$$

$$\therefore x(x-3) = 2(2x-3)$$

$$\therefore x^2 - 3x = 4x - 6$$

$$\therefore x^2 - 7x + 6 = 0$$

$$\therefore (x-1)(x-6) = 0$$

$$\therefore x = 1 \text{ or } 6$$

But $x - 3 > 0$ as lengths must be positive, so x cannot be 1.

$\therefore x = 6$

$\therefore$ the flagpole is 6 m high.

12 Find the value of x:

a

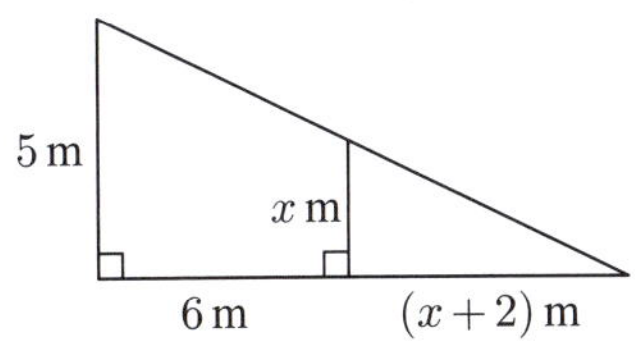

b

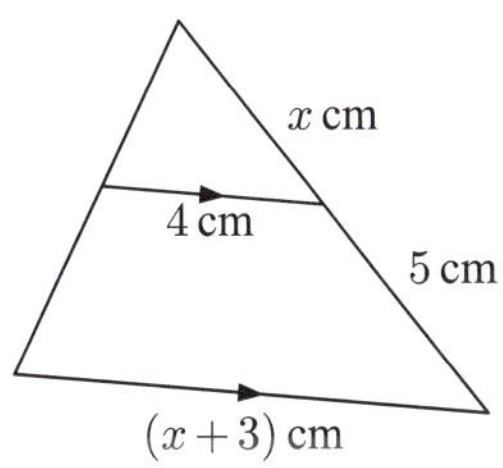

c

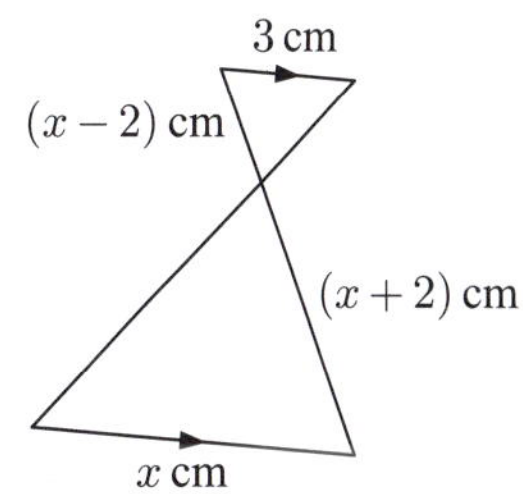

d

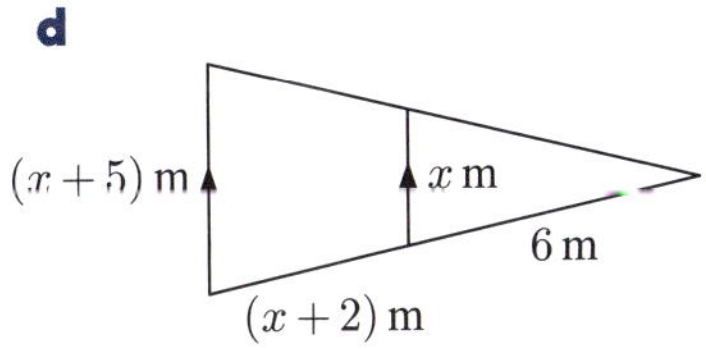

e

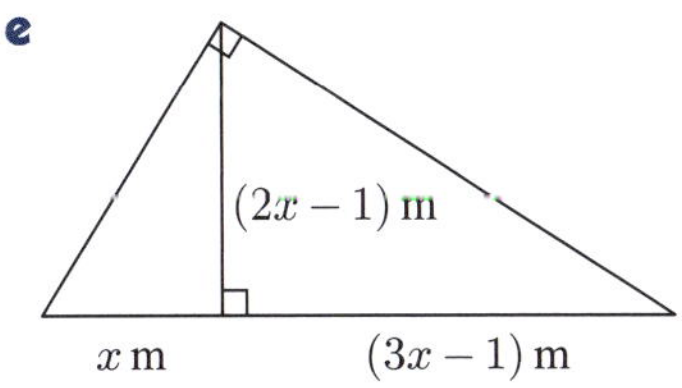

f

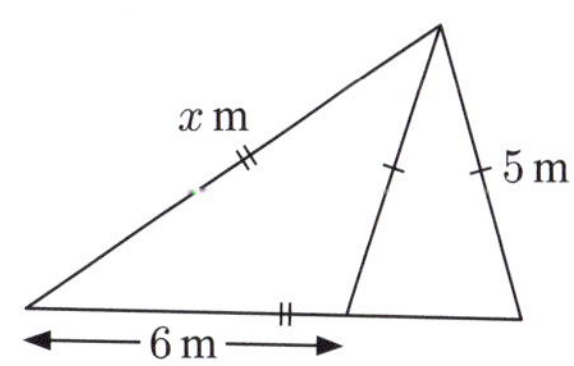

13

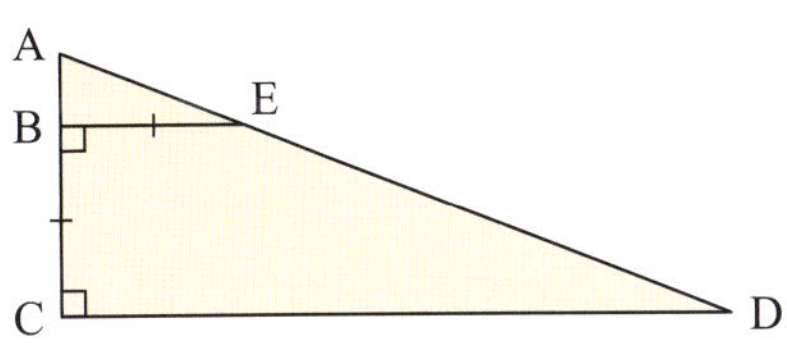

In the figure alongside, AC = 10 cm, BC = BE, and CD is 3 cm more than twice BE.

Find BE.

14 A, B, C, and D are posts on the banks of a 20 m wide canal. A and B are 2 m apart, and OA = CD.

Find the exact distance CD.

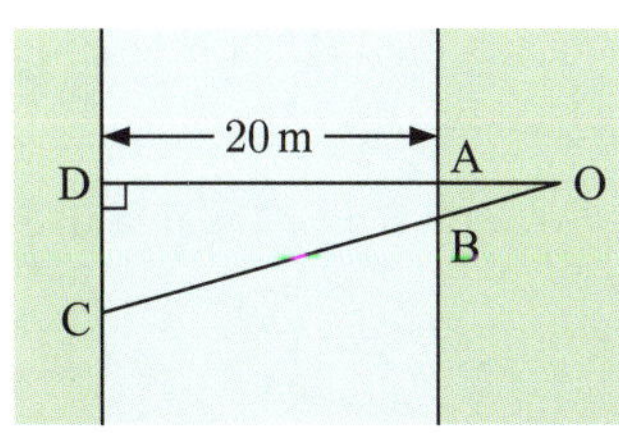

D AREAS AND VOLUMES OF SIMILAR OBJECTS

INVESTIGATION 2

In this Investigation we consider the relationships between areas and volumes of similar objects.

Cylinder **A** had radius r cm and height h cm.

Suppose it is enlarged with scale factor k to produce a *similar* cylinder **B**.

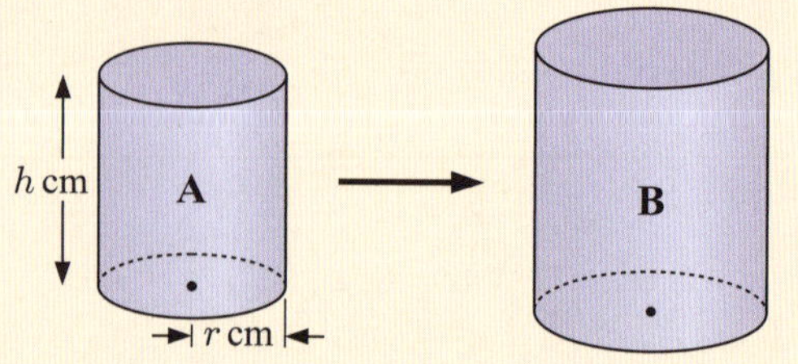

What to do:

1. Write an expression for the height and radius of cylinder **B**.
2. **a** Write an expression for the base area of cylinder: **i** A **ii** B

   b Hence find the ratio $\dfrac{\text{base area of cylinder } \mathbf{B}}{\text{base area of cylinder } \mathbf{A}}$.
3. **a** Write an expression for the volume of cylinder: **i** A **ii** B

   b Hence find the ratio $\dfrac{\text{volume of cylinder } \mathbf{B}}{\text{volume of cylinder } \mathbf{A}}$.

When an object is enlarged with scale factor k to produce a similar object,
- all lengths are multiplied by k
- all areas are multiplied by k^2
- all volumes are multiplied by k^3.

Example 7 **Self Tutor**

For each pair of similar figures, find the value of x:

a

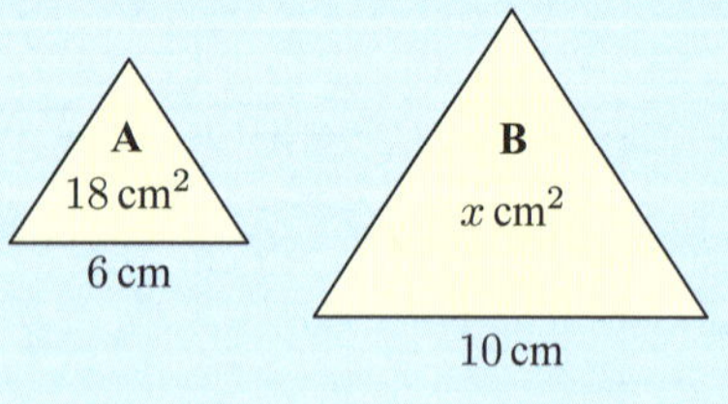

b

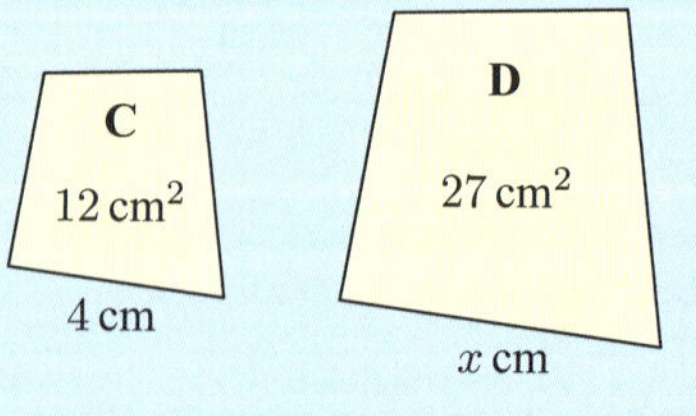

a Suppose figure **A** is enlarged with scale factor k to give figure **B**.

Using the known side lengths,

$k = \frac{10}{6} = \frac{5}{3}$

Area of **B** $= k^2 \times$ area of **A**

$\therefore\ x = \left(\frac{5}{3}\right)^2 \times 18$

$\therefore\ x = 50$

b Suppose figure **C** is enlarged with scale factor k to give figure **D**.

Area of **D** $= k^2 \times$ area of **C**

$\therefore\ 27 = k^2 \times 12$

$\therefore\ \frac{9}{4} = k^2$

$\therefore\ k = \frac{3}{2}$ {as $k > 0$}

Now $x = \frac{3}{2} \times 4$ {corresponding sides}

$\therefore\ x = 6$

EXERCISE 15D

1 For each pair of similar figures, find the value of x:

a

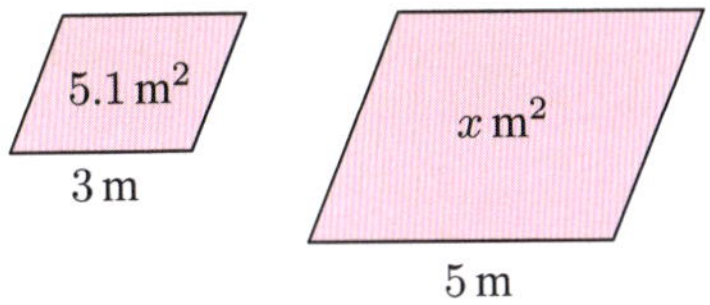

b

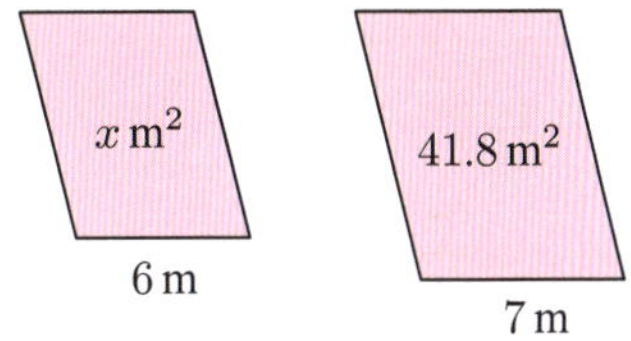

c

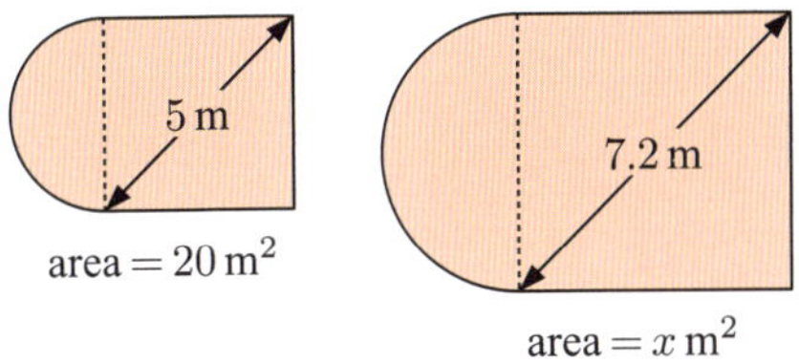

d

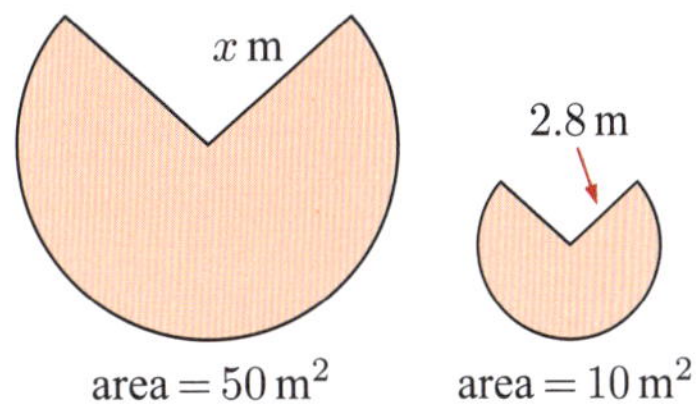

e

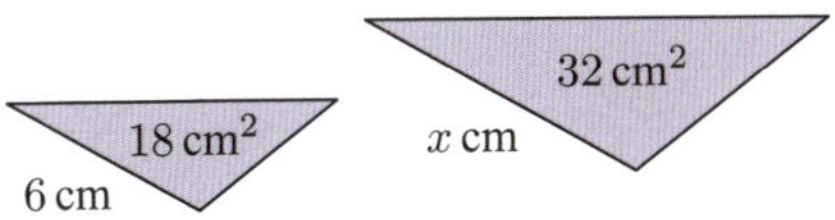

f

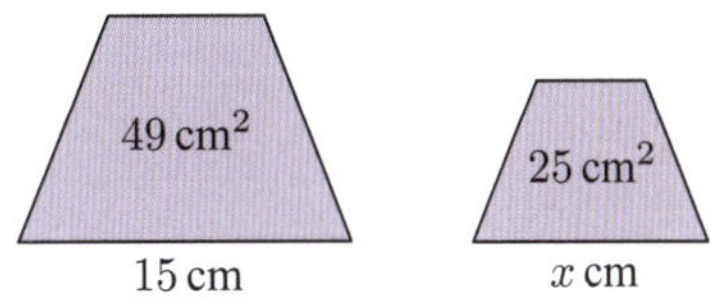

Example 8 — Self Tutor

For each pair of similar solids, find the value of x:

a

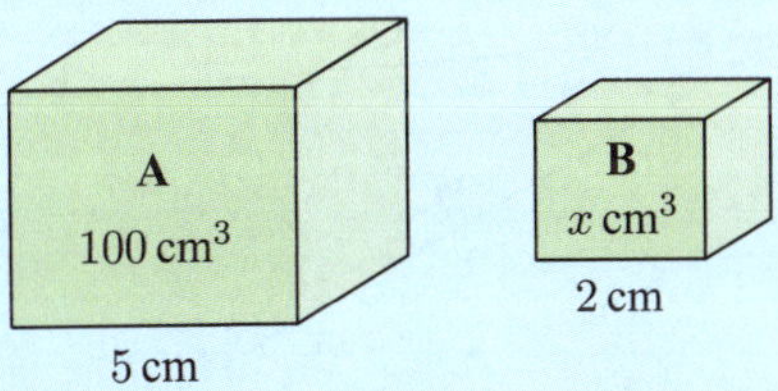

b

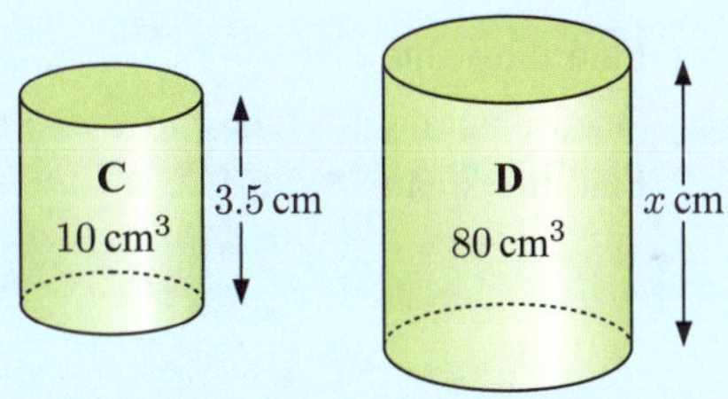

a Suppose box **A** is reduced with scale factor k to give box **B**.

Using the known side lengths, $k = \frac{2}{5}$

Volume of **B** $= k^3 \times$ volume of **A**

$\therefore\ x = (\frac{2}{5})^3 \times 100$

$\therefore\ x = 6.4$

b Suppose cylinder **C** is enlarged with scale factor k to give cylinder **D**.

Volume of **D** $= k^3 \times$ volume of **C**

$\therefore\ 80 = k^3 \times 10$

$\therefore\ 8 = k^3$

$\therefore\ k = 2$

Now $x = 2 \times 3.5$ {corresponding sides}

$\therefore\ x = 7$

2 For each pair of similar solids, find the value of x:

a

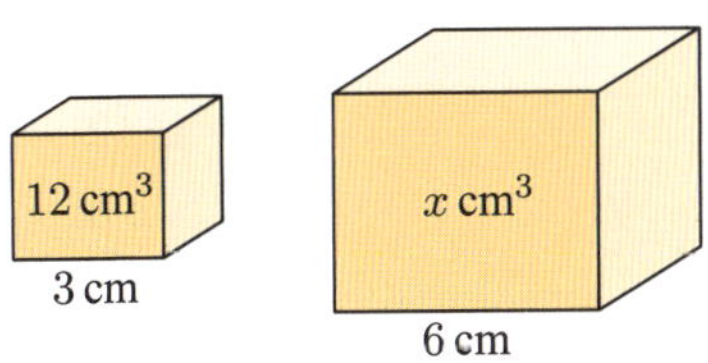

b

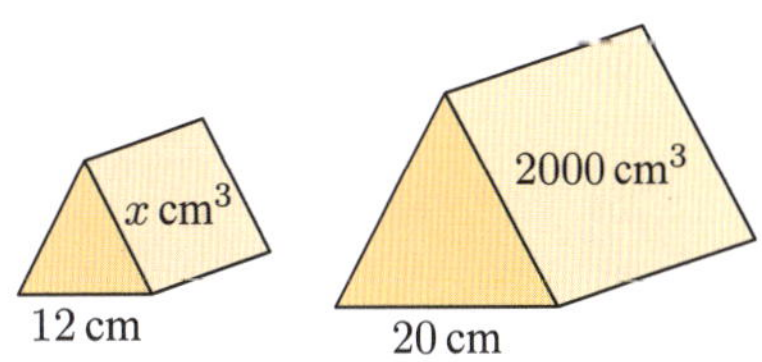

c

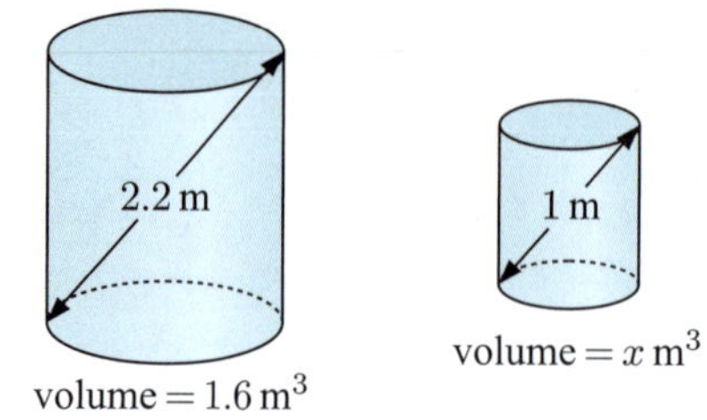

d

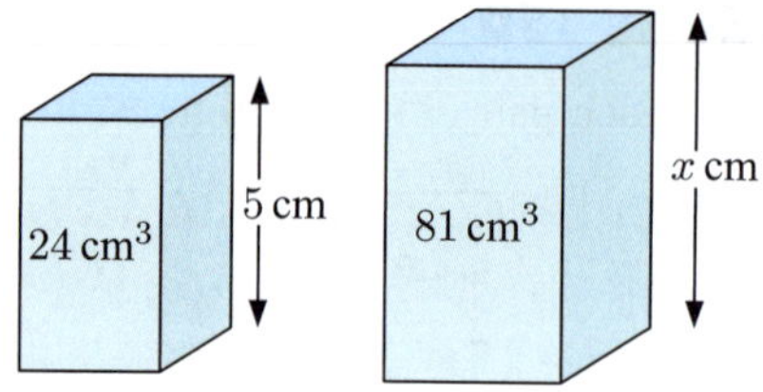

e

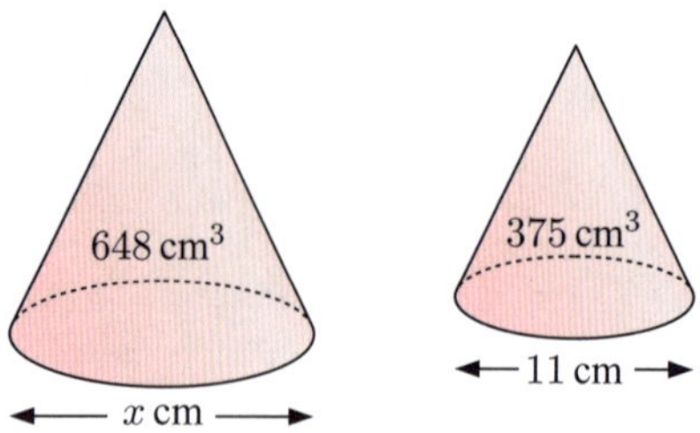

f

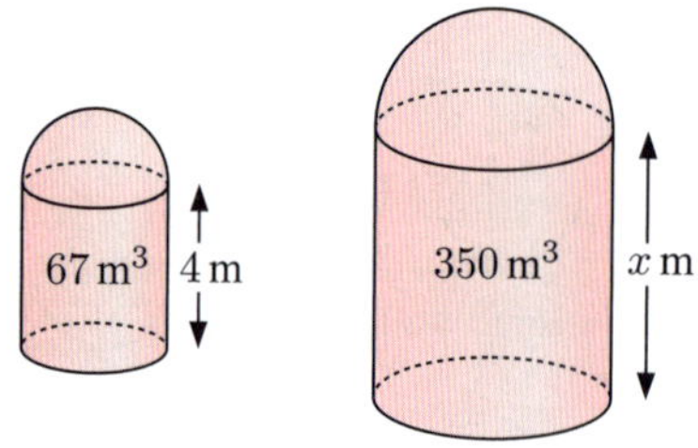

3 The area of $\triangle$BCD is 6.4 cm^2.
Find the area of:

a $\triangle$ACE **b** quadrilateral ABDE.

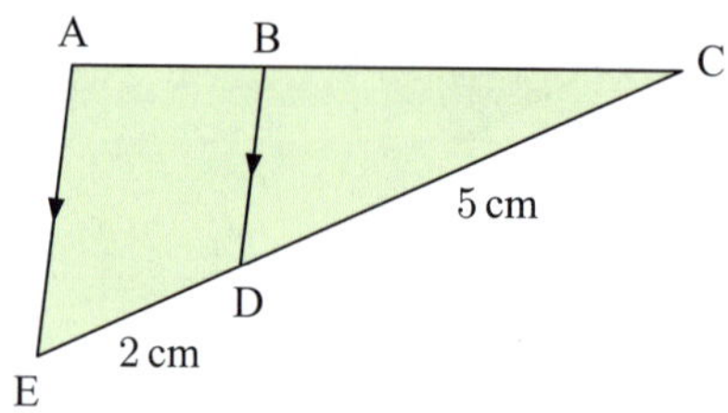

4 **a** Find the value of x.

b Quadrilateral QRST has area 22 m^2.
Find the area of $\triangle$PQT.

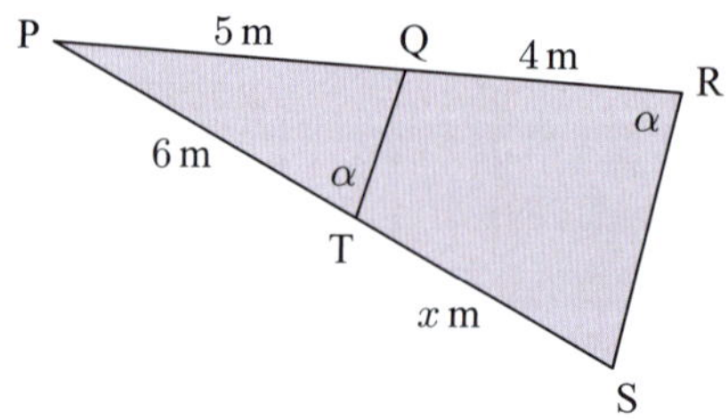

5 A rhombus has sides 13.5 cm long and area 162 cm^2.
A *similar* rhombus has area 50 cm^2. Find its perimeter.

6 The *density* of an object is its ratio of mass to volume. Objects made of the same material have the same density, so their mass is proportional to their volume.

a What happens to the mass of a sphere if its radius is:

i doubled **ii** increased by 20%?

b What happens to the mass of a cylinder if its radius and height are both:

i halved **ii** increased by 50%?

7

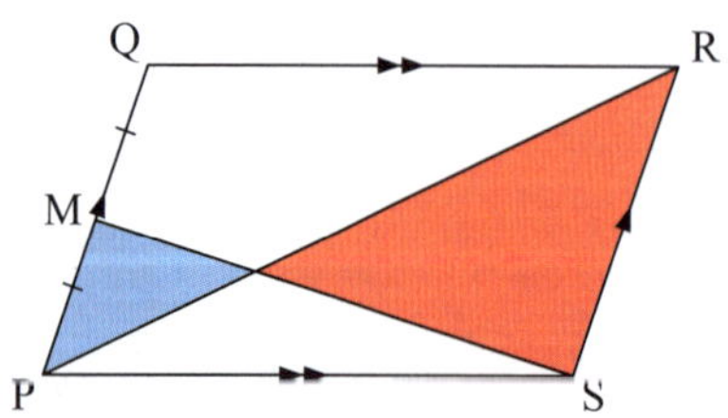

In parallelogram PQRS, M is the midpoint of [PQ].
Show that the red area is 4 times the blue area.

8 Determine whether each pair of objects is similar:

a

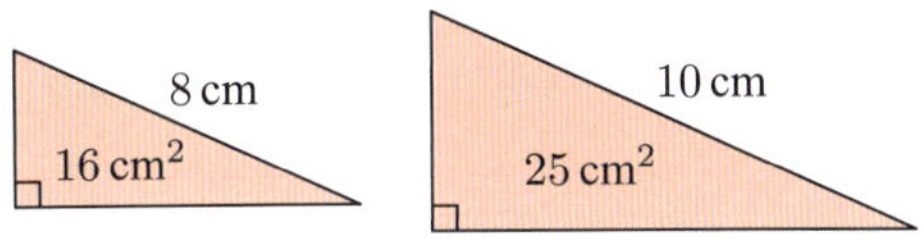

b

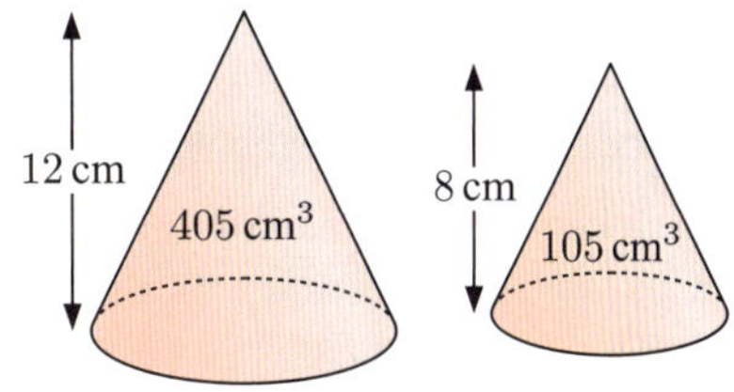

9 A scale model is made of a 300 year old sailing ship. The model is a $1:200$ reduction of the original. Find:

a the height of the mast in the model if the original mast was 20 m high

b the area of a sail in the model if the original sail was 120 m^2

c the height and radius of a keg in the model if the original was 1.2 m high and 0.9 m in diameter

d the capacity of the water tank in the model if the capacity of the original was 10 000 litres.

Example 9 **Self Tutor**

The volumes of two similar boxes are 48 cm^3 and 162 cm^3.

a Find the scale factor of the enlargement from the smaller box to the larger box.

b If the smaller box has length 6 cm, find the length of the larger box.

c If the larger box has surface area 198 cm^2, find the surface area of the smaller box.

a Suppose the smaller box is enlarged with scale factor k to give the larger box.

Volume of larger box $= k^3 \times$ volume of smaller box

$$\therefore\ 162 = k^3 \times 48$$

$$\therefore\ k^3 = \frac{162}{48} = \frac{27}{8}$$

$$\therefore\ k = \sqrt[3]{\frac{27}{8}} = \frac{3}{2}$$

b Length of larger box $= k \times$ length of smaller box

$$= \tfrac{3}{2} \times 6$$

$$= 9 \text{ cm}$$

c Surface area of larger box $= k^2 \times$ surface area of smaller box

$$\therefore\ 198 = \left(\tfrac{3}{2}\right)^2 \times \text{surface area of smaller box}$$

$\therefore$ surface area of smaller box $= 198 \times \frac{4}{9} = 88$ cm^2

10 The surface areas of two similar cylinders are 6 cm^2 and 54 cm^2.

a Find the scale factor of the enlargement from the smaller cylinder to the larger cylinder.

b If the larger cylinder has height 12 cm, find the height of the smaller cylinder.

c If the volume of the smaller cylinder is 24 cm^3, find the volume of the larger cylinder.

11 Two similar cones made from the same material have surface areas 192 cm^2 and 75 cm^2. The volume of the larger cone is 200 cm^3. The mass of the smaller cone is 320 g.

a Find the scale factor of the enlargement from the smaller cone to the larger cone.

b Find the volume of the smaller cone.

c Find the mass of the larger cone.

12 Two buckets are similar in shape. The smaller one is 30 cm tall and the larger one is 45 cm tall. Both buckets contain water to a depth equal to half of their heights. The volume of water in the small bucket is 4400 cm^3.

a What is the scale factor in comparing the smaller bucket to the larger one?

b Find the volume of water in the larger bucket.

c The surface area of the water in the larger bucket is 630 cm^2. Find the surface area of the water in the smaller bucket.

13 Answer the **Opening Problem** on page **236**.

PUZZLE

A glassware company manufactures cylindrical drinking glasses in six different sizes. Their heights and capacities are given below. Which two of the glasses are similar?

A

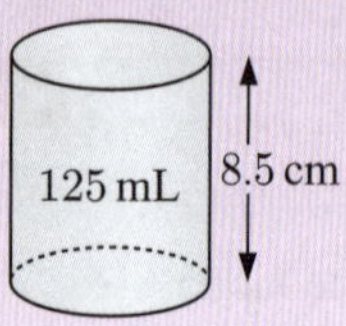

B

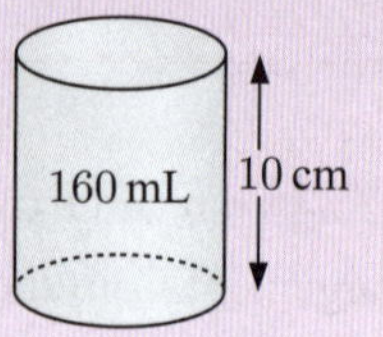

C

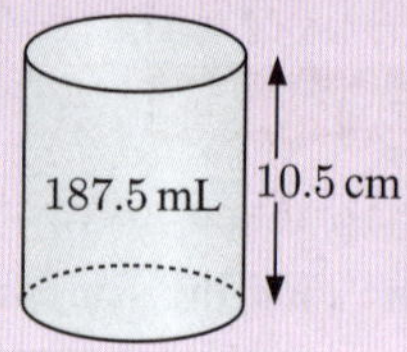

D

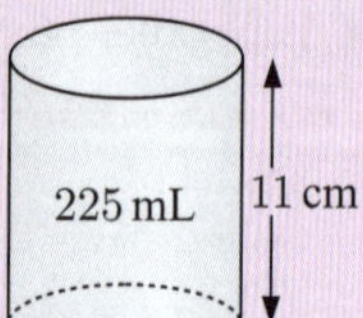

E

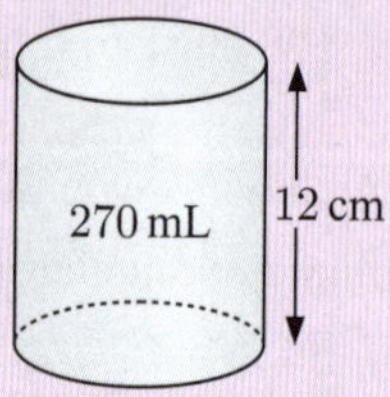

F

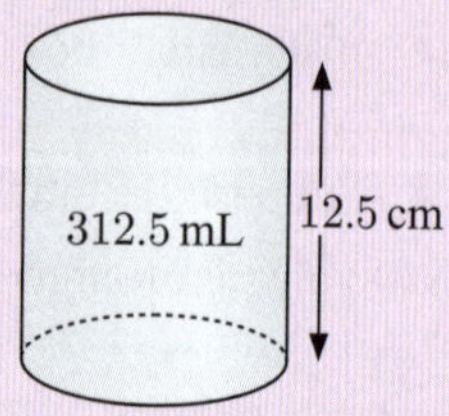

QUICK QUIZ

MULTIPLE CHOICE QUIZ

REVIEW SET 15A

1 Determine whether each pair of triangles is congruent, giving reasons for your answers.

a

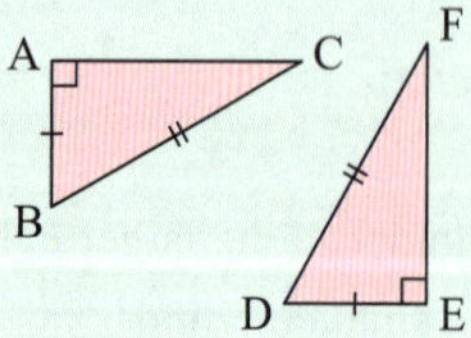

b

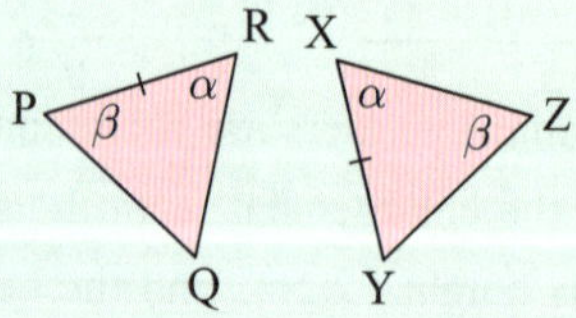

c

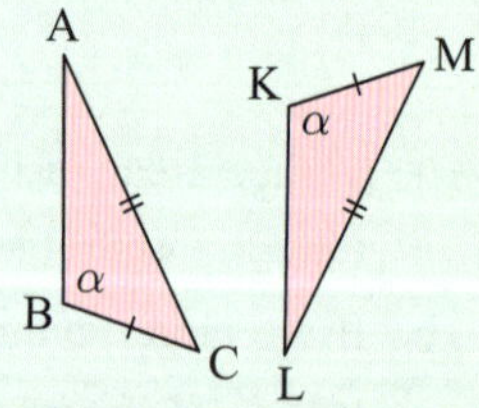

2 Consider the quadrilateral ABCD alongside.

- **a** Show that triangles ABC and CDA are congruent.
- **b** Hence deduce that ABCD is a parallelogram.

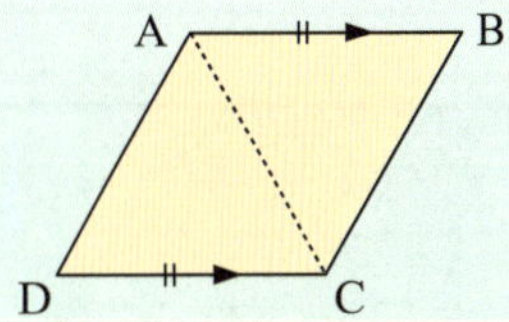

3 The three angle bisectors of a triangle meet at a point called the **incentre** of the triangle.
Prove that the incentre is equidistant from each edge of the triangle.
Hint: Draw a perpendicular line from each edge to the incentre.

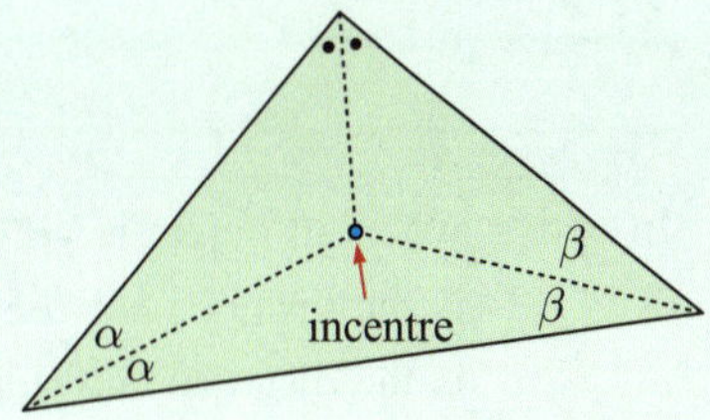

4 Show that these figures possess similar triangles:

a

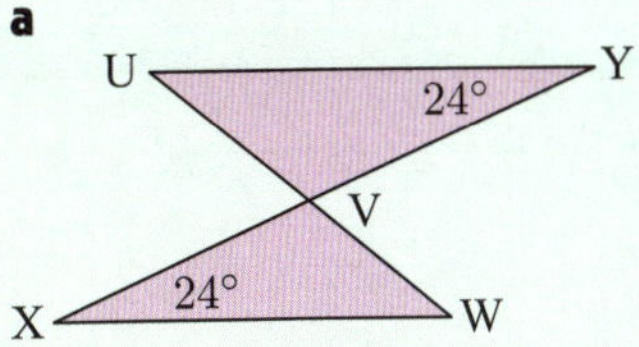

b

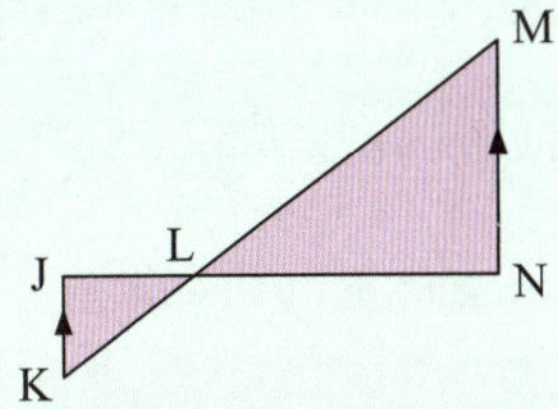

c

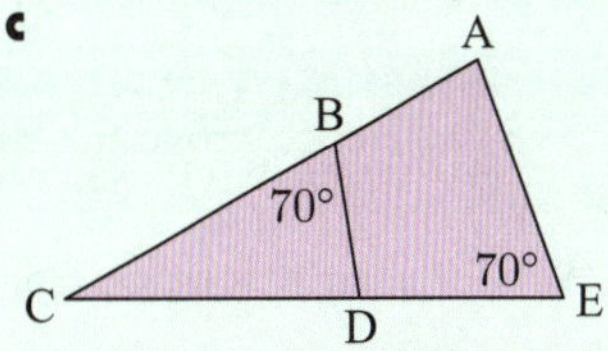

5 In each figure, show that a pair of triangles is similar. Hence find the value of x:

a

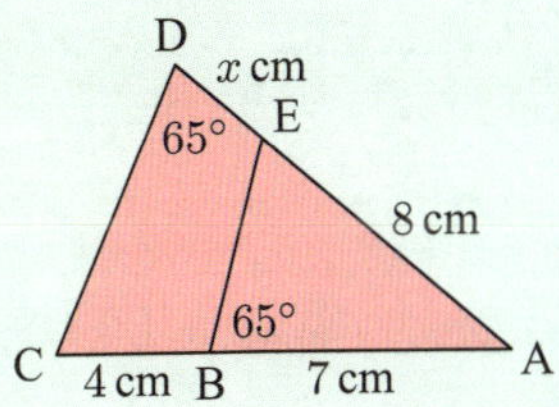

b

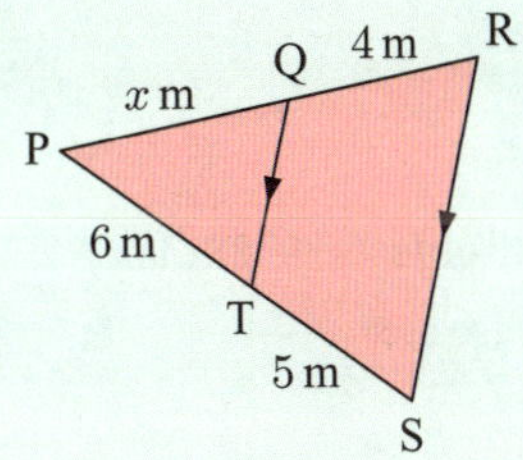

c

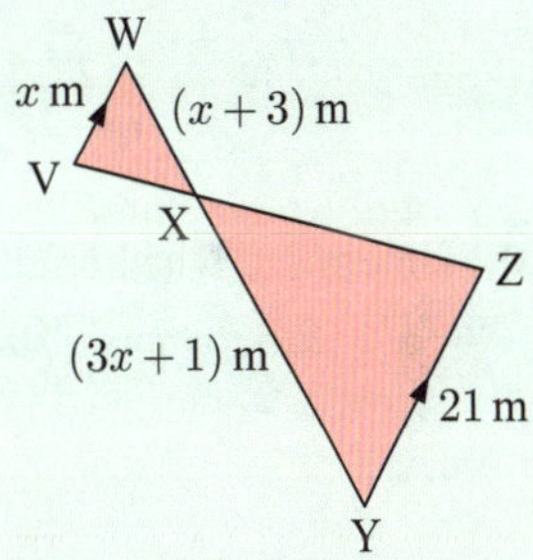

6

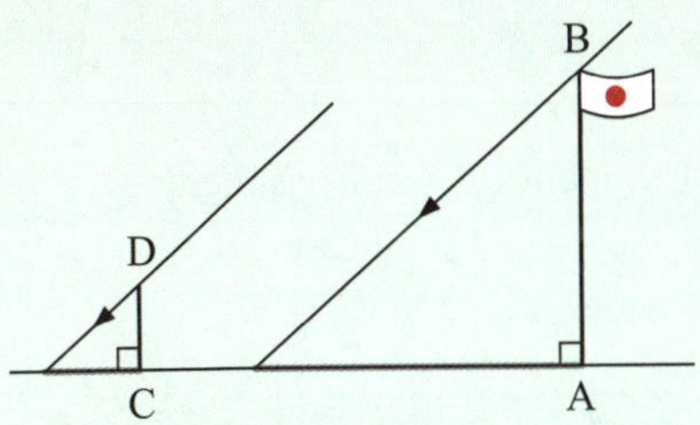

[AB] is a vertical flagpole.
[CD] is a vertical stick 1.4 m long.
When the shadow of the flagpole is 12.3 m long, the shadow of the stick is 1.65 m long.
Find the height of the flagpole.

7
- **a** Show that there are three similar triangles in this figure.
- **b** Explain why $BD^2 = 2\,AB$.
- **c** Use Pythagoras to show that $AB = (1 + \sqrt{5})$ cm.

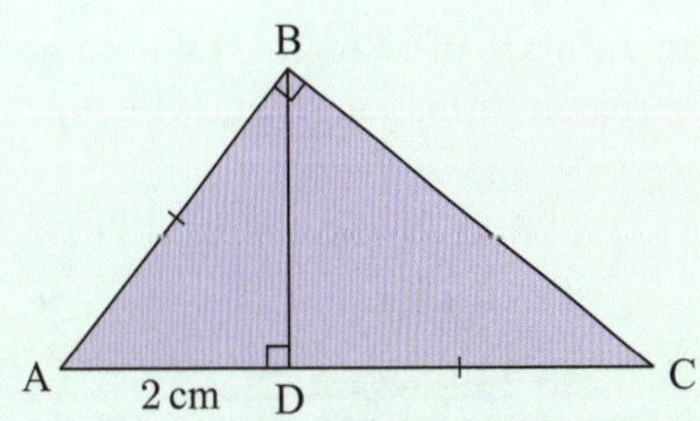

8

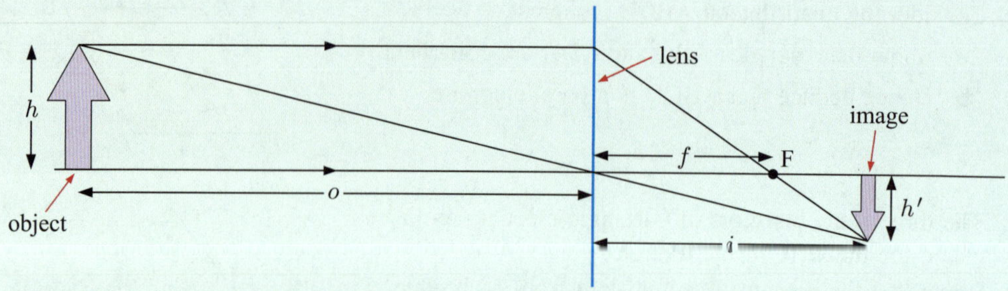

In optics, when an object is viewed through a thin lens:

f is the distance between the lens and the focal point F

o is the distance between the object and the lens

i is the distance between the image and the lens.

a Use similar triangles to prove that: **i** $\dfrac{h'}{h} = \dfrac{i}{o}$ **ii** $\dfrac{h'}{h} = \dfrac{i-f}{f}$

b Hence prove the **thin lens equation** $\dfrac{1}{f} = \dfrac{1}{o} + \dfrac{1}{i}$.

9 For each pair of similar figures, find the value of x:

a

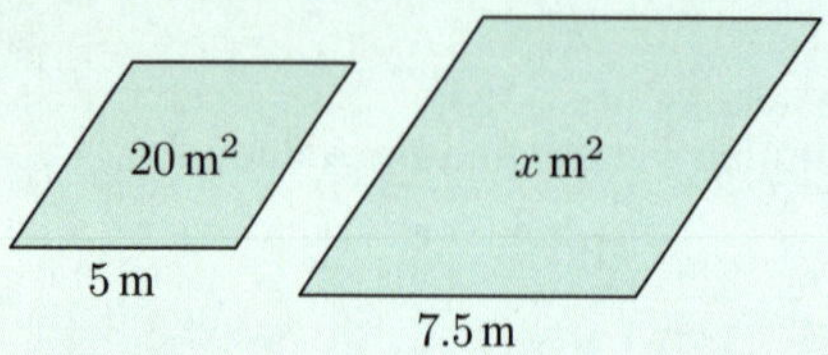

b

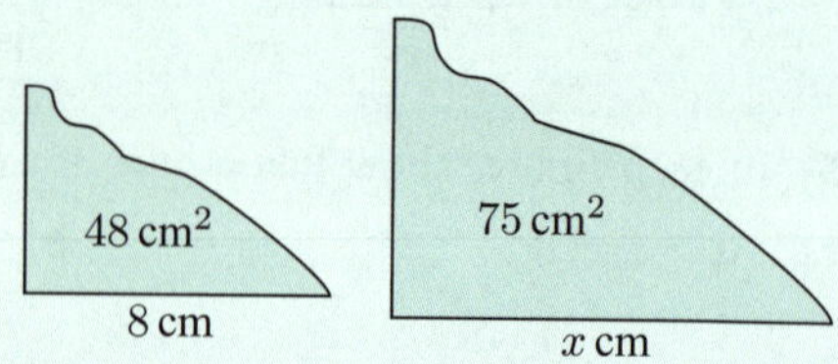

10 Determine whether each pair of objects is similar:

a

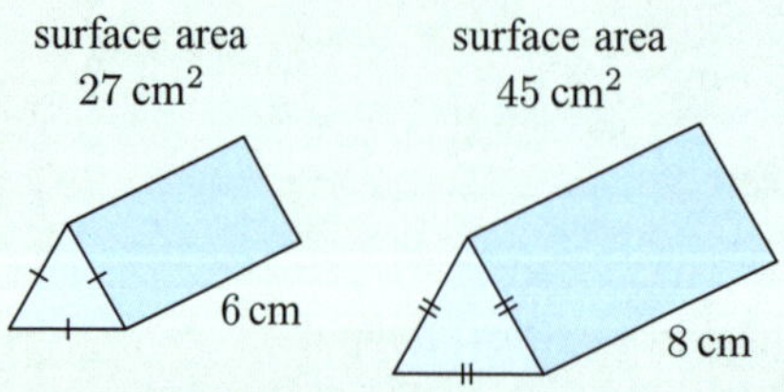

b

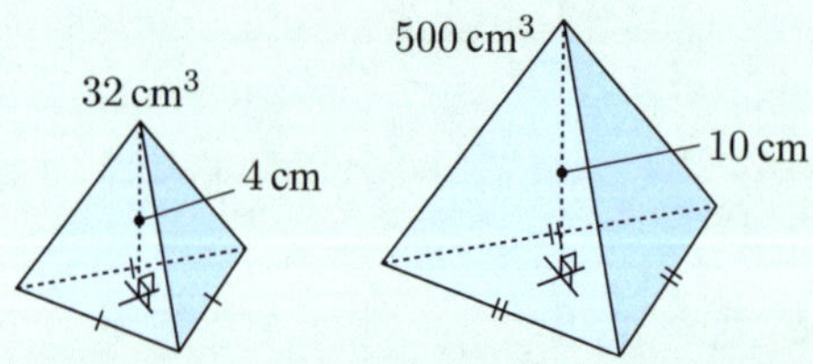

11 A sphere of lead with radius 10 cm is melted into 125 identical smaller spheres. Find the radius of each new sphere.

REVIEW SET 15B

1 In each set of triangles, two are congruent. The triangles are *not* drawn to scale, but contain correct information. State which pair of triangles is congruent, giving a reason for your answer.

a

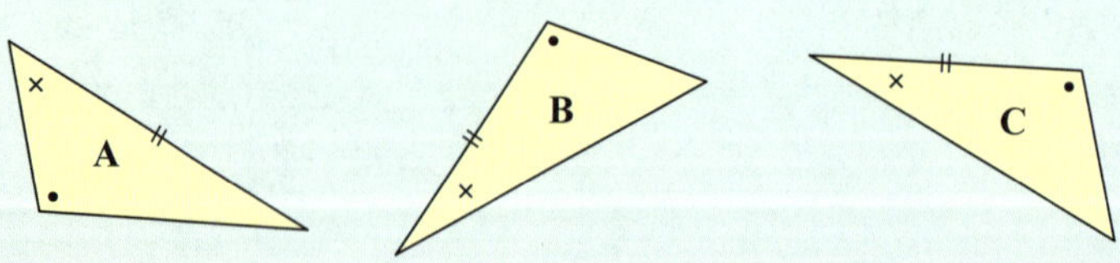

b

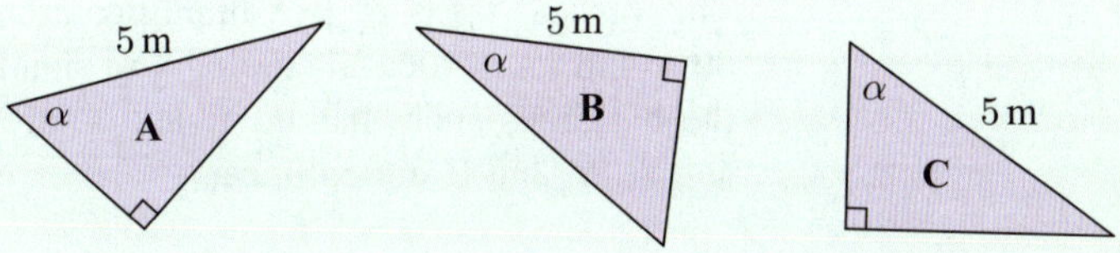

2 PQRS is a kite, with $PQ = PS$ and $QR = SR$.
M and N are the midpoints of [PQ] and [PS] respectively.
Prove that triangle MNR is isosceles.

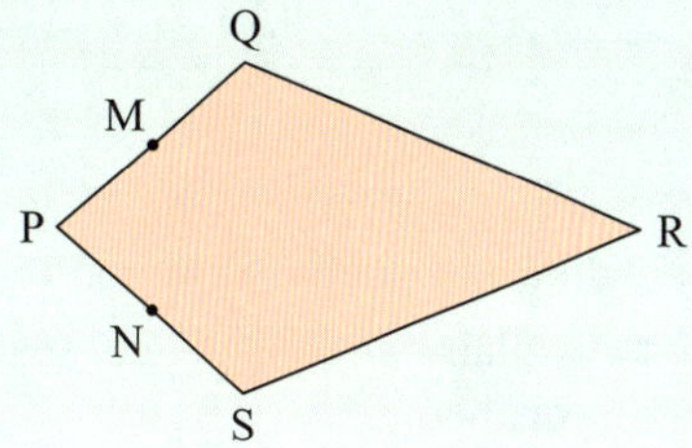

3

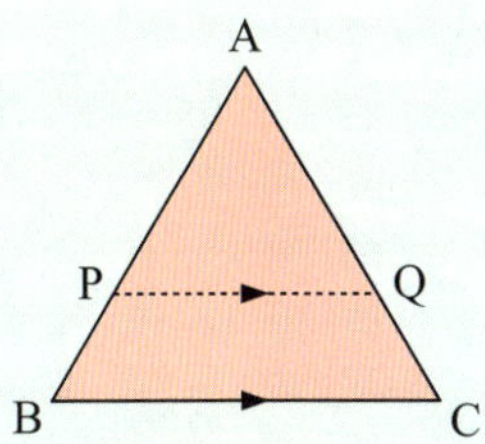

Triangle ABC is isosceles with $AB = AC$.
[PQ] is parallel to [BC].
Prove that $CP = BQ$.

4 In triangle PQR, M is the midpoint of [QR]. [MX] is drawn perpendicular to [PQ], and [MY] is drawn perpendicular to [PR].
Suppose these perpendiculars are equal in length.

a Prove that triangle MQX is congruent to triangle MRY.

b Hence prove that triangle PQR is isosceles.

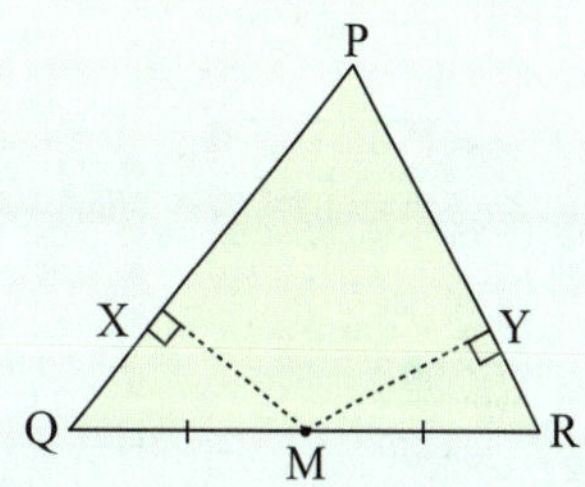

5 In each figure, show that a pair of triangles is similar. Hence find the value of x:

a

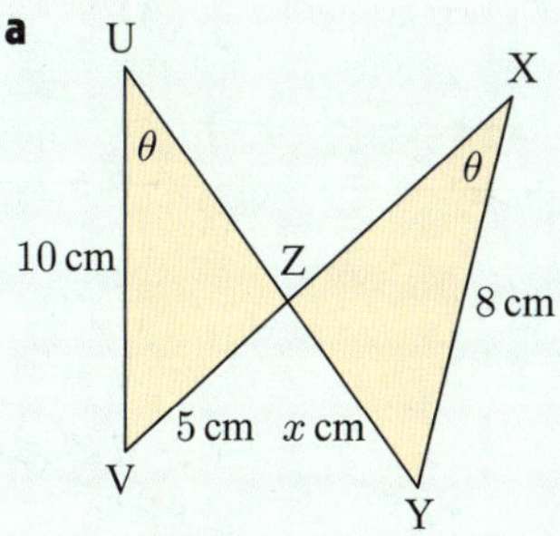

b

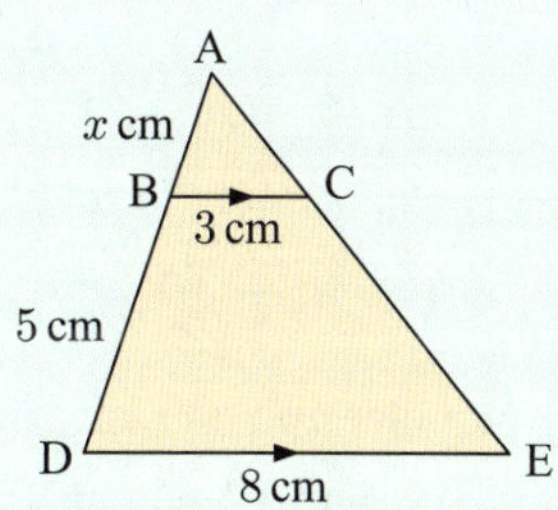

c

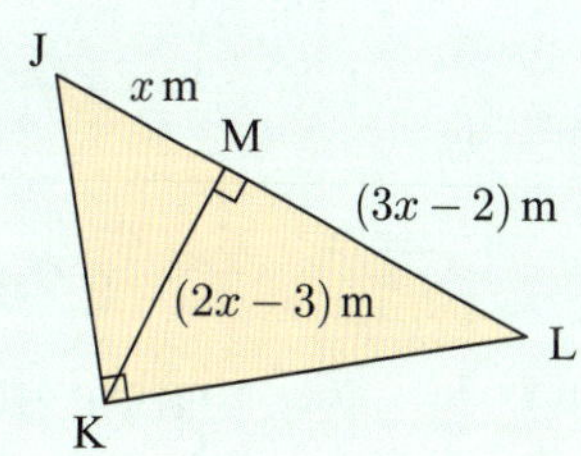

6 A, B, and C are pegs on the bank of a canal which has parallel straight sides. Peg D is directly opposite peg C.
$AB = 30$ m and $BC = 140$ m.
When I walk from A directly away from the bank, I reach a point E, 25 m from A, such that E, B, and D are collinear.
How wide is the canal?

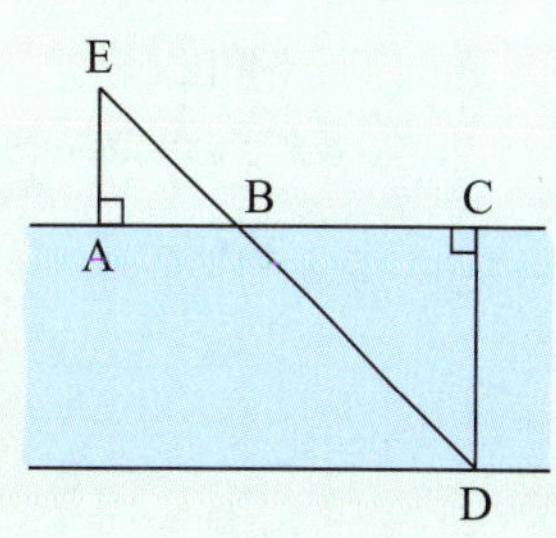

7 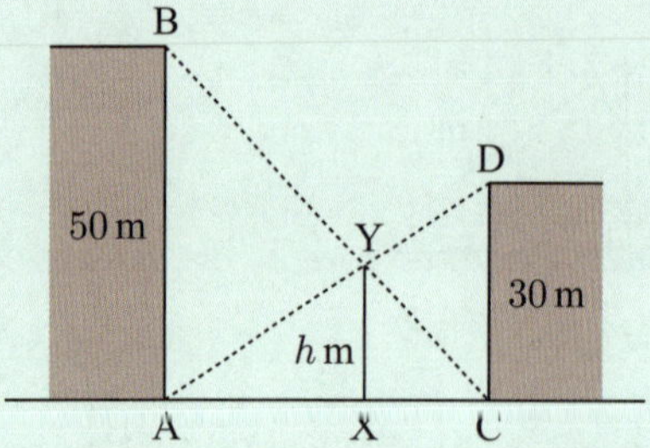

The vertical walls of two buildings are 50 m and 30 m tall. A vertical flagpole [XY] stands between the buildings such that B, Y, and C are collinear, and A, Y, and D are collinear.

a Show that $\frac{h}{50} + \frac{h}{30} = 1$.

b Hence find the height of the flagpole.

8 When the diagonals of the regular pentagon ABCDE are drawn, a smaller pentagon PQRST is formed.

a Explain why all of the angles marked • are of equal size.

b Hence show that the interior angles of PQRST are of equal size.

c Hence show that PQRST is a regular pentagon.

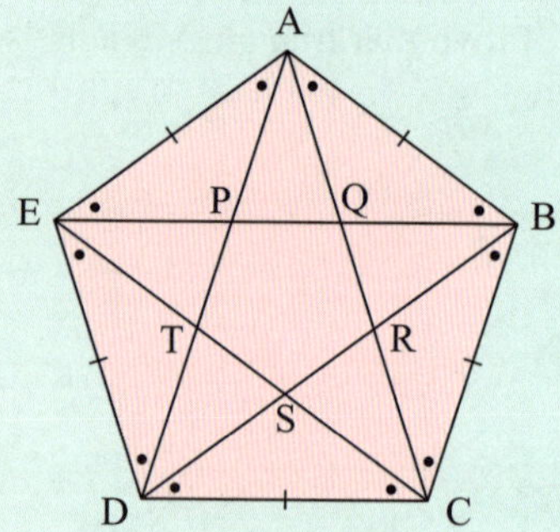

9 For each pair of similar solids, find the value of x:

a

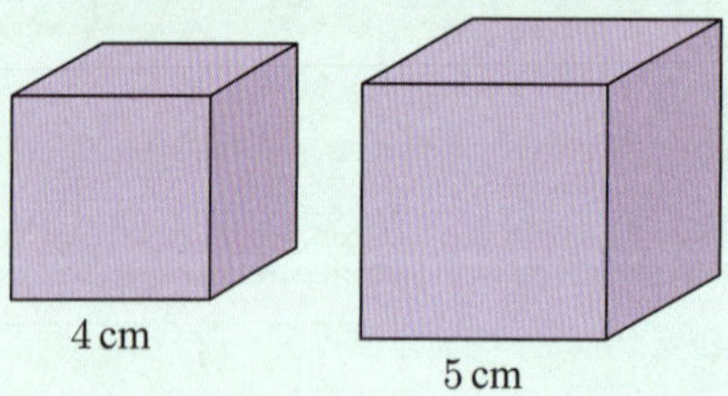

b

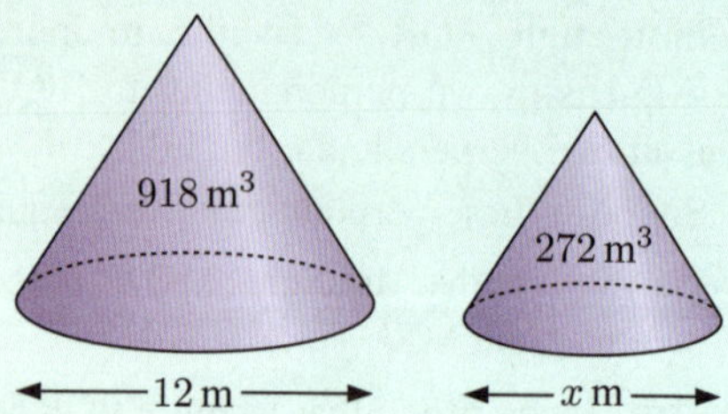

10 Triangle ABC has area 15 cm^2. Find the area of:

a triangle CDE

b quadrilateral PQED.

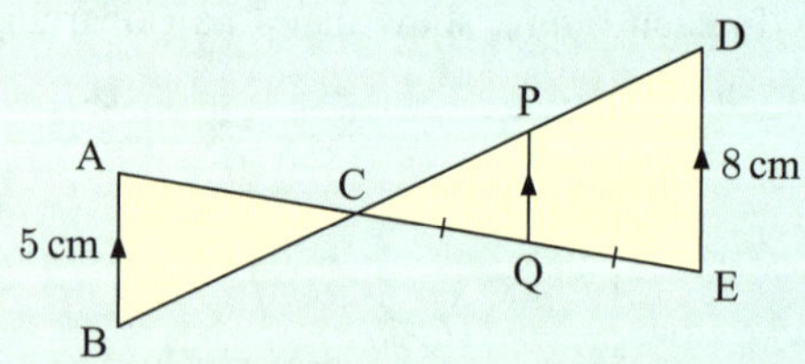

11 In a measuring cup set, the $\frac{1}{2}$ cup measure is 6 cm wide. The set also contains a 1 cup measure and a $\frac{1}{3}$ cup measure, both of which are similar in shape to the $\frac{1}{2}$ cup measure.

Find the width of:

a the 1 cup measure

b the $\frac{1}{3}$ cup measure.

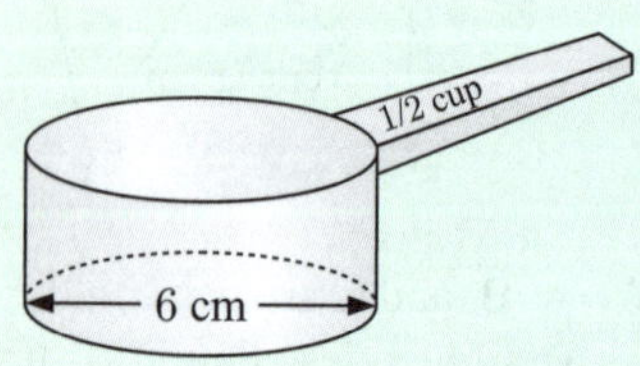

Chapter 16

Circle geometry

Contents:

OPENING PROBLEM

Market gardener Joe has four long straight pipes of different lengths. He places the pipes on the ground, and joins them with rubber hose to form a garden bed in the shape of a quadrilateral. A sprinkler which casts water in a semi-circle with diameter equal to the length of a pipe is placed at the midpoint of each pipe.

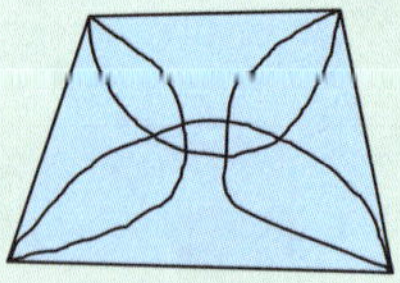

Joe's sketch

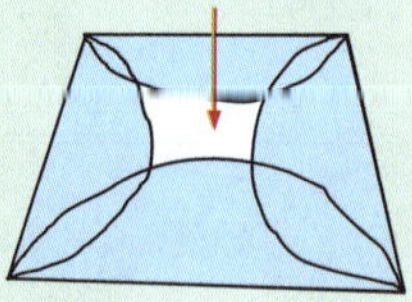

Clinton's sketch

Joe draws a rough sketch of the watering system, and decides that his sprinklers will water the whole of the garden. His son Clinton is sceptical of his father's idea, and draws his own sketch which suggests that there will be an unwatered patch in the centre.

Things to think about:

- **a** By drawing an accurate diagram of this situation, can you determine whether Joe or Clinton is correct?
- **b** Can you *prove* your answer using geometric theorems?

The geometry of triangles, quadrilaterals, and circles has been used for at least 3000 years in art, design, and architecture. Many amazing discoveries have been made by mathematicians and non-mathematicians who were simply drawing figures with rulers and compasses.

A ANGLE IN A SEMI-CIRCLE THEOREM

The **angle in a semi-circle theorem** states that:

The angle in a semi-circle is a right angle.

$$A\widehat{B}C = 90°$$

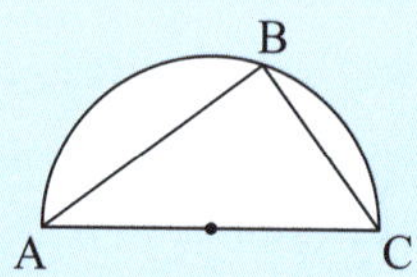

The **converse of the angle in a semi-circle theorem** states that:

If line segment [AB] subtends a right angle at C, then the circle through A, B, and C has diameter [AB].

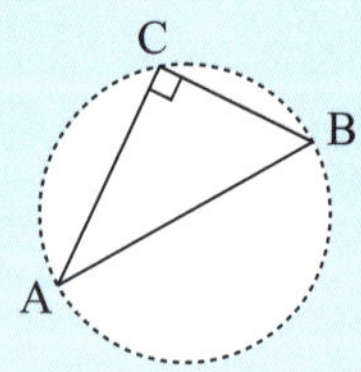

EXERCISE 16A

1 In this question we prove the *angle in a semi-circle* theorem. Suppose [AC] is the diameter of a circle with centre O, and B is any point on the semi-circle.

Let $O\widehat{A}B = x°$ and $O\widehat{C}B = y°$.

- **a** Find $O\widehat{B}A$ and $O\widehat{B}C$ in terms of x and y.
- **b** Hence prove that $A\widehat{B}C = 90°$.

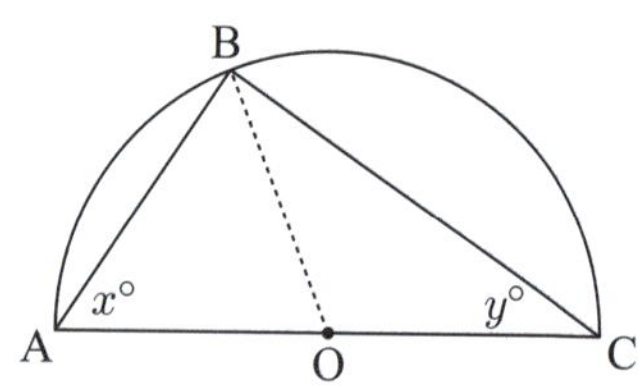

Example 1 — Self Tutor

Find the value of x, giving brief reasons for your answer.

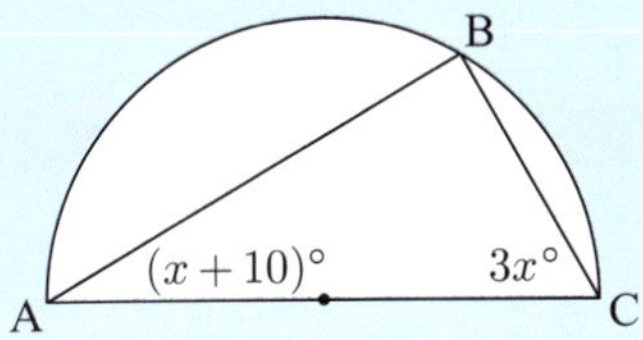

$A\widehat{B}C$ measures $90°$ {angle in a semi-circle}

$\therefore\ (x+10)+3x+90=180$ {angles in a triangle}

$\therefore\ 4x+100=180$

$\therefore\ 4x=80$

$\therefore\ x=20$

2 Find the value of x, giving brief reasons for your answer:

a
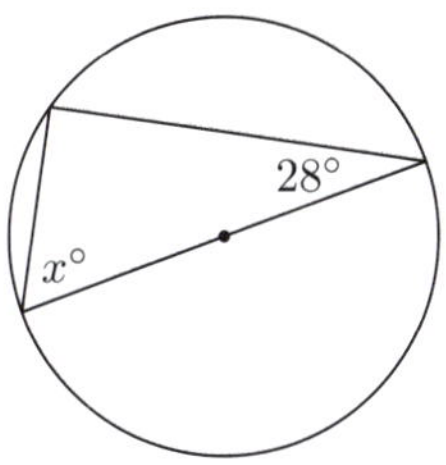

b
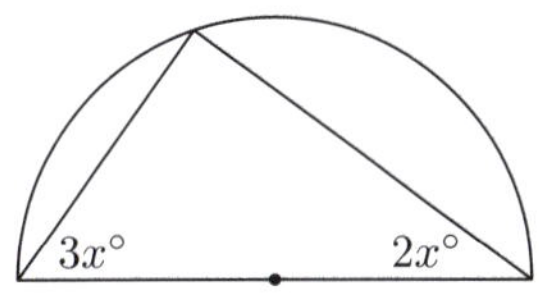

c
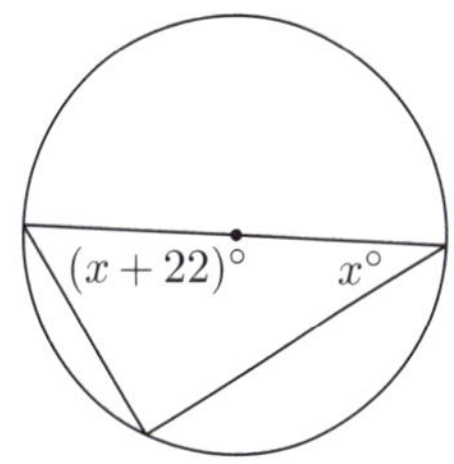

d
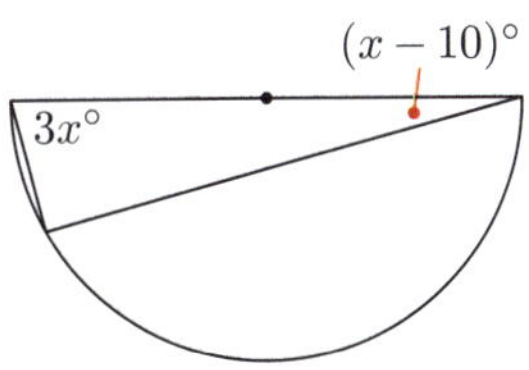

e
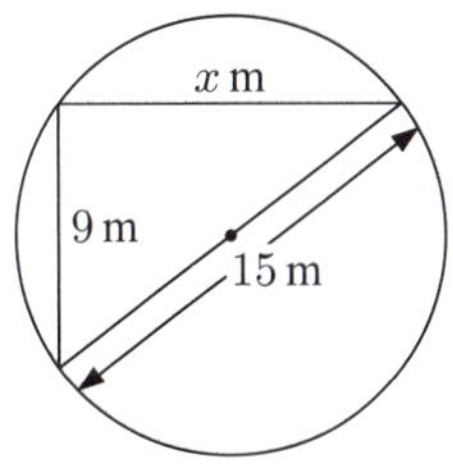

f
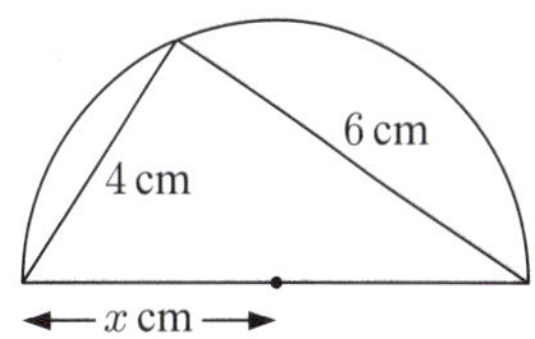

3 A circle has diameter [AB] of length 11 cm. The chord [AC] has length 7 cm. Find the length of the chord [BC].

4 A circle has diameter [AB] of length 20 cm. The point C is placed on the circle such that [AC] is twice the length of [BC]. Find the length of [BC].

5 **a** Find the value of x.

b Hence find the diameter of the circle.

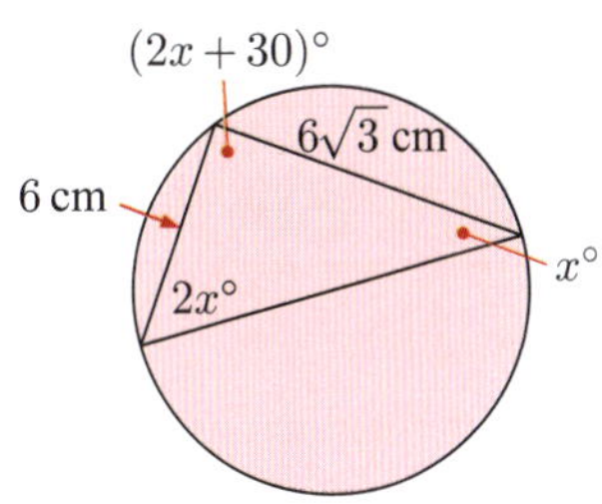

6 Consider the **Opening Problem** on page **258**. The two semi-circles in the figure alongside have diameters [AB] and [BC].

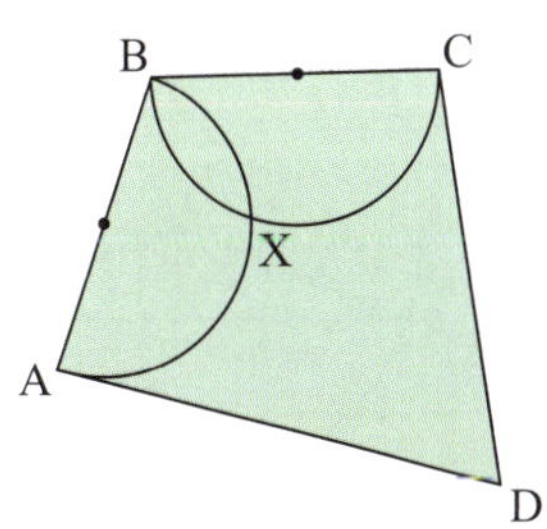

a Determine the measure of $B\widehat{X}A$ and $B\widehat{X}C$.

b What can you conclude about the points A, X, and C?

c Will Joe's four sprinklers water the whole garden? Explain your answer.

7 Show that Area A + Area B = Area C.

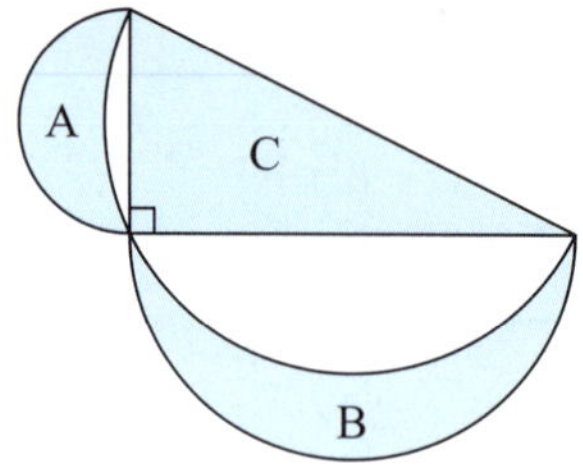

8 The shaded area in this figure is called an *arbelos*, which is a Greek word meaning "shoemaker's knife". It is formed by drawing two smaller semi-circles inside a large semi-circle.

Suppose P is a point on the diameter [AB].

Archimedes showed that if a line is drawn from P perpendicular to [AB], meeting the arbelos again at Q, then the area of the arbelos is equal to the area of the circle with diameter [PQ].

Prove this result.

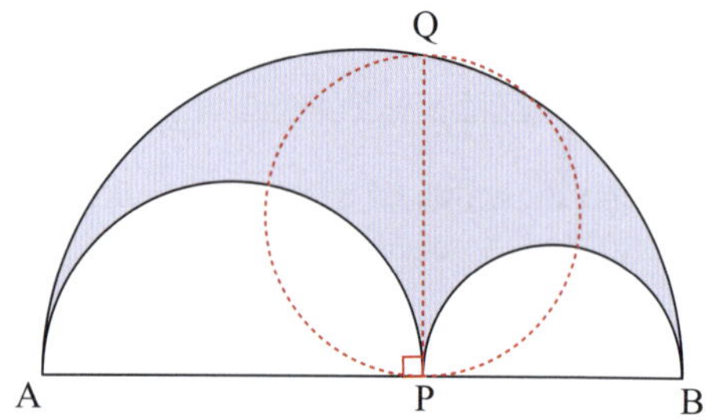

B CHORDS OF A CIRCLE THEOREM

The **chords of a circle theorem** states that:

The perpendicular from the centre of a circle to a chord bisects the chord.

AM = BM

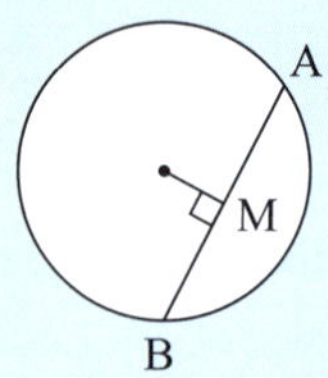

Example 2 Self Tutor

A circle has radius 5 cm. A chord drawn across the circle passes 1.2 cm from its centre. Find, to the nearest mm, the length of the chord.

Suppose the chord has length $2x$ cm.

Let M be the foot of the perpendicular from the centre O to the chord [AB].

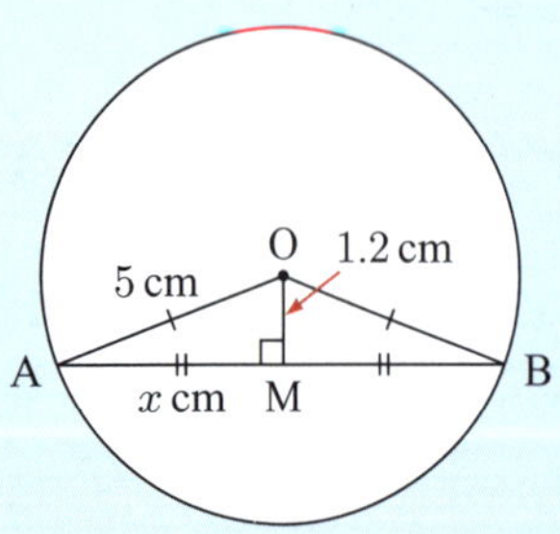

$\therefore$ M is the midpoint of [AB] {chords of a circle}

$\therefore$ AM = x cm

In triangle OAM, $x^2 + 1.2^2 = 5^2$ {Pythagoras}

$\therefore x^2 = 23.56$

$\therefore x \approx 4.854$ {as $x > 0$}

$\therefore 2x \approx 9.71$

The chord is about 9.7 cm long.

The **converse of the chords of a circle theorem** states that:

The perpendicular bisector of a chord of a circle passes through its centre.

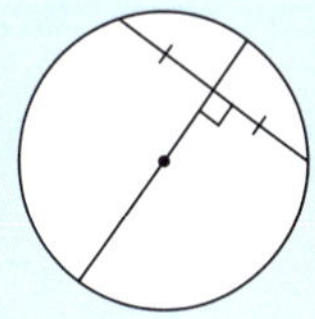

EXERCISE 16B

1 In this question we prove the *chords of a circle* theorem.
Suppose [AB] is a chord of a circle with centre O, and [OM] is perpendicular to [AB].

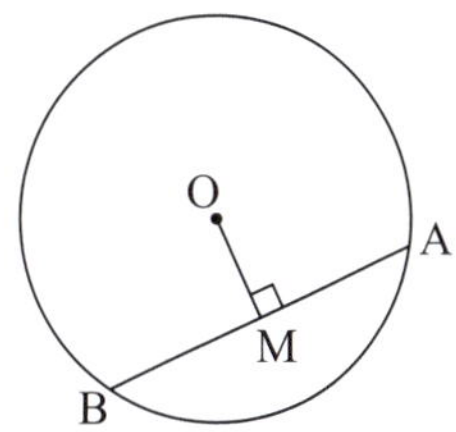

a Show that triangles OAM and OBM are congruent.

b Hence prove that AM = BM.

2 Find the value of x, rounding your answer to 2 decimal places:

a

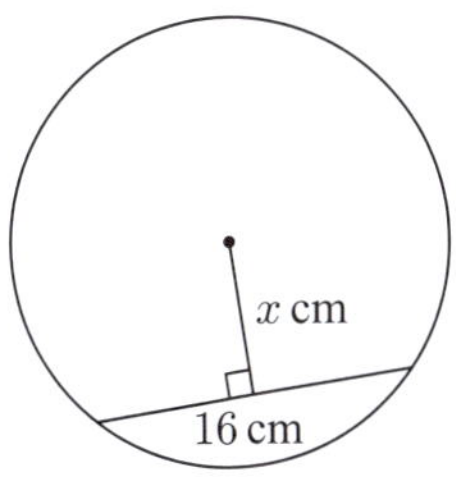

radius = 11 cm

b

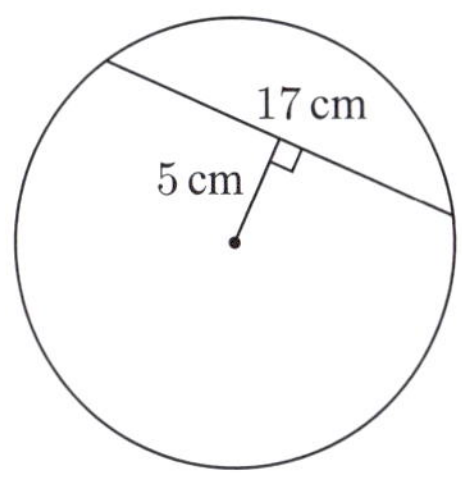

radius = x cm

c

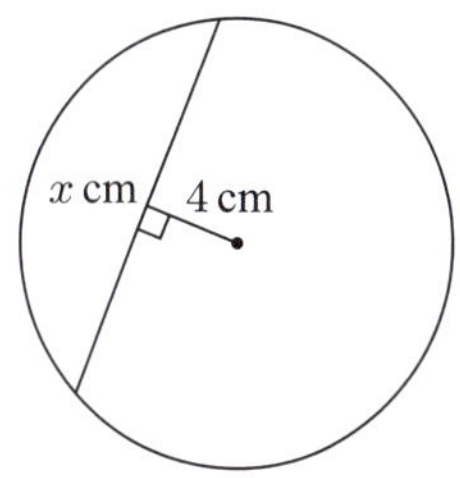

radius = 9 cm

3 A circle has diameter 12 cm. A chord of the circle has length 7 cm. Find, to the nearest mm, the shortest distance from the chord to the circle's centre.

4 The chord [AB] is equal in length to the radius of the circle.
Find the length of [XY].

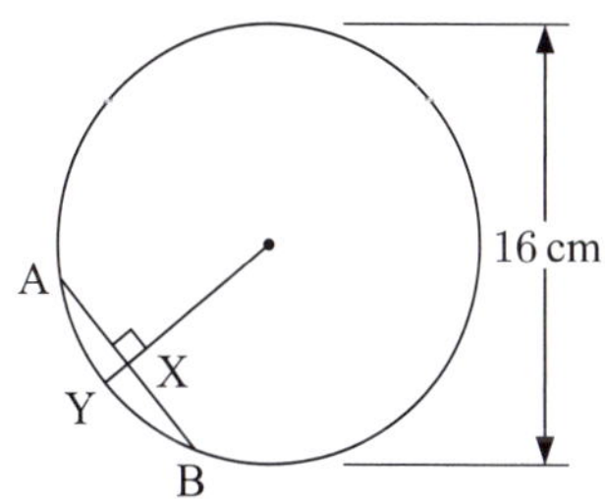

5 [AB] is the perpendicular bisector of the chord [CD].
Find the diameter of the circle.

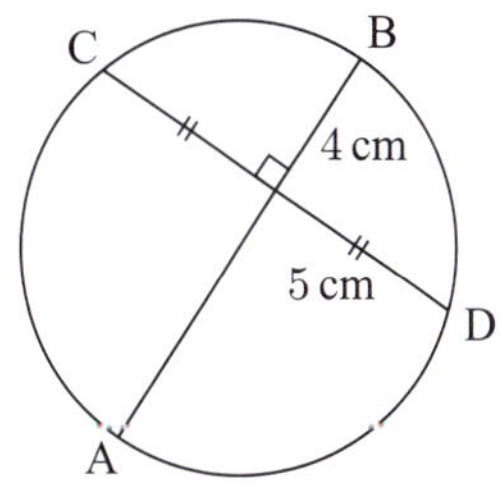

6 Consider three non-collinear points A, B, and C.
Explain how you can use perpendicular bisectors to find the centre of the circle which passes through A, B, and C.

C RADIUS-TANGENT THEOREM

The **radius-tangent theorem** states that:

The tangent to a circle is perpendicular to the radius at the point of contact.

$$\widehat{OAT} = 90^\circ$$

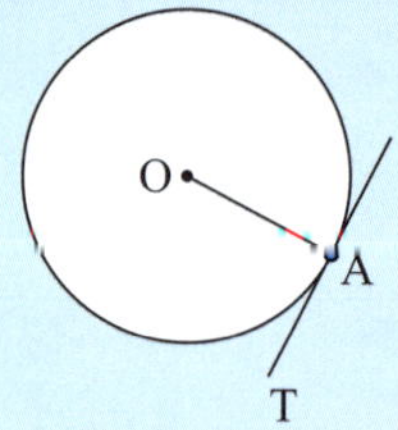

Example 3 — Self Tutor

Find the value of a.

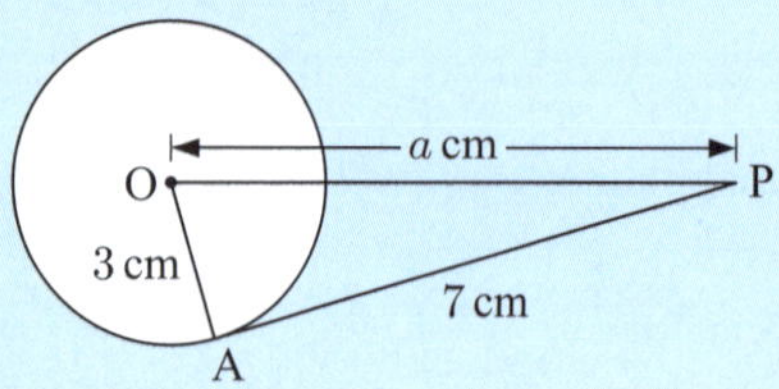

$\widehat{OAP} = 90^\circ$ {radius-tangent}
$\therefore\ a^2 = 3^2 + 7^2$ {Pythagoras}
$\therefore\ a^2 = 58$
$\therefore\ a = \sqrt{58}$ $\{a > 0\}$

EXERCISE 16C

1 Find the values of the unknowns, giving reasons for your answers:

a

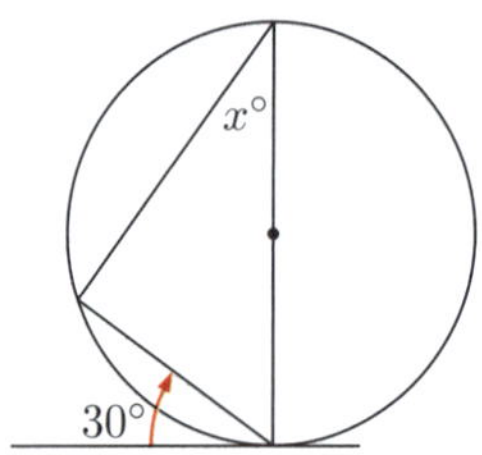

b

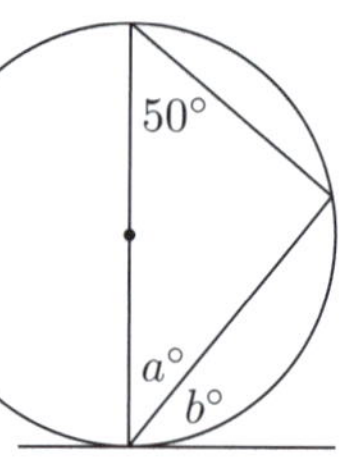

c

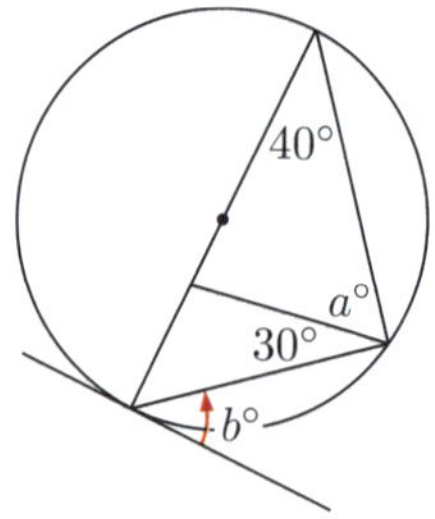

2 [PQ] is the diameter of a circle. Prove that the tangents to the circle at P and Q are parallel.

3 Two circles have a common tangent. The points of contact A and B are 7 cm apart. The radii are 4 cm and 2 cm respectively. Find the distance between the centres.

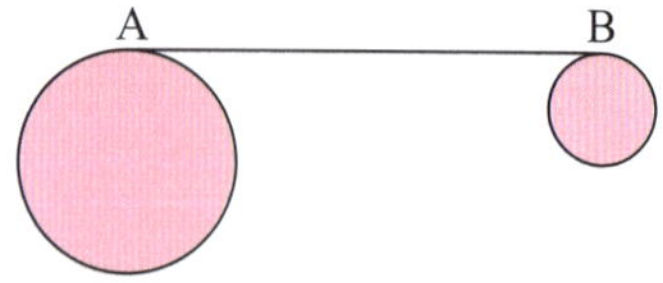

4 In this question we prove the *radius-tangent* theorem.
Suppose the tangent meets the circle at A. Let P be the point on the tangent such that $\widehat{OPT} = 90^\circ$.

a Show that $OP^2 + AP^2 = OA^2$. Hence explain why $OP \leqslant OA$.

b Use the definition of a circle to explain why $OP \geqslant OA$.

c Prove that A and P must be the same point, and hence find the size of $\widehat{OAT}$.

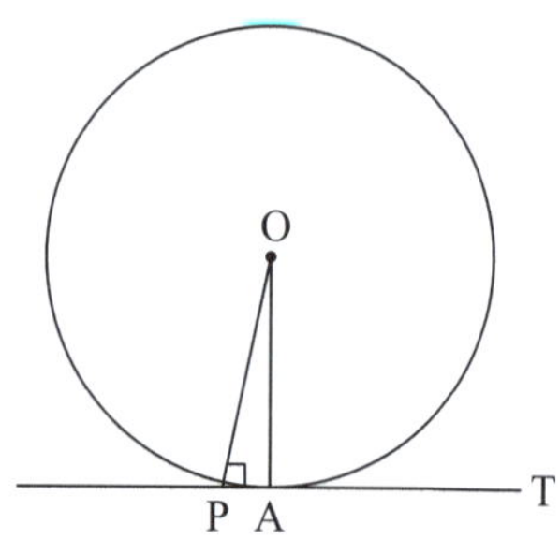

D TANGENTS FROM AN EXTERNAL POINT THEOREM

The **tangents from an external point theorem** states that:

Tangents from an external point are equal in length.

$$AP = BP$$

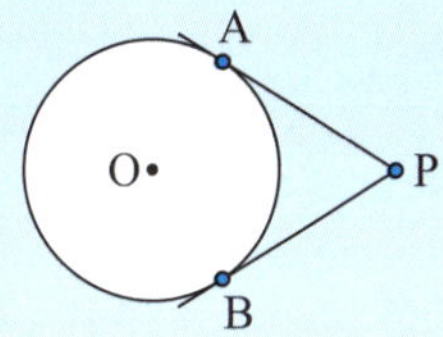

EXERCISE 16D

1 In this question we prove the *tangents from an external point* theorem.
We join [OA], [OB], and [OP].

a Show that triangles AOP and BOP are congruent.

b Hence show that AP = BP.

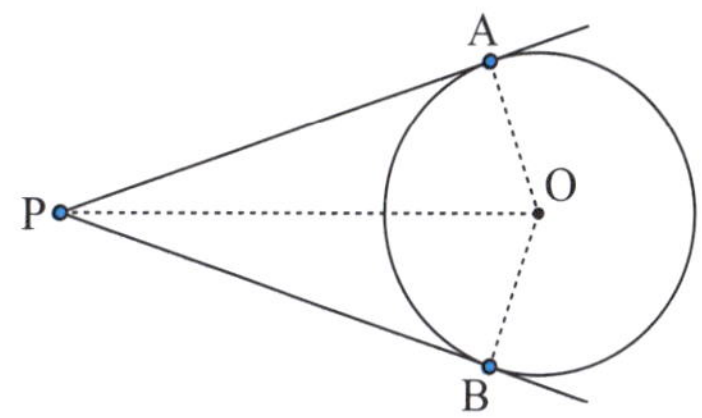

2 In each diagram, O is the centre of the circle, and [TP] and [SP] are tangents. Find the values of the unknowns.

a

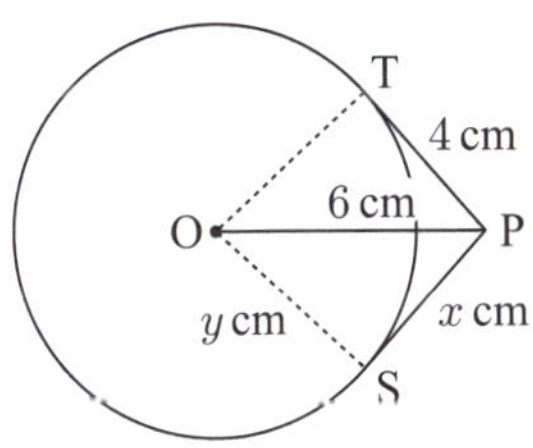

b

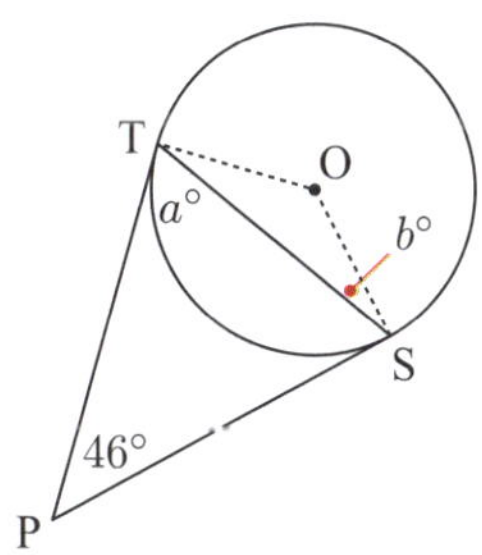

c

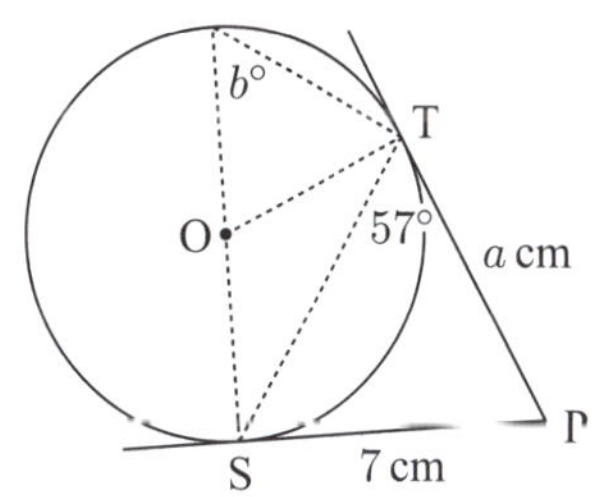

3

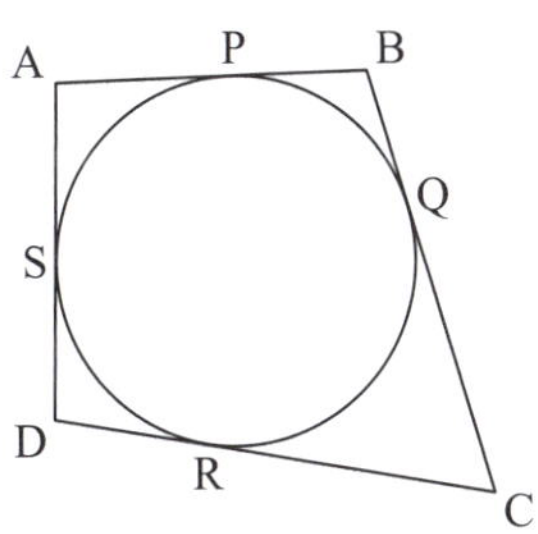

Four tangents to a circle are constructed as shown.
Deduce that AB + CD = BC + AD.

4 A circle is inscribed in a right angled triangle. The radius of the circle is 3 cm, and [BC] has length 8 cm. Find the perimeter of triangle ABC.

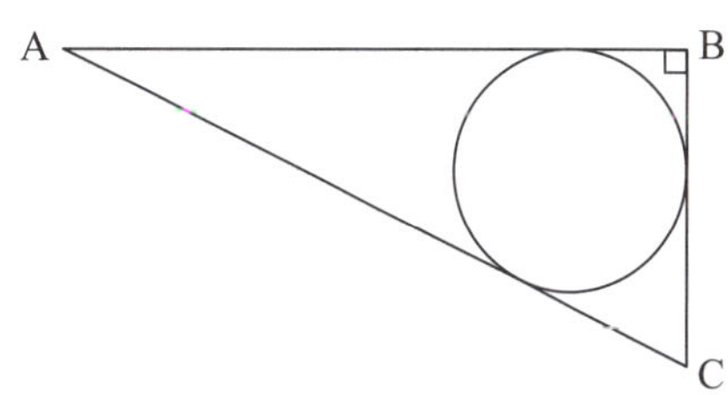

5 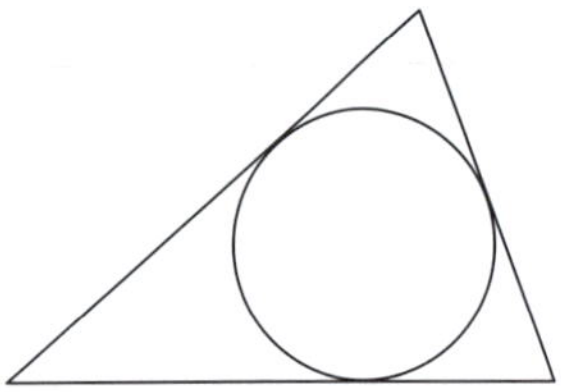

A circle is inscribed in a triangle. The triangle has area A and perimeter P. Show that:

a the radius of the circle is given by $r = \dfrac{2A}{P}$

b the ratio of perimeters of the circle and triangle is the same as the ratio of their areas.

ACTIVITY

This figure consists of three circles of unequal radii. Common external tangents are drawn between each pair of circles and extended until they meet.

Click on the icon to see what interesting fact emerges.

GEOMETRY PACKAGE

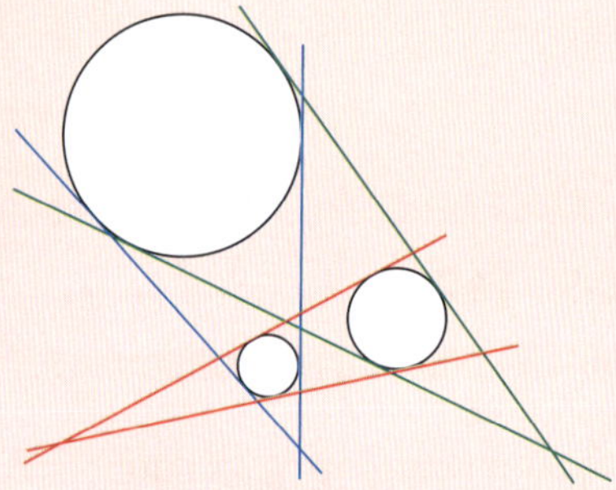

E ANGLE BETWEEN A TANGENT AND A CHORD THEOREM

INVESTIGATION 1 ANGLE BETWEEN A TANGENT AND A CHORD THEOREM

What to do:

GEOMETRY PACKAGE

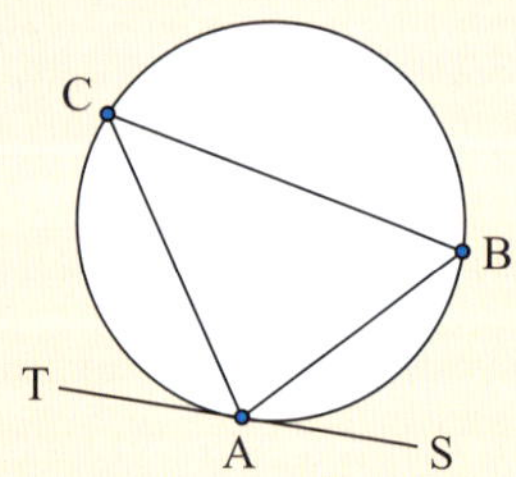

1. Draw a circle. Mark on it points A, B, and C.
2. Draw tangent [TS] meeting the circle at A, and join [AB], [BC], and [CA].
3. Measure $B\widehat{A}S$ and $A\widehat{C}B$. What do you notice?
4. Repeat the above steps with another circle.
5. Copy and complete:

   The angle between a tangent and a chord at the point of contact is ...... to the angle subtended by the chord in the alternate ......

The **angle between a tangent and a chord theorem** states that:

The angle between a tangent and a chord at the point of contact is equal to the angle subtended by the chord in the alternate segment.

$$B\widehat{A}S = A\widehat{C}B$$

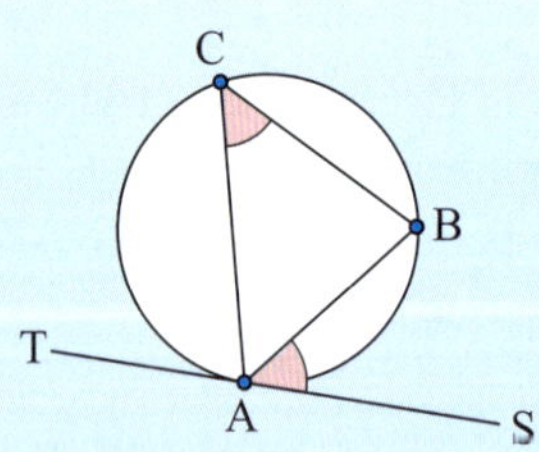

Example 4 **Self Tutor**

[AT] is a tangent to the circle. Find x.

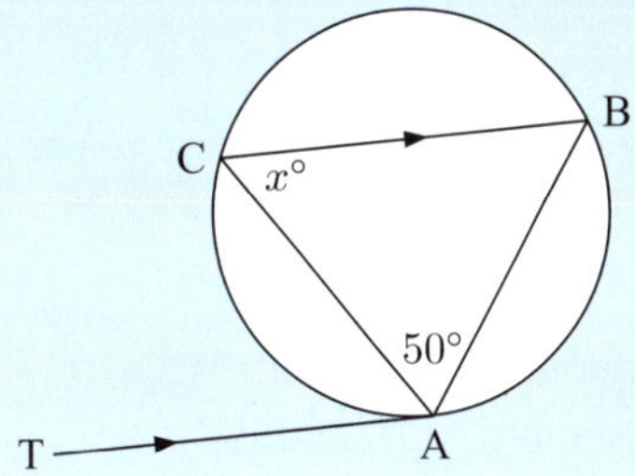

$$\widehat{CAT} = x^\circ \quad \{\text{equal alternate angles}\}$$
$$\therefore \ \widehat{ABC} = \widehat{CAT} = x^\circ \quad \{\text{angle between a tangent and a chord}\}$$
$$\therefore \ x + x + 50 = 180 \quad \{\text{angles in a triangle}\}$$
$$\therefore \ 2x = 130$$
$$\therefore \ x = 65$$

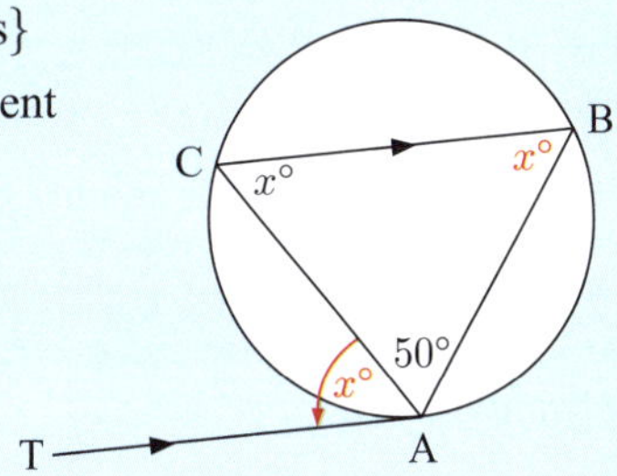

EXERCISE 16E

1 In each diagram, C is the point of contact of tangent [CT]. Find the value of x, giving reasons:

a

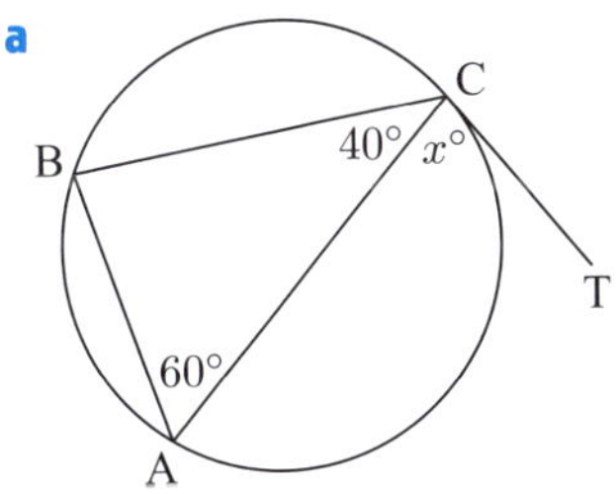

b

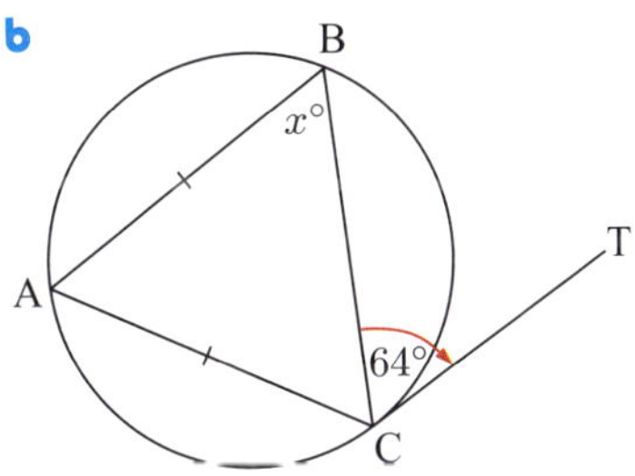

c

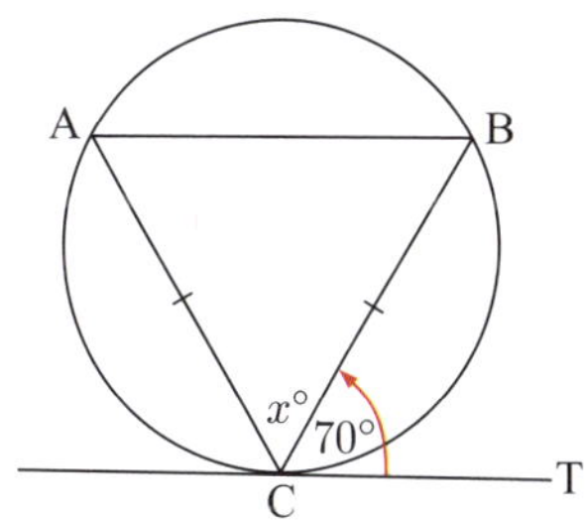

2 ABC is an isosceles triangle which is inscribed in a circle as shown.

[TC] is a tangent to the circle.

Prove that [AC] bisects $\widehat{BCT}$.

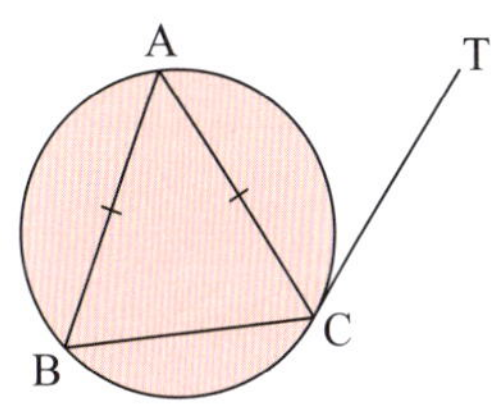

3 [QR] is a chord of a circle which is parallel to the tangent at P. Prove that triangle PQR is isosceles.

4 [BV] and [CU] are tangents to the circle shown.

Find the size of $\widehat{ABC}$.

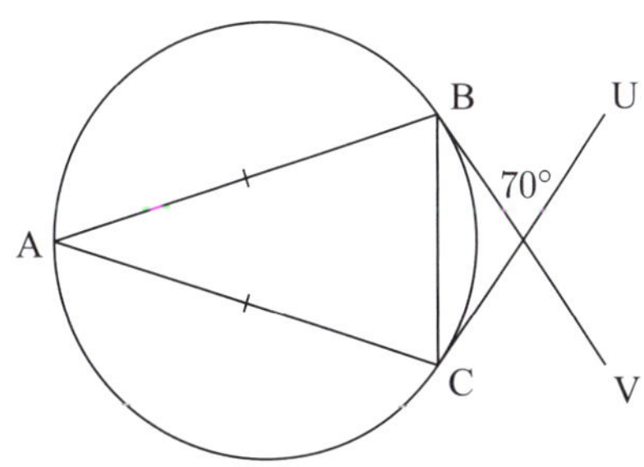

F ANGLE AT THE CENTRE THEOREM

INVESTIGATION 2 ANGLE AT THE CENTRE THEOREM

What to do:

1 Draw a large circle with centre O.
Mark on it points A, B, and P.

2 Join [AO], [BO], [AP], and [BP].

3 Measure $A\widehat{O}B$ and $A\widehat{P}B$.
What do you notice?

4 Repeat the above steps with another circle.

5 Copy and complete:
The angle at the centre of a circle is ...... the angle on the circle subtended by the same arc.

GEOMETRY PACKAGE

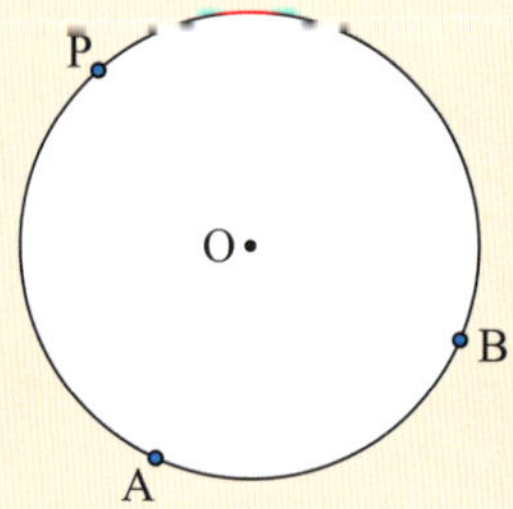

The **angle at the centre theorem** states that:

The angle at the centre of a circle is twice the angle on the circle subtended by the same arc.

$$A\widehat{O}B = 2 \times A\widehat{C}B$$

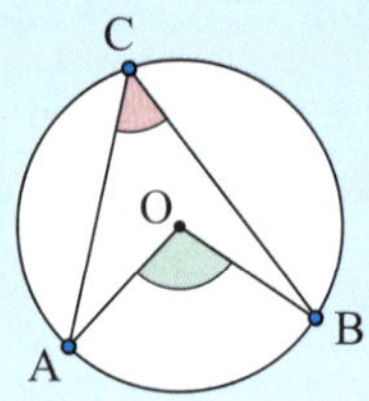

Notice that:

- The following diagrams are all cases of the *angle at the centre theorem*. They can all be explored using the **geometry package**.

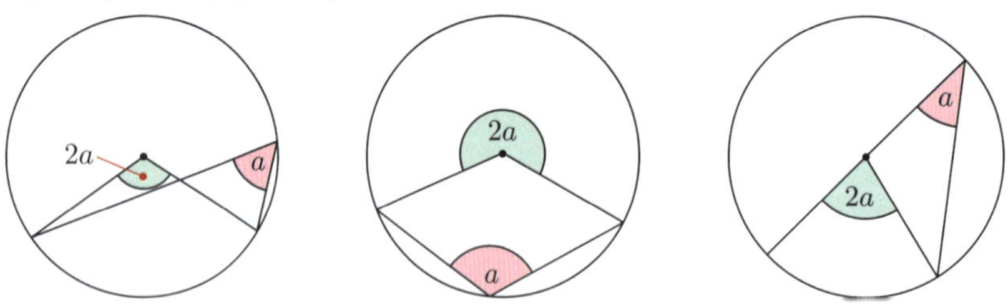

GEOMETRY PACKAGE

- The **angle in a semi-circle theorem** is a special case of the angle at the centre theorem.

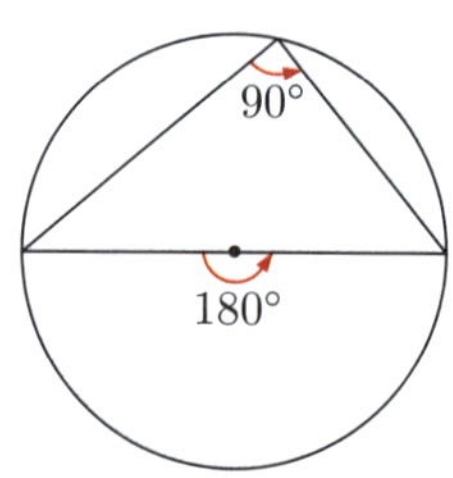

Example 5 — Self Tutor

Find the value of x:

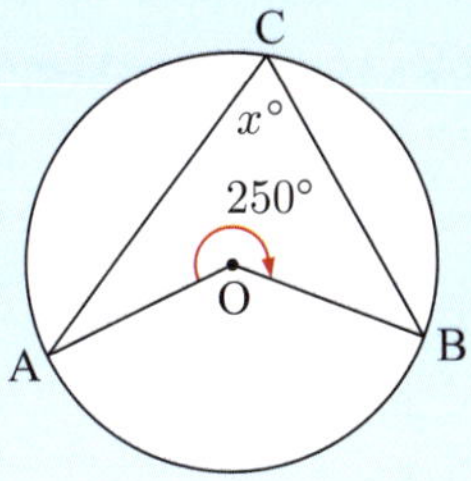

Obtuse $\text{A}\widehat{\text{O}}\text{B} = 360° - 250°$ {angles at a point}

$\therefore$ obtuse $\text{A}\widehat{\text{O}}\text{B} = 110°$

$\therefore$ $2x = 110$ {angle at the centre}

$\therefore$ $x = 55$

EXERCISE 16F

1 Find, giving reasons, the value of x:

a

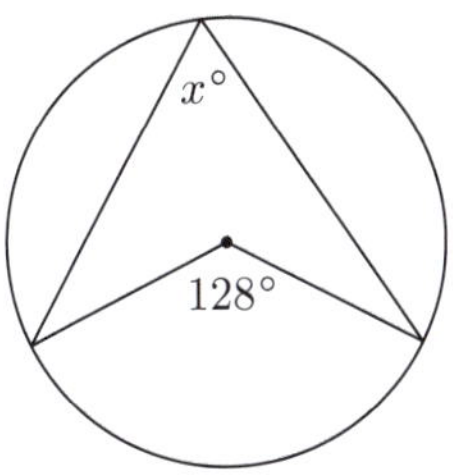

b

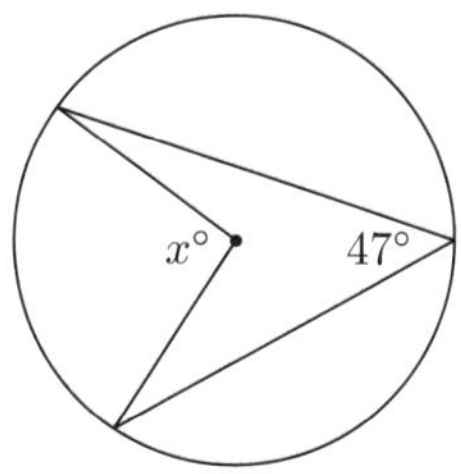

c

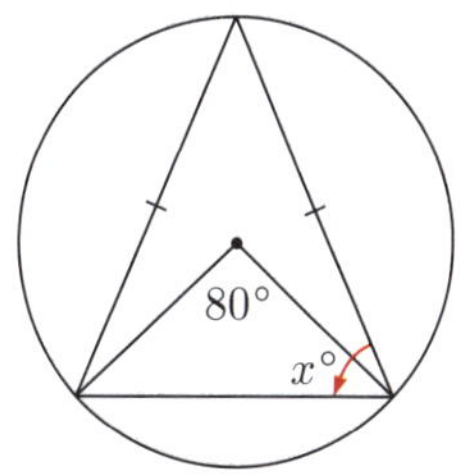

d

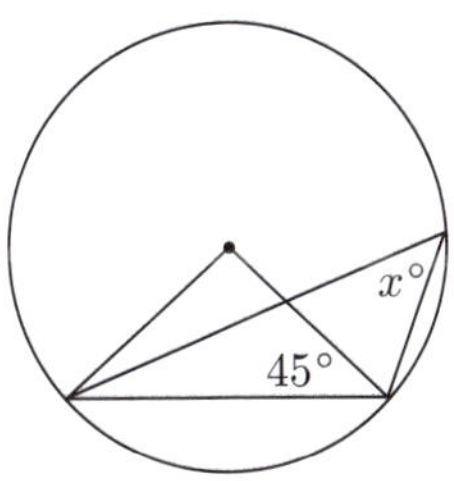

e

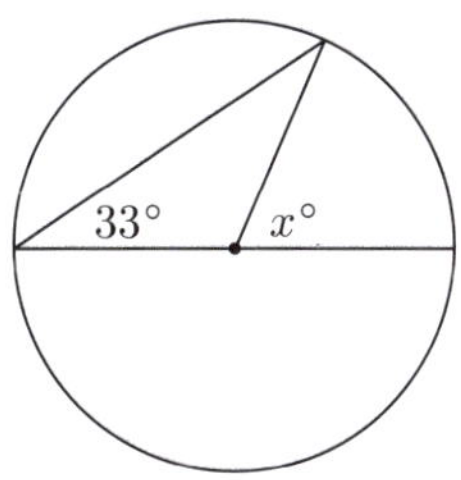

f

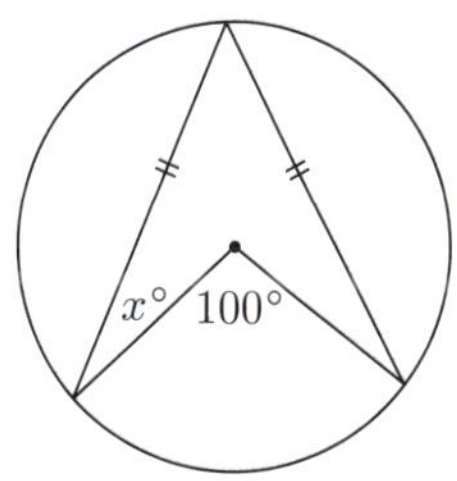

2

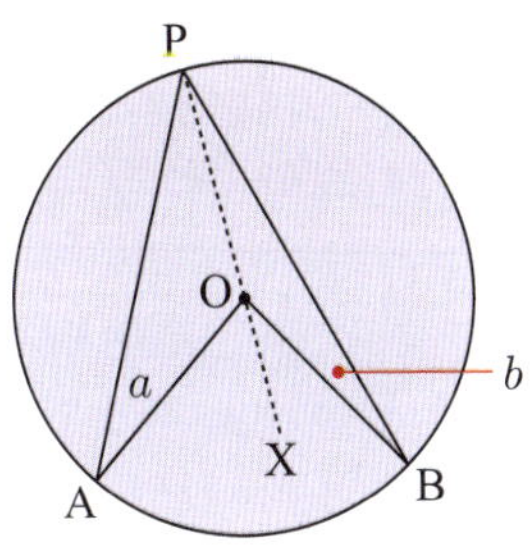

In this question we prove the *angle at the centre* theorem for the case where the centre lies within the triangle formed by the other three points.

a Explain why triangles OAP and OBP are isosceles.

b Write in terms of a and b:

i $\text{A}\widehat{\text{P}}\text{O}$ **ii** $\text{B}\widehat{\text{P}}\text{O}$ **iii** $\text{A}\widehat{\text{O}}\text{X}$

iv $\text{B}\widehat{\text{O}}\text{X}$ **v** $\text{A}\widehat{\text{P}}\text{B}$ **vi** $\text{A}\widehat{\text{O}}\text{B}$

c Hence show that the angle at the centre of the circle is twice the angle on the circle subtended by the same arc.

3 [AB] is the diameter of a circle with centre O. X is a point on the circle, and [AX] is *produced* or extended to Y such that OX = XY.

Let $\text{X}\widehat{\text{Y}}\text{O} = \alpha$.

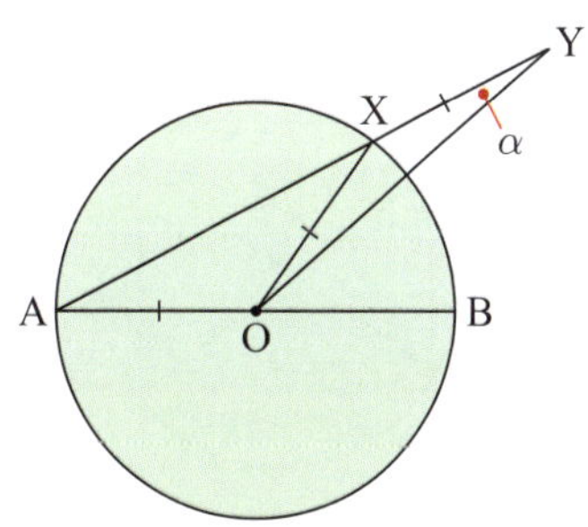

a Write in terms of α:

i $\text{X}\widehat{\text{O}}\text{Y}$ **ii** $\text{A}\widehat{\text{X}}\text{O}$ **iii** $\text{X}\widehat{\text{A}}\text{O}$

iv $\text{X}\widehat{\text{O}}\text{B}$ **v** $\text{B}\widehat{\text{O}}\text{Y}$

b What is the relationship between $\text{B}\widehat{\text{O}}\text{Y}$ and $\text{X}\widehat{\text{O}}\text{Y}$?

G ANGLES SUBTENDED BY THE SAME ARC THEOREM

INVESTIGATION 3 ANGLES SUBTENDED BY THE SAME ARC THEOREM

What to do:

GEOMETRY PACKAGE

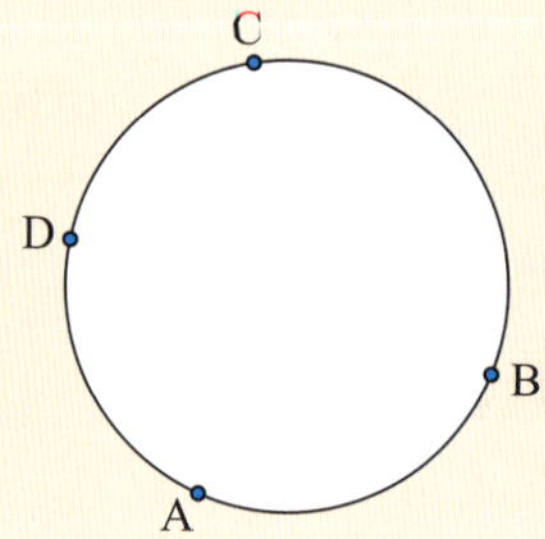

1. Draw a circle.
   Mark on it points A, B, C, and D.
2. Join [AC], [BC], [AD], and [BD].
3. Measure $\widehat{ACB}$ and $\widehat{ADB}$.
   What do you notice?
4. Repeat the above steps with another circle.
5. Copy and complete:
   Angles subtended by the same arc on the circle are ...... in size.

The **angles subtended by the same arc theorem** states that:

Angles subtended by the same arc on a circle are equal in size.

$$\widehat{ADB} = \widehat{ACB}$$

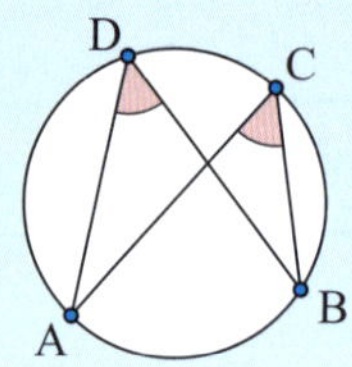

Example 6 — Self Tutor

Find the value of x:

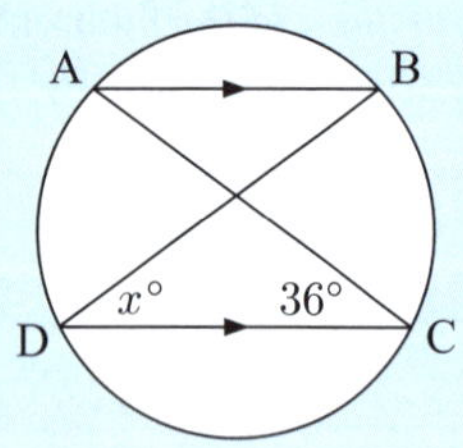

$\widehat{ABD} = 36°$ {angles subtended by the same arc}

and $\widehat{BDC} = \widehat{ABD}$ {equal alternate angles}

$\therefore\ x = 36$

EXERCISE 16G

1 Find, giving reasons, the value of each unknown:

a

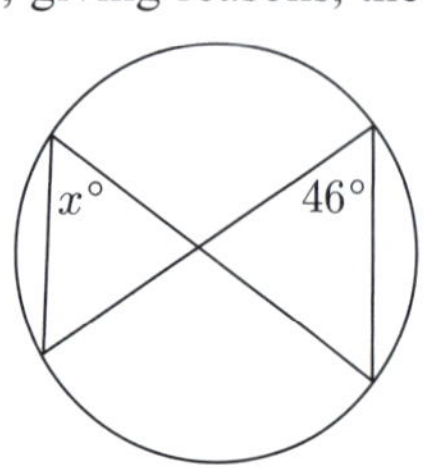

b

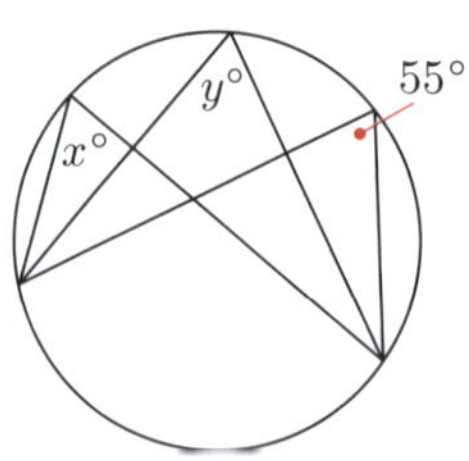

c

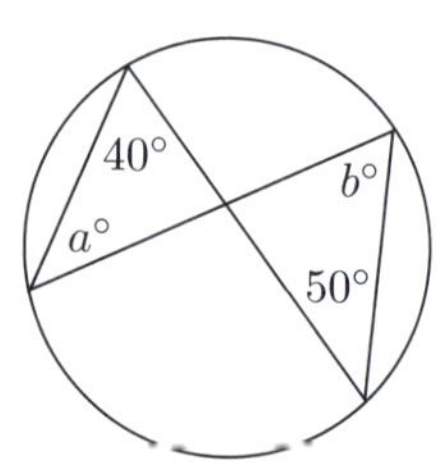

d

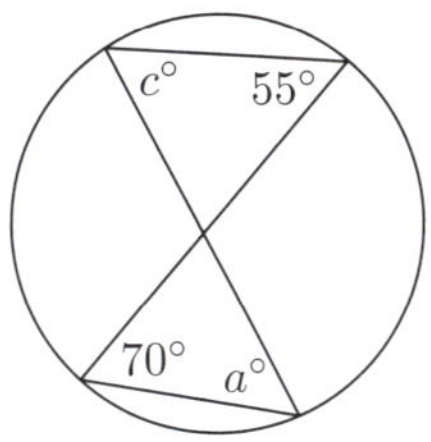

e

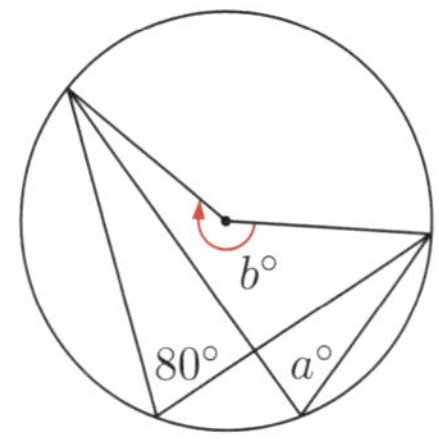

f

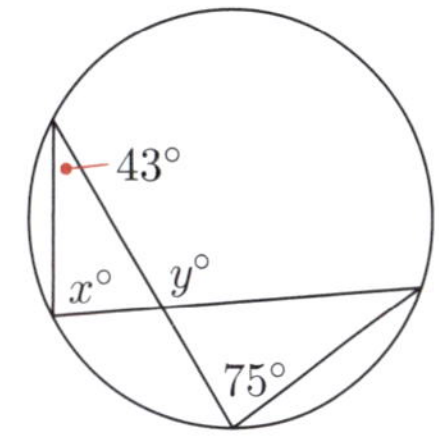

g

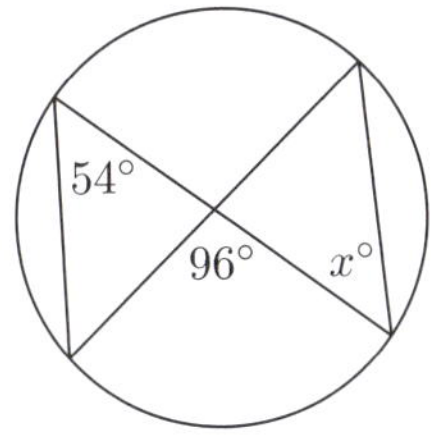

h

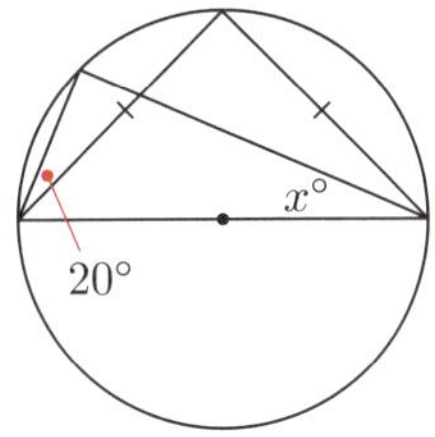

i

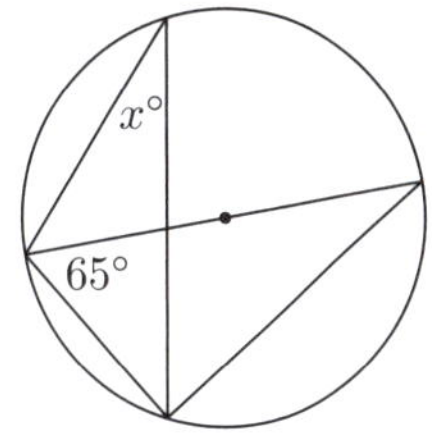

2 In this question we prove the *angles subtended by the same arc* theorem.

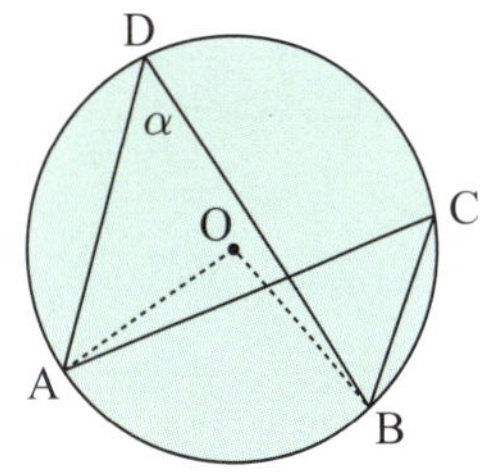

a Use the *angle at the centre* theorem to write $A\widehat{O}B$ in terms of α.

b Hence write $A\widehat{C}B$ in terms of α.

c State the relationship between $A\widehat{D}B$ and $A\widehat{C}B$.

3

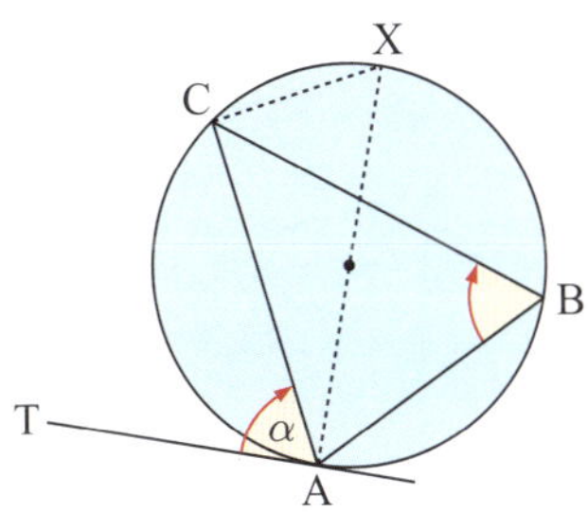

In this question we prove the *angle between a tangent and a chord* theorem for the case where the angle between the tangent and the chord is acute.

a Draw diameter [AX] and join [CX].
Find the size of:

i $T\widehat{A}X$ **ii** $A\widehat{C}X$

b Let $T\widehat{A}C = \alpha$. Write, in terms of α:

i $C\widehat{A}X$ **ii** $C\widehat{X}A$ **iii** $C\widehat{B}A$

c State the relationship between $T\widehat{A}C$ and $C\widehat{B}A$.

4 Two circles intersect at A and B.
Straight lines [PQ] and [XY] are drawn through A to meet the circles as shown.
Show that $X\widehat{B}P = Y\widehat{B}Q$.

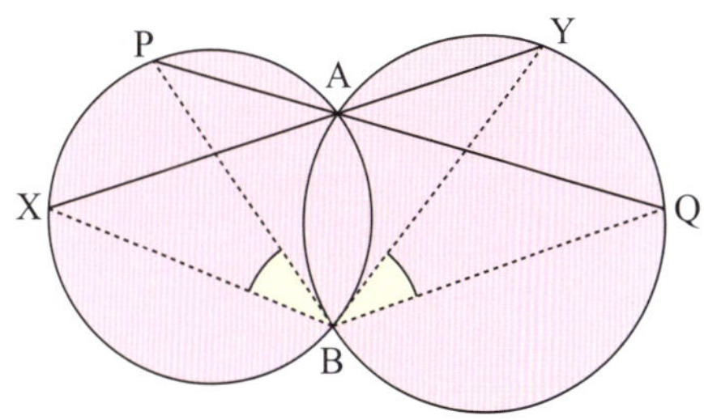

H CYCLIC QUADRILATERALS

I TESTS FOR CYCLIC QUADRILATERALS

Click on the headings to obtain these Sections online.

MULTIPLE CHOICE QUIZ

QUICK QUIZ

REVIEW SET 16A

1 Find the value of a:

a

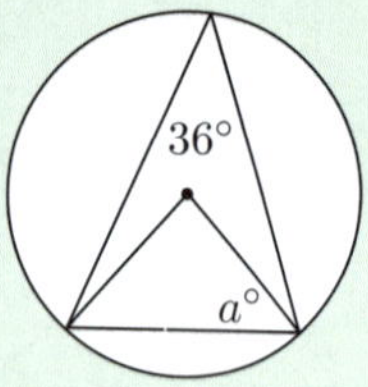

b

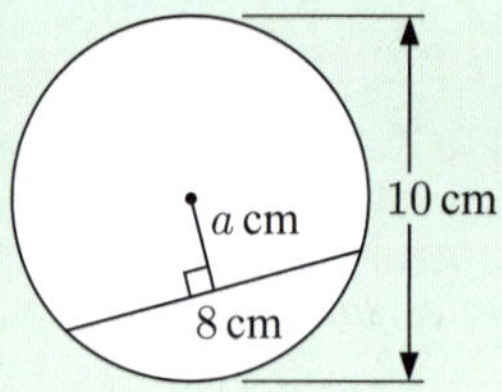

c

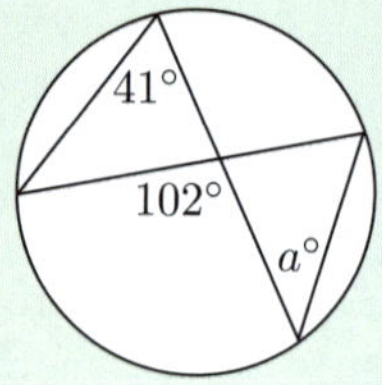

d

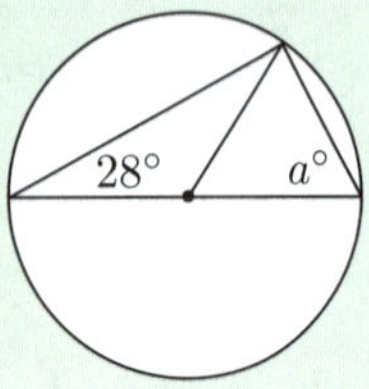

e

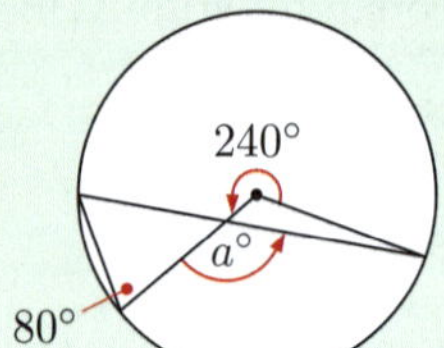

2 [AB] and [AC] are tangents to the circle.
Find an equation connecting α, β, and γ.

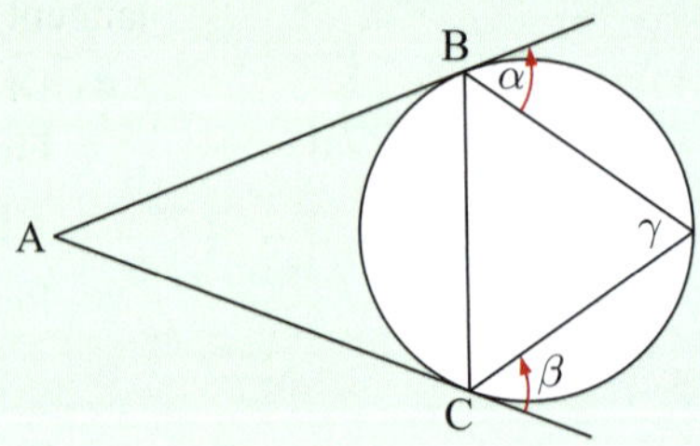

3 Find the value of x:

a

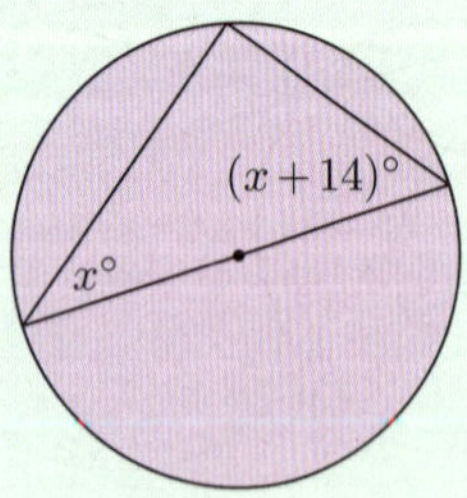

b

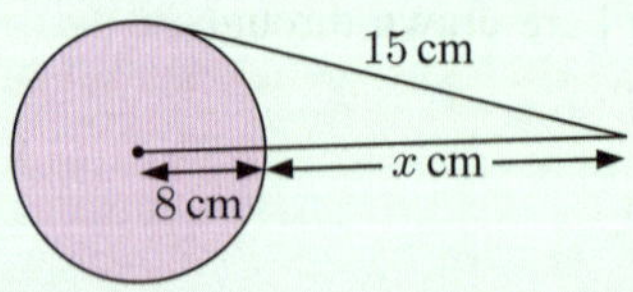

c

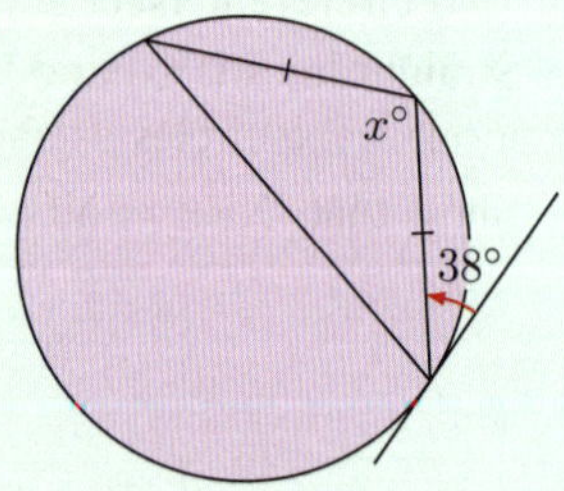

4

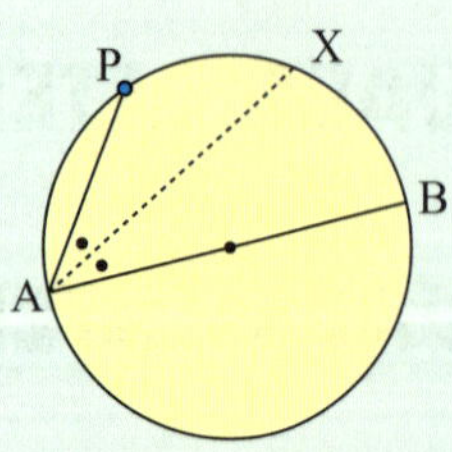

The circle alongside has diameter [AB], and P is another point on the circle. The angle bisector of $\widehat{PAB}$ meets the circle at X.

Show that the tangent at X is parallel to [PB].

5 O is the centre of the circle alongside.
The chords [AC] and [BD] are perpendicular.
Prove that $\alpha+\beta=180°$.

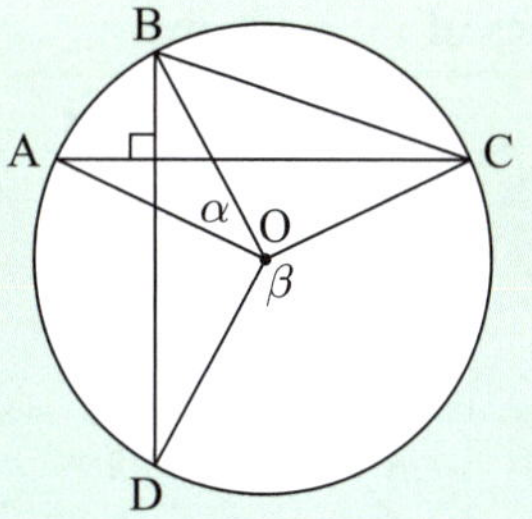

6 [AB] and [AQ] are chords of equal length. Their perpendicular bisectors meet at X. $\text{Q}\widehat{\text{A}}\text{X}=\alpha$ and N is the midpoint of [AQ].

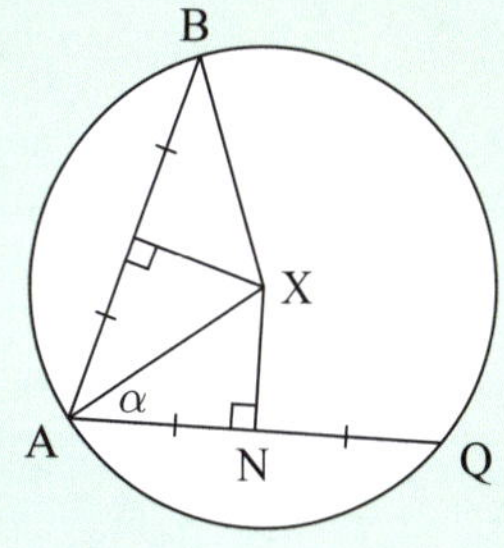

- **a** Explain why X must be the centre of the circle.
- **b** Find the value of α such that:
  - **i** [BQ] is a diameter
  - **ii** B, X, and N are collinear.

7

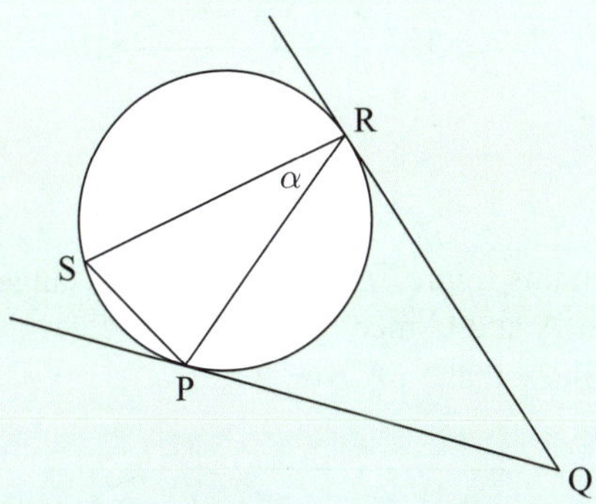

[QP] and [QR] are tangents to a circle. S is a point on the circle such that $\text{P}\widehat{\text{S}}\text{R}$ and $\text{P}\widehat{\text{Q}}\text{R}$ are equal, and both are double $\text{P}\widehat{\text{R}}\text{S}$.
Let $\text{P}\widehat{\text{R}}\text{S}$ be α.

- **a** Show that $\alpha=30°$.
- **b** What can you conclude about [RS]?

REVIEW SET 16B

1 Find the value of x:

a

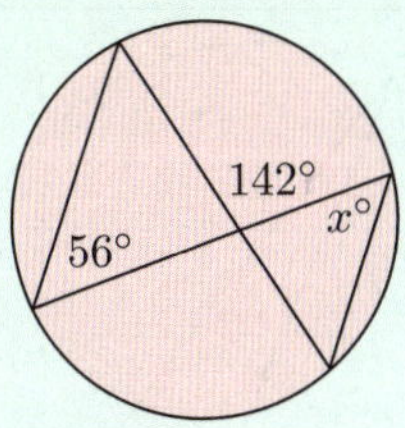

b

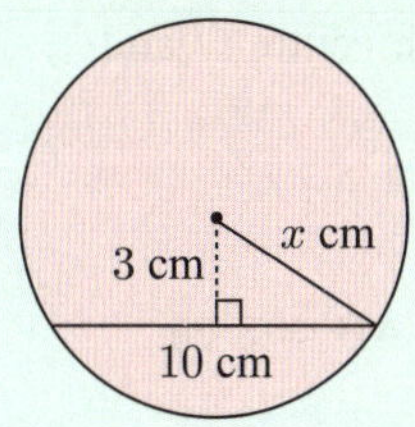

c

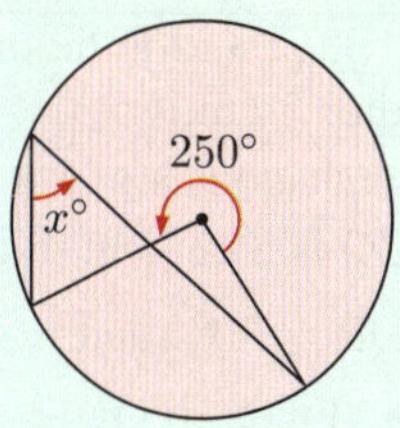

d

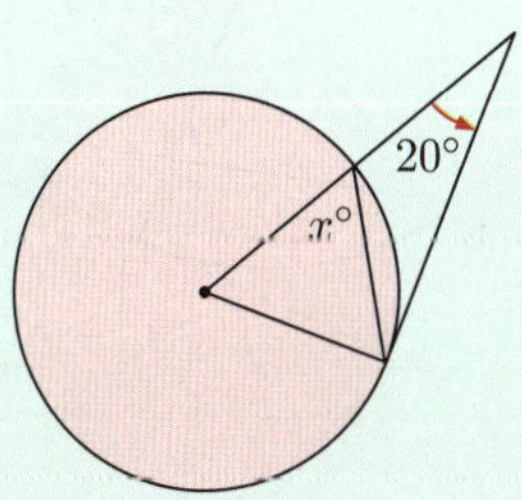

e

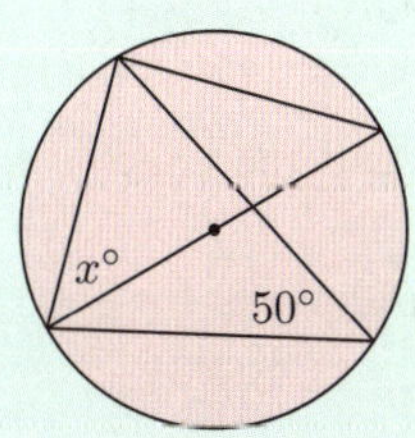

2 Find the values of the unknowns:

a

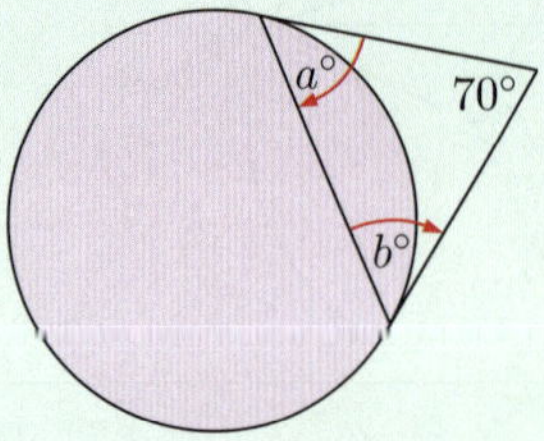

b

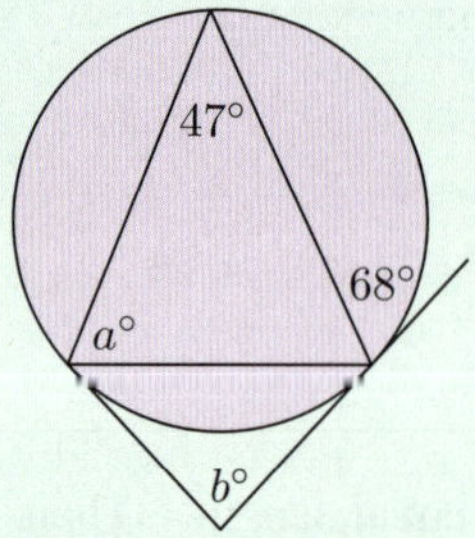

c

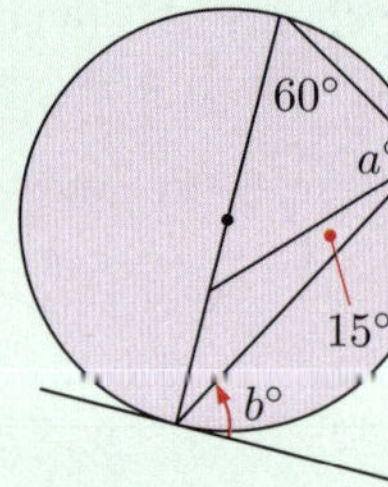

3 Two circles intersect at A and B. [AX] and [AY] are diameters as shown. Prove that X, B, and Y are collinear.

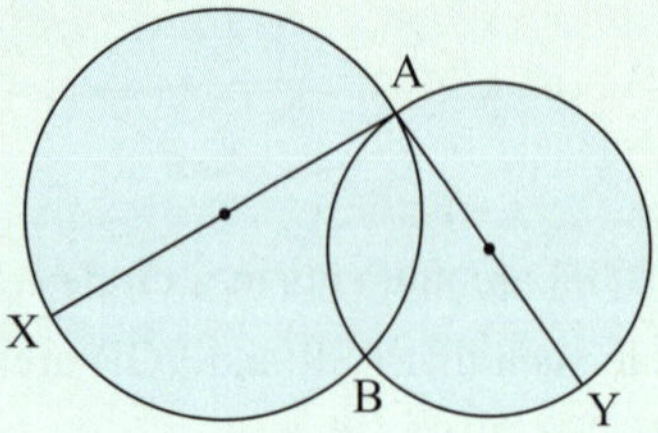

4 [AB] is the diameter of a circle with centre O.
[AC] and [BD] are any two chords.
Show that $\widehat{BDO} = \widehat{ACD}$.

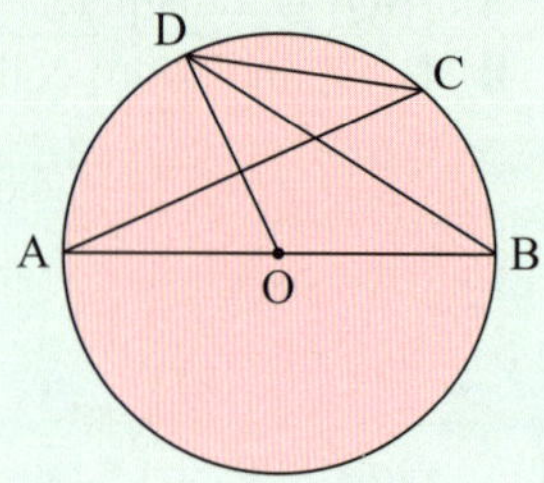

5 In the circle below, $AB = BD$ and [AC] bisects $\widehat{BAD}$.
Show that $\widehat{AXB} = \widehat{ADC}$.

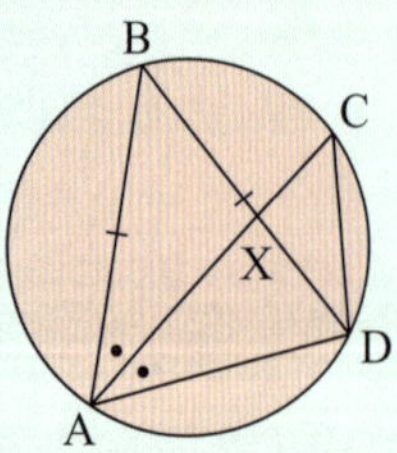

6 In this figure, $AC = BC$, and the tangents at A and C meet at D.
Show that $\alpha + \beta = 90°$.

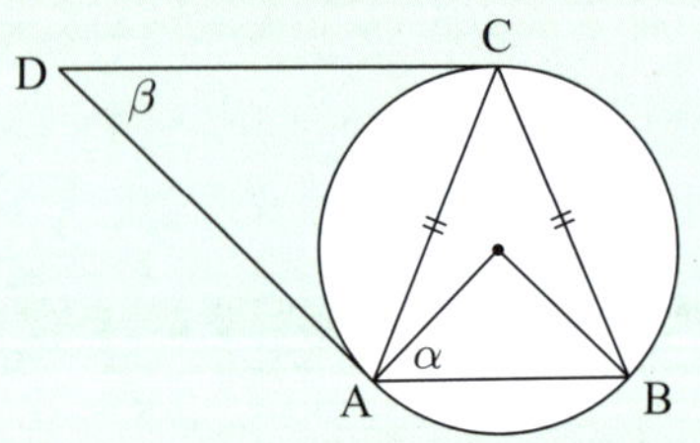

7 A, B, C, and D lie on a circle with centre O and radius 5 cm.
The diagonals of ABCD intersect at X.
(AZ) is a tangent to the circle.
$\widehat{ADB} = 30°$ and $\widehat{CXD} = 105°$. $AB = 5$ cm.

a Find, giving reasons:

i $\widehat{BCA}$ **ii** $\widehat{BOA}$
iii $\widehat{BAZ}$ **iv** $\widehat{CBX}$

b Use Pythagoras' theorem to find the exact length of [BC].

c Find the length of [AX] exactly.

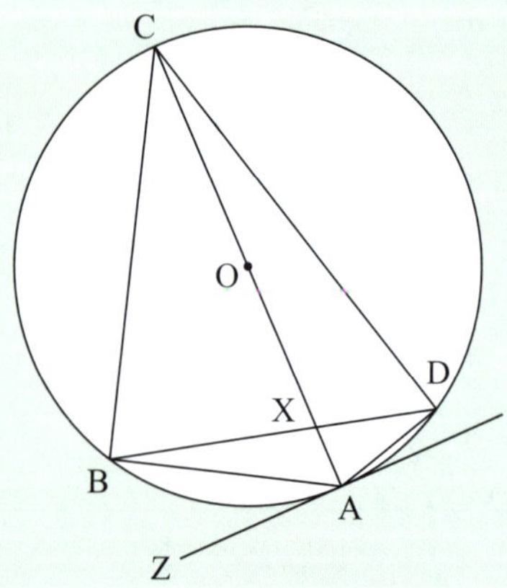

Chapter 17

Trigonometry

Contents:

OPENING PROBLEM

The school gymnasium has the symmetrical cross-section shown.

From a point 30 m from the side wall, the angle of elevation to the top of the side wall is 18.8°.

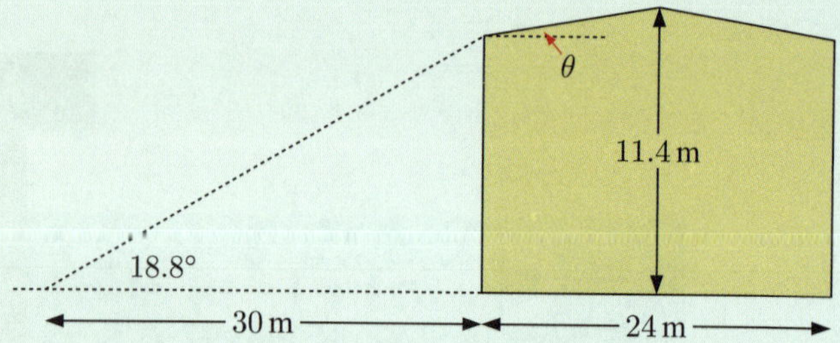

Things to think about:

- **a** How high is the side wall of the gymnasium?
- **b** Can you find the angle θ which is the pitch of the roof?
- **c** Mrs Garrett's eye level is 1.65 m above the ground. How far back from the gymnasium does Mrs Garrett need to stand in order to see the ridgeline?

Trigonometry is the study of the relationships between the side lengths and angles of triangles.

We can apply trigonometry in engineering, astronomy, architecture, navigation, surveying, the building industry, and in many other branches of applied science.

A LABELLING RIGHT ANGLED TRIANGLES

In trigonometry, there is a convention for labelling the sides of right angled triangles.

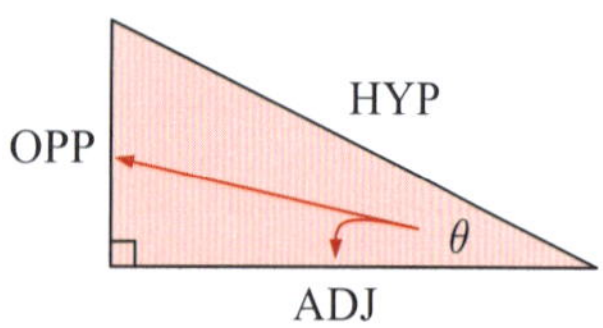

For the right angled triangle with angle θ:

- the **hypotenuse (HYP)** is the longest side
- the **opposite (OPP)** side is opposite θ
- the **adjacent (ADJ)** side is adjacent to θ.

Example 1 — Self Tutor

For the triangle alongside, name the:

- **a** hypotenuse
- **b** side opposite θ
- **c** side adjacent to θ.

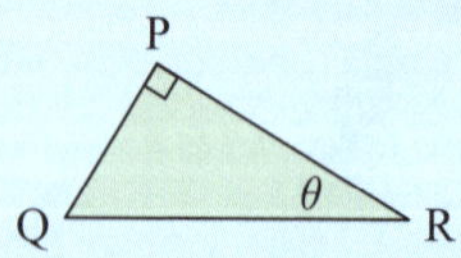

- **a** The hypotenuse is [QR].
- **b** The side opposite θ is [PQ].
- **c** The side adjacent to θ is [PR].

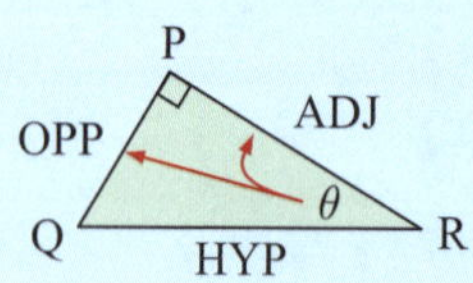

EXERCISE 17A

1 For each triangle, name the:

i hypotenuse **ii** side opposite θ **iii** side adjacent to θ.

a 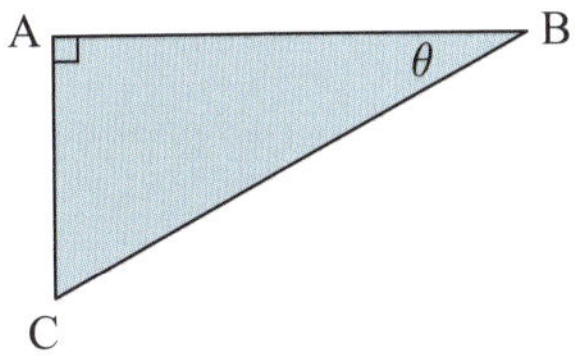

b 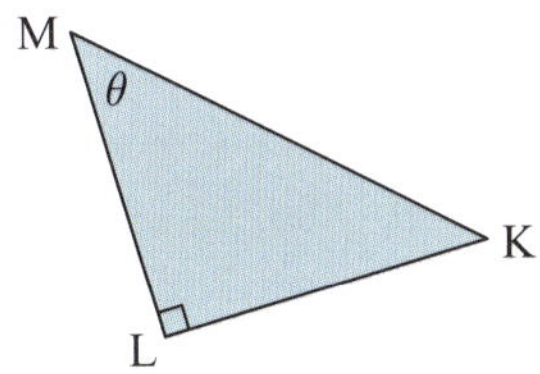

c 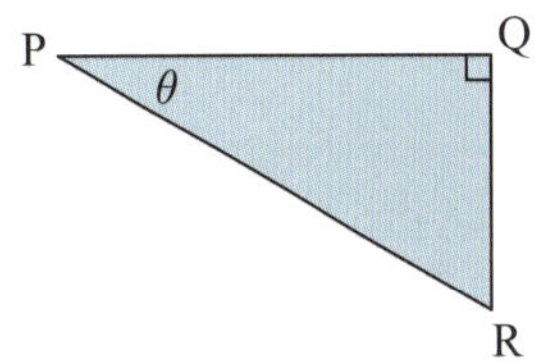

2 For the given triangle, name the:

a hypotenuse

b side opposite α

c side adjacent to α

d side opposite β

e side adjacent to β.

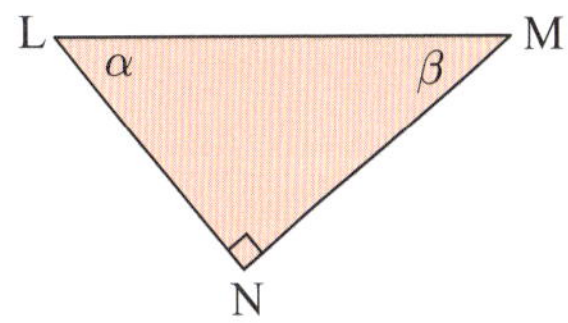

B THE TRIGONOMETRIC RATIOS

INVESTIGATION 1 RATIO OF SIDES OF RIGHT ANGLED TRIANGLES

What to do:

1 Consider the set of all right angled triangles with one acute angle $15°$. For each triangle we label the sides ADJ, OPP, and HYP.

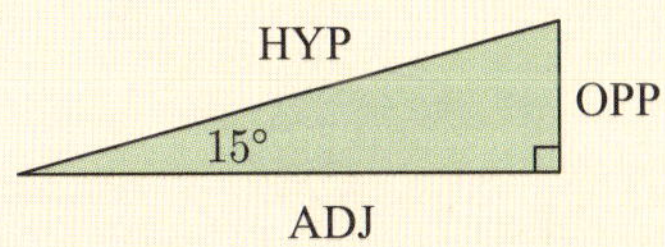

a Explain why all triangles in this set are *similar*.

b What does this tell you about the ratios $\frac{\text{OPP}}{\text{HYP}}$, $\frac{\text{ADJ}}{\text{HYP}}$, and $\frac{\text{OPP}}{\text{ADJ}}$ for these triangles?

2 Use a ruler and protractor to draw a large right angled triangle with one acute angle $15°$.

a Measure each side length to the nearest millimetre. Hence calculate, to 2 decimal places:

i $\frac{\text{OPP}}{\text{HYP}}$ **ii** $\frac{\text{ADJ}}{\text{HYP}}$ **iii** $\frac{\text{OPP}}{\text{ADJ}}$

b Compare your ratio $\frac{\text{OPP}}{\text{ADJ}}$ with $2 - \sqrt{3}$.

In **Investigation 1** you should have recognised that for any set of *similar* right angled triangles, the ratios $\frac{\text{OPP}}{\text{HYP}}$, $\frac{\text{ADJ}}{\text{HYP}}$, and $\frac{\text{OPP}}{\text{ADJ}}$ are constant.

These ratios have the traditional names **sine**, **cosine**, and **tangent** respectively. We abbreviate them to **sin**, **cos**, and **tan**.

In any right angled triangle with one angle θ:

$$\sin\theta = \frac{\text{OPP}}{\text{HYP}}, \quad \cos\theta = \frac{\text{ADJ}}{\text{HYP}}, \quad \tan\theta = \frac{\text{OPP}}{\text{ADJ}}$$

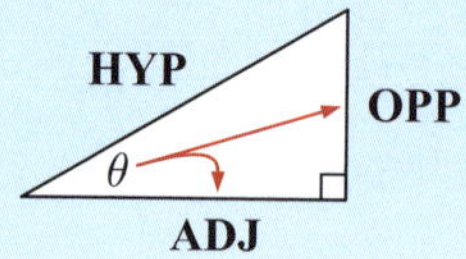

Notice that $\dfrac{\sin\theta}{\cos\theta} = \dfrac{\frac{\text{OPP}}{\text{HYP}}}{\frac{\text{ADJ}}{\text{HYP}}} = \dfrac{\text{OPP}}{\text{ADJ}} = \tan\theta$. So, $\boldsymbol{\tan\theta = \dfrac{\sin\theta}{\cos\theta}}$

In **Investigation 1** you should also have found that:

$$\sin 15^\circ \approx 0.26 \qquad \cos 15^\circ \approx 0.97 \qquad \tan 15^\circ \approx 0.27$$

Trigonometric ratios for any angle can be found using a calculator. You should first check that your calculator is in "degree" mode.

GRAPHICS CALCULATOR INSTRUCTIONS

EXERCISE 17B

1 For each triangle, find: **i** $\sin\theta$ **ii** $\cos\theta$ **iii** $\tan\theta$

a

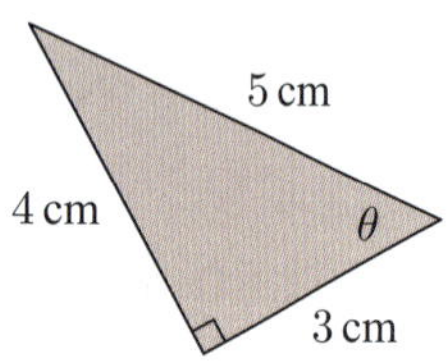

b

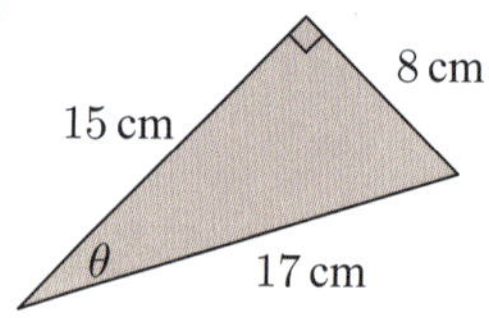

c

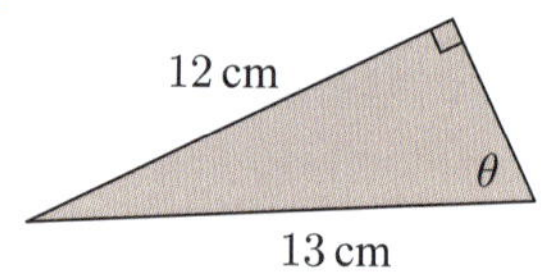

d

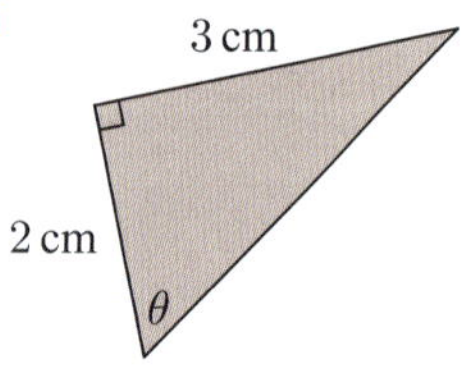

e

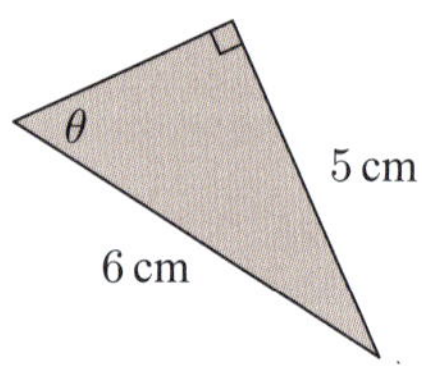

f

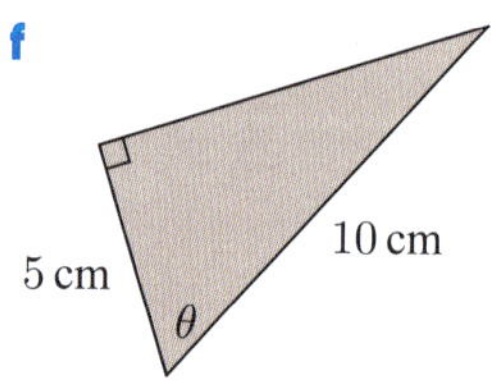

2 **a** Use a ruler to find the length of each side of the triangle, to the nearest millimetre.

b Hence estimate the value of:

i $\sin 57^\circ$ **ii** $\cos 57^\circ$ **iii** $\tan 57^\circ$

c Check your answers using a calculator.

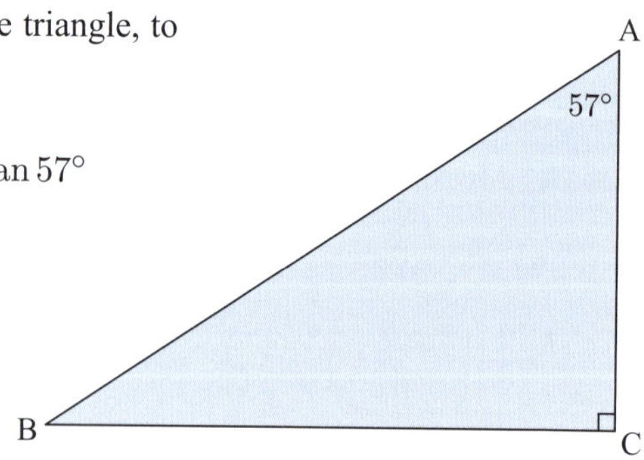

3 Use the triangle alongside to prove that for any acute angle θ:

a $\sin(90^\circ - \theta) = \cos\theta$

b $\cos(90^\circ - \theta) = \sin\theta$

c $\cos^2\theta + \sin^2\theta = 1$

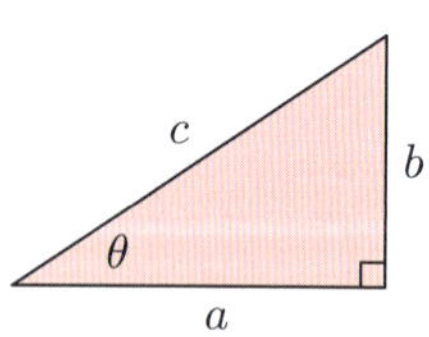

4 Use the triangles alongside to help complete this table. Give *exact* values for each ratio.

θ	$\sin\theta$	$\cos\theta$	$\tan\theta$
$30°$			
$45°$			
$60°$			

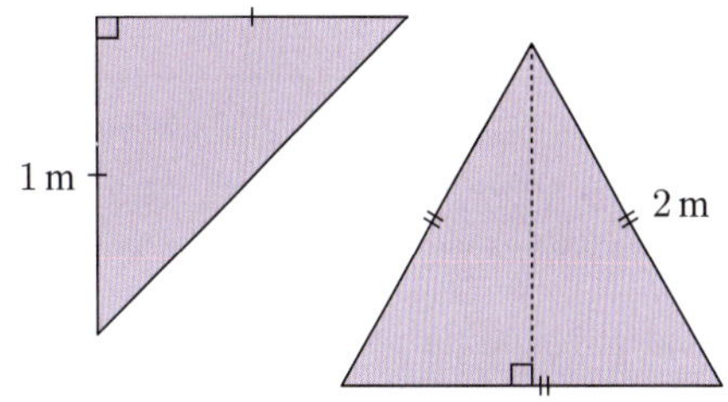

C FINDING SIDE LENGTHS

Suppose we are given the angles of a right angled triangle, and the length of a side. We can use the trigonometric ratios to find the other side lengths.

Example 2 — Self Tutor

Find x, rounding your answer to 2 decimal places:

a

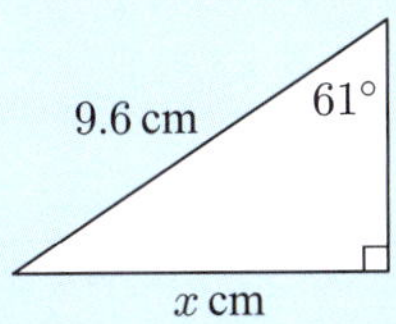

b

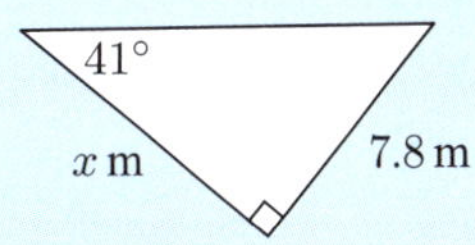

a The relevant sides are OPP and HYP, so we use the *sine* ratio.

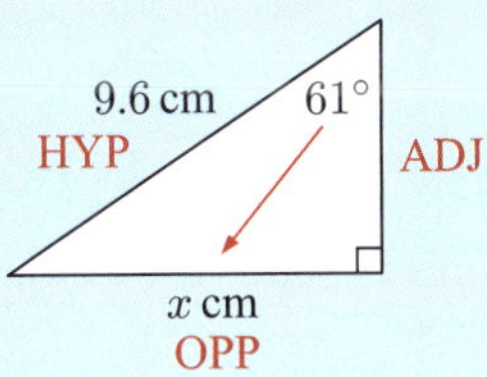

$$\sin\theta = \frac{\text{OPP}}{\text{HYP}}$$

$$\therefore \sin 61° = \frac{x}{9.6}$$

$$\therefore \sin 61° \times 9.6 = x$$

$$\therefore x \approx 8.40$$

b The relevant sides are OPP and ADJ, so we use the *tangent* ratio.

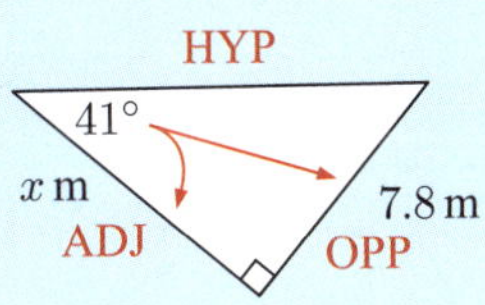

$$\tan\theta = \frac{\text{OPP}}{\text{ADJ}}$$

$$\therefore \tan 41° = \frac{7.8}{x}$$

$$\therefore x \times \tan 41° = 7.8$$

$$\therefore x = \frac{7.8}{\tan 41°}$$

$$\therefore x \approx 8.97$$

EXERCISE 17C

1 Write down a trigonometric equation connecting the angle and the sides given.

a

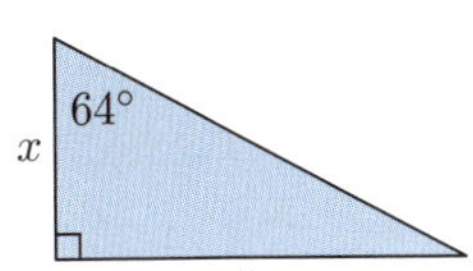

b

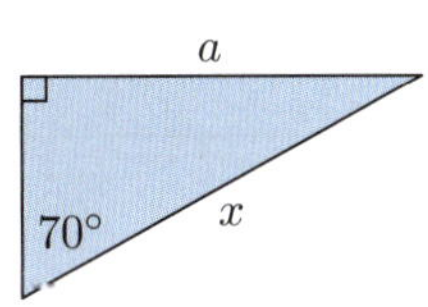

c

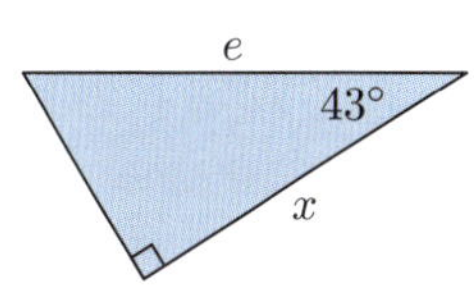

d

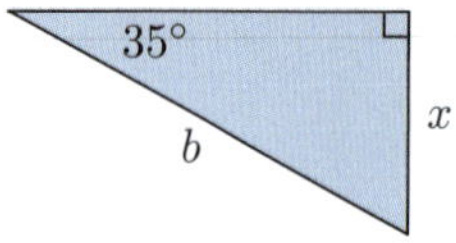

e

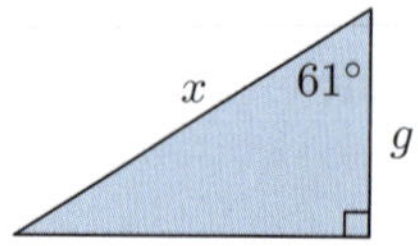

f

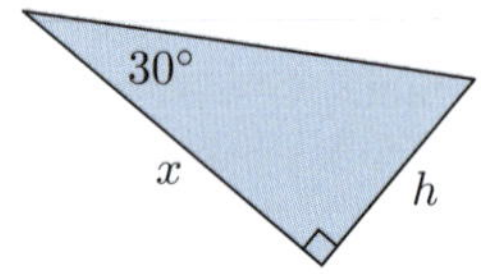

2 Find x, rounding your answer to 2 decimal places:

a

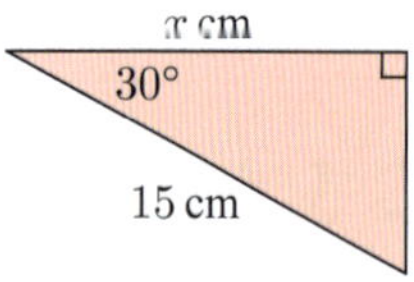

b

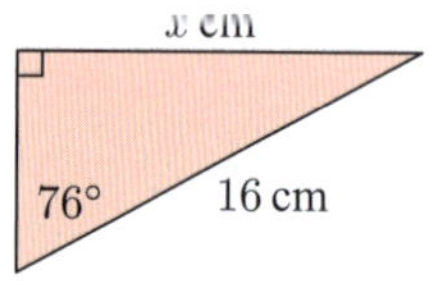

c

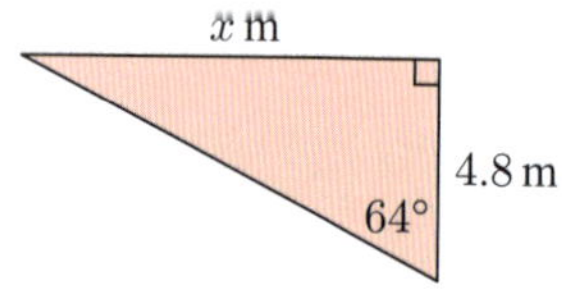

d

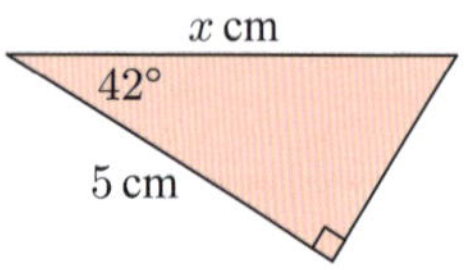

e

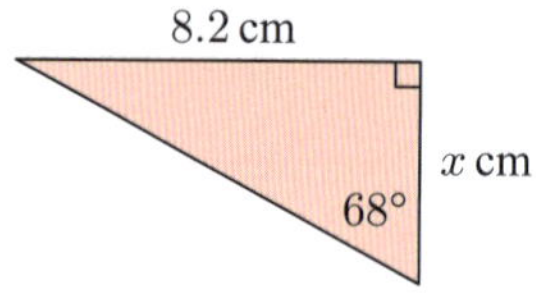

f

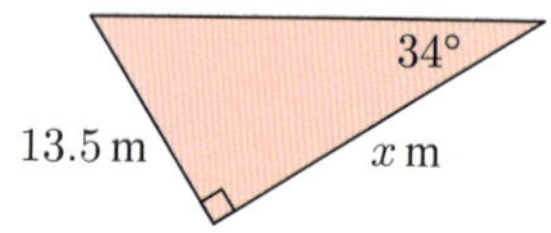

g

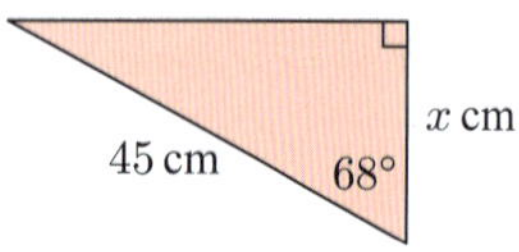

h

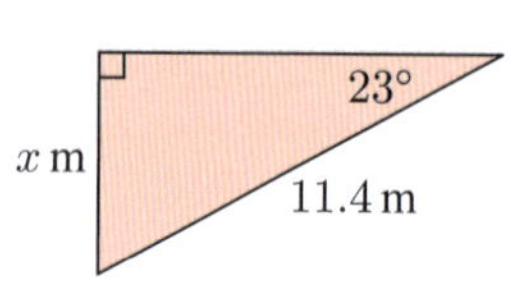

i

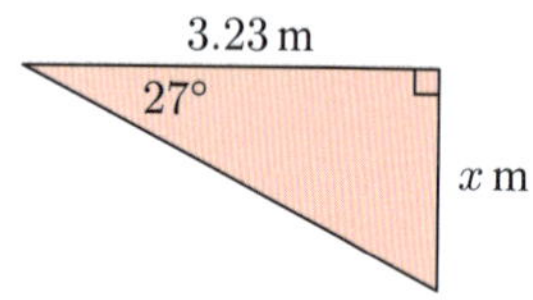

j

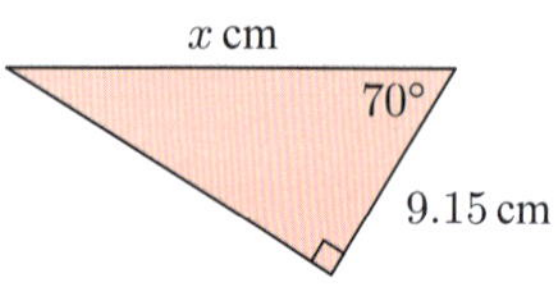

k

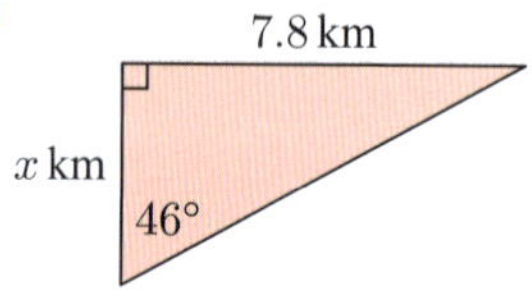

l

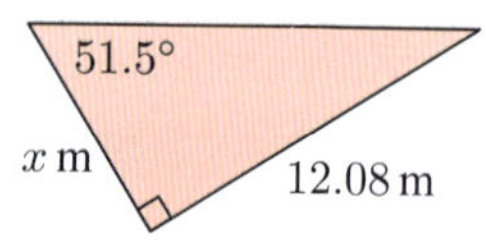

3 Find *all* the unknown angles and sides of:

a

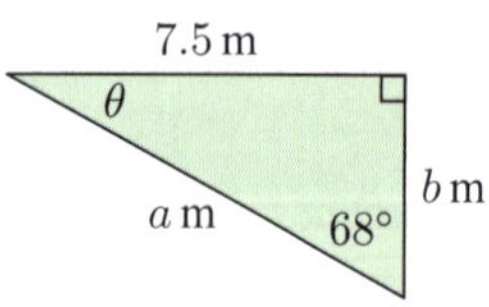

b

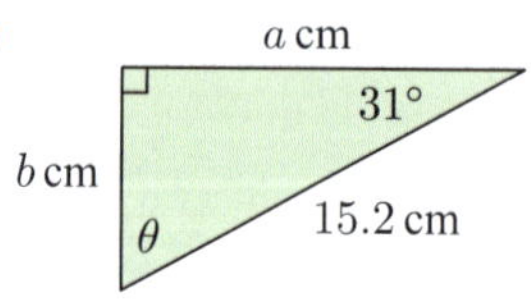

c

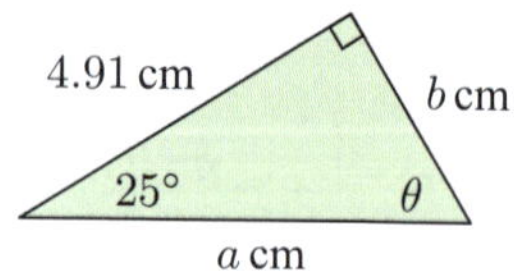

4 Find the area of the triangle:

a

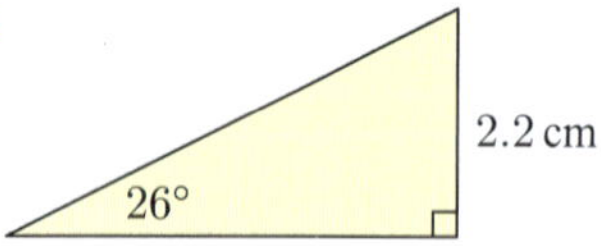

b

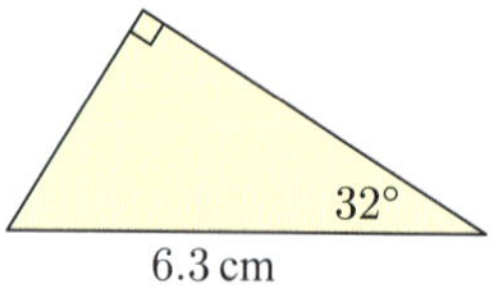

5 **a** Find the length PQ.

b Find the size of $\widehat{PYX}$.

c Hence show that $\tan 75° = 2 + \sqrt{3}$.

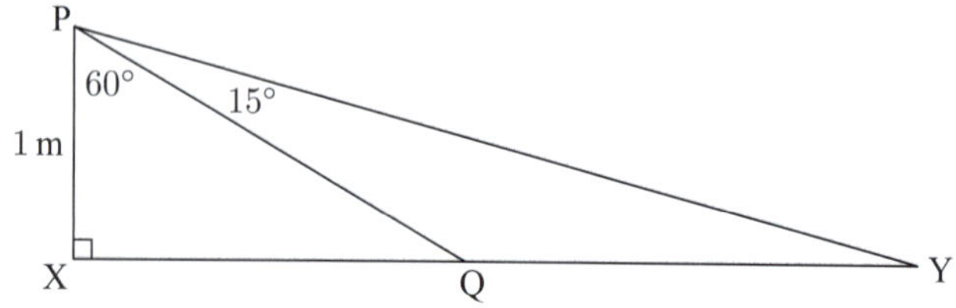

HISTORICAL NOTE

In the 5th century AD, the Hindu mathematician **Aryabhata** constructed a table of the lengths of half-chords of a circle with radius 1 unit.

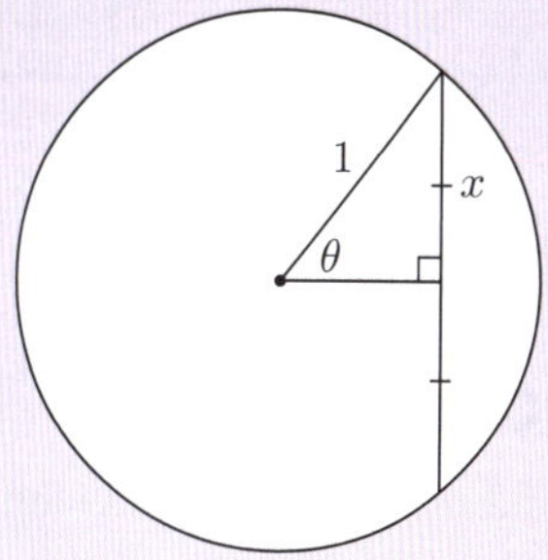

Since $\sin\theta = x$, this was actually the first known table of sine values.

In the late 16th century, **Georg Joachim de Porris**, also known as **Rheticus**, produced comprehensive and remarkably accurate tables of all the trigonometric ratios. This involved a tremendous number of tedious calculations, all without the aid of calculators or computers.

Rheticus helped his tutor **Nicolaus Copernicus** to publish his work *De revolutionibus orbium coelestium*, which means "On the Revolutions of the Heavenly Spheres".

D FINDING ANGLES

If we know two side lengths of a right angled triangle, we can use trigonometry to find the acute angles.

In the right angled triangle shown, $\sin\theta = \frac{3}{5}$.

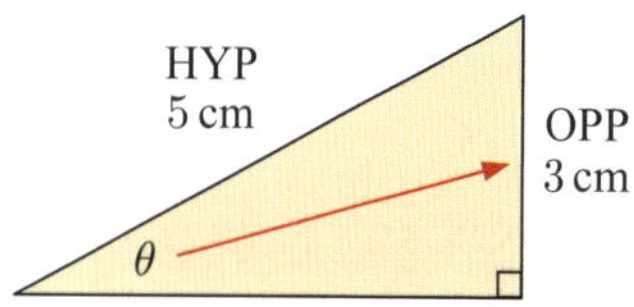

To find θ, we need an angle whose sine is $\frac{3}{5}$.

We say that θ is the **inverse sine** of $\frac{3}{5}$, and write $\theta = \sin^{-1}(\frac{3}{5})$.

We can use a calculator to evaluate inverse sines. Click on the icon for instructions.

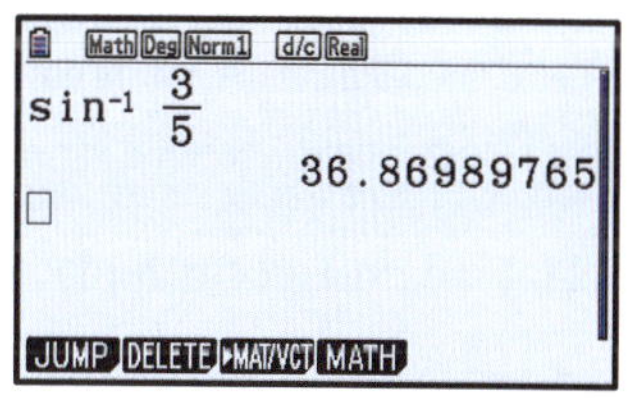

For the right angled triangle above, we find that $\theta \approx 36.9°$.

We define **inverse cosine** and **inverse tangent** in a similar way.

Example 3 — Self Tutor

Find the measure of the angle marked θ:

a

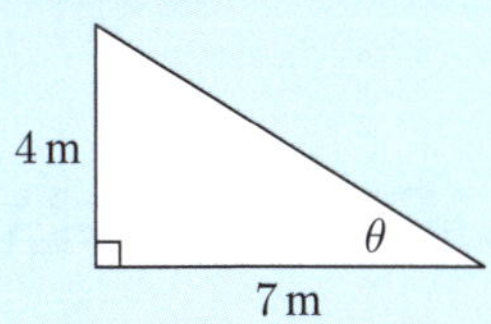

b

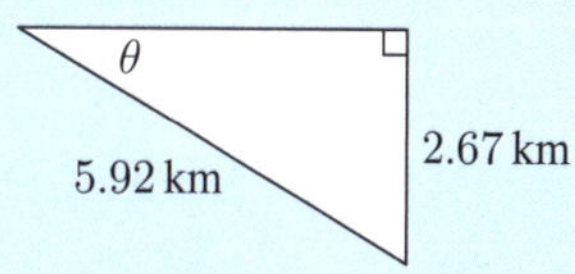

a

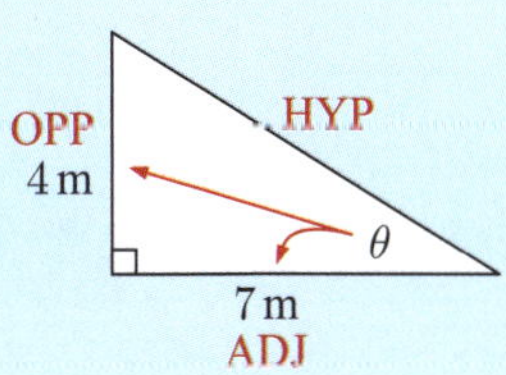

$$\tan\theta = \frac{\text{OPP}}{\text{ADJ}} = \frac{4}{7}$$

$$\therefore\ \theta = \tan^{-1}(\tfrac{4}{7})$$

$$\therefore\ \theta \approx 29.7°$$

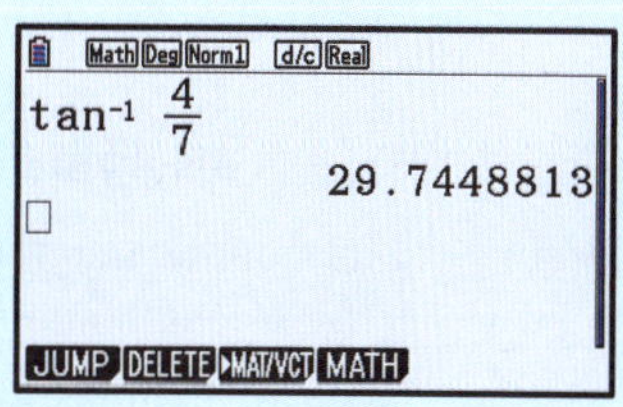

b

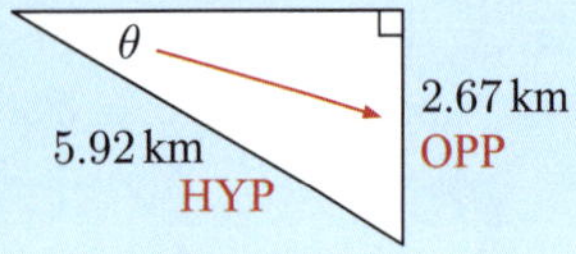

$$\sin\theta = \frac{\text{OPP}}{\text{HYP}} = \frac{2.67}{5.92}$$

$$\therefore\ \theta = \sin^{-1}\left(\frac{2.67}{5.92}\right)$$

$$\therefore\ \theta \approx 26.8°$$

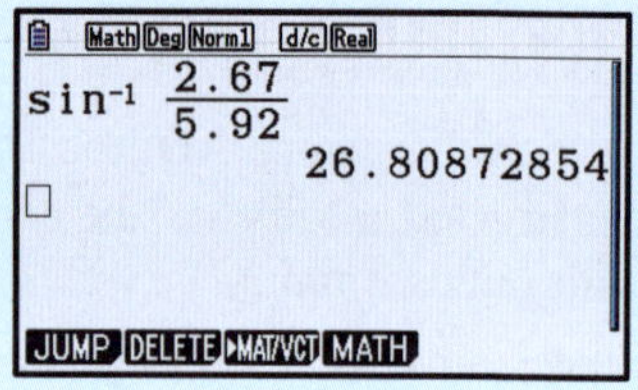

EXERCISE 17D

1 Find the measure of the angle marked θ:

a

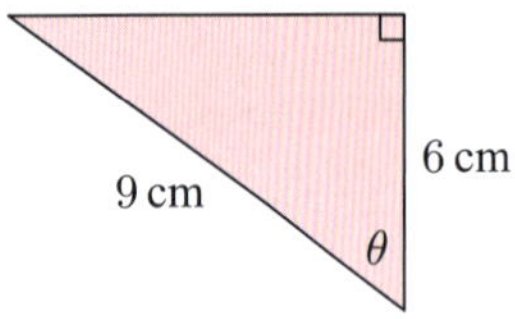

b

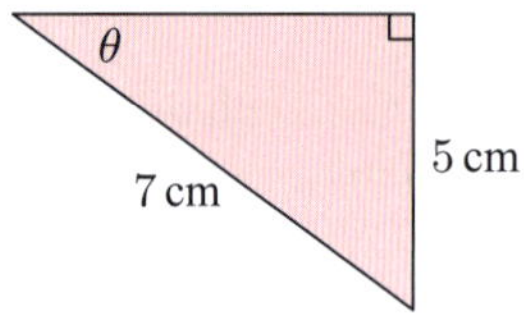

c

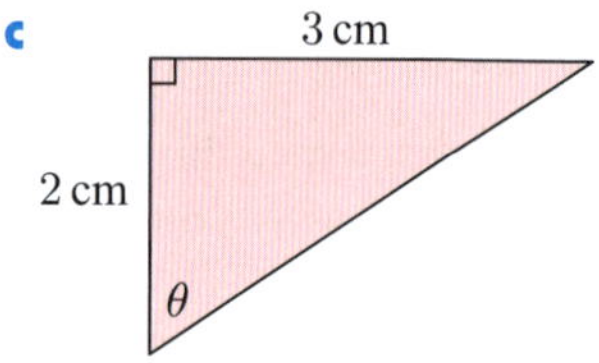

d

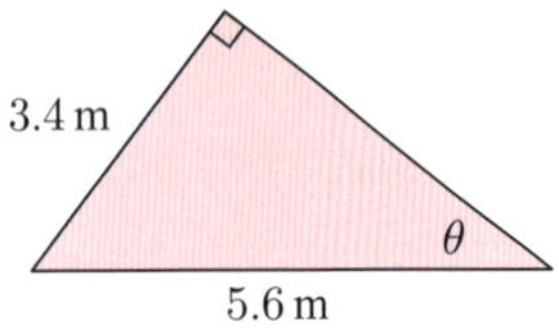

e

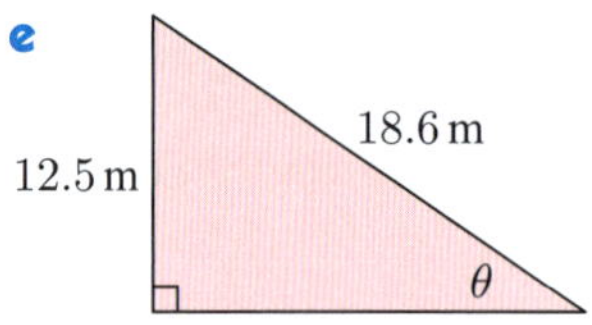

f

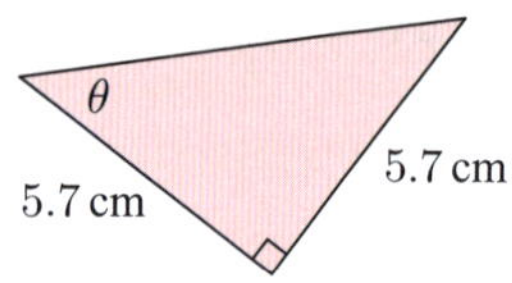

g

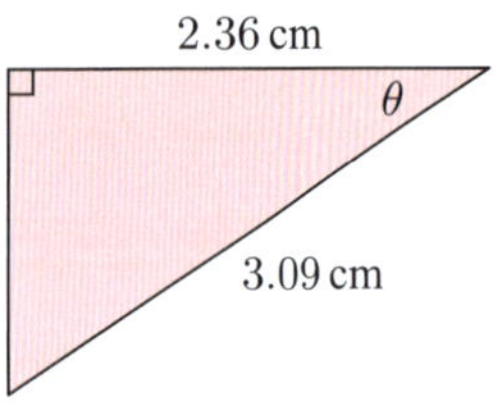

h

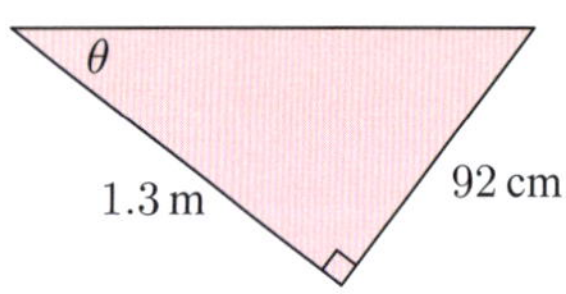

i

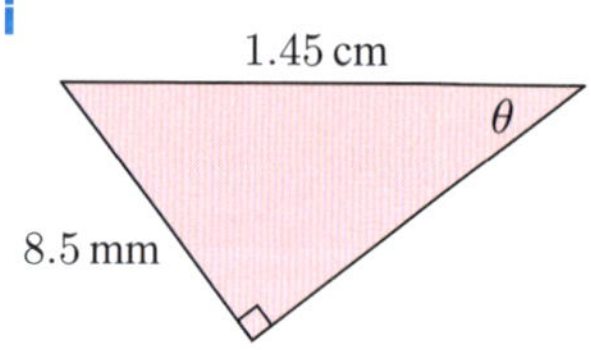

2 Find, to 1 decimal place, *all* the unknown sides and angles:

a

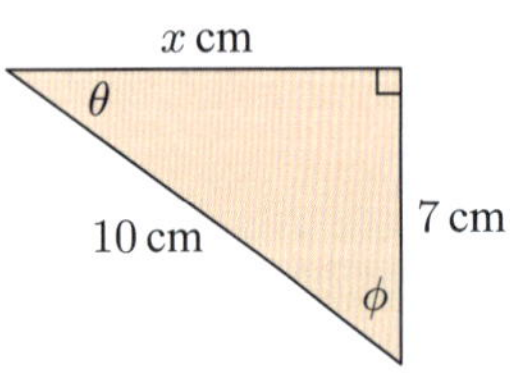

b

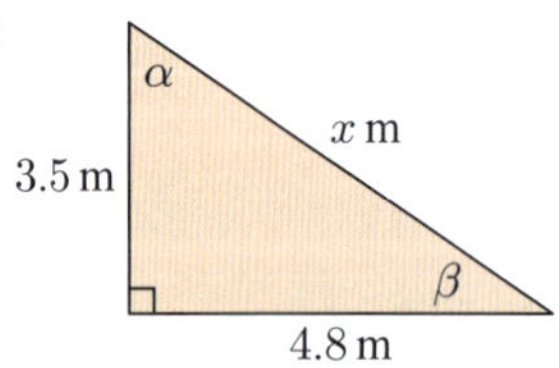

c

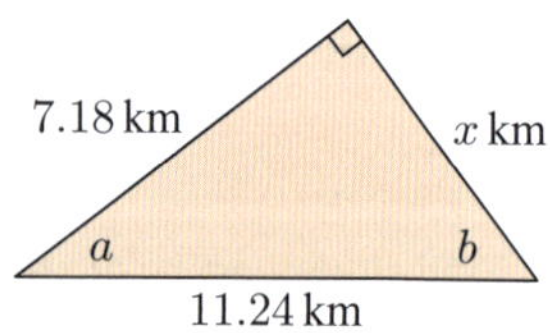

E PROBLEM SOLVING

The trigonometric ratios can be used to solve a wide variety of problems involving right angled triangles. When solving these problems it is important to follow the steps below:

Step 1: Draw a **diagram** to illustrate the situation.

Step 2: Mark on the diagram the **unknown** angle or side that needs to be calculated. We often use x for a length and θ for an angle.

Step 3: Identify a **right angled triangle** in your diagram.

Step 4: Write an **equation** connecting an angle and two sides of the triangle using an appropriate trigonometric ratio.

Step 5: **Solve** the equation to find the unknown.

Step 6: **Write** your answer in a sentence.

Example 4 — Self Tutor

The roof alongside has a pitch of 16°. Find the length of the horizontal beam.

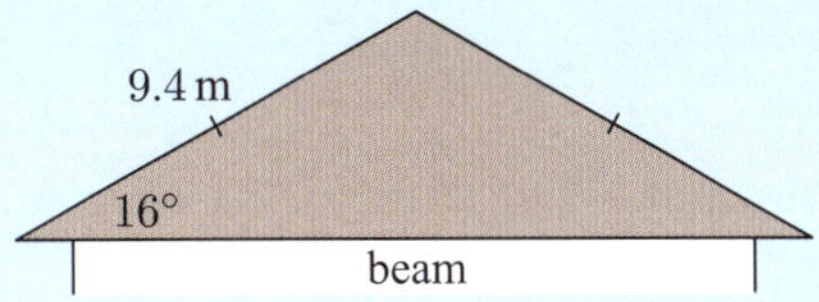

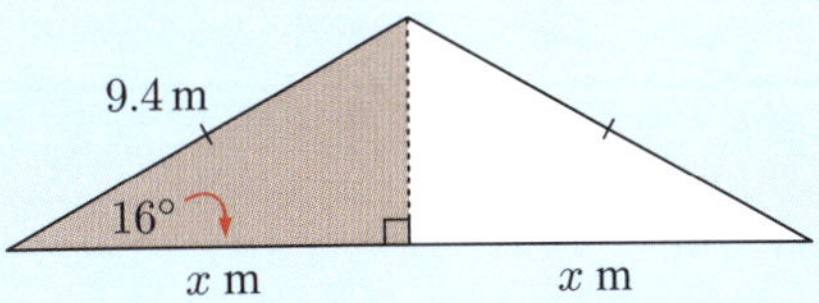

$$\cos\theta = \frac{\text{ADJ}}{\text{HYP}}$$

$$\therefore \cos 16° = \frac{x}{9.4}$$

$$\therefore x = 9.4 \times \cos 16°$$

$$\therefore x \approx 9.036$$

$$\therefore 2x \approx 18.1$$

The beam is about 18.1 m long.

ANGLES OF ELEVATION AND DEPRESSION

When an object is **higher** than an observer, the **angle of elevation** is the angle from the horizontal **up** to the object.

When an object is **lower** than an observer, the **angle of depression** is the angle from the horizontal **down** to the object.

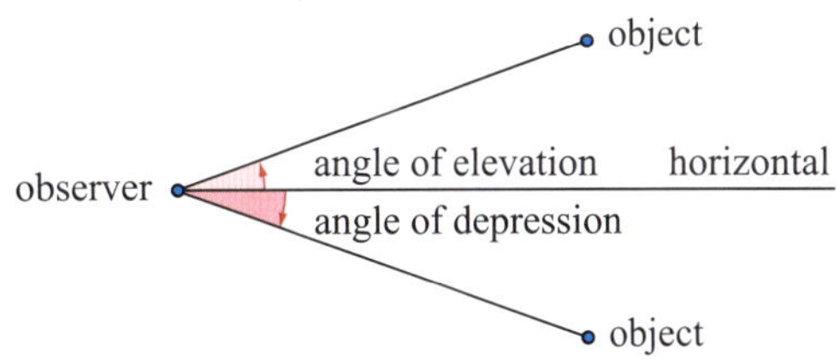

If the angle of elevation from A to B is θ, then the angle of depression from B to A is also θ.

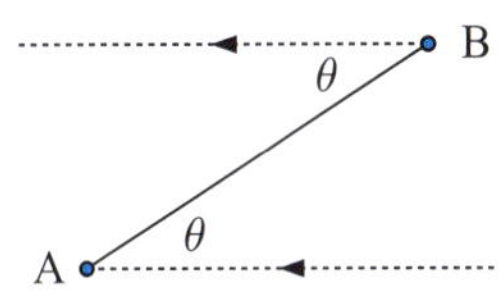

EXERCISE 17E

1 From a point 235 m from the base of a cliff, the angle of elevation to the cliff top is 25°. Find the height of the cliff.

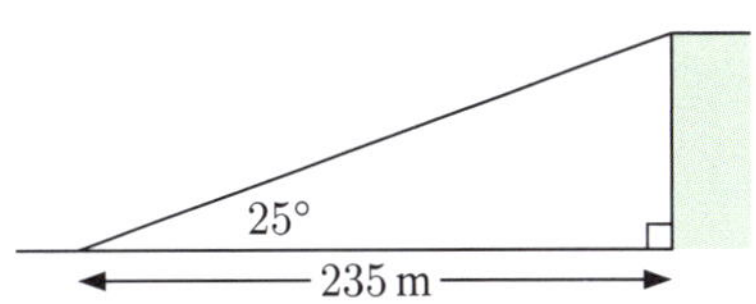

2 A 5 m ladder reaches 4.2 m up a wall. What angle does the ladder make with the wall?

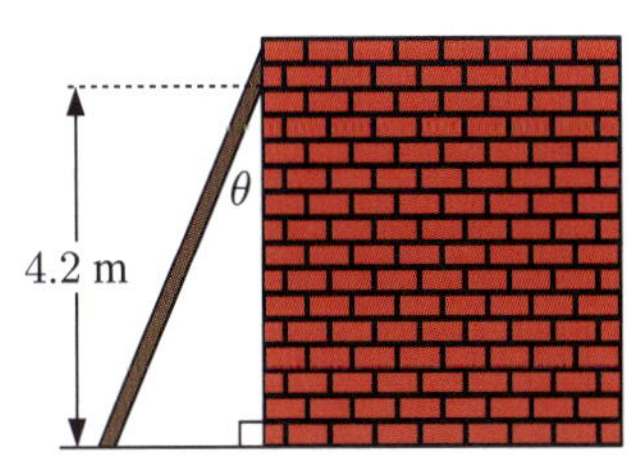

3 The angle of elevation from a row boat to the top of a lighthouse 25 m above sea level is $6°$. Calculate the horizontal distance from the boat to the lighthouse.

4 From a lookout 80 m above sea level, a whale is observed at an angle of depression of $6°$.
Find the distance between the observer at the lookout, and the whale.

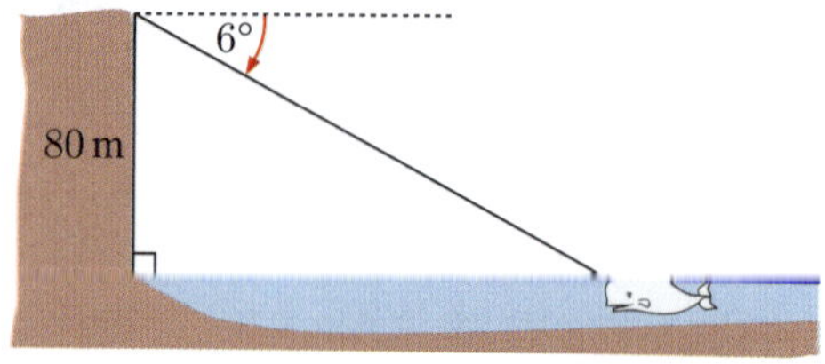

5 A train travelling up an incline of $4°$ travels a horizontal distance of 4 km. How much altitude has the train gained?

6 At the entrance to a building there is a ramp for wheelchair access. The length of the ramp is 5 metres, and it rises 60 cm. Find the angle θ that the ramp makes with the ground.

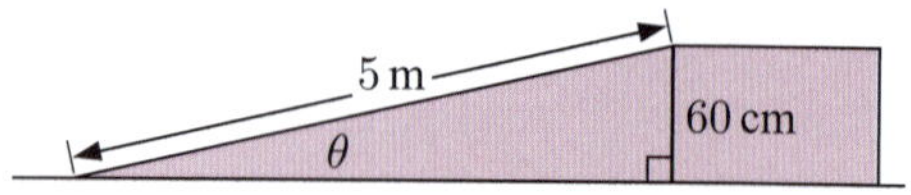

7 A model helicopter takes off from the horizontal ground with a constant vertical speed of 5 m/s. After 10 seconds the angle of elevation from Sam to the helicopter is $62°$. Given that Sam is 1.8 m tall, how far is Sam's head from the helicopter at this time?

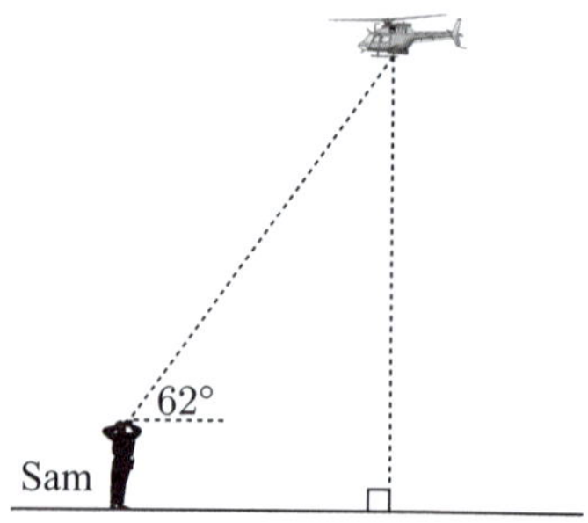

8 A post was hit by lightning and snapped in two. The top of the post is now resting 15 m from its base, at an angle of $25°$. Find the height of the post before it snapped.

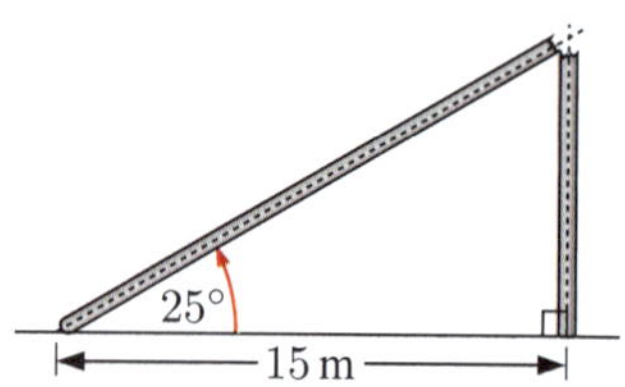

9 A rectangle has length 6 m and width 4 m. Find the acute angle formed where the diagonals intersect.

10 A rhombus has sides of length 10 cm, and the angle between two adjacent sides is $76°$. Find the length of the longer diagonal of the rhombus.

11

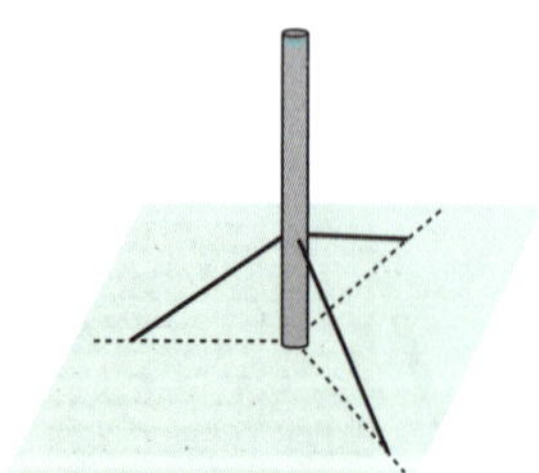

Three strong cables are used to brace a 12 m tall pole at a height 3.4 m from the ground.
Each cable makes an angle of $40°$ with the ground.

a Find the total length of the cables.

b Find the angle of elevation from the point where a cable is pinned to the ground, to the top of the pole.

Example 5 **Self Tutor**

A point P is 9.5 cm from the centre C of a circle with radius 2 cm.
Find the angle between [PC] and the tangent to the circle from P.

$\sin\theta = \dfrac{\text{OPP}}{\text{HYP}}$

$\therefore\ \sin\theta = \dfrac{2}{9.5}$

$\therefore\ \theta = \sin^{-1}\left(\dfrac{2}{9.5}\right)$

$\therefore\ \theta \approx 12.2°$

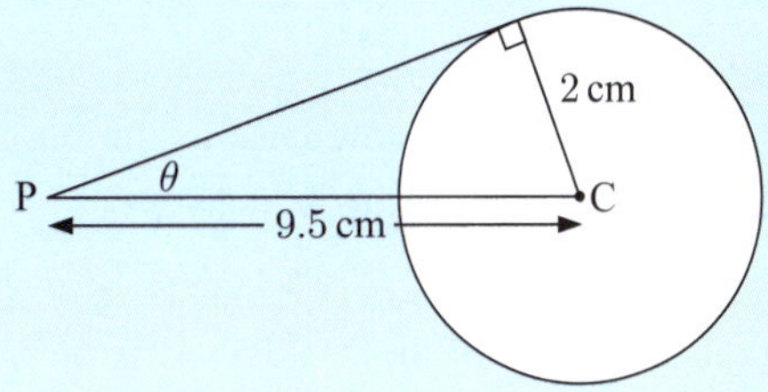

The angle between [PC] and the tangent is approximately 12.2°.

12 A tangent from point P to a circle of radius 4 cm is 10 cm long. Find the angle between the tangent and the line joining P to the centre of the circle.

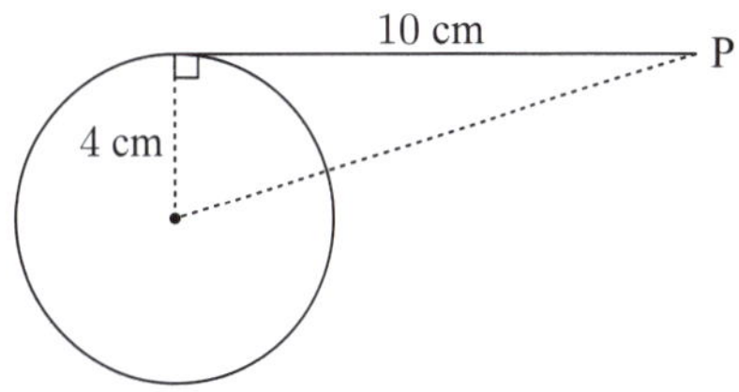

13 Find the values of the unknowns, giving reasons for your answers:

a

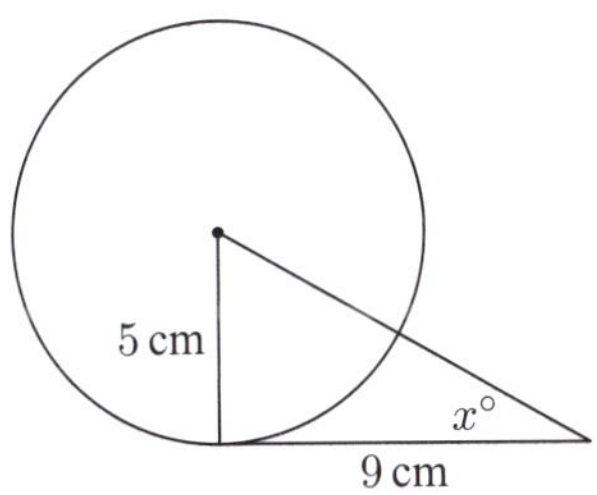

b

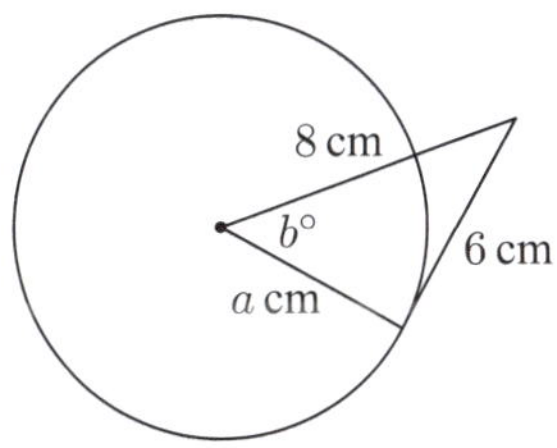

c

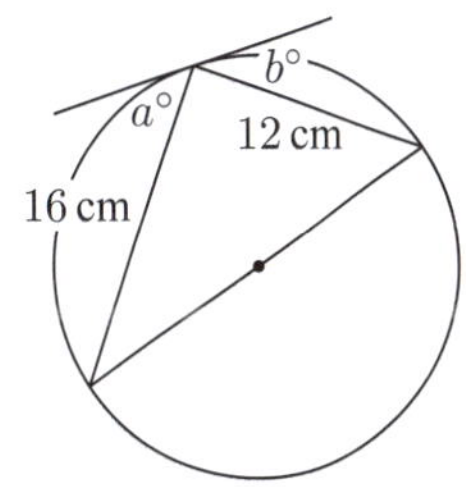

14 Point P is 32 mm from the centre C of a circle. The length of a tangent from P to the circle is 26 mm. Find:

a the radius of the circle

b the angle between the tangent and [CP].

15 [AB] is a chord of a circle with centre O and radius 5 cm. If $AB = 8$ cm, find $A\widehat{O}B$.

16 Find the area of the parallelogram:

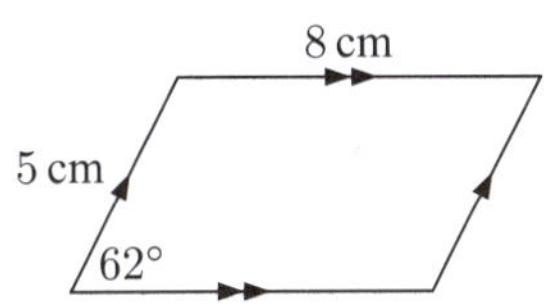

17 In an isosceles triangle, the equal sides are $\frac{2}{3}$ of the length of the base. Determine the measure of the base angles.

18 For the circle given, find:

a the radius of the circle

b the distance between A and B.

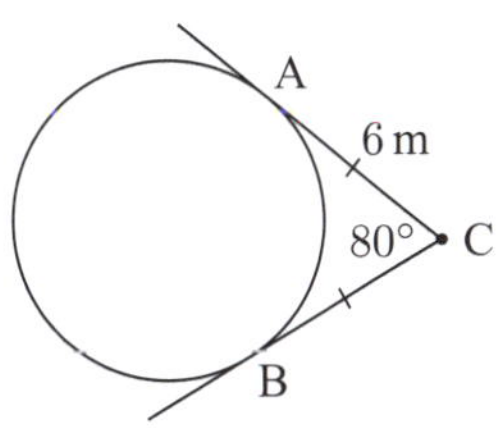

19 In triangle ABC, the angle bisector at A meets [BC] at X.
Is X the midpoint of [BC]? If not, what is the distance between X and the midpoint?

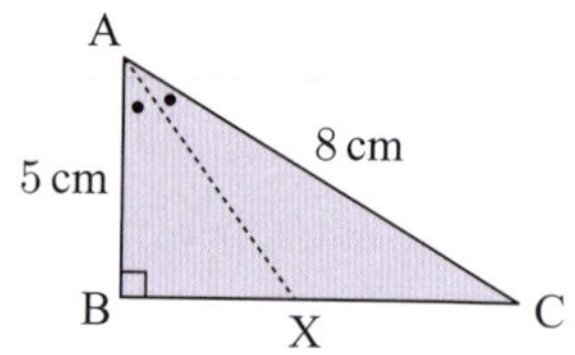

20 An isosceles triangle is drawn with base angles 24° and base 28 cm. Find the base angles of the isosceles triangle with the same base length but with treble the area.

21 Answer the **Opening Problem** on page **274**.

22 From a point A which is 30 m from the base of a building B, the angle of elevation to the top of the building C is 56°, and to the top of the flagpole [CD] is 60°.
Find the height of the flagpole.

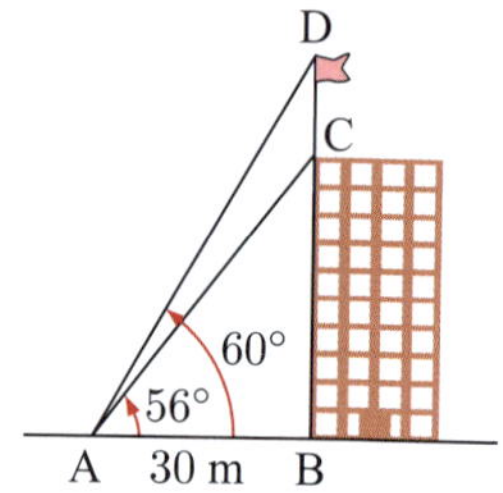

23 The angle of elevation from a marker on level ground to the top of a building 100 m high is 22°.

a Find the distance from the marker to the base of the building.

b How far must the marker be moved towards the building to make the angle of elevation 40°?

24 An observer notices an aeroplane flying directly overhead. Two minutes later the aeroplane is at an angle of elevation of 27°. Assuming the aeroplane is travelling with constant speed and altitude, what will be its angle of elevation after another two minutes?

Example 6 **Self Tutor**

A cube has sides of length 10 cm.

Find the angle between the diagonal [AB] and the edge [BC].

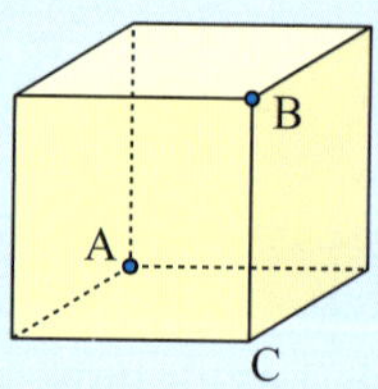

Let AC = x cm.

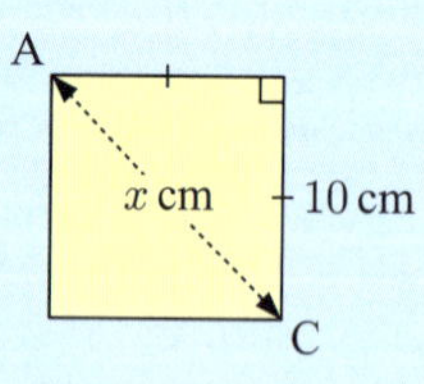

Using Pythagoras, $x^2 = 10^2 + 10^2$

$\therefore\ x^2 = 200$

$\therefore\ x = \sqrt{200}$

Let $\hat{ABC} = \theta$.

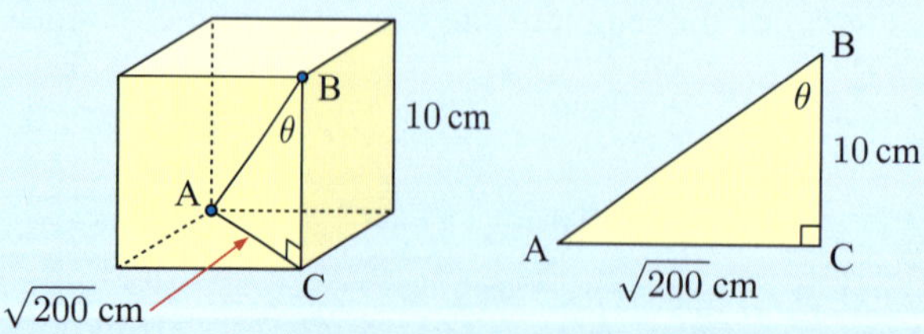

$\tan\theta = \frac{\text{OPP}}{\text{ADJ}} = \frac{\sqrt{200}}{10}$

$\therefore\ \theta = \tan^{-1}\left(\frac{\sqrt{200}}{10}\right)$

$\therefore\ \theta \approx 54.7°$

The angle between the diagonal [AB] and the edge [BC] is about 54.7°.

25 The figure alongside is a cube with sides of length 15 cm. Find:

a EG **b** $A\widehat{G}E$

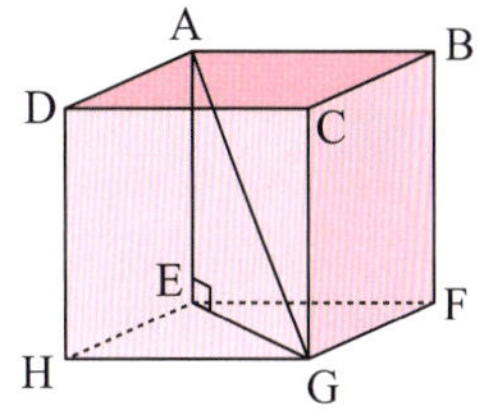

26 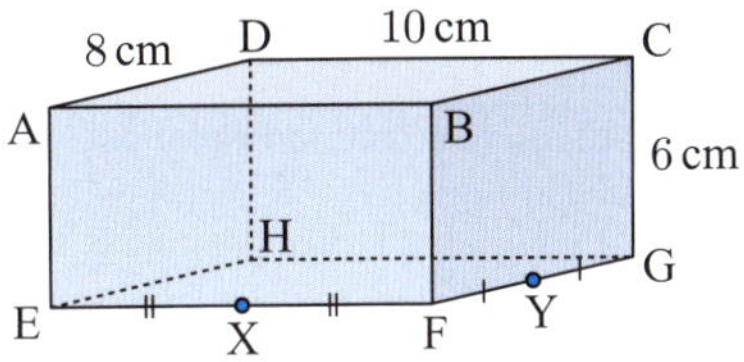

The figure alongside is a rectangular prism. X and Y are the midpoints of [EF] and [FG] respectively. Find:

a HX **b** $D\widehat{X}H$

c HY **d** $D\widehat{Y}H$

27 In the triangular prism alongside, find:

a DF **b** $A\widehat{F}D$

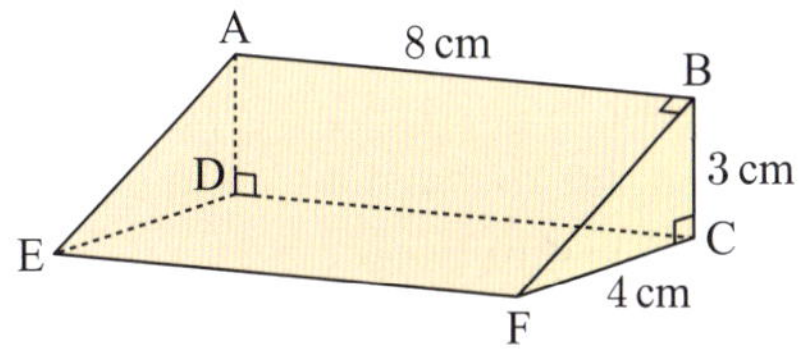

28 Each side of a tent is fixed to the ground by ropes as shown. The peg is 1.5 m from the side of the tent. Find the angle θ between the ropes.

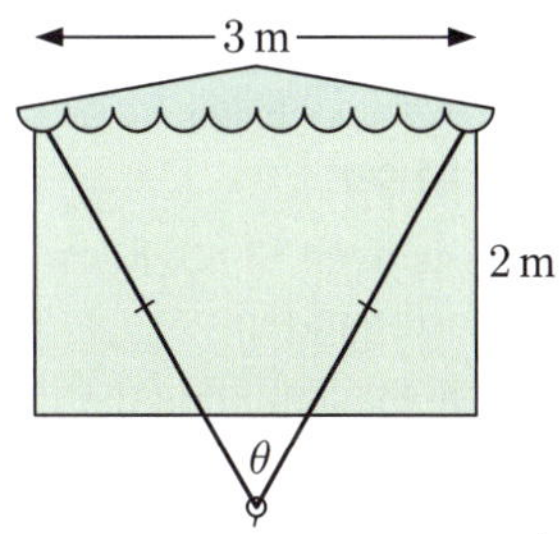

INVESTIGATION 2 THE ANGLE BETWEEN A LINE AND A PLANE

Suppose Jonathon holds his pen at an angle to a piece of paper. If we shine a light from directly above the pen, a shadow will be cast on the paper. This shadow is called the **projection** of the pen onto the paper.

The angle between the pen and the *plane* of the paper is the angle between the pen and its projection.

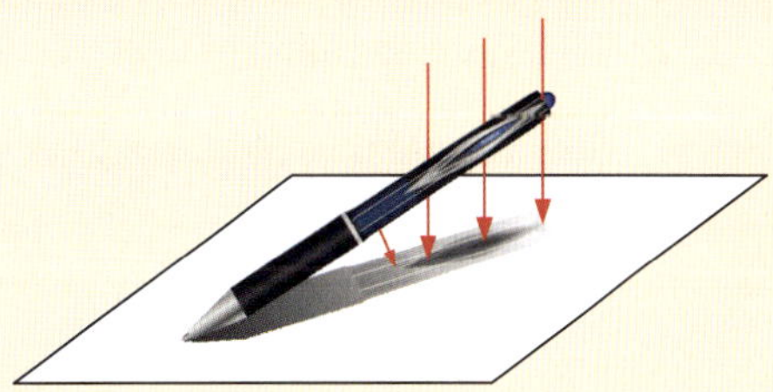

What to do:

1 Jonathon's pen is 16 cm long, and its projection onto the paper is 12 cm long. Find the angle that Jonathon's pen makes with the paper.

DEMO

2 **a** Name the projection of each of the following if a light is shone from directly above the figure:

i [BE] **ii** [CF] **iii** [BH]

b Find the angle that each of these line segments makes with the base plane EFGH:

i [DG] **ii** [AH] **iii** [CE]

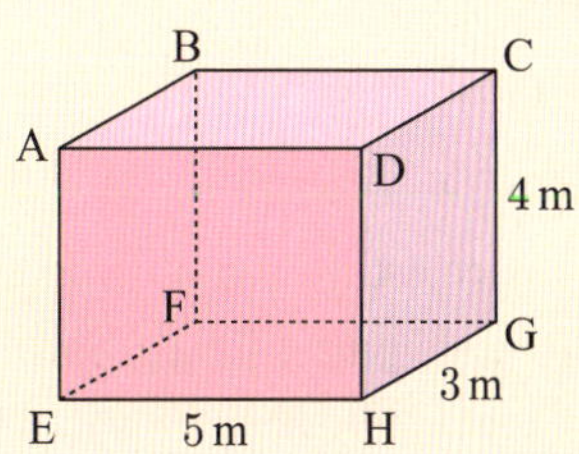

3 A utility pole is 10 m long, but is not quite vertical to the ground. When the Sun is directly above it, the pole casts a 50 cm long shadow on the ground.

   a What angle does the pole make with the ground?

   b How many degrees is the pole from vertical?

INVESTIGATION 3 — HIPPARCHUS AND THE UNIVERSE

Hipparchus of Nicaea was a Greek astronomer and mathematician in the 2nd century BC. He is considered among the greatest astronomers of all time.

Part 1: How Hipparchus measured the distance to the moon

Suppose A and B are two points on the Earth's equator, and that C is the centre of the Earth.

Suppose the moon M is directly above A. From B the moon is just visible, since [MB] is a tangent to the Earth and is therefore perpendicular to [BC].

$\widehat{BCM}$ is the difference in longitude between A and B, which Hipparchus calculated to be about $89°$.

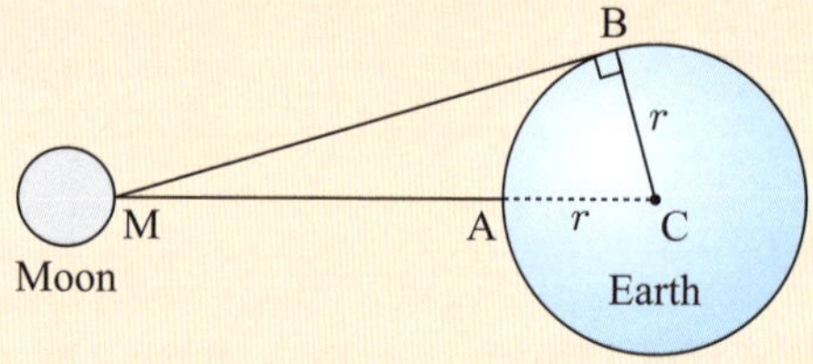

What to do:

1 Given $r \approx 6378$ km and $\widehat{BCM} \approx 89°$, estimate the distance from the centre of the Earth C to the moon.

2 Hence calculate the distance AM *between* the Earth and the moon.

3 In calculating just one distance between the Earth and the moon, Hipparchus was assuming that the orbit of the moon is circular. In fact it is not.
Research the shortest and longest distances to the moon.

- How were these distances determined?
- How do they compare with the distance obtained using Hipparchus' method?

Part 2: How Hipparchus measured the radius of the moon

From point A on the Earth's surface, Hipparchus measured the angle between an imaginary line to the centre of the moon and a tangent to the moon to be about $0.25°$.

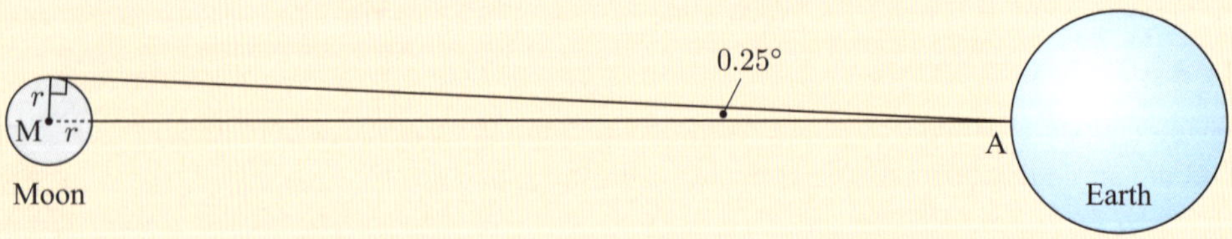

The average distance from the Earth to the moon is about 384 403 km.

What to do:

1 Use the diagram to show that $\sin 0.25° = \dfrac{r}{r + 384\,403}$.

2 Hence estimate r, the radius of the moon.

3 Research the actual radius of the moon, and if possible, find out how it was calculated. Compare your answer in **2** to the actual radius.

F TRUE BEARINGS

We can describe a direction by comparing it with the **true north direction**. We call this a **true bearing**.

Imagine you are standing at point A, facing north. If you turn clockwise through an angle until you face B, the bearing of B from A is the angle through which you have turned.

> The **bearing of B from A** is the clockwise measure of the angle between the "north" line through A, and [AB].

In the diagram alongside, the bearing of B from A is 72° from true north. We write this as 072°.

To find the bearing of A from B, we place ourselves at point B, face north, then turn clockwise until we face A. The true bearing of A from B is 252°.

Note that:

- A true bearing is always written using three digits. For example, we write 072° rather than 72°.
- The bearing of A from B always differs from the bearing of B from A by 180°.

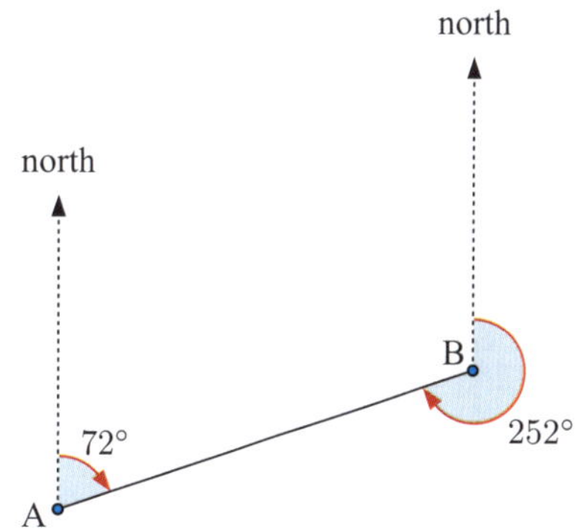

EXERCISE 17F

1 Draw diagrams to represent bearings from O of:

a 136° **b** 240° **c** 051° **d** 327°

2 Find the bearing of Q from P if the bearing of P from Q is:

a 054° **b** 113° **c** 263° **d** 304°

3 Write down the bearing of Q from P:

a
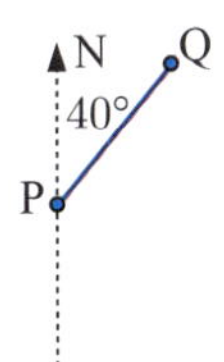

b
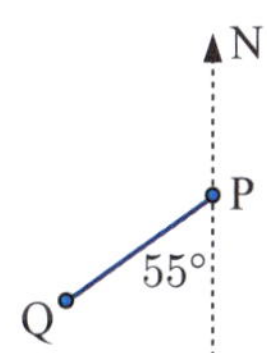

c
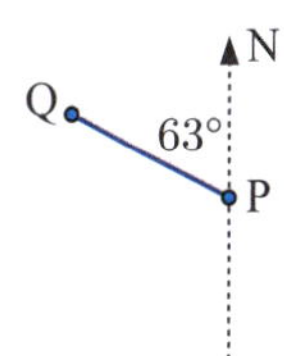

d
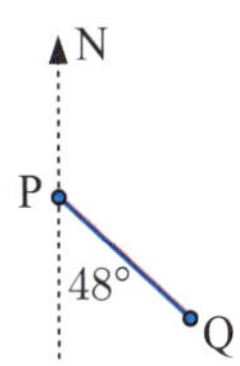

e
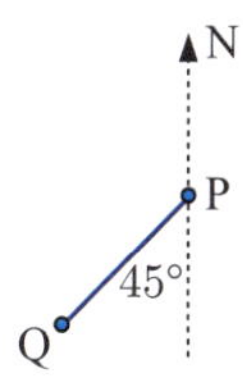

f
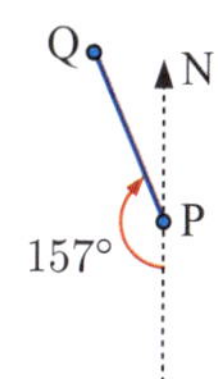

4 Use angle pair properties for parallel lines cut by a transversal to explain why the bearing of A from B always differs from the bearing of B from A by 180°.

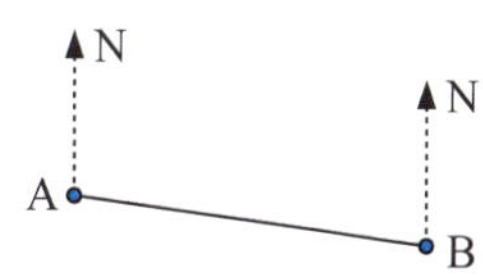

5 A, B, and C are the checkpoints of a triangular orienteering course. Find the bearing of:

a B from A **b** C from B **c** A from C
d B from C **e** C from A **f** A from B.

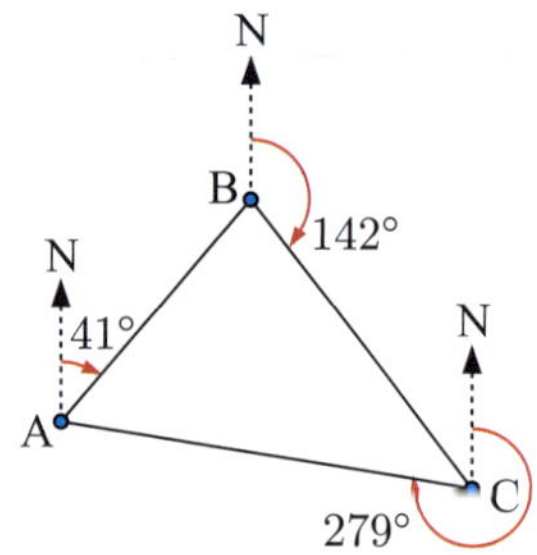

6 Find the bearing of:

a B from A **b** A from B
c C from A **d** A from C
e C from B **f** B from C.

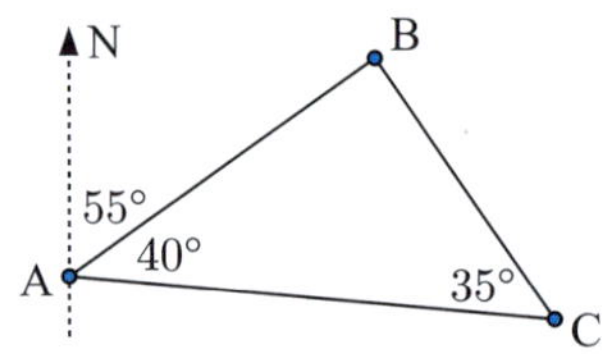

Example 7 **Self Tutor**

A motorbike rider travels to a point 7 km east and 3 km north of his starting position.
Find the bearing back to the starting position.

$\tan\theta = \frac{\text{OPP}}{\text{ADJ}} = \frac{7}{3}$

$\therefore \quad \theta = \tan^{-1}\left(\frac{7}{3}\right)$

$\therefore \quad \theta \approx 66.8°$

The bearing back to the starting position is $180° + 66.8° \approx 247°$.

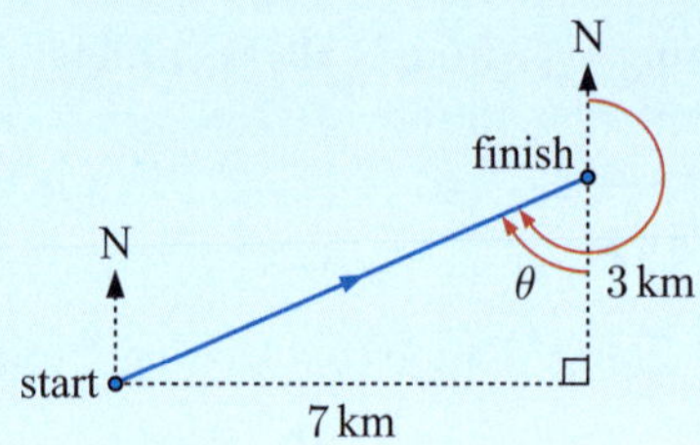

7 A bushwalker hikes to a campsite 5.4 km east and 2.8 km north of her starting point. Find the bearing of:

a the campsite from the starting point **b** the starting point from the campsite.

8 A courier delivers a parcel to a house 10.2 km north and 13.8 km west of its depot.

a How far is the house from the depot?

b On what bearing is the house from the depot?

9 A kayaker paddles to a point 3.1 km south and 1.6 km west of his starting point.

a How far is he from his starting point?

b On what bearing must he paddle to return directly to his starting point?

Example 8 **Self Tutor**

A rally driver travels on the bearing 145° for 2.85 km.
How far east of his starting position is the rally driver?

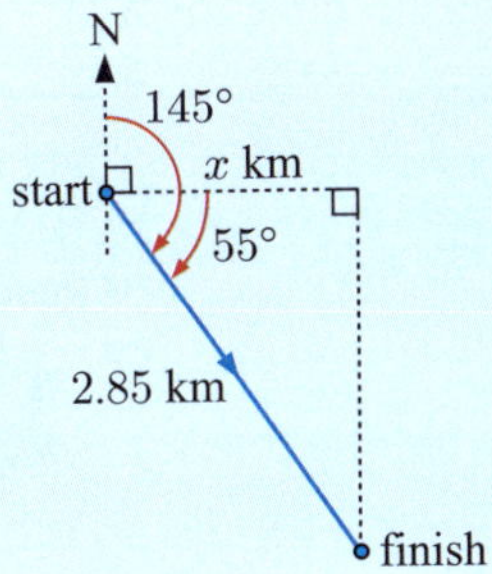

$$\cos\theta = \frac{\text{ADJ}}{\text{HYP}}$$

$$\therefore \cos 55° = \frac{x}{2.85}$$

$$\therefore x = 2.85 \times \cos 55°$$

$$\therefore x \approx 1.63$$

The driver is about 1.63 km east of his starting position.

10 A ship sails for 60 km on the bearing 040°. How far north of its starting point is the ship?

11 An athlete ran for 12 minutes in the direction 164° at 14 km/h.

a Find the distance travelled by the athlete.

b How far east of his starting point is the athlete?

12 According to her GPS, a runner is on the bearing 215° from her starting point, and she is 2 km south of it. How far is she from her starting point?

Example 9 — Self Tutor

A small aeroplane departs A and flies on a 143° bearing for 368 km to B. Having landed and delivered cargo, it departs on a 233° course and flies a further 472 km to C. Find:

a the distance of C from A

b the bearing of C from A.

$A\widehat{B}N = 180° - 143° = 37°$ {co-interior angles}

$\therefore A\widehat{B}C = 360° - 37° - 233°$

$= 90°$ {angles at a point}

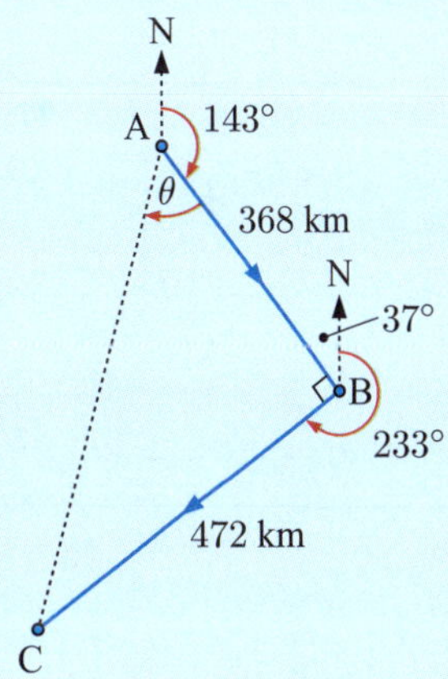

a $AC^2 = 368^2 + 472^2$ {Pythagoras}

$\therefore AC = \sqrt{368^2 + 472^2}$ {as $AC > 0$}

≈ 599

So, C is about 599 km from A.

b To find the bearing of C from A, we first need to find θ.

Now $\tan\theta = \frac{\text{OPP}}{\text{ADJ}} = \frac{472}{368}$

$\therefore \theta = \tan^{-1}\left(\frac{472}{368}\right)$

$\therefore \theta \approx 52.1°$

The bearing of C from A is $143° + 52.1° \approx 195°$.

13 A cyclist departs point R and rides on a straight road for 2.3 km in the direction 197°. She then turns and rides for 1.8 km in the direction 107° to point S.

a Find the distance between R and S.

b Find the bearing of S from R.

14 A fishing trawler sails from port P in the direction 313° for 10 km, then in the direction 043° for 32 km. Find the distance and bearing of the trawler from P.

15 An orienteer runs 2.16 km in the direction 024°, then 1.48 km in the direction 114°. Find the distance and bearing of the finishing point from the starting point.

QUICK QUIZ

MULTIPLE CHOICE QUIZ

REVIEW SET 17A

1 The side lengths in this figure have been rounded to the nearest millimetre.

a Use the diagram to estimate the value of:

i $\cos 42°$ **ii** $\sin 64°$

iii $\tan 48°$ **iv** $\sin 26°$

b Check your answers using a calculator.

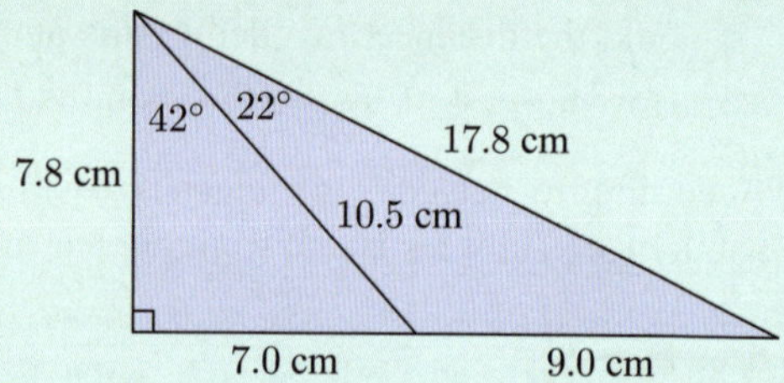

2 Find the value of x:

a

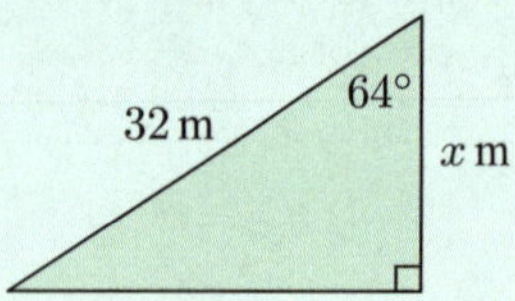

b

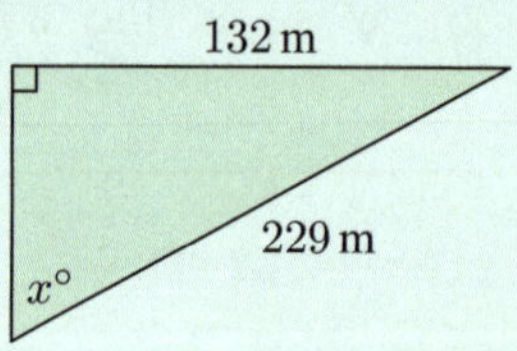

c

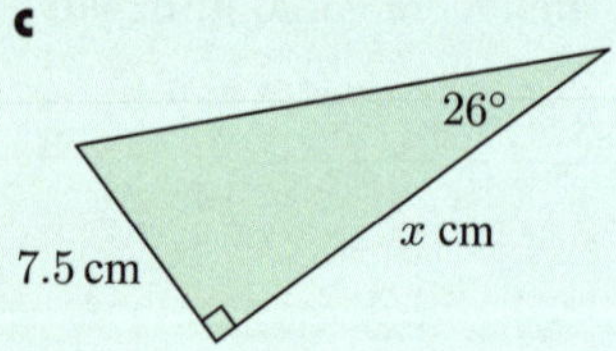

3

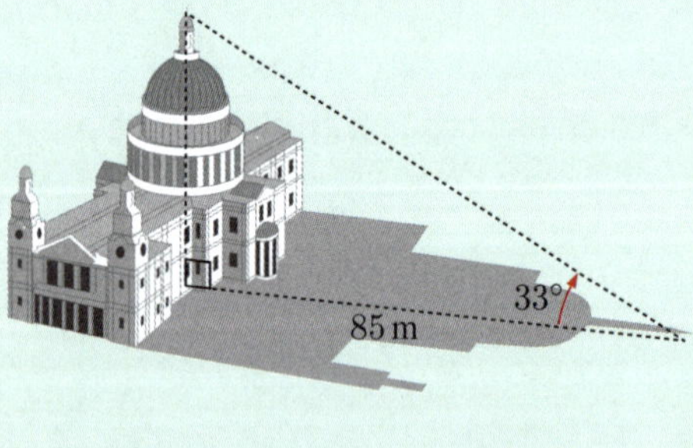

The shadow of a cathedral is 85 m in length. The angle of elevation from the end of the shadow to the top of the steeple is 33°. Find the height of the cathedral.

4 Find all unknown sides and angles of this triangle.

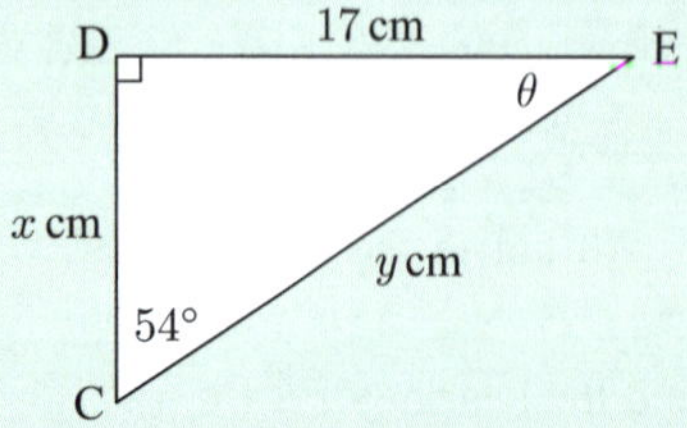

5 Point P is 13 cm from the centre O of a circle. The tangent from P to the circle is 11 cm long. Find the angle between the tangent and [OP].

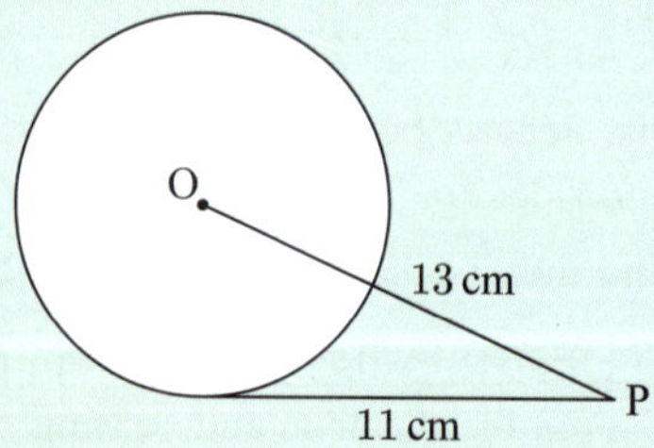

6 Write down the bearing of Q from P:

a

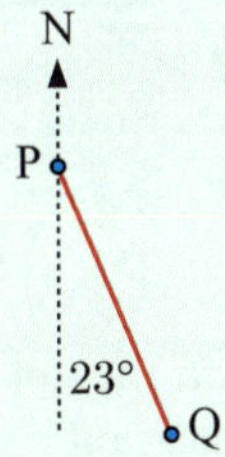

b

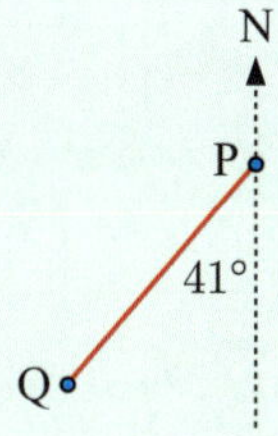

c

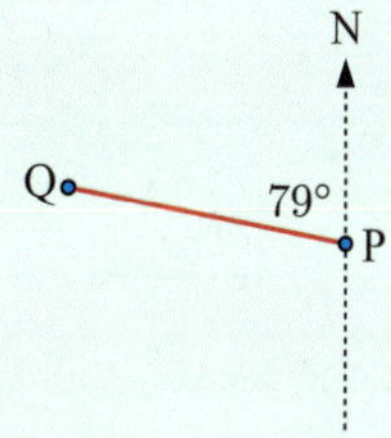

7 A taxi journey takes the passenger to a destination 8 km south and 3 km west of its starting point. Find the distance and bearing of the destination from its starting point.

8 Two aeroplanes leave from an airport at the same time. Aeroplane A flies in the direction 124° at 450 km/h. Aeroplane B flies in the direction 214° at 380 km/h. Find the distance and bearing of B from A after 30 minutes.

9 Find the measure of:

a $B\widehat{G}F$ **b** $A\widehat{G}E$

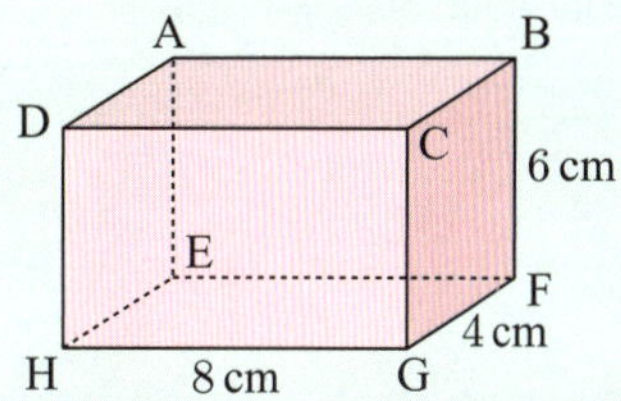

10 Laura is at the top of a scenic lookout. Her friends Ariel and Briannah are at ground level, 50 m apart.
The angle of depression from L to A is 36°, and the angle of depression from L to B is 29°.

a Find the angle of elevation from B to L.

b Find the height of the lookout.

c The girls' car is at C, at an angle of depression of 20° from L. How far is the car from Ariel?

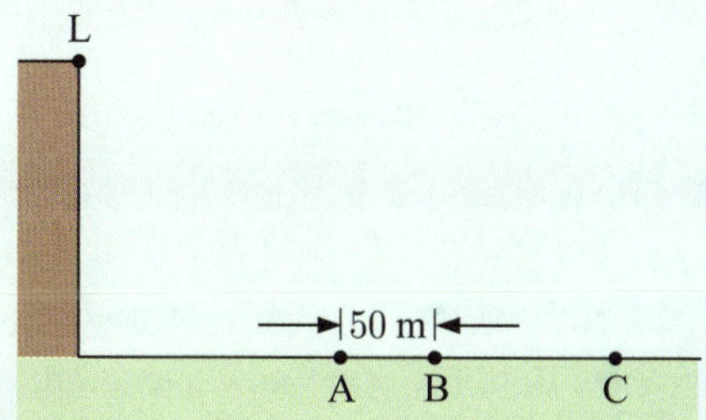

REVIEW SET 17B

1 **a** Find the measure of $A\widehat{B}C$ and $B\widehat{A}C$.

b Hence show that $\tan 67.5° = \sqrt{2} + 1$.

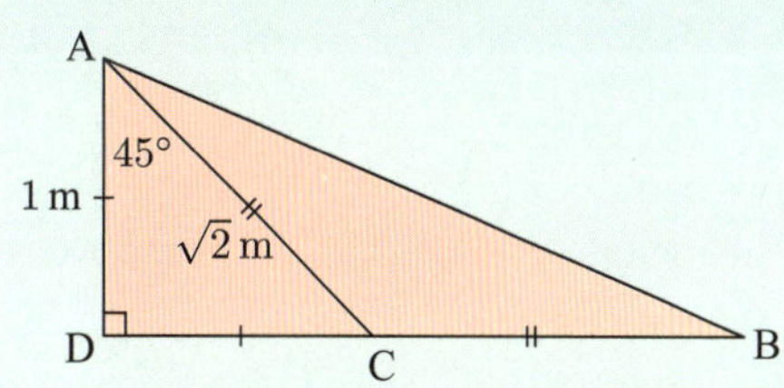

2 Find the measure of the angle marked θ:

a

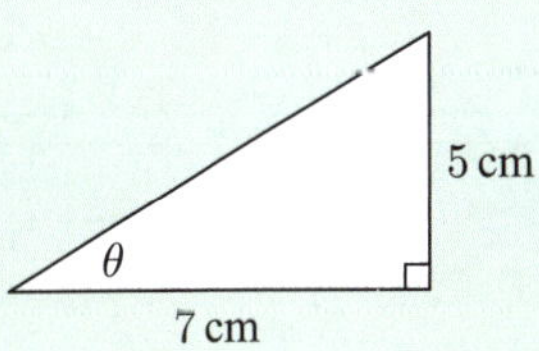

b

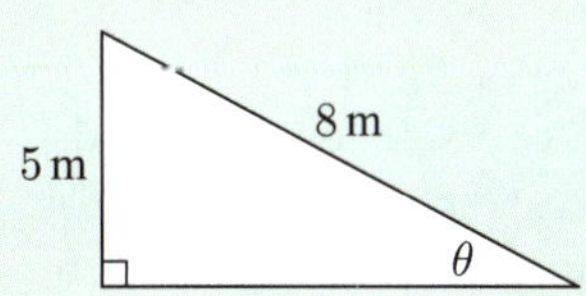

c

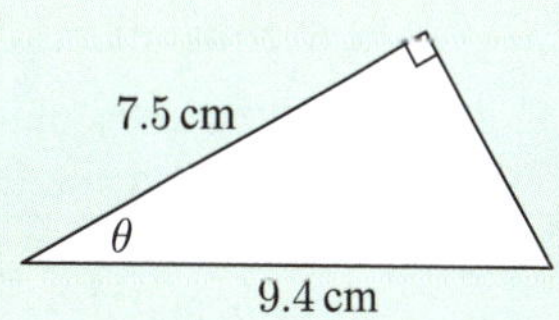

3 Find the value of x, rounding your answer to 2 decimal places:

a

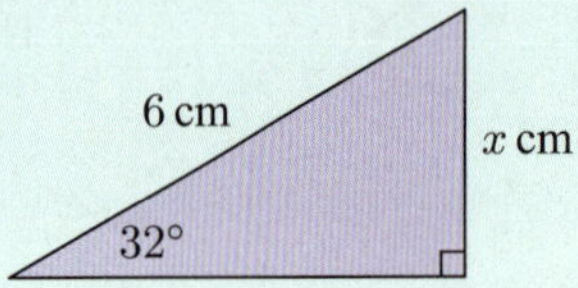

b

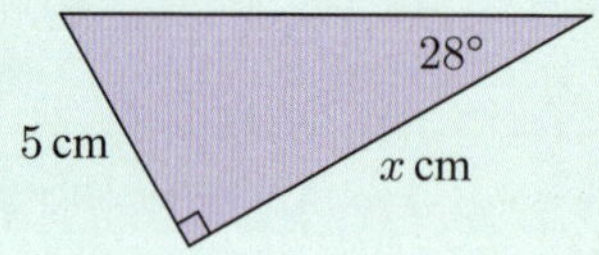

4 Find the area of this triangle.

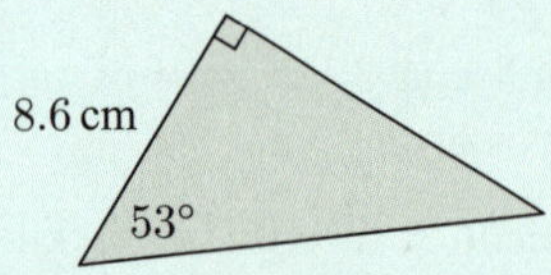

5 Find, to 1 decimal place, *all* the unknown sides and angles in this triangle.

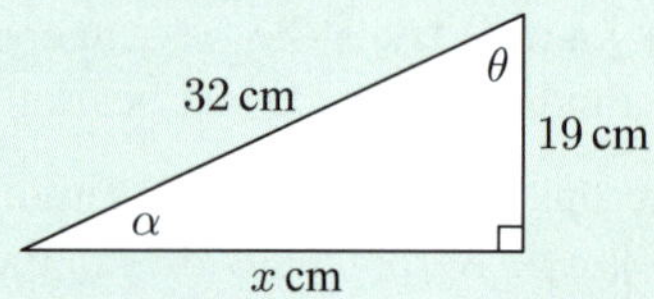

6 Find the diameter of this circle.

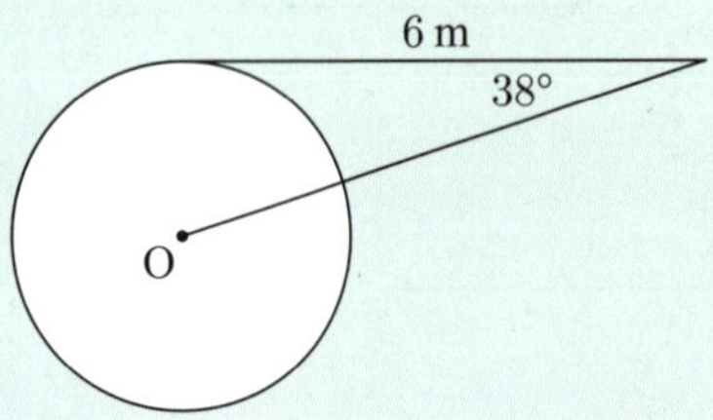

7 Find the bearing of:

a P from Q

b R from Q

c R from P.

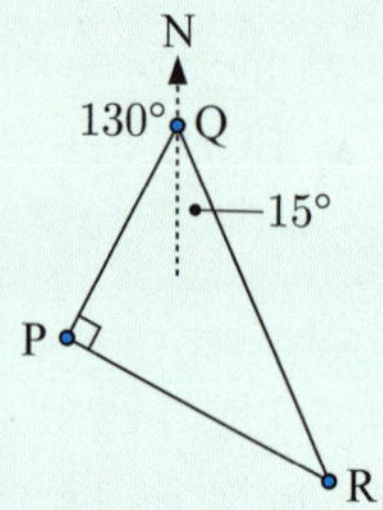

8 A ship sails 40 km on the bearing 156°. How far south is it from its starting point?

9 Boats A and B left port P at the same time. Boat A travelled on the bearing 112° until it was 2 km south of P. Boat B travelled on the bearing 202° until it was 5 km south of P.

a How far did each boat travel?

b Both boats are now anchored.

i Find the distance between the boats.

ii Find the bearing of boat B from boat A.

10 Find the measure of:

a $A\widehat{D}C$ **b** $A\widehat{D}M$

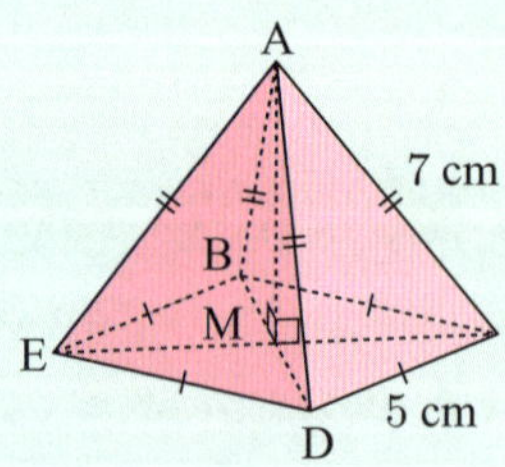

11 In triangle ABC, AB = 7 m, AC = 9 m, and $B\widehat{A}C = 39°$.

a Find the *exact* height h of the triangle.

b Hence find the area of the triangle, rounded to 2 decimal places.

c Suppose a triangle has two side lengths a and b, and an acute angle of size θ between them. Write an expression for the area of the triangle in terms of a, b, and θ.

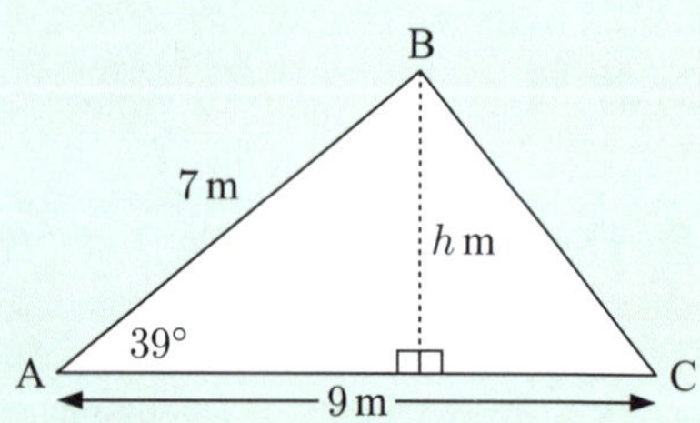

Chapter 18

Non-right angled triangle trigonometry

Contents:

OPENING PROBLEM

Andy is a surveyor who measures the exact lengths of property boundaries.

Andy normally uses a laser measuring device, but today he cannot directly measure the distance from A to B because a number of trees block his vision.

Instead, Andy measures the distance from A to the corner of a shed C to be 124 m. He measures angle ACB to be 130°, and the distance CB to be 105 m.

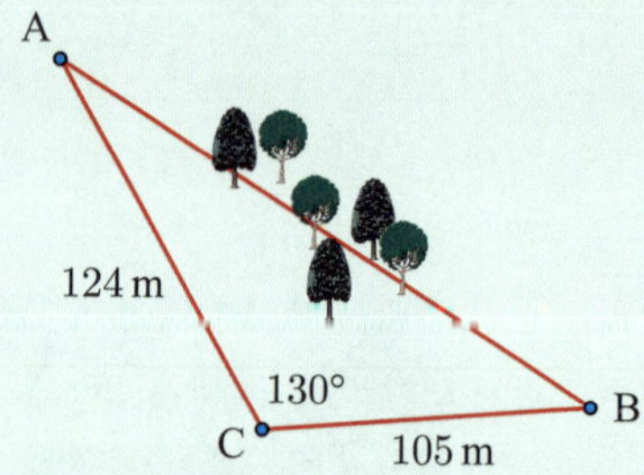

Things to think about:

a The angle ACB is obtuse. Does it make sense to talk about trigonometric ratios such as $\sin 130°$ and $\cos 130°$?

b How can we use trigonometry to find:

i the area of triangle ABC

ii the length of the boundary [AB]?

In our previous study of trigonometry for right angled triangles, we defined the trigonometric ratios sine, cosine, and tangent for acute angles.

If a triangle is not right angled, it *may* contain an obtuse angle. So, before we can study trigonometry for non-right angled triangles, we must first define the trigonometric ratios for obtuse angles.

A TRIGONOMETRY WITH OBTUSE ANGLES

Consider the circle of radius 1 unit with its centre at the origin O. This circle is called the **unit circle**.

Suppose P lies on the circle so that [OP] makes an angle θ with the positive x-axis. θ is always measured in the anticlockwise direction.

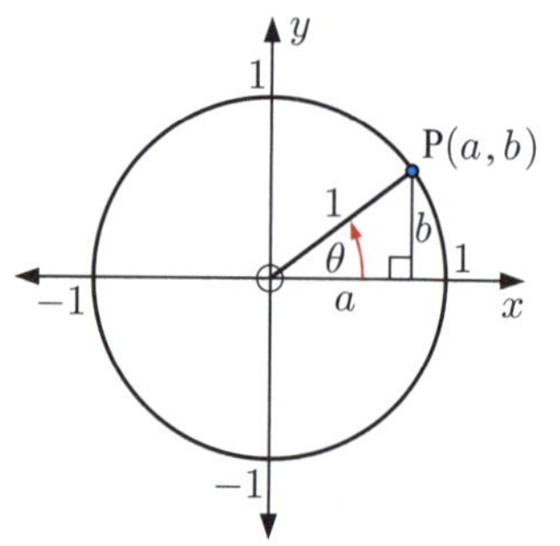

For any acute angle θ, notice that:

$$\cos\theta = \frac{\text{ADJ}}{\text{HYP}} = \frac{a}{1} = a \quad \text{and} \quad \sin\theta = \frac{\text{OPP}}{\text{HYP}} = \frac{b}{1} = b.$$

So, if P is in the first quadrant, then P has coordinates $(\cos\theta, \sin\theta)$.

If we use these coordinates to *define* sine and cosine, we can then use the same definitions to give meaning to the trigonometric ratios for non-acute angles.

If P is *any* point on the unit circle such that [OP] makes an angle θ measured anticlockwise from the positive x-axis:

- $\boldsymbol{\cos\theta}$ is the x-coordinate of P
- $\boldsymbol{\sin\theta}$ is the y-coordinate of P.

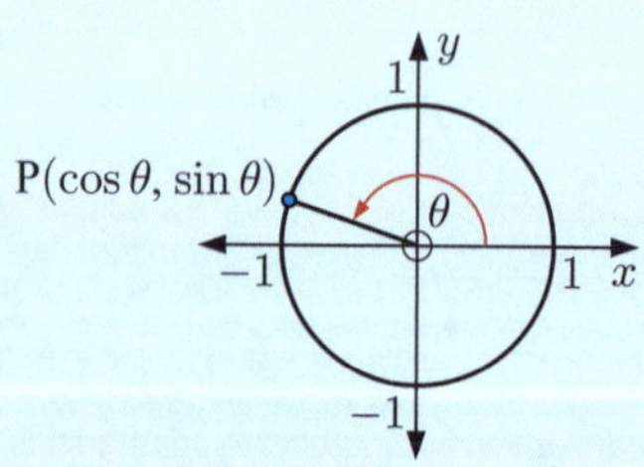

INVESTIGATION — SUPPLEMENTARY ANGLES

In this Investigation we consider the relationships between the trigonometric ratios of supplementary angles.

What to do:

1 Copy and complete the following table, giving your answers correct to 4 decimal places:

θ	$\cos\theta$	$\sin\theta$	$\cos(180° - \theta)$	$\sin(180° - \theta)$
18°				
53°				
80°				
125°				

2 By considering the results in your table, suggest a relationship between:

a $\cos\theta$ and $\cos(180° - \theta)$ **b** $\sin\theta$ and $\sin(180° - \theta)$.

3 Consider a point $P(\cos\theta, \sin\theta)$ in the first quadrant.
We place point P′ on the unit circle in the second quadrant such that [OP′] makes angle θ with the *negative* x-axis.

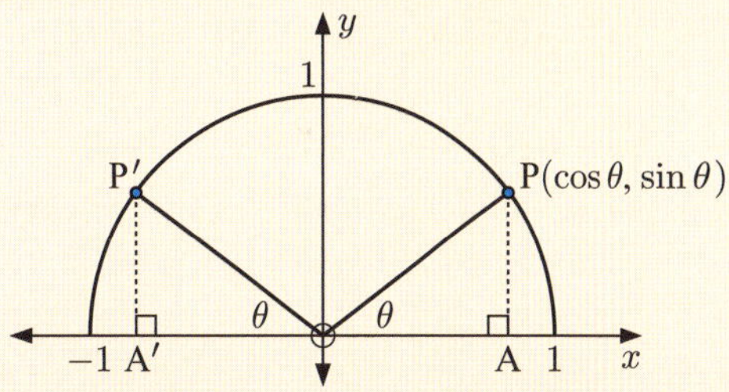

a Find the angle between [OP′] and the *positive* x-axis. *Hence* state the coordinates of P′.

b Suppose A and A′ are points on the x-axis such that $O\widehat{A}P$ and $O\widehat{A'}P'$ are right angles.

i Prove that triangles OAP and OA′P′ are congruent.

ii Hence find alternative expressions for the coordinates of P′.

c Prove your suggestions in **2**.

From the **Investigation**, you should have found that for any angle θ,

$$\cos(180° - \theta) = -\cos\theta$$
$$\sin(180° - \theta) = \sin\theta$$

EXERCISE 18A

1 **a** Write down the exact coordinates of:

i P **ii** Q **iii** R

b Use your calculator to find, to 3 significant figures, the coordinates of:

i P **ii** Q **iii** R

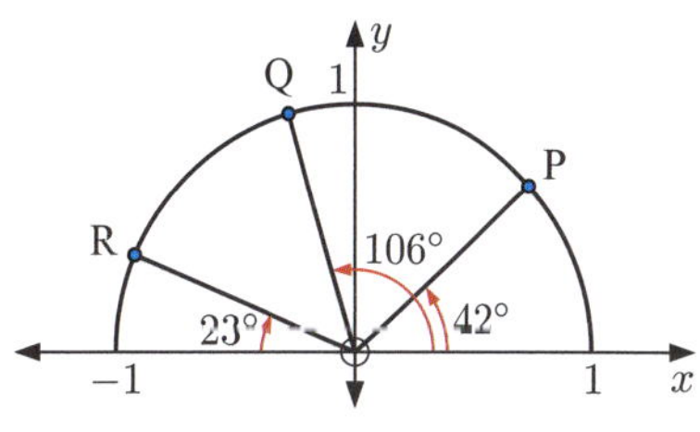

2 Find the obtuse angle which has the same sine as:

a 26° **b** 45° **c** 69° **d** 86°

3 Find the acute angle which has the same sine as:

a $98°$ **b** $127°$ **c** $156°$ **d** $168°$

4 Find the obtuse angle whose cosine is the negative of:

a $\cos 26°$ **b** $\cos 45°$ **c** $\cos 69°$ **d** $\cos 86°$

5 Find the acute angle whose cosine is the negative of:

a $\cos 98°$ **b** $\cos 127°$ **c** $\cos 156°$ **d** $\cos 168°$

Example 1 Self Tutor

Find an acute and an obtuse angle with sine 0.3.

$\sin^{-1}(0.3) \approx 17.46°$

The acute angle with sine 0.3
$\approx 17.5°$

The obtuse angle with sine 0.3
$\approx 180° - 17.46°$
$\approx 162.5°$

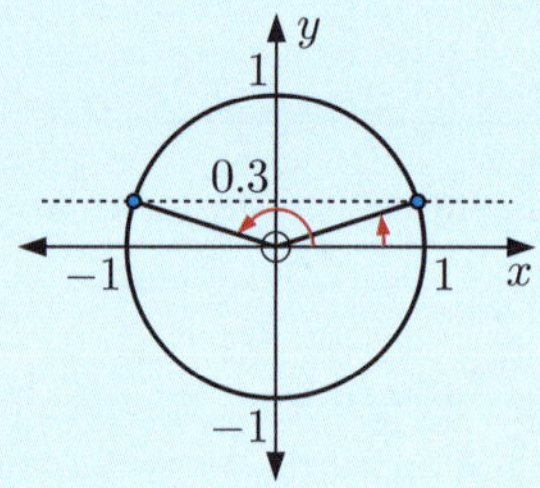

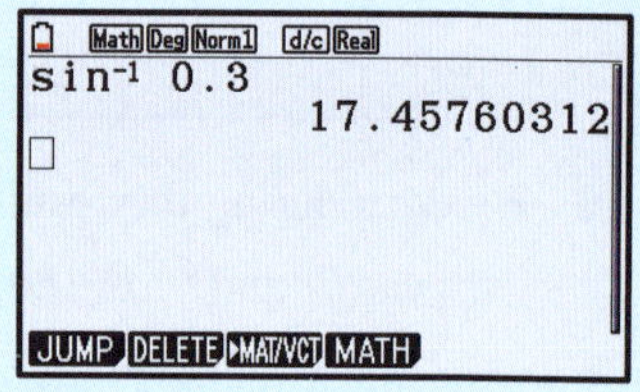

6 Find an acute and an obtuse angle with sine:

a $\frac{1}{2}$ **b** $\frac{\sqrt{3}}{2}$ **c** $\frac{1}{\sqrt{2}}$ **d** 0.1 **e** 0.4 **f** 0.7

B THE AREA OF A TRIANGLE

We can use trigonometry to find the area of a triangle if we are given the lengths of **two sides**, as well as the **included angle** between the sides.

Consider triangle ABC in which the sides opposite angles A, B, and C are labelled a, b, and c respectively.

Suppose N lies on [BC] such that [AN] is perpendicular to [BC].

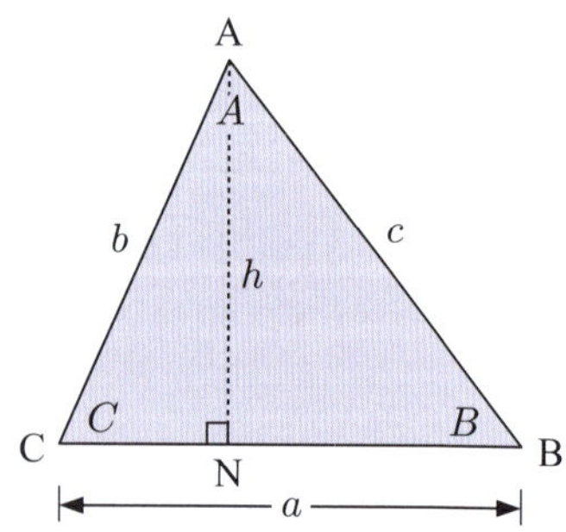

In $\triangle$ANC, $\sin C = \frac{h}{b}$

$\therefore \quad h = b \sin C$

Since the area of $\triangle ABC = \frac{1}{2}ah$, we find:

$$\textbf{Area} = \frac{1}{2}ab\sin C$$

If the altitudes from B and C were drawn, we could also show that the area is $\frac{1}{2}bc\sin A$ or $\frac{1}{2}ac\sin B$.

The **area of a triangle** is *half the product of two sides and the sine of the included angle.*

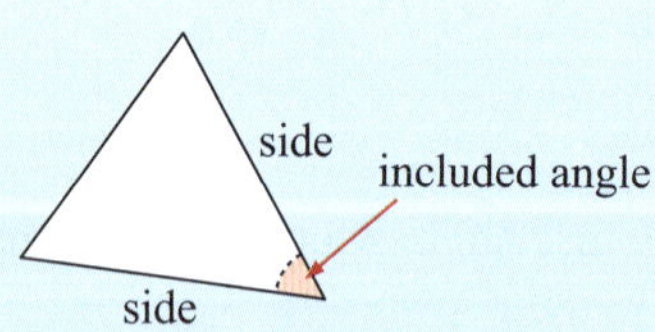

Example 2 — Self Tutor

Find the area of triangle ABC.

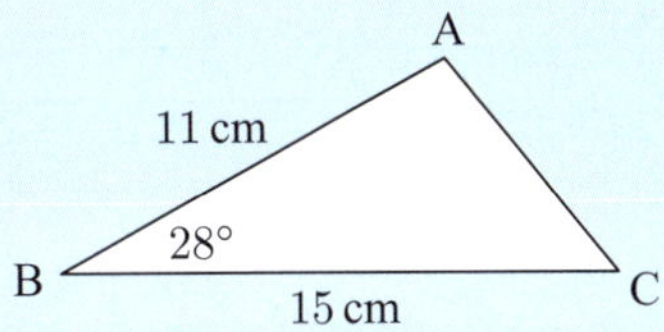

$$\text{Area} = \tfrac{1}{2}ac\sin B$$
$$= \tfrac{1}{2} \times 15 \times 11 \times \sin 28°$$
$$\approx 38.7 \text{ cm}^2$$

EXERCISE 18B

1 Find the area of:

a
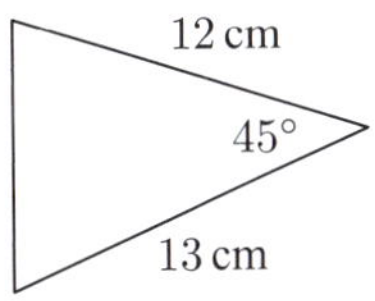

b
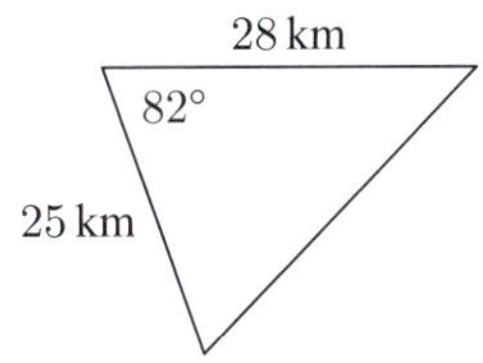

c
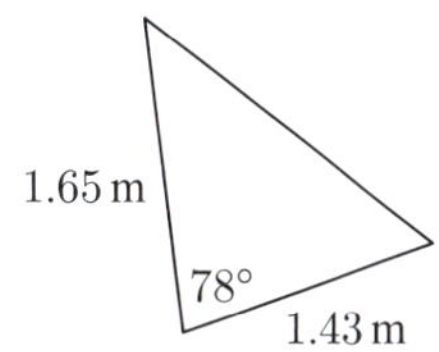

2 **a** Use the unit circle to state the value of $\sin 90°$.

b Prove that the area of a triangle formula is valid for right angled triangles.

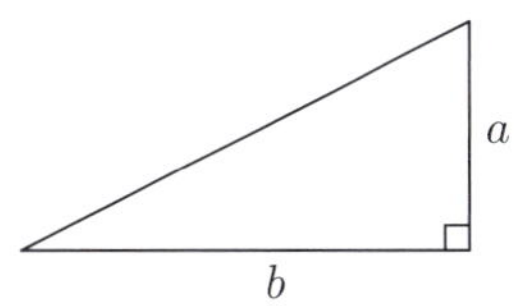

3 In the proof given in the theory on the previous page, we assumed that the included angle C is acute.
We now suppose angle C is obtuse.
Let the altitude from side [BC] have height h as shown.

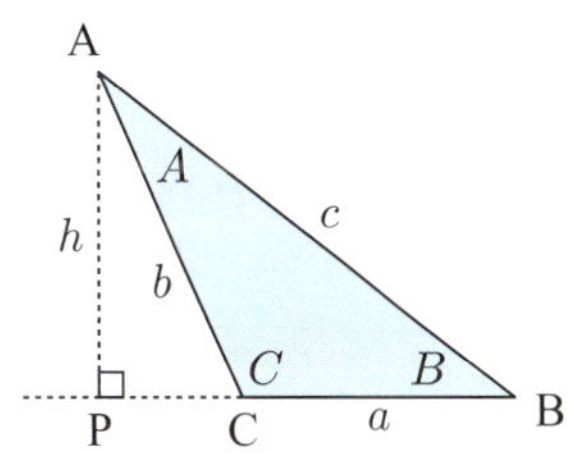

a Write the area of $\triangle$ABC in terms of a and h.

b Use $\triangle$ACP to write an expression for h in terms of b and C.

c Hence prove that the area of $\triangle$ABC is $\frac{1}{2}ab\sin C$.

4 Find the area of:

a
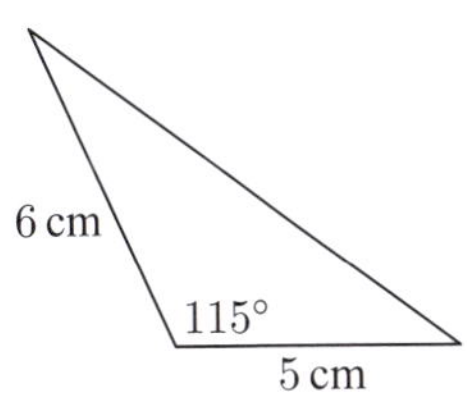

b
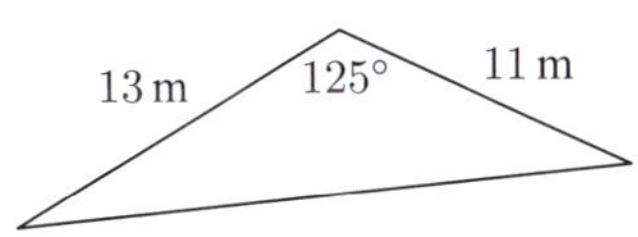

c
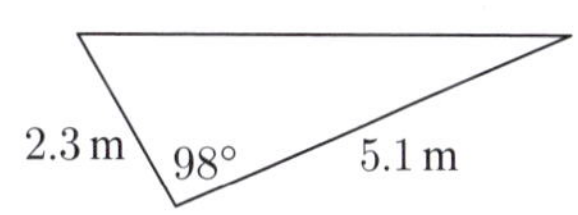

5 Find the area of:

a
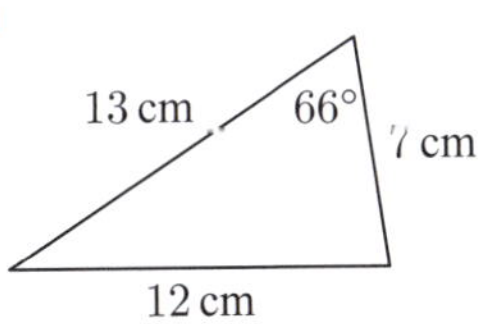

b
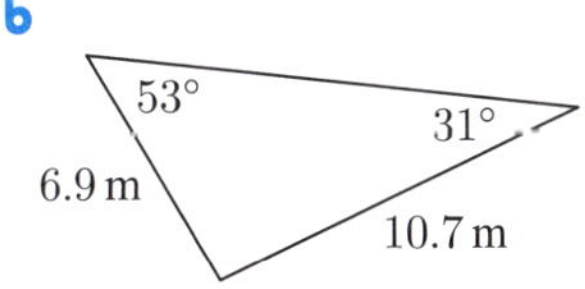

c
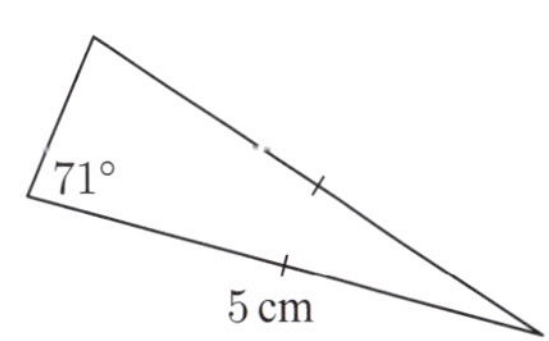

6 Find the area of a parallelogram with sides 6.4 cm and 8.7 cm, and one interior angle 64°.

7 Triangle ABC has area 150 cm^2.
Find the value of x.

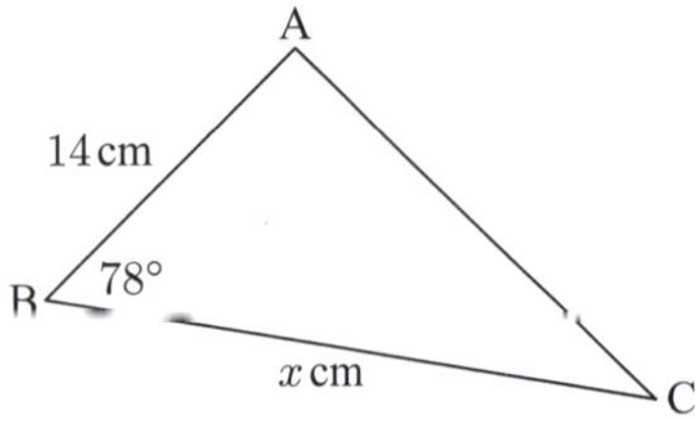

8 Triangle XYZ has area 80 cm^2.
Find the length of the equal sides.

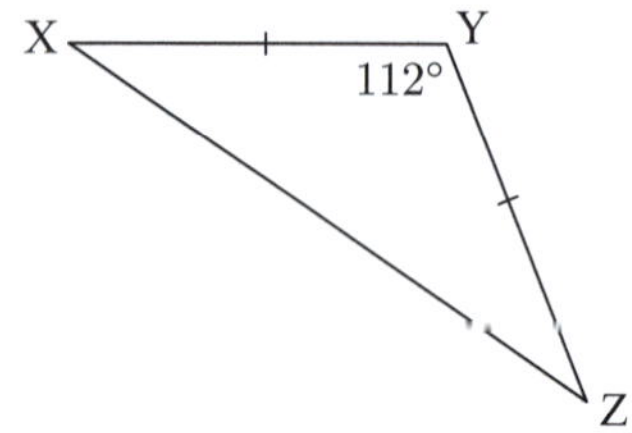

9 C is a point on the semi-circle with diameter [AB], centre O, and radius r.
Show that the shaded triangles have equal area.

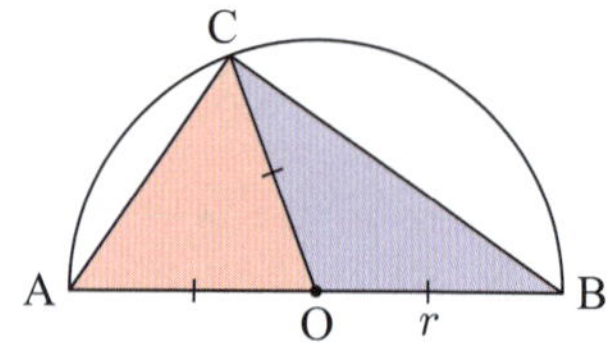

10 Triangle PQR has $\text{P}\widehat{\text{Q}}\text{R} = \theta$, PQ = 10 m, QR = 12 m, and its area is 30 m^2. Find the *two* possible values of θ.

11 Triangle ABC has AB = 13 cm, BC = 17 cm, and its area is 73.4 cm^2. Find the *two* possible sizes of $\text{A}\widehat{\text{B}}\text{C}$.

12 **a** Find the area of triangle ABC using:
i angle A **ii** angle C.

b Hence show that $\dfrac{\sin A}{a} = \dfrac{\sin C}{c}$.

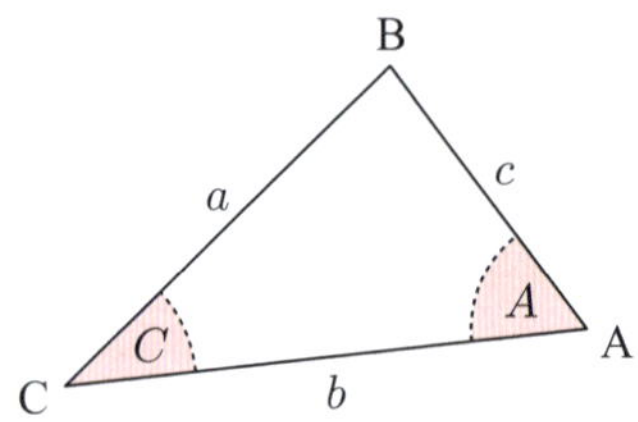

C THE SINE RULE

The **sine rule** is a set of equations which connects the lengths of the sides of any triangle with the sines of the opposite angles.

In any triangle ABC with sides a, b, and c units, and opposite angles A, B, and C respectively,

$$\frac{\sin A}{a} = \frac{\sin B}{b} = \frac{\sin C}{c} \quad \text{or} \quad \frac{a}{\sin A} = \frac{b}{\sin B} = \frac{c}{\sin C}.$$

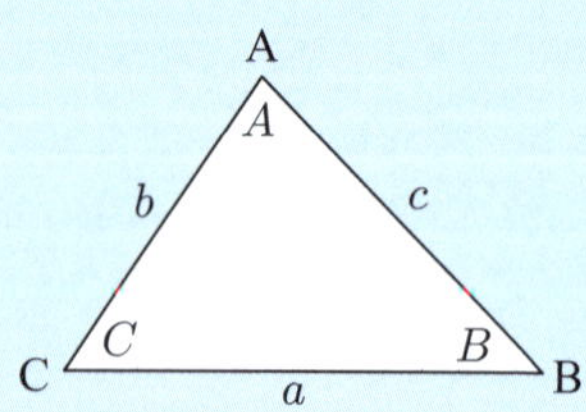

Proof:

The area of any triangle ABC is given by $\frac{1}{2}bc\sin A = \frac{1}{2}ac\sin B = \frac{1}{2}ab\sin C$.

Dividing each expression by $\frac{1}{2}abc$ gives $\dfrac{\sin A}{a} = \dfrac{\sin B}{b} = \dfrac{\sin C}{c}$.

GEOMETRY PACKAGE

FINDING SIDES

If we are given two angles and one side of a triangle, we can use the sine rule to find another side length.

When finding side lengths, we use the form $\frac{a}{\sin A} = \frac{b}{\sin B}$ so that the unknown is in the numerator.

Example 3 — Self Tutor

Find the value of x, rounded to 2 decimal places:

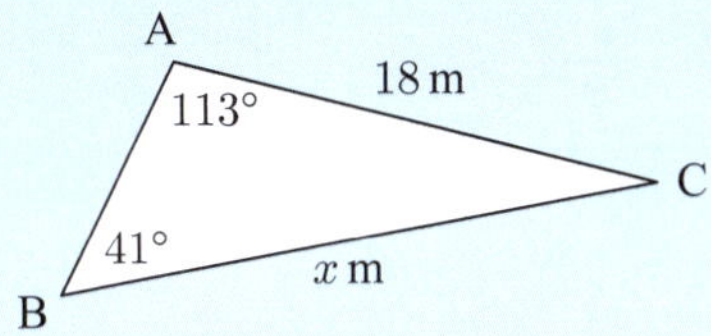

Using the sine rule,

$$\frac{x}{\sin 113°} = \frac{18}{\sin 41°}$$

$$\therefore \quad x = \frac{18 \times \sin 113°}{\sin 41°}$$

$$\therefore \quad x \approx 25.26$$

EXERCISE 18C.1

1 Find the value of x, rounded to 2 decimal places:

a

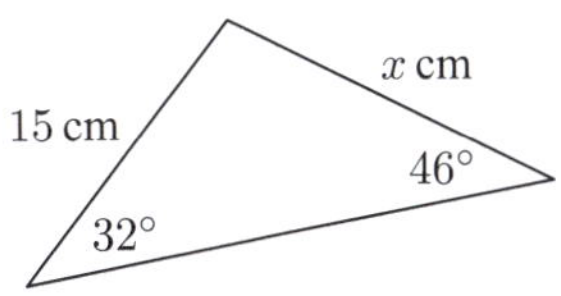

b

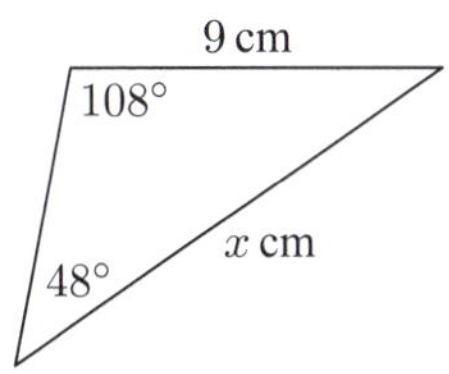

c

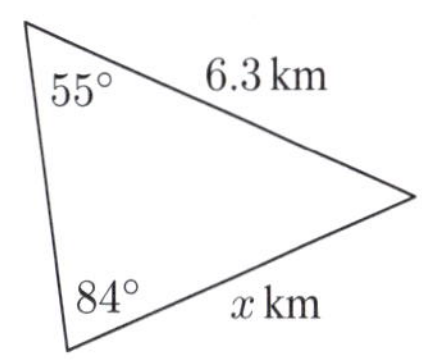

2 Find the value of x:

a

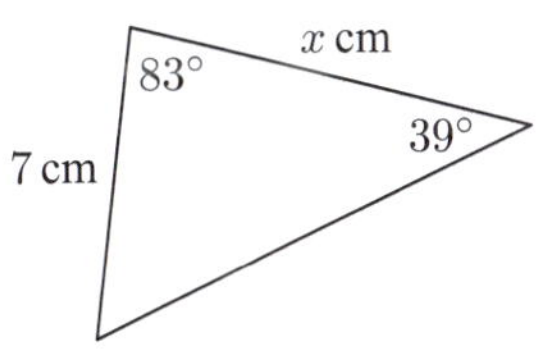

b

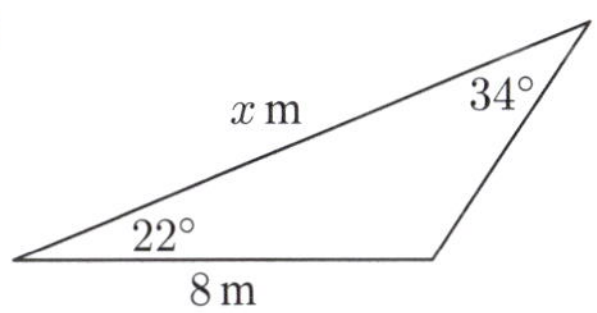

c

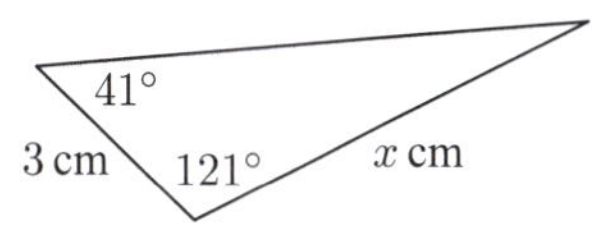

3 Find the area of this triangle.

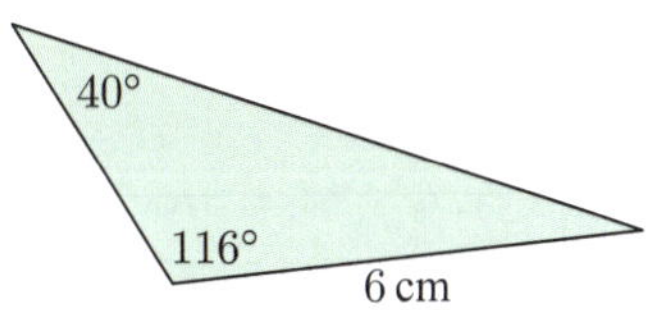

4 Find *all* unknown sides and angles of:

a

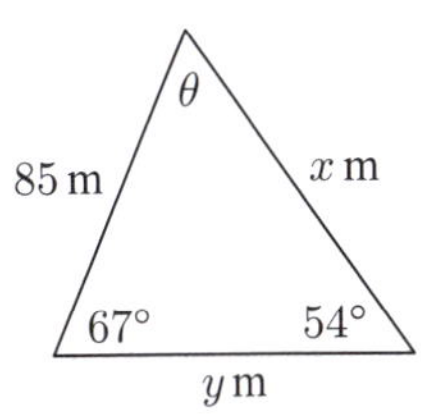

b

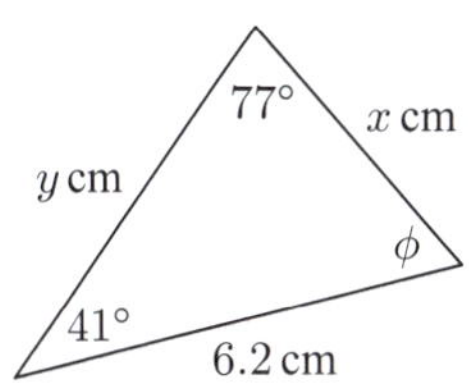

c

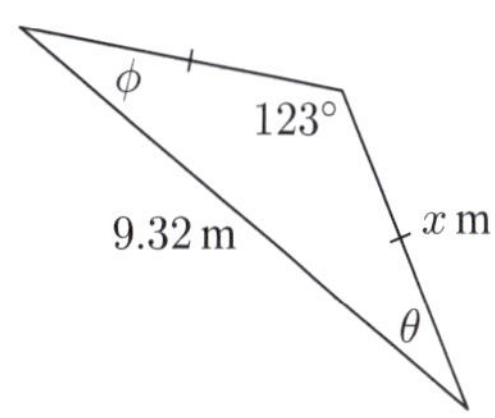

FINDING ANGLES

Finding angles using the sine rule is more complicated than finding sides because there *may* be two possible answers. We must examine both of the possible solutions to see whether each is feasible. Sometimes there is information which enables us to *reject* one of the solutions.

Example 4 **Self Tutor**

In triangle ABC, AB = 12 cm, AC = 8 cm, and angle B measures $28°$.
Find, rounded to 1 decimal place, the measure of angle C.

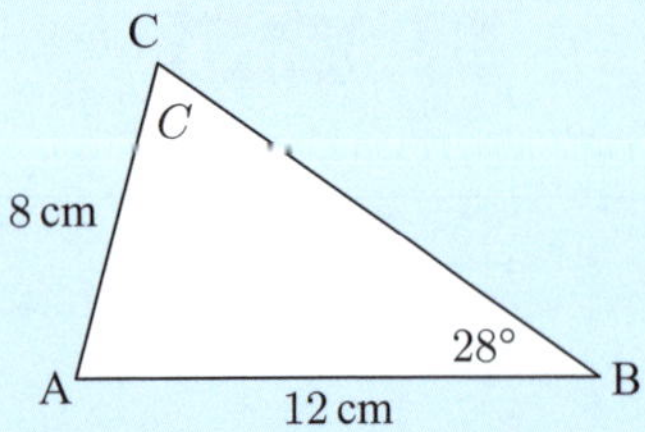

Using the sine rule, $\dfrac{\sin C}{12} = \dfrac{\sin 28°}{8}$

$\therefore \sin C = \dfrac{12 \times \sin 28°}{8}$

Now $\sin^{-1}\left(\dfrac{12 \times \sin 28°}{8}\right) \approx 44.8°$

Since the angle at C could be acute or obtuse, $C \approx 44.8°$ or $(180 - 44.8)°$
$\therefore C \approx 44.8°$ or $135.2°$

In this case there is insufficient information to determine the actual shape of the triangle.

The validity of the two answers in the above **Example** can be demonstrated by construction:

Step 1: Draw [AB] of length 12 cm, and construct an angle of $28°$ at B.

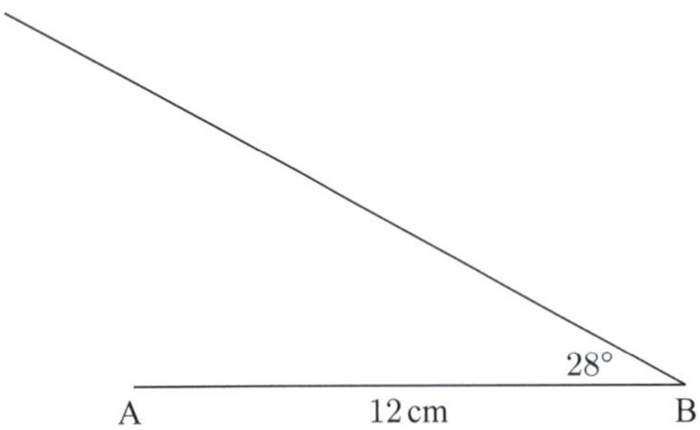

Step 2: From A, draw an arc of radius 8 cm. There are two possible locations of C.

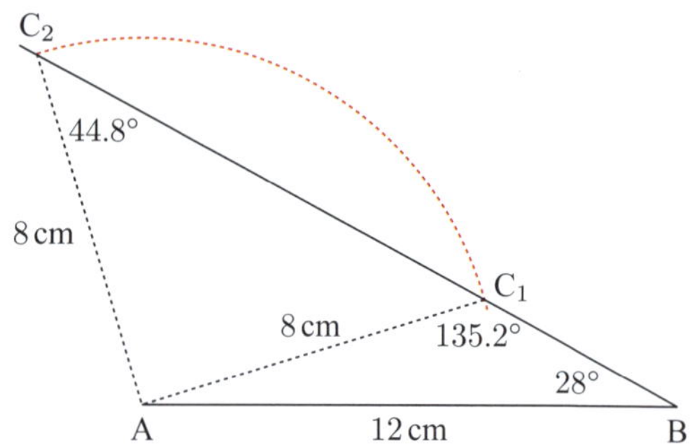

Example 5 **Self Tutor**

In triangle KLM, $L\widehat{K}M$ measures $52°$, LM = 158 m, and KM = 128 m.
Find the measure of angle L.

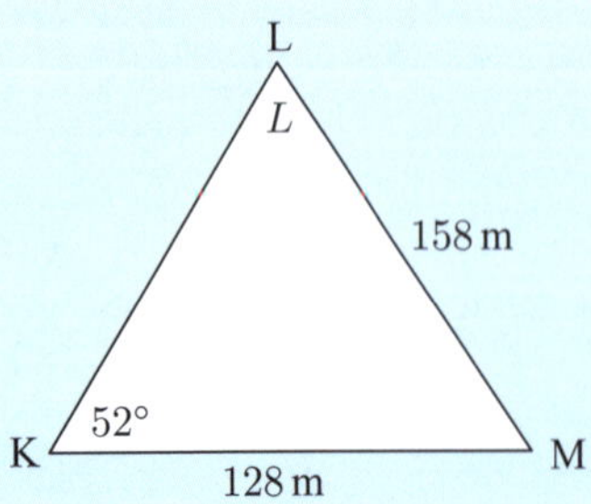

Using the sine rule, $\dfrac{\sin L}{128} = \dfrac{\sin 52°}{158}$

$\therefore \sin L = \dfrac{128 \times \sin 52°}{158}$

Now $\sin^{-1}\left(\dfrac{128 \times \sin 52°}{158}\right) \approx 39.7°$

$\therefore$ since L could be acute or obtuse,
$L \approx 39.7°$ or $(180 - 39.7)° \approx 140.3°$

However, we can reject $L \approx 140.3°$ as $140.3° + 52° > 180°$ which is impossible.
$\therefore$ the angle $L \approx 39.7°$.

EXERCISE 18C.2

1 In triangle ABC, AB = 10 cm, BC = 7 cm, and $\widehat{CAB}$ measures 42°.

- **a** Find the two possible values for $\widehat{ACB}$.
- **b** Draw a diagram to illustrate the two possible triangles.

2 In triangle PQR, PQ = 10 cm, QR = 12 cm, and $\widehat{RPQ}$ measures 42°.

- **a** Show that there is only one possible value for $\widehat{PRQ}$, and state its measure.
- **b** Draw a diagram to demonstrate that only one triangle can be drawn from the information given.

3 The following diagrams are not drawn to scale, but the information on them is correct.
Find the value of θ:

a

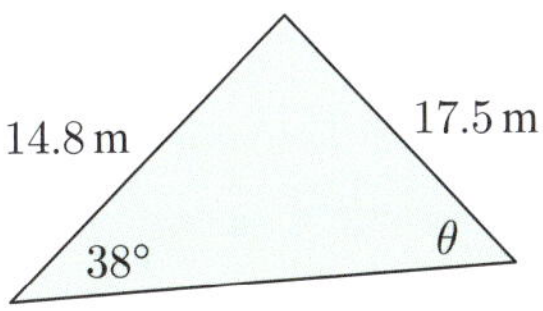

b

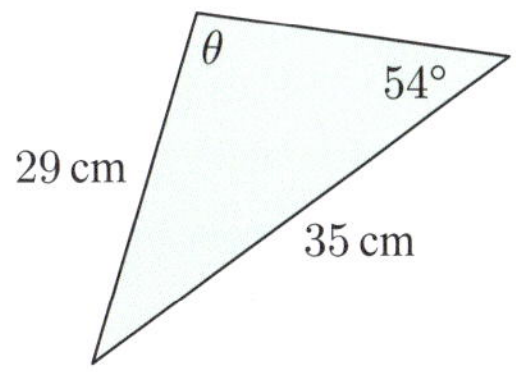

c

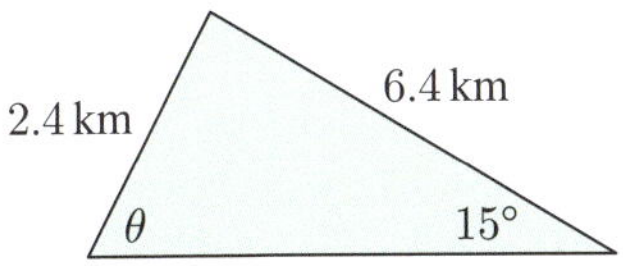

d

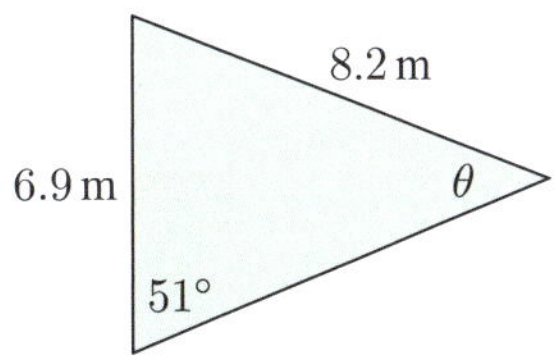

4 Consider triangle ABC. Find the measure of:

- **a** angle A if $a = 12.6$ cm, $b = 15.1$ cm, and $\widehat{ABC} = 65°$
- **b** angle B if $b = 38.4$ cm, $c = 27.6$ cm, and $\widehat{ACB} = 43°$
- **c** angle C if $a = 5.5$ km, $c = 4.1$ km, and $\widehat{BAC} = 71°$.

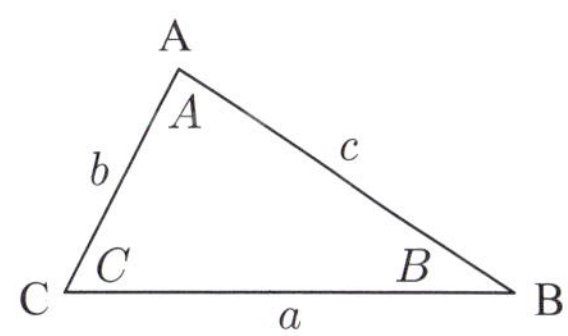

D THE COSINE RULE

The **cosine rule** relates the three sides of a triangle and one of its angles.

Consider an acute angled triangle ABC with side lengths a, b, and c as shown.

Using Pythagoras' theorem in $\triangle$BCN,

$$a^2 = h^2 + (c - x)^2$$
$$\therefore \quad a^2 = h^2 + c^2 - 2cx + x^2$$

In $\triangle$ACN, $b^2 = h^2 + x^2$ and so $h^2 = b^2 - x^2$

Thus, $a^2 = (b^2 - x^2) + c^2 - 2cx + x^2$

$$\therefore \quad a^2 = b^2 + c^2 - 2cx$$

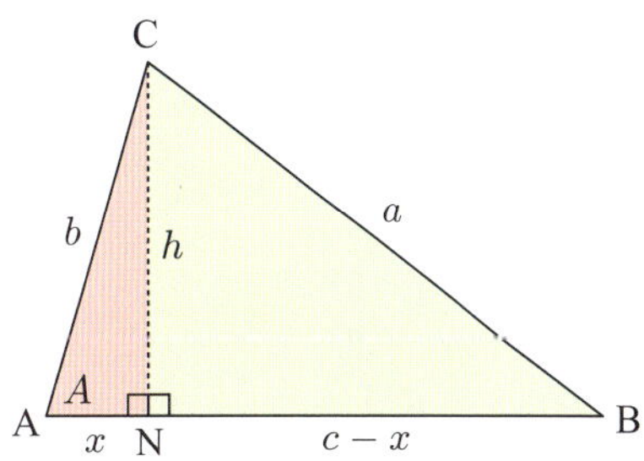

In $\triangle$ACN, $\cos A = \dfrac{x}{b}$ and so $x = b\cos A$

$$\therefore \quad a^2 = b^2 + c^2 - 2bc\cos A$$

In any triangle ABC with sides a, b, and c units, and opposite angles A, B, and C respectively:

$$\mathbf{a^2 = b^2 + c^2 - 2bc\cos A}$$
$$\mathbf{b^2 = a^2 + c^2 - 2ac\cos B}$$
$$\mathbf{c^2 = a^2 + b^2 - 2ab\cos C.}$$

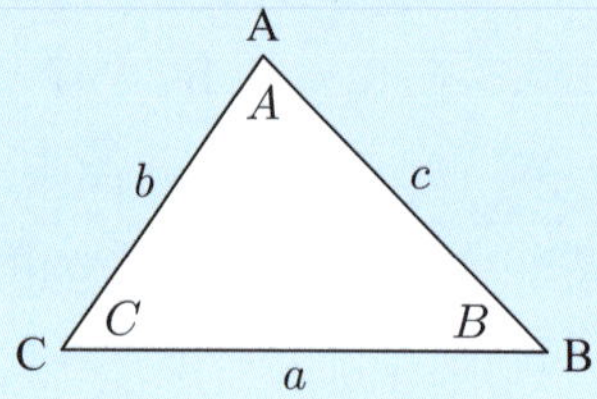

GEOMETRY PACKAGE

If we are given two sides of a triangle and the included angle, we can use the cosine rule to find the third side.

Example 6 **Self Tutor**

Find, correct to 2 decimal places, the length of [BC].

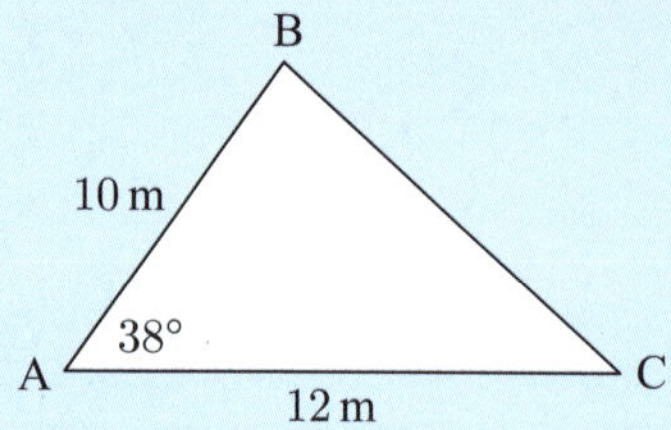

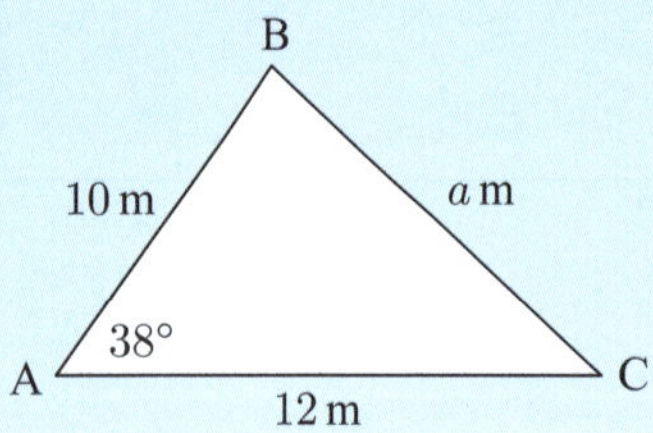

Using the cosine rule,

$a^2 = b^2 + c^2 - 2bc\cos A$

$\therefore\ a = \sqrt{12^2 + 10^2 - 2 \times 12 \times 10 \times \cos 38°}$

$\therefore\ a \approx 7.41$

$\therefore$ [BC] is approximately 7.41 m long.

If we know all three side lengths of a triangle, we can rearrange the cosine rule formulae to find any of the angles.

$$\cos A = \frac{b^2 + c^2 - a^2}{2bc}, \qquad \cos B = \frac{a^2 + c^2 - b^2}{2ac}, \qquad \cos C = \frac{a^2 + b^2 - c^2}{2ab}$$

Example 7 **Self Tutor**

Find the measure of angle C in the given figure.

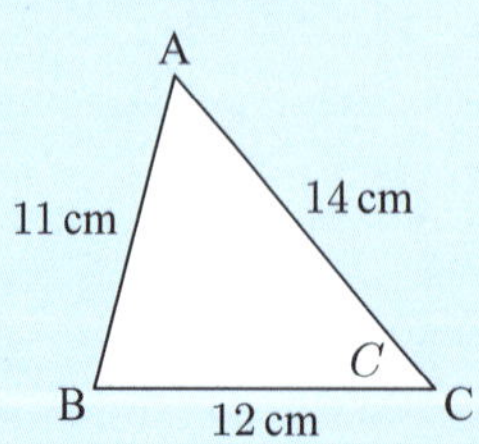

$\cos C = \dfrac{a^2 + b^2 - c^2}{2ab}$

$\therefore\ \cos C = \dfrac{12^2 + 14^2 - 11^2}{2 \times 12 \times 14}$

$\therefore\ \cos C = \dfrac{219}{336}$

$\therefore\ C = \cos^{-1}\left(\dfrac{219}{336}\right)$

$\therefore\ C \approx 49.3°$

EXERCISE 18D

1 Find the length of the remaining side in each triangle:

a
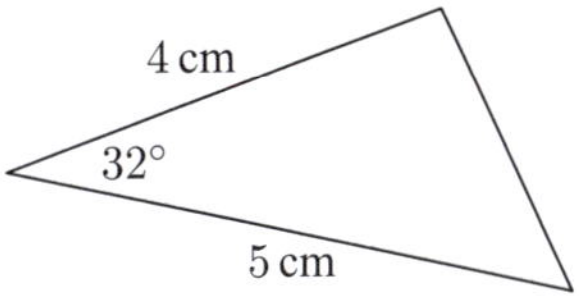

b
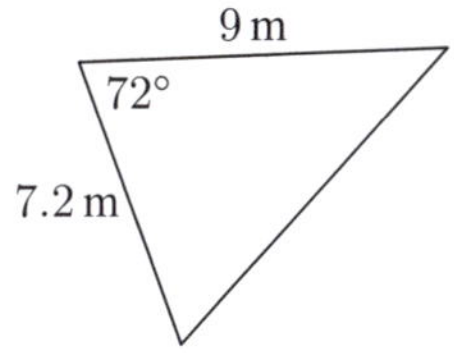

c
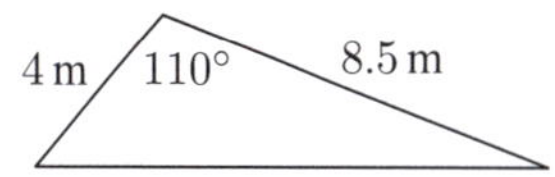

d
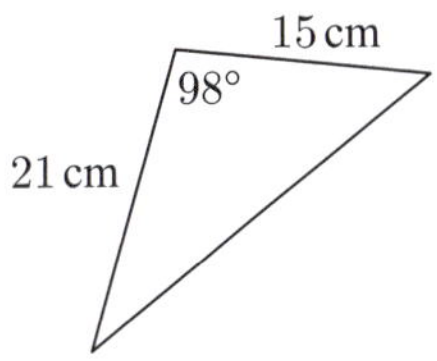

e
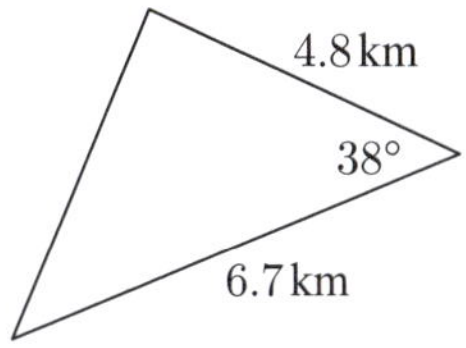

f
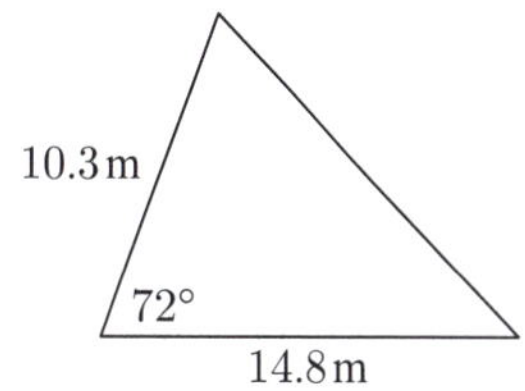

2 Find the value of θ, rounded to 1 decimal place:

a
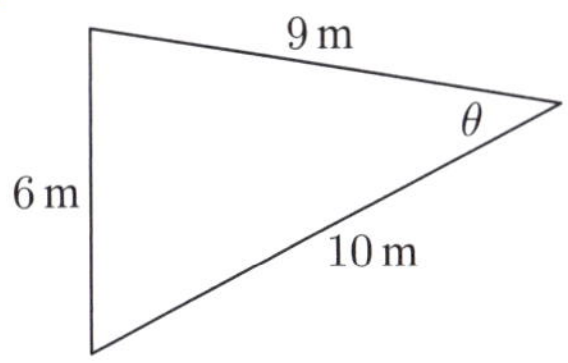

b
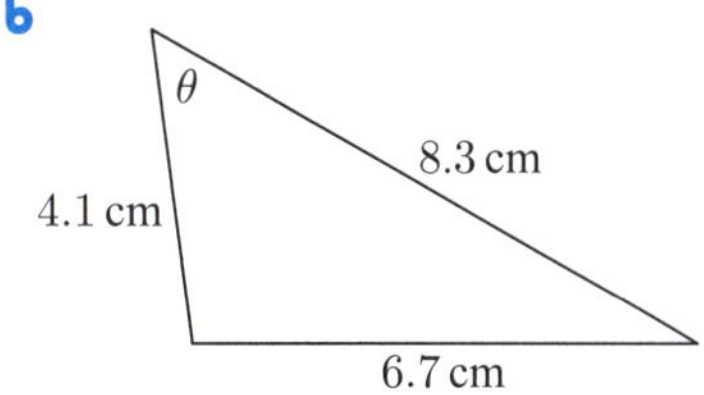

c
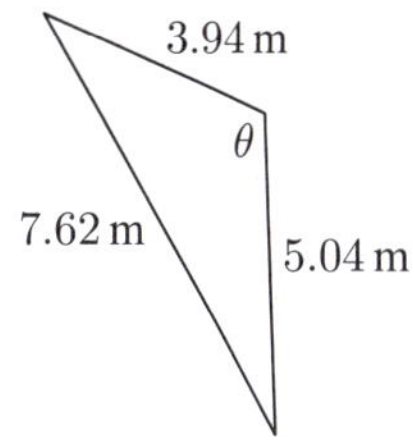

3 Find the measure of all angles in each triangle:

a
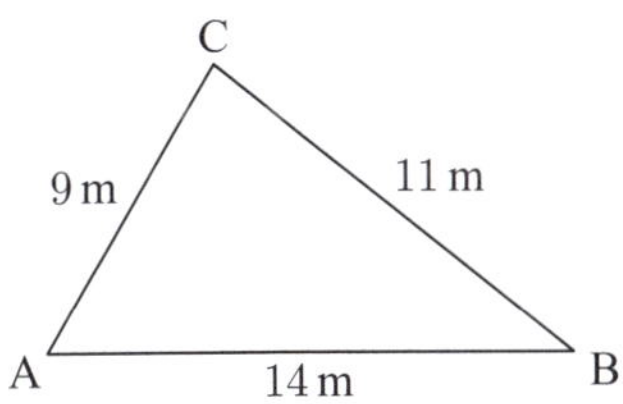

b
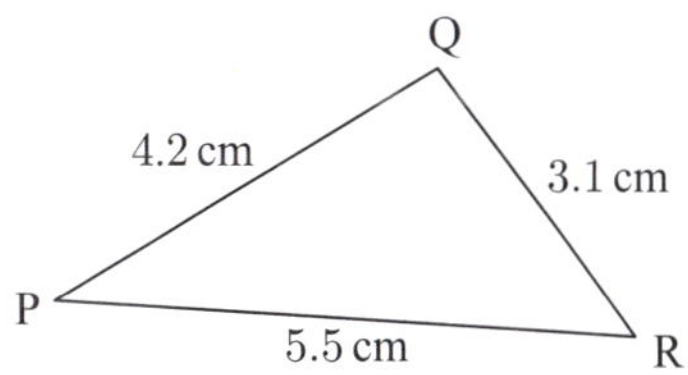

4 Prove the cosine rule $a^2 = b^2 + c^2 - 2bc\cos A$ in the case where A is obtuse.

Hint: $\cos(180° - \theta) = -\cos\theta$

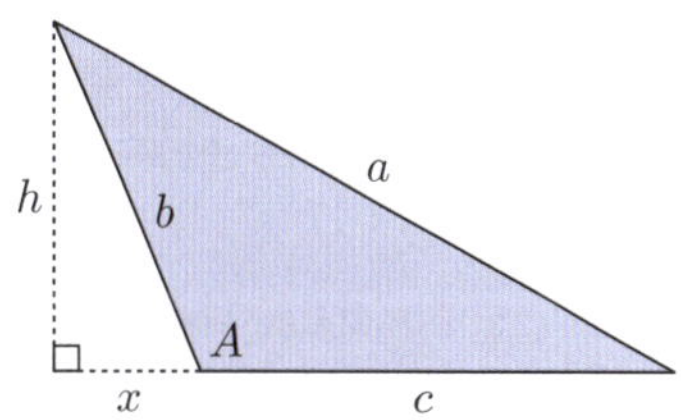

5 Find:

a the smallest angle of a triangle with sides 9 cm, 11 cm, and 13 cm

b the largest angle of a triangle with sides 3 cm, 5 cm, and 7 cm.

6 Find the area of the triangle, correct to 3 significant figures:

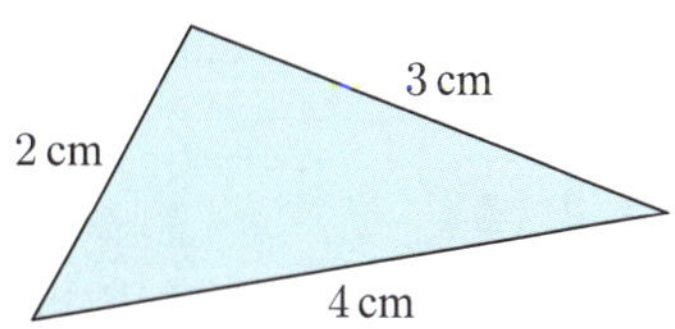

7 Given that ABC is an equilateral triangle, show that $c^2 = a^2 + ab + b^2$.

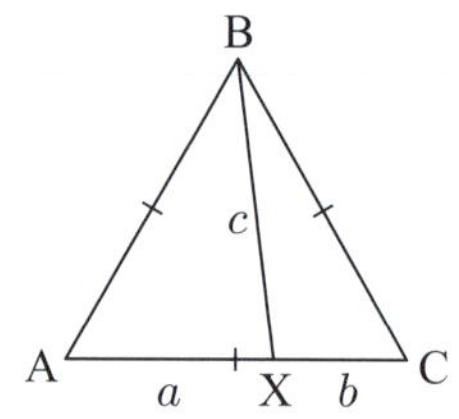

8 In triangle ABC, AB = 10 cm, AC = 9 cm, and $\widehat{ABC} = 60°$. Let BC = x cm.

a Use the cosine rule to show that x is a solution of $x^2 - 10x + 19 = 0$.

b Solve this equation to find the values of x.

c Use a scale diagram and a compass to explain why there are two possible values of x.

E PROBLEM SOLVING

When using trigonometry to solve problems, you should draw a diagram showing the given information. If you can choose between the sine and cosine rules, use the *cosine* rule to avoid the ambiguous case.

Example 8 — Self Tutor

A triangular property is bounded by two roads and a long, straight drain. Find:

a the area of the property in hectares

b the length of the drain boundary

c the angle that the Johns Road boundary makes with the drain.

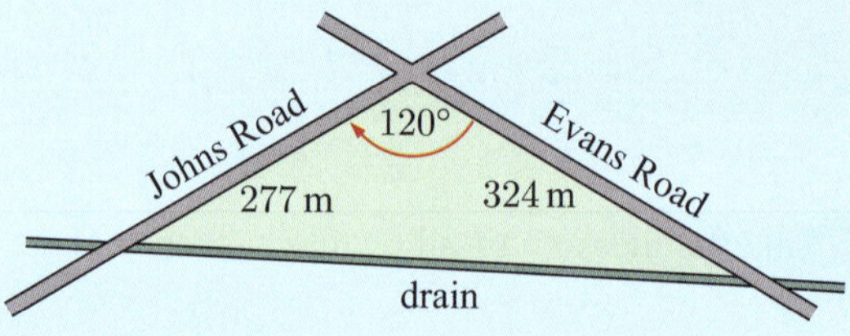

a Area $= \frac{1}{2} \times 277 \times 324 \times \sin 120°$

$\approx 38\,862 \text{ m}^2$

≈ 3.89 ha $\{1 \text{ ha} = 10\,000 \text{ m}^2\}$

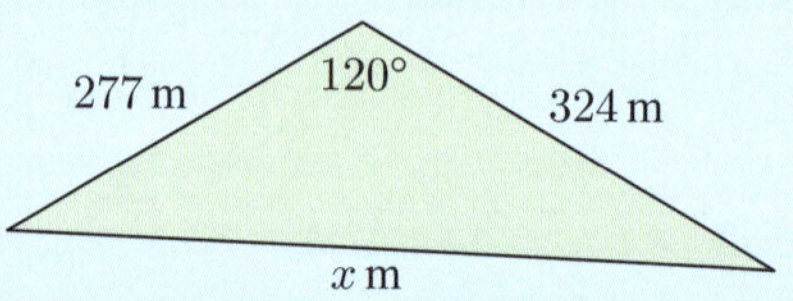

b Let the drain boundary be x m long.

$\therefore\ x^2 = 277^2 + 324^2 - 2 \times 277 \times 324 \times \cos 120°$ {cosine rule}

$\therefore\ x = \sqrt{277^2 + 324^2 - 2 \times 277 \times 324 \times \cos 120°}$

$\therefore\ x \approx 521$

The drain boundary is approximately 521 m long.

c Let the angle be θ.

$\cos\theta \approx \dfrac{277^2 + 521^2 - 324^2}{2 \times 277 \times 521}$ {cosine rule}

≈ 0.843

$\therefore\ \theta \approx \cos^{-1}(0.843) \approx 32.6°$

The Johns Road boundary makes an angle of approximately 32.6° with the drain.

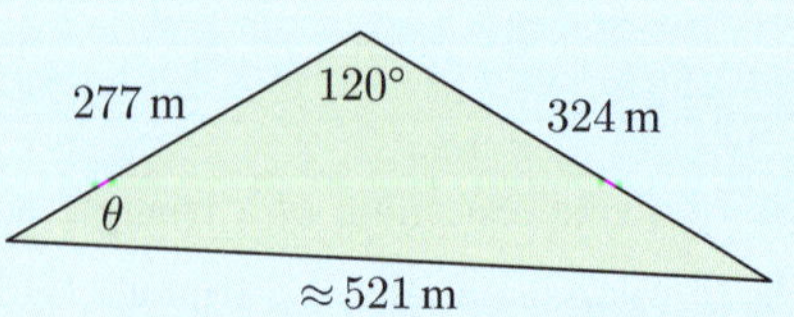

EXERCISE 18E

1 Two farm houses A and B are 10.3 km apart. A third farm house C is located such that $\widehat{BAC} = 83°$ and $\widehat{ABC} = 59°$. How far is C from A?

2 A roadway is horizontal from A to B, then rises up a $12°$ incline from B to C. How far is it directly from A to C?

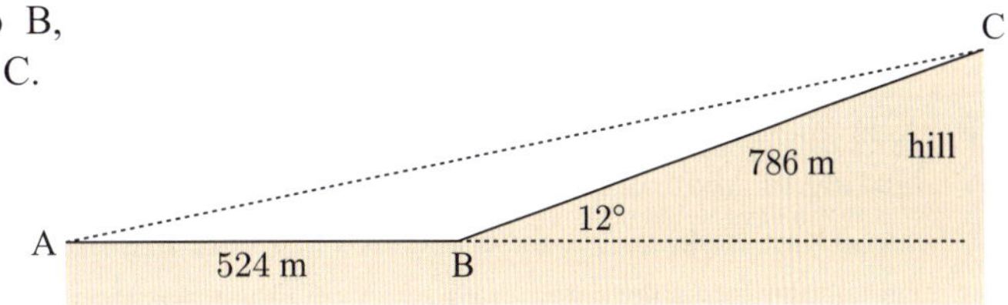

3 Sharon drives a golf ball 253 m from the tee T to point X on the fairway. X is 93 m from the flag, and $\widehat{XFT}$ is $39°$.
Find the angle θ that Sharon's drive was off line.

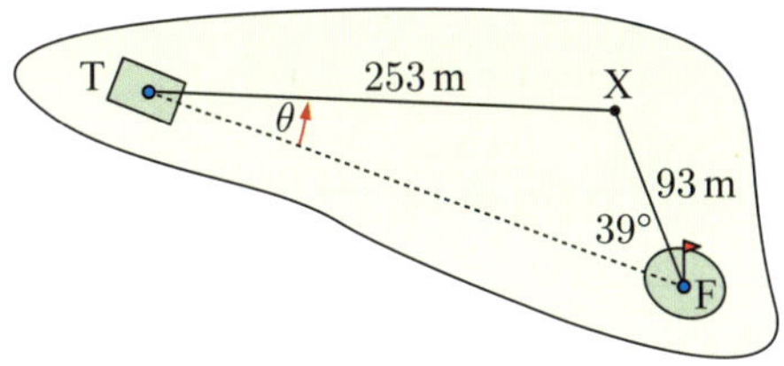

4 Towns A, B, and C are located such that $\widehat{BAC} = 50°$ and B is twice as far from C as A is from C. Find the measure of $\widehat{BCA}$.

5 Hazel's property is triangular with the dimensions shown.

   a Find the measure of the angle at A, rounded to 2 decimal places.

   b Find the area of Hazel's property, to the nearest hectare.

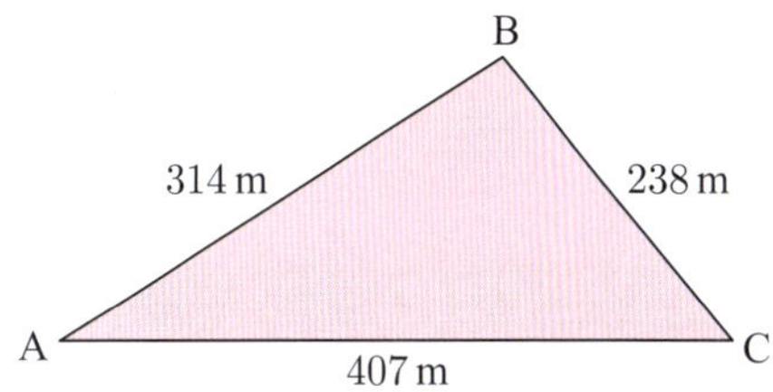

Example 9 — Self Tutor

A ship sails 58 km on the bearing $072°$. Once it has passed a reef, it turns and sails 41 km on the bearing $158°$. How far is the ship from its starting point?

We suppose the ship starts at S, sails to A, then changes direction and sails to F.

$\widehat{SAN} = 180° - 72°$ {co-interior angles}
$= 108°$

$\therefore\ \widehat{SAF} = 360° - 158° - 108°$ {angles at a point}
$= 94°$

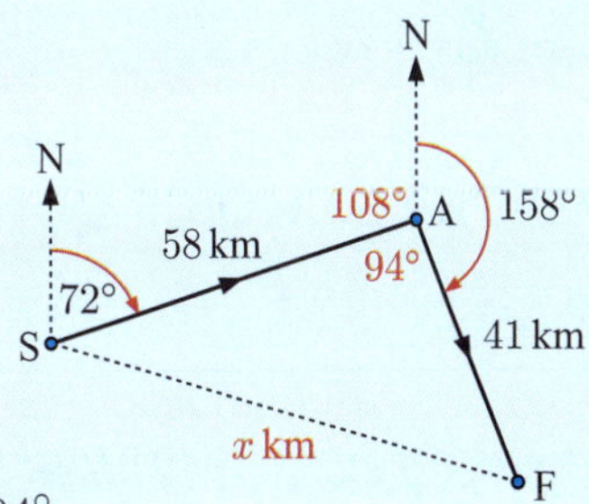

Let $SF = x$ km.

Using the cosine rule, $x^2 = 58^2 + 41^2 - 2 \times 58 \times 41 \times \cos 94°$

$\therefore\ x = \sqrt{58^2 + 41^2 - 2 \times 58 \times 41 \times \cos 94°}$

$\therefore\ x \approx 73.3$

The ship is about 73.3 km from its starting point.

6 Fred walks 83 m in the direction $111°$, and then 78 m in the direction $214°$. How far is Fred from his starting point?

7 A boat travels 13 km in the direction $138°$, and then a further 11 km in the direction $067°$. Find the distance and bearing of the boat from its starting point.

8 Mount X is 9 km from Mount Y, on the bearing $146°$. Mount Z is 14 km from Mount X, and is on the bearing $072°$ from Mount Y. Find the bearing of X from Z.

9 A parallelogram has sides of length 8 cm and 12 cm. Given that one of its angles is $65°$, find the lengths of its diagonals.

10 X is 20 km north of Y. A mobile telephone mast M is to be placed 15 km from Y and with the bearing of M from X being $140°$.

- **a** Draw a sketch to show the two possible positions where the mast could be placed.
- **b** Calculate the distance of each of these positions from X.

11 The area of triangle ABD is 33.6 m^2.
Find the length of [CD].

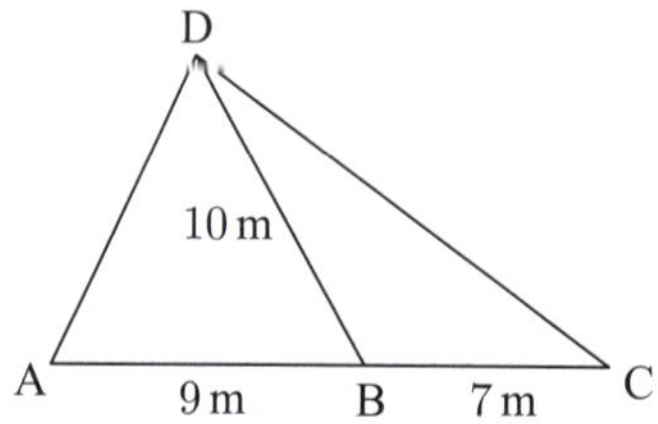

12

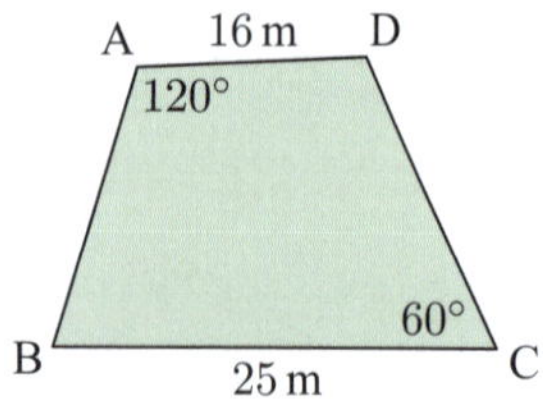

The quadrilateral ABCD represents David's garden plot.
$AD = 16$ m, $BC = 25$ m, and [DC] is 5 m longer than [AB].
A fence runs around the entire boundary of the plot. How long is the fence?

GLOBAL CONTEXT — COORDINATE SYSTEMS

Global context: Scientific and technical innovation
Statement of inquiry: Different coordinate systems can be used to describe the position of a point in space.
Criterion: Communicating

GLOBAL CONTEXT

MULTIPLE CHOICE QUIZ

QUICK QUIZ

REVIEW SET 18A

1
- **a** Write down the exact coordinates of P and Q.
- **b** Use your calculator to state the coordinates of P and Q rounded to 3 significant figures.

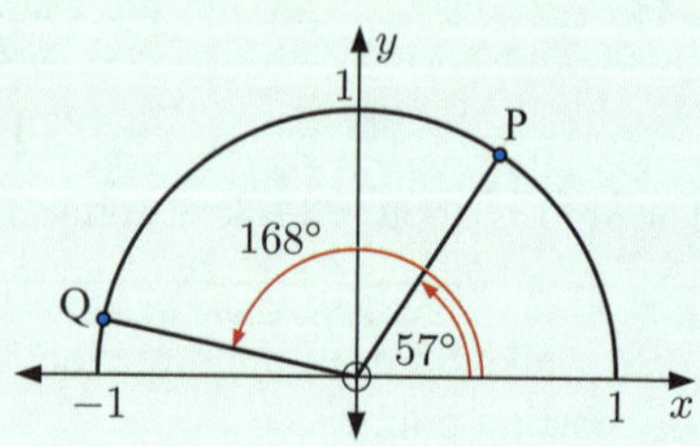

2 The area of triangle ABC is 21 m^2.
Find the value of x.

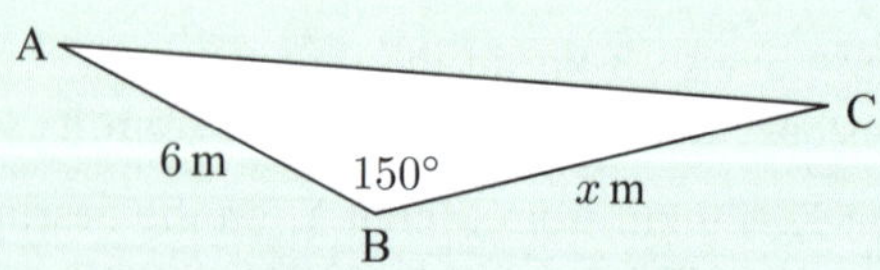

3 Find the area of:

a

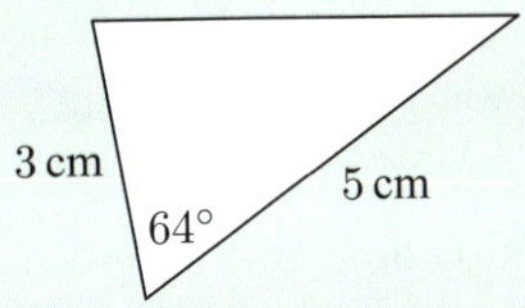

b

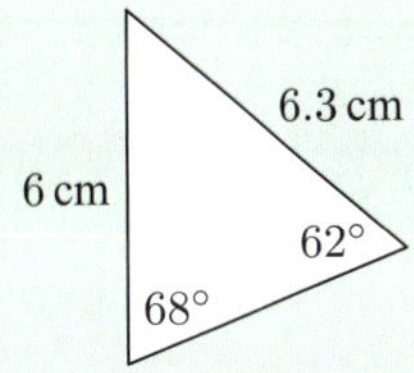

4 Find the value of x, rounded to 2 decimal places:

a

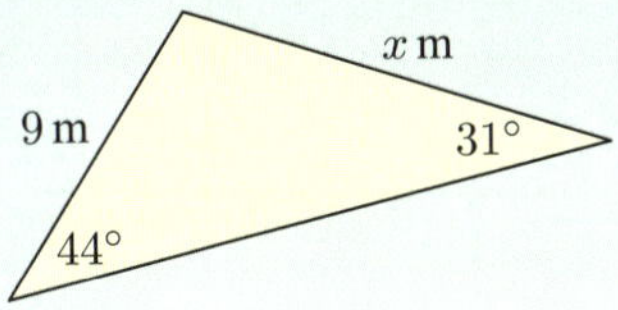

b

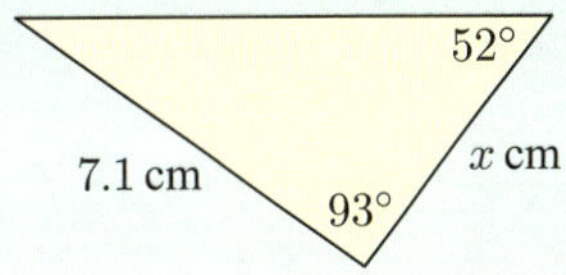

5 Find the value of x:

a

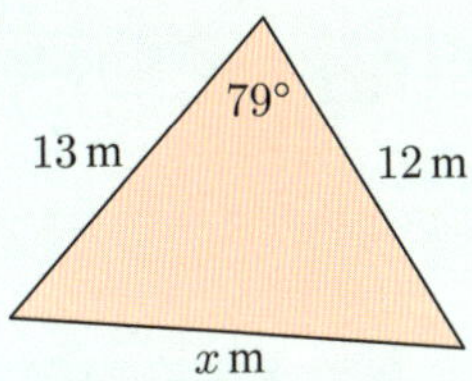

b

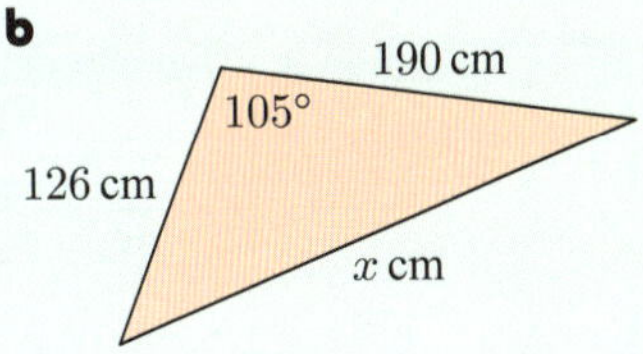

c

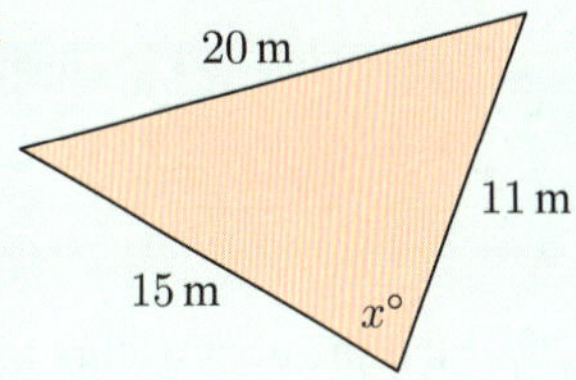

6 Find the smallest angle of a triangle with sides 5 cm, 7 cm, and 10 cm.

7 Stuart swam 200 m in the direction 124°, and then 150 m in the direction 156°. Find the distance and bearing of Stuart from his starting point.

8

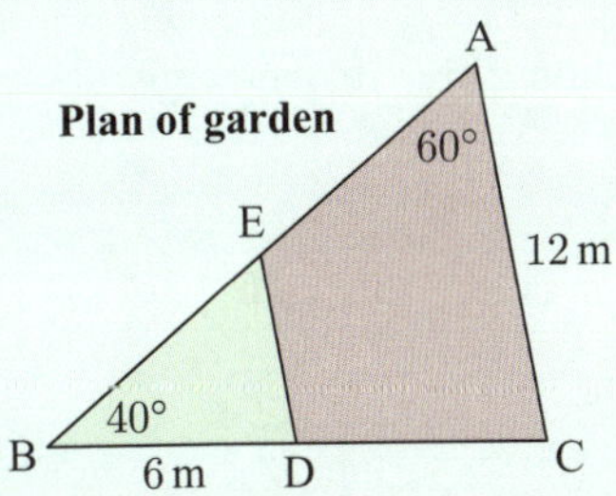

In the given plan view, AC = 12 m, $\widehat{BAC} = 60°$, and $\widehat{ABC} = 40°$.
D is a post 6 m from corner B, E is a tap, and triangle BDE is a lawn with area 13.5 m^2.

a Find the length of:
   i [DC] **ii** [BE]

b Find the area of quadrilateral ACDE.

9 In triangle ABC, AB = 6 cm, BC = 8 cm, and $\widehat{ACB}$ measures 30°.

a Find the two possible values for $\widehat{BAC}$.

b Illustrate the two possible triangles.

c If $\widehat{BAC}$ is obtuse, find:
   i the length of [AC] **ii** the measure of $\widehat{ABC}$ **iii** the area of the triangle.

REVIEW SET 18B

1 Find the area of:

a

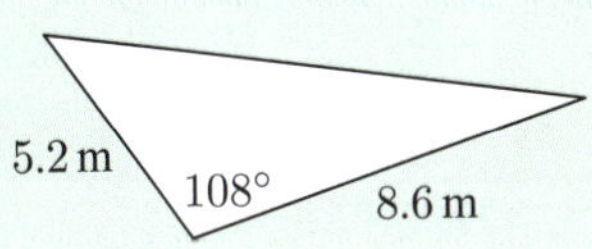

b

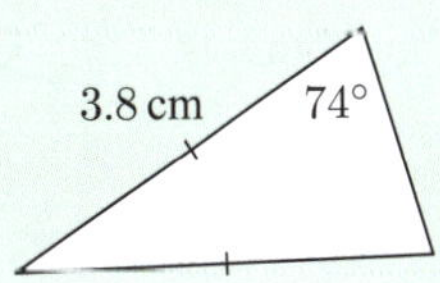

2 Find the obtuse angle:

a which has the same sine as 38°

b whose cosine is the negative of $\cos 52°$.

3 Triangle ABC has $B\widehat{A}C = \theta$, $AB = 12$ cm, $AC = 14$ cm, and its area is 33.4 cm^2. Find the *two* possible values of θ.

4 The following diagrams are not drawn to scale, but the information on them is correct. Find the value of θ:

a

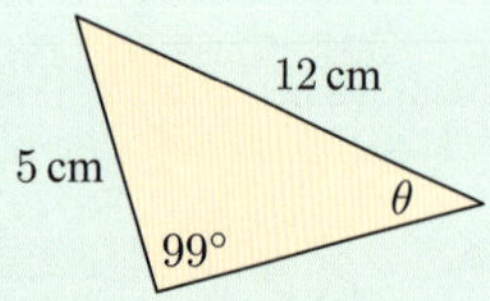

b

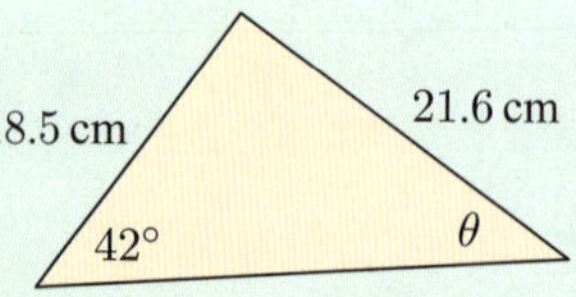

c

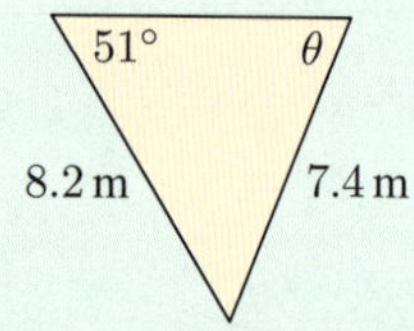

5 Find the value of x:

a

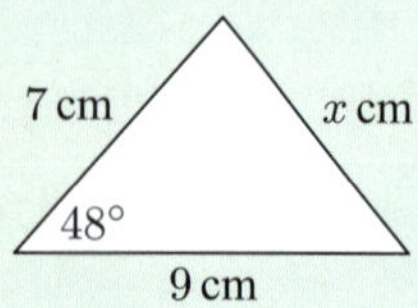

b

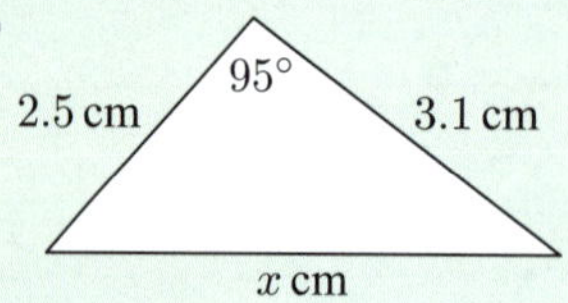

c

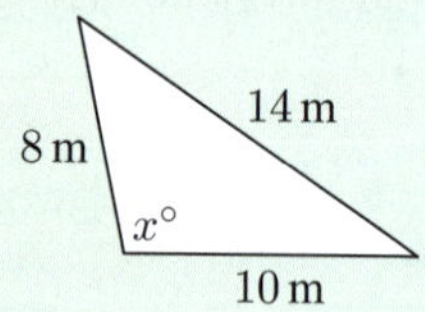

6 Find the area of this triangle.

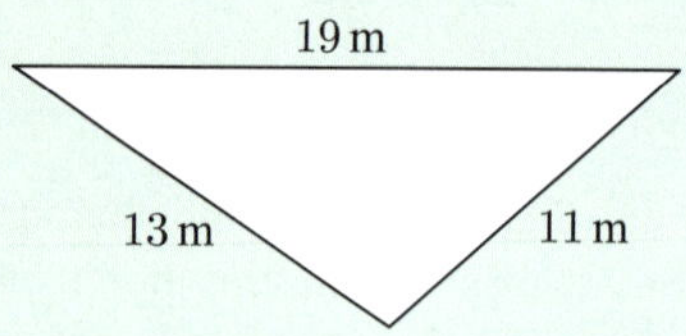

7 A ship leaves port P and travels for 50 km in the direction 181°. It then sails 60 km in the direction 274° to an island port Q.

a How far is Q from P?

b To sail directly back from Q to P, in what direction must the ship sail?

8 Jane and Peter are considering buying a block of land. The land agent has supplied them with the given sketch.
Find the area of the property, giving your answer in:

a m^2

b hectares.

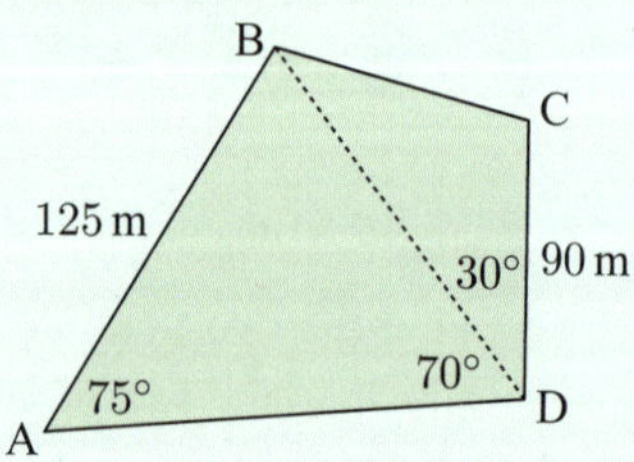

9 Ports P, Q, and R are equally spaced along the coast. A boat B is 4 km from P, and 2 km from Q.

a Show that $\cos B\widehat{Q}P = \frac{13}{20}$.

b Hence find $\cos B\widehat{Q}R$.

c Show that the boat is $\sqrt{42}$ km from port R.

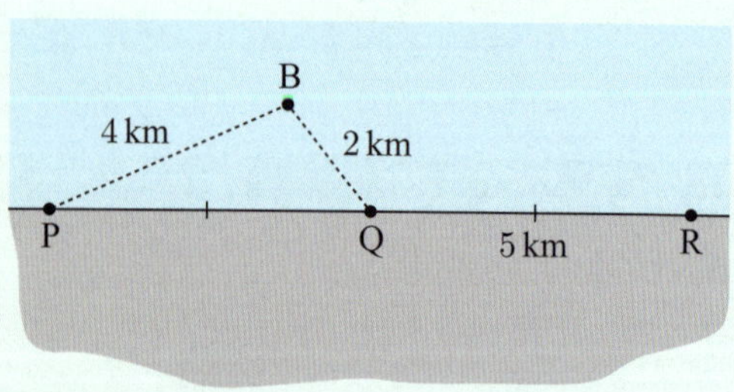

Chapter 19

Probability

Contents:

OPENING PROBLEM

A gift box contains 14 dark chocolates and 18 milk chocolates.

Things to think about:

a Suppose Caroline picks a chocolate at random from the box. What is the probability that the chocolate will be dark?

b Now suppose Caroline picks *three* chocolates at random from the box.

i Can you explain why the probability that all three will be dark is *not* the cube of the probability in **a**?

ii What is the correct probability that all three chocolates will be dark?

The **probability** of an event is the chance or likelihood of it occurring.

An **impossible** event has 0% chance of happening, and is assigned the probability 0.

A **certain** event has 100% chance of happening, and is assigned the probability 1.

All other events have probability between 0 and 1.

We can determine probabilities based on:

- what we theoretically expect to happen
- the results of an experiment.

HISTORICAL NOTE

The Italian **Girolamo Cardano** (1501 - 1576) was the first mathematician to consider probability theory. He studied dice rolling, believing there were scientific principles which governed the occurrence of particular outcomes, rather than luck. He applied his knowledge to gambling, generally with success, but resulting in an addiction which cost him time and reputation.

Cardano's work on chance was written around 1564, but not published until 1663 in *Liber de Ludo Aleae*. In the meantime, **Blaise Pascal** and **Pierre de Fermat** would, through correspondence in 1654, cover much of the same work, including Cardano's dice rolling problem.

Girolamo Cardano

A SAMPLE SPACE AND EVENTS

The **sample space** of an experiment is the set of its possible outcomes.

We usually use U to represent the sample space, as it is the universal set of possible outcomes.

We can represent sample spaces using:

- lists
- 2-dimensional grids
- tree diagrams.

Example 1 🔊 **Self Tutor**

When two coins are tossed, the possible outcomes are:

two heads — head and tail — tail and head — two tails

Represent the sample space for tossing two coins using:

a a list in set notation **b** a 2-dimensional grid **c** a tree diagram.

We let H represent a "head" and T represent a "tail".

a {HH, HT, TH, TT}

b

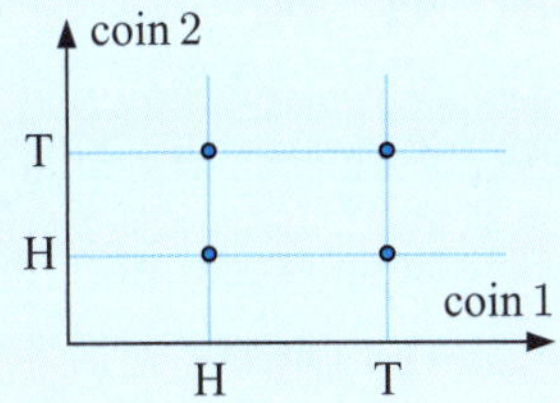

c

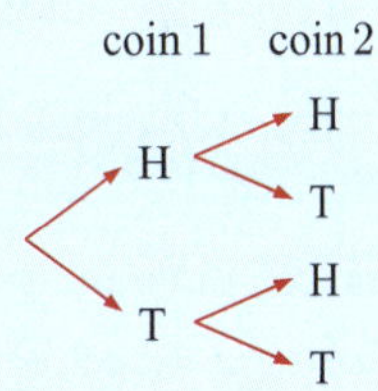

An **event** is a set of outcomes in the sample space that have a particular property.

The **complement** of an event E is E', the event that E does not occur.

Example 2 🔊 **Self Tutor**

Two marbles are drawn from a bag containing red, green, and yellow marbles. Let E be the event that at least one red marble is drawn.

a Draw a tree diagram to illustrate the sample space. Highlight the outcomes corresponding to event E.

b Hence list the outcomes in E using set notation.

a Let R be "drawing a red",
G be "drawing a green",
and Y be "drawing a yellow".

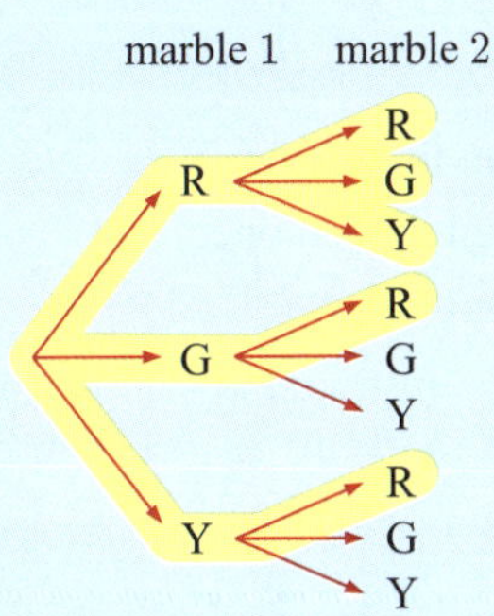

b $E = \{RR, RG, RY, GR, YR\}$

EXERCISE 19A

1 List the sample space for:

- **a** tossing a coin twice
- **b** twirling a square spinner labelled A, B, C, D
- **c** the genders of a 3-child family
- **d** the order in which 4 blocks W, X, Y, and Z can be lined up.

2 Use a 2-dimensional grid to illustrate the sample space for:

- **a** rolling a die and tossing a coin simultaneously
- **b** rolling a die and spinning a spinner with sides A, B, C, D.

3 Two children are asked to pick a camp activity out of hiking, rock climbing, or kayaking. Use a 2-dimensional grid to illustrate the sample space.

4 Use a grid to illustrate the sample space for twirling two square spinners, one labelled A, B, C, D, and the other 1, 2, 3, 4. Circle the outcomes corresponding to the event "a B or an odd number".

5 Draw a tree diagram to illustrate the sample space for:

- **a** tossing a 5 cent and 10 cent coin simultaneously
- **b** tossing a coin and twirling an equilateral triangular spinner labelled A, B, and C
- **c** twirling two equilateral triangular spinners labelled 1, 2, 3, and X, Y, Z respectively
- **d** selecting bag A or bag B, then drawing a ball from that bag.

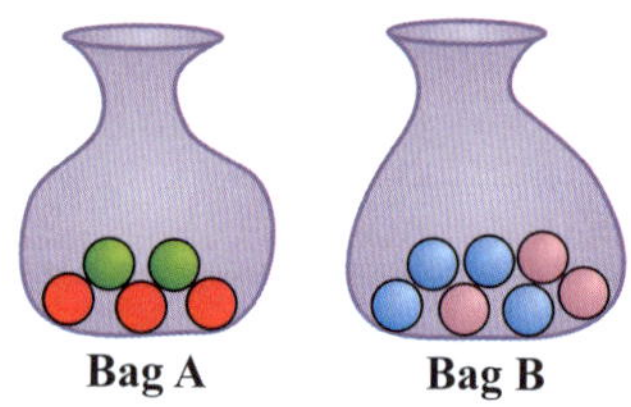

6 Two tickets are drawn from a hat containing pink, blue, and white tickets. Let E be the event that at least one blue ticket is drawn.

- **a** Draw a tree diagram to illustrate the sample space. Highlight the outcomes corresponding to event E.
- **b** Hence list, using set notation, the outcomes of:
  - **i** E
  - **ii** E'

7 The spinner alongside is spun three times.

Let A be the event of spinning green once, and
B be the event of spinning blue at least twice.

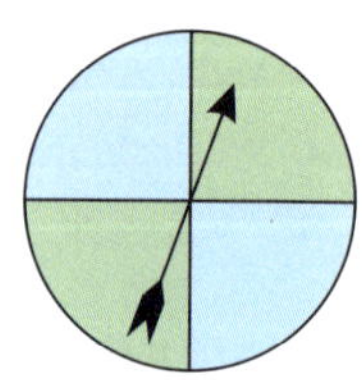

- **a** Draw a tree diagram to illustrate the sample space. Highlight the outcomes corresponding to events A and B in different colours.
- **b** Use set notation to list the outcomes in:
  - **i** the sample space U
  - **ii** A
  - **iii** B
  - **iv** A'
  - **v** B'
- **c** Is it possible to observe A but not B? Explain your answer.
- **d** Illustrate the sample space and the events A and B on a Venn diagram.

B THEORETICAL PROBABILITY

If the possible outcomes of an experiment are equally likely, the theoretical probability of an event E is given by

$$\mathbf{P}(E) = \frac{\textbf{number of outcomes in } E}{\textbf{total number of possible outcomes}}.$$

For any event E, $\mathbf{P}(E') = 1 - \mathbf{P}(E)$.

Example 3 — Self Tutor

Suppose three coins are tossed simultaneously. Find the probability of getting:

a three heads **b** at least one head.

The possible outcomes are: {HHH, HHT, HTH, THH, TTH, THT, HTT, TTT}.

There are 8 possible outcomes.

a $\text{P(three heads)} = \frac{1}{8}$ ← three heads only occurs in the outcome HHH; ← 8 possible outcomes

b $\text{P(at least one head)} = \frac{7}{8}$ ← all outcomes except TTT contain at least one head; ← 8 possible outcomes

EXERCISE 19B

1 The three letters E, N, and T are placed at random in a row. Find the probability of:

- **a** spelling NET
- **b** E appearing first
- **c** E *not* appearing first
- **d** spelling NET or TEN.

2 Determine the probability that a randomly selected 3-child family consists of:

- **a** all boys
- **b** all girls
- **c** boy, then girl, then girl
- **d** two girls and a boy
- **e** a girl for the eldest
- **f** at least one boy.

3 Four friends Alex, Bodi, Claire, and Daniel sit randomly in a row. Determine the probability that:

- **a** Alex is on one of the ends
- **b** Alex is *not* on one of the ends
- **c** Claire and Daniel are on the ends
- **d** Bodi is on an end, and Claire is seated next to him
- **e** Alex and Claire sit next to each other.

4 List the possible outcomes when four coins are tossed simultaneously. Hence determine the probability of getting:

- **a** all heads
- **b** two heads and two tails
- **c** more tails than heads
- **d** at least one tail
- **e** exactly one head.

Example 4 **Self Tutor**

A die has the numbers 0, 0, 1, 1, 4, and 5. It is rolled twice.

a Illustrate the possible outcomes using a 2-dimensional grid.

b Hence find the probability of getting:

i a total of 5 **ii** two numbers which are the same.

a

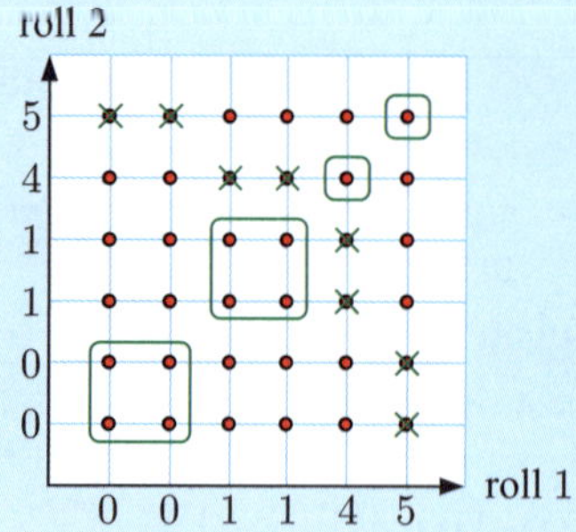

b There are $6 \times 6 = 36$ possible outcomes.

i P(total of 5)
$= \frac{8}{36}$ {×}
$= \frac{2}{9}$

ii P(same numbers)
$= \frac{10}{36}$ {circled points}
$= \frac{5}{18}$

5 **a** Draw a grid to illustrate the sample space when a 10 cent and a 50 cent coin are tossed simultaneously.

b Hence determine the probability of getting:

i two heads **ii** two tails **iii** exactly one head **iv** at least one head.

6 A coin and a pentagonal spinner with sectors 1, 2, 3, 4, and 5 are tossed and spun respectively.

a Draw a grid to illustrate the sample space of possible outcomes.

b Hence determine the chance of getting:

i a head and a 4 **ii** a tail and an odd number **iii** an even number
iv a tail or a 3.

7 **a** Use a grid to display the possible outcomes when a pair of dice is rolled.

b Hence determine the probability of:

i one die showing a 4 and the other a 5 **ii** both dice showing the same number
iii at least one of the dice showing a 3 **iv** either a 4 or 6 being displayed
v both dice showing even numbers **vi** the sum of the numbers being 7.

8 A die has the numbers 1, 2, 2, 2, 5, and 6. It is rolled twice.

a Draw a grid to illustrate the possible outcomes.

b Hence find the probability of getting:

i a 2 and a 5 **ii** a total of 7
iii numbers which are the same **iv** at least one 6
v at least one prime number **vi** numbers whose product is at least 5.

9 The spinners shown are spun once, and the numbers spun are multiplied together.

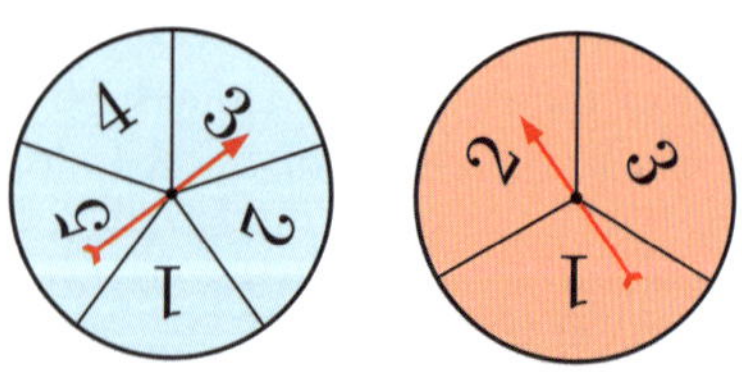

a Find the probability that the result is:

i 9 **ii** 6
iii greater than 5 **iv** prime.

b Is the result more likely to be even or odd?

10 A 52 card pack is well shuffled, then one card is dealt from the top of the pack. Find the probability that the card is:

- **a** a Queen
- **b** the Jack of hearts
- **c** a spade
- **d** a picture card
- **e** a red 7
- **f** a diamond or a club
- **g** a King or a heart
- **h** a Queen and a 3
- **i** not a picture card.

Example 5 — Self Tutor

In a library group of 50 readers, 31 like science fiction, 20 like detective stories, and 12 dislike both.

- **a** Draw a Venn diagram to represent this information.
- **b** If a reader is randomly selected, find the probability that he or she:
  - **i** likes both science fiction and detective stories
  - **ii** likes exactly one of science fiction and detective stories.

a Let S represent readers who like science fiction, and D represent readers who like detective stories.

Let $n(S \cap D) = x$

$\therefore\ n(S \cap D') = 31 - x$ and $n(S' \cap D) = 20 - x$

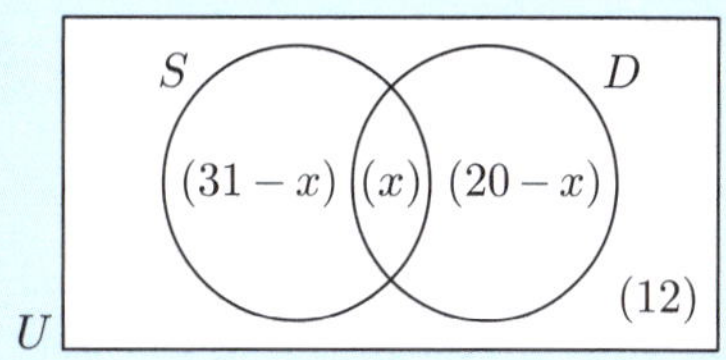

We know 12 dislike both types of book.

Since there are 50 readers in total,

$$(31 - x) + x + (20 - x) + 12 = 50$$
$$\therefore\ 63 - x = 50$$
$$\therefore\ x = 13$$
$$\therefore\ n(S \cap D') = 18$$
$$\text{and}\quad n(S' \cap D) = 7$$

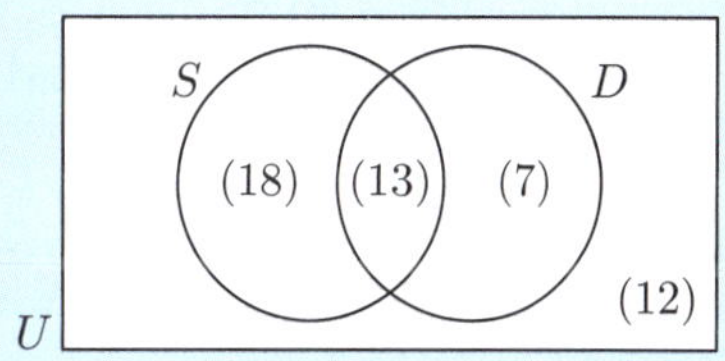

b **i** P(likes both) $= \frac{13}{50}$

ii P(likes exactly one) $= \dfrac{18+7}{50} = \frac{1}{2}$

11 The Venn diagram shows the sports played by the students in a Year 10 class.

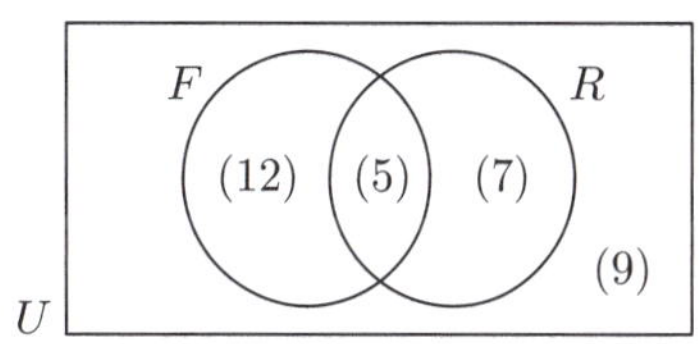

$F \equiv$ football
$R \equiv$ rugby

- **a** How many students are in the class?
- **b** A student is chosen at random.
  Find the probability that the student:
  - **i** plays football
  - **ii** plays both sports
  - **iii** plays football or rugby
  - **iv** plays exactly one of these sports.

12 60 married men were asked whether they gave their spouse flowers or chocolates for their last birthday. It was found that 21 gave flowers, 26 gave chocolates, and 5 gave both flowers and chocolates.

- **a** Draw a Venn diagram to illustrate this information.
- **b** If one of the married men is chosen at random, find the probability that he gave his spouse:
  - **i** flowers but not chocolates
  - **ii** neither flowers nor chocolates
  - **iii** flowers or chocolates.

13 In a class of 24 students, 10 study Biology, 12 study Chemistry, and 5 study neither Biology nor Chemistry. Find the probability that a student picked at random from the class studies:

a Chemistry, but not Biology **b** both Chemistry and Biology.

14 50 tourists went on a "thrill seekers" holiday. 40 went white-water rafting, 21 went paragliding, and each tourist did at least one of these activities.

Find the probability that a randomly selected tourist:

a did both activities **b** went white-water rafting but not paragliding

15 100 people were asked the question: "Which of yesterday's meals did you make yourself?"

51 people made their breakfast, 41 people made their lunch, and 53 people made their dinner. 21 people made breakfast and lunch, 28 people made breakfast and dinner, and 26 people made lunch and dinner. 9 people made all three meals.

Let B represent the people who made their breakfast,

L represent the people who made their lunch,

and D represent the people who made their dinner.

a Copy and complete the Venn diagram alongside.

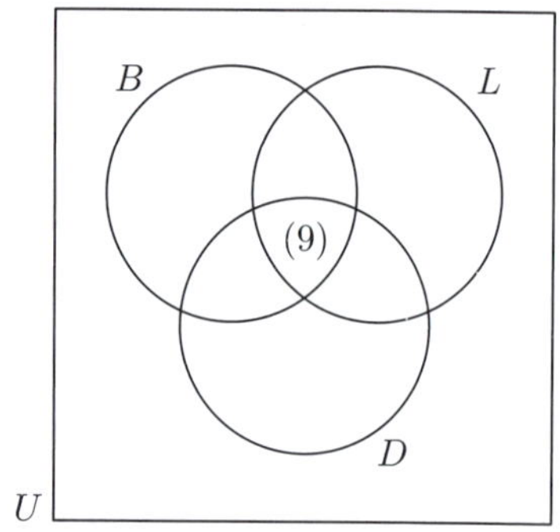

b One of the people surveyed is chosen at random. Find the probability that they made:

i breakfast only

ii none of the meals

iii lunch and dinner only

iv breakfast or lunch, but not both

v exactly two of the three meals.

PUZZLE — THE DICE PROBLEM

Suppose we roll a pair of standard dice, and find the sum of the numbers rolled.

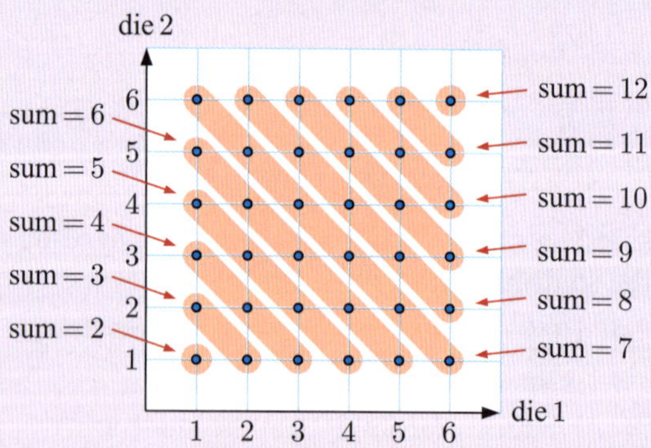

We can use a 2-dimensional grid to find the probability of obtaining each sum.

Sum	2	3	4	5	6	7	8	9	10	11	12
Probability	$\frac{1}{36}$	$\frac{2}{36}$	$\frac{3}{36}$	$\frac{4}{36}$	$\frac{5}{36}$	$\frac{6}{36}$	$\frac{5}{36}$	$\frac{4}{36}$	$\frac{3}{36}$	$\frac{2}{36}$	$\frac{1}{36}$

Can you find *another* way to label two dice with positive whole numbers, so that the sum probabilities are the same as those above?

The same number can appear more than once on a die, and numbers greater than 6 can be used.

C THE ADDITION LAW OF PROBABILITY

We now more carefully consider **compound events** or **combined events** where there is more than one event in our sample space. This may be because the experiment has more than one process, or because we are interested in more than one property of the outcome.

Suppose there are two events A and B in a sample space U. Following set notation:

- The event that **both** A **and** B occur is written $\boldsymbol{A \cap B}$, and read as "A intersection B".
- The event that **either** A **or** B **or both** occur is written $\boldsymbol{A \cup B}$, and read as "A union B".

From the Venn diagram alongside, we know that:

- $n(A) = a + b$
- $n(B) = b + c$
- $n(A \cap B) = b$
- $n(A \cup B) = a + b + c$

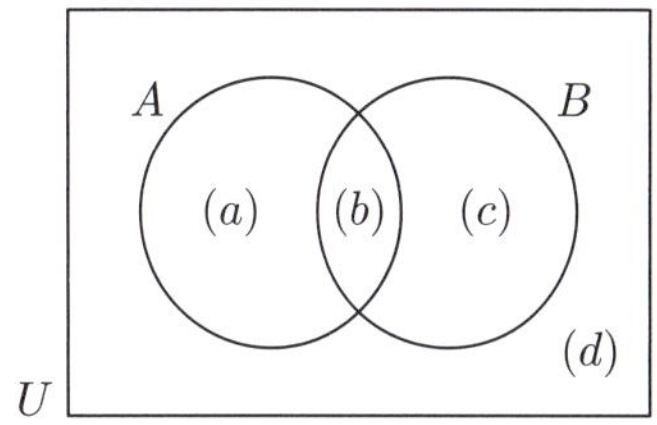

We can hence conclude that for any events A and B,

$$n(A \cup B) = n(A) + n(B) - n(A \cap B).$$

Dividing both sides by $n(U)$,

$$\frac{n(A \cup B)}{n(U)} = \frac{n(A)}{n(U)} + \frac{n(B)}{n(U)} - \frac{n(A \cap B)}{n(U)}$$

$$\therefore \ \text{P}(A \cup B) = \text{P}(A) + \text{P}(B) - \text{P}(A \cap B)$$

This result is described as the **addition law of probability**:

For two events A and B, $\quad \text{P}(A \cup B) = \text{P}(A) + \text{P}(B) - \text{P}(A \cap B)$

which means: $\quad \text{P}(\textbf{either } A \textbf{ or } B \textbf{ or both}) = \text{P}(A) + \text{P}(B) - \text{P}(\textbf{both } A \textbf{ and } B).$

Example 6 **Self Tutor**

If $\text{P}(A) = 0.2$, $\text{P}(A \cup B) = 0.6$, and $\text{P}(A \cap B) = 0.1$, find $\text{P}(B)$.

$\text{P}(A \cup B) = \text{P}(A) + \text{P}(B) - \text{P}(A \cap B)$

$\therefore \ 0.6 = 0.2 + \text{P}(B) - 0.1$

$\therefore \ \text{P}(B) = 0.5$

MUTUALLY EXCLUSIVE EVENTS

Two events are **mutually exclusive** or **disjoint** if they have no common outcomes.

If A and B are mutually exclusive events then $\text{P}(A \cap B) = 0$ and so the addition law becomes $\text{P}(A \cup B) = \text{P}(A) + \text{P}(B)$.

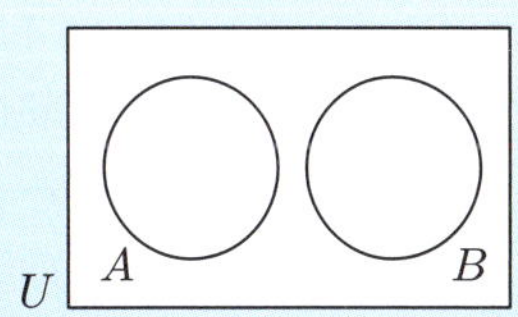

Example 7 **Self Tutor**

A card is selected at random from a normal pack of 52 playing cards.

Let X be the event that the card is a numbered spade, and Y be the event that the card is a red queen.

a Are X and Y mutually exclusive? Explain your answer.

b Illustrate the sample space using a 2-dimensional grid. Clearly indicate the outcomes in X and Y.

c Find: **i** $P(X)$ **ii** $P(Y)$ **iii** $P(X \cap Y)$ **iv** $P(X \cup Y)$

---

a A card cannot be both a spade and red.

$\therefore$ X and Y are mutually exclusive.

b

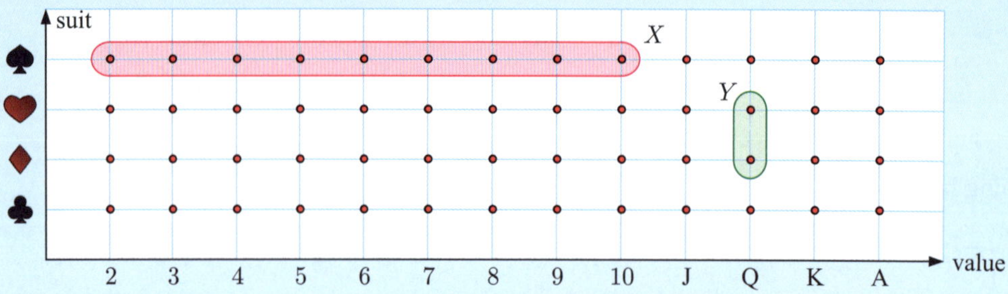

c **i** $P(X) = \frac{9}{52}$

ii $P(Y) = \frac{2}{52} = \frac{1}{26}$

iii $P(X \cap Y) = 0$

iv $P(X \cup Y) = P(X) + P(Y)$

$= \frac{9}{52} + \frac{2}{52}$

$= \frac{11}{52}$

EXERCISE 19C

1 Suppose $P(A) = 0.7$, $P(B) = 0.2$, and $P(A \cap B) = 0.15$.

a Are A and B mutually exclusive? Explain your answer.

b Find $P(A \cup B)$.

2 Suppose $P(B) = 0.4$, $P(A \cap B) = 0.3$, and $P(A \cup B) = 0.5$. Find $P(A)$.

3 Suppose $P(A) = 0.35$, $P(B) = 0.45$, and $P(A \cup B) = 0.65$. Find $P(A \cap B)$.

4 A coin and an ordinary die are tossed and rolled simultaneously.
Let A be the event of tossing a head, and B be the event of rolling a 5.

a Are A and B mutually exclusive? Explain your answer.

b Draw a grid to illustrate the possible outcomes. Clearly indicate the outcomes in A and B.

c Find:

i $P(A)$ **ii** $P(B)$ **iii** $P(A \cap B)$ **iv** $P(A \cup B)$

d Verify that $P(A \cup B) = P(A) + P(B) - P(A \cap B)$.

5 Suppose $P(X) = 0.3$, $P(X \cup Y) = 0.7$, and that X and Y are mutually exclusive. Find $P(Y)$.

6 Suppose $P(C) = 0.6$ and $P(D) = 0.7$. Explain why C and D are not mutually exclusive.

7 Balls numbered 1 to 20 are placed in a box, and a ball is selected at random. Let A be the event of selecting a single digit number, and B be the event of selecting a number greater than 15.

  a Are A and B mutually exclusive? Explain your answer.

  b Find:

   i $P(A)$ **ii** $P(B)$ **iii** $P(A \cap B)$ **iv** $P(A \cup B)$

D INDEPENDENT EVENTS

Consider spinning these spinners simultaneously.

Let A be the event that spinner 1 lands on red, and
B be the event that spinner 2 lands on an even number.

Spinner 1 **Spinner 2**

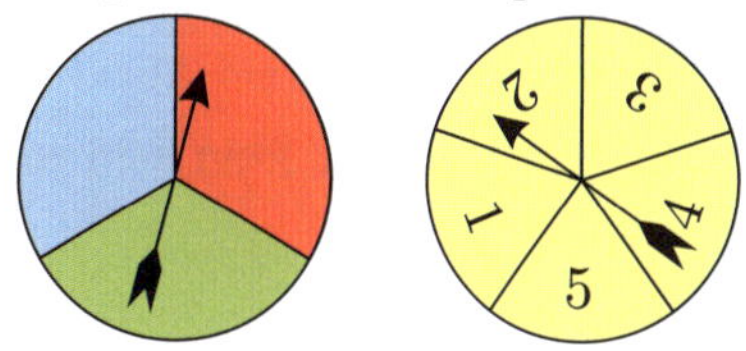

The possible outcomes for the combined event are:

{R1, **R2**, R3, **R4**, R5, G1, G2, G3, G4, G5, B1, B2, B3, B4, B5}

We can see that $P(A \text{ and } B) = \frac{2}{15}$.

For a compound event like this, we can also calculate the probability *without* listing the outcomes.

The outcome of one spinner does not affect the outcome of the other. We say the events are *independent*.

Notice that $P(A) = \frac{1}{3}$ and $P(B) = \frac{2}{5}$, so $P(A) \times P(B) = \frac{2}{15}$.

So, $P(A \textbf{ and } B) = P(A) \times P(B)$.

Two events are **independent** if the occurrence of each event does not affect the occurrence of the other.

If two events A and B are **independent** then $P(A \textbf{ and } B) = P(A \cap B) = P(A) \times P(B)$.

Example 8 **Self Tutor**

A 4-sided die and a 6-sided die are rolled simultaneously.

Find the probability that both numbers rolled will be greater than 1.

The events are independent, since the outcome from the 4-sided die does not affect the outcome from the 6-sided die, and vice versa.

$\therefore$ $P(\text{both numbers} > 1) = P(\text{4-sided die} > 1 \text{ and } \text{6-sided die} > 1)$

$= P(\text{4-sided die} > 1) \times P(\text{6-sided die} > 1)$

$= \frac{\cancel{3}^{1}}{4} \times \frac{5}{\cancel{6}_{2}}$

$= \frac{5}{8}$

This rule can be extended to any number of independent events.

For example: If A, B, and C are all independent events, then
$\mathrm{P}(A \textbf{ and } B \textbf{ and } C) = \mathrm{P}(A \cap B \cap C) = \mathrm{P}(A) \times \mathrm{P}(B) \times \mathrm{P}(C)$.

EXERCISE 19D

1 A die is rolled and the spinner alongside is spun.
Find the probability of obtaining:

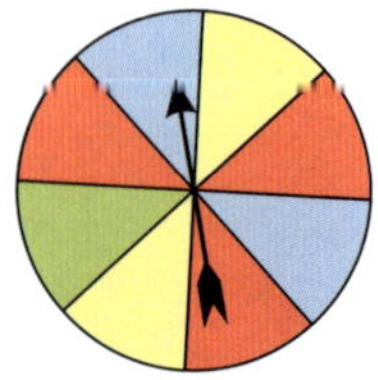

- **a** a 5 and a green
- **b** an even number and not red.

2 A disc is selected from each container alongside.
Find the probability of selecting:

- **a** a red disc from **A** and a green disc from **B**
- **b** a blue disc from both containers.

3 A coin and a 6-sided die are tossed simultaneously. Find the probability of getting:

- **a** a tail and a number larger than 2
- **b** a head and a composite number.

4 A school has two photocopiers. On any one day, machine A has an 8% chance of having a paper jam, and machine B has a 12% chance of having a paper jam.
Find the probability that, on any one day, both machines will:

- **a** have paper jams
- **b** work uninterrupted.

5 A boy and a girl were each asked what day of the week they were born. Find the probability that:

- **a** the boy was born on a Monday and the girl was born on a Wednesday
- **b** the boy was born on a weekend, but the girl was not
- **c** both children were born on a weekday.

6 Each day, Steve attempts the easy, medium, and hard crosswords in the newspaper. He has probability 0.84 of completing the easy crossword, probability 0.59 of completing the medium crossword, and probability 0.11 of completing the hard crossword.
Find the probability that, on any given day, Steve will:

- **a** complete all 3 crosswords
- **b** leave all 3 crosswords incomplete
- **c** complete the easy and medium crosswords, but not the hard crossword
- **d** complete the medium crossword, but not the other two crosswords.

Example 9 Self Tutor

A marble is selected at random from each of the bags alongside.

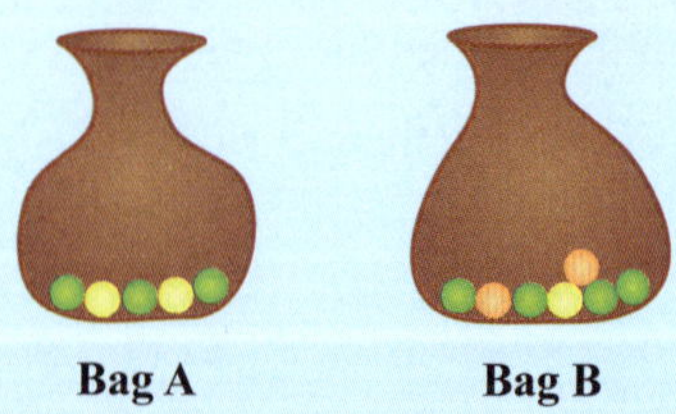

- **a** Draw a tree diagram to display the possible outcomes.
- **b** Find the probability of obtaining a green marble and a yellow marble.

a Let G represent a green marble,
Y represent a yellow marble, and
O represent an orange marble.

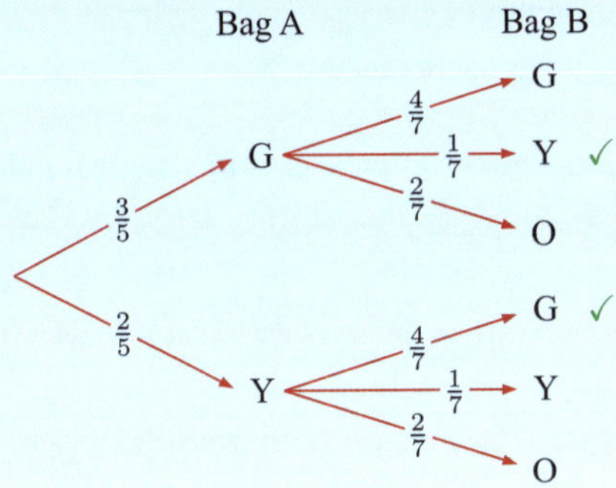

b P(a green and a yellow marble)

$= \text{P(GY or YG)}$ {branches marked ✓}

$= \frac{3}{5} \times \frac{1}{7} + \frac{2}{5} \times \frac{4}{7}$

$= \frac{11}{35}$

7 Each of the spinners alongside is spun once.

a Copy and complete this tree diagram to display the possible outcomes.

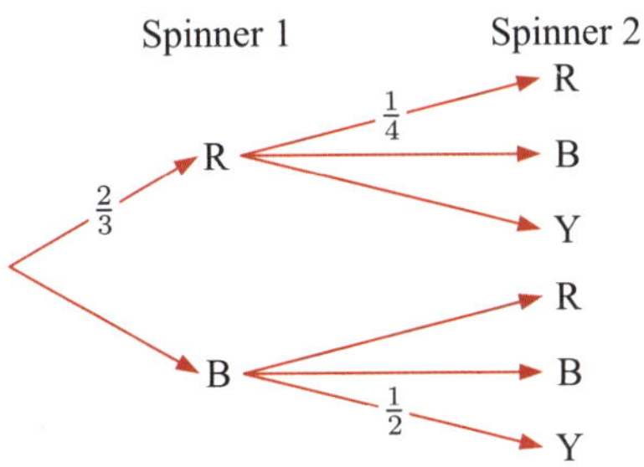

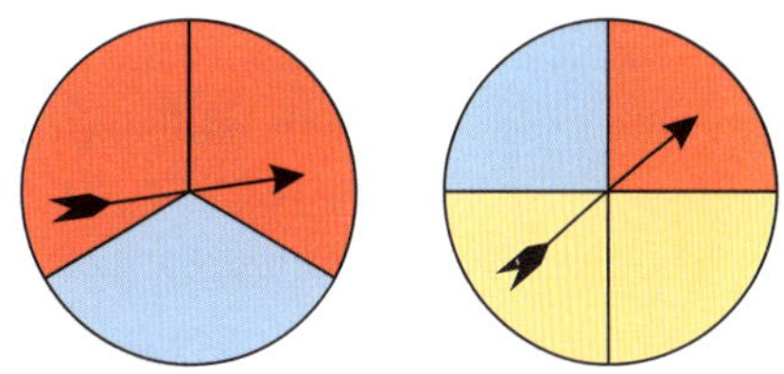
Spinner 1 **Spinner 2**

b Find the probability that:

i the spinners land on the same colour

ii the spinners land on different colours

iii exactly one of the spinners lands on blue.

8 A ticket is selected at random from each of these boxes.

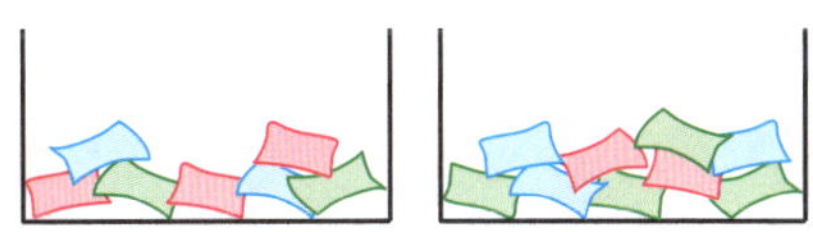
Box A **Box B**

a Draw a tree diagram to display the possible outcomes.

b Find the probability of obtaining:

i a blue ticket and a green ticket

ii two tickets of the same colour

iii exactly one pink ticket.

c Find the sum of the probabilities in **b**. Explain your result.

9 Sharon, Christine, and Keith have agreed to meet at a restaurant for lunch. Sharon has probability 0.8 of being on time, Christine has probability 0.7 of being on time, and Keith has probability 0.4 of being on time. Find the probability that at least 2 people will arrive on time.

10 X and Y are independent events such that $P(X) = 0.5$ and $P(Y) = 0.3$. Find $P(X \cup Y)$.

11 A and B are independent events such that $P(A \cup B) = 0.9$ and $P(A \cap B) = 0.4$. Find $P(A)$ and $P(B)$ given that $P(A) > P(B)$.

E DEPENDENT EVENTS

Two events are **dependent** if they are **not independent**.

The occurrence of one of the events *does* affect the occurrence of the other event.

The rule for finding compound event probabilities for dependent events is different from the rule for independent events.

If A and B are dependent events, then
$\text{P}(A \text{ and } B) = \text{P}(A \cap B) = \text{P}(A) \times \text{P}(B \text{ given that } A \text{ has occurred})$.

Example 10 Self Tutor

A fruit bowl contains 3 apples and 5 oranges. Callan selects a piece of fruit at random, and eats it. His sister Michelle then selects a piece of fruit for herself.

Find the probability that:

a both pieces of fruit selected are apples

b Callan selects an orange and Michelle selects an apple.

a P(both are apples)

= P(Callan selects an apple *and* Michelle selects an apple)

= P(Callan selects an apple) × P(Michelle selects an apple *given* that Callan selects an apple)

$= \frac{3}{8} \times \frac{2}{7}$ ← 2 apples remaining, ← 7 pieces of fruit to choose from

$= \frac{6}{56} = \frac{3}{28}$

b P(Callan selects an orange *and* Michelle selects an apple)

= P(Callan selects an orange) × P(Michelle selects an apple *given* that Callan selects an orange)

$= \frac{5}{8} \times \frac{3}{7}$ ← 3 apples remaining, ← 7 pieces of fruit to choose from

$= \frac{15}{56}$

EXERCISE 19E

1 A pocket in a golf bag contains 6 white and 4 yellow golf balls. Two balls are randomly selected without replacement.

If the first ball is not replaced before the second is chosen, the events are *dependent*.

a Determine the probability that:

  i both are white

  ii the first is white and the second is yellow

  iii one of each colour is selected.

b Explain why your answers in **a** do not add up to 1.

2 In a class of 27 students, 13 students have brothers, 16 have sisters, and 5 have neither.

a Illustrate this information on a Venn diagram.

b Two students are selected at random from the class. Find the probability that both students have only sisters.

3 A box contains 3 red discs and 5 blue discs. Two discs are selected from the box. Find the probability that the discs have the same colour if:

a the first disc is replaced before the second disc is selected

b no replacement occurs.

4 Balls numbered 1 to 10 are placed in a bag. Two of the balls are selected at random. Find the probability that the numbers on the balls are consecutive.

5 Two cards are selected, without replacement, from a standard deck of 52 cards. Find the probability that:

a both of the cards are red

b the first card is a club and the second card is a diamond

c both of the cards are aces.

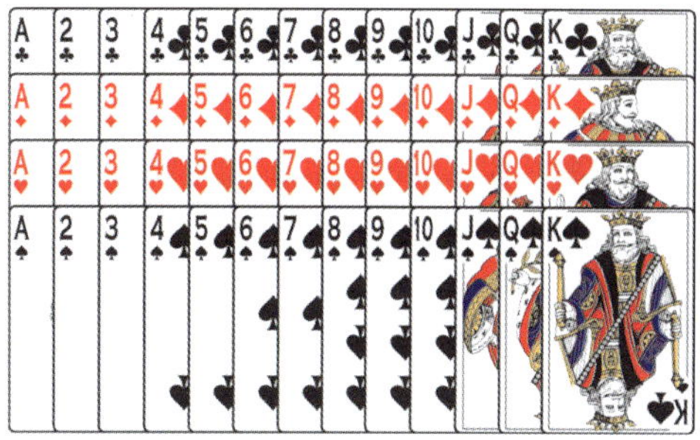

6 Answer the **Opening Problem** on page **310**.

Example 11 — Self Tutor

A container holds 3 banana iceblocks, 4 chocolate iceblocks, and 2 raspberry iceblocks. George randomly selects an iceblock from the container and eats it. His brother Oliver then randomly selects an iceblock from the container.

a Draw a tree diagram to display the possible outcomes.

b Find the probability that the brothers selected the same type of iceblock.

a Let B represent a banana iceblock,
C represent a chocolate iceblock,
and R represent a raspberry iceblock.

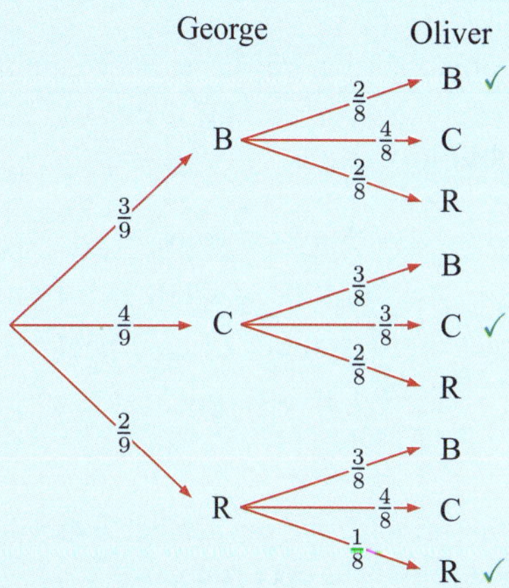

b P(brothers selected the same type of iceblock)

$= \text{P(BB or CC or RR)}$ {branches marked ✓}

$= \frac{3}{9} \times \frac{2}{8} + \frac{4}{9} \times \frac{3}{8} + \frac{2}{9} \times \frac{1}{8}$

$= \frac{20}{72}$

$= \frac{5}{18}$

7 Matt is the best player in his lacrosse team. He has an injured knee, and has only a 60% chance of playing the next game. The team has a 70% chance of winning the next game if Matt plays, but only a 45% chance of winning if he does not play.

a Represent this information on a tree diagram.

b Find the probability that the team will win the next game.

8 In a netball game, the Goal Shooter takes 65% of the team's shots and scores on 70% of her attempts. The Goal Attack takes the remainder of the shots, and scores on 60% of her attempts.
Find the probability that the team will score with their next shot.

9 A drawer contains 5 blue pens, 3 red pens, and 2 green pens. A pen is selected at random from the drawer and put to one side. A second pen is then selected at random from the drawer.
Find the probability that:

a the selected pens are the same colour **b** exactly one of the pens is green.

10 A bag contains 2 blue marbles and 5 red marbles. Jacinta tosses a fair coin. If the result is heads, she selects *one* marble from the bag. If the result is tails, she selects *two* marbles without replacement. Find the probability that Jacinta will select at least one red marble.

11 The spinner below is used to select box A, B, or C. Two discs are then randomly selected from that box, without replacement.

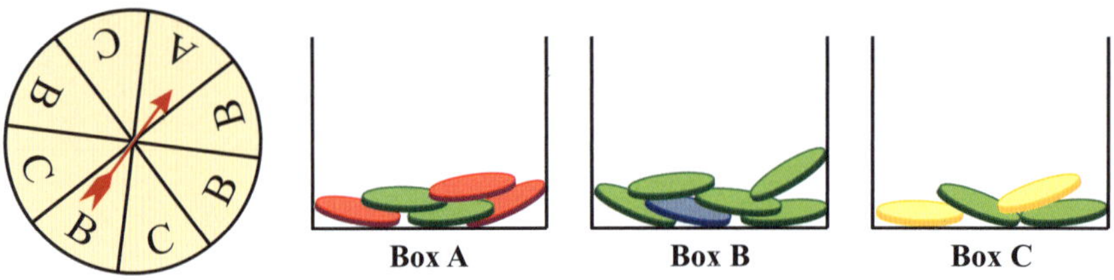

Find the probability that exactly one green disc is selected.

F EXPERIMENTAL PROBABILITY

In situations where we cannot calculate theoretical probabilities, we can *estimate* the probability of an event by performing many trials of the experiment. This estimate is known as an **experimental probability**.

In a probability experiment:

- the **number of trials** is the total number of times the experiment is repeated
- the **frequency** of a particular event is the number of times that this event is observed
- the **relative frequency** of an event is the frequency of that event divided by the total number of trials.

In the absence of any further data, the relative frequency of each outcome is our best estimate of the probability of it occurring.

- The relative frequency is the **experimental probability** of the event.
- The accuracy of an experimental probability is improved by increasing the number of trials.

Example 12 — Self Tutor

Larissa took 43 shots at a netball goal, and scored 29 times.
Estimate the probability that Larissa will miss with her next shot.

From 43 shots, Larissa scored 29 times and missed 14 times.

$\therefore$ P(misses next shot) $\approx \frac{14}{43} \approx 0.326$

EXERCISE 19F

1 Ivy bought some hens to free-range on her property. Of the first 123 eggs that they laid, she found that 11 had double-yolks. Estimate the probability that the next egg laid will have a double-yolk.

2 Richard arrived late to school 8 times during the first 45 days of the school year. Estimate the probability that he will arrive to school on time on the next school day.

3 A biased coin is flipped 200 times. It lands on heads 143 times, and on tails for the remainder. If the coin is flipped 3 times, estimate the probability of getting:

a all heads **b** all tails.

4 Karla tossed a metal nut 200 times, and found that it finished on its end 137 times. Her friend Sam tossed the nut 80 times, and found that it finished on its end 51 times.

a Estimate the probability of the nut finishing on its end using:

i Karla's results **ii** Sam's results.

b Which of the estimates in **a** do you think is the most accurate? Explain your answer.

c Describe how you can obtain an even more accurate estimate. Use your method to estimate the probability of the nut finishing on its end.

Example 13 — Self Tutor

A marketing company surveys 80 randomly selected people to discover what brand of shoe cleaner they use. The results are shown in the table alongside.

Brand	*Frequency*
Shine	27
Brite	22
Cleano	20
No Scuff	11

a Based on these results, estimate the probability that a randomly selected person uses:

i Brite **ii** Cleano.

b How accurate do you think your estimates in **a** will be? Explain your answer.

a We start by calculating the relative frequency for each brand.

Brand	*Frequency*	*Relative Frequency*
Shine	27	0.3375
Brite	22	0.2750
Cleano	20	0.2500
No Scuff	11	0.1375

i P(Brite) ≈ 0.275

ii P(Cleano) ≈ 0.250

b We expect the estimates will not be very accurate, as the sample size is small.

5 A marketing company conducted a survey on the brands of soap people use. The results are shown alongside.

Brand	*Frequency*	*Relative frequency*
Silktouch	62	
Super	53	
Just Soap	46	
Indulgence	39	
Total		

a Copy and complete the table.

b Estimate the probability that the soap used by a randomly selected person is:

i Just Soap **ii** Indulgence **iii** Silktouch.

6 This table shows the flavour of ice creams sold in a café one afternoon.

Flavour	*Frequency*
Chocolate	21
Strawberry	17
Lemon	4
Vanilla	15

a How many ice creams were sold?

b Estimate the probability that the next ice cream sold will be:

i strawberry **ii** chocolate or vanilla.

c Estimate the probability that the next ice cream sold will *not* be lemon.

7 Results from a poll for a local Council election are shown in the table.

Councillor	*Frequency*
Mr Tony Trimboli	216
Mrs Andrea Sims	72
Mrs Sara Chong	238
Mr John Henry	
Total	600

a How many of the people surveyed will vote for John Henry?

b Estimate the probability that a randomly selected person from this electorate will vote for:

i John Henry **ii** a female councillor.

Example 14 Self Tutor

The two-way table alongside shows the language studied by a random sample of students at a language school.

	French	German	Italian
Male	14	15	11
Female	18	9	16

Estimate the probability that a randomly chosen student from the school:

a is female **b** is studying French **c** is male, but not studying German.

We extend the table to include totals.

There are a total of 83 students included in the sample.

	French	German	Italian	*Total*
Male	14	15	11	40
Female	18	9	16	43
Total	32	24	27	83

a P(female) $\approx \frac{43}{83} \approx 0.518$

b P(studying French) $\approx \frac{32}{83} \approx 0.386$

c P(male, but not studying German) $\approx \dfrac{14+11}{83} \approx \frac{25}{83} \approx 0.301$

8 The table alongside gives the age distribution of inmates in a prison on a particular day.
If a new prisoner entered the prison the next day, estimate the probability that the new prisoner would be:

- **a** male
- **b** aged from 17 to 19
- **c** female and 19 or under
- **d** male and aged from 30 to 49.

Age	*Female*	*Male*	*Total*
15	0	6	6
16	5	23	28
17 to 19	26	422	448
20 to 24	41	1124	1165
25 to 29	36	1001	1037
30 to 34	32	751	783
35 to 39	31	520	551
40 to 49	24	593	617
50 to 59	16	234	250
60+	5	148	153
Total	216	4822	5038

9 The passenger numbers for a flight are shown in the two-way table.

- **a** Extend the table to include totals.
- **b** How many passengers are on the flight?
- **c** Estimate the probability that a randomly chosen passenger on the next flight will be:
  - **i** an adult
  - **ii** in business class
  - **iii** a child, but not in economy class.

	Adult	Child	Infant
First Class	11	3	1
Business Class	28	9	2
Economy Class	60	25	11

10 A random selection of hotels in Paris are given a gold star rating for quality, and a green star rating for environmental friendliness.
Estimate the probability that a randomly selected Paris hotel is given:

- **a** 2 gold stars and 4 green stars
- **b** 3 gold stars or higher
- **c** the same number of gold stars as green stars
- **d** more green stars than gold stars.

		Green star			
		★	★★	★★★	★★★★
Gold star	☆	5	2	1	1
	☆☆	4	10	4	3
	☆☆☆	2	8	13	8
	☆☆☆☆	1	7	9	4

G EXPECTATION

The probability of an event can be used to predict the number of times the event will occur in a number of trials.

For example, suppose that from the tennis tournaments last season, the experimental probability that Danielle hits her first serve in is 0.73. If Danielle has 92 first serves in her next match, we would *expect* her to hit $92 \times 0.73 \approx 67$ of them in.

If there are n trials of an experiment, and the probability of an event occurring in each trial is p, then the **expectation** of the occurrence of that event is np.

Example 15 **Self Tutor**

On one day, 75 of the 324 cars which came to a particular T-junction turned left. The next day, 579 cars came to the T-junction. How many of these cars would you expect to turn left?

The experimental probability of a car turning left is $p = \frac{75}{324}$.

We expect $579 \times \frac{75}{324} \approx 134$ of the cars to turn left the next day.

EXERCISE 19G

1 A goalkeeper has probability $\frac{3}{10}$ of saving a penalty kick. How many penalty kicks would he expect to save out of 90 attempts?

2 During the snow season there is probability $\frac{3}{7}$ of snow falling on any particular day. If Dan skis for five weeks, on how many days would he expect to see snow falling?

3 When a drawing pin was tossed 40 times, it landed on its back 14 times. If the drawing pin is tossed 250 more times, how many times would you expect it to land on its back?

4 Suppose you roll two dice simultaneously 180 times. How many times do you expect to get a double?

5 When Annette throws darts at the target shown, she hits the inner circle 17% of the time and the outer circle 72% of the time.

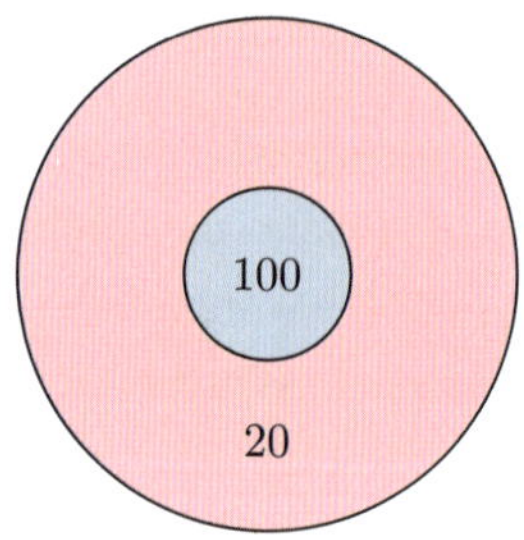

- **a** Estimate the probability that Annette will miss the target completely with her next throw.
- **b** Suppose Annette throws a dart 100 times at the target. She gets 100 points if she hits the inner circle and 20 points if she hits the outer circle. Find:
  - **i** the total number of points you would expect her to get
  - **ii** the average number of points you would expect per throw.

6

The pair of spinners alongside are spun 500 times. How many times do you expect the difference between the spins to equal 1?

7 250 people are asked to vote two people onto a committee from a group of 6 candidates. The group includes Freda and her brother Bjork. If every person votes for two people at random, how many would you expect to vote for neither Freda nor her brother?

8 Each morning, the probability that Eric goes jogging is 0.8 if he hears his alarm, and 0.4 if he does not. The probability that his brother Joel goes jogging is 0.3 if Eric goes jogging, and 0.1 if Eric does not. Given that Eric hears his alarm 60% of the time, on how many mornings in April would you expect exactly one of the brothers to go jogging?

9 Each day, the probabilities of Mandy receiving a text message from her mother, father, and brother are 0.6, 0.5, and 0.2 respectively. During a 365 day year, on how many days would you expect Mandy to receive a text message from:

a her mother

b her father but not her brother

c all three family members

d her brother and exactly one parent?

H CONDITIONAL PROBABILITY

We are often interested in the probability of an event occurring *given* that another event occurs.

For example:

- "Given that it is raining, what is the probability of Newcastle winning?"
- "Given that a randomly selected person likes oranges, what is the probability that they also like mandarins?"
- "Given that Mae has to pick up her son on the way, what is the probability that she will be late?"

For events A and B, we use the notation "$A \mid B$" to represent the event "*A given that B has occurred*".

We call $\text{P}(A \mid B)$ a **conditional probability** because it is the probability of A occurring on the *condition* that B has occurred.

Venn diagrams, two-way tables, and 2-dimensional grids are useful for answering questions involving conditional probabilities.

Example 16 — Self Tutor

In a group of 25 students, 15 like milk (M) and 17 like iced coffee (C). Two students like neither, and 9 students like both. Find the probability that a randomly selected student:

a likes milk

b likes milk given that he or she likes iced coffee.

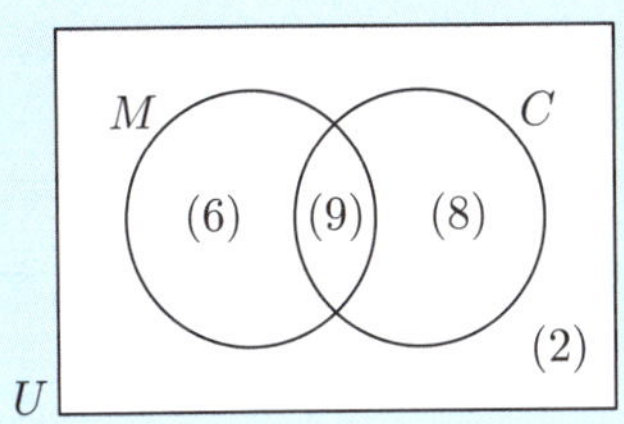

a 15 of the 25 students like milk.

$\therefore \text{P}(M) = \frac{15}{25} = \frac{3}{5}$

b Of the 17 students who like iced coffee, 9 also like milk.

$\therefore \text{P}(M \mid C) = \frac{9}{17}$

EXERCISE 19H

1 In a class of 25 students, 19 have fair hair and 15 have blue eyes. 22 students have fair hair, blue eyes, or both. A student is selected at random. Find the probability that the student has:

a fair hair and blue eyes

b blue eyes, given that the student has fair hair.

2 When 30 students sit for a French examination, 25 pass the aural part, 26 pass the written part, and 3 fail both parts. Find the probability that a randomly selected student:

a passed the written part, given that they passed the aural part

b failed the written part, given that they passed the aural part.

3 28 members of a youth group went hiking. 23 got sunburn, 8 got blisters, and 5 got both sunburn and blisters. Find the probability that a randomly selected person:

a got blisters or sunburn **b** got blisters, given that the person was sunburnt

c was sunburnt, given that the person did not get blisters.

Example 17 **Self Tutor**

A restaurant owner collected the following data about his customers one night:

Amount spent		Dine in	Takeaway	Delivery
	Less than $20	12	17	9
	At least $20	31	15	13

Find the probability that a randomly selected customer:

a spent at least $20

b spent less than $20, given that they had their meal delivered

c dined in, given that they spent less than $20.

Amount spent		Dine in	Takeaway	Delivery	*Total*
	Less than $20	12	17	9	38
	At least $20	31	15	13	59
	Total	43	32	22	97

a $\text{P(spent at least \$20)} = \frac{59}{97}$

b Of the 22 customers who had their meal delivered, 9 spent less than $20.

$\therefore$ $\text{P(spent less than \$20} \mid \text{delivered)} = \frac{9}{22}$

c Of the 38 customers who spent less than $20, 12 dined in.

$\therefore$ $\text{P(dined in} \mid \text{spent less than \$20)} = \frac{12}{38} = \frac{6}{19}$

4 Lee's soccer team played 37 games this season. Their results are shown in the two-way table.
Find the probability that a randomly selected game was:

a won **b** won, given that it was at home

c at home, given that it was won.

	Win	Draw	Lose
Home	11	5	3
Away	6	8	4

5 The Year 10 students at a school were asked how many people and pets lived in their house. The results are shown alongside.
Find the probability that a randomly selected student's house has:

a 2 pets, given that it has 4 people

b 3 people, given that it has at least 2 pets

c at most 1 pet, given that it has at least 4 people.

Number of people

Number of pets		2	3	4	5+
	0	4	6	7	3
	1	4	11	12	6
	2	2	9	11	7
	3+	1	5	7	5

Example 18 **Self Tutor**

A pair of dice is rolled. Given that at least one of the dice shows a 6, find the probability that the sum of the rolls is 10.

Let A be the event "the sum of the rolls is 10".

Let B be the event "at least one of the dice shows a 6".

Of the 11 outcomes in B {highlighted}, 2 of the outcomes are also in A {marked ×}.

$\therefore \quad P(A \mid B) = \frac{2}{11}$

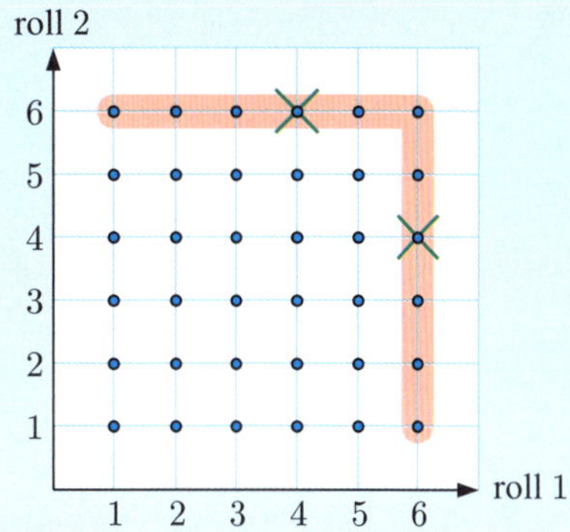

6 A pair of dice is rolled.

a Given that both of the dice show less than 5, find the probability that the sum of the rolls is 6.

b Given that the sum of the rolls is 8, find the probability that one of the dice shows a 3.

7 A card is selected at random from each of the bags shown.

Bag A Bag B

a Find the probability that the card drawn from bag A is the 4.

b If you knew that the sum of the selected cards was 5, would this make it more or less likely that the card drawn from bag A was the 4? Explain your answer.

c Given that the sum of the selected cards is 5, calculate the probability that the card drawn from bag A is the 4.

8 Suppose each spinner alongside is spun once.
Given that exactly one of the spins is green, find the probability that the result of spinner 2 is green.

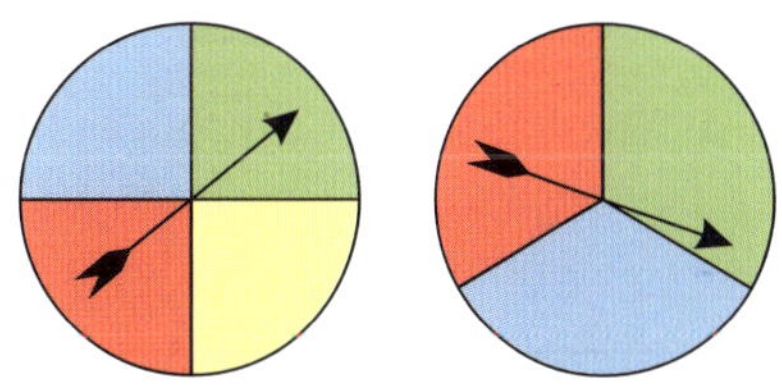

Spinner 1 Spinner 2

ACTIVITY INTERPRETING CONDITIONAL PROBABILITY

In this online Activity, we consider how conditional probabilities are *interpreted*.

In particular, we see how conditional probabilities can be misleading if they are interpreted incorrectly, leading to serious consequences.

ACTIVITY

MULTIPLE CHOICE QUIZ

QUICK QUIZ

REVIEW SET 19A

1 A coin and a pentagonal spinner with sides labelled A, B, C, D, and E are tossed and spun simultaneously. Illustrate the possible outcomes using a 2-dimensional grid.

2 A wheel numbered 1 to 20 is spun.
Find the probability that the result is:

- **a** 13
- **b** a multiple of 3
- **c** greater than 11.

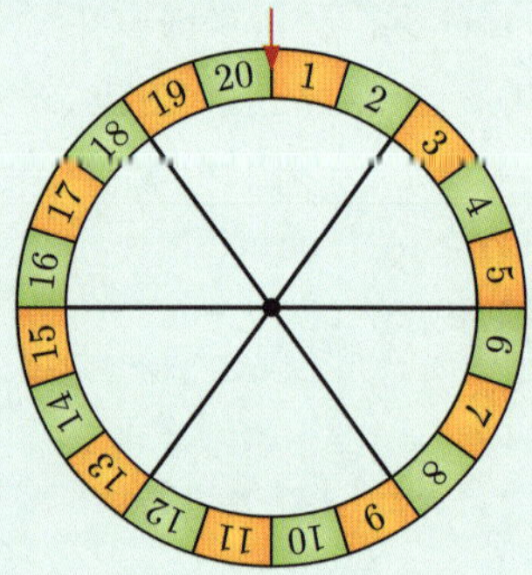

3 At the local girls' school, 65% of the students play netball, 60% play tennis, and 20% play neither sport.

- **a** Display this information on a Venn diagram.
- **b** Hence find the probability that a randomly chosen student plays:
  - **i** netball
  - **ii** netball but not tennis
  - **iii** at least one of these two sports
  - **iv** exactly one of these two sports.

4 Suppose $P(A) = 0.6$, $P(B) = 0.35$, and $P(A \cap B) = 0.25$. Find $P(A \cup B)$.

5 Suppose an ordinary die is rolled. Let X be the event of rolling a 5, and Y be the event of rolling an even number.

- **a** Are X and Y mutually exclusive? Explain your answer.
- **b** Find:
  - **i** $P(X)$
  - **ii** $P(Y)$
  - **iii** $P(X \cap Y)$
  - **iv** $P(X \cup Y)$

6 Brothers Paul, Cameron, and Bruce play in the same rugby team. Their probabilities of getting injured during the season are 0.4, 0.3, and 0.2 respectively. Find the probability that none of the brothers will be injured during the season.

7 A bag contains 4 green and 3 red marbles. Two marbles are randomly selected from the bag, the first being put to one side before the second is drawn. Find the probability that:

- **a** both are green
- **b** they are different in colour.

8 A paper plate is tossed in the air 50 times. It lands face up 37 times, and lands face down 13 times.

- **a** Estimate the probability that the paper plate will land face up the next time it is tossed.
- **b** If the plate is tossed 240 times, how many times would you expect it to land face up?
- **c** If the plate is tossed three times, estimate the probability that the plate will land face down on all three occasions.

9 Each morning when Harold has a shower, there is a 90% chance that the hot water is working. The probability that Harold has a long shower is 80% if the hot water is working, and 10% if the hot water is not working.
During a 365 day year, how many long showers would you expect Harold to have?

10 The two-way table alongside describes the books on Elizabeth's bookshelf.

	Biography	Novel	Reference
Softcover	3	22	6
Hardcover	5	7	4

a How many books are on Elizabeth's bookshelf?

b Find the probability that a randomly selected book is:

i a softcover book **ii** a hardcover reference book

iii a novel, given that it has a hard cover.

11 The spinners alongside are each spun once. Given that the spins are the same colour, find the probability that they are both red.

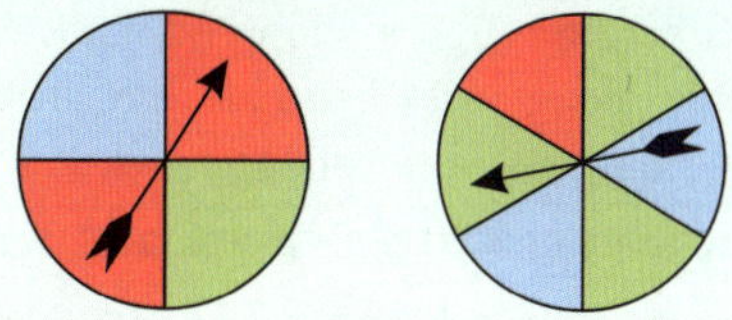

12 Shelley draws 3 cards without replacement from the container alongside. She will win a prize if all 3 cards are the same colour.

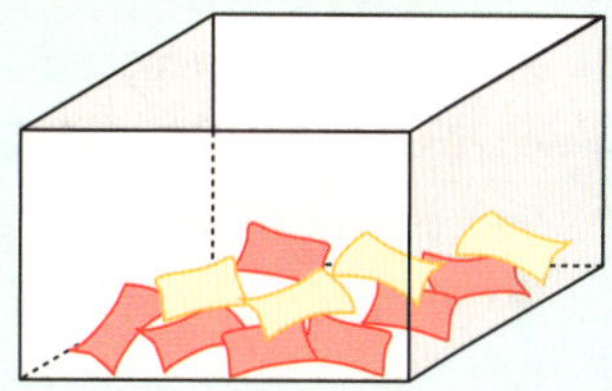

a Find the probability that Shelley will win a prize.

b Suppose the rules change so that Shelley instead draws her cards *with* replacement.

i Will this change increase or decrease her probability of winning? Explain your answer.

ii Find the probability that Shelley will win under the new rules.

REVIEW SET 19B

1 Two balls are drawn from a bag containing blue and yellow balls. Let E be the event that the balls have the same colour.

a Draw a tree diagram to illustrate the sample space. Highlight the outcomes corresponding to event E.

b List the outcomes in E using set notation.

2 The digits 1, 6, and 9 are placed in a random order to create a 3 digit number. Find the probability that this number will be a perfect square.

3 Suppose $P(A) = 0.4$, $P(B) = 0.55$, and $P(A \cup B) = 0.6$. Find $P(A \cap B)$.

4 A knife block contains 6 knives of different sizes. Each knife fits exactly into its specific hole, and will also fit into the holes of larger knives.

Suppose a knife is chosen at random, and a hole in the knife block is chosen at random. Find the probability that the knife will fit in the hole.

5 Alec and Joe sit for an examination in Chemistry. Alec has a 95% chance of passing and Joe has a 25% chance of passing. Determine the probability that:

a both pass **b** both fail

c Joe passes *and* Alec fails **d** Alec passes *and* Joe fails.

6 A married couple own a large car and a small car. Glen uses the small car 30% of the time. When he goes to the shops, the probability that he can park in the car park is 80% if he has the small car, and 60% if he has the large car.

Find the probability that on a given day, Glen is able to park in the shop's car park.

7 A ball is selected from box A and placed in box B. A ball is then selected from box B and placed in box C. A ball is then selected from box C.
Find the probability that the ball is blue.

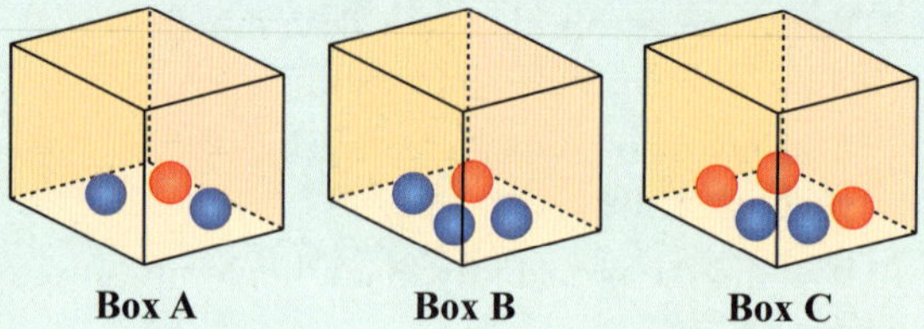

8 Pierre conducted a survey to investigate the ages of people walking through a shopping mall. The results are shown in the table alongside. Estimate the probability that the next person Pierre meets in the shopping mall will be:

Age	*Frequency*
0 to 19	22
20 to 39	43
40 to 59	39
60+	14

a between 20 and 39 years of age
b less than 40 years of age **c** at least 20 years of age.

9 From past experience, a surfer has probability 0.83 of catching a wave. In one week she tries to catch 75 waves. How many waves do you expect her to catch?

10 Suppose a pair of dice is rolled 600 times. How many times do you expect the sum of the rolls to equal 10?

11 University students were asked if they own a car and if they own a computer.
The results are shown in the two-way table.
Estimate the probability that a randomly selected university student has:

	car	no car
computer	124	168
no computer	16	22

a a computer **b** a car **c** a computer and a car
d a car, given that the student does not have a computer.

12 A class consists of 25 students. 15 own a cat, 9 own a dog, and 3 own both a cat and a dog.
a Represent this information on a Venn diagram.
b Hence find the probability that a randomly selected student from the class:
i owns neither a cat nor a dog **ii** owns a cat but not a dog
iii owns a dog, given that he or she owns a cat
iv does not own a dog, given that he or she does not own a cat.

Chapter 20

Statistics

Contents:

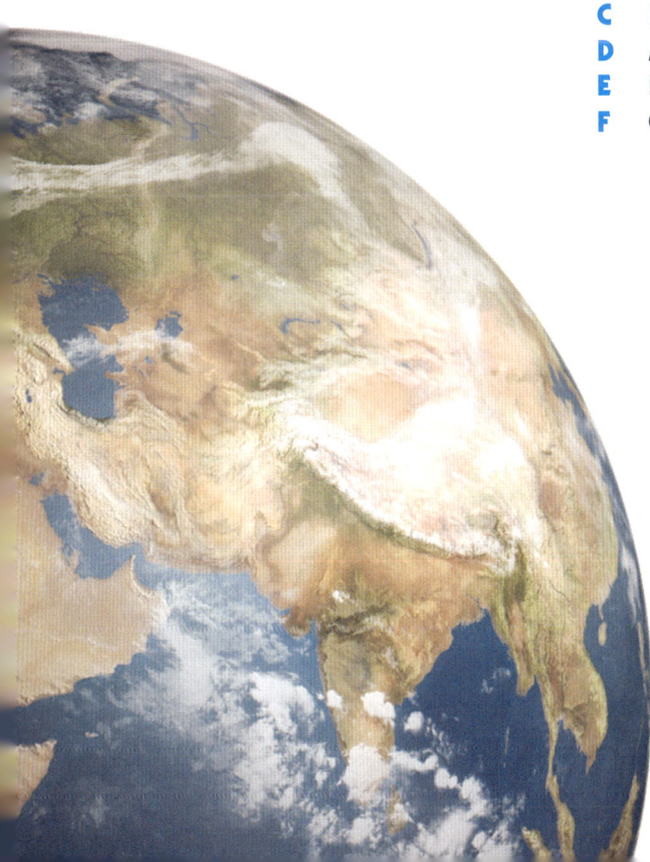

OPENING PROBLEM

Roland owns two hotels, one in New York and one in Miami. He wants to find out whether there is a difference in the number of nights guests stay at the hotels. He therefore inspects the next 40 bookings for each hotel, and records the number of nights the guests are staying.

New York:	2	3	1	2	4	2	6	3	4	5	8	3	1	3	4	2	1	2	4	5
	3	6	2	3	2	1	3	6	2	4	8	1	5	7	2	1	8	5	3	2
Miami:	2	4	4	5	3	6	2	3	1	7	2	3	4	3	5	6	5	2	4	7
	3	2	8	1	7	3	1	2	5	6	4	5	6	4	5	4	8	1	3	7

Things to think about:

a What is the best way to *organise* this data?

b How can we *display* the data?

c What is the *most common* length of stay at each hotel?

d How can Roland best measure:

- **i** the *average* length of stay for each hotel
- **ii** the *spread* of each data set?

e Can Roland draw a reliable conclusion from the data? What factors could affect the reliability of the conclusion?

f How could Roland improve the accuracy of his investigation?

Statistics is the study of collecting and analysing data.

The data we collect describes a particular feature or **variable** which differs between the people or objects we are studying.

The process of **statistical enquiry** or **investigation** includes the following steps:

Step 1: Decide on a topic you wish to investigate. Determine what data you will need to collect and how you will collect it.

Step 2: Collect the data.

Step 3: Organise the data.

Step 4: Summarise and display the data.

Step 5: Analyse the data and make a conclusion in the context of your investigation.

Step 6: Write a report to describe what you have done.

DATA COLLECTION

For any statistical investigation there is a target **population**. This is the group of things or people we are interested in finding information about.

When conducting a statistical investigation, we need to choose whether to collect data from the whole population, or from a **sample** chosen from that population.

A **census** is the process of collecting data from the whole population.

A **survey** is the process of collecting data from a sample.

Conclusions based on data from samples always involve some error. However, if the sample is unbiased and sufficiently large, we can use the properties of the sample to reliably **estimate** the properties of the population.

TYPES OF DATA

When we collect data, we measure or observe a particular feature or **variable** associated with the population.

- A **categorical variable** describes a particular quality or characteristic.
  The data is divided into **categories**, and the information collected is called **categorical data**.
- A **numerical variable** describes a quantity which takes a value which is a number.
  The information collected is called **numerical data**.

A DISCRETE NUMERICAL DATA

A **discrete numerical variable** takes exact numerical values. It is usually a result of counting.

For example, in the **Opening Problem**, the *number of nights stayed* is a discrete numerical variable. It can only take an exact value such as 1, 2, 3, 4, 5, ...

Discrete numerical data can be organised in a **tally and frequency table** and displayed in a **column graph** or **dot plot**.

The most frequently occurring value in a data set is called the **mode**.

For example, for the New York hotel data in the **Opening Problem**, we have:

Tally and frequency table

Nights	*Tally*	*Frequency*
1	卌 |	6
2	卌 卌	10
3	卌 |||	8
4	卌	5
5	||||	4
6	|||	3
7	|	1
8	|||	3

Column graph

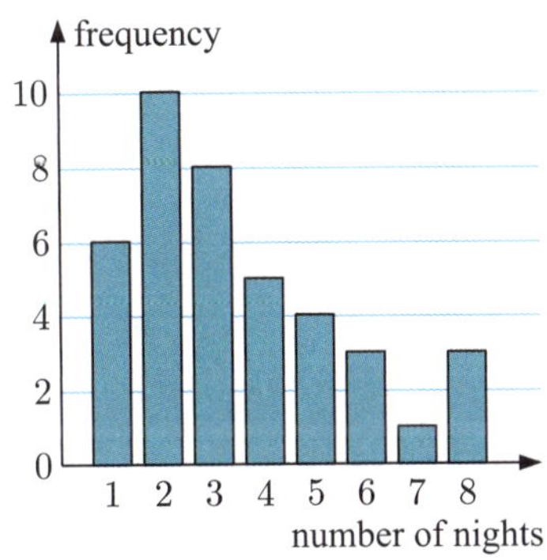

Dot plot

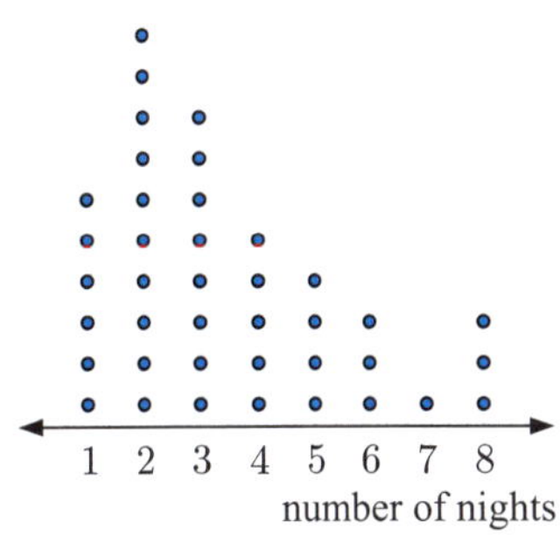

The mode of this data is 2. This means that the most common number of nights that guests have booked to stay for is 2 nights.

GROUPED DISCRETE DATA

If there are a large number of different data values, it may be appropriate to group the data into **class intervals**.

The **modal class** is the class interval with the highest frequency.

Example 1 **Self Tutor**

A popular wedding reception venue hosted receptions for these numbers of guests in the past month:

82 53 68 105 88 56 64 92 67 84 78 69 80 66
82 78 102 56 62 55 84 90 71 62 51 68 89 64

a Organise the data in a tally and frequency table using class intervals of width 10.

b Display the data on a column graph.

c State the modal class for the data.

a

Number of guests	*Tally*	*Frequency*
50 to 59	~~||||~~	5
60 to 69	~~||||~~ ||||	9
70 to 79	|||	3
80 to 89	~~||||~~ ||	7
90 to 99	||	2
100 to 109	||	2

b

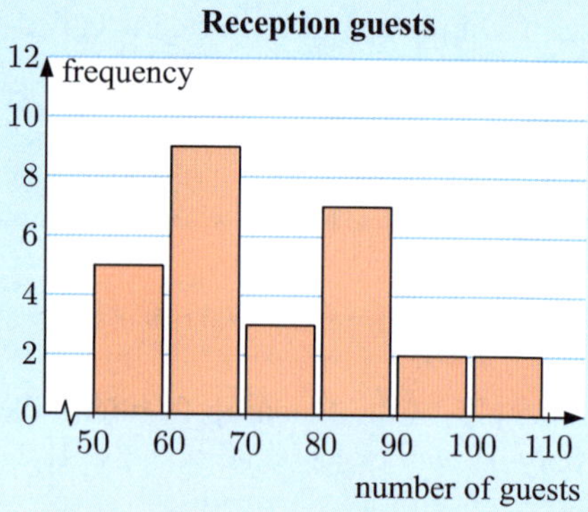

c The modal class is 60 to 69 guests.

EXERCISE 20A

1 The number of salad sandwiches bought from the school canteen each day was recorded for five weeks:

8 11 7 9 8 10 9 11 10 9 12 10 12
11 9 9 10 8 9 12 9 11 13 11 10

a Draw a dot plot to display the data.

b State the mode and explain what it means.

c On what percentage of days were less than 10 salad sandwiches sold?

2 The frequency table alongside summarises the scores when an archer fires 60 arrows at a target.

a Draw a column graph to display the data.

b State the mode and explain what it means.

c On what fraction of attempts did the archer score:

i 10 **ii** less than 9?

Score	*Frequency*
5	1
6	3
7	10
8	16
9	18
10	12

3 Consider the data for the Miami hotel in the **Opening Problem** on page **336**.

a Organise the data in a tally and frequency table.

b Draw a column graph to display the data.

c Compare your column graph with that for the New York hotel on page **337**. In which hotel do guests generally stay longer?

4 This table shows the number of shots in each point during a set of tennis.

Number of shots	*Frequency*
0 to 4	14
5 to 9	15
10 to 14	10
15 to 19	8
20 to 24	7
25 to 29	4

a How many points were played during the set?

b Display the data on a column graph.

c State the modal class for the data.

5 The number of people using a department store elevator was recorded each day for 30 days:

63 52 70 51 66 64 75 82 56 70 69 48 60 71 80
59 72 45 65 86 67 53 44 82 62 50 77 42 67 57

a Organise the data in a tally and frequency table using class intervals of width 10.

b Display the data on a column graph.

c State the modal class for the data.

HISTORICAL NOTE

Florence Nightingale (1820 - 1910) was a British nurse in Turkey during the Crimean War. She worked in very difficult conditions, with overcrowding, poor sanitation, little food, and few basic supplies. With the help of the Turkish Sanitary Commission in 1855, she significantly improved hygiene practices and sanitation. By the time the war ended in 1856, the hospitals were well-run and efficient, with mortality rates no greater than civilian hospitals in England. Nightingale had earned an extraordinary reputation, along with the label "the lady with the lamp".

In addition to her work as a nurse, Nightingale was a pioneer in medical statistics and statistical graphs. Her best known chart shown below was a variation of a pie graph called a **polar area diagram**. It showed the number of deaths each month and their causes. Each month is represented as a sector of the circle, the area of which indicates the number of deaths in that month. Blue indicates death from disease, red indicates death from wounds, and black indicates death from other causes.

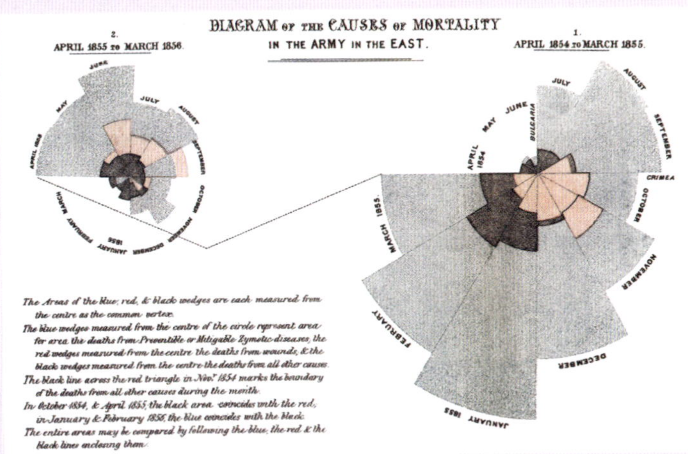

Florence Nightingale

The polar area diagram on the left (1855 - 56) shows a dramatic reduction in death from disease compared with the diagram on the right (1854 - 55), highlighting the impact of Nightingale's work.

Following the war, Nightingale convinced the UK government to completely reform the UK hospital system, with a similar dramatic reduction in rates of infection.

B CONTINUOUS NUMERICAL DATA

A **continuous numerical variable** can take any numerical value within a certain range.
It is usually a result of **measuring**.

For example, a person's *body temperature* is a continuous numerical variable. It could take any numerical value from about 35°C to 41°C.

When we consider a continuous numerical variable, no two data values will be *exactly* the same.

For example, even though we may record the body temperatures of two people as 37.6°C, their body temperatures will not be *exactly* 37.6°C, nor will their temperatures be *exactly* the same.

Since the data values are all different, we organise the data using **class intervals**.

We commonly use a **frequency histogram** or **histogram** to display continuous data. A histogram is similar to a column graph, but because the data is continuous, the columns are joined together.

The **modal class** is the class interval with the highest frequency. On a histogram, the modal class has the highest column.

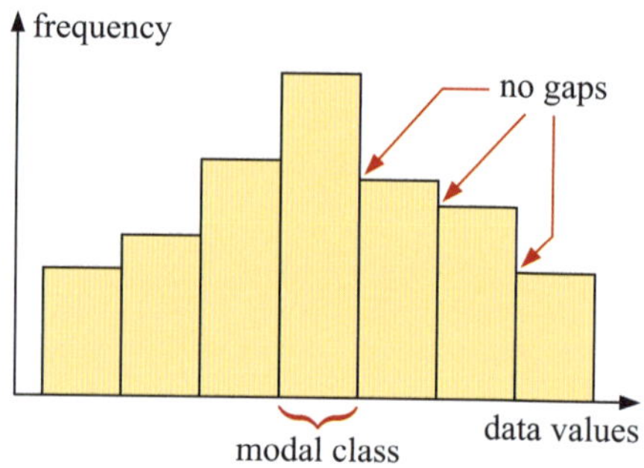

Example 2 **Self Tutor**

The weight w of each parcel sent on a given day from a post office was recorded in kilograms. Each measurement was rounded *down* to the nearest 0.01 kg.

$$\begin{array}{ccccccccc} 5.22 & 4.07 & 1.63 & 3.52 & 2.91 & 3.24 & 3.47 & 5.29 & 4.63 \\ 7.13 & 3.11 & 2.85 & 3.76 & 4.92 & 3.44 & 1.39 & 2.58 & 2.22 \end{array}$$

a Organise the data into class intervals.

b Find the modal class.

c Draw a frequency histogram to display the data.

d Over the next month, 564 parcels are sent from the post office. Estimate the number of parcels that weigh at least 4 kg.

a The lowest weight recorded was 1.39 kg and the highest was 7.13 kg, so we use class intervals of width 1 kg.

Weight (w kg)	*Frequency*
$1 \leqslant w < 2$	2
$2 \leqslant w < 3$	4
$3 \leqslant w < 4$	6
$4 \leqslant w < 5$	3
$5 \leqslant w < 6$	2
$6 \leqslant w < 7$	0
$7 \leqslant w < 8$	1

b The modal class is $3 \leqslant w < 4$ kg.

c **Weights of parcels**

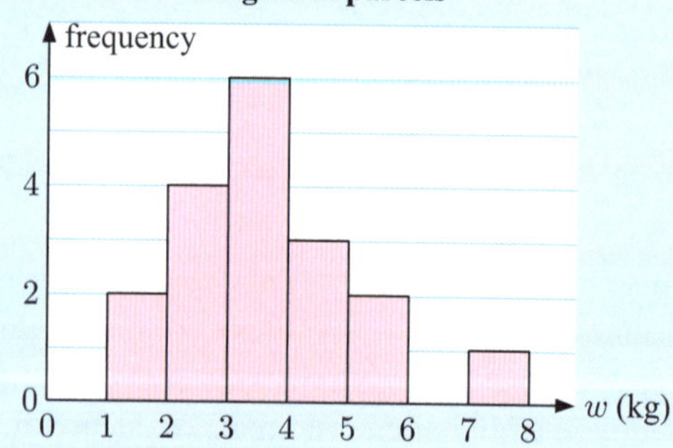

d Of the 18 parcels sent on the given day, $\frac{6}{18} = \frac{1}{3}$ of them weighed at least 4 kg.

$\therefore$ we expect $\frac{1}{3} \times 564 = 188$ parcels to weigh at least 4 kg.

EXERCISE 20B

1 Bianca travels to school by either bicycle or train. This histogram shows the times Bianca took to travel to school each day for 50 days.

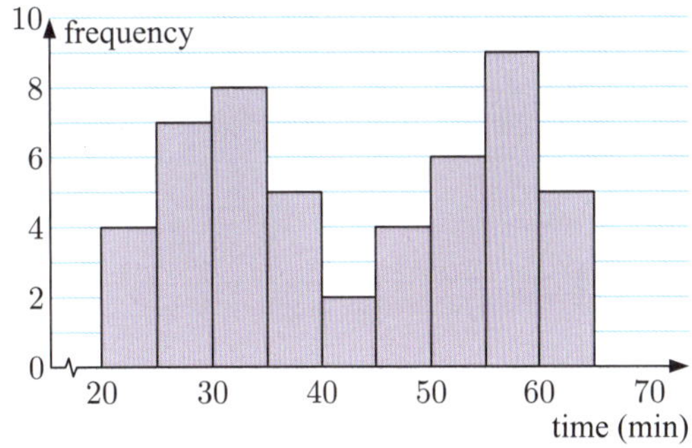

a Explain why the data is continuous.

b State the modal class.

c On what percentage of days did Bianca take less than 40 minutes to travel to school?

2 This table shows the weights of players in a rugby squad.

a Explain why *weight* is a continuous variable.

b Draw a frequency histogram to display the data.

c State the modal class and explain what it means.

d What percentage of the players weigh less than 90 kg?

Weight (w kg)	*Frequency*
$75 \leqslant w < 80$	2
$80 \leqslant w < 85$	5
$85 \leqslant w < 90$	8
$90 \leqslant w < 95$	7
$95 \leqslant w < 100$	5
$100 \leqslant w < 105$	1

3

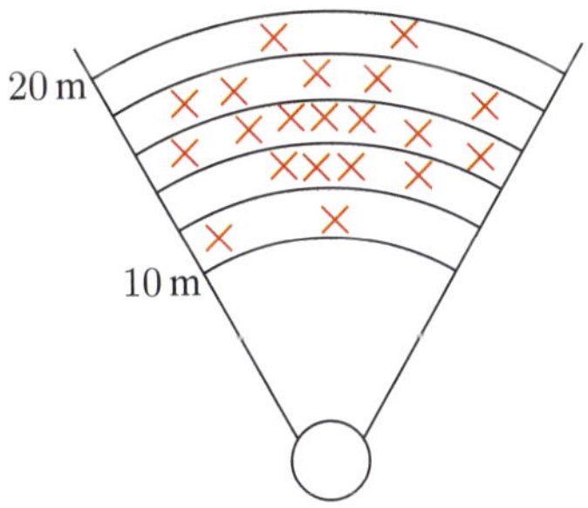

During a training session, Daniel performed 20 throws of the shot put. The results of the throws are shown alongside.

a Organise the data into class intervals.

b Draw a histogram to display the data.

c State the modal class and explain what it means.

d What fraction of Daniel's throws were farther than 14 m?

4 A group of young athletes obtained the following times, in seconds, for running 200 metres:

26.57	25.22	27.09	26.44	24.13	27.83	25.72	26.40
23.12	27.44	24.76	25.09	28.70	26.13	23.94	27.23
26.35	28.91	26.30	27.02	24.19	25.27	27.45	26.45
27.40	27.22	25.88	23.50	26.49	27.19	28.37	25.17
28.08	26.80	28.14	26.82	27.66	25.41	24.89	27.92

a Organise the data into class intervals.

b What percentage of the athletes obtained a time faster than 25 seconds?

c Draw a histogram to display the data.

d Find the modal class.

C DESCRIBING THE DISTRIBUTION OF DATA

When we display numerical data on a graph, the axis which shows the data values is a *number line*. The *shape* of the graph therefore gives us information about the **distribution** of the data.

Placing a curve over the data can help us to describe its distribution.

For example:

- This data is **symmetric** about the mode.

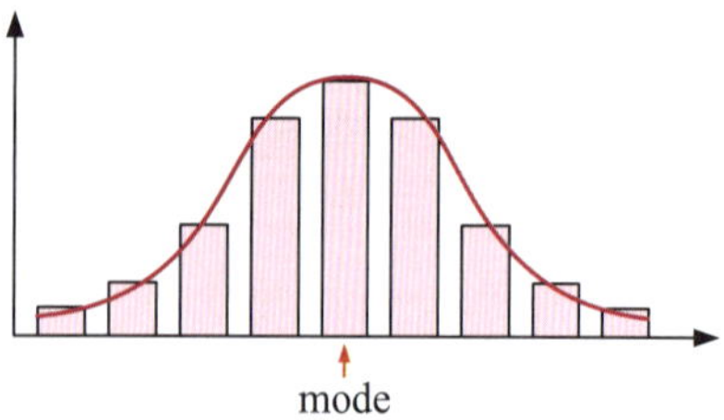

- This data is "stretched" on the right or positive side of the mode. We say the data is **positively skewed**.

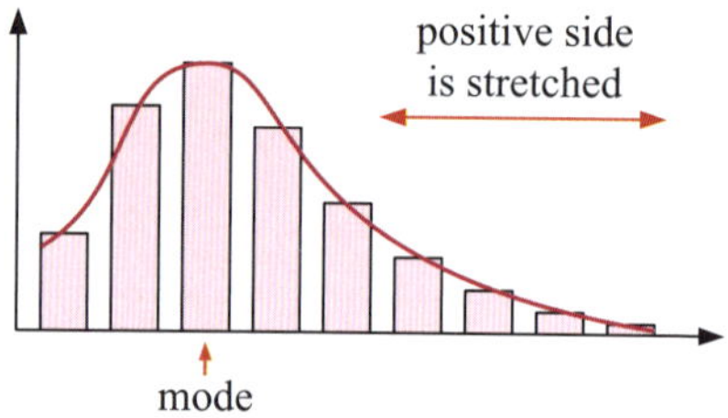

- This data is "stretched" on the left or negative side of the mode. We say the data is **negatively skewed**.

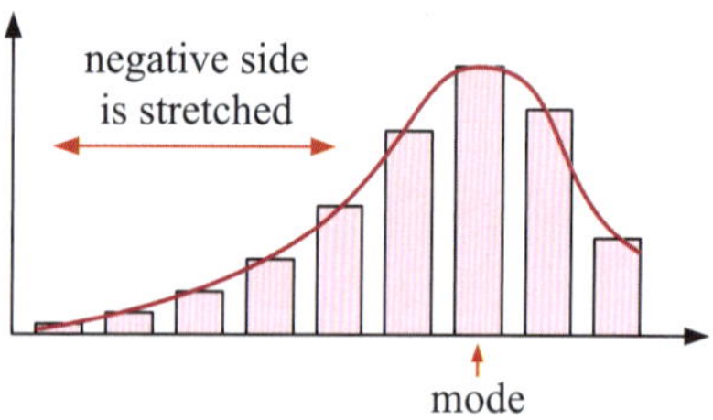

- This data contains two distinct peaks. We say the data is **bimodal**.

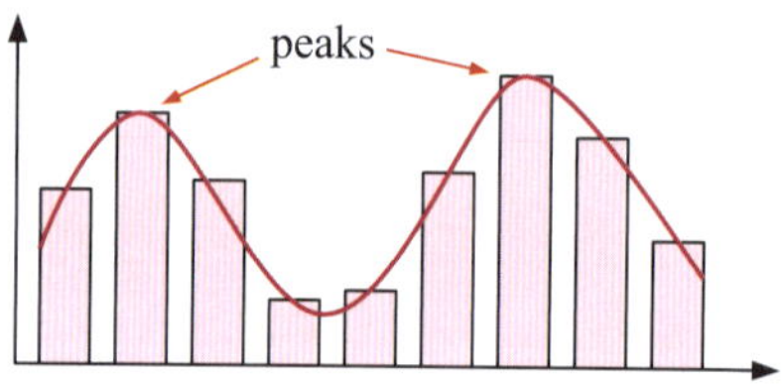

OUTLIERS

Outliers are data values that are either much larger or much smaller than the general body of data.

On a graph, an outlier is separated from the rest of the data.

When describing a distribution, you should mention any outliers.

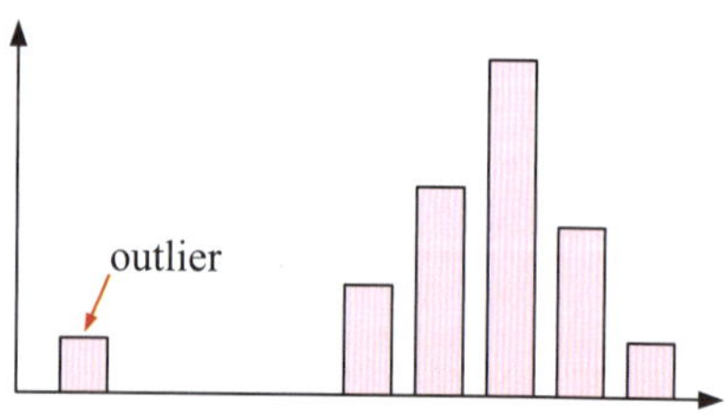

Example 3 — Self Tutor

A random sample of shoppers were asked how many times they visited the supermarket in the past week. The results are shown in the table.

a Draw a column graph to display the data.

b Describe the distribution of the data.

Number of visits	*Frequency*
1	9
2	23
3	18
4	6
5	1
6	0
7	1

a

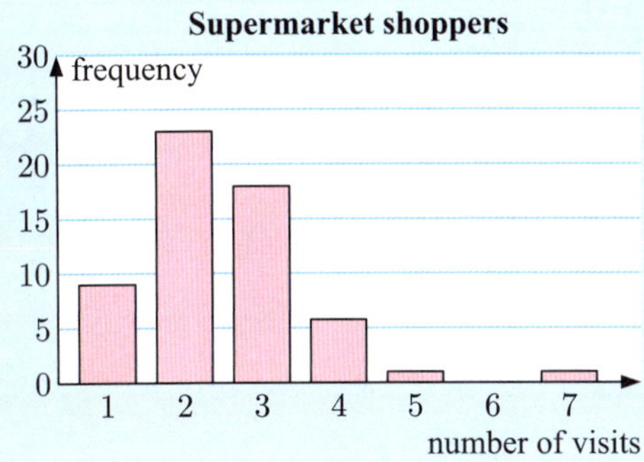

b The data is positively skewed.
The data value "7" is an outlier.

EXERCISE 20C

1 For each graph, describe the distribution of the data:

a

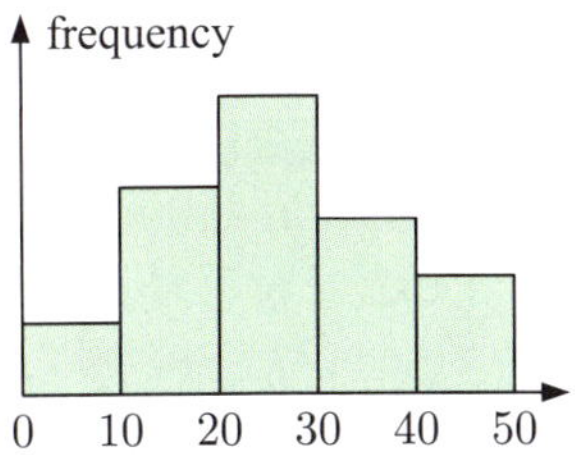

b

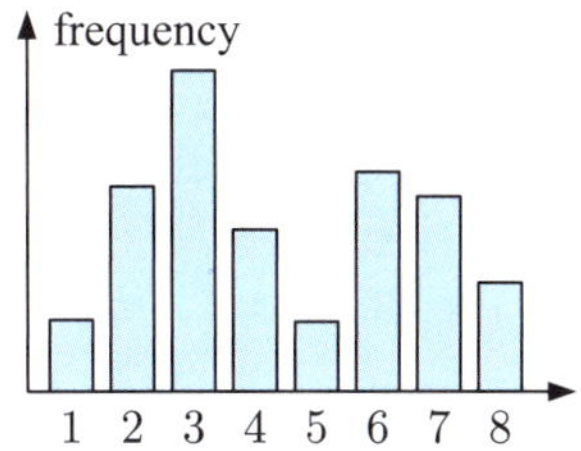

c

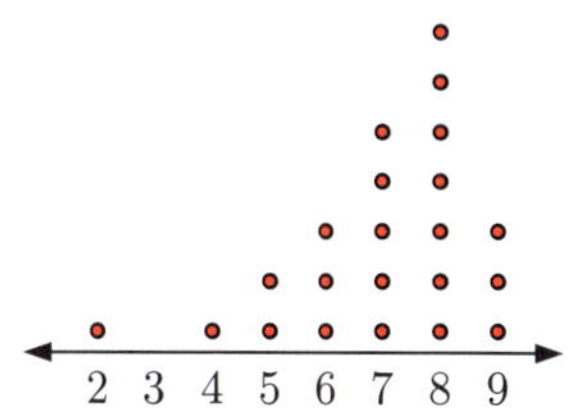

d

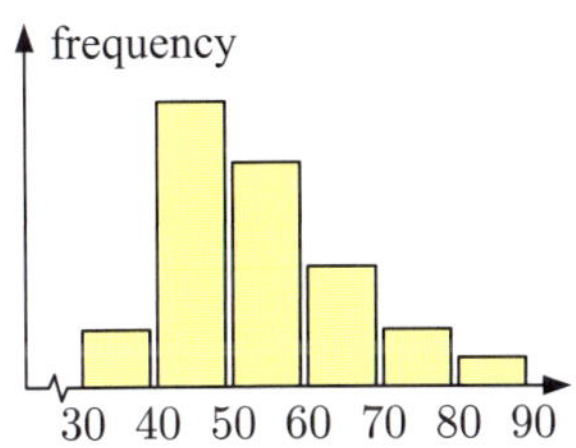

e

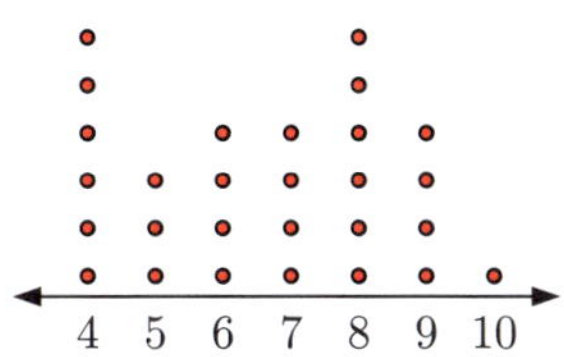

f

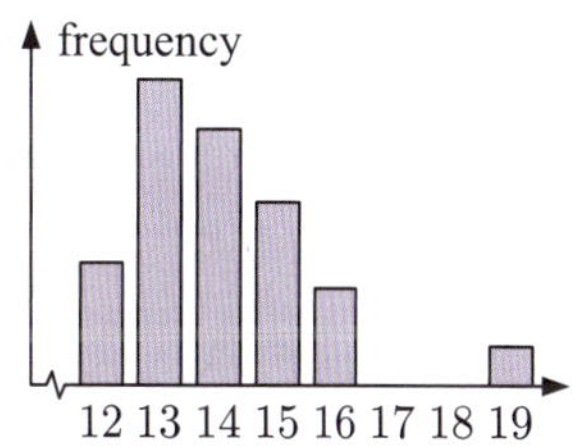

2 Employees of a company were asked how many times they left the office on business appointments during one week. The dot plot alongside was constructed from the data.

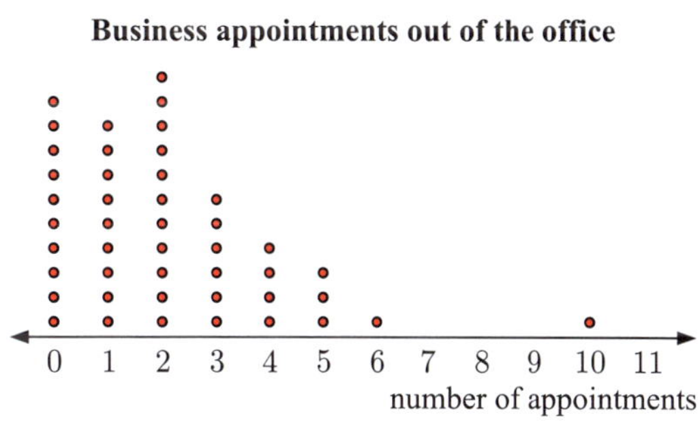

a How many employees did not leave the office?

b What percentage of the employees left the office more than 5 times?

c Describe the distribution of the data.

3 Rhiannon is a ten-pin bowler. She recorded the number of pins she knocked down in each frame during the last 6 games.

a Find the value of x.

b In how many frames did Rhiannon knock down more than 7 pins?

c In what percentage of frames did Rhiannon knock down less than 7 pins?

d Draw a column graph to display the data.

e Describe the distribution of the data.

Number of pins	*Frequency*
4	1
5	3
6	5
7	12
8	x
9	13
10	8
Total	60

4 The number of toothpicks in a box is stated as 50, but the actual number of toothpicks has been found to vary. To investigate this, the number of toothpicks in a box was counted for a sample of 60 boxes. The results were:

50	52	51	50	50	51	52	49	50	48	51	50	45	50	52
48	50	49	51	50	49	50	52	51	50	50	52	50	53	48
50	51	50	50	49	48	51	49	52	50	49	49	50	52	50
51	49	52	52	50	49	50	49	51	50	50	51	50	53	48

a Organise the data in a tally and frequency table.

b Display the data using a column graph.

c Describe the distribution of the data.

d What percentage of the boxes contained exactly 50 toothpicks?

5 A plant inspector takes a random sample of seedlings from a nursery, and measures their height in millimetres.

Height (h mm)	*Frequency*
$20 \leqslant h < 40$	4
$40 \leqslant h < 60$	17
$60 \leqslant h < 80$	15
$80 \leqslant h < 100$	8
$100 \leqslant h < 120$	2
$120 \leqslant h < 140$	4

a State the modal class.

b Draw a frequency histogram to display the data.

c Describe the distribution of the data.

d There are 857 seedlings in the nursery. Estimate the number of seedlings which measure:

i less than 100 mm

ii between 40 mm and 100 mm.

6 The graphs below show the heights from two samples of men aged 20.

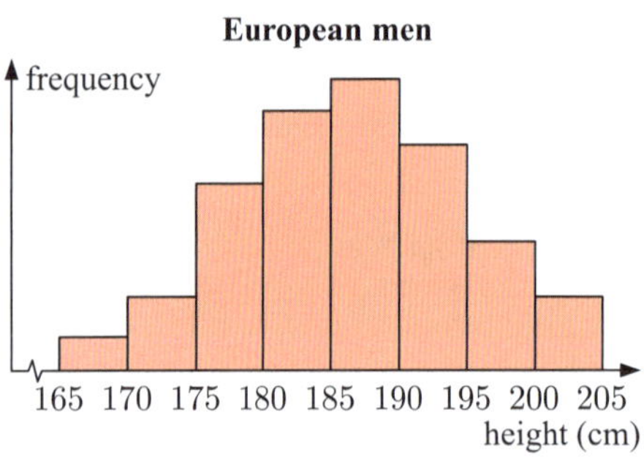

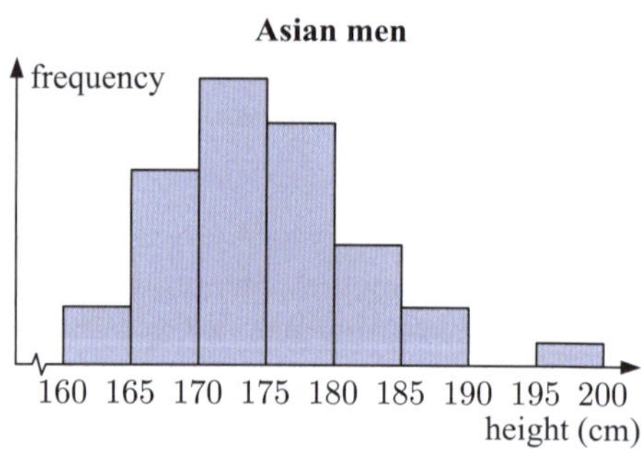

a Describe the distribution of each data set.

b Which group of men is generally taller? Explain your answer.

7 The scores for a group of 56 students in their end of year Physics examination were:

48	73	91	61	75	63	52	84	58	67	19	52	65	71
69	71	52	54	62	58	70	66	78	63	59	51	64	82
86	68	73	56	70	81	57	64	49	65	68	52	65	66
59	77	60	52	74	59	83	62	47	55	69	71	68	53

a Organise the data using class intervals of width 10.

b Draw a column graph to display the data.

c Describe the distribution of the data.

d Students scoring more than 80 were awarded an "A". What percentage of students were awarded an "A"? Give your answer to 1 decimal place.

D MEASURES OF CENTRE

We can gain a better understanding of a data set by locating the **middle** or **centre** of the data, and by measuring its **spread**. Knowing one of these without the other is often of little use.

There are three statistics we commonly use to measure the **centre** of a data set:

- the **mean**
- the **median**
- the **mode**.

THE MEAN

The **mean** $\overline{x}$ of a data set is the *arithmetic average* of the data values.

$$\textbf{mean} = \frac{\textbf{sum of data values}}{\textbf{number of data values}}$$

or $\overline{x} = \dfrac{\sum x}{n}$ where $\sum x$ is the sum of the data values.

$\overline{x}$ is read "x bar".
$\sum$ is read "the sum of all".

The mean is not necessarily a member of the data set.

THE MEDIAN

The **median** of a data set is the *middle* value when the set is written in *order*.

The median splits the data in two halves. Half of the data are less than or equal to the median, and half are greater than or equal to the median.

If there is an *odd* number of data values, there is one middle value which is the median.

If there is an *even* number of data values, there are two middle values. The median is the average of these values.

If there are n data values, the median is the $\left(\dfrac{n+1}{2}\right)$th data value.

For example:

- When $n = 17$, $\dfrac{n+1}{2} = 9$, so the median is the 9th ordered data value.
- When $n = 20$, $\dfrac{n+1}{2} = 10.5$, so the median is the average of the 10th and 11th ordered data values.

THE MODE

We have seen that for ungrouped discrete numerical data:

The **mode** is the most frequently occurring value in the data set.

If there are two data values which are the most frequently occurring, both values are modes and we say the data is **bimodal**.

If there are more than two data values which are the most frequently occurring, we say the mode is **undefined** and do not use it.

Example 4 Self Tutor

The number of peas in a randomly selected sample of pods were:

3 6 5 7 7 4 6 5 6 7 6 8 10 7 8

a For this data set, find the:

i mean **ii** median **iii** mode.

b A 16th pod is randomly selected, and it is found to contain 9 peas. How does this affect the:

i mean **ii** median?

a **i** mean $\overline{x} = \dfrac{3+6+5+ .... +10+7+8}{15}$ ← $\sum x$, ← n

$= \dfrac{95}{15}$

≈ 6.33 peas

ii The ordered data set is:

3 4 5 5 6 6 6 6 7 7 7 7 8 8 10 $\{n = 15, \ \dfrac{n+1}{2} = 8\}$

$\therefore$ the median $= 6$ peas

iii 6 and 7 occur most often, so the data is bimodal with modes 6 peas and 7 peas.

b **i** We expect the mean to rise since the new data value is greater than the old mean.

In fact, the new mean $= \dfrac{95+9}{16}$

$= \dfrac{104}{16}$

$= 6.5$ peas

ii The new ordered data set is:

3 4 5 5 6 6 6 6 7 7 7 7 8 8 9 10 $\{n = 16, \ \dfrac{n+1}{2} = 8.5\}$

$\therefore$ the new median $= \dfrac{6+7}{2} = 6.5$ peas

Given a frequency table, we can add:

- a *product* column to record the sum of scores with each particular value, and help us calculate the mean
- a *cumulative frequency* column to record the number of data values less than or equal to that particular score, and help us calculate the median.

Example 5 Self Tutor

When Caitlin proofread the first chapter of her novel, she recorded the number of errors she found on each page. The results are shown in the table.

For this data set, find the:

a mode **b** median **c** mean.

Number of errors	*Frequency*
0	6
1	9
2	5
3	2
4	1
5	1

Number of errors	*Frequency*	*Product*	*Cumulative frequency*
0	6	$0 \times 6 = 0$	6
1	9	$1 \times 9 = 9$	$6 + 9 = 15$
2	5	$2 \times 5 = 10$	$15 + 5 = 20$
3	2	$3 \times 2 = 6$	$20 + 2 = 22$
4	1	$4 \times 1 = 4$	$22 + 1 = 23$
5	1	$5 \times 1 = 5$	$23 + 1 = 24$
Total	$n = 24$	$\sum x = 34$	

a The most common number of errors was 1, so the mode = 1 error.

b $n = 24$, so $\dfrac{n+1}{2} = \dfrac{25}{2} = 12.5$

The 12th and 13th values are both 1, so the median = 1 error.

c mean $\overline{x} = \dfrac{\sum x}{n} = \dfrac{34}{24} \approx 1.42$ errors

EXERCISE 20D.1

1 Find the mean, median, and mode of each data set:

GRAPHICS CALCULATOR INSTRUCTIONS

STATISTICS PACKAGE

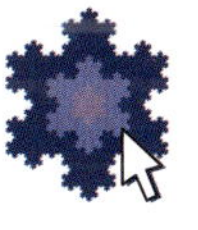

- **a** 12, 17, 20, 24, 30, 30, 42
- **b** 8, 8, 8, 10, 11, 11, 12, 12, 16, 20, 20, 24
- **c** 7.9, 8.5, 9.1, 9.2, 9.9, 10.0, 11.1, 11.2, 11.2, 12.6, 12.9
- **d** 427, 423, 415, 405, 445, 433, 442, 415, 435, 448, 429, 427, 403, 430, 446, 440, 425, 424, 419, 428, 441

Use technology to check your answers.

2 Consider these data sets:

Data set A: 5, 6, 6, 7, 7, 7, 8, 8, 9, 10, 12

Data set B: 5, 6, 6, 7, 7, 7, 8, 8, 9, 10, 20

- **a** Find the mean of both *data set A* and *data set B*.
- **b** Find the median of both *data set A* and *data set B*.
- **c** Explain why the mean of *data set A* is less than the mean of *data set B*.
- **d** Explain why the median of *data set A* is the same as the median of *data set B*.

3 The following raw data is the daily rainfall (to the nearest millimetre) of a city for the month of February 2020:

0, 4, 1, 0, 0, 0, 2, 9, 3, 0, 0, 0, 8, 27, 5, 0, 0, 0, 0, 8, 1, 3, 0, 0, 15, 1, 0, 0, 0

- **a** Find the mean, median, and mode of the data set.
- **b** **i** Identify any outliers in the data set.
  - **ii** If the outliers were excluded, which measure of centre would be most affected?
  - **iii** In the case of rainfall data, would you expect that the outliers would be errors, or is it reasonable that they are genuine data values?

4 A basketball team scored 38, 52, 43, 54, 41, and 36 points in their first six matches.

a Find the mean score for the first six matches.

b What score does the team need in their next match to maintain the same mean score?

c The team scores only 20 points in their seventh match. Find the new mean score.

d The team scores 42 points in their eighth and final match.

i Will their previous mean score increase or decrease? Explain your answer.

ii Find the mean score for all eight matches.

5 Chiara planted 19 seedlings, which she divided into groups A and B. Plants in group A were fertilised and plants in group B were not.
The heights of Chiara's plants in centimetres after one month were:

Group A: 18.8, 21.2, 17.6, 22.8, 20.0, 15.7, 19.7, 24.1, 21.0, 18.2
Group B: 11.8, 14.3, 17.6, 20.6, 15.4, 18.5, 27.9, 18.4, 16.2

a Calculate the mean and median of each data set.

b Do you think the fertiliser is effective? Explain your answer.

6 Three coins were tossed simultaneously 40 times, and the number of heads for each toss was recorded.
For this data, find the:

a mode **b** median

c mean.

Number of heads	*Frequency*
0	6
1	16
2	14
3	4
Total	40

7 The column graph alongside gives the value of donations for an overseas aid organisation, collected in a particular street.

a Construct a frequency table from the column graph.

b Determine the total number of donations.

c Which of the measures of centre can be found easily from the column graph only?

d Find the:

i mode **ii** median **iii** mean.

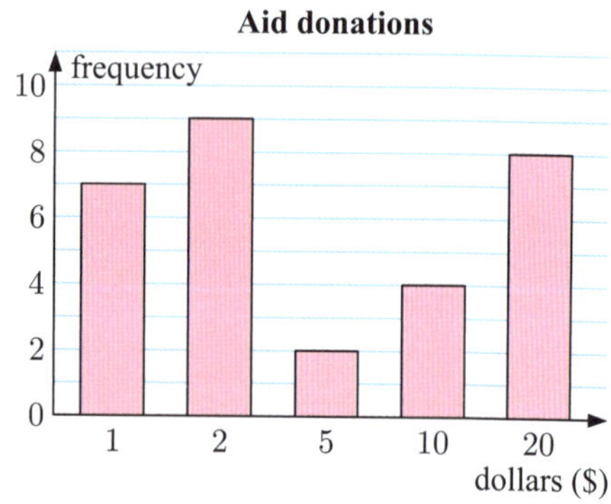

8 Hui breeds ducks. The number of ducklings surviving for each pair after one month is recorded in the table.

a For this data, find the:

i mode **ii** median **iii** mean.

b Describe the distribution of the data.

c How does the skewness of the data affect the measures of the centre of the distribution?

Number of survivors	*Frequency*
0	1
1	2
2	5
3	9
4	20
5	30
6	9
Total	76

9 Grace and Jack sell jewellery online. This side-by-side column graph shows the number of sales each person has made each week for the last 40 weeks.

  a Describe the distribution of each data set.

  b For each data set, find the:

    i mode **ii** median **iii** mean.

  c Who generally sells more jewellery?

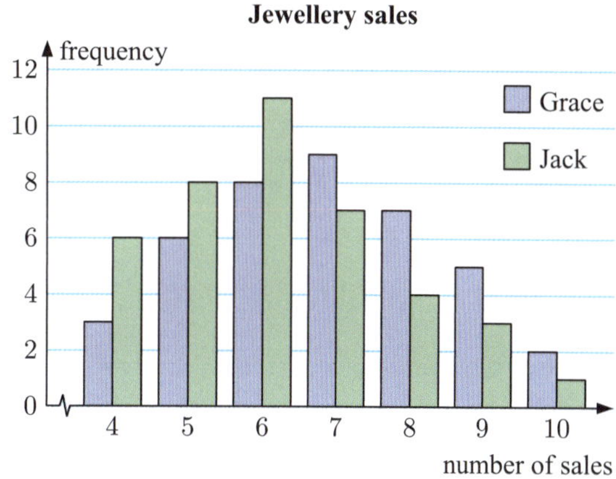

10 Consider the **Opening Problem** on page **336**.

  a For each data set, find the:

    i mean **ii** mode **iii** median.

  b At which hotel do guests generally stay longer?

Example 6 Self Tutor

Linda has taken four Mathematics tests so far this year. Each test has been out of 20 marks, and her average mark has been 15.

What mark does Linda need in the 5th test to raise her average to 16?

Average mark $= \dfrac{\text{sum of marks}}{4} = 15$

$\therefore$ sum of marks $= 15 \times 4 = 60$

Let Linda's mark for the 5th test be x.

$\therefore$ she needs $\dfrac{60+x}{5} = 16$

$\therefore 60 + x = 80$

$\therefore x = 20$

So, Linda needs a mark of 20 in the 5th test.

11 Jackie has played 8 games of football this season, obtaining an average of 17 possessions per game. How many possessions does she need in the next game to increase her average to 20?

12 A sample of 12 measurements has a mean of 8.5, and a sample of 20 measurements has a mean of 7.5. Find the mean of all 32 measurements.

13 Find x given that 7, 15, 6, 10, 4, and x have mean 9.

14 Find a given that 10, a, 15, 20, a, a, 17, 7, and 15 have mean 12.

15 On Saturday, Derek picked pears from 32 trees. He picked an average of 17 pears per tree. On Sunday he picked some more pears, averaging 12 pears per tree. Over the whole weekend he picked an average of 14 pears per tree. From how many trees did Derek pick on Sunday?

16 The mean, median, and mode of seven numbers are 8, 7, and 6 respectively. Two of the numbers are 8 and 10. If the smallest of the numbers is 4, find the largest number.

ESTIMATING THE MEAN OF GROUPED DATA

When data is presented in **class intervals**, either for grouped discrete data or for continuous data, the actual data values are not known. This makes it impossible to calculate the exact mean of the data set.

To *estimate* the mean of the data, we use the **midpoint** of an interval to represent all of the data values within the interval.

For example:

- For the discrete data class interval 50 to 59, the interval midpoint is $\frac{50+59}{2} = 54.5$.
- For the continuous data class interval $80 \leqslant t < 100$, the interval midpoint is $\frac{80+100}{2} = 90$.

Example 7 — Self Tutor

The transport department collected data regarding the distance each of its trams travelled in one day. The results are shown in the table.

Estimate the mean distance travelled by the trams each day.

Distance (d km)	*Frequency*
$50 \leqslant d < 100$	10
$100 \leqslant d < 150$	15
$150 \leqslant d < 200$	16
$200 \leqslant d < 250$	9

Distance (d km)	*Frequency*	*Interval midpoint*	*Product*
$50 \leqslant d < 100$	10	75	750
$100 \leqslant d < 150$	15	125	1875
$150 \leqslant d < 200$	16	175	2800
$200 \leqslant d < 250$	9	225	2025
Total	50		7450

$$\therefore \text{ mean} = \frac{\sum x}{n} \approx \frac{7450}{50} \approx 149 \text{ km}$$

EXERCISE 20D.2

1 Nick served a tennis ball 200 times. The speeds of the serves are summarised in the table alongside.

Speed (s km/h)	*Frequency*
$150 \leqslant s < 155$	18
$155 \leqslant s < 160$	28
$160 \leqslant s < 165$	35
$165 \leqslant s < 170$	43
$170 \leqslant s < 175$	41
$175 \leqslant s < 180$	35

  a Find the modal class of the data.
  b If possible, find the:
    i number of serves faster than 170 km/h
    ii number of serves slower than 162 km/h
    iii percentage of serves between 155 km/h and 175 km/h.
  c Estimate the mean speed of the serves.

2 A zoo conducts bus tours leaving every 15 minutes. This table summarises the number of passengers on each bus tour during one day.

Number of passengers	*Frequency*
10 to 14	7
15 to 19	9
20 to 24	15
25 to 29	6
30 to 34	3

  a State the modal class for the data.
  b Estimate the mean number of passengers on each tour.
  c Estimate the number of bus tours that will have less than 25 passengers in the next 14 days.

3 The Year 10 students at a school were asked how long they slept last night.
The responses are shown alongside, sorted by gender.

Time slept (t hours)	Frequency	
	Male	Female
$4 \leqslant t < 5$	6	4
$5 \leqslant t < 6$	10	8
$6 \leqslant t < 7$	13	10
$7 \leqslant t < 8$	9	11
$8 \leqslant t < 9$	7	8

- **a** Draw a histogram to display each data set.
- **b** Describe the distribution of each data set.
- **c** Estimate the mean of each data set.
- **d** Which group do you think slept longer on average? Comment on the reliability of your answer.

4 The teachers of a Year 7 class and a Year 10 class recorded how many days each of their students was absent from school in the past year.

Number of absent days	Frequency	
	Year 7	Year 10
0 to 4	3	7
5 to 9	5	8
10 to 14	9	5
15 to 19	10	3
20 to 24	6	2

- **a** Estimate the mean for each class.
- **b** Hence predict which class's students had the lower number of absent days on average.
- **c** Use the *lower* value of each interval to calculate the *minimum* possible value for the actual mean of each data set.
- **d** Calculate the *maximum* possible value for the actual mean of each data set.
- **e** Can you be *certain* that your prediction in **b** is correct? Explain your answer.

5 The amounts of petrol bought in one hour by customers at a service station are shown below, rounded down to the nearest 0.01 litres:

41.59	33.09	37.21	58.85	47.20	26.01	31.12	41.11	56.21	43.59
31.77	44.56	23.15	46.67	44.43	58.55	40.09	37.51	43.72	27.56
28.90	36.82	47.19	59.23	39.08	47.81	29.95	55.91	34.11	44.75
46.12	27.09	57.85	33.13	51.05	34.80	56.14	47.33	51.91	57.76
37.10	52.39	48.52	41.08	22.09	49.91	38.10	58.77	25.87	39.21

Suppose the amount of petrol bought by a customer is p litres.

- **a** Find the mean $\overline{p}$ of the data.
- **b** **i** Organise the data into class intervals of width 10.
  ii Hence *estimate* the mean of the data.
- **c** **i** Organise the data into class intervals of width 5.
  ii Hence *estimate* the mean of the data.
- **d** Compare your estimates of the mean in **b** and **c** with the actual mean found in **a**.
- **e** In general, would you expect the use of larger or smaller intervals to produce a more accurate estimate of the mean? Explain your answer.

GLOBAL CONTEXT — DEMOGRAPHICS

Global context: Identities and relationships
Statement of inquiry: Collecting and interpreting data can help us to understand our place in the world.
Criterion: Communicating

GLOBAL CONTEXT

E BOX PLOTS

THE FIVE-NUMBER SUMMARY OF A DATA SET

We have seen how the **median** divides an ordered data set into two halves. By finding the middle value of each half, we divide the original data into *quarters*. These middle values are therefore called **quartiles**.

The middle value of the *lower* half is called the **lower quartile** or $\mathbf{Q_1}$.
One quarter or 25% of the data values are less than or equal to Q_1.

The middle value of the *upper* half is called the **upper quartile** or $\mathbf{Q_3}$.
Three quarters or 75% of the data values are less than or equal to Q_3.

The quartiles form part of the **five-number summary** of a data set, which tells us how each quarter of the data is distributed.

The **five-number summary** of a data set consists of:

- the minimum value (min)
- the lower quartile (Q_1)
- the median (Q_2)
- the upper quartile (Q_3)
- the maximum value (max)

THE RANGE AND INTERQUARTILE RANGE

From the five-number summary of a data set, we can immediately calculate two measures of *spread* of the data:

- The **range** of a data set is the difference between the **maximum** data value and the **minimum** data value.

  range = maximum value − minimum value

- The **interquartile range (IQR)** is the range of the middle half or 50% of the data.

  interquartile range = upper quartile − lower quartile
  or $\mathbf{IQR = Q_3 - Q_1}$

BOX PLOTS

A **box plot** or **box-and-whisker plot** is a visual display of the five-number summary of a data set.

In a box plot:

- The rectangular box represents the middle half of the data set. Its length is the interquartile range.
- The lower whisker represents the 25% of the data with the lowest values.
- The upper whisker represents the 25% of the data with the highest values.
- The distance between the ends of the whiskers is the range.

For example, the following box plot is for a data set with:

- min $= 3$
- $Q_1 = 5$
- median $= 6$
- $Q_3 = 8$
- max $= 11$
- range $= 11 - 3 = 8$
- IQR $= 8 - 5 = 3$

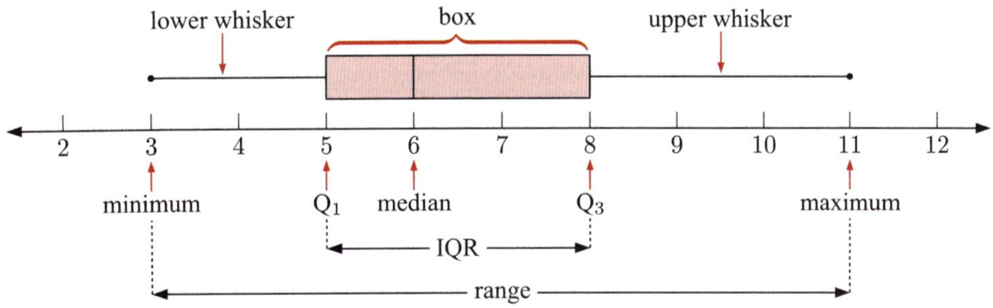

Example 8 — Self Tutor

Consider the data: 6 7 3 7 9 8 5 5 4 6 6 8 7 6 6 5 4 5 6

a Find the five-number summary of the data set.

b Find the range and IQR.

c Draw a box plot for the data.

---

a The ordered data set is:

3 4 4 5 5 5 5 6 6 6 6 6 6 7 7 7 8 8 9

min = 3   median = 6   max = 9

Since $n = 19$, $\dfrac{n+1}{2} = 10$

$\therefore$ the median = 6

Since the median is a data value, we now ignore it and split the remaining data into two:

lower: 3 4 4 5 5 5 5 6 6   6   upper: 6 6 6 7 7 7 8 8 9

$\therefore$ $Q_1 = 5$ and $Q_3 = 7$

The five-number summary is:

min = 3, $Q_1 = 5$, median = 6, $Q_3 = 7$, max = 9

b Range = max − min = 9 − 3 = 6

IQR = $Q_3 - Q_1 = 7 - 5 = 2$

c

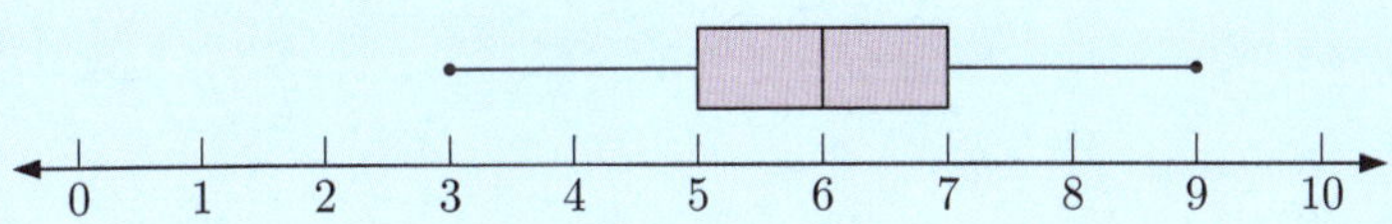

You can use your **calculator** or the **statistics package** to draw the box plot of a data set, and find the range and interquartile range.

STATISTICS PACKAGE

GRAPHICS CALCULATOR INSTRUCTIONS

EXERCISE 20E.1

1 This box plot summarises the goals scored by a netball team in matches during a season.

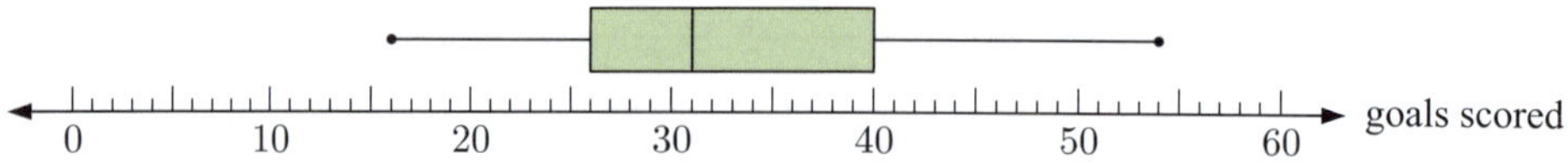

State the value of:

a min
b Q_1
c the median
d Q_3
e max
f the range
g the IQR.

2 This box plot summarises the ratings out of 100 points that a movie received on a website.

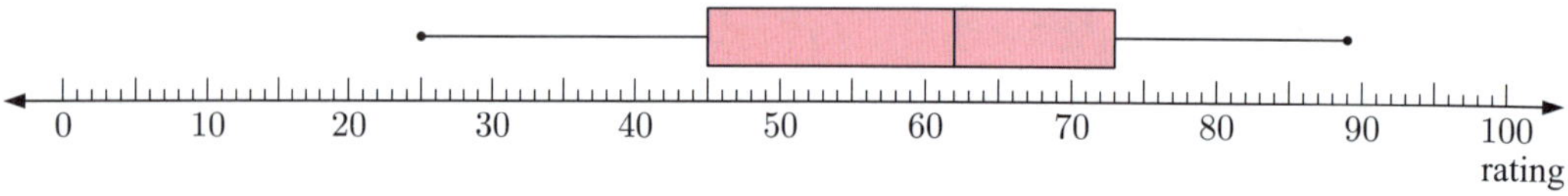

a Copy and complete each statement about their results:
 i The highest rating was ...... points.
 ii The lowest rating was ...... points.
 iii Half of the ratings were greater than or equal to ...... points.
 iv The top 25% of the ratings were at least ...... points.
 v The middle half of the ratings were between ...... and ...... points.

b Find the range of the data.

c Find the interquartile range of the data.

Example 9 — Self Tutor

Consider the data: 9 8 2 3 7 6 5 4 5 4 6 8 9 5 5 5 4 6 6 8

a Find the five-number summary of the data set.
b Find the range and IQR.
c Draw a box plot for the data.

a The ordered data set is:

2 3 4 4 4 5 5 5 5 5 6 6 6 6 7 8 8 8 9 9

min = 2   median = 5.5   max = 9

Since $n = 20$, $\frac{n+1}{2} = 10.5$

$\therefore$ the median $= \frac{5+6}{2} = 5.5$

Since the median is the average of two data values, we split the ordered data into two halves:

lower: 2 3 4 4 4 5 5 5 5 5   upper: 6 6 6 6 7 8 8 8 9 9

$\therefore$ $Q_1 = \frac{4+5}{2} = 4.5$ and $Q_3 = \frac{7+8}{2} = 7.5$

The five-number summary is:

min = 2, $Q_1 = 4.5$, median = 5.5, $Q_3 = 7.5$, max = 9

b Range = max − min = 9 − 2 = 7

IQR = $Q_3 - Q_1 = 7.5 - 4.5 = 3$

c

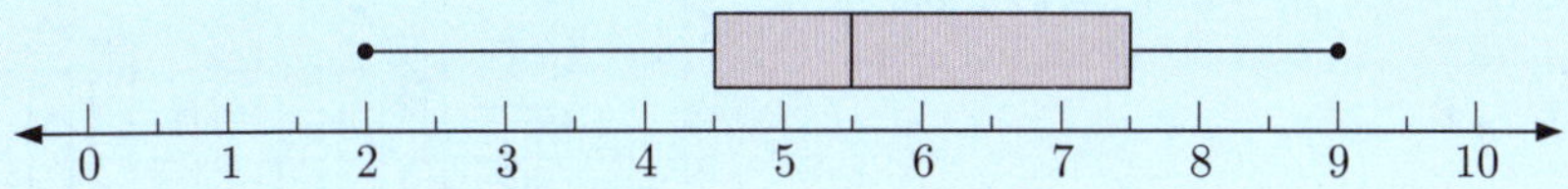

3 A class of 23 Year 6 students sat a spelling test out of 12 marks. Their scores were:

6 8 11 9 7 9 10 12 9 6 8 7
11 9 5 8 11 6 10 9 7 8 12

a Find the five-number summary of the data set. **b** Find the range and IQR.

c Draw a box plot for the data.

4 The production manager at a biscuit factory records the number of workers late for shifts each week for 17 weeks. The results are:

4 3 1 8 5 0 3 4 6 1 3 5 6 4 3 4 7

a Find the five-number summary of the data set. **b** Find the range and IQR.

c Draw a box plot for the data.

5 The scores below are the number of runs scored by a baseball team in matches during a season:

3 4 7 11 2 6 3 14 10 6 9 8

a Find the five-number summary of the data set. **b** Find the range and IQR.

c Draw a box plot for the data.

6 Draw a box plot for each data set:

a 11, 13, 16, 13, 25, 19, 20, 19, 19, 16, 17, 21, 22, 18, 19, 17, 23, 15

b 5, 5, 10, 9, 4, 2, 8, 6, 5, 8, 6, 7, 9, 6, 10, 3, 11

7 The table alongside shows the number of tows performed by a tow truck driver each day over a 45 day period.

a Draw a box plot for the data.

b Find the range and IQR of the data.

Number of tows	*Frequency*
3	2
4	5
5	6
6	9
7	12
8	8
9	3

ACTIVITY

Click on the icon to play a card game for box plots.

GAME

DISCUSSION

- The column graph and box plot of a data set is shown alongside.
  From the column graph, we can see that the data is symmetric. How is this observed on the box plot?

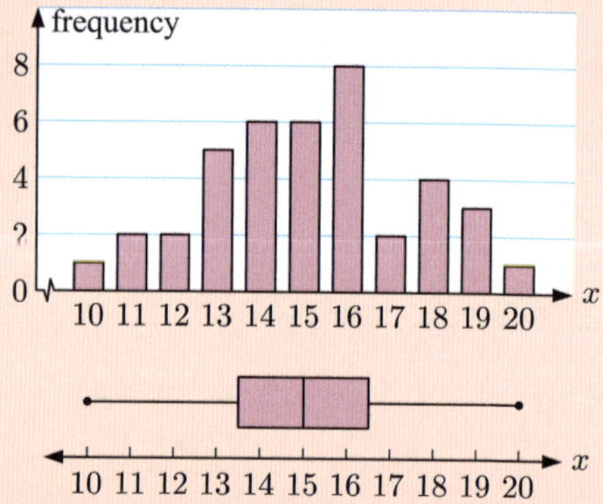

- What features of a box plot would indicate that a data set is:
  - positively skewed
  - negatively skewed?
- For each box plot, discuss whether the data is symmetric, positively skewed, or negatively skewed.

A

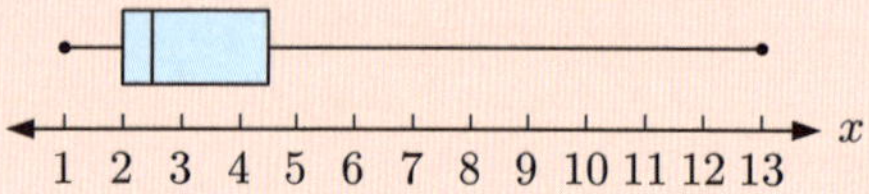

B

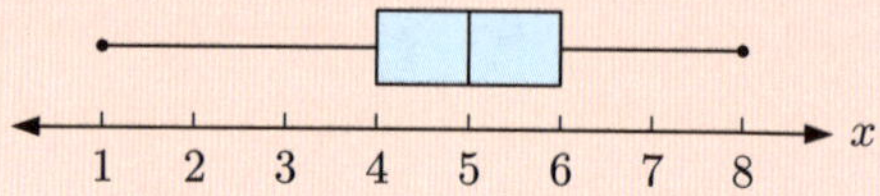

C

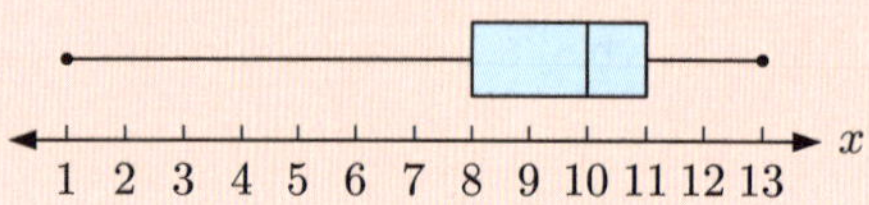

D

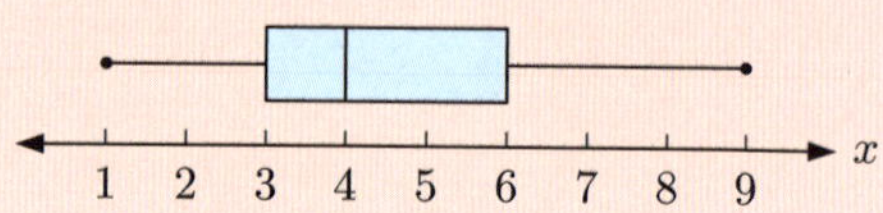

- Is it possible to determine if a data set is *bimodal* from its box plot alone?

PARALLEL BOX PLOTS

To compare two data sets, we can draw a box plot for each data set on the same scale. This is called a **parallel box plot**.

Parallel box plots enable us to make a *visual comparison* of the statistics and distributions of two data sets.

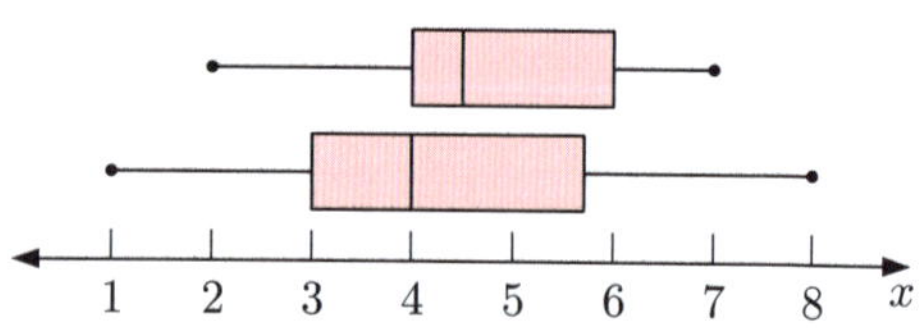

Example 10 **Self Tutor**

An office worker has the choice of travelling to work by car or bus. He wants to know which method of transport is the quickest and which is more reliable.

He recorded the following travel times from recent journeys using both methods. The results are shown in minutes, in ascending order.

Car:	13	14	18	18	19	21	22	22	24	25	27	28	30	33	43
Bus:	16	16	16	17	17	18	18	18	20	20	21	21	23	28	30

a Draw a parallel box plot for the data sets.

b Hence determine which method of transport is: **i** quicker **ii** more reliable.

a We first find the five-number summary of each data set:

Car: min = 13, $Q_1 = 18$, median = 22, $Q_3 = 28$, max = 43

Bus: min = 16, $Q_1 = 17$, median = 18, $Q_3 = 21$, max = 30

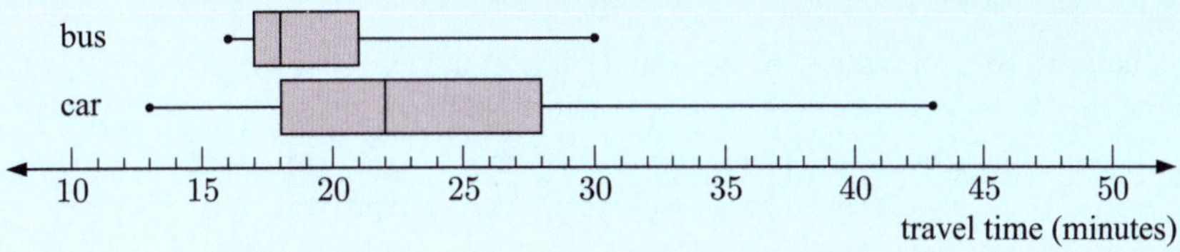

b **i** The bus travel times are generally lower than the car travel times, so the bus is generally quicker.

ii The car travel times have greater spread than the bus travel times, so the bus is also more reliable.

EXERCISE 20E.2

1 This parallel box plot compares the times students in Year 10 and Year 12 spend on homework over a one-week period.

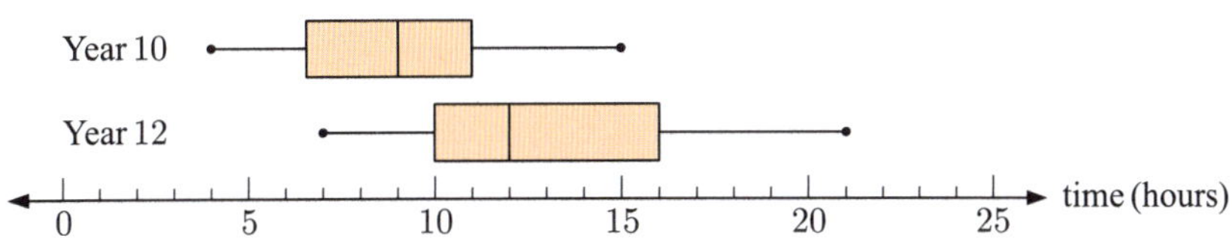

a State the five-number summary of each year group.

b For each group, find the:

i range **ii** interquartile range.

c Which year group generally spends more time on homework?

d For which year group is there greater variability in the amount of time spent on homework?

2 This parallel box plot shows the lengths of barramundi caught in Indonesian and Australian waters.

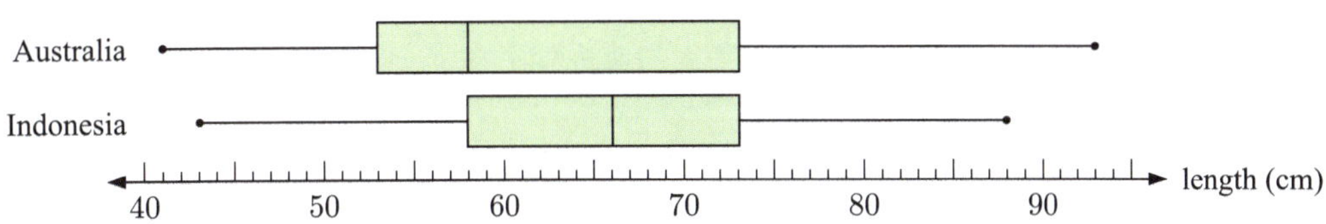

a For each region, find the:

i greatest length **ii** shortest length **iii** range **iv** interquartile range.

b The legal length for a barramundi to be kept is 58 cm. What percentage of fish caught were of legal length in:

i Indonesia **ii** Australia?

c Copy and complete:

i The fish caught in ...... generally have greater length.

ii The fish caught in ...... have greater length variability.

3 Consider the **Opening Problem** on page **336**.

a Find the five-number summary of each data set.

b Draw a parallel box plot to display the data sets.

c Determine the hotel in which the guests generally stay longer.

4 The heights, in centimetres, of boys and girls in a Year 10 class are:

Boys: 165 171 169 169 172 171 171 180 168 168 166 168 170
165 171 173 187 181 175 174 165 167 163 160 169 167
172 174 177 188 177 185 167 160

Girls: 162 171 156 166 168 163 170 171 177 169 168 165 156
159 165 164 154 171 172 166 152 169 170 163 162 165
163 168 155 175 176 170 166

a Find the five-number summary of each data set.

b Draw a parallel box plot to display the data sets.

c Hence compare the data sets.

5 Samples of crayfish were caught in two adjacent sounds in southwest New Zealand. The following data shows the average weights in kilograms for the two sounds over a 20 day catching period.

Sound 1: 2.6 2.5 2.7 2.4 2.9 2.7 2.6 2.7 2.8 2.5
2.7 2.6 2.8 2.6 2.5 2.8 2.5 2.4 2.7 2.3

Sound 2: 2.7 3.0 2.6 2.9 2.7 2.8 2.9 2.6 2.7 2.7
2.9 3.1 2.6 2.7 2.7 2.8 3.2 2.7 2.8 2.8

a Find the five-number summary of each data set.

b Draw a parallel box plot to display the data sets.

c Hence compare the data sets.

F CUMULATIVE FREQUENCY GRAPHS

When we calculated the median from a frequency table, we used a **cumulative frequency** column to record the number of data values less than or equal to a particular value.

A **cumulative frequency graph** shows the cumulative frequency plotted against the data values.

We can use a cumulative frequency graph to estimate:

- the number of data values less than or equal to a particular value
- the value corresponding to a particular cumulative frequency
- the quartiles and median of the data set.

A **percentile** is the score below which a certain percentage of the data lies.

For example, if your score in a test is the 75th percentile, then 75% of the class scored less than you.

Notice that:

- the **lower quartile** (Q_1) is the 25th percentile
- the **median** (Q_2) is the 50th percentile
- the **upper quartile** (Q_3) is the 75th percentile.

Example 11

Self Tutor

The table summarises the weights of 80 male basketball players.

a Add a cumulative frequency column to the table.

b Draw a cumulative frequency graph for the data.

c Use your graph to estimate the:

- **i** median weight
- **ii** number of men weighing less than 83 kg
- **iii** number of men weighing more than 92 kg
- **iv** 85th percentile.

Weight (w kg)	*Frequency*
$70 \leqslant w < 75$	3
$75 \leqslant w < 80$	8
$80 \leqslant w < 85$	16
$85 \leqslant w < 90$	21
$90 \leqslant w < 95$	19
$95 \leqslant w < 100$	8
$100 \leqslant w < 105$	3
$105 \leqslant w < 110$	1
$110 \leqslant w < 115$	1

a

Weight (w kg)	*Frequency*	*Cumulative frequency*
$70 \leqslant w < 75$	3	3
$75 \leqslant w < 80$	8	$3 + 8 = 11$
$80 \leqslant w < 85$	16	$11 + 16 = 27$
$85 \leqslant w < 90$	21	$27 + 21 = 48$
$90 \leqslant w < 95$	19	$48 + 19 = 67$
$95 \leqslant w < 100$	8	$67 + 8 = 75$
$100 \leqslant w < 105$	3	$75 + 3 = 78$
$105 \leqslant w < 110$	1	$78 + 1 = 79$
$110 \leqslant w < 115$	1	$79 + 1 = 80$

48 players weigh less than 90 kg

b **Cumulative frequency graph of basketballers' weights**

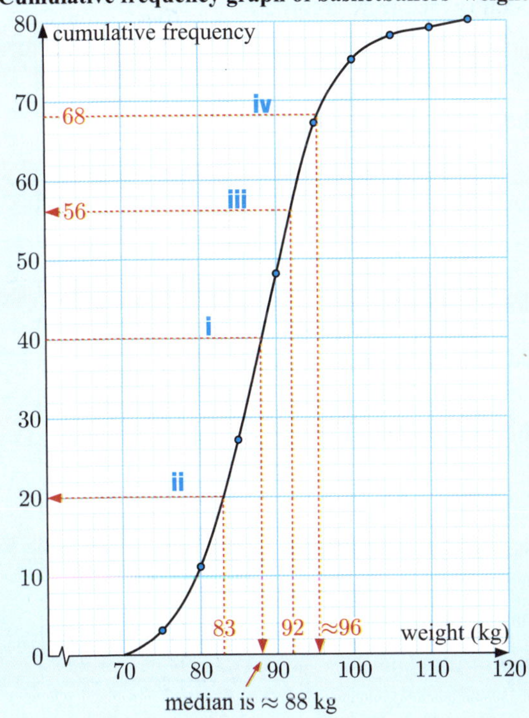

c

- **i** 50% of $80 = 40$, so the median ≈ 88 kg
- **ii** There are 20 men who weigh less than 83 kg.
- **iii** There are $80 - 56 = 24$ men who weigh more than 92 kg.
- **iv** 85% of $80 = 68$, so the 85th percentile ≈ 96 kg.

EXERCISE 20F

1 This cumulative frequency graph shows the weights of Sam's goat herd in kilograms.

- **a** How many goats does Sam have?
- **b** Estimate the median goat weight.
- **c** Find the 60th percentile and explain what it means.
- **d** Find the IQR for the weights of Sam's herd.

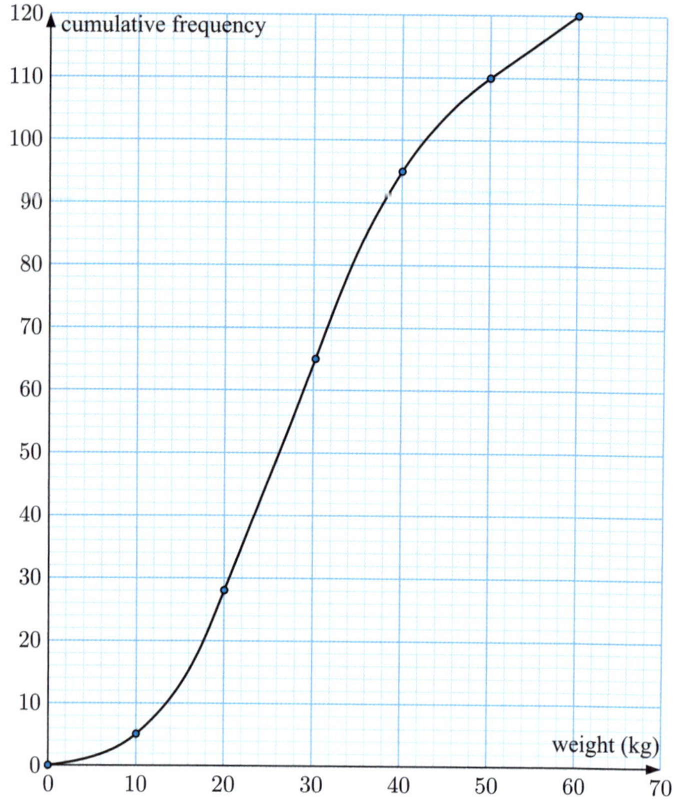

2 The times of 160 competitors in a cross-country race were recorded in the table alongside.

- **a** Add a cumulative frequency column to the table.
- **b** Draw a cumulative frequency graph to display the data.
- **c** Estimate the:
  - **i** median time
  - **ii** number of runners whose time was less than 32 minutes
  - **iii** time in which the fastest 40 runners completed the race
  - **iv** interquartile range.

Time (t min)	*Frequency*
$20 \leqslant t < 25$	18
$25 \leqslant t < 30$	45
$30 \leqslant t < 35$	37
$35 \leqslant t < 40$	33
$40 \leqslant t < 45$	19
$45 \leqslant t < 50$	8

3 The lengths of 30 trout were measured and the results summarised in the table.

- **a** Draw a cumulative frequency graph for the data.
- **b** Estimate the:
  - **i** percentage of trout with length less than 39 cm
  - **ii** median length of trout caught
  - **iii** IQR and explain what this represents
  - **iv** 35th percentile and explain what it means.
- **c** Use a calculator to estimate the mean of the data.

Length (l cm)	*Frequency*
$30 \leqslant l < 32$	1
$32 \leqslant l < 34$	1
$34 \leqslant l < 36$	3
$36 \leqslant l < 38$	7
$38 \leqslant l < 40$	11
$40 \leqslant l < 42$	5
$42 \leqslant l < 44$	2

4 This cumulative frequency graph describes the weights of 40 watermelons.

a Estimate the:

i median weight

ii lower and upper quartiles.

b Use your estimates in **a** to draw a box plot for the data.

c Construct a cumulative frequency table for the data, including a frequency column.

d Estimate the mean weight of the watermelons.

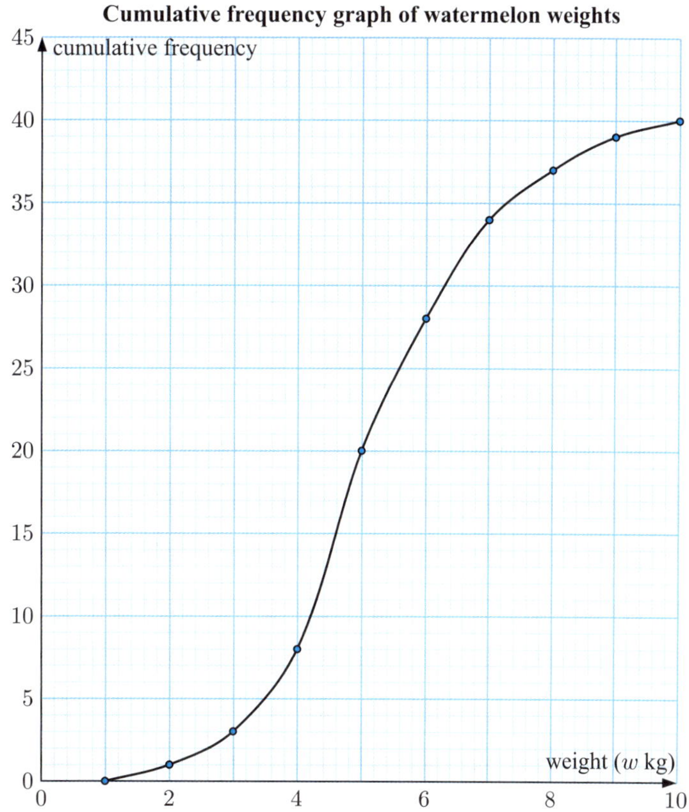

5 This box plot describes the longest distances a baseball was thrown by 60 students.

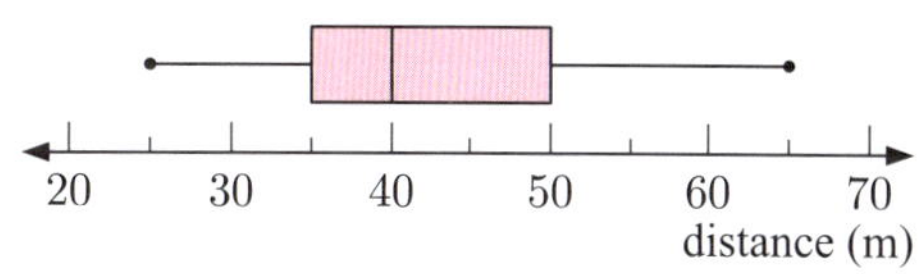

a State the lower quartile of the data.

b In what range of values can you be *certain* that the 60th percentile lies?

c Sketch a cumulative frequency graph for the data.

d Use your graph to *estimate* the 60th percentile of the data.

6 The times taken for hikers to climb Cradle Mountain were recorded in the table alongside.

a Draw a cumulative frequency graph for the hiking times.

b Estimate the median time for the hike.

c Estimate the IQR and explain what it means.

d Guides on the hike say that anyone who completes the hike in 3 hours 40 minutes or less is extremely fit. Estimate the proportion of extremely fit hikers.

Time (t min)	*Frequency*
$200 \leqslant t < 225$	5
$225 \leqslant t < 250$	18
$250 \leqslant t < 275$	37
$275 \leqslant t < 300$	16
$300 \leqslant t < 325$	4

DISCUSSION

What does a cumulative frequency graph look like, for a distribution which is:

a symmetric **b** positively skewed **c** negatively skewed?

QUICK QUIZ

MULTIPLE CHOICE QUIZ

REVIEW SET 20A

1 A class of 20 students was asked "How many bedrooms are there in your house?" The following data was collected:

3 2 3 2 2 4 3 4 2 3 2 1 2 2 3 2 4 2 3 2

a Is the data discrete or continuous?

b Are there any outliers in the data?

c Construct a dot plot to display the data.

d State the mode of the data and explain what it means.

2 The numbers of people at a judo class each week are shown below.

10 8 10 9 7 11 9 11 10 10 9 8 9
9 11 8 10 9 10 11 10 7 9 11 10 8

a Draw a column graph to display the data.

b Describe the distribution of the data.

3 This table summarises the number of laps of a pool Erica swam each day for 40 days.

a On how many days did Erica swim at least 30 laps?

b Estimate the mean number of laps swam per day.

Number of laps	*Frequency*
10 to 19	12
20 to 29	14
30 to 39	9
40 to 49	5

4 A sample of 15 measurements has a mean of 14.2, and a sample of 10 measurements has a mean of 12.6. Find the mean of all 25 measurements.

5 The masses of eggs in a carton marked "50 g eggs" are recorded alongside.

a Draw a frequency histogram to display the data.

b State the modal class.

c Describe the distribution of the data.

d Estimate the mean of the data.

Mass (m g)	*Frequency*
$48 \leqslant m < 49$	1
$49 \leqslant m < 50$	1
$50 \leqslant m < 51$	16
$51 \leqslant m < 52$	4
$52 \leqslant m < 53$	3

6 The scores out of 100 for an exam are displayed in the box plot below.

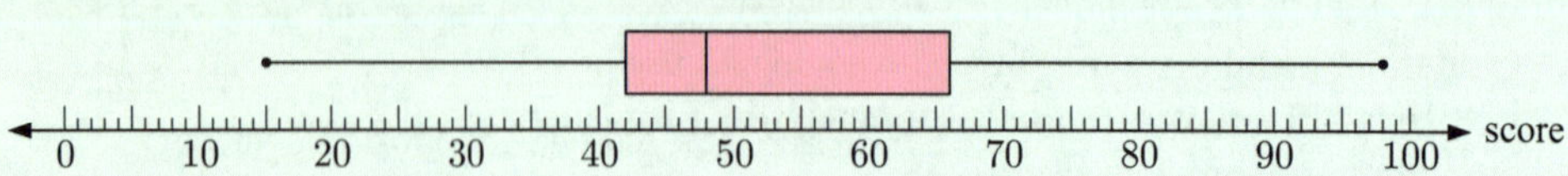

a State the:

i median score **ii** maximum score **iii** minimum score

iv upper quartile **v** lower quartile.

b Find the:

i range **ii** interquartile range.

7 This cumulative frequency graph illustrates the points scored by competitors in a ski aerials competition.

a How many competitors took part in the competition?

b What percentage of competitors scored less than 200 points?

c Estimate the:

i median **ii** lower quartile **iii** upper quartile.

d Hence estimate the interquartile range of the data.

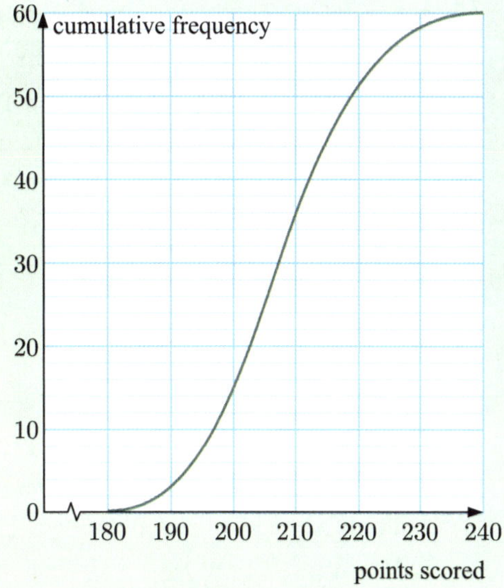

8 The data below gives the number of letters received by a household each week:

6 6 6 6 7 7 7 7 7 7 7 8 8 8 8 8 8 8 8 8 8 9 9 9 9 9 10 10 10

a Find the range of the data.

b Find the five-number summary of the data.

c Draw a box plot for the data.

d Find the IQR of the data.

REVIEW SET 20B

1 For each graph, describe the distribution of the data:

a

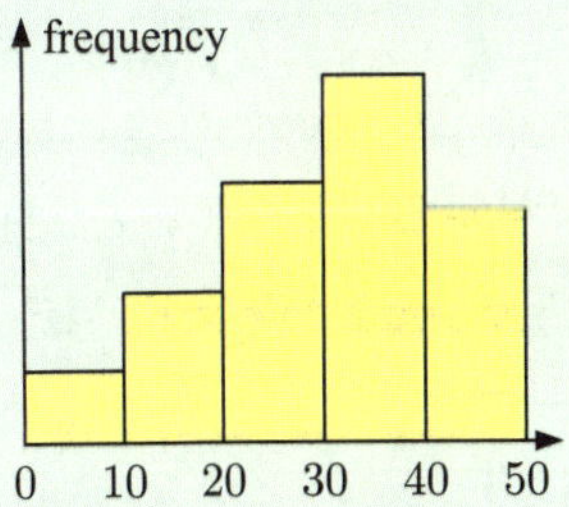

b

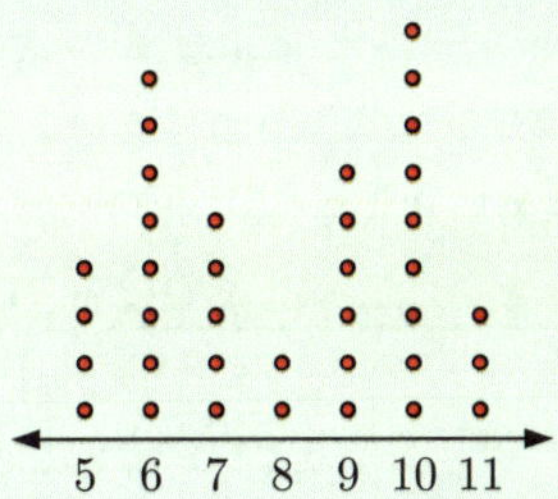

c

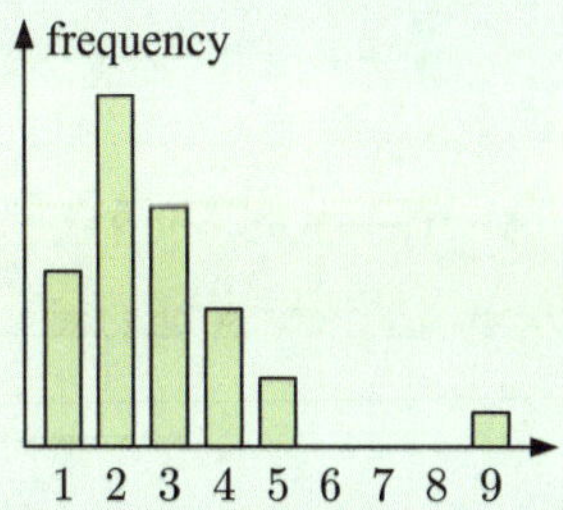

2 As punishment for misbehaving, a class of students had to pick up litter during recess. This dot plot shows the number of litter pieces picked up by the students.

For this data, find the:

a mean **b** median.

Litter pieces picked up by students

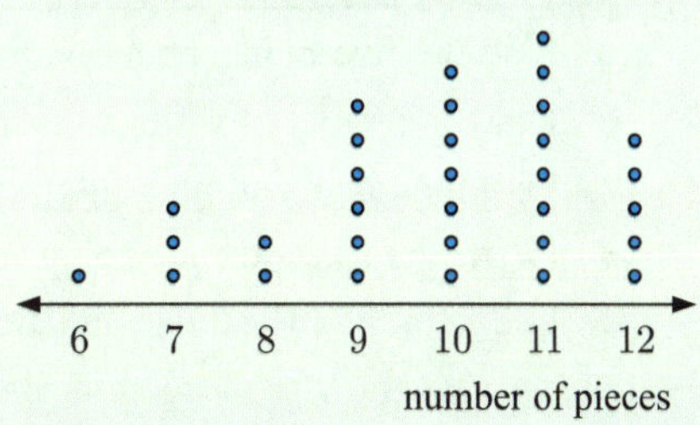

3 Nine scores have an average of 8. Scores of x and $x + 1$ are added, and these scores increase the average to 9. Find x.

4 The data below shows the number of employees in 30 businesses on a street.

13 22 17 15 36 29 36 25 8 40 31 20 26 33 24
22 16 7 28 17 32 26 18 34 25 14 43 9 21 13

a Organise the data in a tally and frequency table using class intervals of width 10.

b Display the data on a column graph.

c State the modal class for the data.

5 Consider the column graph alongside.

a Construct a frequency table for the data.

b Find the total number of:

i games played **ii** goals scored.

c Find the:

i mean **ii** median

iii mode **iv** range.

Goals per game by a hockey player

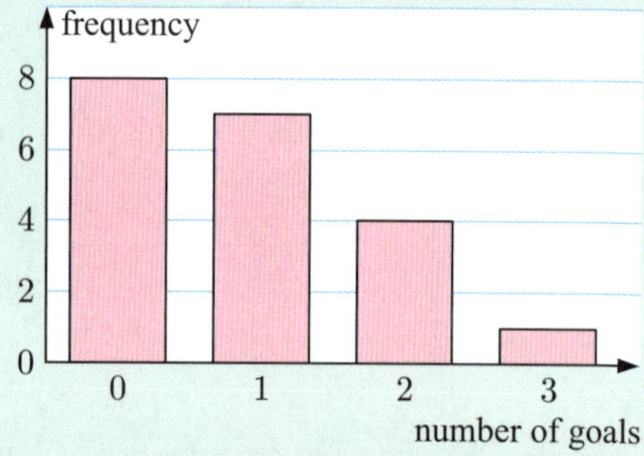

6 The scores for a year group test are summarised in the table.

a Draw a cumulative frequency graph for the data.

b Hence estimate the median and interquartile range.

c Draw a histogram for the data.

d Estimate the mean score.

e Describe the distribution of the data.

Score (s)	*Frequency*
$10 \leqslant s < 20$	4
$20 \leqslant s < 30$	8
$30 \leqslant s < 40$	14
$40 \leqslant s < 50$	20
$50 \leqslant s < 60$	45
$60 \leqslant s < 70$	75
$70 \leqslant s < 80$	60
$80 \leqslant s < 90$	20
$90 \leqslant s < 100$	13

7 The Davis and Douglas families kept a record of the amount they spent at the supermarket each week for 10 weeks:

Davis: $102.50 $115.95 $107.60 $122.15 $131.05
$111.15 $120.50 $127.55 $100.95 $113.40

Douglas: $109.80 $86.75 $94.50 $129.75 $72.05
$133.05 $121.05 $97.60 $73.80 $105.35

a Find the five-number summary of each data set.

b Draw a parallel box plot to display the data sets.

c Find the range and interquartile range for each data set.

d Use technology to find the mean of each data set.

e Use your graph and the statistics you calculated to determine which family:

i generally spent more at the supermarket each week

ii had the greater variation in the amount they spent each week.

Chapter 21

Bivariate statistics

Contents:

OPENING PROBLEM

The relationship between the *height* and *weight* of members of a football team is to be investigated. Data for each player is given below.

Player	1	2	3	4	5	6	7	8	9
Height (cm)	203	189	193	187	186	197	180	186	188
Weight (kg)	106	93	95	86	85	92	78	84	93

Player	10	11	12	13	14	15	16	17	18
Height (cm)	181	179	191	178	178	186	190	189	193
Weight (kg)	84	86	92	80	77	90	86	95	89

Things to think about:

- **a** Which is the *dependent* variable?
- **b** Can you display this data using a scatter graph?
- **c** Does an increase in the *height* of a player generally correspond to an increase or a decrease in their *weight*?
- **d** How can we measure how close the relationship between the variables is to a straight line?
- **e** How can we use this data to estimate the weight of a player who is 200 cm tall? How reliable will this estimate be?

We often want to know how two variables are **associated** or **related**. We want to know whether a change in one variable results in a change in the other. We call this **bivariate statistics** because we are considering *two* variables at the same time.

The data we study is called **bivariate data** as two variables are recorded for each individual.

If we think that a change in one variable results in a change in the other, we try to identify which is the **dependent** variable and which is the **independent** variable. The value of the dependent variable *depends* on the value of the independent variable.

A SCATTER GRAPHS

To analyse the relationship between two **numerical** variables, we usually draw a **scatter graph** to display the data.

- The **independent variable** is placed on the horizontal axis.
- The **dependent variable** is placed on the vertical axis.

Each pair of data values from a particular individual is plotted as an ordered pair on the scatter graph.

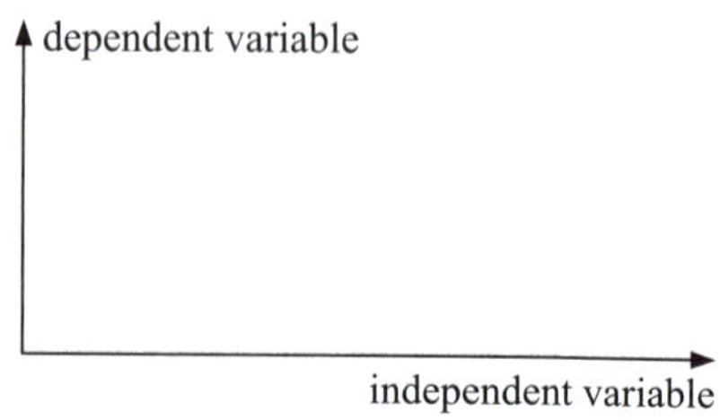

For example, the scatter graph alongside displays the data from the **Opening Problem**.

The red point (187, 86) corresponds to player 4.

GRAPHICS CALCULATOR INSTRUCTIONS

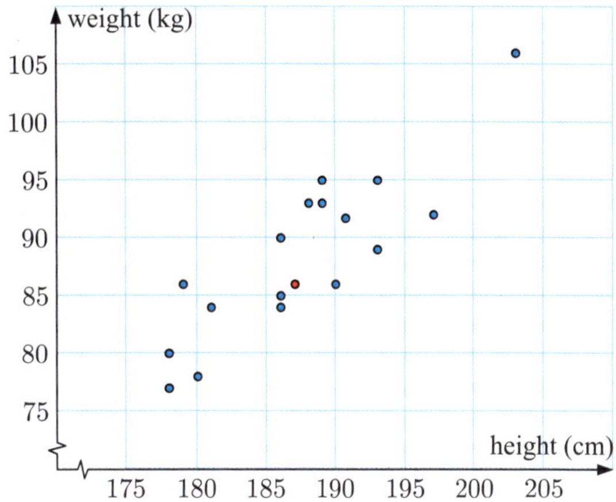

EXERCISE 21A

1 Consider the scatter graph shown alongside.

- **a** Identify the independent variable and the dependent variable.
- **b** Describe what the red point represents.

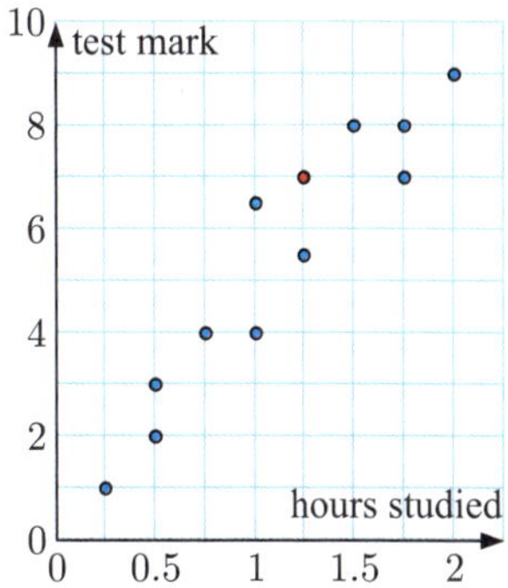

2 This scatter graph shows the productivity of five employees, and the number of hours of training they have received.

- **a** Who has received the most hours of training?
- **b** Who is the least productive?
- **c** Are there employees with equal productivity?
- **d** Do you think that the training given to the employees has been worthwhile? Explain your answer.

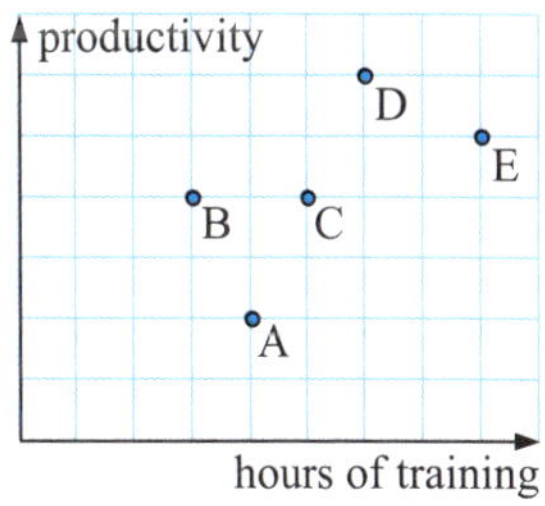

3 This table shows how far 10 students live from school, and how long it takes them to travel there:

Distance to school (km)	8	5	12	6	4	7	6	10	5	8
Travel time (min)	10	5	15	15	7	11	9	11	8	8

- **a** Identify the independent variable and the dependent variable.
- **b** Draw a scatter graph for the data.

4 This table shows the ages of 12 children and the number of baby teeth they have remaining:

Age	6	9	12	6	7	10	12	8	11	7	8	8
Number of baby teeth	10	6	1	9	10	4	0	4	3	8	7	8

- **a** Identify the independent variable and the dependent variable.
- **b** Draw a scatter graph for the data.

5 This table shows the ages of 11 sprinters in an athletics team, and their fastest times for the 100 m:

Age	16	19	14	12	15	17	14	18	15	13	16
Time (s)	12.4	10.8	13.2	15.1	12.6	11.9	13.0	11.2	13.1	14.1	12.6

a Identify the independent variable and the dependent variable.

b Draw a scatter graph for the data.

c Do you think the age of the sprinters affects their fastest times? Explain your answer.

DISCUSSION

Look at the scatter graphs in the previous Exercise.

What words can we use to *describe* the patterns or *trends* in the points plotted?

B CORRELATION

Correlation is a measure of the strength of the relationship or association between two variables.

We can use a scatter graph to describe the correlation between two variables.

Step 1: Look at the scatter graph for any **pattern**.

For a generally *upward* shape, we say that the correlation is **positive**.

As the independent variable increases, the dependent variable generally increases.

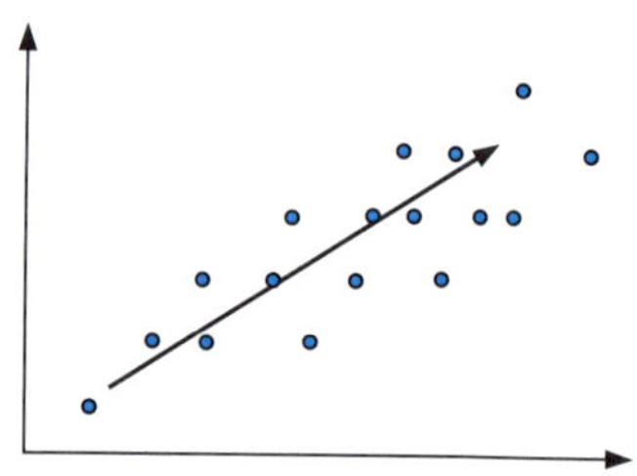

For a generally *downward* shape, we say that the correlation is **negative**.

As the independent variable increases, the dependent variable generally decreases.

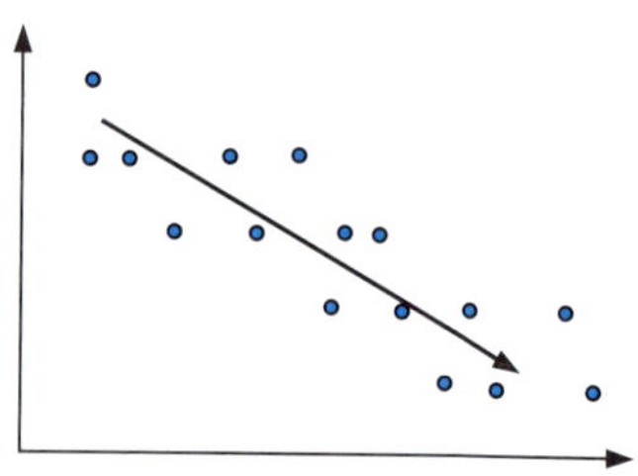

For *randomly scattered* points with no upward or downward trend, we say there is **no correlation**.

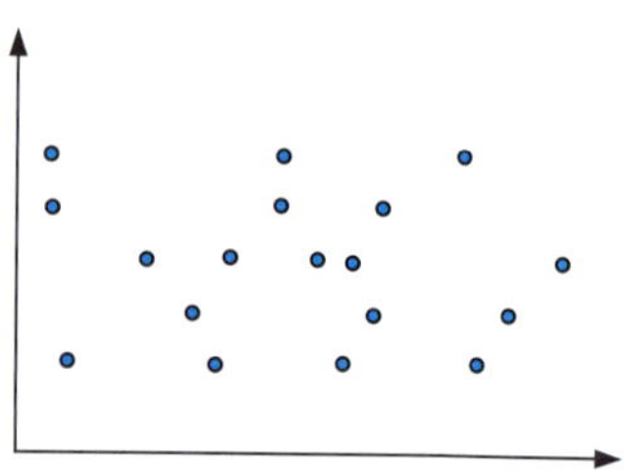

Step 2: Look at the pattern of points to see whether the relationship is **linear**.

The relationship is approximately linear.

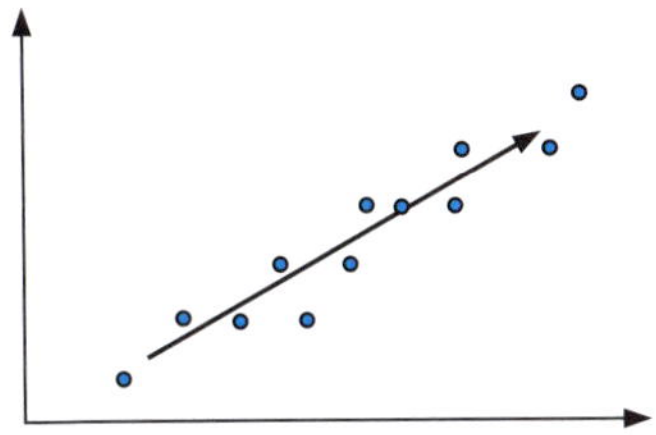

The relationship is not linear.

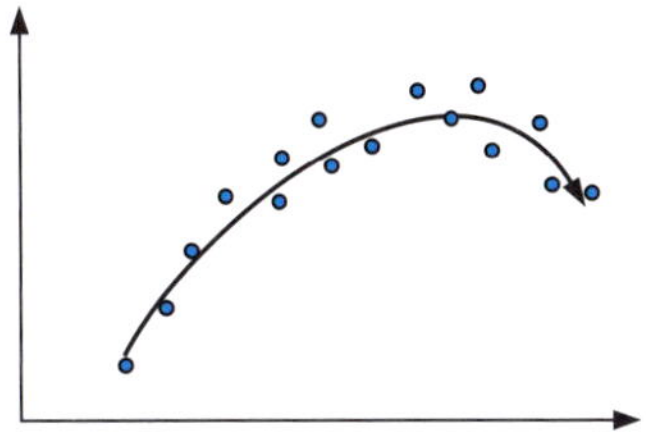

Step 3: Look at the spread of points to judge the **strength** of the correlation.

These scatter graphs show strength classifications for positive relationships:

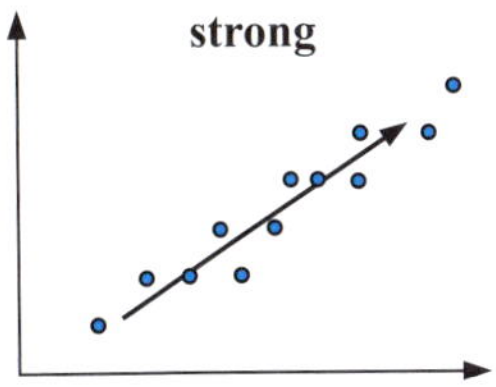

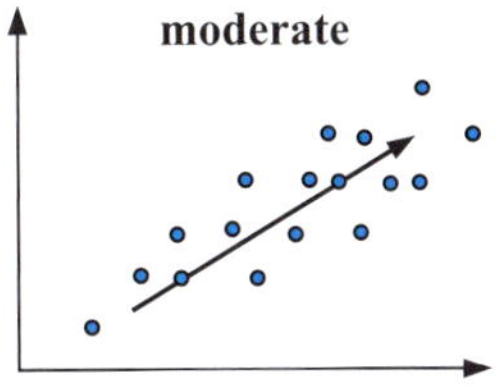

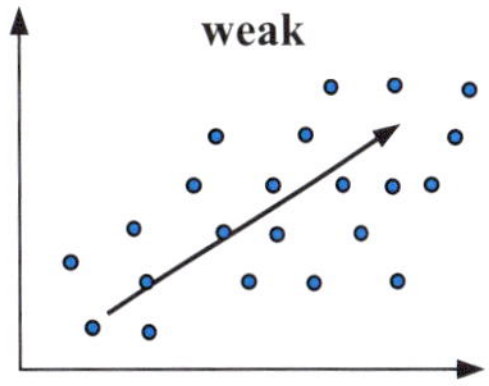

These scatter graphs show strength classifications for negative relationships:

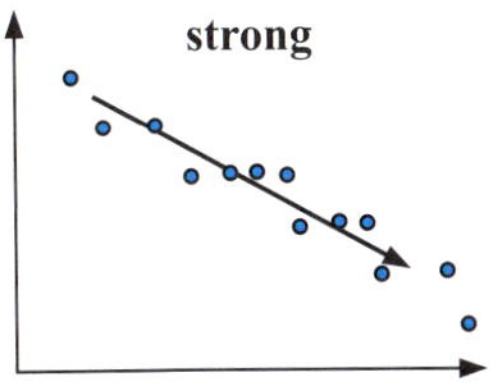

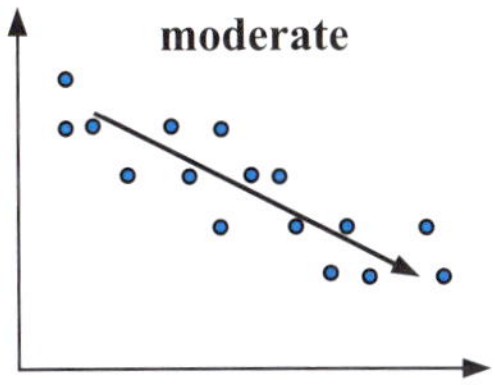

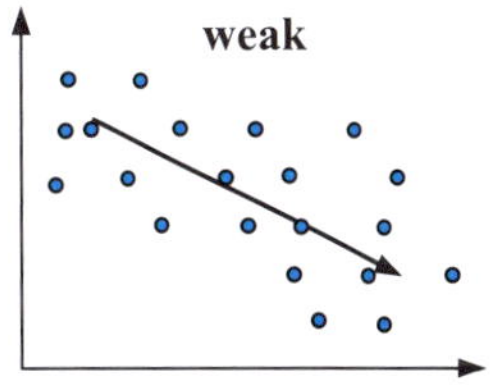

Example 1 **Self Tutor**

Alexander researched the elevation above sea level and mean annual temperature of 12 cities around the world. The results are given in this table.

Elevation (m)	600	850	150	300	100	200	500	450	750	30	300	50
Mean annual temperature (°C)	15	10	16	15	25	20	21	19	9	27	22	28

a Draw a scatter graph for the data.

b Describe the relationship between *elevation* and *mean temperature*.

a

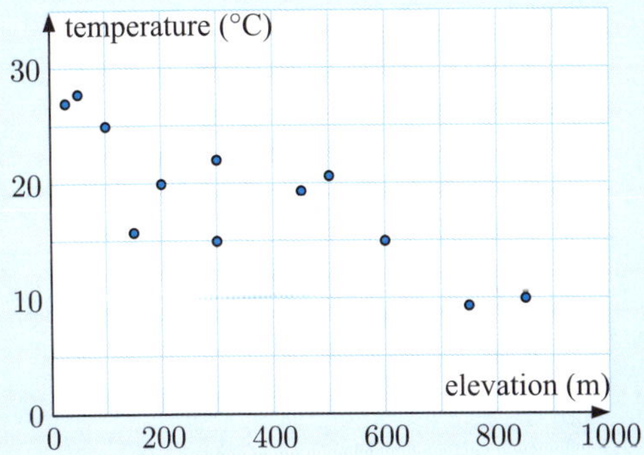

b There is a moderate negative linear correlation between *elevation* and *mean temperature*.

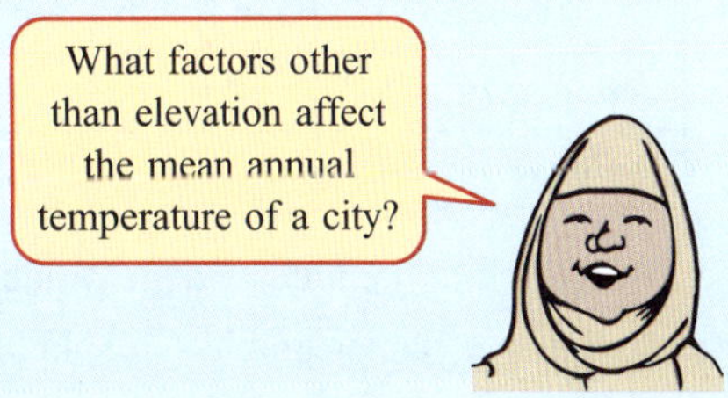

EXERCISE 21B

1 For each scatter graph:

- **i** State whether there is positive, negative, or no association between the variables.
- **ii** Decide whether the relationship between the variables is linear.
- **iii** Describe the strength of the association as zero, weak, moderate, or strong.

a

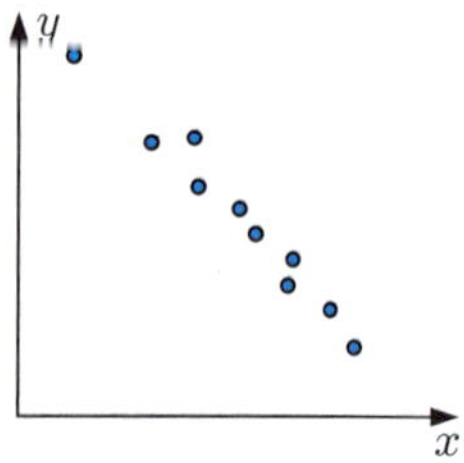

b

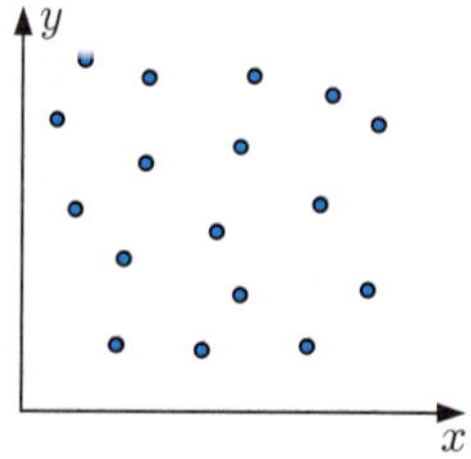

c

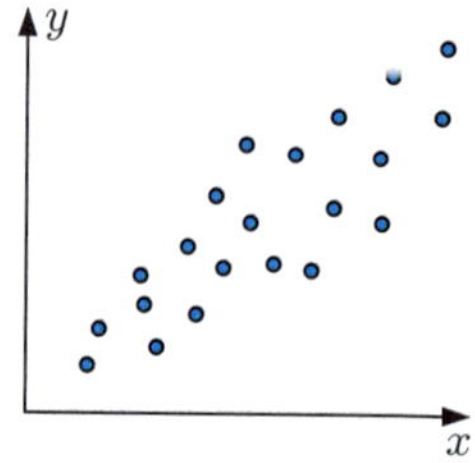

d

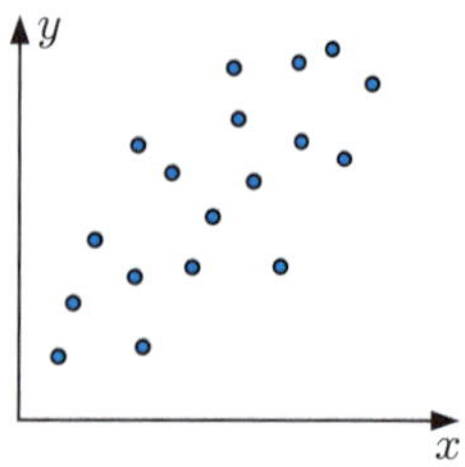

e

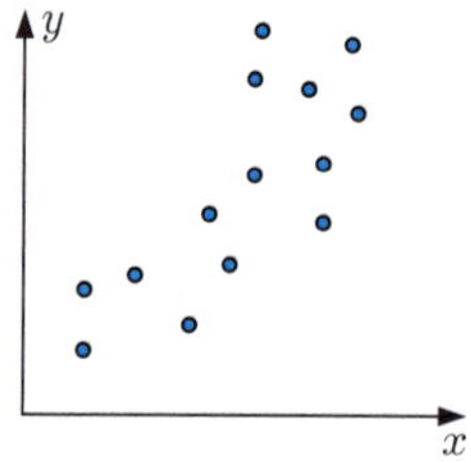

f

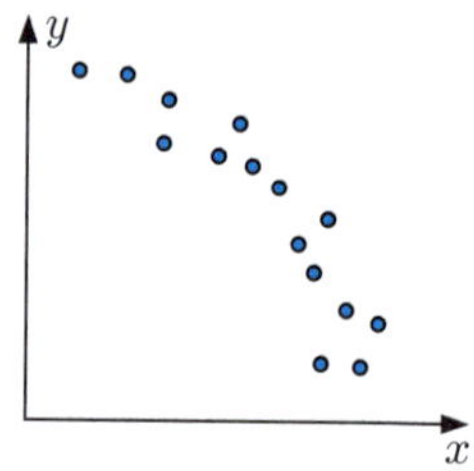

2 **a** There is a positive association between x and y. What happens to y as x increases?

b There is a negative correlation between T and d. What happens to d as T increases?

c Copy and complete: If there is no association between two variables then the points on the scatter graph are ....

3 A class of 15 students was asked how many text messages they had sent and received in the last week. The results are shown below:

Student	A	B	C	D	E	F	G	H	I	J	K	L	M	N	O
Messages sent	5	0	12	9	17	15	10	4	8	18	25	17	0	6	13
Messages received	8	0	15	7	19	11	8	7	12	15	21	16	4	6	16

a Draw a scatter graph to display the data, with *messages sent* on the horizontal axis.

b Describe the relationship between *messages sent* and *messages received.*

4 The following table shows the sales of hot drinks in a popular café each month, along with the average daily temperature for the month.

Month	Jan	Feb	Mar	Apr	May	Jun	Jul	Aug	Sep	Oct	Nov	Dec
Temperature (°C)	6	5	12	13	18	20	21	22	17	14	10	7
Sales (\$ × 1000)	18	22	25	20	15	16	12	12	8	10	15	16

a Which is the independent variable?

b Draw a scatter graph to display the data.

c Comment on the relationship between *sales* and *temperature.*

5 **a** 10 students were asked for their exam marks in Physics and Mathematics. Their percentages are given in the table below.

Student	A	B	C	D	E	F	G	H	I	J
Physics	75	83	45	90	70	78	88	50	55	95
Mathematics	68	70	50	65	60	72	75	40	45	80

i Draw a scatter graph for the data, with the Physics marks on the horizontal axis.

ii Comment on the relationship between the Physics and Mathematics marks.

b The same students were asked for their Art exam results. Their percentages were:

Student	A	B	C	D	E	F	G	H	I	J
Art	75	70	80	85	82	70	60	75	78	65

Draw a scatter graph to see whether there is any relationship between the Physics marks and the Art marks.

6 A group of shoe store customers were asked to record their height in centimetres, and their shoe size. The results are shown in the table below:

Height (cm)	165	155	140	145	158	148	160	164	160	155	150	160
Shoe size	6.5	4.5	4	5.5	6	5.5	6	6.5	5.5	5	5	5.5

a Draw a scatter graph to display the data.

b Comment on the relationship between *height* and *shoe size*.

C PEARSON'S CORRELATION COEFFICIENT r

Simple observation of the points on a scatter graph is a fairly inaccurate way to describe the strength of correlation between two variables. Instead, we can calculate a statistic called **Pearson's correlation coefficient**.

Pearson's correlation coefficient r measures the strength of correlation between two numerical variables.

- The values of r range from -1 to $+1$.
- The **sign** of r indicates the **direction** of the correlation.
  - A positive value for r indicates **positive correlation**.
  - A negative value for r indicates **negative correlation**.
- The **size** of r indicates the **strength** of the correlation.
  - A value of r close to $+1$ or -1 indicates strong correlation between the variables.
  - A value of r close to zero indicates weak correlation between the variables.

The formula for r is quite complicated, so we use technology to calculate it.

Click on the icon for instructions.

GRAPHICS CALCULATOR INSTRUCTIONS

The following table is a guide for describing the strength of linear correlation using r.

Positive correlation

$r = 1$	perfect positive correlation	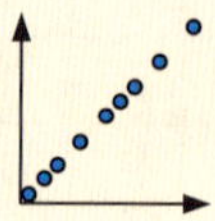
$0.95 \leqslant r < 1$	very strong positive correlation	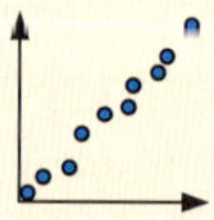
$0.87 \leqslant r < 0.95$	strong positive correlation	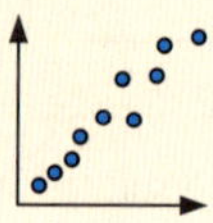
$0.7 \leqslant r < 0.87$	moderate positive correlation	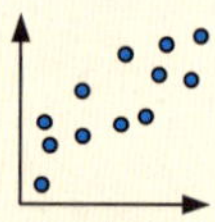
$0.5 \leqslant r < 0.7$	weak positive correlation	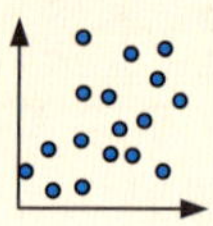
$0 < r < 0.5$	very weak positive correlation	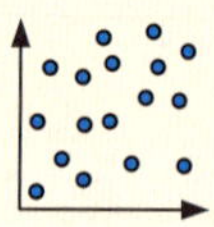

Negative correlation

$r = -1$	perfect negative correlation	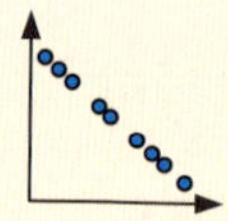
$-1 < r \leqslant -0.95$	very strong negative correlation	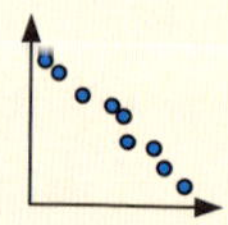
$-0.95 < r \leqslant -0.87$	strong negative correlation	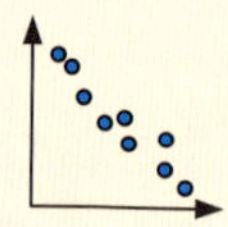
$-0.87 < r \leqslant -0.7$	moderate negative correlation	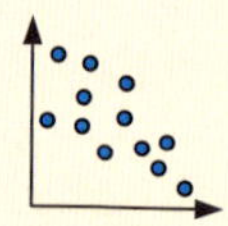
$-0.7 < r \leqslant -0.5$	weak negative correlation	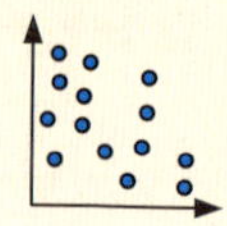
$-0.5 < r < 0$	very weak negative correlation	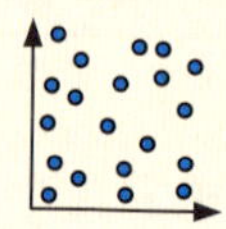

Example 2 Self Tutor

The body mass index (BMI) and resting heart rate of a group of people were measured:

BMI (kg/m^2)	27	31	33	26	33	32	32	38	26	38	28
Heart rate (bpm)	69	72	75	72	73	72	73	78	71	75	70

a Draw a scatter graph for the data.

b Calculate the correlation coefficient r for the data, and interpret its value.

a

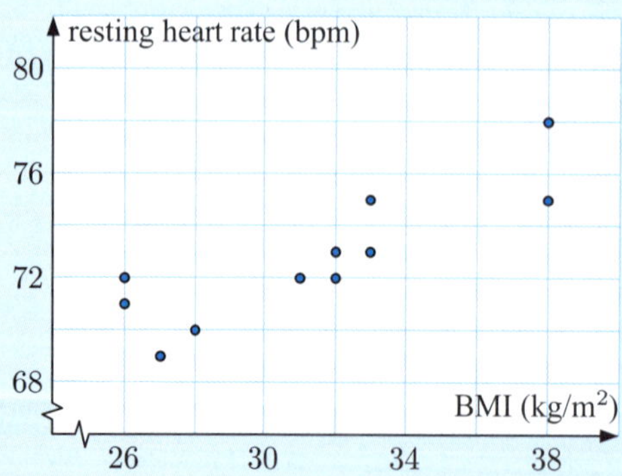

b

```
Deg Norm1 d/c Real
LinearReg(ax+b)
    a =0.50948103
    b =56.7944111
    r =0.85837
    r²=0.73679906
  MSe=1.87696828
y=ax+b
                COPY DRAW
```

Using technology, $r \approx 0.858$.

There is a moderate positive correlation between a person's *BMI* and their *resting heart rate*.

EXERCISE 21C

1 Match each scatter graph with the correct correlation coefficient:

a

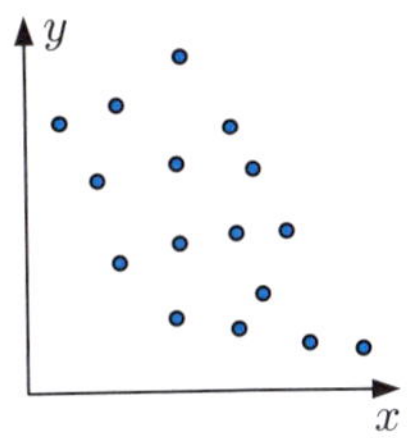

b

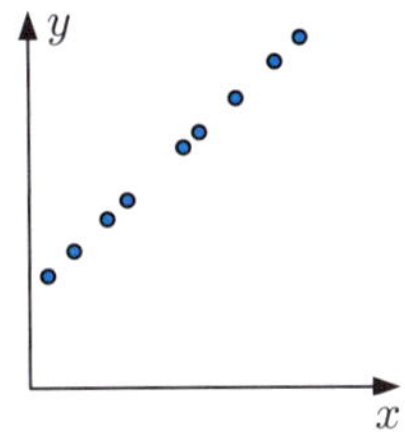

c

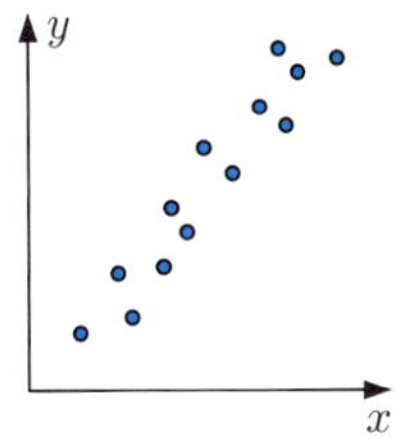

A $r = 1$ **B** $r \approx 0.8$ **C** $r \approx -0.3$

2 The correlation coefficient for a bivariate data set is $r \approx 0.812$.

- **a** Is the correlation between the variables positive or negative? Explain your answer.
- **b** Describe the strength of the correlation between the variables.

3 For each data set:

- **i** Draw a scatter graph for the data.
- **ii** Calculate Pearson's correlation coefficient r.
- **iii** Describe the linear correlation between x and y.

a

x	1	2	3	4	5	6
y	15	14	8	3	3	6

b

x	2	5	15	8	7	11	6	9
y	16	13	3	10	11	7	12	10

c

x	4	5	6	4	3	4	6	4	4	5
y	7	4	7	6	10	10	8	6	10	10

4 A movie review website shows the average critic rating and average audience rating for each movie. The critic and audience ratings for 9 randomly selected movies are shown in the table below:

Movie	A	B	C	D	E	F	G	H	I
Critic rating	18	28	11	92	28	94	61	31	90
Audience rating	31	50	30	82	53	80	84	61	85

- **a** Construct a scatter graph for the data, with *critic rating* on the horizontal axis, and *audience rating* on the vertical axis.
- **b** Calculate the correlation coefficient r.
- **c** Describe the correlation between *critic rating* and *audience rating*.

5 The scores awarded by two judges at a diving competition are shown in the table.

Competitor	P	Q	R	S	T	U	V	W	X	Y
Judge A	6	4	7	5.5	9	7	6.5	8.5	5.5	1.5
Judge B	7	4.5	7	6.5	8.5	7.5	6.5	8	6.5	3

- **a** Construct a scatter graph for the data, with Judge A's scores on the horizontal axis, and Judge B's scores on the vertical axis.
- **b** Calculate the correlation coefficient r.
- **c** Describe the correlation between the judges' scores.

6 Drew wants to investigate the relationship between an *amount* of money and the least *number of coins* required to obtain that amount. For example:

- only one coin is needed to obtain 20 cents
- 3 coins are required to obtain 35 cents.

20 cents

35 cents

a What do you suspect is the strength and direction of correlation between *amount* and *number of coins*? Explain your answer.

b Copy and complete:

Amount (cents)	5	10	15	20	25	30	35	40	45	50	55	....	100
Least number of coins				1			3						

c Draw a scatter graph for the data.

d Calculate the correlation coefficient r.

e Describe the correlation between *amount* and *number of coins*.

DISCUSSION

When describing the relationship between variables, why is it important to look at a scatter graph of the data as well as the correlation coefficient?

ACTIVITY — WHAT'S IN A NAME?

Do people with long surnames generally have long given names as well?

What to do:

1. As a class, predict the nature of the correlation between the *lengths of given names* and the *lengths of surnames*.
2. By considering the original question, decide which is the independent variable.
3. Use the names of each student in your class to create a scatter graph.
4. Calculate the correlation coefficient r.
5. Describe the correlation between the *lengths of given names* and the *lengths of surnames*.

D LINE OF BEST FIT BY EYE

If the scatter graph and correlation coefficient for a data set indicate there is at least moderate linear correlation between two variables, it is reasonable to use a line of best fit to model the data.

For a bivariate data set involving x and y, we use these steps to find the **line of best fit by eye**:

Step 1: Find the means $\overline{x}$ and $\overline{y}$ of the x and y values respectively.

Step 2: Plot the **mean point** $(\overline{x}, \overline{y})$ on a scatter graph of the data.

Step 3: Draw a line through the mean point which fits the trend of the data and which has about as many points above the line as below it.

For the data in the **Opening Problem**, the mean point is approximately $(187, 88)$.

We have added the mean point and the line of best fit by eye on the scatter graph.

As this line is an estimate only, lines drawn by eye will vary from person to person.

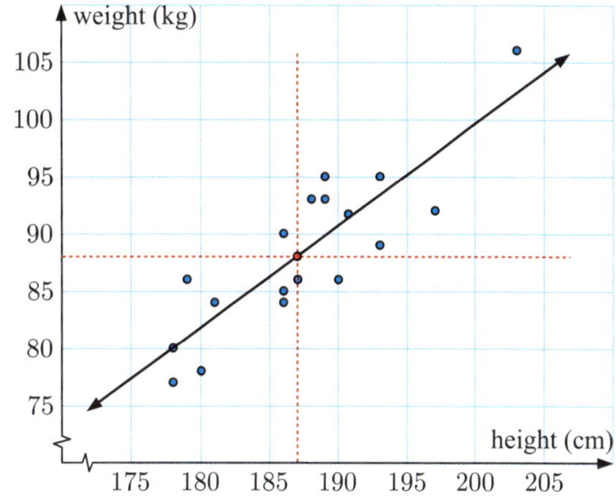

The line of best fit is a **linear model** for the relationship between the variables. We can use it to estimate the value of one variable for any given value of the other.

INTERPOLATION AND EXTRAPOLATION

In a bivariate data set, the data values with the lowest and highest values of x are called the **poles**.

If we use values of x **in between** the poles to estimate y, we say we are **interpolating** between the poles.

If we use values of x **outside** the poles to estimate y, we say we are **extrapolating** outside the poles.

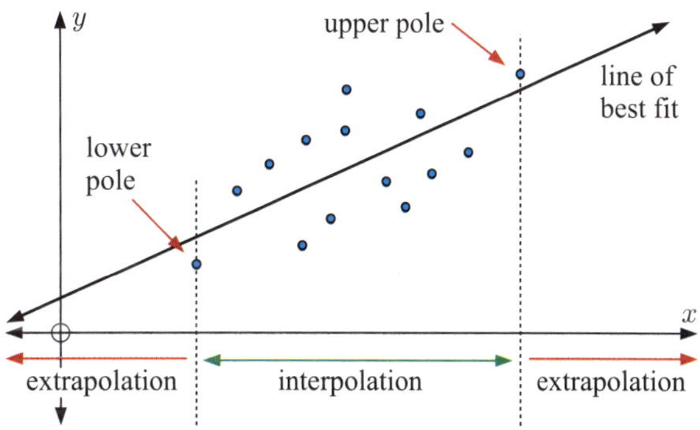

As a general rule, it is reasonable to interpolate between the poles, but unreliable to extrapolate outside them.

Example 3 **Self Tutor**

This table shows the temperature at 9 am and the temperature at 3 pm on the same day in a particular city for 6 randomly selected days of the year:

9 am temperature (x °C)	19	30	16	11	21	11
3 pm temperature (y °C)	26	35	18	15	26	12

a Calculate $\overline{x}$ and $\overline{y}$.

b Draw a scatter graph for the data. Plot the mean point $(\overline{x}, \overline{y})$ on the scatter graph, and draw a line of best fit through this point.

c Predict the temperature at 3 pm for a day the temperature at 9 am is:

i 25°C **ii** 40°C.

Comment on the reliability of your estimates.

a $\overline{x} = \dfrac{19+30+16+11+21+11}{6} = 18, \quad \overline{y} = \dfrac{26+35+18+15+26+12}{6} = 22$

b

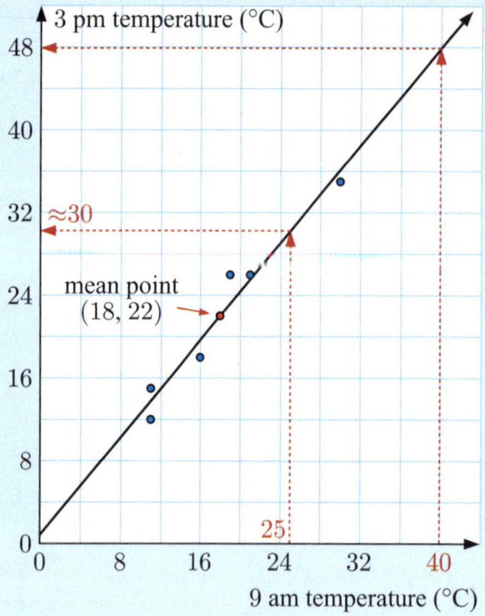

c **i** When $x = 25$, $y \approx 30$.
When the temperature at 9 am is 25°C, the temperature at 3 pm will be approximately 30°C.
This estimate is an interpolation, so we expect it to be reliable.

ii When $x = 40$, $y \approx 48$.
When the temperature at 9 am is 40°C, the temperature at 3 pm will be approximately 48°C.
This estimate is an extrapolation, and therefore may not be reliable. We cannot assume that the linear trend will continue up to a 9 am temperature of 40°C.

EXERCISE 21D

1 Consider the bivariate data alongside.

x	5	10	2	13	6
y	11	3	18	5	13

a Find $\overline{x}$ and $\overline{y}$.
b Draw a scatter graph for the data.
c Does the data appear to be positively correlated or negatively correlated?
d Plot the mean point $(\overline{x}, \overline{y})$ on the scatter graph, and draw a line of best fit through this point.
e Estimate the value of y when $x = 8$.

2 Each month, an opinion poll shows the approval rating of the Prime Minister and the Opposition leader. The approval ratings for the last 10 polls are shown below:

Prime Minister (x%)	55	59	68	61	46	42	38	45	42	44
Opposition (y%)	37	35	31	35	43	40	42	37	41	39

a Calculate $\overline{x}$ and $\overline{y}$.
b Draw a scatter graph for the data. Include the mean point $(\overline{x}, \overline{y})$, and draw a line of best fit through this point.
c In a new opinion poll, the Prime Minister's approval rating is 47%. Estimate the approval rating of the Opposition leader.

3 The table alongside shows the *percentage of unemployed adults* and the *number of major thefts per day* in eight large cities.

City	*Unemployment* (%)	*Thefts*
A	7	113
B	6	67
C	10	117
D	8	88
E	9	120
F	6	38
G	3	61
H	7	76

a Find the mean unemployment percentage $\overline{x}$ and the mean number of thefts $\overline{y}$.
b Draw a scatter graph for the data. Include the mean point $(\overline{x}, \overline{y})$, and draw a line of best fit through this point.
c Estimate the number of major thefts per day for a city with 15% unemployment. Comment on the reliability of your estimate.

4 A café manager believes that during April, the *number of people wanting dinner* is related to the *temperature at noon*. Over a 13 day period, the number of diners and the noon temperature were recorded.

Temperature (x °C)	8	10	13	15	15	12	10	13	17	16	18	14	12
Number of diners (y)	63	70	74	81	77	65	75	87	91	75	96	82	88

a Find the mean point $(\overline{x}, \overline{y})$.

b Draw a scatter graph for the data. Include the mean point $(\overline{x}, \overline{y})$, and draw a line of best fit through this point.

c Estimate the number of diners at the café when the temperature is:

i 11°C **ii** 4°C.

d Comment on the reliability of your estimates in **c**.

5 An ice cube tray is filled with water and placed in a freezer. This table shows the energy lost by the water E kJ in reducing its temperature to t°C.

Temperature (t°C)	28	25	19	13	9	2
Energy (E kJ)	0	6.1	18.7	31.2	39.7	54.3

a What was the temperature of the water when it was placed in the freezer? Explain your answer.

b Find the mean point $(\overline{t}, \overline{E})$.

c Draw a scatter graph for the data. Include the mean point $(\overline{t}, \overline{E})$, and draw a line of best fit through this point.

d Hence estimate the energy lost by the water in reducing its temperature to -4°C.

e The water actually loses 234 kJ of energy in reducing its temperature to -4°C.
Discuss the inaccuracy of your extrapolation in **d**. Can you suggest a reason for this inaccuracy?

E LINEAR REGRESSION

The problem with drawing a line of best fit by eye is that the answer will vary from one person to another, and the line may not be very accurate. Instead, we can use a method called **linear regression**.

Linear regression is a formal method of fitting a line which best fits a set of data.

This line of best fit is called the **least squares regression line**. It is the line which makes the sum of the squares of the distances $d_1^2 + d_2^2 + d_3^2 + ...$ as small as possible.

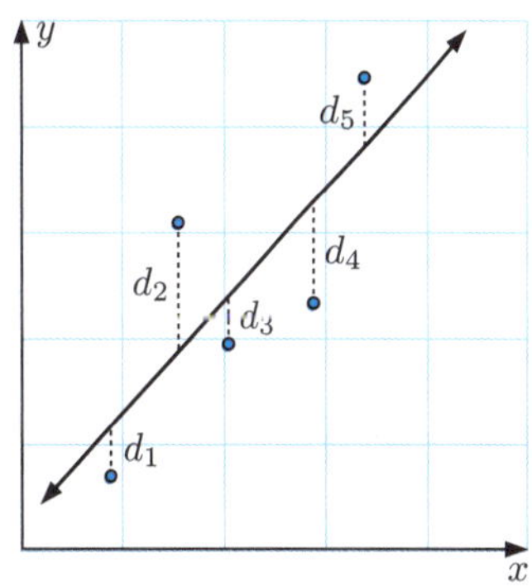

We can use a **graphics calculator** or the **statistics package** to find the equation of the least squares regression line.

Example 4 **Self Tutor**

An ice cream vendor recorded the maximum temperature and the number of ice cream sales each day for 2 weeks during summer:

Maximum temperature (x °C)	34	40	35	32	34	34	31	32	29	33	37	35	31	30
Number of ice creams sold (y)	119	195	131	152	206	169	122	143	90	128	208	155	144	167

a Construct a scatter graph to illustrate the data.

b Use technology to find the least squares regression line.

c Estimate the number of ice creams sold when the maximum temperature is 36°C. Comment on whether this estimate is likely to be reliable.

a

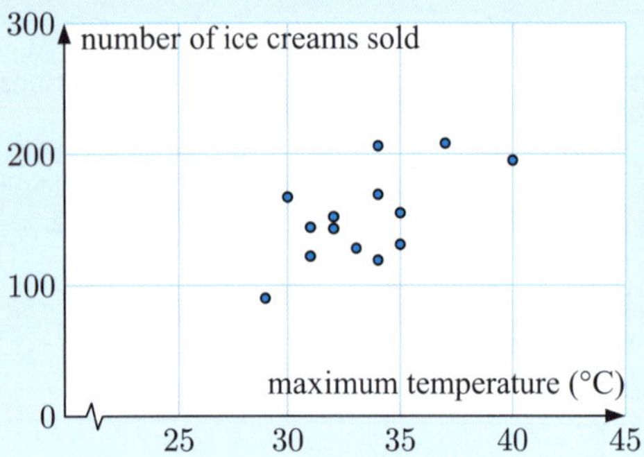

b

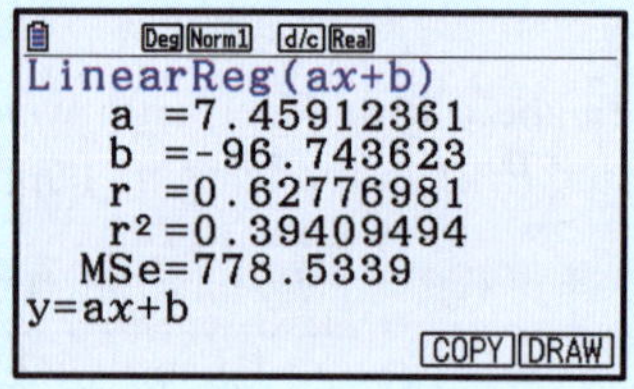

Using technology, the least squares regression line is $y \approx 7.46x - 96.7$.

c When $x = 36$, $y \approx 7.46(36) - 96.7 \approx 172$

So, we expect about 172 ice creams to be sold when the maximum temperature is 36°C. Although this is an interpolation, the correlation is only moderate. So, the estimate may not be reliable.

EXERCISE 21E

1 For each data set:

i Draw a scatter graph for the data.

ii Find the equation of the least squares regression line.

iii Comment on the strength and direction of the linear relationship between the variables.

a

x	3	5	6	9	13	15	18	20
y	4	8	11	17	15	19	25	24

b

x	16	6	11	8	14	24	2	14	19	6
y	15	30	20	26	17	9	35	10	7	27

2 A group of friends competed in a fun-run. The table below shows how long each friend spent training, and the time they recorded for the fun-run.

Training time (x hours)	7	2	11	3	7	15	3	0	5	9	0
Fun-run time (y minutes)	60	75	47	70	52	37	72	75	60	62	80

a Draw a scatter graph for the data.

b Find the correlation coefficient r and discuss the significance of this value.

c Find the equation of the least squares regression line.

d Interpret the gradient of this line.

e Another friend of the group trained for 30 hours.

i Use the least squares regression line to estimate his fun-run time.

ii Comment on the reliability of your estimate.

3 The table below shows the yield of cherries from an orchard over several years with different numbers of frosty mornings.

Frosty mornings (n)	18	29	23	38	35	27
Yield (Y tonnes)	29.4	34.6	32.1	36.9	36.1	32.5

a Draw a scatter graph of Y against n.

b Find the correlation coefficient r and discuss the significance of this value.

c Find the linear model which best fits the data.

d Estimate the yield from the orchard if there are 31 frosty mornings in the year.

4 Carbon dioxide (CO_2) is linked to acid rain and global warming. The concentration of CO_2 in the atmosphere has been recorded over a 40 year period. It is measured in parts per million or ppm found in Law Dome Ice Cores in Antarctica.

Year	1960	1970	1980	1990	2000
CO_2 *concentration* (C ppm)	313	321	329	337	345

Let t be the number of years since 1960.

a Draw a scatter graph of C against t.

b Describe the correlation between C and t.

c Find the linear model which best fits the data.

d Estimate the CO_2 concentration for 1987.

e If CO_2 emission continues at the same rate, estimate the concentration in 2025.

5 The year in which apartment buildings in a city were built and the average floor area of their apartments are shown in the table below:

Year (x)	1980	1985	1965	1975	1994	1976	1988
Area (y m^2)	78.4	80.6	75.2	77.3	81.9	75.8	80.5

Year (x)	1985	2015	1993	1981	1980	1979	2004
Area (y m^2)	75.2	80.4	79.9	77.8	78.7	74.8	82.6

a Draw a scatter graph for the data.

b Describe the relationship between the variables.

c Calculate the correlation coefficient r and interpret its value.

d Find the equation of the least squares regression line.

e Estimate the average floor area of the apartments in a building built in the year:

i 2000

ii 2050.

f Comment on the reliability and reasonableness of your estimates.

6 Answer the **Opening Problem** on page **366**.

7 A class of 25 students was asked to record the amount of time they spent studying for a test marked out of 60. The data below was collected:

Study time (x min)	75	30	35	65	110	60	40	80	56	70	50	110	18
Score (y)	25	31	30	38	55	20	39	47	35	45	32	33	34

Study time (x min)	80	22	30	15	10	85	100	60	55	80	50	75
Score (y)	38	17	38	17	17	26	41	50	30	45	30	23

a Draw a scatter graph for the data.

b Find the equation of the least squares regression line.

c Interpret the y-intercept of this line.

d Comment on the strength and direction of the linear relationship between the variables.

e Use the regression line to predict the test score of a student who spent $3\frac{1}{2}$ hours studying. Comment on whether this prediction is reasonable.

GLOBAL CONTEXT — THE CAPTURE-RECAPTURE METHOD

Global context: Globalisation and sustainability

Statement of inquiry: Collecting and analysing samples can help us model animal populations.

Criterion: Applying mathematics in real-life contexts

GLOBAL CONTEXT

QUICK QUIZ

MULTIPLE CHOICE QUIZ

REVIEW SET 21A

1 Traffic analysts want to find the association between the *average speed* of cars in a city and the *age of drivers*. Devices for measuring average speed were fitted to the cars of drivers participating in a survey. The results are shown in the scatter graph.

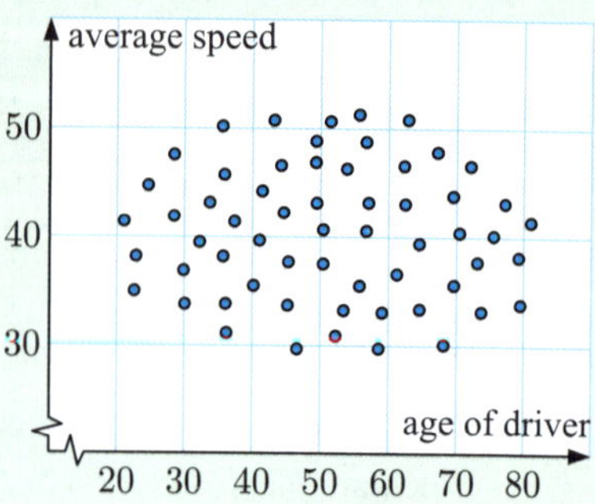

a Which is the independent variable?

b Describe the association between the variables.

c Is it sensible to find the linear regression line for this data? Explain your answer.

2 Sam and Richard competed in a series of 8 chess games. The table below shows how many pieces remained for each player at the end of each game.

Sam	4	2	6	2	4	1	5	7
Richard	2	1	3	7	4	8	3	5

a Construct a scatter graph for the data.

b Calculate the correlation coefficient r and interpret its value.

3 The maximum speed of a Chinese dragonboat with different numbers of paddlers is recorded in the table below.

Number of paddlers (x)	4	6	10	14	18	20
Maximum speed (y km/h)	8	10	12	14	16	17

a Find $\overline{x}$ and $\overline{y}$.

b Draw a scatter graph for the data. Include the mean point and a line of best fit by eye.

c Predict the maximum speed of a dragonboat with:

i 12 paddlers **ii** 24 paddlers.

4 The whorls on a cone shell get wider as you go from the top of the shell towards the bottom. Measurements from a shell are summarised in the following table:

Position of whorl (p)	1	2	3	4	5	6	7	8
Width of whorl (w cm)	0.7	1.2	1.4	2.0	2.0	2.7	2.9	3.5

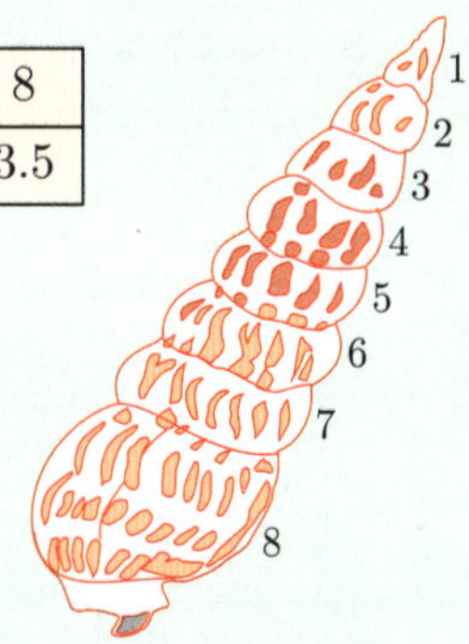

a Draw a scatter graph for this data.

b Find the correlation coefficient r and interpret its value.

c Find the equation of the least squares regression line.

d **i** If a cone shell has 14 whorls, what width do you expect the 14th whorl to have?

ii How reliable is this prediction?

5 The table below shows the number of games won, and the final position on the Premier League ladder, for the Liverpool Football Club from 2009 - 10 to 2020 - 21:

Games won (x)	18	17	14	16	26	18	16	22	21	30	32	20
Position (y)	7	6	8	7	2	6	8	4	4	2	1	3

a Would you expect x and y to be positively or negatively correlated? Explain your answer.

b Draw a scatter graph for the data.

c Describe the correlation between the variables.

d Use technology to find the line of best fit.

e Suppose Liverpool wins 22 games next season. Predict their position on the ladder.

REVIEW SET 21B

1 The scatter graph shows the number of defective items made by each employee of a factory, plotted against the employee's number of weeks of experience.

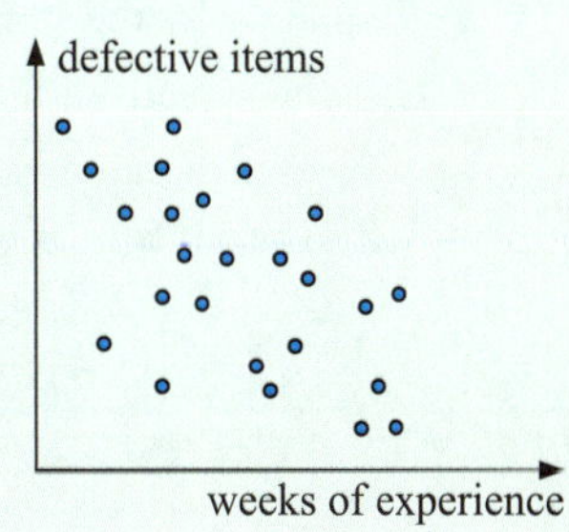

a Identify the independent variable and the dependent variable.

b Describe the correlation between the variables.

2 Consider the relationship between a *number* and the *number of factors* it has.

a Discuss whether you would expect the correlation between these variables to be:

i positive or negative **ii** strong, moderate, or weak.

b Copy and complete:

Number (x)	1	2	3	4	5	....	16	17	18	19	20
Number of factors (y)	1	2	2	3		....				2	6

c Draw a scatter graph for the data.

d Find the correlation coefficient r and interpret its value.

3 Consider the bivariate data alongside.

x	11	7	13	3	12	8
y	17	14	20	5	26	8

a Find $\overline{x}$ and $\overline{y}$.

b Draw a scatter graph for the data. Include the mean point $(\overline{x}, \overline{y})$, and draw a line of best fit through this point.

c Estimate the value of y when $x = 5$. How reliable is this estimate?

4 Paul researched the first 6 games of a Twenty20 cricket tournament. For the team which batted first, he recorded their score after 10 overs, and their final score.

Score after 10 *overs* (x)	76	46	68	51	75	74
Final score (y)	182	123	159	93	191	196

a Draw a scatter graph for the data.

b Describe the correlation between the variables.

c Find the equation of the least squares regression line.

d Predict the final score for a team which scores 60 runs in the first 10 overs. How reliable is your prediction?

5 The following table gives peptic ulcer rates per 1000 people for differing family incomes.

Income ($\$I \times 1000$)	20	25	30	35	40	50	60	80	100
Peptic ulcer rate (R)	8.3	7.7	6.9	7.3	5.9	4.7	3.6	2.6	1.2

a Draw a scatter graph for the data.

b Find the equation of the line of best fit for the data.

c Describe the correlation between the variables.

d Estimate the peptic ulcer rate in families with an income of \$55 000.

e Explain why the model is inadequate for families with an income in excess of \$120 000.

f It is later realised that one of the values was recorded incorrectly.

i Which value is it likely to be? Explain your answer.

ii Repeat **b** to **d** with the incorrect data value removed.

Chapter 22

Relations and functions

Contents:

OPENING PROBLEM

Fernando and Gerard are competing in a cycling race.

This graph shows the *displacement* y m of Fernando relative to Gerard after x minutes.

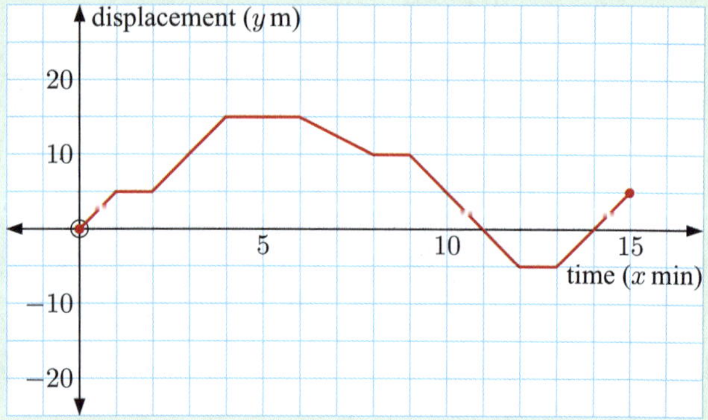

For example:

- If $y = 5$ then Fernando is 5 m ahead of Gerard.
- If $y = -5$ then Fernando is 5 m behind Gerard.

Things to think about:

a What values of *time* are illustrated on the graph?

b What values of *displacement* were recorded during the race?

c What is the value of y when $x = 7$? How can we interpret this value?

d What are the values of x for which $y = 10$? How can we interpret these values?

e How can we *transform* the graph to show the displacement of *Gerard* relative to *Fernando* after x minutes?

In the **Opening Problem** we see a relationship between the two variables *time* and *displacement*.

In this Chapter we explore what it means for the relationship between two variables to be called a **function**. We will then explore properties of functions which will help us work with and understand them.

A RELATIONS AND FUNCTIONS

A **relation** between variables x and y is any set of points in the (x, y) plane. We say that the points *connect* the two variables.

A relation is often expressed as an **equation** connecting x and y.

For example, $y = x - 2$ is an equation of a relation. This equation generates an infinite set of points which form a straight line.

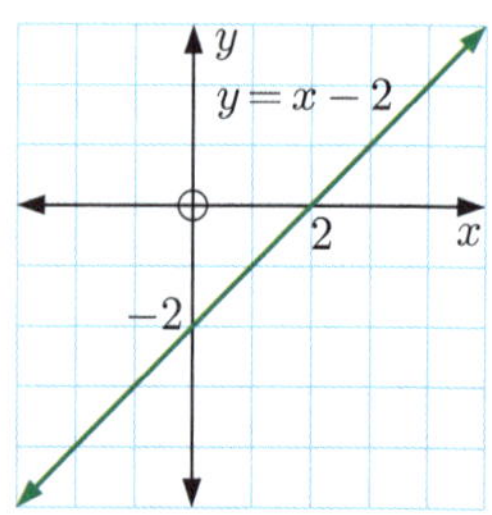

However, not all relations can be defined by an equation.

For example:

- In the **Opening Problem**, the relationship between *displacement* and *time* is described by an infinite set of points, but not an equation.

- These 8 points form a relation.
  It can be described as a finite set of points, but not an equation.

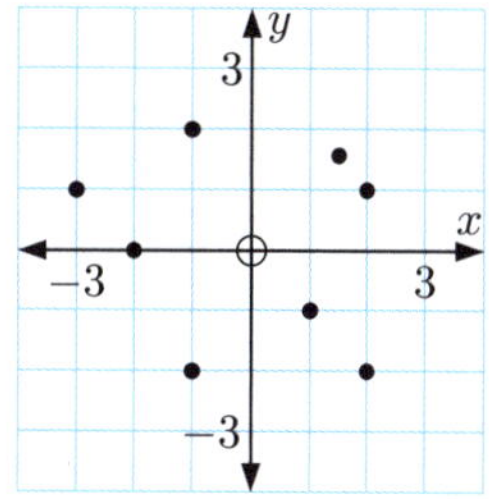

FUNCTIONS

A **function** is a relation in which no two different points have the same x-coordinate or first component.

We can see from this definition that a function is a special type of relation.

Every function is a relation, but not every relation is a function.

For example, in the relation alongside, the two circled points $(-1, -2)$ and $(-1, 3)$ have the same x-coordinate. So, the relation is not a function.

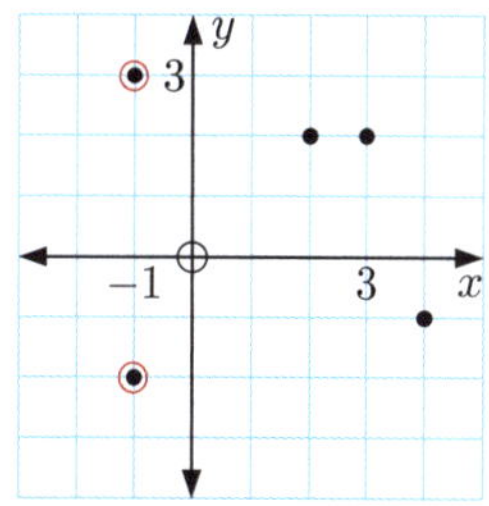

GEOMETRIC TEST OR VERTICAL LINE TEST FOR FUNCTIONS

Suppose we draw all possible vertical lines on the graph of a relation.

- If each line cuts the graph at most once, then the relation is a function.
- If at least one line cuts the graph more than once, then the relation is *not* a function.

Example 1 Self Tutor

Which of these relations are functions?

a

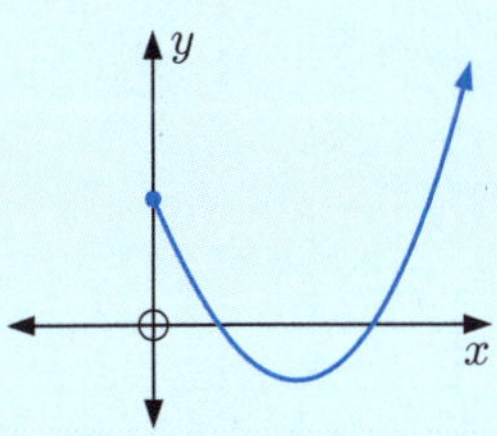

b

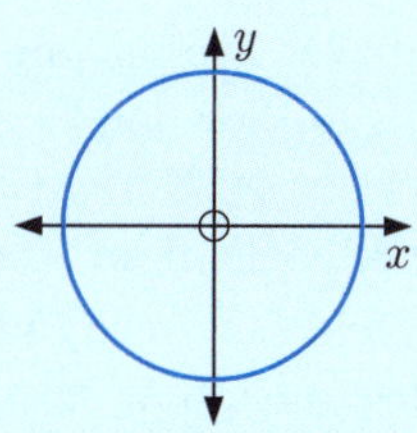

a Every vertical line we could draw cuts the graph only once.
$\therefore$ the relation is a function.

b This vertical line cuts the graph twice.
$\therefore$ the relation is not a function.

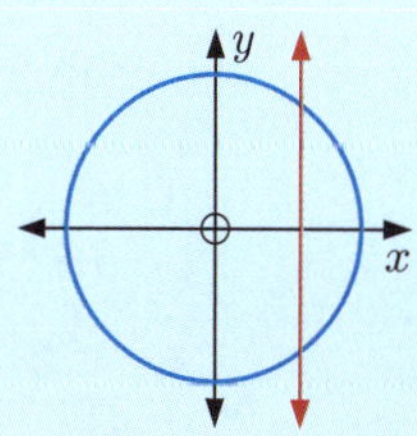

EXERCISE 22A

1 Which of the following sets of points are functions? Explain your answers.

- **a** $\{(1, 1), (2, 2), (3, 3), (4, 4)\}$
- **b** $\{(-1, 2), (-3, 2), (3, 2), (1, 2)\}$
- **c** $\{(2, 5), (-1, 4), (-3, 7), (2, -3)\}$
- **d** $\{(3, -2), (3, 0), (3, 2), (3, 4)\}$
- **e** $\{(-7, 0), (-5, 0), (-3, 0), (-1, 0)\}$
- **f** $\{(0, 5), (0, 1), (2, 1), (2, -5)\}$

2 Use the vertical line test to determine whether each relation is a function:

a

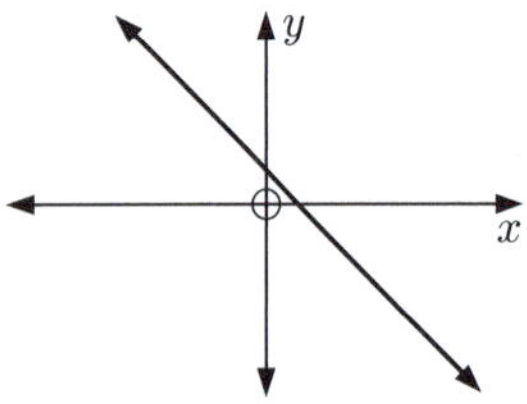

b

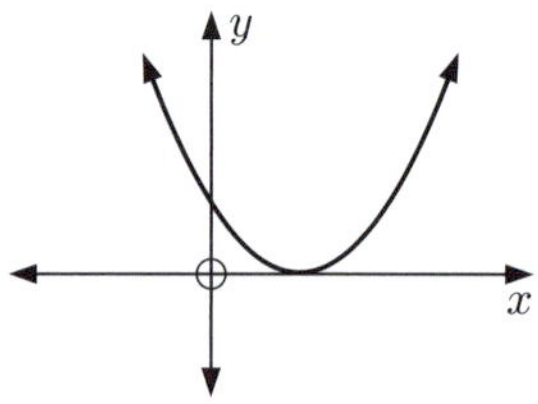

c

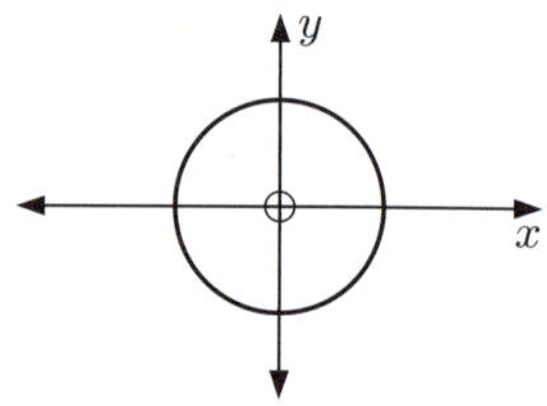

d

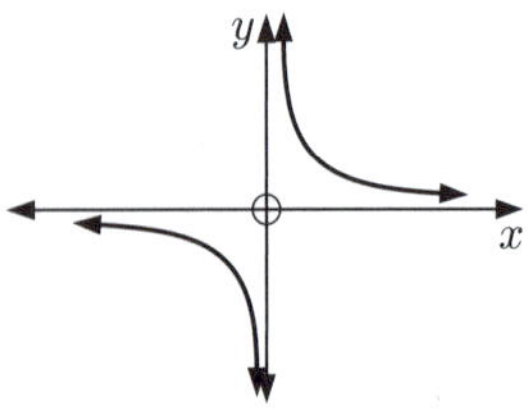

e

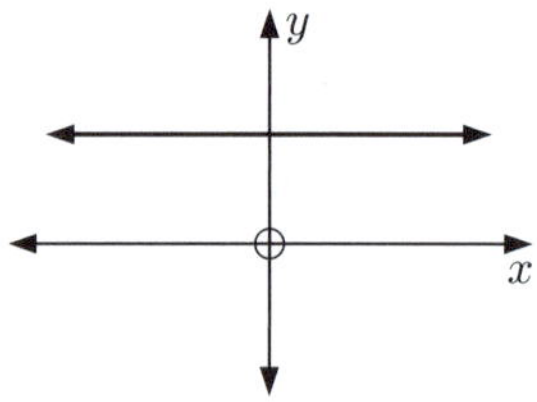

f

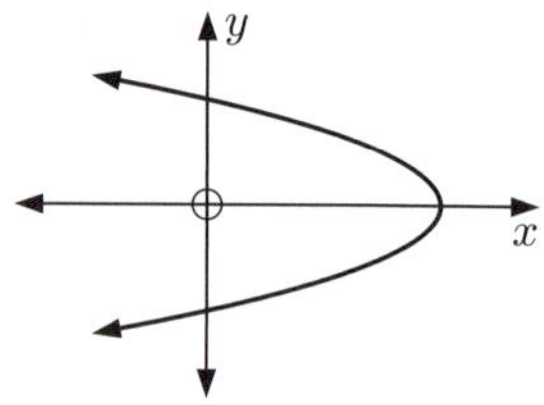

g

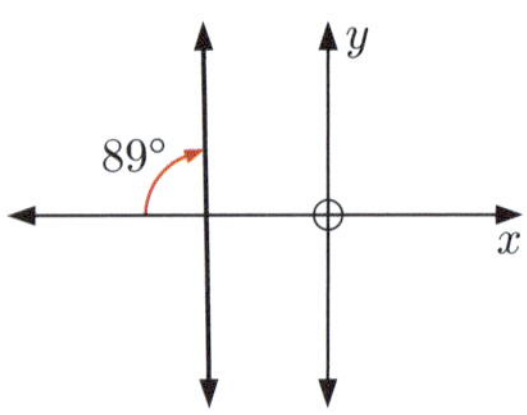

h

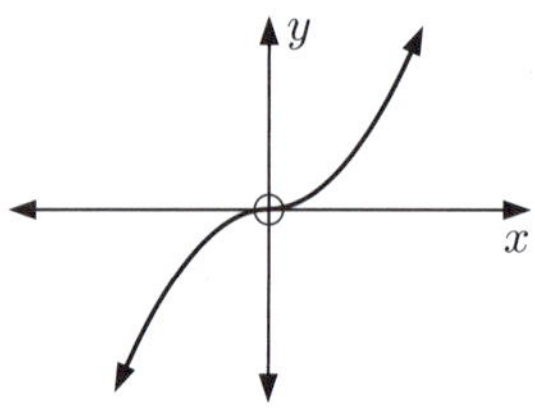

i

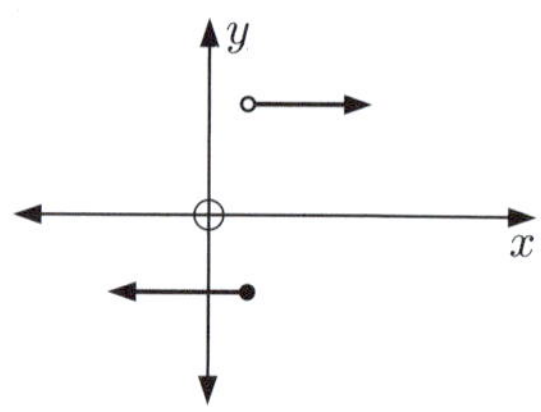

3 From the **Opening Problem** on page **384**, decide whether the relation between *time* and *displacement* is a function.

4 Is the graph of a straight line always a function? Explain your answer.

5 Xavier and Yvonne have a box of 8 biscuits.

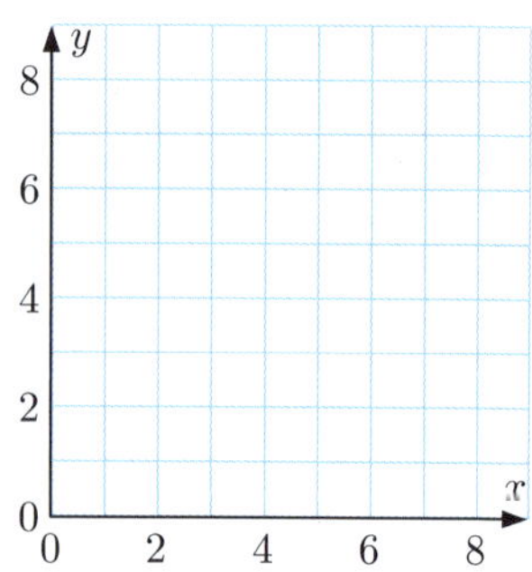

- **a** Suppose Xavier and Yvonne eat *all* of the biscuits between them.
  - **i** On a set of axes like the one alongside, draw a set of points to show the possible combinations of biscuits eaten by Xavier (x) and Yvonne (y).
  - **ii** Do the set of points form a function?
  - **iii** If you are told how many biscuits Xavier ate, can you determine how many biscuits Yvonne ate?
- **b** Now suppose Xavier and Yvonne are allowed to eat all the biscuits, but they do not necessarily do so.
  - **i** Draw a set of points to show the possible combinations of biscuits eaten by Xavier and Yvonne.
  - **ii** Do the set of points form a function?
  - **iii** If you are told how many biscuits Xavier ate, can you necessarily determine how many biscuits Yvonne ate?

B FUNCTION NOTATION

We sometimes use a *function machine* to illustrate how functions behave.

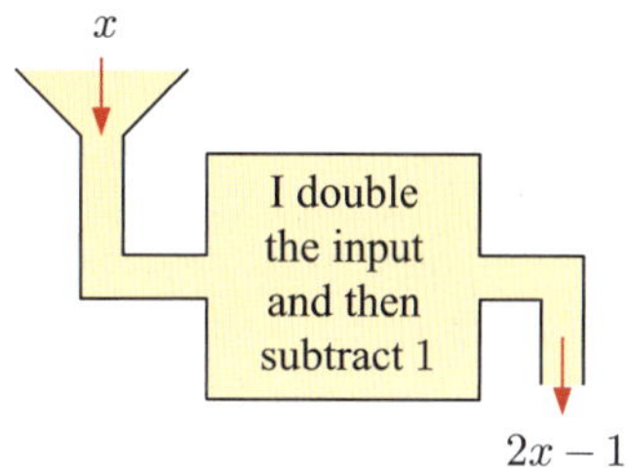

This machine has been programmed to perform the function: "double the number and then subtract 1".

If 3 is the input fed into the machine, the output is $2(3) - 1 = 5$.

If we use f to represent this function, we can write "f is the function that will convert x into $2x - 1$".

So, f would convert 2 into $2(2) - 1 = 3$ and
-4 into $2(-4) - 1 = -9$.

In **function notation**, we write $f(x) = 2x - 1$ or $y = 2x - 1$.

> $f(x)$ is the value of y for a given value of x, so $y = f(x)$.

$y = f(x)$ is sometimes called the **function value** or **image** of x.

For $f(x) = 2x - 1$:

- $f(2) = 2(2) - 1 = 3$
  $\therefore$ the point $(2, 3)$ lies on the graph of the function.
- $f(-4) = 2(-4) - 1 = -9$
  $\therefore$ the point $(-4, -9)$ also lies on the graph.

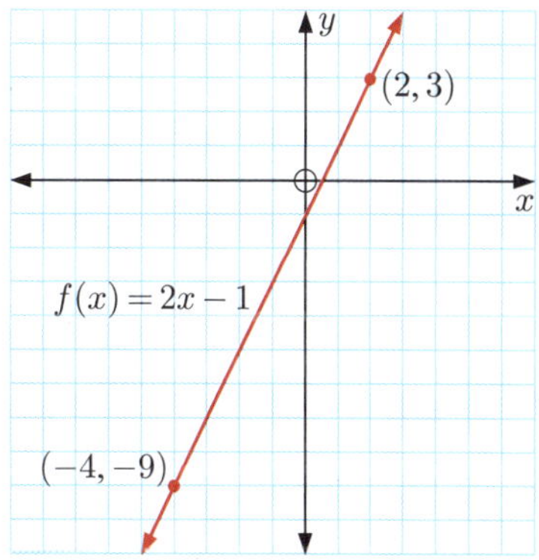

Example 2 — Self Tutor

For $f(x) = 3x^2 - 4x$, find the value of:

a $f(2)$ **b** $f(-5)$

a $f(2)$
$= 3(2)^2 - 4(2)$ {replacing x with (2)}
$= 3 \times 4 - 8$
$= 4$

b $f(-5)$
$= 3(-5)^2 - 4(-5)$ {replacing x with (-5)}
$= 3(25) + 20$
$= 95$

EXERCISE 22B

1 For $f(x) = 2x + 3$, find the value of:

a $f(0)$ **b** $f(2)$ **c** $f(-1)$ **d** $f(-5)$ **e** $f(-\frac{1}{2})$

2 For $E(x) = 2(x - 3)$, find the value of:

a $E(0)$ **b** $E(1)$ **c** $E(5)$ **d** $E(-2)$ **e** $E(\frac{7}{2})$

3 For $h(x) = \dfrac{x}{x-3}$, find the value of:

a $h(2)$ **b** $h(5)$ **c** $h(10)$ **d** $h(0)$ **e** $h(-7)$

4 The graph of $y = f(x)$ is shown alongside.

a Find:

i $f(2)$ **ii** $f(3)$

b Find the value of x such that $f(x) = 4$.

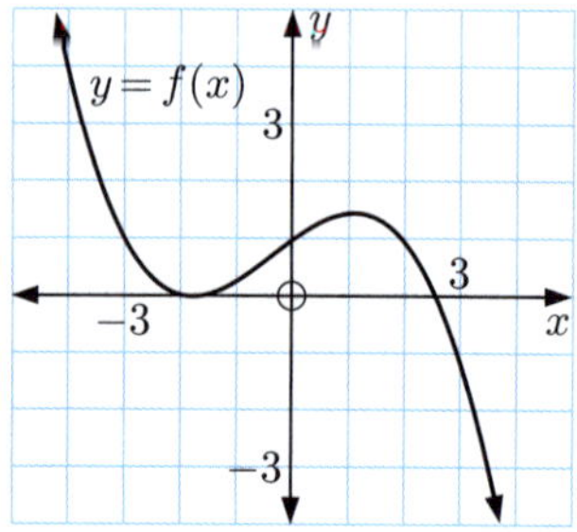

5

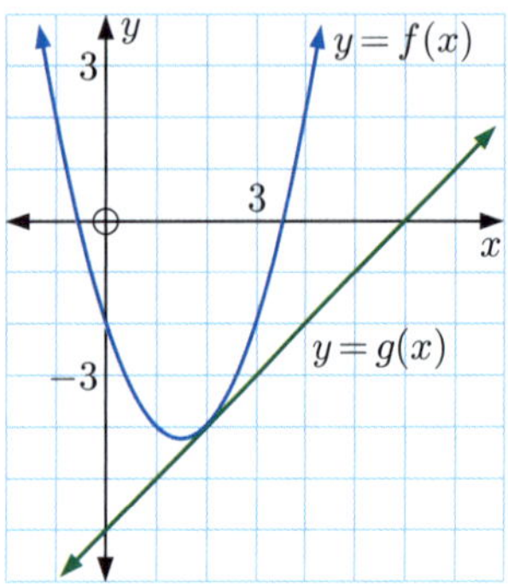

Consider the graphs of $y = f(x)$ and $y = g(x)$.

a Find:

i $f(4)$ **ii** $g(0)$ **iii** $g(5)$

b Find the *two* values of x such that $f(x) = -2$.

c Find the value of x such that $f(x) = g(x)$.

d Show that $g(x) = x - 6$.

6 Suppose $m(x) = \dfrac{x+1}{3}$. Find the value of a for which:

a $m(a) = 2$ **b** $m(a) = -1$ **c** $m(a) = -\frac{7}{6}$

7 Consider the function $R(x) = \dfrac{2x-3}{x+2}$.

a Find the value of:

i $R(0)$ **ii** $R(1)$ **iii** $R(-\frac{1}{2})$

b Find a value of x such that $R(x)$ is undefined.

c Find x such that $R(x) = -5$.

8 Draw a graph of $y = f(x)$ such that $f(-2) = 5$, $f(1) = 0$, and $f(4) = 3$.

9 The value of a car t years after purchase is given by $V(t) = 28\,000 - 4000t$ dollars.

a Find $V(4)$ and state what this value means.

b Find t when $V(t) = 8000$ and explain what this represents.

c Find the original purchase price of the car.

d For what values of t is it reasonable to use this function?

10 The graph of $y = f(x)$ is a straight line passing through $(-3, -5)$ and $(1, 7)$.

a Draw the graph of $y = f(x)$. **b** Find $f(x)$.

Example 3 — Self Tutor

For $f(x) = 4 - 3x - x^2$, find:

a $f(-x)$ **b** $f(x+2)$

a $f(-x) = 4 - 3(-x) - (-x)^2$ {replacing x with $(-x)$}
$= 4 + 3x - x^2$

b $f(x+2) = 4 - 3(x+2) - (x+2)^2$ {replacing x with $(x+2)$}
$= 4 - 3x - 6 - (x^2 + 4x + 4)$
$= -x^2 - 7x - 6$

11 For $g(x) = -5x + 3$, find:

a $g(-x)$ **b** $g(x+4)$ **c** $g(x-1)+2$ **d** $2g\left(\frac{x}{5}\right)$

12 For $f(x) = 2x^2 - 3x + 2$, find:

a $f(-x)$ **b** $f(x+1)$ **c** $3f(x-2)$ **d** $f\left(\frac{x}{2}\right) - 1$

C DOMAIN AND RANGE

We have seen that a relation is a set of points which connects two variables.

The **domain** of a relation is the set of values which the variable on the horizontal axis can take. This variable is usually x.

The **range** of a relation is the set of values which the variable on the vertical axis can take. This variable is usually y.

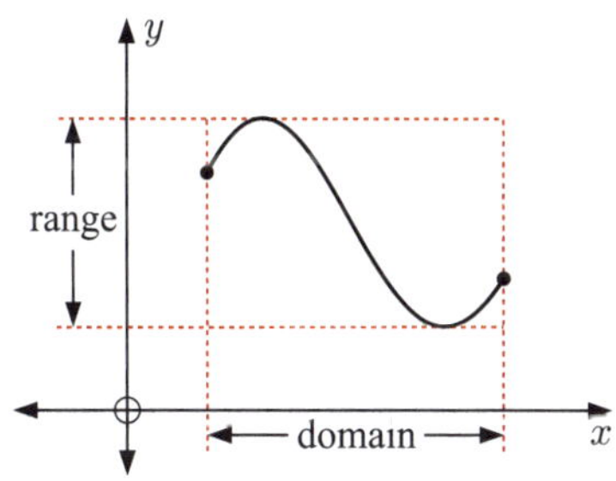

The domain and range of a relation are often described using **interval notation**.

For example:

- This relation is defined for all x such that $1 \leqslant x < 5$, so the domain is $\{x \mid 1 \leqslant x < 5\}$.
  The y values the relation can take satisfy $-2 < y \leqslant 4$, so the range is $\{y \mid -2 < y \leqslant 4\}$.

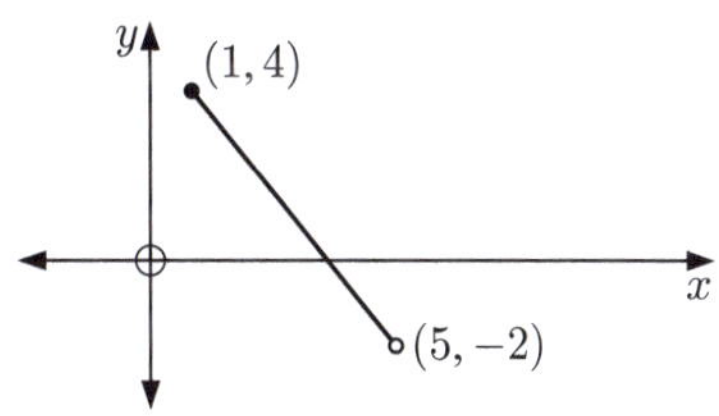

- x can take any real value, so the domain is $\{x \in \mathbb{R}\}$.
  y cannot be less than -3, so the range is $\{y \mid y \geqslant -3\}$.

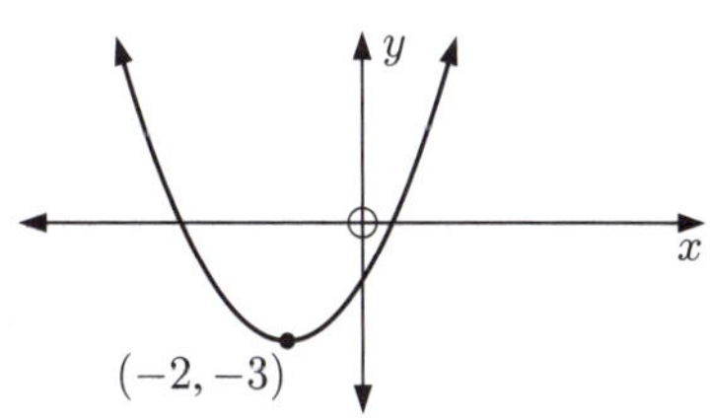

Example 4 **Self Tutor**

For each graph, state the domain and range:

a

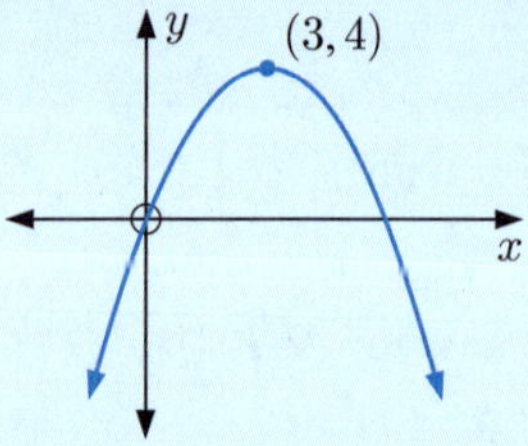

b

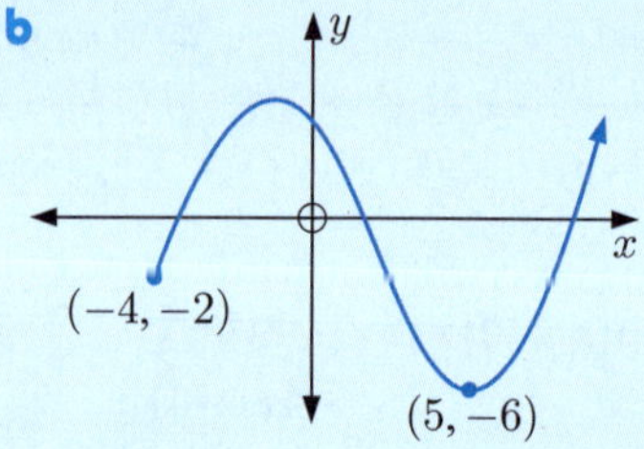

c

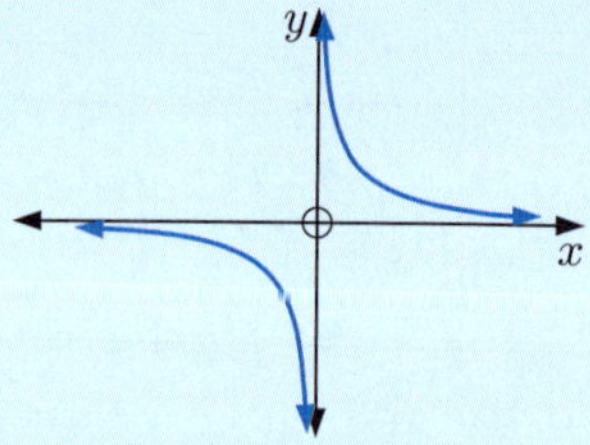

a Domain is $\{x \in \mathbb{R}\}$.
Range is $\{y \mid y \leqslant 4\}$.

b Domain is $\{x \mid x \geqslant -4\}$.
Range is $\{y \mid y \geqslant -6\}$.

c Domain is $\{x \mid x \neq 0\}$.
Range is $\{y \mid y \neq 0\}$.

EXERCISE 22C.1

1 For each graph, state the domain and range:

a

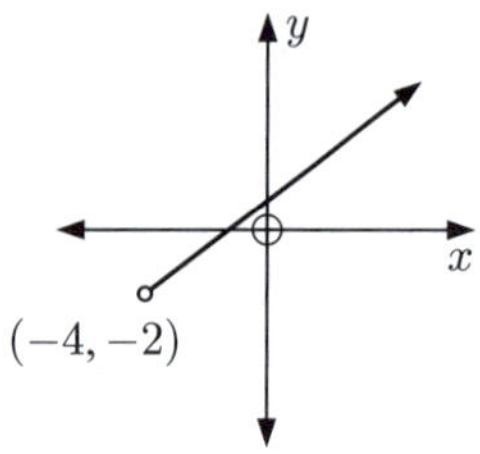

b

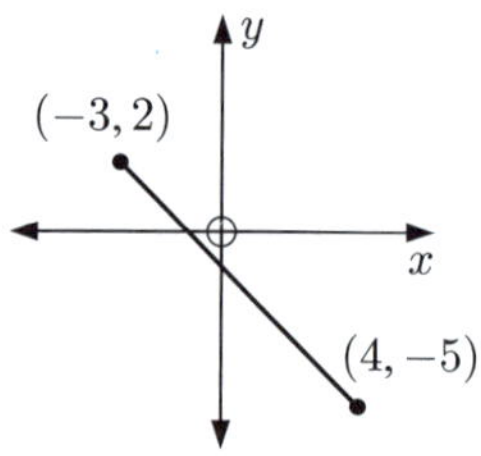

c

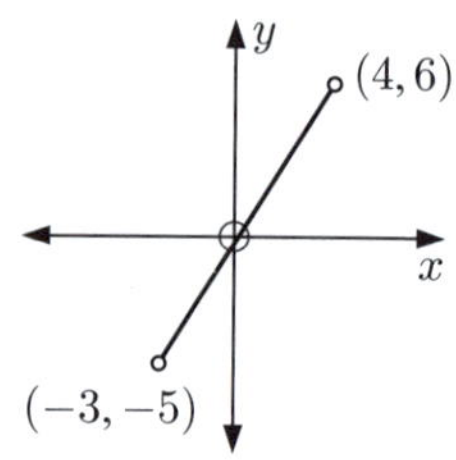

d

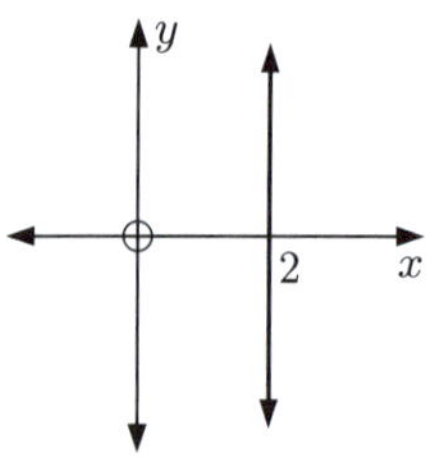

e

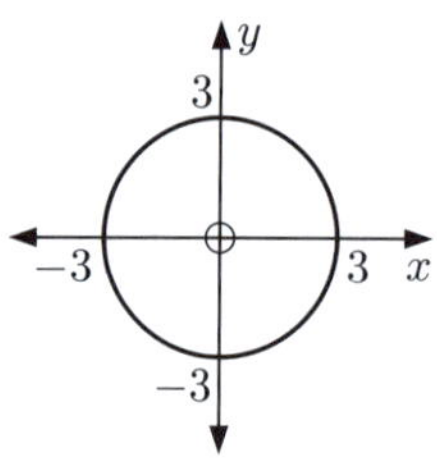

f

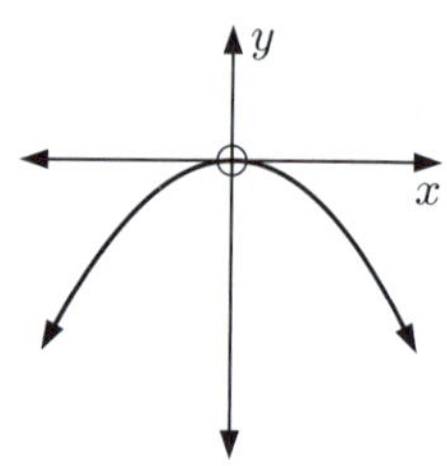

g

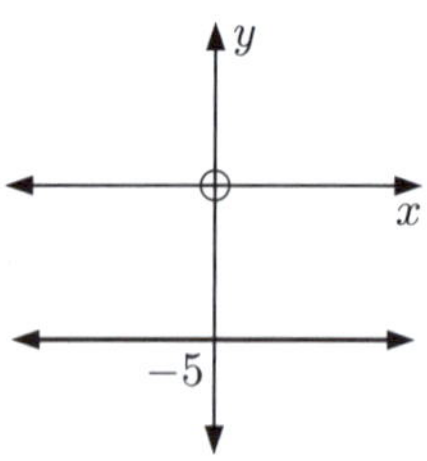

h

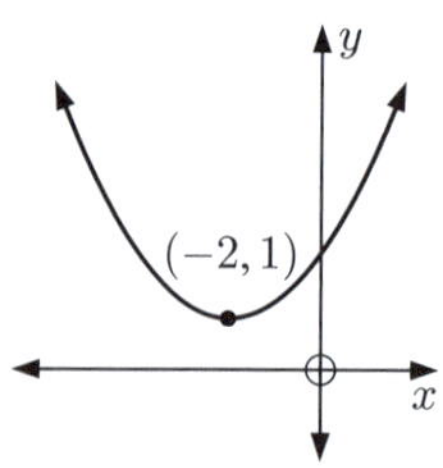

i

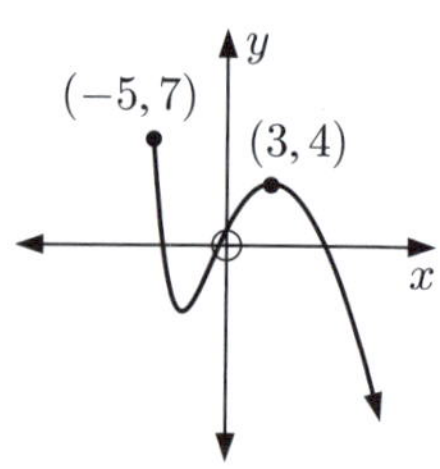

j

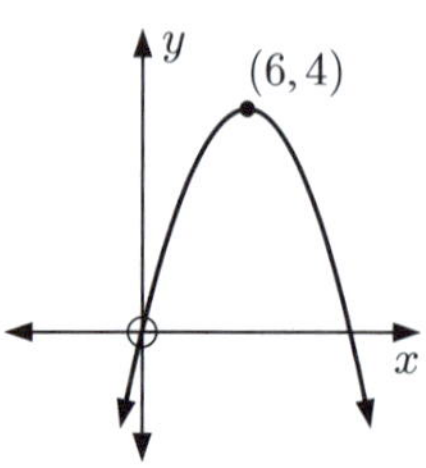

k

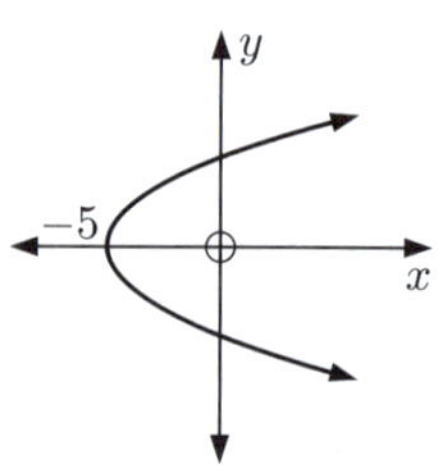

l

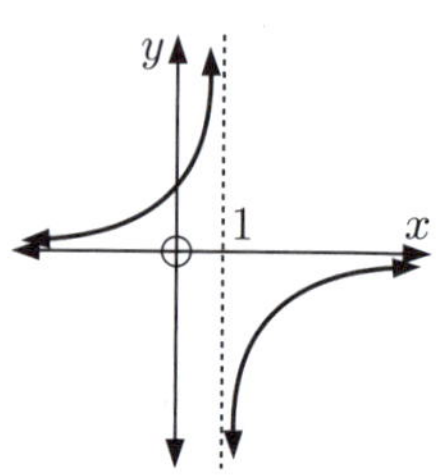

2 Find the domain and range for the graph in the **Opening Problem** on page **384**.

3 This graph shows the speed of a Formula One car during one lap of a race.

   a Explain why a speed graph like this must be a function.

   b Find the domain and range of the function.

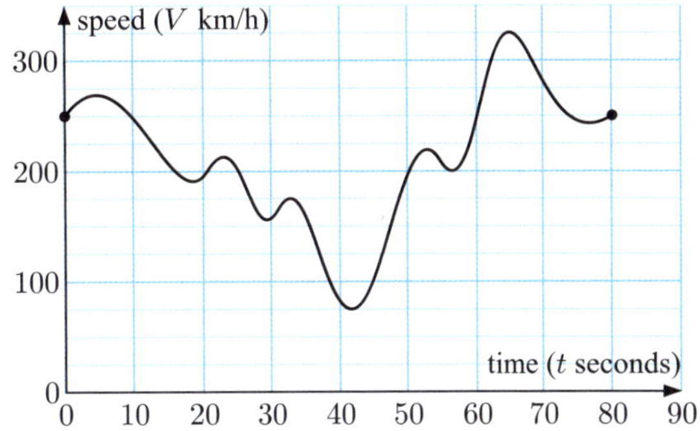

THE NATURAL DOMAIN OF A FUNCTION

To fully describe a function, we need both a rule and a domain.

For example, we can specify $f(x) = \dfrac{12}{x}$ where $x > 0$.

If a domain is not specified, we use the **natural domain**, which is the largest subset of $\mathbb{R}$ for which $f(x)$ is defined.

For example:

- The natural domain of $f(x) = \sqrt{x}$ is $\{x \mid x \geqslant 0\}$, since $\sqrt{x}$ is only defined for $x \geqslant 0$.
- The natural domain of $f(x) = \dfrac{1}{\sqrt{x-1}}$ is $\{x \mid x > 1\}$ since:
  - when $x - 1 = 0$, we are "dividing by zero"
  - when $x - 1 < 0$, $\sqrt{x-1}$ is undefined.

Example 5 **Self Tutor**

Find the natural domain of $f(x) = \dfrac{2}{\sqrt{x+3}}$.

$f(x) = \dfrac{2}{\sqrt{x+3}}$ is defined when $x + 3 > 0$

$\therefore\ x > -3$

$\therefore$ the natural domain of $f(x)$ is $\{x \mid x > -3\}$.

EXERCISE 22C.2

1 Explain why the natural domain of:

   a $f(x) = 4\sqrt{x-2}$ is $\{x \mid x \geqslant 2\}$

   b $f(x) = \dfrac{3}{x-2}$ is $\{x \mid x \neq 2\}$.

2 Find the natural domain of:

   a $f(x) = 2x$

   b $f(x) = \dfrac{1}{x}$

   c $f(x) = \dfrac{1}{x+5}$

   d $f(x) = \dfrac{1}{(x-1)(x+2)}$

   e $f(x) = \dfrac{x}{x^2-9}$

   f $f(x) = \dfrac{3}{x^2-5x+4}$

   Check your answers using the graphing package.

GRAPHING PACKAGE

3 Find the natural domain of:

a $f(x) = \sqrt{x-3}$ **b** $f(x) = \dfrac{1}{\sqrt{x}}$ **c** $f(x) = \sqrt{3-x}$ **d** $f(x) = \dfrac{1}{\sqrt{4-x}}$

Check your answers using the graphing package.

D SIGN DIAGRAMS

A **sign diagram** is a number line which indicates the values of x for which a function is negative, zero, positive, or undefined.

A sign diagram consists of:

- a **horizontal line** which represents the x-axis
- **positive** (+) and **negative** (−) signs indicating where the graph of the function is **above** and **below** the x-axis respectively
- **critical values**, which are the values of x for which the function is zero, and where the function is undefined.

On the number line, we use a solid mark to indicate where the function is zero, and a dashed mark to indicate where the function is undefined.

For example, consider the graph:

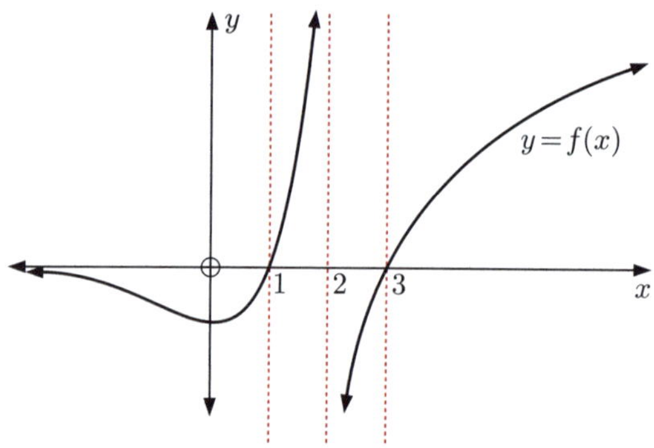

The corresponding sign diagram is:

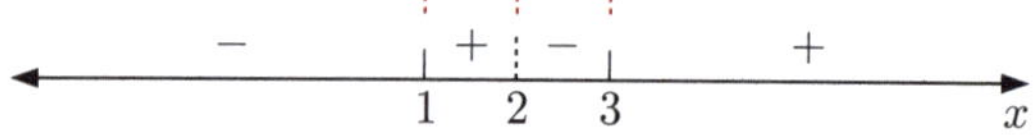

INVESTIGATION 1

What to do:

1 Suppose $f(x) = (x-3)(x+1)$. The graph of $y = f(x)$ is shown alongside.

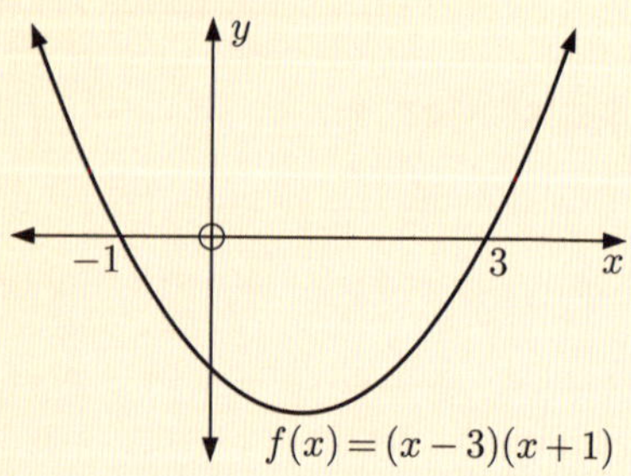

a Explain why -1 and 3 are both x-intercepts of $y = f(x)$.

b Explain why the sign diagram for $y = f(x)$ is

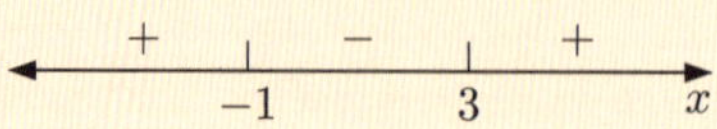

c What happens to the sign of $y = f(x)$ either side of each x-intercept of the graph?

d Are the *powers* of the factors $(x-3)$ and $(x+1)$ odd or even?

2 Suppose $f(x) = 3(x+2)^2$. The graph of $y = f(x)$ is shown alongside.

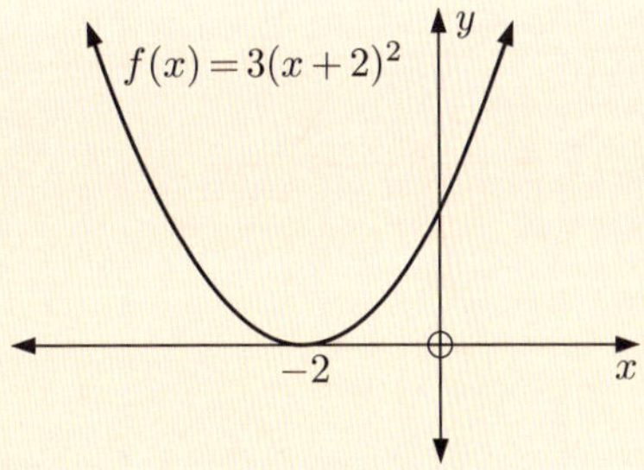

a Explain why -2 is the *only* x-intercept of $y = f(x)$.

b Explain why the sign diagram for $y = f(x)$ is

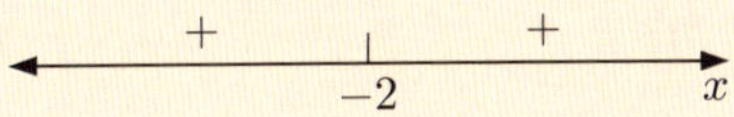

c What happens to the sign of $y = f(x)$ either side of the x-intercept of the graph?

d Is the *power* of the factor $(x+2)$ odd or even?

3 Suppose $f(x) = \dfrac{2}{4-x}$. The graph of $y = f(x)$ is shown alongside.

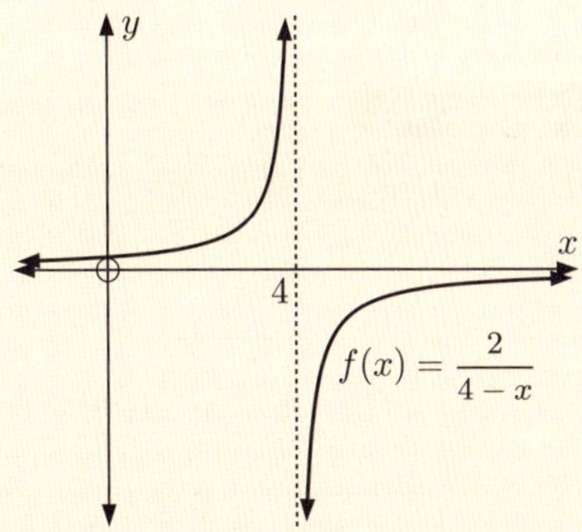

a For what value of x is $f(x)$ undefined?

b Explain why the sign diagram for $y = f(x)$ is

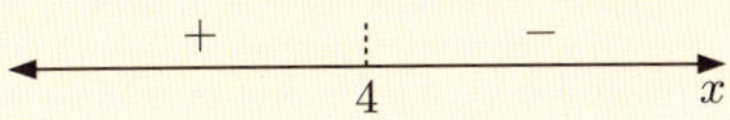

c What happens to the sign of $y = f(x)$ either side of the critical value for which $f(x)$ is undefined?

d **i** Write $f(x)$ in the form $f(x) = 2(4-x)^k$.

ii Is the *power* of the factor $(4-x)$ odd or even?

4 Under what circumstances does the sign of $y = f(x)$ either side of a critical value:

a change **b** stay the same?

From the **Investigation**, you should have concluded that:

When a linear factor of a function has:

- an **odd power**, there is a change of sign about the corresponding critical value
- an **even power**, there is no change of sign about the corresponding critical value.

EXERCISE 22D

1 Draw a sign diagram for each graph:

a

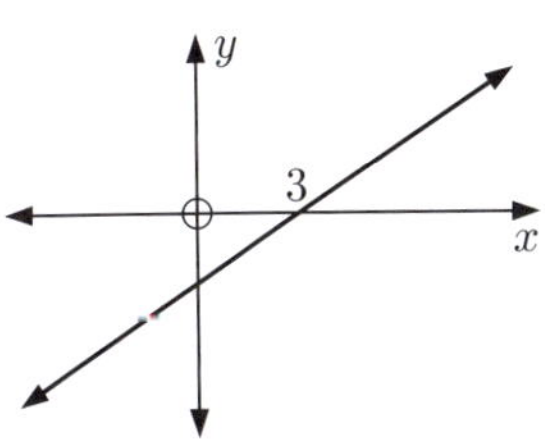

b

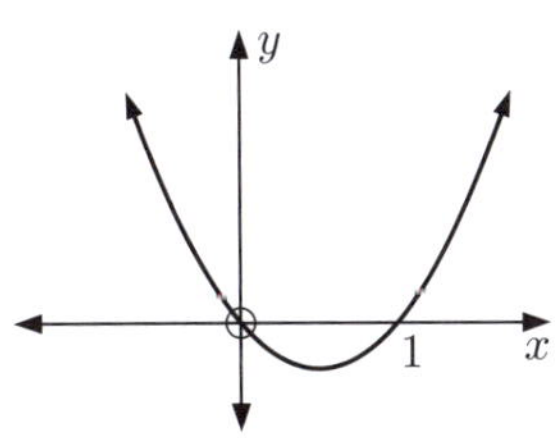

c

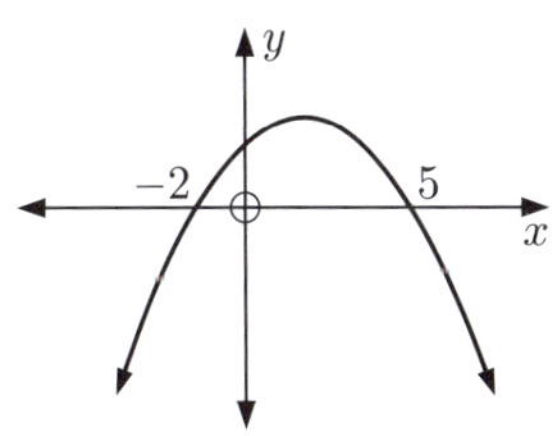

d

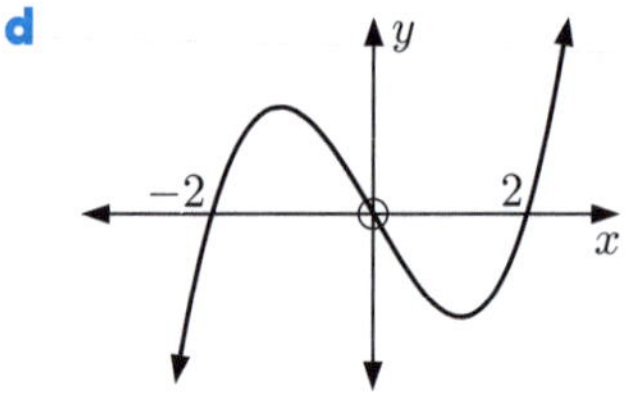

e

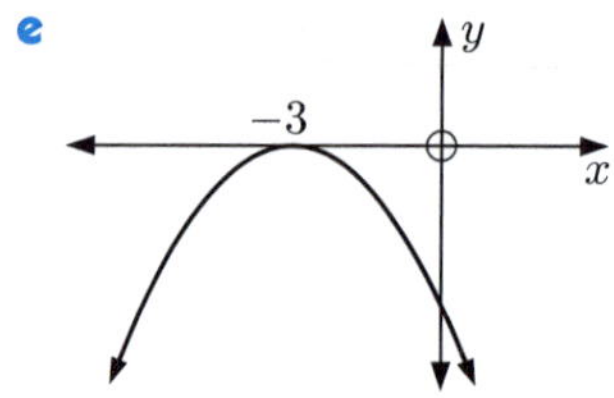

f

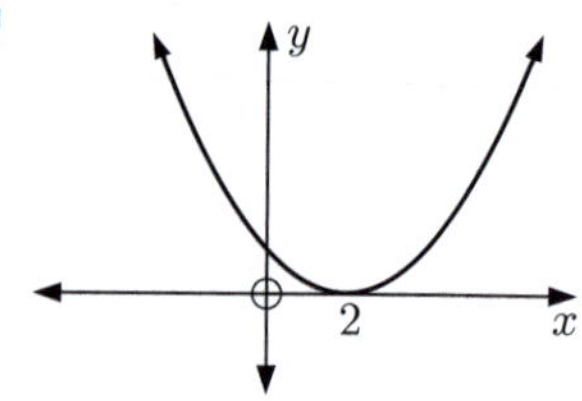

g

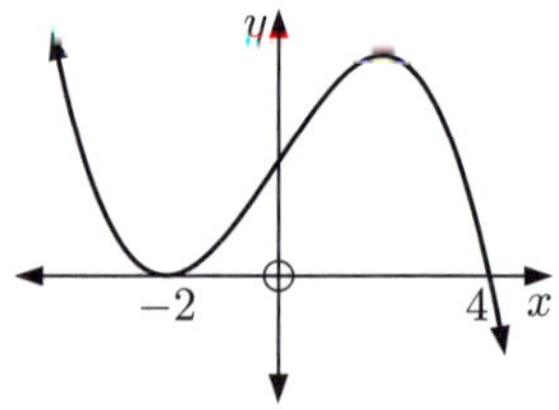

h

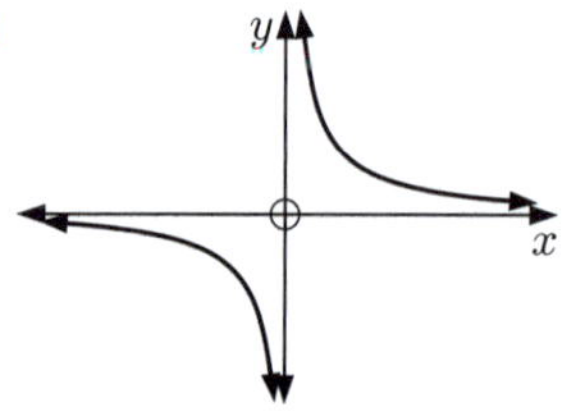

i

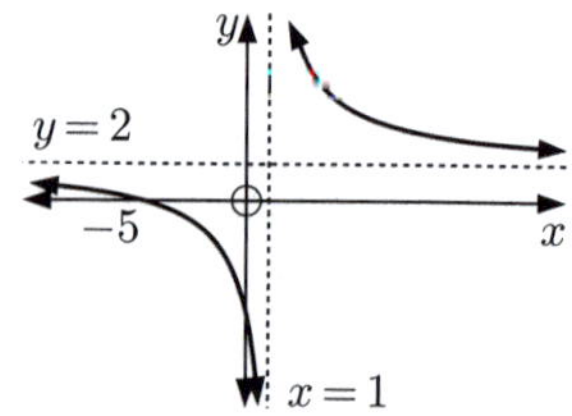

Example 6 **Self Tutor**

Draw a sign diagram for:

a $f(x) = -(x+2)^2$

b $f(x) = \dfrac{x-3}{x+1}$

a $-(x+2)^2$ is zero when $x = -2$.

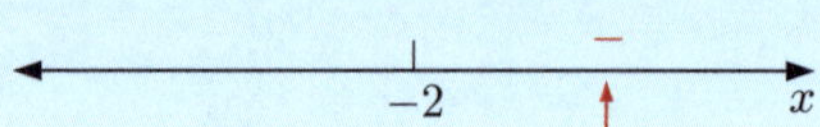

$f(0) = -(2)^2 = -4$

so we put a − sign here.

As the factor has an even power, the sign does not change about the critical value.

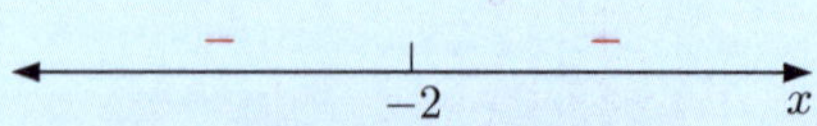

b $\dfrac{x-3}{x+1}$ is zero when $x = 3$

and undefined when $x = -1$.

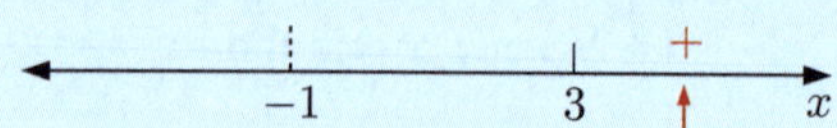

$f(4) = \dfrac{4-3}{4+1} = \dfrac{1}{5}$

so we put a + sign here.

As the factors both have odd powers, the signs change about each critical value.

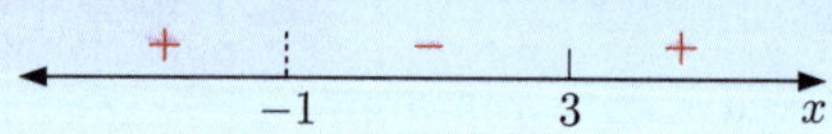

2 Draw a sign diagram for:

a $f(x) = 2x$ **b** $f(x) = 5 - x$ **c** $f(x) = x + 4$

d $f(x) = -2x + 6$ **e** $f(x) = \dfrac{1}{x}$ **f** $f(x) = -x^2$

3 Draw a sign diagram for:

a $f(x) = (x+3)(x-1)$ **b** $f(x) = x(x-4)$ **c** $f(x) = x(x+5)$

d $f(x) = -(x+2)(x-3)$ **e** $f(x) = (3x-1)(4-x)$ **f** $f(x) = (3-x)(1-2x)$

4 Draw a sign diagram for:

a $f(x) = (x-1)^2$ **b** $f(x) = (x+4)^2$ **c** $f(x) = -(x+3)^2$

5 Draw a sign diagram for:

a $f(x) = x^2 - 4x + 3$ **b** $f(x) = x^2 - 16$ **c** $f(x) = 2x - x^2$

d $f(x) = x^2 - 6x + 9$ **e** $f(x) = -x^2 + 2x + 24$ **f** $f(x) = -x^2 + 2x - 1$

6 Draw a sign diagram for:

a $f(x) = \dfrac{x+1}{x-2}$ **b** $f(x) = \dfrac{2x+5}{2-x}$ **c** $f(x) = \dfrac{3x}{(x+3)^2}$ **d** $f(x) = \dfrac{x^2-1}{x}$

E TRANSFORMATIONS OF GRAPHS

Different parts of a function's equation affect the properties of its graph. If we understand what part of the function controls a particular feature, we can learn how to **transform** the function. This is useful because we can often draw the graph of one function from the graph of a related function.

TRANSLATIONS $y = f(x) + k$ AND $y = f(x - h)$

INVESTIGATION 2 TRANSLATIONS

In this Investigation we will explore translations we can apply to $y = f(x)$.

GRAPHING PACKAGE

To help draw the graphs in this Investigation, you may wish to use the **graphing package** or your **graphics calculator**.

What to do:

1 Let $f(x) = x^3$.

a Write down:

i $f(x) + 3$ **ii** $f(x) - 2$

Graph $y = f(x)$, $y = f(x) + 3$, and $y = f(x) - 2$ on the same set of axes.

b Write down:

i $f(x+1)$ **ii** $f(x-3)$

Graph $y = f(x)$, $y = f(x+1)$, and $y = f(x-3)$ on the same set of axes.

c Write down:

i $f(x-1) - 2$ **ii** $f(x+2) + 4$

Graph $y = f(x)$, $y = f(x-1) - 2$, and $y = f(x+2) + 4$ on the same set of axes.

2 What transformation moves:

a $y = f(x)$ to $y = f(x) + k$ **b** $y = f(x)$ to $y = f(x - h)$

c $y = f(x)$ to $y = f(x - h) + k$?

From the **Investigation** you should have discovered that:

- $y = f(x) + k$ is a **vertical translation** of $y = f(x)$ through k units.
  - If $k > 0$, the graph moves **upwards**.
  - If $k < 0$, the graph moves **downwards**.
- $y = f(x - h)$ is a **horizontal translation** of $y = f(x)$ through h units.
  - If $h > 0$, the graph moves to the **right**.
  - If $h < 0$, the graph moves to the **left**.
- $y = f(x - h) + k$ is a **translation** of $y = f(x)$ through h units horizontally and k units vertically.

Example 7 **Self Tutor**

Consider $f(x) = 3 - x$. On separate sets of axes, graph:

a $y = f(x)$ and $y = f(x-2)$ **b** $y = f(x)$ and $y = f(x) - 4$.

a

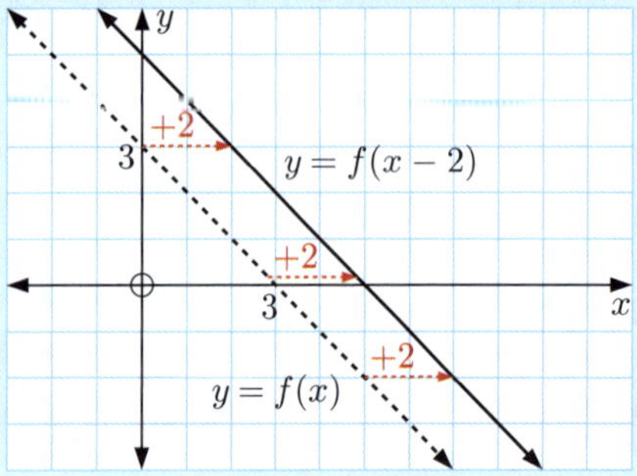

b

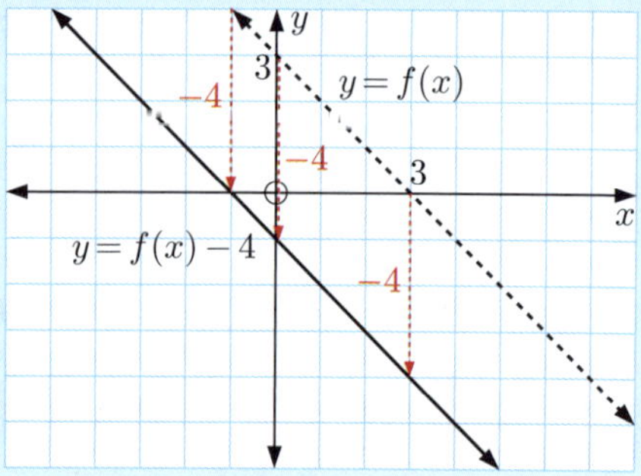

EXERCISE 22E.1

1 Describe the transformation which maps $y = f(x)$ onto:

a $y = f(x) + 5$ **b** $y = f(x-3)$ **c** $y = f(x+6) - 1$

2 Copy the graph of $y = f(x)$ alongside.
On the same set of axes, draw the graphs of $y = f(x) - 2$ and $y = f(x+3)$.

PRINTABLE GRIDS

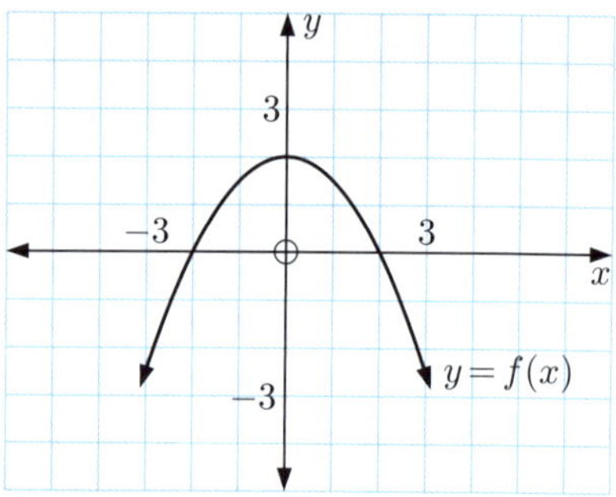

3 Consider $f(x) = 3x - 2$.

a Describe the transformation which maps $y = f(x)$ onto:

i $y = f(x) + 4$ **ii** $y = f(x+4)$

b On the same grid, graph $y = f(x)$, $y = f(x) + 4$, and $y = f(x+4)$. Label each graph.

4 Consider $f(x) = \frac{1}{2}x + 1$. On separate sets of axes, graph:

a $y = f(x)$ and $y = f(x) - 3$ **b** $y = f(x)$ and $y = f(x-1)$.

5 Copy the graph of $y = f(x)$ alongside.
On the same set of axes, draw the graph of $y = f(x-2) - 1$.

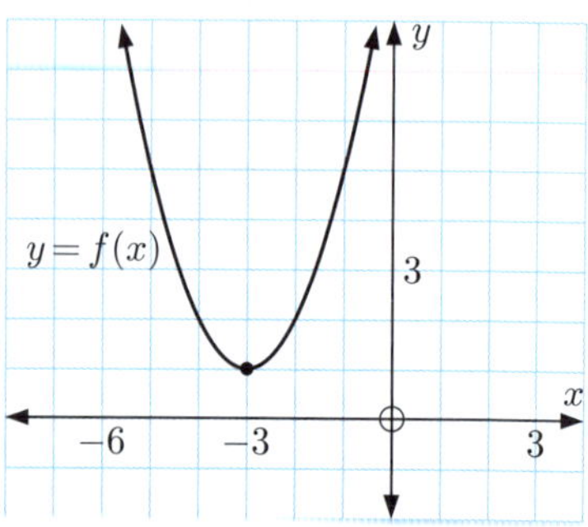

Example 8 **Self Tutor**

Find the equation of the resulting function $g(x)$ when $f(x) = -x^2 + 3$ is translated:

a 3 units to the left **b** 2 units upwards and 3 units to the right.

a
$$\begin{aligned} g(x) &= f(x+3) \\ &= -(x+3)^2 + 3 \\ &= -(x^2 + 6x + 9) + 3 \\ &= -x^2 - 6x - 9 + 3 \\ &= -x^2 - 6x - 6 \end{aligned}$$

b
$$\begin{aligned} g(x) &= f(x-3) + 2 \\ &= -(x-3)^2 + 3 + 2 \\ &= -(x^2 - 6x + 9) + 5 \\ &= -x^2 + 6x - 9 + 5 \\ &= -x^2 + 6x - 4 \end{aligned}$$

6 Find the equation of the resulting function $g(x)$ when:

- **a** $f(x) = 3x + 2$ is translated 5 units downwards
- **b** $f(x) = 4 - 2x$ is translated 1 unit to the right
- **c** $f(x) = x^2 - 3$ is translated 4 units upwards and 2 units to the left
- **d** $f(x) = \dfrac{1}{x}$ is translated 3 units downwards and 4 units to the right.

7 Suppose $f(x)$ has domain $\{x \mid -2 \leqslant x \leqslant 5\}$ and range $\{y \mid y \geqslant 3\}$.
Find the domain and range of:

a $y = f(x) + 2$ **b** $y = f(x+4)$ **c** $y = f(x-1) - 5$

VERTICAL STRETCHES $y = p\,f(x),\ p > 0$

INVESTIGATION 3 VERTICAL STRETCHES

In this Investigation we will explore vertical stretches we can apply to $y = f(x)$.

GRAPHING PACKAGE

To help draw the graphs in this Investigation, you may wish to use the **graphing package** or your **graphics calculator**.

What to do:

1 Let $f(x) = x + 3$.

- **a** Find, in gradient-intercept form:
  - **i** $2\,f(x)$ **ii** $\frac{1}{3}\,f(x)$ **iii** $4\,f(x)$
- **b** Graph all four functions on the same set of axes.

2 What transformation moves $y = f(x)$ to $y = p\,f(x)$, $p > 0$?

3 Copy and complete:

For the transformation $y = p\,f(x)$, $p > 0$, each point becomes ...... times its previous distance from the x axis.

From the **Investigation** you should have found that:

$y = p\,f(x),\ p > 0$ is a **vertical stretch** of $y = f(x)$ with **scale factor** p.

- Each point becomes p times its previous distance from the x-axis.
- If $p > 1$, points move further away from the x-axis.
- If $0 < p < 1$, points move closer to the x-axis.

Example 9 **Self Tutor**

Consider $f(x) = 3 - x$.

Graph $y = f(x)$ and $y = 2\,f(x)$ on the same set of axes.

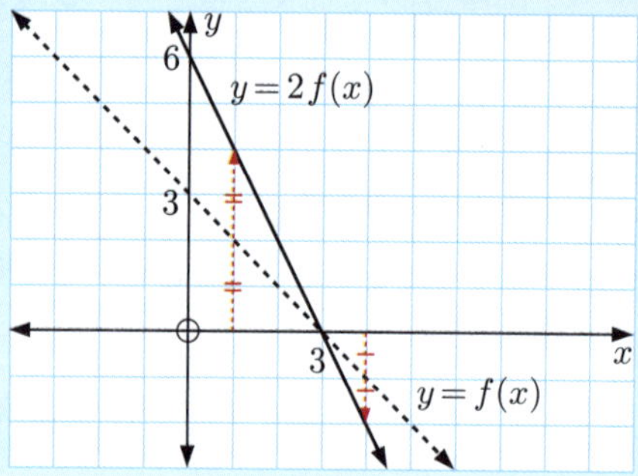

EXERCISE 22E.2

1 Describe the transformation which maps $y = f(x)$ onto:

a $y = 6\,f(x)$ **b** $y = \frac{1}{4}\,f(x)$

2 Copy the graph of $y = f(x)$ alongside.
On the same set of axes, draw the graphs of $y = 2\,f(x)$ and $y = \frac{1}{2}\,f(x)$.

PRINTABLE GRIDS

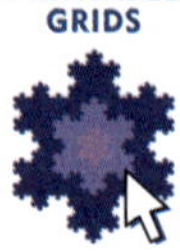

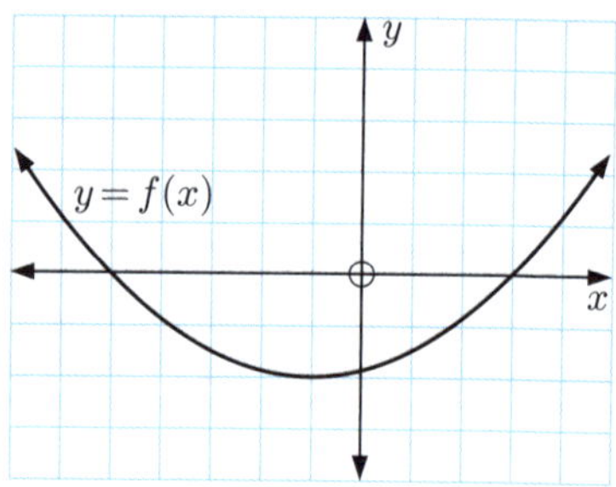

3 Consider $f(x) = 4 - 2x$. On separate sets of axes, graph:

a $y = f(x)$ and $y = 2\,f(x)$ **b** $y = f(x)$ and $y = \frac{1}{4}\,f(x)$.

4 Copy the graph of $y = f(x)$.
On the same set of axes, draw the graph of $y = \frac{1}{2}\,f(x)$.

a

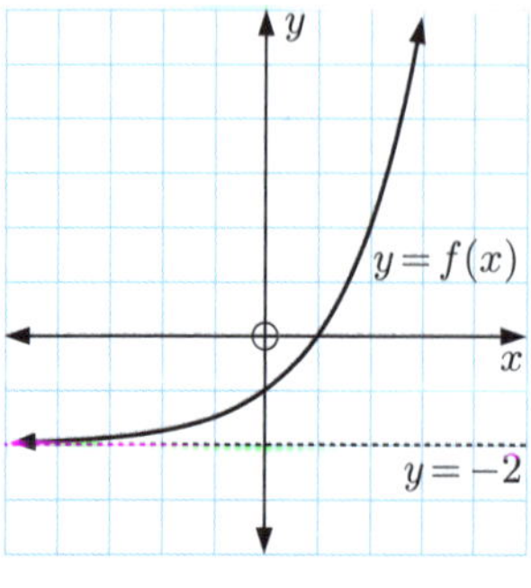

b

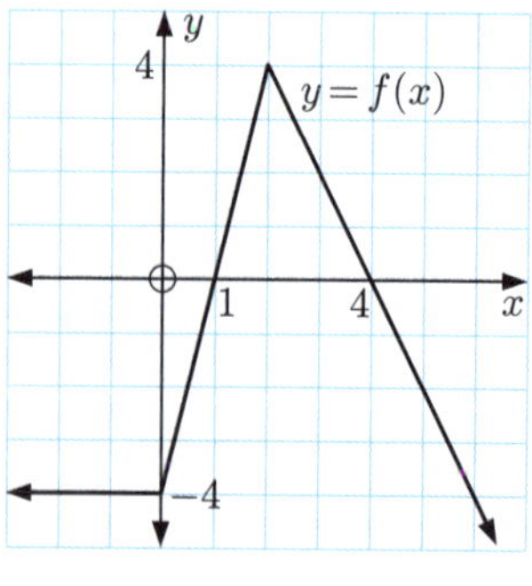

5 Find the equation of the resulting function $g(x)$ when:

a $f(x) = x^2 + 2x$ is stretched vertically with scale factor 3

b $f(x) = 4 - x^2$ is stretched vertically with scale factor $\frac{1}{4}$.

6 Suppose $f(x)$ has domain $\{x \mid -1 \leqslant x \leqslant 6\}$ and range $\{y \mid y \leqslant 4\}$.
Find the domain and range of $y = \frac{1}{2} f(x)$.

HORIZONTAL STRETCHES $y = f(qx),\ q > 0$

INVESTIGATION 4 HORIZONTAL STRETCHES

GRAPHING PACKAGE

In this Investigation we will explore horizontal stretches we can apply to $y = f(x)$.

To help draw the graphs in this Investigation, you may wish to use the **graphing package** or your **graphics calculator**.

What to do:

1 Let $f(x) = x + 3$.

a Find, in gradient-intercept form: **i** $f(2x)$ **ii** $f\left(\frac{1}{2}x\right)$ **iii** $f(3x)$

b Graph all four functions on the same set of axes.

2 What transformation moves $y = f(x)$ to $y = f(qx)$, $q > 0$?

3 Copy and complete:

For the transformation $y = f(qx)$, $q > 0$, each point becomes ...... times its previous distance from the y-axis.

From the **Investigation** you should have found that:

$y = f(qx)$, $q > 0$ is a **horizontal stretch** of $y = f(x)$ with **scale factor** $\frac{1}{q}$.

- Each point becomes $\frac{1}{q}$ times its previous distance from the y-axis.
- If $q > 1$, points move closer to the y-axis.
- If $0 < q < 1$, points move further away from the y-axis.

Example 10 Self Tutor

Copy the graph of $y = f(x)$ below.

On the same set of axes, graph $y = f(2x)$.

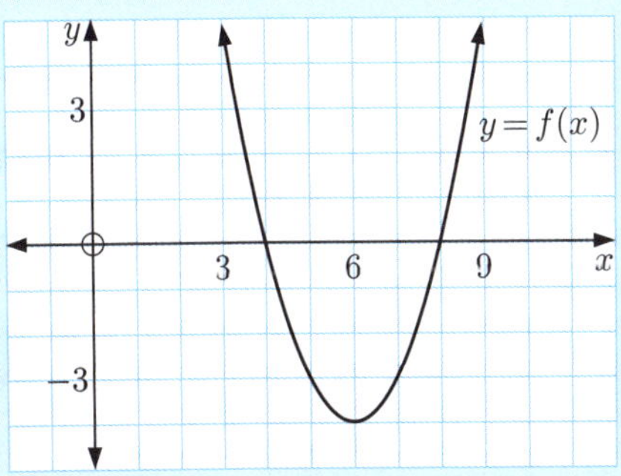

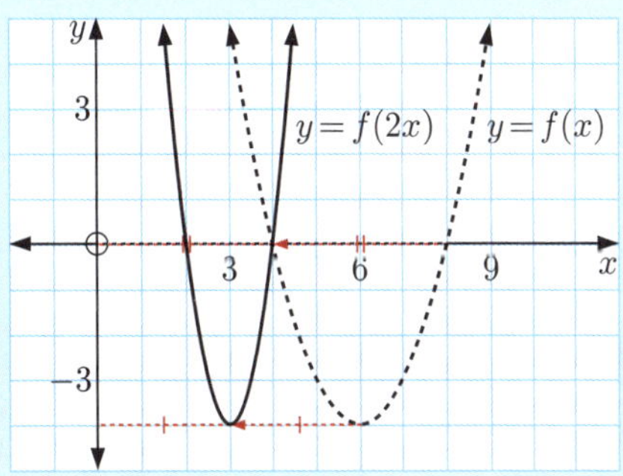

EXERCISE 22E.3

1 Describe the transformation which maps $y = f(x)$ onto:

a $y = f\left(\frac{1}{3}x\right)$ **b** $y = f(4x)$

2 Copy the graph of $y = f(x)$ alongside.
On the same set of axes, draw the graphs of $y = f\left(\frac{1}{2}x\right)$ and $y = f(2x)$.

PRINTABLE GRIDS

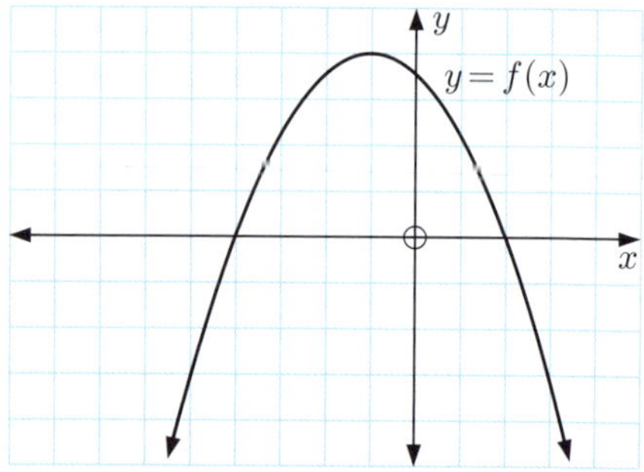

3 Consider $f(x) = x - 1$. On separate sets of axes, graph:

a $y = f(x)$ and $y = f(3x)$ **b** $y = f(x)$ and $y = f\left(\frac{1}{3}x\right)$.

4 Copy the graph of $y = f(x)$.
On the same set of axes, draw the graph of $y = f\left(\frac{1}{2}x\right)$.

a

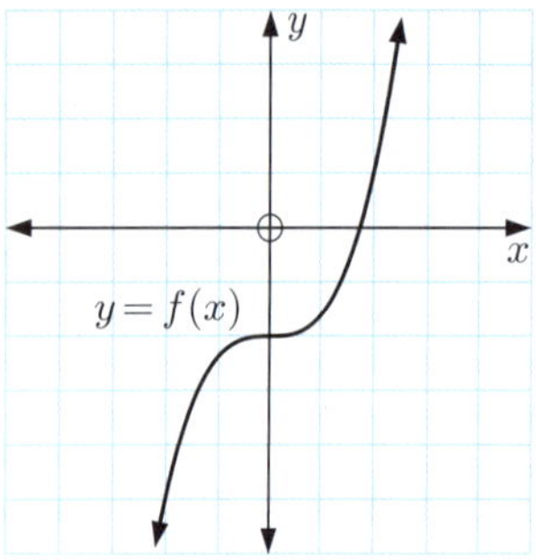

b

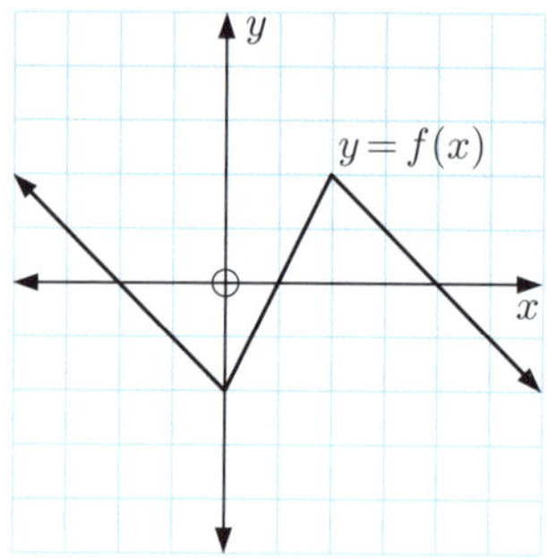

5 Find the equation of the resulting function $g(x)$ when:

a $f(x) = 4 - x^2$ is stretched horizontally with scale factor $\frac{1}{4}$

b $f(x) = 1 - x^2$ is stretched horizontally with scale factor 2.

6 Suppose $f(x)$ has domain $\{x \mid -1 \leqslant x < 3\}$ and range $\{y \mid y \geqslant 4\}$.

Find the domain and range of $y = f\left(\frac{x}{2}\right)$.

REFLECTIONS $y = -f(x)$ AND $y = f(-x)$

INVESTIGATION 5 REFLECTIONS

In this Investigation we will explore reflections we can apply to $y = f(x)$.

To help draw the graphs in this Investigation, you may wish to use the **graphing package** or your **graphics calculator**.

GRAPHING PACKAGE

What to do:

1. Let $f(x) = x - 2$.
   a. Find, in simplest form: i $-f(x)$ ii $f(-x)$
   b. Graph $y = f(x)$, $y = -f(x)$, and $y = f(-x)$ on the same set of axes.
2. Let $f(x) = 3x + 3$.
   a. Find, in simplest form: i $-f(x)$ ii $f(-x)$
   b. Graph $y = f(x)$, $y = -f(x)$, and $y = f(-x)$ on the same set of axes.
3. What transformation moves:
   a. $y = f(x)$ to $y = -f(x)$
   b. $y = f(x)$ to $y = f(-x)$?

From the **Investigation** you should have found that:

- $y = -f(x)$ is a **reflection** of $y = f(x)$ in the ***x*-axis**.
- $y = f(-x)$ is a **reflection** of $y = f(x)$ in the ***y*-axis**.

Example 11 Self Tutor

Consider $f(x) = 3 - x$.

Graph $y = f(x)$ and $y = f(-x)$ on the same set of axes.

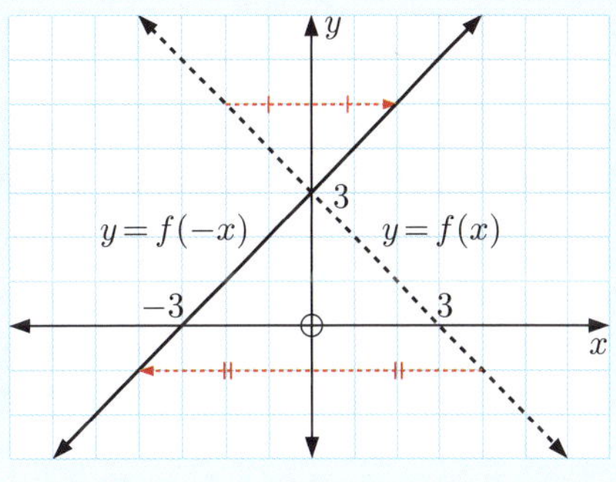

EXERCISE 22E.4

1. Describe the transformation which maps $y = f(x)$ onto:
   a. $y = f(-x)$
   b. $y = -f(x)$
2. Copy the graph of $y = f(x)$ alongside.
   On separate sets of axes, draw the graphs of:
   a. $y = f(x)$ and $y = f(-x)$
   b. $y = f(x)$ and $y = -f(x)$.

PRINTABLE GRIDS

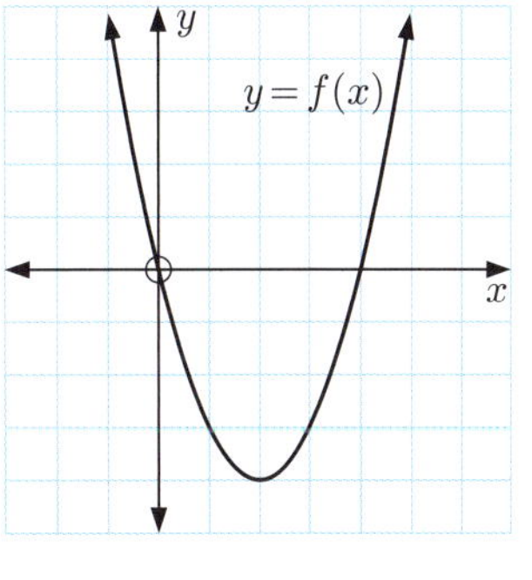

3. Consider $f(x) = 2x + 4$. On separate sets of axes, graph:
   a. $y = f(x)$ and $y = -f(x)$
   b. $y = f(x)$ and $y = f(-x)$.

4 Copy the graph of $y = f(x)$ alongside.
On separate sets of axes, draw the graphs of:

a $y = f(x)$ and $y = f(-x)$

b $y = f(x)$ and $y = -f(x)$.

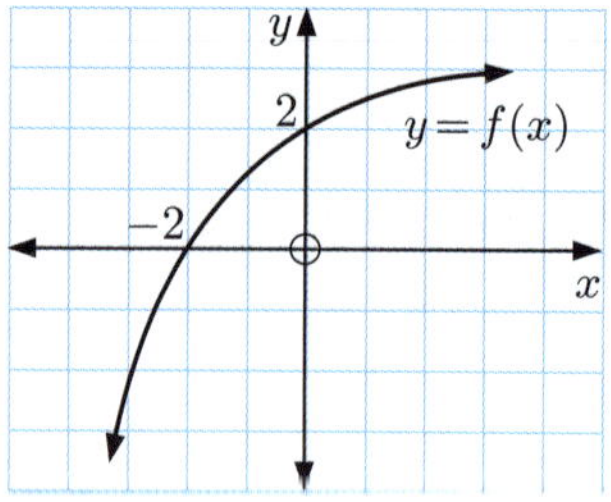

5 Find the equation of the resulting function $g(x)$ when:

a $f(x) = 2x - \frac{3}{x}$ is reflected in the x-axis

b $f(x) = 2x^3 - 5x^2 + 4x - 3$ is reflected in the y-axis.

6 Suppose $f(x)$ has domain $\{x \mid -3 \leqslant x \leqslant 5\}$ and range $\{y \mid y \geqslant -2\}$.
Find the domain and range of: **a** $y = -f(x)$ **b** $y = f(-x)$

7 The function $f(x) = mx + c$ is a straight line. State the gradient of:

a $y = f(x)$ **b** $y = f(x-1)$ **c** $y = f\left(\frac{1}{3}x\right)$

d $y = 2f(x)$ **e** $y = -f(x)$ **f** $y = f(-x)$

DISCUSSION

Invariant points are points which do not move under a transformation.

- Are any points invariant when a function is translated?
- Which points are invariant when a function is:
  - stretched vertically
  - stretched horizontally
  - reflected in the x-axis
  - reflected in the y-axis?

QUICK QUIZ

MULTIPLE CHOICE QUIZ

REVIEW SET 22A

1 Determine whether each relation is a function:

a

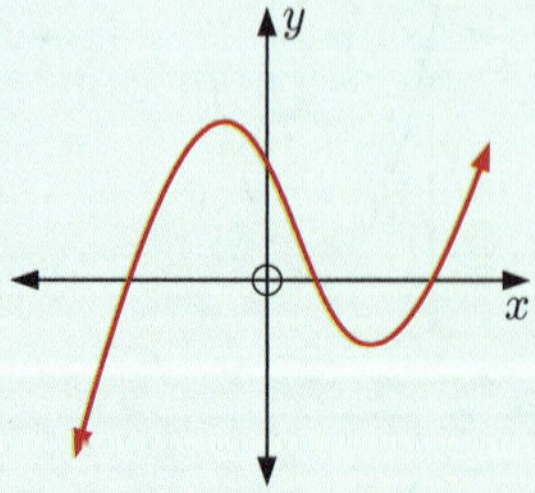

b

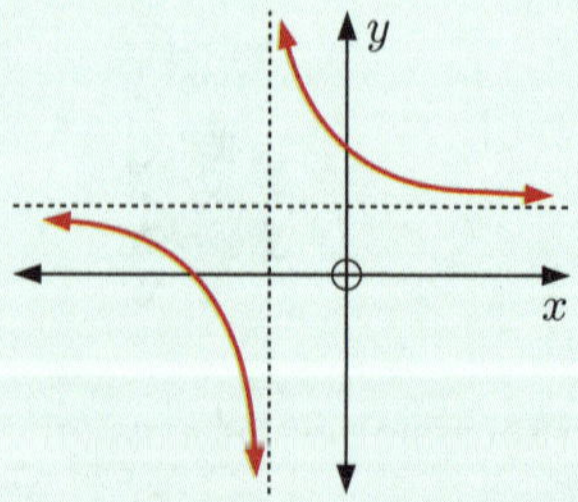

c

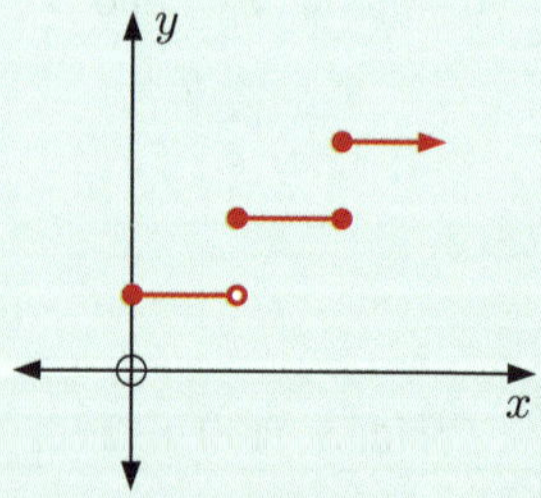

2 For $f(x) = 3x - x^2$, find:

a $f(2)$ **b** $f(-1)$ **c** $f(x-3)$ **d** $f(2x) - 1$

3 For each graph, state the domain and range.

a

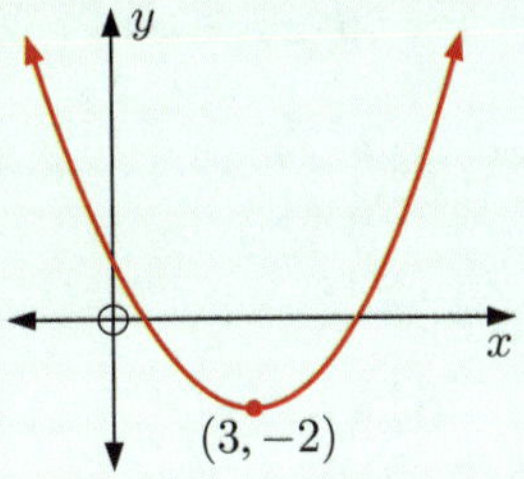

b

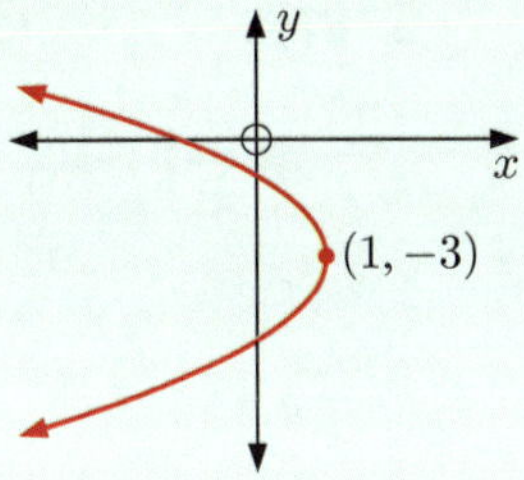

c

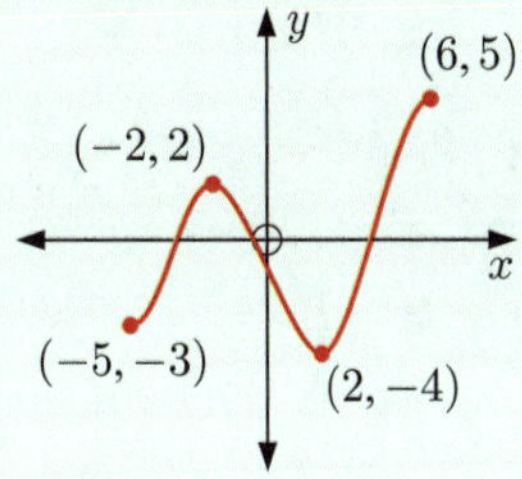

4 Copy the graph alongside. On the same set of axes, draw the graphs of $y = f(x) + 3$ and $y = f(-x)$.

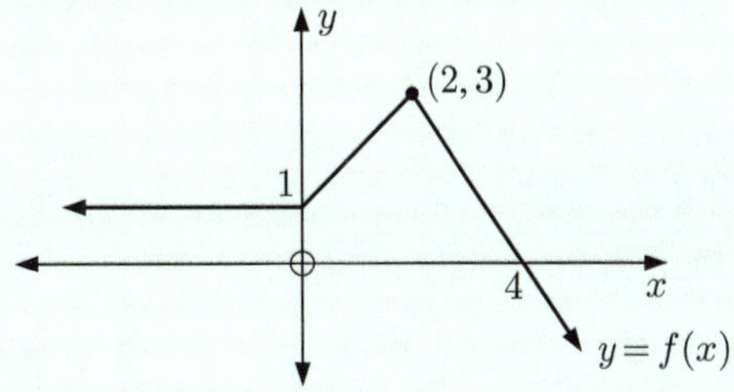

5 Draw a sign diagram for:

a $f(x) = 3x - 2$ **b** $f(x) = (x+4)(2-x)$ **c** $f(x) = \dfrac{2x}{1-x}$

6 Let $f(x) = 2x - 1$. On the same set of axes, graph $y = f(x)$, $y = f(x-2)$, $y = f(3x)$, and $y = -f(x)$.

7 Find the equation of the resulting function $g(x)$ when:

a $f(x) = 4x - 3$ is reflected in the x-axis

b $f(x) = x^2 - x + 3$ is translated 3 units to the left

c $f(x) = \dfrac{4}{x} - 2$ is stretched vertically with scale factor $\frac{1}{2}$.

8 This graph shows the height above ground level $h(t)$ m of a bird t seconds after leaving its nest.

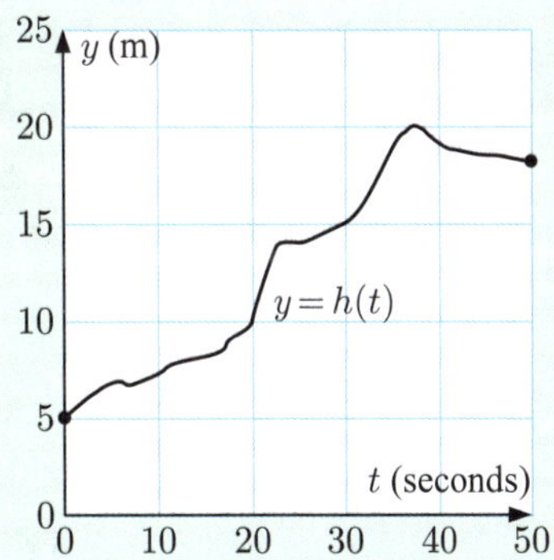

a Find $h(0)$ and interpret your answer.

b Find t such that $h(t) = 10$.

c State the domain and range of the graph.

d Suppose $g(t)$ is the height of the bird above its *nest* after t seconds.

i Write $g(t)$ in terms of $h(t)$.

ii State the domain and range of $g(t)$.

iii Find $g(30)$ and interpret your answer.

REVIEW SET 22B

1 Consider the function $g(x) = x^2 + 2x$. Find:

a $g(2)$ **b** the values of x such that $g(x) = 15$.

2 Consider the function $f(x) = \sqrt{x} + 1$. Find:

a $f(9)$ **b** $f(x-3)$ **c** the value of x such that $f(x) = 6$.

3 Find the natural domain of:

a $f(x) = -\frac{16}{x}$ **b** $f(x) = 3x^2 - 8x$ **c** $f(x) = \sqrt{16 - x}$

4 Find:

a $f(-3)$

b $g(2)$

c the values of x such that $f(x) = g(x)$.

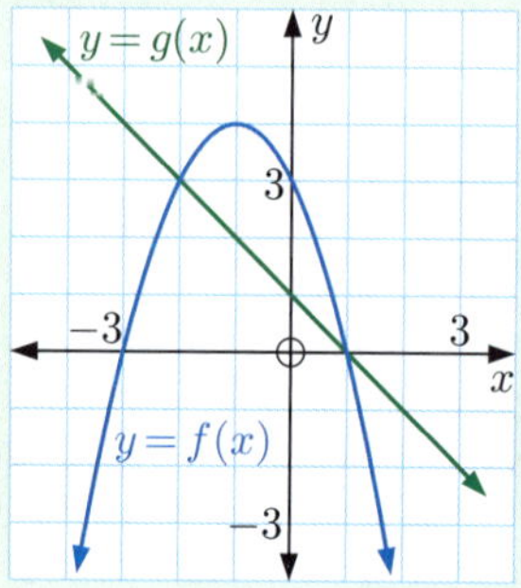

5 Draw a sign diagram for:

a $f(x) = (x+4)(x-1)$ **b** $f(x) = -(x+5)^2$ **c** $f(x) = \frac{2x-1}{(x-4)^2}$

6 Describe the transformation which maps $y = f(x)$ onto:

a $y = f(x+2) - 3$ **b** $y = 5\,f(x)$ **c** $y = f\left(\frac{x}{10}\right)$

7 $y = f(x)$ is mapped onto $y = g(x)$ using a single transformation.

a Describe the transformation fully.

b Write $g(x)$ in terms of $f(x)$.

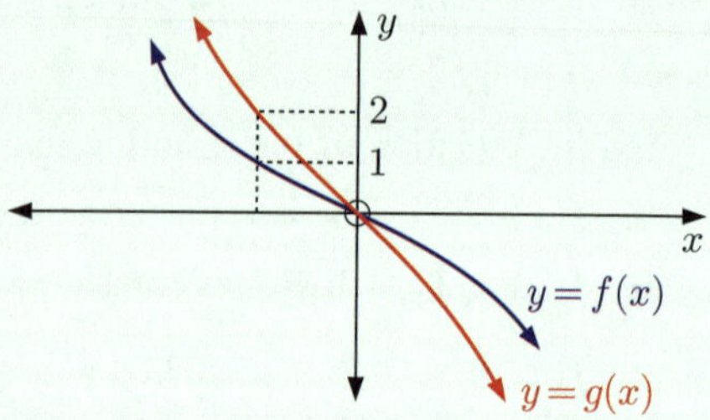

8 Consider the graph of $y = f(x)$ shown.

a State the domain and range of $f(x)$.

b On the same set of axes, draw the graphs of $y = f(x)$, $y = f(x-2) + 1$, and $y = f\left(\frac{1}{2}x\right)$.

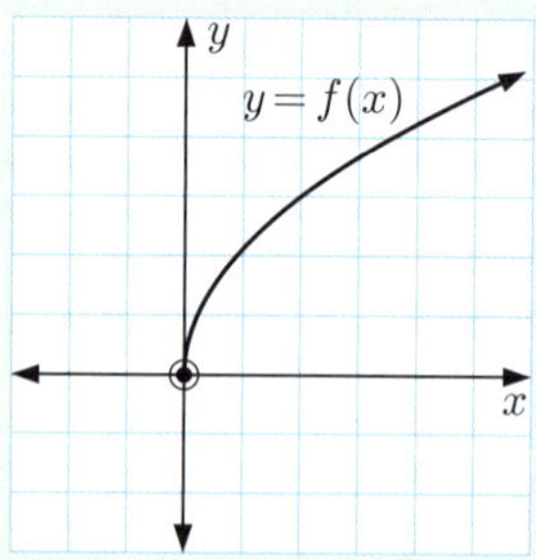

9 Suppose $f(x)$ has domain $\{x \mid x > -2\}$ and range $\{y \mid -4 \leqslant y < 6\}$.
Find the domain and range of:

a $y = f(x-4) + 3$ **b** $y = f(-x)$

Chapter 23

Quadratic functions

Contents:

OPENING PROBLEM

Tennis player Bradley tosses the ball in the air before he serves it. The ball's height above the ground t seconds after it is tossed is given by the function $H(t) = -5t^2 + 6t + 2$ metres.

Things to think about:

a How high was the ball when it was released?

b What was the maximum height reached by the ball?

c Bradley hits the ball on its way down when it is 3 metres above the ground. How long after Bradley releases the ball does he hit it?

In this Chapter we consider functions which are written as a quadratic expression of the independent variable, and the graphs of these functions which are curves called *parabolas*.

A QUADRATIC FUNCTIONS

A **quadratic function** is a relationship between two variables which can be written in the form $y = ax^2 + bx + c$ where a, b, c are constants, $a \neq 0$.

Using function notation, we can write $f(x) = ax^2 + bx + c$.

FINDING y GIVEN x

For any given value of x, the corresponding value of y can be found by substituting into the function.

Example 1 Self Tutor

Suppose $y = 2x^2 + 4x - 5$. Find the value of y when:

a $x = 0$ **b** $x = 3$

a When $x = 0$,
$$y = 2(0)^2 + 4(0) - 5$$
$$= 0 + 0 - 5$$
$$= -5$$

b When $x = 3$,
$$y = 2(3)^2 + 4(3) - 5$$
$$= 18 + 12 - 5$$
$$= 25$$

FINDING x GIVEN y

When we substitute a value for y, we are left with a quadratic equation which we need to solve for x. Since the equation is quadratic, there may be 0, 1, or 2 possible values for x for any given value of y.

Example 2 — Self Tutor

Suppose $y = x^2 - 6x + 8$. Find the value(s) of x for which:

a $y = 15$ **b** $y = -1$

a When $y = 15$,
$$x^2 - 6x + 8 = 15$$
$$\therefore\ x^2 - 6x - 7 = 0$$
$$\therefore\ (x+1)(x-7) = 0$$
$$\therefore\ x = -1 \text{ or } x = 7$$

b When $y = -1$,
$$x^2 - 6x + 8 = -1$$
$$\therefore\ x^2 - 6x + 9 = 0$$
$$\therefore\ (x-3)^2 = 0$$
$$\therefore\ x = 3$$

EXERCISE 23A

1 State whether each function is quadratic. If it is, state the values of a, b, and c. If it is not, explain why.

a $y = 15x - 8$ **b** $y = \frac{1}{3}x^2 + 6$ **c** $y = 5 + 2x - x^2$

d $y = \frac{1}{3}x^2 - 4x$ **e** $y = 15x^3 + 2x - 16$ **f** $3y + 2x^2 - 7 = 0$

2 For the given function and value of x, find the corresponding value of y:

a $y = x^2 + 5x - 14$ when $x = 2$ **b** $y = 2x^2 + 9x$ when $x = -5$

c $y = -2x^2 + 3x - 6$ when $x = 3$ **d** $y = 4x^2 + 7x + 10$ when $x = -2$

3 State whether the function is satisfied by the given ordered pair:

a $f(x) = 6x^2 - 10$ (0, 4) **b** $y = 2x^2 - 5x - 3$ (4, 9)

c $y = -4x^2 + 6x$ $(-\frac{1}{2}, -4)$ **d** $f(x) = -7x^2 + 9x + 11$ (−1, −6)

e $f(x) = 3x^2 - 11x + 20$ (2, −10) **f** $y = -3x^2 + x + 6$ $(\frac{1}{3}, 4)$

4 For the given function and value of y, find the corresponding value(s) of x:

a $y = x^2 + 6x + 10$ when $y = 1$ **b** $y = x^2 + 5x + 8$ when $y = 2$

c $y = x^2 - 5x + 1$ when $y = -3$ **d** $y = 3x^2$ when $y = -3$

5 Find the value(s) of x for which:

a $f(x) = 3x^2 - 3x + 6$ has the value 6 **b** $f(x) = x^2 - 2x - 7$ has the value −4

c $f(x) = -2x^2 - 14x + 3$ has the value −13 **d** $f(x) = 2x^2 - 10x + 1$ has the value −11.

6 The points A(5, a), B(b, 4), and C(1, 2) lie on the curve $y = -x^2 + kx - 2$.

a Find the value of: **i** k **ii** a

b Find the possible values of b.

7 An object is projected into the air. Its height after t seconds is given by the function $h(t) = 80t - 5t^2$ metres.

a Calculate the height of the object after:

i 1 second **ii** 3 seconds **iii** 5 seconds.

b Calculate the time(s) at which the height of the object is:

i 140 m **ii** 0 m.

c Explain your answers to **b**.

B GRAPHS OF QUADRATIC FUNCTIONS

INVESTIGATION 1 PARABOLAS

In geometry, a **parabola** is defined as the *locus* or set of all points which are equidistant from a fixed point called the **focus** and a fixed line called the **directrix**.

What to do:

1. Print the sheet of circular-linear graph paper.
2. Mark the point which is 2 units from both F and the directrix.
3. Mark *two* points which are 3 units from both F and the directrix. Repeat this process for each integer distance up to 10 units.
4. Draw a smooth curve through the points you have marked.

CIRCULAR-LINEAR GRAPH PAPER

In the **Investigation** we used the geometric definition to generate a parabola.

Suppose we choose a set of axes so the focus is $F(0, a)$ and the directrix is the horizontal line $y = -a$.

If $P(x, y)$ lies on the parabola then $NP = FP$ where N has coordinates $(x, -a)$.

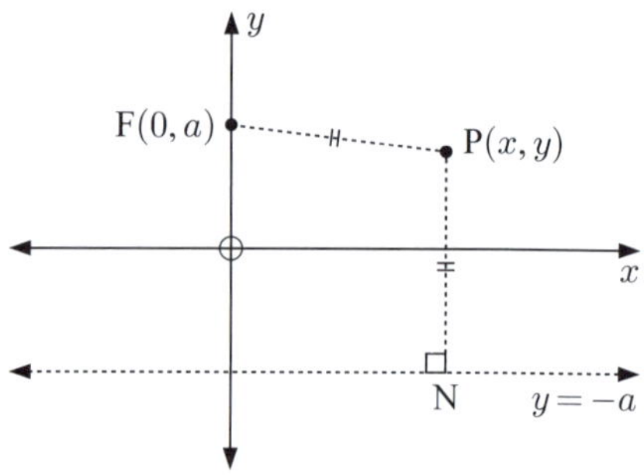

Using the distance formula, $y + a = \sqrt{x^2 + (y-a)^2}$

$$\therefore\ (y+a)^2 = x^2 + (y-a)^2$$
$$\therefore\ (y+a)^2 - (y-a)^2 = x^2$$
$$\therefore\ (y+a+y-a)(y+a-(y-a)) = x^2$$
$$\therefore\ 2y \times 2a = x^2$$
$$\therefore\ y = \frac{x^2}{4a}$$

So, the parabola is described in coordinate geometry by a quadratic function.

The graph of any quadratic function is a parabola.

GRAPHING FROM A TABLE OF VALUES

The simplest quadratic function is $y = x^2$. Its graph can be drawn from a table of values.

x	-3	-2	-1	0	1	2	3
y	9	4	1	0	1	4	9

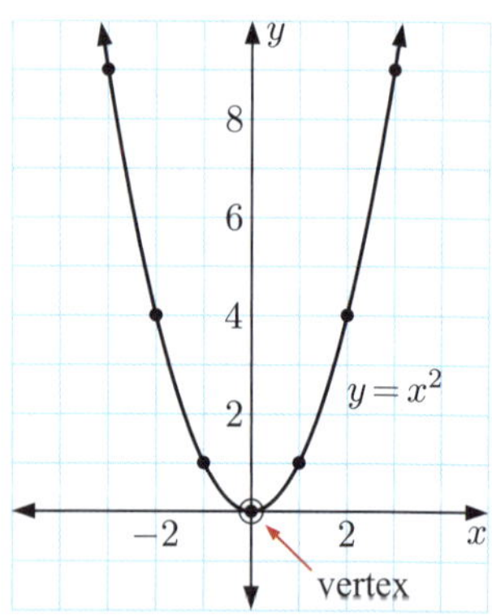

We can see that the graph:

- is symmetrical about the y-axis
- has a minimum turning point at $(0, 0)$ which we call the **vertex** of the parabola.

Example 3 — Self Tutor

Draw the graph of $y = x^2 + 2x - 3$ using a table of values from $x = -3$ to $x = 3$.

When $x = -3$, $y = (-3)^2 + 2(-3) - 3$
$= 9 - 6 - 3$
$= 0$

We complete the table by performing similar calculations for the other values of x:

x	-3	-2	-1	0	1	2	3
y	0	-3	-4	-3	0	5	12

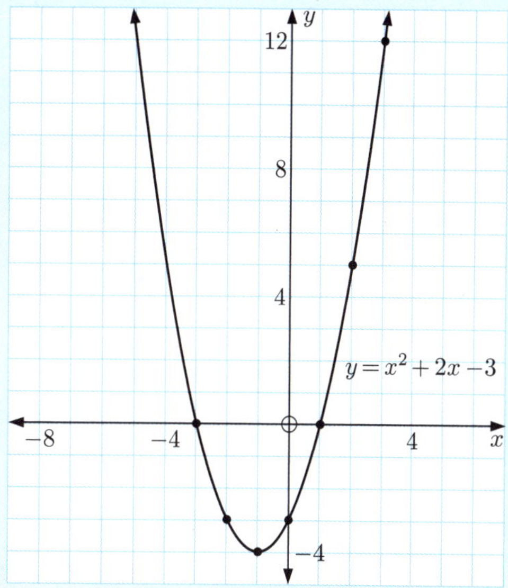

EXERCISE 23B

1 Use a table of values from $x = -3$ to $x = 3$ to draw the graph of:

- **a** $y = x^2 - 2x + 8$
- **b** $f(x) = -x^2 + 2x + 1$
- **c** $y = 2x^2 + 3x$
- **d** $y = -2x^2 + 4$
- **e** $y = x^2 + x + 4$
- **f** $f(x) = -x^2 + 4x - 9$

Use technology to check your graphs.

GRAPHING PACKAGE

2 **a** Use tables of values to graph $y = 2x^2 - x - 3$ and $y = -x^2 + 2x + 3$ on the same set of axes.

b *Hence* find the values of x for which $2x^2 - x - 3 = -x^2 + 2x + 3$.

c Solve algebraically: $2x^2 - x - 3 = -x^2 + 2x + 3$

3 **a** Use tables of values to graph $y = 2x^2 - 5x - 2$ and $y = x^2 - 3x - 3$ on the same set of axes.

b *Hence* find the value of x for which $2x^2 - 5x - 2 = x^2 - 3x - 3$.

c Solve algebraically: $2x^2 - 5x - 2 = x^2 - 3x - 3$.

INVESTIGATION 2 — PARABOLIC MIRRORS

We use parabolas almost every day without realising it. The parabola is the shape used in the mirror surfaces of motor vehicle headlights, torches, and satellite dishes.

In this online Investigation we consider the property of a parabola which makes it suitable for these tasks.

PARABOLIC MIRRORS

C USING TRANSFORMATIONS TO GRAPH QUADRATICS

By observing how a quadratic function is related to $f(x) = x^2$, we can transform the graph of $y = x^2$ to produce the graph of the function.

From the transformations of functions we studied in **Chapter 22**, you should recognise that:

- Graphs of the form $y = (x - h)^2 + k$ have the same shape as the graph of $y = x^2$.
  The graph of $y = x^2$ is translated through h units horizontally and k units vertically to give the graph of $y = (x - h)^2 + k$.
  The vertex is shifted to (h, k).
- $y = -x^2$ is a reflection of $y = x^2$ in the x-axis.

  Since $y = x^2$ opens upwards, $y = -x^2$ opens downwards.

INVESTIGATION 3 GRAPHS OF QUADRATIC FUNCTIONS

What to do:

GRAPHING PACKAGE

1 On the same set of axes, graph:
  a $y = x^2$, $y = 2x^2$, $y = 4x^2$, $y = \frac{1}{2}x^2$
  b $y = -x^2$, $y = -2x^2$, $y = -4x^2$, $y = -\frac{1}{2}x^2$

2 For a function of the form $y = ax^2$, what effect does a have on:
  a the position of the graph
  b the shape of the graph
  c the direction in which the graph opens?

3 Graph the two functions on the same set of axes and label the vertex of each graph.
  a $y = 2x^2$ and $y = 2(x - 1)^2 + 3$
  b $y = -x^2$ and $y = -(x + 2)^2 - 1$
  c $y = \frac{1}{2}x^2$ and $y = \frac{1}{2}(x - 3)^2 - 2$
  d $y = -3x^2$ and $y = -3(x + 1)^2 + 4$

4 Copy and complete:
  The graph of $y = a(x - h)^2 + k$ is found by translating $y = ax^2$ through ......

In the **Investigation** you should have found that:

- If $|a| > 1$, $y = ax^2$ is "thinner" than $y = x^2$.
  If $|a| < 1$, $y = ax^2$ is "wider" than $y = x^2$.
- If $a > 0$, $y = ax^2$ opens upwards.

  If $a < 0$, $y = ax^2$ opens downwards.

- The graph of $y = a(x-h)^2 + k$ has the same shape and opens in the same direction as the graph of $y = ax^2$. It is found by translating $y = ax^2$ through h units horizontally and k units vertically.

Example 4 — Self Tutor

Sketch each function on the same set of axes as $y = x^2$, and label its vertex.

a $y = x^2 + 3$ **b** $y = (x+3)^2$

a We draw $y = x^2$ then translate it 3 units upwards.

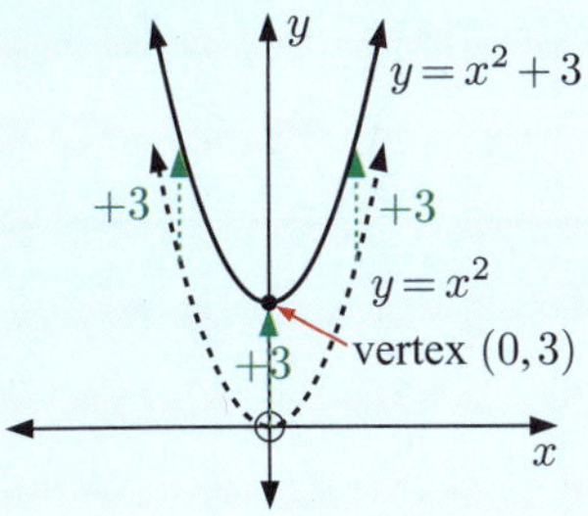

b We draw $y = x^2$ then translate it 3 units to the left.

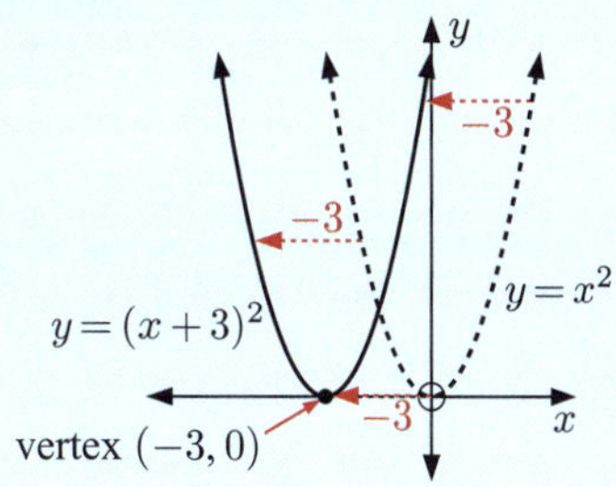

EXERCISE 23C.1

1 Sketch each function on the same set of axes as $y = x^2$, and label its vertex.

a $y = x^2 - 3$ **b** $y = x^2 - 1$ **c** $y = x^2 + 2$

d $y = x^2 - 5$ **e** $y = x^2 + 5$ **f** $y = x^2 - \frac{1}{2}$

2 Sketch each function on the same set of axes as $y = x^2$, and label its vertex.

a $y = (x-3)^2$ **b** $y = (x+1)^2$ **c** $y = (x-2)^2$

d $y = (x-5)^2$ **e** $y = (x+5)^2$ **f** $y = (x - \frac{3}{2})^2$

Example 5 — Self Tutor

Sketch each function on the same set of axes as $y = x^2$, and label its vertex.

a $y = (x-2)^2 + 3$ **b** $y = (x+2)^2 - 5$

a We draw $y = x^2$ then translate it 2 units to the right and 3 units upwards.

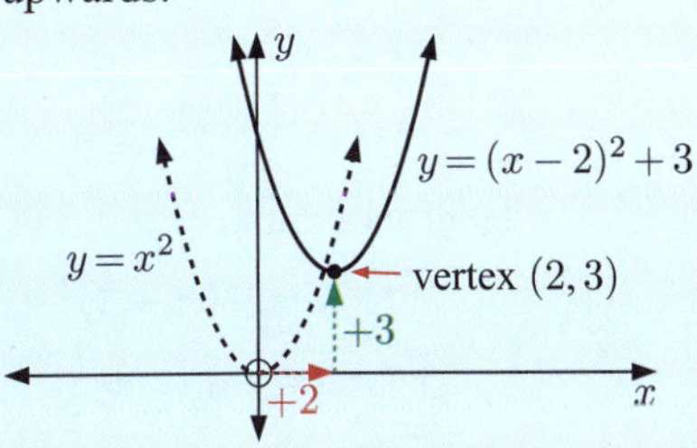

b We draw $y = x^2$ then translate it 2 units to the left and 5 units downwards.

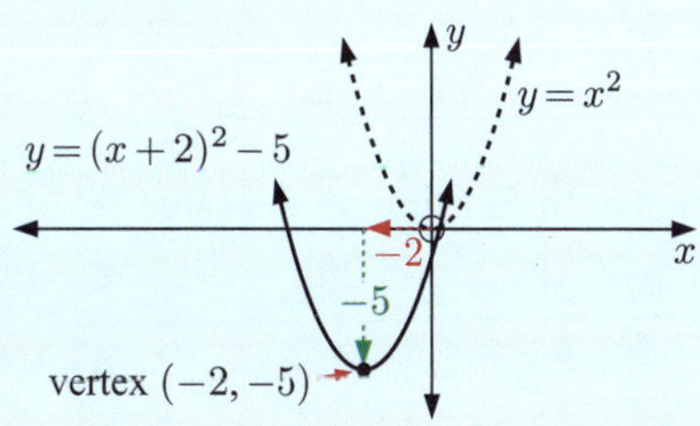

3 Sketch each function on the same set of axes as $y = x^2$, and label its vertex.

a $y = (x-1)^2 + 3$ **b** $y = (x-2)^2 - 1$ **c** $y = (x+1)^2 + 4$

d $y = (x+2)^2 - 3$ **e** $y = (x+3)^2 - 2$ **f** $y = (x-3)^2 + 3$

Example 6 Self Tutor

Sketch $y = x^2$ on a set of axes and hence sketch:

a $y = 3x^2$ **b** $y = -3x^2$

a $y = 3x^2$ is "thinner" than $y = x^2$.

b $y = -3x^2$ has the same shape as $y = 3x^2$, but opens downwards.

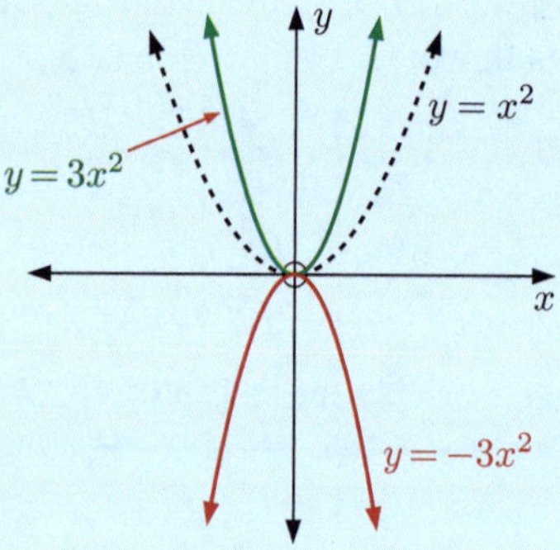

4 Sketch each function on the same set of axes as $y = x^2$. Comment on the shape of the graph, and the direction in which the graph opens.

a $y = 5x^2$ **b** $y = -5x^2$ **c** $y = \frac{1}{3}x^2$

d $y = -\frac{1}{3}x^2$ **e** $y = -4x^2$ **f** $y = \frac{1}{4}x^2$

Example 7 Self Tutor

Sketch the graph of $y = -(x-2)^2 - 3$ from the graph of $y = x^2$, and label its vertex.

$y = -(x-2)^2 - 3$

- $-$: reflect in x-axis
- 2 : horizontal translation 2 units right
- -3 : vertical translation 3 units down

We start with $y = x^2$ then reflect it in the x-axis to give $y = -x^2$.

We then translate $y = -x^2$ 2 units to the right and 3 units down.

The vertex of $y = -(x-2)^2 - 3$ is $(2, -3)$.

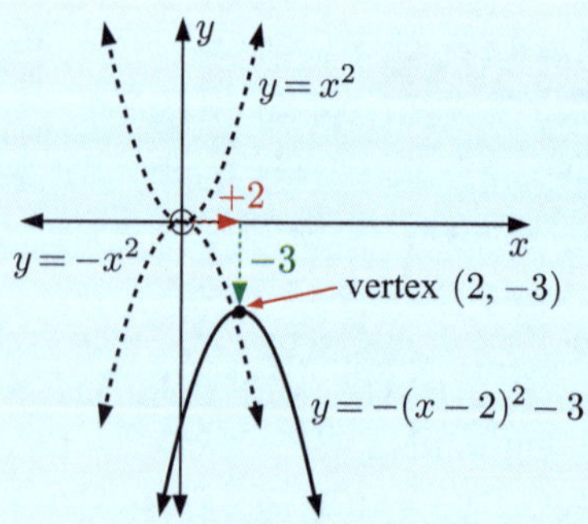

5 Sketch each function on the same set of axes as $y = x^2$, and label its vertex.

a $y = -(x-1)^2 + 3$ **b** $y = 2x^2 + 4$ **c** $y = -(x-2)^2 + 4$

d $y = 3(x+1)^2 - 4$ **e** $y = \frac{1}{2}(x+3)^2$ **f** $y = -\frac{1}{2}(x+3)^2 + 1$

g $y = -2(x+4)^2 + 3$ **h** $y = 2(x-3)^2 + 5$ **i** $y = \frac{1}{2}(x-2)^2 - 1$

6 Match each quadratic function with its graph:

a $y = -(x+2)^2 - 3$

b $y = (x-3)^2 + 2$

c $y = 2(x+3)^2 + 2$

d $y = -(x-3)^2 + 2$

e $y = -\frac{1}{2}(x+2)^2 - 3$

f $y = (x+3)^2 + 2$

g $y = -2(x+3)^2 + 2$

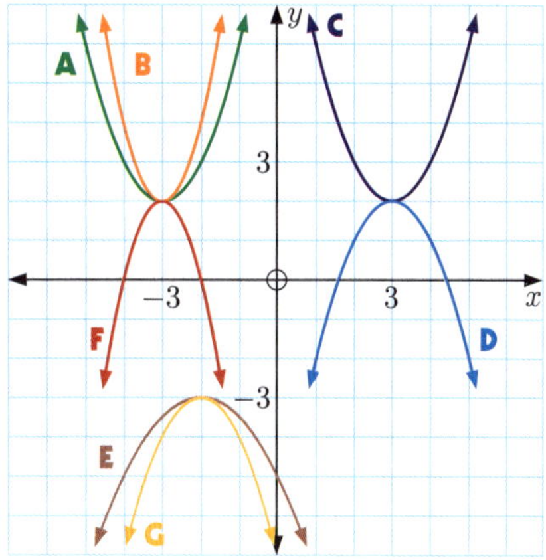

7 **a** Show that $3(x-1)^2 + 2 = 3x^2 - 6x + 5$.

b Hence graph $y = 3x^2$ and $y = 3x^2 - 6x + 5$ on the same set of axes.

8 **a** Show that $-(x+3)^2 + 13 = -x^2 - 6x + 4$.

b Hence graph $y = -x^2$ and $y = -x^2 - 6x + 4$ on the same set of axes.

9 **a** Show that $-2(x - \frac{3}{2})^2 + \frac{9}{2} = -2x^2 + 6x$.

b Hence graph $y = -2x^2$ and $y = -2x^2 + 6x$ on the same set of axes.

COMPLETING THE SQUARE

In **Chapter 12** we used the process of **completing the square** to help solve quadratic equations which did not factorise.

This process can also be used to convert quadratic functions into the **completed square form** $y = a(x-h)^2 + k$, and hence make them easier to graph.

Example 8 Self Tutor

Write $y = x^2 + 2x + 5$ in the form $y = (x-h)^2 + k$.

Hence sketch $y = x^2 + 2x + 5$, including the coordinates of its vertex.

$y = x^2 + 2x + 5$

$y = x^2 + 2x + 1^2 + 5 - 1^2$

$y = (x+1)^2 + 4$

translate 1 unit left; translate 4 units up

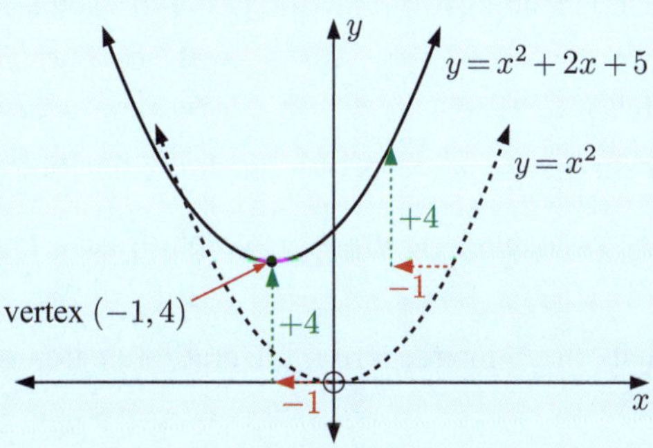

EXERCISE 23C.2

1 Write each function in the form $y = (x-h)^2 + k$. Hence sketch the function, including the coordinates of its vertex. Use technology to check your answers.

GRAPHING PACKAGE

a $y = x^2 + 2x + 4$ **b** $y = x^2 - 6x + 3$ **c** $y = x^2 + 4x - 1$

d $y = x^2 - 2x + 5$ **e** $y = x^2 - 2x$ **f** $y = x^2 + 5x$

g $y = x^2 + 5x - 3$ **h** $y = x^2 - 3x + 3$ **i** $y = x^2 + x - 6$

2 Write each function in the form $y = a(x-h)^2 + k$. Hence sketch the function, including the coordinates of its vertex. Use technology to check your answers.

a $y = 2x^2 + 10x + 8$ **b** $y = -x^2 + x + 6$ **c** $y = -3x^2 - 6x + 2$

D AXES INTERCEPTS

- An **x-intercept** of a function is a value of x where its graph meets the x-axis.
  x-intercepts are found by letting y be 0 in the equation of the function.
- A **y-intercept** of a function is a value of y where its graph meets the y-axis.
  y-intercepts are found by letting x be 0 in the equation of the function.

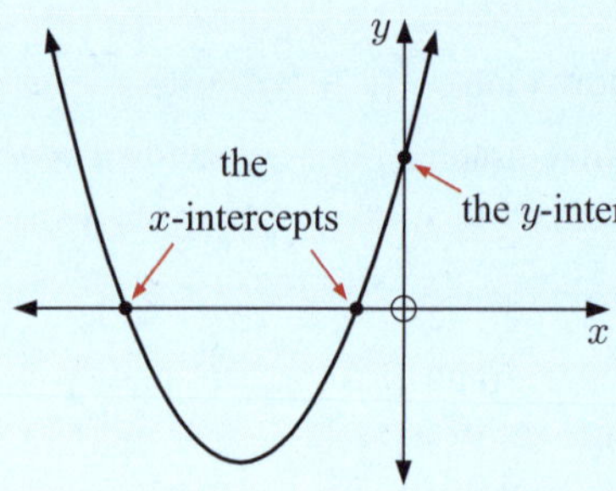

INVESTIGATION 4 AXES INTERCEPTS

What to do:

1 **a** Use technology to draw the graph of each function, including its y-intercept.

GRAPHING PACKAGE

i $y = x^2 - 3x - 4$ **ii** $y = -x^2 + 2x + 8$

iii $y = 2x^2 - 3x$ **iv** $y = -2x^2 + 2x - 3$

v $y = 3x^2 + 2$ **vi** $y = -x^2 - 6$

b State the y-intercept of a quadratic function of the form $y = ax^2 + bx + c$.

2 **a** Use technology to draw the graph of each function, including its x-intercepts.

i $y = (x-2)(x+2)$ **ii** $y = (x-1)(x-3)$ **iii** $y = -(x+2)(x-3)$

iv $y = 2x(x+4)$ **v** $y = -3x(x-4)$ **vi** $y = 3(x+1)(x+4)$

b State the x-intercepts of a quadratic function of the form $y = a(x-\alpha)(x-\beta)$.

3 **a** Use technology to draw the graph of each function, including its x-intercepts.

i $y = (x+4)^2$ **ii** $y = 2(x-2)^2$ **iii** $y = -3(x+1)^2$

b State the x-intercept of a quadratic function of the form $y = a(x-\alpha)^2$.

From the **Investigation** you should have found:

- For a quadratic function in the form $y = ax^2 + bx + c$, the y-intercept is the constant term c.

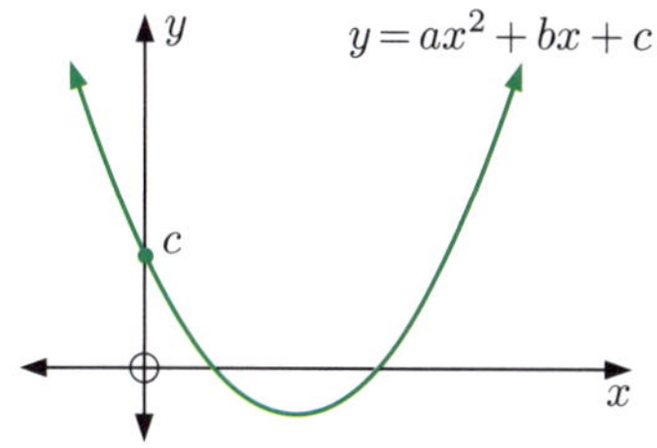

Proof:

Consider $y = ax^2 + bx + c$.
If we let $x = 0$ then $y = a(0)^2 + b(0) + c = c$
$\therefore$ the y-intercept is c.

- A quadratic function with factored form $y = a(x - \alpha)(x - \beta)$ has x-intercepts α and β.

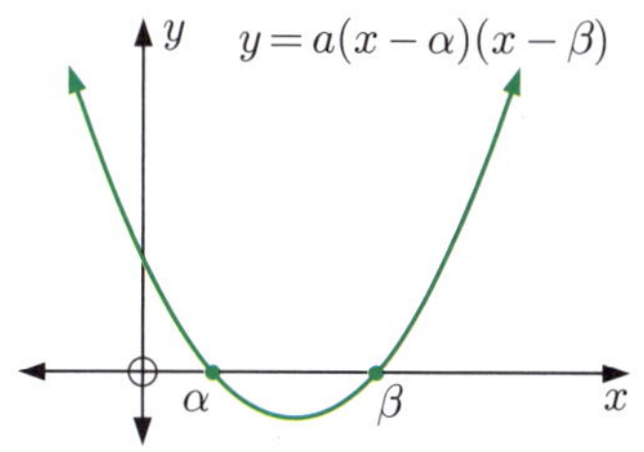

Proof:

Consider $y = a(x - \alpha)(x - \beta)$.
If we let $y = 0$ then $a(x - \alpha)(x - \beta) = 0$
$\therefore x - \alpha = 0$ or $x - \beta = 0$
$\therefore x = \alpha$ or $x = \beta$
$\therefore$ the x-intercepts are α and β.

- A quadratic function with factored form $y = a(x - \alpha)^2$ *touches* the x-axis at $x = \alpha$. The only x-intercept is α.

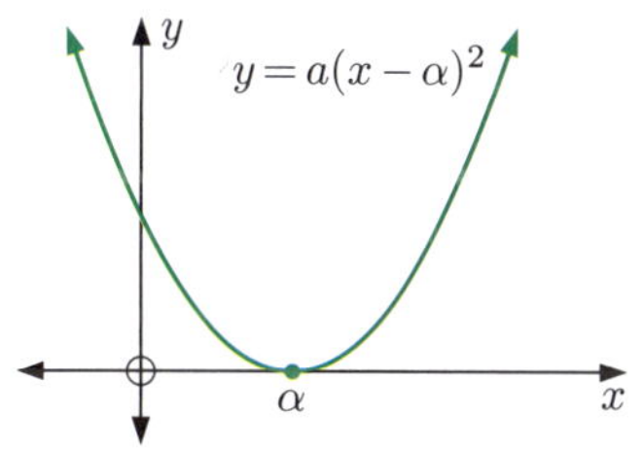

Example 9 Self Tutor

Find the x-intercept(s) of:

a $y = 2(x - 3)(x + 2)$

b $y = -(x - 4)^2$

a When $y = 0$, $2(x - 3)(x + 2) = 0$
$\therefore x = 3$ or $x = -2$
$\therefore$ the x-intercepts are 3 and -2.

b When $y = 0$, $-(x - 4)^2 = 0$
$\therefore x = 4$
$\therefore$ the x-intercept is 4.

FACTORISING TO FIND x-INTERCEPTS

For any quadratic function of the form $y = ax^2 + bx + c$, the x-intercepts can be found by solving the equation $ax^2 + bx + c = 0$.

The quadratic equation may have two, one, or no real solutions. These solutions correspond to the two, one, or no x-intercepts of the graph of the corresponding quadratic function.

Example 10 — Self Tutor

Find the x-intercept(s) of the quadratic function:

a $y = x^2 - 6x + 9$ **b** $y = -x^2 - x + 6$

a When $y = 0$,
$x^2 - 6x + 9 = 0$
$\therefore (x-3)^2 = 0$
$\therefore x = 3$
$\therefore$ the x-intercept is 3.

b When $y = 0$,
$-x^2 - x + 6 = 0$
$\therefore x^2 + x - 6 = 0$
$\therefore (x+3)(x-2) = 0$
$\therefore x = -3$ or $x = 2$
$\therefore$ the x-intercepts are -3 and 2.

EXERCISE 23D

1 State the y-intercept of:

a $y = x^2 + 3x + 3$ **b** $y = x^2 - 5x + 2$ **c** $f(x) = 2x^2 + 7x - 8$

d $y = 3x^2 - x + 1$ **e** $f(x) = -x^2 + 3x + 6$ **f** $y = -2x^2 + 5 - x$

g $y = 6 - x - x^2$ **h** $f(x) = 8 + 2x - 3x^2$ **i** $y = 5x - x^2 - 2$

2 Find the axes intercepts of:

a $y = (x-3)(x+1)$ **b** $f(x) = -(x-2)(x-4)$ **c** $y = 2(x+3)(x+2)$

d $y = -3(x-4)(x-5)$ **e** $y = 2(x+3)^2$ **f** $f(x) = -5(x-1)^2$

3 Find the x-intercept(s) of:

a $y = x^2 - 9$ **b** $y = 25 - x^2$ **c** $y = x^2 - 6x$

d $f(x) = x^2 + 7x + 10$ **e** $y = x^2 + x - 12$ **f** $y = 4x - x^2$

g $y = -x^2 - 6x - 8$ **h** $f(x) = -2x^2 - 4x - 2$ **i** $y = 4x^2 - 24x + 36$

Example 11 — Self Tutor

Graph each quadratic function using its axes intercepts:

a $y = x^2 - 2x - 3$ **b** $y = -2(x+1)(x-2)$ **c** $y = 2(x-3)^2$

a When $x = 0$, $y = -3$
$\therefore$ the y-intercept is -3.
When $y = 0$, $x^2 - 2x - 3 = 0$
$\therefore (x-3)(x+1) = 0$
$\therefore x = 3$ or $x = -1$
$\therefore$ the x-intercepts are 3 and -1.

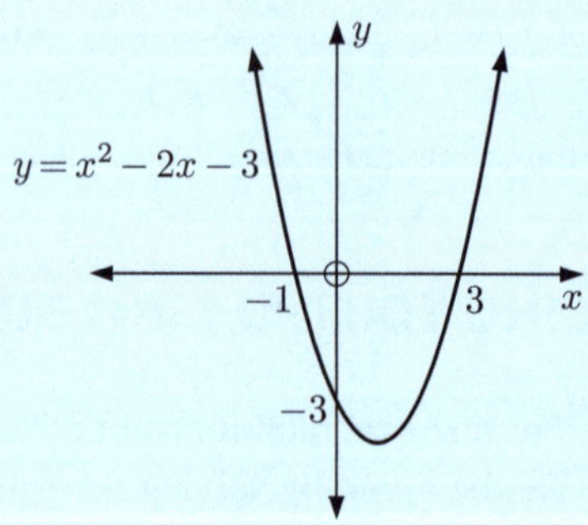

b When $x = 0$, $y = -2(1)(-2) = 4$

$\therefore$ the y-intercept is 4.

When $y = 0$, $-2(x+1)(x-2) = 0$

$\therefore$ $x = -1$ or $x = 2$

$\therefore$ the x-intercepts are -1 and 2.

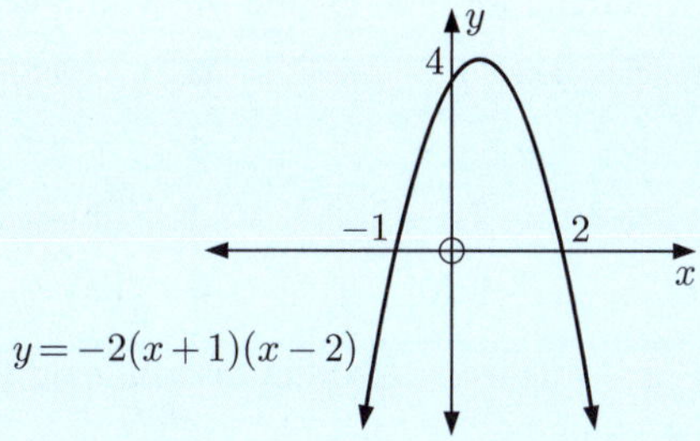

c When $x = 0$, $y = 2(-3)^2 = 18$

$\therefore$ the y-intercept is 18.

When $y = 0$, $2(x-3)^2 = 0$

$\therefore$ $x = 3$

$\therefore$ the x-intercept is 3.

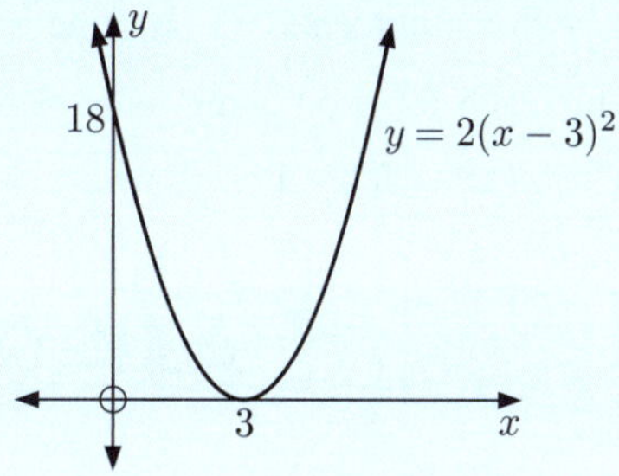

4 Sketch the graph of the quadratic function which has:

a x-intercepts -1 and 1, and y-intercept -1
b x-intercepts -3 and 1, and y-intercept 2
c x-intercepts 2 and 5, and y-intercept -4
d x-intercept 2 and y-intercept 4.

5 Graph each quadratic function using its axes intercepts:

a $f(x) = (x-1)(x+3)$
b $y = 2(x+2)^2$
c $f(x) = -(x-2)(x+1)$
d $y = -3(x+1)^2$
e $y = -3(x-4)(x-1)$
f $y = 2(x+3)(x+1)$
g $y = x^2 - 4x + 4$
h $y = x^2 - x - 6$
i $y = -x^2 + 8x - 12$

Example 12 **Self Tutor**

Use the quadratic formula to find any x-intercepts of the quadratic function:

a $y = x^2 - 2x - 5$
b $y = 2x^2 + 4x + 3$

a When $y = 0$, $x^2 - 2x - 5 = 0$

$\therefore x = \dfrac{2 \pm \sqrt{(-2)^2 - 4(1)(-5)}}{2(1)}$

$\therefore x = \dfrac{2 \pm \sqrt{4 + 20}}{2}$

$\therefore x = \dfrac{2 \pm \sqrt{24}}{2}$

$\therefore x = \dfrac{2 \pm 2\sqrt{6}}{2}$

$\therefore x = 1 \pm \sqrt{6}$

$\therefore$ the x-intercepts are $1 + \sqrt{6}$ and $1 - \sqrt{6}$.

b When $y = 0$, $2x^2 + 4x + 3 = 0$

$\therefore x = \dfrac{-4 \pm \sqrt{4^2 - 4(2)(3)}}{2(2)}$

$\therefore x = \dfrac{-4 \pm \sqrt{16 - 24}}{4}$

$\therefore x = \dfrac{-4 \pm \sqrt{-8}}{4}$

There are no real solutions, so the function has no x-intercepts.

6 Use the quadratic formula to find the x-intercepts of each function:

a $y = x^2 - 4x + 1$ **b** $y = x^2 + 4x - 3$

c $y = -x^2 + 6x - 4$ **d** $y = x^2 + x + 1$

e $f(x) = 3x^2 - 7x - 2$ **f** $f(x) = 2x^2 - x - 5$

g $f(x) = -2x^2 + 5x - 4$ **h** $f(x) = -4x^2 + 9x - 3$

7 Let $y = x^2 - 6x + c$. Use the discriminant to find the value(s) of c for which the graph has:

a two x-intercepts **b** one x-intercept **c** no x-intercepts.

8 Graph each function using its axes intercepts:

a $y = x^2 - 5x + 1$ **b** $y = -2x^2 - 3x + 5$ **c** $f(x) = -x^2 + 8x - 10$

E AXIS OF SYMMETRY OF A QUADRATIC

The graph of any quadratic function is symmetrical about a vertical line passing through the vertex. This line is called the **axis of symmetry**.

If the graph has two x-intercepts, then the axis of symmetry is midway between them.

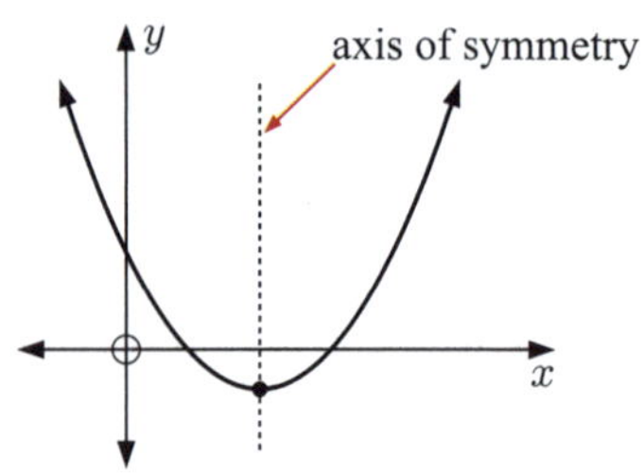

Example 13 **Self Tutor**

Find the equation of the axis of symmetry for this quadratic function:

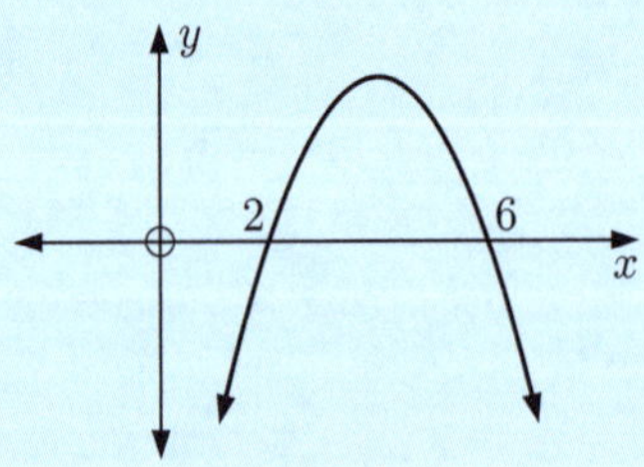

The x-intercepts are 2 and 6.

4 is midway between 2 and 6, so the axis of symmetry is $x = 4$.

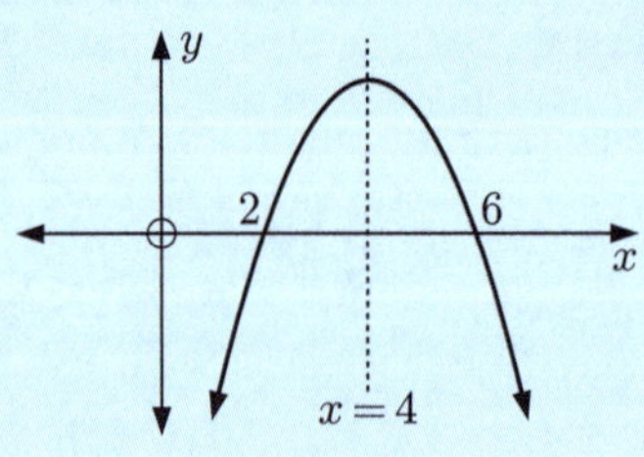

If the quadratic does not have any x-intercepts, or if we do not know the x-intercepts, we can still use the knowledge that the axis of symmetry passes through the vertex.

We know that a function in the completed square form $y = a(x-h)^2 + k$ has vertex (h, k), so its line of symmetry is $x = h$.

Expanding, $$\begin{aligned} y &= a(x-h)^2 + k \\ &= a(x^2 - 2hx + h^2) + k \\ &= ax^2 - 2ahx + [ah^2 + k]. \end{aligned}$$

Comparing the coefficient of x with that from $y = ax^2 + bx + c$, we find $-2ah = b$

$$\therefore\ h = \frac{-b}{2a}$$

The equation of the **axis of symmetry** of $y = ax^2 + bx + c$ is $\boldsymbol{x = \dfrac{-b}{2a}}$.

Example 14 🔊 **Self Tutor**

Find the equation of the axis of symmetry of $y = 2x^2 + 3x + 1$.

$y = 2x^2 + 3x + 1$ has $a = 2$, $b = 3$, $c = 1$.

Now $\dfrac{-b}{2a} = \dfrac{-3}{2 \times 2} = -\dfrac{3}{4}$

$\therefore$ the axis of symmetry has equation $x = -\frac{3}{4}$.

EXERCISE 23E

1 For each graph, find the equation of the axis of symmetry:

a
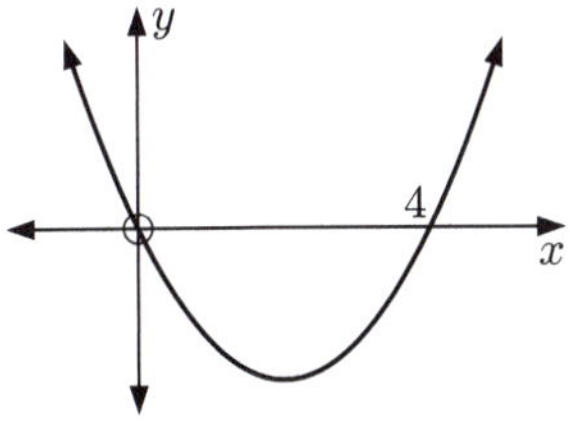

b
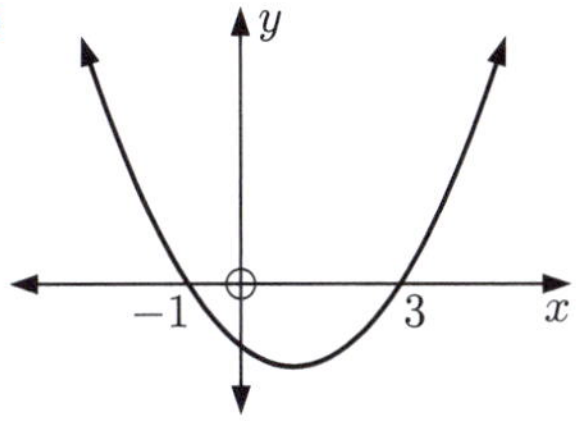

c
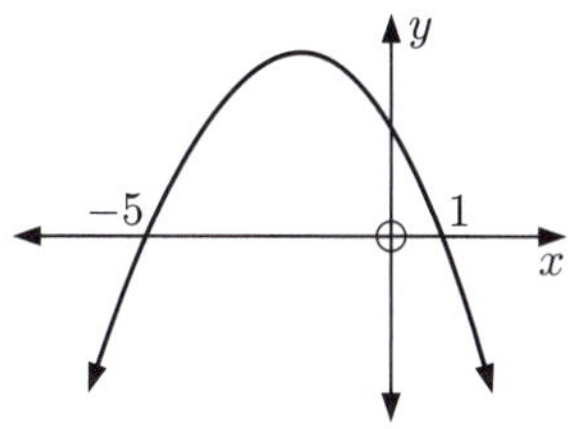

d
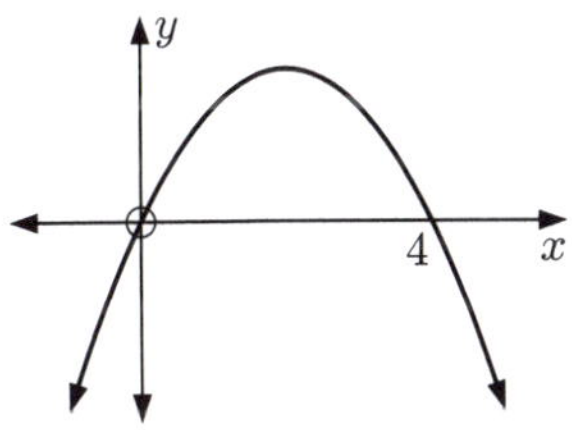

e
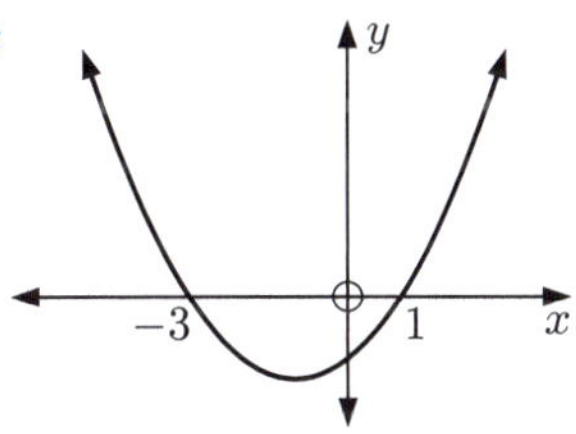

f
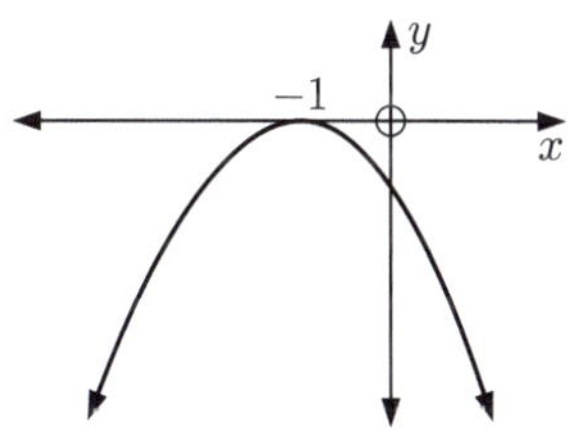

g
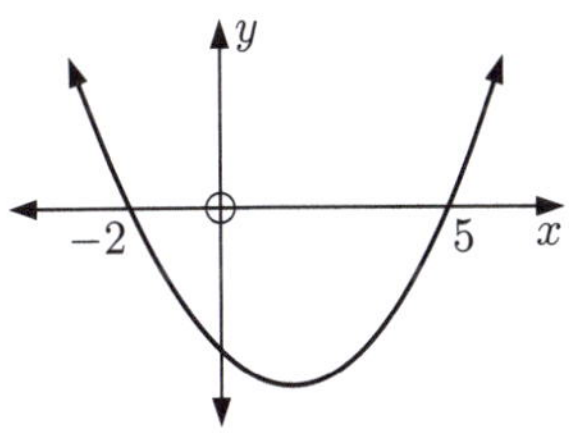

h
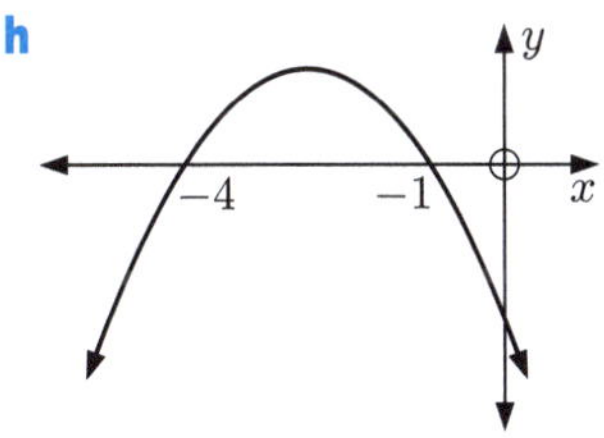

i
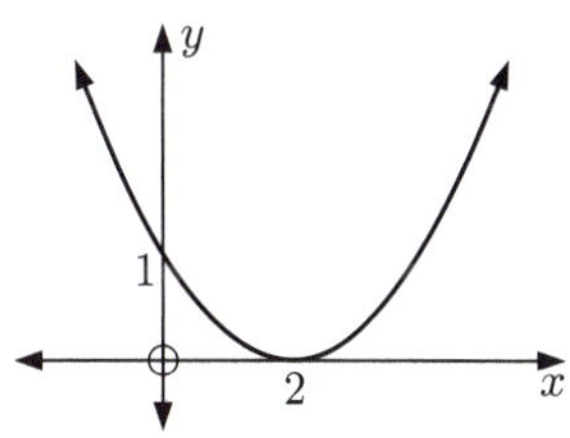

2 Find the equation of the axis of symmetry of each function:

a $y = (x-2)(x-4)$ **b** $y = -(x+1)(x-5)$ **c** $y = 2(x+3)(x-3)$

d $y = x(x+5)$ **e** $y = -3(x+4)^2$ **f** $y = 4(x+6)(x-9)$

3 Find the equation of the axis of symmetry of each function:

a $y = x^2 + 4x + 1$ **b** $y = 2x^2 - 6x + 3$ **c** $f(x) = 3x^2 + 4x - 1$

d $y = -x^2 - 4x + 5$ **e** $y = -2x^2 + 5x + 1$ **f** $f(x) = \frac{1}{2}x^2 - 10x + 2$

g $y = \frac{1}{3}x^2 + 4x$ **h** $f(x) = 100x - 4x^2$ **i** $y = -\frac{1}{10}x^2 + 30x$

4 Find all x-intercepts of the quadratic function which:

a cuts the x-axis at 1 and has axis of symmetry $x = 2$

b cuts the x-axis at -2 and has axis of symmetry $x = 1$

c cuts the x-axis at -1 and has axis of symmetry $x = -\frac{3}{2}$.

5 Sketch the graph, including the axis of symmetry, of the parabola with:

a x-intercepts -3 and 3, and y-intercept 6

b x-intercepts -1 and 2, and y-intercept -3

c only x-intercept -2, and y-intercept 4.

F VERTEX OF A QUADRATIC

The **vertex** or **turning point** of the quadratic function $y = ax^2 + bx + c$ is the point at which the function has:

- a **maximum value** for $a < 0$, *or*
- a **minimum value** for $a > 0$.

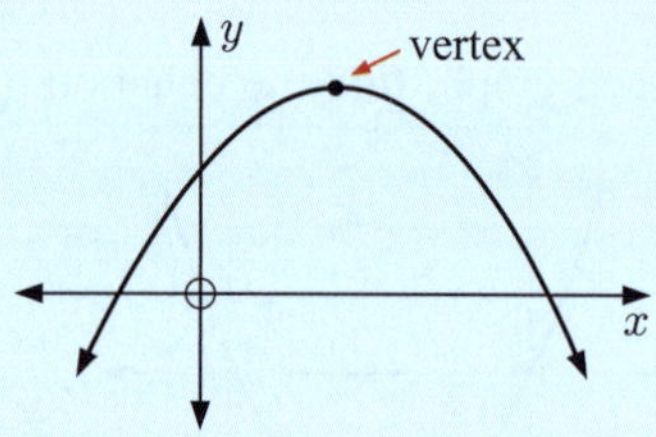

Since the vertex lies on the axis of symmetry, its x-coordinate is $\dfrac{-b}{2a}$.

The y-coordinate is found by substituting this x-coordinate into the function.

Example 15 **Self Tutor**

Consider the quadratic function $y = -x^2 + 2x + 3$.

a Find the axes intercepts.

b Find the equation of the axis of symmetry.

c Find the coordinates of the vertex.

d Hence sketch the function.

a When $x = 0$, $y = 3$

$\therefore$ the y-intercept is 3.

When $y = 0$,

$-x^2 + 2x + 3 = 0$

$\therefore x^2 - 2x - 3 = 0$

$\therefore (x-3)(x+1) = 0$

$\therefore x = 3$ or $x = -1$

$\therefore$ the x-intercepts are 3 and -1.

b $a = -1$, $b = 2$, $c = 3$

$\therefore \dfrac{-b}{2a} = \dfrac{-2}{-2} = 1$

$\therefore$ the axis of symmetry is $x = 1$.

c When $x = 1$, $y = -(1)^2 + 2(1) + 3$

$= -1 + 2 + 3$

$= 4$

$\therefore$ the vertex is $(1, 4)$.

d

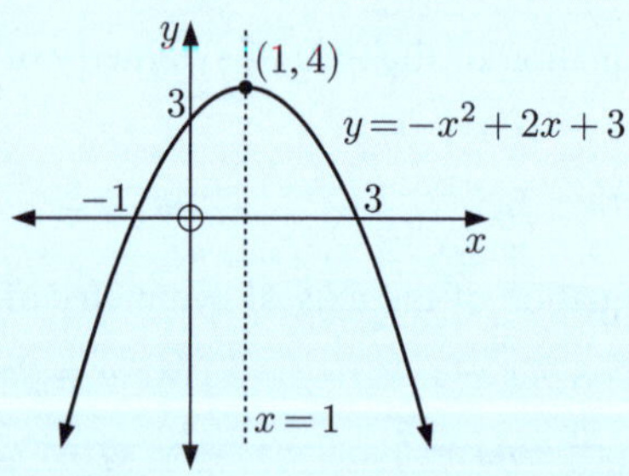

EXERCISE 23F

1 For each quadratic function:

- **i** Find the coordinates of the vertex.
- **ii** Determine whether the vertex is a maximum or minimum turning point.
- **iii** Find the range of the function.

a $y = x^2 - 4x + 2$ **b** $y = x^2 + 2x - 3$ **c** $f(x) = 2x^2 + 4$

d $y = -3x^2 + 1$ **e** $y = -x^2 - 4x - 4$ **f** $y = 2x^2 - 10x + 3$

2 For each quadratic function:

- **i** Find the axes intercepts.
- **ii** Find the equation of the axis of symmetry.
- **iii** Find the coordinates of the vertex.
- **iv** Hence sketch the function.

a $y = x^2 - 2x - 8$ **b** $y = 4x - x^2$ **c** $y = x^2 + 3x$

d $f(x) = x^2 + 4x + 4$ **e** $y = x^2 + 3x - 4$ **f** $y = -x^2 + 2x - 1$

g $y = 2x^2 + 5x - 3$ **h** $f(x) = -3x^2 - 4x + 4$ **i** $y = x^2 - 6x + 3$

Example 16

Self Tutor

Consider the quadratic function $y = 2(x - 2)(x + 4)$.

a Find the axes intercepts. **b** Find the equation of the axis of symmetry.

c Find the coordinates of the vertex. **d** Hence sketch the function.

a When $x = 0$, $y = 2(-2)(4) = -16$

$\therefore$ the y-intercept is -16.

When $y = 0$,

$2(x - 2)(x + 4) = 0$

$\therefore x = 2$ or $x = -4$

$\therefore$ the x-intercepts are 2 and -4.

b The axis of symmetry is halfway between the x-intercepts.

-1 is halfway between 2 and -4, so the axis of symmetry is $x = -1$.

c When $x = -1$, $y = 2(-1 - 2)(-1 + 4)$

$= 2(-3)(3)$

$= -18$

$\therefore$ the vertex is $(-1, -18)$.

d

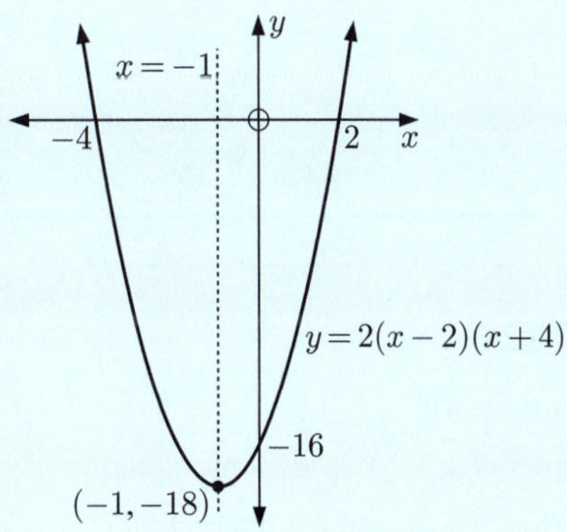

3 For each quadratic function:

- **i** Find the axes intercepts.
- **ii** Find the equation of the axis of symmetry.
- **iii** Find the coordinates of the vertex.
- **iv** Hence sketch the function.

a $f(x) = x(x - 2)$ **b** $y = 2(x - 3)^2$ **c** $y = -(x - 1)(x + 3)$

d $y = -2(x - 1)^2$ **e** $f(x) = -5(x + 2)(x - 2)$ **f** $y = 2(x + 1)(x + 4)$

4 Suppose $f(x) = x^2 + bx + c$.
Find the values of b and c if $y = f(x)$ has vertex:

a $(1, 3)$ **b** $(0, 2)$ **c** $(3, 0)$ **d** $(-3, -2)$ **e** $(\frac{1}{2}, 1)$

G FINDING A QUADRATIC FUNCTION

We can use information about axes intercepts, the vertex, and other points on a graph to establish the equation of a quadratic function.

The form we choose to write the function in is often determined by the information available:

- If a quadratic has vertex (h, k), we use the form $f(x) = a(x - h)^2 + k$.
- If a quadratic has x-intercepts α and β, we use the form $f(x) = a(x - \alpha)(x - \beta)$.
- If a quadratic touches the x-axis at α, we use the form $f(x) = a(x - \alpha)^2$.

Example 17 **Self Tutor**

Find the quadratic function with:

a vertex $(1, 2)$ and y-intercept 3

b vertex $(2, 11)$ and which passes through $(-1, -7)$.

a Using the vertex, $f(x) = a(x - 1)^2 + 2$ for some value of a.
But $f(0) = 3$, so $a + 2 = 3$
$\therefore a = 1$
$\therefore f(x) = (x - 1)^2 + 2$

b Using the vertex, $f(x) = a(x - 2)^2 + 11$ for some value of a.
But $f(-1) = -7$, so $a(-3)^2 + 11 = -7$
$\therefore 9a = -18$
$\therefore a = -2$
$\therefore f(x) = -2(x - 2)^2 + 11$

EXERCISE 23G

1 Find, in the form $f(x) = a(x - h)^2 + k$, the quadratic function with:

a vertex $(2, -5)$ and y-intercept 3
b vertex $(-4, 19)$ and y-intercept 3
c vertex $(1, 8)$ and y-intercept 7
d vertex $(-2, 11)$ and y-intercept 3.

2 Find the quadratic function with the given graph. Give your answers in the form $y = a(x-h)^2 + k$.

a
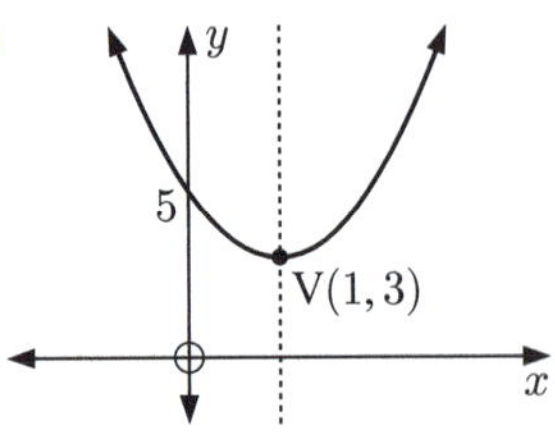

b
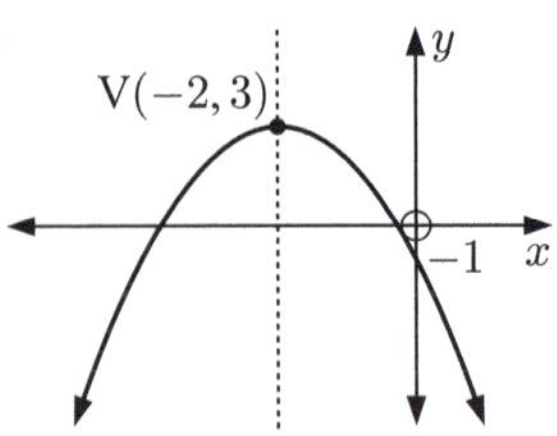

c
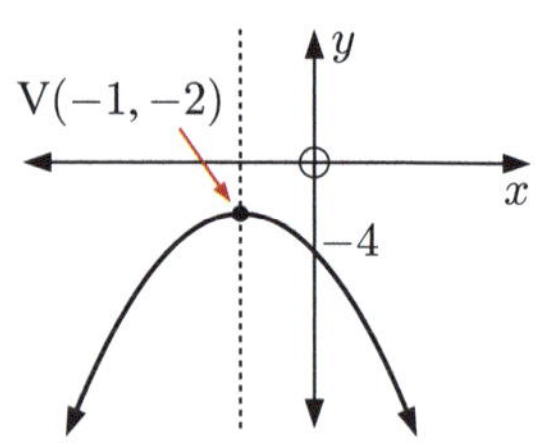

3 Find, in the form $f(x) = a(x-h)^2 + k$, the quadratic function which has:

- **a** vertex $(-2, -5)$ and which passes through $(1, 13)$
- **b** vertex $(3, -19)$ and which passes through $(-2, 31)$.

4 Find, in the form $f(x) = ax^2 + bx + c$, the quadratic function with:

- **a** vertex $(1, -4)$ and y-intercept -7
- **b** vertex $(-3, -5)$ and y-intercept 7
- **c** vertex $(2, -3)$ and which passes through $(-1, 15)$
- **d** vertex $(-1, 5)$ and which passes through $(3, -3)$.

Example 18 — Self Tutor

The graph of a quadratic function has x-intercepts -5 and 1, and it passes through $(3, 32)$. Find the function.

Using the x-intercepts, $f(x) = a(x+5)(x-1)$

But $f(3) = 32$, so $a(8)(2) = 32$

$$\therefore\ 16a = 32$$
$$\therefore\ a = 2$$
$$\therefore\ f(x) = 2(x+5)(x-1)$$

5 Suppose $f(x) = x^2 + bx + c$. Find the values of b and c given $y = f(x)$ has x-intercepts:

- **a** 0 and 2
- **b** -4 and 1
- **c** -5 and 2
- **d** -7 and 0.

6 Find the quadratic function which has x-intercepts:

- **a** -2 and 2, and which passes through $(0, 8)$
- **b** 1 and 4, and which passes through $(0, -12)$
- **c** -2 and 3, and which passes through $(4, 18)$
- **d** -4 and 5, and which passes through $(-1, 36)$
- **e** $1\frac{1}{2}$ and 3, and which passes through $(1, 2)$.

7 Find the quadratic function whose only x-intercept is:

- **a** -2 and has y-intercept 4
- **b** 3 and has y-intercept -3
- **c** 1 and which passes through $(3, 8)$
- **d** -4 and which passes through $(-1, -27)$.

8 Find the quadratic function which passes through $(-3, 4)$, $(-1, 10)$, and $(5, 4)$.

ACTIVITY

Click on the icon to practise matching quadratic functions with their graphs and properties.

MATHS CARDS

H PROBLEM SOLVING

Example 19 **Self Tutor**

The height of a rocket t seconds after it is fired upwards is given by
$H(t) = 100t - 5t^2$ metres, $t \geqslant 0$.

- **a** How long does the rocket take to reach its maximum height?
- **b** Find the maximum height reached by the rocket.
- **c** How long does it take for the rocket to fall back to Earth?

a $H(t) = 100t - 5t^2$
$\therefore H(t) = -5t^2 + 100t$

Now $a = -5$ which is < 0, so the shape of the graph is  .

The maximum height is reached when $t = \dfrac{-b}{2a} = \dfrac{-100}{2(-5)} = 10$

$\therefore$ the maximum height is reached after 10 seconds.

b $H(10) = 100(10) - 5(10)^2 = 500$
$\therefore$ the maximum height reached is 500 m.

c The rocket falls back to Earth when $H(t) = 0$

$$\therefore -5t^2 + 100t = 0$$
$$\therefore -5t(t - 20) = 0$$
$$\therefore t = 0 \text{ or } 20$$

$\therefore$ the rocket falls back to Earth after 20 seconds.

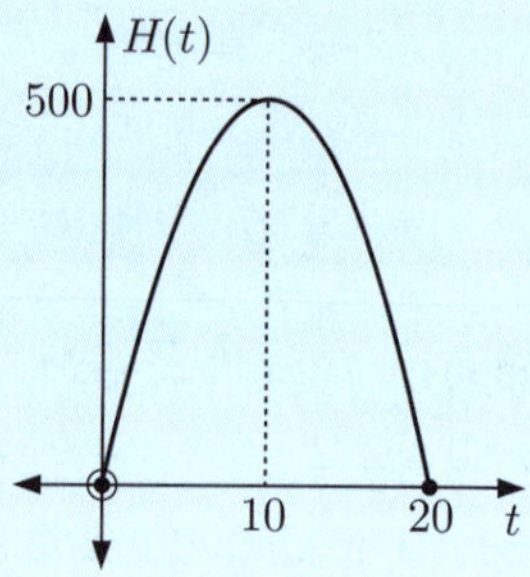

EXERCISE 23H

1 The height of a ball t seconds after it is kicked upwards is given by $H(t) = 20t - 5t^2$ metres.

- **a** How long does the ball take to reach its maximum height?
- **b** Find the maximum height reached by the ball.
- **c** How long does it take for the ball to hit the ground?

2 A manufacturer finds that the profit from assembling x bicycles per day is given by $P(x) = -x^2 + 50x - 200$ dollars.

- **a** How many bicycles should be assembled per day to maximise the profit?
- **b** Find the maximum profit.
- **c** Find the loss made if no bicycles are assembled in a day. Suggest why this loss would be made.

3 The hourly profit obtained from operating a fleet of n taxis is given by $P(n) = 120n - 200 - 2n^2$ dollars.

- **a** What number of taxis gives the maximum hourly profit?
- **b** Find the maximum hourly profit.
- **c** How much money is lost per hour if no taxis are on the road?

4 The temperature in a greenhouse t hours after 7:00 pm is given by $T(t) = \frac{1}{4}t^2 - 6t + 25$ °C for $t \leqslant 20$.

- **a** Find the temperature in the greenhouse at 7:00 pm.
- **b** At what time is the temperature at a minimum?
- **c** Find the minimum temperature in the greenhouse for $0 \leqslant t \leqslant 20$.

5 Answer the **Opening Problem** on page **406**.

6 A car travelling downhill is rapidly accelerating, so the driver applies the brakes. The speed of the car t seconds after the brakes are applied, is given by $s(t) = -6t^2 + 12t + 60$ km/h.

- **a** How fast was the car travelling when the driver applied the brakes?
- **b** After how many seconds does the car reach its maximum speed?
- **c** Find the maximum speed reached.
- **d** The model is valid until the time when the vehicle is stopped. How long does this take?

7 Infinitely many rectangles may be inscribed within the triangle ACE shown. One of them is illustrated.
Suppose $\text{EF} = x$ cm.

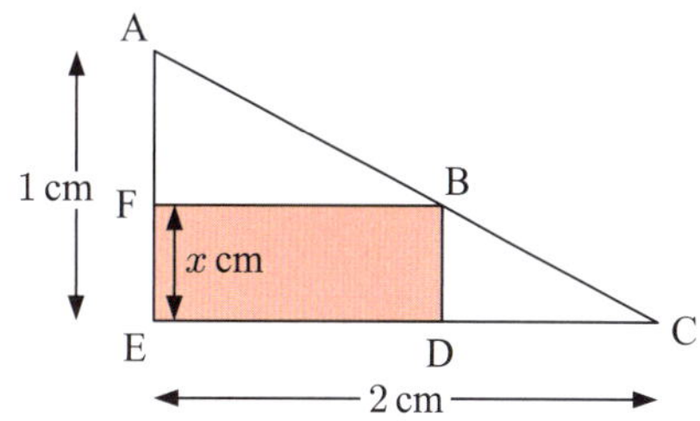

- **a** Show that triangles ABF and ACE are similar.
- **b** Show that $\text{BF} = 2(1-x)$ cm.
- **c** Show that the area of rectangle BDEF is given by $A = -2x^2 + 2x$ cm^2.
- **d** **i** Find the value of x which maximises the area of the rectangle.
  - **ii** Calculate this maximum area.

8 Dennis throws a stone from a bridge 6 m above the water.
The peak of its flight occurs after travelling 9 m horizontally, and it lands in the water after travelling 27 m horizontally.
We choose a set of axes with the origin where the bridge meets the water, as shown.

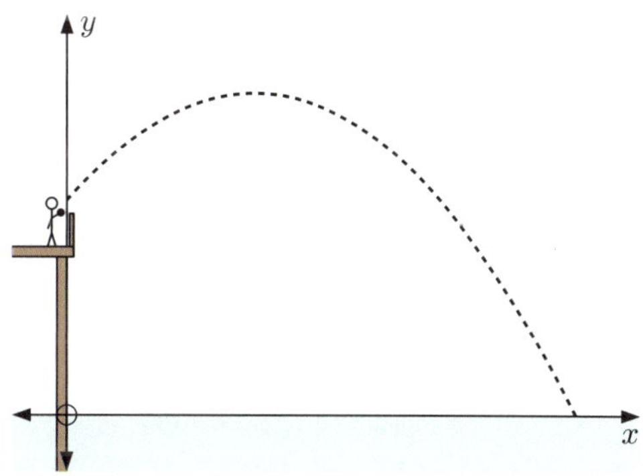

- **a** Write down:
  - **i** the y-intercept
  - **ii** the axis of symmetry.
- **b** Find *both* x-intercepts.
- **c** Write a function for the height of the stone above the water in the form $y = a(x-\alpha)(x-\beta)$.
- **d** Find the maximum height reached by the stone.

GLOBAL CONTEXT — ARCHES

Global context: Personal and cultural expression

Statement of inquiry: Mathematical models can describe the characteristics of the architecture we see around us.

Criterion: Knowing and understanding

GLOBAL CONTEXT

MULTIPLE CHOICE QUIZ

QUICK QUIZ

REVIEW SET 23A

1 For the quadratic function $y = x^2 - 3x - 15$, find:

a the value of y when $x = 4$

b the values of x when $y = 3$.

2 Determine whether the ordered pair $(2, 5)$ satisfies the function $f(x) = x^2 - 3x + 8$.

3 Find the y-intercept of:

a $y = 2x^2 - 7x + 5$

b $y = -\frac{1}{3}(x+1)(5-x)$

4 Sketch each function on the same set of axes as $y = x^2$, and label its vertex.

a $y = x^2 - 4$

b $y = (x-2)^2 + 1$

c $y = -(x+3)^2 - 2$

5 Graph the function $f(x) = x^2 - 4x - 21$ using its axes intercepts.

6 Use the quadratic formula to find the x-intercepts of:

a $y = 3x^2 - x - 5$

b $y = -x^2 + 2x + 6$

7 Find the equation of the axis of symmetry for:

a

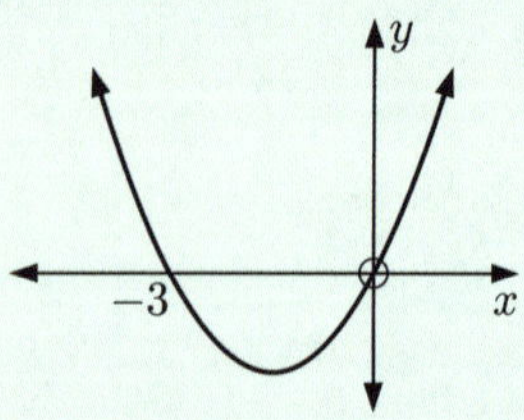

b

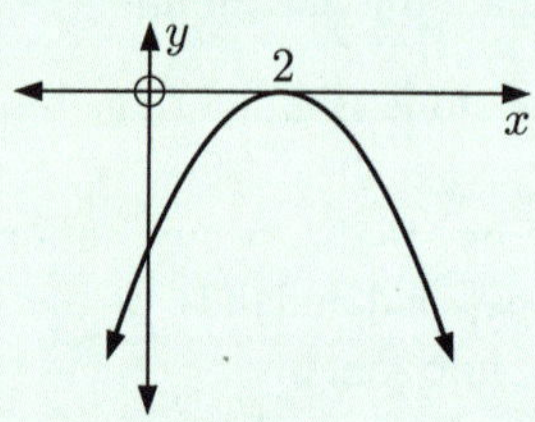

8 Find the vertex of each quadratic function:

a $y = x^2 - 8x - 3$

b $y = -4x^2 + 4x - 3$

9 For each quadratic function:

i Find the axes intercepts.

ii Find the equation of the axis of symmetry.

iii Find the coordinates of the vertex.

iv Hence sketch the function.

a $y = -2(x-1)(x+3)$

b $y = x^2 - 2x - 15$

10 The graph of $f(x) = x^2 + bx + c$ has vertex $(-3, -11)$. Find the values of b and c.

11 Find the quadratic function with vertex $(-1, -5)$ and y-intercept -3. Give your answer in the form $f(x) = a(x-h)^2 + k$.

12 Find, in the form $f(x) = ax^2 + bx + c$, the quadratic function:

a which has x-intercepts -4 and $-\frac{1}{3}$, and which passes through $(-1, -18)$

b with vertex $(6, -2)$ and which passes through $(4, 16)$.

13 A gardener has 40 m of fencing to enclose a rectangular garden plot where one side is an existing brick wall. Suppose the plot is x m wide as shown.

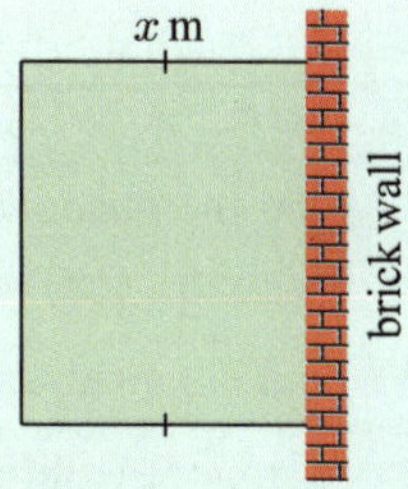

- **a** Show that the area enclosed is given by $A = -2x^2 + 40x$ m^2.
- **b** Find the value of x which maximises the area of the garden.
- **c** Calculate this maximum area.

14 Isaac threw an apple directly upwards from a height of 1.75 m. The apple reached a maximum height of 13 metres after 1.5 seconds.

- **a** Write a quadratic function for the height H m of the apple above ground level after t seconds, giving your answer in the form:
  - **i** $H = a(t-h)^2 + k$
  - **ii** $H = at^2 + bt + c$
- **b** Find the height of the apple after 1 second.
- **c** At what times was the apple 8 metres above ground level?
- **d** Find, rounded to 2 decimal places, the time taken for the apple to fall to the ground.

REVIEW SET 23B

1 Find the values of x for which $f(x) = x^2 + x - 12$ has the value 30.

2 Use a table of values from $x = -3$ to $x = 3$ to draw the graph of $y = -x^2 + x - 2$.

3 Sketch each function on the same set of axes as $y = x^2$, and label its vertex:

a $y = (x+2)^2 + 5$ **b** $y = -\frac{1}{2}x^2$ **c** $y = (x+3)^2 - 9$

4 **a** Write $y = x^2 - 4x + 10$ in the form $y = (x-h)^2 + k$.

b Hence sketch the graph of $y = x^2 - 4x + 10$, including the coordinates of the vertex.

5 Find the axes intercepts of:

a $y = 5x(x+4)$ **b** $y = 2x^2 + 6x - 56$ **c** $y = x^2 - 6x + 9$

6 Graph each equation using its axes intercepts:

a $f(x) = 2(x-3)(x+1)$ **b** $f(x) = -x^2 + 2x + 24$ **c** $f(x) = 3x^2 + x - 2$

7 Find the equation of the axis of symmetry of each function:

a $f(x) = (x+3)(x-5)$ **b** $f(x) = 3x^2 - 5x + 9$

8 **a** Find the vertex of the quadratic function $f(x) = -x^2 + 4x - 7$.

b Hence find the range of the function.

9 Consider the quadratic function $f(x) = (x-1)(x-4)$.

- **a** Find $f(-1)$.
- **b** Find the axes intercepts.
- **c** Find the equation of the axis of symmetry.
- **d** Find the coordinates of the vertex.
- **e** Hence sketch the function.

10 The quadratic function $f(x) = x^2 + bx + c$ has x-intercepts -5 and 3. Find the values of b and c.

11 Find the quadratic function with vertex $(-4, 15)$ and y-intercept -17. Give your answer in the form $f(x) = ax^2 + bx + c$.

12 Find the quadratic function whose only x-intercept is -3 and which passes through $(-5, -2)$.

13 A stone was thrown from the top of a cliff 60 metres above sea level. The height of the stone above sea level t seconds after it was released is given by $H(t) = -5t^2 + 20t + 60$ metres. Find:

a the time taken for the stone to reach its maximum height

b the maximum height above sea level reached by the stone.

14 The graph of the function $y = a(x - h)^2 + k$ is shown alongside.

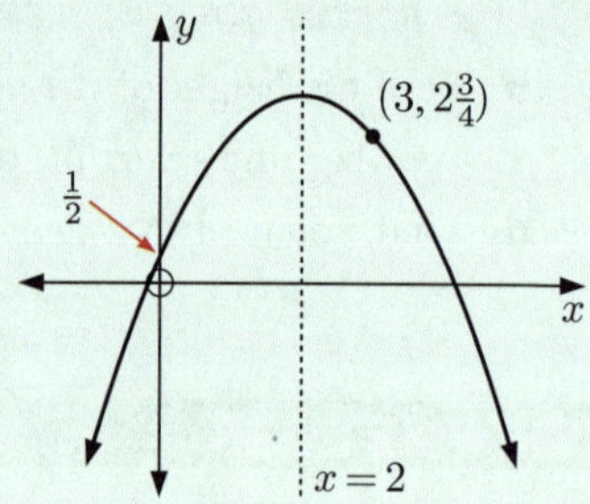

a State the value of h.

b Find the values of a and k by solving simultaneous equations.

c Hence state the coordinates of the vertex.

d Write the equation of the function in the form $y = ax^2 + bx + c$.

e Hence find the x-intercepts of the function. Give your answers in the form $m \pm k\sqrt{n}$, where $m, n \in \mathbb{Z}$, $k \in \mathbb{Q}$.

Chapter 24

Number sequences

Contents:

OPENING PROBLEM

Suppose we draw all the diagonals of an n-sided polygon. The first few cases are:

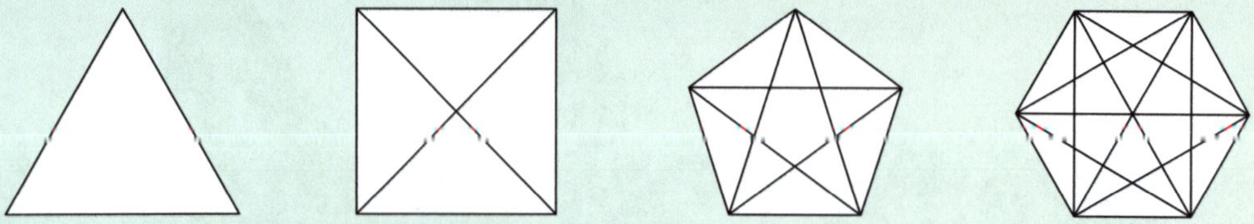

$n=3$	$n=4$	$n=5$	$n=6$
0 diagonals	2 diagonals	5 diagonals	9 diagonals

Things to think about:

a Consider the number of diagonals in each figure:

$$0 \xrightarrow{+\,....} 2 \xrightarrow{+\,....} 5 \xrightarrow{+\,....} 9$$

i Can you *describe* the pattern to get from one number to the next?

ii Can you hence *predict* the number of diagonals in a heptagon (7-gon)? Draw a heptagon and all its diagonals to check your answer.

b Consider the function $f(n) = \dfrac{n(n-3)}{2}$.

What is the value of:

i $f(3)$ **ii** $f(4)$ **iii** $f(5)$ **iv** $f(6)$ **v** $f(7)$?

c Can you write a *formula* for the number of diagonals in an n-sided polygon?

d Can you *explain* why your formula gives the correct number of diagonals?

A NUMBER SEQUENCES

A **number sequence** is an ordered list of numbers defined by a rule.

The numbers in a sequence are called the **terms** of the sequence.

The **nth term** or **general term** is written as u_n, so the sequence is $u_1, u_2, u_3, u_4, ...$

For example, consider this pattern of balls:

The first layer has just one blue ball, so $u_1 = 1$.
The second layer has three pink balls, so $u_2 = 3$.
The third layer has five black balls, so $u_3 = 5$.
The fourth layer has seven green balls, so $u_4 = 7$.

The pattern could be continued forever, generating a **sequence** of numbers.

There are many ways to describe this sequence:

- We can **list the terms** in order: 1, 3, 5, 7, 9, 11, 13, ...
  The string of dots indicates that the pattern continues forever.
- We can write the sequence in **words**: "The sequence starts at 1 and increases by 2 each time."

- We can write an **explicit formula** for the nth term of the sequence in terms of n.
  For this sequence, $u_n = 2n - 1$.
  We can substitute any value of $n \in \mathbb{Z}^+$. For example, $u_{10} = 2(10) - 1 = 19$.
- We can write a **recurrence relation** which connects the nth term of a sequence with one or more preceding terms.
  For this sequence, $u_1 = 1$, $u_n = u_{n-1} + 2$, $n \geqslant 2$.
  For any recurrence relation, at least one initial term must be defined.

DISCUSSION

In the previous example, how is the description in words linked to the recurrence relation?

Can you write a description for the sequence in words which is linked to the explicit formula instead?

Example 1

Self Tutor

Describe the sequence in words, and hence find its next two terms:

a 3, 7, 11, 15, 19, ... **b** 2, 6, 18, 54, ... **c** 1, 1, 2, 3, 5, 8, ...

a Start at 3 and add 4 each time.
$u_6 = 23$ and $u_7 = 27$.

b Start at 2 and multiply by 3 each time.
$u_5 = 54 \times 3 = 162$ and $u_6 = 162 \times 3 = 486$.

c The first two terms are 1 and 1, and each term thereafter is the sum of the previous two terms.
$u_7 = 5 + 8 = 13$ and $u_8 = 8 + 13 = 21$.

EXERCISE 24A

1 Consider the sequence 8, 5, 2, -1, ...

a Describe the sequence in words. **b** State the values of u_2 and u_4.

c Find the values of u_5 and u_6.

2 Write down the first five terms of the sequence:

a starting at 3, and increasing by 4 each time

b starting at 40, and decreasing by 5 each time

c whose nth term is the nth prime number.

3 Describe the sequence 5, 8, 11, 14, 17, 20, ... using:

a words **b** a recurrence relation.

4 Describe the sequence 1, 4, 9, 16, 25, 36, ... using:

a words **b** an explicit formula.

5 Describe the sequence in words, and hence find its next two terms:

a 2, 6, 10, 14, 18, ... **b** 20, 16, 12, 8, 4, ... **c** 8, 19, 30, 41, 52, ...

d 1, 2, 4, 8, 16, ... **e** $\frac{1}{81}, \frac{1}{27}, \frac{1}{9}, \frac{1}{3}, ...$ **f** 405, 135, 45, 15, ...

g 1, 8, 27, 64, ... **h** $1, \frac{1}{4}, \frac{1}{9}, \frac{1}{16}, ...$ **i** 1, 2, 2, 4, 8, ...

Example 2 — Self Tutor

Find the first five terms of the sequence with nth term:

a $u_n = 5n - 3$ **b** $u_n = n(n+2)$ **c** $u_n = 3 \times 2^n$

a $u_1 = 5(1) - 3 = 2$
$u_2 = 5(2) - 3 = 7$
$u_3 = 5(3) - 3 = 12$
$u_4 = 5(4) - 3 = 17$
$u_5 = 5(5) - 3 = 22$

b $u_1 = 1(3) = 3$
$u_2 = 2(4) = 8$
$u_3 = 3(5) = 15$
$u_4 = 4(6) = 24$
$u_5 = 5(7) = 35$

c $u_1 = 3 \times 2^1 = 6$
$u_2 = 3 \times 2^2 = 12$
$u_3 = 3 \times 2^3 = 24$
$u_4 = 3 \times 2^4 = 48$
$u_5 = 3 \times 2^5 = 96$

6 Find the first four terms of the sequence with nth term:

a $u_n = 2n + 3$ **b** $u_n = 3n + 5$ **c** $u_n = 17 - 4n$
d $u_n = n^2 - 1$ **e** $u_n = n^2 + n$ **f** $u_n = n^3 - 1$
g $u_n = 5 \times 2^n$ **h** $u_n = 4 \times 3^{n-1}$ **i** $u_n = 24 \times \left(\frac{1}{2}\right)^{n-1}$

7 The **triangular numbers** are numbers which can be represented by a triangular arrangement of dots.

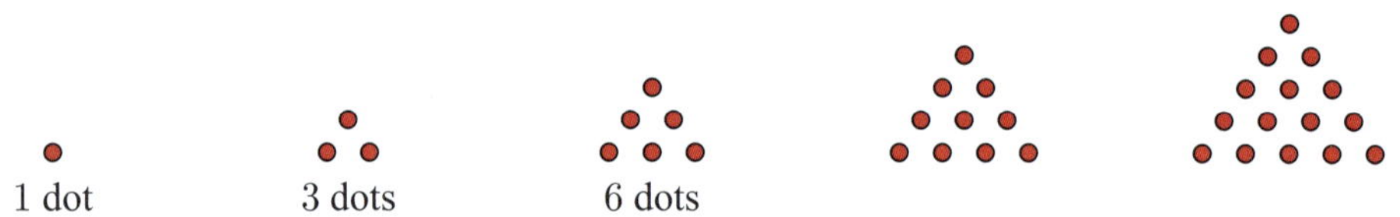

a Find the first five numbers in the sequence of triangular numbers.
b The general term for the sequence of triangular numbers is given by $u_n = \frac{1}{2}n(n+1)$.
  i Verify this rule for the first five triangular numbers.
  ii Use the rule to find the 10th triangular number.

8 Find the first four terms of the sequence with recurrence relation:

a $u_1 = 7, \quad u_n = u_{n-1} + 3, \quad n \geqslant 2$ **b** $u_1 = 25, \quad u_n = u_{n-1} - 4, \quad n \geqslant 2$
c $u_1 = 5, \quad u_n = 3 \times u_{n-1}, \quad n \geqslant 2$ **d** $u_1 = 100, \quad u_n = \frac{u_{n-1}}{2}, \quad n \geqslant 2$
e $u_1 = 3, \quad u_n = 2 \times u_{n-1} - 1, \quad n \geqslant 2$ **f** $u_1 = 4, \quad u_n = 10 - u_{n-1}, \quad n \geqslant 2$
g $u_1 = 3, \quad u_2 = 4, \quad u_n = u_{n-1} \times u_{n-2}, \quad n \geqslant 3$

Example 3 — Self Tutor

a Write a formula for the general term u_n of the sequence 3, 6, 9, 12, 15, ...
b Hence find a formula for the general term u_n of:
  i 4, 7, 10, 13, 16, ... **ii** 1, 4, 7, 10, 13, ... **iii** $\frac{1}{5}, \frac{1}{8}, \frac{1}{11}, \frac{1}{14}, \frac{1}{17}, ...$

a $u_1 = 3 \times 1, \quad u_2 = 3 \times 2, \quad u_3 = 3 \times 3, \quad u_4 = 3 \times 4, \quad u_5 = 3 \times 5$
$\therefore \; u_n = 3 \times n = 3n$

b **i** $u_1 = 3 + 1, \quad u_2 = 6 + 1, \quad u_3 = 9 + 1, \quad u_4 = 12 + 1, \quad u_5 = 15 + 1$
Each term is 1 more than in the sequence in **a**.
$\therefore \; u_n = 3n + 1$

ii $u_1 = 3 - 2, \quad u_2 = 6 - 2, \quad u_3 = 9 - 2, \quad u_4 = 12 - 2, \quad u_5 = 15 - 2$

Each term is 2 less than in the sequence in **a**.

$\therefore \quad u_n = 3n - 2$

iii $u_1 = \frac{1}{3+2}, \quad u_2 = \frac{1}{6+2}, \quad u_3 = \frac{1}{9+2}, \quad u_4 = \frac{1}{12+2}, \quad u_5 = \frac{1}{15+2}$

The denominator of each term is 2 more than in the sequence in **a**.

$\therefore \quad u_n = \frac{1}{3n+2}$

9 **a** Write a formula for the general term u_n of the sequence: 2, 4, 6, 8, 10, ...

b Hence find a formula for the general term u_n of:

i 4, 6, 8, 10, 12, ... **ii** 1, 3, 5, 7, 9, ... **iii** $\frac{1}{2}, \frac{1}{4}, \frac{1}{6}, \frac{1}{8}, \frac{1}{10}, ...$

10 **a** Write a formula for the general term u_n of the sequence: 5, 10, 15, 20, 25, ...

b Hence find a formula for the general term u_n of:

i 6, 11, 16, 21, 26, ... **ii** 3, 8, 13, 18, 23, ... **iii** $\frac{1}{5}, \frac{1}{10}, \frac{1}{15}, \frac{1}{20}, \frac{1}{25}, ...$

11 **a** Write a formula for the general term u_n of the sequence: 2, 4, 8, 16, 32, ...

b Hence find a formula for the general term u_n of:

i 6, 12, 24, 48, 96, ... **ii** $\frac{1}{2}, \frac{1}{4}, \frac{1}{8}, \frac{1}{16}, \frac{1}{32}, ...$

12 Write a formula for the general term u_n of the sequence:

a 1, 2, 3, 4, 5, 6, ... **b** 3, 4, 5, 6, 7, 8, ...

c $\frac{1}{1}, \frac{1}{2}, \frac{1}{3}, \frac{1}{4}, \frac{1}{5}, ...$ **d** $\frac{1}{2}, \frac{1}{3}, \frac{1}{4}, \frac{1}{5}, \frac{1}{6}, ...$

e $\frac{1}{2}, \frac{2}{3}, \frac{3}{4}, \frac{4}{5}, \frac{5}{6}, ...$ **f** $1 \times 2, \; 2 \times 3, \; 3 \times 4, \; 4 \times 5, \; ....$

g $2 \times 3, \; 3 \times 4, \; 4 \times 5, \; 5 \times 6, \; ....$ **h** $\frac{1}{3}, \frac{4}{6}, \frac{7}{9}, \frac{10}{12}, ...$

13 Consider the **Opening Problem** on page **430**. The sequence of the number of diagonals in an n-sided polygon is 0, 2, 5, 9, ... where the first term is u_3.

a Write an explicit formula for the general term u_n.

b Hence predict the number of diagonals in:

i an octagon **ii** a decagon **iii** a 20-gon.

INVESTIGATION — THE FIBONACCI SEQUENCE

The **Fibonacci sequence** is

1, 1, 2, 3, 5, 8, 13, 21, 34, 55, ...

The sequence starts with two 1s, and after that each term is obtained by adding the two terms preceding it:

$$1 + 1 = 2$$
$$1 + 2 = 3$$
$$2 + 3 = 5$$
$$3 + 5 = 8$$
$$\vdots$$

Leonardo Fibonacci

We can hence write down a recurrence relation for the sequence:

$$u_1 = u_2 = 1, \quad u_n = u_{n-1} + u_{n-2} \quad \text{for } n \geqslant 3.$$

Check: If $n = 3$, $u_3 = u_2 + u_1 = 1 + 1 = 2$ ✓

If $n = 4$, $u_4 = u_3 + u_2 = 2 + 1 = 3$ ✓

If $n = 5$, $u_5 = u_4 + u_3 = 3 + 2 = 5$ ✓

The Fibonacci sequence is named after the Italian mathematician **Leonardo Fibonacci** (1170 - 1250). Fibonacci wrote a mathematical puzzle about rabbit populations which involved this sequence, in his book *Liber Abaci*.

The Fibonacci sequence occurs frequently in nature. For example, the number of petals on a flower is often a member of the Fibonacci sequence.

3 petals	lily, iris
5 petals	buttercup
8 petals	delphinium
13 petals	cineraria
21 petals	aster
34 petals	pyrethrum

The Fibonacci sequence can also be observed in:

- the number of leaves arranged about the stem in plants
- the seed patterns of a sunflower
- the seed pattern on a pine cone.

The recurrence relation is very useful for dealing with the Fibonacci sequence because its explicit formula is very complicated. In fact, it is not obvious from the explicit formula

$$u_n = \frac{1}{\sqrt{5}}\left[\left(\frac{1+\sqrt{5}}{2}\right)^n - \left(\frac{1-\sqrt{5}}{2}\right)^n\right]$$

that the terms of the sequence will be integers.

What to do:

1. Show that the explicit formula is correct for $n = 1$ and $n = 2$.
2. Use the recurrence relation to find the next five terms of the Fibonacci sequence after 55.
3. Can you see a pattern of even and odd numbers in the Fibonacci sequence? Do you think the 100th term of the sequence will be even or odd?
4. Research other occurrences of the Fibonacci sequence in nature.

GLOBAL CONTEXT — THE GOLDEN RATIO

GLOBAL CONTEXT

Global context: Personal and cultural expression

Statement of inquiry: Mathematics can help us understand beauty in the world around us.

Criterion: Investigating patterns

B ARITHMETIC SEQUENCES

An **arithmetic sequence** is a sequence in which each term differs from the previous one by the same constant. We call this number the **common difference** d.

A sequence is arithmetic if $u_{n+1} - u_n = d$ for all $n \in \mathbb{Z}^+$.

For example:

- the sequence 1, 5, 9, 13, 17, ... is arithmetic with common difference 4, since

$$\begin{aligned} 5 - 1 &= 4 \\ 9 - 5 &= 4 \\ 13 - 9 &= 4, \quad \text{and so on.} \end{aligned}$$

- the sequence 42, 37, 32, 27, ... is arithmetic with common difference -5, since

$$\begin{aligned} 37 - 42 &= -5 \\ 32 - 37 &= -5 \\ 27 - 32 &= -5, \quad \text{and so on.} \end{aligned}$$

THE GENERAL TERM OF AN ARITHMETIC SEQUENCE

Suppose the first term of an arithmetic sequence is u_1, and the common difference is d.

Notice that:

$$\begin{aligned} u_2 &= u_1 + d \\ u_3 &= u_2 + d = u_1 + 2d \\ u_4 &= u_3 + d = u_1 + 3d \\ &\vdots \\ u_n &= u_{n-1} + d = u_1 + (n-1)d \end{aligned}$$

For an **arithmetic sequence** with **first term** u_1 and **common difference** d, the **general term** or n**th term** is $u_n = u_1 + (n-1)d$.

Example 4 **Self Tutor**

Consider the sequence 3, 9, 15, 21, 27, ...

a Show that the sequence is arithmetic.

b Find a formula for the general term u_n.

c Find the 100th term of the sequence.

d Is **i** 489 **ii** 1592 a term of the sequence?

When we see the string of dots, we can *assume* that the pattern continues.

a $9 - 3 = 6, \quad 15 - 9 = 6, \quad 21 - 15 = 6, \quad 27 - 21 = 6$

The difference between successive terms is constant.

$\therefore$ the sequence is arithmetic with $u_1 = 3$ and $d = 6$.

b

$$\begin{aligned} u_n &= u_1 + (n-1)d \\ \therefore\ u_n &= 3 + 6(n-1) \\ \therefore\ u_n &= 6n - 3 \end{aligned}$$

c

$$\begin{aligned} u_{100} &= 6(100) - 3 \\ &= 597 \end{aligned}$$

d **i** Let $u_n = 489$
$\therefore \quad 6n - 3 = 489$
$\therefore \quad 6n = 492$
$\therefore \quad n = 82$
$\therefore$ 489 is the 82nd term of the sequence.

ii Let $u_n = 1592$
$\therefore \quad 6n - 3 = 1592$
$\therefore \quad 6n = 1595$
$\therefore \quad n = 265\frac{5}{6}$
But n must be an integer, so 1592 is not a term of the sequence.

EXERCISE 24B

1 Determine whether each sequence is arithmetic. If so, state the common difference d.

a 2, 5, 8, 11, 14, ...
b 5, 9, 13, 18, 22, ...
c 29, 23, 16, 10, 4, ...
d 11, 4, −3, −10, −17, ...

2 Find the unknowns given that the following sequences are arithmetic:

a 4, 10, □, 22, 28, ...
b 13, 20, 27, □, 41, ...
c 19, □, 11, 7, △, ...
d 22, □, 4, △, −14, ...

3 Write down the first term u_1 and common difference d for the arithmetic sequence of numbers given by the:

a green arrow
b blue arrow
c purple arrow
d red arrow.

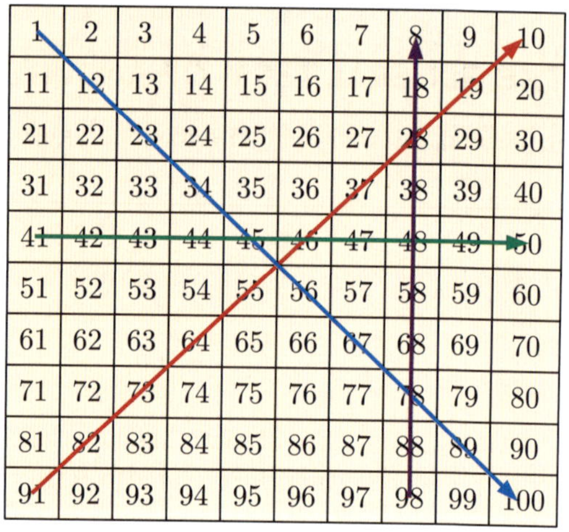

4 Consider the sequence 4, 11, 18, 25, 32, ...

a Show that the sequence is arithmetic.
b Find a formula for the general term u_n.
c Find the 30th term of the sequence.
d Is 340 a member of the sequence?
e Is 738 a member of the sequence?

5 Consider the sequence 67, 63, 59, 55, ...

a Show that the sequence is arithmetic.
b Find a formula for the general term u_n.
c Find the 60th term of the sequence.
d Is −143 a member of the sequence?
e Is −85 a member of the sequence?

6 The general term of a sequence is $u_n = \dfrac{11n - 7}{2}$.

a Find $u_{n+1} - u_n$. Hence show that the sequence is arithmetic.
b Find u_1 and d.
c Find the 20th term of the sequence.
d Find the smallest term of the sequence which is greater than 150.

Example 5 **Self Tutor**

Find k given that $k+5$, -1, and $2k-1$ are consecutive terms of an arithmetic sequence.

Since they are consecutive terms of an arithmetic sequence,

$$-1-(k+5) = (2k-1)-(-1) \quad \{\text{equating common differences}\}$$
$$\therefore \; -1-k-5 = 2k-1+1$$
$$\therefore \; -k-6 = 2k$$
$$\therefore \; -6 = 3k$$
$$\therefore \; k = -2$$

7 Find k given the consecutive terms of an arithmetic sequence:

a $31, k, 13$ **b** $k, 8, k+11$ **c** $k+2, 2k+3, 17$

8 **a** Find k given that $k-1$, 13, and $3k+3$ are consecutive terms of an arithmetic sequence.

b Write down the next two terms of the sequence.

9 Suppose $k-2$, $k+2$, and k^2 are the first three terms of an arithmetic sequence.
Find the possible values of k. List the first three terms of the sequence in each case.

Example 6 **Self Tutor**

Find the general term u_n for the arithmetic sequence with $u_3 = 4$ and $u_7 = -24$.

$u_3 = 4 \quad \therefore \; u_1 + 2d = 4$ .... (1) $\quad \{\text{using } u_n = u_1 + (n-1)d\}$

$u_7 = -24 \quad \therefore \; u_1 + 6d = -24$ .... (2)

We now solve (1) and (2) simultaneously:

$$-u_1 - 2d = -4 \quad \{(1) \times -1\}$$
$$u_1 + 6d = -24 \quad \{(2)\}$$

Adding, $\quad 4d = -28$

$$\therefore \; d = -7$$

Substituting $d = -7$ into (1) gives $u_1 + 2(-7) = 4$

$$\therefore \; u_1 - 14 = 4$$
$$\therefore \; u_1 = 18$$

Now $u_n = u_1 + (n-1)d$

$\therefore \; u_n = 18 - 7(n-1)$

$\therefore \; u_n = 18 - 7n + 7$

$\therefore \; u_n = 25 - 7n$

Check:

$u_3 = 25 - 7(3) = 25 - 21 = 4$ ✓

$u_7 = 25 - 7(7) = 25 - 49 = -24$ ✓

10 Find the general term u_n for the arithmetic sequence with:

a $u_4 = 37$ and $u_{10} = 67$ **b** $u_5 = -10$ and $u_{12} = -38$

c $u_4 = -4$ and $u_{15} = 29$ **d** $u_{10} = -16$ and $u_6 = -13$.

11 An arithmetic sequence has $u_3 = 10$ and $u_{11} = 58$.

a Find a formula for the general term u_n. **b** Find the 30th term of the sequence.

C GEOMETRIC SEQUENCES

A **geometric sequence** is a sequence in which each term can be obtained from the previous one by multiplying by the same non-zero constant. We call this number the **common ratio** r.

A sequence is geometric if $\dfrac{u_{n+1}}{u_n} = r$ for all $n \in \mathbb{Z}^+$.

For example: 2, 6, 18, 54, ... is a geometric sequence as each term can be obtained by multiplying the previous term by 3.

The common ratio is 3 since $\frac{6}{2} = \frac{18}{6} = \frac{54}{18} = 3$.

THE GENERAL TERM OF A GEOMETRIC SEQUENCE

Suppose the first term of a geometric sequence is u_1 and the common ratio is r.

Notice that:

$$\begin{aligned} u_2 &= u_1 r \\ u_3 &= u_2 r = u_1 r^2 \\ u_4 &= u_3 r = u_1 r^3 \\ &\vdots \\ u_n &= u_{n-1} r = u_1 r^{n-1} \end{aligned}$$

For a **geometric sequence** with **first term u_1** and **common ratio r**, the **general term** or **nth term** is $u_n = u_1 r^{n-1}$.

Example 7 — Self Tutor

Consider the sequence 16, 8, 4, 2, 1, ...

a Show that the sequence is geometric.

b Find the general term u_n.

c Find the 10th term as a fraction.

a $\frac{8}{16} = \frac{1}{2}$, $\frac{4}{8} = \frac{1}{2}$, $\frac{2}{4} = \frac{1}{2}$, $\frac{1}{2} = \frac{1}{2}$

Consecutive terms have a common ratio $\frac{1}{2}$.

$\therefore$ the sequence is geometric with $u_1 = 16$ and $r = \frac{1}{2}$.

b $u_n = u_1 r^{n-1}$

$\therefore u_n = 16 \times \left(\frac{1}{2}\right)^{n-1}$

c $u_{10} = 16 \times \left(\frac{1}{2}\right)^9 = \dfrac{2^4}{2^9} = \dfrac{1}{2^5} = \frac{1}{32}$

EXERCISE 24C

1 Write down the common ratio r for these geometric sequences:

a 3, 6, 12, 24, ...
b 60, 30, 15, 7.5, ...
c 4000, -400, 40, -4, ...

2 Determine whether each sequence is geometric. If so, state the common ratio r.

a 5, 10, 20, 40, ...
b 10 000, 1000, 100, 10, ...
c 4, 12, 36, 72, ...
d 45, 15, 5, $\frac{5}{3}$, ...
e 1, -4, 16, 64, ...
f 2, -10, 50, -250, ...

3 For the geometric sequence with first two terms given, find the next two terms:

a 3, 6, ... **b** 8, 2, ... **c** 15, -5, ...

4 Consider the sequence 1, 3, 9, 27, ...

a Show that the sequence is geometric. **b** Find the general term u_n.

c Find the 10th term of the sequence.

5 Consider the sequence 40, -20, 10, -5, ...

a Show that the sequence is geometric. **b** Find the general term u_n.

c Find the 12th term as a fraction.

6 **a** Show that the sequence 16, -4, 1, -0.25, ... is geometric.

b Hence find the 8th term as a decimal.

7 Find the general term u_n of the geometric sequence: 3, $3\sqrt{2}$, 6, $6\sqrt{2}$, ...

Example 8 — Self Tutor

$k-1$, $k+3$, and $2k$ are consecutive terms of a geometric sequence.
Find the possible values of k.

Since they are consecutive terms of a geometric sequence,

$$\frac{2k}{k+3} = \frac{k+3}{k-1} \quad \{\text{equating common ratios}\}$$

$$\therefore\ 2k(k-1) = (k+3)^2$$

$$\therefore\ 2k^2 - 2k = k^2 + 6k + 9$$

$$\therefore\ k^2 - 8k - 9 = 0$$

$$\therefore\ (k+1)(k-9) = 0$$

$$\therefore\ k = -1 \text{ or } 9$$

Check: If $k=-1$, the terms are -2, 2, and -2. ✓ $\{r=-1\}$

If $k=9$, the terms are 8, 12, and 18. ✓ $\{r=\frac{3}{2}\}$

8 Find the value(s) of k given the consecutive terms of a geometric sequence:

a k, 2, 6 **b** 4, 6, k **c** k, $2\sqrt{2}$, k^2

d 3, k, 27 **e** k, $3k$, $10k+7$ **f** k, $k+4$, $8k+2$

9 Suppose k, $k+4$, and $3k-6$ are consecutive terms of a geometric sequence.
Find the possible values of k. In each case, state the common ratio r.

Example 9 — Self Tutor

Find the general term u_n for the geometric sequence with $u_2 = -5$ and $u_5 = 40$.

$u_2 = u_1 r = -5$ .... (1)

and $u_5 = u_1 r^4 = 40$ .... (2)

$$\text{Now } \frac{u_1 r^4}{u_1 r} = \frac{40}{-5} \quad \{(2) \div (1)\}$$

$$\therefore\ r^3 = -8$$

$$\therefore\ r = -2$$

Substituting $r = -2$ into (1) gives $u_1(-2) = -5$

$$\therefore \quad u_1 = \frac{5}{2}$$

So, $u_n = \frac{5}{2} \times (-2)^{n-1}$.

10 Find the general term u_n for a geometric sequence with:

a $u_3 = 16$ and $u_4 = 48$

b $u_3 = 32$ and $u_6 = -4$

c $u_2 = 10$ and $u_4 = 250$

d $u_3 = 3$ and $u_7 = \frac{3}{4}$.

11 A geometric sequence has $u_3 = 48$ and $u_7 = 3$. Find the possible values of u_{12}.

GLOBAL CONTEXT — SPIDER WEBS

Global context: Scientific and technical innovation
Statement of inquiry: Patterns in nature often have mathematical properties.
Criterion: Investigating patterns

D SEQUENCES IN FINANCE

Geometric sequences are observed in financial contexts such as **compound interest** and **depreciation**.

In these contexts we have an **initial condition** corresponding to time zero. This may be an initial investment or an initial value. We therefore allow a "zeroth" term u_0 to begin the sequence.

COMPOUND INTEREST

When you invest money in a bank or financial institution, you will usually be paid **compound interest**. After a certain amount of time called the **period**, the bank pays interest which is calculated as a percentage of the money already in the account.

It is called *compound* interest because the interest generated in one period will itself earn more interest in the next period.

For example, suppose you invest €1000 in a bank account which pays 6% per annum (p.a.) compound interest. The interest is added to your investment each year, so at the end of each year you will have $100\% + 6\% = 106\%$ of the value at the start of the year. This corresponds to a *multiplier* of 1.06.

Let u_n be the value of the investment after n years.

We can model the investment using the recurrence relation $u_0 = 1000, \quad u_n = u_{n-1} \times 1.06, \quad n \geqslant 1$.

Observe that:

$$\begin{aligned}
u_0 &= 1000 && = \text{initial value} \\
u_1 &= u_0 \times 1.06 && = \text{value after 1 year} \\
u_2 &= u_1 \times 1.06 = (u_0 \times 1.06) \times 1.06 = u_0 \times 1.06^2 && = \text{value after 2 years} \\
u_3 &= u_2 \times 1.06 = (u_0 \times 1.06^2) \times 1.06 = u_0 \times 1.06^3 && = \text{value after 3 years} \\
&\vdots \\
u_n &= u_{n-1} \times 1.06 = (u_0 \times 1.06^{n-1}) \times 1.06 = u_0 \times 1.06^n && = \text{value after } n \text{ years}
\end{aligned}$$

Suppose an investment earns compound interest each year.
The value of the investment after n years is

$$u_n = u_0(1+i)^n$$

where u_0 is the initial investment
and i is the interest rate per annum.

The amount of interest earned after n years is given by $u_n - u_0$.

Example 10 — Self Tutor

$20 000 is invested for 5 years at 3% p.a. compound interest.

a What will the investment amount to at the end of this period?

b How much interest has been earned?

a $u_n = u_0 \times (1+i)^n$

$\therefore\ u_5 = 20\,000 \times (1.03)^5 \quad \{3\% = 0.03\}$

$\approx 23\,185.48$

The investment will amount to $23 185.48.

b The interest earned

$= u_5 - u_0$

$= \$23\,185.48 - \$20\,000$

$= \$3185.48$

EXERCISE 24D.1

1 Jane invested £4000 at 5% p.a. compound interest. Find the value of her investment after 6 years.

2 $13 000 is invested for 4 years at 3.5% p.a. compound interest.

a What will the investment amount to at the end of this period?

b How much interest has been earned?

3 If ¥500 000 is invested for 7 years at 2.6% p.a. compound interest, how much interest is earned?

4 **a** Find the interest earned by investing $10 000 at 4% p.a. compound interest for:

i 5 years **ii** 10 years.

b If the duration of the investment doubles, does the total amount of interest earned double? Explain your answer.

Example 11 — Self Tutor

How much does Amy need to invest now at 5.2% p.a. compound interest, to get a maturing value of $16 000 in 6 years' time? Round your answer up to the next dollar.

The initial investment u_0 is unknown.

Now $u_n = u_0 \times (1+i)^n$

$\therefore\ 16\,000 = u_0 \times (1.052)^6 \quad \{5.2\% = 0.052\}$

$\therefore\ u_0 = \dfrac{16\,000}{(1.052)^6}$

$\therefore\ u_0 \approx 11\,803.90$

Amy needs to invest $11 804 now.

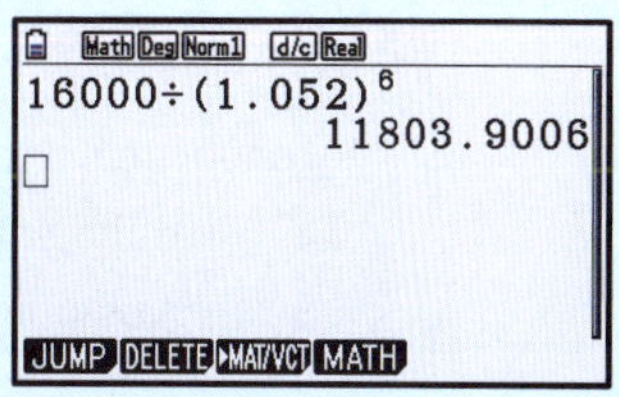

5 How much does Luca need to invest now at 5% p.a. compound interest, to get a maturing value of €12 000 in 4 years' time? Round your answer up to the next euro.

6 Elise invests a sum of money at 3.6% p.a. compound interest. After 5 years, there is $7518.64 in the account. How much did Elise initially invest?

7 **a** How much should I invest now at 4.5% p.a. compound interest, to get a maturing value of $20 000 in 3 years' time?

b How much compound interest will I earn?

DEPRECIATION

Items such as vehicles and electrical appliances lose value as time passes. This is due to wear and tear, technology becoming old, fashions changing, and other reasons.

Depreciation is the loss in value of an item over time.

Suppose a computer with initial value $1500 depreciates by 20% per year. At the end of each year, the computer is only worth $100\% - 20\% = 80\%$ of its value the previous year. This corresponds to a *multiplier* of 0.8.

Let u_n be the value of the computer after n years.

We can model the depreciation of the computer using the recurrence relation
$u_0 = 1500, \quad u_n = u_{n-1} \times 0.8, \quad n \geqslant 1.$

Observe that:

$$\begin{aligned}
u_0 &= 1500 && = \text{initial value}\\
u_1 &= u_0 \times 0.8 && = \text{value after 1 year}\\
u_2 &= u_1 \times 0.8 = (u_0 \times 0.8) \times 0.8 = u_0 \times 0.8^2 && = \text{value after 2 years}\\
u_3 &= u_2 \times 0.8 = (u_0 \times 0.8^2) \times 0.8 = u_0 \times 0.8^3 && = \text{value after 3 years}\\
&\vdots\\
u_n &= u_{n-1} \times 0.8 = (u_0 \times 0.8^{n-1}) \times 0.8 = u_0 \times 0.8^n && = \text{value after } n \text{ years}
\end{aligned}$$

The value of an object after n years is $u_n = u_0(1 - d)^n$
where u_0 is the initial value of the object
and d is the rate of depreciation per annum.
The amount of depreciation after n years is given by $u_0 - u_n$.

Example 12 Self Tutor

A photocopier was purchased for $8500 and depreciated by 12% per year.

a Find the value of the photocopier after 6 years.

b By how much did it depreciate?

a $\quad u_n = u_0 \times (1-d)^n$

$\therefore \quad u_6 = 8500 \times (0.88)^6 \qquad \{12\% = 0.12\}$

≈ 3947.43

After 6 years the value of the photocopier is $3947.43.

b The depreciation $= u_0 - u_6$

$= \$8500 - \$3947.43$

$= \$4552.57$

EXERCISE 24D.2

1 A car was purchased for $32 500 and depreciated by 16% per year. Find the value of the car after 10 years.

2 A refrigerator was purchased for £4500 and depreciated by 25% per year.

- **a** Find the value of the refrigerator after 3 years.
- **b** By how much did it depreciate?

3 Eric bought a laptop for $2500. It depreciated by 22% per year.

- **a** Find the value of the laptop after 5 years.
- **b** By how much did it depreciate?

4 Molly purchased a truck which depreciated by 18% per year. After 4 years, the truck was valued at $16 818.93. How much did Molly pay for the truck originally?

MULTIPLE CHOICE QUIZ

QUICK QUIZ

REVIEW SET 24A

1 Write down the first five terms of the sequence:

- **a** starting at 8, and increasing by 5 each time
- **b** starting at 19, and decreasing by 7 each time
- **c** whose nth term is the nth cubic number.

2 Describe the sequence in words, and hence find its next two terms:

- **a** 6, 10, 14, 18, 22, ...
- **b** 810, 270, 90, 30, ...
- **c** $\frac{1}{4}$, 1, 4, 16, ...

3 Find the first four terms of the sequence with recurrence relation $u_1 = 2, \quad u_n = 2u_{n-1} + 4, \quad n \geqslant 2.$

4 Consider the sequence $-6, -2, 2, 6, ...$

- **a** Show that the sequence is arithmetic.
- **b** Find a formula for the general term u_n.
- **c** Find the 100th term of the sequence.
- **d** Is 200 a member of the sequence?

5 **a** Find k given that $k-2$, $2k-1$, and $4k-6$ are consecutive terms of an arithmetic sequence.

- **b** State the common difference d for the sequence.

6 Write down the common ratio r for these geometric sequences:

a 250, 200, 160, 128, ...

b 4, −12, 36, −108, ...

7 Find the general term u_n of the geometric sequence with $u_4 = 24$ and $u_7 = -192$.

8 £120 000 is invested at 7.4% p.a. compound interest for 5 years.

a What will the investment amount to at the end of this period?

b How much interest has been earned?

9 A caravan was purchased for $15 000 and depreciated at 16% per year.
Find the caravan's value after 8 years.

REVIEW SET 24B

1 Describe the sequence 19, 16, 13, 10, 7, ... using:

a words

b a recurrence relation.

2 Find the first five terms of the sequence with nth term:

a $u_n = 61 - 4n$

b $u_n = n^2 - 10$

3 **a** Find the first five terms of the sequence with recurrence relation
$u_1 = 5, \quad u_n = 12 - u_{n-1}, \quad n \geqslant 2$.

b Hence state the value of u_{100}.

4 **a** Write a formula for the general term u_n of the sequence 4, 8, 12, 16, ...

b *Hence* write a formula for the general term u_n of:

i 1, 5, 9, 13, ...

ii $\frac{1}{3}, \frac{1}{7}, \frac{1}{11}, \frac{1}{15}, ...$

5 Find the unknowns given that the following sequences are arithmetic:

a 9, 17, □, 33, 41, ...

b 27, □, 15, △, 3, ...

6 An arithmetic sequence has $u_3 = 24$ and $u_{11} = -36$.

a Find a formula for the general term u_n.

b Find the 50th term of the sequence.

7 Consider the sequence 64, −32, 16, −8, 4, ...

a Show that the sequence is geometric.

b Find the general term u_n.

c Find the 16th term as a fraction.

8 Suppose $2k + 7$, $1 - k$, and $k - 7$ are consecutive terms of a geometric sequence.
Find the possible values of k. In each case, state the common ratio r.

9 Brayden invests a sum of money at 4.2% p.a. compound interest. After 7 years there is $11 470.24 in the account. How much did Brayden initially invest?

10 Mia bought a television for $3500. It depreciated by 17% per year.

a Find the value of the television after 6 years.

b By how much did it depreciate?

Chapter 25

Exponentials

Contents:

OPENING PROBLEM

A lotus plant initially covers an area of 40 cm^2. The area it covers increases by 20% each week.

Things to think about:

- **a** Does the area covered by the plant increase by a constant amount each week?
- **b** Can you explain why the area covered by the lotus plant after n weeks is given by the function $A(n) = 40 \times 1.2^n$ cm^2?
- **c** What does the graph of $A(n)$ look like?
- **d** What area is covered by the lotus plant after 3 weeks?
- **e** How long will it take for the plant to cover an area of 100 cm^2?

In the **Opening Problem**, the function $A(n) = 40 \times 1.2^n$ includes the variable n in the exponent. Each week, the value of n increases by 1, and the area A is multiplied by 1.2, an increase of 20%. We describe this kind of increase as **exponential growth**.

You might recognise that the compound interest we have just studied is also a form of exponential growth. The difference is that compound interest is added at *discrete* time intervals, whereas the area function $A(n)$ gives the area for any *continuous* value of n, including non-integer values.

A EXPONENTIAL FUNCTIONS

An **exponential function** is a function in which the variable occurs as part of the exponent or index.

Examples of exponential functions are $f(x) = 3^x$, $g(x) = 2^{x-4}$, and $h(x) = 6 + 5^{-x}$.

Example 1 — Self Tutor

For the function $f(x) = 3 - 2^{-x}$, find:

a $f(0)$ **b** $f(3)$ **c** $f(-x)$

a $f(0) = 3 - 2^0$
$= 3 - 1$
$= 2$

b $f(3) = 3 - 2^{-3}$
$= 3 - \frac{1}{8}$
$= 2\frac{7}{8}$

c $f(-x) = 3 - 2^{-(-x)}$
$= 3 - 2^x$

EXERCISE 25A

1 Determine whether each function is an exponential function:

- **a** $f(x) = 7^x$
- **b** $f(x) = x^4$
- **c** $f(x) = 5 - 3^{x-2}$
- **d** $f(x) = 10 \times 2^{\frac{x}{2}}$
- **e** $f(x) = 9x - x^6$
- **f** $f(x) = -2 - 5^{3x}$

2 For the function $f(x) = 3^x + 2$, find:

a $f(0)$ **b** $f(2)$ **c** $f(-1)$ **d** $f(2x)$

3 For the function $f(x) = 5^{-x} - 3$, find:

a $f(0)$ **b** $f(1)$ **c** $f(-2)$ **d** $f(-x)$

4 For the function $g(x) = 3^{x-2}$, find:

a $g(0)$ **b** $g(4)$ **c** $g(-1)$ **d** $g(x+5)$

B GRAPHS OF EXPONENTIAL FUNCTIONS

We can draw graphs of exponential functions using a table of values.

For example, for the function $f(x) = 2^x$ we can construct a table of values and hence draw the graph of $y = f(x)$:

x	-3	-2	-1	0	1	2	3
y	$\frac{1}{8}$	$\frac{1}{4}$	$\frac{1}{2}$	1	2	4	8

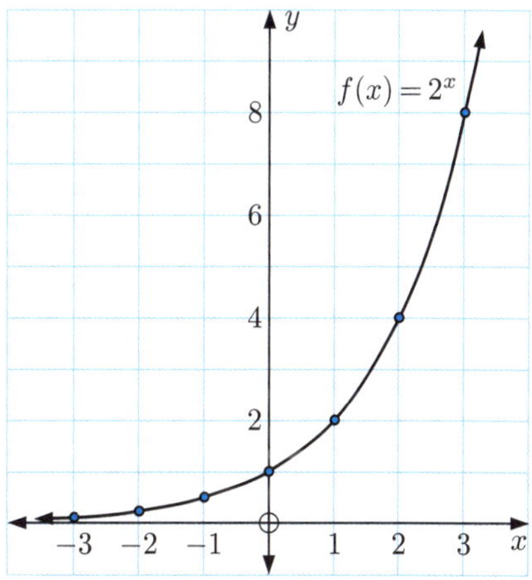

Notice that:

- The y-intercept of the function is 1.
- The graph lies entirely above the x-axis.
- As the values of x get smaller, the values of y get closer and closer to zero, but never actually reach zero. We say that the line $y = 0$ is a **horizontal asymptote** of the function.

INVESTIGATION SIMPLE EXPONENTIAL FUNCTIONS

In this Investigation we will study simple exponential functions of the form $f(x) = a^x$ where $a > 0$, $a \neq 1$.

GRAPHING PACKAGE

GRAPHICS CALCULATOR INSTRUCTIONS

What to do:

1 Use technology to draw the graph of:

a $f(x) = 1.3^x$ **b** $f(x) = 2^x$ **c** $f(x) = 3^x$ **d** $f(x) = 5^x$

e $f(x) = 0.8^x$ **f** $f(x) = \left(\frac{1}{2}\right)^x$ **g** $f(x) = 0.3^x$ **h** $f(x) = 0.1^x$

2 What do you notice about:

a the y-intercept of each graph **b** the horizontal asymptote of each graph?

Explain why this is the case.

3 For the graph of $f(x) = a^x$, what effect does a have on:

a whether the graph is increasing or decreasing **b** the steepness of the graph?

In the **Investigation** you should have found that:

For all exponential functions of the form $f(x) = a^x$, $a > 0$, $a \neq 1$:

- The **y-intercept** is 1, since $f(0) = a^0 = 1$.
- The graph has the **horizontal asymptote** $y = 0$.
- If $a > 1$, the graph is **increasing**.

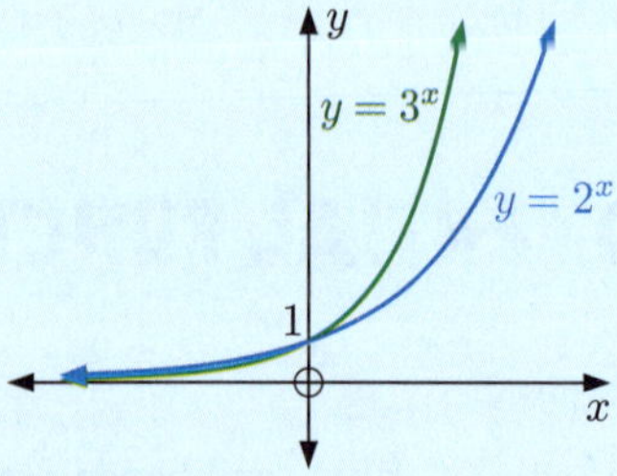

The rise towards infinity is steeper if a is larger.

- If $0 < a < 1$, the graph is **decreasing**.

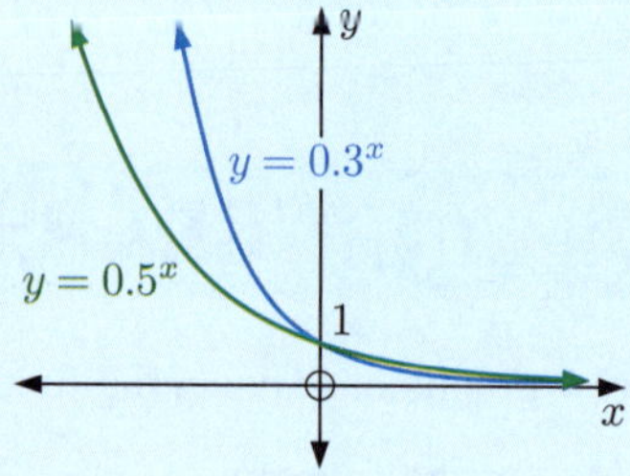

The fall from infinity is steeper if a is smaller.

EXERCISE 25B

1 Use a table of values from $x = -3$ to $x = 3$ to help sketch these exponential functions:

a $f(x) = 3^x$ **b** $f(x) = 4^x$ **c** $f(x) = \left(\frac{1}{2}\right)^x$ **d** $f(x) = \left(\frac{1}{3}\right)^x$

2 Use technology to graph on the same set of axes: $y = 5^x$, $y = 1.8^x$, $y = 0.7^x$, and $y = \left(\frac{2}{5}\right)^x$.

3 Match each function with the correct graph:

a $y = 3.6^x$

b $y = 0.9^x$

c $y = 1.5^x$

d $y = \left(\frac{1}{4}\right)^x$

e $y = \left(\frac{2}{3}\right)^x$

f $y = \left(\frac{5}{2}\right)^x$

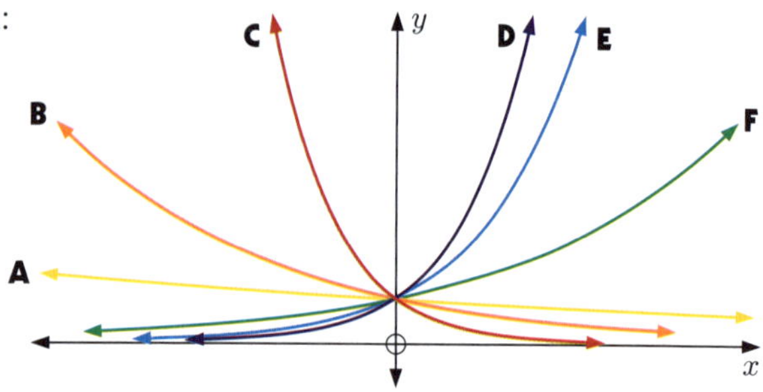

Example 2 **Self Tutor**

Sketch the graph of:

a $y = 2^x + 3$ **b** $y = 2 \times 3^x$

a The graph of $y = 2^x + 3$ is a translation of $y = 2^x$ by 3 units upwards.

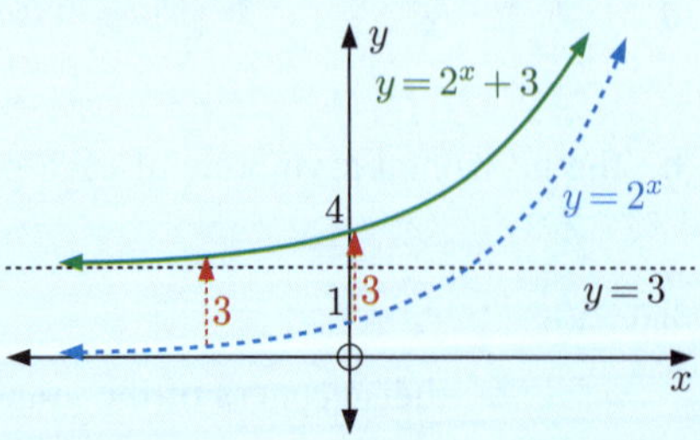

b The graph of $y = 2 \times 3^x$ is a vertical stretch of $y = 3^x$ with scale factor 2.

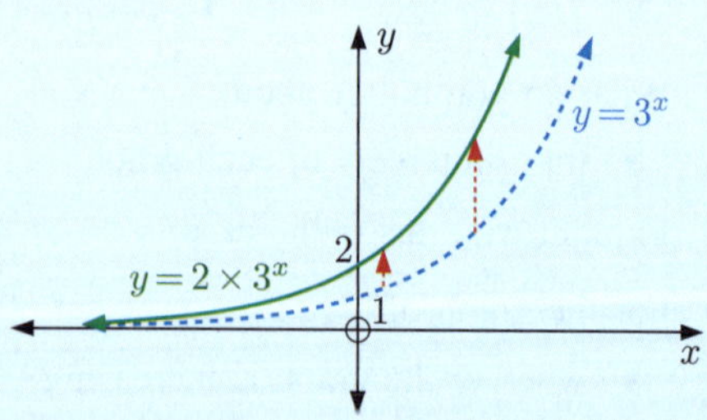

4 **a** State the transformation from $y = f(x)$ to $y = f(x) + k$.

b Sketch the graph of:

i $y = 2^x - 1$ **ii** $y = 3^x + 2$ **iii** $y = \left(\frac{1}{2}\right)^x - 3$

5 **a** State the transformation from $y = f(x)$ to $y = f(x - h)$.

b Sketch the graph of:

i $y = 2^{x-3}$ **ii** $y = 5^{x+2}$ **iii** $y = \left(\frac{1}{3}\right)^{x-1}$

6 **a** State the transformation from $y = f(x)$ to $y = p \times f(x)$, $p > 0$.

b Sketch the graph of:

i $y = 3 \times 2^x$ **ii** $y = \frac{1}{2} \times 5^x$ **iii** $y = 4 \times \left(\frac{1}{2}\right)^x$

7 **a** State the transformation from $y = f(x)$ to $y = f(qx)$, $q > 0$.

b Sketch the graph of:

i $y = 5^{\frac{x}{2}}$ **ii** $y = 2^{3x}$ **iii** $y = \left(\frac{2}{3}\right)^{2x}$

8 **a** State the transformation from $y = f(x)$ to $y = -f(x)$.

b Sketch the graph of:

i $y = -2^x$ **ii** $y = -5^x$ **iii** $y = -\left(\frac{1}{3}\right)^x$

9 **a** State the transformation from $y = f(x)$ to $y = f(-x)$.

b Sketch the graph of:

i $y = 4^{-x}$ **ii** $y = 2^{-x}$ **iii** $y = \left(\frac{1}{3}\right)^{-x}$

10 The following graphs show functions of the form $y = p \times a^x$.
Find the values of a and p for each graph.

a

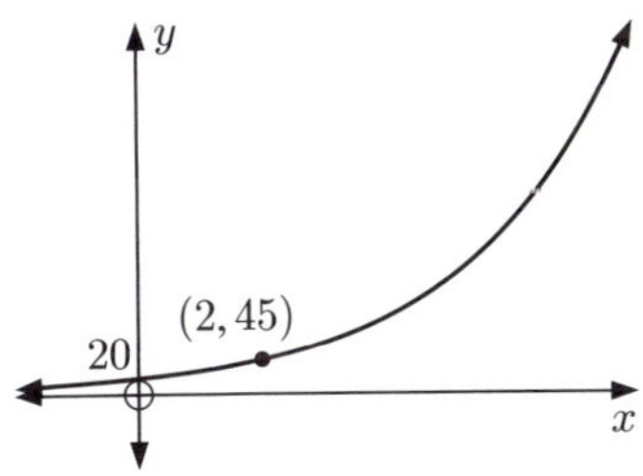

b

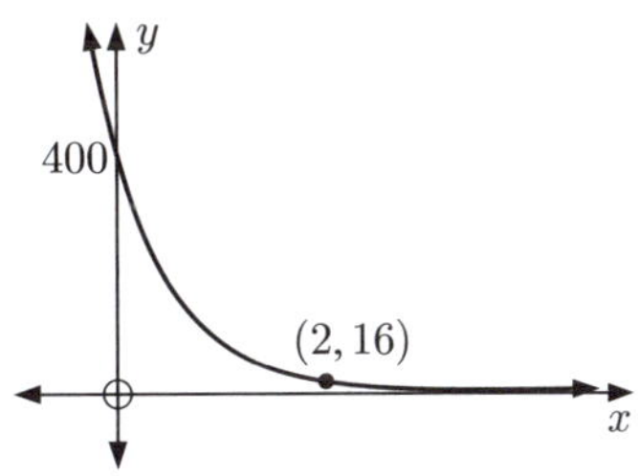

11 Use the exponent laws to explain why $y = 3^{-x}$ and $y = \left(\frac{1}{3}\right)^x$ have the same graph.

12 Write down the function and sketch its graph:

a a translation of $y = 3^x$ by 2 units downwards

b a reflection of $y = 3^x$ in the x-axis

c a vertical stretch of $y = 3^x$ with scale factor $\frac{1}{2}$

d a translation of $y = 3^x$ by 3 units to the left

e a horizontal stretch of $y = 3^x$ with scale factor 5

f a vertical stretch of $y = 3^x$ with scale factor 4

g a horizontal stretch of $y = 3^x$ with scale factor $\frac{1}{3}$.

C EXPONENTIAL EQUATIONS

An **exponential equation** is an equation in which the unknown occurs as part of the exponent or index.

For example: $2^x = 8$ and $30 \times 3^x = 7$ are both exponential equations.

To solve exponential equations, we try to write both sides of the equation with the **same base**. We can then **equate exponents**.

$$\textbf{If } a^x = a^k, \textbf{ then } x = k.$$

Example 3 — Self Tutor

Solve for x:

a $2^x = 32$ **b** $3^{x-2} = \frac{1}{9}$

a
$$2^x = 32$$
$$\therefore 2^x = 2^5$$
$$\therefore x = 5 \quad \{\text{equating exponents}\}$$

b
$$3^{x-2} = \frac{1}{9}$$
$$\therefore 3^{x-2} = 3^{-2}$$
$$\therefore x - 2 = -2 \quad \{\text{equating exponents}\}$$
$$\therefore x = 0$$

Example 4 — Self Tutor

Solve for x:

a $6 \times 3^x = 54$ **b** $4^{x-1} = \left(\frac{1}{2}\right)^{1-3x}$

a
$$6 \times 3^x = 54$$
$$\therefore 3^x = 9$$
$$\therefore 3^x = 3^2$$
$$\therefore x = 2 \quad \{\text{equating exponents}\}$$

b
$$4^{x-1} = \left(\tfrac{1}{2}\right)^{1-3x}$$
$$\therefore (2^2)^{x-1} = (2^{-1})^{1-3x}$$
$$\therefore 2^{2x-2} = 2^{3x-1}$$
$$\therefore 2x - 2 = 3x - 1 \quad \{\text{equating exponents}\}$$
$$\therefore -2 + 1 = 3x - 2x$$
$$\therefore x = -1$$

If we are unable to write both sides of the equation with the same base, we can solve the equation using technology. Click on the icon for instructions for your calculator.

GRAPHICS CALCULATOR INSTRUCTIONS

EXERCISE 25C

1 Solve for x:

a $3^x = 3$ **b** $3^x = 9$ **c** $2^x = 8$

d $5^x = 1$ **e** $3^x = \frac{1}{3}$ **f** $5^x = \frac{1}{5}$

g $2^x = \frac{1}{16}$ **h** $7^x = 343$ **i** $4^x = \frac{1}{256}$

2 Solve for x:

a $5^{x+2}=25$
b $2^{x+2}=\frac{1}{4}$
c $3^{x-1}=\frac{1}{27}$
d $2^{x-1}=32$
e $3^{1-2x}=\frac{1}{27}$
f $4^{2x+1}=\frac{1}{2}$
g $9^{x-3}=3$
h $\left(\frac{1}{2}\right)^{x-1}=2$
i $\left(\frac{1}{3}\right)^{2-x}=9$

3 Solve for x:

a $5\times 2^{x}=40$
b $6\times 2^{x+2}=24$
c $3\times\left(\frac{1}{2}\right)^{x}=12$
d $4\times 5^{x}=500$
e $54\times 3^{x+2}=2$
f $7\times\left(\frac{1}{3}\right)^{x}=63$

4 Solve for x:

a $2^{2-5x}=4^{x}$
b $5^{x-1}=\left(\frac{1}{25}\right)^{x}$
c $9^{x-2}=\left(\frac{1}{3}\right)^{3x-1}$

5 Solve for x:

a $\dfrac{3^{2x+1}}{3^{x}}=9^{x}$
b $\dfrac{25^{x}}{5^{x+4}}=25^{1-x}$
c $\dfrac{4^{x}}{2^{x+2}}=\dfrac{2^{x+1}}{8^{x}}$

6 Use technology to solve for x, giving your answer correct to 3 decimal places:

a $2^{x}=100$
b $2^{x}=0.271$
c $2^{x}=-3$
d $5^{x}=8$
e $7^{-x}=23$
f $9^{x}=10\,000$
g $3^{x}=0.006\,51$
h $5\times 2^{-x}=18$
i $200\times 2^{x}=5800$
j $300\times 2^{-3x}=4.1$
k $25\times 3^{-2x}=0.035$
l $4\times 2^{-0.02x}=0.07$
m $3^{x+1}=4$
n $2^{3x-2}=5$
o $3(2^{x-2})=1$

D EXPONENTIAL GROWTH

Exponential functions can be used to model many real-life situations in science and commerce.

If a quantity increases exponentially over time, we say it exhibits **exponential growth**.

An exponential function of the form $f(x)=p\times a^{x}$ exhibits **exponential growth** if $a>1$.

An example of exponential growth is when a population is able to grow under favourable conditions with abundant resources. Eventually a population will be limited by food, water, space, or predators, at which time an exponential model will no longer be valid.

Example 5 Self Tutor

The population of rabbits on a farm is given by the function $R=50\times 1.07^{n}$ where n is the number of weeks after the rabbit farm was established.

a What was the original rabbit population?
b How many rabbits were present after 15 weeks?
c Sketch the graph of R against n for $n\geqslant 0$.
d How long will it take for the population to reach 500?

a When $n = 0$, $R = 50 \times 1.07^0$
$= 50 \times 1$
$= 50$

$\therefore$ there were 50 rabbits originally.

b When $n = 15$, $R = 50 \times 1.07^{15}$
≈ 137.95

$\therefore$ there were 138 rabbits after 15 weeks.

c

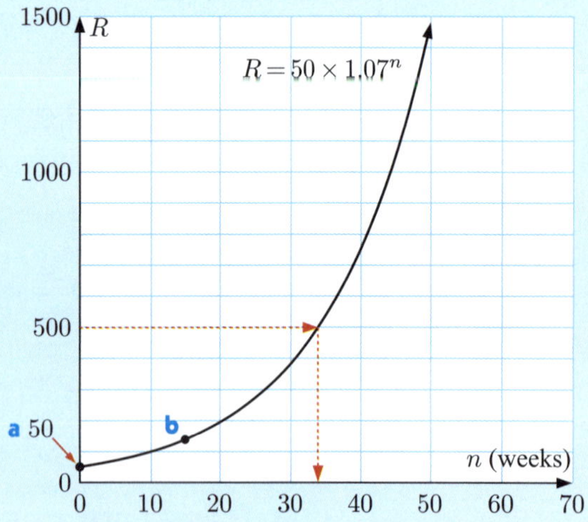

d We use technology to solve $50 \times 1.07^n = 500$.

We find $n \approx 34.03$

$\therefore$ it will take about 34 weeks.

DISCUSSION

By what percentage does the rabbit population grow each week?

How is the rate of exponential growth connected to the base number a in the exponential function?

EXERCISE 25D

1 A local zoo starts a breeding program to ensure the survival of a species of mongoose. Using results from a previous program, the expected population in n years' time is given by $P = 40 \times 1.15^n$.

- **a** What is the initial population purchased by the zoo?
- **b** Find the expected population after:
  - **i** 3 years
  - **ii** 10 years
  - **iii** 30 years.
- **c** Sketch the graph of P against n for $n \geqslant 0$.
- **d** How long will it take for the population to reach 100?

2 In Tasmania, Australia, a reserve is set aside for the breeding of echidnas. The expected population after t years is given by $P = 50 \times 1.26^t$.

- **a** What is the initial breeding colony size?
- **b** Find the expected colony size after:
  - **i** 3 years
  - **ii** 9 years
  - **iii** 20 years.
- **c** Sketch the graph of P against t for $t \geqslant 0$.
- **d** How long will it take for the population to reach 150?

3 In Uganda, the number of breeding females in an endangered population of gorillas is G_0. Biologists predict that the number of breeding females G in n years' time will grow according to $G = G_0 \times 1.01^n$.

  a There are currently 28 breeding females in the colony. Find G_0.

  b Predict the number of breeding females after:

    i 5 years **ii** 10 years **iii** 20 years.

  c Sketch the graph of G against n for $n \geqslant 0$.

  d Estimate how long it will take for the number of breeding females to reach 200.

4 Answer the **Opening Problem** on page **446**.

5 10 g of bacteria is placed in a petri dish. The mass of the bacteria increases by 12% each day.

  a Write an exponential function for the mass M grams of the bacteria after t days.

  b Find the mass of the bacteria after:

    i 2 days **ii** 1 week.

  c Sketch the graph of M against t for $t \geqslant 0$.

  d How long will it take for the mass of the bacteria to reach 30 g?

E EXPONENTIAL DECAY

If a quantity decreases exponentially over time, we say it exhibits **exponential decay**.

An exponential function of the form $f(x) = p \times a^x$ exhibits **exponential decay** if $0 < a < 1$.

Examples of exponential decay include:

- the depreciation in the value of goods over time
- the temperature of an object as it cools
- radioactive decay
- the drop in current when an electrical appliance is turned off.

Example 6 **Self Tutor**

The current I flowing through the electric circuit in a fan, t milliseconds after it is switched off, is given by $I = 320 \times 0.7^t$ milliamps.

  a Find the initial current in the circuit.

  b Find the current after: **i** 4 milliseconds **ii** 10 milliseconds.

  c Sketch the graph of I against t for $t \geqslant 0$.

  d How long will it take for the current to reduce to 40 milliamps?

a When $t = 0$, $I = 320 \times 0.7^0$
$= 320$ milliamps

b **i** When $t = 4$, $I = 320 \times 0.7^4$
≈ 76.8 milliamps

ii When $t = 10$, $I = 320 \times 0.7^{10}$
≈ 9.04 milliamps

c

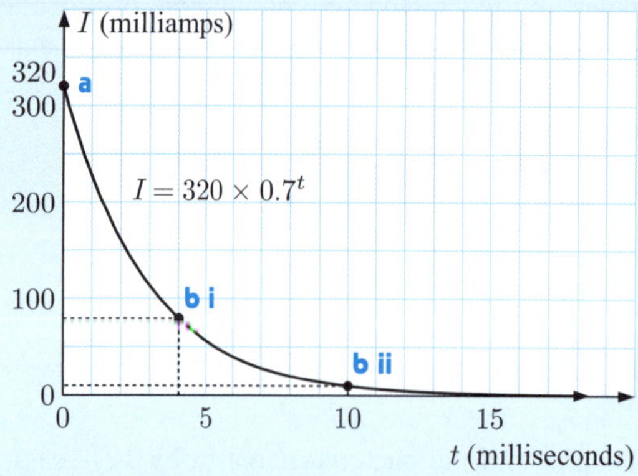

d We use technology to solve
$$320 \times 0.7^t = 40$$
We find $t \approx 5.83$

$\therefore$ it will take about 5.83 milliseconds.

EXERCISE 25E

1 Boiling water is left in a pot to cool. After t minutes, its temperature is given by $T = 100 \times 0.84^t$ °C.

- **a** Find the initial temperature of the water.
- **b** Find the water temperature after:
  - **i** 2 minutes
  - **ii** 10 minutes
  - **iii** 20 minutes.
- **c** Sketch the graph of T against t for $t \geqslant 0$.
- **d** How long will it take for the temperature to fall to 48°C?

2 The weight of radioactive material in an ore sample after t years is given by $W = 2.3 \times 0.96^t$ grams.

- **a** Find the initial weight.
- **b** Find the weight after:
  - **i** 20 years
  - **ii** 40 years
  - **iii** 60 years.
- **c** Sketch the graph of W against t for $t \geqslant 0$.
- **d** Find the percentage weight loss in the first 20 years.

3 After m months, the value of a washing machine is given by $V = 500 \times 0.981^m$ dollars.

- **a** What was the washing machine's original value?
- **b** How much is the washing machine worth after:
  - **i** 6 months
  - **ii** 4 years?
- **c** How long will it take for the washing machine's value to reduce to $150?

4 There are currently 500 cuscus in a section of jungle. As the area is reduced in size for timber, the population is expected to fall by 8% each year.

- **a** Write an exponential function for the expected cuscus population P after t years.
- **b** Find the expected cuscus population after:
  - **i** 2 years
  - **ii** 8 years.
- **c** Sketch the graph of P against t for $t \geqslant 0$.

QUICK QUIZ

MULTIPLE CHOICE QUIZ

REVIEW SET 25A

1 For the function $f(x) = 3^x - 1$, find:

a $f(0)$ **b** $f(3)$ **c** $f(-1)$ **d** $f(2x)$

2 Sketch each graph and label the y-intercept and horizontal asymptote:

a $y = 2^x + 2$ **b** $y = 3^{-x}$ **c** $y = 4^{\frac{x}{2}}$

3 The following graphs show functions of the form $y = p \times a^x$.
Find the values of a and p for each graph:

a

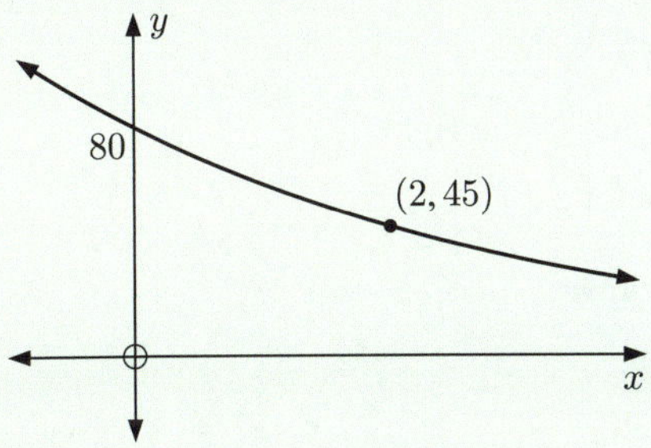

b

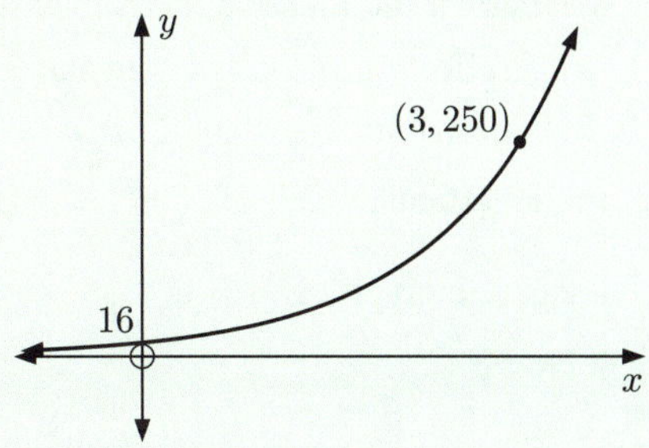

4 Solve for x:

a $3^x = \frac{1}{81}$ **b** $2^{x-2} = \frac{1}{8}$ **c** $25^{4-x} = 1$

d $5 \times \left(\frac{1}{2}\right)^x = 80$ **e** $8^x = 4^{2-x}$ **f** $\dfrac{3^{x+2}}{9^{3-x}} = \dfrac{27^{1-2x}}{3^{2x}}$

5 Suppose $f(x) = 2^x$ and $g(x) = 2^{x-2}$.

a State the transformation from $y = f(x)$ to $y = g(x)$.

b Find the y-intercept and horizontal asymptote of $g(x)$.

c State the domain and range of $g(x)$.

d Sketch $y = f(x)$ and $y = g(x)$ on the same set of axes.

e $f(x)$ can also be transformed to $g(x)$ using a vertical stretch. Find the scale factor for this stretch.

6 Danielle writes an online blog. The number of people following her blog after t weeks is given by $N(t) = 10 \times 1.3^t$.

a How many people are following Danielle's blog:

i initially **ii** after 4 weeks **iii** after 8 weeks?

b Sketch the graph of N against t for $t \geqslant 0$.

c Use technology to find how long it will take for Danielle's blog to have 200 followers.

7 At the zoo, Terry the giant tortoise weighs 130 kg. It is decided to put him on a diet. Terry's weight decreases by 2% each week during the diet.

a Write an exponential function for Terry's weight W kg after t weeks.

b Find Terry's weight after:

i 2 weeks **ii** 5 weeks.

c The keepers would like Terry to lose 20 kg. Estimate how long it will take to achieve this.

REVIEW SET 25B

1 For the function $P(x) = 2 \times 3^{-x}$, find:

a $P(0)$ **b** $P(1)$ **c** $P(-2)$ **d** $P(x+4)$

2 Write down the function and sketch its graph:

a a reflection of $y = 4^x$ in the x-axis

b a horizontal translation of $y = 5^x$ by 3 units to the right

c a vertical stretch of $y = 3^x$ with scale factor $\frac{1}{3}$.

3 Sketch the graph of $y = 2^x - 3$ and state its range.

4 Solve for x:

a $5^{x+1} = \frac{1}{25}$ **b** $2^{1-x} = 128$ **c** $8^x = 4^{5-x}$

d $9^x = 27^{2-2x}$ **e** $16^{x+1} = 32^{2-x}$

5 Use technology to solve for x, giving your answer correct to 2 decimal places:

a $3^x = 20$ **b** $7^{2x} = -1$ **c** $12 \times 5^{-0.01x} = 10$

6 A conservationist is concerned that a local population of sea birds is declining. He finds that the population after t years is given by $P(t) = 500 \times 0.95^t$.

a Find:

i the initial population **ii** the population after 10 years.

b Sketch the graph of P against t for $t \geqslant 0$.

7 The value of a rare banknote has been modelled by $V = 50 \times 1.11^t$ dollars, where t is the time in years since it was printed.

a Find the initial value of the banknote.

b Find the value of the banknote after:

i 10 years **ii** 20 years **iii** 50 years.

c Use your answers from **a** and **b** to sketch the graph of V against t for $t \geqslant 0$.

d Use technology to find how long it will take for the banknote to be worth \$1000.

Chapter 26

Differential calculus

Contents:

CHAPTER

Click on the icon to access this chapter

Chapter 27

Integration

Contents:

CHAPTER

Click on the icon to access this chapter

ANSWERS

EXERCISE 1A

1 **a** 3^7 **b** x^9 **c** x^{5+n} **d** t^{12} **e** 7^4 **f** x^4 **g** t^{6-x} **h** t^{3m-1} **i** 5^6 **j** t^{12} **k** y^{3m} **l** a^{12m}

2 **a** x^{4t+5} **b** k^{2+m-n} **c** c^{5-4x} **d** m^{-1-t}

3 **a** 11^2 **b** 2^5 **c** 3^4 **d** 2^6 **e** 5^4 **f** 3^8

4 **a** 2^{x+1} **b** 7^{t+2} **c** 3^{a-2} **d** 2^{3p-2} **e** 7^2 **f** 3^{2-x} **g** 5^{4t} **h** 2^{3k-12} **i** 2^{2a-b} **j** 2^{3x-4y} **k** 5^{2x} **l** 3^6

5 **a** x^2y^2 **b** a^3b^3 **c** $x^4y^4z^4$ **d** $27b^3$ **e** $625a^4$ **f** $100\,000x^5y^5$ **g** $49p^4q^2$ **h** $81a^{20}b^8$

6 **a** $\frac{p^2}{q^2}$ **b** $\frac{m^3}{n^3}$ **c** $\frac{x^4}{81}$ **d** $\frac{16a^4}{b^4}$ **e** $\frac{p^2}{25q^2}$ **f** $\frac{27x^3}{64y^3}$ **g** $\frac{m^6}{25n^2}$ **h** $\frac{64x^3}{y^9}$

7 **a** $8b^5$ **b** a^2b^2 **c** $6a^4b^2$ **d** $2a^2b^5$ **e** $9p^3q^5$ **f** $\frac{x^2}{3}$ **g** $\frac{a^6}{125b^3}$ **h** $\frac{8r^2}{5}$ **i** $\frac{xy}{3}$ **j** $\frac{5k^2}{16}$ **k** $\frac{1}{4d^3}$ **l** $\frac{125q}{p}$

8 **a** 1 **b** 1 **c** 1 **d** 3 **e** 2 $\{t \neq 0\}$ **f** 1 $\{t \neq 0\}$ **g** 3 **h** $\frac{1}{125}$

9 **a** $\frac{1}{6}$ **b** $\frac{1}{4}$ **c** $\frac{1}{13}$ **d** $\frac{1}{16}$ **e** $\frac{1}{125}$ **f** $\frac{1}{49}$ **g** $\frac{1}{1000}$ **h** $\frac{1}{81}$ **i** $\frac{4}{5}$ **j** $\frac{11}{3}$ **k** $\frac{3}{16}$ **l** $\frac{11}{144}$

10 **a** 1 **b** $\frac{1}{10}$ **c** $\frac{1}{100}$ **d** $\frac{1}{1000}$ **e** $\frac{1}{10\,000}$ **f** $\frac{1}{100\,000}$

11 **a** 5^{-2} **b** 2^{-4} **c** x^{5} **d** y^{-7}

12 **a** $\frac{8}{3}$ **b** $\frac{3}{2}$ **c** 5 **d** 9 **e** $\frac{27}{8}$ **f** $\frac{25}{16}$ **g** $\frac{8}{27}$ **h** $\frac{4}{49}$

13 **a** $\frac{1}{5x}$ **b** $\frac{5}{x}$ **c** $\frac{7}{a}$ **d** $\frac{1}{7a}$ **e** t^2 **f** $\frac{y}{3x}$ **g** $\frac{1}{25t^2}$ **h** $\frac{t^2}{5}$ **i** $\frac{x}{y}$ **j** $\frac{1}{xy}$ **k** $\frac{x}{y^3}$ **l** $\frac{1}{x^3y^3}$ **m** $\frac{1}{3pq}$ **n** $\frac{3}{pq}$ **o** $\frac{3p}{q}$ **p** x^3y^5 **q** $\frac{125y^3}{x^6}$ **r** $\frac{4d^6}{c^2}$ **s** $\frac{r^3t}{3}$ **t** $\frac{25}{4p^2q^4}$

14 **a**
$$\frac{1}{a^{-n}} = \frac{a^0}{a^{-n}} = a^{0-(-n)} = a^{0+n} = a^n$$
b
$$\left(\frac{a}{b}\right)^{-n} = \left[\left(\frac{a}{b}\right)^{-1}\right]^n = \left(\frac{b}{a}\right)^n = \frac{b^n}{a^n}$$

15 **a** $\text{m}\,\text{s}^{-1}$ **b** $\text{m}^3\,\text{h}^{-1}$ **c** $\text{cm}^2\,\text{s}^{-1}$ **d** $\text{cm}^3\,\text{min}^{-1}$ **e** $\text{g}\,\text{s}^{-1}$ **f** $\text{kg}\,\text{m}\,\text{s}^{-1}$ **g** $\text{m}\,\text{s}^{-2}$

16 $2^{125} = (2^5)^{25} = 32^{25}$ and $3^{75} = (3^3)^{25} = 27^{25}$

As $27^{25} < 32^{25}$, 3^{75} is smaller.

17 10^{24}, 5^{36}, 2^{90}, 3^{60}

EXERCISE 1B

1 **a** 2.3×10^2 **b** 5.39×10^4 **c** 3.61×10^{-2} **d** 6.8×10^{-3} **e** 3.26×10^0 **f** 5.821×10^{-1} **g** 3.61×10^8 **h** 1.674×10^{-6}

2 **a** 4×10^6 red blood cells **b** 8×10^{-4} m **c** 4.3252×10^{19} arrangements **d** 3.422×10^{-9}

3 **a** 2300 **b** 0.023 **c** 564 000 **d** 0.000 793 1 **e** 9.97 **f** 60 400 000 **g** 0.4215 **h** 0.000 000 036 21

4 **a** 6 990 000 m **b** 0.018 cm **c** 32 000 000 bacteria **d** 0.0095 g

5 **a** 6×10^{10} **b** 2.8×10^9 **c** 5.6×10^{-8} **d** 5.4×10^{-6} **e** 9×10^{10} **f** 1.6×10^{15} **g** 1.6×10^{-11} **h** 1.25×10^{-7} **i** 1.2×10^0 **j** 2.88×10^7 **k** 2.5×10^{-4} **l** 4×10^6

6 **a** 2×10^3 **b** 3×10^{-2} **c** 2×10^8 **d** 6×10^{-6} **e** 1×10^{-19} **f** 3.2×10^3

7 **a** 10^6 times larger **b** 10^5 times smaller

8 **a** 9×10^8 **b** 5×10^4 **c** 2.7×10^5 **d** 3.77×10^{-4} **e** 5.67×10^4 **f** 2.37×10^8 **g** 3.625×10^{-7} **h** 2.3×10^{-6}

9 **a** $\approx 4.01 \times 10^{13}$ **b** $\approx 2.59 \times 10^{12}$ **c** $\approx 7.08 \times 10^{-9}$ **d** $\approx 4.87 \times 10^{-11}$ **e** $\approx 8.01 \times 10^6$ **f** $\approx 3.55 \times 10^{-9}$ **g** $\approx 1.57 \times 10^{13}$ **h** $\approx 4.23 \times 10^6$ **i** $\approx 6.55 \times 10^{-22}$

10 **a** 2.88×10^8 km **b** $\approx 7.01 \times 10^{10}$ km

11 0.9 s **12** ≈ 23.2 times larger

13 **a** microbe C **b** ≈ 32.9 times heavier

14 **a** 9.1×10^6 km^2
b Quebec, British Columbia, Ontario, Alberta, Saskatchewan, Manitoba, Newfoundland and Labrador, New Brunswick, Nova Scotia, Prince Edward Island
c **i** ≈ 2.55 times **ii** ≈ 333 times **d** $\approx 0.58\%$

15 **a** 1 parsec $\approx$ 31 000 000 000 000 000 m
1 light-year $\approx$ 9 500 000 000 000 000 m
1 astronomical unit $\approx$ 150 000 000 000 m
b Standard form allows us to write these large distances in a way that is shorter and easier to understand.
c $\approx 2.15 \times 10^{16}$ m farther
d $\approx 2.07 \times 10^5$ astronomical units

REVIEW SET 1A

1 **a** k^{5+3m} **b** p^5 **c** m^{48t} **2** 7^{k+1}

3 **a** $9w^2$ **b** $8x^6y^3$ **c** $\frac{a^6}{b^6}$ **d** $\frac{1}{125n^3}$

4 **a** 6 **b** $\frac{1}{7}$ **c** $\frac{1}{16}$ **d** $\frac{99}{100}$

5 **a** $\frac{1}{m^2n^2}$ **b** $\frac{m}{n^2}$ **c** $\frac{125y^6}{x^3}$ **6** $\text{m}^2\,\text{s}^{-1}$

7 **a** 5.9×10^4 **b** 9×10^{-3} **c** 6.085×10^6 **d** 7.71×10^{-6}

8 **a** 1.8×10^{12} **b** 1.5×10^{-8} **c** 4×10^6 **d** 1.6×10^8 **e** 1.25×10^{-8} **f** 1.2×10^{-4}

9 2.6×10^{-3} g

10 **a** **i** $\approx 1.08 \times 10^{12}$ m **ii** $\approx 2.59 \times 10^{13}$ m
iii $\approx 9.47 \times 10^{15}$ m
b $\approx 3.33 \times 10^{-9}$ s **c** $\approx 8.74 \times 10^{5}$ times faster

REVIEW SET 1B

1 **a** x^{a-7} **b** y^{15n} **c** m^{15-pq}

2 **a** x^{7-n} **b** $8t^5$ **c** $\dfrac{ab^4}{5}$

3 **a** $\frac{3}{4}$ **b** $\frac{10}{11}$ **c** $\frac{16}{49}$ **d** $\frac{4}{27}$

4 7^{20} as $2^{60} = (2^3)^{20} = 8^{20}$ and $8^{20} > 7^{20}$

5 z^{-7} **6** **a** $\dfrac{b^2}{a^2}$ **b** $\dfrac{8a^3b^6}{27}$

7 **a** 623 000 **b** 0.000 300 8 **c** 4.597 **d** 0.000 050 6

8 **a** 2.1×10^{15} **b** 4×10^{10} **c** 2.9×10^{11}

9 $\approx 2.49 \times 10^{13}$ red blood cells

10 **a** **i** 1.27×10^4 km **ii** 6.05×10^6 m
b Mercury, Mars, Venus, Earth, Neptune, Uranus, Saturn, Jupiter
c 7.15×10^7 m larger
d **i** ≈ 10.5 times larger **ii** ≈ 20.6 times larger

EXERCISE 2A

1 **a** $2x+12$ **b** $7x-21$ **c** $4x-4y$ **d** $6a+3b$
e $6x-18y$ **f** x^2+4x **g** $6x+15$ **h** $3x^2-2x$
i $4x^2-12x$ **j** $28n+14n^2$ **k** $2x^3-2x$ **l** x^3-3x^2

2 **a** $-2a-8$ **b** $-15+5x$ **c** $-y-2$
d $-3x+3y$ **e** $-6-2x$ **f** $-x^2-x$
g $-4+b^2$ **h** $-3x^2-3xy$ **i** $-x-x^2$
j $-6x^2-21x$ **k** $-10n+5n^2$ **l** a^2b-ab^2

3 **a** $6a-10b+2$ **b** $3x^2-6xy+9x$
c $-2a+6b-10c$ **d** $6x^3+30x^2-60x$
e $-6c^2+9c+21$ **f** $4n^3-8n^2-32n$

4 **a** $3x+5$ **b** $-2x+5$ **c** $4x+3$ **d** $-2x-12$
e a^2 **f** $12m-2m^2$ **g** $3ab+a^2$ **h** $-2b^2+9b$
i $3a^2-12a$ **j** $a+3b$ **k** $-a+5b$ **l** $17n-15$

5 **a** $5x+6$ **b** $-2x+26$ **c** -15
d $x-y$ **e** $a^2+7a-15$ **f** $x^2+2x+12$
g x^2-y^2 **h** $-2x+16$ **i** $9x-9$
j $5x^2-16x$ **k** 0 **l** $9x^2-x^3-12x$

EXERCISE 2B

1 **a** $x^2+7x+10$ **b** x^2-x-12 **c** $x^2+2x-15$
d x^2+2x-8 **e** x^2-2x-3 **f** $x^2-12x+20$
g $3x-2-x^2$ **h** $2x^2-5x-3$ **i** $6x^2-23x+20$

2 **a** $2x^2-xy-y^2$ **b** $-2x^2-7x-3$
c $-x^2-x-2xy-2y$ **d** $9x+18-2x^2$
e $20-4x^2-5x+x^3$ **f** $-x+1-6x^2$
g $-8-14x-5x^2$ **h** $3x^2-2x^3+3-2x$
i $6-x-12x^2$

3 **a** a^2-b^2 **b** $a^2+2ab+b^2$ **c** $a^2-2ab+b^2$

4 **a** x^3+2x^2-3x **b** $-x^3+3x^2+10x$ **c** $12x-3x^3$

5 **a** $x^2+5x-18$ **b** $2x^2+7x-31$ **c** $x^2-8x-12$
d $-7t+2$ **e** $2x^2+3$ **f** $10x^2+53x-96$

EXERCISE 2C

1 **a** x^2-1 **b** b^2-4 **c** t^2-36 **d** x^2-49
e $9-a^2$ **f** $25-x^2$ **g** x^2-y^2 **h** x^2-16
i x^4-1 **j** $4x^2-1$ **k** $16-9y^2$ **l** $9x^2-16z^2$

2 **a** x^2-3 **b** x^2-2 **c** $7-x^2$

3 shaded area $= (a+b)(a-b)$ units^2 and
shaded area $= a^2-ab+ab-b^2$ units^2
$= a^2-b^2$
$\therefore\ (a+b)(a-b) = a^2-b^2$

4 **a** 27 **b** $22p^2+p-4$ **c** $5y^2+8z^2$ **d** $8x^4$

EXERCISE 2D

1 **a** $x^2+10x+25$ **b** $y^2+12y+36$
c $4+4x+x^2$ **d** $9x^2+6x+1$
e $16+24x+9x^2$ **f** $x^2+4xy+4y^2$

2 **a** $x^2-8x+16$ **b** $4-4x+x^2$
c $a^2-14a+49$ **d** $9x^2-6x+1$
e $36-60p+25p^2$ **f** $4x^2-12xy+9y^2$

3 **a** $4x^2+12x+9$ **b** $16a^2-8a+1$
c $9y^2+30y+25$ **d** $9a^2-24a+16$
e $4x^2-28x+49$ **f** $64+48a+9a^2$
g $4+20b+25b^2$ **h** $36+60x+25x^2$
i $16-40y+25y^2$ **j** $25x^2+30xy+9y^2$
k $4x^2-36xy+81y^2$ **l** $1-2xy+x^2y^2$

4

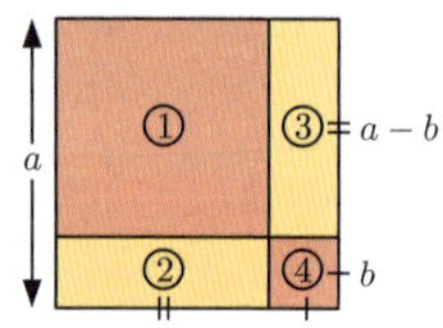

Hint:

Area ①
= total area − 2 × area ②
− area ④

5 **a** x^4+6x^2+9 **b** y^4-4y^2+4
c $4a^4+12a^2+9$ **d** $1-8x^2+16x^4$
e $a^4+2a^2b^2+b^4$ **f** $a^4-2a^2b^2+b^4$
g $25x^4+30x^2y+9y^2$ **h** $25-10x^2+x^4$
i $16x^2-24x^3+9x^4$ **j** $4x^4+16x^3+16x^2$
k $49a^2-42a^3+9a^4$ **l** $x^6+16x^4+64x^2$

6 **a** $x^2+13x+23$ **b** $2x^2+14x+85$ **c** $-2x-17$
d $10x^2-5x-5$ **e** $5x^2+18x-8$ **f** $19x+74$
g $24x^2-4x-8$ **h** $9x^4-10x^2+8x-3$

EXERCISE 2E

1 **a** x^3+3x^2+6x+8 **b** x^3-x^2-9x+9
c x^3+5x^2+7x+3 **d** $2x^3+x^2-6x-5$
e $2x^3+7x^2+8x+3$ **f** $2x^3-9x^2+4x+15$
g $3x^3+14x^2-x+20$ **h** $8x^3-14x^2+7x-1$
i $x^2-6x^3+14x-3$ **j** $-18x^2+18x-4-5x^3$

2 **a** $x^3+6x^2+11x+6$ **b** $x^3+9x^2+26x+24$
c $x^3-x^2-14x+24$ **d** $x^3-10x^2+31x-30$
e $2x^3+x^2-12x+9$ **f** $-3x^3+16x^2-12x-16$
g $2x^3+11x^2+12x-9$ **h** $78x^2-45x^3-44x+8$
i $x^3+6x^2+12x+8$ **j** $x^3-9x^2+27x-27$

3 **a** $2\times 2 = 4$ terms **b** $3\times 2 = 6$ terms
c $2\times 3 = 6$ terms **d** $3\times 3 = 9$ terms
e $2\times 2\times 2 = 8$ terms **f** $3\times 2\times 2 = 12$ terms

EXERCISE 2F

1 **a** x^3+3x^2+3x+1 **b** $x^3+9x^2+27x+27$
c $x^3+15x^2+75x+125$ **d** $x^3+3x^2y+3xy^2+y^3$
e x^3-3x^2+3x-1 **f** $x^3-15x^2+75x-125$
g $x^3-12x^2+48x-64$ **h** $x^3-3x^2y+3xy^2-y^3$
i $8+12y+6y^2+y^3$ **j** $8x^3+12x^2+6x+1$
k $27x^3+27x^2+9x+1$
l $8y^3+36xy^2+54x^2y+27x^3$ **m** $8-12y+6y^2-y^3$
n $8x^3-12x^2+6x-1$ **o** $27x^3-27x^2+9x-1$

3 **a** $x^4+4x^3y+6x^2y^2+4xy^3+y^4$
b $x^4+4x^3+6x^2+4x+1$
c $x^4+8x^3+24x^2+32x+16$
d $x^4+12x^3+54x^2+108x+81$
e $x^4-4x^3y+6x^2y^2-4xy^3+y^4$
f $x^4-4x^3+6x^2-4x+1$
g $x^4-8x^3+24x^2-32x+16$
h $256-256x+96x^2-16x^3+x^4$
i $16x^4-32x^3+24x^2-8x+1$

REVIEW SET 2A

1 **a** $20x-25$ **b** $-4x^2+12x$ **c** $-5x+12+3x^2$
2 **a** x^2+7x+6 **b** x^2-x-30 **c** $5-23x+12x^2$
d $2x^2-2x-1$ **e** $-x^2+18x$
3 **a** t^2-49 **b** $4y^2-25$ **c** $x-36$ **d** $3y^2+8z^2$
4 **a** $k^2+14k+49$ **b** $4m^2-20mn+25n^2$
c $9x^2-12x^3+4x^4$ **d** $17y^2+17$ **e** $11x^2-33x+11$
5 **a** 3 **b** k^4-9k^2+20 **c** x^4-9-x^2-5x
6 **a** $x^3+4x^2-7x-10$ **b** $2x^3+5x^2-8x-6$
c $2x^4+7x^3-5x^2-22x+8$
7 **a** **i** $2\frac{1}{4}$ **ii** $6\frac{1}{4}$ **iii** $12\frac{1}{4}$ **iv** $20\frac{1}{4}$
b In **a**, $2\frac{1}{4}$ is $1\times2+\frac{1}{4}=(1\frac{1}{2})^2$
$6\frac{1}{4}$ is $2\times3+\frac{1}{4}=(2\frac{1}{2})^2$ and so on.
So, we expect $(n+\frac{1}{2})^2=n(n+1)+\frac{1}{4}$.
d **i** $110\frac{1}{4}$ **ii** $380\frac{1}{4}$
8 **a** $216-108x+18x^2-x^3$ **b** $8k^3+36k^2+54k+27$
c $r^3-12r^2t+48rt^2-64t^3$

REVIEW SET 2B

1 **a** $18-12x$ **b** $15a-10b+35$ **c** $x^3+2x-20$
2 **a** $6x^2+13x-5$ **b** $x^2-4x^3+3-12x$ **c** x^2-x+3
3 **a** $4x^2-9$ **b** x^4-16 **c** $10-m$ **d** $3b^2$
4 **a** $36-12n+n^2$ **b** $25x^2-30x+9$
c $9x^4-30x^2+25$ **d** $2x^2+10x-64$ **e** $7-x$
5 **a** x^2-4x-3 **b** $-x-25-x^2$ **c** $7x^2-15x+6$
d $15a^2-4ab-5b^2$
6 **a** x^3+x^2-4x+6 **b** $4x^3-12x^2-15x-4$
c $x^4+x^3+5x^2+5x+12$
7 16 terms
8 **a** $x^4-12x^3+54x^2-108x+81$
b $16x^4+32x^3+24x^2+8x+1$
9 **a** **i** $a^2+2ab+b^2$ **ii** $a^3+3a^2b+3ab^2+b^3$
iii $a^4+4a^3b+6a^2b^2+4ab^3+b^4$
iv $a^5+5a^4b+10a^3b^2+10a^2b^3+5ab^4+b^5$
b **i** 8 **ii** 16 **iii** 32 **c** 2^n
d The sum of the coefficients is found by letting $a=b=1$ in $(a+b)^n$, which gives $(1+1)^n=2^n$.

EXERCISE 3A

1 **a** $2(3x-1)$ **b** $2(2+x)$ **c** $3(a+b)$ **d** $2(3-2x^2)$
e $5(m+2n)$ **f** $n(m-4)$ **g** $2(x-2y)$ **h** $3b(2a+1)$
2 **a** $x(x+7)$ **b** $3x(x+3)$ **c** $x(4+x)$
d $4x(2-x)$ **e** $x(2x-3)$ **f** $5x(1-5x)$
g $3x(2x+5)$ **h** $x^2(x+2)$ **i** $3x^2(1-2x)$
j $x^2(8x-1)$ **k** $xy(x-y)$ **l** $2xy(y+4x)$
3 **a** $x(x^2+x+1)$ **b** $x(2x^2+11x+4)$
c $4(x^2-2x+3)$ **d** $9x(x-5-3x^2)$
e $a(b+c+d)$ **f** $x(y+1-x)$
4 **a** $5(2-a)$ **b** $4(3b-5)$ **c** $8(2d-c)$
d $a(3-b)$ **e** $p(q-1)$ **f** $x(9-2x)$
g $4x(2x-1)$ **h** $4x(3-x)$
5 **a** $-6(x+4)$ **b** $-8(1+x)$ **c** $-5(2p+q)$
d $-a(b+1)$ **e** $-7a(1+3a)$ **f** $-8x(2x+1)$
g $-5x(3+2x)$ **h** $-2x(3x+8)$
6 **a** $(a+2)(a+3)$ **b** $(x-1)(4+x)$ **c** $(x+4)(5-x)$
d $(x-9)(x+1)$ **e** $(c+d)(a-b)$ **f** $(x-3)^2$
g $(r+s)(t-1)$ **h** $(y-6)(y+1)$
7 **a** $(x+4)(x+8)$ **b** $(x+1)(x-1)$
c $(x+2)(x-8)$ **d** $(y-1)(x-y-5)$
e $(n+1)(3n-10)$ **f** $-(n-3)(n+2)$
8 **a** $(x+2)(x-3)$ **b** $(x-1)(x-4)$
c $(x+1)(1-x)$ **d** $(x-2)(3x-7)$
e $(x+1)(3x+2)$ **f** $(x+1)^2(x+2)$
9 **a** $(x+3)(x+4)$ **b** $(x-2)(x+1)$ **c** $(x+4)(x+6)$
d $(x-3)(x-4)$ **e** $(x+4)(x+2)$ **f** $(x-4)(x-9)$
g $2(a+b)(2a+2b-1)$ **h** $-(x^2+2)$

EXERCISE 3B

1 **a** $(x+4)(x-4)$ **b** $(2+x)(2-x)$
c $(x+9)(x-9)$ **d** $(5+x)(5-x)$
e no real linear factors **f** $(8+a)(8-a)$
g no real linear factors **h** $(2x+1)(2x-1)$
i $(3x+4)(3x-4)$ **j** $(2x+3)(2x-3)$
k no real linear factors **l** $(6+7x)(6-7x)$
m $(x+2y)(x-2y)$ **n** $(ab+6)(ab-6)$
o $(4x+3y)(4x-3y)$
2 **a** $3(2+x)(2-x)$ **b** $2(x+3)(x-3)$
c $3(x+3)(x-3)$ **d** $-2(x+2)(x-2)$
e $3(x+5)(x-5)$ **f** $-5(x+1)(x-1)$
g $2(2x+3)(2x-3)$ **h** $-3(3x+5)(3x-5)$
i $3(x+2y)(x-2y)$ **j** $2(1+3x)(1-3x)$
k $x(2+3x)(2-3x)$ **l** $xy(x+y)(x-y)$
3 **a**

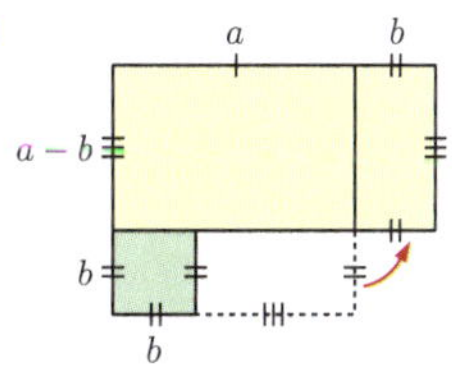

Area of yellow region
$=(a+b)(a-b)$
$\therefore\ a^2-b^2=(a+b)(a-b)$

b no

4 **a** $(x+3)(x-1)$ **b** $4(x+2)(x-1)$ **c** $(x-5)(x+3)$
d $3(x+1)(3-x)$ **e** $(3x+2)(x-2)$
f $(2x+3)(4x-3)$ **g** $(3x-1)(x+3)$
h $8x(x-1)$ **i** $3(2x+1)(2x-3)$

5 $391 = 20^2 - 3^2$
$= (20+3)(20-3)$
$= 23 \times 17$

EXERCISE 3C

1 **a** not a perfect square **b** perfect square
c not a perfect square **d** perfect square
e not a perfect square **f** not a perfect square

2 **a** $(x+3)^2$ **b** $(x+4)^2$ **c** $(x-3)^2$
d $(x-4)^2$ **e** $(x+1)^2$ **f** $(x-5)^2$
g $(y+9)^2$ **h** $(m-10)^2$ **i** $(t+6)^2$

3 **a** $(3x+1)^2$ **b** $(2x-1)^2$ **c** $(3x+2)^2$
d $(5x-1)^2$ **e** $(4x+3)^2$ **f** $(5x-2)^2$
g $(5-3x)^2$ **h** $(2x+7)^2$ **i** $(9-2x)^2$

4 **a** $(x+2y)^2$ **b** $(3x-y)^2$ **c** $(2x+3y)^2$

5 **a** $2(x+1)^2$ **b** $2(x-3)^2$ **c** $4(x-4)^2$
d $-(x-1)^2$ **e** $-2(x+2)^2$ **f** $-3(x+5)^2$
g $3(x+3)^2$ **h** $-(3x-1)^2$ **i** $2(3x+1)^2$

6 **a** $x^2+12x+36 = (x+6)^2 \geqslant 0$ for all real x
$\therefore$ $x^2+12x+36$ is never negative.

b $(x-2)^2 \geqslant 0$
$\therefore$ $x^2-4x+4 \geqslant 0$
$\therefore$ $x^2+4 \geqslant 4x$ for all real x

EXERCISE 3D

1 **a** $(b+2)(a+1)$ **b** $(y+3)(x+1)$ **c** $(a+4)(c+d)$
d $(a+2)(b+3)$ **e** $(m+p)(n+3)$ **f** $(x+3)(x+7)$
g $(x+4)(x+5)$ **h** $(2x+1)(x+3)$ **i** $(3x+2)(x+4)$

2 **a** $(x+5)(x-4)$ **b** $(x+2)(x-7)$ **c** $(x-3)(x-2)$
d $(x-5)(x-3)$ **e** $(x+7)(x-8)$ **f** $(x+2)(x-4)$
g $(2x+1)(x-3)$ **h** $(3x+2)(x-4)$ **i** $(4x-3)(x-2)$
j $(x-1)(9x+2)$ **k** $(3x-4)(x+5)$ **l** $(2y-1)(x+5)$

EXERCISE 3E

1 **a** 1 and 7 **b** 5 and 6 **c** 5 and 12
d -2 and 4 **e** -5 and 4 **f** -8 and -7
g -24 and 2 **h** -10 and 1

2 **a** $(x+2)(x+5)$ **b** $(x+3)(x+11)$ **c** $(x+2)(x+17)$

3 **a** $(x-4)(x-5)$ **b** $(x-4)(x-14)$ **c** $(x-6)(x-7)$

4 **a** $(x-6)(x+1)$ **b** $(x+9)(x-5)$ **c** $(x-9)(x+2)$
d $(x+16)(x-5)$ **e** $(x-5)(x+2)$ **f** $(x+12)(x-6)$

5 **a** $(x+2)(x+3)$ **b** $(x+2)(x-3)$ **c** $(x-2)(x+5)$
d $(x-3)(x+7)$ **e** $(x-4)(x+7)$ **f** $(x-3)(x-8)$
g $(x+4)(x+11)$ **h** $(x+7)(x-8)$ **i** $(x-8)(x+12)$
j $(x-7)(x-5)$ **k** $(x+12)(x+5)$ **l** $(x-13)(x-3)$

6 **a** $2(x+1)(x+4)$ **b** $3(x-1)(x-6)$
c $2(x+3)(x+4)$ **d** $5(x-8)(x+2)$
e $4(x-3)(x+1)$ **f** $3(x-3)(x-11)$
g $2(x-10)(x+9)$ **h** $3(x-4)(x+2)$
i $2(x+4)(x+5)$ **j** $4(x-3)^2$
k $7(x+5)(x-2)$ **l** $x(x-8)(x+1)$

7 **a** $-(x-1)(x-3)$ **b** $(x-2)^2$
c $-(x+3)(x-1)$ **d** $-2(x-10)(x-12)$
e $-3(x-3)(x-7)$ **f** $-5(x-6)(x+2)$

8 **a** $-(x+9)(x-6)$ **b** $2(x+5)(x+2)$
c $-(x+3)(x+7)$ **d** $2(x+8)(x-6)$
e $-3(x-3)^2$ **f** $2(x-7)(x-3)$
g $-2(x-9)(x+7)$ **h** $-2(x-5)^2$
i $-x(x-2)(x+1)$

9 $(x-m)(x-n)$

EXERCISE 3F

1 **a** $7(x-5y)$ **b** $m(m+3p)$ **c** $-5x(x+2)$
d $x(3x+2)$ **e** $2(p^2+4)$ **f** $(x+9)(x-9)$
g $(x-9)(x+1)$ **h** $(d+7)(d-1)$
i $2(x+4)(x-4)$ **j** $(x-3)^2$
k $(x+11)^2$ **l** $3(x+6)(x-6)$
m $5(a-2)(a+1)$ **n** $4t(1+2t)$
o $n^2(n+2)(n-2)$ **p** $2(c-3)(c-1)$
q $x^2(x+1)(x-1)$ **r** $3(b+5)(b-5)$
s $(11+x)(11-x)$ **t** $(5y+4)(5y-4)$
u $x(x+2)^2$ **v** $(x-y)^2$
w $(7y+6z)(7y-6z)$ **x** $(2d+7)^2$
y $a(2b+c)(2b-c)$ **z** $2\pi(R+r)(R-r)$

2 **a** $a(b+c-2)$ **b** $ab(ab-2)$ **c** $2x(3+x)(3-x)$
d $(x+7)^2$ **e** $4a(a+b)(a-b)$
f $xy(x+2)(x-2)$ **g** $4x^2(x+1)(x-1)$
h $(a+3)(a+5)$ **i** $(m-3)^2$

3 **a** $(x-2)(y-z)$ **b** $(x-y)(a+1)$
c $(x+2)(x+3)$ **d** $(x+2)(x-4)$
e $3(x+4)(x-3)$ **f** $5x(x+y-xy)$
g $(y+2)(x+2)$ **h** $(3+2x)(1-2x)$
i $(x+5)(2x+1)$ **j** $(x+1)(x^2+1)$
k $(y+5)(y-9)$ **l** $(4x-1)(2x+5)$

EXERCISE 3G

1 **a** $(x+2)(3x+1)$ **b** $(3x+1)(x+2)$

2 **a** $(2x+3)(x+1)$ **b** $(2x+9)(x+2)$
c $(7x+2)(x+1)$ **d** $(3x+1)(x+4)$
e $(3x+2)(x+2)$ **f** $(3x+7)(x+3)$
g $(4x+1)(2x+3)$ **h** $(7x+1)(3x+2)$
i $(3x+1)(2x+1)$ **j** $(6x+1)(x+3)$
k $(5x+1)(2x+3)$ **l** $(7x+1)(2x+5)$

3 **a** $(2x+3)(2x-1)$ **b** $(2x-1)(2x+3)$

4 **a** $(2x+1)(x-5)$ **b** $(3x-1)(x+2)$
c $(3x+1)(x-2)$ **d** $(2x-1)(x+2)$
e $(2x+5)(x-1)$ **f** $(5x-3)(x-1)$
g $(11x+2)(x-1)$ **h** $(2x+3)(x-3)$
i $(3x-2)(x-5)$ **j** $(5x+2)(x-3)$
k $(3x-2)(x+4)$ **l** $(2x-1)(x+9)$
m $(2x-3)(x+6)$ **n** $(5x+2)(3x-1)$
o $(21x+1)(x-3)$

5 **a** $-(3x+7)(x-2)$ **b** $-(5x-1)(x-2)$
c $-(4x-3)(x+3)$ **d** $-(3x-2)^2$
e $-(4x+1)(2x+3)$ **f** $-(6x+1)(2x-3)$

REVIEW SET 3A

1 a $b(a-5)$ b $3x(x-4)$ c $3x(5-2x)$
d $2x(3-2x)$ e $4a(3-5a)$ f $(x+2)(x-1)$

2 a $(5-3a)(5+3a)$ b $2(x+7)(x-7)$
c $(3x+2)(x+8)$

3 a $(x+12)(x-12)$ b no real linear factors
c $(x-1)(x-7)$

4 a perfect square b not a perfect square

5 a $(m-9)^2$ b $(4x+1)^2$ c $3(x-4)^2$

6 a $(x-1)(5+y)$ b $(3x+7)(1+2b)$ c $(2x+1)(y-z)$

7 a $(x+9)(x+5)$ b $(x-9)(x+4)$ c $(x-7)(x-12)$

8 a $-(x-2)(x+9)$ b $3(x+2)(x-5)$
c $-2(x+4)(x-8)$

9 a $3(t+3)(t-3)$ b $(x-4)(x+8)$ c $(2x-7)^2$
d $4(x+5)^2$ e $2(x+5)(x-2)$ f $-3(x-10)(x+2)$

10 a $(2x+1)(4x+3)$ b $(x-2)(5x-3)$
c $-(3x+1)(3x-2)$

REVIEW SET 3B

1 a $x(6-y)$ b $5b(a+2b)$ c $2m(4m-5)$
d $-6x(4+3x)$ e $(x+5)(x+1)$ f $(x-3)(x-6)$

2 a $(x^2+3y)(x^2-3y)$ b $a(7a+4)(7a-4)$
c $(m+1)(3m-1)$

3 $899 = 30^2 - 1^2$
$= (30+1)(30-1)$
$= 31 \times 29$

4 a $(k+12)^2$ b $(3x-4)^2$ c $(a+3b)^2$

5 $x^2-6x+9=(x-3)^2 \geqslant 0$ for all real x
$\therefore$ x^2-6x+9 is never negative.

6 a $(c+3)(d+3)$ b $(x+3)(2x-5)$ c $(x-2)(x^2+4)$

7 a $(x-11)(x+6)$ b $(x+6)(x+9)$
c $2(x+13)(x-3)$

8 a $(9x+2)(9x-2)$ b $(x-12)(x+5)$
c $4(x-5)(x+3)$ d $(2x+1)^2$
e $3(x-6)^2$ f $5(x-7)(x+2)$

9 a 1111 b $(x+1)(x^2+1)$
c $1111=(10+1)(10^2+1)$ {using **b**}
$=11 \times 101$

10 a $(x-3)(5x+1)$ b $(2x+3)(2x-7)$
c $-(2x-5)(3x+10)$

EXERCISE 4A

1 a $A = \{14, 24, 34, 40, 41, 42, 43, 44, 45, 46, 47, 48, 49, 54, 64, 74, 84, 94\}$
$n(A)=18$
b $B = \{$red, orange, yellow, green, blue, indigo, violet$\}$,
$n(B)=7$

2 a $P=\{10, 25, 50\}$,
$Q = \{6, 8, 10, 14, 15, 21, 22, 26, 27, 33, 34, 35, 38, 39, 46\}$
b i true ii false iii true
c i $n(P)=3$ ii $n(Q)=15$ d no

3 a i $A=\{1, 2, 3, 6\}$ ii finite, $n(A)=4$
b i $B=\{6, 12, 18, 24, ...\}$ ii infinite
c i $C=\{1, 17\}$ ii finite, $n(C)=2$
d i $D=\{1, 3, 5, 7, ...\}$ ii infinite
e i $E=\{2, 3, 5, 7, 11, 13, 17, 19\}$
ii finite, $n(E)=8$
f i $F=\{12, 14, 15, 16, 18, 20, 21, 22, 24, 25, 26, 27, 28\}$
ii finite, $n(F)=13$

4 a yes b no c yes d no

5 a no b yes c no d yes

6 a B contains all the elements in A, and may contain elements which are not in A.
$\therefore$ $n(A) \leqslant n(B)$
b $A=B$ c B is an infinite set

7 a $k=1, 2, 5$, or 10 b $k=10, 20, 30, 40, ...$

EXERCISE 4B

1 a $A'=\{1, 3, 4, 7, 8\}$ b $B'=\{2, 4, 6, 8\}$
c $C'=\{1, 2, 3, 5, 7\}$ d $D'=\{2, 3, 5, 6, 7, 8\}$

2 a $P'=\{$A, B, D, E, G, H, I, K, L, N, O, Q, R, S, T, V, W, X$\}$
b $Q'=\{$A, E, I, O, U$\}$
c $R'=\{$B, C, D, E, F, G, J, K, M, N, Q, R, U, V, W, X, Y, Z$\}$
d $S'=\{$A, B, C, D, E, F, G, H, I, J$\}$

3 a $A=\{1, 2, 3, 5, 6, 10, 15, 30\}$
b $A'=\{4, 7, 8, 9, 11, 12, 13, 14, 16, 17, 18, 19, 20, 21, 22, 23, 24, 25, 26, 27, 28, 29\}$
c $B=\{2, 4, 6, 8, 10, 12, 14, 16, 18, 20, 22, 24, 26, 28, 30\}$
d $B'=\{1, 3, 5, 7, 9, 11, 13, 15, 17, 19, 21, 23, 25, 27, 29\}$

4 $n(A')=n(U)-n(A)$

5 If an element is not in A, it is also not in B.
$\therefore$ $A' \subseteq B'$
$\therefore$ $n(A') \leqslant n(B')$

6 a infinite
b **Note:** Other answers are possible.
i $A=\{1\}$ $(A'=\{2, 3, 4, ...\})$
ii $A=\{3, 4, 5, 6, ...\}$ $(A'=\{1, 2\})$
iii $A=\{$odd numbers$\}$ $(A'=\{$even numbers$\})$
c Every element in U is either in A or in A'. If A and A' are finite, then U must also be finite. This is impossible as U is infinite.

EXERCISE 4C

1 a $A \cap B=\{$d, m$\}$, $A \cup B=\{$b, c, d, j, k, m, p, t$\}$
b $A \cap B=\{11\}$, $A \cup B=\{2, 3, 5, 7, 8, 11, 15, 19\}$
c $A \cap B=\{$red, green, blue$\}$,
$A \cup B=\{$red, pink, green, blue, purple, yellow$\}$

2 a $A=\{1, 2, 3, 4, 6, 9, 12, 18, 36\}$
b $B=\{1, 3, 7, 9, 21, 63\}$ c $A \cap B=\{1, 3, 9\}$
d $A \cup B=\{1, 2, 3, 4, 6, 7, 9, 12, 18, 21, 36, 63\}$

3 a i $A'=\{1, 4, 5, 6, 8\}$ ii $B'=\{1, 2, 4, 5, 7, 8\}$
iii $A \cap B=\{3, 9\}$ iv $A \cup B=\{2, 3, 6, 7, 9\}$
v $A' \cap B'=\{1, 4, 5, 8\}$
vi $A' \cup B'=\{1, 2, 4, 5, 6, 7, 8\}$
b i $(A \cup B)'=\{1, 4, 5, 8\}=A' \cap B'$ ✓
ii $(A \cap B)'=\{1, 2, 4, 5, 6, 7, 8\}=A' \cup B'$ ✓

4 a i $A=\{1, 2, 4, 5, 10, 20\}$
ii $B=\{2, 3, 5, 7, 11, 13, 17, 19\}$
iii $A \cap B=\{2, 5\}$

iv $A \cup B = \{1, 2, 3, 4, 5, 7, 10, 11, 13, 17, 19, 20\}$

b $n(A) + n(B) - n(A \cap B) = 6 + 8 - 2$
$= 12$
$= n(A \cup B)$ ✓

5 **a** **i** $A = \{1, 2, 3, 4, 5, 6, 7, 8, 9\}$
ii $B = \{3, 6, 9, 12, 15, 18\}$ **iii** $A \cap B = \{3, 6, 9\}$
iv $A \cup B = \{1, 2, 3, 4, 5, 6, 7, 8, 9, 12, 15, 18\}$

b $n(A) + n(B) - n(A \cup B) = 9 + 6 - 12$
$= 3$
$= n(A \cap B)$ ✓

6 P and R

7 Every element in A is also in B and so is not in B'.
$\therefore$ A and B' are disjoint.

8 **a** $A = \{$A, L, I, G, T, O, R$\}$, $B = \{$P, I, N, E, A, L$\}$, $C = \{$M, O, U, N, T, A, I$\}$

b **i** $B' = \{$B, C, D, F, G, H, J, K, M, O, Q, R, S, T, U, V, W, X, Y, Z$\}$
This set represents the letters not in PINEAPPLE.

ii $A \cap C = \{$A, I, T, O$\}$
This set represents the letters in both ALLIGATOR and MOUNTAIN.

iii $A \cup B = \{$A, L, I, G, T, O, R, P, N, E$\}$
This set represents the letters in ALLIGATOR or PINEAPPLE or both.

iv $(A \cup B)' = \{$B, C, D, F, H, J, K, M, Q, S, U, V, W, X, Y, Z$\}$
This set represents the letters in neither ALLIGATOR nor PINEAPPLE.

v $A \cap B = \{$A, L, I$\}$
This set represents the letters in both ALLIGATOR and PINEAPPLE.

vi $A \cap C' = \{$L, G, R$\}$
This set represents the letters in ALLIGATOR which are not in MOUNTAIN.

vii $A \cap C \cap B' = \{$T, O$\}$
This set represents the letters in both ALLIGATOR and MOUNTAIN which are not in PINEAPPLE.

viii $A \cap B \cap C = \{$A, I$\}$
This set represents the letters in ALLIGATOR, PINEAPPLE, and MOUNTAIN.

9 **a** infinite **b** cannot be determined **c** infinite **d** finite

EXERCISE 4D

1 **a** true **b** true **c** true **d** true
e false **f** false **g** true **h** true

2 **a** false **b** true **c** true **d** false

3 $\{0\}$ **4** $\mathbb{N}$ **5** $\mathbb{Q}'$

6 **a** rational **b** rational **c** irrational **d** neither
e rational **f** rational **g** neither **h** rational

7 0.527 can be written as $\frac{527}{1000}$, where 527 and 1000 are integers.

8 **Note:** Other answers are possible.

a $\sqrt{2}$ and $-\sqrt{2}$ are irrational, but $\sqrt{2} + (-\sqrt{2}) = 0$ which is rational.

b $\sqrt{2}$ and $\sqrt{50}$ are irrational, but $\sqrt{2} \times \sqrt{50} = \sqrt{100} = 10$ which is rational.

c $\sqrt{3}$ and $2\sqrt{3}$ are irrational, but $\dfrac{2\sqrt{3}}{\sqrt{3}} = 2$ which is rational.

d 0 is rational and $\sqrt{2}$ is irrational, but $0 \times \sqrt{2} = 0$ which is rational.

EXERCISE 4E

1 **a** The set of real numbers less than or equal to 6.
b The set of integers greater than -5.
c The set of natural numbers less than 10.
d The set of real numbers greater than -5 and less than 3.
e The set of integers greater than or equal to -6 and less than 1.
f The set of real numbers less than -4 or greater than or equal to 7.

2 **a** $\{2, 3, 4, 5, 6, 7, 8, 9\}$ **b** $\{1, 2, 3, 4, 5, 6\}$
c $\{0, 1, 2, 3, 4, 5\}$ **d** $\{-7, -6, -5, -4, -3, -2, -1\}$
e $\{...., -7, -6, -5, -4\}$ **f** $\{10, 11, 12, 13, ...\}$

3 **a**
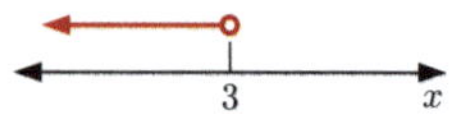

b
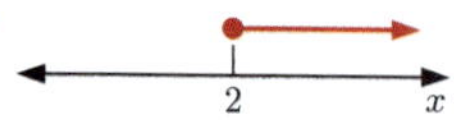

c
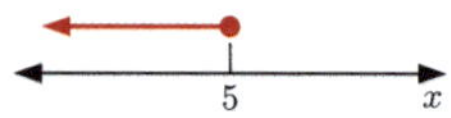

d
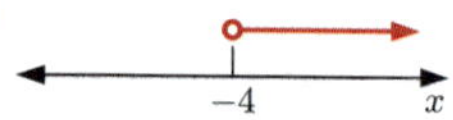

e

f

g

h
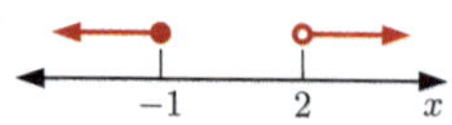

i

j

k
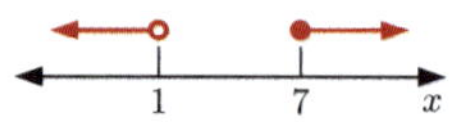

l
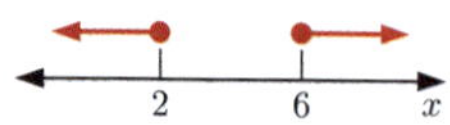

4 **a** $\{x \mid x \geqslant 2\}$ **b** $\{x \mid x < 4\}$ **c** $\{x \mid x > -3\}$
d $\{x \mid x \leqslant \frac{3}{2}\}$ **e** $\{x \mid -1 < x < 5\}$
f $\{x \mid x \leqslant -1 \text{ or } x \geqslant 5\}$ **g** $\{x \mid 3 \leqslant x \leqslant 7\}$
h $\{x \mid x < 3 \text{ or } x > 7\}$ **i** $\{x \mid -2 \leqslant x < 3\}$
j $\{x \mid x < -2 \text{ or } x \geqslant 3\}$ **k** $\{x \mid -\frac{1}{2} < x \leqslant \frac{5}{2}\}$
l $\{x \mid x \leqslant -\frac{1}{2} \text{ or } x > \frac{5}{2}\}$

5 **Note:** Other answers are possible.
a $\{x \in \mathbb{N} \mid 3 \leqslant x \leqslant 7\}$ **b** $\{x \in \mathbb{Z} \mid x \leqslant -1\}$
c $\{x \in \mathbb{N} \mid x \geqslant 11\}$ **d** $\{x \in \mathbb{Z} \mid -1 \leqslant x \leqslant 3\}$
e $\{x \in \mathbb{Z} \mid -3 \leqslant x \leqslant 2\}$

6 **a**
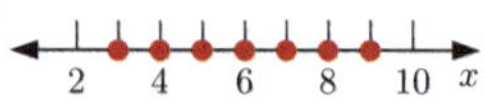

b
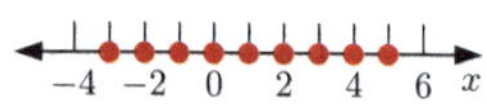

c
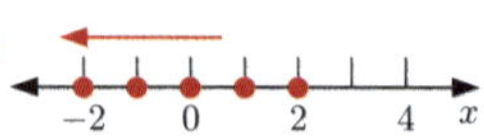

d
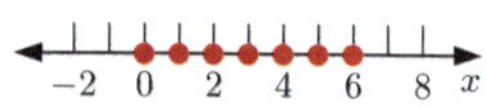

e
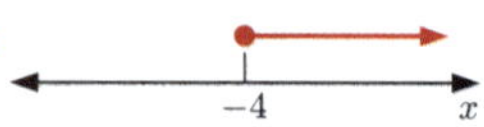

f
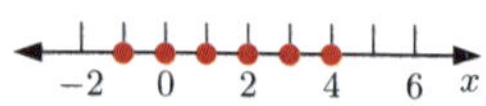

7 **a** **i** The set of integers greater than -2 and less than 5.
ii $B = \{-1, 0, 1, 2, 3, 4\}$ **iii** $n(B) = 6$
iv
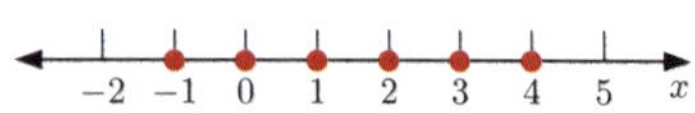

b **i** The set of positive integers greater than or equal to -6 and less than or equal to 3.
ii $B = \{1, 2, 3\}$ **iii** $n(B) = 3$
iv
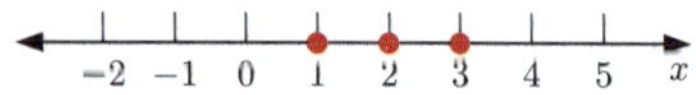

c **i** The set of real numbers greater than 1 and less than 2.
ii not possible **iii** $n(B)$ is infinite
iv 

d **i** The set of real numbers greater than or equal to 4 and less than 7.
ii not possible **iii** $n(B)$ is infinite
iv 

8 **a** $\{x \in \mathbb{Z} \mid 10 < x < 100\}$ **b** $\{x \mid x < 200\}$
c $\{x \mid x \geqslant -8\}$ **d** $\{x \in \mathbb{Q} \mid -1 < x \leqslant 1\}$

9 **a** finite **b** infinite **c** infinite **d** infinite
e finite **f** infinite

10 **a** yes **b** no **c** yes **d** no

11 **a** **i** $A = \{5, 6, 7, 8, 9, 10, 11, 12\}$
ii $A' = \{1, 2, 3, 4, 13, 14, 15\}$ **iii** $B = \{7, 8, 9\}$
iv $B' = \{1, 2, 3, 4, 5, 6, 10, 11, 12, 13, 14, 15\}$
b **i** false **ii** true **iii** true **iv** false

12 **a** **i** $P \cup Q = \{x \mid -5 \leqslant x < 20\}$
ii $Q \cup R = \{x \mid x < 20\}$
iii $P \cap Q = \{x \mid 4 < x < 10\}$
iv $P' = \{x \mid x < -5 \text{ or } x \geqslant 10\}$
b **i** 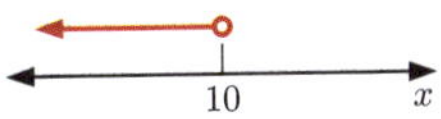

ii 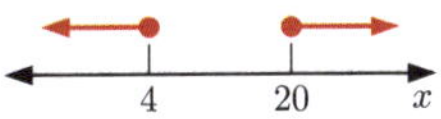

iii 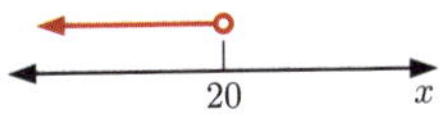

iv 

13 **a** **i** 

ii 

b **i** 

$\{x \mid -1 < x < 9\}$
ii 

$\{x \mid 3 \leqslant x < 4\}$

REVIEW SET 4A

1 **a** $P = \{23, 29, 31, 37\}$ **b** **i** true **ii** false
c $n(P) = 4$

2 **a** no **b** yes **c** yes

3 **a** $A' = \{1, 2, 4, 5, 6, 8, 10\}$ **b** $B' = \{1, 2, 3, 5, 7\}$

4 **a** $P \cup Q = \{13, 18, 23, 28, 30, 31, 32, 33, 34, 35, 36, 37, 38, 39, 43, 48, 53, 58, 63, 68, 73, 78, 80, 81, 82, 83, 84, 85, 86, 87, 88, 89, 93, 98\}$
b $P \cap Q = \{38, 83\}$

5 Each element in A is either in B or in B'.
Now $A \cap B$ is finite as B is finite.
Since A is infinite, $A \cap B'$ must also be infinite.

6 **a** true **b** true **c** true

7 **a** irrational **b** rational **c** rational **d** neither

8 **a** The set of integers greater than or equal to -2 and less than or equal to 7.
b The set of natural numbers less than 8.
c The set of real numbers less than -1 or greater than or equal to 5

9 **a** $\{x \mid x > -3\}$ **b** $\{x \mid 0 < x < 7\}$
c $\{x \mid x < -9 \text{ or } x \geqslant -2\}$

10 **a** infinite **b** finite **c** infinite

11 **a** **i** 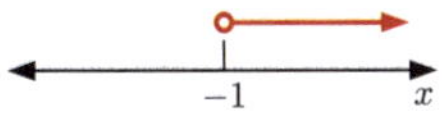

ii 

b **i** 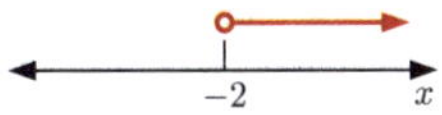

$\{x \mid x > -2\}$
ii 

$\{x \mid -1 < x \leqslant 3\}$

12 **a** The set of integers greater than or equal to -3 and less than or equal to 6.
b $A = \{-3, -2, -1, 0, 1, 2, 3, 4, 5, 6\}$ **c** $n(A) = 10$
d 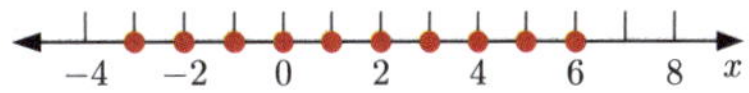

REVIEW SET 4B

1 **a** finite, $n(A) = 8$ **b** infinite **c** finite, $n(C) = 20$

2 $k = 1, 2, 4, 5, 10$, and 20

3 **a** **i** false **ii** true **b** finite **c** infinite **d** yes

4 **a** $B \cup C = \{1, 3, 4, 5, 7, 9, 11, 13, 15, 16, 17, 19, 25, 36\}$
b $A \cap B \cap C = \{9\}$

5 P and R, P and S, Q and S

6 **a** false **b** false **c** true

7 **a** 0.063 can be written as $\frac{63}{1000}$, where 63 and 1000 are integers.
b $\frac{1}{\sqrt{16}} = \frac{1}{4}$ which is rational.

8 **a** 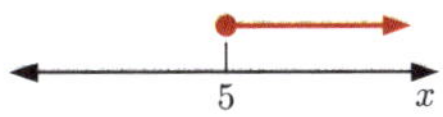

b 

c 

9 **Note:** Other answers are possible.
a $\{x \in \mathbb{Z} \mid x \leqslant -3\}$ **b** $\{x \in \mathbb{Z} \mid -2 \leqslant x \leqslant 5\}$

10 **a** **i** $A \cup B = \{x \mid -4 \leqslant x < 10\}$
ii $A \cap C = \{x \mid -4 \leqslant x \leqslant 3\}$
b **i** 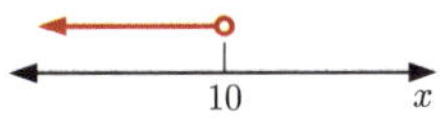

ii 

11 **a** **i** $A = \{2, 4, 6, 8, 10, 12, 14\}$
ii $B = \{1, 2, 3, 5, 6, 10, 15, 30\}$
iii $C = \{1, 2, 3, 4, 5, 6, 7, 8, 9\}$
iv $A \cap B = \{2, 6, 10\}$ **v** $B \cap C = \{1, 2, 3, 5, 6\}$
b $A \cap (B \cap C) = \{2, 6\}$
$(A \cap B) \cap C = \{2, 6\} = A \cap (B \cap C)$ ✓
c Both $A \cap (B \cap C)$ and $(A \cap B) \cap C$ represent the elements in all three sets A, B, and C.

12 **a** $n(A') = 7$, $n(B') = 4$
b There are more elements in B than there are in A, so it is impossible for B to be a subset of A.
c $\{x \in \mathbb{N} \mid 11 \leqslant x \leqslant 15\}$
d 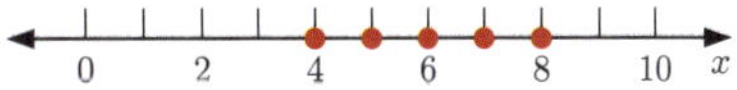

EXERCISE 5A

1 **a** $x = -11$ **b** $x = -6$ **c** $x = -3$ **d** $x = -27$
e $x = 12$ **f** $x = 23$ **g** $x = -8$ **h** $x = 84$

2 **a** $x = -7$ **b** $x = -3$ **c** $x = 9$ **d** $x = -5$
e $x = -2$ **f** $x = 3$ **g** $x = -6$ **h** $x = 11$
i $x = -4$ **j** $x = \frac{7}{2}$ **k** $x = -6$ **l** $x = 11$
m $x = -2$ **n** $x = 4$ **o** $x = -9$

3 **a** $x = -15$ **b** $x = -16$ **c** $x = -12$ **d** $x = 19$
e $x = -11$ **f** $x = 10$ **g** $x = 24$ **h** $x = -32$
i $x = 7$

4 **a** $x = -\frac{11}{2}$ **b** $x = -3$ **c** $x = 17$ **d** $x = -7$
e $x = 3$ **f** $x = \frac{17}{2}$

5 **a** $x = -\frac{3}{5}$ **b** $x = \frac{95}{9}$ **c** $x = -\frac{70}{3}$

6 **a** $x = 2$ **b** $x = -7$ **c** $x = 9$ **d** $x = -2$
e $x = -3$ **f** $x = -3$ **g** $x = 2$ **h** $x = 1$

7 **a** $x = -3$ **b** $x = \frac{6}{5}$ **c** $x = -3$ **d** $x = -\frac{7}{2}$
e $x = 3$ **f** $x = 2$ **g** $x = \frac{13}{2}$ **h** $x = 1$
i $x = \frac{17}{15}$

8 **a** $x = 2$ **b** $x = -1$ **c** $x = 6$ **d** $x = \frac{7}{2}$
e $x = \frac{2}{3}$ **f** $x = \frac{4}{3}$ **g** $x = -\frac{9}{10}$ **h** $x = -\frac{12}{7}$
i $x = -\frac{11}{4}$ **j** $x = \frac{72}{49}$

EXERCISE 5B

1 The number is 8. **2** The integers are 86 and 87.
3 The smallest integer is 35. **4** The number is 18.
5 The number is 4. **6** The number is $4\frac{1}{4}$.
7 The number is $\frac{1}{2}$. **8** Tony made 11 pizzas.
9 120 people attended on Friday. **10** The tour cost was \$10.
11 Marcia used 236 grams of flour for the cake.
12 500 lambs were born.
13 Emma scored 128 points, Alex scored 143 points, and Toni scored 110 points.
14 There are 7 50 cent coins.
15 Five 500 g cans were bought.
16 The corner shop has 15 600 mL cartons.
17 Inka invested \$7200 in mining shares.
18 Bruce has 8 5 cent stamps, 48 10 cent stamps, and 16 20 cent stamps.

19 **a** $0.\overline{7} = \frac{7}{9}$ **b** $0.\overline{41} = \frac{41}{99}$ **c** $0.\overline{324} = \frac{12}{37}$
d $0.0\overline{69} = \frac{23}{330}$

EXERCISE 5C.1

1 **a** $x < 4$ **b** $a \geqslant 2$ **c** $k \leqslant -4$
d $x \geqslant 32$ **e** $x > -5$ **f** $y > -7$
g $a \leqslant -8$ **h** $c \leqslant -42$

2 **a** $x < -\frac{2}{3}$ **b** $a > \frac{9}{5}$ **c** $x \leqslant \frac{1}{3}$
d $x \geqslant -3$ **e** $x < 1$ **f** $x \geqslant 3$
g $x < -1$ **h** $x < 1$ **i** $x \leqslant -\frac{1}{5}$

3 **a** $x > 16$ **b** $x \geqslant -21$ **c** $x \geqslant -36$
d $x \leqslant 9$ **e** $x < -8$ **f** $x \leqslant \frac{35}{2}$
g $x \geqslant -1$ **h** $x > -6$ **i** $x \geqslant 0$

4 **a** $x \leqslant 4$ **b** $x > -5$ **c** $x > -4$
d $x \geqslant \frac{7}{3}$ **e** $x < 1$ **f** $x \leqslant \frac{3}{5}$

5 **a** $x > -\frac{7}{2}$ **b** $x > \frac{4}{3}$ **c** $x \leqslant \frac{1}{3}$
d $x \geqslant 1$ **e** $x < \frac{11}{5}$ **f** $x > -\frac{2}{19}$

6 **a** $x \leqslant 6$ **b** $x < 0$ **c** $x \leqslant -\frac{5}{3}$
d $x > \frac{27}{2}$ **e** $x \geqslant \frac{13}{15}$ **f** $x \leqslant \frac{40}{47}$

EXERCISE 5C.2

1 **a** $2 < x < 5$ **b** $-2 \leqslant a \leqslant 6$ **c** $4 \leqslant x < 13$
d $-5 \leqslant x \leqslant -3$ **e** $-2 < t \leqslant -\frac{1}{4}$ **f** $-6 \leqslant m < 3$

2 **a** $2 \leqslant x \leqslant 6$ **b** $-\frac{4}{5} < x \leqslant 4$ **c** $-9 < x < -4$
d $-6 < x \leqslant \frac{7}{3}$ **e** $16 < x < 24$ **f** $-24 < x \leqslant -8$
g $0 \leqslant x \leqslant 15$ **h** $-30 \leqslant x < -5$ **i** $13 < x \leqslant 19$

j $-43 \leqslant x \leqslant -11$ **k** $-23 < x < 1$ **l** $-11 < x \leqslant 10$

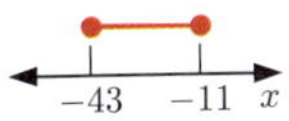

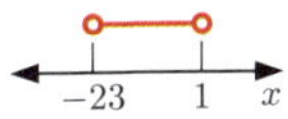

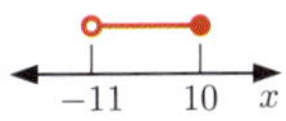

3 **a** $3 \leqslant x < 9$ **b** $\frac{2}{5} < x < \frac{4}{5}$ **c** $-\frac{28}{3} < x \leqslant 8$
d $\frac{13}{3} < x < 9$ **e** $5 \leqslant x \leqslant \frac{25}{2}$ **f** $-\frac{2}{5} < x < \frac{42}{5}$

EXERCISE 5D

1 Jenny's original number was greater than 15.

2 The 15% discount is better when the purchase price is more than \$400.

3 It is cheaper to use Lightning Pace when the weight is less than 10 kg.

4 **a** between 32°F and 212°F **b** less than −40°C

5 more than 10 minutes, but less than or equal to 25 minutes

6 between 45° and 72° **7** between 50° and 80°

8 between 51.5 km and 61.5 km

REVIEW SET 5A

1 **a** $x = -10$ **b** $x = 27$ **c** $x = 19$

2 **a** $x = -\frac{13}{4}$ **b** $x = \frac{1}{2}$ **c** $x = -19$

3 The smallest integer is 20.

4 24 left-handed students attend the school.

5 twelve 5 cent coins

6 **a** $x \leqslant 5$ **b** $x < -21$ **c** $x \geqslant 20$

7 **a** $x < -2$ **b** $x \leqslant \frac{25}{7}$

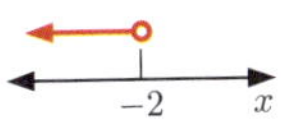

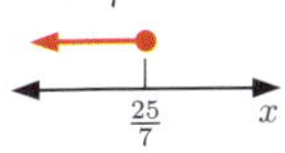

8 **a** $-1 \leqslant m < 12$ **b** $2 \leqslant x < 8$ **c** $-3 \leqslant k \leqslant \frac{11}{3}$

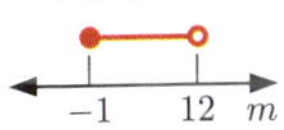

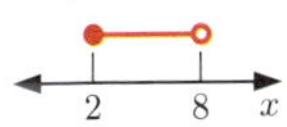

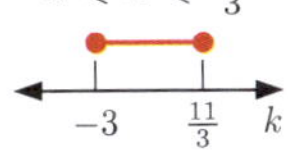

9 For all calls greater than 8 minutes.

10 more than 280 g, but less than or equal to 480 g

11 **a** **i** $6x + 200 = 2000$ **ii** 300 g
b **i** $2000 - 6x \geqslant 300$ **ii** at most ≈ 283 g

REVIEW SET 5B

1 **a** $x = 9$ **b** $x = 6$ **c** $x = -\frac{39}{2}$

2 **a** $x = \frac{1}{6}$ **b** $x = 2$ **c** $x = 3$ **3** 300 mL

4 72 matches **5** $0.\overline{18} = \frac{2}{11}$ **6** 6 writing pads

7 **a** $x > 9$ **b** $x \geqslant -\frac{44}{5}$ **c** $x > 25$

8 **a** $x \geqslant \frac{4}{3}$ **b** $x \leqslant \frac{8}{7}$

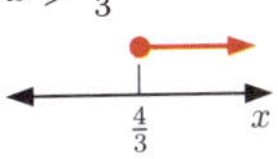

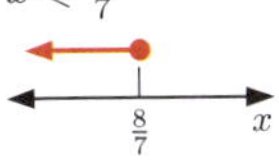

9 **a** $7 \leqslant x < 23$ **b** $-6 < x < 12$

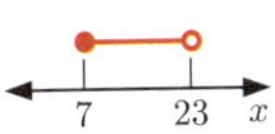

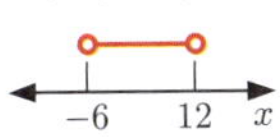

10 Any number greater than 10.

11 **a** **i** 11 dogs, 11 cats, and 6 rabbits **ii** \$3110
b 8 dogs were sold

12 **a** **i** 8 **ii** greater than or equal to 14
iii between 11 and 29

b

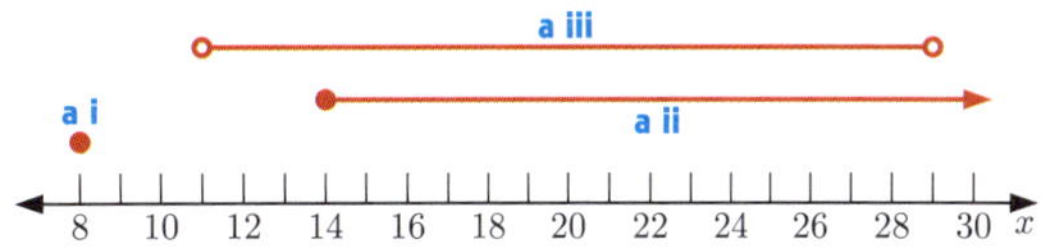

EXERCISE 6A

1 **a** $A = \{2, 3, 5, 7\}$, $A' = \{1, 4, 6, 8\}$ **b**

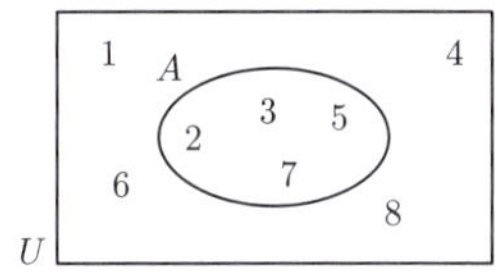

c $n(A) = 4$, $n(A') = 4$

2 **a** $A = \{1, 3, 7, 9\}$ **b** $B = \{1, 2, 5\}$
c $A' = \{2, 4, 5, 6, 8\}$ **d** $B' = \{3, 4, 6, 7, 8, 9\}$
e $A \cap B = \{1\}$ **f** $A \cup B = \{1, 2, 3, 5, 7, 9\}$
g $U = \{1, 2, 3, 4, 5, 6, 7, 8, 9\}$

3 **a**

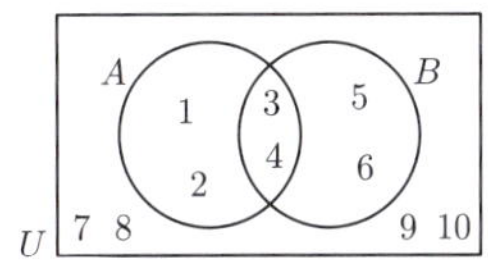

b

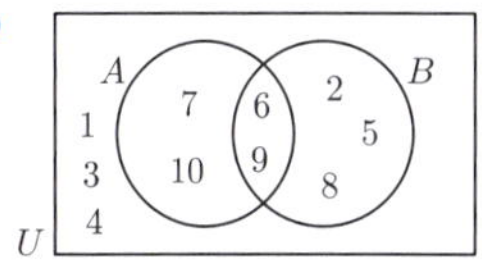

c

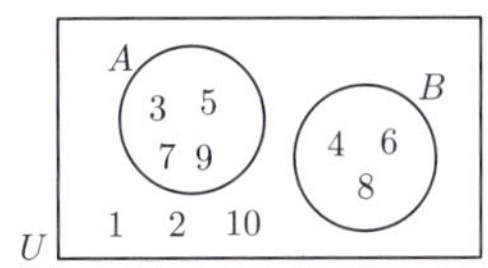

d

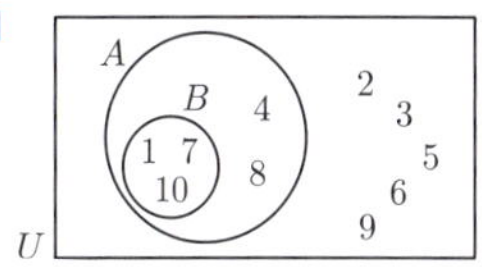

4 **a** **i** $A = \{1, 2, 3, 5, 6, 10, 15, 30\}$
ii $B = \{2, 3, 5, 7, 11, 13, 17, 19, 23, 29\}$
iii $A \cap B = \{2, 3, 5\}$
iv $A \cup B = \{1, 2, 3, 5, 6, 7, 10, 11, 13, 15, 17, 19, 23, 29, 30\}$

b **i** $n(A) = 8$ **ii** $n(B) = 10$
iii $n(A \cap B) = 3$ **iv** $n(A \cup B) = 15$

c $n(A) + n(B) - n(A \cap B) = 8 + 10 - 3 = 15$
$= n(A \cup B)$ ✓

d

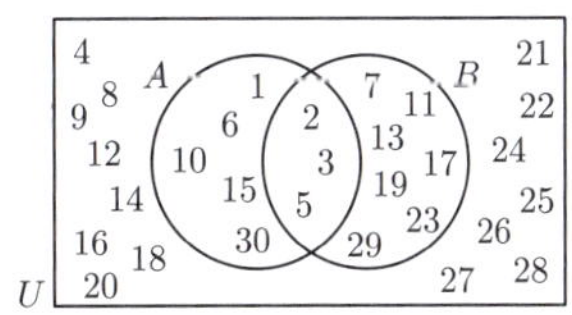

5 **a** $k = 10$ **b** $P = \{3, 6, 9, 12, 15, 18, 21, 24\}$, $Q = \{10, 20\}$

c

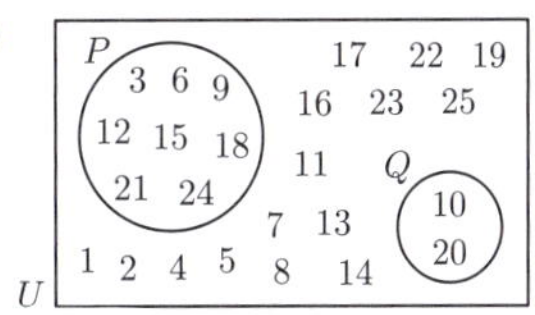

6 **a, c**

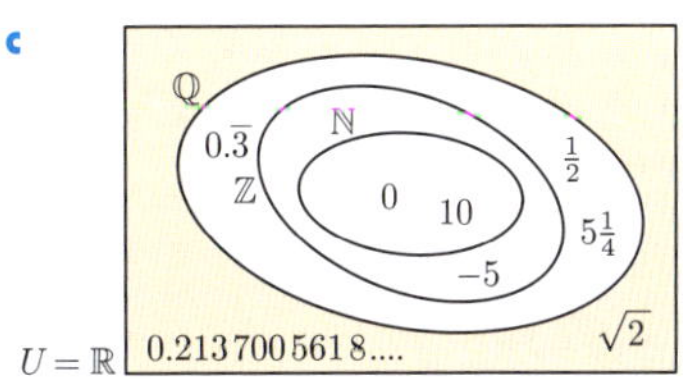

b **i** true
ii true
iii true

7 **a**

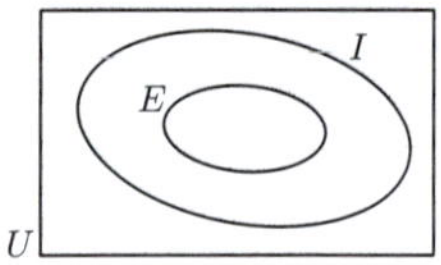

b

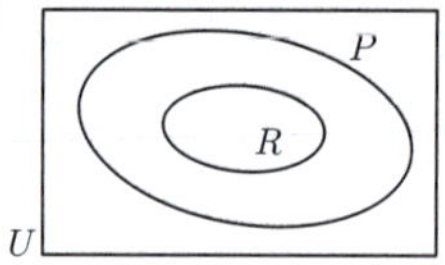

c

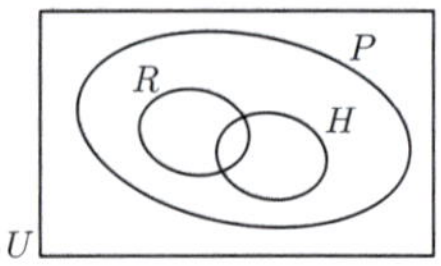

d

e

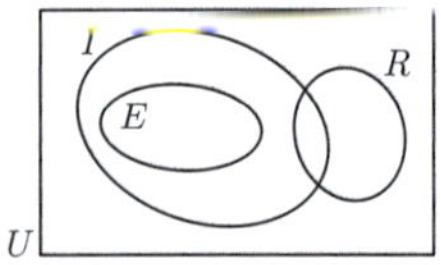

8 **a** **i** $A = \{a, b, f, k, m, q, s, t\}$
ii $B = \{d, k, m, n, o, p, q, r, t, x, z\}$
iii $C = \{a, c, i, k, p, r, t, v, z\}$
iv $A \cap B = \{k, m, q, t\}$
v $B \cup C = \{a, c, d, i, k, m, n, o, p, q, r, t, v, x, z\}$
vi $A \cap C' = \{b, f, m, q, s\}$

b **i** $n(A \cup C) = 14$ **ii** $n(A' \cap B') = 11$
iii $n(A \cap B \cap C) = 2$

9 **a**

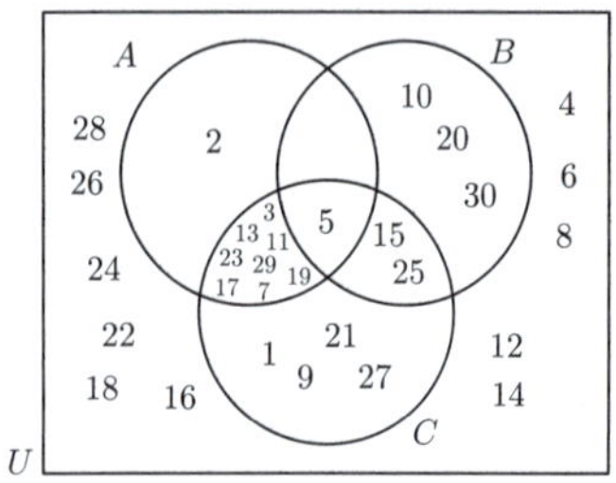

b **i** $\{5, 15, 25\}$ **ii** $\{1, 9, 15, 21, 25, 27\}$
iii $\{3, 7, 11, 13, 17, 19, 23, 29\}$ **iv** $\{5\}$
v $\{4, 6, 8, 12, 14, 16, 18, 22, 24, 26, 28\}$

EXERCISE 6B

1 **a**

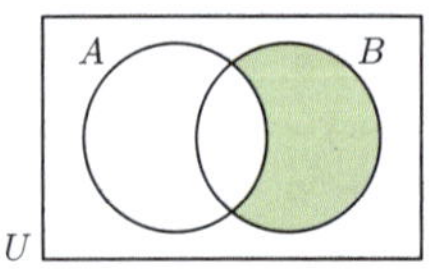

b

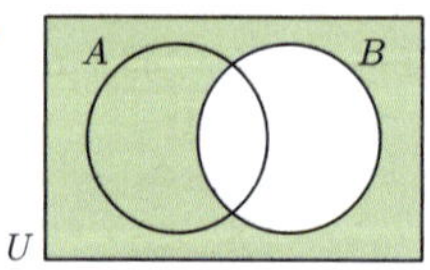

c

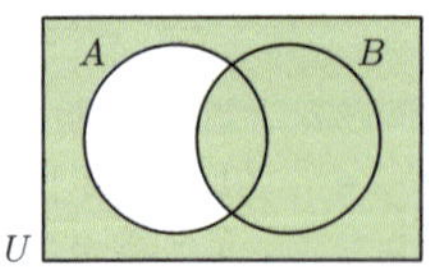

2 **a** not in P **b** in neither P nor Q **c** in P or not in Q

3 **a** **i**

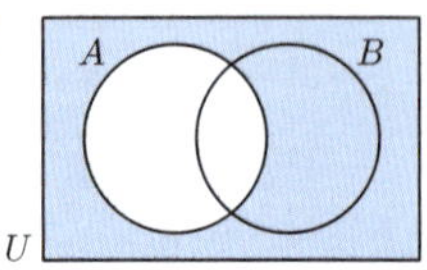

ii

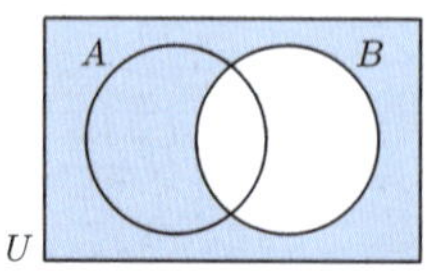

iii

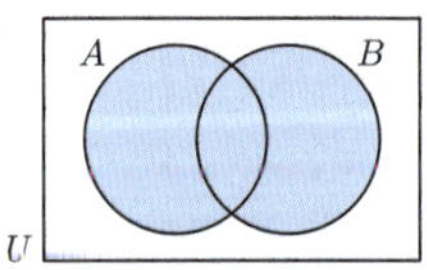

iv

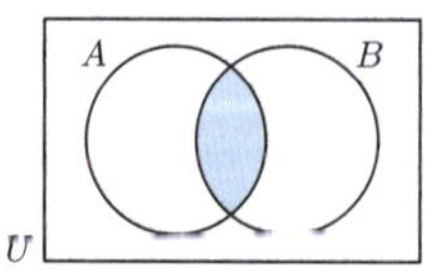

b **i** The shaded region is $A' \cap B'$.
This is the same as the *unshaded* region in **a iii**.
$\therefore \quad (A \cup B)' = A' \cap B'$

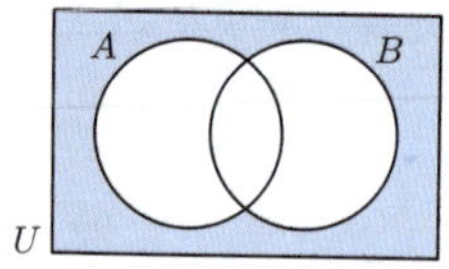

ii The shaded region is $A' \cup B'$.
This is the same as the *unshaded* region in **a iv**.
$\therefore \quad (A \cap B)' = A' \cup B'$

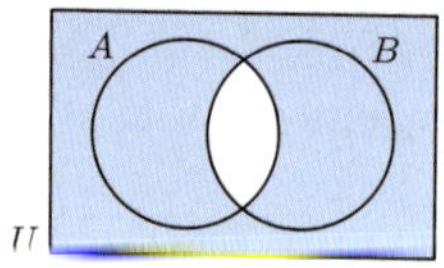

4 **a**

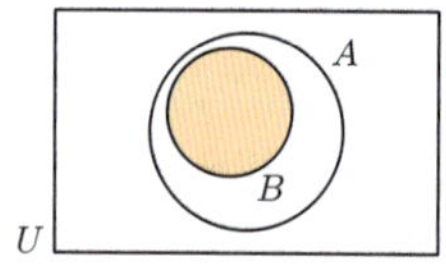

The shaded region is $A \cap B$ which is the same as B.
$\therefore \quad A \cap B = B$

b

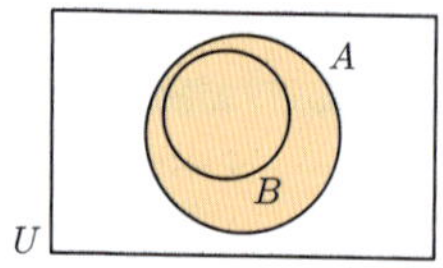

The shaded region is $A \cup B$ which is the same as A.
$\therefore \quad A \cup B = A$

c

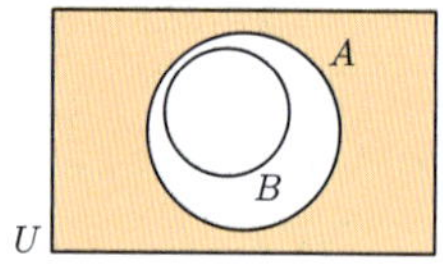

The shaded region is $A' \cap B'$ which is the same as A'.
$\therefore \quad A' \cap B' = A'$

d

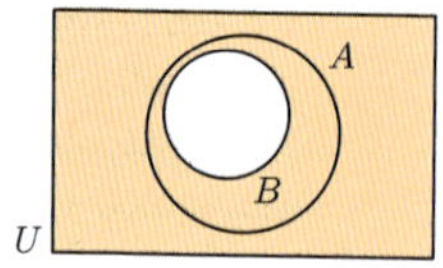

The shaded region is $A' \cup B'$ which is the same as B'.
$\therefore \quad A' \cup B' = B'$

5 **a**

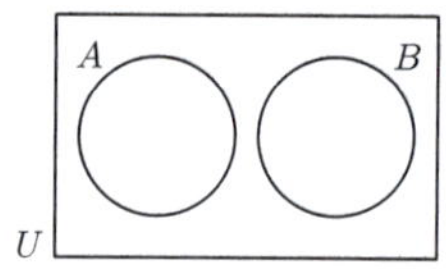

The shaded region is $A \cap B$.
$\therefore \quad (A \cap B)' = U$

b

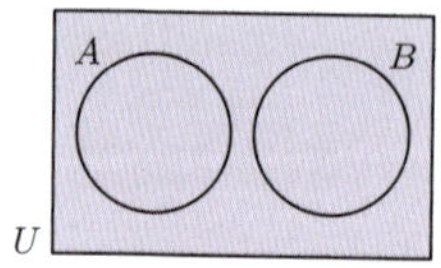

The shaded region is $A' \cup B'$ which is the same as U.
$\therefore \quad A' \cup B' = U$

c

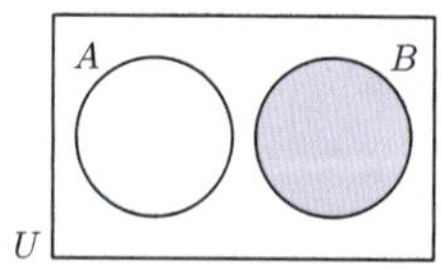

The shaded region is $A' \cap B$ which is the same as B.
$\therefore \quad A' \cap B = B$

6 **a** **i**

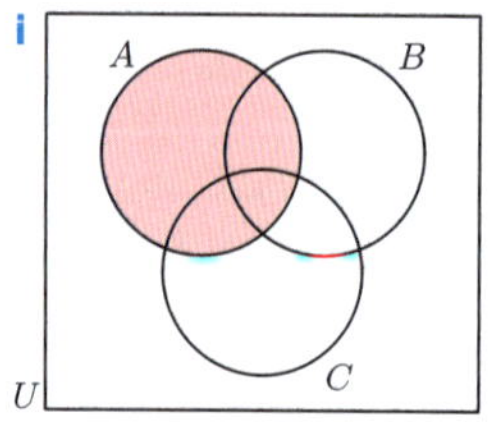

ii

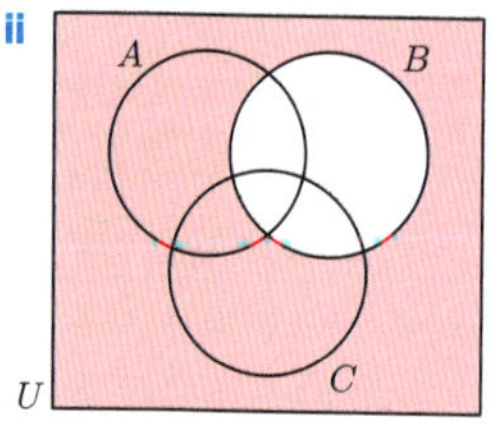

iii

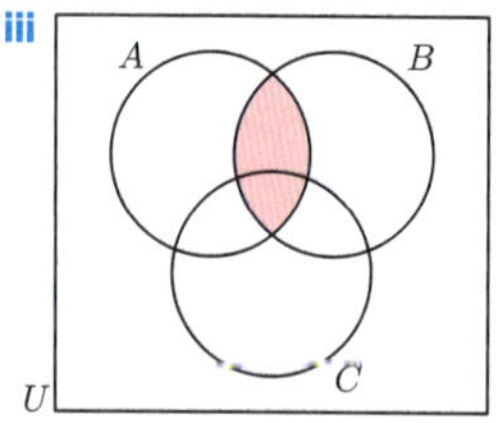

iv

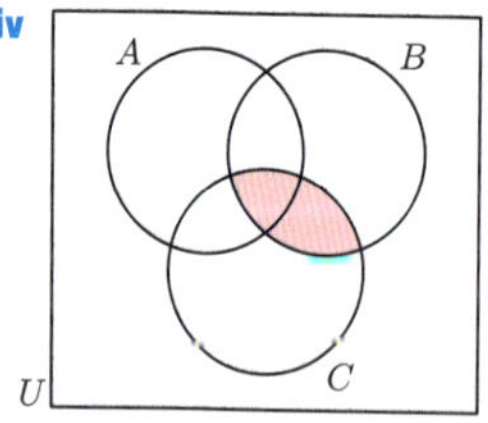

v

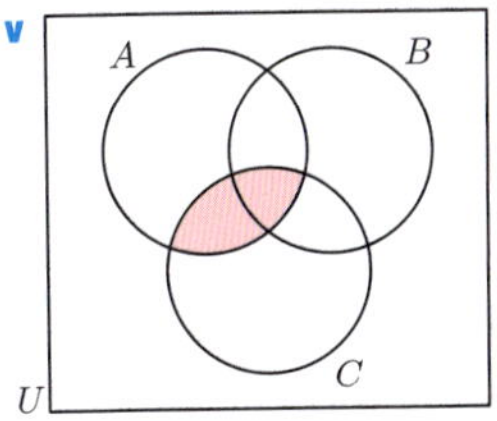

vi

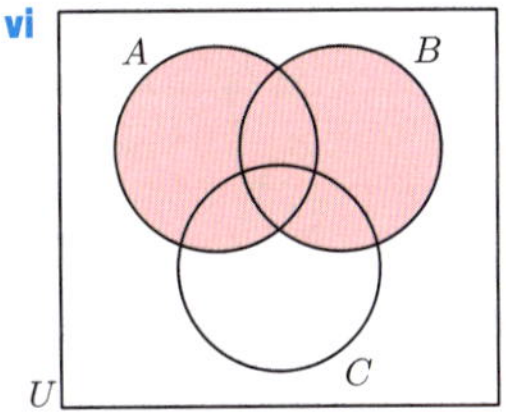

vii

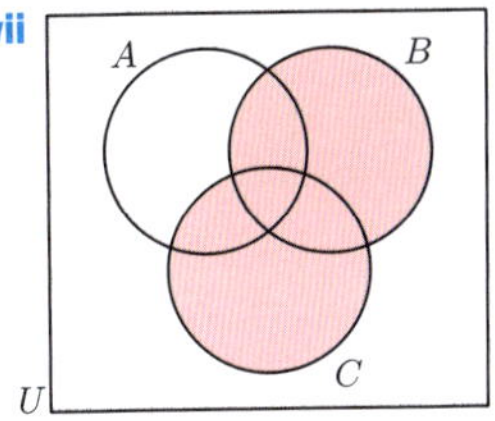

viii

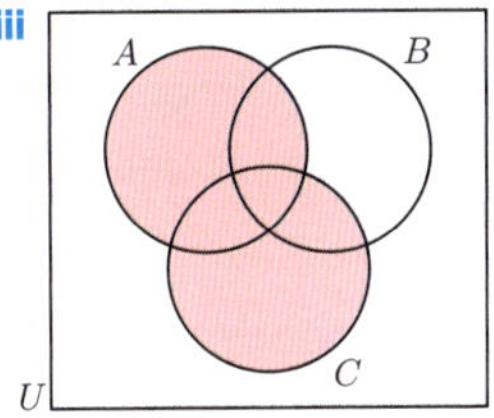

ix

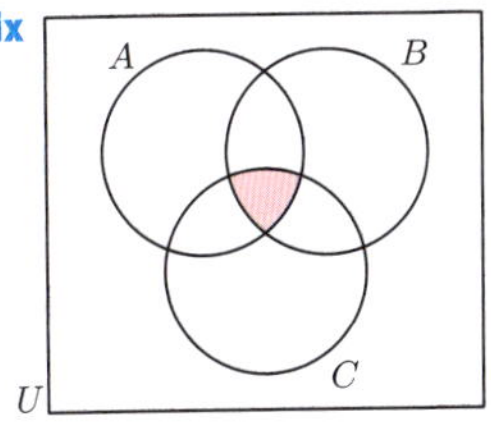

x

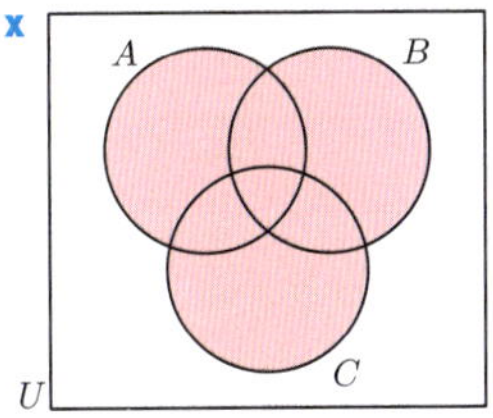

xi

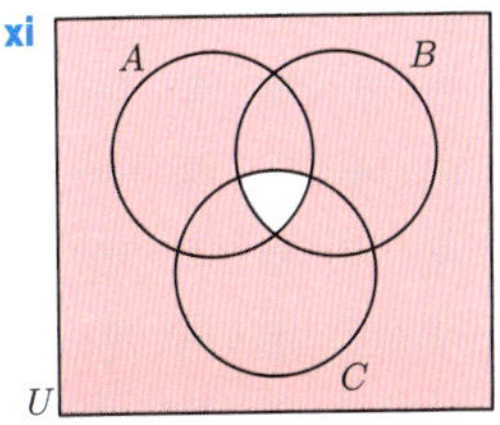

xii

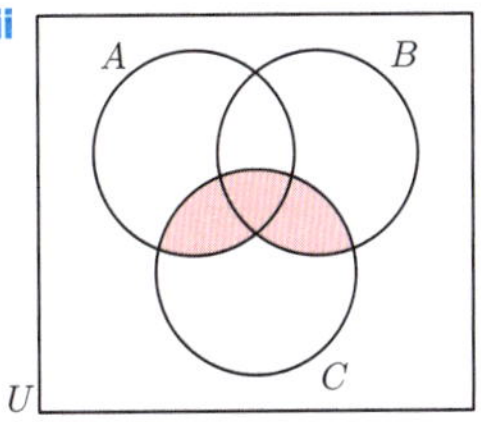

b **i** The shaded region is $A \cup (B \cap C)$ which is the same as $(A \cup B) \cap (A \cup C)$.
$\therefore\ A \cup (B \cap C)$
$= (A \cup B) \cap (A \cup C)$

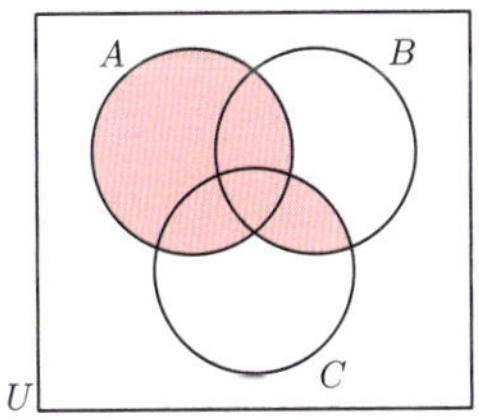

ii The shaded region is $A \cap (B \cup C)$ which is the same as $(A \cap B) \cup (A \cap C)$.
$\therefore\ A \cap (B \cup C)$
$= (A \cap B) \cup (A \cap C)$

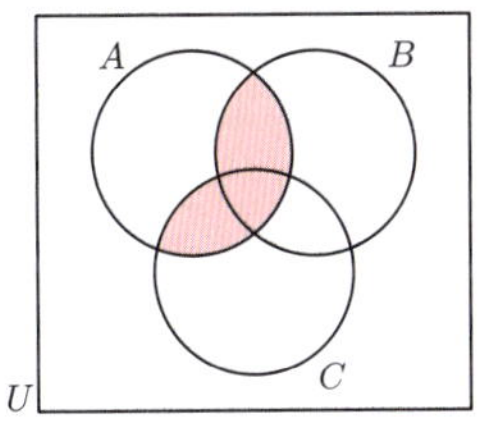

EXERCISE 6C

1 **a** 10 **b** 9 **c** 14 **d** 4 **e** 7 **f** 2

2 **a** **i** 9 **ii** 22 **iii** 13 **iv** 39 **b** 15

3 **a** **i** $n(A) = a + b$ **ii** $n(B) = b + c$
iii $n(A \cap B') = a$ **iv** $n(A \cap B) = b$
v $n(A \cup B) = a + b + c$ **vi** $n(U) = a + b + c + d$

b **i** $n(A) + n(B) - n(A \cup B)$
$= (a + b) + (b + c) - (a + b + c)$
$= a + 2b + c - a - b - c$
$= b$
$= n(A \cap B)$ ✓

ii $n(A) - n(A \cap B) = (a + b) - b$
$= a$
$= n(A \cap B')$ ✓

iii $n(A \cup B') = a + b + d$ and $n(A' \cap B) = c$
Now $n(U) - n(A' \cap B) = (a + b + c + d) - c$
$= a + b + d$
$= n(A \cup B')$ ✓

4 **a**

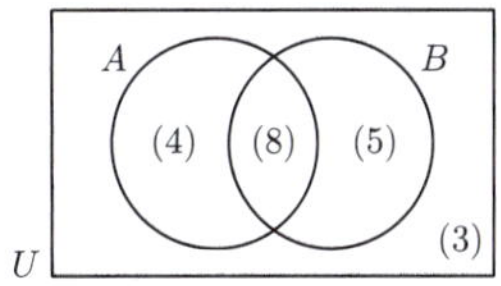

b **i** 17 **ii** 5

5 **a** $n(N) = 7$ **b** $n((M \cup N)') = 10$
c $n(M' \cap N) = 4$

6 **a**

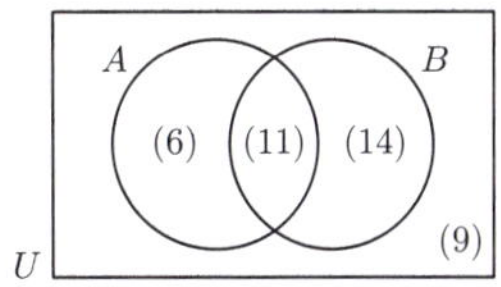

b **i** 11 **ii** 20

7 **a** 13 **b** 19

8 **a**

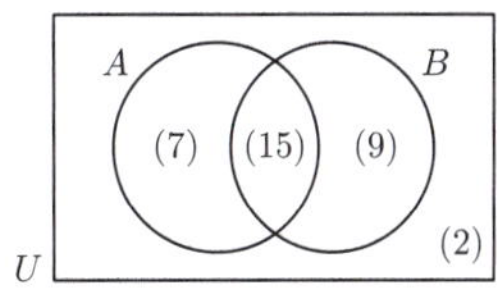

b **i** 31 **ii** 9

9 **a** **i** $n(A \cap B \cap C) = 2$ **ii** $n(A) = 18$

b

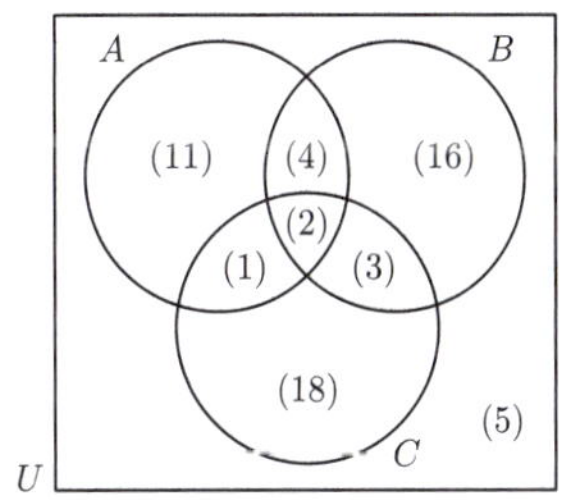

c **i** 24 **ii** 1 **iii** 23 **iv** 6 **v** 8

EXERCISE 6D

1 **a** 75 guests **b** 9 guests **c** 24 guests **d** 42 guests

2 **a**

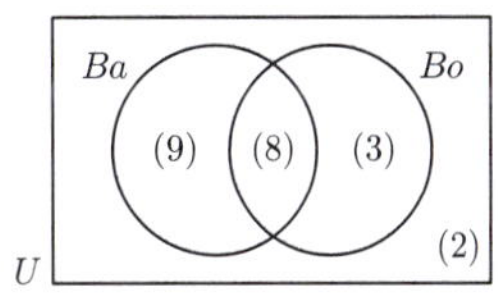

b **i** 3 games **ii** 2 games

3 **a** 8 students **b** 15 students

4 **a**

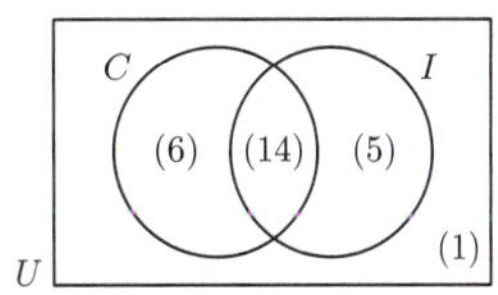

b **i** 14 teenagers **ii** 5 teenagers

5 **a** 27 members **b** 11 members **6** 12%

7 18 cars **8** **a** 37 integers **b** 379 integers

9 **a**

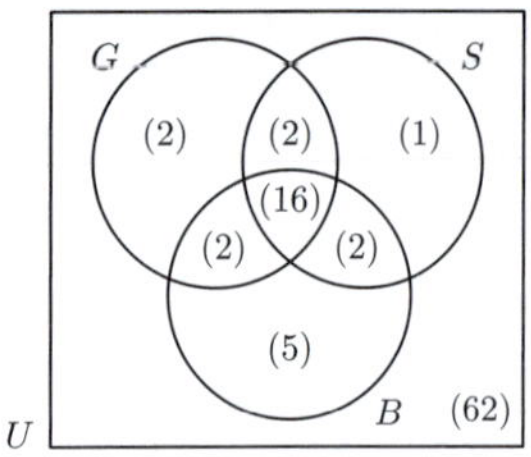

b **i** 62 countries
ii 2 countries
iii 8 countries

10 **a**

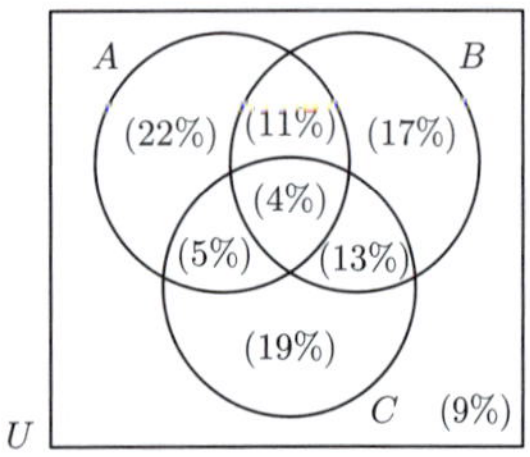

b **i** 9%
ii 91%
iii 58%
iv 72%
v 19%

11 18 students

REVIEW SET 6A

1 **a**

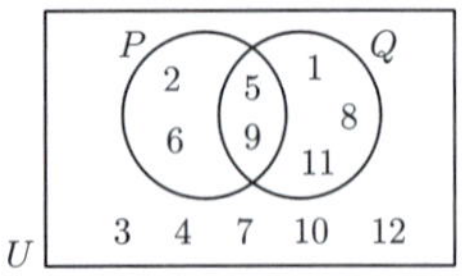

b

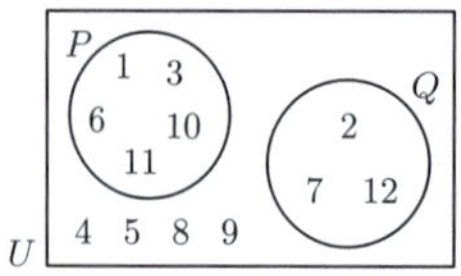

c

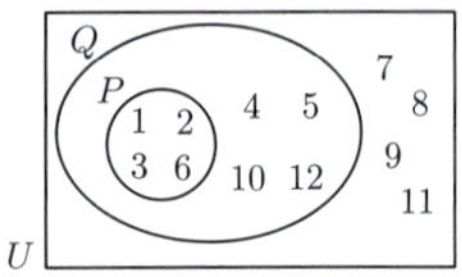

2

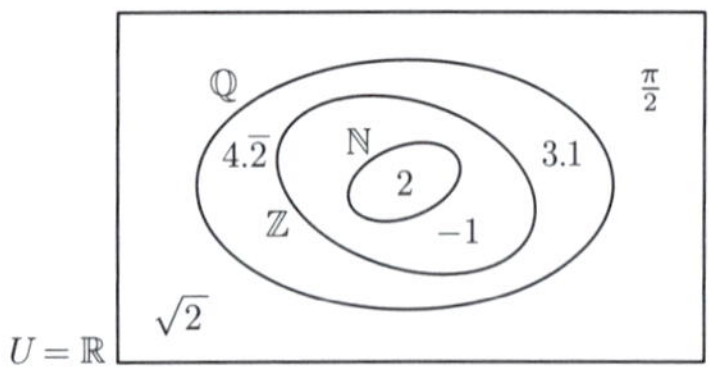

3 **a**

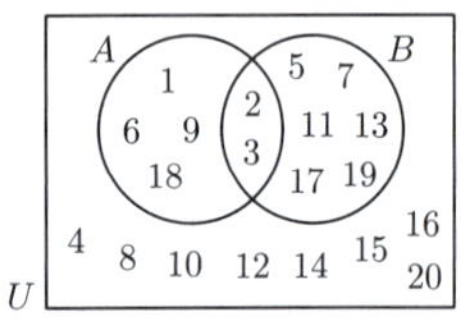

b $n(A \cup B) = 12$

4 **a** in both A and B or in neither A nor B
b in B or C but not A

5 **a** **i**

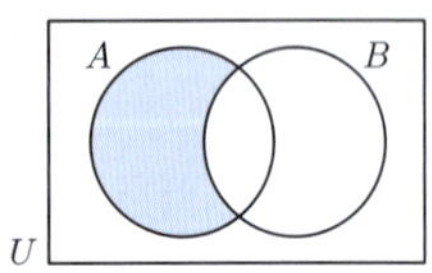

ii

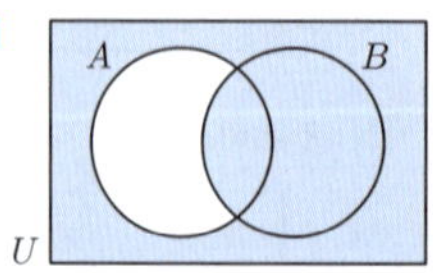

b The shaded region in **a i** is the same as the *unshaded* region in **a ii**.

$\therefore \quad (A' \cup B)' = A \cap B'$

6 **a** 7 **b** 14

7 **a** 6 stalls **b** 37 stalls **8** 12 matches

9 **a**

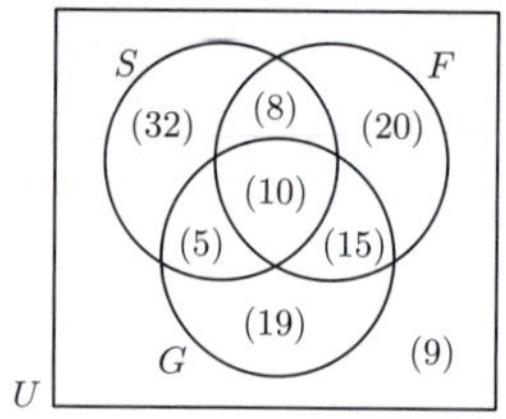

b **i** 118 travellers
ii 28 travellers

REVIEW SET 6B

1 **a**

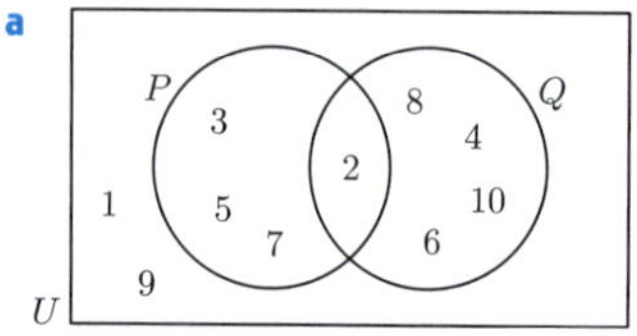

b **i** $n(P') = 6$
ii $n(P \cap Q) = 1$

c yes

2

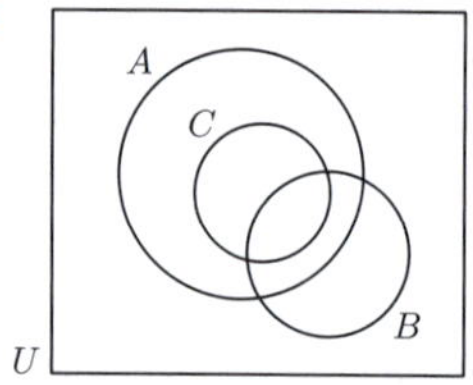

3 **a** 11 **b** 14 **c** 21 **d** 2

4 **a** **i**

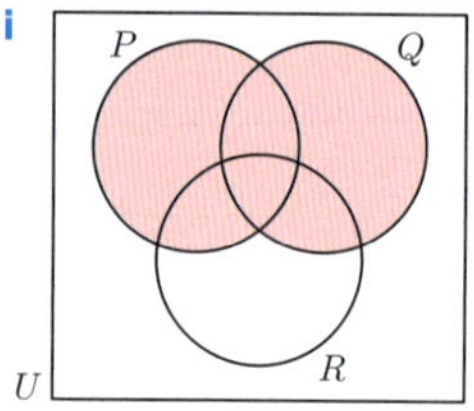

ii

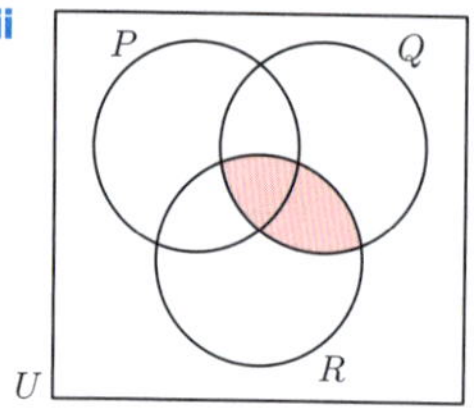

iii

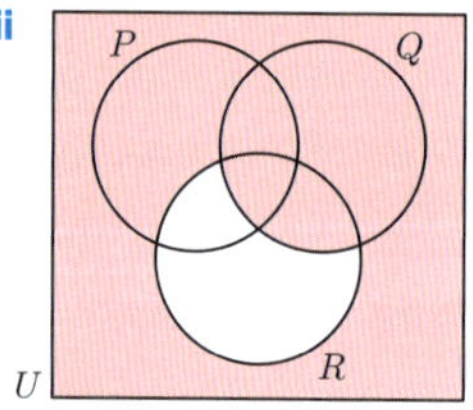

iv

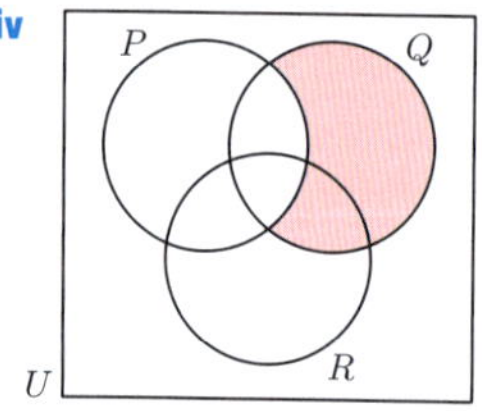

b The shaded region is $(P \cup Q)' \cap R$ which is the same as $P' \cap (Q' \cap R)$.

$\therefore \quad (P \cup Q)' \cap R$
$= P' \cap (Q' \cap R)$

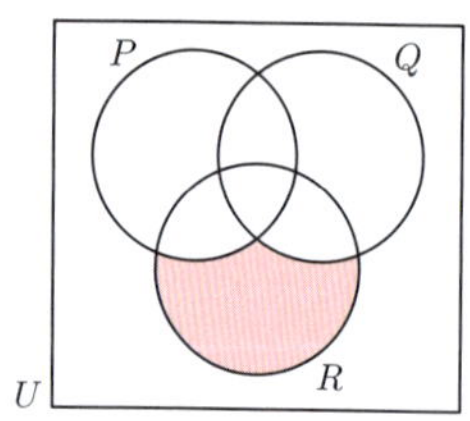

5 **a**

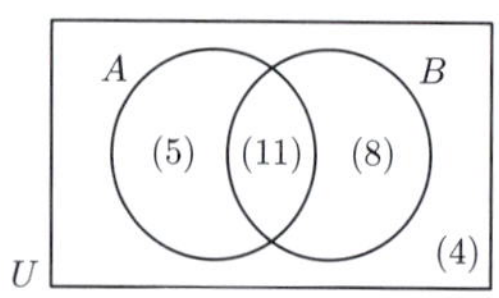

b **i** 24
ii 5

6 **a** 34 families **b** 6 families **7** 19 houses

8 **a** We can represent the percentage of people who read each newspaper on a Venn diagram.

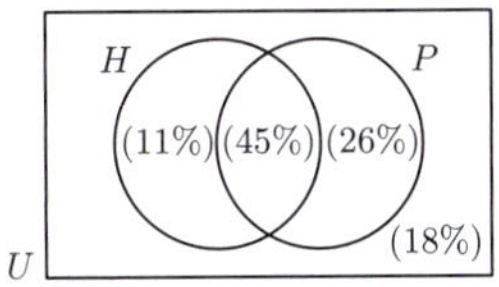

b **i** 45% **ii** 82% **iii** 11% **iv** 37%

9 **a** $A = \{1, 2, 4, 5, 10, 20\}$,
$B = \{2, 4, 6, 8, 10, 12, 14, 16, 18, 20\}$,
$C = \{4, 6, 8, 9, 10, 12, 14, 15, 16, 18, 20\}$

b

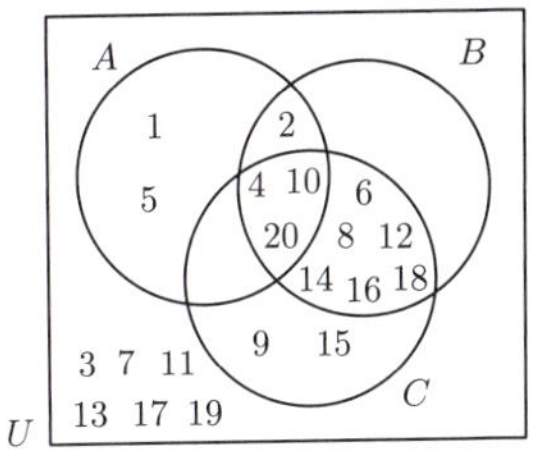

c **i** $n(A \cup B) = 12$
ii $n(A \cap B \cap C) = 3$

d $(B \cup C) \cap A' = \{6, 8, 9, 12, 14, 15, 16, 18\}$
$= A' \cap C$

e The regions $(B \cup C) \cap A'$ and $A' \cap C$ are not the same.
$\therefore$ $(B \cup C) \cap A' = A' \cap C$ does not hold in general, and so is not a set identity.

10 **a**

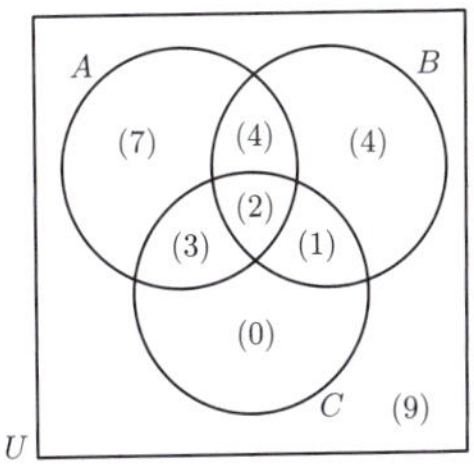

b **i** 17 numbers
ii 8 numbers

c $35 = 5 \times 7$, but no numbers that are a multiple of 5 but not a multiple of 2 and 3 have been selected, so 35 has not been selected.
$60 = 2 \times 2 \times 3 \times 5$ and $30 = 2 \times 3 \times 5$. These are the only numbers from 1 to 75 that are multiples of 2, 3, and 5. From the information two such numbers have been selected, so 60 must have been selected.

d

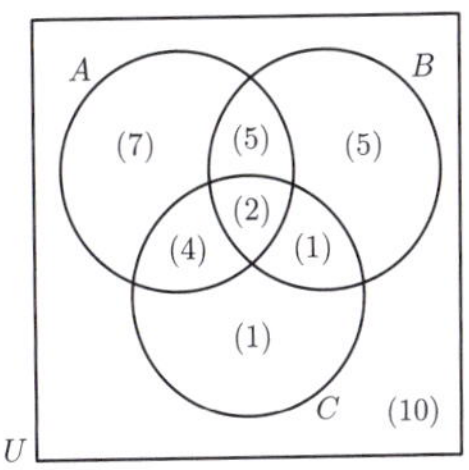

EXERCISE 7A

1 **a** $\sqrt{5}$ is a surd, $\sqrt{5} \approx 2.2361$
b $\sqrt{36} = 6$, $\sqrt{36}$ is not a surd.
c $\sqrt{72}$ is a surd, $\sqrt{72} \approx 8.4853$
d $\sqrt{81} = 9$, $\sqrt{81}$ is not a surd.
e $\sqrt{200}$ is a surd, $\sqrt{200} \approx 14.1421$
f $\sqrt{\frac{1}{10}}$ is a surd, $\sqrt{\frac{1}{10}} \approx 0.3162$
g $\sqrt{\frac{1}{64}} = \frac{1}{8}$, $\sqrt{\frac{1}{64}}$ is not a surd.
h $\sqrt{\frac{3}{14}}$ is a surd, $\sqrt{\frac{3}{14}} \approx 0.4629$

2 **a** D **b** A **c** C **d** E **e** B

3 $\sqrt{1}, \sqrt{4}, \sqrt{9}, \sqrt{16}, ..., \sqrt{961} = 31$ are all rational numbers from the given set.
$\therefore$ $1000 - 31 = 969$ are surds.

4 **a** 3 **b** $3\sqrt{3}$ **c** $9\sqrt{3}$ **d** $\frac{1}{3}$ **e** 5 **f** 25
g 5 **h** 20 **i** $\frac{1}{2}$ **j** $\frac{4}{11}$ **k** $\frac{3}{17}$ **l** $\frac{2}{23}$

5 **a** 8 **b** 32 **c** 12 **d** 20 **e** 45 **f** 28
g 40 **h** 490

6 **a** 24 **b** 30 **c** 70 **d** 32 **e** 147
f -6 **g** 30 **h** -42 **i** -22 **j** $192\sqrt{3}$
k $45\sqrt{5}$ **l** -224

7 **a** 2 **b** -5 **c** $\frac{1}{5}$ **d** 7

EXERCISE 7B

1 **a** $\sqrt{6}$ **b** $\sqrt{34}$ **c** $\sqrt{21}$ **d** $\sqrt{66}$
e $10\sqrt{6}$ **f** $5\sqrt{14}$ **g** $-10\sqrt{14}$ **h** $2\sqrt{30}$
i $2\sqrt{21}$ **j** $3\sqrt{15}$ **k** $80\sqrt{6}$ **l** $162\sqrt{6}$

2 **a** 2 **b** $\frac{1}{3}$ **c** $\sqrt{6}$ **d** $\frac{1}{5}$ **e** $\sqrt{15}$ **f** $\sqrt{\frac{1}{15}}$
g 3 **h** $\frac{1}{4}$ **i** $\sqrt{3}$ **j** 1 **k** 1 **l** $\frac{25}{2}$

3 **a** $\frac{1}{4}$ **b** $\frac{3}{4}$ **c** $\frac{7}{8}$ **d** $\frac{11}{3}$ **e** $\frac{5}{2}$ **f** $\frac{4}{3}$
g $\frac{8}{5}$ **h** $\frac{8}{3}$

4 **a** $2\sqrt{2}$ **b** $3\sqrt{2}$ **c** $5\sqrt{2}$ **d** $7\sqrt{2}$
e $10\sqrt{2}$ **f** $12\sqrt{2}$ **g** $100\sqrt{2}$

5 **a** $2\sqrt{3}$ **b** $3\sqrt{3}$ **c** $5\sqrt{3}$

6 **a** $2\sqrt{5}$ **b** $3\sqrt{5}$ **c** $5\sqrt{5}$

9 $\left(\sqrt{a} + \sqrt{b}\right)^2 = a + 2\sqrt{a}\sqrt{b} + b$
and $\left(\sqrt{a+b}\right)^2 = a + b \neq \left(\sqrt{a} + \sqrt{b}\right)^2$
So in general, $\sqrt{a} + \sqrt{b} \neq \sqrt{a+b}$.

EXERCISE 7C

1 **a** $2\sqrt{3}$ **b** $2\sqrt{7}$ **c** $3\sqrt{6}$ **d** $2\sqrt{15}$
e $3\sqrt{11}$ **f** $2\sqrt{13}$ **g** $2\sqrt{10}$ **h** $3\sqrt{7}$
i $4\sqrt{3}$ **j** $5\sqrt{5}$ **k** $7\sqrt{3}$ **l** $5\sqrt{7}$
m $4\sqrt{11}$ **n** $5\sqrt{6}$ **o** $20\sqrt{5}$

2 **a** $\frac{\sqrt{5}}{3}$ **b** $\frac{3\sqrt{2}}{2}$ **c** $\frac{\sqrt{3}}{2}$ **d** $\frac{5\sqrt{3}}{6}$

3 **a** $2 + \sqrt{2}$ **b** $3 - \sqrt{3}$ **c** $1 + \frac{1}{2}\sqrt{5}$ **d** $2 - \sqrt{2}$
e $2 + \sqrt{2}$ **f** $3 + \frac{1}{2}\sqrt{3}$ **g** $\frac{7}{4} - \frac{1}{4}\sqrt{15}$ **h** $\frac{1}{2} - \sqrt{2}$

EXERCISE 7D

1 **a** $x \approx \pm 3.32$ **b** no real solution **c** $x \approx \pm 8.43$
d $x \approx 9.43$ **e** $x = 2$ **f** $x \approx 2.22$
g $x \approx -2.22$ **h** $x = \pm 3$ **i** no real solution
j $x \approx 1.87$ **k** $x \approx -2.57$ **l** $x \approx 1.84$

2 **a** $x = \pm 2\sqrt{2}$ **b** $x = \pm 6\sqrt{2}$ **c** $x = \pm 4\sqrt{3}$
d $x = \pm 9\sqrt{2}$

3 **a** $x = \pm 4$ **b** $x = \pm 6$ **c** $x = \pm 2\sqrt{15}$
d $x = \pm 3$ **e** no real solution **f** $x = \pm 2$

4 **a** $x = 4$ **b** $x = -2$ **c** $x = \pm 3$

5 **a** $x = \pm 2\sqrt{3}$ **b** $x = \pm 6$ **c** $x = \pm\sqrt{3}$
d $x = \pm 7$ **e** $x = \pm\sqrt{10}$ **f** $x = \pm\sqrt{35}$
g $x = \pm 4$ **h** no real solution

6 **a** $x=4$ **b** $x=81$ **c** $x=0$ **d** $x=7$
e $x=12$ **f** $x=64$ **g** $x=121$ **h** $x=100$

7 $\sqrt{x}$ is defined as the *non-negative* value such that $\sqrt{x}\times\sqrt{x}=x$.

EXERCISE 7E

1 **a** $2\sqrt{5}$ **b** 0 **c** $6\sqrt{3}$ **d** $10\sqrt{2}$ **e** $3\sqrt{3}$
f $-\sqrt{5}$ **g** $\sqrt{10}$ **h** $5\sqrt{6}$ **i** $-6\sqrt{15}$

2 **a** $6\sqrt{2}-2\sqrt{3}$ **b** $7\sqrt{7}+3\sqrt{6}$ **c** $8\sqrt{10}+3\sqrt{5}$
d $-2\sqrt{3}-7\sqrt{13}$ **e** $6\sqrt{7}+5\sqrt{11}$ **f** $-21+6\sqrt{15}$

3 **a** $9\sqrt{2}$ **b** $-8\sqrt{7}$ **c** $28\sqrt{6}$

4 **a** $\sqrt{20}+\sqrt{5}=2\sqrt{5}+\sqrt{5}=3\sqrt{5}=\sqrt{45}$
b $\sqrt{147}-\sqrt{75}=7\sqrt{3}-5\sqrt{3}=2\sqrt{3}=\sqrt{12}$
c $\sqrt{175}-\sqrt{63}=5\sqrt{7}-3\sqrt{7}=2\sqrt{7}=\sqrt{28}$

5 **a** $\frac{7\sqrt{5}}{12}$ **b** $\frac{5\sqrt{6}}{14}$ **c** $\frac{23\sqrt{3}}{24}$ **d** $-\frac{14\sqrt{11}}{45}$
e $\frac{13\sqrt{10}}{12}$ **f** $-\frac{11\sqrt{2}}{28}$

6 **a** $\sqrt{10}+2$ **b** $3\sqrt{2}-2$ **c** $3+\sqrt{3}$
d $\sqrt{3}-3$ **e** $7\sqrt{7}-7$ **f** $2\sqrt{5}-5$
g $22-\sqrt{11}$ **h** $\sqrt{6}-12$ **i** $3+\sqrt{6}-\sqrt{3}$
j $6-2\sqrt{15}$ **k** $6\sqrt{5}-10$ **l** $30+3\sqrt{10}$

7 **a** $2-3\sqrt{2}$ **b** $-2-\sqrt{6}$ **c** $2-4\sqrt{2}$
d $-3-\sqrt{3}$ **e** $-3-2\sqrt{3}$ **f** $-5-2\sqrt{5}$
g $-5+4\sqrt{5}$ **h** $\sqrt{7}-\sqrt{3}$ **i** $4-2\sqrt{2}$
j $9-15\sqrt{3}$ **k** $-14-14\sqrt{3}$ **l** $4-6\sqrt{2}$

8 **a** $4+3\sqrt{2}$ **b** $7+4\sqrt{3}$ **c** $1+\sqrt{3}$
d $10+\sqrt{2}$ **e** -2 **f** $3-3\sqrt{7}$
g $-1-\sqrt{5}$ **h** $1-\sqrt{6}$ **i** $14-7\sqrt{2}$

9 **a** $3+2\sqrt{2}$ **b** $7-4\sqrt{3}$ **c** $7+4\sqrt{3}$
d $5-2\sqrt{6}$ **e** $9+2\sqrt{14}$ **f** $22-8\sqrt{6}$
g $8-4\sqrt{3}$ **h** $13+4\sqrt{10}$ **i** $13-4\sqrt{10}$
j $44+24\sqrt{2}$ **k** $51-10\sqrt{2}$ **l** $19+6\sqrt{2}$

10 **a** 13 **b** 23 **c** 1 **d** -9 **e** 14 **f** 19
g -28 **h** -174 **i** 1 **j** -4 **k** 15 **l** 2

11 **a** $90+34\sqrt{7}$ **b** $9\sqrt{3}-11\sqrt{2}$ **c** $36\sqrt{3}-38\sqrt{2}$

EXERCISE 7F

1 **a** $\frac{\sqrt{5}}{5}$ **b** $\frac{\sqrt{2}}{1}=\sqrt{2}$ **c** $\frac{5\sqrt{3}}{3}$ **d** $\frac{\sqrt{8}}{8}=\frac{\sqrt{2}}{4}$
e $\frac{4\sqrt{7}}{7}$ **f** $\frac{\sqrt{6}}{2}$ **g** $\frac{\sqrt{6}}{3}$ **h** $\frac{\sqrt{21}}{7}$

2 **a** $\frac{\sqrt{3}}{6}$ **b** $\frac{2\sqrt{2}}{3}$ **c** $\frac{\sqrt{15}}{15}$ **d** $\frac{2\sqrt{15}}{15}$
e $\frac{\sqrt{14}}{6}$ **f** $\frac{\sqrt{11}}{2}$ **g** $\frac{3\sqrt{10}}{5}$ **h** $\frac{\sqrt{3}}{5}$

3 **a** $\frac{5\sqrt{3}}{3}$ **b** $\frac{7\sqrt{2}}{2}$ **c** $\frac{\sqrt{3}}{9}$ **d** $\frac{\sqrt{2}}{2}$

REVIEW SET 7A

1 90 numbers **2** **a** 12 **b** 32 **c** 36 **d** -10

3 **a** $\sqrt{77}$ **b** $30\sqrt{15}$ **c** $\sqrt{7}$ **d** $\frac{1}{5}$

4 **a** $3\sqrt{7}$ **b** $6\sqrt{7}$

5 **a** $2\sqrt{11}$ **b** $3\sqrt{10}$ **c** $\frac{3\sqrt{2}}{5}$

6 **a** $x=\pm 2\sqrt{6}$ **b** $x=\pm 5\sqrt{6}$ **c** $x=\pm 4\sqrt{6}$

7 **a** $x=\pm 7$ **b** $x=-2$ **c** $x=\pm\sqrt{21}$ **d** $x=64$

8 **a** $-4\sqrt{7}$ **b** $9\sqrt{6}-3\sqrt{3}$ **c** $\frac{11\sqrt{7}}{30}$

9 **a** $8\sqrt{3}-6$ **b** $16-6\sqrt{7}$ **c** 1
d $\sqrt{5}-4$ **e** $8+5\sqrt{2}$ **f** $79+12\sqrt{14}$

10 **a** $5\sqrt{3}$ **b** $\frac{\sqrt{10}}{10}$ **11** **a** $\frac{27\sqrt{2}}{4}$ **b** $6\sqrt{3}$

REVIEW SET 7B

1 **a** $\frac{9}{10}$ **b** 54 **c** $50\sqrt{2}$

2 **a** $\sqrt{65}$ **b** $20\sqrt{3}$ **c** 2 **d** $\frac{7}{3}$ **3** $4\sqrt{5}$

4 **a** $4+\sqrt{3}$ **b** $3-\frac{3}{4}\sqrt{5}$ **c** $\frac{1}{5}+\frac{2}{5}\sqrt{2}$

5 **a** $x\approx\pm 5.39$ **b** $x\approx 3.08$ **c** $x\approx -3.42$

6 **a** $x=\pm 1$ **b** $x=3$ **c** $x=20$

7 **a** $13\sqrt{3}$ **b** $-\frac{13\sqrt{13}}{60}$ **c** $\frac{29\sqrt{6}}{60}$

8 **a** $\sqrt{54}-\sqrt{24}=3\sqrt{6}-2\sqrt{6}=\sqrt{6}$
b $\sqrt{18}+\sqrt{32}=3\sqrt{2}+4\sqrt{2}=7\sqrt{2}=\sqrt{98}$

9 **a** $6-2\sqrt{3}$ **b** $-2\sqrt{2}+2$ **c** 22
d $9-4\sqrt{5}$ **e** $7\sqrt{2}-9$ **f** 2

10 **a** $7\sqrt{2}$ **b** $\frac{5\sqrt{3}}{6}$

11 **a** 80 cm^2 **b** **i** $\sqrt{80}$ cm **ii** $(12+\sqrt{80})$ cm
c **i** $\frac{\sqrt{80}}{12}$ **ii** $\frac{\sqrt{5}}{3}$

EXERCISE 8A

1 **a** $\sqrt{65}$ cm **b** 10 cm **c** $\sqrt{233}$ km
d $5\sqrt{2}$ cm **e** $2\sqrt{61}$ m **f** $\sqrt{481}$ mm

2 **a** $\sqrt{85}$ cm **b** 8 m **c** $6\sqrt{2}$ cm **d** $3\sqrt{13}$ cm
e $\sqrt{26}$ m **f** 180 cm or 1.8 m

3 **a** ≈ 9.4 cm **b** ≈ 7.1 m **c** ≈ 8.8 cm
d ≈ 5.9 cm **e** ≈ 5.2 m **f** ≈ 16.9 cm

4 **a** $x=\sqrt{11}$ **b** $x=\sqrt{2}$ **c** $x=\sqrt{5}$
d $x=\sqrt{19}$ **e** $x=\frac{1}{\sqrt{2}}$ **f** $x=\frac{1}{2}$

5 **a** $x\approx 3.61$ **b** $x\approx 0.226$ **c** $x\approx 2.49$

6 **a** $x=2\sqrt{2},\ \ y=\sqrt{17}$ **b** $x=3\sqrt{5},\ \ y=\sqrt{29}$
c $x=\sqrt{5},\ \ y=\sqrt{6}$

7 **a** $x=\sqrt{10}$ **b** $x=13$ **c** $x=12$

8 $\text{AC}=\sqrt{39}\text{ m}\approx 6.24$ m

10 **a** $\text{AB}=\sqrt{17}$ cm **b** $\text{AB}=\sqrt{29}$ m **c** $\text{AB}=\sqrt{41}$ m

EXERCISE 8B

1 **a** yes **b** no **c** yes

2 **a** $5^2+7^2=25+49=74\neq 9^2$
b -5 is not a positive integer. **c** $\sqrt{5}$ is not an integer.

3 **a** $k=20$ **b** $k=61$ **c** $k=45$ **d** $k=15$
e $k=21$ **f** $k=11$

4 **b** $\{36, 77, 85\}$ **6** **b** $\{20, 21, 29\}$

EXERCISE 8C

1 ≈ 8.54 cm **2** ≈ 4.59 m **3** ≈ 6.63 cm

4 ≈ 7.07 cm **5** $2\sqrt{445}\text{ km}\approx 42.2$ km

6 $\approx 3.16\text{ cm}\times\approx 9.49$ cm **7** 8 cm **8** ≈ 7.52 m

9 ≈ 2.72 m **10** $\frac{32\sqrt{3}}{3}$ cm

11 **a** Sven: ≈ 115.4 m, Jill: 100 m **b** ≈ 15.4 m

12 **a** ≈ 23.7 km **b** ≈ 2.31 hours or ≈ 2 h 18 min

13 ≈ 2.24 km **14** 6.025 m **15** D is 8 km from B.

16 15 cm

17

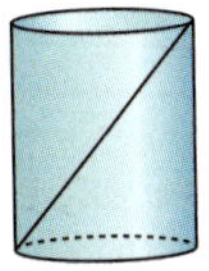

When placed like this, the greatest length a stirrer could be and lie within the glass is about 11.66 cm. So a 12 cm stirrer would go outside the glass.

18 ≈ 3.16 cm **19** $3\sqrt{3}$ cm **20** ≈ 6.80 m

21 ≈ 42.9 m **22** ≈ 71 m

EXERCISE 8D

1 Only **b** is right angled.

2 **a** It is right angled at A. **b** It is not right angled.
c It is right angled at C. **d** It is not right angled.
e It is right angled at A. **f** It is right angled at B.

3 $8^2 + 11.55^2 = 197.4025$ and $14.05^2 = 197.4025$
$\therefore$ is right angled.

4 $40^2 + 42^2 = 3364$, $58^2 = 3364$
$\therefore$ the triangle is right angled.

5 yes

REVIEW SET 8A

1 **a** $x = \sqrt{29}$ **b** $x = 3\sqrt{2}$ **c** $x = \sqrt{11}$
d $x = \sqrt{2}$ **e** $x = 12$

3 **a** $x = \sqrt{113}$, $y = 2\sqrt{14}$ **b** $x = 4\sqrt{5}$, $y = 6$

4 No, $5^2 + 11^2 = 146$ but $13^2 = 169$.

5 **a** $k = 30$ **b** $k = 41$ **c** $k = 60$ **6** ≈ 10.2 cm

7 **a** ≈ 94.1 km **b** ≈ 2.70 hours or ≈ 2 h 42 min

8 AB ≈ 6.55 m

9 Yes, as $4^2 + 1^2 = (\sqrt{17})^2$. The right angle is at A.

10 No, $(\sqrt{41})^2 + (\sqrt{61})^2 \neq 10^2$ **11** 5 cm

REVIEW SET 8B

1 **a** $x \approx 5.74$ **b** $x \approx 4.49$ **c** $x \approx 3.58$
d $x \approx 4.65$ **e** $x = 16$

2 **a** AB $= \sqrt{85}$ m **b** AB $= \sqrt{33}$ cm

3 -3 is not a positive integer. **4** $k = 35$

5 **b** $\{20, 21, 29\}$ **c** $\{40, 96, 104\}$ and $\{24, 32, 40\}$

6 ≈ 1.431 m **7** $4\sqrt{10}$ cm **8** ≈ 55.9 m

9 $2^2 + 5^2 = 29$ and $\left(\sqrt{29}\right)^2 = 29$ $\therefore$ right angled at B.

10 $\widehat{BCD}$ **11** ≈ 80.30 m

EXERCISE 9A

1 **a** 1 **b** $\frac{1}{2}$ **c** 2 **d** $1\frac{3}{5}$ **e** 1 **f** $4\frac{1}{3}$
g 2 **h** -1 **i** 2 **j** 1 **k** 1 **l** $\frac{1}{2}$

2 **a** -2 **b** 3 **c** 1 **d** -1 **e** 5 **f** 1
g $-\frac{1}{2}$ **h** $-\frac{1}{3}$ **i** -1 **j** $-\frac{1}{2}$ **k** $\frac{1}{2}$ **l** 5
m -5 **n** $1\frac{1}{4}$ **o** $-\frac{7}{9}$ **p** $-\frac{1}{3}$

EXERCISE 9B

1 **a** $\frac{a}{2}$ **b** $2m$ **c** 6 **d** 3 **e** $2a$ **f** x^2
g $2x$ **h** 2 **i** $\frac{1}{2a}$ **j** $2m$ **k** 4 **l** $\frac{2}{t}$
m $2d$ **n** $\frac{b}{2}$ **o** $\frac{2b}{3a}$

2 **a** $-\frac{1}{3m}$ **b** $-\frac{m}{2}$ **c** $\frac{1}{3m^2}$ **d** -5
e $\frac{3n}{m}$ **f** $-\frac{2}{m^2}$ **g** $-\frac{n}{2}$ **h** $-\frac{3m}{n}$

3 **a** cannot be simplified **b** n **c** cannot be simplified
d 1 **e** cannot be simplified

4 **a** 4 **b** $2n$ **c** a **d** 1
e a^2 **f** $\frac{3n}{2}$ **g** $\frac{1}{2b^3}$ **h** $\frac{k^4}{2a^3}$

5 **a** $2(x+5)$ **b** $\frac{n+5}{6}$ **c** $\frac{b+2}{2}$ **d** $\frac{3(k-2)}{4}$
e $\frac{5}{t-1}$ **f** $\frac{2}{5(k+4)}$ **g** $\frac{1}{3(x-3)}$ **h** $\frac{5(p+4)}{3}$

6 **a** $\frac{x+2}{9}$ **b** $\frac{12}{a+1}$ **c** $\frac{x+y}{3}$ **d** $x+y$
e $\frac{2}{x+2}$ **f** $\frac{a+5}{3}$ **g** $\frac{y}{3}$ **h** $\frac{y-3}{3}$
i $\frac{x+1}{3}$ **j** $\frac{1}{2(b-4)}$ **k** $\frac{2(p+q)}{3}$ **l** $\frac{8}{5(r-2)}$
m $\frac{x}{x-1}$ **n** $\frac{(x+2)(x+1)}{4}$ **o** $\frac{(x-1)^2}{4x(x+3)}$

7 **a** $x+2$ **b** $x-2$ **c** $\frac{x+2}{2}$ **d** $\frac{x-5}{2}$
e $\frac{3}{y+3}$ **f** $\frac{2}{3(x-5)}$ **g** $\frac{2x+3}{3}$ **h** $\frac{2(2x+3)}{5}$
i $\frac{3(2a-1)}{2}$ **j** $2a-1$ **k** $\frac{2}{3a+1}$ **l** $\frac{3(b+3)}{2}$
m $\frac{b+3}{2}$ **n** $\frac{3}{2(2b-3)}$ **o** $a+b$ **p** $\frac{a(x+1)}{x}$

8 **a** $\frac{3}{4}$ **b** $\frac{5}{3}$ **c** $\frac{2}{5}$ **d** $\frac{4}{5}$ **e** 3 **f** $\frac{1}{2}$ **g** $4x$
h $3x$ **i** $\frac{5x}{7}$ **j** $\frac{3b}{4}$ **k** $\frac{2x}{3}$ **l** $\frac{3(a+b)}{2a^2(2a+b)}$

9 **a** x **b** $\frac{1}{y}$ **c** $\frac{x}{y}$ **d** $\frac{a+b}{c+d}$

10 **a** -2 **b** $-\frac{3}{2}$ **c** -1 **d** $-\frac{1}{2}$ **e** -3
f $-\frac{2}{x}$ **g** $-\frac{ab}{2}$ **h** $-2x$

11 **a** $x+1$ **b** $x-1$ **c** $-(x+1)$ **d** $\frac{1}{x-2}$
e $a-b$ **f** $-(a+b)$ **g** $\frac{2}{x-1}$ **h** $\frac{3+x}{x}$
i $\frac{3(x+y)}{2y}$ **j** $-\frac{2(b+a)}{a}$ **k** $\frac{y}{4x+y}$ **l** $-\frac{4x}{x+4}$

12 **a** $x+1$ **b** $\frac{1}{x-3}$ **c** $\frac{1}{a-5}$ **d** $\frac{2x}{x-5}$
e $\frac{x-2}{x+2}$ **f** $\frac{x+3}{x-1}$ **g** $\frac{x+1}{1-x}$ **h** $\frac{n-5}{n+3}$

13 **a** $\frac{2x-1}{x}$ **b** $\frac{x+2}{x+3}$ **c** $\frac{x+2}{2x-1}$ **d** $\frac{2x+1}{x-1}$
e $\frac{4x-3}{2x+1}$ **f** $\frac{3x+4}{x+2}$

14 $\dfrac{2k^2-7k-15}{k-5} = \dfrac{(2k+3)\cancel{(k-5)}^{\,1}}{\cancel{k-5}_{\,1}} = 2k+3$

For $k \in \mathbb{Z}$, $2k$ is always an even integer, so $(2k+3)$ is always an odd integer.

15 $n = 0, \frac{2}{3}$, or 3

EXERCISE 9C

1 **a** $\dfrac{xy}{10}$ **b** $\frac{3}{2}$ **c** $\dfrac{a^2}{2}$ **d** $\frac{1}{6}$ **e** $\frac{1}{5}$ **f** $\dfrac{c^2}{10}$
g $\dfrac{ac}{bd}$ **h** 1 **i** $\dfrac{1}{2m}$ **j** 2 **k** $\dfrac{a}{b}$ **l** 4
m $\dfrac{3}{m}$ **n** $\dfrac{a^2}{b^2}$ **o** $\dfrac{4}{x^2}$ **p** $\dfrac{1}{c}$

2 **a** $2x^2$ **b** $\dfrac{6}{y^2}$ **c** $\dfrac{2y^2}{x}$ **d** $\dfrac{3}{xy}$

3 **a** $\dfrac{5x}{2(x-2)}$ **b** $\dfrac{4}{3t}$ **c** $\dfrac{4k}{3}$

EXERCISE 9D

1 **a** $\frac{3}{2}$ **b** $\dfrac{3}{a}$ **c** $\dfrac{3x}{16}$ **d** $\frac{3}{4}$ **e** 2 **f** $\dfrac{c}{25}$
g $\frac{1}{5}$ **h** $\dfrac{m^2}{2}$ **i** 2 **j** $\dfrac{n}{m}$ **k** $\dfrac{3}{4g}$ **l** $\dfrac{g}{3}$
m $\dfrac{8}{x^3}$ **n** $\dfrac{x^2}{3}$ **o** $\dfrac{1}{a}$ **p** $\dfrac{3a}{5}$

2 **a** $\dfrac{x^3}{3y}$ **b** $\dfrac{10z^2}{y^2}$ **c** $\dfrac{7a^3}{3b^3}$ **d** $\dfrac{9}{2y^3}$

3 **a** $\dfrac{20}{a}$ **b** $-\dfrac{1}{3x}$ **c** $\dfrac{3m}{2(5m-1)}$

EXERCISE 9E

1 **a** $\dfrac{5a}{6}$ **b** $\dfrac{b}{10}$ **c** $\dfrac{7c}{4}$ **d** $\dfrac{5d-6}{10}$
e $-\dfrac{5x}{14}$ **f** $\dfrac{4a+3b}{12}$ **g** $-\dfrac{2t}{9}$ **h** $\dfrac{5m}{21}$
i $\dfrac{d}{2}$ **j** $\dfrac{11p}{35}$ **k** $\dfrac{22t}{45}$ **l** $\dfrac{19k}{72}$
m m **n** $\dfrac{5a}{12}$ **o** $\dfrac{x}{12}$ **p** $\dfrac{41q}{21}$

2 **a** $\dfrac{x+2}{2}$ **b** $\dfrac{b-12}{4}$ **c** $\dfrac{6+a}{3}$ **d** $\dfrac{y-3}{3}$
e $\dfrac{3a}{2}$ **f** $\dfrac{8-x}{2}$ **g** $\dfrac{4x}{5}$ **h** $\dfrac{2x+1}{x}$
i $\dfrac{5x-2}{x}$ **j** $\dfrac{a^2+2}{a}$ **k** $\dfrac{3+b^2}{b}$ **l** $\dfrac{1-2x^3}{x^2}$

3 **a** $\dfrac{7b+3a}{ab}$ **b** $\dfrac{3c+2a}{ac}$ **c** $\dfrac{4a+15}{3a}$ **d** $\dfrac{2an-am}{mn}$
e $\dfrac{5}{2a}$ **f** $\dfrac{2a+3}{2x}$ **g** $\dfrac{4y-1}{xy}$ **h** $\dfrac{31}{5x}$
i $\dfrac{35}{12z}$ **j** $\dfrac{6+a^2}{2a}$ **k** $\dfrac{3x-2y}{3y}$ **l** $\dfrac{7x}{18y}$
m $-\dfrac{19}{40t}$ **n** $\dfrac{x+1}{x^2}$ **o** $\dfrac{x-2}{x^2}$ **p** $\dfrac{5x+6}{2x^2}$

4 **a** $\dfrac{1+2x}{x^2y}$ **b** $\dfrac{2y^2-1}{xy^2}$ **c** $\dfrac{3-8x}{2x^2}$ **d** $\dfrac{q^2-p}{p^2q}$
e $\dfrac{3s+2r^2}{rs^2}$ **f** $\dfrac{6y-x^2}{2x^2y^2}$

5 **a** $\dfrac{5x+2}{6}$ **b** $\dfrac{-x-1}{4}$ **c** $\dfrac{11x+9}{12}$ **d** $\dfrac{x+5}{6}$
e $\dfrac{-2x-1}{12}$ **f** $\dfrac{10x+3}{6}$ **g** $\dfrac{x-3}{12}$ **h** $\dfrac{23x-6}{30}$
i $\dfrac{5a-b}{6}$ **j** $\dfrac{14x-1}{20}$ **k** $\dfrac{11x-12}{30}$ **l** $\dfrac{3x-4}{20}$

6 **a** $\dfrac{7x+3}{x(x+1)}$ **b** $\dfrac{2x-6}{x(x+2)}$ **c** $\dfrac{x-7}{(x+1)(x-1)}$
d $\dfrac{3x+7}{x+2}$ **e** $\dfrac{5x-4}{x(x-4)}$ **f** $\dfrac{-4x-10}{x+3}$
g $\dfrac{2x^2+x+1}{(x+1)(x-1)}$ **h** $\dfrac{11x-10}{x(x-2)}$
i $\dfrac{-2x-6}{(x+1)(x+2)}$ **j** $\dfrac{2x^2+5x-4}{(x-1)(2x+1)}$
k $\dfrac{x^2+2x+17}{(x-1)(x+3)}$ **l** $\dfrac{-8x}{(x+5)(x-3)}$

7 **a** $\$\left(\dfrac{210x+150}{x(x+1)}\right)$ **b** **i** \$49.50 **ii** \$40

8 **a** $\dfrac{-1}{a(a+1)}$

9 **a** $\dfrac{2+x}{x(x+1)}$ **b** $\dfrac{2x^2+4x+4}{(x+2)(x-3)}$ **c** $\dfrac{x^2-2x+3}{(x-2)(x+3)}$

10 **b** **i** $\frac{1}{2}=\frac{1}{3}+\frac{1}{6}$ **ii** $\frac{1}{5}=\frac{1}{6}+\frac{1}{30}$ **iii** $\frac{1}{9}=\frac{1}{10}+\frac{1}{90}$

EXERCISE 9F

1 **a** $x=8$ **b** $x=-\frac{1}{4}$ **c** $x=3$ **d** $x=-18$
e $x=-5$ **f** $x=-1$ **g** $x=-\frac{1}{5}$ **h** $x=-1$
i $x=-5$

2 **a** $x=\frac{23}{3}$ **b** $x=12$ **c** $x=4$ **d** $x=\frac{5}{3}$
e $x=\frac{25}{9}$ **f** $x=\frac{11}{15}$

3 **a** $x=-25$ **b** $x=2$ **c** $x=\frac{23}{4}$ **d** $x=15$
e $x=\frac{4}{7}$ **f** $x=-16$

4 **a** $x=15$ **b** $x=\frac{9}{2}$ **c** $x=\frac{35}{4}$ **d** $x=-\frac{3}{8}$
e $x=\frac{3}{8}$ **f** $x=-\frac{35}{12}$ **g** $x=\frac{4}{15}$ **h** $x=-\frac{2}{15}$
i $x=8$ **j** $x=\frac{15}{2}$ **k** $x=-\frac{25}{21}$ **l** $x=-\frac{3}{7}$

5 **a** $x=7$ **b** $x=\frac{9}{2}$ **c** $x=-4$ **d** $x=-\frac{3}{8}$
e $x=9$ **f** $x=\frac{5}{7}$

6 **a** $x=2$ **b** $x=-9$ **c** $x=-\frac{4}{3}$ **d** $x=-27$
e $x=-\frac{7}{2}$ **f** $x=-1$

7 **a** $x=13$ **b** $x=-5$ **c** $x=-\frac{7}{2}$ **d** $x=\frac{17}{11}$
e $x=\frac{1}{8}$ **f** $x=-\frac{3}{4}$ **g** $x=-12$ **h** $x=-5$

8 The numerator of the fraction is 20. **9** The number is $\frac{7}{3}$.

10 The fraction is $\frac{15}{20}$. **11** 30 mL of water must be added.

12 Steve needs to add 1.75 L of petrol.

REVIEW SET 9A

1 **a** 4 **b** $\frac{1}{3}$ **c** -1 **2** **a** $\dfrac{ab}{3}$ **b** $\dfrac{2t}{3}$ **c** $\dfrac{x}{3}$

3 **a** $\frac{2}{x+2}$ **b** $\frac{2}{x-3}$ **c** $\frac{8}{3}$ **d** $-\frac{x+4}{2}$
e $\frac{a-3}{a+6}$ **f** $\frac{3x}{3x+1}$

4 **a** $\frac{2m}{n^2}$ **b** $\frac{m}{2}$ **c** $\frac{2}{3xy}$ **d** $\frac{3y}{2(y+2)}$ **e** $-\frac{3}{4x}$

5 **a** $\frac{11x}{12}$ **b** $\frac{14+x}{7}$ **c** $\frac{x-4}{4}$ **d** $\frac{5x}{12}$
e $\frac{5b+9a}{ab}$ **f** $\frac{17}{4x}$

6 **a** $\frac{3x+3}{10}$ **b** $\frac{3x}{(x+1)(x-2)}$ **c** $\frac{4x^2+4x+3}{(x+1)(x+2)}$

7 **a** $x=-2$ **b** $x=\frac{17}{4}$ **c** $x=\frac{1}{15}$ **d** $x=9$
e $x=\frac{5}{6}$ **f** $x=-5$

8 $\frac{72}{48}$ or $\frac{-72}{-48}$

9 **a** $\frac{7x+16}{10}$ goals **b** **i** 10 goals **ii** 17 goals

REVIEW SET 9B

1 **a** -6 **b** 0 **c** 2

2 **a** $\frac{1}{3}$ **b** -2 **c** $2(x-2)$ **d** $-\frac{5}{x}$
e $\frac{x+2}{x}$ **f** $\frac{x-3}{2x+1}$

3 **a** $\frac{k}{3}$ **b** $\frac{3a^2}{b}$ **c** $\frac{t}{24}$

4 **a** $\frac{3}{2n}$ **b** $\frac{x^3}{12y}$ **c** $\frac{7x}{3(x+5)}$

5 **a** $\frac{11}{2x}$ **b** $\frac{6b-ay}{by}$ **c** $\frac{5x}{14}$ **d** $\frac{4x+9}{3x^2}$
e $\frac{3x-y}{x}$ **f** $\frac{6+3x+2y}{6}$

6 **a** $\frac{3x-2}{8}$ **b** $\frac{7-x}{(x-1)(x+2)}$
c $\frac{2x^3+3x^2-10x-12}{x(x+3)(x+2)}$

8 **a** $x=1$ **b** $x=-2$ **c** $x=4$ **d** $x=-7$
e $x=\frac{6}{5}$ **f** $x=-\frac{3}{2}$

9 62.5 mL of water needs to be added.

10 **a** **i** $\frac{1}{2}+\frac{1}{4}$ **ii** $\frac{1}{4}+\frac{1}{6}$ **iii** $\frac{1}{8}+\frac{1}{10}$
b $\frac{1}{x-1}+\frac{1}{x+1}$ {from **a**}

EXERCISE 10A

1 **a** $C=4\times 5$ **b** $C=4x$ **c** $C=mx$

2 **a** $A=5\times 15$ **b** $A=5p$ **c** $A=tp$

3 **a** $A=2000+150\times 8$ **b** $A=2000+150w$
c $A=2000+dw$ **d** $A=P+dw$

4 **a** $C=40+80\times 5$ **b** $C=40+80t$
c $C=40+xt$ **d** $C=F+xt$

5 **a** $P=3\times 10-1(15-10)$ **b** $P=3c-1(20-c)$
c $P=3c-1(a-c)$

6 **a** $D=4\times 6+2\times(4-1)$ **b** $D=pm+b(p-1)$

7 **a** $G=3\times(2-1)+2\times(3-1)$
b $G=5\times(3-1)+3\times(5-1)$
c $G=4\times(4-1)+4\times(4-1)$
d $G=m(n-1)+n(m-1)$

EXERCISE 10B

1 **a** ≈ 71.4 km/h **b** 220 km **c** ≈ 8 h 19 min

2 **a** 19.6 m **b** ≈ 4.52 s **3** **a** ≈ 1.34 s **b** 81 cm

4 **a** 66 line segments **b** 300 line segments

5 **a** ≈ 15.9 km **b** ≈ 49.3 m

6 **a** 160 J **b** ≈ 0.159 kg **c** ≈ 186 m/s

EXERCISE 10C

1 **a** $y=7-x$ **b** $y=x-3$ **c** $y=\frac{1-x}{2}$
d $y=\frac{x-7}{2}$ **e** $y=\frac{10-2x}{5}$ **f** $y=\frac{3x-18}{5}$
g $y=\frac{20-3x}{4}$ **h** $y=\frac{12+2x}{3}$ **i** $y=\frac{20-5x}{2}$

2 **a** $x=r-p$ **b** $x=\frac{z}{y}$ **c** $x=\frac{a-d}{3}$
d $x=5-2y$ **e** $x=\frac{p-by}{a}$ **f** $x=\frac{y-c}{m}$

3 **a** $x=\sqrt{k}$ **b** $x=\frac{\sqrt{M}}{2}$ **c** $x=\sqrt{aN}$
d $x=\sqrt{m-n}$ **e** $x=\sqrt{3b-a}$ **f** $x=\sqrt[3]{\frac{D+y}{3}}$

4 **a** $r=\sqrt{\frac{A}{\pi}}$ **b** $x=\frac{\sqrt{y+7}}{2}$ **c** $b=\sqrt{c^2-a^2}$
d $r=\sqrt[3]{\frac{3V}{4\pi}}$

5 **a** $z=\frac{b}{ac}$ **b** $z=-\frac{pt}{q}$ **c** $z=\frac{2d}{3}$
d $z=\sqrt{2a}$ **e** $z=\sqrt[3]{bn}$ **f** $z=-\sqrt{m(a-b)}$

6 **a** $h=3M-2j$ **b** $j=\frac{3M-h}{2}$

7 **a** $a=d^2n^2$ **b** $a=\sqrt{b^2+c^2}$ **c** $l=\frac{gT^2}{4\pi^2}$
d $b=\frac{16a}{A^2}$

8 **a** $a=\frac{2-bP}{P}$ **b** $q=\frac{Ap-B}{A}$ **c** $x=\frac{3-Ay}{2A}$
d $y=\sqrt{\frac{4-Mx^2}{M}}$

9 **a** $x=3+\frac{2}{y-1}$ **b** $x=-4+\frac{5}{y+2}$
c $x=2-\frac{6}{y+3}$

10 **a** $x=\frac{c-a}{3-b}$ **b** $x=\frac{c}{a+b}$ **c** $x=\frac{a+2}{n-m}$
d $x=-\frac{a}{b+8}$ **e** $x=\frac{a-b}{1-c}$ **f** $x=\frac{e-d}{r+s}$
g $x=\frac{1-4y}{3}$ **h** $x=\frac{5a+b}{b-a}$ **i** $x=\frac{6}{2-3y}$

11 **a** $x = \dfrac{y}{1-y}$ **b** $x = \dfrac{2y+3}{1-y}$ **c** $x = \dfrac{3y+1}{3-y}$
d $x = \dfrac{y-2}{y-5}$ **e** $x = \dfrac{2y+1}{y+4}$ **f** $x = \dfrac{3y-7}{2y+3}$

EXERCISE 10D

1 **a** $P = \dfrac{I}{in}$ **b** $10 000

2 **a** $a = \dfrac{d^2}{2bK}$ **b** **i** $a = \frac{9}{7}$ **ii** $a = \frac{81}{5}$

3 **a** $t = (H-1)^2$
b **i** 1 year **ii** 4 years **iii** $6\frac{1}{4}$ years

4 **a** $v = \sqrt{u^2 + 2as}$ **b** **i** ≈ 20.6 m/s **ii** ≈ 52.9 m/s

5 **a** $r = \sqrt{\dfrac{2P}{\pi \rho v^3 c_p}}$ **b** **i** ≈ 8.64 m **ii** ≈ 3.82 m

6 **a** $\approx 58.8\%$ **b** $w = \dfrac{Pl}{100 - P}$ **c** 9 matches
d 7 consecutive matches

7 **a** $\approx 2.01 \times 10^{20}$ newtons **b** $d = \sqrt{\dfrac{Gm_1m_2}{F}}$
c **i** $\approx 1.50 \times 10^{11}$ m **ii** $\approx 1.43 \times 10^{14}$ m

8 **a** $o = \dfrac{fi}{i-f}$ **b** **i** ≈ 81.8 cm **ii** ≈ 3.48 cm

EXERCISE 10E

1 **a**

Diagram	1	2	3	4
Number of matchsticks	6	11	16	21

b $M = 5n + 1$ **c** 101 matchsticks

2 **a**

Diagram	1	2	3	4
Number of matchsticks	7	10	13	16

b $M = 3n + 4$ **c** 64 matchsticks

3 **a** **i** $S_2 = 4$ **ii** $S_3 = 9$ **iii** $S_4 = 16$ **iv** $S_5 = 25$
b $S_n = n^2$

4 **a** **i** 1 **ii** 2 **iii** 4 **iv** 8 **v** 16 **vi** 32
b **i** 3 **ii** 7 **iii** 15 **iv** 31 **c** $S_n = 2^n - 1$

5 **a** **i** $S_1 = \frac{1}{2}$ **ii** $S_2 = \frac{2}{3}$ **iii** $S_3 = \frac{3}{4}$ **iv** $S_4 = \frac{4}{5}$
b $S_n = \dfrac{n}{n+1}$ **c** $S_{99} = \frac{99}{100}$

6 **b** 338 350

REVIEW SET 10A

1 **a** $B = 15 + 25 \times 5$ **b** $B = s + 25p$ **c** $B = s + mp$

2 **a** ≈ 11.3 g/cm^3 **b** 37.94 g **c** ≈ 9230 cm^3

3 **a** $x = \dfrac{3p - n}{m}$ **b** $x = 3y + 4$ **c** $x = \dfrac{5y}{7}$
d $x = \dfrac{c + 2F}{F}$ **e** $x = \dfrac{a-5}{b-4}$ **f** $x = \dfrac{2y-3}{y-2}$

4 **a** $h = T^2 + l^2$ **b** $h = \dfrac{5 - G^2}{G^2}$

5 **a** $b = \dfrac{a}{a-1}$ **b** $b = \frac{3}{2}$; $3 \times \frac{3}{2} = 3 + \frac{3}{2} = 4\frac{1}{2}$ ✓

6 **a**

Diagram	1	2	3	4
Number of matchsticks	4	7	10	13

b $M = 3n + 1$ **c** 151 matchsticks

7 **a** **i** ≈ 0.816 **ii** ≈ 0.742 **b** $h = C^2H$
c TB: ≈ 2.65 m, GB: ≈ 3.68 m, M: ≈ 2.16 m, SB: ≈ 3.99 m

8 **a** **i** $S_1 = \frac{1}{2}$ **ii** $S_2 = \frac{3}{4}$ **iii** $S_3 = \frac{7}{8}$ **iv** $S_4 = \frac{15}{16}$
b $S_n = \dfrac{2^n - 1}{2^n}$ **c** $S_{10} = \frac{1023}{1024}$

REVIEW SET 10B

1 **a** $E = 2 \times (3-2) + 2 \times (5-2)$
b $E = 2 \times (4-2) + 2 \times (8-2)$
c $E = 2(m-2) + 2(n-2)$

2 **a** 204 800 units **b** ≈ 7.75 cm

3 **a** $x = \dfrac{B+f}{d}$ **b** $x = \dfrac{9Q^2}{t^2}$ **c** $x = \dfrac{2}{2y-5}$
d $x = \dfrac{2y+3}{4-3y}$ **e** $x = -\sqrt{\dfrac{P+r}{2}}$

4 $p = \sqrt{\dfrac{N-1}{N}}$

5 **a** 11 cm **b** $c = \sqrt{L^2 - a^2 - b^2}$ **c** 9 cm

6 **a** 8 amperes **b** $r = \dfrac{E - IR}{I}$ **c** 0.15 ohms

7 **a** $\approx 5.94 \times 10^7$ seconds **b** $r = \sqrt[3]{\dfrac{GMT^2}{4\pi^2}}$
c $\approx 3.16 \times 10^7$ seconds

8 **a** **i** 3 **ii** 6 **iii** 10 **b** **i** 9 **ii** 36 **iii** 100
c $C_n = S_n^2$

EXERCISE 11A

1 **a** 2610 m **b** 4.3 m **c** 8.65 m **d** 70 cm
e 11.5 km **f** 367.5 cm **g** 0.381 m **h** 682 mm
i 567 000 cm **j** 28.6 m **k** 24.3 m **l** 32 800 cm

2 **a** 38.8 cm **b** 17.8 cm **c** 11.9 km **d** 99.6 km
e 19.1 m **f** 30 m

3 30 km **4** 11.1 cm **5** 164 m **6** 18.8 cm

7 **a** ≈ 34.6 cm **b** ≈ 3.39 m

8 **a** ≈ 10.7 m **b** ≈ 441 m **c** ≈ 22.8 cm
d ≈ 75.4 cm **e** ≈ 19.6 mm **f** ≈ 8.09 cm

9 ≈ 127 m **10** ≈ 13.1 cm

11 **a** $P = (3x + 4)$ cm **b** $P = \left(2r + \dfrac{\theta}{180}\pi r\right)$ cm
c $P = (x + 16 + \sqrt{x^2 + 256})$ cm

12 ≈ 20.9 cm **13** ≈ 25.6 cm **14** 155 cm

15 **a** 69.38 m **b** 146.08 m **16** ≈ 2.73 cm

17 **a** ≈ 39.0 m **b** 14 posts **c** $936.16

EXERCISE 11B

1 **a** 3.84 ha **b** 257 000 m^2 **c** 35 000 cm^2
d 50 m^2 **e** 1800 mm^2 **f** 4.6 cm^2
g 9 000 000 m^2 **h** 3500 ha **i** 5 km^2

2 **a** 40 cm^2 **b** 15 cm^2 **c** 42 m^2
d 39 cm^2 **e** 180 m^2 **f** 25 cm^2
g 200 mm^2 **h** ≈ 11.6 cm^2 **i** ≈ 50.3 cm^2
j ≈ 72.4 m^2 **k** ≈ 18.2 cm^2 **l** ≈ 19.6 cm^2

3 **a** ≈ 3.22 cm^2 **b** ≈ 3.67 mm^2 **c** ≈ 70.4 m^2

4 **a** $x = 17$ **b** $x \approx 21.1$ **c** $x = 8$ **5** ≈ 22.2 cm^2

6 $\approx 314\ \text{cm}^2$ **7** **a** ≈ 53.7 cm **b** $160\ \text{cm}^2$
8 17 flights **9** $\approx 50.1\ \text{cm}^2$ **10** ≈ 5.33 cm
11 8 cm **12** **a** $387\ \text{cm}^2$ **b** $36\ \text{cm}^2$ **c** $\approx 6960\ \text{m}^2$
13 **a** 7500 washers **b** $\approx 41.1\%$
14 **a** perimeter ≈ 34.6 cm, area $\approx 77.9\ \text{cm}^2$
b perimeter ≈ 62.8 cm, area $\approx 157\ \text{cm}^2$
c perimeter ≈ 28.3 cm, area $\approx 21.2\ \text{cm}^2$
15 **a** $A = a\sqrt{x^2 - a^2}$ **b** $A = 2ar - \pi r^2$
c $A = 5a^2 + \pi a^2$ **d** $A = \sqrt{3}x^2 + \frac{\pi}{2}x^2$
e $A = \pi(R^2 - r^2)$ **f** $A = a^2 - \dfrac{a}{4}\sqrt{4b^2 - a^2}$

EXERCISE 11C.1

1 **a** $104\ \text{cm}^2$ **b** $4400\ \text{cm}^2$ **c** $41.5\ \text{m}^2$
2 **a** $84\ \text{cm}^2$ **b** $184\ \text{cm}^2$ **c** $3280\ \text{mm}^2$
3 **a** $\approx 5.35\ \text{m}^2$ **b** $\approx 1680\ \text{mm}^2$ **c** $\approx 9.60\ \text{m}^2$
4 **a** $A = 2a^2 + 4ab$ **b** $A = a^2 + 2ab + \sqrt{2}ab$
5 80 cm **6** $167.405\ \text{m}^2$ **7** $\approx 13.9\ \text{cm}^2$

EXERCISE 11C.2

1 **a** $\approx 668\ \text{cm}^2$ **b** $\approx 303\ \text{cm}^2$ **c** $\approx 968\ \text{cm}^2$
2 **a** $\approx 707\ \text{m}^2$ **b** $\approx 145\ \text{km}^2$ **c** $\approx 84.8\ \text{cm}^2$
3 **a** $\approx 320\ \text{cm}^2$ **b** $\approx 1310\ \text{mm}^2$ **c** $\approx 679\ \text{cm}^2$
4 **a** $\approx 452\ \text{cm}^2$ **b** $\approx 942\ \text{m}^2$ **c** $\approx 1640\ \text{cm}^2$
d $\approx 758\ \text{cm}^2$
5 **a** $s \approx 5.39$ **b** $\approx 33.8\ \text{cm}^2$
6 **a** $\approx 251\ \text{m}^2$ **b** 11 cans **c** \$577.50
7 **a** ≈ 5.64 m **b** ≈ 21.8 cm **c** ≈ 25.8 m
8 **a** $\approx 5.15 \times 10^8\ \text{km}^2$ **b** $\approx 3.65 \times 10^8\ \text{km}^2$
c **i** $\approx 1.89\%$ **ii** $\approx 6.50\%$
9 \$1100
10 **a** $A = \frac{3}{4}\pi d^2$ **b** $A = \pi r^2 + \pi r\sqrt{r^2 + h^2}$
c $A = xy + 2xz + 2yz + \frac{\pi}{4}y^2 + \frac{\pi}{2}xy$
11 $\approx 213\ \text{cm}^2$

EXERCISE 11D.1

1 **a** $8650\ \text{mm}^3$ **b** $86\ \text{cm}^3$ **c** $0.3\ \text{m}^3$
d $124\,000\ \text{mm}^3$ **e** $0.3\ \text{cm}^3$ **f** $3\,700\,000\ \text{cm}^3$
2 300 000 erasers **3** 47 400 sinkers

EXERCISE 11D.2

1 **a** $2.464\ \text{m}^3$ **b** $80\ \text{cm}^3$ **c** $525\ \text{mm}^3$
d $\approx 226\ \text{cm}^3$ **e** $\approx 177\ \text{m}^3$ **f** $\approx 353\ \text{cm}^3$
2 **a** $r = \sqrt{\dfrac{V}{\pi h}}$ **b** ≈ 0.865 cm **3** 15 915 handles
4 **a** $\approx 0.0531\ \text{m}^3$ **b** ≈ 27.1 kg
5 **a** $\approx 5.94\ \text{m}^2$ **b** $\approx 3.56\ \text{m}^3$ **6** $\approx 872\ \text{cm}^3$

EXERCISE 11D.3

1 **a** $32\ \text{m}^3$ **b** $\approx 66.0\ \text{cm}^3$ **c** $\approx 233\ \text{m}^3$
d $\approx 738\ \text{cm}^3$ **e** $\approx 42.7\ \text{m}^3$ **f** $\approx 1.89\ \text{m}^3$
2 **a** $\approx 268\ \text{cm}^3$ **b** $\approx 102\ \text{cm}^3$ **c** $\approx 82.3\ \text{cm}^3$
d $\approx 1.07\ \text{m}^3$
3 $4500\ \text{m}^3$ **4** $\approx 2.48\ \text{m}^3$ **5** $\approx 0.905\ \text{m}^3$
6 611 garden stakes **7** ≈ 1.79 m **8** $V = \frac{1}{12}\pi d^3$
9 $\approx 938\ \text{mm}^3$ **10** $V = \pi r^2 h + \frac{2}{3}\pi r^3$ **11** $\approx 0.116\ \text{m}^3$
12 **Hint:** Let r be the radius of the cylinder.

EXERCISE 11E

1 **a** 5600 mL **b** 524 000 L **c** 0.28 L
d 3500 kL **e** 0.743 ML **f** 840 000 mL
2 **a** 24 mL **b** ≈ 23.6 mL **c** ≈ 9.11 mL **d** 300 kL
3 368 bottles **4** ≈ 24.4 mL
5 **a** 53.76 L **b** ≈ 618 km **6** ≈ 817 mL
7 ≈ 12.7 cm **8** $\approx 6.35 \times 10^7$ L **9** ≈ 1410 kL
10 ≈ 0.888 L
11 **a** $1000\ \text{cm}^3$ **b** ≈ 6.20 cm **c** ≈ 39.0 cm
d $\approx 484\ \text{cm}^2$
12 ≈ 0.707 m **13** **a** ≈ 18.1 kL **b** ≈ 0.177 m
14 ≈ 3.07 cm **15** $\frac{91}{108}$ cm

REVIEW SET 11A

1 **a** perimeter $= 42$ m, area $= 60\ \text{m}^2$
b perimeter ≈ 256 mm, area $\approx 3320\ \text{mm}^2$
c perimeter $= 9$ m, area $= 4.2\ \text{m}^2$
2 1273 revolutions
3 **a** $P = (5x + 1)$ m **b** $P = \left(2a + 5 + \sqrt{10a + 25}\right)$ cm
4 $\approx 132\ \text{m}^2$
5 **a** $182.88\ \text{cm}^2$ **b** $\approx 2630\ \text{mm}^2$ **c** $\approx 102\ \text{cm}^2$
6 $\approx 21.7\ \text{m}^2$
7 **a** $\approx 9560\ \text{cm}^3$ **b** $\approx 246\ \text{cm}^3$ **c** $\approx 754\ \text{cm}^3$
8 **a** $\approx 209\ \text{cm}^3$ **b** $\approx 227\ \text{cm}^2$ **9** $\approx 41.8\ \text{m}^3$
10 **a** 5 L **b** ≈ 3.60 kL **c** ≈ 66.7 L
11 **a** **i** $\dfrac{x}{2}$ m **ii** $\dfrac{x}{4}$ m
b **Hint:** Show both areas are $\dfrac{\pi x^2}{4}\ \text{m}^2$.

REVIEW SET 11B

1 **a** perimeter ≈ 20.8 cm, area $= 20\ \text{cm}^2$
b perimeter ≈ 19.8 m, area $= 22.5\ \text{m}^2$
c perimeter ≈ 41.7 cm, area $\approx 37.7\ \text{cm}^2$
2 ≈ 46.6 cm **3** ≈ 7.81 cm
4 **a** $A = bc + \left(\dfrac{a+b}{2}\right)h$ **b** $A = \left(\dfrac{\sqrt{3} + \sqrt{15}}{4}\right)a^2$
5 **a** surface area $= 14\,278\ \text{cm}^2$, volume $= 107\,911\ \text{cm}^3$
b surface area $\approx 28.3\ \text{m}^2$, volume $\approx 14.1\ \text{m}^3$
c surface area $= 496\ \text{mm}^2$, volume $\approx 570\ \text{mm}^3$
6 3 m **7** 614 bollards **8** $\approx 52.4\%$ **9** ≈ 10.9 cm
10 no **11** $\approx 53\,100$ kL
12 **a** $\approx 37.4\ \text{cm}^2$ **b** $15\ \text{cm}^3$
c **i** 13.755 mL **ii** 145 ice blocks

EXERCISE 12A

1 **a** $x = \pm 10$ **b** $x = \pm 5$ **c** $x = \pm 2$
d $x = \pm 3$ **e** no real solutions **f** $x = 0$
g $x = \pm 3$ **h** no real solutions **i** $x = \pm\sqrt{2}$
2 **a** $x = 4$ or -2 **b** $x = 0$ or -8 **c** no real solutions
d $x = 4 \pm \sqrt{5}$ **e** no real solutions **f** $x = -2$
g $x = \frac{5}{2}$ **h** $x = 0$ or $-\frac{4}{3}$ **i** $x = \frac{8}{3}$ or $-\frac{10}{3}$
j $x = -\frac{1}{2} \pm 2\sqrt{3}$ **k** $x = \dfrac{3 \pm \sqrt{7}}{2}$ **l** $x = \dfrac{-3 \pm \sqrt{6}}{2}$
3 **a** $x = \pm 2$ **b** $x = \pm\sqrt{10}$ **c** $x = \pm 4$

EXERCISE 12B

1 **a** $p = 0$ or $q = 0$ **b** $a = 0$ **c** $x = 0$ or $y = 0$ **d** $w = 0$ or $x = 0$

2 **a** $x = 0$ or -3 **b** $x = 0$ or 5 **c** $x = 0$ or 2 **d** $x = 0$ or $-\frac{1}{2}$ **e** $x = 0$ or $\frac{5}{3}$ **f** $x = 0$ or $\frac{6}{5}$

3 **a** $x = 1$ or 3 **b** $x = -5$ or 2 **c** $x = -1$ **d** $x = 5$ or -4 **e** $x = -2$ or 7 **f** $x = 3$ **g** $x = -2$ or $\frac{1}{2}$ **h** $x = -\frac{1}{3}$ **i** $x = 5$ or $-\frac{2}{3}$ **j** $x = -\frac{1}{2}$ **k** $x = -\frac{3}{2}$ or $-\frac{1}{2}$ **l** $x = -\frac{1}{7}$ or $\frac{5}{9}$

4 **a** $x = 0$ or b **b** $x = b$ or c **c** $x = b$ **d** $x = 0$ or $\frac{c}{b}$, $b \neq 0$ **e** $x = b$ or $\frac{d}{c}$, $c \neq 0$

EXERCISE 12C.1

1 **a** $x = 0$ or 7 **b** $x = 0$ or -11 **c** $x = 0$ or 5 **d** $x = 0$ or -2 **e** $x = 0$ or $-\frac{5}{2}$ **f** $x = 0$ or $\frac{3}{4}$

2 **a** $x = 0$ or 8 **b** $x = 0$ or -4 **c** $x = 0$ or $\frac{5}{4}$ **d** $x = 0$ or -3 **e** $x = 0$ or -1 **f** $x = 0$ or $\frac{1}{3}$

3 **a** $x = -5$ or 4 **b** $x = 2$ or -3 **c** $x = -3$ or 1

EXERCISE 12C.2

1 **a** $x = \pm 5$ **b** $x = \pm 8$ **c** no real solutions **d** $x = \pm 3$ **e** $x = \pm 3$ **f** $x = \pm 6$

2 **a** $x = \pm\frac{5}{2}$ **b** $x = \pm\frac{7}{2}$ **c** $x = \pm\frac{4}{3}$ **d** $x = \pm\frac{3}{4}$ **e** $x = \pm\frac{1}{2}$ **f** $x = \pm\frac{2}{3}$

3 **a** LHS $= 4(1)^2 = 4$, RHS $= (1+1)^2 = 2^2 = 4$ ✓

b LHS $= 4(-\frac{1}{3})^2 = \frac{4}{9}$,
RHS $= (-\frac{1}{3}+1)^2 = (\frac{2}{3})^2 = \frac{4}{9}$ ✓

c Jack lost a solution in the third line by not considering the case $2x = -(x+1)$.

d $x = 1$ or $-\frac{1}{3}$

4 **a** $x = \frac{1}{3}$ or -1 **b** $x = -1$ **c** $x = 0$ or 1 **d** $x = \frac{3}{5}$ or 1

EXERCISE 12C.3

1 **a** $x = -1$ or -2 **b** $x = 1$ or 2 **c** $x = 5$ **d** $x = -2$ or -3 **e** $x = 2$ or 3 **f** $x = -1$ or -6 **g** $x = -2$ or -7 **h** $x = -5$ or -6 **i** $x = -5$ or 3 **j** $x = -6$ or 2 **k** $x = 3$ or 8 **l** $x = 7$

2 **a** $x = -3$ or -6 **b** $x = -4$ or -7 **c** $x = -4$ or 2 **d** $x = -4$ or 3 **e** $x = 3$ or 2 **f** $x = 2$ **g** $x = 3$ or -2 **h** $x = -12$ or 5 **i** $x = 10$ or -7

3 **a** $x = -4$ or -3 **b** $x = -3$ or 1 **c** $x = \pm 3$ **d** $x = 11$ or -3 **e** $x = -\frac{1}{2}$ **f** $x = 3$ or -5

4 **a** $x = 3$ or -7 **b** $x = -2$ or 1 **c** $x = -6$ or 2 **d** $x = 2$ or -1 **e** $x = 4$ or -1 **f** $x = \frac{1}{3}$

5 **a** $x = 4$ or -1 **b** $x = 15$ or 2 **c** $x = -9$ or -2 **d** $x = 0$ or $\pm\sqrt{11}$ **e** $x = 0$ or -3

6 **a** $x = \frac{1}{2}$ or 2 **b** $x = -3$ or $\frac{1}{3}$ **c** $x = -4$ or $-\frac{5}{3}$ **d** $x = \frac{1}{2}$ or -3 **e** $x = \frac{1}{2}$ or 5 **f** $x = -1$ or $-\frac{5}{2}$ **g** $x = -\frac{1}{3}$ or -4 **h** $x = -\frac{2}{5}$ or 3 **i** $x = \frac{1}{2}$ or 0 **j** $x = -1$ or $\frac{5}{2}$ **k** $x = \frac{4}{3}$ or -2 **l** $x = \frac{3}{2}$ or -6

m $x = -\frac{1}{6}$ or 3 **n** $x = \frac{3}{2}$ or -4 **o** $x = \frac{3}{2}$ or $\frac{1}{3}$ **p** $x = -\frac{8}{3}$ or 2 **q** $x = \frac{4}{7}$ or $-\frac{1}{2}$ **r** $x = \frac{1}{4}$ or $-\frac{4}{3}$

7 **a** $x = -1$ or $\frac{2}{3}$ **b** $x = \frac{5}{2}$ or 4 **c** $x = -\frac{3}{4}$ or $\frac{5}{2}$ **d** $x = 1$ or $-\frac{1}{3}$ **e** $x = \frac{1}{2}$ or -1 **f** $x = 3$ or $-\frac{7}{2}$ **g** $x = 4$ or $\frac{4}{3}$ **h** $x = 1$ or $\frac{6}{5}$

8 $x = 2$ **9** $x = \pm 1$ or ± 2

EXERCISE 12D

1 **a** **i** 1^2 **ii** $(x+1)^2 = 6$
b **i** 1^2 **ii** $(x-1)^2 = -6$
c **i** 3^2 **ii** $(x+3)^2 = 11$
d **i** 3^2 **ii** $(x-3)^2 = 6$
e **i** 5^2 **ii** $(x+5)^2 = 26$
f **i** 4^2 **ii** $(x-4)^2 = 21$
g **i** 6^2 **ii** $(x+6)^2 = 49$
h **i** $(\frac{5}{2})^2$ **ii** $(x+\frac{5}{2})^2 = \frac{17}{4}$
i **i** $(\frac{7}{2})^2$ **ii** $(x-\frac{7}{2})^2 = \frac{65}{4}$

2 **a** $x = 2 \pm \sqrt{3}$ **b** $x = 1 \pm \sqrt{3}$ **c** $x = 2 \pm \sqrt{7}$ **d** $x = -1 \pm \sqrt{2}$ **e** $x = -2 \pm \sqrt{3}$ **f** $x = -3 \pm \sqrt{6}$ **g** $x = -1$ or -2 **h** $x = -4 \pm \sqrt{2}$ **i** $x = \frac{3 \pm \sqrt{13}}{2}$

3 **a** no real solutions **b** $x = 3$ or 2 **c** no real solutions **d** $x = \frac{-1 \pm \sqrt{5}}{2}$ **e** $x = \frac{-5 \pm \sqrt{33}}{2}$ **f** no real solutions

4 **a** $x = \frac{-b \pm \sqrt{b^2 - 4c}}{2}$

b **i** $b^2 - 4c > 0$ **ii** $b^2 - 4c = 0$ **iii** $b^2 - 4c < 0$

5 **a** $x = -1 \pm \sqrt{\frac{3}{2}}$ **b** $x = 2 \pm \sqrt{\frac{5}{3}}$ **c** $x = 1 \pm \sqrt{\frac{2}{5}}$ **d** $x = -3 \pm \sqrt{\frac{23}{2}}$ **e** no real solutions **f** $x = -\frac{1}{2} \pm \sqrt{\frac{7}{12}}$

EXERCISE 12E

1 **a** **i, ii** $x = -2$ or -4 **b** **i, ii** $x = 5$ **c** **i, ii** $x = -\frac{2}{3}$ or 3

2 **a** $x = \frac{-1 \pm \sqrt{21}}{2}$ **b** $x = \frac{5 \pm \sqrt{5}}{2}$ **c** $x = 2 \pm \sqrt{5}$ **d** $x = \frac{3 \pm \sqrt{5}}{2}$ **e** $x = \frac{1 \pm \sqrt{7}}{2}$ **f** $x = \frac{1 \pm \sqrt{2}}{3}$

3 **a** $x \approx -1.85$ or 0.18 **b** $x \approx -1.64$ or 2.14 **c** $x \approx 0.14$ or 1.46 **d** $x \approx -0.16$ or 0.88 **e** $x \approx -1.22$ or 0.55 **f** $x \approx 0.05$ or 0.75

4 **a** $\Delta = -39$ which is < 0 $\therefore$ no real solutions
b $\Delta = -12$ which is < 0 $\therefore$ no real solutions
c $\Delta = -7$ which is < 0 $\therefore$ no real solutions

5 **a** $k < \frac{25}{12}$ **b** $k = \frac{25}{12}$ **c** $k > \frac{25}{12}$

6 **a** no real solutions **b** $x = 5$ or -1 **c** no real solutions **d** no real solutions **e** $x = \frac{3 \pm \sqrt{19}}{2}$ **f** $x = \frac{-1 \pm \sqrt{17}}{4}$

7 **a** $x=\dfrac{-1\pm\sqrt{29}}{2}$ **b** $x=\dfrac{-1\pm\sqrt{5}}{2}$ **c** no real solutions

EXERCISE 12F

1 The number is -11 or 10. **2** The number is -3 or 8.
3 The numbers are 3 and 6.
4 The numbers are -2 and 5, or 2 and -5. **5** 8 cm
6 10 m **7** **a** $x=6$ **b** $x=5$ **c** $x=12$
8 **a** $(30-x)$ m **b** 18 m by 12 m **9** 5 cm
10

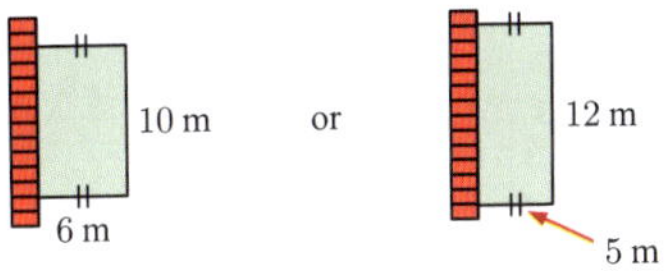

11 $BC = 16$ cm or 5 cm **12** $n=6$ **13** $\frac{2}{5}$ or $\frac{-9}{-6}$
14 40 oranges
15 **a** The radius of each quarter circle is $\dfrac{16-2x}{2}=(8-x)$ m.
b Total area of yellow flowers $= 9\times$ area of red flowers
$\therefore\ \pi\times(8-x)^2 = 9\times\pi\times x^2$
$\therefore\ (8-x)^2 = 9x^2$
c $-8x^2-16x+64=0$ or $x^2+2x-8=0$
d $x=2$ or -4, the equation is easier to solve when written in the form $ax^2+bx+c=0$.
e No, $x=-4$ is not a valid solution because lengths cannot be negative.
f 2 m
16 $x=5$ **17** The number is $\frac{4}{3}$ or $\frac{3}{4}$.
18 The numbers are $\dfrac{4+\sqrt{14}}{2}$ and $\dfrac{4-\sqrt{14}}{2}$.
19 Cut out squares with sides 2 cm.
20 3.2 cm **21** **c** 1.5 m

REVIEW SET 12A

1 **a** $x=\pm\sqrt{11}$ **b** $x=\pm 4$ **c** $x=6$ or $-\frac{2}{3}$
2 **a** $x=9$ or -1 **b** $x=0$ or -1 **c** $x=\pm 6$
3 **a** $x=0$ or 5 **b** $x=0$ or $\frac{6}{5}$ **c** $x=1$ or -5
4 **a** $x=7$ or -3 **b** $x=2$ or -15 **c** $x=9$ or -2
5 **a** $x=\pm\frac{9}{4}$ **b** $x=-3$ **c** $x=\frac{1}{2}$
6 **a** $x=5$ or -3 **b** $x=-6$ or 9
7 **a** $x=2$ or $-\frac{3}{2}$ **b** $x=\frac{1}{2}$ or $-\frac{2}{5}$ **c** $x=1$ or $-\frac{10}{3}$
8 **a** $x=4\pm3\sqrt{2}$ **b** $x=-3\pm\sqrt{5}$ **c** $x=\dfrac{-1\pm\sqrt{97}}{2}$
9 **a** $x=2$ or $-\frac{1}{2}$ **b** $x=\dfrac{-2\pm\sqrt{19}}{3}$ **c** $x=\dfrac{13\pm\sqrt{105}}{8}$
10 The numbers are 5 and 8. **11** $x=12$
12 12 people, \$40 each **13** $x=7$
14 The number is $\frac{2}{3}$ or $\frac{3}{2}$. **15** $\dfrac{15+30\sqrt{2}}{7}$ cm

REVIEW SET 12B

1 **a** $x=\pm\sqrt{6}$ **b** $x=\pm 1$ **c** $x=4$ or $-\frac{7}{3}$
2 **a** $x=-3\pm\sqrt{19}$ **b** $x=\frac{1}{2}$ or $\frac{5}{2}$ **c** $x=\dfrac{1\pm\sqrt{17}}{3}$
3 **a** $x=\pm 12$ **b** $x=+\frac{5}{3}$ **c** $x=-6$ or $\frac{4}{5}$
4 **a** $x=11$ or -3 **b** $x=9$ or -5 **c** $x=6$ or 10
5 **a** $x=2$ or -1 **b** $x=5$ or -1 **c** $x=-\frac{1}{3}$
6 **a** $x=\pm\sqrt{\frac{3}{2}}$ **b** $x=-3$ or 1
7 **a** $x=\frac{2}{3}$ or $-\frac{1}{2}$ **b** $x=-\frac{2}{5}$ or $\frac{3}{2}$ **c** $x=5$ or $-\frac{3}{2}$
8 **a** $x=1\pm\sqrt{101}$ **b** $x=-\frac{7}{2}\pm\sqrt{13}$ **c** $x=3\pm\sqrt{17}$
9 **a** $x=\dfrac{-1\pm\sqrt{3}}{2}$ **b** $x=-5$ or $\frac{2}{3}$ **c** $x=\dfrac{2\pm\sqrt{2}}{2}$
10 **a** $k<\frac{9}{8}$ **b** $k=\frac{9}{8}$ **c** $k>\frac{9}{8}$ **11** 12 mm
12 10 cm **13** $\frac{2}{8}$ or $\frac{20}{-10}$ **14** The number is $\frac{4}{5}$ or -1.
15

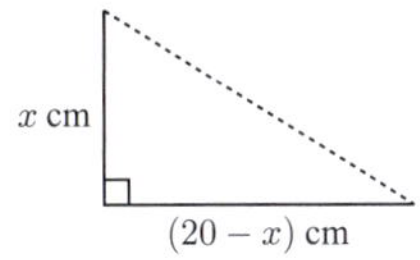

a **i** $(10+2\sqrt{10})$ cm, $(10-2\sqrt{10})$ cm, and $2\sqrt{70}$ cm
ii $(20+2\sqrt{70})$ cm
b No, $\frac{1}{2}x(20-x)=51$ has no real solutions.

EXERCISE 13A

1 **a** $2\sqrt{2}$ units **b** 7 units **c** $2\sqrt{5}$ units
d $3\sqrt{5}$ units **e** 7 units **f** $\sqrt{5}$ units
g $\sqrt{10}$ units **h** $\frac{\sqrt{5}}{2}$ units **i** $\sqrt{13}$ units
2 **a** $10\sqrt{2}\approx 14.1$ km **b** $20\sqrt{5}\approx 44.7$ km
c $10\sqrt{26}\approx 51.0$ km
3 **a** isosceles with $AB = AC = \sqrt{85}$ units **b** scalene
c isosceles (and right angled at B) with $AB = BC$
d isosceles with $BC = AC = \sqrt{7}$ units
e equilateral, all sides $2\sqrt{3}$ units
f isosceles $(AC = BC)$ if $b\neq 2\pm a\sqrt{3}$, $a\neq 0$, $b\neq 2$
equilateral if $b=2\pm a\sqrt{3}$, $a\neq 0$
4 **a** right angled at B **b** not right angled
c right angled at A **d** not right angled
5 ≈ 13.8 units **6** a rhombus
7 **a** $a=2$

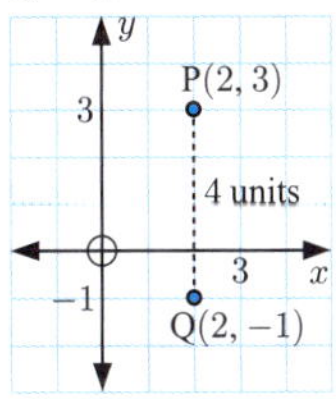

b $a=-5$ or 3

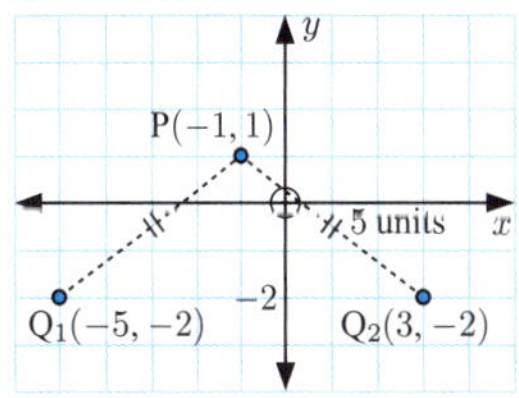

c $a=2$ or 6

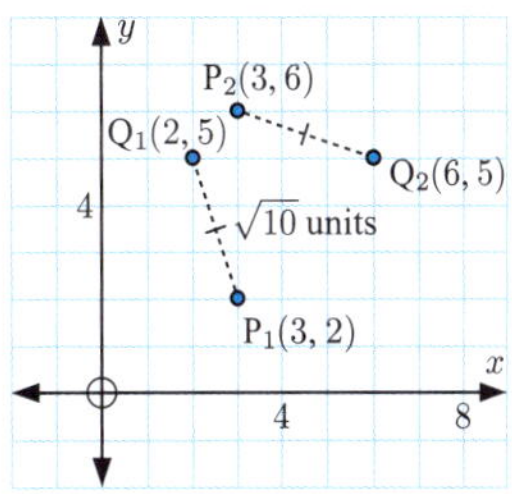

d $a=\pm 2$

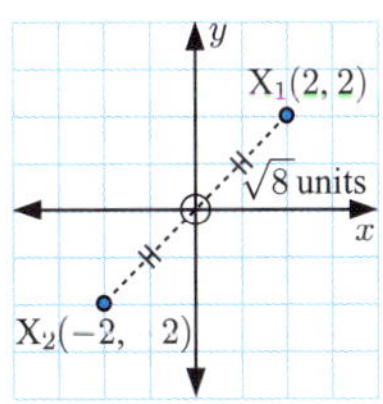

e $a=-1$

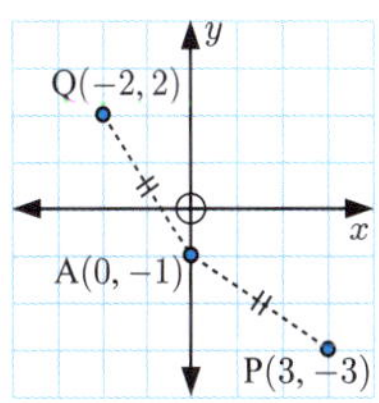

8 **a** **i** $x^2 + y^2 = 9$ **ii** $(x-1)^2 + (y-3)^2 = 4$

b 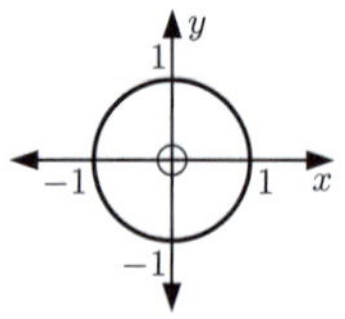

This set represents a circle with centre (0, 0) and radius 1 unit.

c $(x-a)^2 + (y-b)^2 = r^2$

EXERCISE 13B

1 **a** M(5, 3) **b** M(1, −1) **c** M($1\frac{1}{2}$, 3) **d** M(0, 4)
e M(2, $-1\frac{1}{2}$) **f** M(1, $2\frac{1}{2}$) **g** M(1, $a-1$) **h** M(1, 0)

2 **a** **i** $(\frac{1}{2}, 1)$ **ii** $(\frac{1}{2}, 1)$
b Since the midpoints of [AC] and [BD] coincide, the diagonals of ABCD bisect each other.
$\therefore$ ABCD is a parallelogram.

3 **a** B(0, −6) **b** B(5, −2) **c** B(0, 6) **d** B(0, 7)
e B(−7, 3) **f** B(−3, 0)

4 C(1, −3) **5** P(7, −3) **6** $\frac{\sqrt{89}}{2} \approx 4.72$ units

7 **a** **i** C(1, 3) **ii** D(4, −1)
b **i** 5 units **ii** 5 units **c** ABCD is a rhombus.

EXERCISE 13C.1

1 **a** $\frac{1}{3}$ **b** 0 **c** −3 **d** $\frac{2}{3}$ **e** $-\frac{3}{4}$
f undefined **g** −4 **h** $-\frac{2}{5}$

2 **a** 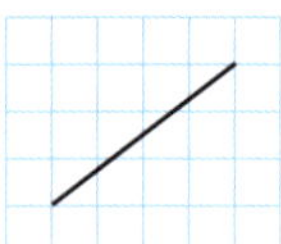 **b** 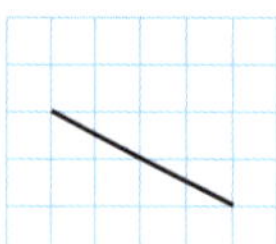 **c** 
d 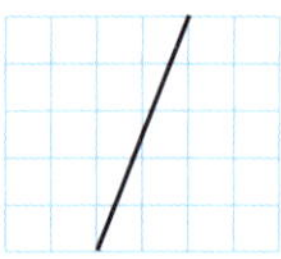 **e** 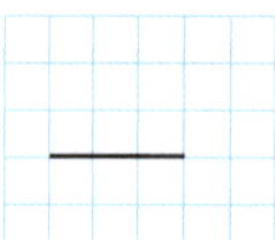 **f** 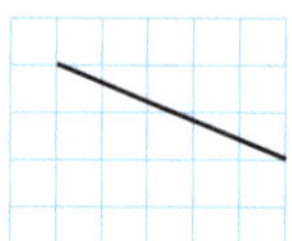

3 **a** 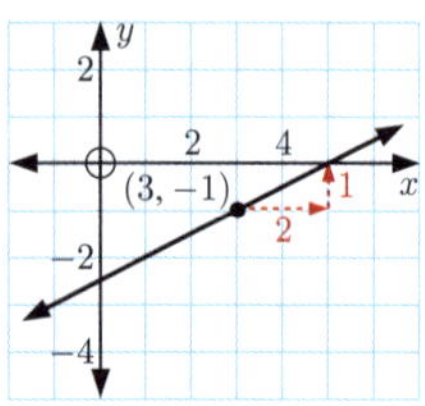

b 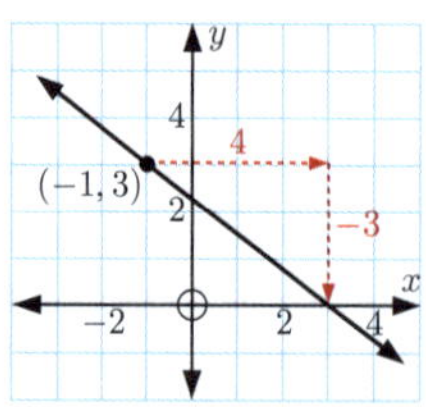

4 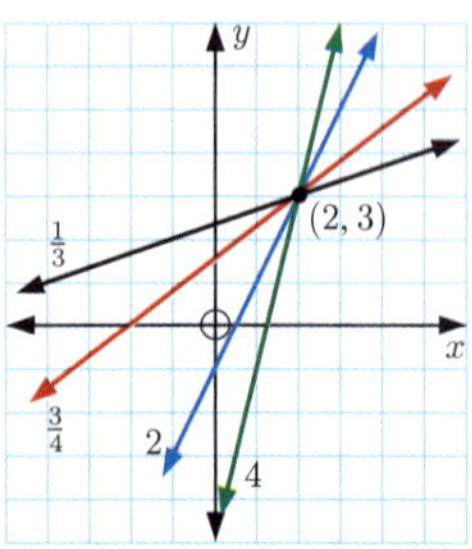

5 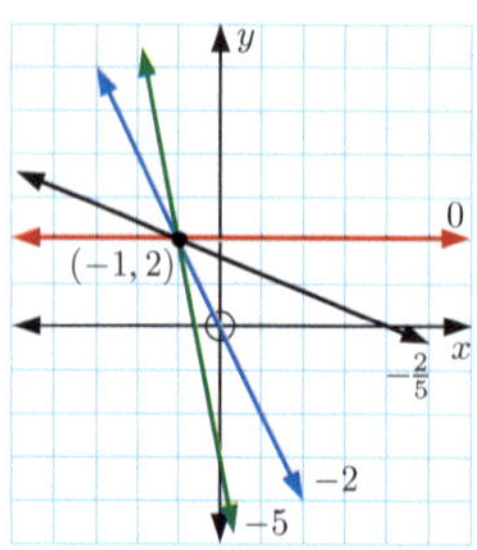

EXERCISE 13C.2

1 **a** $-\frac{1}{3}$ **b** $\frac{4}{3}$ **c** $\frac{b}{a}$

2 $\frac{4}{7}$ 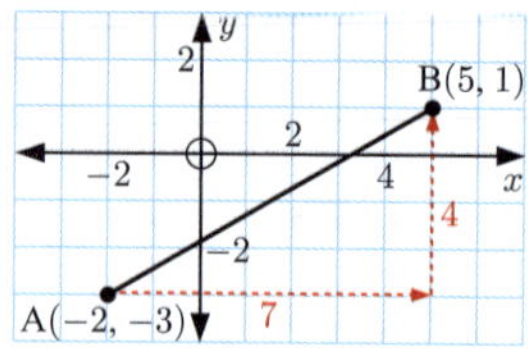

3 **a** $\frac{1}{5}$ **b** $\frac{1}{4}$ **c** 4 **d** 0 **e** undefined
f $\frac{2}{7}$ **g** $-\frac{2}{7}$ **h** 1 **i** $-\frac{4}{3}$

4 **a** $t = 19$ **b** $t = \frac{29}{2}$ **c** $t = \frac{13}{3}$ **d** $t = \frac{2}{3}$
e $t = 9$ **f** $t = 5$

5 R$(\frac{28}{3}, 0)$

EXERCISE 13D

1 **a** 2 **b** −4 **c** $\frac{1}{3}$ **d** $-\frac{3}{4}$

2 **a** −2 **b** $-\frac{5}{2}$ **c** $-\frac{1}{3}$ **d** $-\frac{1}{7}$ **e** $\frac{5}{2}$ **f** $\frac{2}{7}$
g $\frac{3}{4}$ **h** 1

3 The line pairs in **c**, **d**, **f**, and **h** are perpendicular. **4** 0

5 **a** [AB]: $\frac{3}{5}$, [BC]: −2, [CD]: $\frac{9}{2}$, [DE]: $-\frac{1}{2}$,
[EF]: −2, [FA]: 2
b **i** [BC] ∥ [EF] **ii** [DE] ⊥ [FA]

6 **a** $a = 9$ **b** $a = 1$ **7** **a** $t = \frac{1}{5}$ **b** $t = 5$

8 **a** $t = 4$ **b** $t = 4$ **c** $t = 14$ **d** $t = 3\frac{1}{7}$ **9** $k = -2$

10 **a** not collinear **b** collinear **c** not collinear
d collinear

11 **a** $c = 3$ **b** $c = -5$

EXERCISE 13E

1 **a** PQ = PR = $2\sqrt{5}$ units
b M(4, 4)
c gradient of [PM] = $-\frac{1}{3}$,
gradient of [QR] = 3,
and their product is −1
$\therefore$ [PM] ⊥ [QR]

d 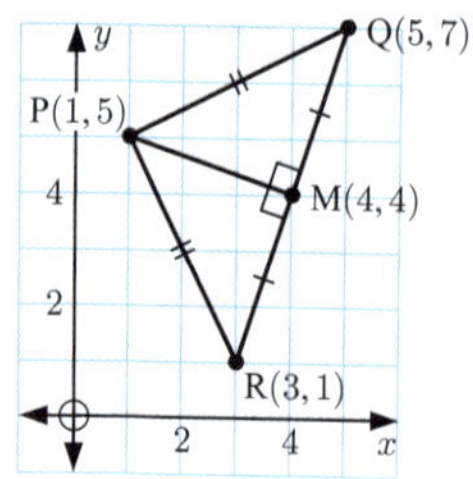

2 M(0, 3) and N(3, 3)
gradient of [MN] = 0, gradient of [AC] = 0
$\therefore$ [MN] ∥ [AC]
MN = 3 units and AC = 6 units $\therefore$ MN = $\frac{1}{2}$AC

3 **a** M(1, −4)
b gradient of [AM] = gradient of [MC] = 1, $\therefore$ collinear
c gradient of [AC] = 1, gradient of [BD] = −1,
and $1 \times -1 = -1$
$\therefore$ [AC] ⊥ [BD]

4 **a** All sides have length 5 units. $\therefore$ ABCD is a rhombus.
b (2, 1) and (2, 1)
c gradient of [AC] = −2,
gradient of [BD] = $\frac{1}{2}$,
and their product is −1
$\therefore$ [AC] ⊥ [BD]

d 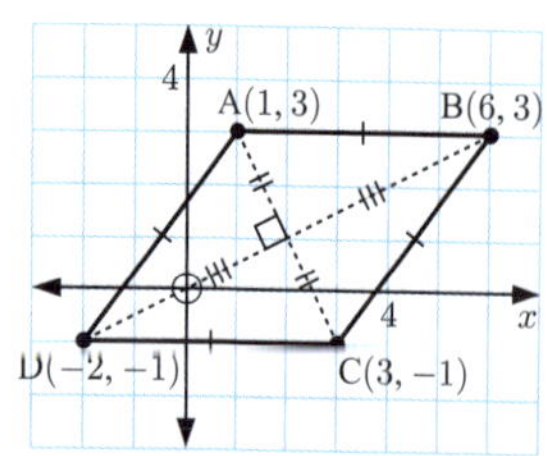

5 **a** **i** P(0, 5) **ii** Q($4\frac{1}{2}$, 2) **iii** R($\frac{1}{2}$, $-2\frac{1}{2}$)
iv S(-4, $\frac{1}{2}$)
b **i** $-\frac{2}{3}$ **ii** $\frac{9}{8}$ **iii** $-\frac{2}{3}$ **iv** $\frac{9}{8}$
c PQRS is a parallelogram.

6 **a** $s = 6$ **b** **i** $\frac{1}{2}$ **ii** -2
c gradient of [PS] $\times$ gradient of [SQ] $= -1$
$\therefore$ $\widehat{PSQ} = 90°$

7 **a** B($a+b$, c) **b** midpoint of [AC] is $\left(\dfrac{a+b}{2}, \dfrac{c}{2}\right)$
midpoint of [OB] is $\left(\dfrac{a+b}{2}, \dfrac{c}{2}\right)$
c The diagonals of a parallelogram bisect each other.

EXERCISE 13F

1 **a** gradient is 2, y-intercept is 5
b gradient is -6, y-intercept is -1
c gradient is $\frac{3}{5}$, y-intercept is $-\frac{2}{5}$
d gradient is -1, y-intercept is 10
e gradient is -4, y-intercept is 1
f gradient is $-\frac{7}{4}$, y-intercept is $\frac{5}{4}$
g gradient is $\frac{3}{7}$, y-intercept is $\frac{1}{7}$
h gradient is $\frac{1}{2}$, y-intercept is $-\frac{5}{8}$
i gradient is $-\frac{7}{9}$, y-intercept is $\frac{2}{3}$

2 **a** $x - y = 7$ **b** $x + y = 4$ **c** $x - 2y = 10$
d $3x + y = 2$ **e** $3x - 4y = -2$ **f** $5x + 2y = -3$

3 **a** $y = 2x + 1$ **b** $y = -3x + 3$ **c** $y = \frac{1}{4}x + 2$
d $y = -\frac{5}{2}x + 10$ **e** $y = \frac{1}{3}x - \frac{7}{6}$ **f** $y = -\frac{7}{2}x - 14$

4 **a** $2x - 3y = 6$ has gradient $\frac{2}{3}$,
$6x + ky = 4$ has gradient $-\dfrac{6}{k}$
b **i** $k = -9$ **ii** $k = 4$

5 **a** yes **b** no **c** yes **d** yes

6 **a** yes **b** no **c** no **d** yes

7 **a** $c = 4$ **b** $m = \frac{1}{5}$ **c** $a = -2$ **d** $k = \frac{15}{4}$

8 **a** $k = 11$ **b** $k = 19$ **c** $k = -4$ **d** $k = 10$

9 **a** -2 **b** 5 **c** 5 **d** 2 **e** 6 **f** 9
g -3 **h** $\frac{7}{2}$ **i** $-\frac{3}{4}$

10 **a** $\dfrac{d}{b}$ given $b \neq 0$
The condition $b \neq 0$ means that the line is not vertical.
b $\dfrac{d}{a}$ given $a \neq 0$
The condition $a \neq 0$ means that the line is not horizontal.
c $-\dfrac{a}{b}$ given $b \neq 0$
d **i** $-\frac{1}{3}$ **ii** 2 **iii** -4 **iv** $\frac{3}{2}$ **v** $-\frac{1}{5}$ **vi** $\frac{4}{3}$
e **i** Lines parallel to $3x + 5y = 2$ have gradient $-\frac{3}{5}$.
$\therefore$ they have the form $3x + 5y = d$, where d is a constant.
ii Lines perpendicular to $3x + 5y = 2$ have gradient $\frac{5}{3}$.
$\therefore$ they have the form $5x - 3y = d$, where d is a constant.

EXERCISE 13G

1 **a**
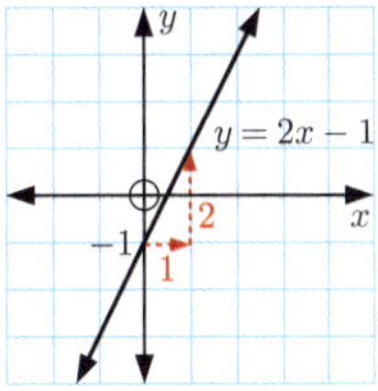

b
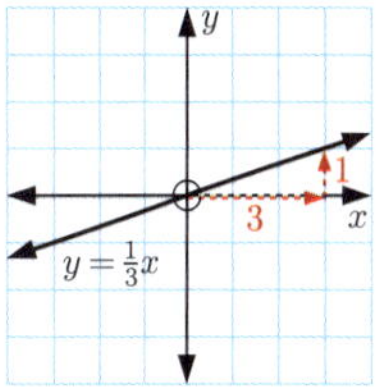

c
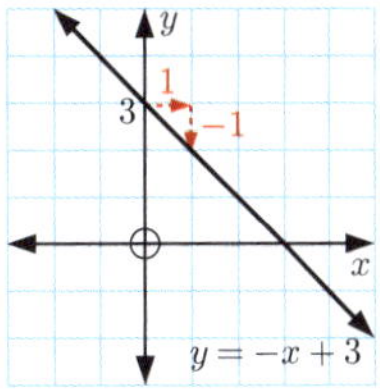

d
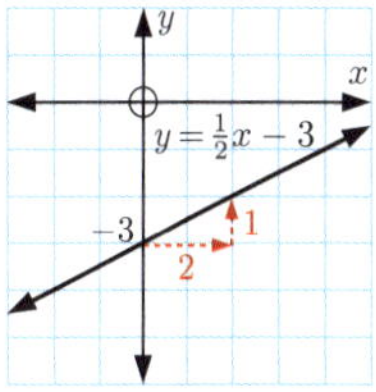

e
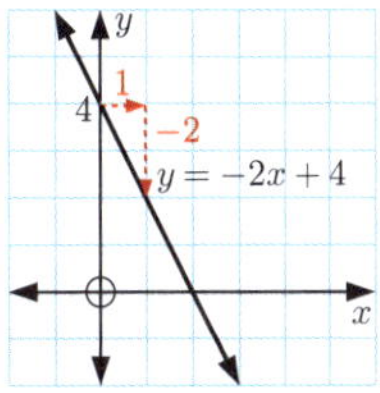

f
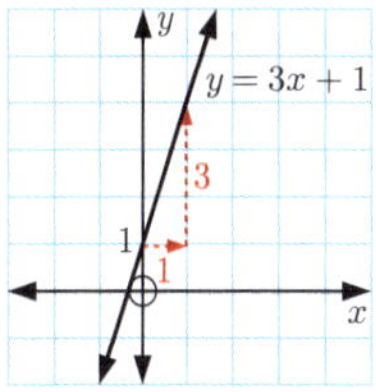

g
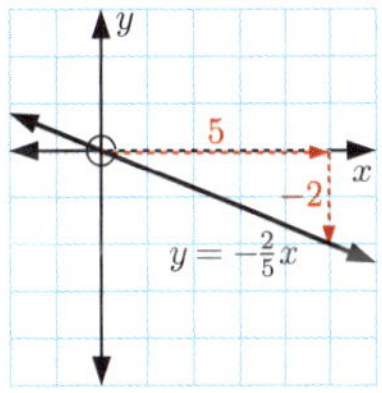

h
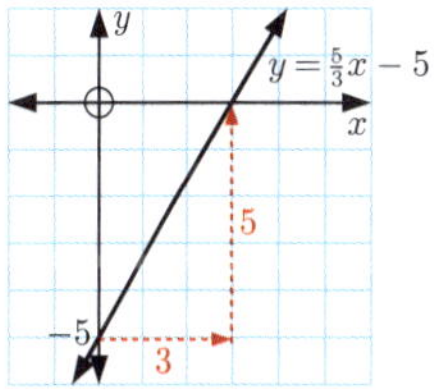

i
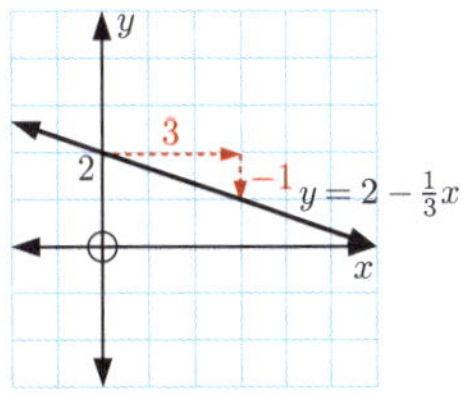

2 **a**
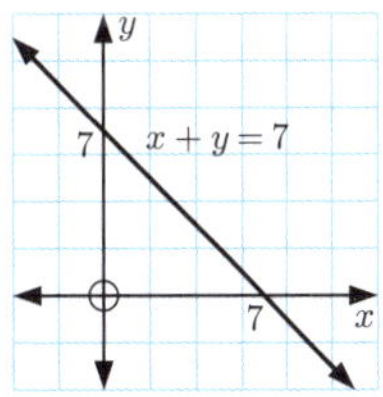

b
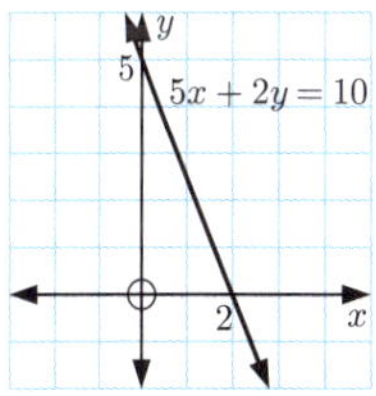

c
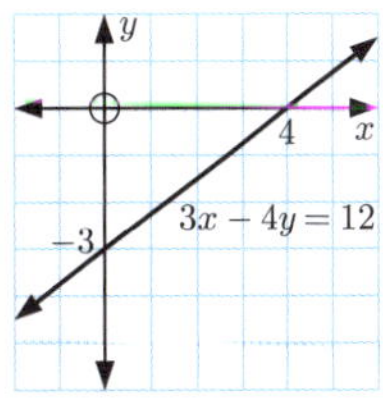

d
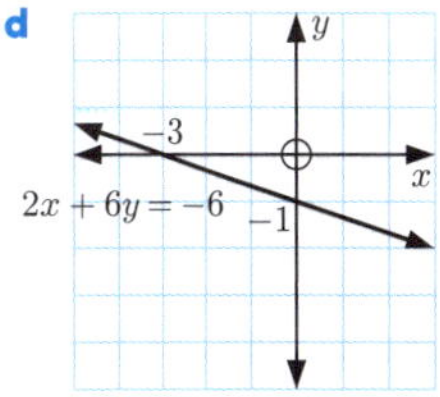

e

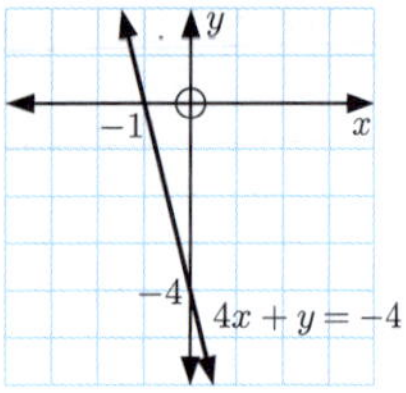

f

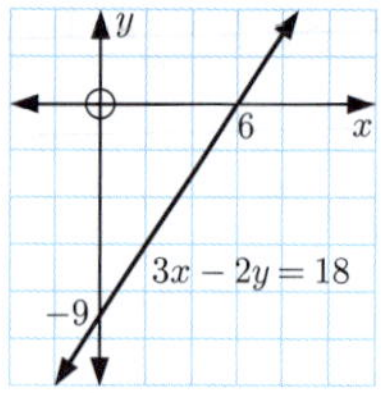

g

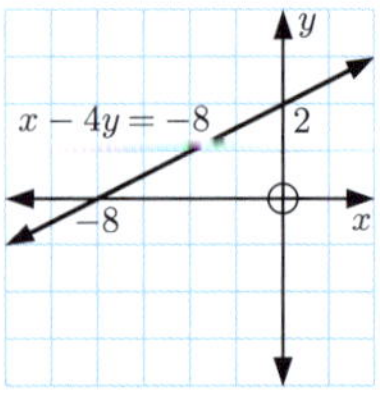

h

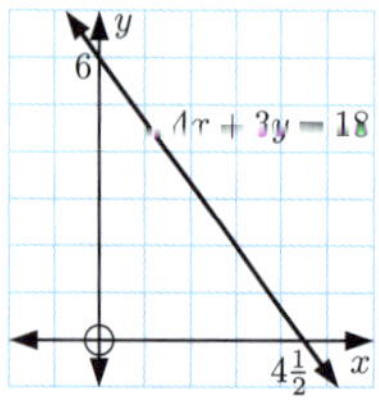

i

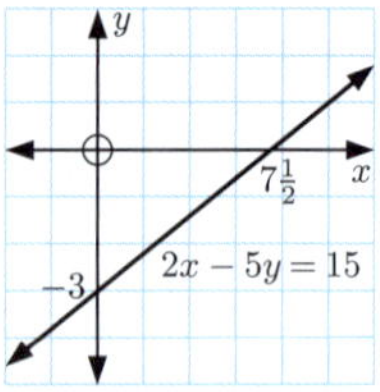

3 **a**

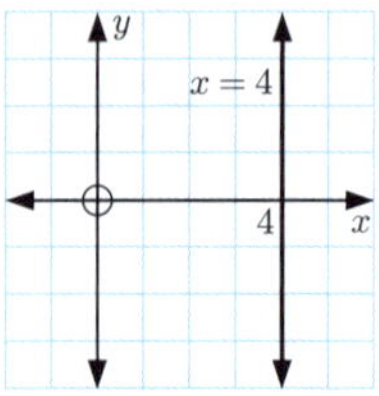

b

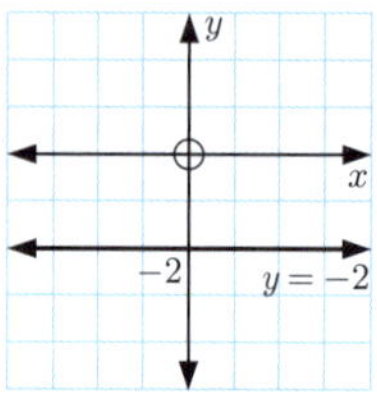

c

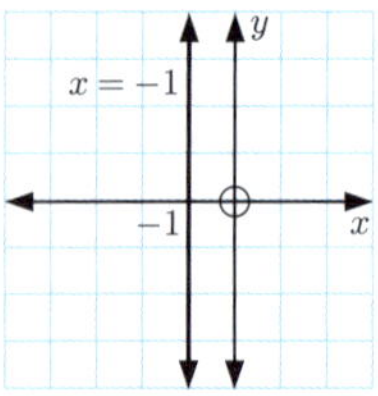

d

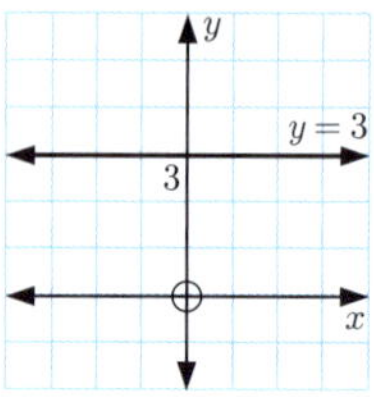

4 **a**

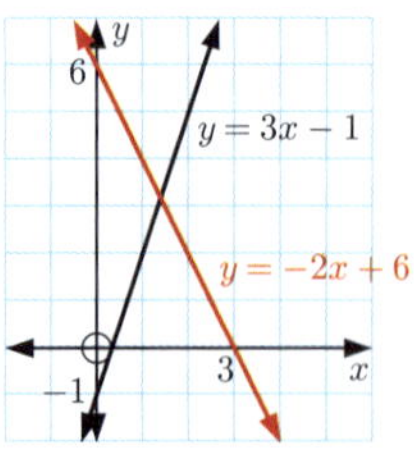

b

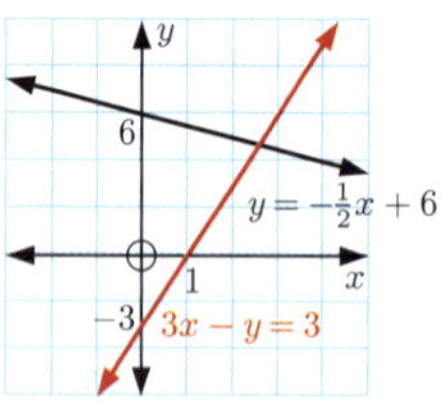

c

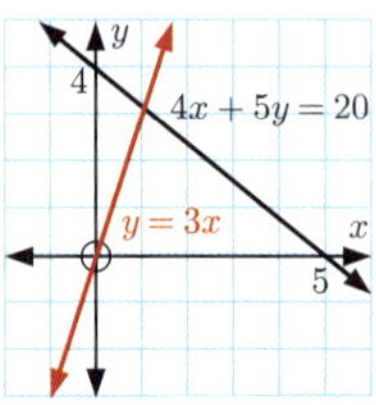

d

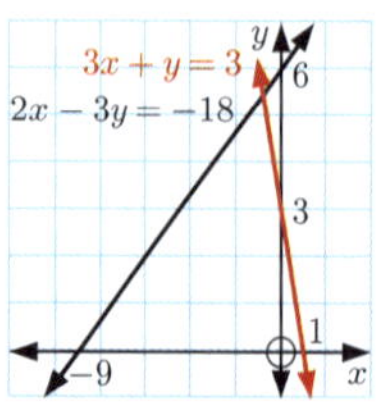

e

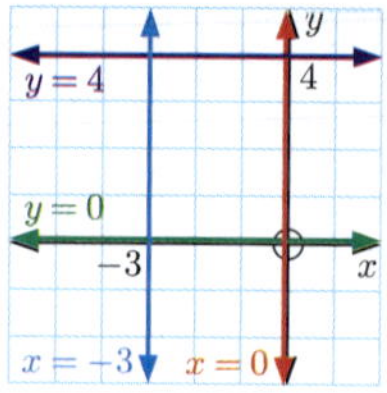

5 **a** $C = 39x + 95$ **b**

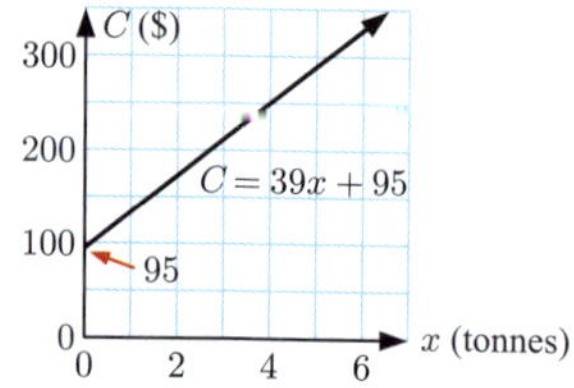

c gradient is 39; the cost of purchasing sand is $39 per tonne. C-intercept is 95; the fixed delivery fee is $95.

6 **a** The total cost of x skipping ropes at $12 each, and y hoops at $16 each is $528.

$\therefore \quad 12x + 16y = 528$

b

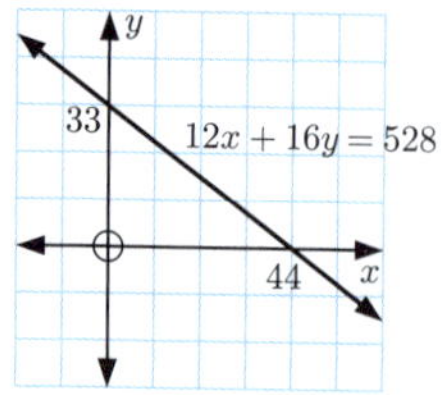

c **i** x-intercept is 44; this is the number of skipping ropes bought if no hoops were bought.

ii y-intercept is 33; this is the number of hoops bought if no skipping ropes were bought.

iii The points (x, y) on the line for which $x, y \in \mathbb{N}$ represent the possible combinations of skipping ropes and hoops Belinda could buy.

EXERCISE 13H

1 **a** $y = 2x + 4$ **b** $y = -3x - 2$ **c** $y = \frac{1}{2}x + \frac{1}{4}$
d $y = -\frac{1}{3}x$

2 **a** $3x - 4y = -8$ **b** $2x + 3y = -9$

3 **a** $y = -\frac{2}{3}x + 5$ **b** $\frac{15}{2}$

4 **a** **i** $y = \frac{2}{3}x - 2$ **ii** $2x - 3y = 6$
b **i** $y = -2x + 5$ **ii** $2x + y = 5$
c **i** $y = \frac{5}{3}x + 3$ **ii** $5x - 3y = -9$

5 **a** $y = 2x + 1$ **b** $y = -x + 1$ **c** $y = \frac{2}{3}x + 3$
d $y = -\frac{4}{5}x + \frac{6}{5}$

6 **a** $y = x - 4$ **b** $y = -x + 4$ **c** $y = 6x + 16$
d $y = \frac{1}{2}x + 8$ **e** $y = -\frac{1}{3}x - \frac{7}{3}$ **f** $y = -\frac{3}{5}x - \frac{14}{5}$

7 **a** $4x - y = 7$ **b** $3x + 5y = -1$ **c** $x - 3y = -11$
d $2x - 7y = 25$

8 **a** $3x - y = -1$ **b** $2x + y = 3$ **c** $y = -3$

9 **a** $4x + 5y = 13$ **b** $4(12) + 5(-7) = 48 - 35 = 13$ ✓

10 **a** $x + 4y = 1$ **b** $5x - 3y = 10$ **c** $x = 2$

11 **a** $y = 2x + 6$ **b** $4x + 3y = 20$

12 $-\frac{5}{3}$ **13** $x + 3y = 13$

14 **a** **i** $y = -\frac{1}{2}x + 2$ **ii** $y = 2x - 3$ **iii** $y = -\frac{1}{2}x + \frac{19}{2}$

EXERCISE 13I

1 **a** $(4, 4)$ **b** $-\frac{2}{3}$ **c** $\frac{3}{2}$ **d** $3x - 2y = 4$

2 **a** $x - y = 4$ **b** $2x - y = -6$ **c** $12x - 10y = -35$

3 **a** $2x + 3y = -1$ **b** $2(-5) + 3(3) = -1$ ✓

c $AC = BC = \sqrt{65}$ units

4 $2x - 3y = -5$

5 **a** $2x - 5y = -32$ **b**

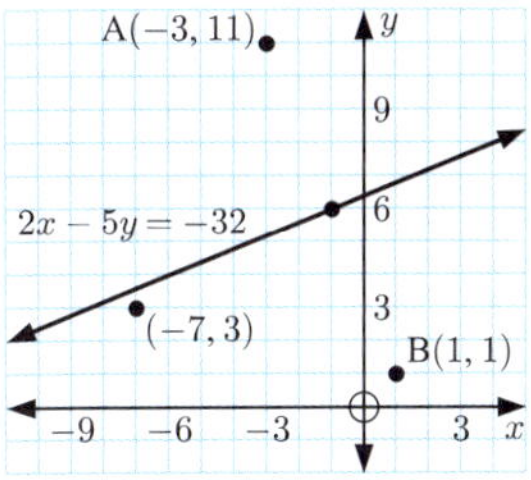

c hospital B

6 **a** $AB = AD = \sqrt{10}$ units, $BC = CD = 5$ units

There are two pairs of equal adjacent sides.

$\therefore$ ABCD is a kite.

b $x - 2y = -2$

c $0 - 2(1) = -2$ and $6 - 2(4) = -2$ ✓

7 **a** ≈ 10.2 km

b The boundaries are the perpendicular bisectors of each pair of points.

c $x - 5y = -3$, $3x + 4y = 2$, $4x - y = -1$

d station B

e $3(-6) + 4(5) = -18 + 20 = 2$ ✓

$\therefore$ $(-6, 5)$ lies on $3x+4y = 2$ which is the perpendicular bisector of [AC].

$\therefore$ $(-6, 5)$ is equally closest to stations A and C.

EXERCISE 13J

1 **a**

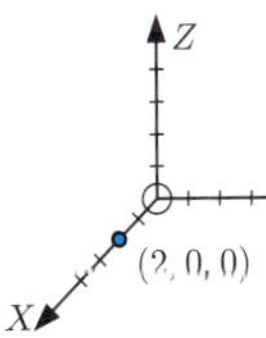

b

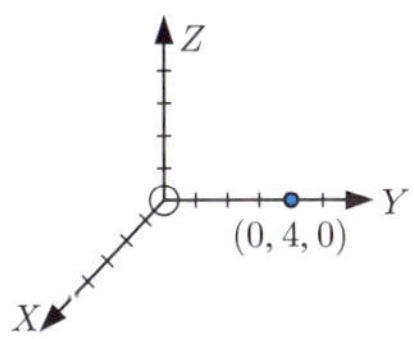

c

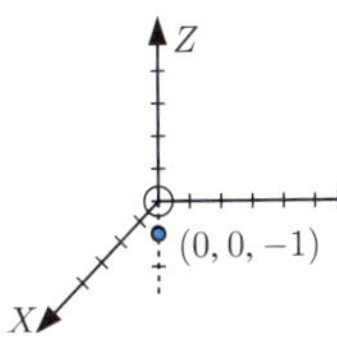

d

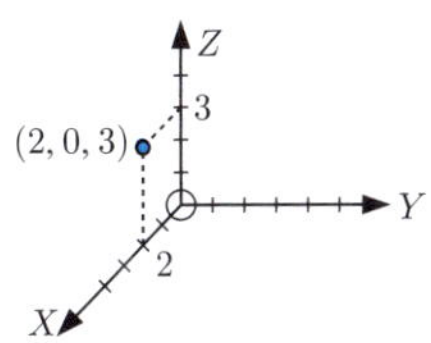

e

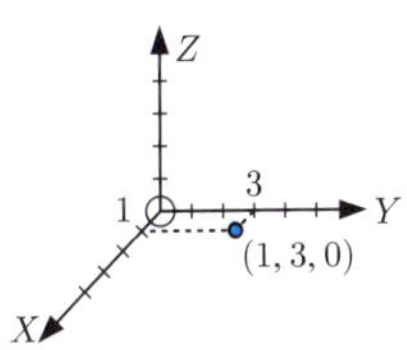

f

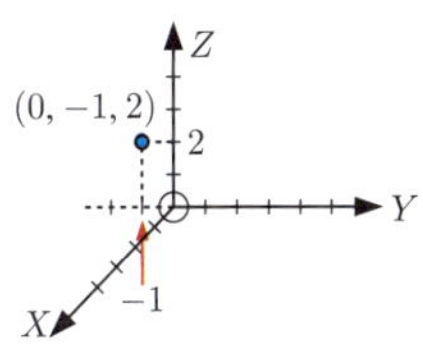

g

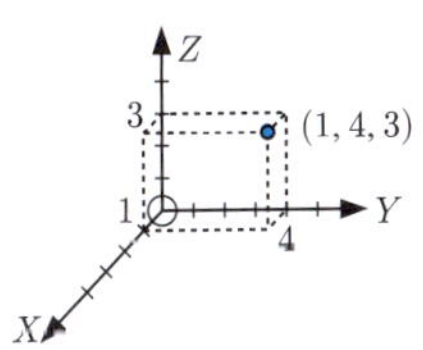

h

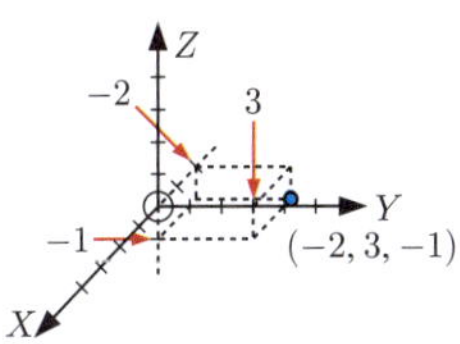

i

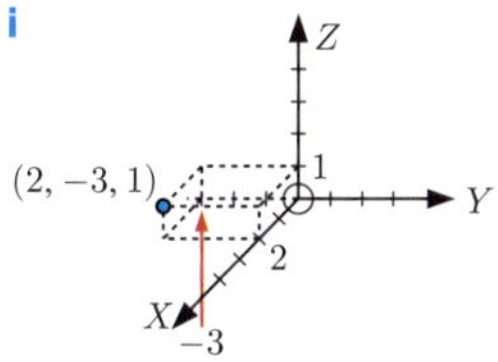

j

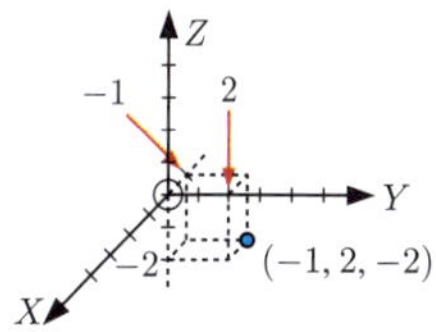

k

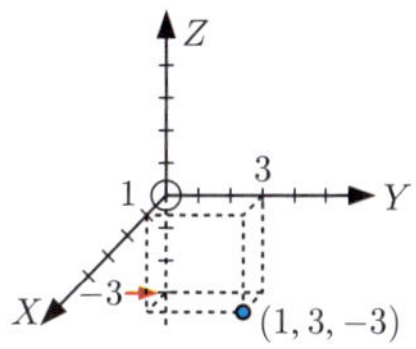

l

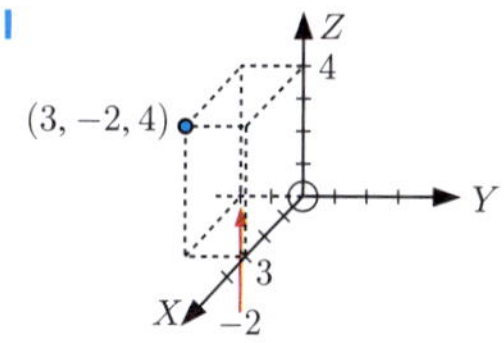

2 $d = \sqrt{a^2 + b^2 + c^2}$ units

3 **a** **i** $\sqrt{11}$ units **ii** $(2\frac{1}{2}, -\frac{1}{2}, -\frac{1}{2})$

b **i** 3 units **ii** $(1, -\frac{1}{2}, 3)$

c **i** $\sqrt{17}$ units **ii** $(4, -1, -1\frac{1}{2})$

d **i** $\sqrt{34}$ units **ii** $(-\frac{1}{2}, -1, 3\frac{1}{2})$

4 **a** isosceles $\{AC = BC = \sqrt{14}$ units$\}$

b right angled at B $\{AB^2 + BC^2 = AC^2 = 54\}$

5 $k = 1 \pm \sqrt{19}$

6 $x^2 + y^2 + z^2 = 4$ which is the equation of a sphere, centre $(0, 0, 0)$, radius 2 units.

REVIEW SET 13A

1 **a** 5 units **b** $\sqrt{13}$ units **2** $m = 6$ or -2

3 **a** $(5, -1)$ **b** $(-5\frac{1}{2}, 6\frac{1}{2})$

4 **a** $D(4, -3)$ **b** 5 units

5

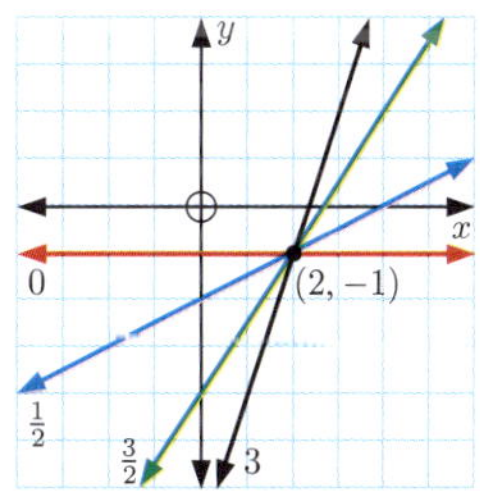

6 **a** 2 **b** $-\frac{2}{7}$

7 $-\frac{3}{10}$

8 [PS] and [QR]

9 **a** $AB = BC = 5$ units,

gradient of $[AB] = \frac{3}{4}$, gradient of $[BC] = -\frac{4}{3}$,

and $\frac{3}{4} \times -\frac{4}{3} = -1$

$\therefore$ $\triangle ABC$ is right angled at B and isosceles.

b $X(\frac{1}{2}, \frac{1}{2})$

c gradient of $[BX] = 7$, gradient of $[AC] = -\frac{1}{7}$

and $7 \times -\frac{1}{7} = -1$, $\therefore$ $[BX] \perp [AC]$.

10 gradient of [AB] = gradient of [BC] = 2 and B is common

11 **a** [AB]: $\frac{1}{5}$, [BC]: -2, [CD]: $\frac{1}{5}$, [AD]: -2

$\therefore$ ABCD is a parallelogram.

b $(\frac{1}{2}, \frac{1}{2})$ for both diagonals.

$\therefore$ the diagonals of the parallelogram bisect each other.

c **i** [AC]: $-\frac{3}{7}$, [BD]: $\frac{5}{3}$

ii product $= -\frac{5}{7}$ $\therefore$ ABCD is not a rhombus.

12 **a** gradient is 6, y-intercept is -4
b gradient is $-\frac{2}{5}$, y-intercept is 3
c gradient is $\frac{2}{3}$, y-intercept is $-\frac{7}{6}$

13 **a** $3x+4y=24$ has gradient $-\frac{3}{4}$,

$6x+ky=12$ has gradient $-\dfrac{6}{k}$

b **i** $k=8$ **ii** $k=-\frac{9}{2}$

14 **a** no **b** yes

15 **a**

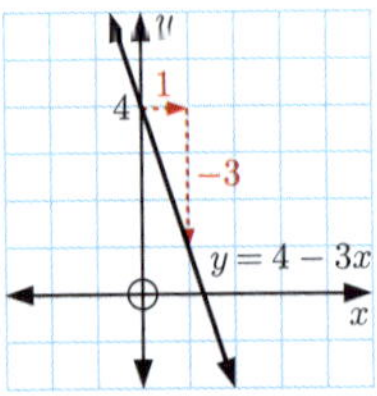

b

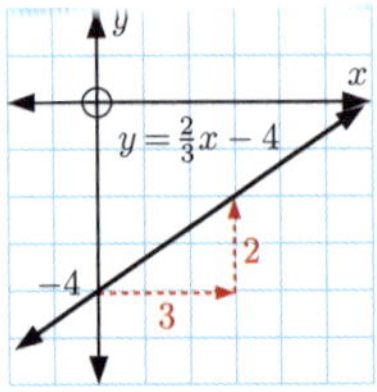

c

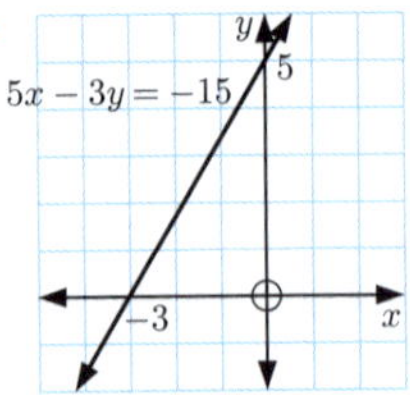

16 **a** $y=-2x+7$ **b** $y=-\frac{2}{3}x+\frac{7}{3}$

17 **a** $x-3y=6$ **b** $7x-4y=-12$ **c** $9x+5y=7$

18 $4x-5y=2$ **19** **a** $16x-2y=-17$ **b** $c=-\frac{15}{2}$

20 **a** $PQ=QR=\sqrt{50}$ units, $PR=\sqrt{40}$ units
b $x-3y=-7$ **c** $5-3(4)=-7$ ✓

21 **a** $PQ=2\sqrt{14}$ units **b** $(0, 0, 0)$

REVIEW SET 13B

1 **a** $ST=\sqrt{73}$ units **b** $(3, -\frac{1}{2})$ **2** $a=-1$ or -3

3 $B(5, -4)$

4 **a** isosceles $\{AB=BC=\sqrt{65}$ units$\}$ **b** not right angled

5

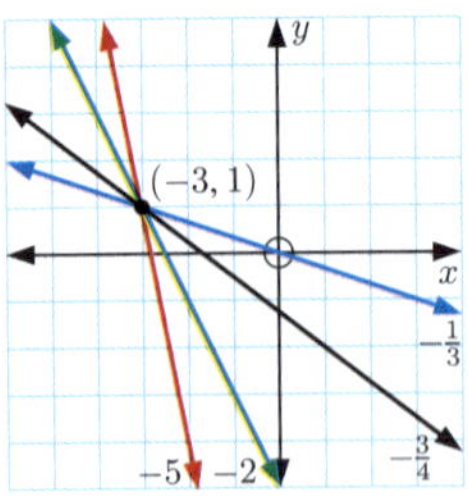

6 **a** -1
b undefined
c $\frac{3}{2}$

7 **a** $k=-4\frac{1}{2}$
b $k=-15$

8

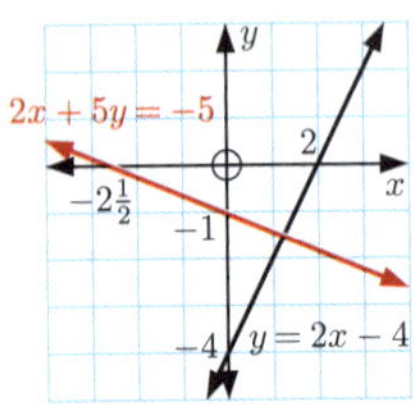

9 $a=4$

10 $b=-3$

11 **a** **i** $-\frac{3}{5}$ **ii** $\frac{7}{11}$ **iii** $\frac{5}{3}$
b $-\frac{3}{5}\times\frac{5}{3}=-1$ $\therefore$ $[PQ]\perp[PR]$ **c** 34 units^2
$\therefore$ $Q\widehat{P}R$ is a right angle.

12 **a** $y=\frac{4}{3}x-\frac{2}{3}$ **b** $\frac{4}{3}$ **13** $3x+y=5$

14 **a** $k=\frac{9}{4}$ **b** $k=39$

15 **a**

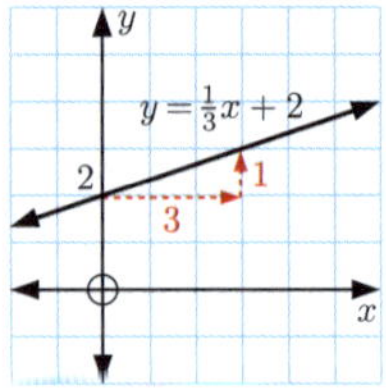

b

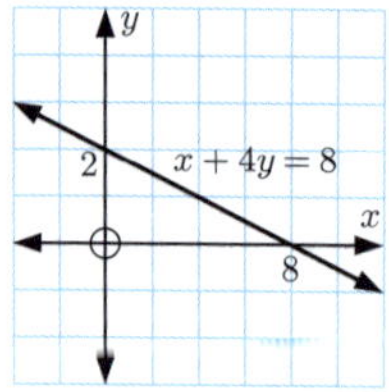

c

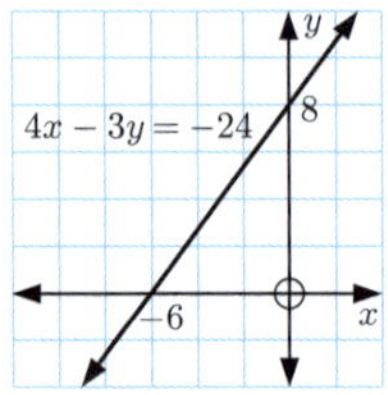

16 **a** $3x-y=-7$ **b** $x+2y=1$

17 **a** $C=15t+50$ **b**

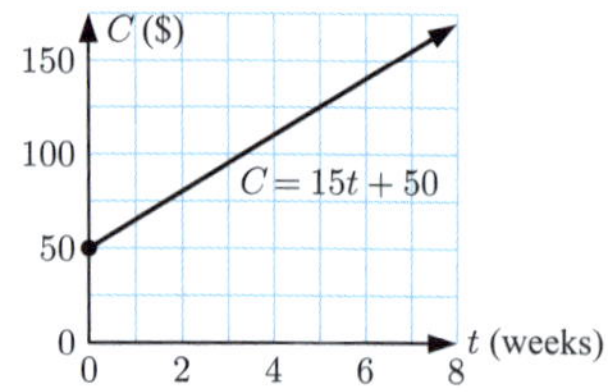

c gradient is 15; the cost of hiring a storage locker is $15 per week.
C-intercept is 50; the fixed reservation fee is $50.

18 $x-2y=-5$

19 **a** $3x-y=-1$, $x+3y=-7$, $x+y=-3$
b restaurant Q
c $3(-1)-(-2)=-3+2=-1$ ✓
$-1+3(-2)=-1-6=-7$ ✓
$-1+(-2)=-1-2=-3$ ✓
$\therefore$ $(-1, -2)$ satisfies all of the boundary equations.
$\therefore$ $(-1, -2)$ is equidistant from all three restaurants.

20 $\sqrt{30}$ units **21** $k=4$ or -6

EXERCISE 14A.1

1 **a** $x=-1$, $y=-5$ **b** $x=1$, $y=3$
c $x=-2$, $y=0$

2 **a** $x=5$, $y=2$ **b** $x=3$, $y=-4$
c $x=-4$, $y=-6$

3 **a** no solutions **b** infinitely many solutions

EXERCISE 14A.2

1 **a** $x=0.75$, $y=1.75$ **b** $x=1.4$, $y=-1.8$
c $x=-0.25$, $y=-4.75$ **d** $x=1.8$, $y=4.2$
e $x\approx1.27$, $y\approx-4.64$ **f** $x\approx-2.71$, $y\approx-1.53$

2 **a** infinitely many solutions **b** no solutions
c no solutions

EXERCISE 14B

1 **a** $x=7$, $y=4$ **b** $x=-1$, $y=-5$
c $x=3$, $y=-7$ **d** $x=5$, $y=-2$
e $x=-\frac{1}{2}$, $y=\frac{5}{2}$ **f** $x=\frac{20}{47}$, $y=\frac{1}{47}$

2 **a** $x=4,\ y=-5$ **b** $x=-2,\ y=-3$
c $x=-2,\ y=\frac{5}{2}$ **d** $x=\frac{1}{3},\ y=-\frac{4}{3}$
e $x=-\frac{3}{4},\ y=\frac{5}{8}$ **f** $x=\frac{23}{31},\ y=-\frac{29}{31}$

3 **a** $x=5,\ y=\frac{1}{2}$ **b** $x=-\frac{4}{3},\ y=2$
c $x=3,\ y=-\frac{1}{4}$ **d** $x=\frac{2}{5},\ y=-\frac{6}{5}$
e $x=\frac{3}{7},\ y=-\frac{5}{7}$ **f** $x=\frac{7}{2},\ y=-\frac{8}{5}$

4 **a** Substituting $y=3x+1$ into $9x-3y=5$ gives $-3=5$, which is absurd.
$\therefore$ there are no solutions.
b Substituting $y=3x+1$ into $6x-2y=-2$ gives $-2=-2$, which is true for all values of x.
$\therefore$ there are infinitely many solutions.

5 **a** $y=\dfrac{-4x-4}{5}$ **b** $x=-\frac{17}{7},\ y=\frac{8}{7}$

EXERCISE 14C

1 **a** $11x=11$ **b** $4y=12$ **c** $9x=9$ **d** $9x=6$
e $-y=11$ **f** $-11y=-11$

2 **a** $x=2,\ y=6$ **b** $x=1,\ y=-2$
c $x=-1,\ y=-3$ **d** $x=1,\ y=-5$
e $x=2,\ y=-2$ **f** $x=-\frac{17}{4},\ y=\frac{5}{2}$

3 **a** $10x+25y=5$ **b** $-3x+y=-4$
c $3x-21y=24$ **d** $-10x-8y=-18$
e $-18x-12y=12$ **f** $-16x+8y=-12$

4 **a** $x=5,\ y=-2$ **b** $x=1,\ y=2$
c $x=1,\ y=-6$ **d** $x=\frac{1}{2},\ y=-3$
e $x=1,\ y=-\frac{2}{3}$ **f** $x=-\frac{3}{4},\ y=3$

5 **a** $x=5,\ y=1$ **b** $x=1,\ y=1$
c $x=-3,\ y=4$ **d** $x=2,\ y=-3$
e $x=-7,\ y=-2$ **f** $x=-\frac{2}{3},\ y=\frac{5}{3}$
g $x=\frac{8}{5},\ y=\frac{22}{5}$ **h** $x=\frac{15}{2},\ y=-\frac{5}{2}$
i $x=-\frac{10}{3},\ y=-\frac{11}{3}$

6 **a** The system reduces to $0=0$, which is true for all x.
$\therefore$ there are infinitely many solutions.
b The system reduces to $0=-5$, which is absurd.
$\therefore$ there are no solutions.

EXERCISE 14D

1 The numbers are 16 and 56. **2** The numbers are 8 and 22.
3 The numbers are 17 and 39.

4 **a** no
b **i** Hafiz spent RM 12.50 buying 3 Ramly burgers and 2 satay sticks.
$\therefore\ 3x+2y=12.5$
ii Ismail spent RM 6.50 buying 1 Ramly burger and 3 satay sticks.
$\therefore\ x+3y=6.5$
c $x=3.5,\ y=1$ **d** yes

5 geranium: \$8, marigold: \$3 **6** \$23
7 fish: \$2.20, chips: \$0.75
8 coffee: \$2.90, muffin: \$1.25 **9** 33 bottles **10** 11 cm
11 before 3 pm: \$2.50 per h, after 3 pm: \$4 per h
12 6 small sets and 5 large sets **13** R(3, 9) **14** 15 units

EXERCISE 14E

1 **a** $x=-3,\ y=20$ and $x=2,\ y=5$
b $x=2,\ y=13$ and $x=-9,\ y=35$
c $x=-3,\ y=24$ and $x=1,\ y=4$
d $x=-2,\ y=18$ and $x=5,\ y=46$
e $x=-2,\ y=1$ and $x=6,\ y=73$
f $x=-3,\ y=19$

2 **a** $x=-5,\ y=-4$ and $x=4,\ y=5$
b $x=4,\ y=8$
c $x=-\frac{5}{3},\ y=-3$ and $x=\frac{1}{2},\ y=10$

3 **a** no real solutions
b $x=-2-\sqrt{10},\ y=2-\sqrt{10}$ and $x=-2+\sqrt{10},\ y=2+\sqrt{10}$
c no real solutions

5 B is $(1+\sqrt{6},\ 2+3\sqrt{6})$ or $(1-\sqrt{6},\ 2-3\sqrt{6})$

REVIEW SET 14A

1 **a** $x=-3,\ y=-13$ **b** $x=6,\ y=-2$
2 **a** $x=3.25,\ y=5.75$ **b** $x=0.2,\ y=-0.6$
3 **a** $x=2,\ y=2$ **b** $x=5,\ y=-3$
4 **a, b** Substituting $y=4x-2$ into $8x-2y=4$ gives $4=4$, which is true for all values of x.
$\therefore$ there are infinitely many solutions.
5 **a** $6x=-6$ **b** $-4y=40$
6 **a** $x=-3,\ y=1$ **b** $x=\frac{84}{11},\ y=\frac{38}{11}$
c $x=\frac{3}{2},\ y=-\frac{5}{2}$
7 adult: \$18, student: \$12
8 14 fifty cent coins and 7 twenty cent coins **9** 17 cm
10 **a** $x=-3,\ y=16$ and $x=5,\ y=0$
b $x=8,\ y=-1$ and $x=\frac{1}{2},\ y=-16$

REVIEW SET 14B

1 **a** $x=3,\ y=1$ **b** $x=4,\ y=-3$
2 infinitely many solutions
3 **a** $x=\frac{23}{9},\ y=\frac{29}{9}$ **b** $x=-\frac{2}{9},\ y=\frac{17}{9}$
4 **a, b** Substituting $y=2x-5$ into $4x-2y=3$ gives $10=3$, which is absurd.
$\therefore$ there are no solutions.
5 $-6x+9y=12$
6 **a** $x=3,\ y=-4$ **b** $x=2,\ y=-5$
c $x=-43,\ y=-31$
7 7 points **8** 11 packs
9 **a** **i** The total shaded region occupies $\frac{1}{4}$ of the square.
$\therefore\ A+B=\frac{1}{4}$
ii One semi-circular region occupies $\frac{\pi}{8}$ of the square.
$\therefore\ 2A+B=\frac{\pi}{8}$
b **i** $\dfrac{\pi-2}{8}$ **ii** $\dfrac{4-\pi}{8}$
c $\dfrac{\pi-2}{8}-\dfrac{4-\pi}{8}=\dfrac{2\pi-6}{8}$
Since $\pi>3,\ 2\pi-6>0$, so $\dfrac{\pi-2}{8}-\dfrac{4-\pi}{8}>0$.
So, the green region is larger.
10 **a** $x=5,\ y=75$
b $x=-6,\ y=-6$ and $x=3,\ y=12$

EXERCISE 15A

1 a $\triangle ABC \cong \triangle FED$ {AAcorS}
b $\triangle PQR \cong \triangle ZYX$ {SAS}
c $\triangle ABC \cong \triangle EDF$ {AAcorS}
d $\triangle ABC \cong \triangle LKM$ {SSS} e $\triangle XYZ \cong \triangle FED$ {RHS}
f not enough information g not enough information
h not enough information i $\triangle ABC \cong \triangle PQR$ {SSS}
j $\triangle ABC \cong \triangle FED$ {AAcorS}

2 a **B** and **D** {SAS} b **A** and **D** {RHS}
c **B** and **C** {AAcorS} d **A** and **D** {SSS}

3 **Hint:** Use AAcorS.

4 This is *not* SAS since the angle α is included in $\triangle ABC$ but not $\triangle CDA$.
$\therefore$ there is not sufficient evidence to conclude that the triangles are congruent.

EXERCISE 15B

1 a **Hint:** Use SSS. b i $47°$ ii $51°$

2 a **Hint:** Use SSS.
b $\widehat{WZX} = \widehat{YXZ}$ so [WZ] ∥ [XY]
and $\widehat{WXZ} = \widehat{YZX}$ so [WX] ∥ [ZY]

3 a **Hint:** Join [AC] and show $\triangle ABC \cong \triangle CDA$
b **Hint:** Join [DB] and let the diagonals meet at M.
Show $\triangle AMD \cong \triangle CMB$

4 **Hint:** Show $\triangle APB \cong \triangle AQC$

5 **Hint:** Join [AX], [BX], and [CX].
Show $\triangle AQX \cong \triangle BQX$ and $\triangle APX \cong \triangle CPX$

6 **Hint:** Join [WZ], [ZY], [YX], and [XW].
Show $\triangle WAZ \cong \triangle YDZ \cong \triangle YCX \cong \triangle WBX$

EXERCISE 15C

1 a $\widehat{TPR} = \widehat{SQR}$ {given}, angle R is common
b $\widehat{ABC} = \widehat{DEC}$ {given}
$\widehat{ACB} = \widehat{DCE}$ {vertically opposite angles}
c $\widehat{AEC} = \widehat{DBC}$ {given}, angle C is common
d $\widehat{WVX} = \widehat{ZYX}$ {given}
$\widehat{VXW} = \widehat{YXZ}$ {vertically opposite angles}
e $\widehat{JNK} = \widehat{MLK}$ {given}
$\widehat{JKN} = \widehat{MKL}$ {vertically opposite angles}
f $\widehat{ABC} = \widehat{BCD}$ {equal base angles of an isosceles triangle}
$\widehat{ACB} = \widehat{BDC}$ {equal base angles of an isosceles triangle}

2 a angle A is common,
$\widehat{ABE} = \widehat{ACD}$ {equal corresponding angles}; $x = 1.2$
b $\widehat{PTR} = \widehat{QSR}$ {given}, angle R is common; $x = \frac{10}{3}$
c $\widehat{CAB} = \widehat{CDE}$ {given}
$\widehat{ACB} = \widehat{DCE}$ {vertically opposite angles};
$x = \frac{20}{7}$
d angle S is common,
$\widehat{PQS} = \widehat{TRS}$ {equal corresponding angles}; $x = 7.5$
e $\widehat{BEA} = \widehat{DCA}$ {given}, angle A is common; $x = \frac{4}{3}$
f $\widehat{YXZ} = \widehat{ZWV}$ {given}
$\widehat{YZX} = \widehat{ZVW}$ {exterior angle of a triangle};
$x = 4.8$
g $\widehat{BEA} = \widehat{CDA}$ {given}, angle A is common; $x = \frac{35}{3}$
h angle P is common,
$\widehat{PTQ} = \widehat{PSR}$ {equal corresponding angles}; $x = 10.8$
i $\widehat{UVW} = \widehat{XYW}$ {given}
$\widehat{UWV} = \widehat{XWY}$ {vertically opposite angles};
$x = \frac{8}{3}$
j $\widehat{EDA} = \widehat{CBA}$ {given}, angle A is common; $x = 12$
k $\widehat{ABC} = \widehat{EDC}$ {given}
$\widehat{BCA} = \widehat{DCE}$ {vertically opposite angles};
$x = 7$
l $\widehat{CBA} = \widehat{EDA}$ {given}, angle A is common; $x = 10$

3 5.25 m 4 ≈ 1.22 m 5 6.375 km
6 ≈ 651 m 7 8 cm

8 She is correct as the viewing region is always 20 m long no matter where the camera is located on [AB].

9 1.6 m 10 ≈ 35.7 m 11 ≈ 5.56 cm

12 a $x = 2$ b $x = 5$ c $x = 6$
d $x = \sqrt{31} - 1$ e $x = \dfrac{3+\sqrt{5}}{2}$ f $x = 3 + \sqrt{34}$

13 BE = 6 cm 14 CD = $(1 + \sqrt{41})$ m

EXERCISE 15D

1 a $x \approx 14.2$ b $x \approx 30.7$ c $x \approx 41.5$
d $x \approx 6.26$ e $x = 8$ f $x \approx 10.7$

2 a $x = 96$ b $x = 432$ c $x \approx 0.150$
d $x = 7.5$ e $x = 13.2$ f $x \approx 6.94$

3 a 12.544 cm^2 b 6.144 cm^2

4 a $x = 1.5$ b 17.6 m^2 5 30 cm

6 a i It is multiplied by 8. ii It is increased by 72.8%.
b i It is divided by 8. ii It is increased by 237.5%.

7 **Hint:** The red triangle is an enlargement of the blue triangle with scale factor 2.

8 a similar b not similar

9 a 10 cm b 30 cm^2
c height 6 mm, radius 2.25 mm d 1.25 mL

10 a 3 b 4 cm c 648 cm^3

11 a $\frac{8}{5}$ b ≈ 48.8 cm^3 c 1310.72 g

12 a $\frac{2}{3}$ b 14 850 cm^3 c 280 cm^2

13 a The dimensions are in the same ratio.
b i $6 : 10 : 15$ ii $\sqrt[3]{6} : \sqrt[3]{10} : \sqrt[3]{15}$
iii $(\sqrt[3]{6})^2 : (\sqrt[3]{10})^2 : (\sqrt[3]{15})^2$
c i ≈ 8.43 cm ii ≈ 4.58 cm iii ≈ 215 cm^2

REVIEW SET 15A

1 a $\triangle ABC \cong \triangle EDF$ {RHS}
b not enough information c not enough information

2 a **Hint:** Use SAS.
b $\widehat{DAC} = \widehat{BCA}$ {$\triangle ABC \cong \triangle CDA$}
$\therefore$ [AD] ∥ [BC] {equal alternate angles}
$\therefore$ ABCD is a parallelogram.

4 a $\widehat{UYV} = \widehat{WXV}$ {given}
$\widehat{UVY} = \widehat{WVX}$ {vertically opposite angles}
b $\widehat{JLK} = \widehat{NLM}$ {vertically opposite angles}
$\widehat{JKL} = \widehat{NML}$ {equal alternate angles}

c $\widehat{CBD} = \widehat{CEA}$ {given}, angle C is common.

5 **a** $\widehat{CDA} = \widehat{EBA}$ {given}, angle A is common; $x = 1.625$

b angle P is common,

$\widehat{PQT} = \widehat{PRS}$ {equal corresponding angles}; $x = 4.8$

c $\widehat{WXV} = \widehat{YXZ}$ {vertically opposite angles}

$\widehat{VWX} = \widehat{ZYX}$ {equal alternate angles};

$x = 9$

6 ≈ 10.4 m

7 **a** **Hint:** Show that triangles ABD, BCD, and ACB are similar.

9 **a** $x = 45$ **b** $x = 10$

10 **a** not similar **b** similar **11** 2 cm

REVIEW SET 15B

1 **a** **B** and **C** {AAcorS} **b** **A** and **C** {AAcorS}

2 **Hint:** Join [MR] and [NR], and show $\triangle MQR \cong \triangle NSR$.

3 **Hint:** Join [BQ] and [CP], and show that $\triangle ABQ \cong \triangle ACP$.

4 **a** **Hint:** Use RHS.

b $\widehat{MQX} = \widehat{MRY}$ {congruence}

$\therefore$ $\triangle PQR$ is isosceles {equal base angles}

5 **a** $\widehat{VUZ} = \widehat{YXZ}$ {given}

$\widehat{UZV} = \widehat{XZY}$ {vertically opposite angles};

$x = 4$

b angle A is common,

$\widehat{ACB} = \widehat{AED}$ {equal corresponding angles}; $x = 3$

c $\widehat{JMK} = \widehat{KML}$ {angles on a straight line}

$\widehat{KJM} = 90° - \widehat{JKM}$ {angles in a triangle}

$= \widehat{LKM}$ {angles in a right angle};

$x = 9$

6 ≈ 117 m

7 **a** **Hint:** Show that $\dfrac{h}{30} = \dfrac{AX}{AC}$ and $\dfrac{h}{50} = \dfrac{XC}{AC}$.

b 18.75 m

8 **a** **Hint:** Show $\triangle ABC$ and $\triangle BCD$ are congruent isosceles triangles, for example.

9 **a** $x = 156.25$ **b** $x = 8$

10 **a** 38.4 cm^2 **b** 28.8 cm^2

11 **a** ≈ 7.56 cm **b** ≈ 5.24 cm

EXERCISE 16A

1 **a** $\widehat{OBA} = x°$, $\widehat{OBC} = y°$

b $2x + 2y = 180$ {angles in a triangle}

$\therefore$ $x + y = 90$

and $\widehat{ABC} = (x+y)° = 90°$

2 **a** $x = 62$ {angle in a semi-circle, angles in a triangle}

b $x = 18$ {angle in a semi-circle, angles in a triangle}

c $x = 34$ {angle in a semi-circle, angles in a triangle}

d $x = 25$ {angle in a semi-circle, angles in a triangle}

e $x = 12$ {angle in a semi-circle, Pythagoras}

f $x = \sqrt{13}$ {angle in a semi-circle, Pythagoras}

3 $6\sqrt{2} \approx 8.49$ cm **4** $4\sqrt{5} \approx 8.94$ cm

5 **a** $x = 30$ **b** 12 cm

6 **a** $\widehat{BXA} = \widehat{BXC} = 90°$ {angles in a semi-circle}

b A, X, and C are collinear.

c Yes, the illustrated sprinklers water one side of the diagonal [AC]. We similarly draw semi-circles on sides [CD] and [AD] to conclude that the remaining sprinklers water the other side of the diagonal [AC].

8 **Hint:** Join [AQ] and [BQ], then show that triangles APQ and QPB are similar.

EXERCISE 16B

1 **a** **Hint:** Use RHS.

2 **a** $x \approx 7.55$ **b** $x \approx 9.86$ **c** $x \approx 16.12$

3 ≈ 4.9 cm **4** ≈ 1.07 cm **5** 10.25 cm

6 The centre of the circle passing through A, B, and C is the intersection of the perpendicular bisectors of [AB], [AC], and [BC].

EXERCISE 16C

1 **a** $x = 30$ {radius-tangent, angle in a semi-circle, angles in a triangle}

b $a = 40$ {angle in a semi-circle, angles in a triangle}

$b = 50$ {radius-tangent}

c $a = 60$ {angle in a semi-circle}

$b = 40$ {angles in a triangle, radius-tangent}

2 **Hint:** Show that co-interior angles are supplementary.

3 $\sqrt{53} \approx 7.28$ cm

4 **a** Using Pythagoras in $\triangle OAP$, $OP^2 + AP^2 = OA^2$.

Since [OA] is the hypotenuse, $OP \leqslant OA$.

b No point on the tangent lies inside the circle, which has radius OA, so $OP \geqslant OA$.

c From **a** and **b**, $OP = OA$.

$\therefore$ A and P must be the same point.

$\therefore$ $\widehat{OAT} = \widehat{OPT} = 90°$

EXERCISE 16D

1 **a** **Hint:** Use RHS.

2 **a** $x = 4$, $y = 2\sqrt{5}$ **b** $a = 67$, $b = 23$

c $a = 7$, $b = 57$

4 40 cm

5 **a, b** **Hint:** Show that $A = r(a + b + c)$.

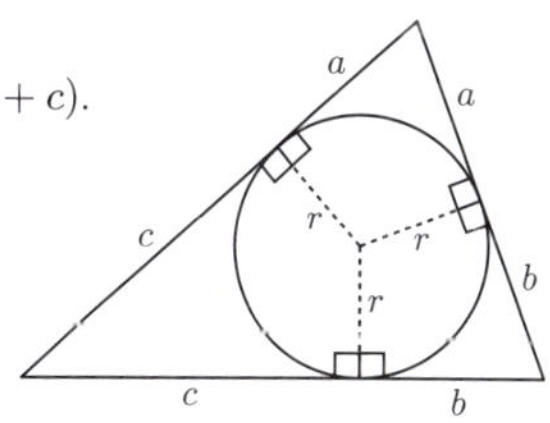

EXERCISE 16E

1 **a** $x = 80$ {angles in a triangle, angle between a tangent and a chord}

b $x = 58$ {angle between a tangent and a chord, angles in an isosceles triangle}

c $x = 40$ {angle between a tangent and a chord, angles in an isosceles triangle}

2 **Hint:** Let $\widehat{ACT} = \alpha$. **4** 72.5°

EXERCISE 16F

1 **a** $x = 64$ {angle at the centre}

b $x = 94$ {angle at the centre}

c $x = 70$ {angle at the centre, angles in an isosceles triangle}

d $x = 45$ {angles in an isosceles triangle, angle at the centre}

e $x = 66$ {angle at the centre}

f $x = 25$ {angle at the centre, base angles of isosceles triangle}

2 **a** OA, OP, and OB are radii of the circle. These lengths are equal, hence $\triangle$OAP and $\triangle$OBP are isosceles.

b **i** $\widehat{APO} = a$ **ii** $\widehat{BPO} = b$ **iii** $\widehat{AOX} = 2a$
iv $\widehat{BOX} = 2b$ **v** $\widehat{APB} = a + b$ **vi** $\widehat{AOB} = 2a + 2b$

c $\widehat{AOB} = 2(a+b) = 2\,\widehat{APB}$ ✓

3 **a** **i** $\widehat{XOY} = \alpha$ **ii** $\widehat{AXO} = 2\alpha$ **iii** $\widehat{XAO} = 2\alpha$
iv $\widehat{XOB} = 4\alpha$ **v** $\widehat{BOY} = 3\alpha$

b $\widehat{BOY} = 3\,\widehat{XOY}$

EXERCISE 16G

1 **a** $x = 46$ {angles subtended by the same arc}
b $x = y = 55$ {angles subtended by the same arc}
c $a = 50,\ b = 40$ {angles subtended by the same arc}
d $a = 55,\ c = 70$ {angles subtended by the same arc}
e $a = 80$ {angles subtended by the same arc}
$b = 200$ {angles at the centre, angles at a point}
f $x = 75$ {angles subtended by the same arc}
$y = 118$ {exterior angle of a triangle}
g $x = 42$ {angles subtended by the same arc, exterior angle of a triangle}
h $x = 25$ {angles in a semi-circle, base angles of isosceles triangle, angles subtended by the same arc}
i $x = 25$ {angle in a semi-circle, angles in a triangle, angles subtended by the same arc}

2 **a** $\widehat{AOB} = 2\alpha$ **b** $\widehat{ACB} = \alpha$ **c** $\widehat{ADB} = \widehat{ACB}$

3 **a** **i** $\widehat{TAX} = 90°$ **ii** $\widehat{ACX} = 90°$
b **i** $\widehat{CAX} = 90° - \alpha$ **ii** $\widehat{CXA} = \alpha$ **iii** $\widehat{CBA} = \alpha$
c $\widehat{TAC} = \widehat{CBA}$

REVIEW SET 16A

1 **a** $a = 54$ **b** $a = 3$ **c** $a = 61$ **d** $a = 62$
e $a = 140$

2 $\alpha + \beta + \gamma = 180°$

3 **a** $x = 38$ **b** $x = 9$ **c** $x = 104$

4 **Hint:** Join [PB] and [BX].

6 **a** The perpendicular bisectors of [AB] and [AQ] pass through the centre of the circle.
b **i** $\alpha = 45°$ **ii** $\alpha = 30°$

7 **b** [RS] is a diameter of the circle.

REVIEW SET 16B

1 **a** $x = 86$ **b** $x \approx 5.83$ **c** $x = 55$ **d** $x = 55$
e $x = 40$

2 **a** $a = b = 55$ **b** $a = 68,\ b = 86$ **c** $a = 75,\ b = 60$

3 **Hint:** Use the angle in a semi-circle theorem.

4 **Hint:** Use the properties of isosceles triangles and the angles subtended by the same arc theorem.

6 **Hint:** Use the angle between a tangent and a chord theorem.

7 **a** **i** $\widehat{BCA} = 30°$ {angles subtended by the same arc}
ii $\widehat{BOA} = 60°$ {angle at the centre}
iii $\widehat{BAZ} = 30°$ {angle between a tangent and a chord}
iv $\widehat{CBX} = 75°$ {exterior angle of a triangle}

b $5\sqrt{3}$ cm **c** $(10 - 5\sqrt{3})$ cm

EXERCISE 17A

1 **a** **i** [BC] **ii** [AC] **iii** [AB]
b **i** [KM] **ii** [KL] **iii** [LM]
c **i** [PR] **ii** [QR] **iii** [PQ]

2 **a** [LM] **b** [MN] **c** [LN] **d** [LN] **e** [MN]

EXERCISE 17B

1 **a** **i** $\frac{4}{5}$ **ii** $\frac{3}{5}$ **iii** $\frac{4}{3}$
b **i** $\frac{8}{17}$ **ii** $\frac{15}{17}$ **iii** $\frac{8}{15}$
c **i** $\frac{12}{13}$ **ii** $\frac{5}{13}$ **iii** $\frac{12}{5}$
d **i** $\frac{3}{\sqrt{13}}$ **ii** $\frac{2}{\sqrt{13}}$ **iii** $\frac{3}{2}$
e **i** $\frac{5}{6}$ **ii** $\frac{\sqrt{11}}{6}$ **iii** $\frac{5}{\sqrt{11}}$
f **i** $\frac{\sqrt{3}}{2}$ **ii** $\frac{1}{2}$ **iii** $\sqrt{3}$

2 **a** AB $\approx$ 6.8 cm, BC $\approx$ 5.7 cm, AC $\approx$ 3.7 cm
b **i** ≈ 0.838 **ii** ≈ 0.544 **iii** ≈ 1.54
c **i** ≈ 0.839 **ii** ≈ 0.545 **iii** ≈ 1.54

4

θ	$\sin\theta$	$\cos\theta$	$\tan\theta$
30°	$\frac{1}{2}$	$\frac{\sqrt{3}}{2}$	$\frac{1}{\sqrt{3}}$
45°	$\frac{1}{\sqrt{2}}$	$\frac{1}{\sqrt{2}}$	1
60°	$\frac{\sqrt{3}}{2}$	$\frac{1}{2}$	$\sqrt{3}$

EXERCISE 17C

1 **a** $\tan 64° = \frac{c}{x}$ **b** $\sin 70° = \frac{a}{x}$ **c** $\cos 43° = \frac{x}{e}$
d $\sin 35° = \frac{x}{b}$ **e** $\cos 61° = \frac{g}{x}$ **f** $\tan 30° = \frac{h}{x}$

2 **a** $x \approx 12.99$ **b** $x \approx 15.52$ **c** $x \approx 9.84$ **d** $x \approx 6.73$
e $x \approx 3.31$ **f** $x \approx 20.01$ **g** $x \approx 16.86$ **h** $x \approx 4.45$
i $x \approx 1.65$ **j** $x \approx 26.75$ **k** $x \approx 7.53$ **l** $x \approx 9.61$

3 **a** $\theta = 22°,\ a \approx 8.09,\ b \approx 3.03$
b $\theta = 59°,\ a \approx 13.0,\ b \approx 7.83$
c $\theta = 65°,\ a \approx 5.42,\ b \approx 2.29$

4 **a** ≈ 4.96 cm^2 **b** ≈ 8.92 cm^2

5 **a** PQ = 2 m **b** $\widehat{PYX} = 15°$
c **Hint:** Find the length of [XQ] and [QY].

EXERCISE 17D

1 **a** $\theta \approx 48.2°$ **b** $\theta \approx 45.6°$ **c** $\theta \approx 56.3°$
d $\theta \approx 37.4°$ **e** $\theta \approx 42.2°$ **f** $\theta = 45°$
g $\theta \approx 40.2°$ **h** $\theta \approx 35.3°$ **i** $\theta \approx 35.9°$

2 **a** $\phi \approx 45.6°,\ \theta \approx 44.4°,\ x \approx 7.1$
b $\alpha \approx 53.9°,\ \beta \approx 36.1°,\ x \approx 5.9$
c $a \approx 50.3°,\ b \approx 39.7°,\ x \approx 8.6$

EXERCISE 17E

1 ≈ 110 m **2** $\approx 32.9°$ **3** ≈ 238 m **4** ≈ 765 m
5 ≈ 280 m **6** $\approx 6.89°$ **7** ≈ 54.6 m **8** ≈ 23.5 m
9 $\approx 67.4°$ **10** ≈ 15.8 cm

11 **a** ≈ 15.9 m **b** $\approx 71.3°$ **12** $\approx 21.8°$

13 **a** $x \approx 29.1$ {radius-tangent}
b $a \approx 5.29$ {radius-tangent, Pythagoras}
$b \approx 48.6$ {radius-tangent}

c $a \approx 53.1$ {angle between a tangent and a chord, angle in a semi-circle}
$b \approx 36.9$ {angles on a straight line}

14 **a** ≈ 18.7 mm **b** $\approx 35.7°$
15 $\approx 106°$ **16** ≈ 35.3 cm^2 **17** $\approx 41.4°$
18 **a** ≈ 5.03 m **b** AB ≈ 7.71 m **19** No, ≈ 0.721 cm
20 $\approx 53.2°$ **21** **a** ≈ 10.2 m **b** $\approx 5.65°$ **c** ≈ 86.6 m
22 ≈ 7.48 m **23** **a** ≈ 248 m **b** ≈ 128 m
24 $\approx 14.3°$ **25** **a** ≈ 21.2 cm **b** $\approx 35.3°$
26 **a** ≈ 9.43 cm **b** $\approx 32.5°$ **c** ≈ 10.8 cm **d** $\approx 29.1°$
27 **a** ≈ 8.94 cm **b** $\approx 18.5°$ **28** $\theta \approx 61.9°$

EXERCISE 17F

1 **a**
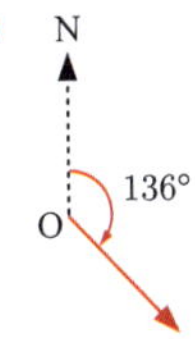

b
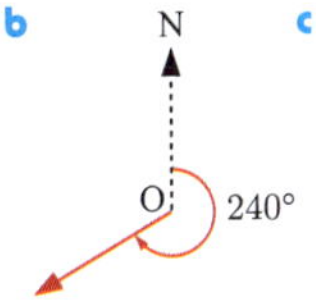

c
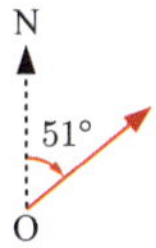

d
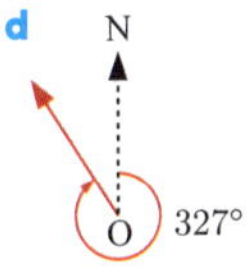

2 **a** 234° **b** 293° **c** 083° **d** 124°
3 **a** 040° **b** 235° **c** 297° **d** 132° **e** 225°
f 337°

4
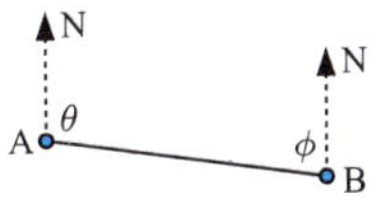

Let θ be the bearing of B from A.
$\therefore$ $\phi = 180° - \theta$ {supplementary co-interior angles}
$\therefore$ bearing of A from B
$= 360° - (180° - \theta)$ {angles at a point}
$= 180° + \theta$
$\therefore$ the difference between the bearings is always 180°.

5 **a** 041° **b** 142° **c** 279° **d** 322° **e** 099°
f 221°
6 **a** 055° **b** 235° **c** 095° **d** 275° **e** 130°
f 310°
7 **a** $\approx 062.6°$ **b** $\approx 243°$ **8** **a** ≈ 17.2 km **b** $\approx 306°$
9 **a** ≈ 3.49 km **b** $\approx 027.3°$ **10** ≈ 46.0 km
11 **a** 2.8 km **b** ≈ 772 m **12** ≈ 2.44 km
13 **a** ≈ 2.92 km **b** $\approx 159°$
14 ≈ 33.5 km, on bearing $\approx 025.6°$
15 ≈ 2.62 km, on bearing $\approx 058.4°$

REVIEW SET 17A

1 **a** **i** $\cos 42° \approx 0.743$ **ii** $\sin 64° \approx 0.899$
iii $\tan 48° \approx 1.114$ **iv** $\sin 26° \approx 0.438$
b **i** $\cos 42° \approx 0.743$ **ii** $\sin 64° \approx 0.899$
iii $\tan 48° \approx 1.111$ **iv** $\sin 26° \approx 0.438$
2 **a** $x \approx 14.0$ **b** $x \approx 35.2$ **c** $x \approx 15.4$
3 ≈ 55.2 m **4** $\theta = 36°$, $x \approx 12.4$, $y \approx 21.0$
5 $\approx 32.2°$ **6** **a** 157° **b** 221° **c** 281°
7 ≈ 8.54 km, on bearing $\approx 201°$
8 ≈ 294 km, on bearing $\approx 264°$ **9** **a** $\approx 56.3°$ **b** $\approx 33.9°$
10 **a** 29° **b** ≈ 117 m **c** ≈ 160 m

REVIEW SET 17B

1 **a** $\widehat{ABC} = \widehat{BAC} = 22.5°$
2 **a** $\theta \approx 35.5°$ **b** $\theta \approx 38.7°$ **c** $\theta \approx 37.1°$
3 **a** $x \approx 3.18$ **b** $x \approx 9.40$ **4** ≈ 49.1 cm^2
5 $\alpha \approx 36.4°$, $\theta \approx 53.6°$, $x \approx 25.7$ **6** ≈ 9.38 m
7 **a** 230° **b** 165° **c** 140° **8** ≈ 36.5 km
9 **a** Boat A: ≈ 5.34 km, Boat B: ≈ 5.39 km
b **i** ≈ 7.59 km **ii** $\approx 247°$
10 **a** $\approx 69.1°$ **b** $\approx 59.7°$
11 **a** $h = 7\sin 39°$ **b** ≈ 19.82 m^2 **c** $A = \frac{1}{2}ab\sin\theta$

EXERCISE 18A

1 **a** **i** P(cos 42°, sin 42°) **ii** Q(cos 106°, sin 106°)
iii R(cos 157°, sin 157°)
b **i** P(0.743, 0.669) **ii** Q(−0.276, 0.961)
iii R(−0.921, 0.391)
2 **a** 154° **b** 135° **c** 111° **d** 94°
3 **a** 82° **b** 53° **c** 24° **d** 12°
4 **a** 154° **b** 135° **c** 111° **d** 94°
5 **a** 82° **b** 53° **c** 24° **d** 12°
6 **a** 30° and 150° **b** 60° and 120° **c** 45° and 135°
d $\approx 5.7°$ and $\approx 174.3°$ **e** $\approx 23.6°$ and $\approx 156.4°$
f $\approx 44.4°$ and $\approx 135.6°$

EXERCISE 18B

1 **a** ≈ 55.2 cm^2 **b** ≈ 347 km^2 **c** ≈ 1.15 m^2
2 **a** $\sin 90° = 1$ **b** Area $= \frac{1}{2}ab\sin 90° = \frac{1}{2}ab$ ✓
3 **a** Area $= \frac{1}{2}ah$ **b** $h = b\sin(180° - C)$
c **Hint:** $\sin(180° - C) = \sin C$
4 **a** ≈ 13.6 cm^2 **b** ≈ 58.6 m^2 **c** ≈ 5.81 m^2
5 **a** ≈ 41.6 cm^2 **b** ≈ 36.7 m^2 **c** ≈ 7.70 cm^2
6 ≈ 50.0 cm^2 **7** $x \approx 21.9$ **8** ≈ 13.1 cm
9 **Hint:** $\widehat{AOC} = 180° - \widehat{BOC}$
10 $\theta = 30°$ or 150° **11** $\widehat{ABC} \approx 41.6°$ or $\approx 138.4°$
12 **a** **i** Area $= \frac{1}{2}bc\sin A$ **ii** Area $= \frac{1}{2}ab\sin C$
b From **a**, $\frac{1}{2}bc\sin A = \frac{1}{2}ab\sin C$
$\therefore \dfrac{\sin A}{a} = \dfrac{\sin C}{c}$ $\{\div \frac{1}{2}abc$ both sides$\}$

EXERCISE 18C.1

1 **a** $x \approx 11.05$ **b** $x \approx 11.52$ **c** $x \approx 5.19$
2 **a** $x \approx 9.43$ **b** $x \approx 11.9$ **c** $x \approx 6.37$
3 ≈ 10.2 cm^2
4 **a** $\theta = 59°$, $x \approx 96.7$, $y \approx 90.1$
b $\phi = 62°$, $x \approx 4.17$, $y \approx 5.62$
c $\theta = \phi = 28.5°$, $x \approx 5.30$

EXERCISE 18C.2

1 **a** $\widehat{ACB} \approx 72.9°$ or $\approx 107.1°$ **b**
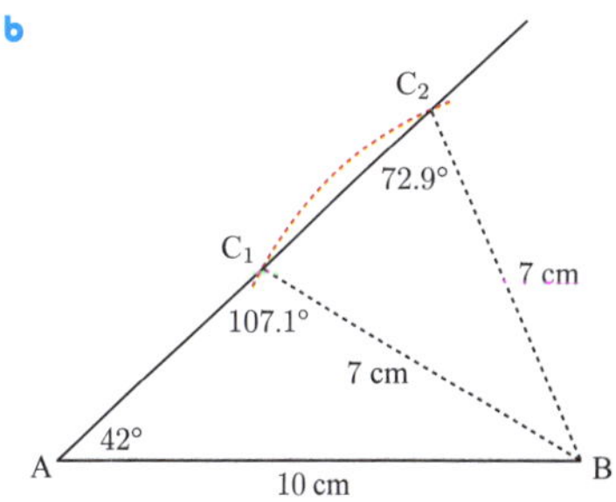

2 **a** Using the sine rule, $\widehat{PRQ} \approx 33.9°$ or $\approx 146.1°$.
However, $42° + 146.1° > 180°$.
$\therefore \widehat{PRQ} \approx 33.9°$

b

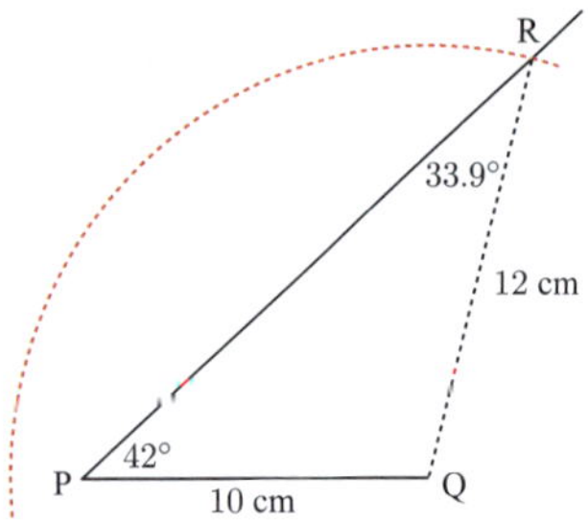

3 **a** $\theta \approx 31.4°$ **b** $\theta \approx 77.5°$ or $\approx 102.5°$
c $\theta \approx 43.6°$ or $\approx 136.4°$ **d** $\theta \approx 40.8°$

4 **a** $A \approx 49.1°$ **b** $B \approx 71.6°$ or $\approx 108.4°$ **c** $C \approx 44.8°$

EXERCISE 18D

1 **a** ≈ 2.66 cm **b** ≈ 9.63 m **c** ≈ 10.6 m
d ≈ 27.5 cm **e** ≈ 4.15 km **f** ≈ 15.2 m

2 **a** $\theta \approx 36.3°$ **b** $\theta \approx 53.2°$ **c** $\theta \approx 115.6°$

3 **a** $\widehat{BAC} \approx 51.8°$, $\widehat{ABC} \approx 40.0°$, $\widehat{ACB} \approx 88.3°$
b $\widehat{QPR} \approx 34.0°$, $\widehat{PQR} \approx 96.6°$, $\widehat{PRQ} \approx 49.3°$

5 **a** $\approx 43.0°$ **b** $120°$ **6** ≈ 2.90 cm^2

7 **Hint:** Use the cosine rule in $\triangle ABX$, and $\widehat{BAX} = 60°$.

8 **b** $x = 5 \pm \sqrt{6}$
c

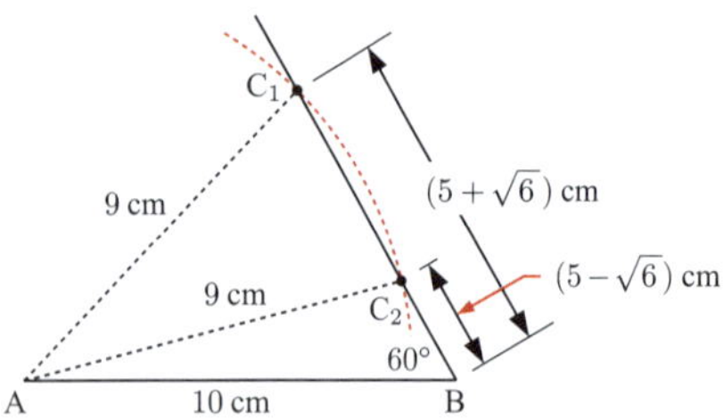

EXERCISE 18E

1 ≈ 14.3 km **2** ≈ 1300 m **3** $\theta \approx 13.4°$
4 $\approx 107°$ **5** **a** $\approx 35.69°$ **b** ≈ 4 ha **6** ≈ 100 m
7 ≈ 19.6 km, on bearing $\approx 106°$ **8** $\approx 214°$
9 ≈ 11.3 cm and ≈ 17.0 cm

10 **a**

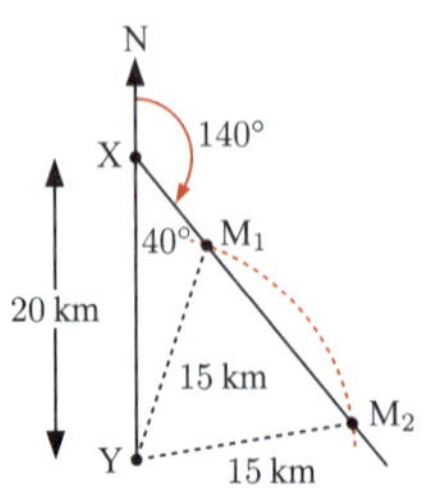

b $XM_1 \approx 7.59$ km
$XM_2 \approx 23.0$ km

11 ≈ 15.6 m **12** ≈ 63.4 m

REVIEW SET 18A

1 **a** P(cos 57°, sin 57°), Q(cos 168°, sin 168°)
b P(0.545, 0.839), Q(−0.978, 0.208)

2 $x = 14$ **3** **a** ≈ 6.74 cm^2 **b** ≈ 14.5 cm^2

4 **a** $x \approx 12.14$ **b** $x \approx 5.17$

5 **a** $x \approx 15.9$ **b** $x \approx 254$ **c** $x \approx 99.4$

6 $\approx 27.7°$ **7** ≈ 337 m, on bearing $\approx 138°$

8 **a** **i** ≈ 10.2 m **ii** ≈ 7.00 m **b** ≈ 82.0 m^2

9 **a** $\widehat{BAC} \approx 41.8°$ or $\approx 138.2°$
b

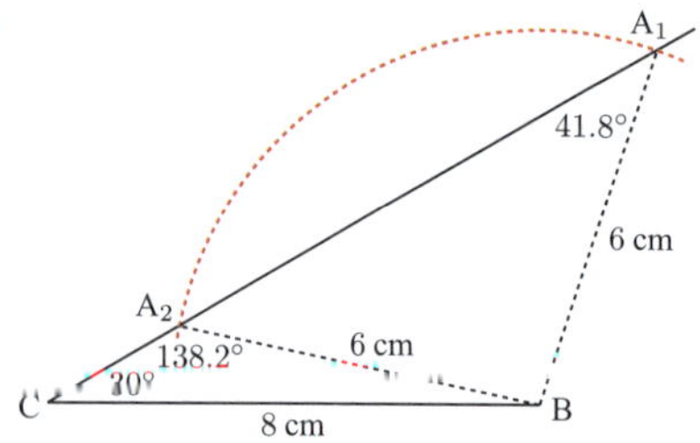

c **i** ≈ 2.46 cm **ii** $\approx 11.8°$ **iii** ≈ 4.91 cm^2

REVIEW SET 18B

1 **a** ≈ 21.3 m^2 **b** ≈ 3.83 cm^2

2 **a** $142°$ **b** $128°$ **3** $\theta \approx 23.4°$ or $\approx 156.6°$

4 **a** $\theta \approx 24.3°$ **b** $\theta \approx 35.0°$ **c** $\theta \approx 59.4°$ or $\approx 120.6°$

5 **a** $x \approx 6.76$ **b** $x \approx 4.15$ **c** $x \approx 102$

6 ≈ 69.3 m^2 **7** **a** ≈ 76.1 km **b** $\approx 053.0°$

8 **a** ≈ 7500 m^2 **b** ≈ 0.750 ha

9 **a** **Hint:** Use the cosine rule in $\triangle BPQ$.
b $\cos \widehat{BQR} = -\frac{13}{20}$

EXERCISE 19A

1 **a** {HH, HT, TH, TT} **b** {A, B, C, D}
c {GGG, GGB, GBG, BGG, GBB, BGB, BBG, BBB}
d {WXYZ, WXZY, WYXZ, WYZX, WZXY, WZYX, XWYZ, XWZY, XYWZ, XYZW, XZWY, XZYW, YWXZ, YWZX, YXWZ, YXZW, YZWX, YZXW, ZWXY, ZWYX, ZXWY, ZXYW, ZYWX, ZYXW}

2 **a**

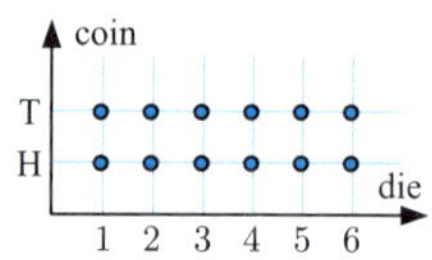

b

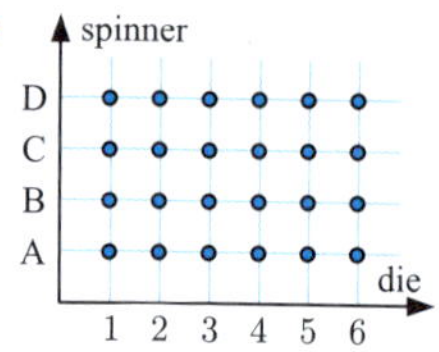

3

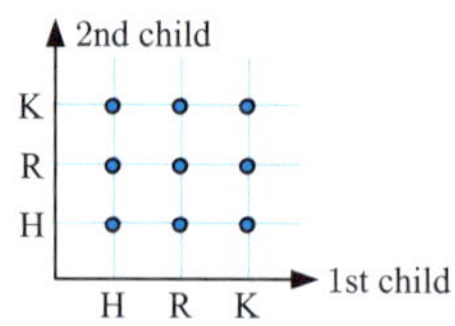

4

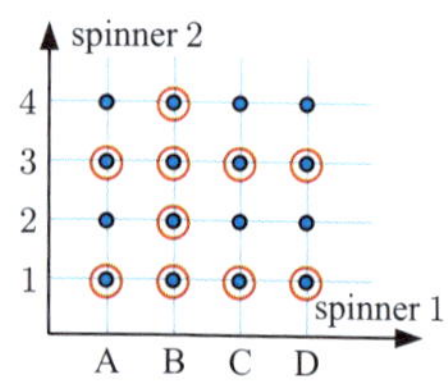

5 **a**

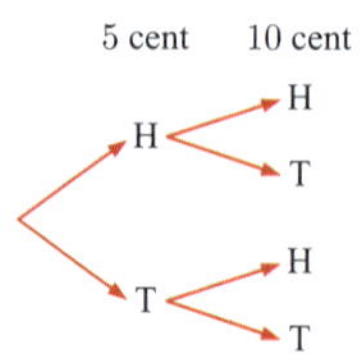

b

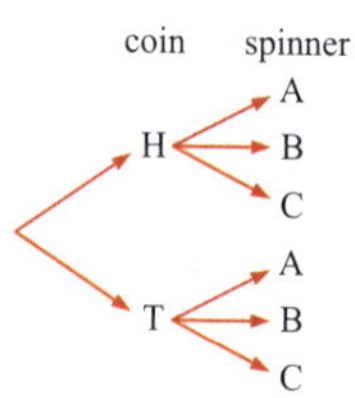

c

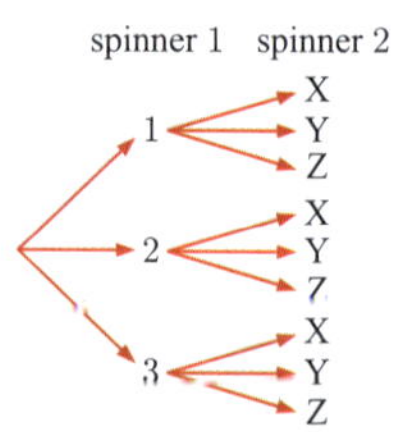

d

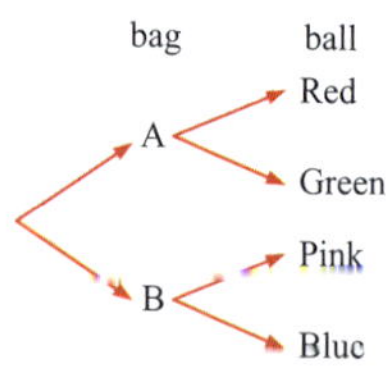

6 **a**

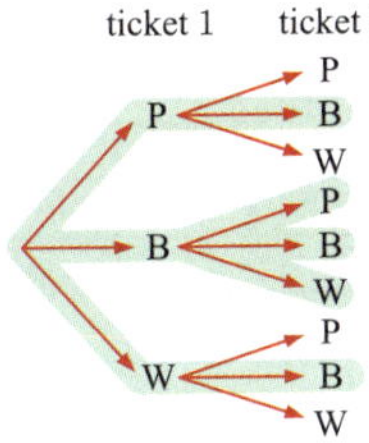

b **i** $E = \{PB, BP, BB, BW, WB\}$
ii $E' = \{PP, PW, WP, WW\}$

7 **a**

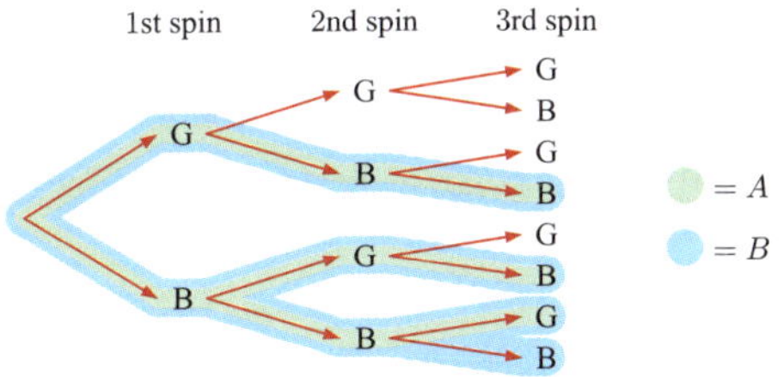

b **i** $U = \{GGG, GGB, GBG, GBB, BGG, BGB, BBG, BBB\}$
ii $A = \{GBB, BGB, BBG\}$
iii $B = \{GBB, BGB, BBG, BBB\}$
iv $A' = \{GGG, GGB, GBG, BGG, BBB\}$
v $B' = \{GGG, GGB, GBG, BGG\}$

c No, $A \subseteq B$. **d**

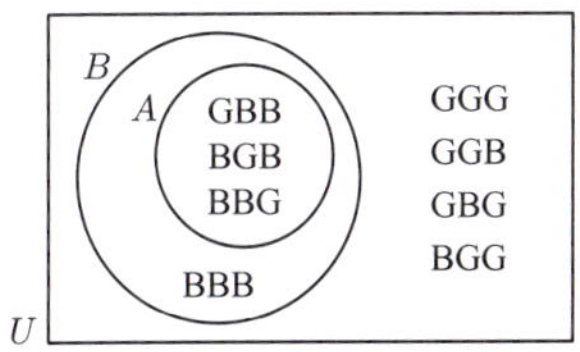

EXERCISE 19B

1 **a** $\frac{1}{6}$ **b** $\frac{1}{3}$ **c** $\frac{2}{3}$ **d** $\frac{1}{3}$

2 **a** $\frac{1}{8}$ **b** $\frac{1}{8}$ **c** $\frac{1}{8}$ **d** $\frac{3}{8}$ **e** $\frac{1}{2}$ **f** $\frac{7}{8}$

3 **a** $\frac{1}{2}$ **b** $\frac{1}{2}$ **c** $\frac{1}{6}$ **d** $\frac{1}{6}$ **e** $\frac{1}{2}$

4 {HHHH, HHHT, HHTH, HTHH, THHH, HHTT, HTHT, HTTH, THTH, TTHH, THHT, TTTH, TTHT, THTT, HTTT, TTTT}

a $\frac{1}{16}$ **b** $\frac{3}{8}$ **c** $\frac{5}{16}$ **d** $\frac{15}{16}$ **e** $\frac{1}{4}$

5 **a**

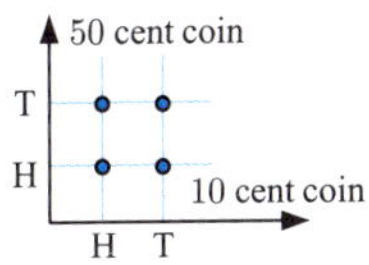

b **i** $\frac{1}{4}$ **ii** $\frac{1}{4}$ **iii** $\frac{1}{2}$ **iv** $\frac{3}{4}$

6 **a**

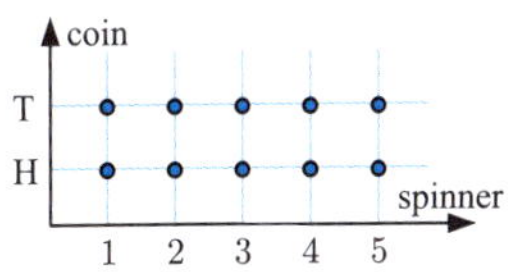

b **i** $\frac{1}{10}$ **ii** $\frac{3}{10}$ **iii** $\frac{2}{5}$ **iv** $\frac{3}{5}$

7 **a**

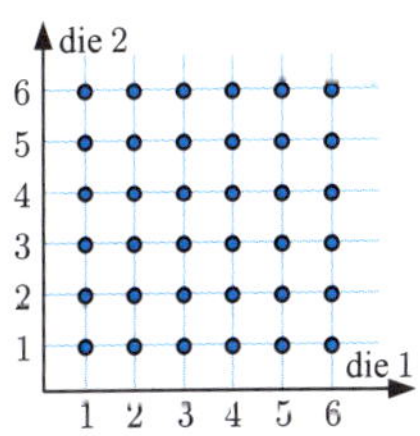

b **i** $\frac{1}{18}$ **ii** $\frac{1}{6}$ **iii** $\frac{11}{36}$ **iv** $\frac{5}{9}$ **v** $\frac{1}{4}$ **vi** $\frac{1}{6}$

8 **a**

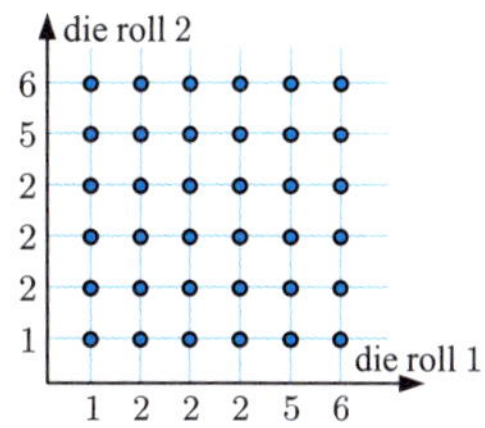

b **i** $\frac{1}{6}$ **ii** $\frac{2}{9}$ **iii** $\frac{1}{3}$ **iv** $\frac{11}{36}$ **v** $\frac{8}{9}$ **vi** $\frac{5}{9}$

9 **a** **i** $\frac{1}{15}$ **ii** $\frac{2}{15}$ **iii** $\frac{7}{15}$ **iv** $\frac{1}{3}$

b $P(\text{odd}) = \frac{6}{15}$, $P(\text{even}) = \frac{9}{15}$ $\therefore$ even

10 **a** $\frac{1}{13}$ **b** $\frac{1}{52}$ **c** $\frac{1}{4}$ **d** $\frac{3}{13}$ **e** $\frac{1}{26}$ **f** $\frac{1}{2}$ **g** $\frac{4}{13}$ **h** 0 **i** $\frac{10}{13}$

11 **a** 33 students **b** **i** $\frac{17}{33}$ **ii** $\frac{5}{33}$ **iii** $\frac{8}{11}$ **iv** $\frac{19}{33}$

12 **a**

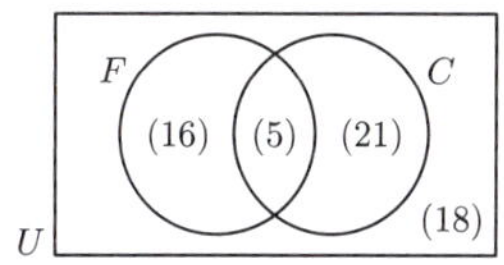

b **i** $\frac{4}{15}$ **ii** $\frac{3}{10}$ **iii** $\frac{7}{10}$

13 **a** $\frac{3}{8}$ **b** $\frac{1}{8}$ **14** **a** $\frac{11}{50}$ **b** $\frac{29}{50}$

15 **a**

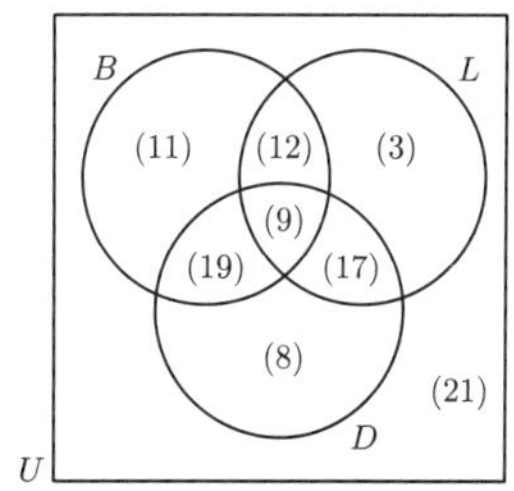

b **i** $\frac{11}{100}$ **ii** $\frac{21}{100}$ **iii** $\frac{17}{100}$ **iv** $\frac{1}{2}$ **v** $\frac{12}{25}$

EXERCISE 19C

1 **a** No, since $P(A \cap B) \neq 0$ **b** 0.75 **2** 0.4 **3** 0.15

4 **a** No, it is possible to "toss a head" *and* "roll a 5". **b**

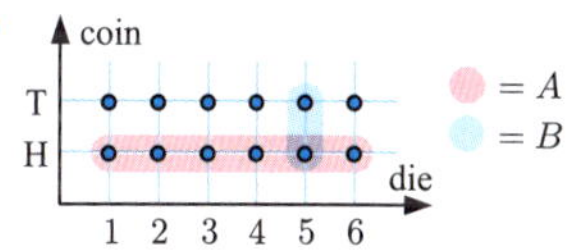

c **i** $\frac{1}{2}$ **ii** $\frac{1}{6}$ **iii** $\frac{1}{12}$ **iv** $\frac{7}{12}$

d $P(A) + P(B) - P(A \cap B) = \frac{1}{2} + \frac{1}{6} - \frac{1}{12}$
$= \frac{7}{12} = P(A \cup B)$ ✓

5 0.4 **6** $P(C) + P(D) > 1$

7 **a** Yes, the numbers greater than 15 all have 2 digits.

b **i** $\frac{9}{20}$ **ii** $\frac{1}{4}$ **iii** 0 **iv** $\frac{7}{10}$

EXERCISE 19D

1 **a** $\frac{1}{48}$ **b** $\frac{5}{16}$ **2** **a** $\frac{2}{9}$ **b** $\frac{4}{15}$ **3** **a** $\frac{1}{3}$ **b** $\frac{1}{6}$

4 **a** 0.0096 **b** 0.8096

5 **a** $\frac{1}{49}$ **b** $\frac{10}{49}$ **c** $\frac{25}{49}$

6 **a** ≈ 0.0545 **b** ≈ 0.0584 **c** ≈ 0.441 **d** ≈ 0.0840

7 **a**

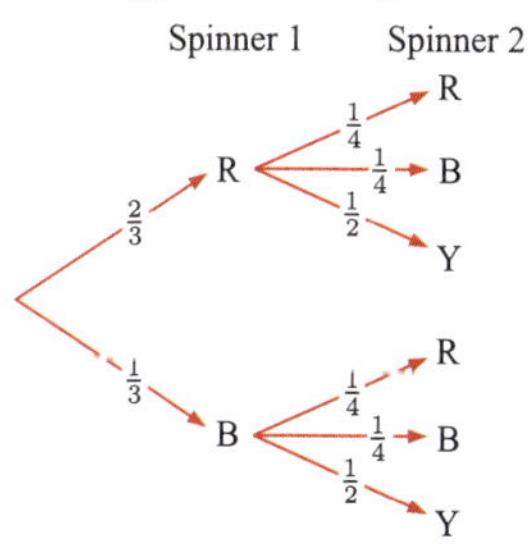

b **i** $\frac{1}{4}$ **ii** $\frac{3}{4}$ **iii** $\frac{5}{12}$

8 **a**

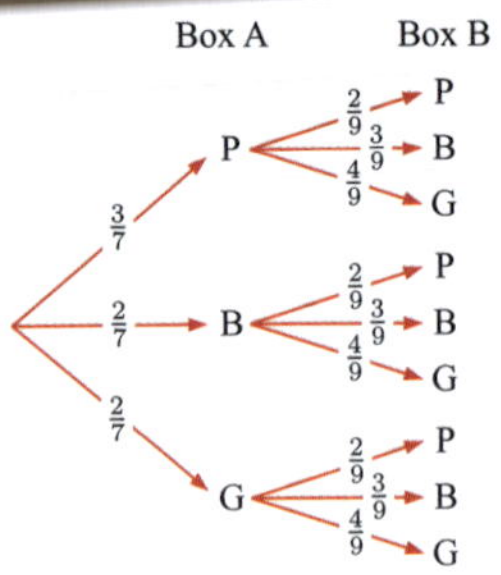

b **i** $\frac{2}{9}$ **ii** $\frac{20}{63}$ **iii** $\frac{29}{63}$

c 1; this is because every possible outcome is covered by these three events.

9 0.712 **10** 0.65 **11** $P(A) = 0.8$, $P(B) = 0.5$

EXERCISE 19E

1 **a** **i** $\frac{1}{3}$ **ii** $\frac{4}{15}$ **iii** $\frac{8}{15}$

b The possibilities are: WW, WY, YW, YY.
The 3 events do not cover all these possibilities.
So, the probability sum will not necessarily be 1.

2 **a**

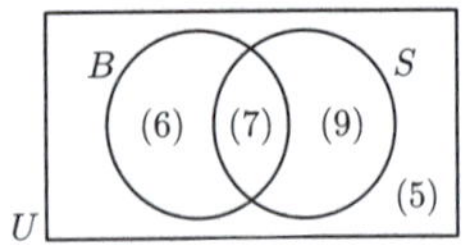

b $\frac{4}{39}$

3 **a** $\frac{17}{32}$ **b** $\frac{13}{28}$ **4** $\frac{1}{5}$ **5** **a** $\frac{25}{102}$ **b** $\frac{13}{204}$ **c** $\frac{1}{221}$

6 **a** $\frac{7}{16}$

b **i** The probability that each successive chocolate will be dark will not stay the same because it depends on what types of chocolate were previously drawn.

ii $\frac{91}{1240}$

7 **a** M is the event of Matt playing in a game.
W is the event of his team winning.

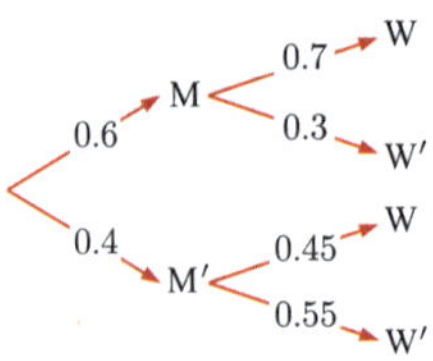

b 0.6

8 0.665 **9** **a** $\frac{14}{45}$ **b** $\frac{16}{45}$ **10** $\frac{5}{6}$ **11** $\frac{59}{120}$

EXERCISE 19F

1 ≈ 0.0894 **2** ≈ 0.822 **3** **a** ≈ 0.366 **b** ≈ 0.0231

4 **a** **i** ≈ 0.685 **ii** ≈ 0.638

b The estimate in **a i** is likely to be more accurate because it uses a larger sample size.

c combine Karla and Sam's results, ≈ 0.671

5 **a**

Brand	*Frequency*	*Relative Frequency*
Silktouch	62	0.31
Super	53	0.265
Just Soap	46	0.23
Indulgence	39	0.195
Total	200	1

b **i** ≈ 0.23 **ii** ≈ 0.195 **iii** ≈ 0.31

6 **a** 57 ice creams **b** **i** ≈ 0.298 **ii** ≈ 0.632
c ≈ 0.930

7 **a** 74 people **b** **i** ≈ 0.123 **ii** ≈ 0.517

8 **a** ≈ 0.957 **b** ≈ 0.0889 **c** $\approx 0.006\,15$ **d** ≈ 0.370

9 **a**

	Adult	Child	Infant	*Total*
First Class	11	3	1	15
Business Class	28	9	2	39
Economy Class	60	25	11	96
Total	99	37	14	150

b 150 passengers **c** **i** ≈ 0.66 **ii** ≈ 0.26 **iii** ≈ 0.08

10 **a** ≈ 0.0366 **b** ≈ 0.634 **c** ≈ 0.390 **d** ≈ 0.232

EXERCISE 19G

1 27 kicks **2** 15 days **3** ≈ 88 times **4** 30 times

5 **a** 0.11 **b** **i** 3140 points **ii** 31.4 points

6 175 times **7** 100 people **8** ≈ 15 mornings

9 **a** 219 days **b** 146 days **c** ≈ 22 days **d** ≈ 37 days

EXERCISE 19H

1 **a** $\frac{12}{25}$ **b** $\frac{12}{19}$ **2** **a** $\frac{24}{25}$ **b** $\frac{1}{25}$

3 **a** $\frac{13}{14}$ **b** $\frac{5}{23}$ **c** $\frac{9}{10}$ **4** **a** $\frac{17}{37}$ **b** $\frac{11}{19}$ **c** $\frac{11}{17}$

5 **a** $\frac{11}{37}$ **b** $\frac{14}{47}$ **c** $\frac{14}{29}$ **6** **a** $\frac{3}{16}$ **b** $\frac{2}{5}$

7 **a** $\frac{1}{4}$

b It increases the probability. In order for the sum to be 5, the card picked from bag A cannot be the 1 or the 2.

c $\frac{1}{2}$

8 $\frac{3}{5}$

REVIEW SET 19A

1

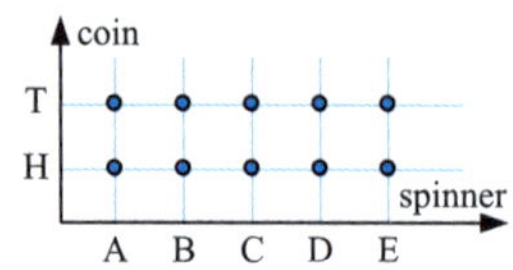

2 **a** $\frac{1}{20}$ **b** $\frac{3}{10}$ **c** $\frac{9}{20}$

3 **a**

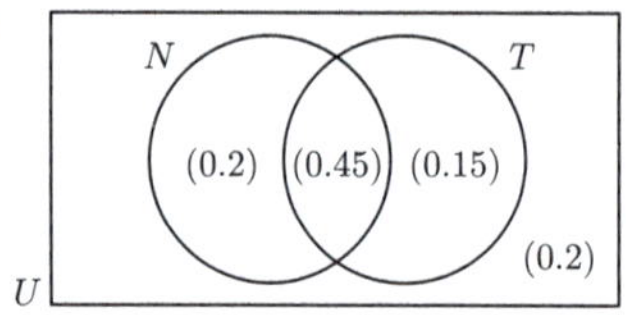

b **i** 0.65 **ii** 0.2 **iii** 0.8 **iv** 0.35

4 0.7

5 **a** Yes, 5 is an odd number.

b **i** $\frac{1}{6}$ **ii** $\frac{1}{2}$ **iii** 0 **iv** $\frac{2}{3}$

6 0.336 **7** **a** $\frac{2}{7}$ **b** $\frac{4}{7}$

8 **a** ≈ 0.74 **b** ≈ 178 times **c** ≈ 0.0176

9 ≈ 266 long showers

10 **a** 47 books **b** **i** $\frac{31}{47}$ **ii** $\frac{4}{47}$ **iii** $\frac{7}{16}$ **11** $\frac{2}{7}$

12 **a** $\frac{13}{55} \approx 0.236$

b **i** Increases, as the number of tickets of the desired colour does not decrease with each draw.

ii $\frac{37}{121} \approx 0.306$

REVIEW SET 19B

1 **a**

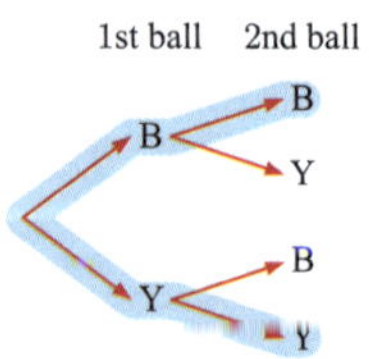

b $E = \{BB, YY\}$

2 $\frac{1}{2}$ **3** 0.35 **4** $\frac{7}{12}$

5 **a** 0.2375 **b** 0.0375 **c** 0.0125 **d** 0.7125

6 0.66 **7** $\frac{41}{90}$

8 **a** ≈ 0.364 **b** ≈ 0.551 **c** ≈ 0.814

9 ≈ 62 waves **10** 50 times

11 **a** ≈ 0.885 **b** ≈ 0.424 **c** ≈ 0.376 **d** ≈ 0.421

12 **a**

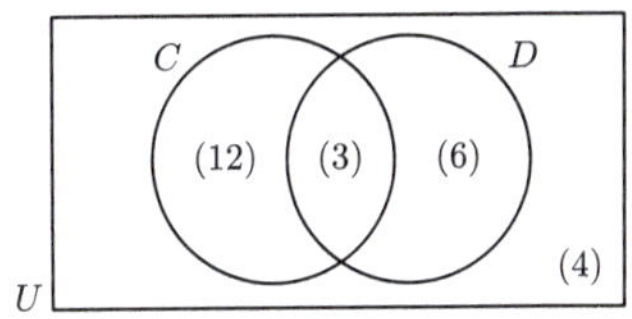

b **i** $\frac{4}{25}$ **ii** $\frac{12}{25}$ **iii** $\frac{1}{5}$ **iv** $\frac{2}{5}$

EXERCISE 20A

1 **a**

Salad sandwiches bought

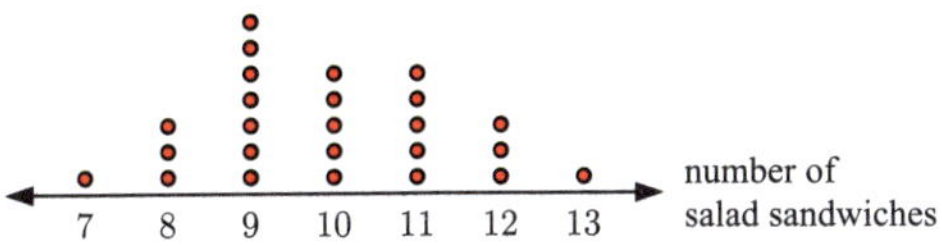

b 9 salad sandwiches, it is the most common number of salad sandwiches bought in one day.

c 44%

2 **a**

Archery scores

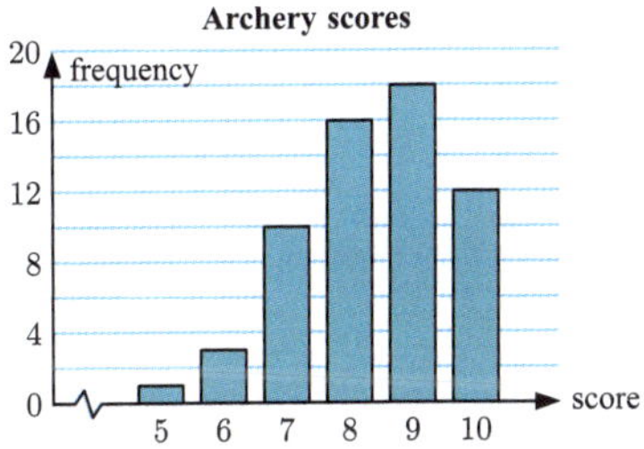

b 9, it is the most common score an archer achieves from one arrow.

c **i** $\frac{1}{5}$ **ii** $\frac{1}{2}$

3 **a**

Number of nights	*Tally*	*Frequency*
1	||||	4
2	卌 |	6
3	卌 ||	7
4	卌 ||	7
5	卌 |	6
6	||||	4
7	||||	4
8	||	2
	Total	40

b

Miami hotel data

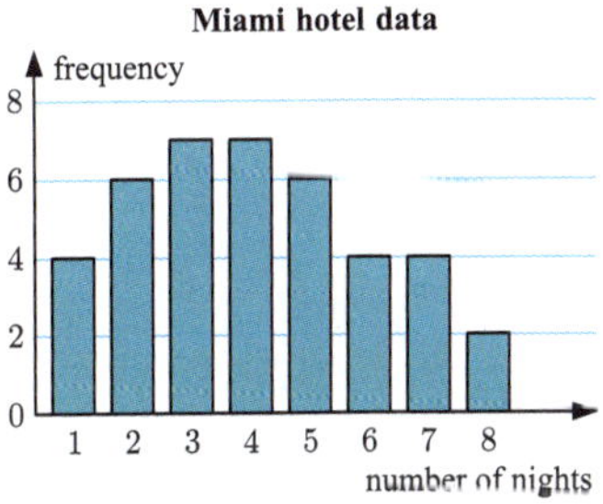

c the Miami hotel

4 **a** 58 points

b

Tennis set data

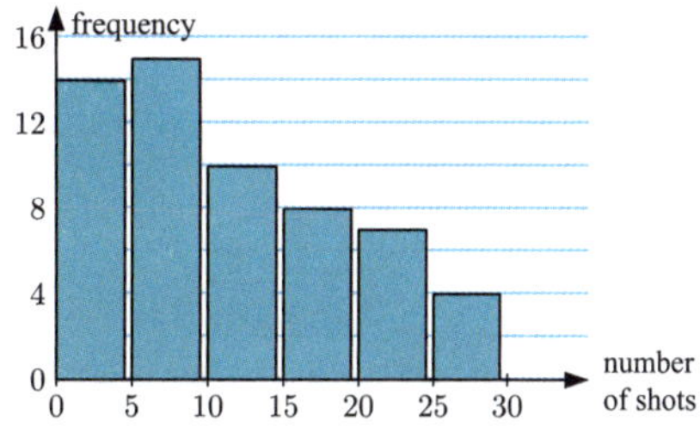

c 5 to 9 shots

5 **a**

Number of people	*Tally*	*Frequency*
40 to 49	||||	4
50 to 59	卌 ||	7
60 to 69	卌 ||||	9
70 to 79	卌 |	6
80 to 89	||||	4
	Total	30

b

Elevator use

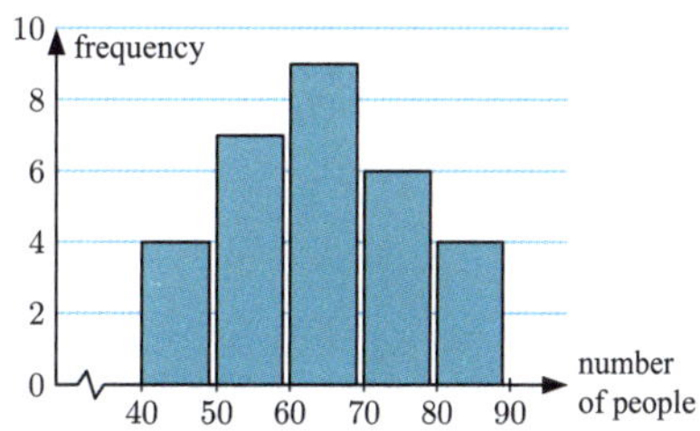

c 60 to 69 people

EXERCISE 20B

1 **a** The travel time can take any value in the given range, and is measured, not counted.

b $55 \leqslant t < 60$ min **c** 48%

2 **a** *Weight* can take any value in the given range, and is measured, not counted.

b

Weights of rugby squad

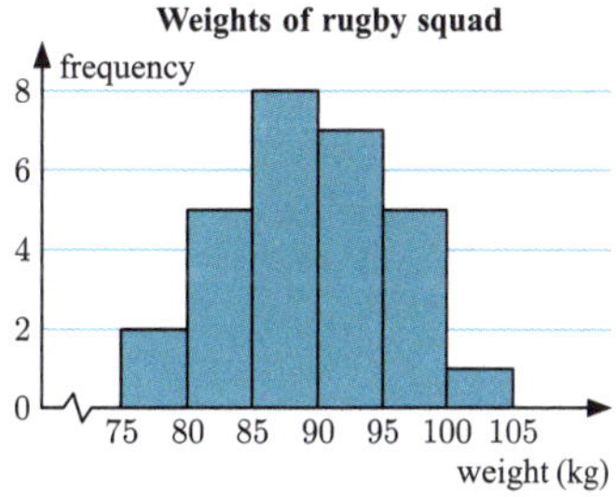

c $85 \leqslant w < 90$ kg. More people in the rugby squad have weights between 85 and 90 kg than in any other interval.

d $\approx 53.6\%$

3 **a**

Distance (d m)	*Frequency*
$10 \leqslant d < 12$	2
$12 \leqslant d < 14$	4
$14 \leqslant d < 16$	7
$16 \leqslant d < 18$	5
$18 \leqslant d < 20$	2

b **Shotput distances thrown**

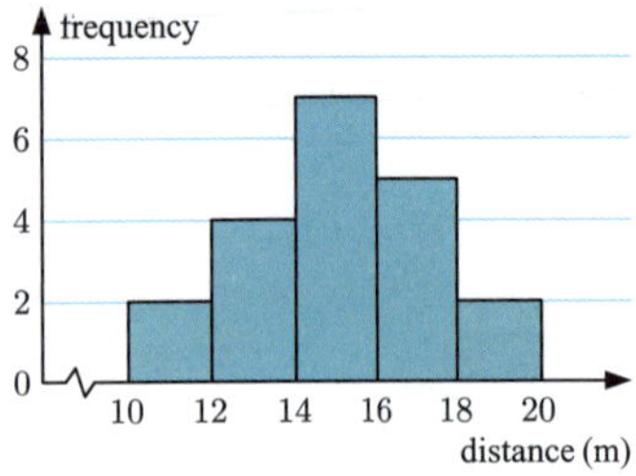

c $14 \leqslant d < 16$ m; Daniel threw the shotput between 14 m and 16 m more times than any other interval.

d $\frac{7}{10}$

4 **a**

Time (t s)	*Frequency*
$23 \leqslant t < 24$	3
$24 \leqslant t < 25$	4
$25 \leqslant t < 26$	7
$26 \leqslant t < 27$	10
$27 \leqslant t < 28$	11
$28 \leqslant t < 29$	5

b 17.5%

c **200 m times run by young athletes**

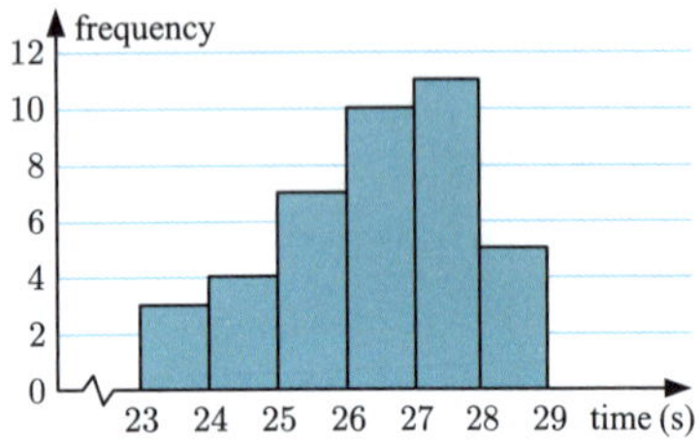

d $27 \leqslant t < 28$ s

EXERCISE 20C

1 **a** symmetric **b** bimodal
c negatively skewed, "2" is an outlier
d positively skewed **e** bimodal
f positively skewed, "19" is an outlier

2 **a** 10 employees **b** $\approx 4.44\%$
c positively skewed, "10" is an outlier

3 **a** $x = 18$ **b** 39 frames **c** 15%

d **Ten-pin bowling data**

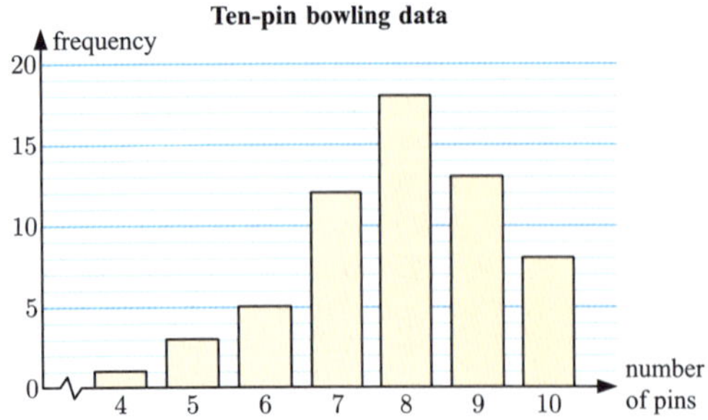

e negatively skewed

4 **a**

Number of toothpicks	*Tally*	*Frequency*
45	|	1
46		0
47		0
48	𝍸	5
49	𝍸 𝍸	10
50	𝍸 𝍸 𝍸 𝍸 |||	23
51	𝍸 𝍸	10
52	𝍸 ||||	9
53	||	2
	Total	60

b **Number of toothpicks in boxes**

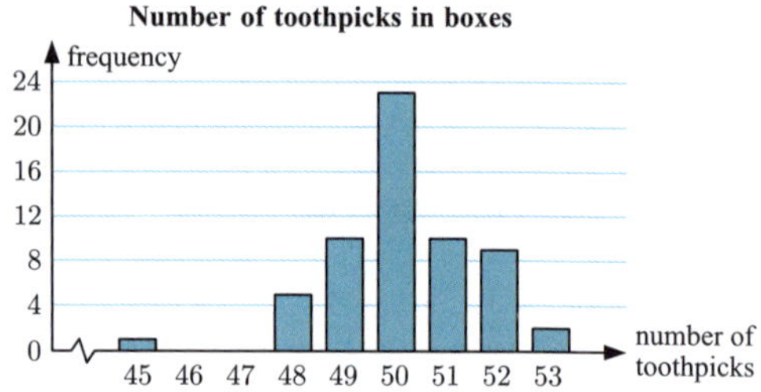

c approximately symmetric, "45" is an outlier **d** $\approx 38.3\%$

5 **a** $40 \leqslant h < 60$ mm

b **Heights of seedlings in a nursery**

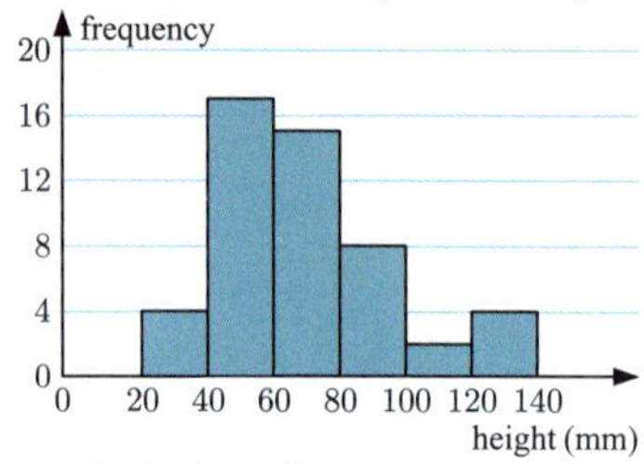

c positively skewed

d **i** ≈ 754 seedlings **ii** ≈ 686 seedlings

6 **a** European men: approximately symmetric,
Asian men: slightly positively skewed,
$195 \leqslant h < 200$ is an outlier.

b European men, the centre of the distribution is higher for European men than Asian men.

7 **a**

Score	*Frequency*
10 to 19	1
20 to 29	0
30 to 39	0
40 to 49	3
50 to 59	16
60 to 69	19
70 to 79	11
80 to 89	5
90 to 99	1

b **Physics examination scores**

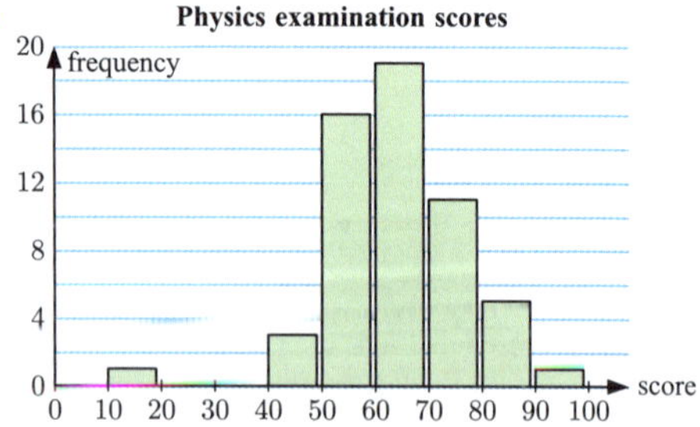

c approximately symmetric, "19" is an outlier **d** $\approx 10.7\%$

EXERCISE 20D.1

1 **a** mean $= 25$, median $= 24$, mode $= 30$
b mean ≈ 13.3, median $= 11.5$, mode $= 8$
c mean ≈ 10.3, median $= 10$, mode $= 11.2$
d mean ≈ 429, median $= 428$, mode $= 415, 427$

2 **a** *data set A*: ≈ 7.73, *data set B*: ≈ 8.45
b *data set A*: 7, *data set B*: 7
c The data sets are the same except for the last value. Since the last value of A is less than the last value of B, the mean of A is less than the mean of B.
d The middle value of both data sets is the same, so the median is the same.

3 **a** mean $= 3$ mm, median $= 0$ mm, mode $= 0$ mm
b **i** "15" and "27" are outliers **ii** the mean
iii The outliers are reasonable values for rainfall, so we expect that they are not errors and are genuine data values.

4 **a** 44 points **b** 44 points **c** ≈ 40.6 points
d **i** Increase, as 42 is greater than the mean calculated in **c**.
ii 40.75 points

5 **a** *Group A*: mean $= 19.91$ cm, median $= 19.85$ cm
Group B: mean ≈ 17.9 cm, median $= 17.6$ cm
b The mean and median are higher for the plants which were fertilised than for those which were not. This suggests that the fertiliser is effective.

6 **a** 1 head **b** 1 head **c** 1.4 heads

7 **a**

Donation (\$)	*Frequency*
1	7
2	9
5	2
10	4
20	8
Total	30

b 30 donations
c the mode
d **i** \$2
ii \$2
iii $\approx$ \$7.83

8 **a** **i** 5 ducklings **ii** 5 ducklings **iii** 4.25 ducklings
b negatively skewed
c The mean is lower than the mode and median.

9 **a** Grace: approximately symmetric,
Jack: slightly positively skewed
b **i** Grace: 7 sales, Jack: 6 sales
ii Grace: 7 sales, Jack: 6 sales
iii Grace: 6.85 sales, Jack: 6.2 sales
c Grace

10 **a** **i** New York: 3.475 nights, Miami: 4.075 nights
ii New York: 2 nights, Miami: 3 nights and 4 nights
iii New York: 3 nights, Miami: 4 nights
b the Miami hotel

11 44 possessions **12** 7.875 **13** $x = 12$
14 $a = 8$ **15** 48 trees **16** 15

EXERCISE 20D.2

1 **a** $165 \leqslant s < 170$ km/h
b **i** 76 serves **ii** not possible **iii** 73.5%
c ≈ 167 km/h

2 **a** 20 to 24 passengers **b** ≈ 20.6 passengers
c 434 bus tours

3 **a**

Male

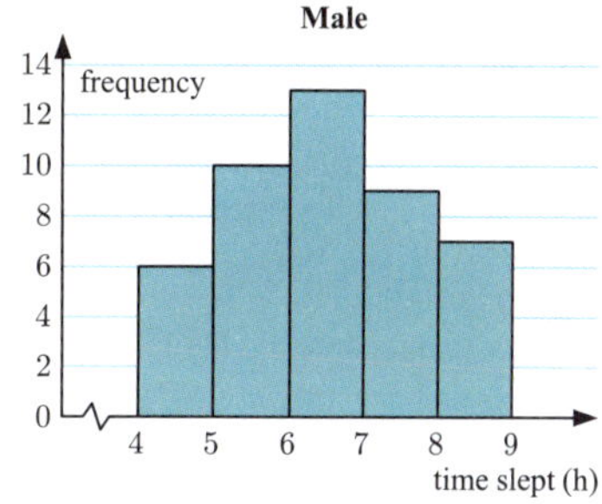

Female

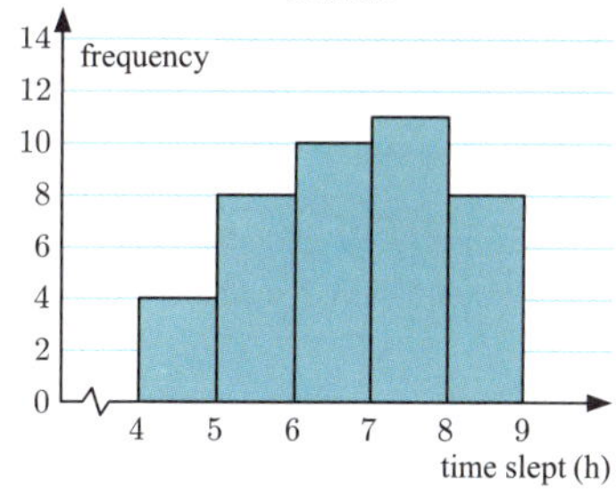

b Male: approximately symmetric,
Female: negatively skewed
c Male: ≈ 6.52 hours, Female: ≈ 6.77 hours
d The female students. However the means are only estimates, and the difference in means is small, so the answer may not be reliable.

4 **a** Year 7: ≈ 13.7 days, Year 10: ≈ 9 days **b** Year 10
c Year 7: ≈ 11.7 days, Year 10: 7 days
d Year 7: ≈ 15.7 days, Year 10: 11 days
e Yes, the maximum possible mean for the Year 10 class is still lower than the minimum possible mean for the Year 7 class.

5 **a** $\overline{p} \approx 42.28$ litres
b **i**

Petrol bought (p L)	*Frequency*
$20 \leqslant p < 30$	8
$30 \leqslant p < 40$	13
$40 \leqslant p < 50$	17
$50 \leqslant p < 60$	12

ii ≈ 41.6 litres

c **i**

Petrol bought (p L)	*Frequency*
$20 \leqslant p < 25$	2
$25 \leqslant p < 30$	6
$30 \leqslant p < 35$	6
$35 \leqslant p < 40$	7
$40 \leqslant p < 45$	9
$45 \leqslant p < 50$	8
$50 \leqslant p < 55$	3
$55 \leqslant p < 60$	9

ii ≈ 42.1 litres

d The estimate found in **c** is closer to the actual mean found in **a**. However, both estimates are reasonable approximations.
e Using smaller intervals should produce a more accurate estimate of the mean. This is because as the intervals get smaller, the values in those intervals get closer to the interval midpoint used in estimating the mean.

EXERCISE 20E.1

1 **a** 16 goals **b** 26 goals **c** 31 goals **d** 40 goals
e 54 goals **f** 38 **g** 14

2 **a** **i** The highest rating was 89 points.
ii The lowest rating was 25 points.
iii Half of the ratings were greater than or equal to 62 points.

iv The top 25% of the ratings were at least 73 points.
v The middle half of the ratings were between 45 and 73 points.

b 64 c 28

3 a min = 5 marks, $Q_1 = 7$ marks, median = 9 marks, $Q_3 = 10$ marks, max = 12 marks
b range = 7, IQR = 3
c

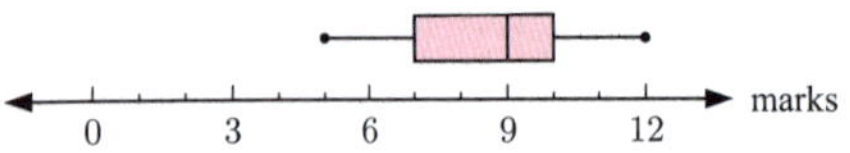

4 a min = 0 workers, $Q_1 = 3$ workers, median = 4 workers, $Q_3 = 5.5$ workers, max = 8 workers
b range = 8, IQR = 2.5
c

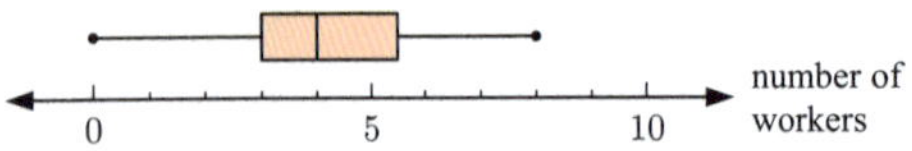

5 a min = 2 runs, $Q_1 = 3.5$ runs, median = 6.5 runs, $Q_3 = 9.5$ runs, max = 14 runs
b range = 12, IQR = 6
c

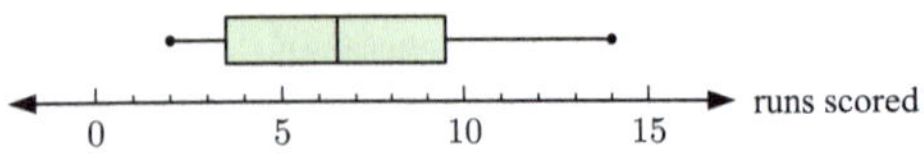

6 a

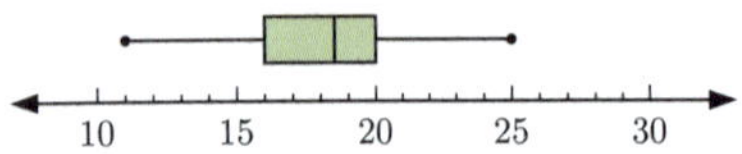

b

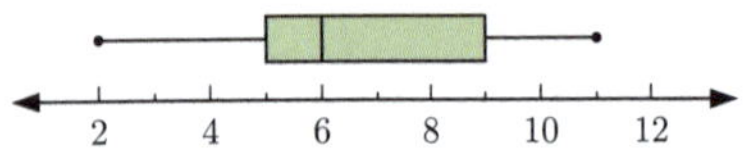

7 a

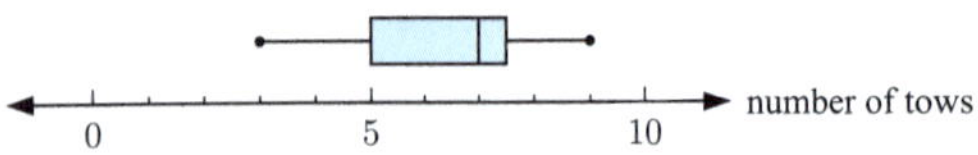

b range = 6, IQR = 2.5

EXERCISE 20E.2

1 a Year 10: min = 4 hours, $Q_1 = 6.5$ hours, median = 9 hours, $Q_3 = 11$ hours, max = 15 hours
Year 12: min = 7 hours, $Q_1 = 10$ hours, median = 12 hours, $Q_3 = 16$ hours, max = 21 hours
b i Year 10: 11 hours, Year 12: 14 hours
ii Year 10: 4.5 hours, Year 12: 6 hours
c Year 12 d Year 12

2 a i Indonesia: 88 cm, Australia: 93 cm
ii Indonesia: 43 cm, Australia: 41 cm
iii Indonesia: 45 cm, Australia: 52 cm
iv Indonesia: 15 cm, Australia: 20 cm
b i 75% ii 50%
c i The fish caught in Indonesia generally have greater length.
ii The fish caught in Australia have greater length variability.

3 a New York: min = 1 night, $Q_1 = 2$ nights, median = 3 nights, $Q_3 = 5$ nights, max = 8 nights
Miami: min = 1 night, $Q_1 = 2.5$ nights, median = 4 nights, $Q_3 = 5.5$ nights, max = 8 nights
b

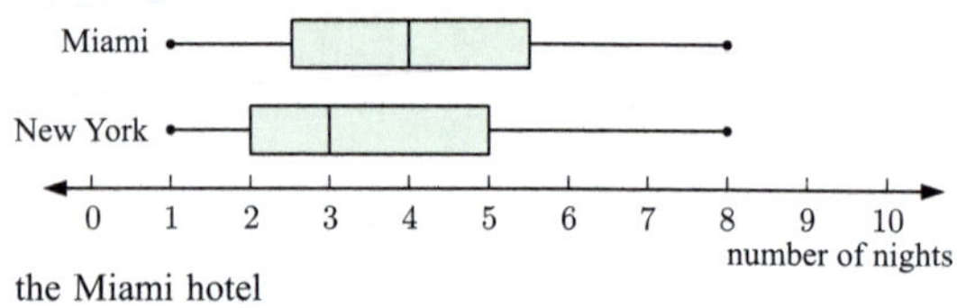

c the Miami hotel

4 a Boys: min = 160 cm, $Q_1 = 167$ cm, median = 170.5 cm, $Q_3 = 174$ cm, max = 188 cm
Girls: min = 152 cm, $Q_1 = 162.5$ cm, median = 166 cm, $Q_3 = 170$ cm, max = 177 cm
b

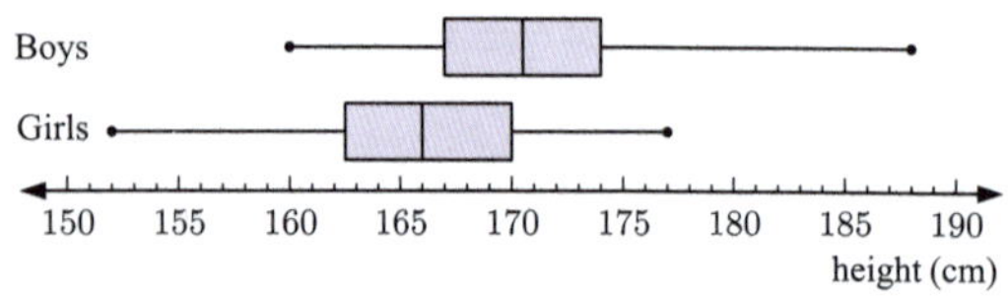

c The distributions show that in general, the boys are taller than the girls.

5 a Sound 1: min = 2.3 kg, $Q_1 = 2.5$ kg, median = 2.6 kg, $Q_3 = 2.7$ kg, max = 2.9 kg
Sound 2: min = 2.6 kg, $Q_1 = 2.7$ kg, median = 2.75 kg, $Q_3 = 2.9$ kg, max = 3.2 kg
b

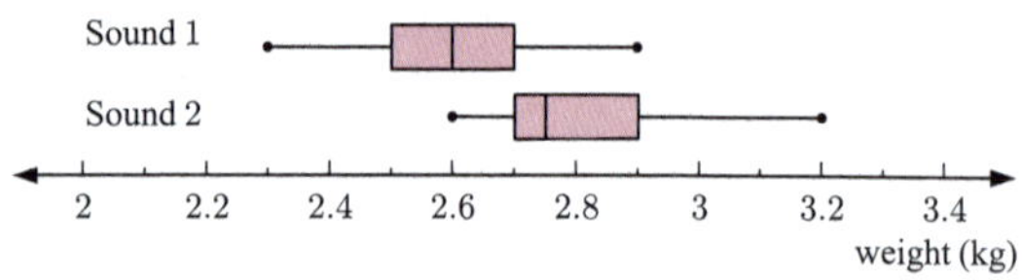

c The distributions show that in general, crayfish in sound 2 weigh more than those in sound 1. Both sounds have similar variability in weights.

EXERCISE 20F

1 a 120 goats b ≈ 29 kg
c ≈ 32 kg, 60% of the goats have weight less than or equal to 32 kg.
d ≈ 17.5 kg

2 a

Time (t min)	Frequency	Cumulative frequency
$20 \leqslant t < 25$	18	18
$25 \leqslant t < 30$	45	63
$30 \leqslant t < 35$	37	100
$35 \leqslant t < 40$	33	133
$40 \leqslant t < 45$	19	152
$45 \leqslant t < 50$	8	160

b

Cumulative frequency graph of race data

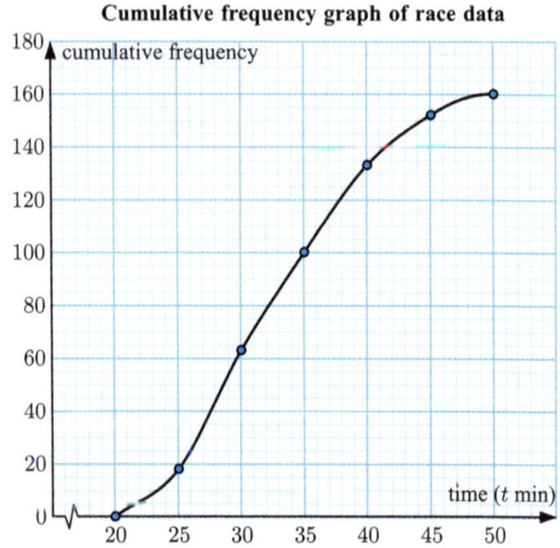

c **i** ≈ 32 min **ii** ≈ 80 runners **iii** ≈ 27.5 min
iv ≈ 10.5 min

3 **a** **Cumulative frequency graph of trout length**

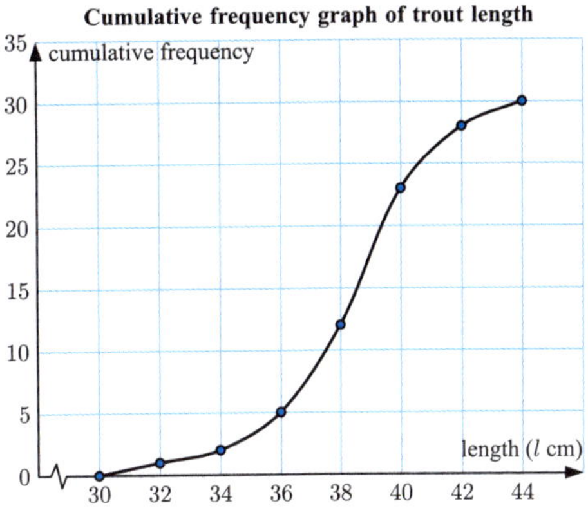

b **i** $\approx 58\%$ **ii** ≈ 38.6 cm
iii ≈ 3.0 cm, this is the range in the lengths of the middle 50% of the population of trout.
iv ≈ 37.6 cm. 35% of the trout have lengths less than or equal to 37.6 cm.

c ≈ 38.3 cm

4 **a** **i** ≈ 5 kg **ii** $Q_1 \approx 4.2$ kg, $Q_3 \approx 6.3$ kg

b

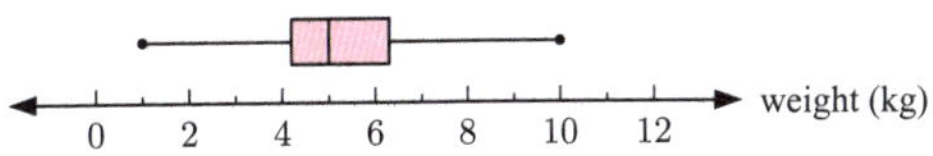

c

Weight (w kg)	Frequency	Cumulative frequency
$1 \leqslant w < 2$	1	1
$2 \leqslant w < 3$	2	3
$3 \leqslant w < 4$	5	8
$4 \leqslant w < 5$	12	20
$5 \leqslant w < 6$	8	28
$6 \leqslant w < 7$	6	34
$7 \leqslant w < 8$	3	37
$8 \leqslant w < 9$	2	39
$9 \leqslant w < 10$	1	40

d ≈ 5.25 kg

5 **a** 35 m **b** 40 to 50 m

c **Cumulative frequency graph of baseball throws**

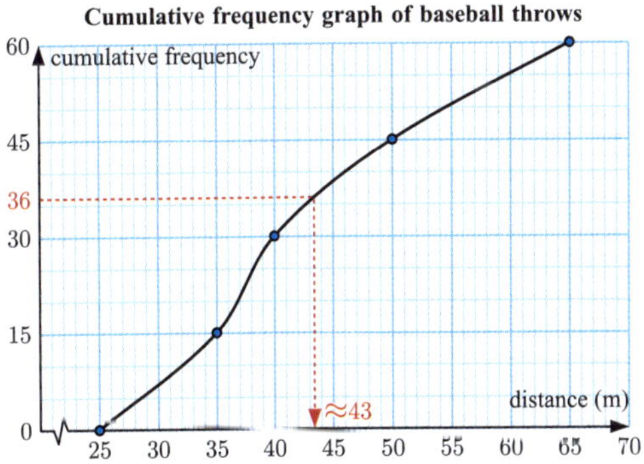

d ≈ 43 m

6 **a** **Cumulative frequency graph of Cradle Mountain climb data**

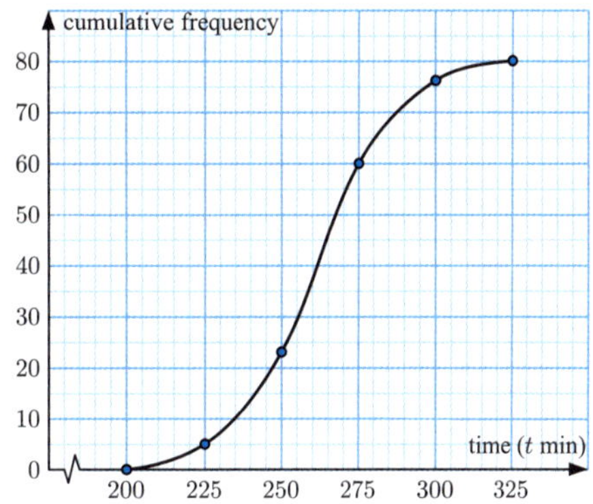

b ≈ 262 min
c ≈ 28 min. This is the range in times for the middle 50% of the data.
d $\approx 5\%$

REVIEW SET 20A

1 **a** discrete
b no
c **Number of bedrooms in students' houses**

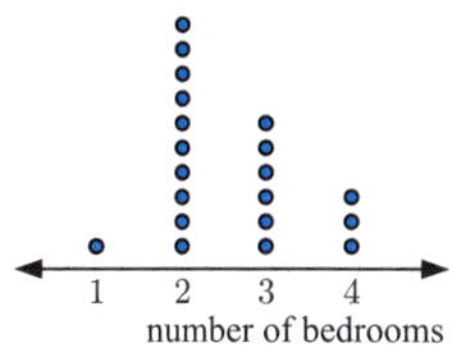

d 2 bedrooms, it is the most common number of bedrooms.

2 **a** **Number of people at judo class**

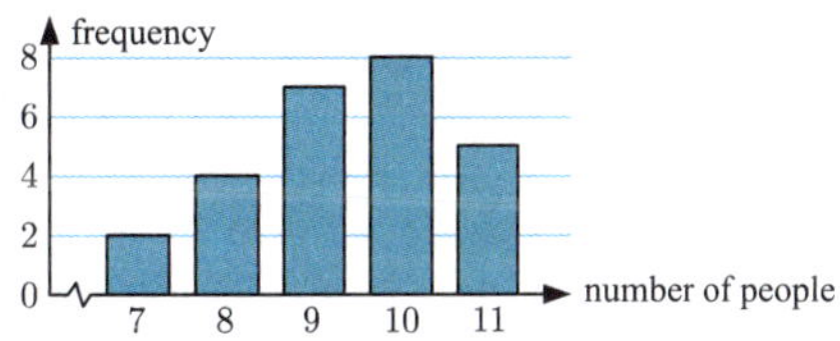

b negatively skewed

3 **a** 14 days **b** ≈ 26.3 laps

4 13.56

5 **a** **Masses of eggs in a carton**

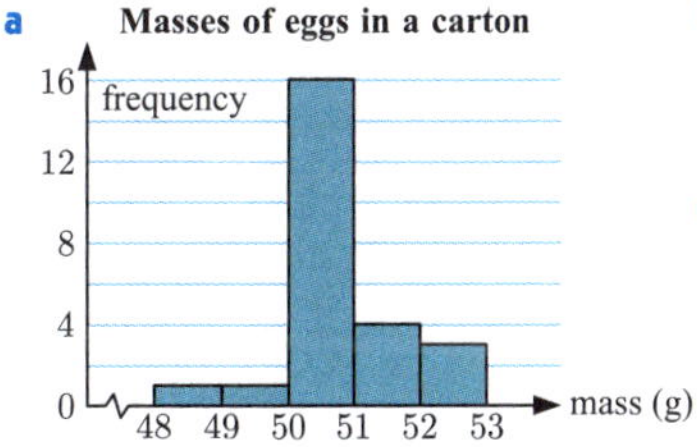

b $50 \leqslant m < 51$ g
c approximately symmetric
d ≈ 50.8 g

6 **a** **i** 48 **ii** 98 **iii** 15 **iv** 66 **v** 42
b **i** 83 **ii** 24

7 **a** 60 competitors **b** $\approx 25\%$
c **i** ≈ 208 points **ii** ≈ 200 points **iii** ≈ 215 points
d ≈ 15 points

8 **a** 4 letters
b min $= 6$ letters, $Q_1 = 7$ letters, median $= 8$ letters, $Q_3 = 9$ letters, max $= 10$ letters
c

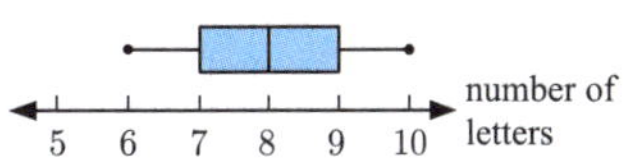

d 2 letters

REVIEW SET 20B

1 **a** negatively skewed **b** bimodal
c positively skewed, "9" is an outlier

2 **a** ≈ 9.84 pieces **b** 10 pieces **3** $x = 13$

4 **a**

Number of employees	*Tally*	*Frequency*
0 to 9	|||	3
10 to 19	~~||||~~ |||	8
20 to 29	~~||||~~ ~~||||~~ |	11
30 to 39	~~||||~~ |	6
40 to 49	||	2

b

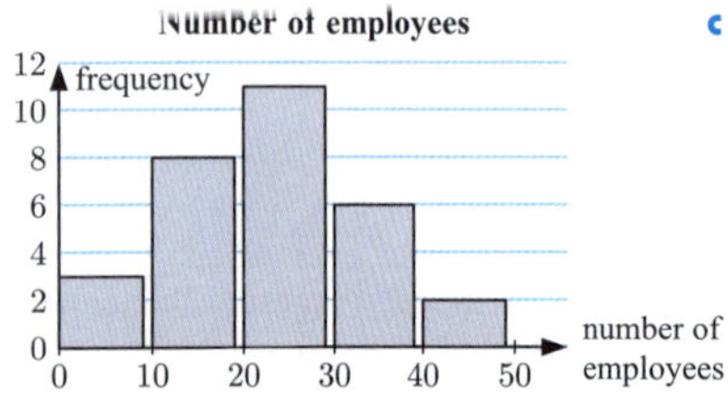

c 20 to 29 employees

5 **a**

Number of goals	*Frequency*
0	8
1	7
2	4
3	1
Total	20

b **i** 20 games **ii** 18 goals
c **i** 0.9 goals **ii** 1 goal **iii** 0 goals **iv** 3

6 **a**

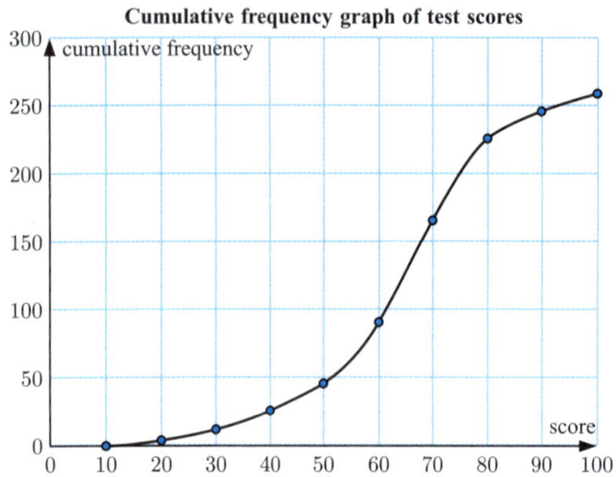

b median ≈ 65, IQR ≈ 18

c

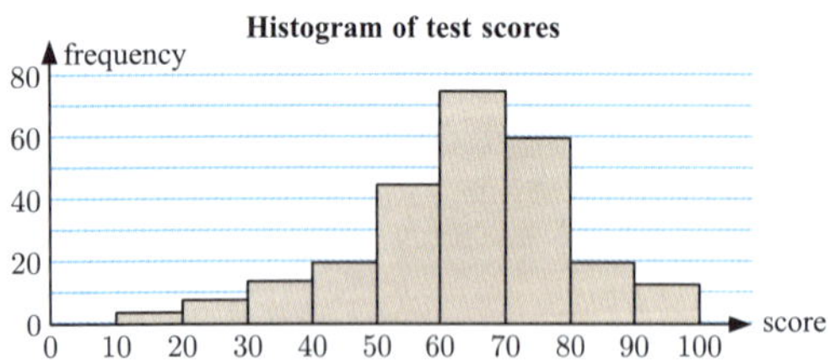

d ≈ 63.5 **e** negatively skewed

7 **a** *Davis*: min = \$100.95, Q_1 = \$107.60, median $\approx$ \$114.68, Q_3 = \$122.15, max = \$131.05
Douglas: min = \$72.05, Q_1 = \$86.75, median $\approx$ \$101.48, Q_3 = \$121.05, max = \$133.05

b

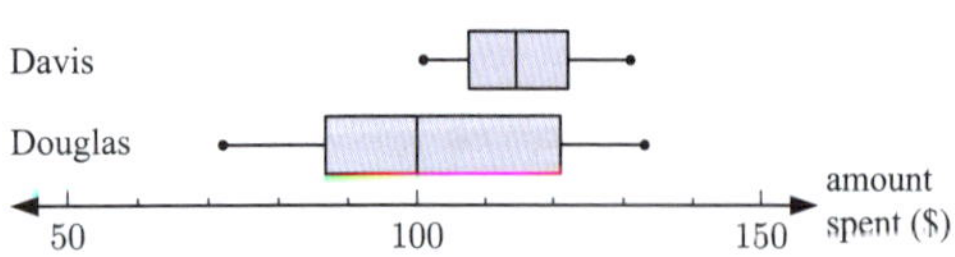

c *Davis*: range = \$30.10, IQR = \$14.55;
Douglas: range = \$61, IQR = \$34.30

d *Davis*: \$115.28; *Douglas*: \$102.37

e **i** the Davis family **ii** the Douglas family

EXERCISE 21A

1 **a** independent variable: *hours studied*,
dependent variable: *test mark*
b a person who studied for 1.25 hours and obtained a test mark of 7

2 **a** E **b** A **c** yes, B and C
d Yes, generally more hours of training results in higher productivity (from the graph).

3 **a** independent variable: *distance to school*,
dependent variable: *travel time*

b

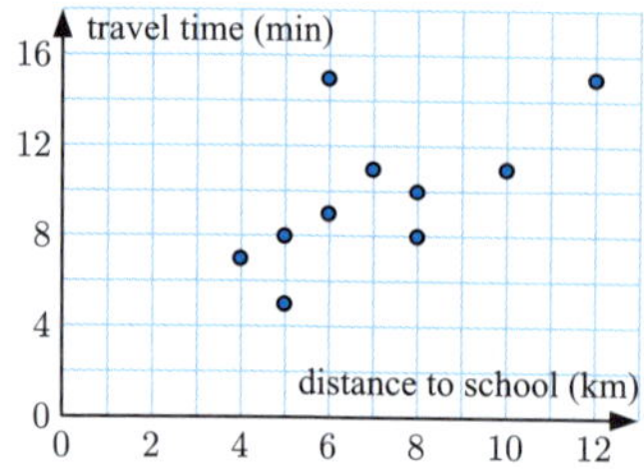

4 **a** independent variable: *age*,
dependent variable: *number of baby teeth*

b

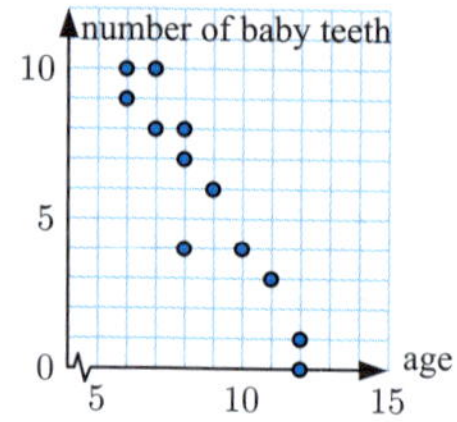

5 **a** independent variable: *age*,
dependent variable: *time*

b

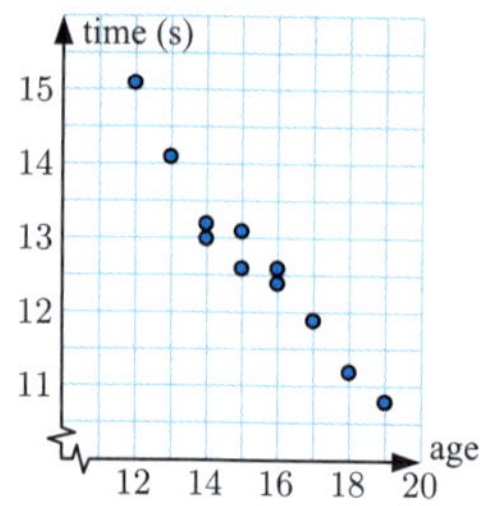

c Yes, older sprinters generally have lower fastest times.

EXERCISE 21B

1

a	**i**	negative association	**ii**	linear	**iii**	strong
b	**i**	no association	**ii**	not linear	**iii**	zero
c	**i**	positive association	**ii**	linear	**iii**	moderate
d	**i**	positive association	**ii**	linear	**iii**	weak
e	**i**	positive association	**ii**	not linear	**iii**	moderate
f	**i**	negative association	**ii**	not linear	**iii**	strong

2 **a** as x increases, y increases **b** as T increases, d decreases
c If there is no association between two variables then the points on the scatter graph are randomly scattered.

3 **a**

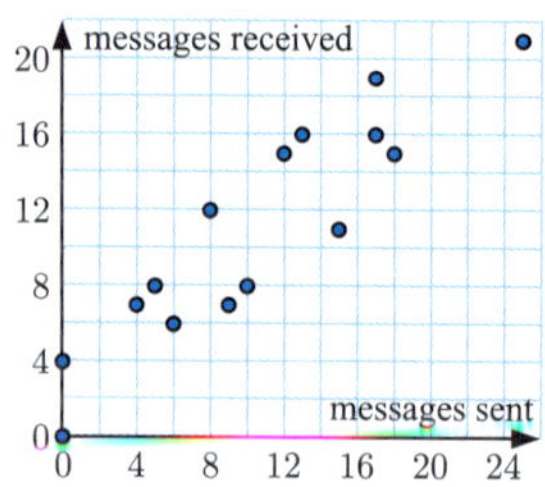

b A moderate, positive, linear correlation.

4 **a** *temperature*

b 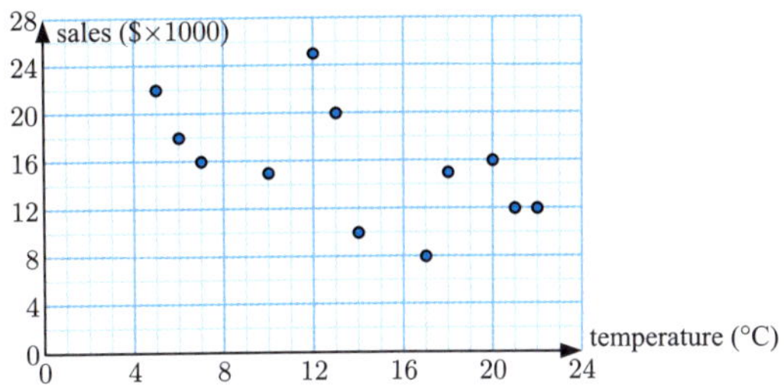

c There is a weak, negative, linear correlation between *sales* and *temperature*.

5 **a** **i** 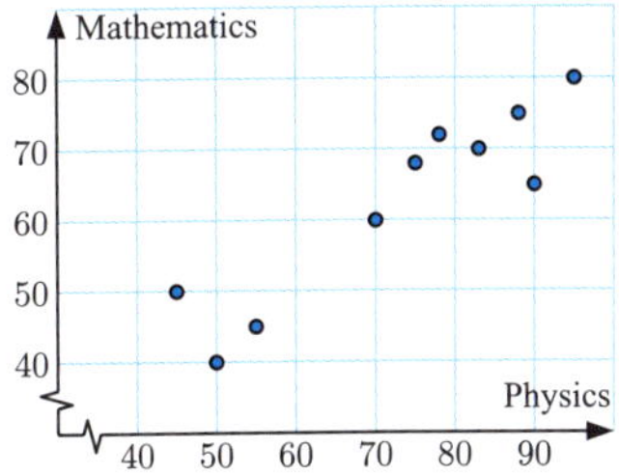

ii A moderate, positive, linear correlation.

b 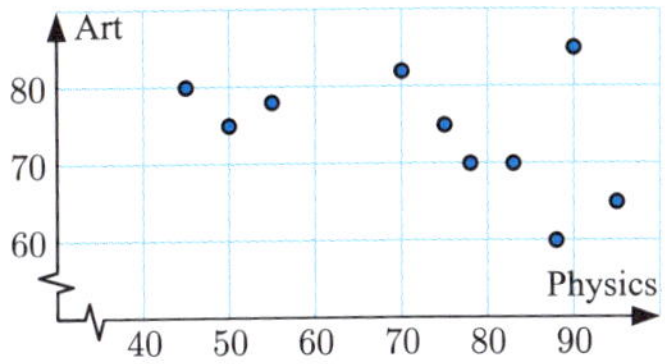

A weak, negative, linear correlation (virtually zero correlation).

6 **a** 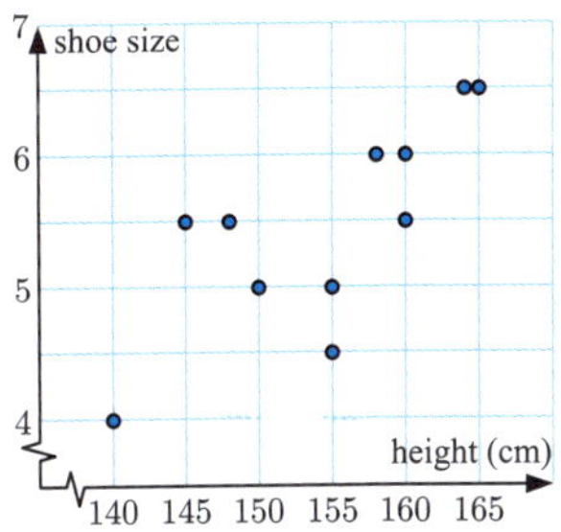

b A weak, positive, linear correlation exists between *height* and *shoe size*.

EXERCISE 21C

1 **a** C **b** A **c** B

2 **a** positive correlation as $r > 0$

b moderate positive linear correlation

3 **a** **i** 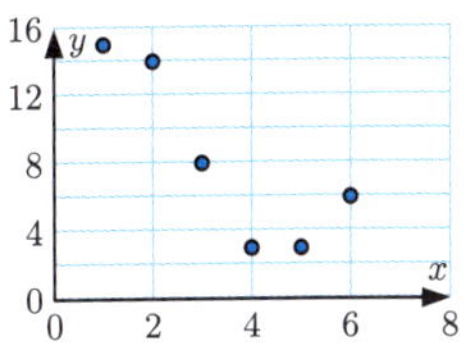

ii $r \approx -0.842$

iii moderate negative linear correlation

b **i** 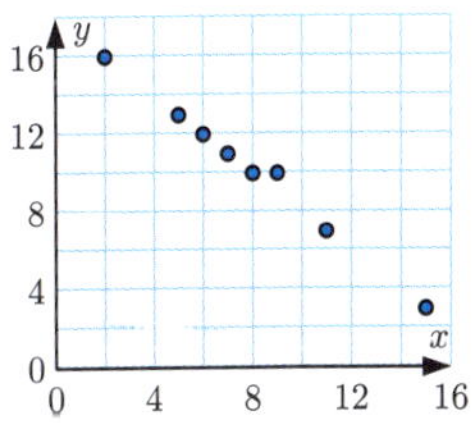

ii $r \approx -0.996$

iii very strong negative linear correlation

c **i** 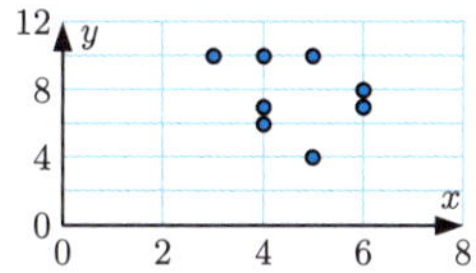

ii $r \approx -0.266$

iii very weak negative linear correlation

4 **a** 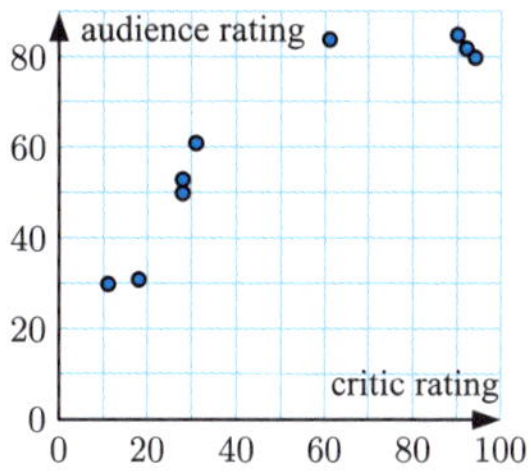

b $r \approx 0.911$

c There is a strong positive linear correlation between the *critic rating* and the *audience rating*.

5 **a** 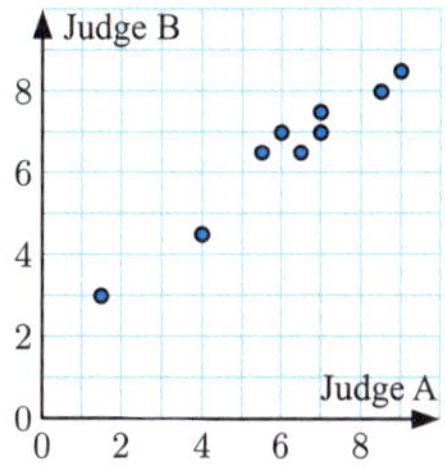

b $r \approx 0.974$

c A very strong positive linear correlation exists between Judge A's scores and Judge B's scores.

6 **a** A weak positive correlation is suspected. In general, the higher the amount, the more coins needed to make that amount. However, only one coin is needed for certain higher values.

b

Amount (cents)	5	10	15	20	25	30	35
Least number of coins	1	1	2	1	2	2	3

Amount (cents)	40	45	50	55	60	65	70
Least number of coins	2	3	1	2	2	3	2

Amount (cents)	75	80	85	90	95	100
Least number of coins	3	3	4	3	4	1

c 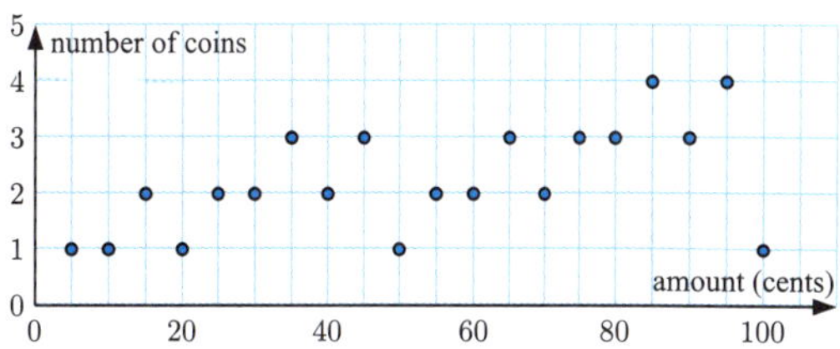

d $r \approx 0.557$

e A weak positive linear correlation exists between *amount* and *number of coins*.

EXERCISE 21D

1 **a** $\overline{x} = 7.2$, $\overline{y} = 10$

b, d 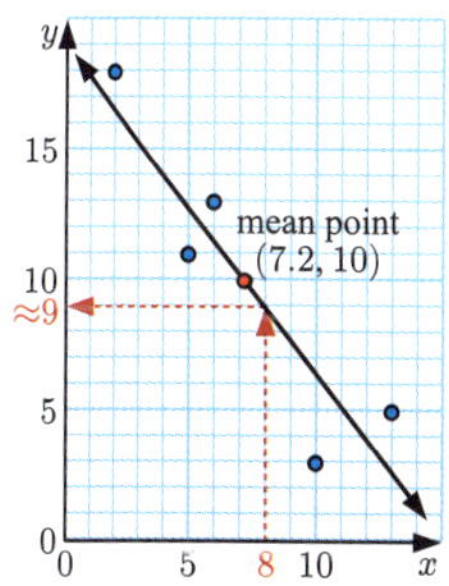

c negatively correlated

e when $x = 8$, $y \approx 9$

2 **a** $\overline{x} = 50, \quad \overline{y} = 38$

b

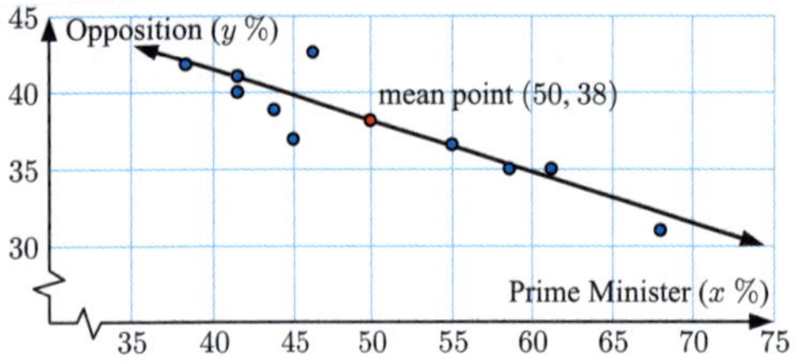

c $\approx 39\%$

3 **a** $\overline{x} = 7, \quad \overline{y} = 85$

b

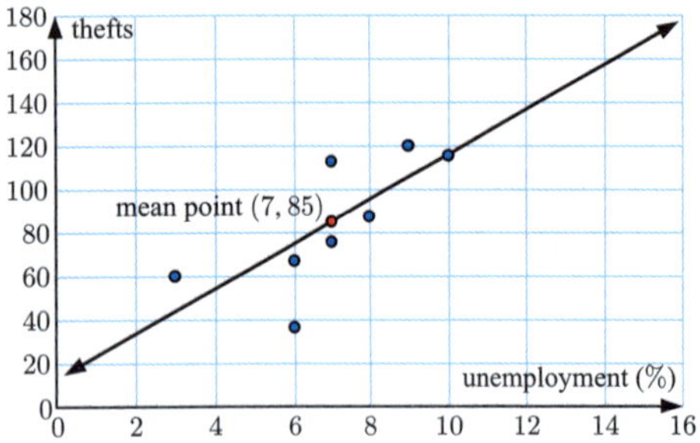

c $\approx$ 170 thefts, this estimate is unreliable as it is an extrapolation well beyond the upper pole.

4 **a** (13.3, 78.8)

b

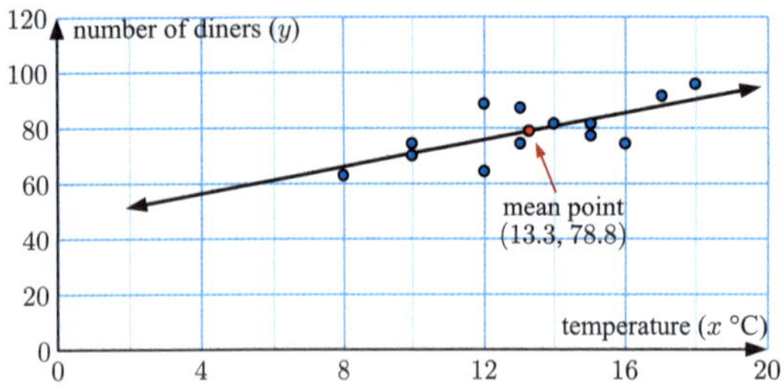

c **i** ≈ 73 diners **ii** ≈ 56 diners

d We expect the first estimate to be reliable as it is an interpolation, but the second may not be reliable as it is an extrapolation.

5 **a** 28°C, at this temperature no energy had been lost yet.

b (16, 25)

c

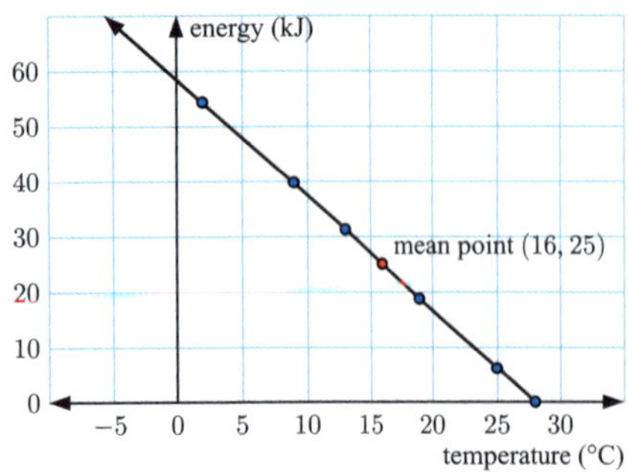

d ≈ 67 kJ

e The extrapolation in **d** is a significant underestimate of the energy lost. This may be because extra energy is lost when the water freezes at 0°C.

EXERCISE 21E

1 **a** **i**

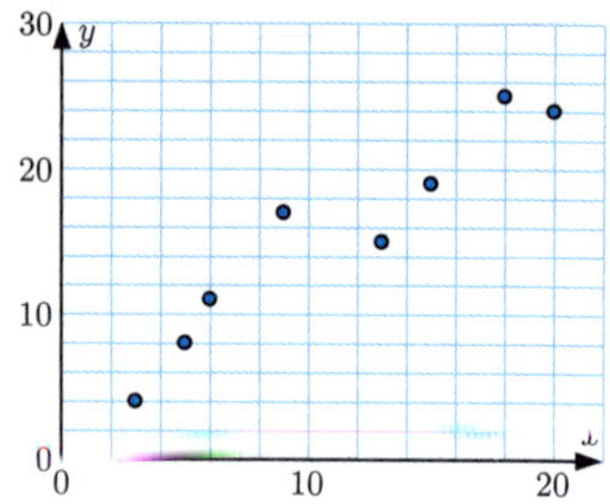

ii $y \approx 1.12x + 2.90$

iii $r \approx 0.953$, there is a very strong positive correlation between the variables.

b **i**

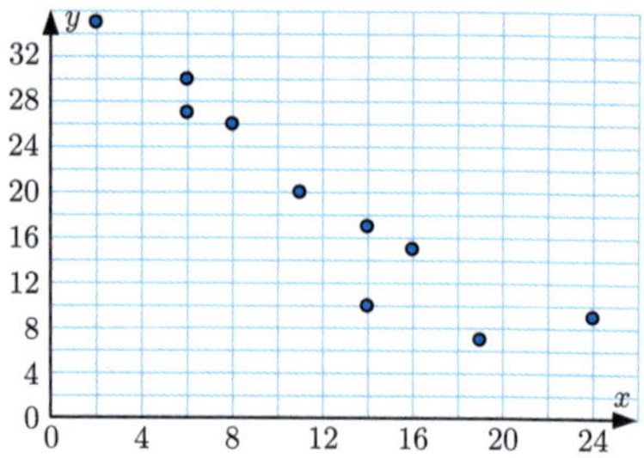

ii $y \approx -1.34x + 35.7$

iii $r \approx -0.937$, there is a strong negative correlation between the variables.

2 **a**

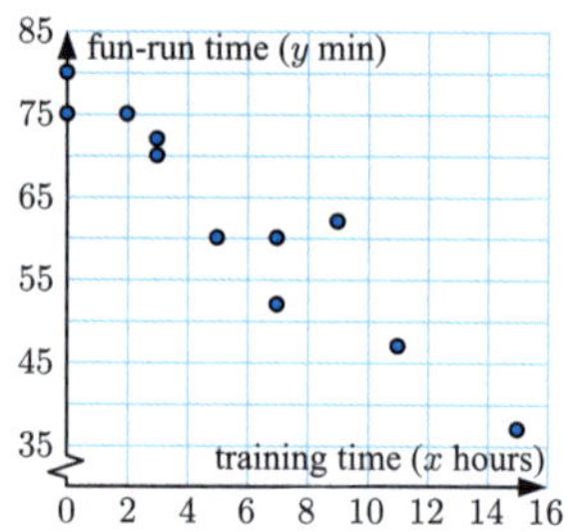

b $r \approx -0.952$ there is a very strong negative correlation between the variables.

c $y \approx -2.69x + 77.9$

d The gradient indicates that an increase in training time of 1 hour will decrease the fun-run time by ≈ 2.69 minutes.

e **i** ≈ -2.86 minutes

ii This value is clearly absurd as one cannot record a negative time for a fun-run. It is very unreliable as it is an extrapolation well beyond the upper pole.

3 **a**

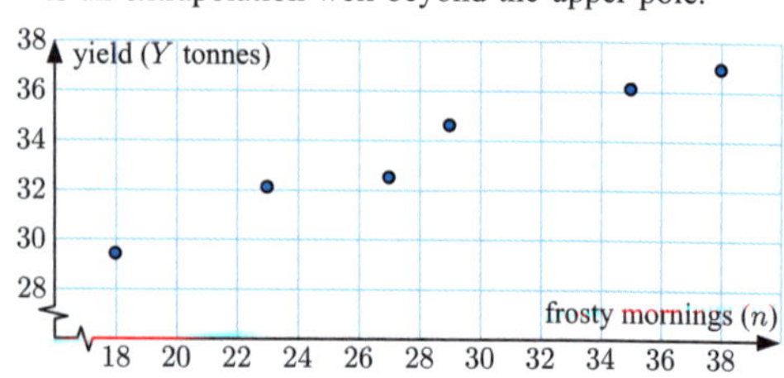

b $r \approx 0.982$, there is a very strong positive correlation between the variables.

c $Y \approx 0.371n + 23.1$ **d** ≈ 34.6 tonnes

4 **a**

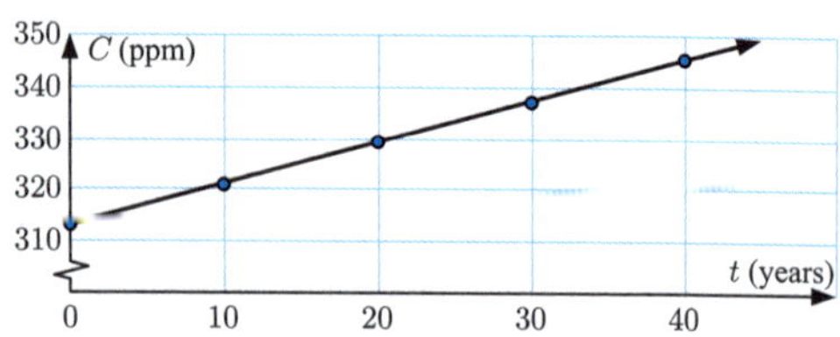

b $r = 1$, perfect positive linear correlation

c $C = 0.8t + 313$ **d** 334.6 parts per million

e 365 parts per million

5 **a**

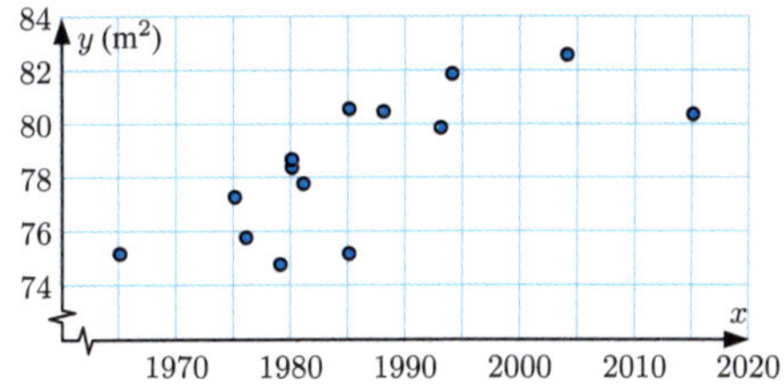

b As *year* increases, the average floor *area* of an apartment generally increases also.

c $r \approx 0.725$, there is a moderate positive correlation between the variables.

d $y \approx 0.148x - 215$

e **i** ≈ 80.6 m^2 **ii** ≈ 88.0 m^2

f The estimate in **e i** is likely to be reliable because it is an interpolation. The estimated area is also close to the actual average floor areas of apartment buildings built in similar years. So this estimate is also reasonable.

The estimate in **e ii** may be unreliable because it is an extrapolation. While it is unreasonable to assume that the trend will continue indefinitely, the estimated average floor area is not significantly higher than that of the most recently built apartment buildings. So this estimate is still reasonable.

6 **a** weight

b

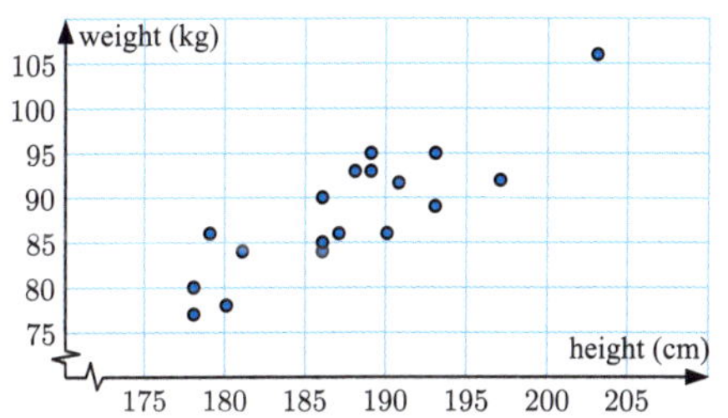

c An increase in height generally corresponds to an increase in weight.

d By calculating Pearson's correlation coefficient r, $r \approx 0.864$.

e We can make estimations using the equation of the least squares regression line. We estimate that a 200 cm tall player weighs ≈ 99.7 kg. As the estimate is an interpolation, we expect it to be reliable.

7 **a**

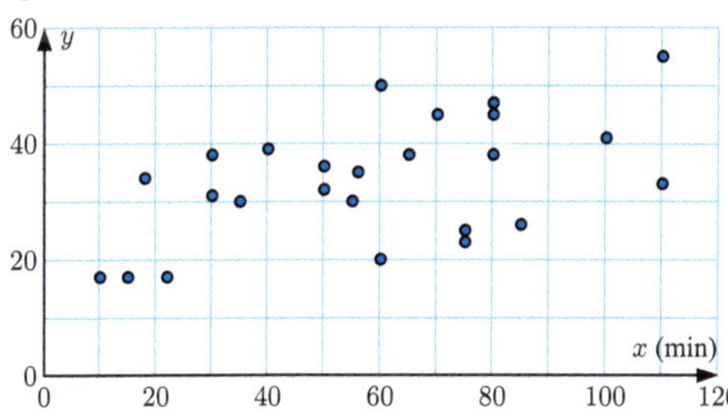

b $y \approx 0.188x + 22.7$

c The y-intercept indicates that a student is expected to obtain a score of 22.7 when they do not study for the test.

d As $r \approx 0.516$, there is a weak positive correlation between the variables.

e ≈ 62.1, this estimate is unreasonable because the maximum possible mark is 60.

REVIEW SET 21A

1 **a** The independent variable is *age*. **b** No association exists.

c It is not sensible to find it as the variables are not linearly related.

2 **a**

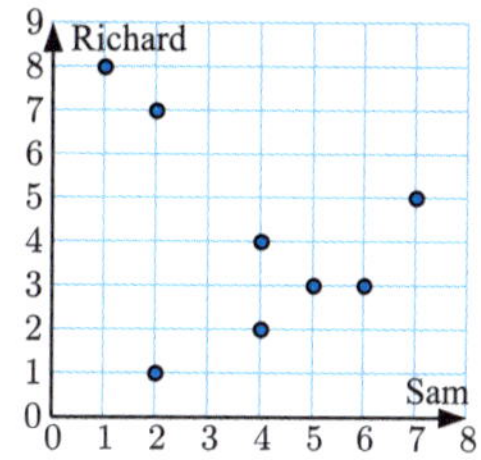

b $r \approx -0.334$, there is a very weak negative correlation between the variables.

3 **a** $\overline{x} = 12$, $\overline{y} \approx 12.8$

b

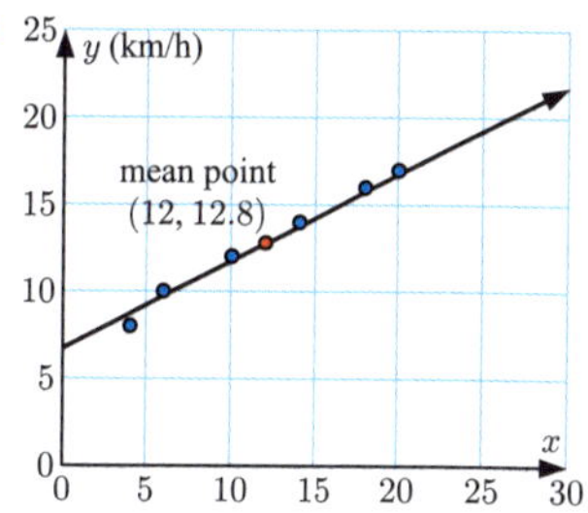

c **i** ≈ 13 km/h **ii** ≈ 19 km/h

4 **a**

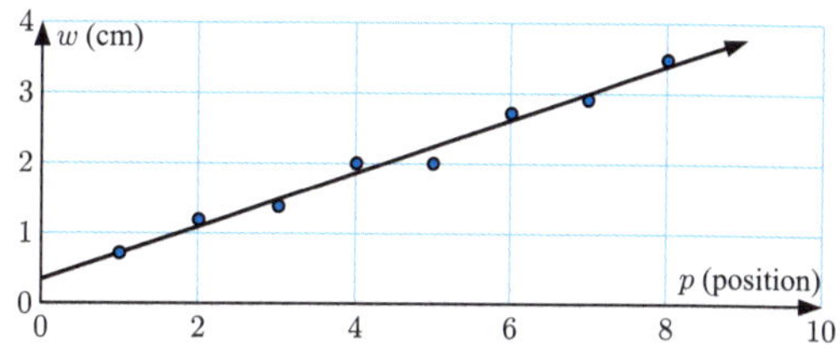

b $r \approx 0.990$, there is a very strong positive correlation between the variables.

c $w \approx 0.381p + 0.336$

d **i** ≈ 5.7 cm

ii As $p = 14$ is outside the poles, this prediction could be unreliable.

5 **a** Negatively correlated. The more games won, the higher the team's position on the ladder (and so the value for *Position* is smaller).

b

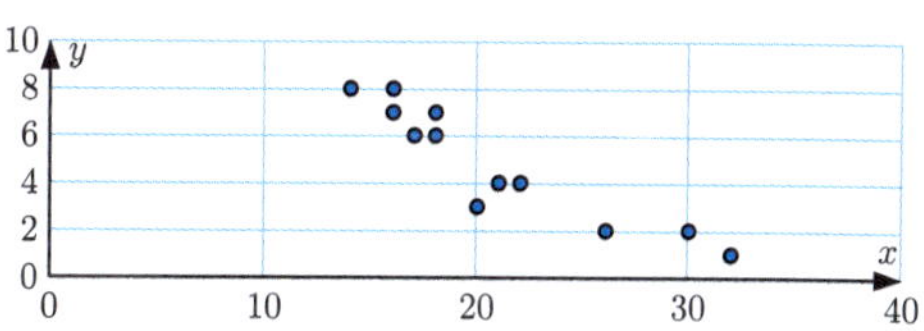

c $r \approx -0.923$, there is a strong negative correlation between the variables.

d $y \approx -0.399x + 13.1$ **e** 4th

REVIEW SET 21B

1 **a** independent: *weeks of experience*, dependent: *defective items*

b There is a weak, negative, linear correlation between the variables.

2 **a** **i** Positive, as numbers get larger they have more possible factors.

ii Weak, even among large numbers there will be numbers such as primes that have very few factors.

b

Number	1	2	3	4	5	6	7	8	9	10
Number of factors	1	2	2	3	2	4	2	4	3	4

Number	11	12	13	14	15	16	17	18	19	20
Number of factors	2	6	2	4	4	5	2	6	2	6

c

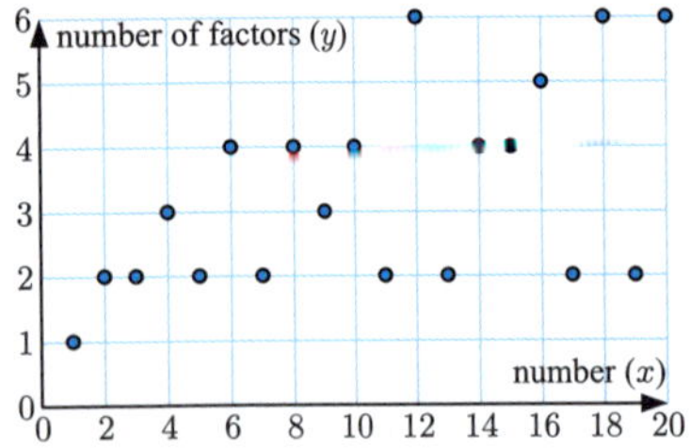

d $r \approx 0.531$, there is a weak positive linear correlation between the variables.

3 **a** $\overline{x} = 9$, $\overline{y} = 15$ **b**

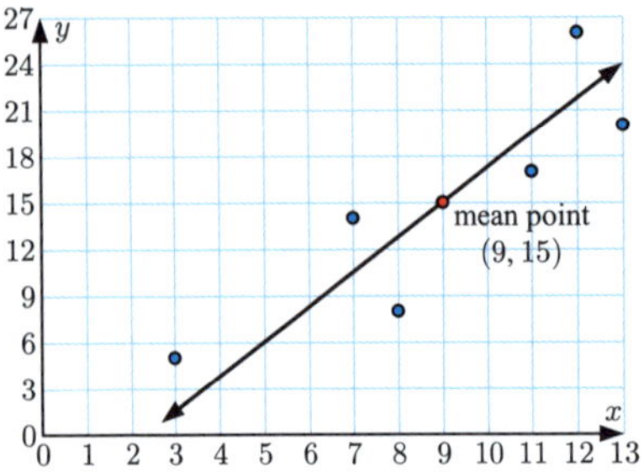

c $y \approx 6$. This estimate is likely to be reliable.

4 **a**

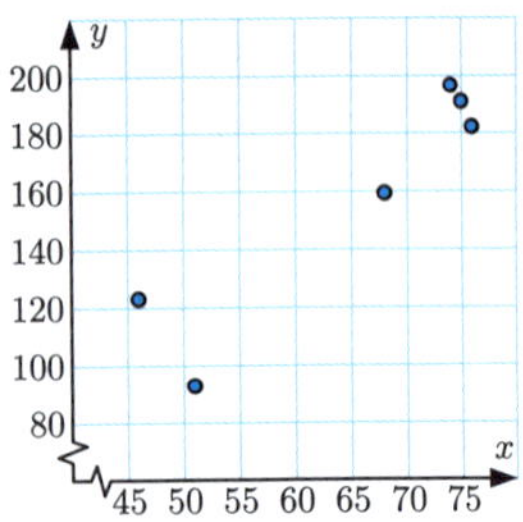

b strong positive linear correlation

c $y \approx 2.90x - 30.9$

d ≈ 143 runs
This should be fairly reliable as it is an interpolation.

5 **a**

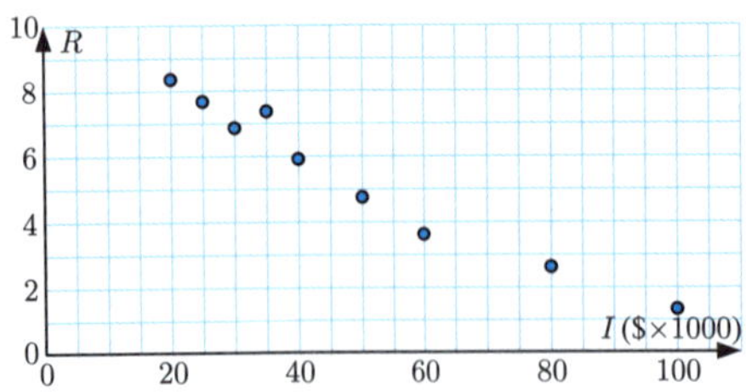

b $R \approx -0.0907I + 9.79$

c There is a very strong negative correlation between the variables.

d when $I = 55$, $R \approx 4.8$

e \$120 000 gives a rate of -1.09 which is meaningless.
So, \$120 000 is outside the data range of this model.

f **i** The point (35, 7.3), as it does not follow the general trend of the data in that area.

ii $R \approx -0.0887I + 9.61$ There is a very strong negative correlation between the variables.
When $I = 55$, $R \approx 4.7$

EXERCISE 22A

1 **a**, **b**, and **e** are functions as no two points have the same x-coordinate.

2 **a**, **b**, **d**, **e**, **g**, **h**, and **i** are functions.

3 The relation between *time* and *displacement* is a function as any vertical line cuts the relation at most once.

4 No, a vertical line is not a function as each point on the line has the same x-coordinate.

5 **a** **i**

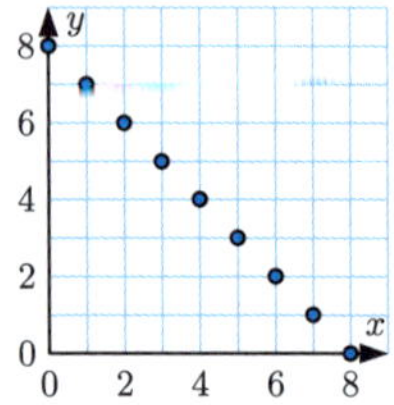

ii yes **iii** yes

b **i**

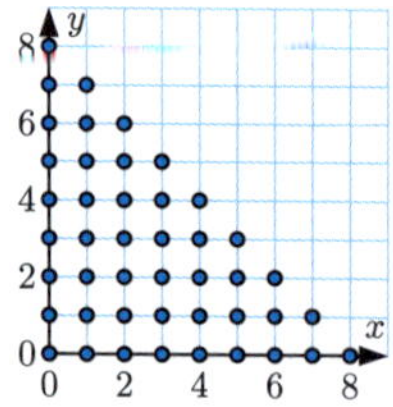

ii no **iii** no

EXERCISE 22B

1 **a** 3 **b** 7 **c** 1 **d** -7 **e** 2

2 **a** -6 **b** -4 **c** 4 **d** -10 **e** 1

3 **a** -2 **b** $\frac{5}{2}$ **c** $\frac{10}{7}$ **d** 0 **e** $\frac{7}{10}$

4 **a** **i** $f(2) = 1$ **ii** $f(3) = -1$ **b** $x = -4$

5 **a** **i** $f(4) = 2$ **ii** $g(0) = -6$ **iii** $g(5) = -1$
b $x = 0$ and $x = 3$ **c** $x = 2$
d g has gradient 1 and y-intercept -6.

6 **a** $a = 5$ **b** $a = -4$ **c** $a = -\frac{9}{2}$

7 **a** **i** $-\frac{3}{2}$ **ii** $-\frac{1}{3}$ **iii** $-\frac{8}{3}$ **b** $x = -2$ **c** $x = -1$

8

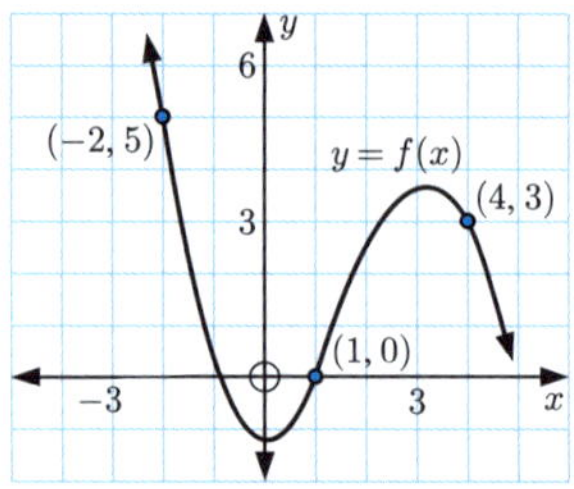

Note: Other graphs are possible.

9 **a** $V(4) = 12\,000$. The value of the car after 4 years is \$12 000.
b $t = 5$, 5 years after purchase the value of the car is \$8000.
c \$28 000 **d** $0 \leqslant t \leqslant 7$

10 **a**

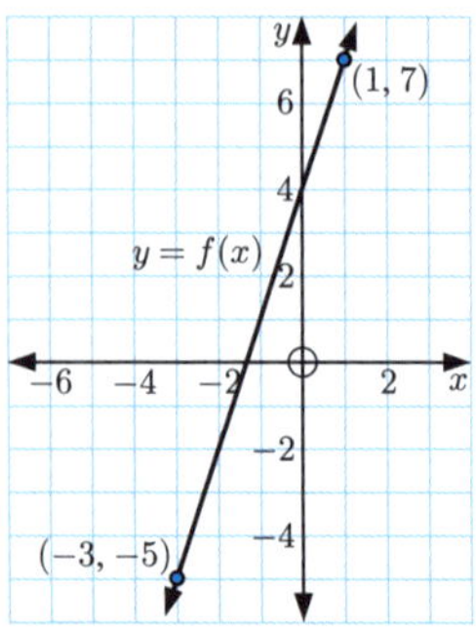

b $f(x) = 3x + 4$

11 **a** $g(-x) = 5x + 3$ **b** $g(x+4) = -5x - 17$
c $g(x-1) + 2 = -5x + 10$ **d** $2g\left(\frac{x}{5}\right) = -2x + 6$

12 **a** $f(-x) = 2x^2 + 3x + 2$ **b** $f(x+1) = 2x^2 + x + 1$

c $3f(x-2) = 6x^2 - 33x + 48$

d $f\left(\dfrac{x}{2}\right) - 1 = \frac{1}{2}x^2 - \frac{3}{2}x + 1$

EXERCISE 22C.1

1 **a** Domain is $\{x \mid x > -4\}$. Range is $\{y \mid y > -2\}$.
b Domain is $\{x \mid -3 \leqslant x \leqslant 4\}$. Range is $\{y \mid -5 \leqslant y \leqslant 2\}$.
c Domain is $\{x \mid -3 < x < 4\}$. Range is $\{y \mid -5 < y < 6\}$.
d Domain is $\{x \mid x = 2\}$. Range is $\{y \in \mathbb{R}\}$.
e Domain is $\{x \mid -3 \leqslant x \leqslant 3\}$. Range is $\{y \mid -3 \leqslant y \leqslant 3\}$.
f Domain is $\{x \in \mathbb{R}\}$. Range is $\{y \mid y \leqslant 0\}$.
g Domain is $\{x \in \mathbb{R}\}$. Range is $\{y \mid y = -5\}$.
h Domain is $\{x \in \mathbb{R}\}$. Range is $\{y \mid y \geqslant 1\}$.
i Domain is $\{x \mid x \geqslant -5\}$. Range is $\{y \mid y \leqslant 7\}$.
j Domain is $\{x \in \mathbb{R}\}$. Range is $\{y \mid y \leqslant 4\}$.
k Domain is $\{x \mid x \geqslant -5\}$. Range is $\{y \in \mathbb{R}\}$.
l Domain is $\{x \mid x \neq 1\}$. Range is $\{y \mid y \neq 0\}$.

2 Domain is $\{x \mid 0 \leqslant x \leqslant 15\}$. Range is $\{y \mid -5 \leqslant y \leqslant 15\}$.

3 **a** At any point in time, the speed of the car can only take one value.
$\therefore$ the speed graph must be a function.
b Domain is $\{t \mid 0 \leqslant t \leqslant 80\}$.
Range is $\{V \mid 75 \leqslant V \leqslant 325\}$.

EXERCISE 22C.2

1 **a** $f(x) = 4\sqrt{x-2}$ is defined when $x - 2 \geqslant 0$
$\therefore x \geqslant 2$
$\therefore$ the natural domain of $f(x)$ is $\{x \mid x \geqslant 2\}$.

b $f(x) = \dfrac{3}{x-2}$ is defined when $x - 2 \neq 0$
$\therefore x \neq 2$
$\therefore$ the natural domain of $f(x)$ is $\{x \mid x \neq 2\}$.

2 **a** $\{x \in \mathbb{R}\}$ **b** $\{x \mid x \neq 0\}$ **c** $\{x \mid x \neq -5\}$
d $\{x \mid x \neq -2 \text{ and } x \neq 1\}$ **e** $\{x \mid x \neq 3 \text{ and } x \neq -3\}$
f $\{x \mid x \neq 1 \text{ and } x \neq 4\}$

3 **a** $\{x \mid x \geqslant 3\}$ **b** $\{x \mid x > 0\}$ **c** $\{x \mid x \leqslant 3\}$
d $\{x \mid x < 4\}$

EXERCISE 22D

1 **a** 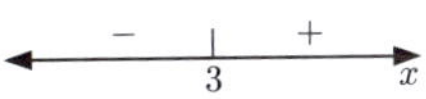

b 

c 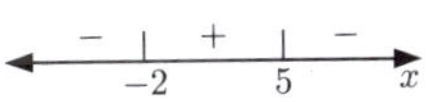

d 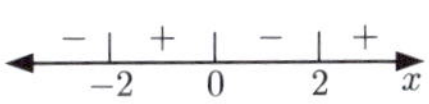

e 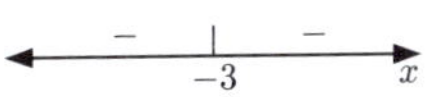

f 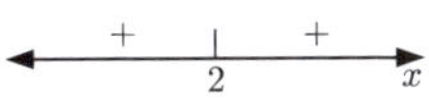

g 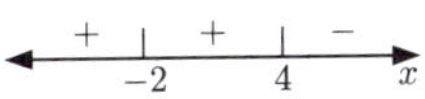

h 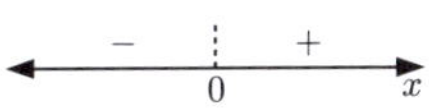

i 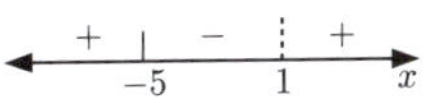

2 **a** 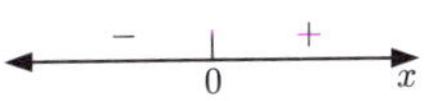

b 

c 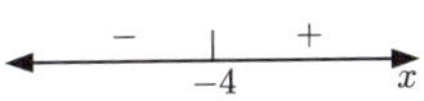

d 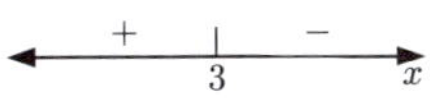

e 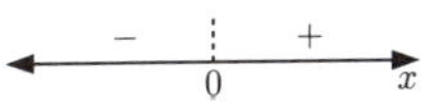

f 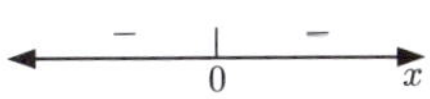

3 **a** 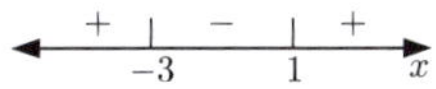

b 

c 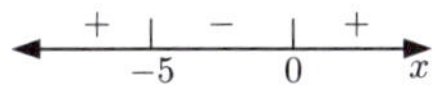

d 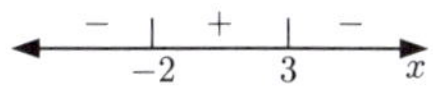

e 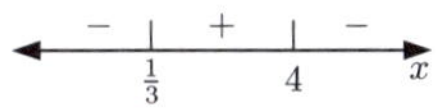

f 

4 **a** 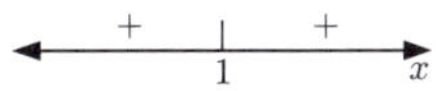

b 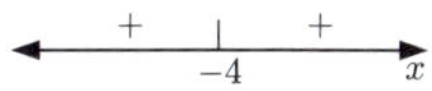

c 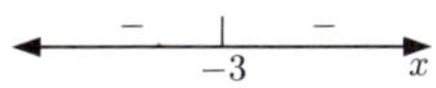

5 **a** 

b 

c 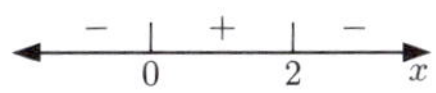

d 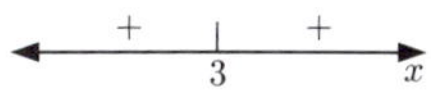

e 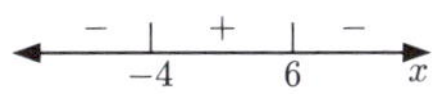

f 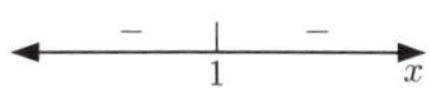

6 **a** 

b 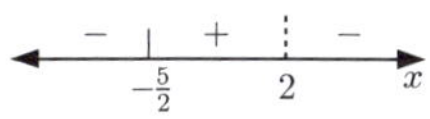

c 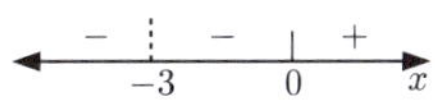

d 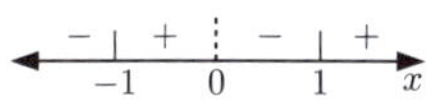

EXERCISE 22E.1

1 **a** A translation 5 units upwards.
b A translation 3 units to the right.
c A translation 6 units to the left, and 1 unit downwards.

2 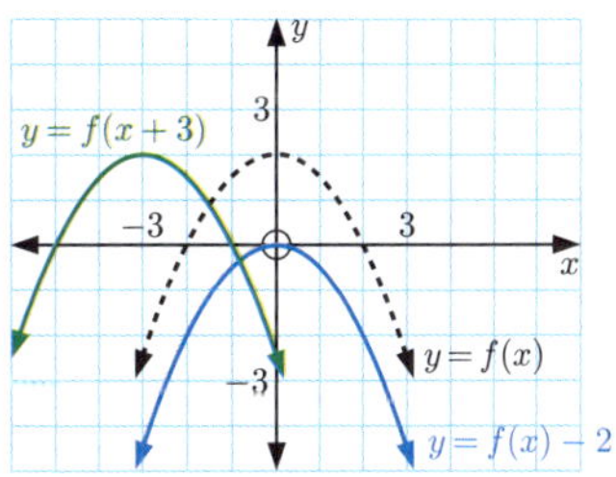

3 **a** **i** A translation 4 units upwards.
ii A translation 4 units to the left.

b 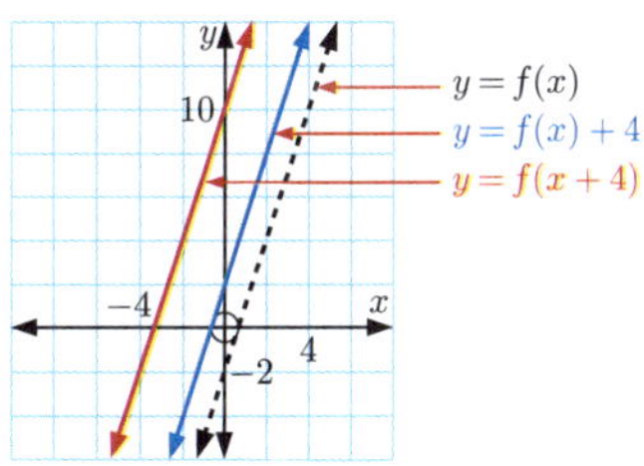

4 **a** 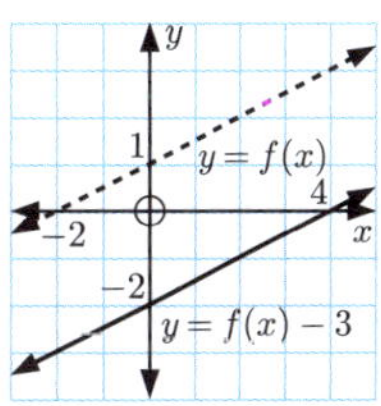

b 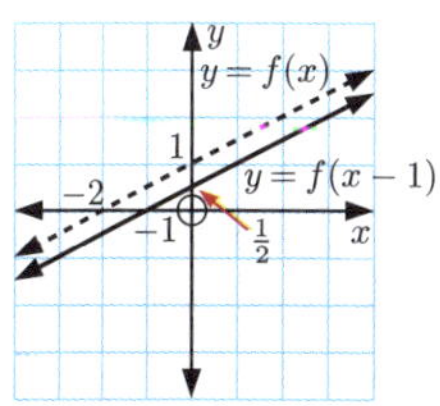

5

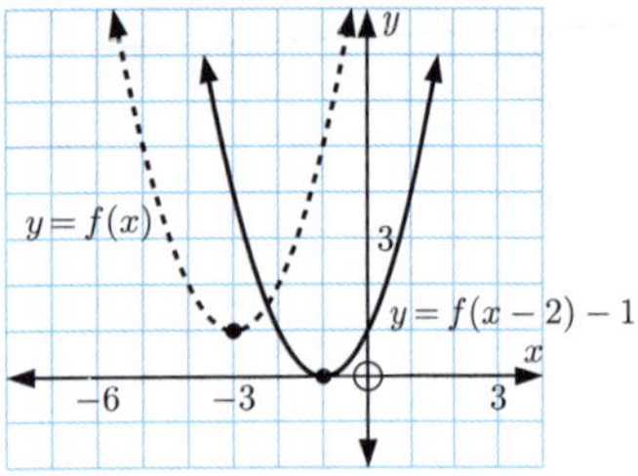

6 **a** $g(x) = 3x - 3$ **b** $g(x) = 6 - 2x$

c $g(x) = x^2 + 4x + 5$ **d** $g(x) = \dfrac{13 - 3x}{x - 4}$

7 **a** Domain is $\{x \mid -2 \leqslant x \leqslant 5\}$. Range is $\{y \mid y \geqslant 5\}$.

b Domain is $\{x \mid -6 \leqslant x \leqslant 1\}$. Range is $\{y \mid y \geqslant 3\}$.

c Domain is $\{x \mid -1 \leqslant x \leqslant 6\}$. Range is $\{y \mid y \geqslant -2\}$.

EXERCISE 22E.2

1 **a** A vertical stretch with scale factor 6.

b A vertical stretch with scale factor $\frac{1}{4}$.

2

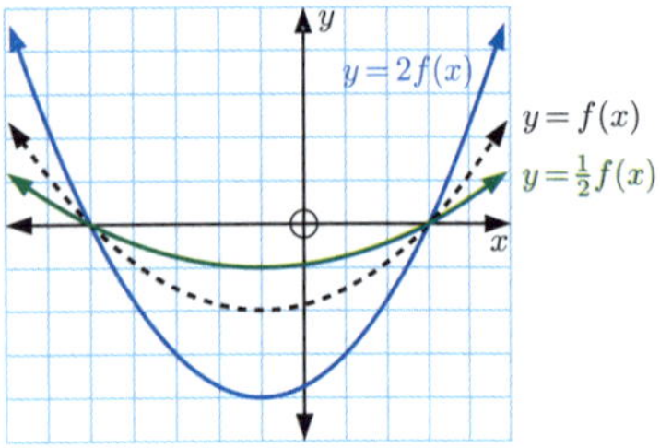

3 **a**

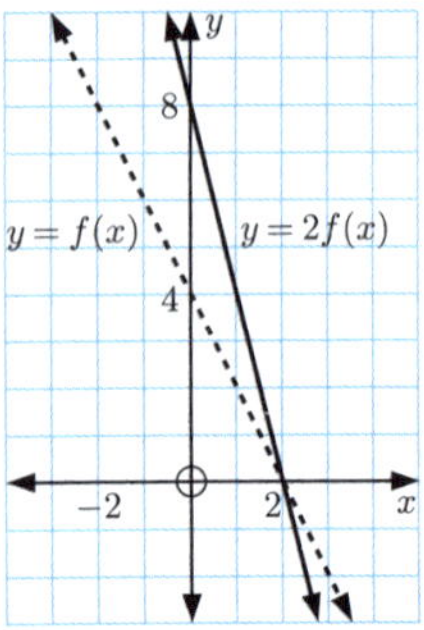

b

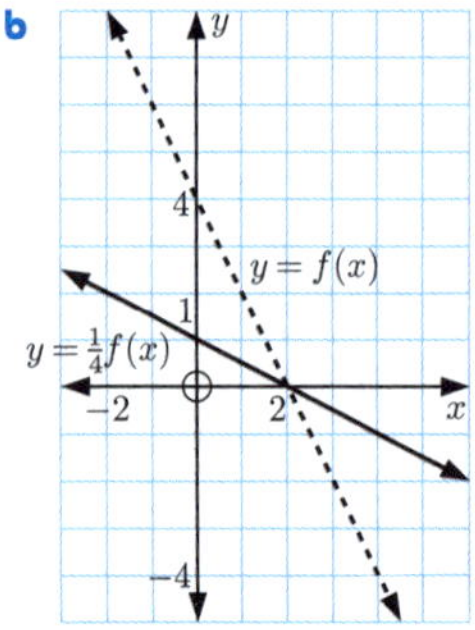

4 **a**

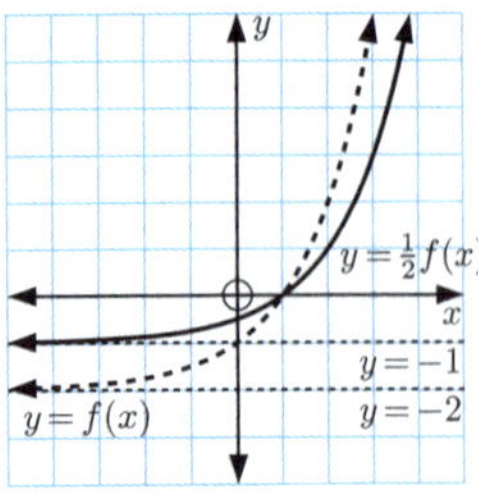

b

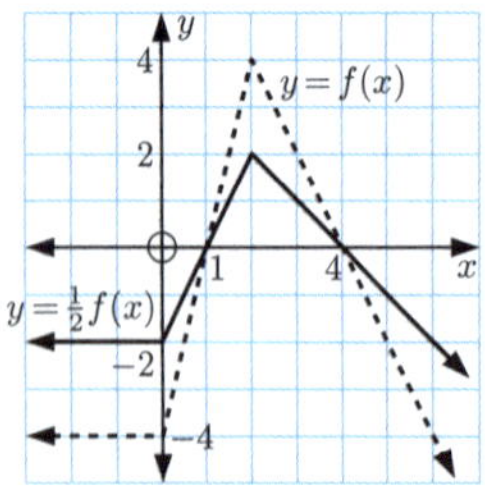

5 **a** $g(x) = 3x^2 + 6x$ **b** $g(x) = 1 - \frac{1}{4}x^2$

6 Domain is $\{x \mid -1 \leqslant x \leqslant 6\}$. Range is $\{y \mid y \leqslant 2\}$.

EXERCISE 22E.3

1 **a** A horizontal stretch with scale factor 3.

b A horizontal stretch with scale factor $\frac{1}{4}$.

2

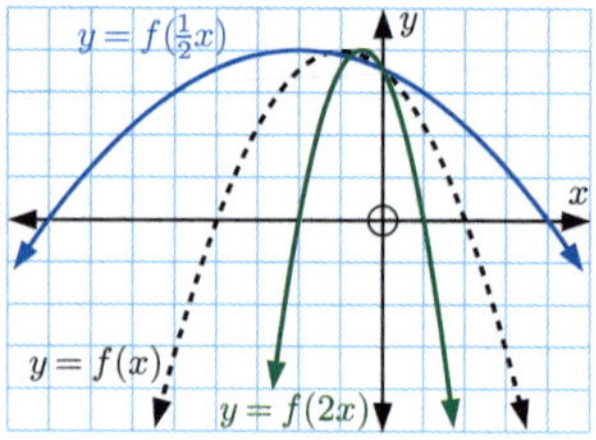

3 **a**

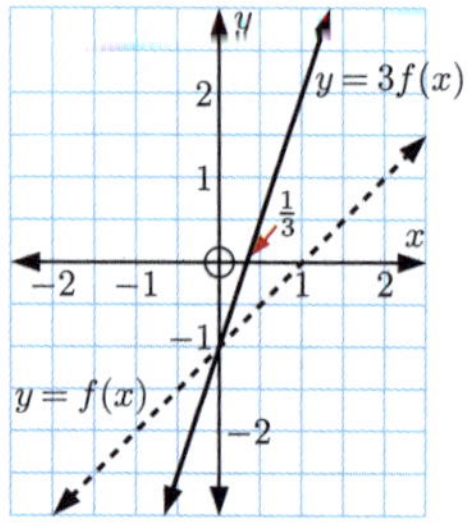

b

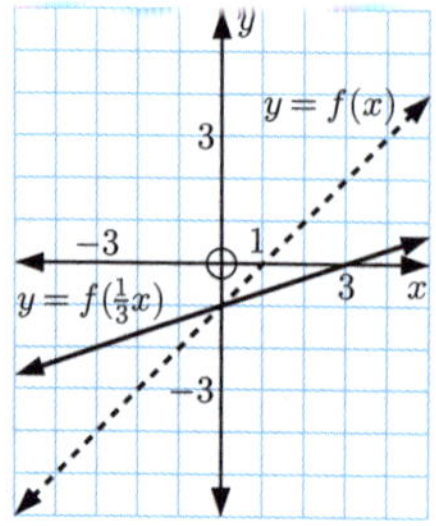

4 **a**

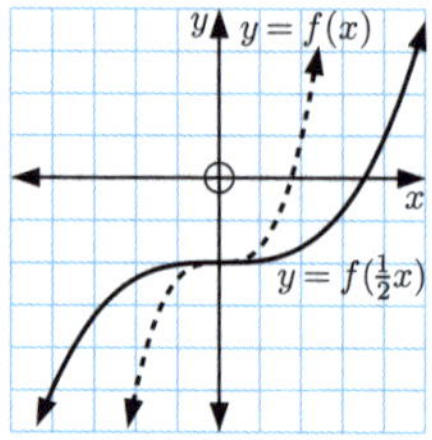

b

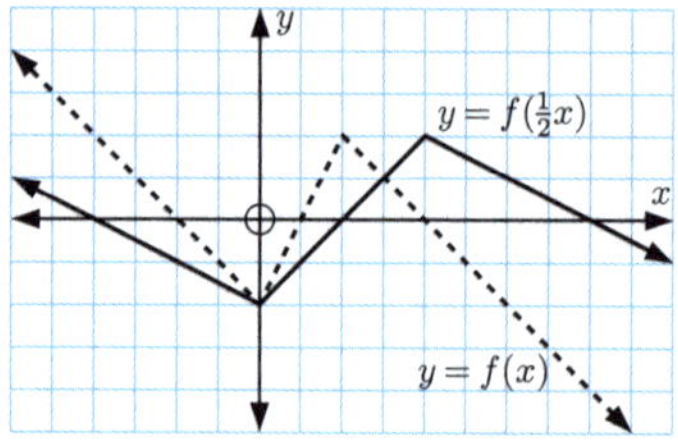

5 **a** $g(x) = 4 - 16x^2$ **b** $g(x) = 1 - \frac{1}{4}x^2$

6 Domain is $\{x \mid -2 \leqslant x < 6\}$. Range is $\{y \mid y \geqslant 4\}$.

EXERCISE 22E.4

1 **a** A reflection in the y-axis. **b** A reflection in the x-axis.

2 **a**

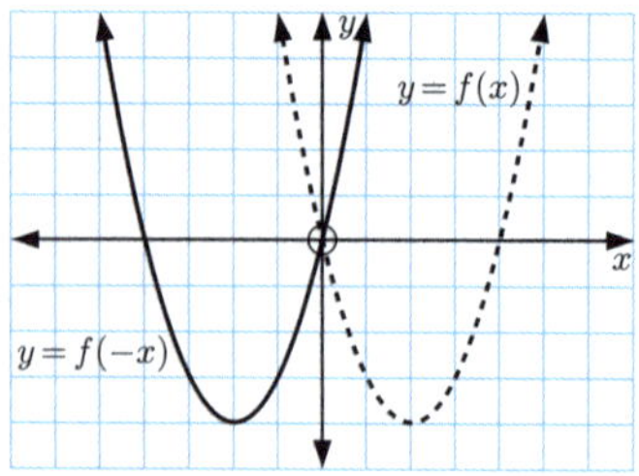

b

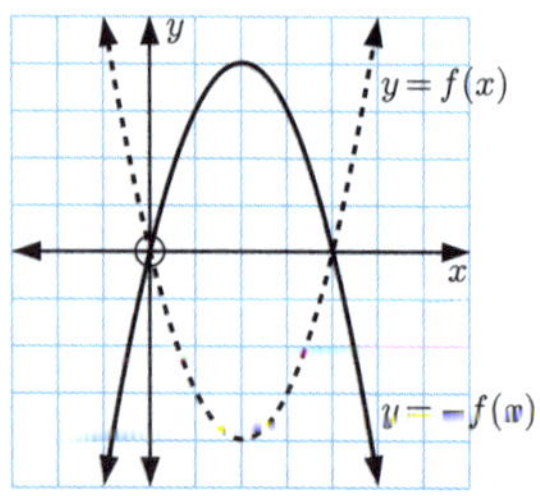

3 **a**

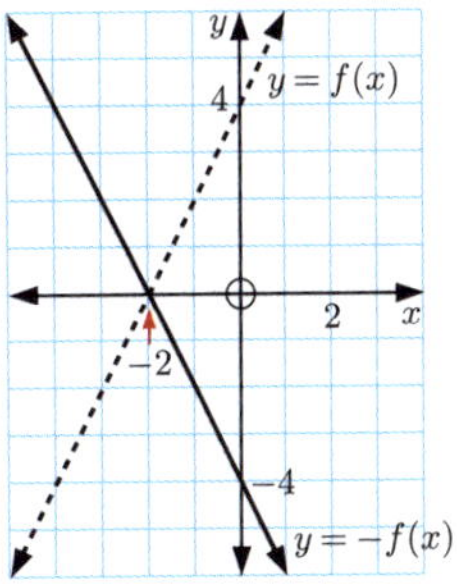

b

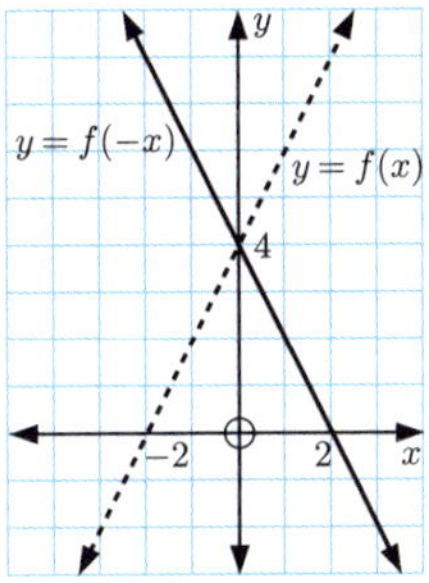

4 **a**

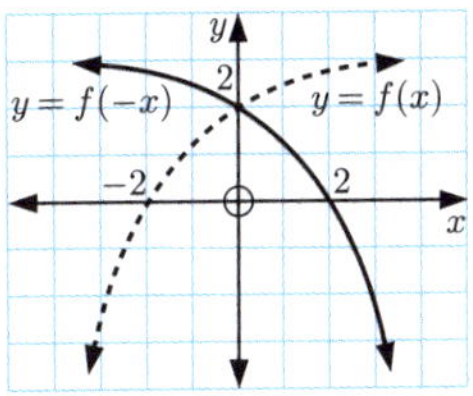

b

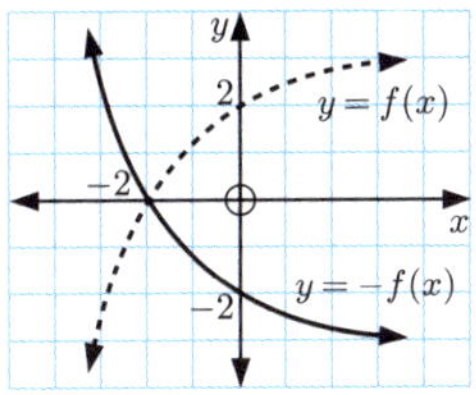

5 **a** $g(x) = \dfrac{3}{x} - 2x$ **b** $g(x) = -2x^3 - 5x^2 - 4x - 3$

6 **a** Domain is $\{x \mid -3 \leqslant x \leqslant 5\}$. Range is $\{y \mid y \leqslant 2\}$.
b Domain is $\{x \mid -5 \leqslant x \leqslant 3\}$. Range is $\{y \mid y \geqslant -2\}$.

7 **a** m **b** m **c** $\dfrac{m}{3}$ **d** $2m$ **e** $-m$ **f** $-m$

REVIEW SET 22A

1 **a** function **b** function **c** not a function

2 **a** $f(2) = 2$ **b** $f(-1) = -4$
c $f(x-3) = -x^2 + 9x - 18$
d $f(2x) - 1 = -4x^2 + 6x - 1$

3 **a** Domain is $\{x \in \mathbb{R}\}$. Range is $\{y \mid y \geqslant -2\}$.
b Domain is $\{x \mid x \leqslant 1\}$. Range is $\{y \in \mathbb{R}\}$.
c Domain is $\{x \mid -5 \leqslant x \leqslant 6\}$. Range is $\{y \mid -4 \leqslant y \leqslant 5\}$.

4

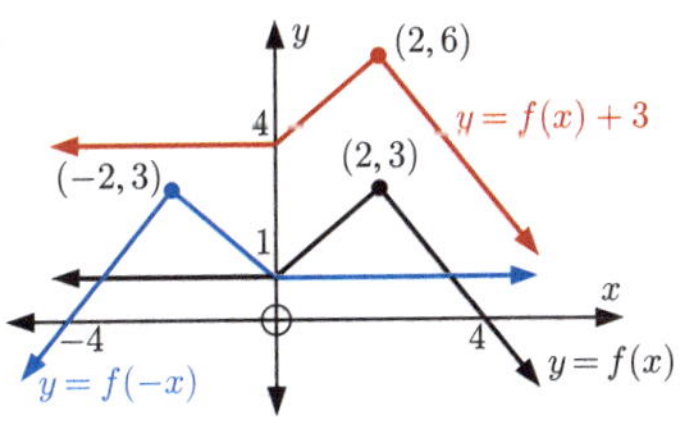

5 **a**

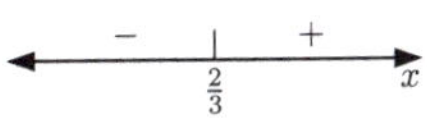

b

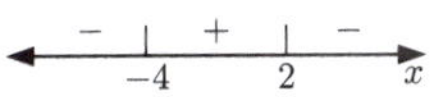

c

6

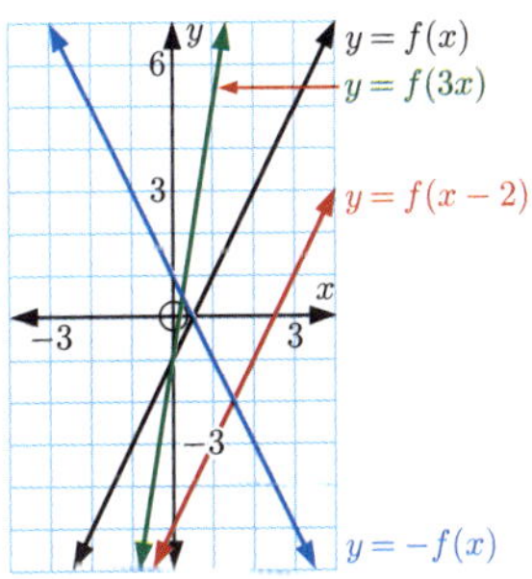

7 **a** $g(x) = 3 - 4x$
b $g(x) = x^2 + 5x + 9$
c $g(x) = \dfrac{2}{x} - 1$

8 **a** $h(0) = 5$. The nest is 5 m above ground level.
b $t = 20$
c Domain is $\{t \mid 0 \leqslant t \leqslant 50\}$. Range is $\{y \mid 5 \leqslant y \leqslant 20\}$.
d **i** $g(t) = h(t) - 5$
ii Domain is $\{t \mid 0 \leqslant t \leqslant 50\}$.
Range is $\{y \mid 0 \leqslant y \leqslant 15\}$.
iii $g(30) = 10$. The bird is 10 m above its nest after 30 seconds.

REVIEW SET 22B

1 **a** $g(2) = 8$ **b** $x = -5$ or 3

2 **a** $f(9) = 4$ **b** $f(x-3) = \sqrt{x-3} + 1$ **c** $x = 25$

3 **a** $\{x \mid x \neq 0\}$ **b** $\{x \in \mathbb{R}\}$ **c** $\{x \mid x \leqslant 16\}$

4 **a** $f(-3) = 0$ **b** $g(2) = -1$ **c** $x = -2$ and 1

5 **a**

b

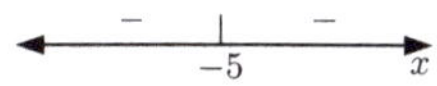

c

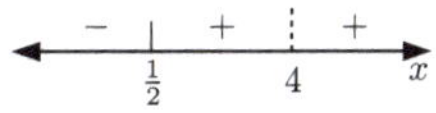

6 **a** A translation 2 units to the left and 3 units downwards.
b A vertical stretch with scale factor 5.
c A horizontal stretch with scale factor 10.

7 **a** A vertical stretch with scale factor 2. **b** $g(x) = 2\,f(x)$

8 **a** Domain is $\{x \mid x \geqslant 0\}$.
Range is $\{y \mid y \geqslant 0\}$.
b

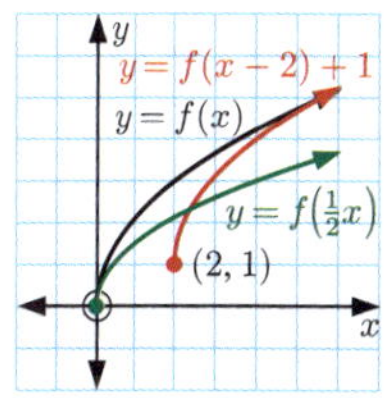

9 **a** Domain is $\{x \mid x > 2\}$. Range is $\{y \mid -1 \leqslant y < 9\}$.
b Domain is $\{x \mid x < 2\}$. Range is $\{y \mid -4 \leqslant y < 6\}$.

EXERCISE 23A

1 **a** not a quadratic function; no x^2 term
b quadratic function; $a = \frac{1}{3}$, $b = 0$, $c = 6$
c quadratic function; $a = -1$, $b = 2$, $c = 5$
d quadratic function; $a = \frac{1}{3}$, $b = -4$, $c = 0$
e not a quadratic function; contains x^3 term
f quadratic function; $a = -\frac{2}{3}$, $b = 0$, $c = \frac{7}{3}$

2 **a** $y = 0$ **b** $y = 5$ **c** $y = -15$ **d** $y = 12$

3 **a** no **b** yes **c** yes **d** no **e** no **f** no

4 **a** $x = -3$ **b** $x = -2$ or -3 **c** $x = 1$ or 4
d no real solution

5 **a** $x = 0$ or 1 **b** $x = -1$ or 3 **c** $x = 1$ or -8
d $x = 2$ or 3

6 **a** **i** $k = 5$ **ii** $a = -2$ **b** $b = 2$ or 3

7 **a** **i** 75 m **ii** 195 m **iii** 275 m
b **i** At $t = 2$ s and $t = 14$ s.
ii At $t = 0$ s and $t = 16$ s.
c The object leaves the ground at $t = 0$ s.
At $t = 2$ it is rising and at $t = 14$ it is falling.
Height 0 m is ground level and the time of flight is 16 s.

EXERCISE 23B

1 a

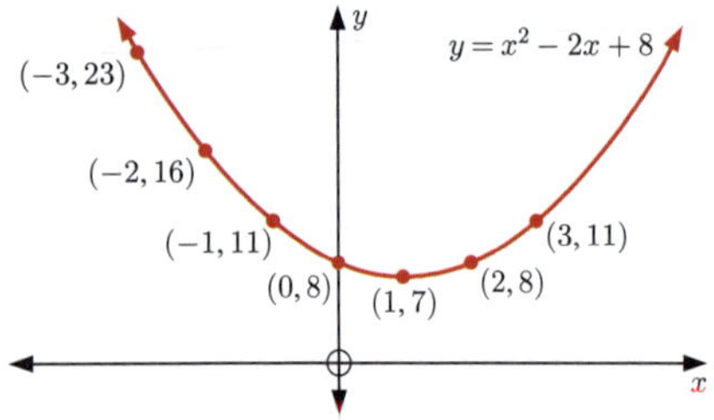

b

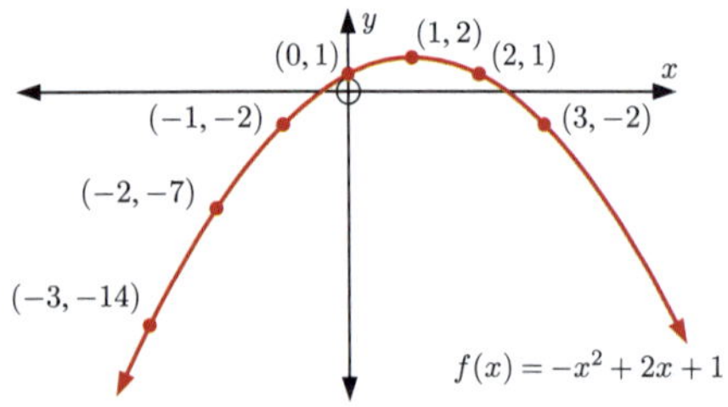

c

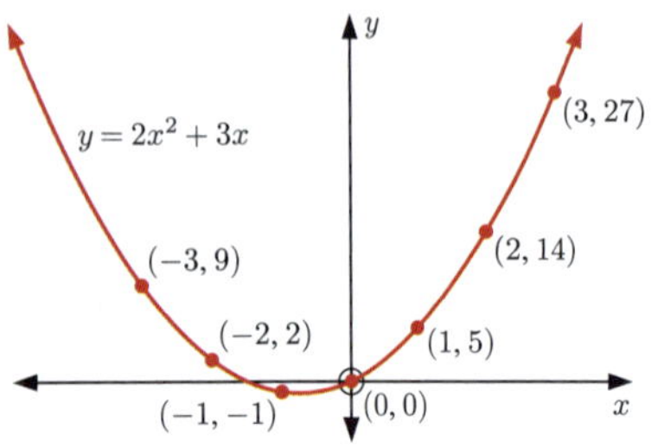

d

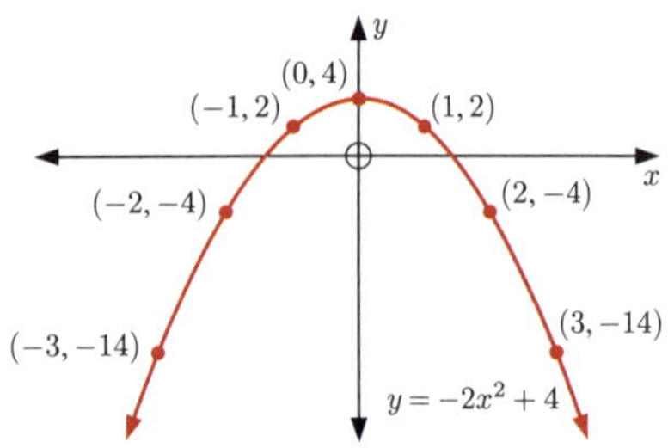

e

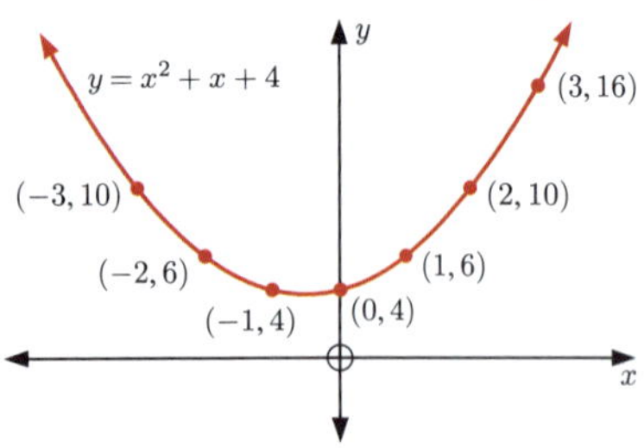

f

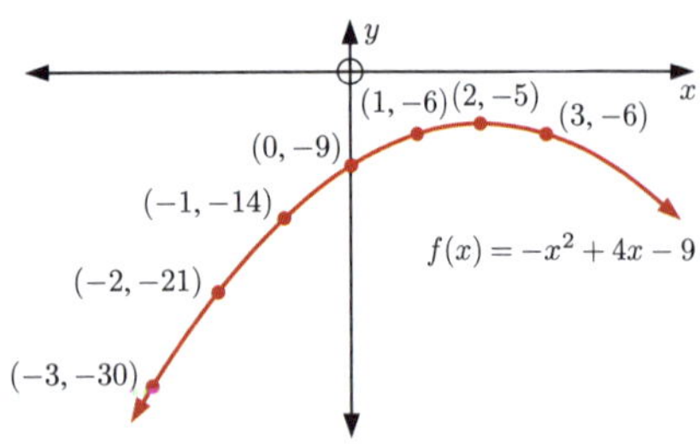

2 a

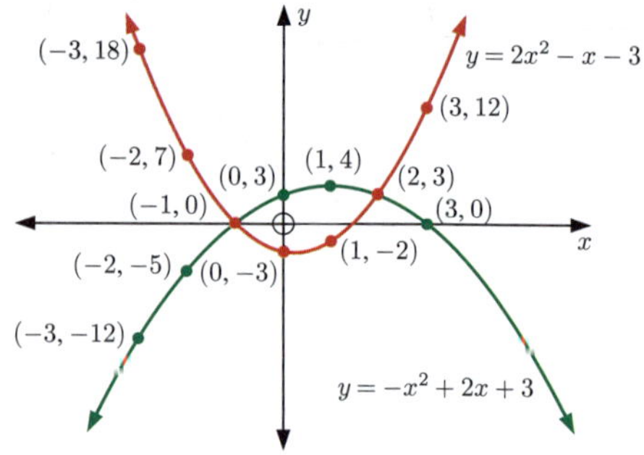

b $x = -1$ or 2 (where they meet) **c** $x = -1$ or 2

3 a

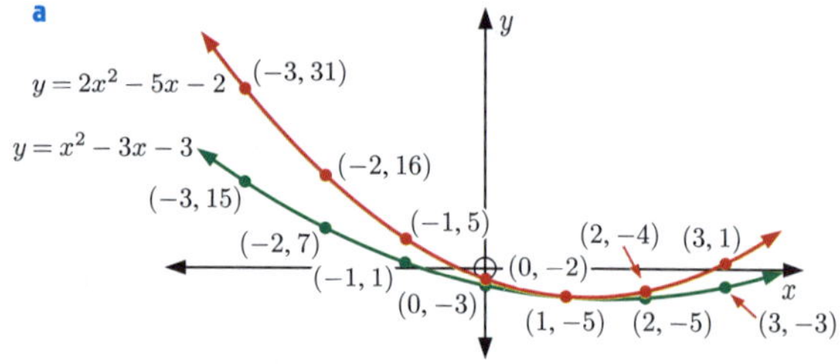

b $x = 1$ (where they meet) **c** $x = 1$

EXERCISE 23C.1

1 a

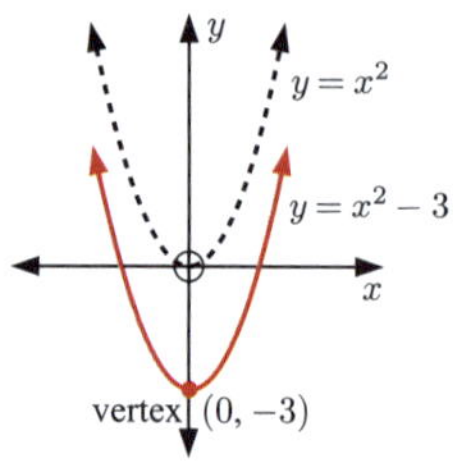

b

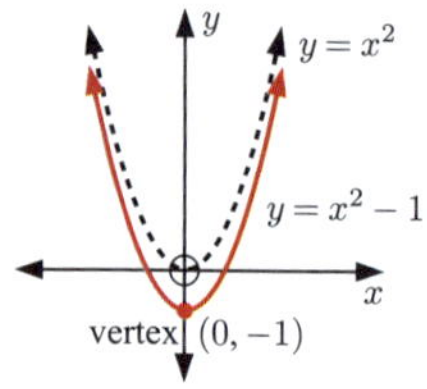

c

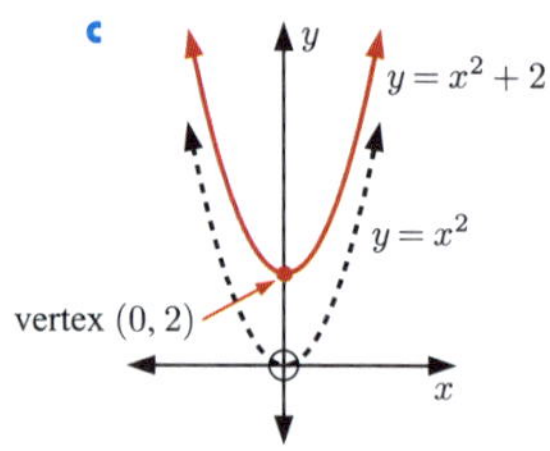

d

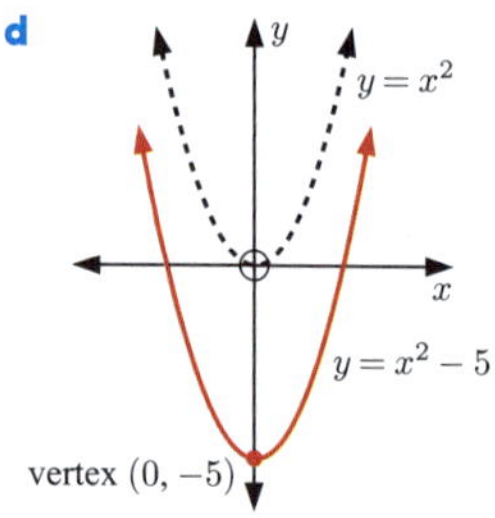

e

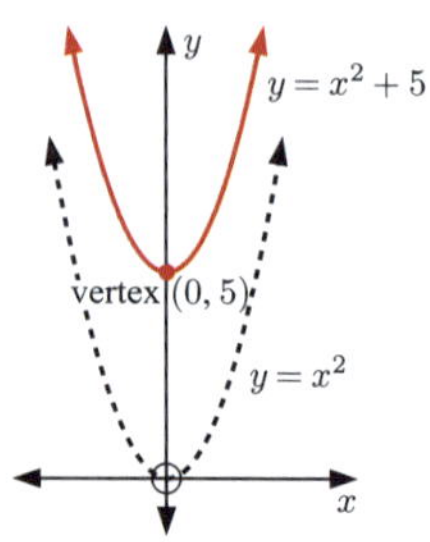

f

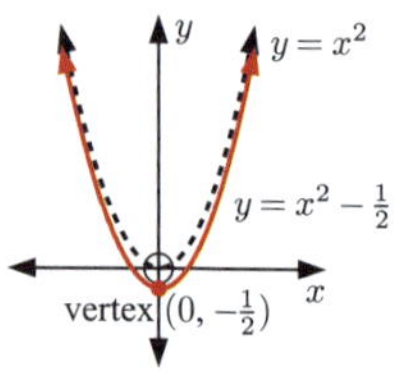

2 **a**

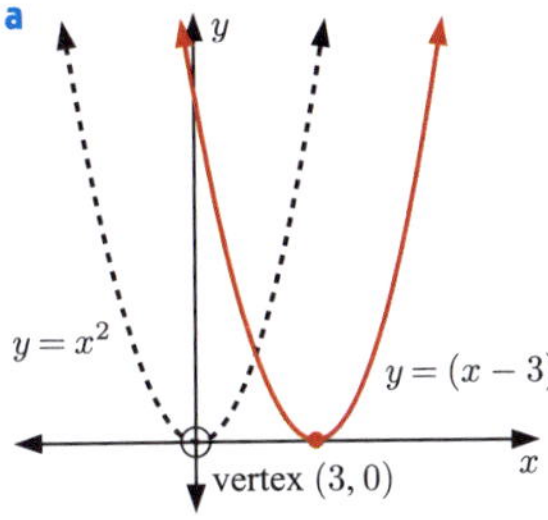

b

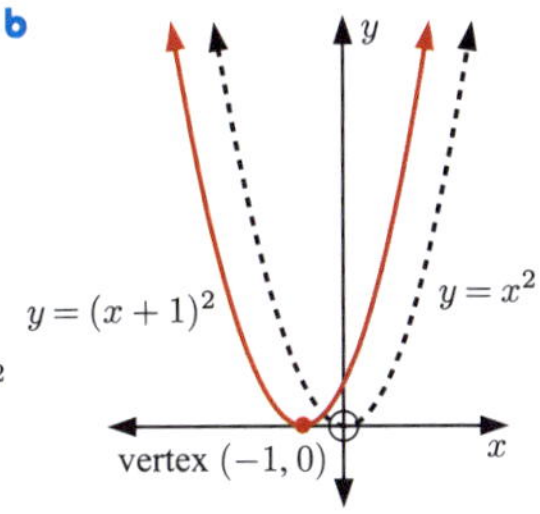

c

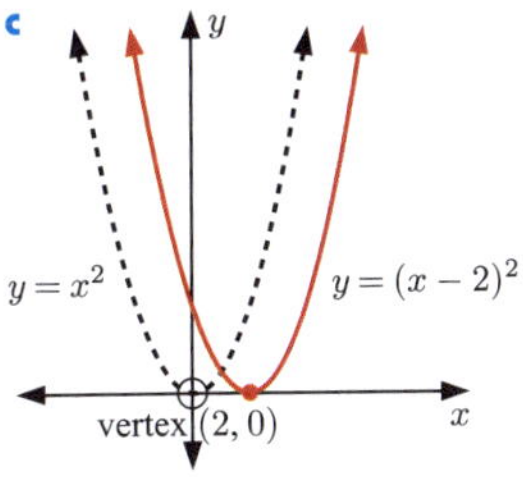

d

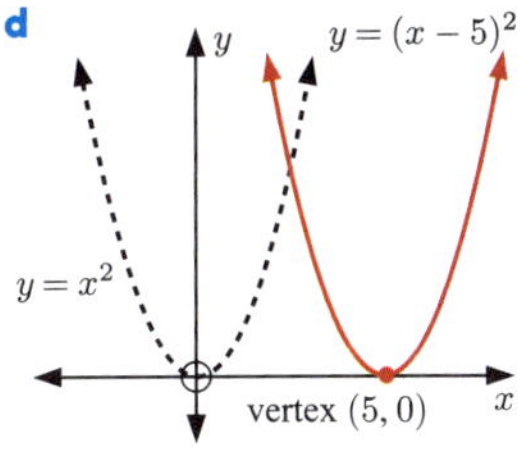

e

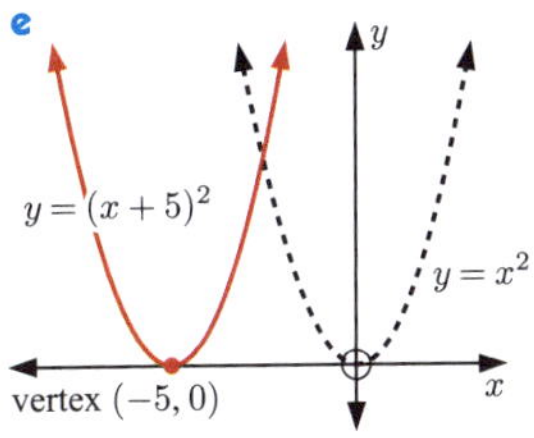

f

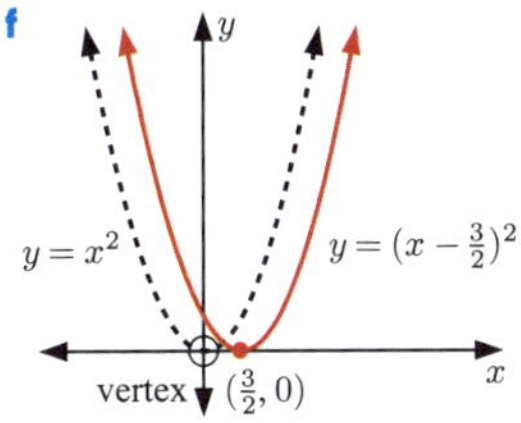

3 **a** $y = (x-1)^2 + 3$

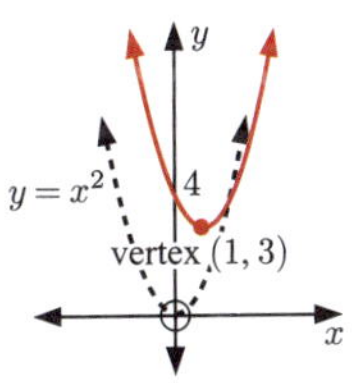

b $y = (x-2)^2 - 1$

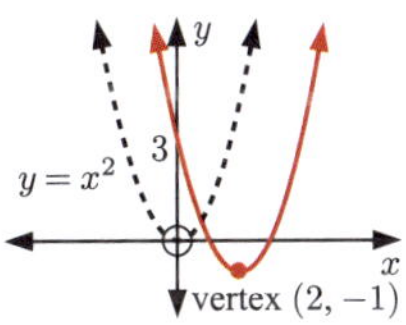

c $y = (x+1)^2 + 4$

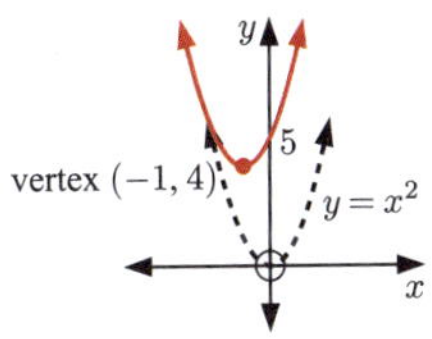

d $y = (x+2)^2 - 3$

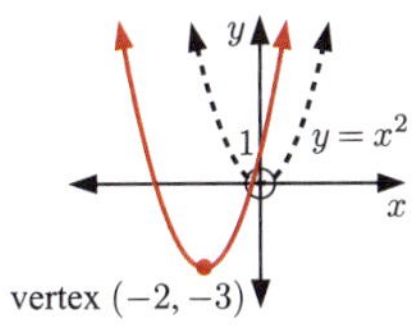

e $y = (x+3)^2 - 2$

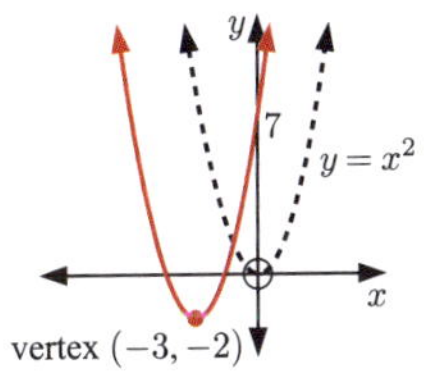

f $y = (x-3)^2 + 3$

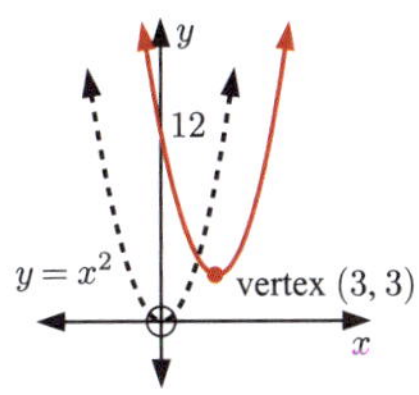

4 **a**

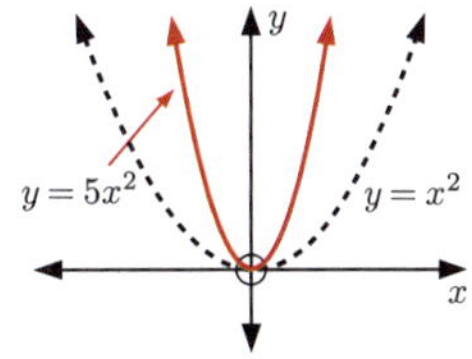

$y = 5x^2$ is "thinner" than $y = x^2$ and the graph opens upwards.

b

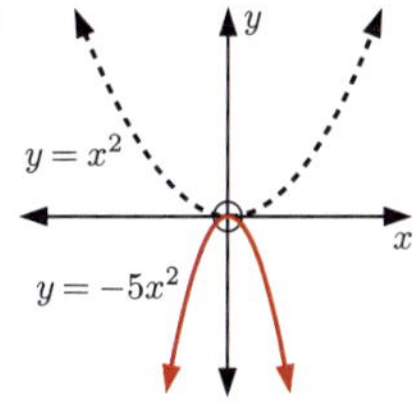

$y = -5x^2$ is "thinner" than $y = x^2$ and the graph opens downwards.

c

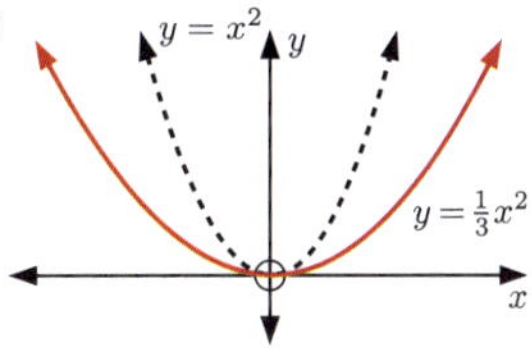

$y = \frac{1}{3}x^2$ is "wider" than $y = x^2$ and the graph opens upwards.

d

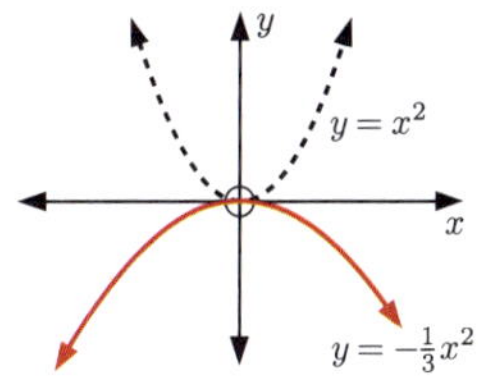

$y = -\frac{1}{3}x^2$ is "wider" than $y = x^2$ and the graph opens downwards.

e

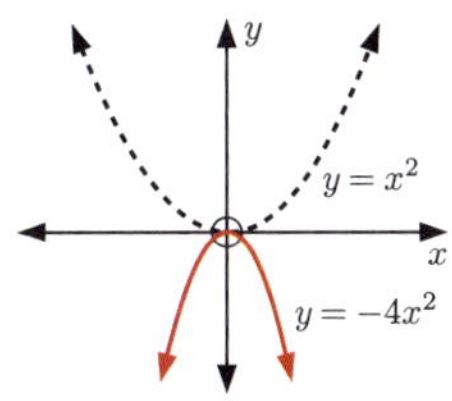

$y = -4x^2$ is "thinner" than $y = x^2$ and the graph opens downwards.

f

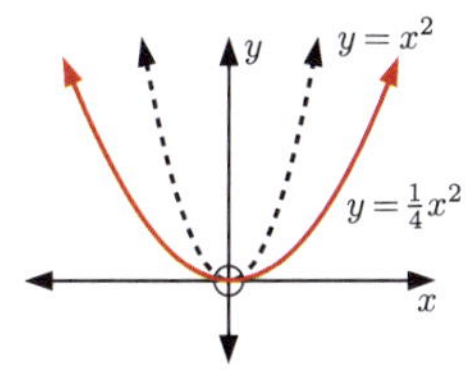

$y = \frac{1}{4}x^2$ is "wider" than $y = x^2$ and the graph opens upwards.

5 **a**

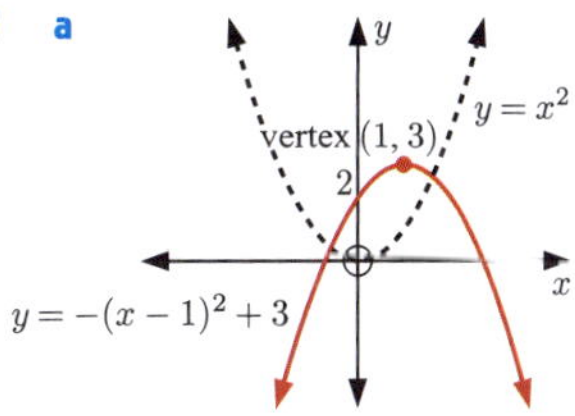

b

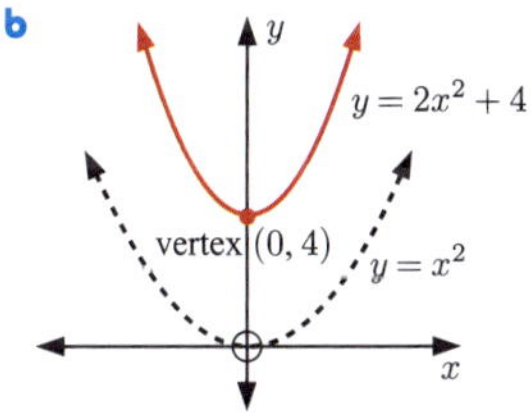

c

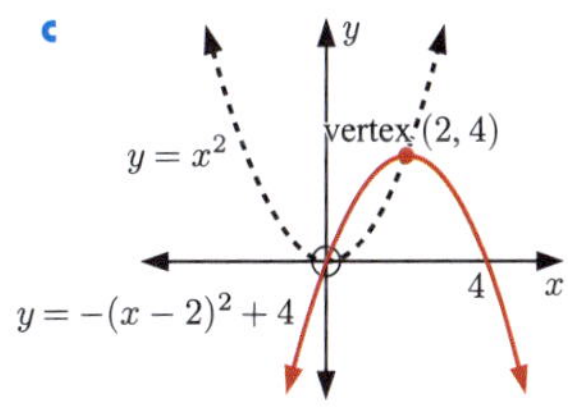

d

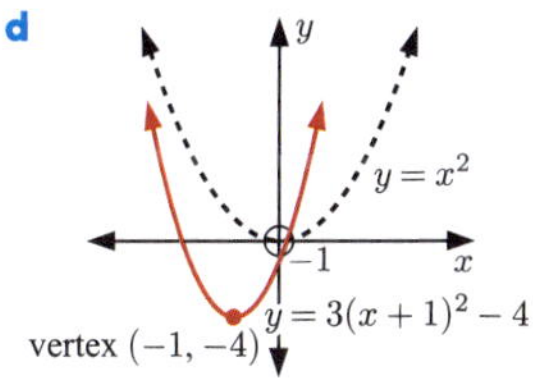

e

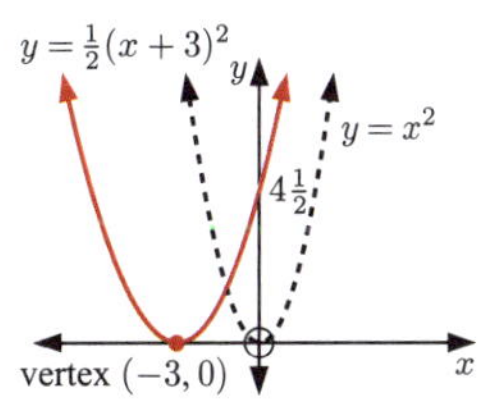

f

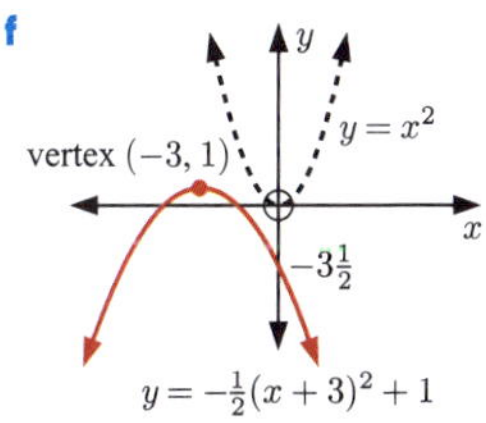

g

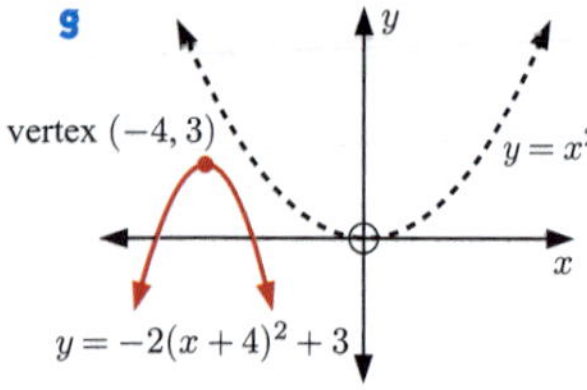

h

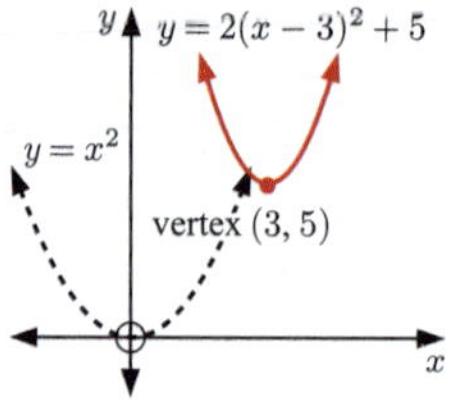

i

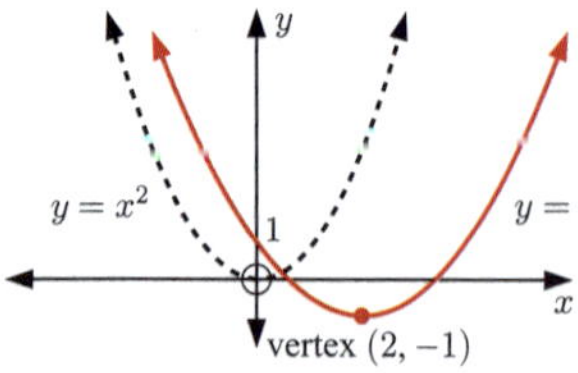

6 **a** G **b** C **c** B **d** D **e** E **f** A **g** F

7 **b**

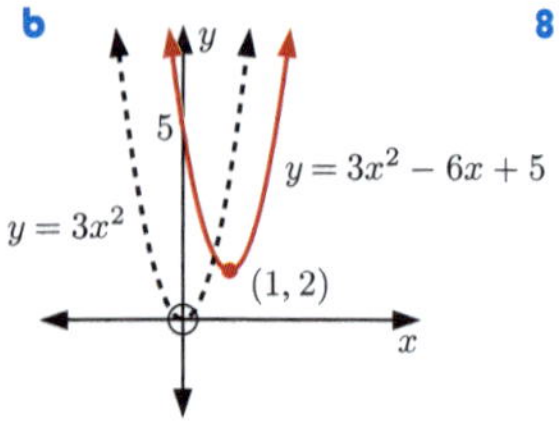

8 **b**

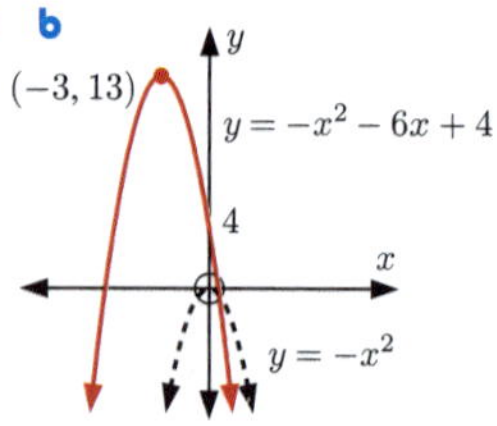

9 **b**

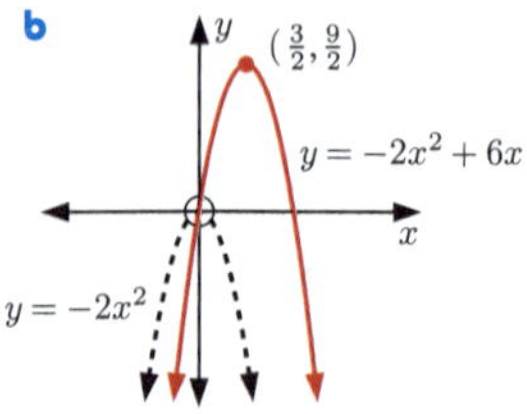

EXERCISE 23C.2

1 **a** $y = (x+1)^2 + 3$

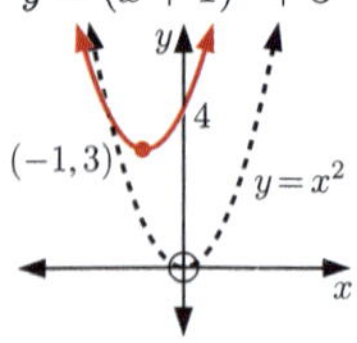

b $y = (x-3)^2 - 6$

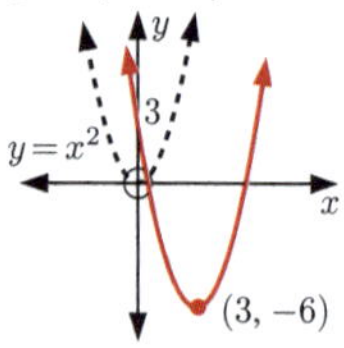

c $y = (x+2)^2 - 5$

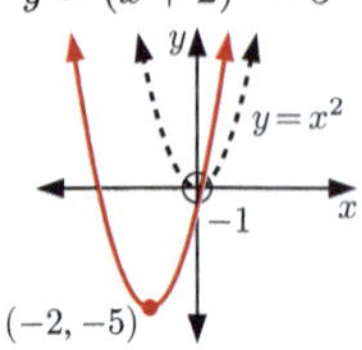

d $y = (x-1)^2 + 4$

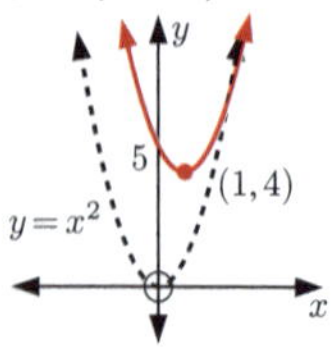

e $y = (x-1)^2 - 1$

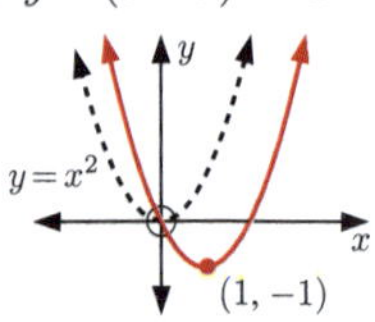

f $y = (x+\frac{5}{2})^2 - 6\frac{1}{4}$

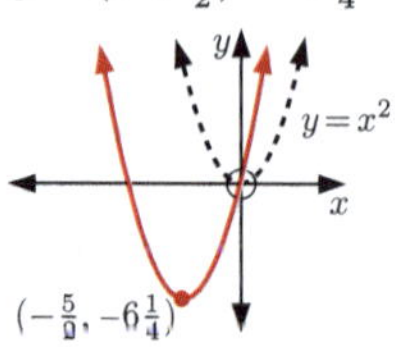

g $y = (x+\frac{5}{2})^2 - 9\frac{1}{4}$

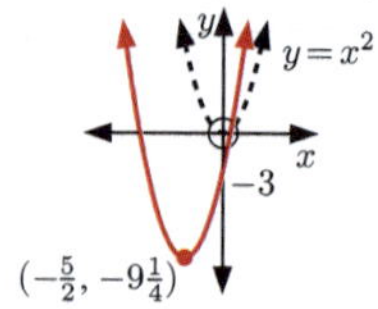

h $y = (x-\frac{3}{2})^2 + \frac{3}{4}$

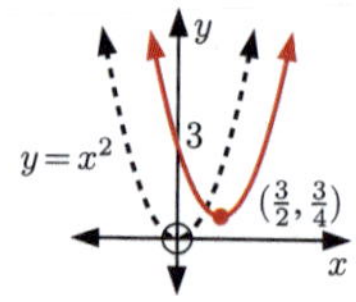

i $y = (x+\frac{1}{2})^2 - 6\frac{1}{4}$

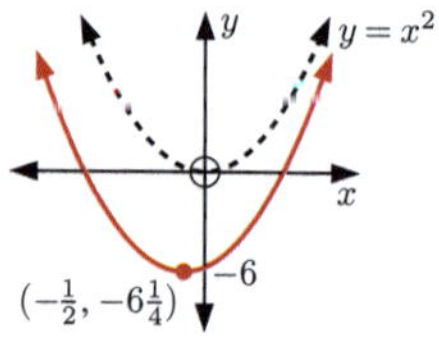

2 **a** $y = 2(x+\frac{5}{2})^2 - \frac{9}{2}$

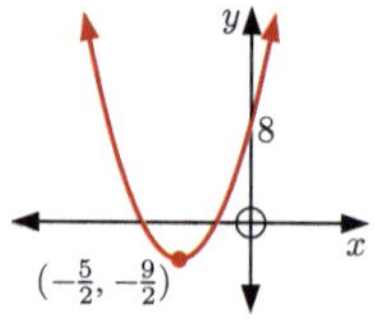

b $y = -(x-\frac{1}{2})^2 + 6\frac{1}{4}$

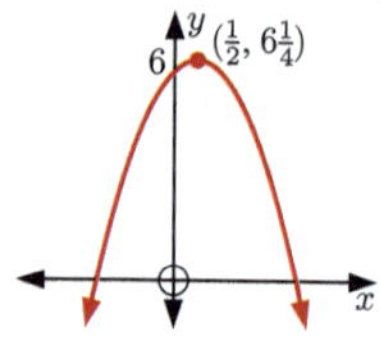

c $y = -3(x+1)^2 + 5$

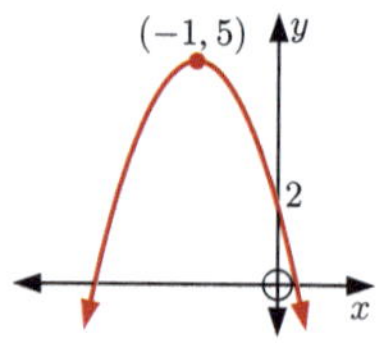

EXERCISE 23D

1 **a** 3 **b** 2 **c** −8 **d** 1 **e** 6 **f** 5 **g** 6 **h** 8 **i** −2

2 **a** x-intercepts 3 and −1, y-intercept −3
b x-intercepts 2 and 4, y-intercept −8
c x-intercepts −3 and −2, y-intercept 12
d x-intercepts 4 and 5, y-intercept −60
e x-intercept −3, y-intercept 18
f x-intercept 1, y-intercept −5

3 **a** −3 and 3 **b** −5 and 5 **c** 0 and 6
d −5 and −2 **e** −4 and 3 **f** 0 and 4
g −2 and −4 **h** −1 **i** 3

4 **a**

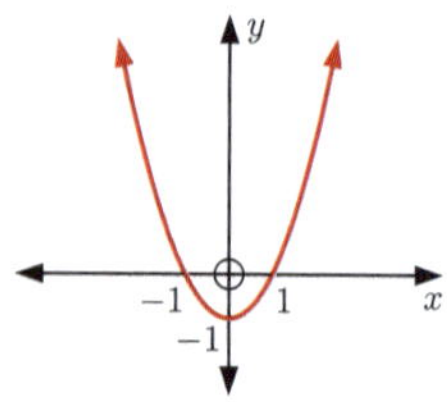

b

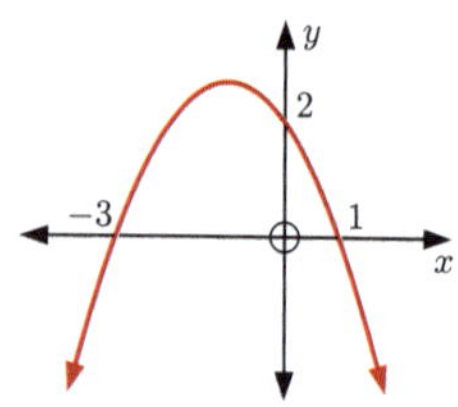

c

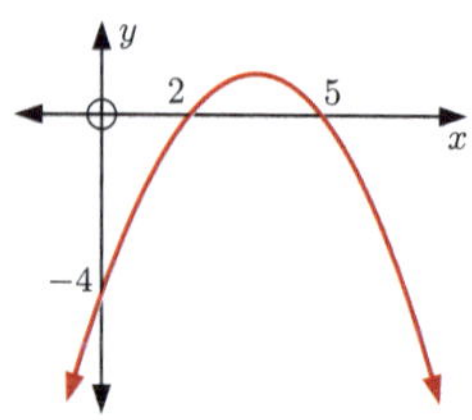

d

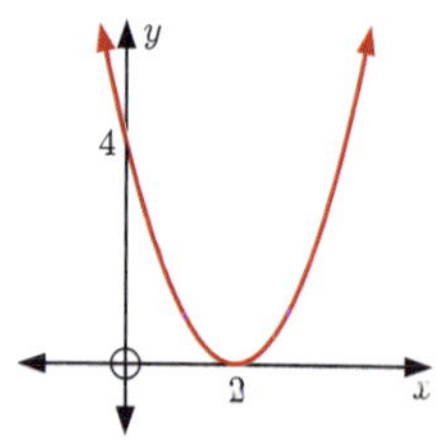

5 **a**
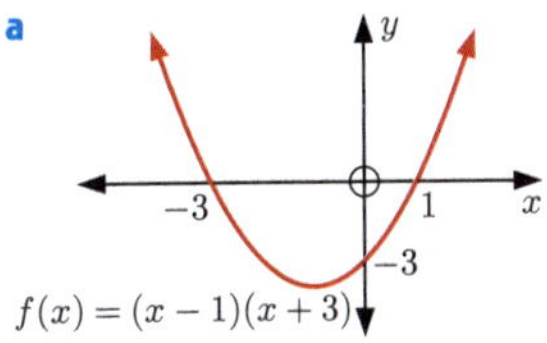

b
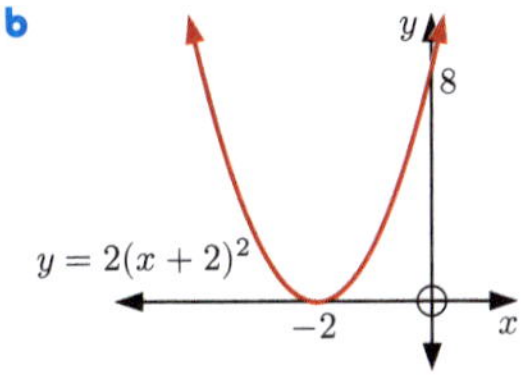

c
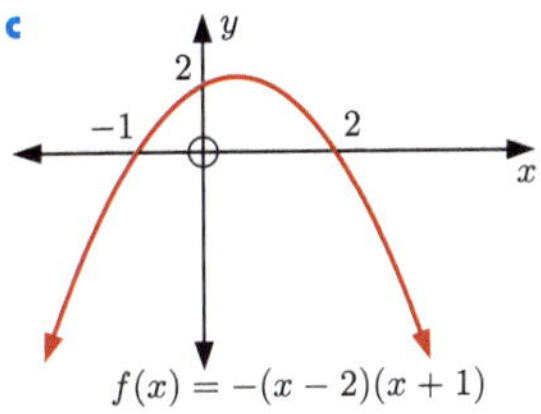

d
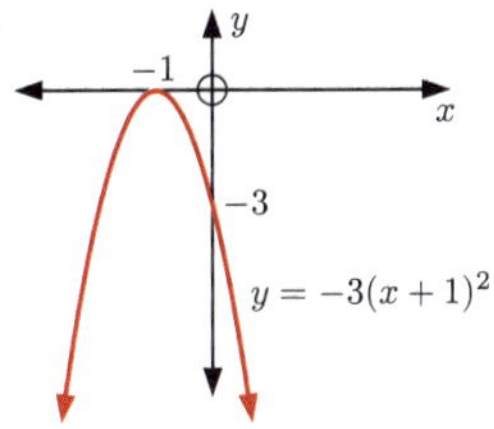

e
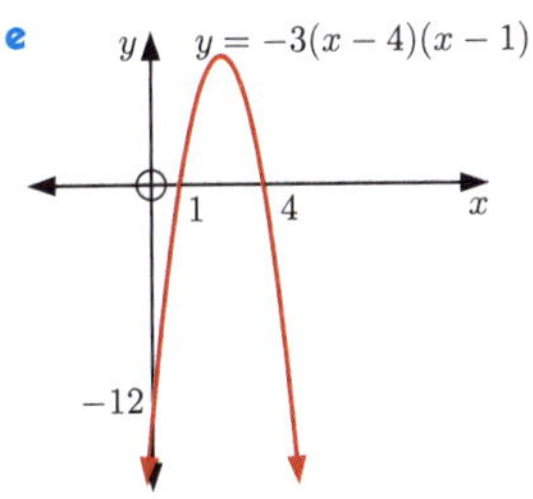

f
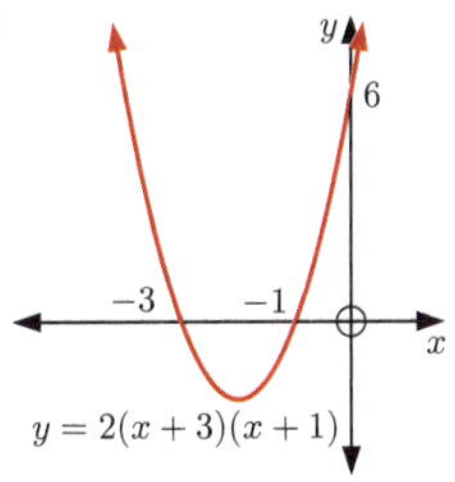

g
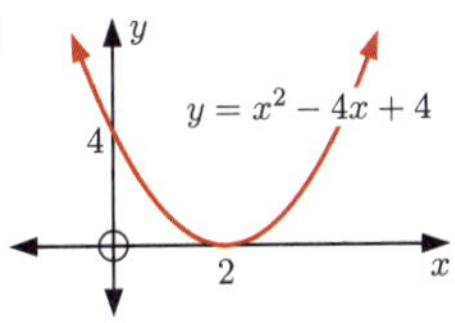

h
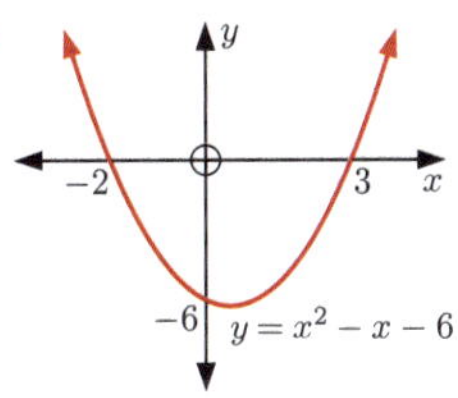

i
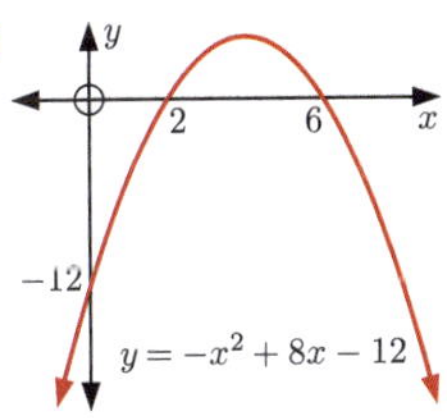

6 **a** $2 \pm \sqrt{3}$ **b** $-2 \pm \sqrt{7}$ **c** $3 \pm \sqrt{5}$

d no x-intercepts **e** $\dfrac{7 \pm \sqrt{73}}{6}$ **f** $\dfrac{1 \pm \sqrt{41}}{4}$

g no x-intercepts **h** $\dfrac{9 \pm \sqrt{33}}{8}$

7 **a** $c < 9$ **b** $c = 9$ **c** $c > 9$

8 **a**
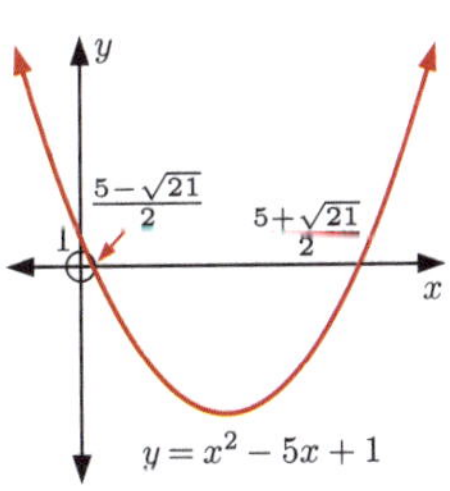

b
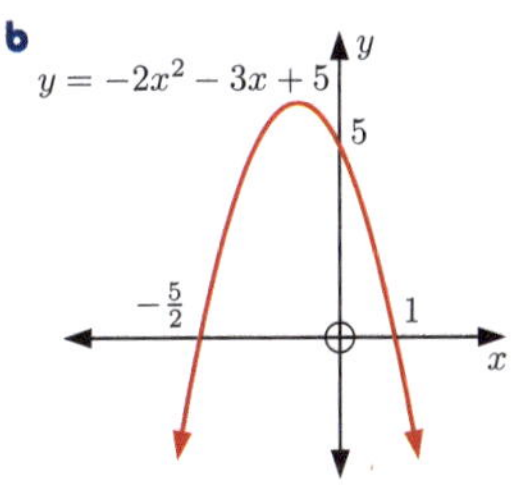

c
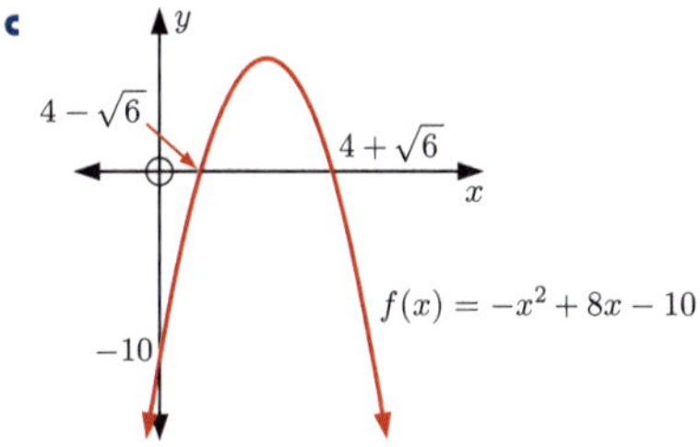

EXERCISE 23E

1 **a** $x = 2$ **b** $x = 1$ **c** $x = -2$ **d** $x = 2$
e $x = -1$ **f** $x = -1$ **g** $x = \frac{3}{2}$ **h** $x = -\frac{5}{2}$
i $x = 2$

2 **a** $x = 3$ **b** $x = 2$ **c** $x = 0$ **d** $x = -\frac{5}{2}$
e $x = -4$ **f** $x = \frac{3}{2}$

3 **a** $x = -2$ **b** $x = \frac{3}{2}$ **c** $x = -\frac{2}{3}$ **d** $x = -2$
e $x = \frac{5}{4}$ **f** $x = 10$ **g** $x = -6$ **h** $x = 12\frac{1}{2}$
i $x = 150$

4 **a** x-intercepts 1 and 3 **b** x-intercepts -2 and 4
c x-intercepts -1 and -2

5 **a**
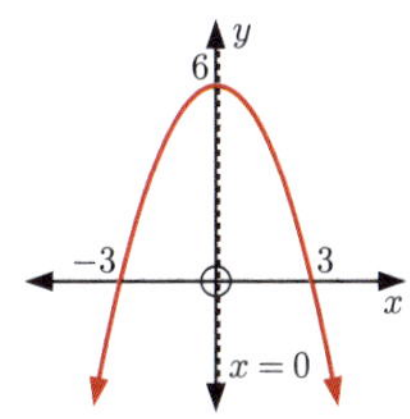

b
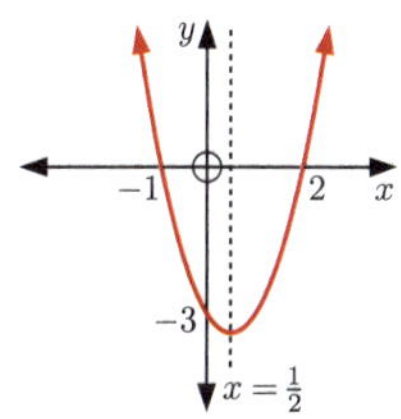

c
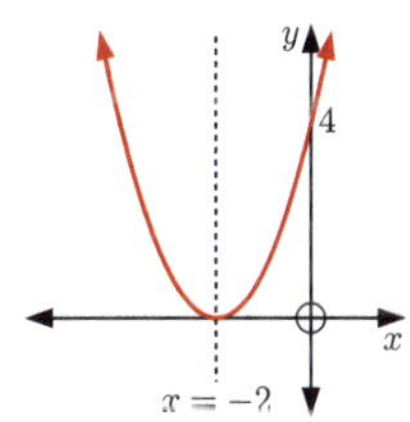

EXERCISE 23F

1 **a** **i** $(2, -2)$ **ii** minimum **iii** $\{y \mid y \geqslant -2\}$
b **i** $(-1, -4)$ **ii** minimum **iii** $\{y \mid y \geqslant -4\}$
c **i** $(0, 4)$ **ii** minimum **iii** $\{y \mid y \geqslant 4\}$
d **i** $(0, 1)$ **ii** maximum **iii** $\{y \mid y \leqslant 1\}$
e **i** $(-2, 0)$ **ii** maximum **iii** $\{y \mid y \leqslant 0\}$
f **i** $(\frac{5}{2}, -\frac{19}{2})$ **ii** minimum **iii** $\{y \mid y \geqslant -\frac{19}{2}\}$

2 **a** **i** x-intercepts -2 and 4, y-intercept -8
ii $x = 1$ **iii** $(1, -9)$
iv
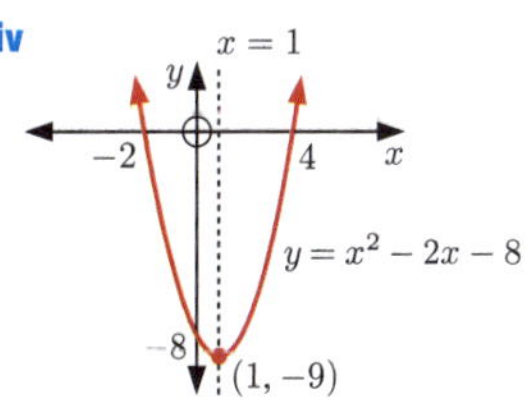

b **i** x-intercepts 0 and 4, y-intercept 0
ii $x = 2$ **iii** $(2, 4)$
iv
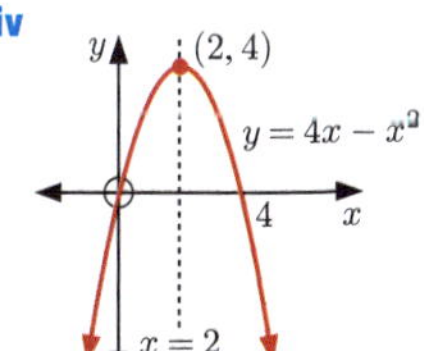

c **i** x-intercepts 0 and -3, y-intercept 0
ii $x = -\frac{3}{2}$
iii $(-\frac{3}{2}, -\frac{9}{4})$
iv

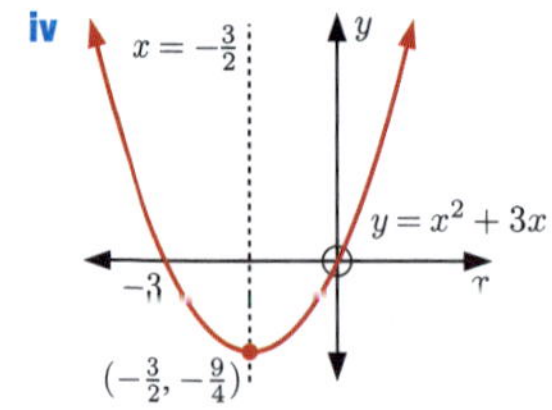

d **i** x-intercept -2, y-intercept 4
ii $x = -2$
iii $(-2, 0)$
iv

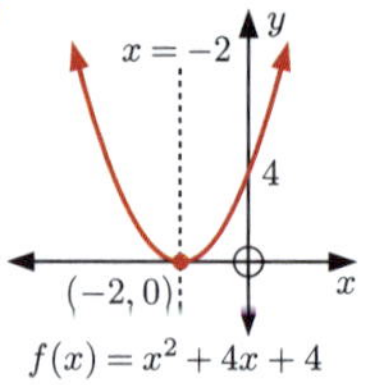

e **i** x-intercepts -4 and 1, y-intercept -4
ii $x = -\frac{3}{2}$
iii $(-\frac{3}{2}, -\frac{25}{4})$
iv

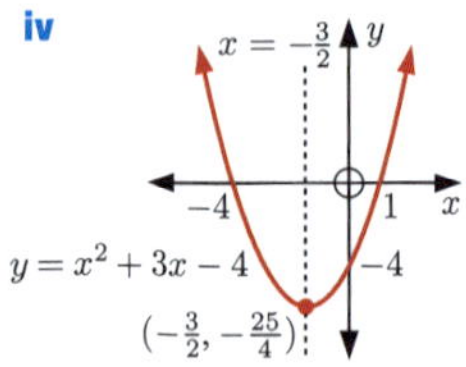

f **i** x-intercept 1, y-intercept -1
ii $x = 1$
iii $(1, 0)$
iv

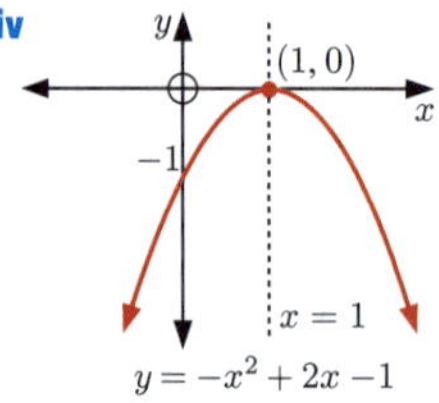

g **i** x-intercepts -3 and $\frac{1}{2}$, y-intercept -3
ii $x = -\frac{5}{4}$
iii $(-\frac{5}{4}, -\frac{49}{8})$
iv

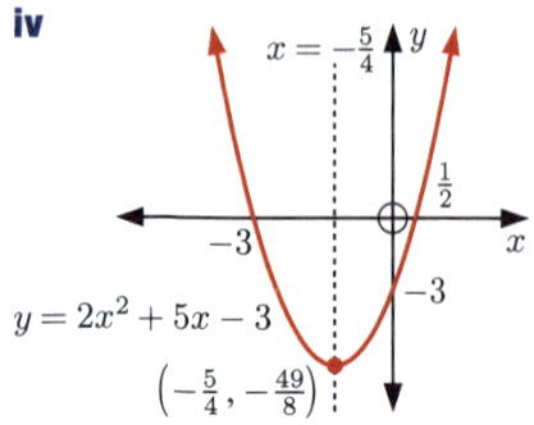

h **i** x-intercepts -2 and $\frac{2}{3}$, y-intercept 4
ii $x = -\frac{2}{3}$
iii $(-\frac{2}{3}, \frac{16}{3})$
iv

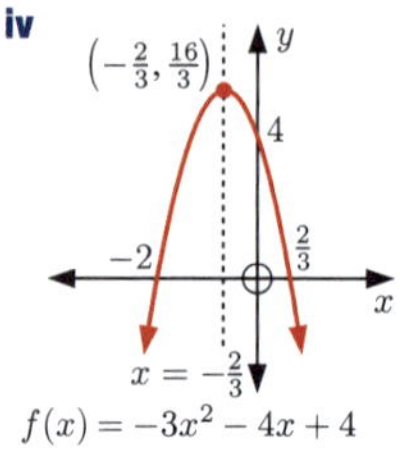

i **i** x-intercepts $3 \pm \sqrt{6}$, y-intercept 3
ii $x = 3$
iii $(3, -6)$

iv

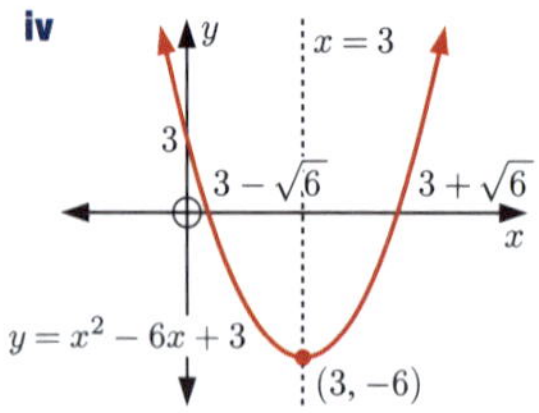

3 **a** **i** x-intercepts 0 and 2, y-intercept 0
ii $x = 1$ **iii** $(1, -1)$
iv

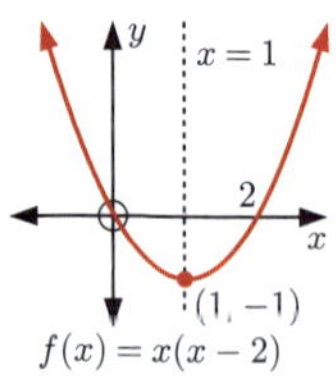

b **i** x-intercept 3, y-intercept 18
ii $x = 3$ **iii** $(3, 0)$
iv

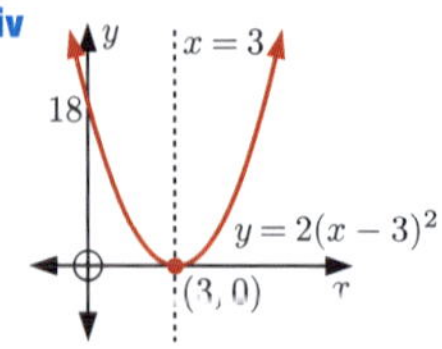

c **i** x-intercepts -3 and 1, y-intercept 3
ii $x = -1$ **iii** $(-1, 4)$
iv

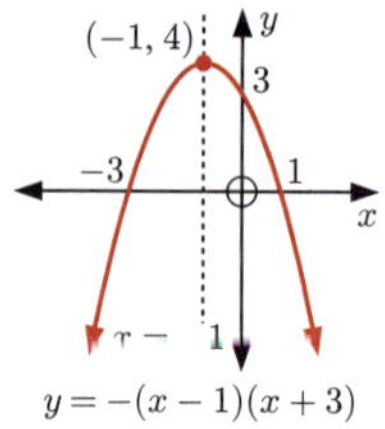

d **i** x-intercept 1, y-intercept -2
ii $x = 1$ **iii** $(1, 0)$
iv

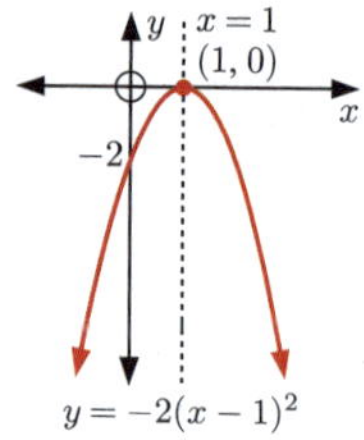

e **i** x-intercepts -2 and 2, y-intercept 20
ii $x = 0$
iii $(0, 20)$
iv

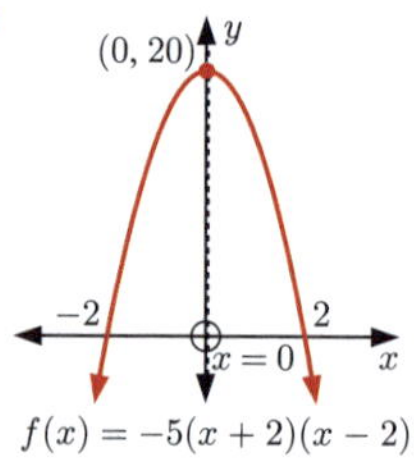

f **i** x-intercepts -1 and -4, y-intercept 8
ii $x = -\frac{5}{2}$
iii $(-\frac{5}{2}, -\frac{9}{2})$
iv

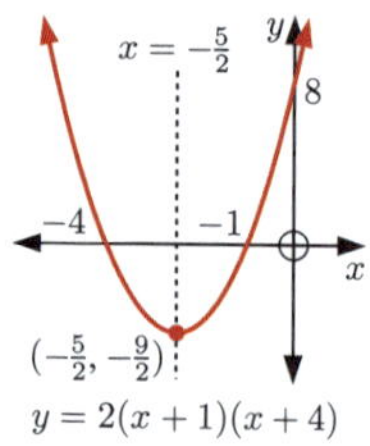

4 **a** $b = -2$, $c = 4$ **b** $b = 0$, $c = 2$ **c** $b = -6$, $c = 9$
d $b = 6$, $c = 7$ **e** $b = -1$, $c = \frac{5}{4}$

EXERCISE 23G

1 **a** $f(x) = 2(x - 2)^2 - 5$ **b** $f(x) = -(x + 4)^2 + 19$
c $f(x) = -(x - 1)^2 + 8$ **d** $f(x) = -2(x + 2)^2 + 11$

2 **a** $y = 2(x - 1)^2 + 3$ **b** $y = -(x + 2)^2 + 3$
c $y = -2(x + 1)^2 - 2$

3 **a** $f(x) = 2(x + 2)^2 - 5$ **b** $f(x) = 2(x - 3)^2 - 19$

4 **a** $f(x) = -3x^2 + 6x - 7$ **b** $f(x) = \frac{4}{3}x^2 + 8x + 7$
c $f(x) = 2x^2 - 8x + 5$ **d** $f(x) = -\frac{1}{2}x^2 - x + \frac{9}{2}$

5 **a** $b = -2$, $c = 0$ **b** $b = 3$, $c = -4$
c $b = 3$, $c = -10$ **d** $b = 7$, $c = 0$

6 **a** $f(x) = -2(x + 2)(x - 2)$ **b** $f(x) = -3(x - 1)(x - 4)$
c $f(x) = 3(x + 2)(x - 3)$ **d** $f(x) = -2(x + 4)(x - 5)$
e $f(x) = (2x - 3)(x - 3)$

7 **a** $f(x) = (x + 2)^2$ **b** $f(x) = -\frac{1}{3}(x - 3)^2$
c $f(x) = 2(x - 1)^2$ **d** $f(x) = -3(x + 4)^2$

8 $f(x) = -\frac{1}{2}(x - 1)^2 + 12$

EXERCISE 23H

1 **a** 2 seconds **b** 20 metres **c** 4 seconds

2 **a** 25 bicycles **b** \$425
c \$200 (Due to fixed daily costs such as wages and electricity.)

3 **a** 30 taxis **b** \$1600/hour **c** \$200

4 **a** 25°C **b** 7:00 am the next day **c** -11°C

5 **a** 2 m **b** 3.8 m **c** 1 s

6 **a** 60 km/h (when $t = 0$) **b** 1 s **c** 66 km/h
d ≈ 4.32 s

7 **d** **i** $x = \frac{1}{2}$ **ii** $\frac{1}{2}$ cm^2

8 **a** **i** 6 **ii** $x = 9$ **b** 27 and -9

c $y = -\frac{2}{81}(x-27)(x+9)$ **d** 8 m

REVIEW SET 23A

1 **a** $y = -11$ **b** $x = 6$ or -3 **2** no

3 **a** 5 **b** $-\frac{5}{3}$

4 **a**

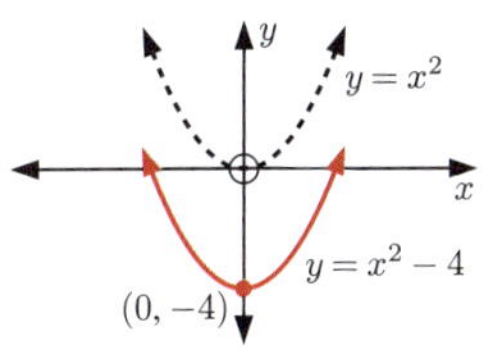

b

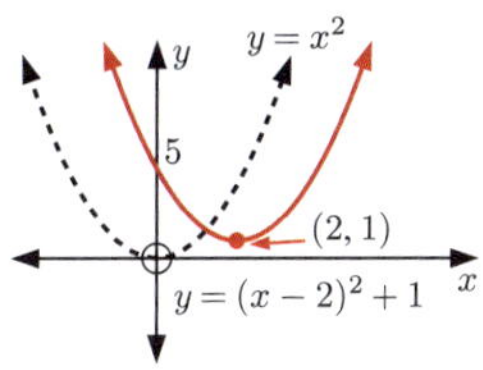

c

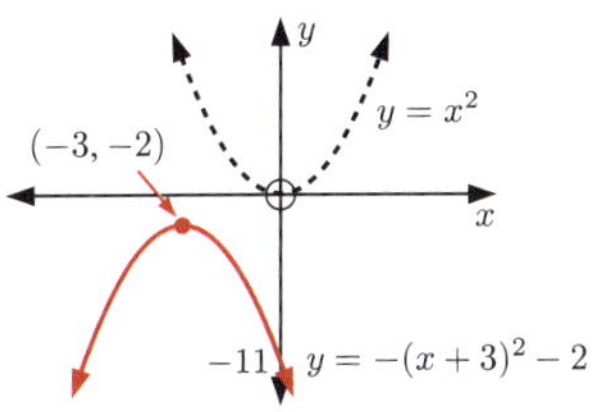

5

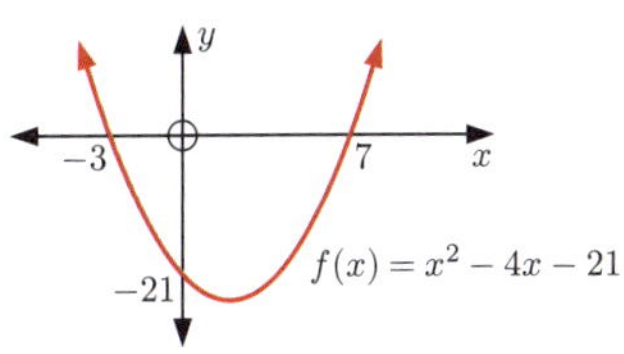

6 **a** $\dfrac{1 \pm \sqrt{61}}{6}$ **b** $1 \pm \sqrt{7}$

7 **a** $x = -\frac{3}{2}$ **b** $x = 2$ **8** **a** $(4, -19)$ **b** $(\frac{1}{2}, -2)$

9 **a** **i** x-intercepts -3 and 1, y-intercept 6
ii $x = -1$
iii $(-1, 8)$
iv

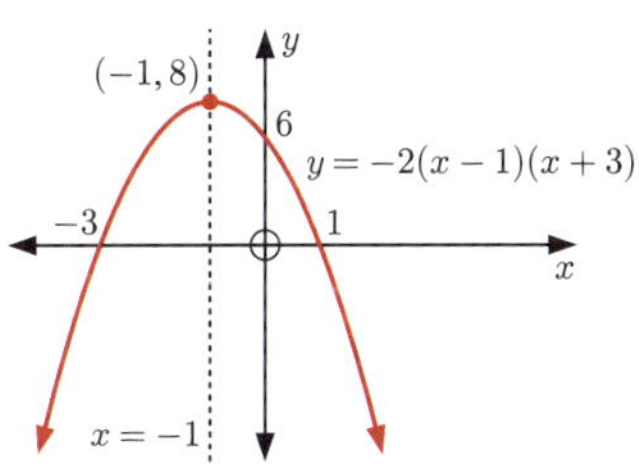

b **i** x-intercepts -3 and 5, y-intercept -15
ii $x = 1$
iii $(1, -16)$
iv

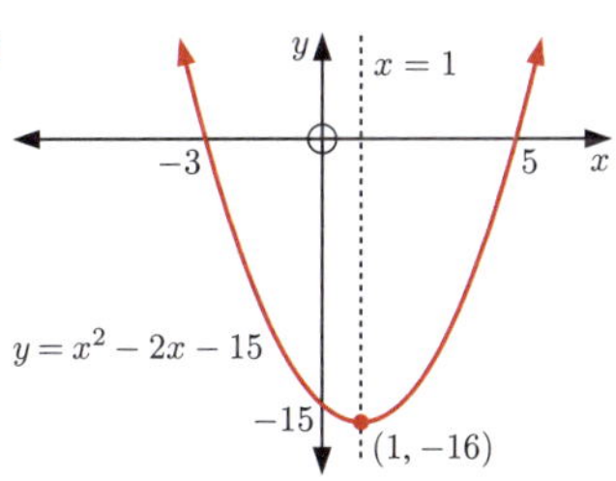

10 $b = 6$, $c = -2$ **11** $f(x) = 2(x+1)^2 - 5$

12 **a** $f(x) = 9x^2 + 39x + 12$ **b** $f(x) = \frac{9}{2}x^2 - 54x + 160$

13 **b** $x = 10$ **c** 200 m^2

14 **a** **i** $H = -5(t - 1.5)^2 + 13$
ii $H = -5t^2 + 15t + 1.75$

b 11.75 m **c** after 0.5 s and 2.5 s

d ≈ 3.11 seconds

REVIEW SET 23B

1 $x = -7$ or 6

2

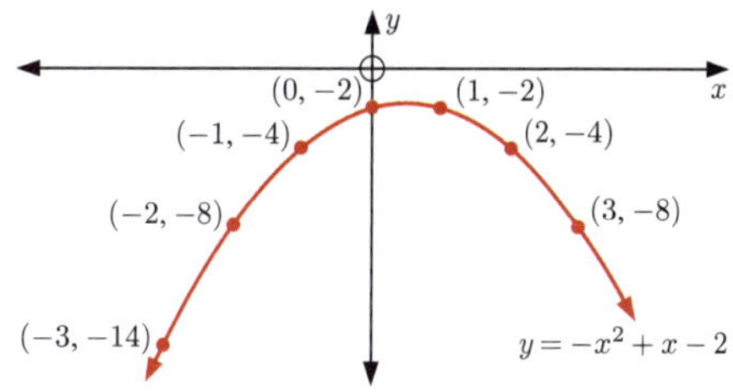

3 **a**

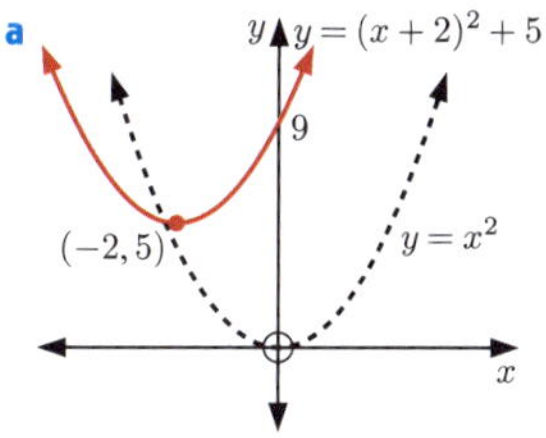

b

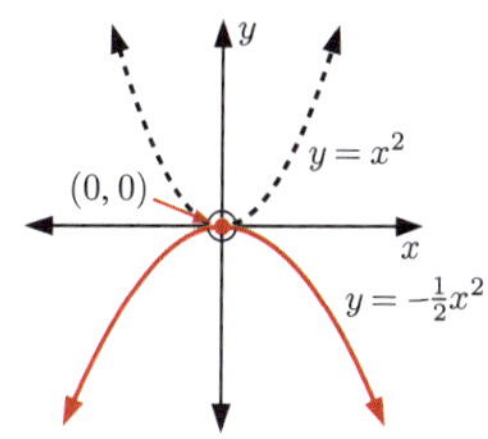

c

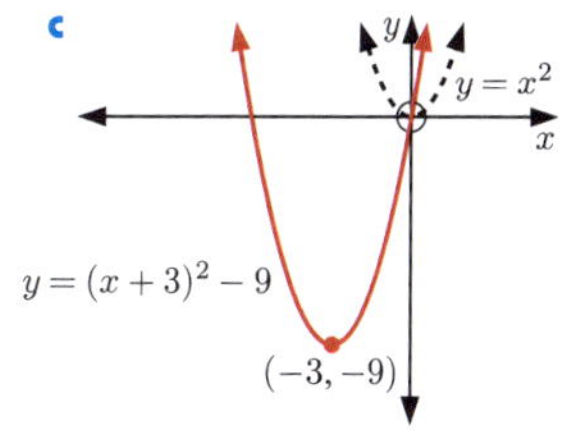

4 **a** $y = x^2 - 4x + 10 = (x-2)^2 + 6$

b

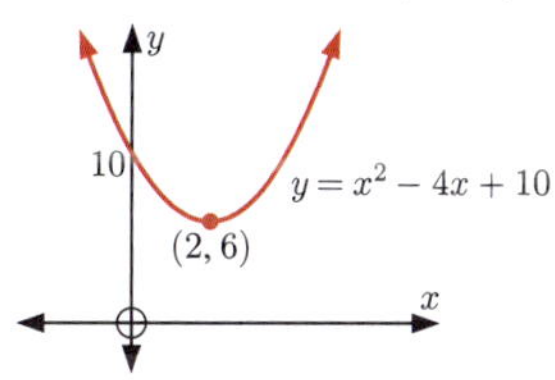

5 **a** x-intercepts 0 and -4, y-intercept 0
b x-intercepts -7 and 4, y-intercept -56
c x-intercept 3, y-intercept 9

6 **a**

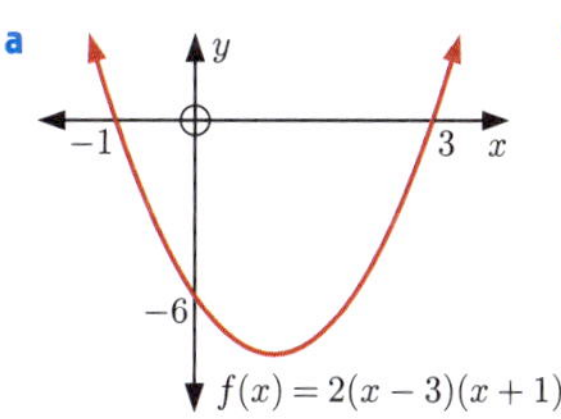

b

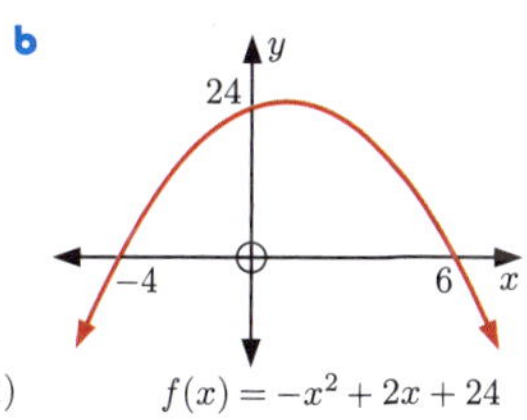

c

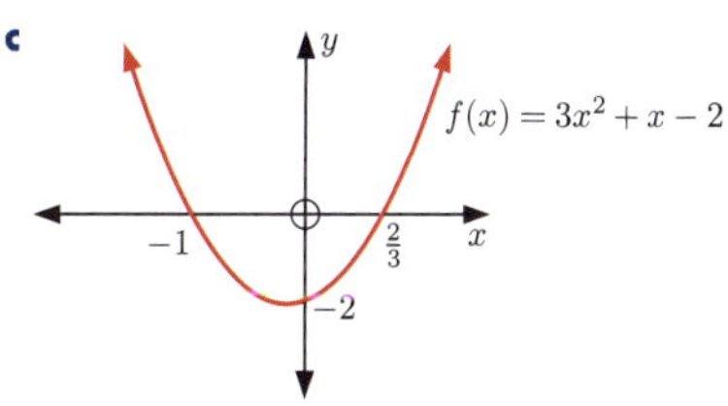

7 **a** $x = 1$ **b** $x = \frac{5}{6}$

8 **a** $(2, -3)$ **b** $\{y \mid y \leqslant -3\}$

9 **a** $f(-1) = 10$
b x-intercepts 1 and 4, y-intercept 4
c $x = \frac{5}{2}$
d $(\frac{5}{2}, -\frac{9}{4})$
e

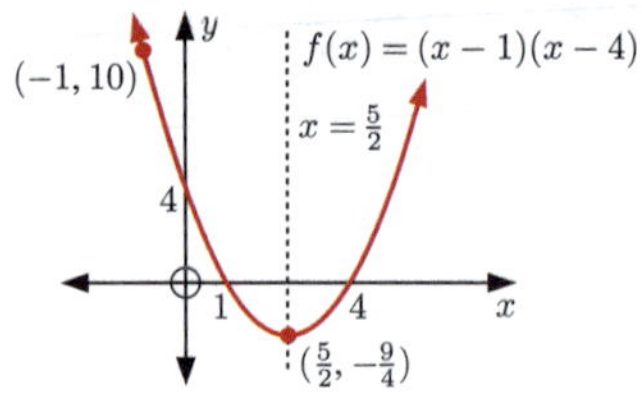

10 $b = 2,\ c = -15$ **11** $f(x) = -2x^2 - 16x - 17$

12 $f(x) = -\frac{1}{2}(x+3)^2$ **13** **a** 2 seconds **b** 80 m

14 **a** $h = 2$ **b** $a = -\frac{3}{4},\ k = \frac{7}{2}$ **c** $(2, \frac{7}{2})$
d $y = -\frac{3}{4}x^2 + 3x + \frac{1}{2}$ **e** $2 \pm \frac{1}{3}\sqrt{42}$

EXERCISE 24A

1 **a** Start at 8 and subtract 3 each time.
b $u_2 = 5,\ u_4 = -1$ **c** $u_5 = -4,\ u_6 = -7$

2 **a** 3, 7, 11, 15, 19 **b** 40, 35, 30, 25, 20 **c** 2, 3, 5, 7, 11

3 **a** Start at 5 and add 3 each time.
b $u_1 = 5,\ \ u_n = u_{n-1} + 3$ for $n \geqslant 2$

4 **a** The nth term of the sequence is the nth square number.
b $u_n = n^2$

5 **a** Start at 2 and add 4 each time. $u_6 = 22,\ u_7 = 26$
b Start at 20 and subtract 4 each time. $u_6 = 0,\ u_7 = -4$
c Start at 8 and add 11 each time. $u_6 = 63,\ u_7 = 74$
d Start at 1 and multiply by 2 each time. $u_6 = 32,\ u_7 = 64$
e Start at $\frac{1}{81}$ and multiply by 3 each time. $u_5 = 1,\ u_6 = 3$
f Start at 405 and divide by 3 each time. $u_5 = 5,\ u_6 = \frac{5}{3}$
g The nth term is the nth cubic number. $u_5 = 125$, $u_6 = 216$
h The nth term is the reciprocal of the nth square number. $u_5 = \frac{1}{25},\ u_6 = \frac{1}{36}$
i The first two terms are 1 and 2, and each term thereafter is the product of the previous two terms. $u_6 = 32,\ u_7 = 256$

6 **a** 5, 7, 9, 11 **b** 8, 11, 14, 17 **c** 13, 9, 5, 1
d 0, 3, 8, 15 **e** 2, 6, 12, 20 **f** 0, 7, 26, 63
g 10, 20, 40, 80 **h** 4, 12, 36, 108 **i** 24, 12, 6, 3

7 **a** 1, 3, 6, 10, 15 **b** **ii** 55

8 **a** 7, 10, 13, 16 **b** 25, 21, 17, 13 **c** 5, 15, 45, 135
d 100, 50, 25, 12.5 **e** 3, 5, 9, 17
f 4, 6, 4, 6 **g** 3, 4, 12, 48

9 **a** $u_n = 2n$
b **i** $u_n = 2n + 2$ **ii** $u_n = 2n - 1$ **iii** $u_n = \dfrac{1}{2n}$

10 **a** $u_n = 5n$
b **i** $u_n = 5n + 1$ **ii** $u_n = 5n - 2$ **iii** $u_n = \dfrac{1}{5n}$

11 **a** $u_n = 2^n$ **b** **i** $u_n = 3 \times 2^n$ **ii** $u_n = \dfrac{1}{2^n}$

12 **a** $u_n = n$ **b** $u_n = n + 2$ **c** $u_n = \dfrac{1}{n}$
d $u_n = \dfrac{1}{n+1}$ **e** $u_n = \dfrac{n}{n+1}$ **f** $u_n = n(n+1)$
g $u_n = (n+1)(n+2)$ **h** $u_n = \dfrac{3n-2}{3n}$

13 **a** $u_n = \frac{1}{2}n(n-3),\ \ n \geqslant 3$
b **i** 20 diagonals **ii** 35 diagonals **iii** 170 diagonals

EXERCISE 24B

1 **a** arithmetic, $d = 3$ **b** not arithmetic
c not arithmetic **d** arithmetic, $d = -7$

2 **a** $\square = 16$ **b** $\square = 34$ **c** $\square = 15,\ \triangle = 3$
d $\square = 13,\ \triangle = -5$

3 **a** $u_1 = 41,\ d = 1$ **b** $u_1 = 1,\ d = 11$
c $u_1 = 98,\ d = -10$ **d** $u_1 = 91,\ d = -9$

4 **a** common difference is 7 **b** $u_n = 7n - 3$
c $u_{30} = 207$ **d** yes, the 49th term **e** no

5 **a** common difference is -4 **b** $u_n = 71 - 4n$
c $u_{60} = -169$ **d** no **e** yes, the 39th term

6 **a** $u_{n+1} - u_n = \frac{11}{2}$, which is constant.
$\therefore$ the sequence is arithmetic.
b $u_1 = 2,\ d = \frac{11}{2}$ **c** $u_{20} = \frac{213}{2} = 106\frac{1}{2}$
d $u_{28} = \frac{301}{2} = 150\frac{1}{2}$

7 **a** $k = 22$ **b** $k = \frac{5}{2}$ **c** $k = \frac{13}{3}$

8 **a** $k = 6$ **b** 29, 37

9 $k = -2\ \ \{-4, 0, 4\}$ or $k = 3\ \ \{1, 5, 9\}$

10 **a** $u_n = 5n + 17$ **b** $u_n = 10 - 4n$ **c** $u_n = 3n - 16$
d $u_n = -\frac{17}{2} - \frac{3}{4}n$

11 **a** $u_n = 6n - 8$ **b** $u_{30} = 172$

EXERCISE 24C

1 **a** $r = 2$ **b** $r = \frac{1}{2}$ **c** $r = -\frac{1}{10}$

2 **a** geometric, $r = 2$ **b** geometric, $r = \frac{1}{10}$
c not geometric **d** geometric, $r = \frac{1}{3}$
e not geometric **f** geometric, $r = -5$

3 **a** 12, 24 **b** $\frac{1}{2}, \frac{1}{8}$ **c** $\frac{5}{3}, -\frac{5}{9}$

4 **a** $r = 3$ **b** $u_n = 3^{n-1}$ **c** $u_{10} = 19\,683$

5 **a** $r = -\frac{1}{2}$ **b** $u_n = 40 \times (-\frac{1}{2})^{n-1}$ **c** $u_{12} = -\frac{5}{256}$

6 **a** $r = -\frac{1}{4}$ **b** $-0.000\,976\,562\,5$ **7** $u_n = 3\left(\sqrt{2}\right)^{n-1}$

8 **a** $k = \frac{2}{3}$ **b** $k = 9$ **c** $k = 2$ **d** $k = \pm 9$
e $k = -7$ **f** $k = -\frac{8}{7}$ or 2

9 $k = -1,\ r = -3$ or $k = 8,\ r = \frac{3}{2}$

10 **a** $u_n = \frac{16}{9} \times 3^{n-1}$ **b** $u_n = 128 \times (-\frac{1}{2})^{n-1}$
c $u_n = 2 \times 5^{n-1}$ or $u_n = (-2) \times (-5)^{n-1}$
d $u_n = 6 \times (\pm\frac{1}{\sqrt{2}})^{n-1}$

11 $u_{12} = \pm\frac{3}{32}$

EXERCISE 24D.1

1 £5360.38 **2** **a** $14 917.80 **b** $1917.80 **3** ¥98 414

4 **a** **i** $2166.53 **ii** $4802.44
b No, more interest is earned in the latter half of the investment period because the value of the investment is greater.

5 €9873 **6** $6300 **7** **a** $17 525.94 **b** $2474.06

EXERCISE 24D.2

1 $5684.29 **2** **a** £1898.44 **b** £2601.56

3 **a** $721.79 **b** $1778.21 **4** $37 200

REVIEW SET 24A

1 **a** 8, 13, 18, 23, 28 **b** 19, 12, 5, −2, −9
c 1, 8, 27, 64, 125

2 **a** Start at 6 and add 4 each time. $u_6 = 26$, $u_7 = 30$
b Start at 810 and divide by 3 each time. $u_5 = 10$, $u_6 = \frac{10}{3}$
c Start at $\frac{1}{4}$ and multiply by 4 each time. $u_5 = 64$, $u_6 = 256$

3 2, 8, 20, 44

4 **a** $d = 4$ **b** $u_n = 4n - 10$ **c** $u_{100} = 390$ **d** no

5 **a** $k = 6$ **b** $d = 7$ **6** **a** $r = \frac{4}{5}$ **b** $r = -3$

7 $u_n = -3 \times (-2)^{n-1}$

8 **a** £171 475.73 **b** £51 475.73 **9** $3718.14

REVIEW SET 24B

1 **a** Start at 19 and subtract 3 each time.
b $u_1 = 19$, $u_n = u_{n-1} - 3$, $n \geqslant 2$

2 **a** 57, 53, 49, 45, 41 **b** −9, −6, −1, 6, 15

3 **a** 5, 7, 5, 7, 5 **b** $u_{100} = 7$

4 **a** $u_n = 4n$ **b** **i** $u_n = 4n - 3$ **ii** $u_n = \dfrac{1}{4n-1}$

5 **a** $\square = 25$ **b** $\square = 21$, $\triangle = 9$

6 **a** $u_n = \frac{93}{2} - \frac{15}{2}n$ **b** $u_{50} = -\frac{657}{2}$

7 **a** $r = -\frac{1}{2}$ **b** $u_n = 64 \times (-\frac{1}{2})^{n-1}$ **c** $u_{16} = -\frac{1}{512}$

8 $k = -5$, $r = -2$ or $k = 10$, $r = -\frac{1}{3}$ **9** $8600

10 **a** $1144.29 **b** $2355.71

EXERCISE 25A

1 **a**, **c**, **d**, and **f** are exponential functions.

2 **a** 3 **b** 11 **c** $2\frac{1}{3}$ **d** $3^{2x} + 2$

3 **a** −2 **b** $-2\frac{4}{5}$ **c** 22 **d** $5^x - 3$

4 **a** $\frac{1}{9}$ **b** 9 **c** $\frac{1}{27}$ **d** 3^{x+3}

EXERCISE 25B

1 **a**

x	−3	−2	−1	0	1	2	3
y	$\frac{1}{27}$	$\frac{1}{9}$	$\frac{1}{3}$	1	3	9	27

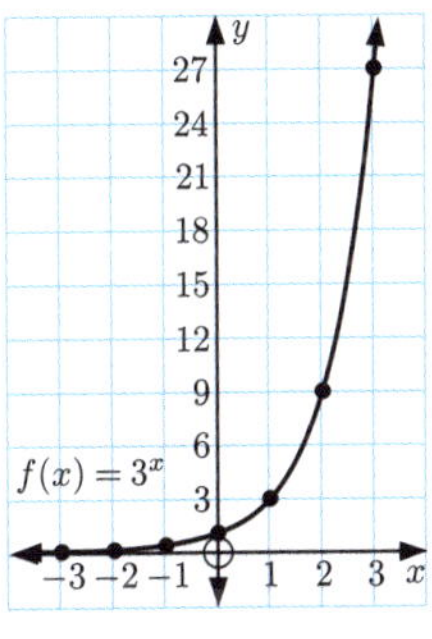

b

x	−3	−2	−1	0	1	2	3
y	$\frac{1}{64}$	$\frac{1}{16}$	$\frac{1}{4}$	1	4	16	64

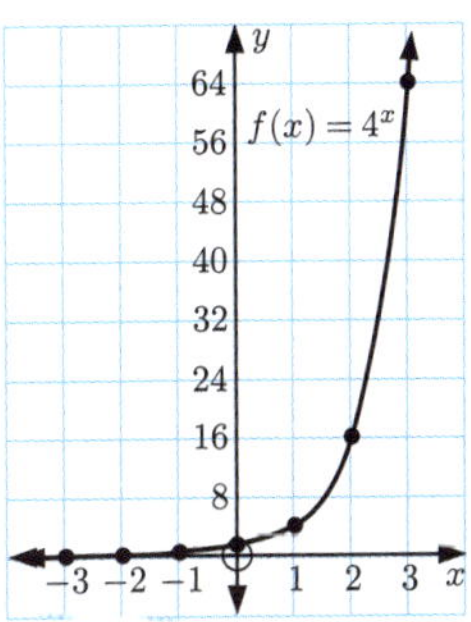

c

x	−3	−2	−1	0	1	2	3
y	8	4	2	1	$\frac{1}{2}$	$\frac{1}{4}$	$\frac{1}{8}$

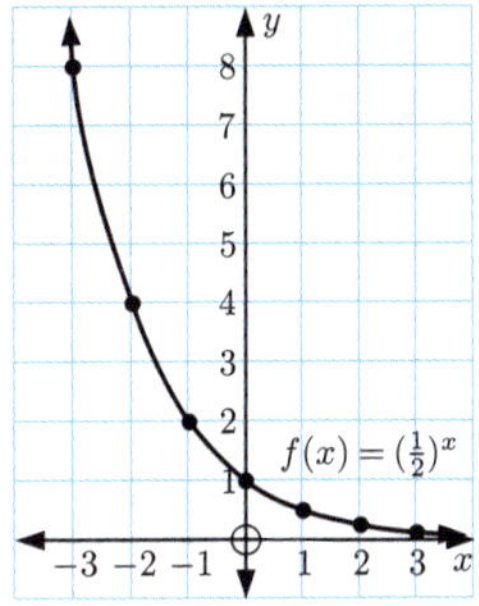

d

x	−3	−2	−1	0	1	2	3
y	27	9	3	1	$\frac{1}{3}$	$\frac{1}{9}$	$\frac{1}{27}$

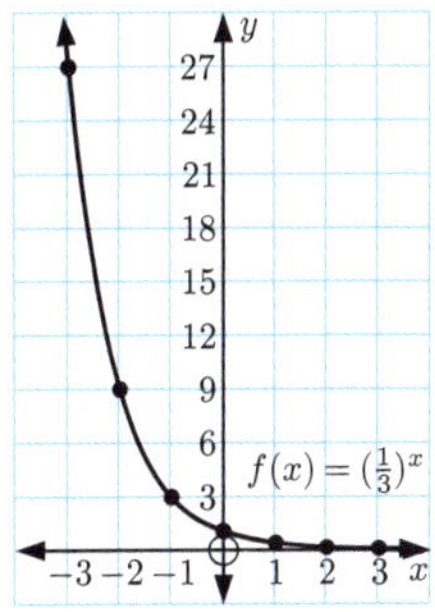

2

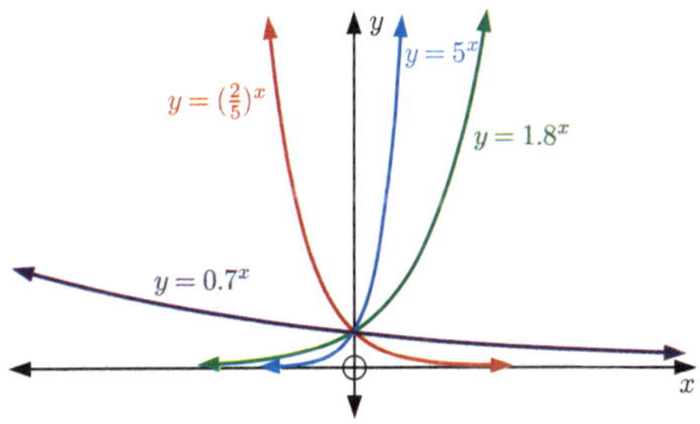

3 **a** D **b** A **c** F **d** C **e** B **f** E

4 **a** A vertical translation of $y = f(x)$ through k units.

b **i**

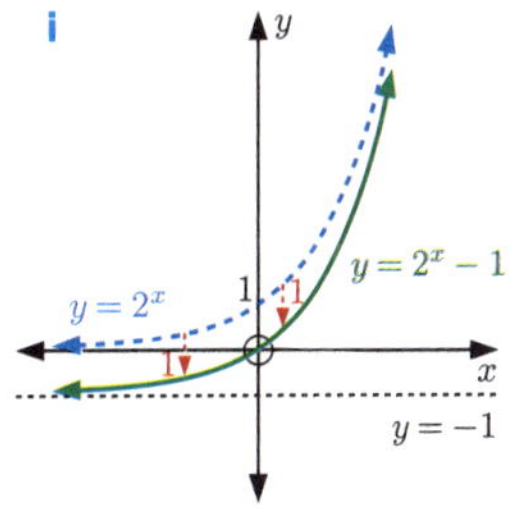

ii

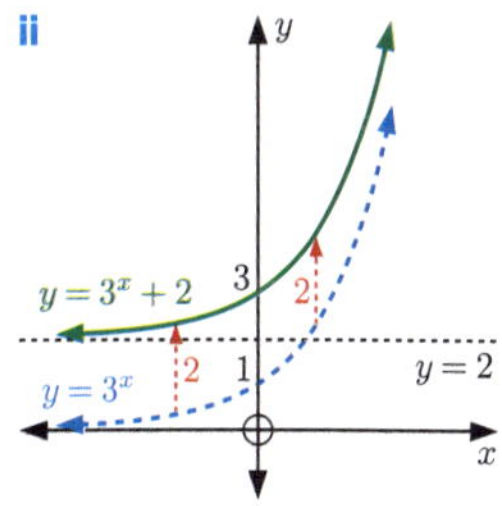

iii

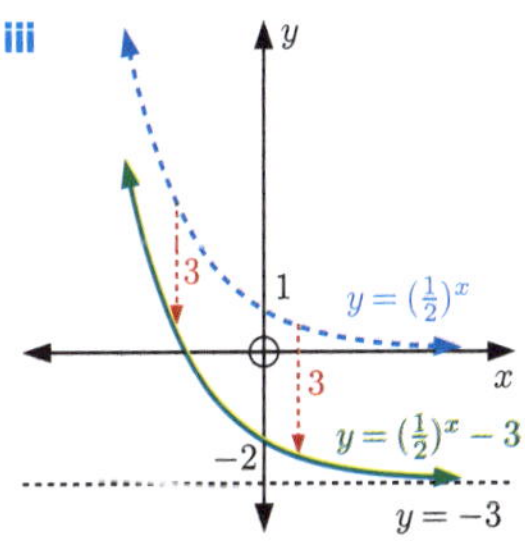

5 **a** A horizontal translation of $y = f(x)$ through h units.

b **i**

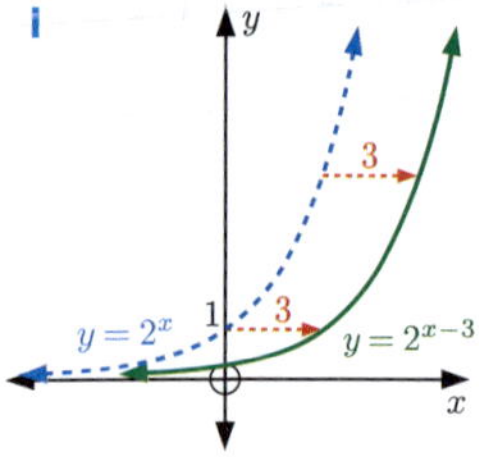

ii

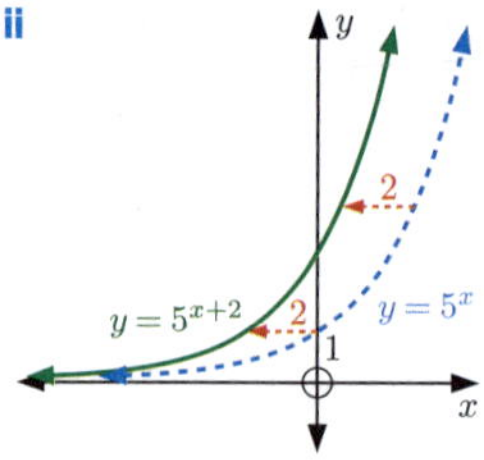

iii

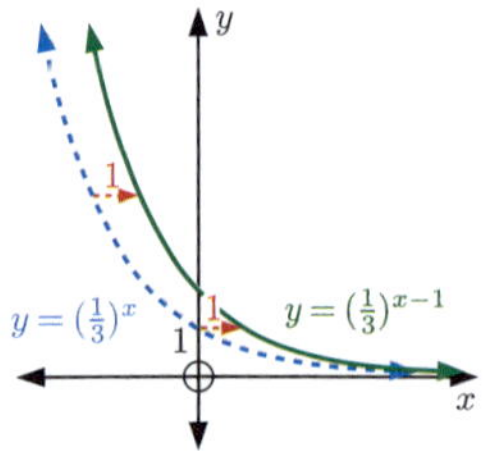

6 **a** A vertical stretch with scale factor p.

b **i**

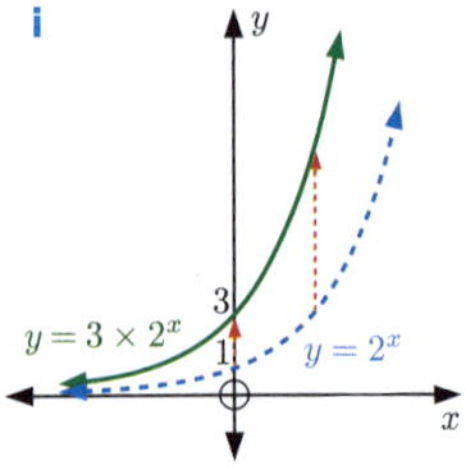

ii

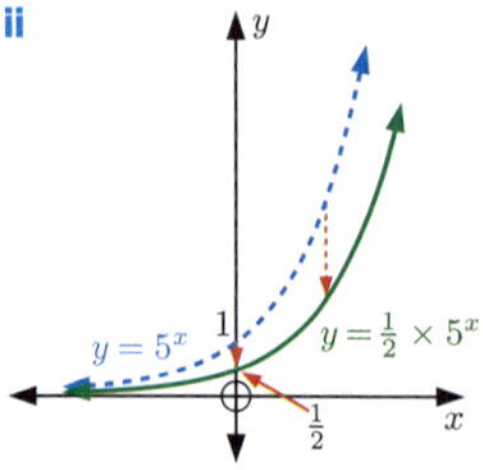

iii

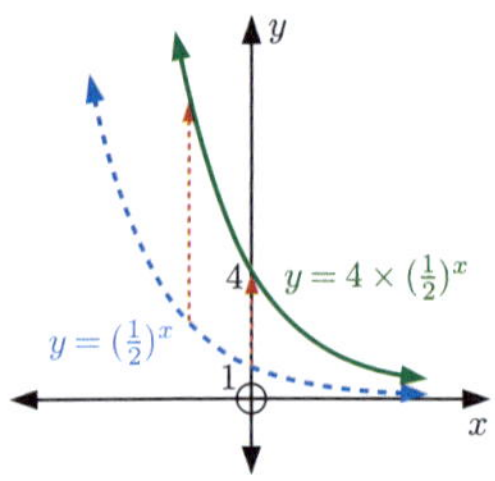

7 **a** A horizontal stretch with scale factor $\dfrac{1}{q}$.

b **i**

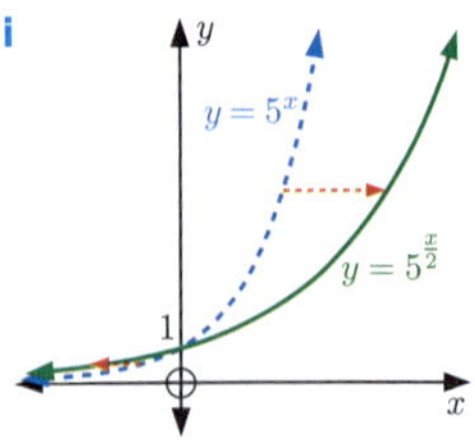

ii

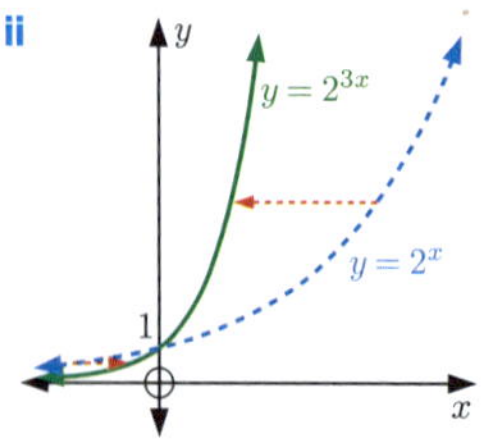

iii

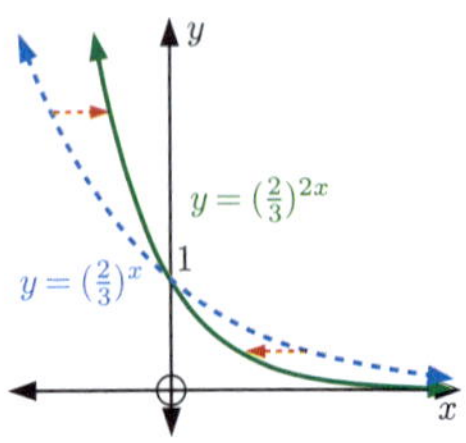

8 **a** A reflection of $y = f(x)$ in the x-axis.

b **i**

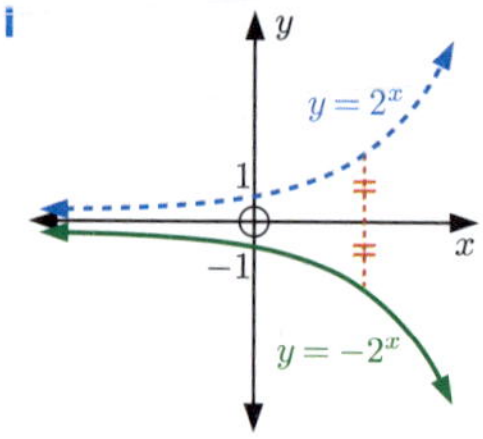

ii

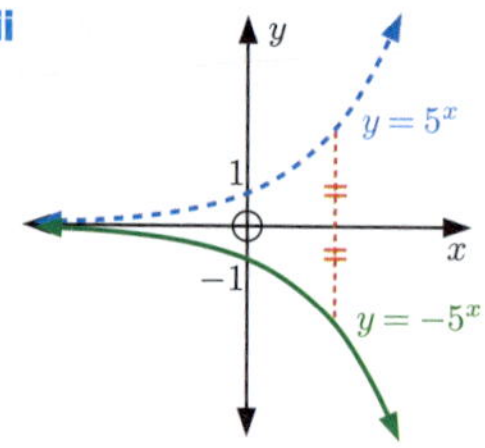

iii

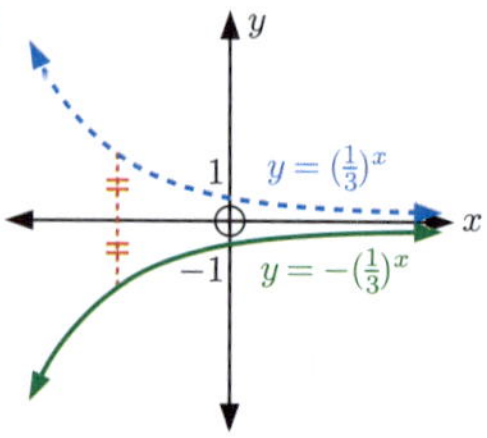

9 **a** A reflection of $y = f(x)$ in the y-axis.

b **i**

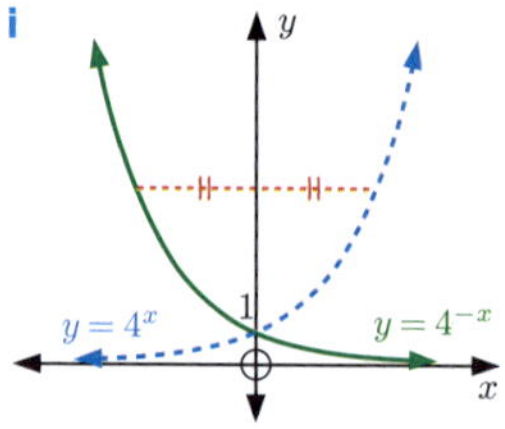

ii

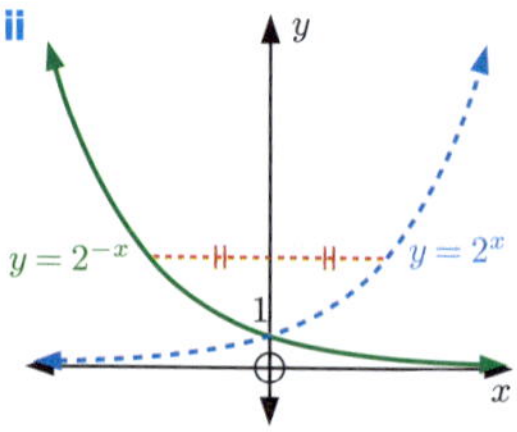

iii

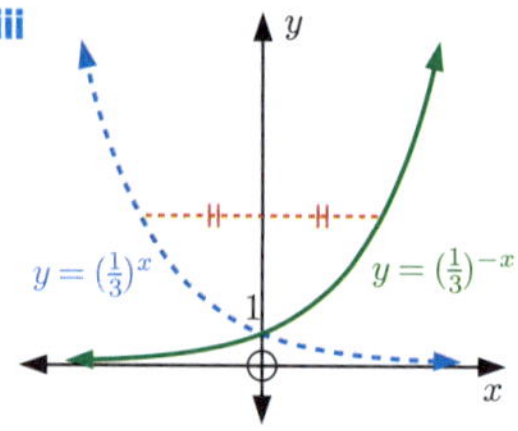

10 **a** $a = \frac{3}{2}$, $p = 20$ **b** $a = \frac{1}{5}$, $p = 400$

11 $3^{-x} = \dfrac{1}{3^x} = \left(\frac{1}{3}\right)^x$ for all x

$\therefore$ $y = 3^{-x}$ and $y = \left(\frac{1}{3}\right)^x$ have the same graph.

12 **a** $y = 3^x - 2$

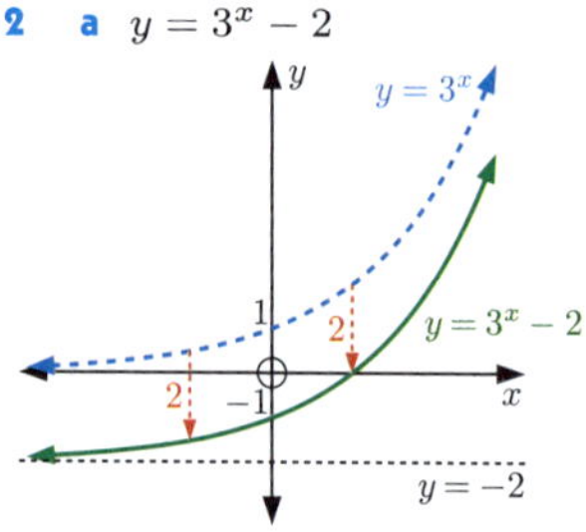

b $y = -3^x$

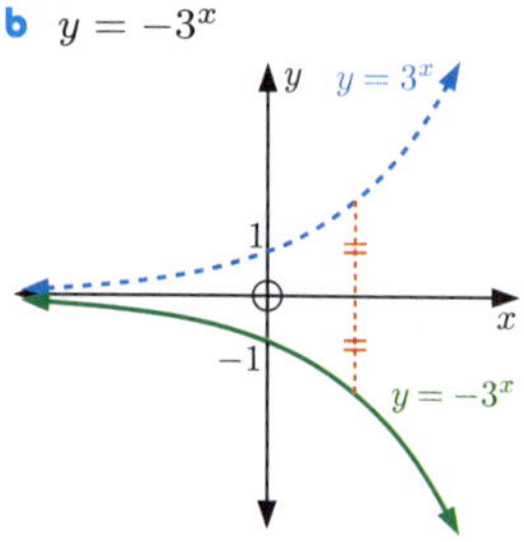

c $y = \frac{1}{2} \times 3^x$

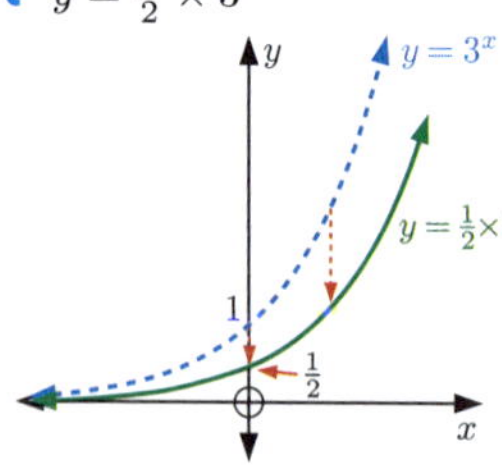

d $y = 3^{x+3}$

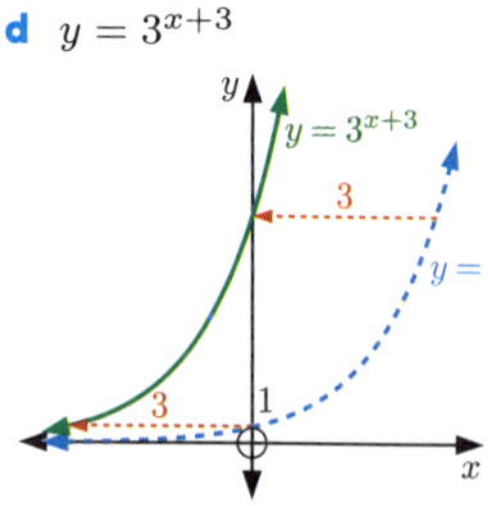

e $y = 3^{\frac{x}{5}}$

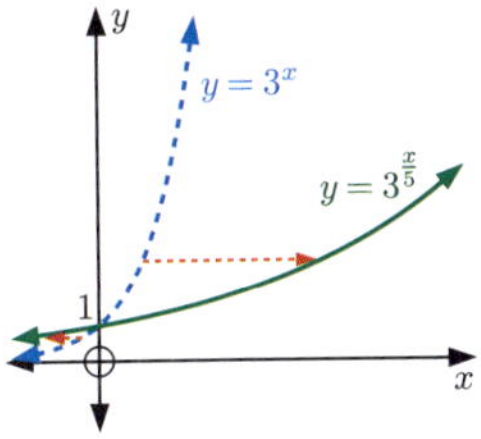

f $y = 4 \times 3^x$

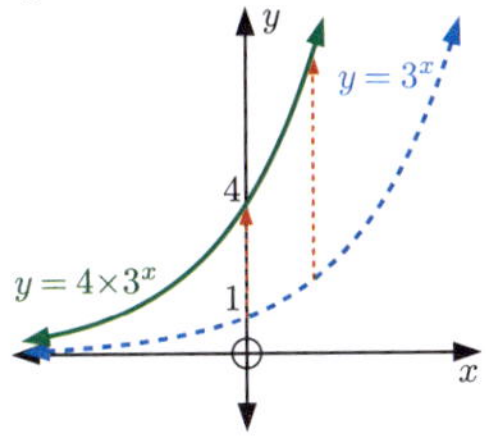

g $y = 3^{3x}$

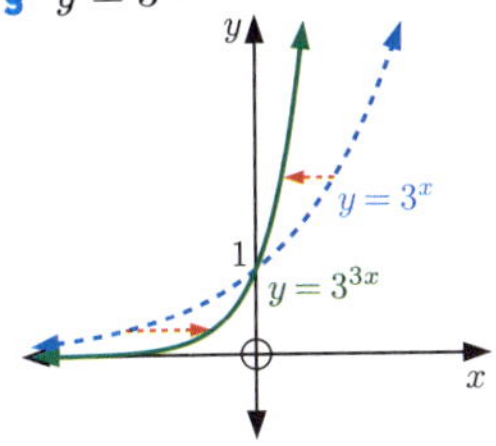

EXERCISE 25C

1 **a** $x = 1$ **b** $x = 2$ **c** $x = 3$ **d** $x = 0$
e $x = -1$ **f** $x = -1$ **g** $x = -4$ **h** $x = 3$
i $x = -4$

2 **a** $x = 0$ **b** $x = -4$ **c** $x = -2$ **d** $x = 6$
e $x = 2$ **f** $x = -\frac{3}{4}$ **g** $x = \frac{7}{2}$ **h** $x = 0$
i $x = 4$

3 **a** $x = 3$ **b** $x = 0$ **c** $x = -2$ **d** $x = 3$
e $x = -5$ **f** $x = -2$

4 **a** $x = \frac{2}{7}$ **b** $x = \frac{1}{3}$ **c** $x = 1$

5 **a** $x = 1$ **b** $x = 2$ **c** $x = 1$

6 **a** $x \approx 6.644$ **b** $x \approx -1.884$ **c** no solution
d $x \approx 1.292$ **e** $x \approx -1.611$ **f** $x \approx 4.192$
g $x \approx -4.583$ **h** $x \approx -1.848$ **i** $x \approx 4.858$
j $x \approx 2.064$ **k** $x \approx 2.991$ **l** $x \approx 291.825$
m $x \approx 0.262$ **n** $x \approx 1.441$ **o** $x \approx 0.415$

EXERCISE 25D

1 **a** 40 mongooses
b **i** $\approx$ 61 mongooses **ii** $\approx$ 162 mongooses
iii $\approx$ 2650 mongooses
c

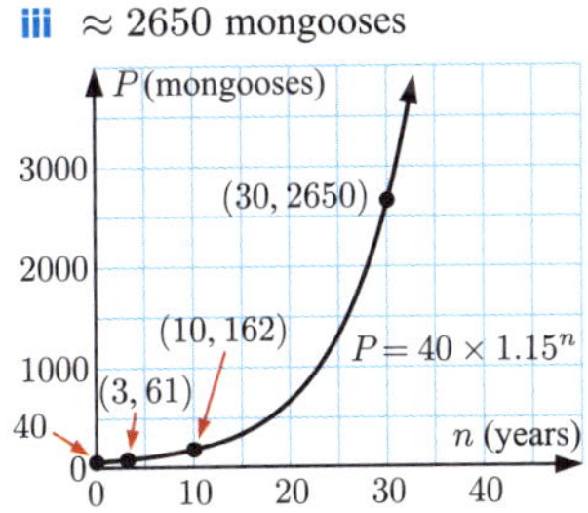

d $\approx$ 6.56 years (6 years 7 months)

2 **a** 50 echidnas
b **i** $\approx$ 100 echidnas **ii** $\approx$ 400 echidnas
iii $\approx$ 5090 echidnas
c

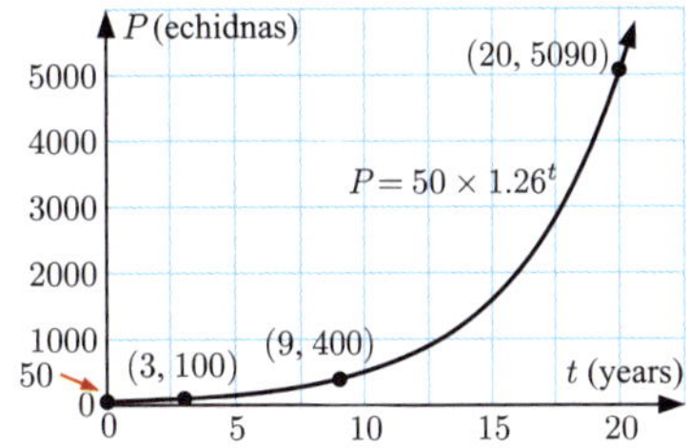

d $\approx$ 4.75 years (4 years 9 months)

3 **a** $G_0 = 28$
b **i** ≈ 29
ii ≈ 31
iii ≈ 34
c

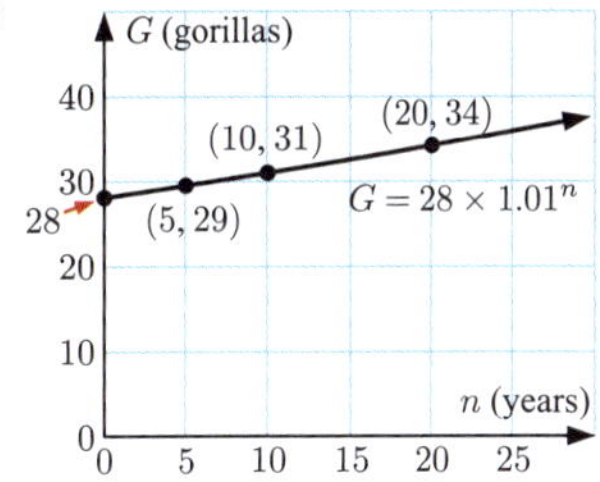

d $\approx$ 198 years

4 **a** no
b The initial area covered is 40 cm^2, and the area covered each week is 1.2 times greater than the area covered in the previous week.
c

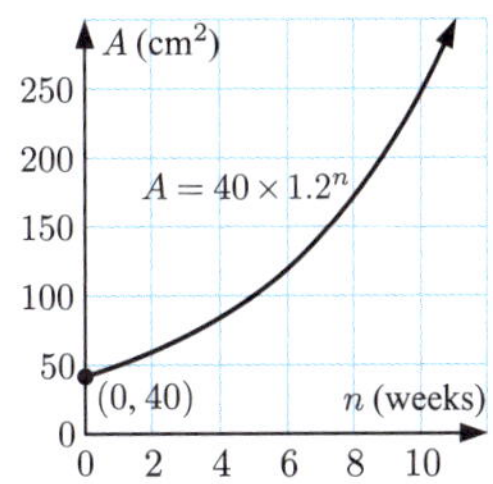

d 69.12 cm^2
e $\approx$ 5.03 weeks

5 **a** $M = 10 \times 1.12^t$ grams
b **i** $\approx$ 12.5 grams **ii** $\approx$ 22.1 grams
c

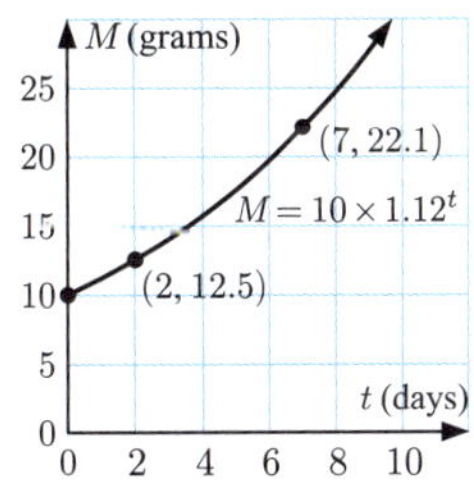

d $\approx$ 9.69 days

EXERCISE 25E

1 **a** 100°C
b **i** $\approx$ 70.6°C
ii $\approx$ 17.5°C
iii $\approx$ 3.06°C
c

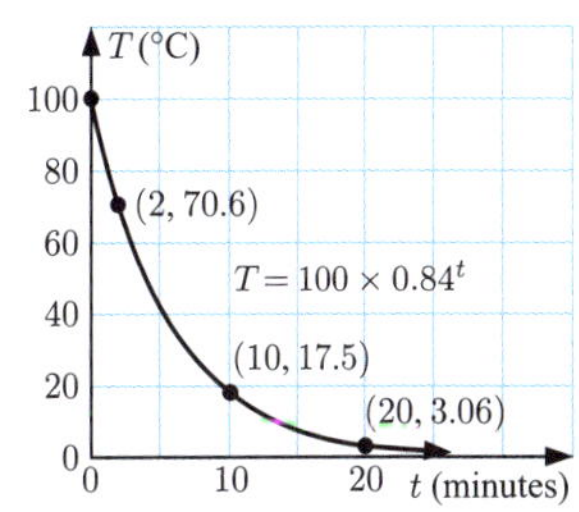

d $\approx$ 4.21 min
(4 min 13 s)

2 **a** 2.3 g

b **i** ≈ 1.02 g

ii ≈ 0.449 g

iii ≈ 0.199 g

c

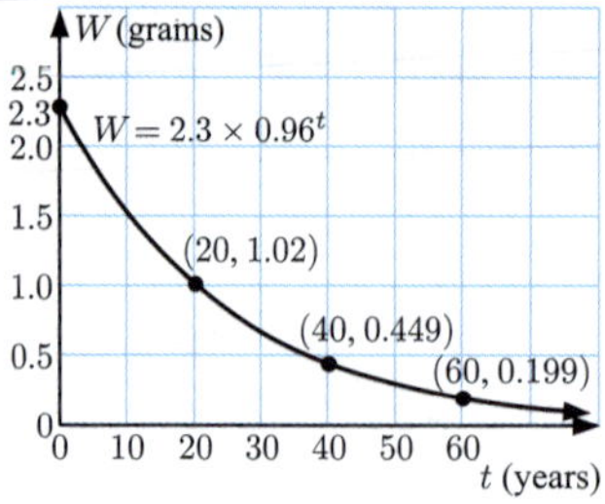

d $\approx 55.8\%$ loss

3 **a** \$500 **b** **i** \$445.64 **ii** \$199.11

c ≈ 62.8 months (5 years 3 months)

4 **a** $P = 500 \times 0.92^t$

b **i** ≈ 423

ii ≈ 257

c

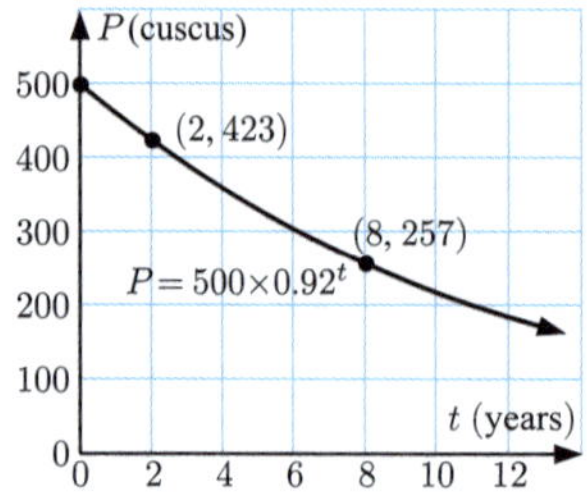

REVIEW SET 25A

1 **a** 0 **b** 26 **c** $-\frac{2}{3}$ **d** $3^{2x} - 1$

2 **a**

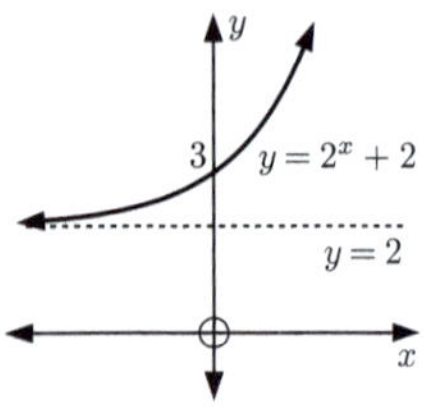

b

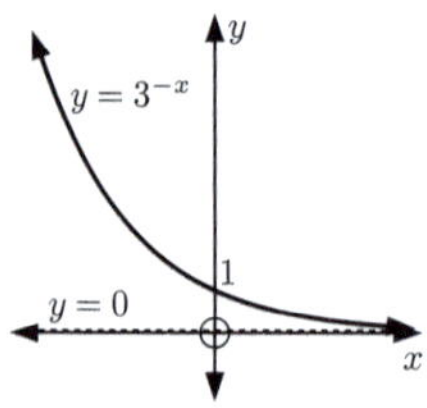

c

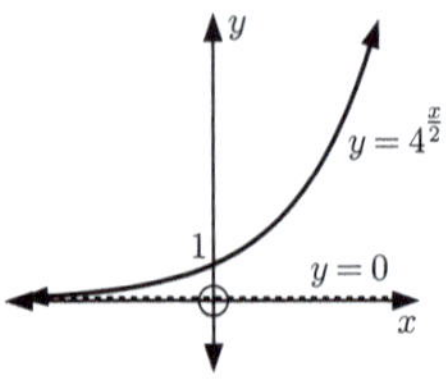

3 **a** $a = \frac{3}{4},\ p = 80$ **b** $a = \frac{5}{2},\ p = 16$

4 **a** $x = -4$ **b** $x = -1$ **c** $x = 4$ **d** $x = -4$

e $x = \frac{4}{5}$ **f** $x = \frac{7}{11}$

5 **a** A horizontal translation 2 units to the right.

b y-intercept: $\frac{1}{4}$, horizontal asymptote: $y = 0$

c domain $= \{x \in \mathbb{R}\}$,

range $= \{y \mid y > 0\}$

d

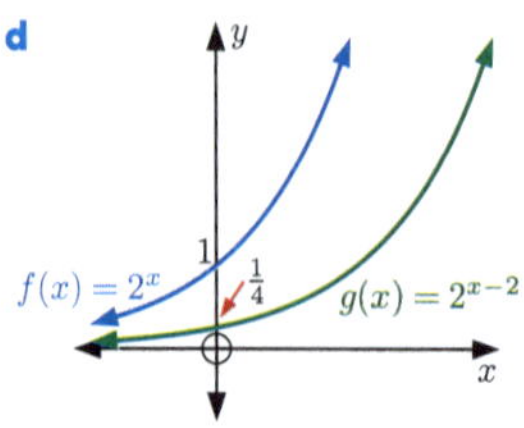

e $\frac{1}{4}$

6 **a** **i** 10 people **ii** ≈ 29 people **iii** ≈ 82 people

b

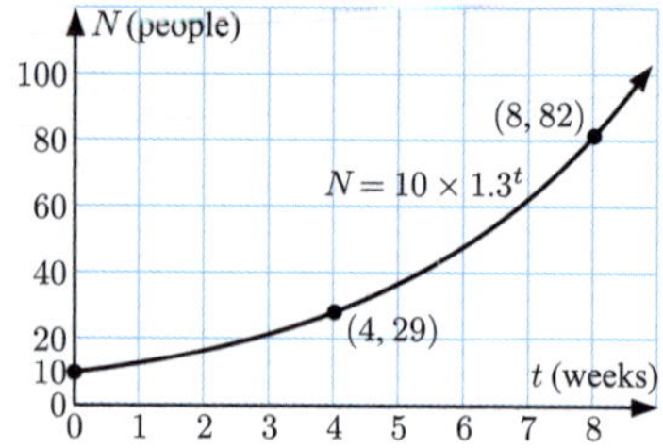

c ≈ 11.4 weeks

7 **a** $W = 130 \times 0.98^t$

b **i** ≈ 125 kg **ii** ≈ 118 kg **c** ≈ 8.27 weeks

REVIEW SET 25B

1 **a** 2 **b** $\frac{2}{3}$ **c** 18 **d** $2 \times 3^{-x-4}$

2 **a** $y = -4^x$

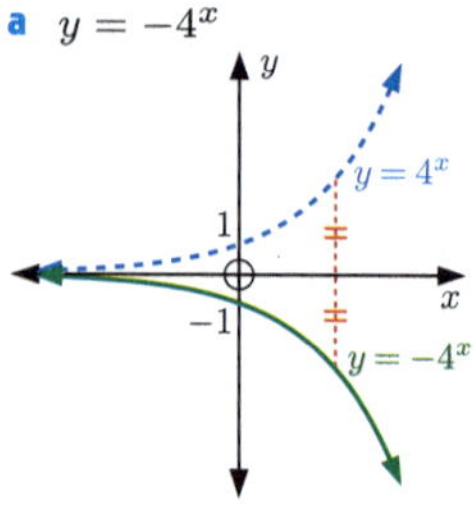

b $y = 5^{x-3}$

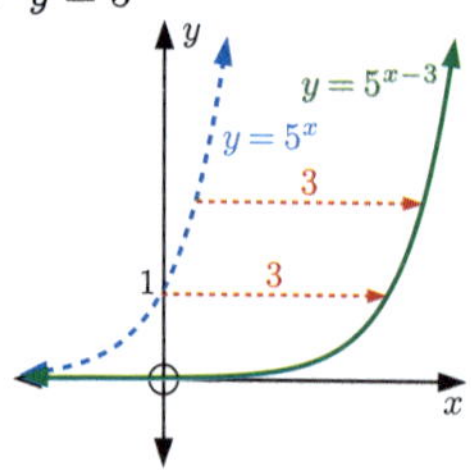

c $y = \frac{1}{3} \times 3^x$

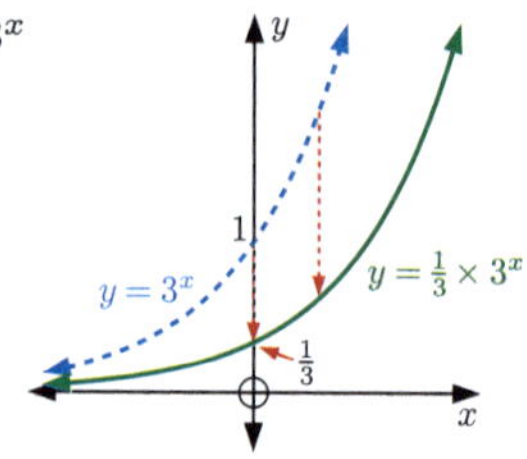

3

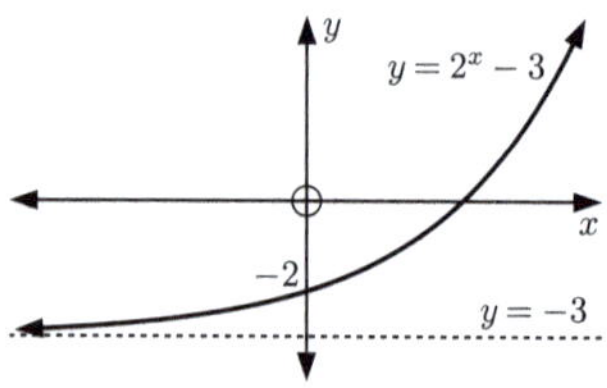

range $= \{y \mid y > -3\}$

4 **a** $x = -3$ **b** $x = -6$ **c** $x = 2$ **d** $x = \frac{3}{4}$

e $x = \frac{2}{3}$

5 **a** $x \approx 2.73$ **b** no solutions **c** $x \approx 11.33$

6 **a** **i** 500 birds **ii** ≈ 299 birds

b

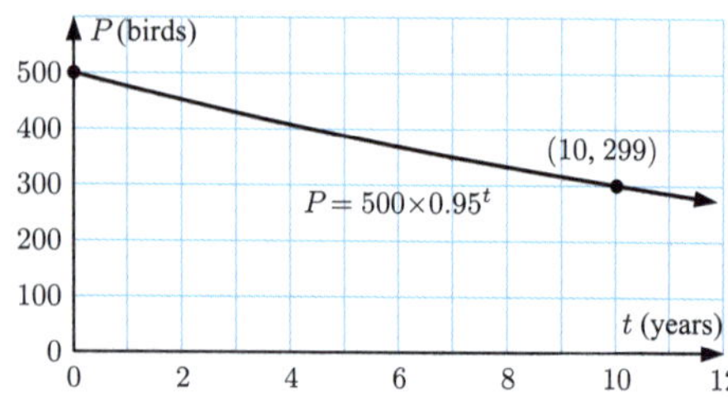

7 **a** \$50

b **i** \$141.97 **ii** \$403.12 **iii** \$9228.24

c

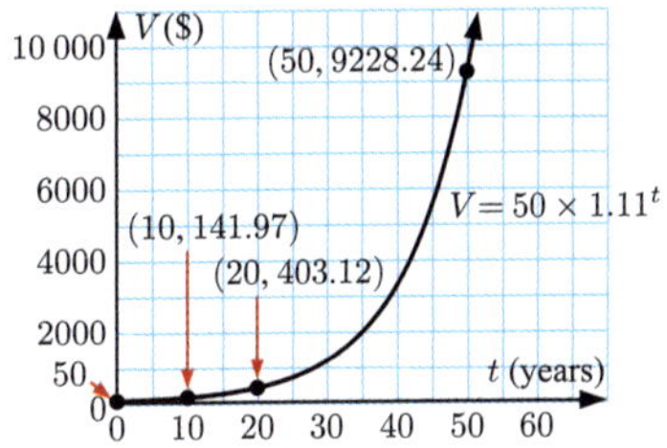

d ≈ 28.7 years (28 years 8 months)

INDEX